Inositol Phospholipid Metabolism
and Phosphat...

LABORATORY TECHNIQUES IN BIOCHEMISTRY AND MOLECULAR BIOLOGY

Edited by

P.C. van der Vliet — *Department for Physiological Chemistry, University of Utrecht, Utrecht, Netherlands*
and

S. Pillai — *MGH Cancer Center, Boston, Massachusetts, USA*

Volume 30

ELSEVIER

AMSTERDAM–BOSTON–HEIDELBERG–LONDON–NEW YORK–OXFORD–PARIS–SAN DIEGO
SAN FRANCISCO–SINGAPORE–SYDNEY–TOKYO

INOSITOL PHOSPHOLIPID METABOLISM AND PHOSPHATIDYL INOSITOL KINASES

A. Kuksis

Banting and Best Department of Medical Research
University of Toronto
112 College Street, Toronto
Ontario M5G 1L6
Canada

2003
ELSEVIER

AMSTERDAM−BOSTON−HEIDELBERG−LONDON−NEW YORK−OXFORD−PARIS−SAN DIEGO
SAN FRANCISCO−SINGAPORE−SYDNEY−TOKYO

Elsevier
The Boulevard, Langford Lane, Kidlington, Oxford OX5 1GB, UK
Radarweg 29, PO Box 211, 1000 AE Amsterdam, The Netherlands

First edition 2003

Library of Congress Cataloging in Publication Data
A catalog record from the Library has been applied for.

British Library Cataloguing in Publication Data
A catalogue record from the British Library has been applied for.

ISBN: 0-444-51321-3 (library edition)
ISBN: 0-444-51304-3 (pocket edition)
ISBN: 0-7204-4200-1 (series)

Transferred to digital print, 2007
Printed and bound by CPI Antony Rowe, Eastbourne

Preface

The writing of this book was undertaken as an opportunity to examine the analytical validity of the biochemical transformations that constitute the basis of the lipid signaling pathways. It is hoped that this review of past triumphs and failures will facilitate the understanding and research in the area of inositol phosphate and phosphatide metabolism and will build confidence in the methodology that is subject to continued improvement and correction. The development of advanced methods for the isolation, identification and quantification of old and new inositol lipids and inositol phosphates from natural and synthetic systems has been a major advancing force in phosphoinositol research.

In the past progress in studies of inositol lipid and inositol phosphate metabolism has been frequently hampered by inadequate methods of analysis of both substrates and products of metabolic transformation of the various lipid and water-soluble inositol derivatives. The early failure to recognize that stimulated phospholipid breakdown occurs at the level of phosphatidylinositol bisphosphate rather than at phosphatidylinositol was directly due to the inability of the lipid extraction methods to recover phosphatidylinositol phosphates, and the absence of techniques for the separation of the phosphatides in the acidic extracts, which might have contained them.

Later, much improved conventional methods failed to separate the novel isomers resulting from the 3-kinase activity, from the well known and more abundant phosphatidylinositol phosphates. Specifically, phosphatidylinositol 3-phosphate and phosphatidylinositol 4-phosphate co-migrated in nearly all one- and two-dimensional thin-layer chromatographic systems used, while phosphatidylinositol 3,4-bisphosphate and phosphatidylinositol 4,5-bisphosphate co-migrated on conventional thin-layer chromatographic systems. In addition the salt gradients commonly employed to separate the deacylated phosphatidylinositol phosphates could not resolve the various isomers of deacylated phosphatidylinositol polyphosphates. In other instances the biased belief that naturally occurring inositols are limited to D-*myo*-inositol has resulted in neglecting other inositol isomers. Difficulties have also arisen from the loss of symmetry and change of locant numbering upon substitution of the inositol ring.

Detection of novel phosphatidylinositol phosphates *in vivo* has been hampered by the fact that these lipids are present in relatively low abundance in all cells. Therefore, to detect the inositol phospholipids in intact cells, it has been necessary to incorporate high levels of radio-isotope into the total cellular phospholipids. However, the $4'$ and $5'$-positions of the phosphatidylinositol phosphates are rapidly labeled with $^{32}P_i$ *via* phosphatidylinositol kinases and the labels may be removed by phosphatases. Furthermore, the phosphate labels in the 4- and 5-positions tend to equilibrate quickly with that in the γ-phosphate of ATP and, thus, to reflect more the amounts of phosphatidylinositol and phosphatidylinositol bisphosphate in a preparation than their natural biosynthetic labeling. Whenever these difficulties have been overcome, the progress was rapid. Clearly, the advances in inositol phosphate and phosphatide research have been propelled by developments in methodology and advanced methodology remains the basis of new discovery.

The introduction of methods of molecular biology for the investigation of metabolic signaling in the late 90s has spurred further advances in the field and new discoveries have been announced in rapid succession. At present, the methodology of molecular biology already assumes a major role in the elucidation of the signaling pathways by providing for the first-time access to sufficient mass of the enzyme, carrier and receptor proteins for the investigation of their structure and function. In this process, chromatography along with mass spectrometry remain the separation methods of choice, as they have been in the past, except that their role is enhanced by the availability of better defined products of biochemical transformations for separation, identification and quantification.

Despite much progress in the field, much remains to be discovered. While there is evidence that all eight phosphatidylinositol phosphates have distinct cellular functions, the possibility that all 63 inositol phosphates would also have individual cellular functions remains to be demonstrated. Allowance must also be made for discovery of new structures and functions in this group of compounds of seemingly endless variety. Further complexity arises from the known natural occurrence of pyrophosphate derivatives and more than one inositol isomer.

Finally, many biological systems that posses only a limited variety inositol phosphates and phosphatides contain a multitude of inositol phosphate kinases and phosphatases, which attaches great importance to the location of the substrates and the metabolizing enzymes. The full exploration of the catalytic flexibility and function of the phosphatidylinositol and inositol phosphate kinases and phosphatases, as well as of the phosphatidylinositol phospholipases remains in the hands of the molecular biologists. It is hoped that in these efforts the highly sensitive quantitative chromatographic and mass spectrometric methods discussed in this book will be of practical help.

I am very grateful to the Editors of Elsevier for the original suggestion to include the 3-kinases in this book. This led to a much more extensive review of the field than originally intended and to the recognition that the kinases could not be discussed without considering the phosphatases and phospholipases. Fortunately, modern indexing and library service has kept the work up to date. I regret, that, despite this effort, limitations of time and space prevented coverage of all promising methodologies and the examination of all original discoveries in the detail that they deserve. Finally, I wish to express my appreciation to the Editorial Board and staff of Elsevier for their patience during the lengthy preparation and continued updating of the text required by each new advance and new publication on the subject.

The three dimensional representations of phosphatidylinositol and its mono- and bis-phosphates were generously supplied by Avanti Polar Lipids, Inc., Alabaster, AL, USA.

March 25, 2003
Arnis Kuksis

Contents

Preface   v

Abbreviations   xxiii

Chapter 1.   General Introduction   1

1.1.   Introduction   1
1.2.   Natural occurrence   6
   1.2.1.   Phosphatidylinositol   6
   1.2.2.   Phosphatidylinositol phosphates   7
   1.2.3.   Inositol phosphates   7
   1.2.4.   Glycosyl phosphatidylinositols   8
1.3.   Biological significance   9
   1.3.1.   Structural components   9
   1.3.2.   Signaling molecules   10
   1.3.3.   Metabolic intermediates   13
1.4.   Nomenclature   14
   1.4.1.   Inositol glycerophospholipids   15
      1.4.1.1.   Lipid components   15
      1.4.1.2.   Stereoisomers of glycerolipids   16
      1.4.1.3.   Positional isomers   19
   1.4.2.   Inositol and phosphatidylinositol phosphates   19
      1.4.2.1.   Stereoisomers of inositols   20
      1.4.2.2.   Numbering of phosphate groups   24
   1.4.3.   Inositol pyrophosphates   30
   1.4.4.   Kinases, phosphatases and phospholipases   31
   1.4.5.   Other conventions   33
1.5.   Scope of the book   33

Chapter 2.   Phosphatidylinositols   37

2.1.   Isolation   37
   2.1.1.   Solvent extraction   37
      2.1.1.1.   Chloroform/methanol   38
      2.1.1.2.   Other solvents   40

2.1.2. Solid phase extraction 42
    2.1.2.1. TLC and HPTLC 42
    2.1.2.2. HPLC 45
    2.1.2.3. Cartridge extraction 47
2.2. Identification 48
  2.2.1. Determination of structure 49
    2.2.1.1. Chemical methods 50
    2.2.1.2. Enzymic methods 54
    2.2.1.3. Mass spectrometric methods 59
    2.2.1.4. NMR 62
  2.2.2. Composition of fatty acids 66
    2.2.2.1. Total 66
    2.2.2.2. sn-1-position 66
    2.2.2.3. sn-2-position 68
  2.2.3. Determination of molecular species 70
    2.2.3.1. Intact PtdIns 70
    2.2.3.2. PtdOH 79
    2.2.3.3. Diradylglycerols 83
    2.2.3.4. Inositol isomers 95
2.3. Quantification 96
  2.3.1. Intact PtdIns 98
  2.3.2. Diacylglycerols 99
  2.3.3. Fatty acids 101
  2.3.4. Inositol 101
  2.3.5. Phosphorus 105

Chapter 3.   *Phosphatidylinositol Phosphates* 107

3.1. Isolation 107
  3.1.1. Solvent extraction 107
  3.1.2. Solid phase extraction (cartridge) 111
    3.1.2.1. TLC and HPTLC 111
    3.1.2.2. HPLC 121
    3.1.2.3. Ion exchange 126
    3.1.2.4. Affinity chromatography 127
3.2. Determination of structure 128
  3.2.1. Chemical methods 129
    3.2.1.1. General assays 129
    3.2.1.2. Deacylation 130
    3.2.1.3. Deglyceration 130
    3.2.1.4. Dephosphorylation 132

3.2.2.   Enzymic methods   132
         3.2.2.1.   Phospholipases   132
         3.2.2.2.   Phosphatases   133
         3.2.2.3.   Kinases   133
         3.2.2.4.   Polyol dehydrogenase   134
3.2.3.   Chromatographic methods   135
         3.2.3.1.   Resolution of intact PtdInsPs   135
         3.2.3.2.   Resolution of GroPInsPs   136
         3.2.3.3.   Resolution of InsPs   141
         3.2.3.4.   Resolution of isomeric Ins   141
3.2.4.   Mass spectrometric methods   141
         3.2.4.1.   FAB/MS   142
         3.2.4.2.   ES/MS/MS   143
         3.2.4.3.   MALDI-TOF/MS   148
3.3.   Resolution of molecular species   151
       3.3.1.   Intact PtdInsPs and PtdIns   153
       3.3.2.   Phosphatidic acids   158
       3.3.3.   Diacylglycerols   158
       3.3.4.   Positional distribution of fatty acids   159
       3.3.5.   Inositol isomers   162
3.4.   Quantification   162
       3.4.1.   MS/MS   163
       3.4.2.   Autoradiography   165
       3.4.3.   Radioreceptor assay   167
       3.4.4.   Other methods   171

Chapter 4.   *Inositol Phosphates*   175

4.1.   Introduction   175
4.2.   Isolation   176
       4.2.1.   Acid extraction   176
       4.2.2.   Removal of nucleotides   178
       4.2.3.   Neutralization of acid extracts   178
4.3.   Resolution of chemical families   179
       4.3.1.   TLC   180
       4.3.2.   Anion exchange cartridge chromatography   182
       4.3.3.   HPLC analysis   185
       4.3.4.   Other   190
4.4.   Resolution of positional isomers   191
       4.4.1.   Inositol monophosphates ($InsP_1s$)   194

4.4.2.   Inositol bisphosphates (InsP$_2$s)   195
4.4.3.   Inositol trisphosphates (InsP$_3$s)   196
4.4.4.   Inositol tetrakisphosphates (InsP$_4$s)   198
4.4.5.   Inositol pentakisphosphates (InsP$_5$s)   200
4.4.6.   Inositol hexakis and pyrophosphates (InsP$_6$s, InsP$_7$s
         and InsP$_8$s)   205
4.5.   Determination of chemical structure   207
4.5.1.   Chemical analysis   207
4.5.2.   Nuclear magnetic resonance spectroscopy (NMR)   212
4.5.3.   Mass spectrometry   217
4.6.   Determination of stereochemical structure   223
4.6.1.   Optical rotation   224
4.6.2.   Polyol dehydrogenases   224
4.6.3.   Inositol phosphate kinases   232
4.6.4.   Inositol phosphate phosphatases   233
4.7.   Quantification   235
4.7.1.   Radio-isotope assays   236
4.7.2.   Radio-receptor assays   239
4.7.3.   Mass assays of inositol phosphates   243

Chapter 5.   Glycosylphosphatidylinositols   253

5.1.   Introduction   253
5.2.   Natural occurrence and isolation   254
5.2.1.   Isolation of free glycosyl PtdIns-anchors   255
         5.2.1.1.   Extraction   256
         5.2.1.2.   Purification   258
5.2.2.   Isolation of glycosyl PtdIns-anchored proteins   259
         5.2.2.1.   Metabolic labeling   260
         5.2.2.2.   Chemical labeling   262
5.2.3.   Release of bound glycosyl PtdIns-anchors   263
         5.2.3.1.   Proteolysis   264
         5.2.3.2.   Nitrous acid deamination   266
         5.2.3.3.   Hydrofluoric acid   267
5.3.   Determination of chemical structure   267
5.3.1.   Protein-ethanolamine bridge   271
5.3.2.   Glucosamine-inositol linkage   279
5.3.3.   Glycan moiety   280
5.3.4.   PtdIns moiety   288
5.3.5.   Diradylglycerol moiety   294

　　　　5.3.6.　Composition of fatty chains 302
　　　　5.3.7.　Inositol moiety 307
　　5.4.　Determination of molecular species 307
　　　　5.4.1.　Mass spectrometry 308
　　　　　　5.4.1.1.　Intact protein 308
　　　　　　5.4.1.2.　Intact glycosyl PtdIns 310
　　　　　　5.4.1.3.　PtdIns moiety 311
　　　　5.4.2.　Chromatography/mass spectrometry 311
　　　　　　5.4.2.1.　Intact glycosyl PtdIns 311
　　　　　　5.4.2.2.　Diradyl GroPIns 312
　　　　　　5.4.2.3.　Diradyl GroPOH 313
　　　　　　5.4.2.4.　Diradylglycerols 315
　　5.5.　Quantification 322
　　　　5.5.1.　Mass spectrometry 323
　　　　5.5.2.　Chromatography 324
　　　　5.5.3.　Chemical analysis 326
　　5.6.　Related structures 326
　　　　5.6.1.　Glycosylinositol phosphoceramides 326
　　　　5.6.2.　Lipophosphoglycans and glycoinositolphospholipids 330
　　　　5.6.3.　Lipoarabinomannans and arabinogalactans 331

Chapter 6.　Biosynthesis of Inositol Phospholipids 335

　　6.1.　Introduction 335
　　6.2.　Phosphatidylinositols 336
　　　　6.2.1.　De novo synthesis 336
　　　　　　6.2.1.1.　Synthesis of CPD-DAG via CTP 341
　　　　　　6.2.1.2.　Synthesis of PtdIns via CPD-DAG 343
　　　　　　6.2.1.3.　Synthesis of PtdIns via CPD-DAG in plants . . . 344
　　　　6.2.2.　Remodeling 345
　　　　　　6.2.2.1.　Acyl group exchange 345
　　　　　　6.2.2.2.　Head-group exchange 348
　　6.3.　Phosphatidylinositol phosphates 352
　　　　6.3.1.　PtdIns monophosphates 353
　　　　　　6.3.1.1.　PtdIns(4)P 354
　　　　　　6.3.1.2.　PtdIns(3)P 359
　　　　　　6.3.1.3.　PtdIns(5)P 363
　　　　6.3.2.　PtdIns bisphosphates 365
　　　　　　6.3.2.1.　PtdIns(4,5)P_2 366
　　　　　　6.3.2.2.　PtdIns(3,4)P_2 369
　　　　　　6.3.2.3.　PtdIns(3,5)P_2 371

6.3.3. PtdIns trisphosphates 373

        6.3.3.1. PtdIns(3,4,5)P$_3$ 373

6.4. Glycosyl phosphatidylinositols 376

    6.4.1. GPtdIns protein anchors 377

        6.4.1.1. GlcNAc-PtdIns 379

        6.4.1.2. GlcN-PtdIns 382

        6.4.1.3. GlcN-PtdIns acyl esters 384

        6.4.1.4. GlcN-PtdIns mannosides 385

        6.4.1.5. GlcN-PtdIns(Man)$_3$EtnP 388

        6.4.1.6. Transfer to protein 390

        6.4.1.7. Remodeling 391

    6.4.2. Protein-free GPtdIns 398

        6.4.2.1. Protozoal GPtdIns-related structures 399

        6.4.2.2. Lipophosphoglycans 400

        6.4.2.3. Glycoinositolphospholipids (GIPLs) 401

        6.4.2.4. Glucosyl PtdIns glycolipids 401

Chapter 7.   *Phosphatidylinositol and Inositol Phosphate Kinases* 403

7.1. Introduction 403

7.2. Phosphatidylinositol kinases 404

    7.2.1. PtdIns 3-kinase 407

        7.2.1.1. Assay of immunoprecipitates 408

        7.2.1.2. Assay of cell lysates 411

    7.2.2. PtdIns 4-kinases 412

        7.2.2.1. Assays in micelles 416

        7.2.2.2. Assays of recombinant proteins 417

        7.2.2.3. Plant tissue assays 418

    7.2.3. PtdIns 5-kinases 419

        7.2.3.1. Assays of cellular proteins 420

        7.2.3.2. Assays of recombinant protein 422

7.3. Phosphatidylinositol monophosphate kinases 423

    7.3.1. PtdInsP 3-kinases 424

        7.3.1.1. PtdIns(4)P 3-kinase 425

        7.3.1.2. PtdIns(5)P 3-kinase 426

    7.3.2. PtdInsP 4-kinases 427

        7.3.2.1. PtdIns(3)P 4-kinase 427

        7.3.2.2. PtdIns(5)P 4-kinase 428

7.3.3.   PtdInsP 5-kinases 430
      7.3.3.1.   PtdIns(3)P 5-kinase 430
      7.3.3.2.   PtdIns(4)P 5-kinase 432
7.4.   Phosphatidylinositol bisphosphate kinases 434
    7.4.1.   PtdIns $(4,5)P_2$ 3-kinase 434
    7.4.2.   PtdIns$(3,5)P_2$ 4-kinase 437
    7.4.3.   PtdIns$(3,4)P_2$ 5-kinase 437
7.5.   Inositol monophosphate kinases 439
    7.5.1.   InsP 1-kinases 439
      7.5.1.1.   Ins(3)P 1-kinase 439
    7.5.2.   InsP 6-kinases 440
      7.5.2.1.   Ins(3)P 6-kinases 440
      7.5.2.2.   Ins(2)P 6-kinases 440
7.6.   Inositol bisphosphate kinases 440
    7.6.1.   InsP$_2$ 4-kinases 441
      7.6.1.1.   Ins$(3,6)P_2$ 4-kinases 441
    7.6.2.   InsP$_2$ 5-kinases 441
      7.6.2.1.   Ins$(1,3)P_2$ 5-kinase 441
      7.6.2.2..   Ins$(2,6)P_2$ 5-kinase 441
    7.6.3.   InsP$_2$ 6/3-kinase 441
7.7.   Inositol trisphosphate kinases 442
    7.7.1.   InsP$_3$ 1-kinases 442
      7.7.1.1.   Ins$(3,4,6)P_3$ 1-kinase 442
    7.7.2.   InsP$_3$ 3-kinases 442
      7.7.2.1.   Ins$(1,4,5)P_3$ 3-kinase 442
      7.7.2.2.   Ins$(2,5,6)P_3$ 3-kinase 445
    7.7.3.   InsP$_3$ 6/3 kinases 445
      7.7.3.1.   Ins$(1,4,5)P_3$ 6/3-kinase 445
    7.7.4.   InsP$_3$ 6/5 kinase 448
      7.7.4.1.   Ins$(1,3,4)P_3$ 6/5-kinase 448
    7.7.5.   Ins$(1,3,5)P_3$ 6-kinase 450
7.8.   Inositol tetrakisphosphate kinases 450
    7.8.1.   InsP$_4$ 1-kinases 450
      7.8.1.1.   Ins$(3,4,5,6)P_4$ 1-kinase 450
    7.8.2.   InsP$_4$ 2/4 kinases 452
      7.8.2.1.   Ins$(1,3,5,6)P_4$ 2/4-kinase 452
    7.8.3.   InsP$_4$ 3-kinases 452
    7.8.4.   InsP$_4$ 4-kinases 452
      7.8.4.1.   Ins$(2,3,5,6)P_4$ 4-kinase 452
    7.8.5.   InsP$_4$ 5-kinases 452
      7.8.5.1.   Ins$(1,3,4,6)P_4$ 5-kinase 452

7.9. Inositol pentakisphosphate kinases 453
    7.9.1. InsP$_5$ 1-kinases 453
            7.9.1.1. Ins(2,3,4,5,6)P$_5$ 1-kinase 453
    7.9.2. InsP$_5$ 2-kinases 454
            7.9.2.1. Ins(1,3,4,5,6)P$_5$ 2-kinase 454
    7.9.3. InsP$_5$ 3-kinases 457
            7.9.3.1. Ins(1,2,4,5,6)P$_5$ 3-kinase 457
    7.9.4. InsP$_5$ 5/6 kinases 457
            7.9.4.1. D/L-Ins(1,2,3,4,6)P$_5$ 5/6-kinase
                      (Stephens et al., 1991) 457
    7.9.5. InsP$_5$ pyrophosphokinase 457
            7.9.5.1. Ins(1,3,4,5,6)P$_5$ pyrophosphokinase 457
7.10. Inositol hexakisphosphate kinases 457
    7.10.1. InsP$_6$ 5-kinase 458
    7.10.2. PP-InsP$_4$ kinase 461
    7.10.3. bis-PP-InsP$_4$ 462

Chapter 8.   *Phosphatidylinositol Phosphate and Inositol
              Phosphate Phosphatases* 463

8.1. Introduction 463
8.2. PtdIns phosphate phosphatases 465
    8.2.1. PtdIns monophosphate phosphatases 465
            8.2.1.1. PtdIns (3)P 3-phosphatase 465
            8.2.1.2. PtdIns(4)P 4-phosphatase 472
            8.2.1.3. PtdIns(5)P 5-phosphatase 474
    8.2.2. PtdIns bisphosphate phosphatases 474
            8.2.2.1. PtdInsP$_2$ 3-phosphatases 474
            8.2.2.2. PtdInsP$_2$ 4-phosphatases 475
            8.2.2.3. PtdInsP$_2$ 5-phosphatases 478
    8.2.3. PtdIns trisphosphate phosphatases 483
            8.2.3.1. PtdInsP$_3$ 3-phosphatases 484
            8.2.3.2. PtdInsP$_3$ 4-phosphatases 489
            8.2.3.3. PtdInsP$_3$ 5-phosphatases 490
8.3. Inositol phosphate phosphatases 497
    8.3.1. Inositol monophosphate phosphatases 497
            8.3.1.1. Ins(1)P phosphatases 497
            8.3.1.2. Ins(3)P phosphatases 502
    8.3.2. Inositol bisphosphate phosphatases 504
            8.3.2.1. InsP$_2$ 1-phosphatase 505

8.3.2.2. InsP$_2$ 3-phosphatases 506
8.3.2.3. InsP$_2$ 4-phosphatase 508
8.3.2.4. InsP$_2$ 5-phosphatases 510
8.3.3. Inositol trisphosphate phosphatases 510
8.3.3.1. InsP$_3$ 1-phosphatase 511
8.3.3.2. InsP$_3$ 3-phosphatases 513
8.3.3.3. InsP$_3$ 4-phosphatases 514
8.3.3.4. InsP$_3$ 5-phosphatases 516
8.3.4. Inositol tetrakisphosphate phosphatases 526
8.3.4.1. InsP$_4$ 3-phosphatases 526
8.3.4.2. InsP$_4$ 4-phosphatases 526
8.3.4.3. InsP$_4$ 5-phosphatases 527
8.3.5. Inositol pentakisphosphate phosphatases 528
8.3.5.1. InsP$_5$ 1-phosphatases 528
8.3.5.2. InsP$_5$ 3-phosphatases 529
8.3.5.3. Ins(1,3,4,5,6)P$_5$ 6-phosphatases 531
8.3.6. Inositol hexakisphosphate phosphatases 531
8.3.6.1. InsP$_6$ 6-phytase 532
8.3.6.2. InsP$_6$ 3-phytase 532
8.3.6.3. InsP$_6$ 5-phytase 533
8.3.7. Inositol pyrophosphate phosphatases 534
8.3.7.1. (PP)$_2$InsP$_4$ pyrophosphatase 534
8.3.7.2. PPInsP$_5$ pyrophosphatases 536
8.4. Other phosphatases and phytases 537
8.4.1. Alkaline phosphatases 537
8.4.2. Acid phosphatases 539
8.4.3. *Bacillus* phytases 540

Chapter 9.   *Phosphatidylinositol Phospholipases* 541

9.1. Introduction 541
9.2. PtdIns-nonspecific phospholipases 542
9.2.1. Phospholipase A$_1$ 543
9.2.1.1. Substrate specificity 544
9.2.1.2. Methods of assay 547
9.2.2. Phospholipase A$_2$ 550
9.2.2.1. Substrate specificity 553
9.2.2.2. Methods of assay 556
9.2.3. Phospholipase C 561
9.2.3.1. Substrate specificity 561

        9.2.3.2. Methods of assay 565
    9.2.4. Phospholipase D 568
        9.2.4.1. Substrate specificity 570
        9.2.4.2. Methods of assay 572
9.3. PtdIns and GPtdIns-specific phospholipases 575
    9.3.1. Phospholipase A_1 and A_2 576
        9.3.1.1. Substrate specificity 577
        9.3.1.2. Methods of assay 577
    9.3.2. Phospholipase C 579
        9.3.2.1. Substrate specificity 582
        9.3.2.2. Methods of assay 591
    9.3.3. Phospholipase D 599
        9.3.3.1. Substrate specificity 603
        9.3.3.2. Methods of assay 604
9.4. PtdIns(4,5)P_2 activated phospholipases 606
    9.4.1. Phospholipase A_2 606
        9.4.1.1. Affinity assay 607
        9.4.1.2. Catalytic assay 608
    9.4.2. Phospholipase C 609
        9.4.2.1. Affinity assays 609
        9.4.2.2. Catalytic assay 611
    9.4.3. Phospholipase D 612
        9.4.3.1. In vitro assay 619
        9.4.3.2. In vivo assay 624
9.5. PtdIns(3,4,5)P_3 activated phospholipases 626
    9.5.1. Phospholipase A_2 626
    9.5.2. Phospholipase C 627
    9.5.3. Phospholipase D 629

Chapter 10.   Preparation of Standards 633

10.1. Introduction 633
10.2. Phosphatidylinositol phosphates 634
    10.2.1. PtdIns monophosphates 635
        10.2.1.1. PtdIns(4)P 635
        10.2.1.2. PtdIns(3)P 639
        10.2.1.3. PtdIns(5)P 644
    10.2.2. PtdIns bis-phosphates 645
        10.2.2.1. PtdIns(4,5)P_2 645
        10.2.2.2. PtdIns(3,5)P_2 648

    10.2.2.3. PtdIns(3,4)P_2 649
  10.2.3. PtdIns trisphosphates 651
    10.2.3.1. PtdIns(3,4,5)P_3 651
10.3. Inositol phosphates 653
  10.3.1. Inositol monophosphates (InsP) 654
    10.3.1.1. Ins(1)P 654
    10.3.1.2. Ins(2)P 654
    10.3.1.3. Ins(3)P 655
    10.3.1.4. Ins(4)P 655
    10.3.1.5. Ins(5)P 655
  10.3.2. Ins bisphosphates 655
    10.3.2.1. Ins(1,2)P_2 656
    10.3.2.2. Ins(1,3)P_2 and Ins(3,4)P_2 656
    10.3.2.3. Ins(1,4)P_2 658
    10.3.2.4. Ins(1,5)P_2 659
    10.3.2.5. Ins(3,5)P_2 659
    10.3.2.6. Ins(4,5)-P_2 659
  10.3.3. InsP_3 660
    10.3.3.1. Ins(1,3,4)P_3 660
    10.3.3.2. Ins(1,3,5)P_3 662
    10.3.3.3. Ins(1,3,6)P_3 662
    10.3.3.4. Ins(1,4,5)P_3 662
    10.3.3.5. Ins(1,4,6)P_3 666
    10.3.3.6. Ins(3,4,5)P_3 666
    10.3.3.7. Ins(3,4,6)P_3 666
    10.3.3.8. Ins(4,5,6)P_3 667
    10.3.3.9. Ins(1,2cyc4,5)P_3 667
    10.3.3.10.. Ins(1,2,4)P_3 667
  10.3.4. InsP_4 668
    10.3.4.1. D/L [^{3}H]-or [^{32}P]Ins(1,2,3,4)P_4 668
    10.3.4.2. Ins(1,2,4,5)P_4 669
    10.3.4.3. Ins(1,3,4,5)P_4 669
    10.3.4.4. Ins(1,3,4,6)P_4 671
    10.3.4.5. Ins(1,4,5,6)P_4 673
    10.3.4.6. Ins(3,4,5,6)P_4 674
  10.3.5. InsP_5s 676
    10.3.5.1. Ins(1,2,3,4,5)P_5 and Ins(1,2,4,5,6)P_5 679
    10.3.5.2. D-Ins(1,2,3,5,6)P_5 680
    10.3.5.3. D-Ins(1,2,4,5,6)P_5 681
    10.3.5.4. Ins(1,3,4,5,6)P_5 681
    10.3.5.5. Ins(1,2,3,4,6)P_5 682

10.3.6. $InsP_6$ 682
      10.3.6.1. $InsP_6$ 682
      10.3.6.2. $[^3H]InsP_6$ and $[^{32}P]InsP_6$ 683
10.4. Inositol pyrophosphates 683
    10.4.1. $PP\text{-}InsP_5$ 684
    10.4.2. $PP\text{-}[^3H]InsP_4\text{-}PP$ 685
10.5. Structural analogues 685
    10.5.1. Deoxy-*myo*-InsPs 685
    10.5.2. Membrane permeators 686
    10.5.3. Fluorescent derivatives 687
    10.5.4. Affinity reagents 688

Chapter 11.   *InsPs and PtdInsPs as Signaling Molecules*   689

11.1. Introduction 689
11.2. InsPs as cellular signals 691
    11.2.1. Evidence for signaling 691
        11.2.1.1. Analytical 691
        11.2.1.2. Metabolic......................... 693
        11.2.1.3. Molecular 695
    11.2.2. InsP-protein interaction 696
        11.2.2.1. $Ins(1,4,5)P_3$ receptors 697
        11.2.2.2. $Ins(1,3,4,5)P_4$ receptors 699
        11.2.2.3. $Ins(3,4,5,6)P_4$ receptors 700
        11.2.2.4. $InsP_6$ receptors 702
    11.2.3. Regulation of InsP kinases 702
        11.2.3.1. $Ins(1,4,5)P_3/Ins(1,3,4,5)P_4$ cycle 703
        11.2.3.2. $Ins(1,3,4,5,6)P_5/Ins(3,4,5,6)P_4$ cycle 705
        11.2.3.3. $Ins(1,3,4,5,6)P_5/Ins(1,4,5,6)P_4$ cycle 707
        11.2.3.4. $InsP_6/InsP_5$ cycle 708
        11.2.3.5. Pyrophosphate cycle 710
    11.2.4. Specific biological effects 711
        11.2.4.1. Ion channel physiology 711
        11.2.4.2. Membrane dynamics 715
        11.2.4.3. Nuclear signaling 718
11.3. PtdInsPs as cellular signals 719
    11.3.1. Evidence for signaling 720
        11.3.1.1. Analytical 720
        11.3.1.2. Metabolic......................... 722
        11.3.1.3. Molecular 723

11.3.2.   PtdInsPs-protein interactions 725
          11.3.2.1.   PtdInsP₃ binding 726
          11.3.2.2.   PtdInsP₂ binding 729
          11.3.2.3.   PtdInsP₁ binding 732
11.3.3.   Regulation of PtdInsP kinases 734
          11.3.3.1.   PtdIns 3-kinases 735
          11.3.3.2.   PtdIns-4-kinases 737
          11.3.3.3.   PtdIns 5-kinases 738
11.3.4.   Specific biological effects 740
          11.3.4.1.   Membrane and vesicular trafficking 741
          11.3.4.2.   Cell growth and differentiation 747
          11.3.4.3.   Cytoskeletal organization and cell motility ... 752
          11.3.4.4.   Apoptosis 756
          11.3.4.5.   DNA synthesis 762

References 765

Index 951

Abbreviations

A23187, ionophore
ABA, (RS)-2-*cis*,4-*trans*-abscisic acid
AcChoE, acetylcholine esterase
AEBSF, 4-(2-aminoethyl)-benzene-
  sulfonyl fluoride
AHM, anhydromannitol
AKT/PKB, protein kinase B
1-AlkyllysoGroPIns, plasmanic acids
AM, heptakis(acetoxy-methyl) esters
AP, alkaline phosphatase
AP-2, clathrin adapter protein
APAM, *Aspergillus saito* α-manno-
  sidase
APCI, atmospheric pressure chemical
  ionization
AppA2, acid phosphatase gene
ARF, ADP ribosylation factor
β-ARK, β-adrenergic receptor kinase
ARNO, Arf nucleotide-binding site
  opener
ASAP-1, PH domain of centaurin β4
ATM, *Ataxia telangiectasia* mutated
ATR, Ataxia and Rad-related

BAD, Bcl-2/Bcl-X1-antagonist
BHT, butylated hydroxytoluene
BSA, bovine serum albumin
Btk, Bruton's tyrosine kinase

C1, protein kinase C homology-1
C2, protein kinase C homology-2
C2A/C2B-deletion mutant,
CABP, 4-carboxy-D-arabinitol-1,5-
  bisphosphate
CaCC, Ca^{2+} activated Cl^- conductance
CAD, collision associated dissociation
CALM-N, clathrin adaptor protein

CaM, calmodulin
CaMKII, Ca^{2+}/calmodiuml-dependent
  protein kinase II
CAP, Cbl-associated protein
Cbl, adaptor protein
CBAG, coffee bean alpha-galactosidase
CBR1, calcium binding regions-1
CBR2, calcium binding region-2
CD, circular dichroism
CD52-I, a GPtdIns anchored glyco-
  peptide
CD52-II, a GPtdIns anchored glyco-
  peptide
CDK1, cyclin dependent kinase
CDP, cytidine disphosphate
CDP-DAG, CDP-diacylglycerol
CDTA, *trans*-1,2-diaminocyclo
  hexane-N,N,N',N'-tetraacetic acid
CGS93453B, a calmodulin antagonist
CHAPS, 3-[3-cholamidopropyl)
  dimethylammonio]-1-propane-
  sulfonate
CHO, Chinese hamster ovary
Chx, cycloheximide
CID, collision induced dissociation
CI-MS, chemical ionization-mass
  spectromeery
CF-LSI-MS, continuous flow liquid
  secondary ion mass spectrometry
Ck, cyclin-dependent kinase
CL, cardiolipin
CRMV, collagenase-released matrix
  vesicle
CVVQ, putative C-terminal isopreny-
  lation site
Cyc, cyclic

DAF, decay accelerating factor
DAG, diacylglycerol
Dbi,
DEAE, diethylaminoethyl
DEP, dishevelled-glin-pleckstrin homology
DFP, diisopropylfluorophosphate
DG, diacylglycerol (see DAG)
DH, Dbi homology
DHB, 2,5-dihydroxybenzic acid
DHS, dihydrosphingosine.
DIPP, diphosphoinositol polyphosphate phosphohydrolase
hDIPP, human DIPP
DMEM, Delbecco's modified essential medium
DMF, N,N-dimethylformamide
DMG, 3,3-dimethylglutaric acid
DNPH, dintrophenylhydrazone.
DNPU, dinitrophenylurethane
Dol, dolichol
DPG, diphosphoglycerol
DPilcIp,diC7 GroPIns,
DRM, detergent-resistant membrane
DSP, dual-specificity phosphatase
DTT, dithiothreitol

E^{bo} AcChoE, bovine erythrocyte acetylcholinesterase
EDTA, ethylenediamine tetraacetic acid
E^{hu} AChE, see E^{hu} AChoE
E^{hu} AcChoE, human erythrocyte acetylcholinesterase
EEA1, early endosome antigen 1
EF, elongation factor
EGF, epidermal growth factor
EGTA, ethyleneglycol-bis-(β-amino-ethyl ether)N,N,N',N'-tetraacetic acid
ELISA, enzyme-linked immunosorbent assay

ENTH, epsin amino-terminal homology
ER, endoplasmic reticulum
ERK, extracellular signal-regulated kinase
ESI, electrospray ionization
EthN(Me)$_2$, N,N-dimethylethanol-amine
Etn, ethanolamine
ETNH, epsin amino-terminal homology
EtnP, ethanolamine phosphate

FAB, fast atom bombardment
FAB1, yeast homologue of PtdInsP 5-kinase gene
Fab1p, yeast PtdInsP 5-kinase homo-logue
FACS, fluorescence-activated cell sorter
FCS, fetal calf serum
$F_c\epsilon$ RI, high affinity Fc receptor for immunoglobulin E
FERM, four-point-one-ezrin-radixin-monesin
fMLP, formylmethionylleucylphenyla-lanine
FID, flame ionization detection
FITB, fill in the blank domain
FPLC, fast protein liquid chromato-graphy
FYVE, Fab1-YOTP-Vac 1-EEA1, PtdIns(3)P-binding module

G proteins, hetrotrimeric GTP-binding proteins
Gal, galactose,
Gal-Nac, N-acetylgalactosamine
GAP, GTPase-activating protein
GAP1^{IP4BP}, Ins(1,3,4,5)P$_4$-binding protein
GATA-1, megakaryocyte transcription factor

GEF, guanine nucleiotide exchange factor

GFP, green fluorescent protein

GlcN, glucosamine

GlcNAc, N-acetylglucosamine

Glc6Pase, glucose-6-phosphatase

GLC, gas liquid chromatography

GIPLs, glycoinositolphospholipids

GIPs, GAP domain-containing PtdInsP 5-phosphatases

GLUT4, glucose transporter 4

Glycolipid A, fatty acid-labeled P2

Glycolipid C, fatty acid labeled P3

Gpi1, a *S. cerevisiae* involved in first step of glycosylPtdIns biosynthesis

GPI, glycosylated PtdIns

GPI-PLD, glycosyl PtdIns specific PLD

GPMI-P_2, L-α-glycerophospho-D myo-Ins(4,5)P_2

GRB2, guanine nucleotide releasing binding protein

Gro, glycerol

GroPIns, glycerophosphoinositol

Grp1, general receptor for PtdInsPs-1, cytohesin-1

GST, glutathione S-transferase

GTPγS, guanidine trisphosphate γ-sulfate

GTP, guanine triphosphate

HD-motif, C-terminal His-Asp motif

HEES, N-2-hydroxyethylpiperazine N'-2-ethanesulfonic acid

HEPES, 4-(2-hydroxyethyl)-1-piperazine-ethanesulfonic acid

HF, hydrofluoric acid

HFB$_6$-Ins, perheptafluorobutyryl-inositol

HFBA, heptafluorobutyric anhydride

HiPER, histidine phosphatase of endoplasmic reticulum

HPLC, high performance liquid chromatography

HPTLC, high peformance thin-layer chromatography

Hrs, *Drosophila* protein

IDH, inositol dehydrogenase

IGF-1, insulin growth factor-1

IGF, insulin-like growth factor

Ins, inositol

InsP, inositol phosphatate

InsPs, inositol phosphates

InsP$_7$, PP-InsP$_5$

InsP$_8$, [PP]$_2$InsP$_4$

Ipmk, inositol polyphosphate multikinase

IPP5C, inositol 5-phosphatase catalytic domain

IRS-1, insulin receptor substrate 1,

ISC1, Ins phosphosphingolipid-PLC in yeast

JAKs, Janus protein kinases

JBAM, jack bean α-mannosidase

K1AA0371, human myotubularin related protein

LC, liquid chromatography

LPG, lipophosphoglycan

LPPGs, lipopeptidophosphoglycans

LPG, lipophosphoglycans

LSIMS, liquid secondary ion mass spectrometry

LUV, large unilamellar vesicles

LY294002, PtdIns 3-kinase inhibitor

MAG, monoacylglycerol

MALDI, matrix-assisted laser desorption ionization

Man, mannose

MAP, mitogen activated protein
MAPK, mitogen-activated protein kinase
MAPKK, MAP kinase kinase
MBP, myelin basic protein
MDD, metal-dye detector
MDH, malonyl dehydrogenase
MES, 2-(N-morpholino)ethanesulfonic
MIKES, mass analyzed ion kinetic spectra
Minpp1, ER based MIPP
MIPP, multiple inositol polyphosphate phosphatase
Mops, 3-[N-morpholino]propanesulfonic acid
MPLAc-1, membrane-bound PLC1
MS, mass spectrometry
MS/MS, mass spectrometry-mass spectrometry (tandem)
MSS4, yeast homologue of PtdIns(4)P 5-kinase gene
MTMR3, myotubularin-related protein 3
m/z, mass/charge

NANA, N-acetylneuraminic acid
NBD, N-4(nitrobenzo-2-oxa-1,3-diazole)
NEU, naphthylethylurethane
NICI, negative chemical ionization
NIH, Image software
NL, neutral lipid
NMR, nuclear magnetic resonance
NOE, nuclear Overhauser effect
NSF, N-ethylmaleimide-sensitive factor
OCRL, oculocerebrorenal syndrome protein
nOG, n-octylglucopyranoside

PAF, platelet activating factor
3-PAP, human 3-phosphatase adapter subunit

PAR, 4-(2-pyridylazo) resourcinol
PARP, procyclic acidic repetitive protein
PBS, phosphate-buffered saline
PCA, perchloric acid
PCI, positive chemical ionization
PDGF, platelet derived growth factor
PDK, PtdIns-dependent protein kinase
PDK1, D3-PtdInsP-dependent PK-1
PDZ, central organizer of protein complexes at plasma membrane
PEI, polyethyleneimine
PFP, pentafluoropropionic
PGE_1, prostaglandin E_1
PH, pleckstrin homology
Pharbin, proline-rich InsP phosphatase
PhyC, *Bacillus* phytase
PI, PtdIns
PI-PLC, PtdIns specific PLC
PIG-C, Takahashi et al, 1996
PIG-CP3-(4-azidoanilido)uridine 5'-trispVSG,
PIKfyve, mammalian PtdIns kinase, p235
PIPKINS, PtdInsP kinases
PIPs, proline-rich InsPs 5-phosphatase
PIPP, proline-rich Ins polyphosphate 5-phosphatase
PKA, protein kinase A
PKB, also known as Akt, protein kinase B
PKB/Akt, protein kinase B related to AKR mouse T cell lymphoma-derived oncogenic product
PKC, protein kinase C (Ca^{2+}-dependent)
PKI, protein kinase 1
PLA_1, phospholipase A_1
PLA_2, phospholipase A_2
$cPLA_2$, cytosolic PLA_2

iPLA$_2$, Ca^{2+}-independent PLA$_2$
sPLA$_2$, secretory PLA$_2$
srPLA$_2$, PLA$_2$, which interacts with
   v-Src oncoprotein
PLAP, placental alkaline phosphatase
PLB, phospholipase B
PLC, phospholipase C
PLC$_{BC}$, PLC from *Bacillus cereus*
PLD$_1$, phospholipase D$_1$
PMSF, phenylmethylsulfonyl floride
POH, Garcia et al, 1995, Lemon et al,
   1995
PP-InsP$_4$, diphosphoinositol tetra-
   kisphosphate
PP-InsP$_5$, diphosphoinositol penta-
   kisphosphate
[PP]$_2$-InsP$_4$, bis-diphosphoinositol
   tetrakisphosphate
PP1, protein phosphatase-1
PPBM, putative PtdIns(3,4,5)P$_3$
   binding motif
PtdBt, phosphatidylbutanol
PtdCho, phosphatidylcholine
PtdDB, phosphatidyldimethylbutanol
PtdEt, phosphatidylethanol
PtdEtn, phosphatidylethanolamine
PtdIns, phosphatidylinositol
PtdInsPs, phosphatidylinositol phos-
   phates
PtdInsPK1, PtdInsP kinase type 1
PtdSer, phosphatidylserine
PTEN, phosphates and tensin
   homologue delayed on
   chromosome 10
PTEN/MMAC1, phosphatase and
   tensin homologue or mutated in
   multiple advanced cancers
PTK, protein-tyrosine kinase
PTP, protein tyrosine phosphatase
PX, phox homology

RA, Ras associating
Rac, primary regulator of actin
   remodeling
Radyl, any hydrocarbon sidechain
Rap2B, Ras-related GTPase
Ras, low-molecular mass G-protein
Rho, GTP binding protein
RSALP, retrieval sequence-alkliane
   phosphatse
Rsd1, yeast protein Sac1

SAC, protein domain of phosphatases,
Sac1p, suppressor of actin mutations in
   yeast, phosphatase domain
SAX, strong anion exchange resin
Sbf1, myotubularin-related protein
SCIPs, Sac1p-containing InsP 5-phos-
   phatases,
scIpk1, yeast InsP$_5$ kinase-1
SDS, sodium dodecyl sulfate
SDS-PAGE, SDS-polyacrylamide gel
   electrophoresis
SET, binding domain of myotubularin-
   related proteins
SH2, sarc homology 2
Shc, an SH-2-containing adopter
   protein
SHIP, SH2-domain containing InsP
   5-phosphatase
SHIP1, leukocyte specific InsP
   5-phosphatase
SIM, single ion monitoring
SIP-110, InsPs 5-phosphatase
SKIP, skeletal muscle and kidney
   enriched inositolphosphatase
Slc1p, 1-acyl-*sn*-Gro(3)P acyltrans-
   ferase
SM, sphingomyelin
SNAP, soluble NSF attachment protein
SNARE, soluble NSF (N-ethylmalea-
   mide-sensitive factor) receptor

SOC, store operated channel
SoP, *Salmonella* virulence protein,
SopB, InsPs 3-phosphatase
SPH, sphingosine
Src, a domain of approximately 100 amino acid residues located N terminal to tyrosine kinase
STA2, 9,11-Epithio-11,12-methano-thromboxane A_2
STT4, staurosporine- and temperature-sensitive gene product homologous to PtdIns 4-kinase
SUV, small unilamellar vesicles

TAG, triacylglycerol
TBAHS, tetrabutylammonium hydrogen sulfate
TBDMS, *tert*-butyldimethylsilyl
TCA, trichloroacetic acid
TEAB, triethylammonium bicarbonate
TID, 3-(trifluoromethyl)-3-(m-iodo-phenyl)diazirine)
TLC, thin layer chromatography
TMCS, trimethylchlorosilane
TMS, trimethylsilyl
TM4SF, transmembrane 4 superfamily
TNE, mixed buffer containing Tris, NaCl, EDTA and EGTA
TNF, tumor necrosis factor
TNM, mixed buffer containing Tris, NaCl and MagCl$_2$
TOF, time of flight
TOF-MS, time of flight mass spectrometry

TPIP, TPTE and PTEN homologous inositol lipid phosphatase
TPIPα, human homologue of PTEN

TPTE, transmembrane phosphatase with tensin,
Tris, trishydroxy-methyl)amino-methane.
TRP, transient receptor potential
TSI, thermospray ionization
TT4, PtdIns 4-kinase from yeast
TUBBY, mouse mutant with maturity onset obesity
TULP, tubby-like protein
TXA2, thromboxane A_2

UDP, uridine diphosphate
UV, ultraviolet absorption

VAMP, vesicle-associated membrane protein
VHS, domain containing Vps27p, Hrs and STAM
Vitride, sodium bis(2-ethoxy-ethoxy) aluminium hydride,
Vps27p, vacuolar protein sorting mutant
Vps34p, vacuolar protein sorting mutant
VSG, variant surface glycoprotein

WASP, Wiskott-Aldrich syndrome protein
N-Wasp, neural WASP
WAX, weak anion exchange

X-Y domains, catalytic regions of PLCδ 1

YW3548, inhibitor of GPtdIns biosynthesis

CHAPTER 1

General introduction

1.1. Introduction

The early work on the inositol containing phospholipids is described in detail by Hanahan (1960) and Hawthorne (1960, 1982). Phosphatidylinositol (PtdIns) was first isolated by Faure and Morolec-Coulon from wheat germ and heart muscle (1953–1954). By analysing PtdIns and its hydrolysis products, it was shown that it contained fatty acids, glycerol, *myo*-inositol and phosphoric acid in molar proportions of 2:1:1:1. The work on the phosphatidylinositol phosphates (PtdInsPs) first centered on the brain phospholipid preparation described by Folch (1949) as PtdInsP, which was later found to include PtdIns and PtdIns(4,5)P$_2$ (see Brockerhoff and Ballou, 1961). The chemical structures of PtdIns, PdIns(4)P, and PtdIns(4,5)P$_2$ were determined by Ballou and colleagues between 1959 and 1961. Dawson and Dittmer (1961) succeeded in isolating from the Folch fraction pure PtdInsP$_2$, which Brown and Stewart (1966) characterized as the 1,2-diacyl-*sn*-glycero-3-phosphoryl-1-*myo*-inositol-4,5-bis-phosphate (Fig. 1.1). The structures of PtdIns, PtdInsP and PtdInsP$_2$ have since been confirmed by modern analytical methods in many laboratories. Subsequent to the discovery of 3-kinase, it was shown that it synthesized PtdIns(3)P (Whitman et al., 1988) and PtdIns(3,4,5)P$_3$ (Traynor-Kaplan et al., 1988).

The existence of inositol phosphates (InsPs) has been known for over 80 years (Posternak, 1919, 1921), although the

1

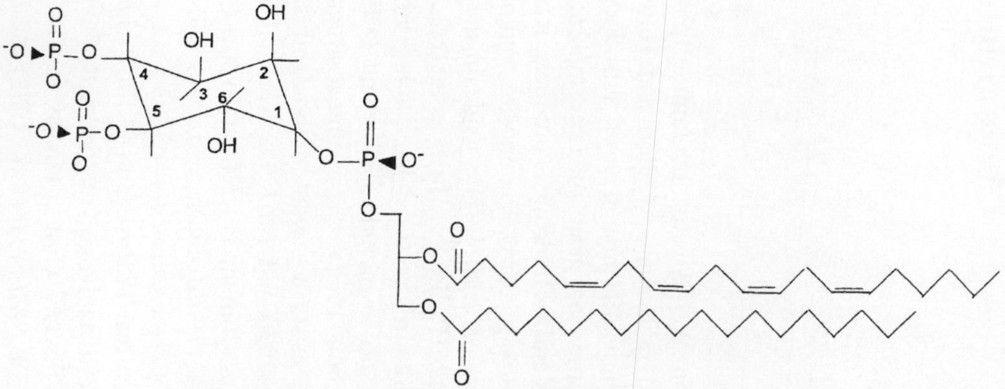

Fig. 1.1. Structure of PtdIns(4,5)P_2 including numbering of inositol ring reproduced from Tolias and Carpenter (2000) with permission of the publisher.

existence in seeds of a highly phosphorylated substance later identified as *myo*-inositol hexaphosphate or phytic acid had been isolated much earlier (Pfeiffer, 1872). The InsPs were subsequently shown to arise as partial hydrolysis products of phytic acid and of PtdInsPs, and as products of phosphorylation of InsP (Chapter 5).

During the 1950s it was observed that stimulation of several tissues with cholinergic antagonists promoted the hydrolysis of PtdIns(4,5)P_2 to Ins(1,4,5)P_3 and diacylglycerol by phospholipase C, as well as the selective incorporation of $[^{32}P]_i$ into PtdIns (Hokin and Hokin, 1953; Hokin, 1985, 1996). It was later observed that cholinergic agonists, which stimulated PtdIns turnover also activated calcium dependent processes inside the cell (Michell, 1975). During the 1980s, attention became focused on InsPs after the discovery that Ins(1,4,5)P_3 is a Ca^{2+} mobilizing second messenger (Streb et al., 1983; Berridge and Irvine, 1989). It is now known that the metabolism of PtdIns and InsPs is much more complex and that many more PtdInsP derivatives are involved in cellular signaling and metabolic regulation (Berridge, 1993; Fruman et al., 1998; Vanhaesebroeck et al., 1997, 2001). The variety of InsPs now includes (the pyrophosphate derivatives, which are widely distributed. Three types of diphosphorylated InsPs are presently known (Safrany et al., 1999): diphosphoinositol tetrakisphosphate (PP-InsP$_4$), diphosphoinositol pentakisphosphate (PP-InsP$_5$, also known as InsP$_7$) and bis-diphosphoinositol tetrakisphosphate ([PP]$_2$-InsP$_4$ also known as 'InsP$_8$') (Fig. 1.2).

In addition, glycosyl PtdIns have been discovered covalently linked to eukaryotic cell surface glycoproteins (Englund, 1993; Medof et al., 1996). These structures constitute a diverse family of lipid molecules also found in prokaryotes. The members of this family are characterized by the sequence EtN-PO$_4$-Manα1-44GlcNα1-6-*myo*-inositol-1-PO$_4$-diradylglycerol (Fig. 1.3). The *myo*-inositol residue in some instances may be palmitoylated (Roberts et al., 1988). Ceramide-based lipids, however, have been

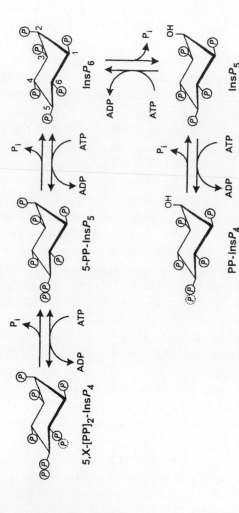

Fig. 1.2. Structure of diphosphatidylinositol polyphosphates in mammalian cells. The location of the diphosphate group of mammalian PP-InsP₅ at the 5-carbon is based on the work of Albert et al. (1997). A broken outline enclosing the symbol for a phosphate group (P) indicates a tentative location on the inositol ring. The suggestion that the InsP₆ kinase also phosphorylates InsP₅ is based on the work of Menniti et al. (1993). Reproduced from Safrany et al. (1999) with permission of the publisher.

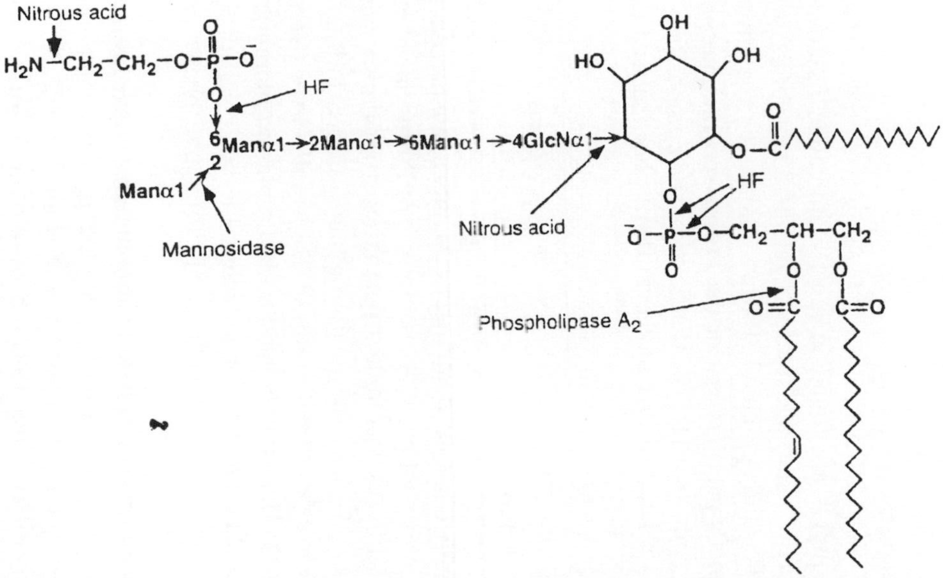

Fig. 1.3. Structure of glycosyl PtdIns from *Plasmodium falciparum* showing the enzymatic and chemical cleavage sites for preparation of glycosyl PtdIns components. Treatment of glycosyl PtdIns with α-mannosidase gives EtN-P-Man₃-GlcN-acyl-InsP-diacylglycerol. Treatment with phospholipase A₂ gives EtN-P-Man₄-GlcN-acyl-InsP-monoacylglycerol. Treatment with hydrofluoric acid (HF) gives Man₄-GlcN-acyl-InsP, Man₄-GlcN-acyl-Ins, and diacylglycerol, while treatment with nitrous acid forms OHC–CH₂–P-Man₄-AHM and acyl-InsP-diacylglycerol. Reproduced from Vijaykumar et al. (2001) with permission of the publisher.

found in yeast (Conzelman et al., 1992; Fankhauser et al., 1993) and *Dictyostelium discoideum* (Stadler et al., 1989; Haynes et al., 1993).

1.2. Natural occurrence

Inositol phospholipids are found in most mammalian tissues, as well as in a variety of plants and microorganisms. They comprise of PtdIns, PtdInsPs and glycosyl PtdIns anchors. The tissue distribution of InsPs parallels that of the PtdIns-specific phospholipases C (PLC) and D (PLD). The distribution and composition of the InsPs is also related to that of the PtdInsP phosphatases and kinases.

1.2.1. Phosphatidylinositol

PtdIns is universally distributed but makes up only a few percent of the total cellular phospholipid. Early estimates of PtdIns have been frequently inflated by its incomplete resolution from PtdSer with which it overlaps in many chromatographic systems. PtdIns is present in rat liver at a concentration of 1.7 μmol/g liver (Creba et al., 1983). The PtdIns is associated with the plasma membrane and usually significant amounts of it occupy the inner half of the phospholipid bilayer. It is the precursor of PtdIns phosphates and glycosyl PtdIns lipid anchors.

The studies of PtdIns metabolism in animal tissues have become further complicated by the discovery of the presence of *chiro*-inositol in the glycerophospholipids isolated from H35 cells and rat liver membranes (Mato et al., 1987), bovine liver (Larner et al., 1988; Pak and Larner, 1992), and diabetic urine (Ostlund et al., 1993), and the presence of *scyllo*-inositol, in addition to the commonly occurring *myo*-inositol, in plant PtdIns (Kinnard et al., 1995).

1.2.2. Phosphatidylinositol phosphates

An important feature of natural PtdIns is its occurrence in various phosphorylated forms. The $PtdInsP_2$ and $PtdInsP_3$, which were originally thought to occur only in the brain and nervous tissues, are now known to occur in small amounts (1–3% of PtdIns) in other tissues where they are involved in cell to cell signaling pathways via a variety of PtdIns kinases (Fruman et al., 1998). Usually significant amounts of the PtdInsPs occupy the outer half of the plasma membrane (Gascard et al., 1991). In vivo, PtdIns phosphates are interrelated by a complex web of reactions (Majerus et al., 1988), and a rapid dynamic equilibrium exists among them. To date, the following PdInsPs have been identified in cells: $PtdIns(3)P$, $PtdIns(4)P$, $PtdIns(5)P$, $PtdIns(3,4)P_2$, $PtdIns(3,5)P_2$, $PtdIns(4,5)P_2$, and $PtdIns(3,4,5)P_3$.

The PtdIns kinases are defined as phosphokinases that transfer a phosphate group from ATP or other high energy phosphate to a hydroxyl group in PtdIns. The PtdInsP phosphatases remove a phosphate group from an appropriate substrate by transferring it to water. The tissue distribution of the kinases, phosphatases and phospholipases that are specifically involved in inositol phosphatide metabolism and cellular signaling are closely linked to the distribution of PtdIns and PtdInsPs. The concerted actions of a series of phosphatases and kinases convert $Ins(1,4,5)P_3$ to a number of other InsPs. At least some of the intermediates produced have their own physiological activities.

1.2.3. Inositol phosphates

The natural occurrence of InsPs follows that of PtdInsPs and phytic acid and is closely associated with the tissue distribution of PtdIns-specific phospholipases C and D, although a clear distinction between PtdIns-specific and non-specific phospholipases is not always possible. Numerous InsP phosphatases have been discovered and characterized in recent years. In many cases, the same enzymes hydrolyze phosphate from both water-soluble InsPs and the corresponding lipid-soluble PtdInsPs with the same arrangement of phosphate groups (Majerus et al., 1999).

Phytases catalyze the hydrolysis of phytate (*myo*-inositol hexakisphosphate), thereby releasing various InsPs and inorganic phosphate. In extracts of intact cells, it is not always possible to be certain about the origin of a specific InsP.

In addition to the simple InsPs, diphosphoinositol polyphosphates (pyrophosphates) are widely distributed. They have been identified in mammals, yeasts, slime molds, plants and free living amoebae (for a minireview, see Safrany et al., 1999). The pyrophosphates of $InsP_6$ have been claimed to represent the final frontier of InsP research and have led to a renewed interest in the water-soluble derivatives of InsPs (Irvine and Schell, 2001), although their function remains obscure.

1.2.4. Glycosyl phosphatidylinositols

Glycosyl PtdIns constitute a diverse family of lipid molecules found in prokaryotes and eukaryotes. The members of this family are characterized by the sequence $EtN-PO_4-Man\alpha1-44GlcN\alpha1-6$-*myo*-inositol-1-$PO_4$-diradylglycerol. These structures were first discovered covalently linked to eukaryotic cell surface glycoproteins and recognized to be an important alternative mechanism for anchoring proteins to cell membrane. Over 100 glycosyl PtdIns-anchored proteins have been described to date (Englund, 1993; Medof et al., 1996).

In glycosyl PtdIns anchors, the lipid moiety released by PtdIns-specific phospholipase C is alkyl/acylglycerol or diacylglycerol. Ceramide-based lipids, however, have been found in yeast (Conzelman et al., 1992; Fankhauser et al., 1993) and *Dictyostelium discoideum* (Stadler et al., 1989; Haynes et al., 1993).

In mammalian cells and yeast, many glycosyl PtdIns-anchored molecules associate with sphingolipid and cholesterol-rich detergent-resistant membranes, known as rafts (Hooper, 1999). Denny et al. (2001) have recently shown that glycosyl PtdIns-anchored parasite macromolecules are present in *Leishmania* and *Trypanosoma brucei* and have suggested that they occur throughout the *Kinetoplastida*. These lipid rafts are enriched in inositol phosphosphingolipid and sterol, in addition to glycosyl PtdIns-anchored glycoconjugate and glycoprotein.

1.3. Biological significance

The discovery that the inositol pospholipids of plasma membrane mediate cellular responses to external signals has led to tremendous interest in the structure and metabolism of PtdInsPs and InsPs. However, PtdIns and PtdInsPs also provide structural components to cell membranes and membrane receptors. Glycosyl PtdIns serve as lipid anchors of membrane proteins.

1.3.1. Structural components

PtdIns provides a negatively charged building block to biological membranes where it usually occupies the inner leaflet of the plasma membrane. The diacylglycerol moiety of PtdIns contains a major species with stearic acid in the *sn*-1-position and arachidonate in the *sn*-2-position. PtdIns serves as a precursor of the glycosylated PtdIns glycans known as the lipid anchors of many membrane associated proteins (Englund, 1993; Medof et al., 1996). Among the glycosyl PtdIns anchor structures, the glycosylinositolphosphoceramides must also be included, where the alkyl/acyl or diacylglycerol moiety has been replaced by a ceramide (Azzouz et al., 1998). These features bestow special physicochemical and biochemical properties to the glycosylated inositol phospholipids.

Azzouz et al. (1995) have isolated from the free living protozoan *Paramaecium primaurelia*, a glycosylinositolphosphoceramide, in which the core glycan has been modified by mannosyl phosphate.

Reggiori et al. (1997) have shown that in yeast, the glycosyl PtdIns anchors contain three types of ceramides which were identified as dihydrosphingosine-C_{26}, phytosphingosine-C_{26} and phytosphingosine-C_{26}-OH. The ceramides were introduced by lipid exchange, which occurred in both in the ER and to a lesser extent in the Golgi.

More recently, Loureiro y Penha et al. (2001) have characterized novel structures of mannosylinositolphosphoryl-ceramides from the yeast forms of *Sporotrix schenkii*. The lipid portion was identified as a ceramide composed of C-18 phytosphingosine N-acylated by either 2-hydroxylignoceric (80%), lignoceric (15%) or 2,3-dihydroxylignoceric (5%) acids. The ceramide was linked through a phosphodiester to *myo*-inositol which is substituted on position O-6 by an oligomannose chain. Novel glycosyl inositol-phosphosphingolipids have been isolated from the edible mushrooms, the basidiomycetes *Agaricus* by Jennemann et al. (1999). These lipids were found to consist of a novel carbohydrate-homologous series of four glyco-inositol-phospho-sphingolipids, designated basidiolipids: Manpβ1-2inositol1-phospho-ceramide, Galpα-6[Fucpα-2]Galpβ-6Manβ-2-inositol1-phospho-ceramide, Galpα-6Galpα-6[Fucpα-2]Galpβ-6Manβ-2inositol1-phospho-ceramide, and Galpα-6Galpα-6[Fucpα-2]Galpβ-6Manpβ-2inositol1-phospho-ceramide. All four glycolipids contained a ceramide which was composed of phytosphingosine and predominantly α-hydroxy-behenic and α-hydroxy-lignoceric acid.

1.3.2. Signaling molecules

The observation that the inositol phospholipids of plasma membrane moderate cellular response to external signals has led to tremendous interest in the structure, metabolism and biological function of PtdInsPs (Berridge, 1993).

A sequential phosphorylation of a PtdIns to PtdIns(4)P and then to PtdIns(4,5)P_2 yields a substrate for PtdIns-specific PLC, which is under the control of cell surface receptors via the G-protein coupled receptor system. Activated PLC cleaves PtdInsP$_2$ to give a water-soluble Ins(1,4,5)P_3 and a lipid-soluble sn-1,2-diacylglycerol (DAG), both of which act as second messengers inside the target cell, Ins(1,4,5)P_3 acting to increase calcium concentration, while DAG activates protein kinase C, promoting the phosphorylation of target proteins in the cell (Nishizuka, 1984, 1995). PtdInsP$_2$ thus serves as a reservoir of sn-1,2-diacylgycerols, the second messengers and activators of protein kinase C (Singer et al., 1997). Since the brain PtdInsPs are rich in 18:0/20:4 species (Holub et al., 1970), the DAG can be further metabolized to produce prostaglandins.

Following the identification of Ins(1,4,5)P_3 as a second messenger, many other InsPs were found in rapid succession, which was accompanied by an active search of their synthetic pathways and possible function. With an explosion in the interest of PtdInsPs, the studies of InsPs appeared to lag until the role of InsPs became discovered in such cell functions as the ion transport, membrane dynamics and nuclear signaling. The Ins(1,4,5)P_3 calcium mobilizing signal was found to be terminated by the metabolism of Ins(1,4,5)P_3 via specific kinases and phosphatases. While the action of 5-phosphatase lead to Ins(1,4)P_2, which is inactive, the phosphorylation by 3-kinase resulted in the formation of Ins(1,3,4,5)P_4, which shows a variety of effects in stimulated cells. Neurotransmitter functions for Ins(1,3,4,5,6)P_5 and InsP$_6$ (Vallejo et al., 1988) have been proposed. The ability of Ins(1,3,4,5,6)P_5 to modulate the affinity of hemoglobin for oxygen is widely accepted (Isaacs and Harkness, 1980). The possibility that other inositol phosphates found in cells may perform cellular functions has heightened interest in the structure and metabolism of inositol phosphates (Menniti et al., 1993; Irvine and Schell, 2001).

The PtdIns signaling pathway employs a host of kinases and phosphatases that form and degrade many signaling molecules that are involved in this system. The system is present in all eukaryotic cells in one form or another. Numerous kinases (Fruman et al., 1998) that phosphorylate PtdIns at specific positions of the inositol ring, phosphatases (Norris and Majerus, 1994; Woscholski et al., 1995; Munday et al., 1999), which remove phosphates from specific positions of the inositol ring, as well as specific phospholipases C and D (Singer et al., 1997) have been recognized and their substrate requirements established. The realization that the metabolic transformations of the PtdInsPs constitute the basis of cellular signaling (Rameh and Cantley, 1999) has been responsible for the great current interest in the inositol phospholipids.

PtdIns 3-kinase is a ubiquitous enzyme that has been shown to be an important mediator of intracellular signaling in mammalian cells. To date, the expanding family of mammalian PtdIns 3-kinase consists of three members, each containing a different p110 catalytic subunit (Ptasznik et al., 1997). Upon activation, PtdIns 3-kinase phosphorylates PtdIns and PtdInsPs at the D-3 position of the inositol ring to generate lipid messengers, such as PtdIns(3)P, PtdIns $(3,4)P_2$ and PtdIns$(3,4,5)P_3$.

A number of recent reviews (Rincon and Boss, 1990; Cote and Crain, 1993; Hetherington and Drobak, 1992; Drobak, 1992, 1993) and other chapters in this book contain information of the structure and metabolism of inositol phospholipids in animal cells. The known cellular functions of the individual inositol phospholipids have been repeatedly reviewed in detail (Divecha and Irvine, 1995; Rittenhouse, 1996; Toker and Cantley, 1997; Vanhaesebroeck et al., 1997; Vanhaesebroeck et al., 2001). A full understanding of cellular regulation of PtdInsPs and InsPs metabolism will require the powerful techniques of modern molecular biology, which is able to focus on individual enzymes and regulatory proteins, leading to a clarification of previously suggested functions and the discovery of new ones.

In addition to its role in anchoring membrane proteins to the plasma membrane, the glycosyl PtdIns structures appear to participate in signal transduction, and in cell targeting (Field and Menon, 1993; 2000). The possibility that the lipid anchors also have roles in the regulation of the immune response is also possible as their degradation mediated by a specific phospholipase C generates the putative second messengers: alkyl/acyl and diacylglycerols (Liscovitch and Cantley, 1994) and ceramides (Prieschl and Baumruker, 2000).

1.3.3. Metabolic intermediates

Not all of the large number of inositol glycerophospholipids and inositol phosphates serve as specific signaling molecules. Many of them clearly are intermediates and precursors of the signaling molecules as well as inactivation products.

There is now overwhelming evidence that a major function of the inositol phospholipids in cells is to store the precursors of second messengers that activate specific processes in the cell. PtdIns is a precursor of the PtdInsPs, and indirectly of the InsPs and the eicosanoids, which were the first phospholipid derived second messengers identified. They have since been joined by diacylglycerols, ceramides, sphingosine, phosphatidic acid, and inositol phosphates. Except for ceramides and sphingosine, PtdIns and PtdInsPs can contribute to the production of all of the above cellular messengers. The PtdIns serves as a reservoir of sn-1,2-diacylglycerols, the secondary messengers and activators of protein kinase C (Singer et al., 1997), as well as of arachidonate, the precursor of the prostaglandins.

The further metabolism of Ins(1,4,5)P_3 derived from PtdInsP$_2$ by hydrolysis with PtdIns-specific PLC, gives rise to three isomeric inositol monophosphates, Ins(4)P, Ins(1)P and Ins(3)P. A single phosphatase is responsible for the hydrolysis of all these compounds to free inositol, which is then used for resynthesis of PtdInsP$_2$.

PtdIns serves as a precursor of the glycosylated PtdIns glycans known as the lipid anchors of many membrane associated proteins (Englund, 1993; Medof et al., 1996).

In addition to the true precursors and products of inositol glycerophospholipid biosynthesis, organic solvent extracts of tissues also include the lipid metabolites and autoxidation products of inositol glycerophospholipids.

The modern metabolic pathways of the PtdInsPs have been worked out largely on the basis of chromatographic analyses of radioactive substrates. Due to the small amount of precursors and products involved, mass analyses have seldom been performed. Therefore, the metabolic relationships among the acylglycerol moieties and the polyunsaturates remain yet to be elucidated.

InsPs are products of metabolism of PtdIns and PtdInsPs, and phytic acid. InsP and InsPs are precursors of inositol polyphosphates. Although InsPs are insoluble in organic solvents, they must be considered along with the lipid precursors when reviewing the structure and function of inositol glycerophospholipids. Furthermore, detailed understanding of the structure of the PtdInsPs requires both deacylation and deglyceration of the inositol phospholipids, which are chemical intermediates frequently employed for the purpose of structural identification and determination of metabolic activity.

1.4. Nomenclature

The term inositol phospholipids or inositol phosphatides is poorly defined and variously understood. The Nomenclature Commission of the International Union of Biochemistry (IUB) and the International Union of Pure and Applied Chemistry (IUPAC) has issued Tentative Rules for the naming of fatty acid esters and other lipids (Anonymous, 1977). The observance of the rules put forward for systematic naming of glycerophospholipids by the IUB, can be of great help for designating inositol phospholipids, but this nomenclature is not rigorously followed by the authors or enforced by the editors of the scientific journals. The majority of publications in scientific journals utilizes ambiguous abbreviations for glycerophospholipids and related compounds. A common error is committed by a reference to diacyl GroPIns a diacylPtdIns inadvertently implying the presence of four acyl groups in the PtdIns molecule.

Furthermore, the numbering of inositol derivatives has led to confusion when following a given inositol derivative through a metabolic pathway. This has led to a relaxation of the lowest-locant rule in a new IUPAC proposal (Anonymous, 1989), which suggests that since all natural occurring inositol phosphates are of the D-configuration, they may all be numbered as the D-inositol derivatives, rather than by strict chemical sequence rules.

1.4.1. Inositol glycerophospholipids

Since inositol glycerophospholipids are glyccrolipids, the rules of naming glycerolipids apply.

1.4.1.1. Lipid components

According to the Nomenclature Commission (Anonymous, 1977) the use of the old terms monoglyceride and diglyceride as they apply to the neutral lipids is discouraged in favor of monoacylglycerol and diacylglycerol, respectively. Likewise, the use of the old terms phosphatide, phosphoglyceride and phosphoinositide are no longer recommended because they do not convey the intended meaning. Especially confusing is the use of the term "phosphoinositide" to include both inositol phospholipids and inositol phosphates. The acylglycerol system conveys the intended meaning more accurately. A glycerophospholipid is any derivative of glycerophosphoric acid that contains at least one O-acyl, O-alkyl or O-alk-1-enyl group attached to the glycerol residue. A glycerophosphoric acid containing two fatty acids attached to the glycerol residue is known as phosphatidic acid. All common glyccrophospholipids are named as derivatives of phosphatidic acid, e.g. 3-*sn*-phosphatidylcholine. The systematic name is 1,2-diacyl-*sn*-glycero-3-phosphocholine. Phosphatidic acid esters of inositol are therefore phosphatidylinositols, or 1,2-diacyl-*sn*-glycero-3-phosphoinositols.

The structural features of fatty acid components of fats are described by the classical alkanoic acid system in which the location of the double bond is indicated by the Δ-position, which is numbered from the carboxyl end of the fatty chain, and the configuration of the double bond indicated by t or c, for *trans* and *cis*, respectively, e.g. oleic acid, Δ^{9c}-octadecenoic and elaidic acid, Δ^{9t}-octadecenoic acid (Anonymous, 1977). A common designation for fatty acids involves the indication of the acyl carbon number along with the number of double bonds and, in case of the unsaturated acids, the distance of the last double bond from the methyl terminal (Farquhar et al., 1959). Thus, palmitic acid is designated as 16:0, oleic as 18:1n-9, linoleic as 18:2n-6 and arachidonic as 20:4n-6. Alternatively, the symbol ω is used in place of n, e.g. arachidonic acid as 20:4ω6. A complication arises from the presence of *trans*- and *cis*-unsaturated as well as cyclic-fatty acids.

Table 1.1 gives the numerical symbols, partial structures, systematic and trivial names of fatty acids along with the three letter symbols proposed by the Nomenclature Commission (Anonymous, 1977) for the more common natural fatty acids. The three letter abbreviations for fatty acids proposed by the Commission are not used very often. The number of known fatty acids exceeds 1000 with the PtdIns containing less than 50 different acids (Kuksis and Myher, 1990).

1.4.1.2. Stereoisomers of glycerolipids

Due to the pro-chiral nature of carbon 2, glycerolipids can occur in two enantiomeric forms. The stereospecific nomenclature of the

TABLE 1.1

Names and symbols for common long-chain fatty acids

Numerical symbol	Structure $H_3C–(R)–CO_2H$	Systematic name[a]	Trivial name[b]	Letter symbol
10:0	$–(CH_2)_8–$	Decanoic	Capric[c]	Dec
12:0	$–(CH_2)_{10}–$	Dodecanoic	Lauric	Lau
14:0	$–(CH_2)_{12}–$	Tetradecanoic	Myristic	Myr
16:0	$–(CH_2)_{14}–$	Hexadecanoic	Palmitic	Pam
16:1	$–(CH_2)_5CH=CH(CH_2)_7–$	9-Hexadecenoic	Palmitoleic	ΔPam
18:0	$–(CH_2)_{16}–$	Octadecanoic	Stearic	Ste
18:1(9)	$–(CH_2)_7CH=CH(CH_2)_7–$	Cis-9-octadecenoic	Oleic	Ole
18:1(11)	$–(CH_2)_5CH=CH(CH_2)_9–$	11-Octadecenoic	Vaccenic	Vac
18:2(9,12)	$–(CH_2)_3(CH_2CH=CH)_2(CH_2)_7–$	Cis, cis-9,12-octadecadienoic	Linoleic	Lin
18:3(9,12,15)	$–(CH_2CH=CH)_3(CH_2)_7–$	9,12,15-Octadecatrienoic	(9,12,15)-Linolenic	αLnn
18:3(6,9,12)	$–(CH_2)_3(CH_2CH=CH)_3(CH_2)_4–$	6,9,12-Octadecatrienoic	(6,9,12)-Linolenic	γLnn
18:3(6,11,13)	$–(CH_2)_3(CH=CH)_3(CH_2)_7–$	9,11,13-Octadecatrienoic	Eleostearic	eSte
20:0	$–(CH_2)_{18}–$	Icosanoic	Arachidic	Ach
20:2(8,11)	$–(CH_2)_6(CH_2CH=CH)_2(CH_2)_6–$	8,11-Icosadienoic[d]		Δ₂Ach
20:3(5,8,11)	$–(CH_2)_6(CH_2CH=CH)_3(CH_2)_3–$	5,8,11-Icosatrienoic[d]		Δ₃Ach
20:4(5,8,11,14)	$–(CH_2)_3(CH_2CH=CH)_4(CH_2)_3–$	5,8,11,14-Icosatetraenoic[d]	Arachidonic	Δ₄Ach
22:0	$–(CH_2)_{20}–$	Docosanoic	Behenic	Beh
24:0	$–(CH_2)_{22}–$	Tetracosanoic	Lignoceric	Lig
24:1	$–(CH_2)_7CH=CH(CH_2)_{13}–$	Cis-15-tetracosenoic	Nervonic	Ner
26:0	$–(CH_2)_{24}–$	Hexacosanoic	Cerotic	Crt
28:0	$–(CH_2)_{26}–$	Octacosanoic	Montanic	Mon

Reproduced from Anonymous (1977) with permission of the publisher.

[a] Ending in '-ic acid,' 'ate,' 'yl', for acid, salt or ester, acyl radical, respectively.

[b] Ending in '-ic acid,' 'ate,' '-oyl' for acid, salt or ester, or acyl radical, respectively.

[c] Not recommended because of confusion with caproic (hexanoic) and caprylic (octanoic) acids. Decanoic is preferred.

[d] Formerly eicosa (changed by IUPAC Commission on Nomenclature of Organic Chemistry, 1975).

glycerophospholipids places phosphate at the *sn*-3-position, e.g. *sn*-glycero-3-phosphoinositol; *sn* designates stereospecific numbering. In order to simplify discussion and promote comprehension, abbreviations and shorthand designation are extensively employed in the lipid field. Table 1.2 lists the symbols of the

TABLE 1.2

Symbols recommended for various constituents of lipids

Name of constituent	Symbol[a]
Alkyl radicals[b]	R
Methyl, ethyl,...,decyl	Me, Et, Pr, Bu, Pe, Hx, Hp, Oc, Nn, Dec
Carboxylic acids[b]	Acyl, RCO–
Formyl, acetic, propionyl, butyryl, valeryl	Fo (or CHO), Ac, Pp, Bu, Vl
Hexanoyl, heptanoyl, octanoyl	Hxo, Hpo, Oco
Glycerol and its oxidation products[c]	
Glycerol, glyceraldehyde	Gro, Gra
Glycosyl	Ose
Glucose, galactose, fucose	Glc[d], Gal, Fuc
Gluconic acid, glucuronic acid	GlcA, GlcU[e]
Glucosamine[f], N-acetylglucosamine	GlcN, GlcNAc
Neuraminic, sialic, muramic acids	Neu, Sia, Mur
N-acetylneuraminic, N-glycolylneuraminic acids	NeuAc, NeuGc
Dexy	d
Phospholipid moieties	
Ceramide, choline, ethanolamine	Cer, Cho, Etn
Inositol, serine	Ins, Ser
Phosphatidyl, phosphoric acids	Ptd, P
Sphingosine, sphingoid	Sph, Spd

[a]These symbols have been constructed in analogy to those already in use for amino acids and saccharides.

[b]Systematic and recommended trivial names of unbranched, acyclic compounds only. Other forms are created by prefixes (e.g. 'iso', 'tert', 'cyclo').

[c]These symbols are a self-consistent series for a group of closely related compounds.

[d]Not Glu (glutamic acid) or G (non-specific).

[e]Recommended in place of Glc UA.

[f]Approved name trivial name for 2-amino-2-deoxyglucose.

common constituents of complex lipids as recommended by the Commission (Anonymous, 1977). The abbreviation for glycerol is Gro and not g; thus, GroPIns, glycerophosphoinositol. The radical of phosphatidic acid is named phosphatidyl (Ptd), which designates 1,2-diacyl-*sn*-glycero-3-phospho group. The major natural glycerophospholipids are phosphatidylcholine (PtdCho), phosphatidylethanolamine (PtdEtn), phosphatidylserine (PtdSer) and phosphatidylinositol (PtdIns). A mixture of 1,2-diacyl-GroPIns, 1-alkenyl-2-acyl-GroPIns, and 1-alkyl-2-acylGroPIns is correctly described as diradyl glycerophosphoinositols. Radyl is any hydrocarbon side chain, such as acyl, alk-1-enyl, or alkyl and the above diradyl-GroPIns could also be referred to as inositol glycerophospholipids.

1.4.1.3. Positional isomers

The diradylglycerol moieties of the PtdIns are now known to contain alkylacyl, alkenylacyl and diacylglycerol species although the diacylglycerol species predominate. The *sn*-2-position of the diradylglycerol moiety is usually occupied by an unsaturated fatty acid, while the *sn*-1-position contains a saturated fatty acid or the alkyl or alkenyl ether group. There is evidence that a similar arrangement of fatty chains is present in the PtdIns phosphates, although this has not been fully investigated. This characteristic composition of the molecular species of the phosphatidylinositols does not extend to that of the glycosyl phosphatidylinositols (Roberts et al., 1988; Kuksis and Myher, 1990; Butikofer et al., 1990; Lee et al., 1991).

1.4.2. Inositol and phosphatidylinositol phosphates

The large variety of phosphate esters that make up the natural inositol glycerophospholipids has made individual naming impractical and has led to the introduction of systematic nomenclature, which has proven as unwieldy as the common names initially employed. The book follows the relaxed D-numbering system of naming the substituted inositols.

1.4.2.1. Stereoisomers of inositols

There are nine possible stereoisomeric forms of inositol: *cis-*, *epi-*, *allo-*, *myo-*, *muco-*, *neo-*, *(+)-chiro-*, *(−)-chiro-*, and *scyllo-*inositols, and their structures are given in Fig. 1.4 (Billington, 1993).

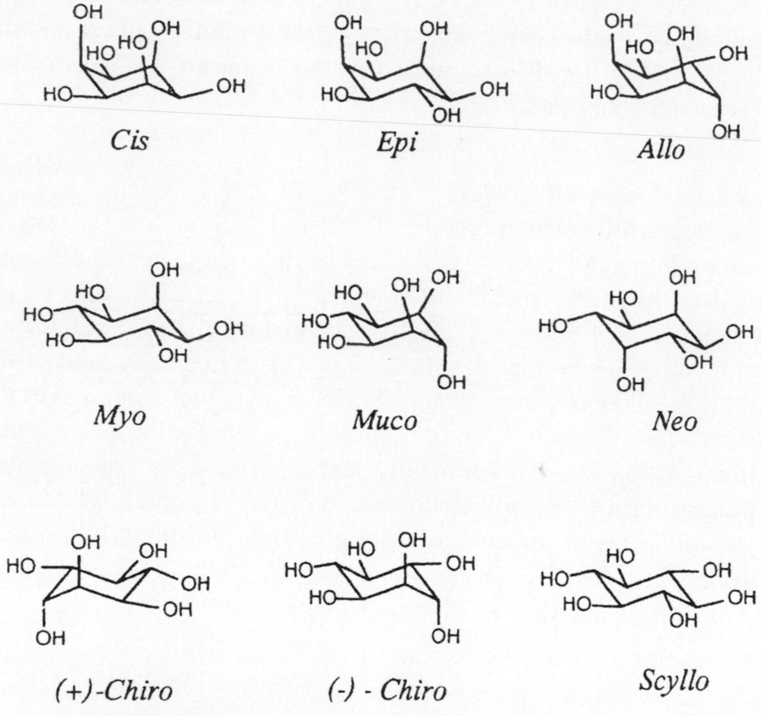

Fig. 1.4. Structures of inositol stereoisomers together with their trivial names. Of these, seven are optically inactive *meso* compounds (*allo, cis, epi, muco, myo, neo* and *scyllo*). The remaining two isomers lack this symmetry and form an enantiomeric pair ((+) and (−) *chiro*). Reproduced from Billington (1993) with permission of the publisher.

Seven are optically inactive or *meso* compounds (*allo, cis, epi, muco, myo, neo* and *scyllo*) because they have an internal plane of symmetry. The remaining two isomers have no such symmetry and form an enantiomeric pair ((+) and (−) *chiro*). The prevalent natural form is *myo*-inositol, but *chiro*- and *scyllo*-inositols are also of interest. *myo*-Inositol was first isolated from heart muscle and initially was known as *meso*-inositol, without realizing that six other optically inactive stereoisomeric inositols would be eventually recognized as equally deserving of this prefix. The inositols are readily resolved by chromatography (Pak and Larner, 1992; Kinnard et al., 1995) on the basis of the number of the axial hydroxyls: *scyllo*-isomer (no axial hydroxyls), *myo*-isomer (one axial hydroxyl), *neo*-, *epi*- and *chiro*- (two axial hydroxyls) and *muco*-, *allo*- and *cis*- (three axial hydroxyls). The only optically active isomer, *chiro*-inositol, has two axial hydroxyl groups in positions 1 and 2 (D), respectively, positions 1 and 6 (L), whereas *scyllo*-inositol carries exclusively equatorial hydroxyl groups.

To completely define *myo*-inositol, a *meso* compound, or its derivatives, which may be *chiral* or *meso*, both the absolute configuration (D or L) and the positional numbers have to be specified. According to the original recommendation of cyclitol numbering (Anonymous, 1974), phosphorylation on O-1 of *myo*-inositol leads to 1-L-*myo*-inositol 1-phosphate, whereas phosphorylation on O-3 alters the numbering of the carbon atoms, reversing C-1 to C-3, and so leads to 1-D-*myo*-inositol 1-phosphate. This is so because the old rule allocates the lowest possible locant to the substituted one, so that both phosphates become 1-phosphate. A relaxation of this rule permits alternative designation to be used, to bring out stereospecific relationships (Anonymous, 1989). It was further recommended (Anonymous, 1989) that the symbol Ins be taken to mean *myo*-inositol with numbering in the 1D configuration, unless 1L is explicitly added.

As an aid to remember the numbering of *myo*-inositol, Agranoff (1978) has suggested the image of a turtle. The head of the turtle represents the axial hydroxyl at the 2-position, while the four limbs and the tail represent the five equatorial hydroxyl groups. The right front limb of the turtle is the D-1 position, and, proceeding counterclockwise, the head is the D-2, etc. The left front limb is the 1-L or 5 D-3 position. In inositol glycerophospholipids, the right front limb, D-1, is esterified to diacylglycerol. Before 1968, in accordance with the rules of carbohydrate nomenclature, the highest locant, C-6, specified the configuration, D, or L (Anonymous, 1977). Because hydroxyls at C-1 and C-6 are *trans* to each other, the compounds assigned configurations D and L circa 1968 are now assigned 1L and 1D, respectively.

The turtle model facilitates identification of the molecule in different orientations. In PtdIns, the right front leg is always phosphodiesterified to DAG. The left front flipper is number 1, and because the head is still number 2, the limbs will be numbered clockwise starting at the front left flipper, which is why the alternative name for Ins(3,4,5,6)P$_4$ is L-Ins(1,4,5,6,)P$_4$.

Figure 1.5 depicts the structures of *myo-*, *chiro-* and *scyllo-*inositol in their chair conformations and in the corresponding two-dimensional projections (Schultz et al., 1966). The two-dimensional projection is from-top view onto the molecule in which all hydroxyl groups, except the one in the 2-position, are roughly positioned in the plane of the paper. The hydroxyl groups in positions 1, 3 and 5 are directed slightly toward, and the hydroxyl groups in position 4 and 6 slightly away from, the observer, each at an angle of 30 from the plane of the paper. The single axial hydroxyl group, by convention in position 2, is positioned vertically towards the observer. A plane of symmetry runs through C-2 and C-5 of *myo*-inositol. In the absence of elements of symmetry, each of the enantiomers is understood as D-configured, according to IUPAC recommendations (Anonymous, 1989). This is indicated by means of a counterclockwise numbering of the ring carbons. The prefix D is commonly left out for *myo*-inositol derivatives, though D and L are used for the not so common *chiro*-inositols.

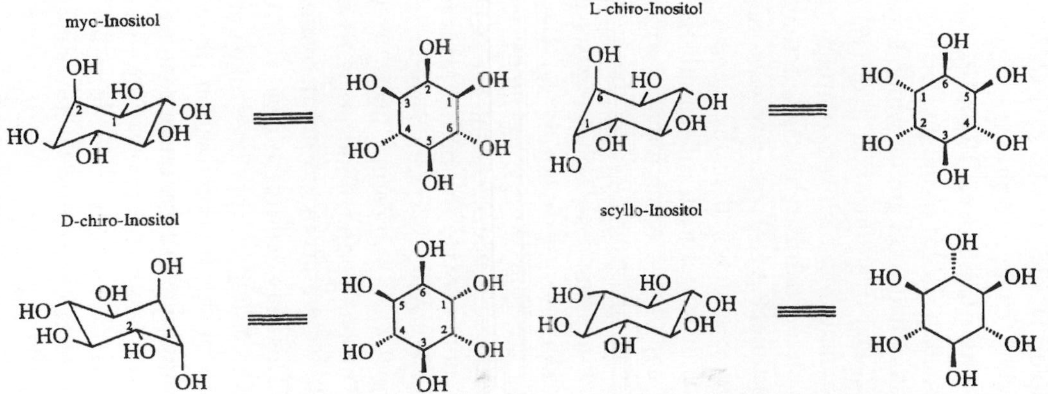

Fig. 1.5. The structures of *myo*-, D- and L-*chiro*- and *scyllo*-inositol in their pseudo-three dimensional chair conformations and the corresponding two-dimensional projections. In the absence of symmetry, each of the enantiomers is understood as D-configured according to IUPAC recommendations (Anonymous, 1989). This is indicated by means of a counterclockwise numbering of the ring carbons. Reproduced from Schultz et al. (1966) with permission of the publisher.

1.4.2.2. Numbering of phosphate groups

The naming and numbering of the inositol phosphates has presented many complex problems, which arise from the loss of plane of symmetry and a change in priority of numbering upon substitution of the ring (Cosgrove, 1980; Parthasarathy and Eisenberg, 1986; Billington, 1993; Irvine and Schell, 2001). *myo*-Inositol has a single axial hydroxyl group, which by convention is assigned to position 2, and five equatorial hydroxyl groups. This leads to a plane of symmetry which runs through C-2 and C-5, resulting in the *meso*-stereochemistry discussed above (Fig. 1.4). Incorporation of a substituent at C-2 or C-5 of *myo*-inositol leads to an optically inactive *meso*-compound, as the plane of symmetry is retained. In contrast, incorporation of a substituent at C-1 (enantiotopic to C-3) and/or C-4 (enantiotopic to C-6) leads to a pair of enantiomers (plane of symmetry lost due to substitution). Figure 1.6 shows that inositol 1-phosphate exists as a pair of enantiomers (**I**) and (**II**), inositol 2-phosphate (**III**) and inositol 1,3-bis-phosphate (**IV**) are both *meso* compounds, while inositol 1,4,5-trisphosphate exists as a pair of enantiomers (**V**) and (**VI**) (Billington, 1993). If anticlockwise numbering leads to the lowest number count, then, by IUPAC rules the derivative is assigned to the prefix D, and clockwise numbering leads to the L-prefix (Anonymous, 1974). This numbering of inositol derivatives has led to confusion when following a given inositol derivative through a metabolic pathway. This has led to a new IUPAC proposal (Anonymous, 1989), which suggests that since all natural occurring inositol phosphates are of the D-configuration, they may all be numbered as the D-inositol derivatives, rather than by strict chemical sequence rules. As a result, the two enantiomers of inositol 1-phosphate (**I**) and (**II**) are named as D-inositol 1-phosphate and D-inositol 3-phosphate, rather than as D- and L-inositol 1-phosphate (Billington, 1993).

The application of the relaxed convention may be illustrated using the enzymatic hydrolysis of D-*myo*-inositol 1,3,4,5-tetrakis-phosphate by the 1-phosphatase as an example. The product is

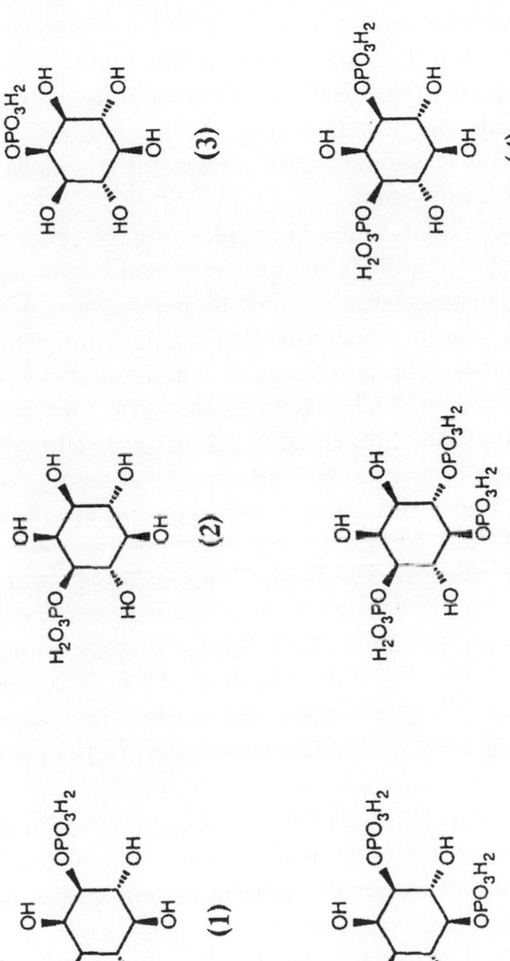

Fig. 1.6. Effect of incorporation of a substituent at C-1 (enantiotopic to C-3) and/or C-4 (enantiotopic to C-6) into *myo*-inositol. Ins(1)P exists as a pair of enantiomers (I) and (II), Ins(2)P (III) and Ins(1,3)P₂ (IV) are both *meso* compounds, while Ins(1,4,5)P₃ exists as a pair of enantiomers (V) and (VI). According to the revised rules of numbering, D-inositol 1-phosphate becomes D-inositol 3-phosphate. Reproduced from Billington (1993) with permission of the publisher.

called D-*myo*-inositol 3,4,5-trisphosphate rather than L-*myo*-inositol 1,5,6-trisphosphate. The IUB also recommends the use of Ins as an abbreviation for D-*myo*-inositol. This recommendation is also followed without further comments in this book. In current biological journals, and in this book, the D numbering is used, but in chemical journals the L numbering of the inositol ring is frequently found. This is understandable as long as it is made clear which numbering is being used.

The turtle analogy can also help in appreciating the enantiomerism in InsPs. A turtle has a plane of symmetry running through its head and tail, so distinguishing a left-flippered form a right-flippered turtle will require a technique that can discriminate D and L configuration, that is, a chiral analysis or separation method. For example, two prominent InsP$_4$ isomers are Ins(3,4,5)P$_4$ and Ins(1,4,5,6)P$_4$, both of which have an unphosphorylated hydroxyl in the 2-position (the head), so they differ only in having vacant either the 1- or the 3-positions. They are therefore an enantiomeric pair, because they can be converted from one to the other by reflection in the plane of symmetry. The standard separation techniques used in inositol phosphate analyses cannot distinguish between enantiomers, so these two InsP$_4$s co-chromatograph exactly. Quantifying Ins(3,4,5,6)P$_4$ and Ins(1,4,5,6)P$_4$ separately is possible only by using enantiomeric-specific enzyme based analyses (Vajanaphanich et al., 1994; Stephens et al., 1988; Carew et al., 2000).

Inositol phosphates are abbreviated according to IUPAC nomenclature (Anonymous, 1989), e.g. InsP, inositol monophosphate; InsP$_2$, inositol bis-phosphate; InsP$_3$, inositol tris-phosphate; InsP$_4$, inositol tetrakisphosphate; InsP$_5$, inositol pentakisphosphate; InsP$_6$, inositol hexakisphosphate; Ins(1)P, inositol 1-phosphate; Ins(2)P, inositol 2-phosphate; Ins(3)P, inositol 3-phosphate; Ins(4)P, inositol 4-phosphate; Ins(5)P, inositol 5-phosphate; Ins(6)P, inositol 6-phosphate; Ins(1,2)P$_2$, inositol (1,2)-bis-phosphate; Is(1,4,5)P$_3$, inositol (1,4,5)-trisphosphate; Ins(3,4,5,6), *myo*-inositol 3,4,5,6,-tetrakisphosphate. Alternatively, Ins(n)P$_1$, inositol monophosphate; $Ins(n)P_2$, inositol bis-phosphate; $Ins(n,n,n)P_3$, inositol trisphosphate; $Ins(n,n,n,n)P_4$, inositol tetrakisphosphate; $Ins(n,n,n,n,n)P_5$, inositol pentakinsphosphate; $Ins(n,n,n,n,n,n)P_6$, inositol hexakisphosphate.

The inositol pyrophosphates are named as follows: $PP\text{-}InsP_4$, diphospho $InsP_4$; $PP\text{-}InsP_5$, diphospho $InsP_5$; $PP_2\text{-}InsP_4$, bis (diphospho)-$InsP_4$; more specifically, 5-PP-*myo*-$InsP_5$, 5-diphospho-*myo*-inositol pentakisphosphate; 6-PP-*myo*-$InsP_5$, 6-diphospho-*myo*-inositol pentakisphosphate; 3,5-bis-PP-*myo*-InsP4, 3,5-bis-diphospho-*myo*-inositol tetrakisphosphate; 5,6-bis-PP-*myo*-$InsP_4$, 5,6-bis-diphospho-*myo*-inositol tetrakisphosphate. Pyrophosphates are derivatives of $InsP_6$ and so far have not involved stereoisomers.

In summary, in order to avoid confusion when reading older literature, the following points should be kept in mind. Between 1968 and 1986, *meso* compounds were numbered in the L designation, and therefore, if no chirality is specified, assuming the L designation is a safer choice. In the literature published before 1968, compounds of the L designation are of the current D designation, and vice versa. For example, before 1968, compound III would have been labeled L-*myo*-inositol-1-monophosphate (carbon 5 used to assign configuration). After 1968, it would be labeled 1D-*myo*-inositol-1-monophosphate or *myo*-inositol-3-monophosphate (unspecified configuration is taken to mean L). Now, III may be labeled $Ins(1)P_1$. Thus, the overall metabolic pathway (Shears et al., 1987; Irvine et al., 1987) can be described as follows: 1D-*myo*-inositol $(1,4,5)P_3$—1D-*myo*-inositol$(1,3,4,5)P_4$—1D-*myo*-inositol$(1,3,4)P_3$—1D-*myo*-inositol$(3,4)P_2$ and not 1L-*myo*-inositol$(1,5)P_2$, as would have been required by the older convention.

It should be noted that cyclic-monophosphoinositol is known to arise from enzymatic hydrolyses of phosphate diesters. The cyclic phosphates of the PtdInsP are referred to as PtdIns cycP.

The phosphorylation products of phosphatidylinositol are phosphatidylinositol phosphates. The phosphorylated derivatives of 1(3-*sn*-phosphatidyl)inositol should be called 1-phosphatidyl-inositol 4-phosphate and 1-phosphatidylinositol 3,4-bis-phosphate, respectively. The use of 'diphosphoinositide' and 'triphosphoinositide' to include all phosphorylated derivatives of PtdIns is discouraged, as these names do not convey the chemical structures of the compounds and can be misleading. For these reasons, the use of the term 'inositide' has been avoided in this book. Thus, Ins refers to D-*myo*-inositol and locants within parentheses refer to phosphorylation sites at OH-groups on the inositol ring. Unlike the case for, e.g. adenosine phosphates, there is only one type of phosphorylatable structure (the OH species of the ring) in phosphatidylinositol, therefore PtdIns(3,4)P_2 is adequate. Whereas PtdIns(3',4')P_2 is not necessary. The term 'plasmalogen' may be used as a generic term for glycerophospholipids in which the glycerol moiety bears an 1-alkenyl ether group.

Figure 1.1 (see above) shows the structure of PtdInsP$_2$ and the site of action of the lipid kinases, phosphatases and PLC that use it as a substrate. GroPIns(4)P, glycerophosphoinositol 4-monophosphate; GroPIns(4,5)P_2, glycerophosphoinositol 4,5-bis-phosphate. Other recommended abbreviations are as follows: PtdInsP, PtdIns phosphate; PtdInsP$_2$, PtdIns bis-phosphate; PtdInsP$_3$, PtdIns trisphosphate; PtdIns(3)P, PtdIns 3-phosphate; PtdIns(4)P, PtdIns 4-phosphate; PtdIns(5)P, PtdIns 5-phosphate; PtdIns(3,4)P_2, PtdIns (3,4)-bis-phosphate; PtdIns(4,5)P_2, PtdIns 4,5-bis-phosphate; PtdIns(3,4,5)P_3, PtdIns (3,4,5)-trisphosphate. Alternatively, PtdIns(*n*)P_1, phosphatidylinositol monophosphate; PtdIns(*n,n*)P_2, phosphatidylinositol bis-phosphate; PtdIns(*n,n,n*)P_3, phosphatidyl-inositol trisphosphate (isomeric form unspecified).

Figure 1.7 shows the structures of natural PtdIns containing the *myo*- and *chiro*-inositol configuration. The Ptd group is attached to the analogous hydroxyl group in *myo*- and *chiro*-PtdIns despite different numbering of the inositol ring. The stereochemical designation and numbering follow IUPAC–IUB recommendations (Anonymous, 1976a,b). The numbering of *myo*-inositol starts from the carbon atom (C-1) adjacent to the carbon atom bearing the axial

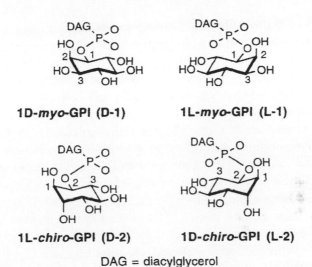

1D-*myo*-GPI (D-1)   **1L-*myo*-GPI (L-1)**

1L-*chiro*-GPI (D-2)   **1D-*chiro*-GPI (L-2)**

DAG = diacylglycerol

Fig. 1.7. The structures of natural PtdIns containing *myo*- and *chiro*-inositol configuration. The Ptd group is attached to the analogous hydroxyl group in *myo*- and *chiro*-PtdIns despite different numbering of the inositol ring. Reproduced from Bruzik et al. (1994) with permission of the publisher.

hydroxyl group (C-2) and increases passing the latter carbon first. The carbon atom C-1 is designated as having the 1D-configuration if numbering increases counterclockwise with the axial hydroxyl group oriented toward the viewer (clockwise in 1L). Numbering of *chiro*-inositol starts from the carbon atom bearing the axial hydroxyl group (C-1) and continues passing the carbon with an equatorial hydroxyl group (C-2). In 1D-*chiro*-inositol numbering increases counterclockwise with the hydroxyl group at C-1 oriented towards the viewer (clockwise 1L). Note that in *chiro*-inositol the following positions are equivalent: 1 and 6, 2 and 5, and 3 and 4, due to C2 symmetry. The *scyllo*-isomer having no axial hydroxyls differs from *myo*-isomer (one axial hydroxyl) and *chiro*-inositol (two axial hydroxyls).

Glycosyl PtdIns stands for a PtdIns tetrasaccharide linked to *myo*-inositol in α1-6 linkage. The glycosyl PtdIns tetrasaccharide

(Manα1-2Manα1-6Manα1-4GlcN) represents a glycan backbone that can be substituted with additional Etn-P groups or with other sugars such as mannose (Man), galactose (Gal), N-acetylgalactosamine (GalNAc), N-acetylglucosamine (GlcNAc) and N-acetylneuraminic acid (NANA) (McConville and Ferguson, 1993).

Glycosyl PtdIns palmitate refers to the glycosyl PtdIns containing a palmitic acid esterified to one of the inositol hydroxyl groups. The glycerolipid moiety of the glycosyl PtdIns may contain either a diacyl, alkylacyl or dialkylglycerol, which is referred to as diradylglycerol. Figure 1.3 (see above) shows the structure for the glycated Ptdns from *T. brucei*, which is representative of glycosyl PtdIns from several sources. The *myo*-inositol residue in some instances may be palmitoylated (Roberts et al., 1988).

Their nomenclature is based on that just described for the diradylglycero phosphoinositols. Finally, it should be noted that ceramide phosphoinositols as well as inositol sphingophospholipids have been isolated and identified from various natural sources (Azzouz et al., 1998). The inositol sphingophospholipids, however, are discussed only in passing. A systematic shorthand nomenclature has yet to be developed for the glycated PtdIns.

The fatty acid composition of the diradylglycerol moieties of the PtdIns is subject to remodeling, which must be clearly distinguished from other types of lipid exchange, e.g. remodeling by which ceramide is introduced into glycosyl PtdIns in exchange for the diradylglycerol moiety (Reggiori et al., 1997).

1.4.3. Inositol pyrophosphates

The diphosphoinositol polyphosphates comprise a group of highly phosphorylated compounds, which have rapid rate of metabolic turnover through tightly regulated kinase/phosphorylase substrate cycles (Safrany et al., 1999), and the authors have wondered if it constitutes the final frontier for inositol phosphate research. It should be noted that these pyrophosphates are referred to as diphosphates in keeping with the practice of naming adenosine pyrophosphates as diphosphates and triphosphates. These diphosphates are to be distinguished from the bis-phosphates, which are inositol residues carrying two is two monophosphate esters on the same inositol molecule. This convention is extended to the trisphosphates indicating that three phosphate monoesters are present in the same inositol residue. However, the higher inositol phosphates are referred to as inositol tetrakisphosphates (four monophosphate residues), pentakis- (five monophosphate residues) and hexakisphosphates (six monophosphate residues). In this instance, the prefix kis stands for polysubstitution.

In addition, the diphosphoinositol polyphosphates are widely distributed. They have been identified in mammals, yeasts, slime molds, plants and free living amoebae (for a minireview, see Safrany et al., 1999). Three types of diphosphorylated inositides are presently known: diphosphoinositol tetrakisphosphate (PP-InsP$_4$), diphosphoinositol pentakisphosphate (PP-InsP$_5$, also known as InsP$_7$) and bis-diphosphoinositol tetrakisphosphate ([PP]$_2$-InsP$_4$ (also known as 'InsP$_8$') (Fig. 1.2). Martin et al. (2000) have reexamined the structure of inositol phosphates present in trophozoites of the parasitic amoeba *Entamoeba histolytica* and have shown that, rather than being *myo*-inositol derivatives (Martin et al., 1993), these compounds belong to a new class of inositol phosphates in which the cyclitol isomer is *neo*-inositol. No evidence for the co-existence of their *myo*-inositol counterparts has been found. Preliminary data indicated that large amounts of the same *neo*-inositol phosphate and diphosphate esters were also present in another primitive amoeba, *Phreatamoeba balamuthi*. Hence. *neo*-InsP$_6$, *neo*-inositol hexakisphosphate; 2-PP-*neo*-InsP$_5$, 2-diphospho-*neo*-inositol pentakisphosphate; 2,5-bis-PP-*neo*-InsP$_4$, 2,5-bis-diphospho-*neo*-inositol tetrakisphosphate.

1.4.4. Kinases, phosphatases and phospholipases

Abbreviations are also used for the enzymes involved in InsPs biosynthesis (phosphokinases) and degradation (phosphatases).

The kinases are named by the substrate as follows: PtdInsPKI, PtdInsP kinase type I; PtdInsPKII, PtdInsP kinase type II; PtdIns3K, PtdIns 3-kinase; PtdIns4K, PtdIns 4-kinase; PtdIns5K, PtdIns 5-kinase; PtdIns 3'-kinase, phosphatidylinositol 3'-kinase.

Hinchcliffe et al. (1998) have introduced the abbreviation PIPkin (an abbreviation they have used in their laboratory for years) because the term PtdIns(4)P 5-kinase used previously no longer applied due to the extensive revision and uncertainty as to the substrates and products that were to be discussed in their review, as only general term such as PtdInsP kinase would suffice. Furthermore, the recent evidence that type II PtdInsP kinase is subject to multiple phosphorylations by protein kinases (Hinchcliffe et al., 1998), which would necessitate to speak of PtdInsP kinase kinases. Therefore, using PtdInsPkins for PtdInsP kinases (and thus PtdInsPkin kinases for the protein kinases that phosphorylate them) would be a preferred term for general use, especially, if the PI was replaced by PtdIns in the proposed abbreviations. Since this book does not deal with protein kinases to any significant extent, the above terminology has not been used.

Ins, D-*myo*-inositol; PtdIns3K, 1-(3-*sn*-phosphatidyl)-1D-*myo*-inositol 3 phosphate-kinase; PtdIns4K, 1-(3-*sn*-phosphatidyl)-1D-*myo*-inositol 4 phosphate-kinase.

A few of the enzymes have received systematic numbers: 1-(3-*sn*-phosphatidyl)-1D-*myo*-inositol 3 phosphate kinase (EC 2.7.1137); 1-(3-*sn*-phosphatidyl)-1D-*myo*-inositol 4-phosphate-kinase (EC 2.71.67); others only provisional naming: InsP(5)P-3K, Inositol 5-phosphate 3-kinase; PtdInsDK1, PtdIns-dependent kinase 1; phosphatidylinositol PLC, phospholipase C; PtdIns-PLC, phosphatidylinositol dependent phospholipase C, rather than phosphoinositidase C; PtdIns 3K, PtdIns 3-kinase or PtdIns 3-hydroxykinase.

In this book, the recommendations of the IUB (1989) for naming inositol glycerophospholipids have been adopted. The inositol isomer has been identified only when other than *myo*-inositol. As a result, the term will include substances ranging from fatty acid esters of simple glycerophosphoinositols to their polyphosphates and complex glycosylated phosphatidylinositol anchors of membrane proteins.

1.4.5. Other conventions

Other conventions and abbreviations are used to name special domains of the enzyme proteins, receptors, channels, etc. which are defined in appropriate chapters of the book. Still other conventions and abbreviations are reserved for the genes and gene products. However, there is a lack of consistency in naming of both the genes and the gene products. As a result, various isomeric kinases, phosphatases and receptor proteins are being given whimsical names, which are whimsically inconsistent (see also Editorial, *Nature* **389** (1997) 1). The progress initially made on basis of common domains has given rise to a variety of isozymes containing additional structural and functional elements, which have been subject to the most undisciplined subnumbering and naming. It will require more effort at characterization to allow systematic classification.

1.5. Scope of the book

The following chapters summarize the basic chemical and biochemical evidence for the structure of phosphatidylinositols and their phosphates along with modern methods for the mass analyses of the molecular species as well as both radio and mass assays of the kinases, phosphatases and specific phospholipases. The book concludes with a collection of methods for the preparation of selected phosphatidylinositol standards, which are not available commercially or cannot be readily stored.

This book reviews the experimental methodology employed for the extraction, isolation, identification and quantification of the phosphatidylinositol, phosphatidylinositol phosphate and inositol phosphate in current use, along with assays of the kinases, phosphatases, and phospholipases involved in their metabolism. Wherever appropriate, the emphasis has been placed on the proof of structure, even when only radioactivity has been measured. The methods have been selected largely based on their frequency of application, which has been taken as an indication of their extensive testing and satisfactory performance. However, new methods of exceptional promise have also been included. The data from intact cells that are referenced in this book are biased towards the most recent publications, since these studies have separately analyzed the largest number of metabolites.

The book opens with a general introduction of the subject, including the natural occurrence and biological significance of inositol phospholipids and inositol polyphosphates (Chapter 1). This chapter includes a brief review of the recommended nomenclature of glycerophospholipids and of the stereochemistry and numbering of the positions in inositols and inositol derivatives. It continues with an extensive survey of modern methods for the isolation, chemical and stereochemical characterization and quantification of all major molecular species of phosphatidyl-inositols (Chapter 2), phosphatidylinositol phosphates (Chapter 3), inositol phosphates (Chapter 4), and glycosyl phosphatidylinositol (Chapter 5). A separate chapter is devoted to assays of enzymes involved in phosphatidylinositol biosynthesis (Chapter 6). It updates and extends the coverage of the subject in earlier reviews (Duckworth and Cantley, 1996; Martin, 1998; Vanhaesebroeck et al., 1997; Vanhaesebroeck et al., 2001). The 3-kinase and other PtdIns and inositol phosphate kinases have been further considered in a special chapter (Chapter 7) as are the corresponding PtdIns phosphate and inositol phosphate phosphatases (Chapter 8) and the phospholipases (Chapter 9). The latter chapter also includes the glycosyl phosphatidylinositol phospholipases. Selected methods of preparation of reference inositol phosphates and inositol phospholipids have also been discussed (Chapter 10). This chapter constitutes an updated and somewhat expanded account of the chemical synthesis described in earlier published texts

(Reitz, 1991; Billington, 1993). The methods of radiolabeling of potential cellular substrates, and the preparation of deacylation and deglyceration products of the inositol phosphatides have been included in the appropriate earlier chapters. The book concludes with a discussion of the signaling properties of the lipid products of the various inositol phosphates, phosphatidylinositol and phosphatidylinositol phosphate kinases, phosphatases and phospholipases (Chapter 11). This chapter provides brief descriptions of the better known signaling events along with the supporting analytical basis. It summarizes the pathways put forward in numerous recent reviews, which have been updated within the limits of the time and space available. The relationships between different enzymes and the extent to which these are isoenzymes have been discussed under the appropriate titles in the preceding chapters.

Throughout the book the emphasis has been placed upon the methodology employed, its improvement and validation, which has proven critical for the advances made in the isolation and identification of the various inositol phosphatides, inositol phosphates, and inositol isomers. In this compilation of material, I am much indebted to outstanding previous reviews and leading papers on modern methodology of inositol phosphatide and inositol phosphate chromatography and radio-chromatography. In the selection of specific methodologies for inclusion in various chapters I have been guided by both the past success and future promise of the methodology. Furthermore, non-chromatographic methods were frequently excluded for lack of general applicability although they may have provided the best choice for a specific application. Whenever available, information has been included to qualify the homeostatic mechanisms controlling the polyphosphoinositol and polyphosphoinositol lipid levels, including the presence of stimulae, inhibitors and compromising kinase and phosphatase activities. Despite resulting redundancy, the analytical details are repeated ad nauseum to provide complete documentation of the applicability of each assay.

Although molecular biologists have already made a significant impact on the field by purifying and sequencing various phosphatases and kinases and by determining their action and regulation, a separate chapter has not been devoted to these efforts.

CHAPTER 2

Phosphatidylinositols

2.1. Isolation

Phosphatidylinositol (PtdIns) is the simplest form of the inositol phospholipids and provides the parent molecule for the PtdIns phosphates and PtdIns glycans. Along with phosphatidylserine (PtdSer) it constitutes the minor acidic phospholipids commonly isolated from cell membranes by organic solvent extraction. The PtdIns make up only a few percent of the total cellular phospholipid. They are associated with the cell membranes and usually occupy the inner half of phospholipid bilayer.

The PtdIns are readily isolated from most tissues by conventional extraction with neutral organic solvents, but most isolations are performed under conditions that would also allow recovery of the PtdIns phosphates, which require acidification of the extraction medium. Solid phase cartridge extraction and affinity chromatography, as well as thin-layer chromatography (TLC) and high performance liquid chromatography (HPLC) have also been employed for this purpose as required.

Prior to the solvent extraction, subcellular fractionation may be carried out according to a variety of procedures (Sun et al., 1988).

2.1.1. Solvent extraction

Extraction with organic solvents provides a simple method of isolating PtdIns regardless of its fatty acid composition. Depending on the nature of the tissue, the extraction may need to be repeated several times. Because both methanol and chloroform are toxic, attempts have been made to replace them with less toxic solvents. Eder et al. (1993) have published a very detailed report on the use of different solutions and extraction times to obtain optimum recovery of different phospholipid from the erythrocyte membranes of rats, while Wu et al. (2002) have dealt with improvements for extraction of the acidic phospholipids from mineralized tissues.

Attempts to improve current methods of lipid extraction have led to experimentation with automated extraction equipment, micro-wave irradiation and supercritical fluid extraction (Carrapiso and Garcia, 2000). These techniques, however, have not yet been critically applied to PtdIns isolation. Wu et al. (2002) have recently reported changes in phospholipid extractability and composition accompanying tissue mineralization, with acidic phospholipids becoming fully extractable only after demineralization.

2.1.1.1. *Chloroform/methanol*
PtdIns is readily isolated along with other common glycerophospholipids and sphingomyelins by extraction with chloroform/methanol (2:1, v/v) as described by Folch et al. (1957) or chloroform/methanol (1:1, v/v) as proposed by Bligh and Dyer (1959), or some variation (Wuthier, 1968) of these methods. A popular variant utilizes chloroform/methanol/0.05% HCl (2:1:0.05, by vol.) as discussed by Shaikh (1984). The tissue (brain) is suspended in 10 volumes of ice-cold 0.32 M sucrose containing 50 mM Tris–HCl (pH 7.4) and 1 mM EDTA and homogenized with a motor-driven glass homogenizer equipped with a Teflon pestle. Since a direct extraction of tissues with acidic solvents leads to degradation of the plasmalogens and the release of lysophospholipids, the isolation of the phospholipids is best carried out in two steps. To avoid plasmalogen destruction, the tissue is first extracted with neutral chloroform/methanol or chloroform/methanol/water mixture followed by extraction with the acidified solvent (Shaikh, 1984; Sun and Lin, 1990).

According to Shaikh (1984) the tissue homogenate is extracted with 4 volumes of chloroform/methanol (2:1, v/v). The volumes of the organic and aqueous phases are proportionally adjusted depending on the amount of tissue. After vortexing and brief centrifugation, the lower organic phase is carefully removed and transferred to another test tube. The aqueous phase is further extracted with 2 volumes of chloroform/methanol/12 M HCl (2:1:0.012, by vol.). The second acidic extraction is important for a more complete recovery of the acidic phospholipids, especially the PtdInsPs (see below). The organic layer obtained from the acidic chloroform/methanol extract is subsequently neutralized with one drop of 4 M NH$_4$OH before being combined with the first organic extract. The neutralization step is needed to protect the plasmalogens, which are sensitive to strong acids. The combined organic solvents are evaporated to dryness in a rotary evaporator. The lipids are resuspended in chloroform/methanol (2:1, v/v) and stored at − 70°C until used. The organic layer containing the PtdIns and other phospholipids may be dried by passing through a short column (Pasteur pipette) containing 0.5 g of anhydrous sodium sulfate and the dried solvents evaporated under nitrogen and saved for further analysis.

Wu et al. (2002) have recently revised the methodology of acidic glycerophospholipid extraction following the observation that changes in phospholipid extractability accompany mineralization of chicken growth plate cartilage matrix vesicles. The initial extract (Extract 1) is still being made with chloroform/methanol (2:1, v/v) using a ratio of 20 ml solvent per ml of aqueous medium followed by sonication (Wuthier, 1968). The tubes are then centrifuged at 3000 rpm for 12 min to sediment the insoluble residue. After collecting the lipid containing supernatant, a second extraction with chloroform/methanol/HCl (200/100/1, by vol.) was performed, assuming that any mineral-complexed lipids would be extractable. However, subsequent work revealed that significant amounts of the acidic phospholipids remained in the residue. Therefore, Wu et al. (2002) now recommend that after the initial extraction, the collagenase-released matrix vesicle (CRMV) pellets be demineralized with 0.5 M sodium salt of EDTA for 20 min at room temperature and sedimented by centrifugation at 3000 rpm for 12 min. After removal of the supernatant, the decalcified residue is re-extracted using chloroform/methanol/HCl (200/100/1, by vol.) (Extract 2), which was found to quantitatively remove the remaining lipids. The crude extracts were dried under N_2 and partitioned through a Sephadex G-25 column to remove non-lipid contaminants (Wuthier, 1966).

Finally, a note on extraction of PtdIns from yeast and neurospora. According to Nasuhoglu et al. (2002) about 250 mg of cells are pelleted at 1000g for 5 min. The cells are then added to a 4-ml tube containing 0.5 ml of ice-cold, acid-washed glass beads and 1.5 ml of chloroform/methanol (1:1, v/v) with 0.5 M HCl and 2 mM AlCl₃. The contents are vortexed vigorously three times for 20 s each, with the tubes placed on dry ice between vortex cycles. The supernatant is then transferred to a 2 ml microfuge tube, 0.4 ml chloroform and 0.4 ml water added, and the mixture is vortexed vigorously, followed by centrifugation at 1000g at 4°C for 10 min. The lower phase is transferred to a new microfuge tube and washed with 1 ml of cold methanol: 2 mM oxalic acid (1:0.9, v/v). The mixture is again vortexed and centrifuged at 1000g at 4°C for 10 min, and the lower phase transferred to a new tube and dried with a stream of nitrogen.

2.1.1.2. Other solvents

Kolarovic and Fournier (1986) have compared the efficiency of four classical methods (chloroform/methanol, hexane/isopropanol, n-butanol and single-phase chloroform/methanol) in extracting phospholipids from animal tissues. After the extraction, total lipids were separated quantitatively by DEAE-Sephadex chromatography into their acidic and non-acidic fractions. The two fractions were then further analyzed by gradient saturation HPTLC combined with scanning photodensitometry after staining with copper acetate. The chromogenic response of acidic phospholipids, including PtdIns, using biphasic solvent systems was shown to be lower by 10–35% in comparison to the single-phase method of Christiansen (1975), who had described the use of a single phase chloroform/methanol solvent system to obtain an unwashed lipid extract, which after concentration to dryness, was converted into a less polar chloroform extract and a more polar ethanol-based extract, thus recovering all radioactive lipids including acyl-CoA.

Hara and Radin (1978) had demonstrated that lipids can be successfully extracted from rat or mouse brains using a single-phase hexane-isopropanol (3:2, v/v) solvent system. In addition to such advantages as freedom from reactive impurities, ultraviolet transparency, and solvent density suitable for sample centrifugation, the use of isopropanol for lipid extraction inhibits the action of lipases, which are responsible for accumulation of phosphatidic acid in plant tissues (Kates, 1972).

Kolarovic and Fournier (1986) have reported a modification of the procedure of Hara and Radin (1978) for a single-phase extraction of lipids from 200 μl of a rat heart microsomal suspension without any washing procedure. According to Kolarovic and Fournier (1986), a 20-ml Pyrex test tube fitted with a polytetrafluoroethylene-line screw cap, containing 200 μl rat heart microsomal suspension (3.6 mg protein) and 200 μl distilled water, was placed in a 50°C water-bath sonicator (100 W). As soon as the sample was dispersed (1–5 min), 10 ml hexane/isopropanol (3:2, v/v) was pipetted into the test tube during sonication. Sample sonication was stopped 1 min later. After centrifugation for 1 min at 2500*g*, the clear liquid phase was decanted and then concentrated to dryness using a Speed Vac vacuum evaporator. As a result the initial radioactivity resulting from the presence of 2-[^{3}H]glycerol 3-phosphate dropped from 100 to 2.7–3% after centrifugation and it was further decreased to the level of 0.5–0.7% (non-acidic fraction) and 0% (acidic fraction), respectively, after DEAE-Sephadex chromatography.

Regardless of the method of extraction, the isolated lipid will contain some non-lipid contaminants, which may be removed by Sephadex bead columns. According to Wurhier (1966), the crude lipid extracts of tissues is taken up in an appropriate amount of the Folch lower phase, which is transferred to a medium porosity sintered glass funnel (to remove precipitated proteins) and filtered directly onto the Sephadex columns soaked overnight and washed in Folch upper phase. The purified lipid is eluted from the column with sufficient Folch lower phase.

2.1.2. Solid phase extraction

The method of solid phase extraction has emerged as an efficient procedure for the rapid isolation of phospholipids and resolution of phospholipid classes using limited amounts of solvents. It can be employed in the form of TLC plates or commercially available adsorption cartridges.

2.1.2.1. TLC and HPTLC

TLC provides the simplest method for the isolation, identification and purification of PtdIns. It provides the means of monitoring all other methods of purification and isolation of PtdIns and PtdIns phosphates. PtdIns is resolved from all other glycerophospholipids and sphingomyelin by conventional one-dimensional TLC (Skipski et al., 1964; Jolles et al., 1981) but two-dimensional systems (Rouser et al., 1969) and multiple developments have also been used. The one-dimensional system of Skipski et al. (1964) uses chloroform/methanol/acetic acid/water (25:15:4:2, by vol.) and silica gel H (without calcium sulfate binder) and has been widely adopted for the resolution of the major phospholipid classes including PtdIns. However, the system presents difficulties in the separation of the anionic phospholipids and frequently results in an overlap between PtdIns and PtdSer. Small differences in the properties of the silica gel and in humidity greatly affect the separation of these phospholipids. Mahadevappa and Holub (1987a) have overcome the problem by altering the proportions of acetic acid and water in the solvent system. Figure 2.1 shows

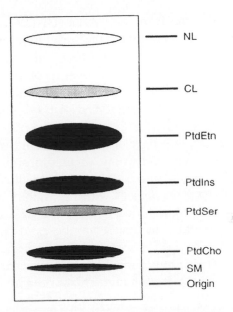

Fig. 2.1. Separation of PtdIns and the major cellular lipids of human platelets by one-dimensional TLC on pre-coated silica gel H plates (Merck). Abbreviations: O, origin; SPH, sphingomyelin; PC, PtdCho; PS, PtdSer; PI, PtdIns; PE, PtdEtn; CL, cardiolipin; NL, neutral lipid. Solvent system: chloroform/methanol/acetic acid/water (50:37.5:3.5:2, by vol.). TLC bands were located by spraying with 2,7-dichlorofluorescein (1% in methanol). Redrawn from Mahadevappa and Holub (1987) with permission of publisher.

a preparative separation of PtdIns and other major cellular phospholipids by one-dimensional TLC on precoated silica gel H plates (Merck and Co.) using chloroform/methanol/acetic acid/water (50:37.5:3.5:2, by vol.). The positions of PtdIns, PtdSer, PtdCho and PtdEtn, SPH and (Ptd)$_2$Gro were located by comparison with phospholipid standards following spraying the plate with 2′,7′-dichlorofluorescein (1% in methanol) or Rhodamine 6G and viewing under ultraviolet light. The system provides an excellent resolution of PtdIns and PtdSer from each other and from the major cellular phospholipids.

More recently, the invention of the HPTLC variant of TLC analysis has greatly improved the resolution of PtdIns from phospholipid mixtures. Sun and Lin (1990) have described a detailed protocol for a two-dimensional resolution of brain phospholipids using HPTLC plates. The first solvent system comprises chloroform/methanol/acetone/16 M NH$_4$OH (70:45:10:10, by vol.), which is allowed to advance for 9 cm up the plates. The HPTLC plates are subsequently developed in the same direction with a second solvent system consisting of chloroform/methanol/16 M NH$_4$OH/water (36:28:2:6, by vol.), which is allowed to advance up to 7 cm. After development with the second solvent system, the plates are thoroughly dried by blowing with a hair drier for 6–8 min. At this time, the plates may be exposed to HCl fumes for 3 min to destroy any plasmalogens. Immediately after HCl exposure, excess HCl fumes should be removed from the plates by blowing with a hair drier for 8 min. This procedure is carried out in fume hood. Finally, the plates are inverted through 90° and placed in a third solvent system consisting of chloroform/methanol/acetone/acetic acid/0.1 M ammonium acetate (140:60:55:4.5:10, by vol.). After development, the plates are dried briefly with the hair drier and the phospholipids are located by spraying with fluorescein or by charring with 3% copper acetate in 8% aqueous phosphoric acid and heating for 20 min at 140°C.

Wilson and Sargent (1993) have described a comparable separation of PtdIns and PtdSer from each other and from other cellular phospholipids by two-dimensional HPTLC using methyl acetate/propanol/chloroform/methanol/0.25% aqueous KCl (25:25:25:10:9, by vol.) in one direction and hexane/diethyl ether/acetic acid (80:20:2 by vol.) in the other direction. The dried plates are sprayed with color reagents, including 2′,7′-dichorofluorescein (0.1%) in methanol/water (7:3, v/v), which exhibits fluorescence with most phospholipids; a mixture containing 37% formaldehyde and conc. sulfuric acid (3:97, v/v), which is sprayed on the plate and then heated at 180°C for 30 min and causes the lipid compounds to appear as black charred spots, or a mixture containing 3% copper acetate in 8% phosphoric acid, which makes phospholipids to appear as dark brown spots. Singh and Jiang (1995) have reviewed the application of TLC to the resolution of PtdIns and PtdInsPs and have concluded that the recoveries are only about 60% of the applied material.

Zhang et al. (1998) have modified the method of Touchstone et al. (1980) for improved separation of PtdIns from PtdSer and other phospholipids in HDL as follows. Briefly, Whatman LK5D silica gel plates (15 nm) were pre-run with chloroform/methanol (1:1, v/v). The plates were then dried in an oven and the lipid samples applied. The plates were then developed with chloroform/methanol/water/triethylamine/boric acid (30:35:4.5:35:1 g, by vol/wt.). The boric acid was dissolved in methanol/water before addition of chloroform and triethylamine. After development, the plates were dried and the radioactivity was quantified on the PhosphorImager (Molecular Dynamics, Sunnyvale, CA).

Recently, Narasimhan et al. (1997) and Carstensen et al. (1999) have reported excellent separation of *scyllo*-PtdIns, *myo*-PtdIns, PtdCho, PtdSer and PtdEtn from each other and from PtdInsPs, which remain at the origin. The separations were performed by HPTLC using chloroform/methanol/ammonium hydroxide (90:90:20, by vol.) as the developing solvent. The HPTLC methods, however, are not well suited for the isolation of the PtdIns for subsequent work-up using methodology less sensitive than HPTLC.

Table 2.1 lists other popular solvent systems for separation of PtdIns from cellular phospholipids.

2.1.2.2. HPLC

Conventional HPLC on normal phase columns yields extensive separation of the common phospholipid classes, but PtdIns and PtdCho may overlap unless the solvent systems have been appropriately selected. One normal phase HPLC system that has permitted effective separation of PtdIns from other phospholipids

TABLE 2.1

Separation of PtdIns and other phospholipids by TLC

Solvent system	Comments	References
Chloroform/methanol/2-propanol/ triethylamine/0.25% KCl (15:4:12.5:9:3, by vol.)	Separation of PtdIns from PtdSer and other phospholipids	Touchstone et al. (1980)
Chloroform/methanol/water/NH$_4$OH (48:40:7:5) in first direction followed by chloroform/methanol/ formic acid (55:25:5, by vol.) in second direction	Separation of PtdIns from PtdInsPs and other glycerophospholipids	Mitchell et al. (1986)
Chloroform/ethanol/water/triethylamine (30:35:6:35, by vol.)	TLC plates prepared with boric acid in silica gel	Leray and Pelletier (1987)
Chloroform/methanol/NH$_4$OH/water (86:76:6:16, by vol.)	TLC plates prepared with 2.1% potassium oxalate	Cho et al. (1992)
Ethyl acetate/propanol/chloroform/methanol/0.25% KCl (25:25:20:15:9 by vol.) (up to 5 cm) after drying for 15 min hexane/diethyl ether/acetic acid (75:21:4 by vol.) (full length)	Whatman LHP-K HPTLC plates (resolves PtdIns from PtdSer and other phospholipids)	Wu et al. (1993)
Chloroform/methanol/28% ammonium hydroxide (65:25:5 by vol.) in first direction; chloroform/ acetone/methanol/acetic acid/water (3:4:1:1:0.5 by vol.) in second direction	Humidity of spotting critical for separation of PtdIns from PtdSer	Singh and Jiang (1995)
Chloroform/methanol/acetic acid/water (75:45:3:1 by vol.)	Resolves PtdIns from other phospholipids	Singh and Jiang (1995)
Chloroform/methanol/NH$_4$OH 90:90:20, by vol.)	HPTLC, resolution of *myo*- and *scyllo*-PtdIns	Carstensen et al. (1999)

utilizes a gradient of hexane/2-propanol (3:3, v/v) (Solvent A) and water/Solvent A (27.5:72.5, v/v). This solvent system separated neutral lipids from acidic phospholipids but did not resolve PtdIns and individual PtdInsPs (Dugan et al., 1986). This solvent is nearly transparent in the 200 nm UV range, which can be used for quantification. Alternatively, the lipids can be detected by light scattering detector or by mass spectrometry, which can be used with most solvent systems. Deacylation of PtdInsPs before analysis may improve the analysis of PtdInsPs. The mass spectrometric detector also provides an indication of the composition of molecular species (see below). Utilization of a reversed phase HPLC system with most solvent systems and all detectors provides an indication of the resolution of molecular species (see below).

Rouser et al. (1969) have reported elaborate protocols for the stepwise elution of different lipids from DEAE or TEAE cellulose columns. The lipid fractions are recovered from the DEAE column in the following order: neutral lipids (chloroform); free fatty acids, ceramides, and choline-containing phospholipids (chloroform/methanol, 9:1 with 0.02% acetic acid); PtdEtn (chloroform/methanol, 7:3, v/v); salts (methanol); PtdOH, PtdIns and PtdGro (chloroform/methanol/ammonium salt). The individual fractions are concentrated under nitrogen and the residue purified by TLC or HPLC. Since HPLC on conventional columns is time consuming and requires large amounts of solvents, cartridge separations have been proposed.

2.1.2.3. Cartridge extraction

Cartridge type of solid-phase extraction methods using pre-packed silica cartridges and various elution solvents have been developed and evaluated as chromatographic means to enrich biological lipid extracts for PtdIns. Both normal and reversed phase HPLC separation can be performed using cartridges. Janero and Burghardt (1990) have optimized a procedure that selectively removes the major tissue neutral lipids and non-choline-containing phospholipids from complex lipid mixtures and yields an acidic phospholipid fraction made up of PtdIns, PtdEtn, PtdGro, (Ptd)$_2$ Gro and about 62% of PtdSer.

According to Janero and Burghardt (1990), lipid samples in about 250 μl chloroform are loaded onto silica Sep-Pak cartridges (600 mg silica; 1.5 ml void volume, Waters, Milford, MA, USA) by using a micropipette whose tip was placed at the upper surface of the adsorbent bed. After adsorption of the sample, the columns were eluted manually at a flow rate of about 30 ml/min at ambient temperature (22°C). The column capacity was 5 mg total lipid. Fraction 1 containing neutral lipid (99.8%) was eluted in 16 ml chloroform/acetic acid (100:1, v/v); fraction 2 containing 97–99% PtdIns, PtdEtn, PtdGro, cardiolipin and 62% PtdSer were eluted in 5 ml methanol/chloroform (2:1, v/v), while fraction 3 was eluted in 10 ml methanol/chloroform/water (2:1:0.8, v/v/v). Alternatively, PtdIns along with other tissue phospholipids free of neutral lipids can be eluted with 30 ml of methanol following an initial wash with 20 ml chloroform. None of these elution systems will recover the PtdInsPs, which are irreversibly adsorbed by the silica gel.

Singh (1992) has shown that reversed phase Sep-Pak cartridges (Accel Plus QMA, Millipore, Bedford, MA, USA) provide excellent separation of PtdCho, PtdIns, PtdInsP and PtdInsP$_2$ from a DEAE eluate as shown in Fig. 2.2.

2.2. Identification

Modern chromatographic methods allow both isolation and preliminary identification of the structure of the PtdIns. Characteristic retention times have been recorded for intact PtdIns using both normal and reversed phase TLC and HPTLC as described above. Further characterization of the chromatographic fractions has been obtained by direct mass spectrometry and NMR, as well as by chemical and enzymatic degradation in combination with further chromatographic, mass spectrometric and NMR analysis.

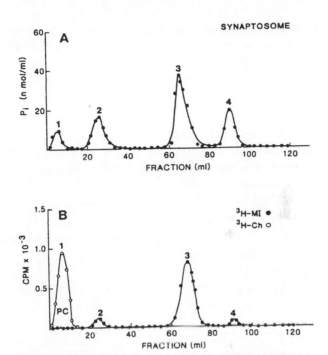

Fig. 2.2. Fractionation of PtdCho, PtdIns, PtdIns(4)P and PtdIns(4,5)P$_2$ by Sep-Pak column. Synaptosomes were labelled with [^{3}H]MI and the loaded samples were extracted for total lipid. The lipid tract was concentrated under nitrogen and then applied onto a DEAE-cellulose column and different phospholipids were eluted from the column. The fractions containing PtdCho and inositol phospholipids were pooled and concentrated under nitrogen. The concentrate was applied onto the Sep-Pak column and eluted with a gradient of 0.5–99.7% of ammonium acetate. A clear separation of PtdCho, PtdIns, PtdIns(4)P and PtdIns(4,5)P$_2$ was obtained. Reproduced from Singh (1992) with permission of publisher.

2.2.1. Determination of structure

Current methods of identification of PtdIns in a lipid extract rely largely on chromatographic migration rates in relation to reference standards, but MS/MS methods are also being utilized as they also provide, in addition to the molecular mass of the PtdIns, characteristic fragment ions needed for unequivocal identification of the component molecular species. Additional characterization of the chromatographic fractions is performed using both chemical and enzymatic degradation in combination with further chromatographic, mass spectrometric and NMR analyses. The general structure of PtdIns has been well established and conventional methods of identification do not require complete structure proof. For completeness, brief reference to the measurement of the various chemical components of the PtdIns molecule is nevertheless made below.

2.2.1.1. Chemical methods

The chemistry of inositol phospholipids has been extensively reviewed (Hawthorne, 1960; Ansell and Hawthorne, 1964; Hawthorne, 1982; Irvine, 1990).

Modern chemical characterization of an unknown PtdIns (Leondaritis and Galanopoulou, 2000) is likely to start with an alkaline hydrolysis of a 2-D TLC-purified [³H]inositol-labeled PtdIns, which is performed according to Dittmer and Wells (1969). Samples are suspended in 0.5 ml chloroform/methanol (1:4, v/v) and, after addition of 0.050 ml 1N NaOH in methanol/water (1:1, v/v), the lipids are hydrolyzed at 37°C for 10 min. Isobutanol is included in the subsequent partitioning in order to enhance the recovery of the lysoPtdIns in the organic phase. The resulting water-soluble products are chromatographed on Dowex columns, and the products are eluted stepwise with increasing concentrations of ammonium formate in formic acid (Downes and Michell, 1981). GroPIns is eluted with 60 mM ammonium formate/5 mM sodium borate, and the glycerol-derivatives of PtdInsP and PtdInsP₂, if any, are eluted with 12–14 ml of 0.4 M ammonium formate/0.1 M formic acid and 10–12 ml of 1 M ammonium formate/0.1 M formic acid, respectively. Alternatively, the total [³H]labeled *Tetrahymena* lipids were deacylated with methanolic NaOH as described by Downes and Michell (1981), and the water-soluble products were analyzed as above. Deacylated standard phosphatidyl[2-^{3}H]inositol 4,5-bisphosphate was used for the calibration of the columns.

The *phosphorus* content is usually determined using the method of Bartlett (1959) or one of the routines described by Ames (1966). Mitchell et al. (1986) have described a microadaptation of the method of Bartlett (1959) for assay of phosphorus in TLC spots (see below under quantification). However, the method described by Stull and Buss (1977) for protein-bound phosphate may be preferred (Pak and Larner, 1992). According to Stull and Buss (1977) the sample is dissolved in 25 μl of 10% MgNO$_3$, 6H$_2$O in 95% ethanol and gently heated to dryness. The samples are then ashed over an intense flame. After dissolving the residue in 150 μl of 1.2N HCl, 50 μl of filtered phosphate reagent is added, and after 5 min, the developed color is measured at 660 nm. The phosphate reagent contained 1 volume of 10% (NH$_4$)$_6$MoO$_{24}$, 4H$_2$O in 4N HCl to 3 volumes of 0.2% Malachite green. Phosphate standards (0.2–1.5 nmol KH$_2$PO$_4$) were also ashed. The variation on multiple assays of the same sample was 5% or less.

The *glycerol* content of lipids can now be readily determined by various commercial kits and by automated equipment. The commercial kits are based on enzymatic oxidation of free glycerol, which is released from the PtdIns by alkaline hydrolysis, as described for release of InsPs (see below).

The *fatty acid content* of the PtdIns samples is determined by GLC following alkaline or acid transmethylation (see also below).

According to Myher and Kuksis (1984a) the silica gel in the zones corresponding to PtdIns is scraped into tubes and then treated for 2 h at 80°C with 1.5 ml of methanol/sulfuric acid (94:6, v/v). After cooling the tubes, 4 ml of hexane and 0.5 ml of chilled 3 M aqueous ammonia are added. After vortexing the extraction mixture, the tubes are centrifuged at 800g for a few minutes. The hexane extracts are then passed through a short column of anhydrous sodium sulfate in a Pasteur pipette containing a small plug of glass wool. (These columns must be prewashed with 2 ml chloroform followed by 2 ml of hexane.) The hexane extract is evaporated under a stream of dry nitrogen gas. It is important with small samples to stop the evaporation process as soon as the solvent is no longer visible in the tube, otherwise it is possible to distort the composition by preferential evaporation of the shorter chain fatty acid methyl esters. The fatty acid methyl esters are quantified by GLC after addition of 50 μg of methyl heptadecanoate as internal standard. The methyl esters were resolved by GLC on a 15 m RTx 2330 polar capillary column (Restek Corp, Port Matilda, PA) using a 7:1 split injection system. The carrier gas was hydrogen at 2 ψ, and the column was programmed from 80 to 130°C at 20°C/min and then to 220°C at 5°C/min.

Originally, *inositol* was determined by a microbiological assay with an inositol-requiring yeast. A specific and sensitive procedure using a strain of *Schizosaccharomyces pombe* has been described by Norris and Darbre (1956). More recently, a variety of enzymatic and physico-chemical methods have been employed for mass assays, but none have gained wide acceptance (for a brief review, see Maslanski and Busa, 1990).

According to Pak and Larner (1992) the PtdIns purified by TLC is hydrolyzed in sealed vials with 6N HCl at 110°C for 48 h. The acid hydrolysates are analyzed for inositol by HPAE-Dionex HPLC with Aminex HPX-87C column (Bio-Rad). GC/MS analysis of the hydrolysates was performed using DB-5 column (0.32 mm × 30 m) after preparation of the hexatrimethylsilyl derivatives. The inositol isomers are identified by reference to standards.

Ostlund et al. (1993) have described a chiral phase GLC method for the separation of enantiomeric *chiro*-inositols in urine and plasma.

Briefly, fasting EDTA-plasma samples are collected on ice. To each sample (0.5 ml) is added 1.25 nmol of deuterated DL-*chiro*-inositol and 17.5 nmol of deuterated *myo*-inositol. The samples are then deproteinized by $ZnSO_4$ and diluted until their conductivity is <2.5 pS and then passed over 1 ml column of prewashed AG501-X8(D) 20–50 mesh (Bio-Rad) mixed-bed resin to remove ionic substances. The column is washed twice with 0.6 ml of water and to the combined eluates are added 1 g (wet weight) of prewashed Amberlite IRA-400 hydroxide resin. The samples are shaken for 30 min, the supernatant removed, the resin washed with 0.5 ml of water, and the samples are lyophilized. The processed samples are converted to hexakispentafluoropropionic (PFP) esters by reaction overnight at 65°C with 150 μl of a 1:1 (v/v) solution of PFP anhydride and acetonitrile. The excess derivatizing reagent is removed prior to GLC. Plasma samples of 10–60 μl are reduced to a few μl under a stream of nitrogen and taken up in 7–15 μl of acetonitrile and the samples analyzed immediately by GC/MS.

Alternatively, the hydrolytic products are separated by descending paper chromatography on Whatman 1 MM filter paper (13 cm × 38 cm) in one of the following three solvent mixtures (Kinnard et al., 1995): solvent mixture 1, isopropanol/pyridine/acetic acid/water (8:8:1:4 by vol.); solvent mixture 2, isopropanol/acetic acid/water (3:1:1 by vol.); solvent mixture 3, phenol/water (3:1 by wt.). Typically papers were developed for 24 h in solvent mixture 1, for 24 h in solvent mixture 2, and 56 h in solvent mixture 3. Labeled products were separated by chromatography in parallel with authentic standards which were visualized by first spraying with 0.1N AgNO$_3$/5N NH$_3$/2N NaOH (1:1:2 by vol.) and subsequently exposing the wet paper to steam. Solvent mixture 2 provided an effective separation of *chiro-*, *epi* and *muco*-inositols from *scyllo-*, *myo-* and *neo-*inositols which overlapped. The *scyllo-*, *neo-* and *myo-*inositols were effectively resolved by solvent mixture 3 (Kinnard et al., 1995).

Bruzik et al. (1994) have synthesized two diastereomers of the analog of PtdIns containing 1D- and 1L-*chiro-*inositol in 12 steps starting from 1D-2,3,4,5-O-tetrakis(methoxymethylene)-*myo-*inositol by the inversion of the hydroxyl group at the 1-position of inositol followed by several protection/deprotection and phosphorylation steps. Both diastereomers were subjected to cleavage by PtdIns-PLC from *Bacillus thuringiensis* as described below. The results suggested that the natural *chiro-*inositol derivatives should have the 1L-configuration if they are produced by PtdIns-PLC, which is in contrast to the 1D-configuration reported by others (Larner et al., 1988; Ostlund et al., 1993). Bruzik et al.(1994) isolated *chiro*-inositol from total bovine liver lipid and showed it to possess the 1L-configuration by ^{13}P-NMR. This result is consistent with stereochemical similarity between 1D-*myo*- and 1L-*chiro*-PtdIns.

In principle, D- and L-*chiro*-inositol phospholipids may be differentiated by PtdIns-specific phospholipase C (*B. thuringiensis*) as reported by Bruzik et al. (1994), although a practical execution of the assay with small amounts of non-radioactive substrate may present problems.

2.2.1.2. Enzymic methods
The structure and composition of PtdIns can be further characterized by hydrolysis with various phospholipases (Chapter 9).

Phospholipase A₁. Phospholipases A_1 attack the *sn*-1-position of glycerophospholipids, including PtdIns. Many of these enzymes, however, also attack the *sn*-2-position (Gassama-Diagne et al., 1991). Of special interest is the phospholipase A_1 activity secreted from guinea pig pancreas, which lacks the classical secretory phospholipase A_2 activity (Gassama-Diagne et al., 1991). A detergent-resistant phospholipase A_1 from the membranes of an *Escherichia coli* strain has been used for large-scale production of the enzyme, which has been purified and partially characterized (Nakagawa et al., 1991). The neutral *sn*-1-lipase from *Rhizopus arrhizus* appears to provide the most specific attack on the *sn*-1-position in comparison to the *sn*-2-position of glycerophospholipids, but it also attacks the *sn*-3-position in the unsaturated enantiomer (Slotboom et al., 1970).

A scaled-down version of the method of Slotboom et al. (1970) consists of hydrolysis of 1 mg or less of a glycerophospholipid or diacylglycerol dispersed in 0.3 mg of sodium deoxycholate and 0.5 mg bovine serum albumin in 1 ml of 0.1 M borate buffer ([Ca^{2+}] $= 5 \times 10^{-3}$ M) at pH 6.5. The purified *Rhyzobium* enzyme was added in amounts of 0.1–0.5 mg and incubations were carried out at 30°C for 0.5–6 h with vigorous vibrating. The degree of hydrolysis was followed by TLC.

Phospholipase A_2. Phospholipase A_2 is the most often utilized phospholipase for the structural characterization of PtdIns. It allows the confirmation of the enantiomeric nature of the *PtdIns* as well as the positional distribution of its fatty acids. For this purpose, the bacterial enzymes (e.g. *Crotalus adamanteus*) are utilized most commonly. These enzymes exhibit preference for glycerophospholipids with choline and ethanolamine headgroups and PtdIns is attacked at a slower rate. The substrate preference toward fatty acid chains in the *sn*-2-position of PtdCho is: 18:1 > 16:0 > 18:1 > 20:4.

According to Kuksis and Myher (1990) the purified PtdIns (0.5 mg) is vortexed for 3 h at 37°C in a mixture made up of 2 ml of buffer [17.5 mM tris(hydroxy-methyl)aminomethane, adjusted to pH 7.3 with dilute HCl, 1.0 mM $CaCl_2$], 2 ml diethyl ether, 50 μg of butylated hydroxytoluene (BHT) and 50 units of phospholipase A_2. After acidification with one drop of conc. HCl, the mixture is extracted first with five parts chloroform/methanol (2:1, v/v), then with two parts chloroform/methanol (4:1, v/v). The two portions of the extract are then resolved by TLC on Silica Gel H. The plates are first developed to a height of 11 cm using chloroform/methanol/ acetic acid/water (100:35:10:3, by vol.), dried under nitrogen gas for 10 min, and then developed to a height of 15 cm using heptane/ isopropyl ether/acetic acid (60:40:4, by vol.). The silica gel in the zones corresponding to free fatty acids, residual PtdIns and lysoPtdIns are scraped into tubes and then treated for 2 h at 80°C with 1.5 ml of methanol/sulfuric acid (94:6, v/v). The methyl esters are recovered as described above for the total fatty acids of PtdIns. The *sn*-configuration of the PtdIns preparation is determined from the relationship of the lysoPtdIns and the composition of the released free fatty acids.

Phospholipase C. PtdIns is readily hydrolyzed by phospholipase C to yield *sn*-1,2-diacylglycerol and a water-soluble InsP.

Crude preparations of phospholipase C from *Bacillus cereus* (such as type III and some batches of type V, Sigma Chemical Co.) have activity against PtdIns as well as against PtdCho, PtdEtn and PtdSer. Purer preparations (such as type XIII, Sigma Chemical Co.) have no phospholipase C activity against PtdIns. The hydrolysis is illustrated with phospholipase C from *B. cereus* (type III or type V, Sigma Chemical Co.), which yields *sn*-1,2-diacylglycerol and a water-soluble InsP.

According to Myher and Kuksis (1984a) the purified PtdIns (approximately 500 μg) plus 50 μg of BHT as antioxidant is suspended in 1.5 ml of peroxide-free diethyl ether and 1.5 ml of buffer [17.5 mM tris(hydroxymethyl)aminomethane adjusted to pH 7.3 with HCl, 1.0 mM $CaCl_2$], along with 10 units of the enzyme. The mixture is vortexed for 10 s and then shaken at 37°C on Buchler rotary Evapo-Mix for 3 h. The diacylglycerols are extracted two times with 2 ml of diethyl ether. The extracts are combined and passed through two successive Pasteur pipettes containing anhydrous sodium sulfate. After reduction of volume, the extent of hydrolysis is checked by spotting the sample in a narrow lane (about 1 cm) beside a 25 μg of undigested phospholipid on a silica gel plate, followed by developing in chloroform/methanol/aqueous ammonia (65:25:4, by vol.). After it has been sprayed with dichlorofluorescein, the amount of any residual phospholipid is estimated by a visual comparison with 25 μg of reference phospholipid standard.

Alternatively, PtdIns could be digested with commercially available PtdIns-specific phospholipase C from *B. cereus* (Boehringer Mannheim Co.) or from *B. thuringiensis* (ICN Co.). The enzyme obtained from a recombinant strain of *B. subtilis* after transfection with the *B. thuringiensis* PtdIns-PLC gene for overproduction of PtdIns-PLC enzyme catalyzes the hydrolysis of PtdIns in discrete steps. First, an intramolecular phosphotransferase reaction yields Ins(1,2cyc)P, which is converted by a cyclic phosphodiesterase activity to Ins(1)P (Zhou et al., 1997a). The activity of the PtdIns-specific enzymes is limited to PtdIns and lyso-PtdIns. These enzymes do not attack PtdGro, SM, PtdCho, PtdEtn or PtdSer.

Bruzik et al. (1994) have synthesized both 1D- and 1L-*chiro*-PtdIns in diastereomerically pure forms and have determined their substrate properties toward bacterial PtdIns-phospholipase C using ^{31}P-NMR. The results indicated that 1L-*chiro*-PtdIns was accepted by PtdIns-PLC at a reduced rate whereas the 1D-diasteromer was not a substrate. It was concluded that the naturally occurring *chiro*-inositol derivatives should have the 1L-configuration if they are derived from PtdIns by PtdIns-PLC. Bruzik et al. (1994) isolated and characterized the *chiro*-inositol from bovine liver lipid and showed that the natural isomer was indeed the 1L-enantiomer.

Hondal et al. (1997) have reported the use of dipalmitoyl thiophosphate analog of PtdIns, D-thio-dipalmitoyl GroPIns, for the assay of PtdIns-PLC. In mixed micelles with diheptanoyl GroPCho, it gave only about half of the activity of sonicated dispersions of D-thio-dimyristoyl GroPIns with PtdIns-PLC. Hendrickson et al. (1997) have reported that L-thio-1,2-dimyristoyloxypropane-3-thio-phospho(1L-1-*myo*-inositol) is an ideal neutral diluent with which to study the kinetics of PtdIns-PLC. Kinetic parameters were derived from the rate as a function of bulk lipid concentration at a constant saturating surface concentration of substrate (A), and as a function of surface concentration of substrate at a constant saturating bulk concentration of lipid (B). The substrate, D-thio-dimyristoylGro-PIns, was diluted with L-thio-dimyristoyl GroPIns or dimyristoyl-GroPMe. Dimyristoyl GroPMe caused enzyme inhibition in case of saturated surface concentration A but no inhibition in case of a constant saturating bulk concentration of lipid.

Phospholipase D. Phospholipase D catalyzes the hydrolysis of glycerophospholipids to produce PtdOH and their respective polar head groups and is distinct from the phospholipase D activity that hydrolyzes glycosyl PtdIns (see below). At least two distinct isoenzymes of phospholipase D are expressed in mammalian cells. One is membrane associated and prefers PtdCho as substrate, whereas the other is a cytosolic enzyme that appears to hydrolyze preferentially PtdEtn or PtdIns. Two major classes of mammalian phospholipases D have been identified, a $PtdInsP_2$-dependent class (PLD 1 and PLD 2) and a fatty acid (oleate)-sensitive form (Singer et al., 1997).

PtdIns-specific phospholipase D liberates free Ins and PtdOH. PtdIns is also subject to transphosphatidylation by non-specific phospholipase D. The action of phospholipase D is often coupled to a PtdOH phosphatase that produces diacylglycerol and PO_4. Therefore, it is difficult in intact cells to differentiate between phospholipase D plus phosphatase and phospholipase C.

$PtdInsP_2$-independent phospholipase D may be assayed as follows (Li and Fleming, 1999). The reaction mixture contained 100 mM MES (pH 7.0), 1 mM SDS, 1% ethanol, 2.5 μg cabbage phospholipase D and 0.2 mM dipalmitoyl-[2-palmitoyl-9,10-^{3}H]GroPCho with varying concentrations of $CaCl_2$ at the mM level. To assay phospholipase activity with PtdIns as substrate, PtdCho was replaced by unlabelled PtdIns, plus 1-palmitoyl-2-arachidonoyl [arachidonoyl-1-^{14}C]PtdEtn in the assay system. The final concentration of PtdEtn was 0.4 mM. The assays were conducted at 37°C for 20 min in a total assay volume of 60 μl.

Li and Fleming (1999) have demonstrated that aluminum fluoride inhibits phospholipase D purified from cabbage in both $PtdInsP_2$-dependent and $PtdInsP_2$-independent assays, consistent with its previously observed effect on mammalian phospholipase D.

The $PtdInsP_2$-dependent phospholipase D may be assayed in a cell free system containing phospholipid vesicles (PtdEtn/PtdIns-P_2/PtdCho in a molar ratio 16:1.4:1) with a final concentration of 8.7 μM PtdCho. The assay is performed with 10 μl lipid vesicles added to the assay system (50 mM HEPES, 80 mM KCl, 3 mM $MgCl_2$, 3 mM DDT, pH 7.0) which contained 2.5 μg cabbage PLD. The free Ca^{2+} concentration was EGTA-buffered to 7.8 μM Ca^{2+} unless otherwise indicated. PLD activity was assayed either by transphosphatidylation reaction in the presence of 1% ethanol, with dipalmitoyl-[2-palmitoyl-9,10-^{3}H]GroPCho as substrate, or by radiolabeled Cho-release with PtdPCho-[methyl-^{3}H]Cho as substrate. The assays were conducted at 37°C for 20 min in a total assay volume of 60 μl.

Bruzik and Tsai (1991) have used phospholipids chirally labeled at phosphorus to determine the stereospecificity at phosphorus for PLA$_2$, PLC and PLD, as well as PtdIns-specific PLC. The study required the use of phosphothioates.

2.2.1.3. Mass spectrometric methods

Fast atom bombardment/mass spectrometry (FAB/MS) was recognized early as being a useful method for establishing the structure of PtdIns. This is due to the unique chemical behavior of this compound having a highly lipophilic region, which enables the molecule to orient on the surface of FAB-matrices, as well as polar functionalities, which readily accept negative charge sites in the gas phase (Murphy and Harrison, 1994). FAB/MS can provide an effective method for characterization of the molecular structure of PtdIns including the molecular ion and characteristic fragment ions. The negative ion FAB spectrum of PtdIns (Fig. 2.3a) contains [M − H]$^-$ ions in low abundance consistent with the loss of [Ins-H$_2$O], such as m/z 671. This ion has the structure identical to [M − H]$^-$ anion of PtdOH. Carboxylate anions from the fatty acyl groups at sn-1 and sn-2 are also observed in the negative ion FAB mass spectrum. Collision induced decomposition of [M + H]$^+$ from PtdIns yields a neutral loss of 260 u, corresponding to the loss of InsP. Collision induced decomposition (CID) of [M − H]$^-$ results in abundant carboxylate anions at sn-1 and sn-2, while precise assignment of the position requires decomposition of the PtdOH anion (m/z 671.5). Minor ions at m/z 595 and 577 in Fig. 2.3b arise from loss of palmitic acid and its ketene analog, respectively, from the [M − H]$^-$ anion. The ions at m/z 553 and 571 result from neutral loss of linoleic acid and its ketene analog, respectively. The ions at m/z 391 and 409 result from the losses of the sn-2 substituent from PtdOH ion formed by loss of inositol.

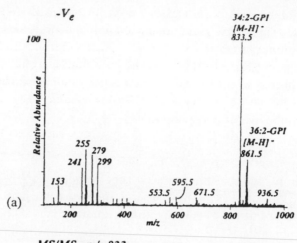

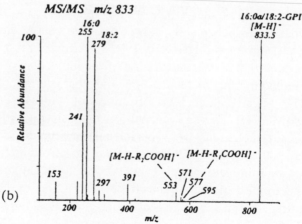

Fig. 2.3. Mass spectrometry of PtdIns derived from soybeans using liquid SIMS ionization. (a) Negative ions; (b) product ion scan of the major molecular species at m/z 833.5 corresponding to 16:0/18:2 GroPIns. Reproduced from Murphy and Harrison (1994) with permission of publisher.

More recently, ESI/MS has provided the molecular masses of PtdIns in the negative ion mode (Han and Gross, 1994; Li et al., 1999; Hsu and Turk, 2000). In the negative ion mode, PtdIns yields abundant $[M - H]^-$ ions by ESI. Following CID, the $[M - H]^-$

ions yield three major series of fragment ions that reflect neutral loss of free fatty acid ($M - H-R_xCO_2H]^-$), neutral loss of ketene ($M - H-R'_xCH=C=O]^-$), and fatty carboxylate anions ($[R_xCO_2]^-$), where $x = 1,2$, $R_x = R'_xCH_2$. Ions reflecting an inositol polar head group are also abundant. Samples in chloroform/methanol (1:4, v/v) at final concentration of 2 pmol/μl were infused (1 μl/min) into ESI source with a Harvard syringe pump. The intensities of the ions arising neutral loss of the sn-2-substituent as a free fatty acid ($[M - R_2CO_2H]^-$) or a ketene ($[M - H-R'_2CH=C=O]^-$) are greater than those of ions reflecting corresponding losses of the sn-1-substituent, and these features have been used to determine the positional distribution of fatty acids.

Hsu and Turk (2000) have presented a mechanistic study of the mass spectrometric characterization of PtdIns, PtdInsP and PtdInsP$_2$ by ESI/MS/MS. In negative ion mode, the major fragmentation pathways under low energy CID for PtdIns arise from neutral loss of free fatty acid substituents ($[M - H-R_xCO_2H]^-$) and neutral loss of the corresponding ketenes ($[M - H R'_xCH=C=O]^-$), followed by consecutive loss of the inositol head group. The intensities of the ions arising from neutral loss of the sn-2-substituent as a free fatty acid ($[M - H-R_2CO_2H]^-$) or as ketene ($[M - H-R'_2CH=C=O]^-$) are greater than those ions reflecting corresponding losses of the sn-1-substituent. This is consistent with a recent finding of Hsu and Turk (2000) that ions reflecting these losses arise from charge-driven processes that occur preferentially at the sn-2-position. Nucleophilic attack of the anionic phosphate onto the C-1 or the C-2 of the glycerol to which the fatty acids are attached expels sn-1-($[R_1CO_2]^-$) or sn-2 ($[R_2CO_2]^-$) carboxylate anion, respectively. This pathway is sterically more favorable at the sn-2 than at sn-1. According to Hsu and Turk (2000) further dissociations of $[M - H-R_xCO_2H-Ins]^-$. $[M - H-R_xCO_2H]^-$, and $[M - H-R'_xCH=C=O]^-$ precursor ions also yield $[R_xCO_2]^-$ ions, whose abundance are affected by the collision energy applied. Therefore, relative intensities of the $[R_xCO_2]^-$ ions in the spectrum do not reflect their positions on the glycerol backbone and determination of their regiospecificities based on their ion intensities is not reliable.

Schiller et al. (1999) and Petkovic et al. (2000) have applied matrix-assisted laser desorption and ionization time-of-flight-mass spectrometry (MALDI-TOF-MS) to the structural characterization of phospholipids, including PtdIns and PtdInsPs. These techniques, however, have not been perfected to the same extent as the older FAB/MS and ESI/MS methods discussed above. All of the MS/MS methods provide diacylglycerol fragmentations from which the molecular species of the PtdIns can be identified (see below).

Modern MS analysis of PtdIns starts with chromatography on a scale sufficient to isolate pure chemical compounds for subsequent structural analyses. The chromatographic methods used for the initial isolation of PtdIns may also serve well for the preliminary characterization of unknown PtdIns. A comparison of relative retention times on TLC plates and of elution times on HPLC, when combined with judicious use of standards, is frequently accepted as evidence of identity of PtdIns. Chromatographic methods are most efficient when combined with on-line MS, but chromatographic fractions collected from the column effluent for subsequent MS analyses have also proven useful.

Karlsson et al. (1996) have used normal-phase LC/ESI/MS and LC/APCI/MS for class separation and identification of glycerophospholipids, including PtdIns. With APCI the response was lower than with other glycerophospholipid classes. With HPLC class separation, the tedious task of correctly identifying a phospholipid class is avoided. If the unknown sample consisted of only a few phospholipid classes, MS/MS could be employed directly without HPLC separation, bearing in mind, however, that suppression effects in the ion source can occur and make quantification difficult.

2.2.1.4. NMR
^{1}H NMR, ^{2}H NMR, ^{13}C NMR, and ^{31}P NMR all have been utilized for the characterization of PtdIns and its derivatives with various

success. PtdIns isolated from natural sources is heterogenous in chain length and unsaturation, while synthetic long chain PtdIns requires solubilization in organic solvents or dispersion in water, which complicates interpretation of spectra. While a ^{2}H NMR study provides possible structures, there are not enough constraints to orient the inositol ring with respect to the glycerol backbone or acyl chains.

Volwerk et al. (1990) have used ^{31}P NMR to analyze the cleavage products of PtdIns-specific phospholipase C from *B. cereus*. ^{31}P NMR spectroscopy could distinguish between the InsP species and PtdIns. The ^{31}P NMR spectra (146.18 MHz) of authentic samples of PtdIns gave -0.411 ppm, D,L-Ins(1:2cyc)P 16.304 ppm, and D,L-Ins (1)P 3.623 ppm. Spectra were recorded at 35°C under the following conditions: PtdIns, 5.65 mM in 50 mM sodium borate/HCl (pH 7.5) containing 12.5 mM sodium deoxycholate and 50% D_2O (by volume); D,L-Ins(1:2cyc)P, 2.20 mM in 50 mM Tris/acetate (pH 7.5) containing 12.5 mM Triton X-100 and 50% D_2O; D,L Ins(1)P, 5.47 mM in the same buffer as for D,L-Ins(1:2cyc)P. Chemical shift values (with reference to phosphoric acid) observed are 0.41, 3.63, 4.45 and 16.30 ppm for PtdIns, *myo*-Ins(1)P, *myo*-Ins(2)P, and *myo*-Ins(1,2cyc)P, respectively.

Computational modeling has identified six possible 'tilted' orientations all of which are consistent with the ^{2}H NMR data (Shibata et al., 1984; Hansbro et al., 1992). Figure 2.4 shows the high resolution ^{1}H-NMR spectra in DMSO-d_6 of synthetic dimyristoyl GroPIns-d_0 (A) and dimyristoyl GroPIns-d_6 (B) (Hansbro et al., 1992). The appropriate resonances are shown in the figure.

In order to use the full power of high resolution NMR for structural analyses, a monomeric or micellar PtdIns is required. Lewis et al. (1993) have synthesized the D- and L-Ins isomers of short chain GroPIns and PtdIns analogs including 2-methoxy derivatives and have used ^{31}P NMR to establish the structural requirements of PtdIns-specific PLC. The ^{31}P NMR parameters were based on those previously used by Volwerk et al. (1990).

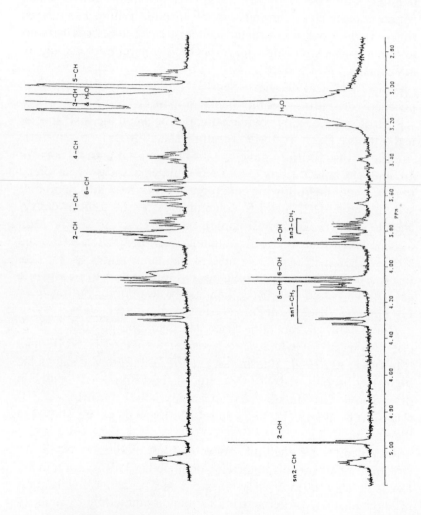

Zhou et al. (1997b) have presented a detailed NMR analysis of the conformation of these short-chain PtdIns molecules in both monomer and micellar states. Coupling constant analysis and nuclear Overhauser effect data addressed the differences between D- and L-Ins isomers as well as the similarity of 2-methoxy-PtdIns to diheptanoyl GrosPIns. QUANTA computational modeling provided a likely orientation of the Ins ring with respect to the interface. ^{13}C-relaxation times characterized the internal motion of the PdIns molecule and highlighted differences between monomers and aggregated PdIns. These results were analyzed in light of PtdIns-specific and non-specific PLC activities.

Leondaritis and Galanopoulou (2000) used ^{1}H NMR analysis to identify PtdIns in *Tetrahymena*. For this purpose the *Tetrahymena* PtdIns and standard PtdIns (approx. 15 µg of lipid phosphorus) was dissolved in 0.8 ml methanol-d_4/chloroform d_1 (2:1, v/v) and transferred to 5 mm NMR tubes. ^{1}H NMR spectra were recorded using Bruker DRX400 and AM 500 NMR spectrometers at 20°C in the Fourier transform (FT) mode with 32 K data points, using a 45° detection pulse and 4.0-s acquisition time. ^{1}H NMR analysis of *Tetrahymena* PtdIns purified by two-dimensional TLC revealed the presence of an Ins moiety; the ^{1}H NMR spectrum of *Tetrahymena* PtdIns was identical to that of standard PtdIns. Characteristic chemical shifts corresponding to the six protons of the Ins ring were observed at about 3.6 (H-4), 3.8 (H-6, a triplet), 4.0 (H-1), and 4.2 ppm (H-2) and at 3.2 and 3.4 ppm (H-5 and H-3, respectively). Peaks at 4–4.2 ppm were partially overlapped by the glycerol C-1 and C-3 proton multiplets. The signal at about 4.4 ppm (double doublets) was assigned to the glycerol C-1 proton downfield resonance (Casu et al., 1991).

Fig. 2.4. MHz ^{1}H-NMR spectra in DMSO-d_6, showing the 3.5–5.5 ppm region of synthetic dimyristoyl GroPIns-d_0 (upper) and dimyristoyl GroPns-d_6 (lower). Reproduced from Hansbro et al. (1992) with permission of publisher.

2.2.2. Composition of fatty acids

The positional distribution of fatty acids in PtdIns is determined by hydrolysis with enzymes, which are specific for either the *sn*-1- or *sn*-2-position of the phosphatide. As noted above, the fatty acids in the *sn*-1- and *sn*-2-position of the PtdIns may be distinguished on basis of characteristic fragment ions produced in the mass spectrometer. The presence of both alkyl and alkenyl ether groups in PtdIns preparations complicates the analysis of the fatty chain composition. Recently, it has become possible to resolve the reverse isomers of the diradylglycerol moieties of glycerophospholipids (Itabashi and Kuksis, 2001), which can provide the positional distribution of both ether and ester fatty chains within a known subclass of species, as illustrated in Fig. 2.5.

2.2.2.1. Total
Early workers recognized that PtdIns contained phosphorus, inositol, glycerol and fatty acids in the molar ratio of 1:1:1:2. Analyses of the fatty acids by GLC later revealed that the bulk of the acids was made up of stearic (18:0) and arachidonic (20:4) acids in the PtdIns of many mammalian tissues. In plants and microorganisms the saturated and monounsaturated fatty acids tend to dominate. The fatty acid composition is usually determined by GLC analysis of the fatty acids as methyl esters. However, HPLC with UV absorbing esters of fatty acids has also been employed as has been HPLC with light scattering and MS detection. Since some PtdIns contain a few percent of ether chains, detailed analyses of the fatty chain composition of PtdIns are best performed following isolation and segregation of the diradylglycerol moieties into the alkylacyl, alkenylacyl and diacylglycerol subclasses, as described in Chapter 5.

2.2.2.2. sn-1-position
The fatty acid composition of the *sn*-1-position may be obtained in a variety of ways. It can be obtained by determining the fatty acids

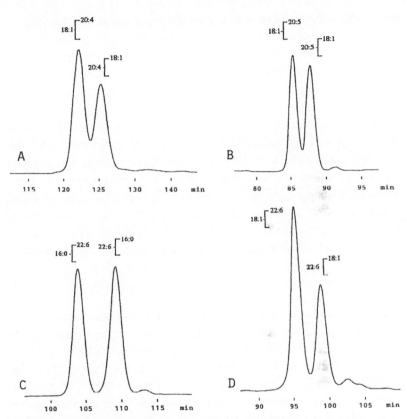

Fig. 2.5. Reversed phase HPLC resolution of the reverse isomers of 1,2-diacyl-*rac*-glycerols containing polyunsaturated acyl chains as the 3,5-dinitrophenyl-urethanes. Peak identification: A, 1-arachidonoyl-2-oleoyl-*rac*-glycerol and 1-oleoyl-2-arachidonoyl-*rac*-glycerol; B, 1-eicosapentaenoyl-2-oleoyl-*rac*-gly-cerol and 1-oleoyl-2-eicosapentaenoyl-*rac*-glycerol; C, 1-docosahexaenoyl-2-palmitoyl-*rac*-glycerol and 1-palmitoyl-2-docosahexaenoyl-*rac*-glycerol; D, 1-docosahexaenoyl-2-oleoyl-*rac*-glycerol and 1-oleoyl-2-docosahexaenoyl-*rac*-glycerol. HPLC conditions as given by Itabashi et al. (2000).

recovered in the lysoPtdIns released by phospholipase A_2, or by subtraction of the fatty acids in the *sn*-2-position (released by phospholipase A_2) from the total. The positional distribution of fatty acids may also be determined by lipase hydrolysis of the *sn*-1,2-diacylglycerol moieties released from the PtdIns by PtdIns-specific phospholipase C. The latter type of positional analysis is best carried out following acetylation of the diacylglycerols, which prevents isomerization. The composition of the fatty acids in the *sn*-1-position can be directly determined by hydrolysis of the phosphatide with the lipase from *Rhyzopus arrhizus*, which releases specifically the fatty acid from the primary (*sn*-1-) position of the glycerol molecule. This assay is best performed with the diacylglycerol moiety of the PdIns. The fatty acid composition of the *sn*-1-position may also be obtained by subtraction of the fatty acids in the *sn*-2-monoacylglycerol from the total. The reaction with *R. arrhizus* lipase may be performed as described by Slotboom et al. (1970) (see above). The composition of the alkyl and alkenyl groups in the *sn*-1-position of PtdIns is determined as described in Chapter 5.

2.2.2.3. sn-2-position

Likewise, the fatty acid composition of the *sn*-2-position may be obtained in a variety of ways. The fatty acid composition of the *sn*-2-position of PtdIns can be obtained directly by isolating the free fatty acids released by phospholipase A_2 from PtdIns. For this purpose the snake venom enzymes are usually employed, but bee venom has also been used. This reaction can be performed on the total PtdIns or a chromatographic fraction thereof (see above). The lipid extract from transmethylation of the *sn*-1-lyso PtdIns would also contain the dimethylacetals resulting from any alkenylacyl-GroPIns and 1-alkylglycerols resulting from any alkylacylGroPIns (see Chapter 5).

A detailed positional distribution of the fatty chains in the PtdIns molecules is only infrequently determined, most investigators relying on the results of the early reports, which dealt exclusively with the positional distribution of the acyl chains. Table 2.2 compares the distribution of fatty acids in the *sn*-1- and *sn*-2-positions of the PtdIns of rat liver (Holub and Kuksis, 1978) and the soybean (Myher and Kuksis, 1984a). The fatty acids of lysoPtdIns

TABLE 2.2

Composition and positional distribution of fatty acids in rat liver and soybean PtdIns [a,b]

Fatty acids	Liver (Holub and Kuksis, 1978)		Soybean (Myher and Kuksis, 1984)	
	Sn-1	Sn-2	Sn-1	Sn-2
16:0	6.6 ± 0.3	2.8 ± 0.2	66.4 ± 1.0	37.0 ± 0.3
16:1	trace	0.5 ± 0.1	0.8 ± 0.1	0.5 ± 0.1
18:0	90.6 ± 1.5	5.3 ± 0.3	15.8 ± 0.4	8.7 ± 0.3
18:1	2.2 ± 0.1	2.0 ± 0.1	6.1 ± 0.1	5.3 ± 0.1
18:2	0.6 ± 0.1	2.5 ± 0.1	7.7 ± 0.1	41.7 ± 0.3
18:3		trace	0.8 ± 0.1	4.7 ± 0.1
20:3		4.3 ± 0.1		
20:4		74.2 ± 1.5		
20:5		1.3 ± 0.1		
22:3		1.4 ± 0.1		
22:5		2.1 ± 0.1		
22:6		3.6 ± 0.1		
Other			1.4 ± 0.2	0.8 ± 0.2

[a]Determined from fatty acid compositon of liberated lysoPtdIns (sn-1-position) and free fatty acids (sn-2-position) following phospholipase A_2 hydrolysis.
[b]All values expressed as averages ± range/2 (mole %).

(sn-1-position) and of free fatty acids (sn-2-position) were obtained by digestion with phospholipase A_2. In both instances the sn-1-position is occupied largely by saturated C_{16} and C_{18} fatty acids but smaller amounts of the mono- and diunsaturated fatty acids are also detected. The animal and plant PtdIns differ greatly in the fatty acid composition of the sn-2-position, which is occupied largely by arachidonic acid in the liver PtdIns and by linoleic acid in the soybean PtdIns.

Interestingly, the soybean PtdIns has been reported to be toxic to tumor cells in culture without affecting normal cell lines (Jett and Alwing, 1983).

Likewise, the distribution of fatty acids in the sn-1- and sn-2-positions of the PtdIns of human red blood cells has been obtained by digestion with phospholipase A_2 (Myher et al., 1998). The primary position contains largely saturated and monounsaturated fatty acids but small amounts of polyunsaturates are also found. The presence of small amounts of polyunsaturates in the *sn*-1-position has been confirmed by lipolysis with the *R. arrhizus* enzyme, which is specific for the primary position (see above). The PtdIns of human and bovine erythrocytes have been shown to contain small amounts of alkylacylGroPIns, but the positional location of the alkyl group has not been specifically determined. It has been assumed to occupy the *sn*-1-position (Lee et al., 1991; Butikofer et al., 1992). Recently, Nair et al. (1999) have reported the presence of dimethylacetals among the transmethylation products of PtdIns isolated from porcine cardiomyocytes. A preferential, but not absolute, location of the arachidonic acid in the *sn*-2-position of PtdIns has been demonstrated by analysis of the reverse isomers of the diradylglycerol moieties (Itabashi and Kuksis, 2001).

2.2.3. Determination of molecular species

The full details of the PtdIns structure are revealed by analysis of molecular species. This analysis can be performed with the intact molecules or appropriate degradation products thereof, and their derivatives. The methods vary greatly in the ease of application. Direct mass spectrometric examination of PtdIns extracts provides the most rapid means of determining the molecular species composition, but quantification presents difficulty and the overall results do not provide the detail obtained by the more time consuming and labor intensive chromatographic examination of the derived diradylglycerol moieties.

2.2.3.1. Intact PtdIns

Although less efficient and more complicated, analyses of molecular species of intact PtdIns are possible and necessary when assessing the molecular association of the radyl chains with radio- or stable-isotope labeled polar head groups. Such analyses are best performed with reversed or normal phase HPLC with mass spectrometry, but flow ES/MS and MS/MS has been more frequently employed.

It may be noted, that attempts to develop methods for the determination of molecular species of PtdIns, which would preserve both polar and non-polar parts of the molecule, were reported long before the arrival of FAB/MS (Luthra and Sheltawy, 1972a,b). A comparison of the molecular species of PtdIns from three non-myelinic and two myelinic fractions of ox cerebral hemispheres showed that the trienoic species constituted 20% of the PtdIns from the two myelinic fractions (large and small) but less than 14% of the non-myelinic fractions. Arachidonic acid was shown to be paired exclusively with stearic acid, while oleic acid in the monoenes was paired with palmitic and stearic acids. The nuclear and nerve ending plus mitochondrial fractions were enriched in the monoenoic and tetraenoic species of PtdIns when compared to myelinic fractions. Only the post-mitochondrial supernatant contained significant quantities of fully saturated species of this glycerophospholipid.

FAB/MS. Mass spectrometric analyses of the molecular species of intact PtdIns were first made with FAB/MS (Jensen et al., 1987). One of the more complete analyses was reported by Cronholm et al. (1992), who determined the molecular species of PtdIns, PtdIns(4)P and PtdIns(4,5)P_2 from the pancreas of rats fed an ethanol-containing liquid diet for 24 h and from corresponding pair-fed animals. The compositions of PtdIns(4)P and PtdIns(4,5)P_2 were similar to that of PtdIns, with the stearoyl/arachidonoyl species constituting about 32% of the total, compared with 38% in PtdIns. The PtdIns species having an arachidonoyl group were about half as abundant in the ethanol-treated as in the control rats. For changes in the molecular species of PtdInsPs (see below). The PtdIns samples were dissolved in 20 μl of chloroform/methanol/water (86:14:1, by vol.) and one tenth of the sample was applied under a gentle stream of N_2 to the FAB target, already treated with triethanolamine. Spectra of negative ions were taken in the negative ion mode and the molecular species were identified based on molecular weight and chromatography. The chromatographic separations were performed on Lipidex-DEAP and PtdIns and PtdSer were eluted with chloroform as described by Schacht (1981).

Flow injection ES/MS. Flow injection ES/MS with MS/MS offers a rapid method of determining molecular species of PtdIns without prior chromatographic separation. Han et al. (1996) have identified and quantified (by means of an internal standard) the five major species of PtdIns in total phospholipid extracts from human platelets. ES/MS of chloroform extracts of human platelets in the negative ion mode demonstrated the presence of multiple individual molecular species of PtdIns, e.g. *m/z* 836 (16:0/18:1 GroPIns); 858 (16:0/20:4 GroPIns); 864 (18:0/18:1 GroPIns); 884 (181/20:4 GroPIns); and 886 (18:0/20:4 GroPIns), which were similar to those previously described, revealing a high proportion of arachidonic acid containing species.

Atmospheric pressure chemical ionization (APCI) MS has also been employed for molecular species analysis of complex lipids in the flow injection mode, but it is not very well suited for the analysis of glycerophospholipids in general (Byrdwell, 2001).

Koivusalo et al. (2001) have concluded that quantitative determination of molecular species of phospholipids, including PtdIns, can be obtained by ESI/MS only if proper attention is paid to experimental details, particularly the choice of internal standards. The molecular species of PtdIns were measured in BHK-21 cells using an ion trap instrument (Esquire-LC, Bruker-Franzen Analytik, Bremen, Germany) and a triple quadrupole instrument (Perkin–Elmer Sciex API 300). The lipids were dissolved in chloroform/methanol 1:2 (v/v) with or without NH_4OH (0.25–1%) or 0.1 mM NaCl, and were infused to the electrospray source via a 50 μm i.d. fused silica capillary using a syringe pump at the flow rate of 5 μl/min. With the ion trap, nitrogen was used as the nebulizing (at 5–6 ψ) and a drying gas (5–7 l/min at 200°C). The potentials of the spray needle, capillary exit, and skimmer 1 were set to ±4000 90–150 and 25–50 V, respectively. With the triple quadrupole instrument, the spary capillary voltage was 4000 V and orifice voltage 25 V in positive and negative ion ESI. Synthetic air was used as the nebulizing gas, and nitrogen was used as the curtain and collision gas. The most important novel finding of the study was that unsaturation of phospholipid acyl chains can have a major effect on instrument response, which was found to vary by as much as 40% depending on the total number of double bonds and the lipid concentration. The structure of the polar head groups was also found to have a major effect on the instrument response. In the negative ion mode, the acidic phospholipids including PtdIns and PtdOH, gave a much higher response than PtdEtn and PtdCho. The variation in instrument response depending on phospholipid acyl chain length had been observed previously (Brugger et al., 1999; Fridriksson et al., 1999).

Normal phase HPLC and LC/MS. Normal phase HPLC is capable of resolving the PtdIns from other glycerophospholipids and therefore does not require prior isolation PtdIns. It is especially convenient to combine normal phase HPLC with on-line ES/MS, which provides the molecular weights of the species. A collision induced dissociation (CID) of the parent ion peaks yields fragment ions from which the corresponding molecular species can be identified. Karlsson et al. (1996) used normal phase LC/ESI/MS to analyze the major molecular species of PtdIns of gastric juice (Fig. 2.6). The parent ions of the major species submitted to MS/MS analysis revealed combinations of 16:0, 18:0, 18:1, 18:2 and 20:4 fatty acids.

Hvattum et al. (2000) used normal-phase HPLC coupled with negative ion ES/MS/MS to determine individual molecular species of PtdIns from Atlantic salmon head kidney along with the response of the distribution to feeding of capelin and soybean oil. In all instances 18:0/20:4 species made up 36 to 55% of total while

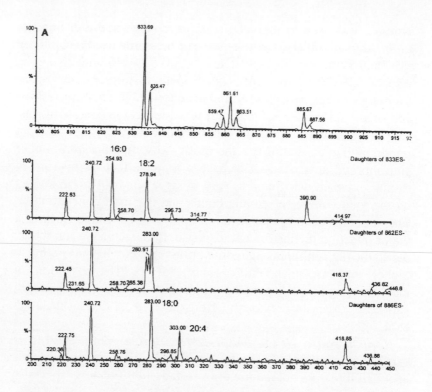

Fig. 2.6. Mass spectra of PtdIns from a gastric juice sample. A, Mass spectrum of PtdIns; B, C and D, MS/MS daughter-ion spectra of major PtdIns species. B, daughters of 833; C, daughters of 862; D, daughters of 886. Major fatty acids are indicated in the figure. Reproduced from Karlsson et al. (1996) with permission of publisher.

18:0/22:6, 16:0/22:6 and 16:0/20:4 provided the other major molecular species and did not exceed 10% each. The intensity ratio of *sn*-1/*sn*-2 fragment ions increased with increasing number of double bonds in the *sn*-2 acyl chain but was not affected by increasing number of double bonds in the *sn*-1-acyl chain of the species examined. The relative distribution of molecular species was determined by multiple reaction monitoring of the carboxylate anion fragment from the *sn*-1- position. The assignment of the fatty acid to the *sn*-1- and *sn*-2-positions for PtdIns was easier than for other phospholipid classes. All the species identified contained fatty acids with carbon chain length 16 or 18 combined with carbon chain lengths of 20 or 22. The *sn*-1 to *sn*-2 abundance ratio of the 16:0/20:4 GroPIns species was approximately 1. Table 2.3 shows the effect of diet on the molecular species composition of the salmon kidney PtdIns, as determined by LC/ES/MS (Hvattum et al., 2000). Table 2.3 emphasizes the high content of 18:0/20:4 species in the PtdIns and its resistance to dietary influences.

Nair et al. (1999) have shown that adult porcine cardiomyocytes grown in media supplemented with 20:4, 20:5 and 22:6 acids become preferentially incorporate 20:5 and 22:6 fatty acids in the PtdIns fraction, as shown by analysis of fatty acids recovered from different phospholipid classes.

TABLE 2.3

Molecular species of PtdIns from extracted salmon kidney after three months on soybean (SO) and fish oil (FO) diets as determined by normal phase HPLC with negative electrospray tandem mass spectrometry (mole %)

	$[M - H]^-$ m/z	50SO–50FO Mean (SD)	SO Mean (SD)	FO Mean (SD)
16:0/20:5	856	$3.1^b \pm 0.6$	$1.2^c \pm 0.1$	$5.4^a \pm 0.5$
16:0/20:4	858	$7.1^b \pm 0.8$	$5.9^c \pm 0.6$	$10.0^a \pm 0.3$
16:0/20:3	860	$2.4^a \pm 0.2$	$1.8^b \pm 0.2$	$1.1^c \pm 0.1$
16:0/22:6	882	$6.7^b \pm 0.3$	$4.9^c \pm 0.2$	$8.6^a \pm 0.2$
18:1/20:5	882	$2.3^b \pm 0.4$	$0.88^c \pm 0.14$	$3.9^a \pm 0.2$
18:0/20:5	884	$9.4^b \pm 0.4$	$4.4^c \pm 0.8$	$10.6^a \pm 0.6$
18:1/20:4	884	$5.0^b \pm 0.4$	$5.8^b \pm 0.7$	$7.0^a \pm 0.2$
18:0/20:4	886	$42.3^b \pm 0.8$	$55.3^a \pm 1.1$	$36.0^c \pm 1.0$
18:1/20:3	886	$2.1^a \pm 0.3$	$2.0^a \pm 0.3$	$0.88^b \pm 0.19$
18:0/20:3	888	$5.5^a \pm 0.4$	$6.5^a \pm 1.5$	$1.6^b \pm 0.1$
18:1/22:6	908	$3.5^b \pm 0.1$	$2.0^c \pm 0.1$	$4.6^a \pm 0.2$
18:0/22:6	910	$9.7^a \pm 0.7$	$8.6^a \pm 0.7$	$9.2^a \pm 0.3$
18:0/22:5	912	$0.92^{a,b} \pm 0.18$	$0.68^b \pm 0.09$	$0.99^a \pm 0.15$

Values marked with different superscript letters (*a*, *b* and *c*) are significantly different, based on 95% confidence interval. The mass of all ions was rounded to nearest integer. Reproduced from Hvattum et al. (2000) with permission of the publisher. SD, standard deviation.

Kurvinen et al. (2000) used normal phase HPLC in combination with on-line electrospray ionization and single quadrupole spectrometer to determine the effect of feeding n-3 and n-6 fatty acid esters on the molecular species of phosphatidylinositol of guinea pig brain using a previously described method (Ravandi et al., 1995). Dietary deficiency of n-3 fatty acids mainly affected the molecular species of alkenylacyl and diacyl GroPEtn and GroPSer, while PtdCho was affected less and PtdIns not at all. Uran et al. (2001) reported quantitative analysis of PtdIns species in human blood using normal phase liquid chromatography coupled with electrospray ionization ion-trap tandem mass spectrometry. Chromatographic baseline separation was obtained between PtdIns and other glycerophospholipids on an HPLC diol column and a gradient of chloroform and methanol with 0.1% formic acid, titrated to pH 5.3 with ammonia and added with 0.05% triethylamine. The species from each phospholipid class were identified using MS^2 or MS^3, which forms characteristic lyso-fragments.

Reversed phase HPLC and LC/MS. Reversed phase HPLC resolution of molecular species of PtdIns requires a prior isolation of pure PtdIns. This is usually accomplished by normal phase TLC or by affinity column chromatography. Reversed phase HPLC of intact PtdIns can be performed, however, without prior derivatization of the PtdIns molecule, although it complicates detection and quantification (Patton et al., 1982; Myher and Kuksis, 1984a). Thus, excellent separations of the molecular species of PtdIns from rat liver have been obtained by Patton et al. (1982) on a 250 mm × 4.6 mm column of 5 μm particle size (Ultrasphere OS, supplier). The column is eluted with 20 mM choline chloride in methanol/water/acetonitrile (90:7:2.5, by vol.). The eluate is monitored at 205 nm. For quantification, however, it is necessary to calibrate the UV response, which varies greatly with unsaturation of the molecular species. The molecular species are resolved in order of increasing partition number. The major species of rat liver

PtdIns are eluted in the following order: 16:0/18:1; 18:0/20:4; 18:0/ 18:2; 16:0/22:5; 18:0/20:3; 18:0/22:4 and 18:0/18:1.

An improved method for separating and quantifying molecular species of intact PtdIns has been reported by Wiley et al. (1992). PtdIns from rat heart and liver was isolated by conventional solvent extraction but was resolved from other glycerophospholipids including PtdSer by silica gel (LiChrosorb Si 100) chromatography by eluting the lipids with isopropanol/hexane/ethanol/1 mM ammonium phosphate/acetic acid (490:370:1165:25:0.4 or 490:370:60:80:0.4 by vol.). For separation of molecular species, the PtdIns fraction was redissolved in 0.5 to 2 ml methanol/water/ acetonitrile (90.5:7.0:2.5, by vol.) per gram wet weight of liver. A portion of the resulting solution (100–500 µl) was injected onto a 4.5×250 mm^2 column packed with octadecyl silica (5 µm particle size Ultrasphere ODS; Rainin Instrument Co., Woburn, MA). The species were resolved by eluting the column with methanol/286 mM aqueous choline chloride/acetonitrile (90.5: 7.0:2.5, by vol.) using a flow rate of 1 ml/min. The molecular species of PtdIns from rat liver contained 18:0/20:4 as a major component accompanied by smaller amounts of 16:0/20:4, 16:0/ 18:2, 18:1/18:2, 18:0/22:6, 18:0/22:5, 18:0/20:3 (Fig. 2.7). The molecular species were identified on basis of retention time of standards and the fatty acid composition of each peak.

Earlier Kim and Salem (1987) had resolved the molecular species of intact PtdIns from rat liver by a reversed phase LC/MS with thermospray interface (Fig. 2.8). The column was eluted with methanol/hexane/0.1 M ammonium acetate in H$_2$O (500:25:25, by vol.) with a flow rate of 1 ml/min of 2 ml/min. Under the thermospray conditions, the spectrum of PtdIns species was dominated by monoacyl-([RO + 74]$^+$), and diacylglycerol-type fragments. There was also an ion due to the ammonium adduct of Ins at m/z 198.

More recently, Li et al. (1999) determined intact PtdIns molecular species including the arachidonoyl-containing phospholipid species using capillary HPLC/continuous flow liquid secondary

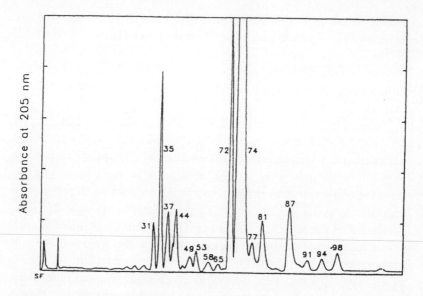

Fig. 2.7. Reversed phase HPLC resolution of molecular species of rat liver PtdIns. Major molecular species: **31**, 16:0/22:6; **35**, 16:0/20:4; **37**, 16:0/18:2; **44**, 18:1/ 18:2; **72**, 18:0/22:6; **74**, 18:0/20:4 **81**, 18:0/22:5; **87**, 18:0/20:3; **98**, 18:0/22:4. Chromatographic conditions: column, packed octadecyl silica (Dynamax-300A). Elution conditions: methanol/286 mM aqueous choline chloride/acetonitrile (90.5:7.0:2.5 by vol.) using a flow rate of 1 ml/min, which was critical for optimum resolution of peaks. Absorbance of eluate was measured at 205 nm and recorded simultaneously. Reproduced from Wiley et al. (1992) with permission of publisher.

ion mass spectrometry (CF-LSIMS). CF-LSIMS analyses were made on a JEOL HX110A double focusing mass spectrometer. The HPLC isolated phospholipid classes were dried under vacuum and dissolved in 1 ml of methanol-propan-2-ol (80:20, v/v) containing 1.5% glycerol. The capillary HPLC flow was supplied by splitting the main flow (1.0 ml/min) from a Waters 600MS HPLC pump to 3 μl/min using an open split at a VALCO tee. The flow directed through a Valco injector with 10 μl loop and then through a KAPPA Hypersil BDS C_{18} capillary column (100 mm × 0.30 mm ID).

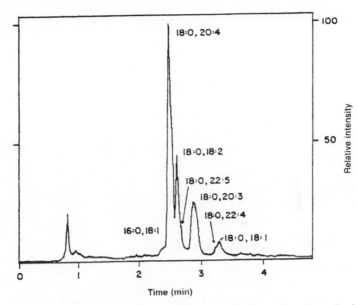

Fig. 2.8. Reversed phase LC/MS profile of the PtdIns of rat liver. Peaks are identified as indicated in the figure. Instrumentation: Beckman Model 114 M Liquid Chromatograph equipped with a Du Pont Zorbax C-18 column (5 μm, 4.6 mm × 25 cm) and connected to an Extrel 400-2 quadrupole mass spectrometer via a Vestec thermospray interface (Vestec, Houston, TX). Total ion chromatogram (mass range 105–1005 Da) obtained for 20 g of PtdIns from bovine liver using methanol/hexane/0.1 M NH₄OAc (500:25:25, by vol.) as developing solvent. Redrawn from Kim and Salem (1987) with permission of publisher.

The mass spectrometer was operated in the LSI/MS mode. Ion was produced by bombardment with a beam of Cs^+ ions (10 keV for the positive ion mode and 15 keV for the negative ion made), with the ion source accelerating voltage at 10 kV (Fig. 2.9).

2.2.3.2. PtdOH

The PtdOH is released from PtdIns by PtdIns-specific phospho-lipase D. PtdOH may also occur free in the tissues from which they are isolated along with other phospholipids. Since PtdOH readily

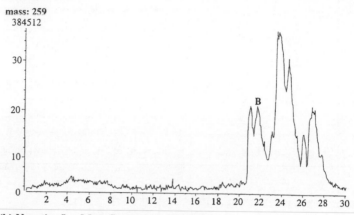

(a) Negative Ion

(b) Negative Ion Mass Spectrum of Peak B

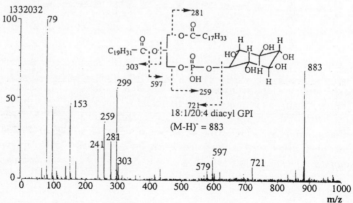

Fig. 2.9. Negative ion CF-LSIMS of PtdIns isolated from human dU937 cells. (a) Reconstructed ion chromatogram of phosphoinositol anion (*m/z* 259); (b) negative ion mass spectrum of 18:1/20:4 GroPIns. A capillary HPLC flow was supplied by splitting the main flow from a Waters 600MS HPLC pump to 3 µl/min using an open slit at a Valco tee. The flow was directed through a Valco injector with 10 µl loop and then through a KAPPA Hypersil BDS C18 capillary column (100 mm × 0.30 mm i.d.) to the frit of the JEOL mass spectrometer.
Reproduced from Li et al. (1999) with permission of publisher.

associates with salts, great care must be used to avoid contamination with metal ions, which leads to adsorption and peak tailing.

When extremely low levels of the PtdIns are suspected, the phospholipase D digestion and the PtdOH isolation may be performed in presence of 25 μg each of myristic acid, dimyristoylglycerol, and dimyristoylGroPOH as carriers (Deeg and Verchere, 1997). The samples are spotted on silica gel 60 TLC plates (Merck and Co., Gibbstown, NJ) and the plates are developed with chloroform/methanol/0.25% KCl (55:45:5, by vol.). The PtdOH spots on the plates were located by iodine vapors. Radioactive peaks were also recognized that migrated with myristic acid and dimyristoylglycerol. PtdOH, the expected product from phospholipase D hydrolysis of PtdIns or other sources of PtdIns accounted for 95% of the total radioactive products. This activity makes PLD_2 a useful reagent for comparing molecular species of all glycerophospholipid classes, including PtdInsPs and glycosylPtdIns, through the formation of PtdOH from each (Holbrook et al., 1992).

PtdOH has been subjected to molecular species analysis by MS using a variety of methods of sample ionization. Murphy and Harrison (1994) have reviewed the FAB/MS of PtdOH. FAB/MS of molecular species readily produces a protonated molecule $[M + H]^+$, provided the liquid matrix is acidified with a strong acid such as HCl. Facile fragmentation with loss of neutral phosphoric acid is a characteristic feature of the FAB (as well as liquid SIMS) spectra of PtdOH. However, negative ions readily form during FAB ionization of this acidic glycerophospholipid because of the very low pK_a of the phosphoric acid residue. Negative ion FAB fragment ions include carboxylate anions corresponding to the fatty acyl groups from the sn-1- and sn-2-positions. MS/MS of the $[M + H]^+$ and $[M - H]^-$ molecular ion reveals the molecular weight of the species. Following CID, $[M + H]^+$ ions decompose to a characteristic ion corresponding to the loss of 98 u (phosphoric acid). A neutral loss scan for 98 mass units can be used to identify the $[M + H]^+$ ions of all the PtdOH species.

Holbrook et al. (1992) has compared the FAB/MS spectra of the PtdOH moieties derived from PtdCho, PtdEtn, PtdSer and PtdIns extracted from PC12 cells and hydrolyzed with phospholipase D. The molecular species of PtdOH can also be readily recognized and quantified by FAB/MS/MS. This method disposes the analytical column, or reduces its service to that of a convenient injection port. The identification of the species is based on MS/MS fragmentation of the parent molecules. Both free PtdOH and the dimethyl esters of PtdOH are equally well suited for this purpose.

When ethanol is present in an *in vitro* incubation along with phospholipase D, phospholipids are converted into PtdEt rather than PtdOH by transphosphatidylation. Holbrook et al. (1992) used this reaction in the formation of PtdEt as a marker for endogenous phospholipase D substrate. The resultant PtdEt was analyzed by FAB/MS (see Chapter 9).

Reversed phase HPLC and LC/MS. The molecular species of PtdOH may be resolved by reversed phase HPLC following methylation. The methyl esters are prepared by reacting the acid form of PtdOH directly with diazomethane. The dimethyl PtdOH is almost pure as shown by TLC on silica gel with diethyl ether as solvent (Luthra and Sheltawy, 1976).

The molecular species of the dimethyl esters of PtdOH may be determined by combining reversed phase HPLC with TS/MS or ES/MS. The dimethyl esters are eluted from the reversed phase HPLC column in order of decreasing polarity. The peaks are detected and quantified by UV at 205 nm. As noted above, this method of detection, however, lacks sensitivity and specificity and requires extensive calibration of response for quantitative analysis.

Normal phase HPLC and LC/MS. Normal phase HPLC on its own is incapable of resolving any molecular species of PtdOH. However, in combination with ES/MS, it can provide a rapid and detailed account of most molecular species as the PtdOH dimethyl esters.

TLC and AgNO₃-TLC. For the resolution of the molecular species of PtdOH, the dimethyl esters of PtdOH are subjected to TLC on silica gel G impregnated with $AgNO_3$ (10%, w/w) (Luthra and Sheltawy, 1976). The plates are spotted and developed at 4°C with $CHCl_3/CH_3OH/H_2O$ (120:20:1, by vol.) as the developing solvent. The developed plates are dried and sprayed with 2′,7′-dichlorofluorescein (0.02%, w/w, in ethanol) and viewed under UV light. The molecular species are resolved into saturates, monoenes, dienes, trienes, tetraenes, pentaenes and hexaenes. The identity of the TLC bands is established by transmethylating the various fractions and determining the fatty acid composition by GLC (Luthra and Sheltawy, 1976).

2.2.3.3. Diradylglycerols

The diradylglycerols may be released from the parent PtdIns by hydrolysis with non-specific phospholipase C (Holub et al., 1970). The hydrolysis is illustrated with phospholipase C from *B. cereus* (Sigma Chemical Co., type III or type V), which yields an *sn*-1,2-diacylglycerol and a water-soluble InsP (Kuksis and Myher, 1990).

According to Kuksis and Myher (1990) purified PtdIns (approximately 500 μg) and 50μg of BHT as antioxidant are suspended in 1.5 ml of peroxide-free diethyl ether and 1.5 ml of buffer (17.5 mM trishydroxy-methyl)aminomethane adjusted to pH 7.3 with HCl, 1.0 mM $CaCl_2$), containing 10 units of phospholipase C (*B. cereus*, Sigma, Type III). The mixture is agitated using a vortex mixer for 10 s and then shaken at 37°C on a Buchler rotary Evapo-Mix for 3 h. The diradylglycerols are extracted two times with 2 ml of hexane. The extracts are combined and passed through two successive Pasteur pipettes containing anhydrous Na_2SO_4. After reduction in volume, the extent of hydrolysis is checked by spotting the sample in a narrow lane (about 1 cm) beside 25 μg of undigested PtdIns on a silica gel plate, followed by developing in chloroform/methanol/aqueous ammonia (65:25:4, by vol.). After the plate has been sprayed with dichlorofluorescein, the amount of any residual PtdIns may be estimated by visual comparison with 25 μg of reference PtdIns standard.

The released diradylglycerols are purified by borate TLC (to remove any sn-1,3-isomers formed by isomerization during the enzyme digestion) and after recovery from the TLC plate are immediately converted into TMS or TBDMS ethers by reaction with the corresponding trialkylchlorosilane under conditions that do not lead to isomerization during the reaction.

Thus, the TMS ethers are prepared (Myher and Kuksis, 1975) by reaction with 100 μl of pyridine/hexamethyldisilazane/trimethylchlorosilane (15:5:2, by vol.) for 30 min at room temperature. The reagents are then evaporated under nitrogen gas and the products dissolved in 2 ml of hexane. After a brief centrifugation (1000g), the supernatant is evaporated under nitrogen gas and the sample dissolved in an appropriate volume of hexane. The centrifugation step can be avoided by using pyridine/bis(trimethylsilyl)-trifluoroacetamide/trimethyl chlorosilane (50:49:1, by vol.) as the silylating reagent, which yields volatile by-products.

The TBDMS ethers are prepared (Myher et al., 1978) by reacting the diradylglycerols with 150 μl of a solution of 1 M t-butyl-dimethylchlorosilane and 2.5 M imidazole in dimethylformamide for 20 min at 80°C. After the reaction mixture is dissolved in 5 ml of hexane, the products are washed three times with H_2O and then dried by passage through a small column (Pasteur pipette) containing anhydrous sodium sulphate. The diacyl-, alkylacyl and alkenylacyl subfractions of the diacylglycerols may be resolved as the TMS or the t-BDMS ethers by normal phase HPLC (Chapter 5).

Non-polar capillary GLC and GC/MS. Non-polar capillary columns provide a highly sensitive and reliable separation and quantification of molecular species within each diradylglycerol subclass on the basis of molecular weight or carbon number. The TMS ethers of the diradylglycerols are resolved on an 8 m fused silica capillary column coated with a non-polar permanently bonded

SE-54 liquid phase. The samples are injected on-column and the oven temperature is programmed from 40 to 150°C at 30°C/min, then to 230°C at 20°C/min and to 340°C at 5°C/min (Myher and Kuksis, 1984a). Hydrogen at 6 ψ (1 ψ = 6.9 kPa) head pressure was used as a carrier gas. This separation may be combined with a preliminary resolution of the diacylglycerols by AgNO$_3$-TLC, AgNO$_3$-HPLC or reversed phase HPLC (Myher and Kuksis, 1982).

   Polar capillary GLC and GC/MS. GLC on polar capillary columns provides the most extensive resolution of molecular species yet achieved by any method within each class of diradylglycerols. The molecular species are resolved on the basis of both carbon and double bond number by GLC at 250°C on a 10 m capillary column coated with an SP-2330 polar liquid phase (68% cyanopropyl, 32% phenylsiloxane) (Myher and Kuksis, 1982). The temperatures of the injector and detector are maintained at 270 and 300°C, respectively. The samples are introduced with a split injector (split ratio 7:1) and hydrogen is used as the carrier gas. More recently (Myher et al., 1989a; Myher and Kuksis, 1989), diradylglycerol TMS ethers have been separated on fused silica columns (15 m × 0.32 mm) coated with cross-bonded RTx 2330 (Restek Corp., Port Matilda, PA) or with stabilized SP 2330 (Supelco) (Myher and Kuksis, 1989; Myher et al., 1989b). The carrier gas was hydrogen at 3 ψ head pressure and the column temperature isothermal at 250°C. The sample was admitted by a split injector (split ratio 7:1). Determinations of major components (greater than 10%) had coefficients of variation (CV) of 2% or less, whereas minor components (less than 1%) had CV of approximately 10%. Figure 2.10 shows the resolution of molecular species of the diacylglycerol moieties of human red blood cell PtdIns (Kuksis and Myher, 1990) on polar and non-polar capillary columns. The polar capillary column (a) resolves 10 times more components than the non-polar GLC column (b), which yields only a few. A similar resolution has been demonstrated elsewhere for PtdIns of the soybean (Myher and Kuksis, 1984a) and of human plasma (Myher et al., 1989a,b). The analyses of the red cell PtdIns, however, did not

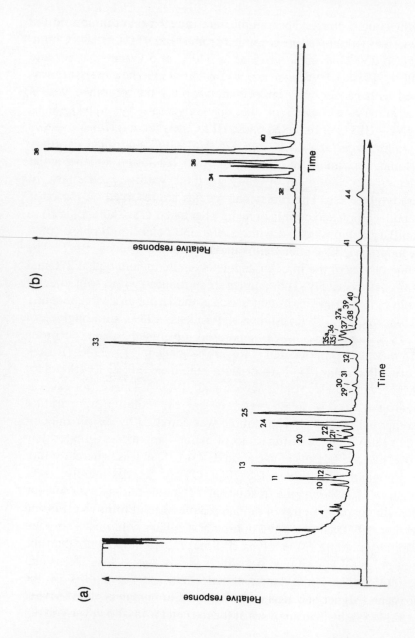

include the minor proportion of the akylacyl GroPIns later reported by Lee et al. (1991) and Butikofer et al. (1992).

The diacylglycerol peaks in the chromatograms are usually identified in relation to the retention times of either genuine standards or well-characterized natural samples (Myher and Kuksis, 1989). When these cannot be measured, which may be the case for very small or extremely complex mixtures, relative retention times are not sufficient for reliable peak identification. In such instances combinations of GLC or HPLC with mass spectrometry are helpful. Since polar capillary columns have low loading capacity and give too much bleed, thermally stable non-polar columns are usually employed for carbon number resolution and on-line GC/MS identification of molecular species of diacylglycerols.

Thus, the TBDMS ethers of diacylglycerols are identified by GC/MS using a non-polar column (Myher et al., 1978). The species within each GLC peak or carbon number are then identified on the basis of the pseudomolecular $[M - 57]^+$ ion, the $[acyl + 74]^+$ ions, and the fragment ions formed by loss of one of the acyl chains $[R - RCOO]^+$. The $[M - 57]^+$ ions arise from a facile loss of the t-butyl group which conveys a special advantage for the identification of molecular species of diradylglycerols by GC/MS. The more efficient capillary columns allow resolution of the molecular species within each carbon number.

Fig. 2.10. GLC resolution of diacylglycerol moieties of human red blood cell PtdIns on (a) polar and (b) non-polar capillary columns. Instrumentation: Hewlett-Packard Model 5880 gas chromatograph with flame ionization detector. Polar column, RTx2330 (15 m × 0.32 mm i.d.) operated isothermally (250°C); Non-polar column, 5% phenylmethyl silicone (8 m × 0.32 mm i.d.), temperature programmed from 40 to 350°C. Carrier gas, H_2 at 3 and 6 ψ, respectively. Major peaks: (a) **11**, 16:0/18:1; **13**, 16:0/18:2; **20**, 18:0/18:1; **24**, 18:0/18:2; **25**, 16:0/20:4; **33**, 18:0/20:4; and (b) peaks identified by total number of acyl carbons.
Reproduced from Kuksis and Myher (1990) with permission of publisher.

HPLC and LC/MS. Both normal and reversed phase HPLC are commonly employed for the purification and resolution of the diacylglycerol moieties of glycerophospholipids. The resolution of molecular species of fluorescent or UV absorbing derivatives of diacylglycerols is usually performed by reversed phase HPLC. However, there are specific applications where normal phase HPLC is required, e.g. resolution of diastereomers of diacylglycerol naphthylethyl urethanes. The reverse isomers of enantiomeric diacylglycerols are resolved by chiral phase HPLC.

Although normal phase HPLC can provide a limited resolution of the molecular species of diradylglycerol derivatives, it is usually neglected in favor of reversed phase HPLC. Nevertheless, it could be shown that arachidonic acid was paired exclusively with stearic acid in this tissue, while oleic acid in the monoenes was paired with palmitic and stearic acids. A comparison of the molecular species of PtdIns from three non-myelinic and two myelinic fractions of ox cerebral hemispheres showed that the trienoic species constituted 20% of the PtdIns from the two myelinic fractions (large and small) but less than 14% of the non-myelinic fractions (Luthra and Sheltawy, 1972b). The nuclear and nerve ending plus mitochondrial fractions were enriched in the monoenoic and tetraenoic species of PtdIns when compared to myelinic fractions. Only the post-mitochondrial supernatant contained significant quantities of fully saturated species of this glycerophospholipid. Thus, there are differences in the distribution and composition of the PtdIns among the tissues and among the subcellular fractions of the tissue. The PtdInsPs are found in the cytosol, along with the free PtdIns anchors.

Normal phase HPLC provides special advantages for the resolution of diastereomeric diradylglycerol naphthylethyl urethanes (Laakso and Christie, 1990). These derivatives yield excellent electrospray spectra in the form of strong signals for the parent ions, which are adequate for the identification of molecular species of the resolved *sn*-1,2-diradylglycerol derivatives (Agren and Kuksis, 2002).

The naphthylethyl urethanes are prepared by reacting the rac-diacylglycerols (<0.5 mg) in dry toluene (0.3 ml) containing 10 μl of (R)-(−)- or (S)-(+)-1-(1-naphthyl)ethyl isocyanate to which 4 mg of 4-pyrrolidinopyridine were added. The mixture was heated at 50°C overnight. After evaporation of solvents under a stream of nitrogen, the reaction products were dissolved in methanol/water (9:5, v/v) and applied to Sep-Pak C_{18} column, which had been solvated with the same solvent. A further 15 ml of this solvent was passed through the column and the naphthylethylurethanes were eluted with 10 ml of acetone (Agren and Kuksis, 2002).

Reversed phase HPLC is less efficient for resolution of molecular species of diacylglycerols than GLC on polar capillary columns, but offers advantages for the separation of polyunsaturated species. HPLC also allows peak collection. For HPLC analysis with UV detection, the diradylglycerols are converted into the benzoates (Blank et al., 1984), dinitrobenzoates (Takamura et al., 1986, 1989) or pentafluorobenzoates (Kuksis et al., 1991) by reaction with the corresponding acid anhydride or acid chloride and pyridine for 10–15 min at 60–80°C (Takamura et al., 1986). The method preparation of the UV absorbing derivatives is illustrated by synthesis of benzoates.

According to Blank et al. (1984) up to 2 mg of diradylglycerols are dissolved in 0.3 ml of benzene containing 10 mg of benzoic anhydride and 4 mg of 4-dimethylaminopyridine and are allowed to stand for 1 h. The samples are placed in an ice bath and 2 ml of 0.1 M NaOH added slowly. The diradylglycerobenzoates produced are extracted three times with hexane. When the amount of diradylglycerols reaches 10 mg, the benzoylation reagents are increased four-fold.

For preparation of dinitrobenzoates (Takamura et al., 1986) 2 mg of diacylglycerol and 25 mg of dinitrobenzoyl chloride are dried for 30 min in vacuo prior to use. The mixture is dissolved in 0.5 ml of dry pyridine and heated in a sealed vial at 60°C for 10 min. After cooling, the reaction is stopped by adding 2 ml of 80% (v/v) methanol and 2 ml of H_2O. Then the reaction mixture is applied to a SepPak C_{18} cartridge (Waters Associates, Walton, MA) and the cartridge is washed with 25 ml of 80% methanol to purify the sample. The product is recovered by eluting with 30 ml of methanol. After evaporation, the residue is dissolved in hexane and washed with H_2O. Diethyl ether and methanol contained 0.01% BHT. A prolonged reaction time and/or elevated temperature may lead to isomerization of the diradylglycerols, which may result in confusion about the true identity of the double peaks observed with the UV absorbing derivatives (Butikofer et al., 1992).

Butikofer et al. (1990) used the benzoates to separate the molecular species of the diradylglycerol moieties of membrane PtdIns of Torpedo marmorate, which, like the glycosyl PtdIns anchor of the acetylcholinesterase, were made up exclusively of the diacyl species (See also Chapter 5). Lee et al. (1991) prepared the benzoate derivatives of the diradyl GroPIns of bovine red blood cells and after a prefractionation into diacyl, alkylacyl and alk-1-enylacyl types, identified 2–2.4% alkenylacyl, 4.8–9.5% alkylacyl and 93–88% diacyl GroPIns as subfractions of prepared from inositol-containing phospholipids of bovine erythrocytes. It was suggested that the alkyacylGroPIns could represent precursors of the diaradyl GroPIns found in glycosyl PtdIns. Using a similar analysis of the diradylglycerolbenzoate derivatives, Butikofer et al. (1992) detected 1.5–3.5% alkylacyl GroPIns in human and 2.5–4.8% alkylacyl GroPIns in bovine erythrocytes. The alkenyl-acyl molecular species of GroPIns, however, were not detected. Table 2.4 gives the molecular species composition of the PtdIns of human and bovine erythrocytes (Butikofer et al., 1992). The table shows a heterogeneous species composition for both human and bovine erythrocyte alkylacyl GroPIns. Their compositions are distinctly different from those of human and bovine erythrocyte acetylcholinesterase glycosyl PtdIns anchors (see also Chapter 5). For HPLC analysis with fluorescence detection, the diradyl-glycerols are converted into naphthylurethanes (Rabe et al., 1989).

TABLE 2.4

Molecular species composition of membrane PtdIns from human and bovine erythrocytes

Molecular species	Composition of membrane PtdIns (mole %)						
	Human			Bovine			
	diacyl		alkylacyl	Diacyl		alkylacyl	
18:1/22:6			3.6				
16:0/22:6	0.3		13.3				
18:1/20:4	1.5		9.1	1.4		2.6	
18:0/20:4	9.1		9.2	1.1			
18:0/20:5			8.2	2.4			
16:1/18:1						3.8	
18:1/18:2		75					
18:0/22:6	4.0	25	5.7				
16:0/18:2	14.6	86		5.5		3.4	
18:0/18:3 + 16:0/20:3		14		4.8		1.0	
16:0/22:4			4.7				
18:0/22:5	2.0		5.8				
18:0/20:4	31.3		14.6	7.5		7.0	
18:1/18:1	1.7			2.5		8.7	
16:0/18:1		44			37		22
18:0/18:2	24.3	53		32.5	54	32.5	78
18:0/22:4		2	14.4				
18:0/20:3					9		
16:0/16:0	1.5			1.0		1.8	
17:0/18:1				3.3		1.5	
18:0/18:1	5.3			30.8		13.1	
16:0/18:0	1.7		3.5	0.7		5.0	
18:0/18:0	1.9			0.7			
Other	0.7		7.8	3.6		8.9	

Reproduced from Butikofer et al. (1992) with permission of publisher.

The preparation of these derivatives requires somewhat higher temperatures and longer times of reaction than the preparation of the trialkylsilyl ethers.

Haroldsen and Murphy (1987) developed a rapid sensitive procedure for analyzing diacylglycerodinitrobenzoates by electron capture negative CI/MS. The diacylglycerols released from the glycerophospholipids were derivatized with 3,5-dinitroben-zoylchloride. The dinitrobenzoates were resolved by reverse phase HPLC and the collected peaks were subjected to direct probe MS, which gave mass spectra characterized by an intense molecular anion. From this molecular anion the total carbon number and degree of unsaturation of the fatty chains could be determined. Analysis of fatty aid content of the molecular species allowed unequivocal assignment of structure of the diacylglycerols. Likewise, Kuypers et al. (1991) analyzed the benzoate derivatives of diacylglycerols released from glycerophospholipids by phospholipase C using reversed phase HPLC in combination with TS/MS. Molecular species showed as base peaks the salt adducts of the molecular ion, which permitted easy deduction of the overall fatty acyl composition. In addition, the diacylglycerol fragment of each species was found at $[M - 122]^-$ and two fragments formed by the loss of the fatty acyl groups (R) in the sn-1 and sn-2-position were found at $[M - R1]^-$ and $[M - R2]^-$, respectively. Since preferential release of either fatty acyl group was observed in positional isomers, the ratio of the intensity of these fragments gave information on the position of the fatty acyl groups in individual HPLC peaks.

Zhu and Eichberg (1993) used the benzoate derivatives of diacylglycerols derived from PtdIns of nerves of normal and experimentally diabetic rats. sn-1,2-Diacylglycerol moieties of purified PtdIns were released by phospholipase C from *B. cereus* and converted into benzoates by reaction with benzoyl chloride as described by Blank et al. (1984). Molecular species of diacylglycerobenzoates were separated on a C_{18} 4.6 × 250 mm Microsorb reversed phase column using a linear gradient of acetonitrile/isopropanol from 80:20 to 44:56 (v/v) from 0 to 70 min at a flow rate of 1 ml/min. The eluted benzoates were detected by a UV detector at 230 nm. The molecular species were provisionally identified on the basis of their retention times in comparison to reference standards. A more definite identification was made by collecting the each diacylglycerolbenzoate peak and determining its fatty acid composition following transmethylation with boron trifluoride-methanol. The most prominent molecular species of PtdIns was 18:0/20:4. The PtdIns molecular species pattern most closely resembled that of free diacylglycerol fraction. The principle difference was that there was a considerably greater proportion of the 18:0/20:4 molecular species in PtdIns than in the diacylglycerol fraction. The analyses demonstrated that the content of 18:0/20:4 GroPIns failed to decline at the significant rate shown for other species of the diacylglycerophospholipids analyzed.

Han et al. (1996) have claimed that reversed phase LC/MS of the benzoates of the diacylglycerol moieties of PtdCho and PtdEtn do not yield an accurate correspondence to the values obtained by flow injection MS/MS. An examination of their results, however, suggests that the HPLC data may have been obtained under less than optimum conditions. In the authors' laboratory an excellent agreement has been obtained between HPLC and LC/MS data (Ravandi et al., 1995).

The TBDMS ethers of the diradylglycerols are stable to moisture and can be effectively employed in reversed phase liquid chromatography (LC/CI/MS), where the facile loss of the t-BDMS group $[M - 132]^-$ together with the formation of prominent $[MH - RCOOH]^-$ ions provides the information necessary for peak identification of the molecular species (Pind et al., 1984). The pentafluorobenzoates are also readily detected in the UV and yield intense negative ions in CI/MS and are well suited for combined LC/MS applications (Kuksis et al., 1991). The pentafluorobenzoates yield a very strong signal for the molecular ion in NCI. The signal for the $[M]^-$ ion in the NCI s 500–1000 times stronger than any signal in positive chemical ionization (PCI). The NCI spectrum, however, does not yield any information about the structural components of the molecular species of the diradylglycerols, which must be obtained from the relative elution times and positive CI, where the $[MH - RCOOH]^+$ ions predominate, although at much reduced overall sensitivity when compared to the $[M]^-$ ion in NCI.

The knowledge of the fatty acid composition of the sn-1- and sn-2-positions of PtdIns, allows the calculation of the fatty acid pairing in the phosphatide molecule, including the reverse isomer composition by assuming the 1-random 2-random distribution. The calculated fatty acid pairing can be compared to that determined directly by molecular species analysis using a polar capillary GLC column, or a reversed phase HPLC column. The reverse isomer content, however, can be directly determined by reverse phase HPLC of the corresponding sn-1,2-diacylglycerol moieties of PtdIns using appropriate derivatives. Itabashi et al. (2000) have obtained excellent resolution of the reverse isomers of mixed chain diacylglycerols as the 3,5-dinitrophenylurethanes on both reversed phase and chiral phase columns.

According to Itabashi and Kuksis (2001) the diacylglycerols released from PtdIns by phospholipase C (see above) are purified by borate TLC on Silica Gel G (Merck, Darmstadt, Germany) impregnated with boric acid, using chloroform/acetone (98:2, v/v) as the developing solvent (Myher and Kuksis, 1984a,b). The purified diacylglycerols are converted into the 3,5-DNPU derivatives by reacting with 3,5-DNPI in dry toluene in the presence of pyridine at room temperature as described by Itabashi et al. (1990). The reverse phase HPLC of the reverse isomers of the sn-1,2-diacylglycerol DNPU derivatives was performed using an ODS column (Supersphere RP-18e, 250×4.6 mm i.d., 4 μm particle size, Merck) with a LiChrosorb RP-18 guard column. The analysis was done isocratically at 18°C using acetonitrile as the mobile phase at a constant flow-rate of 0.5 ml/min, which produced a pump pressure of about 55 kg/cm^2. Reversed phase HPLC with on-line ESI/MS of the reverse isomers was performed on a LCQ ion trap MS (Thermo Separation Products, San Jose, CA) using the same ODS column and mobile phase as those described above. The negative ES/MS spectra of the DNPU derivatives, which was essentially the same for both reverse isomers, gave prominent pseudomolecular ion $[M - 1]^-$.

Itabashi et al. (2000) have shown that it is possible to resolve the reverse isomers of the 1,2-diacyl-diacylglycerol moieties of glycerophospholipids containing polyunsaturated acyl chains as the 3,5-dinitrophenylurethanes using the reversed phase HOLC columns. Thus, the reverse isomer pair 20:4/18:1 and 18:1/20:4 gave a separation equivalent to that of 20:4/18:0 and 18:0/20:4, as well as of 20:4/16:0 and 16:0/20:4. Similar observations were obtained for other mixed acid polyunsaturated diacylglycerols when run as the DNPU derivatives. The new method thus far has not been applied to the resolution of the reverse isomers of the diacylglycerol moieties of natural PtdIns.

2.2.3.4. Inositol isomers

The molecular species of a PtdIns preparation may contain isomeric inositols, e.g. *myo-* and *chiro*-inositol in animal tissues, or *myo-* and *scyllo*-inositols in plant tissues. These isomers may be resolved by a variety of conventional chromatographic methods, as described above for total inositol preparations. The D- and L-inositols may be resolved as the PFP esters on a Chirasil-Val fused capillary columns as described above (Ostlund et al., 1993).

The chromatographic methods offer the opportunity to assay more than one inositol isomer at a time. According to Pak and Larner (1992) the PtdIns purified by TLC is hydrolyzed in sealed vials with 6N HCl at 110°C for 48 h. The acid hydrolysates are analyzed for inositol by HPAA-Dionex HPLC with Aminex HPX-87C column (Bio-Rad). GC/MS analysis of the hydrolysates is performed using a DB-5 column (0.32 mm × 30 m) after preparation of the hexatrimethylsilyl derivatives.

Ostlund et al. (1993) have described a chiral phase GLC method for the separation and quantification of enantiomeric D- and L-*chiro*-inositols in urine and plasma (see above). According Ostlund et al. (1993), to each sample (0.5 ml) was added 1.25 nmol of deuterated DL-chiro-inositol and 17.5 nmol of deuterated *myo*-inositol. The separation was achieved on a Chirasil-Val fused-silica capillary column (25 m × 0.32 mm i.d., Alltech) by modification of a published procedure (Leavitt and Sherman, 1982). For GLC, the inositol samples were converted into the hexakis[pentafluoropropionic] esters by reaction overnight at 65°C with 150 μl of a 1:1 (v/v) solution of pentafluoropropionic acid anhydride and acetonitrile. Most of the derivatizing reagent was removed prior to GLC analysis in order to avoid damage to the column.

Figure 2.11 shows the NICI GC/MS profile of a diabetic patient plasma with simultaneous monitoring of the PFP derivatives of natural *chiro*-inositol at *m/z* 1036 (Upper) and internal standard *chiro*-inositols at *m/z* 1041 (Lower). The slightly earlier elution time of the internal standard is characteristic of deuterium-labeled *chiro*-inositol pentafluoropropyl esters (Ostlund et al., 1993). In principle, D- and L-*chiro*-PtdIns may be differentiated by PtdIns-specific phospholipase C (*B. thuringiensis*) as reported by Bruzik et al. (1994), although a practical execution of the assay with small amounts of non-radioactive substrate may present problems.

2.3. Quantification

Several types of quantification must be distinguished in the quantitative analysis of PtdIns. First, the total amount of PtdIns may be measured in relation to its habitat, the wet tissue weight, or as mg% of plasma or plasma lipoprotein. This can be best accomplished by determining the organic phosphorus. Other techniques may be employed to determine the quantitative relationship among different parts of the PtdIns molecule and of the total molecule, as measured by phosphorus. Some of the quantitative measurements are better suited than others for determination of the quantitative composition of molecular species. MS with stable isotope labeled internal standards constitutes the definitive method for quantification of molecular species of PtdIns as well as for total PtdIns (Han and Gross, 1994). Aside from mass analyses, quantitative analyses may be made of the radioactivity,

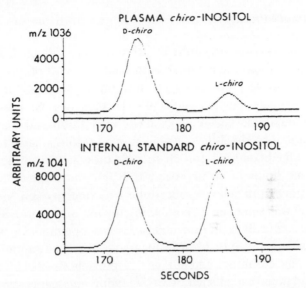

Fig. 2.11. NICI GC/MS of a diabetic patient plasma with simultaneous monitoring of the PFP derivatives of natural *chiro*-inositols at *m/z* 1036 (Upper) and internal standard *chiro*-inositols at *m/z* 1041 (Lower). Instrumentation: Hewlett-Packard 5988A gas chromatograph with on-line electropsray mass spectrometer. Column: Chiasil-Val fused-silica capillary column (25 m × 0.32 mm i.d., Alltech). Typical chromatographic conditions: injector, 189C; pressure, 5 ψ (helium); injector split, 12:1; temp. programmed stepwise from 100 to 185°C. Elution times: as shown in chromatogram. The slightly earlier elution time of the internal standard is characteristic of deuterium-labeled *chiro*-inositol PFP esters. Reproduced from Ostlund et al. (1993) with permission of publisher.

fluorescence, UV absorption and simple calorimetric assays. PtdIns in impure samples is quantified by methods that include some chemical or chromatographic resolution. MS methods can recognize the characteristic masses of PtdIns species and fragment ions and require the least purification. In all instances either an external or internal standard is necessary for reliable quantification.

2.3.1. Intact PtdIns

Intact PtdIns is most difficult to quantify. Direct ES/MS or LC/ES/MS is suitable for work with small amounts of material and does not require preliminary purification. Either a stable isotope labeled or nano-labeled molecular species can be used as an internal standard. The method needs calibration and may suffer from poor reproducibility. Brugger et al. (1997) have extended previous methods for a full characterization and quantification of membrane lipids in an unprocessed Bligh and Dyer (1959) total lipid extract from cells or subcellular structures by ESI/MS alone. This goal was achieved by combining the use of a nano-ESI source allowing the analysis of picomole or even subpicomole amounts of polar lipids, the use of MS/MS techniques such as parent ion and neutral loss scanning, and the use of phospholipid class-specific fragmentations established with synthetic analogues with non-natural fatty acid structures as internal standards. This approach was claimed to allow the specific detection of PtdIns and PtdInsPs, along with PtdCho, SM, PtdEtn, PtdSer, PtdOH and PtdGro.Uran et al. (2001), however, have pointed out that because of the ^{13}C isotope effect it is very difficult or impossible to calculate minor molecular species with fatty chains exceeding $C_{17:0}$ in mixtures of glycerophospholipids. Prior chromatographic separation of lipid classes avoids or minimizes the problem, which can be further eliminated by a further prefractionation of the molecular species by reversed phase HPLC.

Fridriksson et al. (1999) have used positive and negative mode electropsray Fourier transform ion cycloron resonance MS to compare quantitatively the phospholipid composition of isolated detergent-resistant membranes of RBL-2H3 mast cells. Using the relative intensity of PtdIns in positive and negative ion modes to normalize the spectra with respect to each other, the relative sensitivities of the standards were as follows. For the positive ion mode, 1.00:0.75:0.67 for PtdCho/PtdEtn/SM; and for the negative ion mode 1.00:1.14:0.46;0.52:1.2 for PtdIns/PtdGro/PtdSer/PtdOH/PtdEtn.

PtdIns exhibits characteristic features in both ^{31}P NMR and in ^{1}H NMR and the NMR methods can be used for quantification of PtdIns. The NMR methods, however, are most accurate with relatively large amounts of pure sample. Nevertheless, both methods have been successfully employed for this purpose in the past (Volwerk et al., 1990; Hansbro et al., 1992; Leondaritis and Galanopoulou, 2000).

2.3.2. Diacylglycerols

The diacylglycerol moieties released from PtdIns by PLC are readily quantified by high temperature GLC following trimethylsilylation using tridecanoylglycerol as an internal standard. The peak areas of the molecular species, which are resolved on a non-polar capillary column according to carbon number are summed and related to the peak area of a known amount of the internal standard and the results expressed on a weight or mole basis. Appropriate correction factors for incomplete recoveries from the column or differences in the detector response may be included in the calculations as described in detail elsewhere (Kuksis and Myher, 1990).

Table 2.5 compares the relative recoveries of the different molecular species of PtdIns of human plasma by normal phase LC/ES/MS and by capillary GLC using polar columns (Kuksis and Hegele, 2001, unpublished). Earlier close agreement was obtained between the estimates of molecular species of diacylglycerols by GLC/FID and LC/MS with direct liquid inlet interface when analyzed as the TBDMS ethers (Pind et al., 1984).

Dobson et al. (1998) have described an LC/MS method, using a particle beam interface to obtain EI/MS of the nicotinate derivatives of diacylglycerol mixtures, thereby obtaining maximum structural information on natural glycerophospholipids.

TABLE 2.5

Molecular species of PtdIns of human plasma lipoproteins as determined by normal phase LC/ES/MS and by GLC/FID of the TMS ethers of the derived of diacylglycerol moieties (mol %)

| Molecular species | Normal phase LC/ES/MS[a] | | Polar capillary GLC/FID[b] Ave. ± range/2 (Myher et al., 1989a) | Normal phase LC/ES/MS[c] Ave. ± range/2 (Uran et al., 2001) |
	[M − H]⁻	Mean (SD) (Kuksis et al., 2000)		
16:0/18:1	835	10.69 ± 1.0	4.2 ± 0.5	3.0
16:0/18:2	833		4.2 ± 0.5	3.5
18:0/18:0	865		0.2 ± 0.1	0.4
18:0/18:1	863	12.7 ± 1.0	5.5 ± 0.5	4.4
18:1/18:1	861		2.5 ± 0.5	
18:0/18:2	861	16.9 ± 1.5	11.6 ± 1.0	11.0
16:0/20:4	857	0.0 ± 0.0	4.0 ± 0.5	4.8
16:0/20:3	859		3.5 ± 0.5	3.0
18:0/20:4	885	47.59 ± 2.5	50.0 ± 2.5	44.2
16:0/22:4	885		3.9 ± 0.5	
18:0/22:4	913		1.1 ± 0.1	1.3
16:0/22:6	881			
18:0/22:5	911			2.6
18:0/20:3	887	3.75 ± 0.5		3.2
18:0/22:6	909			3.0
18:1/20:4	883	0.5 ± 0.1	2.8 ± 0.5	3.0

[a]Normal phase LC/ES/MS of intact PtdIns.

[b]Polar capillary GLC of TMS ethers of diacylglycerols derived from PtdIns by hydrolysis with phospholipase C (Myher et al., 1989a).

[c]Normal phase LC/ES/MS/MS of intact human blood PtdIns.

A useful feature of these derivatives is that in addition to the size and degree of unsaturation of the acyl chains, the positions of the double bonds and other functional groups can also be located. Dobson and Deighton (2001) have subsequently adopted this method for quantifying phospholipid molecular species as nicotinate derivatives by reversed phase HPLC in conjunction with UV detection and the APCI/MS areas of [MH-123]$^{+}$ ions.

2.3.3. Fatty acids

In principle, PtdIns can be quantified based on its fatty acid content. GLC analysis of the fatty acid content of a purified PtdIns sample in the presence of an internal standard provides one of the simplest mass measurements (see above). The transmethylation of total sample can be performed in the presence of known amount of fatty acid ester internal standard or the esterified standard can be added to the transesterification mixture. A minimum of 10% of the total mass added as internal standard yields readily measured GLC peaks for all major components. This method has been extensively utilized for estimating PtdIns resolved by TLC from other phospholipids. It is most important to avoid contamination of the sample with foreign fatty acids or esters present in solvents, equipment and glassware.

2.3.4. Inositol

The inositol content of a purified PtdIns preparation can also provide a good indication of the mass of the sample. Originally, inositol analysis was carried out mainly by microbial or paper chromatographic techniques. Later a variety of GLC methods offered the advantages of rapidity of analysis, high resolution, and sensitivity when using the flame ionization detector. GLC would appear to provide the simplest quantitative analysis, although sample preparation and derivatization requires care. Modern GLC methods of inositol quantification also provide evidence of the isomer content.

Sherman et al. (1977) and Leavitt and Sherman (1982) adopted GLC analysis of *myo*-inositol to the picomole level and applied it for the study of single cells and histologically defined tissue samples. DaTorre et al. (1990) have further improved this method for direct measurement of the mass of inositol from biological samples of inositol phosphates.

Myo-Ins for quantitative analysis by GC/MS methods may be released from PtdIns by saponification, and the resulting InsP is dephosphorylate with alkaline phosphatase. Heathers et al. (1989) dephosphorylated the isomeric InsP fractions obtained by concentration of the HPLC column eluates. To dephosphorylate the samples (5–10 nmol), $MgCl_2$ (5 mM) and alkaline phosphatase (Sigma, type VIIS, 35–50 units) were added to each tube. Samples were incubated overnight at 37°C, followed by boiling for 3 min to terminate the reaction. Following dephosphorylation, 1 nmol of *chiro*-inositol was added to each sample to serve as an internal standard for GLC analysis.

Figure 2.12 shows the standard curve of HFB_6-*myo*-inositol measured by negative-ion chemical ionization monitoring using *m/z* 1336 for HFB_6-*myo*-inositol and *m/z* 1341 for HFB_6-d_6-*myo*-inositol. Ratios reflect integrated peak areas. The measured amount of *myo*-inositol was linear from 10 fmol (0.4 pmol/40 μl sample, see inset) to 1 pmol per microliter injected onto the column.

Fig. 2.12. Standard curve of HFB_6-*myo*-inositol measured by negative-ion chemical ionization monitoring *m/z* 1336 for HFB_6-*myo*-inositol and *m/z* 1341 for HFB_6-d_6-*myo*-inositol. Ratios reflect integrated peak areas. The measured amount of *myo*-inositol was linear from 10 fmol (0.4 mol/40 μl sample, see inset) to 1 pmol per microliter injected onto the column. Reproduced from Rubin et al. (1993) with permission of the publisher.

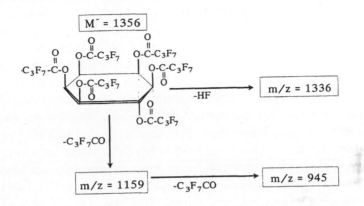

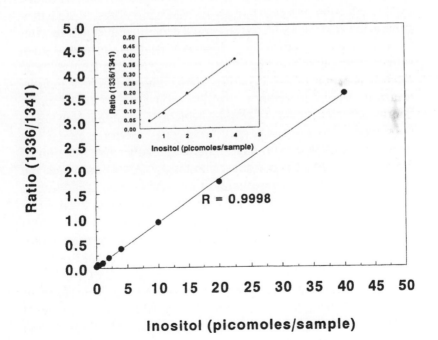

Nevertheless, the colorimetric methods provide the attraction of simplicity. MacGregor and Matschinsky (1984) had developed an enzymatic assay, which was based on the oxidation of *myo*-inositol and reduction of NAD^+ by *myo*-inositol dehydrogenase coupled with the scavenge reaction catalyzed by malate dehydrogenase to reoxidize the NADH produced in the first reaction. The resulting malate, which is stoichiometric with *myo*-inositol, is measured fluorimetrically. The method is highly sensitive but it gives a high blank in the measurement of serum and cells due to endogenous malate. Maslanski and Busa (1990) have described an improved method for mass analysis of Ins in InsPs, which does not require advanced instrumentation yet provides specificity and sensitivity comparable to that achieved by GC/MS. The assay of Maslanski and Busa (1990) is based on two coupled enzyme reactions linked by the reduction and reoxidation of NAD. *myo*-Inositol dehydrogenase, in the presence of NAD, is used to oxidize free *myo*-inositol to *scyllo*-inosose and NADH. The NADH thus formed is then stoichiometrically reoxidized by the enzyme diaphorase. The electron acceptor in the second reaction is the non-fluorescent dye resazurin. Upon reduction, resazurin is converted to the intensely fluorescent compound resorufin, which is quantified using a fluorometer. The method is specific for *myo*-inositol.

Ashizawa et al. (2000) have reported a further modification of the method of MacGregor and Matschinsky (1984), in which the second reaction is the conversion of iodonitrotetrazolium chloride to formazan by diaphorase. This results in low blank even when measurements are made in rat serum. Ashizawa et al. (2000) increased the spectrometric response by substituting triethanolamine hydrochloride $-K_2HPO_4-KOH$ buffer (pH 8.6) for the Tris$-$HCl buffer (pH 9.0) used previously (MacGregor and Matschinsky, 1984; Shayman et al., 1987) and by increasing the concentration of β-NAD in the reaction mixture to 4.8 mM. Although the modified assay is less sensitive than the method of MacGregor and Matschinsky (1984), it is adequately sensitive for measuring *myo*-inositol in serum or tissue weighing 1 mg of more.

2.3.5. Phosphorus

The phosphorus content may be determined using the method of Bartlett (1959). Mitchell et al. (1986) have described a micro-adaptation of the method of Bartlett (1959) for assay of phosphorus in TLC spots. According to Mitchell et al. (1986), the TLC plate is moistened with distilled water and the spots are scraped into 12×85 mm^2 glass test tubes, Sulfuric acid (5 M, 75 μl) is added to each tube, and the tubes are heated at 150°C for 90 min to char the phospholipids. The samples are bleached with 30% H_2O_2 (30 μl or more, if needed) at 150°C. Samples are maximally bleached within 30 min. To ensure complete decomposition of the peroxide yet avoid recharring, the samples are checked periodically with a moistened strip of peroxide test paper held above each tube. Tubes were removed from the heat once the vapors tested negative. The degraded, bleached samples were incubated with ammonium molybdate (2.2 g/l, 690 μl) and SubbaRow reagent (150 mg/ml NaHSO$_3$, 10 mg/ml Na$_2$SO$_3$, 5 mg/ml 1-amino-2-naphthol-4-sulfonic acid, 30μl) for 7 min at 100°C to allow blue color development. The supernatants were transferred to a 1.5 ml centrifuge tubes and centrifuged for 3 min at 8800g to pellet any residual silica gel, and their absorbance were read in a spectrophotometer at 830 nm. A phosphate free region of the TLC plate provided a background value.

However, the method described by Stull and Buss (1977) for protein-bound phosphate may be preferred in other instances (Pak and Larner, 1992).

According to Stull and Buss (1977) the sample is dissolved in 25 μl of 10% MgNO$_3$·6H$_2$O in 95% ethanol and gently heated to dryness. The samples are then ashed over an intense flame. After dissolving the residue in 150 μl of 1.2N HCl, 50 μl of filtered phosphate reagent is added, and after 5 min, the developed color is measured at 660 nm. The phosphate reagent contains 1 volume of 10% $(NH_4)_6MoO_{24} \cdot 4H_2O$ in 4N HCl to three volumes of 0.2% Malachite Green. Phosphate standards (0.2 to 1.5 nmol KH$_2$PO$_4$)

are also ashed. The variation on multiple assays of the same sample is 5% or less.

Others (Heathers et al., 1989; Singh, 1992) have employed the method of Lanzetta et al. (1979) for quantitative determination of organic phosphate in PtdIns or PtdInsPs. It should be noted that Rouser et al. (1970) have described a method specifically adopted for quantification of phosphorus in silica gel scrapings from TLC plates.

CHAPTER 3

Phosphatidylinositol phosphates

3.1. Isolation

PtdIns is the major inositol lipid in mammalian tissues, but appreciable quantities of the PtdIns(4)P and PtdIns(4,5)P$_2$ as well as PtdIns(3,4,5)P$_3$ are also found in nervous tissue, adrenal medulla and kidney (Abdel-Latif, 1986; Irvine, 1986; Holub, 1986). More recent studies have recognized the universal occurrence of the PtdInsPs. The PtdInsPs are of very low abundance in most lipid extracts of cells, but these species have attracted a great deal of interest because they serve as precursors for InsP$_3$, a messenger of signal transduction (Berry and Nishizuka, 1990; Berridge, 1993). Since tissue levels of PtdInsPs are relatively low and post-mortem hydrolysis of PtdInsPs is rapid, accurate analyses of the higher inositol phosphatides are difficult to obtain. Other analytical complications arise from the high polarity of PtdInsPs and the lack of readily prepared derivatives.

3.1.1. Solvent extraction

Because PtdInsPs and their hydrolysis products are degraded very rapidly, it is essential to either extract the samples immediately after collection or freeze the samples in situ after collection and store, if necessary, at $-70°C$. In situ fixation of the samples by freeze-clamping or by microwave irradiation may be helpful in arresting the degradation of the phosphate esters. The recovery efficiency can be determined by means of radiolabeled PtdIns or PtdIns phosphate external standards. Recent studies have demonstrated 50–90% recoveries of PtdIns and InsPs from brain samples (Singh, 1992).

According to Mitchell et al. (1986) each aqueous cell or cell membrane suspension is dissolved in 15 vol. ice-cold methanol/chloroform/conc. HCl (100:50:1, by vol.) and the extraction is completed in the usual way. The acid is needed to ensure extraction of PtdInsPs bound to proteins, but it also accelerates hydrolysis of certain phospholipids (plasmalogens). A quantitative extraction of the PtdInsPs requires acidification of the extraction medium, which is best performed following an initial removal of the neutral phospholipids with neutral solvents (Shaikh, 1986). Christensen (1986) has pointed out that the extraction of PtdIns(4)P and, in particular of PtdIns(4,5)P$_2$, requires an acid solvent system that, when applied to erythrocytes, causes the extraction of considerable amounts of haem products, which interfere in the subsequent separation of the PtdInsPs. Some 95% of the haem co-extracted with the lipids could be removed by precipitation with chloroform, with maximal loss of PtdInsPs of less than 3% (see also Chapter 2).

Low (1990) describes the extraction of PtdIns, PtdInsP and PtdInsP$_2$ from human platelets and the procedure should be adaptable to PtdInsPs from other tissues and cell types. Platelets are prepared from 60 ml of fresh human blood and suspended in 3 ml of 140 mM NaCl, 5 mM Cl, 0.005 mM CaCl$_2$, 0.1 mM MgCl$_2$, 0.1 mM MgCl$_2$, 16.5 mM glucose, 0.1 mg serum albumin/ml, 15 mM Hepes (pH 7.4), containing 2–6 mCi of [^{32}P]Pi (carrier free) for 2 h at 37°C with occasional gentle shaking. Prostaglandin E$_1$ (PGE$_1$) (2.8 μM) is also added, since it prevents activation and increases the yield of [^{32}P]PtdInsPs. The lipids are extracted with 11.25 ml of chloroform/methanol/HCl (50:100:1.3, by vol.) for 30 min at 25°C. Chloroform (3.75 ml) and 0.1 M HCl (3.75 ml) are then added to separate the phases. The lower phase is washed twice with 5 ml of methanol, 4.5 ml of 2 M NaCl, and 0.5 ml of 100 mM

EDTA–NaOH (pH 7.4) and then evaporated to dryness in a clean tube with approximately 1.5 μmol of acid-washed, methanol-precipitated bovine brain PtdInsPs to act as carrier.

Low (1990) describes a procedure for breaking up the Ca^{2+}/Mg^{2+} chelate of $PtdInsP_2$ and converting it to a Na salt using an intermediate treatment with HCl. The method starts with a bovine brain Folch fraction I (1 g) dissolved in 12 ml of chloroform. The PtdInsPs are precipitated by mixing with 22 ml of methanol followed by centrifugation at 1000g for 5 min. The supernatant is discarded, and the pellet redissolved in 12 ml of chloroform. This precipitation with methanol is done a total of six times. The final methanol precipitate is dissolved in a mixture of 15 ml of chloroform, 15 ml of methanol, and 5 ml 1 M HCl. Additional 1 M HCl (8.5 ml) is added to give two phases. After centrifugation the upper phase is discarded and, to the lower phase, 15 ml of methanol and 5 ml of 2 M NaCl are added and the sample mixed until it becomes clear. The phases are separated by the addition of 2.5 ml of 2 M NaCl and 6 ml of H_2O. The upper phase is removed and the lower phase washed once more with methanol and 2 M NaCl, as above (at 0–4°C). The lower phase is transferred to a clean tube, evaporated to dryness under nitrogen gas, the lipids redissolved in 2 ml of chloroform and stored at $-20°C$ until required. The acid wash should be completed as rapidly as possible to minimize acid hydrolysis. One gram of starting material yields 150–200 μl of precipitated phospholipid (as determined by organic phosphorus).

For the isolation of PtdInsPs from cultured cells for in vivo detection of isomeric PtdInsPs, it is necessary to prelabel the cells with inorganic [^{32}P]orthophosphate of myo-[^{3}H]inositol. According to Auger et al. (1990), cells are plated in 10 cm tissue culture plates in standard growth medium. After cells have attached to the substratum, 5–6 ml of inositol-free Dubelco's Modified Eagle's Medium supplemented to contain 10% (v/v) dialyzed fetal bovine serum and 10–20 μCi/ml of myo-[^{3}H]inositol, 15 Ci/mmol) are added to each plate. Cells are incubated for 24–96 h in this medium at 37°C and then placed in serum-free condition: 5 ml of inositol-free medium and 10–20 μCi/ml of [^{3}H]Ins. Cells are incubated for an additional 12–72 h under serum-free conditions in the presence of [^{3}H]Ins to induce quiescence. Cells are labeled with inorganic [^{32}P]orthophosphate in a manner similar to that described for [^{3}H]Ins labeling. For mitogen stimulation, recombinant or purified growth factor is added to the cell culture medium at an optimal mitogenic concentration for a specific period of time prior to the labeled cells being washed and harvested and the lipids extracted.

For the extraction of lipids, individual plates of cells are washed three times with ice-cold PBS to remove unincorporated radioisotope. Cells are harvested with a cell scraper in 750 μl of methanol/1 M HCl (1:1, v/v) and placed in a 1.5 ml microfuge tube. After vigorous vortexing, 380 μl of chloroform is added and the tube is vortexed and incubated on a rocker platform for 15 min at room temperature. The samples are centrifuged to separate the phases. The upper phase is carefully removed, placed in a clean microfuge tube, and then dried by roto-evaporation in vacuo. Approximately 1 ml of distilled, deionized water is then added to the tube and the contents are dried again, and the dried sample stored at −70°C until further use. The organic phase and the material at the interface are extracted two times with an equal volume of MeOH/0.1 M EDTA (1:0.9, v/v) to remove traces of divalent cations. The entire organic phase is carefully removed, placed in a clean microfuge tube, and stored at −70°C under nitrogen gas until deacylation or TLC analysis. Auger et al. (1990) have observed that a load of a few million dpm of [^{3}H]Ins-labeled deacylated lipids is necessary in order to detect the novel minor isomers as they constitute only a small fraction of the total PtdInsP$_2$ in the cell. Furthermore, PtdIns(3,4)P$_2$ and PtdIns(3,4,5)P$_3$ are not present in most cells until they are induced by mitogen or when the cells are proliferating.

According to Van der Kaay et al. (1997) the PtdInsPs are extracted in 0.75 ml of CHCl$_3$/MeOH/HCl (40:80:1, by vol.) for

20 min, and phases were then split by the addition of 0.25 ml of CHCl₃ and 0.45 ml of 0.1 M HCl. The lower phase, obtained after 1 min centrifugation at 13 000g, was transferred to a screw cap tube, and the upper phase was reextracted once with 0.45 ml of the synthetic lower phase. The lower phases were pooled and dried down. The extracts were subjected to alkaline hydrolysis and the water soluble phosphates used for isotope dilution assay.

Wu et al. (2002) have recently revised the methodology of acidic glycerophospholipid extraction following the observation that changes in phospholipid extractability occur during mineralization of chicken growth plate cartilage matric vesicles. The authors recommend demineralization of the tissue for complete glycerophospholipid extraction. The extraction of PtdInsPs from mineralized tissues was not investigated.

Shaikh (1994) recommends that prior to the extraction with acidic the neutral phospholipids, including plasmalogens, be removed with neutral solvents.

Vickers (1995) compared a two-step with a common single-step procedure for extraction of acidic PtdInsPs and concluded that the qualitative results would likely be the same.

3.1.2. Solid phase extraction (cartridge)

As already noted in Chapter 2, solid phase extraction saves solvents, time and labor. It is especially valuable for extraction of PtdInsPs using affinity columns.

3.1.2.1. TLC and HPTLC

TLC provides the simplest method for the isolation, identification and purification of PtdInsPs. It provides the means of monitoring all other methods of purification and isolation of PtdInsPs. The PtdInsPs may be resolved by conventional one-dimensional (Gonzalez-Sastre and Folch-Pi, 1968; Jolles et al., 1981) or two-dimensional TLC (Abdel-Latif et al., 1977; Ferrell and Huestis, 1984) or by HPTLC on silica gel impregnated with oxalic acid

(Akhtar et al., 1983) or boric acid (Fine and Sprecher, 1982; Hegewald, 1996).

A one-dimensional technique (Jolles et al., 1981) yielded chemically pure PtdIns(4,5)P_2 in a single step. Although this separation also produces radiochemically pure PtdIns(4)P, the phosphate-containing compound(s) co-chromatographed sufficiently closely together to make assay of total PtdIns(4)P impossible. Consequently, the samples were chromatographed twice to obtain uncontaminated PtdIns(4)P. Firstly, they were resolved in two-dimensional system (Ferrell and Huestis, 1984), leaving the PtdInsPs on the origin. Secondly, the origin was scraped from the plate and the PtdInsPs were eluted with chloroform/ methanol/HCl (20:40:1, by vol.) containing 5% water, applied to a new plate, and PtdIns(4)P was separated by the one-dimensional separation method described above. Recoveries were of the order of 85%. Some ^{32}P-labeled PtdIns(4,5)P_2 spots were scraped off plates and the silica gel was placed in the deacylating mixture of Clark and Dawson (1981). The resulting GroPIns(4,5)P_2 was purified by the procedure of Clark and Dawson (1981).

Palmer (1981) separated PtdIns, PtdInsP and PtdInsP$_2$ by one-dimensional TLC using precoated silica gel HR plates (Analtech, Newar, NJ) which were dipped in 1% (w/v) potassium oxalate, dried, and reactivated by heating at 120°C for 40 min. These plates were developed with chloroform/acetone/methanol/acetic acid/ water (40:15:13:12:8, by vol.) and yielded excellent separations of the PtdInsPs. The phospholipids were located with the molybdate spray reagent (Gatelli et al., 1978) or by charring with 3% cupric acetate in 8% phosphoric acid (Fewster et al., 1969).

A simple, rapid two-dimensional TLC system which resolves the four PtdIns cycle phospholipids as well as all commonly encountered major and minor phospholipids has been described by Mitchell et al. (1986). Ca^{2+}-free lipid samples are loaded onto silica gel H plates and developed first in chloroform/methanol/ water/conc. ammonia (48:40:7:5, by vol.) and then at an angle of 90° in chloroform/methanol/formic acid (55:25:5, by vol.).

Figure 3.1 shows a separation of phospholipid standards including PtdIns, PtdInsP and PtdInsP$_2$ by two-dimensional TLC (Mitchell et al., 1986). Development with a basic system in the first dimension and an acidic system in the second dimension resulted in an effective resolution of all major and minor phospholipid classes. After an appropriate staining and outlining the spots, the areas could be scraped and the resolved compounds recovered for further analysis. PtdInsP and PtdInsP$_2$ migrate as two distinct spots well separated from the origin. The method was successfully applied to human erythrocytes, human platelets, and BL/L3 murine lymphoma cells. Scored $20 \times 10 \, \text{cm}^2$, 250 µm silica gel H plates (Analtech) were purchased and snapped into $10 \times 10 \, \text{cm}^2$ plates. The plates were used without activation or prewashing.

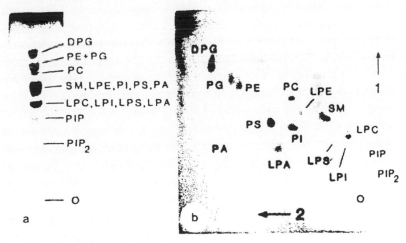

Fig. 3.1. Separation of phospholipid standards, including PtdIns, PtdInsP and PtdInsP$_2$ by one- and two-dimensional TLC. A, One-dimensional TLC on silica gel HL with chloroform/methanol/water/conc. ammonia (48:40:7:5, by vol.); B, TLC on silica gel HL as in A for first dimension, then chloroform/methanol/-formic acid (55:25:5, by vol.) for second dimension at 90°. 3–10 µg of each phospholipid was applied, and the developed plate was stained with iodine vapor.
    Reproduced from Mitchell et al. (1986) with permission of publisher.

The PtdInsP [PtdIns(3)P and PtdIns(4)P], the PtdInsP$_2$ [PtdIns(3,4)P$_2$ and PtdIns(4,5)P$_2$] and PtdInsP$_3$ can clearly be separated by TLC, as noted above, but the resolution of the D-3/D-4 isomers of PtdInsP and PtdInsP$_2$ is critical (Varticovski et al., 1994; Traynor-Kaplan et al., 1988). A few protocols on this subject have been published. Some of them provide separation of PtdIns(3)P from PtdIns(4)P (Walsh et al., 1991; Okada et al., 1994) and others of PtdIns(3,4)P$_2$ from PtdIns(4,5)P$_2$ (Pignataro and Ascoli, 1990; Gaudette et al., 1993; Munnik et al., 1994).

Sun and Lin (1990) have provided a detailed description of the separation of phospholipids by HPTLC using a brain lipid extract equivalent to 1–5 mg protein as an example. The silica gel 60 HPTLC plates (10 cm × 10 cm, Merck, Darmstadt, Germany) are impregnated by dipping in 1% (w/v) potassium oxalate and then dried overnight before use. The oxalate solution consists of 1% (w/v) potassium oxalate, 2 mM Na$_2$EDTA, which are mixed with methanol in the ratio oxalate/methanol 2:3 (v/v). PdInsPs standards (100 μg) are added to the sample as a carrier for visualization purposes. The lipid sample is applied to the lower left corner of the HPTLC plate using a disposable pipette and the plate dried by blowing over it dry nitrogen to remove any residual moisture. The plate is then developed using chloroform/methanol/acetone/16 M NH$_4$OH (70:45:10:10, by vol.) to a height of 9 cm. After development, the plate is dried under a stream of nitrogen for 5 min. The first system displaces all phospholipids from the origin except PtdInsPs. The HPTLC plates are subsequently developed in the same direction with a second solvent system consisting of chloroform/ethanol/16 M NH$_4$/H$_2$O (36:28:2:6, by vol.). The solvent is allowed to move up to about 7 cm. The second solvent is used to separate the PtdInsP and PtdInsP$_2$ from the origin. After development with the second solvent system, the HPTLC plates are thoroughly dried by blowing dry nitrogen over it for 6–8 min. The plates are exposed to HCl fumes for 3 min and the excess HCl fumes removed from the plates by blowing with dry nitrogen for 8 min. The plates are then inverted through 90° and placed in a third solvent system consisting of chloroform/methanol/acetone/acetic acid/1 M ammonium acetate (140:60:55:4.5:10, by vol.). After the development, plates are dried briefly with dry nitrogen. The spots are visualized by spraying the plate with $2',7'$-dichlorofluorescein (1% in methanol) and viewing under an ultraviolet lamp. The fluorescent dye procedure is used to recover the PtdInsPs for analysis of molecular species and fatty acids by GLC. For measuring radioactivity, the lipid spots are identified by exposing the plate to iodine vapor. The brown lipid spots are scraped into counting vials for measurement of radioactivity. Sun and Lin (1990) have demonstrated effective separation of PtdInsPs of a brain homogenate and myelin preparation by the two-dimensional HPTLC. The brain samples were spiked with PtdInsPs standards (100 μg) prior to spotting the plates. After development, the lipids were visualized by charring with a solution containing 3% (v/v) copper acetate in 8% (v/v) aqueous phosphoric acid and subsequent heating for 20 min at 140°C. Chromatograms to the right depict the autoradiograms of the ^{32}P-labeled phospholipids in (c) brain homogenates and (d) myelin. For radiochromatography, an unstained plate was exposed to Kodak OMATAR film at −70°C overnight. It is possible to observe radioactive spots that cannot be visualized by the charring procedure.

The discovery of PtdIns 3-kinase has led to the need to distinguish between isomeric PtdInsP₂s and PtdInsP₃ (Varticovski et al., 1994). The so-called PtdIns 3-kinase phosphorylates PtdIns, PtdIns(4)P and PtdIns(4,5)P₂ to form PtdIns(3)P, PtdIns(3,4)P₂, and PtdIns(3,4,5)P₃, respectively. The simple two-dimensional TLC system used to resolve the commonly encountered major and minor phospholipids are not able to resolve them. The detection of the novel inositol phospholipids in vivo is hampered by their presence in relatively low abundance although they have been found to occur in all cells thus far investigated.

In the literature, TLC separation of the D-3/D-4 isomers is generally performed with alkaline mixtures of chloroform and methanol which result in a slightly lower migration of PtdIns(3)P

(Okada et al., 1994) or of PtdIns(3,4)P$_2$ (Pignataro and Ascoli, 1990; Gaudette et al., 1993; Munnik et al., 1994), relative to the respective 4-D isomer. Other workers have shown that inositol lipids can be separated by methods based on the ability of boric acid to form complexes with 2,3-*cis* diols present in the inositol ring of PtdIns (Hegewald, 1996). In the 3-D isomers the 3-OH is substituted by the phosphate group, thus inhibiting complex formation. The chromatographic behavior of PtdIns(4,5)P$_2$ seems to be unaffected by boric acid, thus retaining its slower migration relative to PtdIns(3,4)P$_2$ present in the boric acid-free TLC system (see Table 3.1).

Hegewald (1996) has described the first protocol for the separation of D-3 and D-4 PtdInsP lipids. The chromatography is performed on two different types of HPTLC plates under otherwise identical conditions and depends on the ability of the D-4, but not D-3, isomers to form complexes with boric acid. For this purpose, precoated HPTLC plates of silica gel 60F254, Art 1.056641, and HPTLC plates NH$_2$, Art 1.12572, were obtained from Merck. Reference standards, ^{32}P-labeled PtdIns(4)P, PtdIns(4,5)P$_2$ and PtdOH were produced by in vitro incubation of human erythrocyte membranes with Mg-[γ-^{32}P]ATP, and the lipids were extracted with hexane/2-propanol/HCl mixtures, essentially as described (Hegewald et al., 1987). ^{32}P-labeled PtdIns(3)P, PtdIns(3,4)P$_2$, and PtdIns(3,4,5)P$_3$ were synthesized by the incubation of a sonicated mixture of the erythrocyte lipids with Mg-[γ-^{32}P]ATP and the recombinant PtdIns-3-kinase γ. Cultured A431 cells were incubated for 24 h with [^{32}P]I and the lipids were extracted as described (Hegewald et al., 1987).

Hegewald (1996) impregnated the HPTLC plates by dipping them upside down into 5% boric acid (w/v) solution in methanol and then drying them for 5 min in an air current. Samples containing about 0.01–0.1 mg lipid were streaked with a CAMAG-Linomat IV (Muttenz, Switzerland) as 10 mm lanes 10 mm above the bottom edge of the plate. The plates, which needed not to be activated, were then developed in 1-propyl acetate/2-propanol/

TABLE 3.1

R_f values of lipids separated on two different types of boric acid impregnated HPTLC plates

Phospholipid	R_f values	
	TLC system A	TLC system B
PtdIns(3,4,5)P$_3$	0.07	0.04
LysoPtdIns(3,4)P$_2$	0.21	0.10
LysoPtdIns(4,5)P$_2$	0.24	0.10
PtdIns(3,4)P$_2$	0.31	0.17
PtdIns(4,5)P$_2$	0.34	0.17
LysoPtdIns(4)P	0.41	0.27
LysoPtdIns(3)P	0.42	0.29
PtdIns(4)P	0.49	0.38
PtdIns(3)P	0.50	0.41
LysoPtdIns	0.54	0.47
PtdIns	0.63	0.56
LysoPtdCho	0.64	0.66
SM	0.66	0.70
PtdOH	0.66	0.60
PtdCho	0.70	0.73
PtdSer	0.72	0.62
PtdEtn	0.82	0.75
Cholesterol	0.93	0.92

[32]P-labeled human erythrocyte phospholipids were incubated with PtdIns 3-kinase γ and Mg-[^{32}P]ATP and extracted with hexane/2-propanol/HCl mixtures and subsequently separated by one-dimensional LC (system A or B) with standard lipids in parallel, and visualized with charring reagent and by phosphor screen autoradiography. TLC system A, HPTLC silica gel 60; TL system B, HPTLC NH$_2$. Reproduced from Hegewald (1996) with permission of publisher.

absolute ethanol/6% aqueous ammonia (3:9:3:9, by vol.), in a well equilibrated paper-lined twin-trough chamber (CAMAG). The mobile phase was allowed to reach the top of the plates (about 2 h). Following chromatography the dried plates were dipped in charring reagent (5% CuSO$_4$ solution in 8% aqueous H$_3$PO$_4$) and heated at 180°C for 15 min. The radioactivity of the ^{32}P containing phospholipids was visualized and quantified using the GS-250 Molecular Imager System (Bio-Rad Laboratories, Hercules, CA).

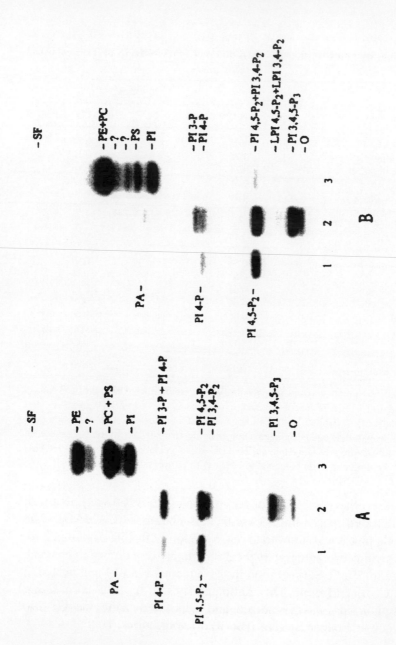

Occasionally the optical densities of the charred lipid bands were scanned at 366 nm in the reflectance mode, with a DESAGA CD 60 densitometer. Figure 3.2A and B show typical separations obtained with the two types of boric acid-impregnated HPTLC plates. The phosphor-screen autoradiograph of the ^{32}P-labeled erythrocyte phospholipids (see lane 1 of Fig. 3.2A and B) generally revealed three radioactive bands with R_f values corresponding to those of commercial PtdIns(4,5)P$_2$, PtdIns(4)P, and PtdOH. Sometimes weakly radioactive bands were detected below PtdIns(4,5)P$_2$ and PtdIns(4)P, and PtdOH. Treatment of sonicated ^{32}P-labeled erythrocyte lipids with phospholipase A$_2$ from *Vipera ammodytes* resulted in three additional radioactive bands with R_f values identical to the above mentioned weak bands (results not shown). Thus, they are thought to consist of the *lyso*-compounds of the 4-D PtdInsPs and of PtdOH. TLC separation of erythrocyte phospholipids labeled additionally by means of Mg-[γ-^{32}P]ATP in the presence of the PtdIns-3-kinase γ is shown in lane 2 of Fig. 3.2A and B. The incubation procedure led to the appearance of at least three new radioactive bands. The identity of the bands was revealed by co-migration of radiolabeled PtdIns(3)P, PtdIns(3,4)P$_2$ and

Fig. 3.2. Separation of D-3 and D-4 PtdInsPs by two one-dimensional HPTLC systems. A, Phosphor-screen autoradiography of lipid compounds separated on boric acid-impregnated HPTLC plates silica gel 60 (TLC system A). Development in propyl acetate/2-propanol/absolute ethanol/6% aqueous ammonia (3:9:3:9, by vol.). ^{32}P-labeled erythrocyte phospholipids (lane 1), ^{32}P-labeled erythrocyte phospholipids incubated additionally with PtdIns 3-kinase γ and Mg-[γ-^{32}P]ATP (lane 2), and ^{32}P-labeled A431 cell phospholipids (lane 3). SF, solvent front. Other abbreviations as in Table 3.1. B, Phosphor screen autoradiography of lipid compounds separated on boric acid-impregnated HPTLC plates NH$_2$ (TLC system B). Development in 1-propyl acetate/2-propanol/absolute ethanol/6% aqueous ammonia (3:9:3:9, by vol.). ^{32}P-labeled erythrocyte phospholipids (lane 1). ^{32}P-labeled erythrocyte phospholipids incubated additionally with PtdIns 3-kinase γ and Mg-[γ-^{32}P]ATP (lane 2), and ^{32}P-labeled A431 cell phospholipids (lane 3). Abbreviations are as described in legend to A. Reproduced from Hegewald (1996) with permission of publisher.

PtdIns(3,4,5)P$_3$, which were synthesized by incubation of the corresponding precursor PtdIns, PtdIns(4)P, or PtdIns(4,5)P$_2$ with PtdIns 3-kinase γ and Mg-[γ-^{32}P]ATP (not shown). Also the lyso compounds of the 3-D inositides were sometimes detectable. Table 3.1 also gives the relative migrations of all major cellular phospholipids separated by two different types of boric acid-impregnated HPTLC plates (Hegewald, 1996).

The TLC System A separates PtdInsP$_3$, lysoPtdIns(3,4)P$_2$, lyso-PtdIns(4,5)P$_2$, PtdIns(3,4)P$_2$, PtdIns(4,5)P$_2$, lysoPtdInsP, PtdIns-, lysoPtdIns, and PtdEtn (Hegewald, 1996). TLC System B is able to separate lysoPtdInsP$_2$, PtdInsP$_2$, lysoPtdIns(4)P, lysoPt-dIns(3)P, PtdIns(4)P, PtdIns (3)P, lysoPtdIns, PtdIns, PtdOH, and PtdSer. SM co-migrates with cardiolipin (not shown). The above HPTLC methods have been designed to separate the inositol phospholipids that exhibit phosphorylation at 4-OH position of the inositol ring (such as PtdIns(4)P and PtdIns(4,5)P$_2$). However, it is now known that PtdIns(3)P, PtdIns(3,4)P$_2$ and PtdIns(3,4,5)P$_3$ also occur in natural systems although at low levels. Detection of the PtdIns(3)P, PtdIns(3,4)P$_2$ and PtdIns(3,4,5)P$_3$ in vivo has been difficult due to the fact that these lipids are present in relatively low abundance in almost all cells. To detect and identify these novel phospholipids in intact cells, it is necessary to incorporate high levels of radioisotope into the total cellular PtdInsPs. Further complications have arisen from the failure of conventional methods to separate these novel isomers from the well-known and more abundant PtdInsPs. Thus, PtdIns(3)P and PtdIns(4)P co-migrate on nearly all one- and two-dimensional TLC systems thus far investigated. PtdIns(3,4)P$_2$ and PtdIns(4,5)P$_2$ also co-migrate on conventional TLC systems.

These PtdIns(3)P, PtdIns(3,4)P$_2$ and PtdIns (3,4,5)P$_3$ are usually recovered from incubations with PtdIns 3-kinase (see Chapter 5). According to Auger et al. (1990), the PtdIns 3-kinase reaction may be performed in a total volume of 50 μl and is initiated by the addition of 10–50 μCi of [γ-^{32}P]ATP, 3000 mCi/mmol in a carrier of 50 μM unlabeled ATP, 10 mM Mg^{2+} and 20 mM Hepes (pH 7.5) to the washed immune complexes that have been pre-incubated at room temperature for 5–10 min. After stopping the reaction with 80 μl of 1 M HCl and extracting the lipids with 160 μl of chloroform/methanol (1:1, v/v), the ^{32}P-labeled phospholipid products in the bottom organic phase are collected after a brief centrifugation and stored at −70°C until further use. It has been found that such samples can be best analyzed and resolved on TLC plates (silica gel G, 0.2 mm thickness, Merck) that have been precoated with 1% (w/v) potassium oxalate and baked at 100°C for 30 min to 1 h immediately before use. Unlabeled phospholipid standards are run in parallel to monitor lipid migration and are visualized by exposure to iodine vapor. The highly phosphorylated PtdIns(3,4,5)P$_3$ is separated from the radioactivity remaining near the origin and any [γ-^{32}P]ATP and [^{32}P]phosphate carried over with the organic phase during lipid extraction of incubation mixtures, by using an acidic solvent system of *n*-propanol/2.0 M acetic acid (13:7, v/v) instead of the more commonly used CHCl$_3$/MeOH/2.5 M NH$_4$OH (9:7:2, by vol.) solvent system. To achieve maximum resolution of each of the phospholipids from each other and from the material close to the origin, the solvent is allowed to migrate nearly to the top of a 20 cm TLC plate, a process that is routinely accomplished within 5–6 h (Auger et al., 1990).

3.1.2.2. HPLC
HPLC using acetonitrile- or hexane-based mobile phases allows the direct detection of phospholipids without prior cleanup of the crude extract, thus making these methods simple and rapid. However, these methods do not effectively separate the different inositol phospholipids. Nakamura et al. (1989) have developed an HPLC procedure that specifically analyzes PtdIns and PtdInsP$_2$ in the brain. In this method, the lipid extract is first derivatized with 9-anthryldiazomethane to produce (9-anthryl) PtdInsP and di(9-anthryl) PtdInsP$_2$ and then the derivatized sample is separated

by HPLC with UV detection at 245 nm. PtdCho, PtdIns, PtdSer, PtdEtn and SM are not derivatized with this reagent and, therefore, do not interfere with the assay (Yamada et al., 1988). This method is more sensitive than the underivatized UV detection method for the analysis of the inositol phospholipids.

Low (1990) has described the use of DEAE cellulose columns for preparative HPLC of brain lipids. The brain phospholipids dissolved in 50 ml of Solvent A (chloroform/methanol/H_2O, 20:9:1, by vol.) and applied to the cellulose column (DEAE cellulose, Sigma), which is washed with alkali and acid to neutrality. The washed DEAE cellulose is packed into a glass column (2 cm × 38 cm) with solvent resistant fittings. The column is then eluted at a flow rate of 100 ml/h with 1 l linear gradient from 100% Solvent A to 100% Solvent B (Solvent A containing 0.6 M ammonium acetate) and 13–15 ml fractions are collected and assayed for phosphorus (Fig. 3.3).

Alternatively, Low (1990) performed the chromatographic step by preparative HPLC using an amino column. The brain phospholipids were dissolved in 5 ml of Solvent A and were applied at a flow rate of 2.5 ml/min to a 10 mm × 250 mm amino-NP column (5 μm spherical silica with n-propylamine binded phase, IBM Instruments, Danbury, CT) with a 4.5 mm × 50 mm guard column equilibrated with Solvent A. The column is eluted at a flow rate of 2.5 ml/min with 25 ml of Solvent A, a 125 ml gradient of 100% Solvent A to 100% Solvent B, and 100 ml of 100% Solvent B; 5 ml fractions being collected. The pooled fractions are concentrated by adjusting to 18 ml with Solvent A and the phospholipids extracted. The average recovery of organic phosphorus is 80%. A complete separation of PtdIns (90 ml), PtdInsP (150 ml) and PtdInsP$_2$ (160–200 ml) was obtained (chromatogram not shown). The PtdIns, however, overlaps with PtdSer. The composition of the peaks was determined with TLC of pooled fractions.

Low (1990) has also described an HPLC purification of the [^{32}P]-labeled PtdIns(4)P and PtdIns(4,5)P$_2$ from human platelets. The lipid extract from platelets was prepared from 60 ml of fresh

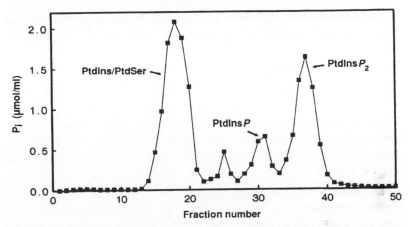

Fig. 3.3. Isolation of PtdInsPs from bovine brain by a DEAE-cellulose column chromatography. Acid washed methanol-precipitated Folch Fraction was eluted from a DEAE-cellulose column with a gradient of 100% Solvent A (chloroform/methanol/H$_2$O, 20:9:1, by vol.) to 100% Solvent B (Solvent A containing 0.6 M ammonium acetate) and 13–15 ml fractions collected. The fractions were assayed for organic phosphorus. Average recovery of applied organic phosphorus was approximately 90%. The peak eluting at fraction number. Twenty-five was reproducibly present but was not identified. The composition of the peaks was determined by TLC analysis of pooled fractions. Peaks are identified as shown in the figure. Reproduced from Low (1990) with permission of publisher.

human blood in the presence of methanol precipitated bovine brain PtdInsPs as a carrier, is finally redissolved in 0.5 l of Solvent A and applied at a flow rate of 1 ml/min to a 4.5 mm × 250 mm amino-NP column equilibrated in Solvent A. The column is eluted with 5 ml of Solvent A (a 25 ml linear gradient from 100% Solvent A to 100% Solvent B), and 20 ml of Solvent B. One milliliter fractions are collected and radioactivity determined by liquid scintillation counting. Average recovery of applied radioactivity is approximately 70%. There is an excellent resolution of the three inositol phosphatides but the PtdIns overlaps with PtdOH. Therefore, this HPLC procedure is not suitable for the preparation of [^{32}P]PtdIns since PtdOH, which in most cell types is more rapidly labeled by [^{32}P]Pi, is not resolved from it. It is likely that other anionic phospholipids such as PtdSer will be poorly resolved from PtdIns by the amino-NP column, as observed with the DEAE-cellulose procedure.

As discussed above, none of the above methods is effective in separating the individual PtdInsPs in pure form from crude tissue extracts. Therefore, the lipid extracts must be processed by several chromatographic steps. For this purpose, the extract is first subjected to Sephadex chromatography for separation of the lipids from the nonlipids in the extract. The lipid fraction is then poured onto a DEAE column that is eluted with different solvents. The PtdInsPs are recovered by elution with chloroform/methanol/ ammonium salt as the final elution step. The individual fractions are concentrated to 100 μl at 45°C under nitrogen and then the concentrated residue is further separated by cartridge-column chromatography, TLC, HPLC or other chromatographic procedures.

Singh (1992) has described the extraction and analysis of PtdIns, PtdIns(4)P and PtdIns(4,5)P$_2$ using a Sep-Pak cartridge. A total lipid extract (Folch et al., 1957) of rat brain synaptosomes and microvessels was first prepared and the first fraction generated (Fraction 1) was collected and further extracted to separate phospholipids and glycolipids as described by Dugan et al. (1986). The phospholipid fraction was subjected to DEAE-cellulose chromatography (separation sequence 4) as described by Rouser et al. (1969) to purify further the inositol lipids. The purified extract was transferred to a Sep-Pak cartridge. Various inositol lipids are eluted from the column with a linear gradient of ammonium acetate and water from 0.5% of ammonium acetate (200 mM) and 99.5% of water to 75% of ammonium acetate and 25% of water in 60 min at a flow rate of 0.5 ml/min (see Chapter 2). The column eluate for 120 min was collected in 1-ml fractions. Different peaks were identified by using PtdIns, PtdIns(4)P and PtdIns(4,5)P$_2$ peaks as reference points. The fractions containing individual inositol lipids were pooled and analyzed by HPLC as described by Geurts van Kessel et al. (1977). Individual phospholipids were detected at 206 nm by using a diode array detector programmed to scan a 200–350 nm range. Singh (1992) has demonstrated 50–90% recovery of PtdIns and InsPs from brain samples. $InsP_3$ and $InsP_4$ exhibited the lowest recovery because of poor quantification rather than a poor extraction method. The cartridge methods have not been applied for the isolation of the PtdIns(3)P, PtdIns(3,4)P_2 and PtdIns (3,4,5)P_3.

Cronholm et al. (1992) have described the isolation of PtdInsPs from rat pancreas. Each pancreas was immediately homogenized in 3 ml of chloroform/methanol (1:2, v/v), and about 3 MBq of ^{3}H-labeled PtdIns(4)P or PtdIns(4,5)P_2 were added together with 0.6 ml of 1.2 aq. HCl prior to extraction with chloroform, as described by Schacht (1981). Methanol was added to the pooled and washed extracts to yield chloroform/methanol (3:2, v/v). The solution was applied to the Lipidex-DEAE column, and most of the lipids were eluted with chloroform/methanol (3:2, v/v) or chloroform/methanol/water (3:6:1 by vol.). PtdIns and PtdSer were eluted together with 0.03 M H_3PO_4 and unidentified material with 0.1 M H_3PO_4. PtdIns(4)P was eluted with 0.25 M H_3PO_4 and PtdIns(4,5)P_2 with 0.5 M H_3PO_4 as indicated by the eluted radioactivity. All fractions were extracted with 0.4 ml of water/ml eluate. The upper phase was back-extracted with 0.1 vol. of chloroform. The pooled chloroform phases were washed with 0.4 vol. of 0.5 M HCl in 50% aq. methanol. The fractions were used for determination of molecular species by FAB/MS (see below).

Gunnarsson et al. (1997) used normal phase HPLC with evaporative light scattering detection for the analysis of commercial samples of PtdInsPs along with other phospholipids. Using a linear gradient of **A**, chloroform/methanol/ammonia (50:45:3, by vol.) and **B**, chloroform/methanol/water/ammonia (25:55:17:3, by vol.) well resolved late emerging peaks for PtdIns(4)P and PtdIns(4,5)P_2 are obtained. The dose response curves for PtdIns(4)P and PtdIns(4,5)P_2 were not linear, which is a characteristic of the light scattering detector (results not shown). For most purposes the HPLC technique is superior in resolution, speed and overall convenience to the DEAE-cellulose procedure and is the method of choice if HPLC equipment is available. DEAE-cellulose is cheap and large scale purification might require the use of this material. The HPLC procedure is also superior to some or in all respects to alternative procedures for purification of polyphosphoinositides, e.g. preparative TLC or neomycin affinity chromatography.

3.1.2.3. Ion exchange

The HPLC procedures that are commonly used for the separation of deacylated PtdInsPs typically utilize salt gradients that cannot resolve the various deacylated phospholipid isomers. However, HPLC systems with a strong anion exchange (SAX) column and a shallow salt gradient can yield baseline separations of the GroPInsPs derived from the PtdIns(3), PtdIns(3,4)P$_2$ and PtdIns(3,4,5)P$_3$ along with those derived from the more common PtdIns(4)P and PtdIns(4,5)P$_2$.

Auger et al. (1990) have described an ion-exchange HPLC method for the identification of PtdIns, PtdIns(3)P, PtdIns(3,4)P$_2$, and PtdIns(3,4,5)P$_3$. The method requires prior deacylation of the phospholipids because the HPLC method separated the phospholipids based on the structural differences in the inositol head group. The phospholipids were deacylated with methylamine reagent as previously described (Whitman et al., 1988). Briefly, the lipids extracted in chloroform were washed with an equal volume of methanol/0.1 M EDTA (1:0.9, v/v) and were dried under N$_2$. Methylamine reagent (42.8% of 25% methylamine in H$_2$O, 45.7% methanol, 11.4% n-butanol) was added to the dried lipids and incubated at 53°C for 50 min. The sample was dried in vacuo, washed with H$_2$O and dried again. The samples were resuspended in H$_2$O and extracted two times with an equal volume of n-butanol/light petroleum ether/ethyl formate (20:4:1 by vol.). The aqueous phase was then dried in vacuo and stored at −70°C until further use.

Deacylated lipids were deglycerated with mild $NaIO_4$ and dimethylhydrazine as previously described (Hawkins et al., 1984). Briefly, the deacylated lipids were resuspended in 1 mM $NaIO_4$ and incubated in the dark for 75 min at room temperature. Ethylene glycol (0.1%) was added, and the incubation was continued for an additional 10 min. A solution of 1,1-dimethylhydrazine (1%, pH adjusted to 4.0 with formic acid) was added and allowed to incubate at room temperature for 1 h. The solution was then added to a 1 ml slurry of Dowex-50 anion exchange resin (+form). Unbound material was collected and dried by roto-evaporation, resuspended in 10 M $(NH_4)_2HPO_4$ (pH 3.8) and analyzed by HPLC.

Jones et al. (1999) have separated the deacylated PtdInsPs by HPLC on a Partisphere SAX column (4.6 × 235 mm^2, 5 μm, Whatman) with a gradient of 0–1 M ammonium phosphate, pH 3.75, over 120 min. The gradient consisted of 0–10 min 100% for pump A, 10–70 min linear rise to 25% for pump B, 70–120 min steep linear rise to 100% for pump B. Radiolabeled deacylated GroPIns(3,5)P$_2$, GroPIns(3,4)P$_2$ and GroPIns(4,5)P$_2$ were eluted in that order and were detected by on-line radiochemical monitoring (Beckman Instruments, Fullerton, CA). Peak-associated radioactivity was expressed as percentage of total radioactivity detected to eliminate any inter-sample variation. 3-Phosphorylated PtdInsPs HPLC standards ([^{32}P]GroPIns(3)P, [^{32}P]GroPIns(3,4)P$_2$, and [^{32}P]GroPIns(3,4,5)P$_3$) were prepared by the action of PtdIns 3-kinase (immunoprecipitated from 10^8 CTLL-2 cells using an anti-p85 antibody) on the substrates PtdIns, PtdIns(4)P and PtdIns(4,5)P$_2$ in the presence of [γ-^{32}P]ATP followed by deacylation.

3.1.2.4. Affinity chromatography

The neutral solvents also extract PtdIns and some PtdInsP, which may be recovered by selective adsorption on neomycin columns as originally described by Schacht (1978, 1981) and refined by Palmer (1981). The antibiotic neomycin exhibits specific affinity for

PtdInsP and PtdInsP$_2$, presumably due to a strong ionic interaction between the cluster of six primary amino groups on the neomycin and the several negatively charged phosphate groups on the lipid molecules. The separation is performed by means of columns of immobilized neomycin (reductively coupled to porous glass beads). All anionic lipids present in the chloroform/methanol extracts, which had been washed first with acid and then with neutral salt solutions, were adsorbed from a chloroform/methanol 1:1 (v/v) solution. All non-acidic lipids, pigments, and the acidic lipids (PtdSer and PtdOH) were eluted with chloroform/methanol/water (5:10:2) (two column volumes). The different PtdInsPs are eluted from the neomycin column by increasing the concentration of ammonium formate. Letcher et al. (1990) have found that simply using the eluants described by Palmer (1981) results in cross-contamination between the three principal inositol lipids. Using a stepwise elution from 200 to 500 mM in 100 mM steps and then taking only the PtdInsP revealed them to be pure by TLC analysis (Irvine, 1990). Each batch of glass beads varies, however.

These PtdIns(3)Ps differ from PtdIns(4)Ps since the former phospholipids do not act as substrates for PLC (Auger et al., 1990). The PtdIns structures derived by chromatographic separations require confirmation by chemical, enzymic and/or mass spectrometric methods.

3.2. Determination of structure

The determination of structure of PtdInsPs has evolved from purely chemical to largely mass spectrometric methods with TLC and HPLC playing major facilitating roles. Although chiral chromatographic methods have been developed for the resolution of stereochemical isomers, the enantiomeric nature of the PtdInsPs must be determined by specific enzyme reactions and by the binding selectivity of receptor proteins.

3.2.1. Chemical methods

Folch (1949) isolated a fraction from ox brain which contained fatty acid, glycerol, phosphorus and inositol in a molar ratio of 1:1:2:1 and proposed a PtdInsP structure. It was later recognized to include PtdIns and PtdInsP$_2$ (Grado and Ballou, 1961), who also proposed the locations of the phosphate groups in the PtdInsP and PtdInsP$_2$ molecules. Subsequent work in several laboratories (Dawson and Dittmer, 1961; Brockerhoff and Ballou, 1961) confirmed the structures (Fig. 1.1, Chapter 1) as 1-phosphatidyl-L-*myo*-inositol (PtdIns), 1-phosphatidyl-L-*myo*-inositol 4-phosphate [PtdIns(4)P], and 1-phosphatidyl-L-*myo*-inositol 4,5-bisphosphate [PtdIns(4,5)P$_2$] (Ansell and Hawthorne, 1964).

3.2.1.1. General assays

Early chemical analyses established that inositol is linked through its 1-hydroxyl to a phosphatidyl group. Mild alkaline hydrolysis yielded GroPIns, which was shown to be optically active and to have glycerol in *sn*-3-linkage to phosphate. Acid hydrolysis under appropriate conditions released inositol phosphate and diacylglycerol. Work from the laboratories of Ballou and Brown (for a review, see Hawthorne and Kemp, 1964) established that the inositol phosphate from soya, liver and brain preparations contained D-*myo*-inositol–1-phosphate structures. Subsequently, chemical synthesis of PtdIns containing a phosphatidyl group linked to the 1-position of inositol completed the chemical proof (Klyashchitskii et al., 1969).

The PtdInsPs have the same structure as PtdIns with the exception that the inositol has an additional phosphate group at the 4-position (PtdIns(4)P) and at the 4,5-positions [PtdIns(4,5)P$_2$]. The structures were originally established by acid migration studies and treatment with periodate, which established the structures as Ins(1,4)P$_2$ and Ins(1,4,5)P$_3$ as components of the Folch 'phosphoinositide' complex.

3.2.1.2. Deacylation

Examination of the products released by mild alkali (as done for PtdIns in Chapter 2) confirmed a D-*myo*-inositol – 1-phosphate structure for the phospholipids that gave rise to the above $InsP_2$ and $InsP_3$. Dawson and Dittmer (1961) independently derived a similar structure for a highly purified $InsP_3$. Chemical degradation of PtdInsPs also yields the fatty acid composition, which, however, was not an early part of the characterization of the PtdInsPs. The fatty acids were determined later by GLC following preparation of methyl esters (see Chapter 2).

Deacylation of PtdInsPs serves two functions. It provides evidence of the composition and general structure of the phospholipid molecules as well as provides derivatives that permit effective separation of isomeric PtdIns, including the novel 3-OH phosphates.

According to Jones et al. (1999) the deacylation is best carried out when 500 μl of freshly prepared methylamine reagent (1-butanol, methanol, 25% aqueous methylamine, 11.5:45.7:42.8, v/v/v), containing 1 mM EDTA is added to the glass vials before heating at 53°C for 1 h. After cooling to room temperature, the contents of the vials were transferred to an Eppendorf tube and vacuum dried at room temperature. After resuspending the mixture in 500 μl of H_2O, fatty acids and any undeacylated lipids were removed by washing twice with 500 μl of the freshly prepared solvent mixture consisting of 1-butanol, petroleum ether, ethyl formate (20:4:1, by vol.). More than 95% of the radioactivity routinely partitioned into the aqueous phase, indicating almost complete phospholipid deacylation.

3.2.1.3. Deglyceration

In order to further characterize the GroPInsPs, the glycerol moiety is removed by mild periodate oxidation (Brown and Stewart, 1966). Both [^{32}P]phosphate and [^{14}C]inositol labeled GroPInsPs can be used. (Letcher et al. (1990) used a modification of the method by Brown and Stewart (1966) to remove the glycerol moiety from the [5-^{32}P]GroPIns(4,5)P$_2$. To the dry [5-^{32}P]GroPIns(4,5)P$_2$ was added 1 ml of 50 mM sodium periodate. After mixing to dissolve the [5-^{32}P]GroPIns(4,5)P$_2$, the tube was placed in the dark for 30 min at room temperature. If the oxidation is allowed to continue for longer the periodate will start to oxidize the inositol ring. Ethylene glycol (150 μl of 10%, v/v) is added to stop the reaction and, after 15 min, 0.5 ml of 1% 1,1-dimethylhydrazine/formic acid (pH 4) is added, and the mixture is allowed to stand for further 60 min. The mixture is passed through 30 ml of a cation exchange resin (BioRad 50WX8, 200–400 mesh in aid form) in a 1.6 cm diameter glass column. The column is washed with 30 ml of H$_2$O, the pooled eluates being adjusted to pH 6 with dilute KOH and evaporated to dryness on a rotary evaporator. The [^{32}P]Ins(1,4,5)P$_3$ is now dissolved in a small volume of H$_2$O for application to HPLC. Letcher et al. (1990) remove the glycerol moiety from the [^{14}C]GroPIns(4)P exactly as described for[5-^{32}P]GroPIns(4,5)P$_2$. After passing through acid Dowex and having been neutralized with KOH, the [^{14}C]Ins(1,4)P$_2$ is dissolved in 2 ml of H$_2$O ready for HPLC. The InsPs are resolved by HPLC and identified as described below (see also Chapter 6).

Ballou and Grado (1961) and Tomlinson and Ballou (1961) devised a strategy for the identification of the structure of InsPs based on a process of oxidation with periodate, reduction and dephosphorylation of the unknown inositol phosphate. The chirality of the final product was established by measuring the specific optical rotation. The discovery of the suitability of yeast polyol dehydrogenase for the stereoselective oxidation of the polyols, has reestablished the periodate oxidation as an effective means of establishing the chirality of the InsPs derived from PtdInsPs. Stephens (1990) has described a scaled down adoptation of the periodate oxidation method in combination with polyol dehydrogenase assay for the analysis of the structure of trace quantities of [^{3}H]InsPs (see Chapter 6). Hirvonen et al. (1988) have reported that InsPs are readily detected by GLC as TMS derivatives with capillary SE-30 column and flame ionization at picomolar range.

3.2.1.4. Dephosphorylation

The chemical characterization of the InsPs derived by deacylation and deglyceration of PtdInsPs may be continued through to the formation of *myo*-inositol and its isomers, if present. This may be accomplished by chemical or by enzymatic dephosphorylation (see below). The chemical dephosphorylation of InsPs is accomplished by alkaline hydrolysis, which leads to an intermediate formation of a cyclic triester, provided a free hydroxyl group is available. The cyclic triester is then rapidly broken down to isomeric α- and β-phosphates. The isomerization of the phosphates can easily invalidate structure conclusions. The confusion introduced by alkaline hydrolysis is eliminated by enzymatic hydrolysis of the InsPs (see Chapter 4).

3.2.2. Enzymic methods

Enzymatic hydrolysis of PtdInsPs has provided characteristic degradation products for structural identification and quantification. Enzymic degradation has been particularly important in positional and stereospecific analyses which acid or alkaline degradation has not been able to provide.

3.2.2.1. Phospholipases

The structure of the PtdInsPs can be characterized by the catabolic enzymes, phospholipase and phosphomonoesterase. A phospholipase C specific for PtdIns has been purified from *Bacillus cereus* (Ikezawa et al., 1976) and *Staphylococcus aureus* (Low and Finean, 1977). Thus, the PtdInsP$_2$ yields diacylglycerol and inositol trisphosphate, while PtdInsP yields diacylglycerol and InsP$_2$. The PtdInsP$_2$ yields PtdInsP and inorganic phosphate, while the PtdInsP yields PtdIns and inorganic phosphate. A specific phospholipase C present in high concentration in aqueous extracts from acetone powders of ox brain was used by Holub et al. (1970) to release the diacylglycerol moieties from PtdIns, PtdInsP and PtdInsP$_2$ of ox brain. The phosphomonoesterase has been purified 430-fold from rat brain (Nijjar and Hawthorne, 1977). Guinea pig pancreas contains a phospholipase A$_1$, which attacks the sn-1-position of PtdIns, while brain and other tissues have a phospholipase A$_2$, which attacks the sn-2-position of PtdIns (Baker and Thompson, 1972, 1973; Shum et al., 1979). Two forms of PtdIns-specific PLC have since been purified from bovine brains (Ryu et al., 1986).

The positional distribution of fatty acids in the diradylglycerol moieties of inositol phospholipids may be determined by hydrolysis with phospholipase A$_2$, which yields the composition of the sn-2-position as free fatty acids and that of the sn-1-position as the corresponding lyso-phosphatide derivative. More specific enzymatic characterization of the positional distribution of the fatty acids in the inositol phosphatides may be obtained following analysis of the derived PtdIns, PtdOH, or sn-1,2-diacylglycerols as shown below.

3.2.2.2. Phosphatases

Alkaline phosphatase preparations provide effective means of dephosphorylation of PtdInsPs as well as GroPInsPs. Hydrolysis with alkaline phosphatase is critical for the dephosphorylation of InsPs without isomerization. Carter et al. (1994) have provided a flow chart of analysis of ^{32}P-labeled PtdIns(3,4,5)P$_3$, starting with the deacylated and deglycerated product and utilizing phospho-monoesterase and radiochromatography (see also Chapter 5).

3.2.2.3. Kinases

More recent work has shown that cellular kinases upon appropriate stimulation yield novel inositol phospholipids in intact cells. A PtdIns 3-kinase phosphorylates the D-3 position of the inositol ring of PtdIns to produce a PtdIns(3)P that is distinct from PtdIns(4)P, the predominant monophosphate from cellular PtdIns

(Whitman et al., 1988). In addition, the PtdIns 3-kinase also phosphorylates PtdIns(4)P and PtdIns(4,5)P$_2$ to generate two, additional novel phospholipids: PtdIns(3,4)P$_2$ and PtdIns(3,4,5)P$_3$, respectively. Using appropriate radio-labeled substrates, these kinases also have been utilized for the characterization of PtdIns and PtdInsPs.

According to Divecha and Irvine (1990) satisfactory purification of PtdIns(4)P kinase can be achieved in a single step by using a DEAE-Sepharose column. The purified enzyme can be then used for the characterization and quantification of PtdIns(4)P by the incorporation of ^{32}P into PtdInsP$_2$ by incubation of the PtdIns(4)P preparation with the enzyme under appropriate reaction conditions. In a model reaction, 1 nmol PtdIns(4)P was sonicated into 80 μl of PIPKIN buffer (50 mM Tris-acetate, 80 mM KCl, 10 mM magnesium acetate, 2 mM EDTA (pH 7.4). A 20 μl portion of the fraction to be analyzed was added and the reaction started by the addition of 100 μl of PIPKIN buffer containing ATP and 0.5 μCi [γ-^{32}P]ATP such that the final concentration in the reaction is 5 μM. The reaction was continued at 37°C for 30 min and stopped by the addition of 750 μl of chloroform/methanol/HCl (80:160:1, by vol.), followed by 250 μl of chloroform containing 0.5 μg P of Folch-fraction inositide mixture and finally 250 μl of 0.9% (w/v) NaCl. The mixture was vortexed, centrifuged and finally lipid extraction completed. The bottom phase was dried and counted.

3.2.2.4. Polyol dehydrogenase

Another enzyme activity of great importance for the determination of the structure of the InsPs is polyol dehydrogenase, which possesses total D versus L selectivity. Polyols recognized as substrates are oxidized to ketones in a β-NAD$^+$-linked reaction. The progress of the reaction is monitored in a spectrophotometer by measuring the accumulation of NADH or by detecting the formation of a ^{3}H-labeled ketone. A commercially available preparation of a yeast-derived L-iditol dehydrogenase (Sigma) oxidizes substrates with a specificity that is consistent with the rules set out by McCorkindale and Eson (1954) (see Chapter 4).

3.2.3. Chromatographic methods

The technically difficult chromatographic separation of PtdInsPs is usually approached in one or two ways. PtdInsPs can be separated by TLC, either in one, two or three dimensions, or by HPLC. Alternatively, the PtdInsPs can be hydrolyzed and the resulting head groups analyzed by anion exchange column chromatography or HPLC. Good separations of PtdInsPs from each other are usually obtained only by sacrificing the resolution of one or more other phospholipids.

3.2.3.1. Resolution of intact PtdInsPs

An early separation of PtdIns, PtdInsP and PtdInsP$_2$ by TLC was reported by Gonzalez-Sastre and Folch-Pi (1968). Glass plates were coated from a slurry of 30 g of Silica Gel H (Merck) in 80 ml of a 1% aqueous solution of potassium oxalate. After drying at room temperature the plates were activated at 110°C for 30 min. The samples obtained from ox brain according to Hendrickson and Ballou (1964) were applied to the plates in amounts ranging from 0.3 to 3 µg of total phosphorus, in water or in chloroform/methanol/water (75:25:2, by vol.). The plates were developed upward with n-propanol/4N NH$_4$OH (2:1, v/v) or with chloroform/methanol/water (9:7:2 by vol.). Spots were revealed either by exposure to iodine vapor followed by spraying with 1% starch solution, or by spraying with Kaggi–Misher reagents (glacial acetic/conc. sulfuric acid/-anisaldehyde 100:1:0.4) and heating in an oven at 120°C for 15 min. Excellent separations were obtained with both solvent systems (chromatograms not shown).

Stein et al. (1990) have described a method for acetylation of PtdIns(4)P and PtdIns(4,5)P$_2$ with [^{3}H]acetic anhydride and for separation of the products from each other and from unchanged starting material. The principle of the method is to acetylate the lipid and then isolate the lipid acetate from unchanged lipid by TLC. The addition, before acetylation, of a sample of lipid labeled with a second isotope (^{14}C or ^{32}P is convenient) allows a ^{3}H/^{14}C or ^{3}H/^{32}P ratio to be obtained for the acetylated lipid by comparison with the ratio obtained for a set of lipid standards treated in an identical fashion, the amount of lipid in the unknown sample can be calculated. This value (a ratio) is not affected by losses in extraction and processing, counting efficiencies, or exactness of specific activity because the addition of a known number of counts of the ^{14}C- or ^{32}P-labeled lipid to both unknown and standard samples at the start of the procedure means that the initial concentration of the lipid present may be determined. Thus, the acetylation may be performed using ^{32}P-labeled PtdInsPs, having approximately 10 000 cpm and approximately 5 µg of pure PtdCho. To thoroughly dried samples, DMF (40 µl) containing 0.5 mg 4-dimethylaminopyridine/ml, diisopropylethylamine (20 µl), and unlabeled acetic anhydride (10 µl) were added. The samples were then sonicated and incubated at 50°C for 0.2 and 6 h. The acetylated products were extracted and separated by TLC using Polygram Sil N-HR precoated plastic sheets soaked in a solution of 1.3% (w/v) potassium oxalate (pH 7.5), 1 mM EGTA, in water/methanol (3:2, v/v), and air dried. Before use the plates are activated by heating at 110°C for 30 min. The plates are developed with chloroform/methanol/4 M ammonia (60:40:10, by vol.). The spots are detected by autoradiography of TLC plates. The spots are cut out and counted in a scintillation counter. A Berthold automatic TLC-linear analyzer may also be used. This measures total (^{32}P or ^{14}C + ^{3}H) radioactivity when the aperture is on open window setting, and ^{32}P, or ^{14}C, radioactivity with the aperture reduced using a polyethylene sheet.

3.2.3.2. Resolution of GroPInsPs

Auger et al. (1989) resolved GroPInsPs by anion exchange HPLC with a Partisphere SAX column (Whatman) using a method modified from Whitman et al. (1988). Figure 3.4 shows the HPLC separation of deacylated GroPInsPs (Auger et al., 1990). [^{32}P]GroPInsPs (solid circles), generated in vitro from anti-phosphotyrosine immunoprecipitates of PDGF-stimulated cells were deacylated and analyzed as described. [^{3}H]Ins-labeled standards are shown as open circles. Baseline separations of all of the GroPInsPs were achieved using an HPLC high resolution 5 μm Partisphere SAX column (Whatman) and a shallow, discontinuous salt gradient (Auger et al., 1990). Dried, deacylated samples are resuspended in 0.1–0.5 ml of 10 mM $(NH_4)_2HPO_4$ (adjusted to pH 3.8 with H_2PO_4) and applied to the column for analysis. The HPLC column is equilibrated with H_2O prior to sample loading and is eluted with a discontinuous gradient of up to 1 M $(NH_4)_2HPO_4{\cdot}H_2PO_4$ (pH 3.8) at a flow rate of 1 ml/min. The gradient is established from 0–1.0 M, $(NH_4)_2HPO_4{\cdot}H_2PO_4$ over 115 min. To develop the shallow gradient required for these separations, dual pumps are used. Pump A contains H_2O; Pump B contains 1.0 M $(NH_4)_2HPO_4{\cdot}H_2PO_4$. The common technique for this purpose consists of the deacylation of the lipids and separation of the resulting GroPInsPs by HPLC (Serunian et al., 1991). Until recently no TLC system was available for the simultaneous resolution of all known isomers of the inositol phosphatides and PtdInsP$_3$ (Hegewald, 1996).

Jones et al. (1999) have been able to achieve base-line separation of the three radiolabeled deacylated isomers of PtdInsP$_2$ ([^{32}P]GroPIns(3,5)P$_2$, [^{32}P]GroIns(3,4)P$_2$, and [^{3}H]GroPIns (4,5)P$_2$) known at this time. Figure 3.5 shows the separation of the standards on the SAX-HPLC. Separation of all the deacylated phospholipids was performed by HPLC employing a Partisphere SAX column (4.6 × 235 mm^2, 5 μm, Whatman) with a gradient of 0–1 M ammonium phosphate, pH 3.75, over 120 min. The gradient consisted of 0–10 min 100% for pump A, 10–70 min linear rise to 2% for pump B, 70–120 min steep linear rise to 100% for pump B. The pump and column washout was from 120 to 130 min with 100% for pump A (H_2O). Radiolabeled deacylated phospholipids

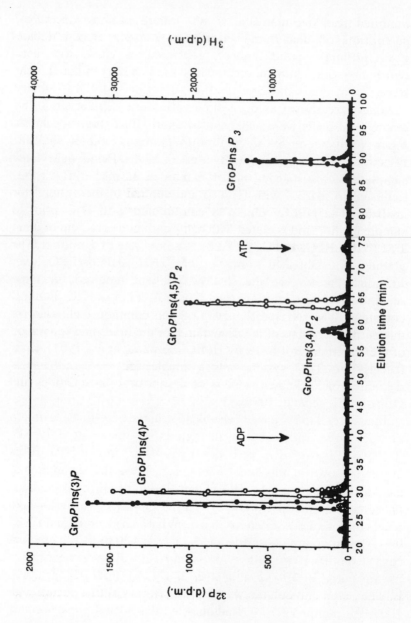

were detected by on-line radiochemical monitoring (Beckman Instruments, Fullerton, CA and E. G. Berthold, Bad Wildbad, Germany). Peak associated radioactivity was expressed as per cent of total radioactivity detected to eliminate any inter-sample variation.

At least four metabolic pools of PtdIns(4)P and PtdIns(4,5)P$_2$ have been distinguished in the human erythrocyte plasma membrane (King et al., 1987). The mechanisms by which multiple non-mixing metabolic pools of PtdOH, PtdIns(4)P and PtdIns(4,5)P$_2$ are sustained over many hours in the plasma membranes presumably involves compartmenting between different leaflets of the lipid bilayer, or there could be inhibitory long-term molecular associations between particular lipid molecules and other components within one leaflet of the bilayer.

Nasuhoglu et al. (2002) have described the quantification of deacylated PtdInsPs by anion HPLC with suppressed conductivity measurements. The major anionic head groups could be quantified in single runs with practical detection limits of about 100 pmol, and the D-3 isoforms of PtdInsP and PtdInsP$_2$ were detected as shoulder peaks. The HPLC system and columns were from Dionex Corporation (Sunnyvale, CA). The NaOH gradient was optimized to separate the major anionic phospholipid head groups in single runs. After injection, the elution was carried out in four stages:

Fig. 3.4. HPLC separation of deacylated PtdInsPs. [^{32}P]PtdInsPs generated in vitro from anti-phosphotyrosine immunoprecipitates of PDGF-stimulated cells were deacylated and analyzed on high resolution 5 μM Partisphere SAX column using a shallow, discontinuous salt gradient. Closed circles, unknowns; open circles, standards. ADP and ATP were added to the samples prior to analysis. The solvent gradient was as follows: Solvent A, 100% water; B, 1.0 M (NH$_4$)$_2$-HPO$_4$·H$_2$PO$_4$. At start, B is run at 0% for 5 min; to 15% B with a duration of 55 min, isocratic elution for 20 min, and then to 100% B over 25 min. The fractions, 100–500 μl are collected at 5–30 s intervals over the entire gradient or the region of interest. Reproduced from Auger et al. (1990) with permission of publisher.

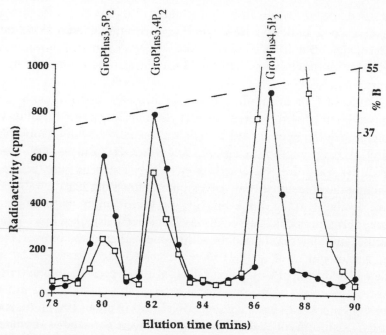

Fig. 3.5. Identification of three PtdInsPs isomers in CTLL-2 cells using SAX-HPLC. Arrested cells were radiolabeled with [^{32}P]orthophosphate followed by phospholipid extraction, deacylation, and analysis by SAX-HPLC. Radioactivity associated with the fractions was detected on-line with a flow detector open squares. In addition, the SAX-HPLC profile of three standard deacylated isomers of PtdInsP$_2$, [^{3}H]GroPIns(3,5)P$_2$, [^{32}P]GroPIns(3,4)P$_2$, and [^{3}H]GroPIns(4,5)P$_2$, is overlaid (closed circles). Reproduced from Jones et al. (1999) with permission of publisher.

(1) 1.5–4 mM NaOH (0–7 min); (2) 5–16 mM NaOH (7–12 min); (3) 16–86 mM NaOH (12–30 min); and finally (4) isocratic 86 mM NaOH (10 min). In HeLa, Hek 293 and COS cells, as well as in intact heart, PtdInsP$_2$ amounted to 0.5–1.5% of total anionic phospholipid (10–30 μmol/l cell water or 0.15–0.45 nmol/mg protein). The new method permitted non-radioactive analysis of PtdInsPs and other anionic phospholipids in cell cultures and tissues.

3.2.3.3. Resolution of InsPs

The chromatographic separation and quantification of various InsPs is described in detail in Chapter 6. For the present purpose, it is sufficient to indicate that two basic methods exist. The first method determines directly any particular InsPs isomer in an extract without initial chromatographic separation from other isomers, e.g. radioligand competition assay employing highly specific high-affinity binding protein or enzymatic assays. However, only a few such assays have been developed to sufficient reliability, e.g. for myo-Ins(1,4,5)P$_3$, that employs a binding protein from adrenal cortex (see Chapter 4). In the second technique, InsPs isomers are first separated chromatographically, most commonly by anion exchange HPLC. Subsequently their masses are determined by different methods, e.g. phosphorus or inositol assays, which are laborious.

Mayr (1989, 1990) has developed an on-line post column metal–dye detection system, in which a transition metal and a metal–specific dye act as reported substances for InsPs and other (oxy)anions eluting from the HPLC column. As shown elsewhere (Chapter 4) InsPs ranging from InsP to Ins(1,3,4,5,6)P$_5$ and InsP$_6$ are separated by one of several elution protocols and quantified by the dye–metal binding method.

3.2.3.4. Resolution of isomeric Ins

Isomeric Ins can be resolved by GLC following trimethylsilylation and quantified by flame ionization detection. The stereoisomers may be resolved on chiral GLC columns. The Ins peaks can be identified by mass spectrometry as shown in Chapter 2.

3.2.4. Mass spectrometric methods

Mass spectrometric techniques are especially useful due to their high sensitivity. Effective applicability of mass spectrometry, however, depends on the proper choice of the corresponding MS technique. Only three of the different ionization techniques have played a major role in the field of phospholipid analyses: FAB, ESI and MALDI.

3.2.4.1. FAB/MS

FAB/MS was the first MS technique successfully applied for the detection of PtdInsPs. Sherman et al. (1985) studied both positive and negative ions generated from PtdIns(4)P and PtdIns(4,5)P$_2$. The PtdIns(4)P and PtdIns(4,5)P$_2$ were shown to behave in FAB similarly to PtdIns (see Chapter 2). GroPIns lipid species behaved similarly to GroPOH lipids since both are acidic glycerophospholipids. While positive ions corresponding to [M + H]$^+$ can be observed, most FAB ions are negative ions. Collision induced decomposition of [M + H]$^+$ from GroPIns lipids yields a neutral loss of 260 u, corresponding to the loss of InsP. This neutral loss is the characteristic transition reported to identify an ion as uniquely derived from GroPIns lipids (Muenster et al., 1986; Murphy and Harrison, 1994). The negative ion FAB spectrum of GroPIns lipids contains [M − H]$^-$ ions in low abundance consistent with the loss of [inositol–H$_2$O], such as m/z 671. This ion has a structure identical to the [M − H]$^-$ anion of GroPOH lipids. Carboxylate anions from the fatty acyl groups at sn-1 and sn-2 are also observed in the negative ion FAB mass spectrum. Collision induced decomposition of [M − H]$^-$ results in abundant carboxylate anions at sn-1- and sn-2-, while precise assignment of the position requires decomposition of the PtdOH anion (m/z 671.5). Minor ions at m/z 595 and 577 arise from loss of palmitic acid and its ketene analog, respectively. The ions at m/z 392 and 409 likewise result from the losses of the sn-2 substituent from the GroPOH ion formed by loss of inositol.

Sherman et al. (1985) also reported several specific GroPIns-related negative ions at m/z 259 (inositol phosphate), 241 (inositol–phosphate–H$_2$O), and 297 (diacylglycerophosphoinositol–R$_1$COOH–R$_2$COOH), which are unique for this class of phospholipid.

The PtdIns and its phosphates are acidic lipids and therefore produce primarily negative ions in FAB experiments (Jensen et al., 1987). An abundant $[M - H]^-$ ion indicates the molecular weight of the compound as well as the acyl substituents. When sufficient sodium is present in the sample, positive ions can be observed corresponding to $[MH]^+$ and $[M + 2Na-H]^+$ (Sherman et al., 1985). CID of the $[M - H]^-$ ions from diradyl GroPIns yield abundant carboxylate anions as well as several phosphoinositol ions at m/z 21, 259, and 299. Jensen et al. (1987) also reported a series of charge remote decomposition ions seen from $[M - H]^-$ from diradyl-GP.

The use of 'hard' ionization methods like FAB requires experience in the analysis of molecule fragmentation patterns, since the molecular ion is often not detectable. Therefore, FAB is limited towards its use in mixture analysis.

3.2.4.2. ES/MS/MS

ES/MS has the advantage that no derivatization of the sample is required and that the ionization is 'soft'. ES/MS can be very easily coupled with HPLC to separate sample components prior to their mass spectrometric analysis. This however leads to ionization of different phospholipid classes at different solvent compositions, which may affect the response. Michelsen et al. (1995) have demonstrated that PtdInsP and PtdInsP$_2$ give prominent singly and doubly negatively charged deprotonated molecules in ES/MS. These ions can be used for quantification of PtdInsP and PtdInsP$_2$ in the low picomole range, without prior chromatographic separation, using selected ion monitoring and consecutive measurements of the signals from the deprotonated singly charged molecules. The dose response curves for both compounds are linear. In a complex matrix consisting of polar lipids (Folch extract) PtdIns and PtdInsP$_2$ monitored at m/z 965.4 and 1045.5 (stearoyl and arachidonoyl) were determined in the low picomole range, at a flow rate of 100 µl/min. Collision-induced decomposition of PtdInsP and PtdInsP$_2$ using a mixture of xenon and argon

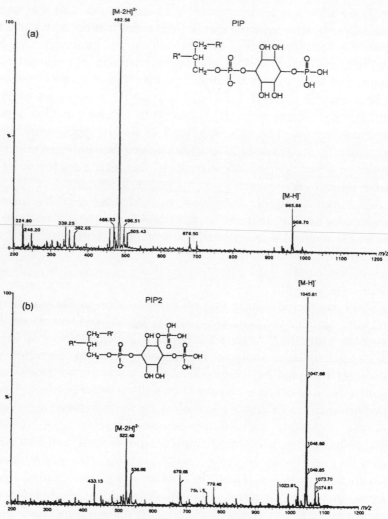

Fig. 3.6. Negative-ion electrospray mass spectrum of PtdIns(4)P (a) and PtdIns(4,5)P$_2$ (b) isolated from bovine brain. Mass spectra were recorded on a VG Quattor 1 instrument equipped with a pneumatically assisted electrospray ion source and a VG mass data handling system. PtdInsP gives rise to a prominent peak at m/z 965.7 corresponding to 18:0/20:4 species. The base peak is due to a doubly charged molecular related ions at m/z 482.6. In the spectrum of PtdInsP$_2$,

at 25 eV afforded identical high mass ions formed by loss of a molecule of water from PtdInsP and a phosphate group and a molecule of water from PtdInsP$_2$. The results indicate that PtdInsPs, and biologically relevant changes in their concentrations, can be quantified directly in cells and cellular membranes by selected ion monitoring with ES/MS. Figure 3.6a and b show the full negative-ion spectra of PtdInsP and PtdnsP$_2$, respectively, isolated from bovine brain. PtdInsP gives rise to a prominent peak at m/z 965.7 (Fig. 3.6a), a value which tallies with that of deprotonated molecules of PtdInsP containing stearoyl and arachidonoyl residues. Interestingly, the base peak is formed from doubly charged molecular-related ions at m/z 682.6. Low intensity ions in the molecular-ion range indicate the presence also of small amounts of other acyl residues. In the negative-ion full spectrum of PtdInsP$_2$ (Fig. 3.6b), deprotonated molecules form the base peak at m/z 1045.6 indicating the presence of the same two acyl residues as in PtdInsP. The doubly charged ions of m/z 522.5 represent less than half of the intensity of the single charged ions. As in the case of PtdInsP, small amounts of acyl residues of different chain lengths and unsaturation were observed in the molecular weight range. Figure 3.7 gives the high mass range negative-ion mass spectrum of phospholipids in a bovine brain extract (Michelsen et al., 1995). Optimal proof with respect to the identity of the analyte may be obtained using CID of the molecular related ions observed in the full mass spectrum. Hereby, fragments used in SIM are correlated with a high degree of certainty to ions from the parent ions. Figure 3.7 shows the high mass range negative-ion mass spectrum obtained from a mixture of the total polar lipids (30 μg/μl) in a bovine brain extract (Michelsen et al., 1995). The deprotonated molecules of PtdIns, PtdInsP, and

deprotonated molecules form base peak at m/z 1045.6 again indicating the presence of 18:0/20:4 species. The doubly charge ions at m/z 522.5 represent less than half of the intensity of the singly charged ions. Reproduced from Michelsen et al. (1995) with permission of publisher.

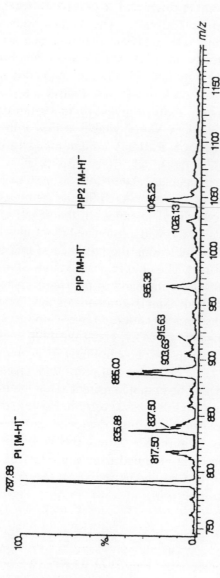

Fig. 3.7. High mass range negative ion mass spectrum of phospholipids of bovine brain. Mass spectrum recorded as in Fig. 3.6. The deprotonated molecules of PtdIns, PtdInsP and PtdInsP$_2$ are clearly recognized at m/z 787.9, 965.4, and 1045.3, respectively (acyl residues 18:0 and 20:4). Here consecutive measurements of the same sample diluted with an equal volume of mobile phase produced signals in SIM at 965.4 and 1045.3, corresponding to a concentration of PtdInsP and PtdInsP$_2$ of 7.4 ± 0.2 and 7.3 ± 0.2 pmol/μl, respectively. Reproduced from Michelsen et al. (1995) with permission of publisher.

$PtdInsP_2$ are clearly recognized at m/z 787.9, 965.4, and 1045.3, respectively (acyl residues 18:0 and 20:4) (chromatogram not shown). In the negative-ion spectrum of PtdInsP, the base peak was formed using a doubly charged molecular-related ions at m/z 482.6. Low intensity ions in the molecular weight range indicated the presence of small amounts of other acyl residues. In the negative-ion full spectrum of $PtdInsP_2$ the doubly charged ions at m/z 522.5 represented less than half of the intensity of the singly charged ions. Michelsen et al. (1995) have quantified the PtdInsP and $PtdInsP_2$ subclasses by single ion monitoring and construction of response curves (see below).

Gunnarsson et al. (1997) combined normal-phase HPLC separation of PtdInsPs with ES/MS. For this purpose, a Gilson HPLC system was combined with a Quattro II mass spectrometer (Micromass) equipped with pneumatically assisted electrospray and atmospheric pressure chemical ionization sources. The HPLC columns (100 × 2 mm id) were packed in the laboratory with silica 5μ (Spherosorb, CT) at 950 bar. The columns were developed with solvent mixtures consisting of A, chloroform/methanol/ammonia (50:45:3, by vol.) and B, chloroform/methanol/water/ammonia (25:55:17:3, by vol.). A linear gradient was run. The HPLC effluent entered the mass spectrometer through an electrospray capillary set at -2.6 kV and 120°C. The PtdInsPs in the HPLC eluate were directly admitted to the ES/MS system. The mass spectra of Folch type brain extract gave mass chromatograms for the ions of m/e 885, 965 and 1045 representing PtdIns, PtdIns(4)P, and PtdIns(4,5)P_2, respectively (chromatograms not shown). The area under each peak in the mass chromatograms reflected the relative abundance of the PtdInsPs in the mixture applied, as was found earlier (Michelsen et al., 1995).

Hsu and Turk (2000) have reported detailed structural characterization of PtdIns, PtdIns(4)P, and PtdIns(4,5)P_2 by collisionally activated dissociation ES/MS/MS. In negative ion mode, the major fragmentation pathways under low energy CAD for PtdIns arise from neutral loss of free fatty acid substituents and neutral loss of the corresponding ketenes, followed by consecutive loss of the inositol head group (see Chapter 2). In negative ion mode, the PtdInsP and PtdInsP$_2$ tandem spectra contain the ions m/z 259, 241 and 223, reflecting the inositol head group, in addition to the ions at m/z 321 and 303, reflecting the doubly phosphorylated inositol ions. The PtdInsP$_2$ also contains unique ions at m/z 401 and 383 that reflect the triply phosphorylated inositol ions. The $[M - H]^-$ ions of PtdInsP and PtdInsP$_2$ undergo fragmentation pathways similar to that of PtdIns upon CAD. However, the doubly charged ($[M - 2H]^{2-}$) molecular ions undergo fragmentation pathways that are typical of the $[M - H]^-$ ions of glycerophosphoethanolamine, which are basic. These results suggest that the further deprotonated gaseous $[M - 2H]^{2-}$ ions of PtdInsP and PtdInsP$_2$ are basic precursors.

3.2.4.3. MALDI-TOF/MS

Like ES/MS, MALDI/MS does not require sample derivatization. Unlike ES/MS, MALDI/MS cannot be easily combined with HPLC. However, MALDI-TOF/MS has the advantage that all measurements can be performed in a single organic phase, since both the lipid and the matrix (e.g. DHB) are readily soluble in organic solvents. This provides extremely homogeneous matrix/analyte crystals and leads to an excellent reproducibility. MALDI is more sensitive than ES and is not affected by impurities to such a high extent. Schiller et al. (1999) have demonstrated that these properties allow the efficient and convenient detection of lipids by MALDI-TOF/MS. Harvey (1995) and Marto et al. (1995) used this technique successfully for the analysis of lipids in model membranes, while Schiller and Arnold (2000) used it in the analysis of organic solvent extracts of cells. PtdInsPs differ from other phospholipids by their headgroup structure containing the inositol ring system as well as by their charge state. While PtdIns bears one negative charge at the dissociated phosphodiester group, additional phosphomonoester groups at the inositol ring lead to higher negative charges for PtdInsP ($z = -3$) and PtdInsP$_2$ ($z = -5$) at pH 7.4 (Toner et al., 1988; Gabev et al., 1989). The charge of PtdInsP$_3$ is still higher ($z = -7$) at pH 7.4. Furthermore, PtdInsPs possess a higher molecular mass than all other known phospholipids containing two acyl chains (more complex lipid structures like cardiolipin are not considered here). The biologically relevant mass region of PtdInsPs ranges from about 880 to about 1120 Da. Mueller et al. (2001) have recently investigated the detectability of PtdInsPs by MALDI-TOF/MS. Using the signal-to-noise-ratio to describe the quality of a given spectrum and the detectability of a certain PtdInsPs, they defined the minimum amount of analyte necessary to obtain a reasonable signal-to-noise-ratio as detection threshold. Figure 3.8 shows the negative ion MALDI-TOF mass spectra of selected PtdInsPs with different degrees of phosphorylation at the inositol ring (Mueller et al., 2001). The phospholipids were from bovine brain: PtdIns (**a**), PtdInsP (**b**) PtdInsP$_2$ (**c**), dipalmitoyl GroPInsP$_3$ (**d**) and stearoyl/arachidonoyl GroP-InsP$_3$ (**e**). The spectra were acquired on a Voyager Biospectrometry DE workstation (PerSeptive Biosystems, Framington, MA). The ion chamber pressure was held between 1×10^{-7} and 4×10^{-7} Torr. All measurements were made under delayed extraction with an extraction voltage of 20 kV. The samples were prepared in a matrix of dihydroxybenzoic acid (0.5 M in methanol). The lipid amounts of the plate were 113 pmol of PtdIns; 261.8 pmol for PtdInsP; 478 pmol of PtdInsP$_2$, 572 pmol for dipalmitoyl GroPInsP$_3$, and 444.3 pmol for stearoyl/arachidonoyl GroPInsP$_3$. Spectra are the average of 128 laser shots. The bovine liver PtdIns (**a**) yielded the isotopically resolved molecular ion peak at m/z 885.5 Da. With an increasing number of phosphate groups the peaks of the corresponding molecular ion were shifted to higher m/z ratios. These changes could be explained by the substitution of a hydroxyl group by a phosphate group, and by the required compensation of the corresponding negative charge of the lipid by the addition of protons or sodium ions. Thus, two protons were involved in the case of PtdInsP (m/z 965.5, **b**), four in

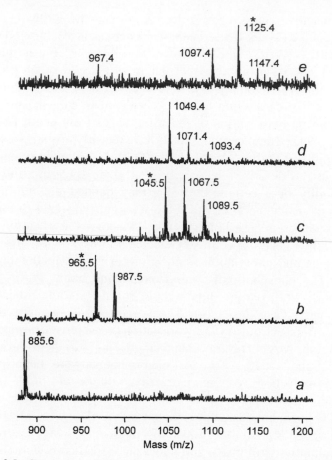

Fig. 3.8. Comparison of the positive and negative ion MALDI-TOF mass spectra of (a) PtdInsP from bovine liver; (b) PtdInsP and PtdInsP$_2$ (c) from bovine brain; (d) dipalmitoyl GroPInsP$_3$ and (e) stearoyl-arachidonoyl GroPInsP$_3$. For all sample a 0.5 M 2,5-dihydroxybenzoic acid (DHB) solution in methanol containing 0.1% trifluoro-acetic acid was used. All lipid solutions were directly applied on the sample plate as 1-μl droplets, followed by the addition of a drop of 1 μl matrix solution. Samples were allowed to crystallize at room temperature. All MALDI spectra were acquired on a Voyager Biospectrometry workstation (Perseptive Biosystems, Framingham, M). Reproduced from Mueller et al. (2001) with permission of publisher.

PtdInsP$_2$ (m/z 1045.5, **c**), and six in PtdInsP$_3$ (m/z 1049.9, for dipalmitoyl GroPInsP$_3$, **d**, and m/z 1125.4 for stearoylarachidonoyl GroPInsP$_3$, **e**). The lipid signals were observed to decrease in lipid mixtures. Thus the detectability of PtdInsP$_2$ was reduced by the presence of PtdCho. While PtdInsP$_2$ alone is well detectable above 123.4 pmol, the detection limit is 156 pmol in the presence of 100 ng (132 pmol) PtdCho. According to Mueller et al. (2001) MALDI-TOF/MS spectra cannot be analyzed quantitatively, if only the measured signal intensity is taken into account, because the signal intensity is influenced by factors that cannot be effectively standardized. However, internal standards of the same chemical nature as the substance of interest can be used to compensate for the external influences. Mueller et al. (2001) have introduced the use of the signal-to-noise ratio as a quantitative measure. It can be done semi-automatically using the standard software packages available for commercial MALDI-TOF mass spectrometers.

PtdInsPs possess higher molecular weights than other phospholipids and a high phosphorylation-dependent negative charge. Both features affect the MALDI detection limits expressed as the minimum of analyte on the sample plate resulting in a signal-to-noise ratio of S/N = 5. Using 2,5-dihydroxybenzoic acid as matrix the detection limit for PtdIns is seven times higher than for PtdCho and further increases with increasing phosphorylation or in mixtures with other well-detectable phospholipids. For PtdInsP$_3$ in a mixture with PtdCho, the limit is about 20 times higher than for PtdIns. It is therefore advisable to pre separate PtdInsPs from biological lipid mixtures prior to the application of MALDI-TOF/MS.

3.3. Resolution of molecular species

Each PtdInsPs subclass like the parent PtdIns is made up of several molecular species, which contain saturated fatty acids in the *sn*-1- and polyunsaturated fatty acids in the *sn*-2-position of the glycerol moiety. In addition, specific saturated fatty acids are found to be paired with specific polyunsaturated fatty acids in a highly characteristic manner. Early studies of the molecular species of PtdIns, PtdIns-P and PtdInsP$_2$ from beef brain showed a similar pattern for each with the 1-stearoyl-2-arachidonoyl-*sn*-glycerol species representing 40% of the total (Holub et al., 1970, Holub and Kuksis, 1978). The major molecular species in each phospholipid class were identified and quantitatively estimated by laborious thin-layer and gas-liquid chromatography of the component diacylglycerols, which were released by hydrolysis with a specific brain phosphodiesterase. More recent chromatographic and mass spectrometric analyses permit much more rapid and more sensitive assays of the molecular species of the inositol phospholipids but only a few original analyses have been performed. In most instances it has been assumed that the PtdInsPs possess the same molecular species as the corresponding PtdIns, the molecular species composition of which is usually not determined but assumed from early literature. Since more recent studies have demonstrated the presence of small amounts of both alkylacyl (Lee et al., 1991; Butikofer et al., 1992) and alkenylacyl (Lee et al., 1991) in addition to the diacylglycerol moieties of PtdIns of human and bovine erythrocytes, it is of great interest to establish whether or not the ether linked fatty chains are present also in the PtdInsPs of the erythrocytes, which are presumably derived from the PtdIns pool. The presence of both alkylacyl and alkenylacylglycerol moieties have been well established to be present among the molecular species of the glycosyl PtdIns anchor of the *Torpedo marmorata* acetylcholinesterase (Butikofer et al., 1990). The presence of alkenylacyl or alkylacylglycerol containing species thus far have not been reported in the PtdInsPs despite occasional thorough GLC analysis of component fatty acids or MS/MS of purified PtdInsPs.

3.3.1. Intact PtdInsPs and PtdIns

Chromatographic analyses of intact polyphosphoinositides usually do not permit resolution of molecular species as the high polarity of these compounds dominates the separation characteristics. However, direct mass spectrometry or mass spectrometry in combination with chromatography has yielded detailed information about the molecular species composition of both isolated inositol phosphatides and in mixtures containing other inositol phosphatides and phospholipids.

It was already pointed out that PtdIns(4)P and PtdIns(4,5)P_2 behave in FAB quite similar to GroPIns (Sherman et al., 1985). Cronholm et al. (1992) also used FAB/MS to analyze the InsP, GroPIns(4)P and GroPInsP(4,5)P_2 lipids isolated from rat pancreas. The molecular species composition was found to be quite similar for the three GroPIns phosphorylated subclasses, except for more fully saturated molecular species appearing in Ins(4,5)P_2. Cronholm et al. (1992) has compared the molecular species composition of the PtdInsPs from the pancreas of rats fed an ethanol-containing diet (E) and from the corresponding pair-fed controls (C) as determined by FAB/MS (Table 3.2). In control rats, the composition of PtdIns(4) was similar to that of PtdIns. The only significant differences noted were a higher proportion of the species containing a palmitoyl and a stearoyl group and a lower proportion of the species containing a stearoyl and a linoleoyl group in PtdIns(4)P than in PtdIns. In addition, PtdIns(4,5)P_2 had a lower content of the arachidonoyl-containing species than PtdIns. The results showed highly significant differences between the ethanol-fed and control rats. Thus, PtdIns species containing an arachidonoyl group were half as abundant in the ethanol-fed rats, while species containing a stearoyl and no arachidonoyl group were twice as abundant in these animals. The percentage of PtdIns containing two palmitoyl residues was also lower in the ethanol-fed rats. The differences in composition of PtdIns(4)P and PtdIns(4,5)P_2 between ethanol-fed and control rats were of the

TABLE 3.2

Molecular species composition of PtdIns from pancreas of rats fed an ethanol-containing liquid diet (E) and from the corresponding pair-fed controls (C), as determined by FAB/MS

Acyl groups	PtdIns		PtdIns(4)P		PtdIns(4,5)P$_2$	
	C	E	C	E	C	E
16:0/16:0	2.9 ± 0.3	1.8 ± 0.4^a	9.6 ± 5.5	8.8 ± 5.1	6.0 ± 0.5	$3.2 \pm 0.7++$
16:0/18:2	3.5 ± 0.4	2.9 ± 0.4	4.1 ± 1.8	4.3 ± 1.6	3.2 ± 0.3	2.8 ± 0.5
16:0/18:1	7.3 ± 0.5	7.0 ± 0.4	8.0 ± 0.5	6.6 ± 0.9^b	8.2 ± 0.7	6.9 ± 0.9
16:0/18:0	6.6 ± 0.5	13.6 ± 1.1^c	10.9 ± 0.8	13.2 ± 3.3	14.9 ± 1.7	22.9 ± 4.0^b
16:0/20:4	6.0 ± 0.1	2.4 ± 0.4^c	5.8 ± 0.7	5.5 ± 1.6	4.5 ± 0.3	2.6 ± 0.7^a
18:1/18:2	4.1 ± 0.2	4.8 ± 0.3^b	4.2 ± 0.6	5.2 ± 0.5^b	3.6 ± 0.7	4.5 ± 0.8
18:0/18:2	13.4 ± 0.4	21.4 ± 1.4^c	9.1 ± 1.6	11.3 ± 2.5	9.5 ± 0.3	12.0 ± 1.6^b
18:0/18:1	10.4 ± 0.7	23.3 ± 3.1^c	8.7 ± 1.9	14.5 ± 2.3^c	11.8 ± 1.6	16.5 ± 1.8^a
18:1/20:4	7.8 ± 0.3	3.7 ± 0.4^c	8.0 ± 0.8	7.0 ± 0.2	6.7 ± 0.5	5.3 ± 1.1
18:0/20:4	37.9 ± 1.9	19.2 ± 1.3^c	31.6 ± 5.3	23.6 ± 4.8	31.5 ± 2.1	23.2 ± 3.6^a

Assigned structures are based on molecular masses and most likely acyl groups as judged from previous results and gas chromatography after transesterification. Values are mean $\pm$ SD ($n = 4$).
Significant differences: [a]$P < 0.05$.
[b]$P < 0.01$.
[c]$P < 0.001$ in comparison to controls and ethanol-fed rats. Reproduced, with permission, from Cronholm, T., Viestam-Rains, M. and Sjovail, J. (1992) Biochemical Journal 287, 925–928. © The Biochemical Society and the Medical Research Society.

same type as for PtdIns, although less marked. No evidence was reported for the presence of alkylacyl or alkenylacyl species in the PtdInsPs thus far examined.

Jensen et al. (1987) studied the negative ion FAB mass spectra of PtdIns isolated from soybeans. Charge remote fragmentation of the carboxylate anion for linoleic acid esterified to GroPIns revealed the double bond locations at C_9-C_{10} and C_7-C_{13} following collision-induced dissociation in the triple sector mass spectrometer (See also Murphy and Harrison, 1994).

Kayganich and Murphy (1994) have determined the time-dependent increase in incorporation of $[^{13}C_{17}]$arachidonic acid into the arachidonate-containing glycerophospholipids. Isotope incorporation values were obtained from the ratio of ion intensity for the transitions of the major anions (A- to m/z 303) compared to the resultant labeled species $[^{13}C_{17}]$A- to m/z 320. The apparent level of isotope incorporation was different for each of the major phospholipid classes.

Schiller et al. (1999) have recently reported the application of MALDI-TOF/MS for the analysis of PtdInsPs. It is shown that in a matrix of 2,5-dihydroxybenzoic acid the molecular ions (M + 1) of the different PtdInsPs are easily detected even in complex mixtures, and thus, detailed data on the fatty acid composition are provided. Fragmentation reactions of fatty acids on the double bonds and on the polar lipid head group are observed to a minor extent in the spectra of all investigated lipids. The inositol phospholipids were obtained from different suppliers: PtdIns (ammonium salt) was from Avanti Polar Lipids Inc. (Alabama) (20 mg/ml $CHCl_3$ solution), whereas PtdIns(4)P (sodium salt) was obtained from Sigma as a 10 mg/ml solution in chloroform (both from bovine liver). The fatty acid composition of both substances was indicated to be mainly stearic acid and arachidonic acid. PtdIns(4,5)P_2 (triammonium salt from bovine brain) was obtained from Calbiochem (Bad Soden, Germany) as a lyophilized powder. It was dissolved in chloroform in a concentration of 30 g/ml (about 30 μM). Two microliter of 1 M HCl was added to a total volume of 10 ml PtdIns(4,5)P_2 solution to give a final concentration of 0.2 mM hydrochloric acid. All solvents (chloroform, methanol and ethyl acetate), the matrix (2,5-dihydroxybenzic acid) and trifluoroacetic acid were obtained in highest commercially available form from FLUKA.

Cells were stimulated for 2 min with the chemotactic tripeptide N-formyl-Met-Leu-Phe (fMLP) and at sampling time treated with 0.5 ml of chloroform/methanol (2:1, v/v). The chloroform layer was used for MALDI-TOF/MS analyses. For all samples a 0.5 M 2,5-dihydroxybenzoic acid solution in methanol containing 0.1% trifluoroacetic acid was used (acetoacetate was also found to be suitable as solvent for the matrix compound). All lipid samples were directly applied on the sample plate as 1 μl droplets, followed by the addition of a drop of 1 μl matrix solution. Drying the samples with a moderate, warm stream of air extremely improved the homogeneity of crystallization.

Schiller et al. (1999) employed a Voyager Biospectrometry workstation (PerSeptive Biosystems, Framington, MA). The system utilizes a pulsed nitrogen laser, emitting at 337 nm. The extraction voltage used was 20 kV. Pressure in the ion chamber was maintained between 1×10^{-7} and 4×10^{-7} Torr. All spectra were acquired in a DHB matrix using a low-mass gate of 4000 Da. To enhance the signal to noise ratio, 128 single shots from the nitrogen laser (337 nm) were averaged for each mass spectrum. All these preparations were commercially available. Due to difficulties in determining the exact amount of inositol phospholipids, all samples were used as supplied. For all lipids under investigation, the spectra recorded in the negative mode contain a lower number of peaks and all peaks are shifted by one (PtdIns) or two mass units (PtdInsP and PtdInsP$_2$) to lower masses in comparison to the positive ion mass spectra. Spectra of negative ions also exhibit an enhanced signal-to-noise ratio for PtdInsP and PtdInsP$_2$, whereas in the case of PtdIns the positive mode spectrum gave better signal-to-noise ratio. The differences between negative and positive ion mass spectra can be explained by different numbers of hydrogen or sodium ions added to compensate for the negative charges. PtdInsP$_2$ contains five negative charges in total. Assuming the appearance of one positive charge under the conditions of laser irradiation in the mass spectrometer, three of five positively charged ions (H$^+$ or Na$^+$) are necessary for charge compensation to obtain ions with one negative or positive charge, respectively. Peaks at 1046 (negative mode) or 1048 (positive mode) obviously correspond to charge compensation solely by hydrogen ions. With an increasing number of sodium ions additional peaks appear with a mass difference of 22 Da. In a similar manner the spectra of PtdInsP can be explained.

Assuming the same mechanism of charge compensation for PtdIns, one should expect only a mass spectrum in the positive mode with two peaks at 886 (+H) and 908 (+Na). However, there is one noticeable peak at 885 in the negative mode. This peak corresponds to the exact molecular mass of the PtdIns ion, bearing one stearic and one arachidonic acid residue. Since there are no further peaks of PtdIns in this region of the mass spectrum (very low intense peaks between 850 and 860 are caused by the matrix, data not shown). Schiller et al. (1999) speculate that the negative PtdIns ion leads to this peak, whereby neither cationization by laser irradiation nor addition of cations for charge compensation occurs. In contrast, the addition of two sodium ions to the PtdIns ion would explain the peak at 931 in the positive mode.

In summary, the MALDI mass spectra of PtdIns and PtdInsPs can easily be interpreted considering the need to compensate a different amount of negative charges for the negative and the positive ion mode. As already noted with other lipid classes, fragmentation reactions do not occur to a large extent (Schiller et al., 2000).

Prior to this time Marto et al. (1995) had shown that different species of phospholipids can easily be investigated by MALDI mass spectrometry, using highly sophisticated experimental methods (Fourier transform ion cyclotron resonance spectrometry).

Since the PtdInsPs can be deliberately dephosphorylated by phosphatases or converted into the corresponding PtdIns during a metabolic transformation, the molecular species of the parent phosphatide could be determined by any of the methods described for the assay of the molecular species of PtdIns (see Chapter 2). In view of the much smaller amounts of mass available for the analyses, the use of carriers would be necessary. Mixtures of PtdIns can be isolated from appropriate sources for determination of the molecular species of radio-labeled (phosphate, glycerol, inositol or fatty acid) samples, while mixtures of deuterium labeled carriers could be employed for a mass spectrometric assessment of the phosphatide species.

3.3.2. *Phosphatidic acids*

Since the PtdInsPs are subject to hydrolysis by phospholipase D, the phosphatidic acids generated by deliberate digestion or transphosphatidylation with phospholipase D (methyl or ethyl phosphatides) or released during the course of a metabolic transformation are amenable to analysis of molecular species by the methods described for the phosphatidic acids released from PtdIns (see Chapter 2). In view of the much smaller amounts of mass available for the analyses, the use of carriers would be necessary.

3.3.3. *Diacylglycerols*

Holub et al. (1970) have described conditions for hydrolysis of PtdInsP and PtdInsP$_2$ with PtdIns-specific phospholipase C from bovine brain. In a scaled down version, the phosphatide (3 mg) was incubated in 5 ml of 0.05 M Tris–HCl buffer (pH 7.2) containing brain extract (1 mg of protein) and 0.02% (w/v) acetylmethylammonium bromide at 37°C for 60 min. This enzyme also hydrolyses brain PtdIns. The liberated diacylglycerols are immediately removed from the incubation medium by extraction with diethyl ether and derivatized. The diradylglycerol acetates prepared were analyzed by silver ion TLC and GLC on nonpolar columns to obtain the molecular species composition of the released diradylglycerols.

The positional distribution of the fatty acids was determined by treating the diacylglycerol acetates with pancreatic lipase. Over 27 molecular species were identified, and these accounted for about 95% of each PtdInsP class, but the 1-stearate-2-arachidonate derivative contributed more than 40% of the total in each class. The other molecular species also were qualitatively and quantitatively similar in the three PtdInsP subclasses. Table 3.3 compares the molecular species composition of the PtdIns, PtdIns-(4)P and PtdIns(3,4)P$_2$ from ox brain (Holub et al., 1970). The brain enzyme used in this study has since been purified from particulate (Katan and Parker, 1987) and cytosol (Ryu et al., 1986) fractions.

It was shown in Chapter 2 that the common molecular species of diacylglycerols are most extensively resolved by GLC on polar capillary columns (Kuksis and Myher, 1990, 1995). Reversed phase HPLC is less efficient, but offers advantages for the separation of polyunsaturated species (Zhu and Eichberg, 1993). HPLC also allows peak collection. Recent work has shown that the certain reverse isomers of the sn-1,2-diacylglycerols may also be obtained by HPLC (Itabashi et al., 1999). Finally, appropriate derivatives of the sn-1,2-diacylglycerol moieties of the PtdInsPs may be subjected to GC/MS and LC/MS analyses as discussed in Chapter 2. Because of the smaller amounts of mass likely to be encountered, more extensive use of carrier molecules may have to be utilized in the isolation and analysis of these diacylglycerols.

The purified PtdIns-specific phospholipase C from *B. cereus* is commercially available (Boehringer Mannheim Co.). It is not known, however, if this enzyme effects a representative release of sn-1,2-diacylglycerols from all PtdInsPs.

3.3.4. Positional distribution of fatty acids

A number of recent improvements in the isolation and purification of the inositol phospholipids have permitted the isolation of pure subclasses of the PtdInsPs, making it possible to determine the fatty

TABLE 3.3

Major molecular species of ox brain PtdIns and PtdInsPs (mol%)

Chemical classes	Molecular species	PtdIns	PtdIns(4)P	PtdIns(4,5)P$_2$
Monoenes	14:0/18:1	0.10	0.10	0.32
	16:0/16:1	0.21	–	–
	16:0/18:1	5.13	4.46	2.83
	16:0/20:1	0.42	0.51	0.06
	18:0/18:1	3.97	4.66	2.83
	18:0/20:1	0.21	0.20	0.26
Dienes	16:0/18:2	0.61	0.08	0.33
	18:1/16:1	0.12	0.33	0.11
	18:1/18;1	1.22	3.42	4.34
	18:0/18:2	1.22	0.56	0.67
	16:0/20:2	0.49	0.16	0.33
	18:0/20:2	1.88	2.93	4.12
	18:1/20:1	0.36		
	20:1/20:1	0.18	0.65	1.23
Trienes	16:0/20:3	0.83	0.78	0.90
	18:1/18:2		0.20	0.23
	18:0/20:3	8.35	17.16	19.18
	18:1/22:2		1.37	2.26
Tetraenes	16:0/20:4	5.74	2.92	2.01
	18:0/20:4	59.58	43.15	41.73
	18:1/20:3	0.72	2.92	1.51
	18:1/22:3	5.74	9.33	5.03
Pentaenes	18:1/20:4	1.68	2.81	7.66
	18:2/22:3:	0.83	0.31	1.05
	20:2/22:3		0.47	0.38
	20:1/22:4		0.31	0.48
Hexaenes	16:0/22:6	Trace		
	18:0/22:6	Trace		
	18:2/20:4	Trace		

Each molecular species is reported as a percentage of the total phosphatide. The positional distribution of fatty acids was determined by pancreatic lipase hydrolysis of the corresponding diacylglycerol moieties. Reproduced from Holub et al. (1970) with permission of the publisher.

acid composition of the individual subclasses of the PtdInsPs. However, such analyses have been reported only infrequently (Holub et al., 1970; Cronholm et al., 1992) and the uniformity of the molecular species composition of PtdIns and PtdInsPs generally assumed remains to be established. Table 3.4 gives the positional distribution of fatty acids in the PtdIns, PtdIns(4)P and PtdIns(4,5)P$_2$ fractions of ox brain as determined by pancreatic lipase hydrolysis of the derived diacylglycerol moieties. Table 3.4 demonstrates the usual preferential association of the saturated fatty acids with the primary and the polyunsaturated

TABLE 3.4

Positional distribution of fatty acids in PtdIns, PdInsP and PtdInsP$_2$ of ox brain (mol%)

Fatty acids	PtdIns		PtdInsP		PtdInsP$_2$	
	sn-1-	sn-2-	sn-1-	sn-2-	sn-1-	sn-2-
14:0	0.1		0.1		0.3	
16:0	14.6		8.6		6.6	
16:1		0.8		0.8		0.2
18:0	74.4		69.3		68.9	
18:1	9.9	10.4	20.2	12.7	21.1	10.3
18:2	0.8	1.8	0.3	0.8	1.0	1.2
20:1§	0.2	1.2	1.0	1.1	1.7	1.6
20:2§		2.4	0.5	3.1	0.4	2.4
20:3 (n − 9)		4.7		9.8		10.4
20:3 (n − 6)		5.2		11.0		11.6
20:4 (n − 6)		67.0		48.9		51.5
22:2§				1.4		2.2
22:3		6.6		10.1		6.5
22:4				0.3		0.5
22:5 (n − 5)						
22:6 (n − 3)		Tr		Tr		Tr

§, tentative identity. Free fatty acids (sn-1-position) released by pancreatic lipase; fatty acids from monoacylglycerols formed by action of pancreatic lipase on the diacylglycerol. Reproduced, with permission, from Cronholm, T., Viestam-Rains, M. and Sjovail, J. (1992) Biochemical Journal 287, 925–928. © The Biochemical Society and the Medical Research Society.

fatty acids with the secondary position of the glycerol molecule. The PtdIns, PtdInsP and PtdInsP$_2$, however, differ significantly in the fatty acid composition of the *sn*-1- as well as the *sn*-2-position, as already noted from the analyses of the molecular species composition (Cronholm et al., 1992). These discrepancies in the fatty acid composition could be accounted for by a selective phosphorylation of specific molecular species of the precursor inositide, or by an independent turnover and remodeling of the molecular species subsequent to phosphorylation. The positional distribution of fatty acids may also be obtained by hydrolysis of the PtdIns, PtdInsP and PtdInsP$_2$ by a lipase, which attacks the primary position of the intact phospholipids (Slotboom et al., 1970).

The positional distribution of the fatty acids in individual intact molecular species of the PtdInsPs could be determined by hydrolysis with bee venom (Thompson, 1969) phospholipase A$_2$, which releases the fatty acid from the *sn*-2-position. Alternatively, the fatty acids in the *sn*-1-position can be specifically released by a fungal lipase, which specifically attacks the primary position of the acylglycerol molecule (Itabashi et al., 2000). Because of the small amounts of mass likely to be involved, such analyses are not practical unless radioactive markers are involved and carriers used.

3.3.5. Inositol isomers

Since at least *myo*- and *chiro*-inositols would be anticipated to be associated with the PtdIns phosphate fractions, it would be of interest to determine the isomer content of individual subclasses and major molecular species of the PtdInsPs. These assays could be performed along the lines described for PtdIns (Chapter 2) and PtdIns glycans (Chapter 5).

3.4. Quantification

Quantification of specific PtdInsPs is best performed by means of tandem mass spectrometry, which in addition to the estimate of a total PtdInsP, PtdInsP$_2$, and PtdInsP$_3$, also yields PtdInsPs subclasses. Alternatively, each subclass of the PtdInsPs can be isolated chromatographically and the molecular species composition of each PtdInsPs subclass determined by further chromatography following partial degradation to the sn-1,2-diacylglycerol or PtdOH moiety as described elsewhere (Kuksis and Myher, 1990; Singh and Jiang, 1995). In those instances, where an effective separation of PtdIns subclasses cannot be directly obtained, quantification of the overlapping subclasses may be obtained following deacylation and/or deglyceration of the sample. In such a case, the information about the molecular species composition is lost. Information about the composition of molecular species is also lost when the subclasses of the PtdInsPs are quantified by receptor binding.

3.4.1. MS/MS

Although much of the mass spectrometric methodology described above for the resolution of the molecular species of PtdInsPs is suitable for quantification, only a few such attempts have been documented. Nakamura et al. (1989) reported FAB studies of the 9-anthryl derivatives of natural PtdInsPs isolated with preparative TLC. The mass spectra showed di- and tri-sodiated molecules of the anthryl derivatives. Cronholm et al. (1992) determined the quantitative composition of the molecular species of the PtdInsPs of rat pancreas, but the quantitative proportions of the PtdInsP subclasses. Michelsen et al. (1995) identified the major negative singly and doubly charged deprotonated ions in standard PtdInsP and PtdInsP$_2$ by loop injection ES/MS. In a complex matrix consisting of polar lipids (Folch extract) PtdInsP and PtdInsP$_2$ monitored at m/z 965.4 and 1045.5 (stearoyl and arachidonoyl) were determined in the low picomole range, at a flow rate of 100 μl/min CID of PtdInsP and PtdInsP$_2$ using a mixture of xenon and argon at 25 eV afforded identical high mass ions formed by loss of a molecule of water from PtdInsP and a phosphate group and a molecule of water from PtdInsP$_2$. These ions were used for quantification of PtdInsP and PtdInsP$_2$ in the low picomole range, without prior chromatographic separation, using selected ion monitoring and consecutive measurements of the signals from the deprotonated singly charged molecules. The dose response curves for both compounds were linear. Michelsen et al. (1995) also investigated the suitability of LC/MS for the quantification of the major PtdInsP and PtdInsP$_2$. Three consecutive chromatograms of the brain extract obtained using chloroform/methanol/water (10:20:8, by vol.) at a flow rate of 100 μl/min, produced signals in SIM at m/z 965.4 and 1045.3 corresponding to concentrations of PtdIns and PtdInsP$_2$ of 7.4 ± 0.2 and 7.2 ± 0.2 pmol/μl, respectively. The results indicate that PtdInsPs, and biologically relevant changes in their concentrations, and be quantified directly in cells and cellular membranes by SIM with ES/MS. Michelsen et al. (1995) have found that the Folch extract used contains 10–20% of PtdInsPs of which about 1% constitutes PtdInsP and PtdInsP$_2$. The results show that PtdInsP and PtdInsP$_2$ are present in nearly equal amounts and that each of the analytes can also be quantified when they constitute only ca. 0.05% of the total phospholipids in a complex mixture.

Gunnarsson et al. (1997) have compared the relative response of PtdIns, PtdIns(4)P and PtdIns(4,5)P$_2$ of a Folch type I brain extract by ES/MS and HPLC with light scattering detection. The area under each peak in the mass chromatograms (m/z 885, 965 and 1045), representing PtdIns, PtdIns(4)P and PtdIns(4,5)P$_2$ reflected the relative abundance of the PtdInsPs in the mixture in agreement with an earlier report (Michelsen et al., 1995). Approximately equal amounts of each phosphatide were present in the synthetic PtdInsPs preparation, while the Folch extract contained very little PtdIns(4)P and PtdIns(4,5)P$_2$ relative to PtdIns. Using selected ion monitoring allowed the quantification of less than 200 pmol of PtdInsPs (Michelsen et al., 1995).

The signal intensity in MALDI-TOF/MS is influenced by sample preparation, inhomogeneities of the laser intensity and ionization and desorption efficiency of different molecules. However, different approaches can be used to overcome the limitations for obtaining quantitative information from MALDI-TOF/MS, e.g. use of internal standards of the same chemical nature as the substance of interest and estimation of relative concentrations of the individual components in a mixture from the respective relative peak intensities (Petkovic et al., 2001). Mueller et al. (2001) have introduced the signal-to-noise ratio as a quantitative measure. It can be done semiautomatically using standard software packages available for commercial MALDI-TOF mass spectrometers. The calculated S/N ratio increases continuously over a wide concentration range with the sample concentration, which also provides an exact criterion for the estimation of the detection limit. Mueller et al. (2001) have observed that the detection limits for PtdInsPs are about 20 times higher than for PtdCho, while Petkovic et al. (2001) have noted that the presence of other easily ionized substances in the mixture may alter the MS response to PtdInsPs. It is therefore advisable to preseparate PtdInsPs from biological lipid mixtures prior to the application of MALDI-TOF/MS. Berrie et al. (2002) have recently reported a novel LC/MS/MS method for direct measurement of GroPInsPs in cell extracts.

3.4.2. Autoradiography

PtdInsPs may be quantified following radiolabeling of the experimental material, but this requires the knowledge of the specific labeling. Furthermore, labeling to equilibrium may be difficult to obtain as there may be more than one PtdInsPs pool in cells having different turnover rates. Nevertheless, autoradiography of PtdInsPs has proven adequate for certain types of quantitative comparisons. An illustration of this method is found in the report of Traynor-Kaplan et al. (1989) where autoradiography was used to demonstrate the transient increase in PtdIns(3,4)P_2 and PtdInsP$_3$ during activation of human neutrophils. Precoated TLC plates (Silica Gel-60, 20×20, 250 μm thick) were impregnated with potassium oxalate by developing in a TLC tank overnight with a solvent system containing 1.2% potassium oxalate in ethanol/water (2:3, v/v). After activation (15 min at 110°C), the plates were spotted with the ^{32}PO$_4$ (500 μCi/ml) prelabeled phospholipids in 100 μl of chloroform/methanol (2:1, v/v). TLC plates were then developed in one dimension in chloroform/acetone/methanol/ acetic acid/water (80:30:26:24:14, by vol.) for an additional 15 min after the solvent reached the top of the TLC plate. Radioactive spots were detected with autoradiography using Kodak X-Omat X-ray film and with AMBIS β-scanning system (San Diego, CA). Spots were also quantified using the conventional technique of scraping and counting in a Beckman liquid scintillation counter. The two methods yielded comparable ratios of peaks, as shown by an autoradiogram of a TLC separation comparing N-formyl-norleu-Leu-Phe-norleu-Tyr-lysine-fluor-escein (FLPEP)-stimulated changes in ^{32}PO$_4$ incorporation into phospholipids from islet-activating protein-pretreated neutrophils with co-incubated controls.

Ptasznik et al. (1996) have described an improved one-dimensional TLC separation and quantification of PtdInsP, while Ptasznik et al. (1997) have reported effective application of this separation to the analysis of radioactive PtdIns, PtdInsP, PtdInsP$_2$ and PtdInsP$_3$ in fetal islet like cell clusters (Fig. 3.9).

A variation of the autoradiographic step involves the separation of the deacylated radiolabeled PtdInsPs by HPLC using a Partisphere SAX column and a gradient based on buffers A (water)/B [1.25 M (NH$_4$)$_2$HPO$_4$]. The data are collected as values for ^{32}P and ^{3}H radioactivity (dpm) per 0.5 ml fractions of eluate (measured using a calibrated ^{3}H/^{32}P dual-label program on a scintillation counter). The determination allowed to investigate the synthesis of 3-phosphorylated PtdInsPs in human astrocytoma cells (Stephens et al., 1989) and in growth factor stimulated Swiss

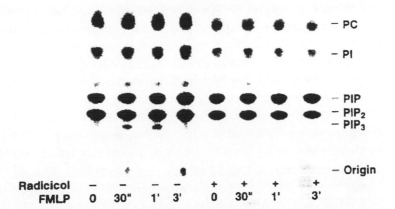

Fig. 3.9. Autoradiogram of a thin-layer chromatographic separation of total neutrophil phospholipids showing the time course of PtdInsP$_3$ formation in neutrophils treated in the presence or the absence of radicicol (200 ng/ml, 2.5 h) and subsequently stimulated with 1 μM fMLP (formylmethionylleucylphenylalanine) for the indicated times. The experiment shows PtdInsP$_3$ formation in fMLP-stimulated neutrophils is blocked by tyrosine kinase inhibitors; analogous results were obtained with 100 μM genistein. PC, PtdCho; PI, PtdIns; PIP, PtdIns(4)P; PIP$_2$, PtdIns(4,5)P$_2$. Reproduced from Ptasznik et al. (1996) with permission of publisher.

3T3 cells (Jackson et al., 1992). Although radiotracer labeling has been extensively used to assess the metabolic interrelationships of the PtdInsPs, the methods suffer from the inevitable assumption that all pools of PtdInsPs have been labeled to equilibrium. Furthermore, they are complex, time-consuming, and relatively insensitive.

3.4.3. Radioreceptor assay

Radioreceptor assay provides an alternative method for mass measurement of PtdInsPs in the intact form or as the derived InsPs (Challiss et al., 1988; Donie and Reiser, 1989).

Chilvers et al. (1991) determined the mass changes in PtdIns(4,5)P$_2$ by assaying for Ins(1,4,5)P$_3$ according to Challis et al. (1988). Thus, dried lipid extracts prepared by routine extraction were dissolved in 250 μl of 1 M KOH and incubated in capped tubes at 100°C for 30 min. Samples were cooled on ice and excess alkali was removed by filtration through columns [0.5 m of 50% (w/v) slurry] of Dowex 50 (200–400 mesh; H$^+$ form). After a column wash with 2.5 ml of water, the combined eluates (pH 3–4) were extracted twice with n-butanol/light petroleum (b.p. 40–60°C) (5:1, v/v) as described by Clark and Dawson (1981). Samples of 1 ml of the resultant lower phase were freeze dried and redissolved in 2 ml of water and portions were assayed for Ins(1,4,5)P$_3$ as described by Challis et al. (1988). Briefly, assays were performed at 4°C in a final volume of 120 μl. To 30 μl of sample or 30 μl containing known amount of D-Ins(1,4,5)P$_3$ (12–36 000 fmol) or DL-Ins(1,4,5)P$_3$ (0.3 nmol, to define non-specific binding) was added 30 μl 100 mM Tris–HCl, 4 mM EDTA, pH 8.0, and 30 μl [^{3}H]Ins(1,4,5)P$_3$ (20 000 dpm/assay) or [^{32}P]Ins(1,4,5)P$_3$ (40 000 cpm/assay). Some 30 μl (0.6 mg protein) of the adrenal cortical binding protein preparation was added and samples were intermittently vortex mixed for 30 min. Separation of bound and free D-Ins(1,4,5)P$_3$ was achieved by centrifugation at 12 000g for 4 min. After aspiration of the supernatant, the pellet was dissolved in 100 μl Lumasolve, scintillant added, and radioactivity determined by scintillation counting.

Divecha et al. (1991) have miniaturized the chemical procedures whereby fatty acids and the glycerol are removed sequentially from the lipid and then assayed the resulting Ins(1,4,5)P$_3$ by a well established radioreceptor assay (Palmer and Wakelam, 1990). In this assay the Ins(1,4,5)P$_3$ present in cell extract competes with a fixed quantity of high specific activity [^{3}H]Ins(1,4,5)P$_3$ (although [^{32}P]Ins(1,4,5)P$_3$ could be used instead) for the Ins(1,4,5)P$_3$-specific binding sites of bovine adrenocortical microsomes. A standard curve using known amounts of unlabeled Ins(1,4,5)P$_3$ is prepared in parallel. Thus, the quantity of $Ins(1,4,5)P_3$ in the cell extract can be calculated (see Chapter 5).

Interestingly, Rivera et al. (1998) have described a radioreceptor assay for mass measurement of $Ins(1,4,5)P_3$ using saponin-permeabilized outdated human platelets. Obtaining large batches of the $Ins(1,4,5)P_3$ binding protein by treating outdated platelets with saponin is simple and quick. The binding of $Ins(1,4,5)P_3$ to saponin-permeabilized blood bank-outdated human platelets, 6 days old, was characterized ($d = 3.8$ nM; $B_{max} = 1.7$ pml/mg protein) and used to develop a novel radioreceptor assay which allows the measurement of the $Ins(1,4,5)P_3$ content in resting or agonist-stimulated cells. This assay is as sensitive (0.25 pmol in 0.25 ml volume), specific, and reproducible as previously proposed methods (see Chapter 4).

Van der Kaay et al. (1997) have reported a simple, reproducible isotope dilution assay which detects $PtdIns(3,4,5)P_3$ at subpico-mole sensitivity that is suitable for measurements of both basal and stimulated levels of $PtdIns(3,4,5)P_3$ obtained from samples containing approximately 1 mg of cellular protein. Total lipid extracts, containing $PtdIns(3,4,5)P_3$, are first subjected to alkaline hydrolysis which results in the release of the polar head group $Ins(1,3,4,5)P_4$. The latter is measured by its ability to displace $[^{32}P]Ins(1,3,4,5)P_4$ from a highly specific binding protein present in cerebellar membrane preparations. Van der Kaay et al. (1997) have shown that this assay solely detects $PtdIns(3,4,5)P_3$ and does not suffer from interference by other compounds generated after alkaline hydrolysis of total cellular lipids. Specifically, the $Ins(1,3,4,5)P_4$ isotope dilution assay was performed as follows (Van der Kaay et al., 1997). The $Ins(1,3,4,5)P_4$ binding protein was obtained from sheep cerebellum. Cerebella were homogenized in ice-cold buffer (20 mM $NaHCO_3$, pH 8.0, 1 mM dithiothreitol, 2 mM EDTA) and centrifuged for 10 min at $5000g$. The pellet was re-extracted once, and the pooled supernatants were centrifuged for 20 min at 38 000g. The pellet was washed twice and resuspended in homogenization buffer at a final protein concentration of

10–20 mg/ml. Assay of 320 μl comprised 80 μl of assay buffer (0.1 M NaAc, 0.1 M KH_2PO_4, pH 5.0, 4 mM EDTA, 80 μl of 3×10^5 dpm of [^{32}P]Ins(1,3,4,5)P_4), 80 μl of sample, and 80 μl of binding protein. Both standard Ins(1,3,4,5)P_4 and samples were in 0.5 M KOH/acetic acid, pH 5.0) and were assayed directly. After addition of binding protein, samples were incubated on ice for 30 min and subsequently subjected to rapid filtration using GF/C filters. Filters were washed twice with 5 ml of ice-cold buffer (25 mM NaAc, 25 mM KH_2PO_4, pH 5.0, 1 mM EDTA and 5 mM $NaHCO_3$). Radioactivity was determined after the filters were extracted for 12 h in 4 ml of scintillant.

According to Van der Kaay et al. (1997) measurements in a wide range of cells, including rat-1 fibroblasts, 1321N1 astrocytoma cells, HEK 293 cells, and rat adipocytes, show wortmannin-sensitive increased levels of PtdIns(3,4,5)P_3 upon stimulation with appropriate agonists. The enhanced utility of this procedure is further demonstrated by measurements of PtdIns(3,4,5)P_3 levels in tissue derived from whole animals. Other InsPs binding assays are described in Chapter 4.

Subsequently, Van der Kaay et al. (1999) have performed the mass assay for Ptdns(3,4,5)P_3 as described by Van der Kaay et al. (1998), which is modified from Van der Kaay et al. (1997) by the use of a different Ins(1,3,4,5)P_4-binding protein that had essentially the same characteristics. Recombinant GST-GP1^{IP4BP} (expressed in *E. coli* and purified on glutathione-agarose beads) replaced crude cerebellar membrane from sheep. Total cellular lipids were isolated from confluent 6-well plates (Van der Kaay et al., 1999).

Maroun et al. (1999) have described such an assay for PtdIns(3,4,5)P_3 using the Gab 1 pleckstrin homology domain as the receptor. GST fusion proteins were generated by cloning wild type Gab1 and A/A and C/C Gab1 mutants as BamHI-EcoRI fragment into pGEX-5X-2 (Amersham Pharmacia Biotech) and expression in *E. coli* BL21 strain. GST fusion proteins were purified on glutathione-Sepharose and the purified proteins (10 pmol) were bound to 30 μl of glutathione-Sephadex beads in

30 mM Hepes, pH 7, 100 mM NaCl, 1 mM EDTA, 0.025% Nonidet P-40 (HNE). Labeled phospholipids were sonicated in the same buffer containing 20 μg/ml Ser(P). Unlabeled competitor lipids (dipalmitoyl GroPIns(3,4)P_2 or PtdIns(3,4,5)P_3 were soni-cated at 100 μM with Ser(P) and 25 μM PtdIns(3,4)P_2 and diluted with Ser(P)/HNE, and constant amounts of labeled [32]PtdInsP$_3$ (800 000 cpm) were added. Beads with bound fusion proteins were incubated with 40 μl of labeled phospholipid with or without competitor for 1 h at 22°C, and the beads were quickly washed twice with 1 ml of HNE, 0.1% Nonidet P-40. Beads were extracted with 60 μl of CHCl$_3$/CH$_3$OH (1:1, v/v) and added directly to scintillation fluid for quantification. Data were corrected for nonspecific binding by subtracting the counts bound to GST. In comparison to MS, the highly specific isotope dilution assays have the disadvantage of assaying only one PtdInsPs at a time.

Gray et al. (1999) have pointed out that the pleckstrin homology domains of protein kinase B and GrP1 are sensitive and selective probes for the cellular detection of PtdInsP$_2$ and PtdInsP$_3$ in vivo. However, only a few of the pleckstrin homology (PH) domains have been exploited for quantitative assay of PtdInsPs or InsPs. Bottomley et al. (1998) has tabulated the ligands and affinities characterizing the interactions between the PH domains and PtdInsPs or InsPs.

3.4.4. Other methods

Matuoka et al. (1988) have prepared monoclonal antibodies specific for PtdIns(4)P and PtdIns(4,5)P_2 and have utilized them to develop an enzyme linked immunosorbent assay to measure mass changes of these lipids in Swiss mouse 3T3 cells after stimulation with growth factors. This method was demonstrated to be sensitive down to picomolar levels.

Divecha and Irvine (1990) and Divecha et al. (1991) have described a PtdIns(4)P mass assay by coupling the activities of a phospholipase C and the DAG kinase (measuring incorporation of radioactivity from [γ-^{32}P]ATP into PtdOH). This assay is sensitive to less than 10 pmol and is, under these conditions, entirely specific for PtdIns [as opposed to PtdIns(4)P, PtdIns(4,5)P$_2$, or other glycerophospholipids]. The enzymes are purified from bovine and rat brain, respectively (see Chapter 7).

Gaudette et al. (1993) measured the mass of PtdInsPs in unstimulated and U46619-stimulated human platelets by TLC separation of PtdIns, PtdInsP, PtdIns(4,5)P$_2$ and PtdIns(3,4)P$_2$ and quantification of the fatty acid components derived from them by transmethylation in the presence of an internal standard. For the analysis of the PtdInsPs, the lipid extracts from four replicate 1-ml aliquots of platelet suspension were pooled. Several (6–12, depending on platelet yield) 2-cm-wide bands were applied to the TLC plates and the lipids resolved in the chloroform/methanol/conc. NH$_4$OH/H$_2$O (45:35:7:5 by vol.) solvents system. Lipids were visualized under UV light following spraying the TLC plates with 8-anilino-1-naphthalenesulfonic acid. The system gave good separation of PtdIns(4)P from other lipids. Although PtdIns(4,5)P$_2$ appeared to be resolved from PtdIns(3,4)P$_2$, HPLC analysis of [^{3}H]Ins-labeled PtdInsPs revealed incomplete purification of PtdIns(3,4)P$_2$ by the single TLC run. The PtdIns(3,4)P$_2$ was purified by rechromatography of the TLC fractions. The recovery of the PtdInsP$_2$ fractions from the silica gel was facilitated by sonication. For quantification the PtdInsPs fractions were transmethylated for 2.5 h at 85°C using 2 ml of acetyl chloride/methanol (5:50, v/v) in the presence of an appropriate quantity of internal standard (heptadecanoic acid) and the fatty acids analyzed by GLC. Appropriate blank areas of the TLC plate were also transmethylated and the values obtained subtracted as a background from the corresponding sample peaks.

Stein et al. (1990, 1991) have described a method for acetylation of PtdIns(4)P and PtdIns(4,5)P$_2$ with [^{3}H]acetic anhydride and for separation of the products from each other and also from unchanged starting material. The addition, before the acetylation, of a sample of lipid labeled with a second isotope (^{14}C or ^{32}P is convenient) allows a ^{3}H/^{14}C or ^{3}H/^{32}P ratio to be obtained for the acetylated lipid. By comparison with the ratio obtained for a set of lipid standards treated in identical fashion, the amount of lipid in the unknown sample can be calculated. As an example, the samples to be acetylated contained ^{32}P-labeled PtdInsPs, approximately 10 000 cpm and about 5 μg of pure PtdCho. After thorough drying, the samples were dissolved in DMF (40 μl) containing 0.5 mg 4-dimethylaminopyridine/ml, diisopropylethanolamine (20 μl), and unlabeled acetic anhydride (10 μl) were added. The samples were then briefly sonicated and then incubated at 50°C for 6 h. The acetylated products were extracted after addition of 5–10 μg of unlabeled carrier PtdInsPs acetates to each sample to aid extraction and TLC of the lipids. TLC is performed on oxalate-impregnated plates, activated at 110°C for 30 min before use. The solvent is chloroform/methanol/4 M ammonia (60:40:10, by vol.). Auto-radiography of TLC plates is carried out to locate the labeled lipids and then the spots cut out and counted in a scintillation counter. Plastic backed TLC plates are used and counting is done in the presence of the plastic. The counting efficiency is improved if the samples are left to stand in the scintillation mixture for about 24 h before counting. Alternatively, a Berthold automatic TLC-scanner analyzer may be used.

The specific protein binding assays of PtdInsPs and InsPs have been further discussed by Prestwich et al. (1998) and Ferguson et al. (2000). Wang et al. (1999) have discussed the differential association of the pleckstrin homology domains of PLC with lipid bilayers and the heterotrimeric G proteins.

CHAPTER 4

Inositol phosphates

4.1. Introduction

Interest in understanding the signal transduction mechanism involving turnover of PtdIns phosphates (PtdInsPs) and generation of diacylglycerols and inositol phosphates (InsPs) has created an urgent need to develop new methods for analysis of these compounds in biological systems (Downes and McPhee, 1990; Murphy et al., 1996; Irvine et al., 1999). The inositol head group of PtdIns contains five free hydroxyl groups with the potential to become phosphorylated. To date, most of the mono-, di-, and polyphosphates of inositol have been claimed to participate in one or more metabolic functions. However, the isolation and analysis of inositol phosphates has posed special problems because of their low concentrations in biological samples and lack of an identifiable chromophore. As a result, analysis of InsPs has relied mainly on sensitive radiotracer techniques, although the mass of the compounds is not determined by such methods. Since these compounds are metabolically very active, information on specific radioactivity is essential for studies on metabolic turnover.

The ubiquitous occurrence of $InsP_{n>1}$, such as $InsP_5$ and $InsP_6$, in eukaryotic cells, and the complex pathways concerned with their synthesis and degradation, suggests that these compounds are functionally significant, but their cellular roles remain obscure at present. Experimental evidence, ranging from the first observations of $Ins(1,4,5)P_3$-stimulated Ca^{2+} mobilization, cDNA sequencing and functional reconstitution of the $Ins(1,4,5)P_3$ receptor, provide unequivocal evidence that $Ins(1,4,5)P_3$ is the second messenger responsible for hormone-stimulated intracellular Ca^{2+} release. Although $Ins(1,3,4,5)P_4$ has been implicated in Ca^{2+} signaling (see below) and $Ins(1,4)P_2$ apparently can activate a low-affinity form of DNA polymerase α (Sylvia et al., 1988), it seems unlikely that many of the rapidly formed metabolites of $Ins(1,4,5)P_3$ serve distinctive functions other than to act as intermediates in the recovery of Ins (Joseph and Williamson, 1989; Taylor, 2002; Dawson et al., 2003).

4.2. Isolation

The methods commonly used to isolate the InsPs associated with the signaling pathways, involve deproteinization of the samples with trichloroacetic acid (TCA) or perchloric acid (PCA), neutralization of the acid and stepwise elution from anion exchange resin (Bio-Rad AG1 X8) of InsP, $InsP_2$, $InsP_3$ and $InsP_4$ with increasing concentration of ammonium formate in dilute formic acid (Singh and Jiang, 1995). The higher forms of InsPs, $InsP_5$ and $InsP_6$, whose metabolism is separate from that of the 'signaling' InsPs, by charcoal extraction (Stephens, 1990), while the $InsP_7$ and $InsP_8$ pyrophosphates are isolated by chromatography on anion exchange resin columns (Q-Sepharose Fast Flow and Resource Q high resolution) (Laussmann et al., 1996) using 250–375 mM HCl as the eluant.

4.2.1. Acid extraction

The use of acidic reagents displaces the InsPs from their salts as well as terminates enzymatic activity. PCA or TCA are most often used. After removal of the denatured proteins by centrifugation, the extracts are neutralized. PCA is removed by precipitation as its potassium salt, while TCA is extracted with water-saturated diethyl ether. Prior to acid extraction, neutral glycerophospholipids may be recovered with neutral solvents (Folch et al., 1957). The acid conditions, also cause hydrolysis of the cyclic phosphates and may isomerize other phosphates. Therefore, the extracts should be neutralized as rapidly as possible. The extraction method should also be checked for other losses of the phosphate of interest (e.g. non-specific binding by filter material or vessel walls).

The isolation of mixed InsPs may be illustrated with a preparation of a sample from rat brain (Sun et al., 1990). One-month-old rats are decapitated and the heads immediately dropped into liquid nitrogen. After immersion of 50 s the skulls are opened and cerebral cortices are removed and weighed while frozen. The brain tissue (about 1 g) is then homogenized at room temperature in 20 ml of $CHCl_3/CH_3OH/12$ M HCl (2:1:0.012, by vol.), extracted with 5 ml of deionized H_2O (HPLC grade), and centrifuged at low speed to separate the phases. The upper aqueous phase is transferred to another test tube and freeze-dried. Samples prepared by this extraction method are subsequently passed through a maxi-clean $IC-H^+$ cartridge (Alltech Associates, Deerfield, IL) to reduce the background conductivity. Following this treatment the samples are ready for ion chromatography.

Maslanski and Busa (1990) described a sensitive procedure for the isolation of InsPs suitable for cell cultures. The method calls for the use of plasticware, because InsPs readily stick to glass plates. The tissue to be assayed is placed in a 4 ml conical polypropylene tube (Sarstedt n. 57.512) and quickly frozen in a solid $CO_2/MeOH$ bath. The tissue is then homogenized, on ice, in 100 μl of 7.5% (w/v) ice-cold PCA using a Teflon pestle. The homogenates are kept on ice for 20 min. The precipitate that forms is pelleted in a Sorval centrifuge at 10 000g, at 4°C for 10 min. The supernatant is treated with 50 μl of 10 mM EDTA, pH 7.0, with KOH. PCA is removed from solution as the potassium salt by the addition of ice-cold 10 M KOH until the pH is greater than 7.0. After removing the potassium perchlorate by centrifugation, the supernatant is withdrawn and the pH of each is brought between 6 and 9 by HCl. The extract is now ready for the isolation of the InsPs by chromatography.

4.2.2. Removal of nucleotides

Negatively charged molecules other than InsPs (e.g. nucleotides) are removed by treatment of the extracts with charcoal (Mayr, 1988; Meek, 1986). Charcoal extraction (using Norit GSX or Darco-G60) may be incorporated into the purification procedure to remove the bulk of $[^{32}P]ATP$ from cell extracts or assays. The charcoal should be thoroughly acid-washed (five times with ice-cold 5% PCA). It is recommended that the charcoal be mixed only with InsPs extracts under acidic conditions (e.g. in the presence of 4% ice-cold PCA (Stephens and Downes, 1990) or at high salt concentrations (Mayr, 1988)). Stephens (1990) recommends that the charcoal is added as a 25 mg/ml slurry in 4% PCA with an air displacement pipette (100 μl is used with an extract from 0.3 ml of packed cells). The charcoal is removed after 15 min (on ice) by adding 40 μl of 10% (w/v) in water BSA, vortexing, and then centrifuging the samples.

Hatzack and Rasmussen (1999) have described a simple inexpensive HPTLC method for the analysis of $InsP_1$ to $InsP_6$ on cellulose precoated plates. This method is suitable for monitoring the nucleotide removal by charcoal treatment of InsPs with similar R_f values.

4.2.3. Neutralization of acid extracts

Using PCA as the deproteinizing agent has the advantage that a number of techniques can be used that simultaneously remove the acid and neutralize the extracts. The two neutralization strategies that are frequently employed are KOH and tri-octylamine/Freon (1:1, v/v) (Sharpes and McCarl, 1982). KOH has the advantage of the relatively low solubility of potassium perchlorate. Typically, the PCA-precipitated suspension is vortexed and the debris pelleted by centrifugation at 0–4°C. The supernatant above the acid-insoluble material is removed and added directly to a previously established fixed volume of 2 M KOH, 0.1 M Mes buffer (pK_a 6) and 10–20 mM EDTA, such that the final pH is about 6.0. Under these conditions, non-specific losses of InsPs are not encountered (Stephens, 1990). When trioctylamine/Freon is used as the neutralizing agent, the ice-cold supernatant from the centrifugation of the PCA-quenched samples is added to 1.1 volumes of tri-n-octylamine/Freon (freshly prepared at room temperature) into which an appropriate portion of Na_2EDTA (e.g. 50 µl of 0.1 M) solution has been placed. The sample is then very well mixed, and centrifuged in a microfuge for 5–10 min at room temperature. A three-layered system should result: an upper aqueous phase, that should be pH 5–6, a viscous middle layer of tri-n-octylamine perchloroate, and a lower phase containing Freon and any excess tri-n-octylamine. The advantage of this approach is that it is simple, rapid, and retains low concentration of perchlorate in the upper phase. There may occur, however, non-specific losses of InsPs (poly), which tend to diffuse into the tri-octylamine/perchlorate residue (Stephens, 1990).

Elution of the phosphates with lithium chloride leads to the lithium salts of the phosphates. Lithium chloride is soluble in ethanol and can be removed by extraction (Shayman et al., 1987). InsPs eluted with phosphate buffers must be desalted by anion exchange. The inorganic phosphate is eluted with triethylammonium formate (0.2 M), while a 10-fold increase of the formate concentration eluted the InsPs (Stephens et al., 1988a), which were lyophilized to remove the triethylammonium formate. Bird et al. (1989) has used a combination of anion- and cation-exchange processes to change elution buffers to the more easily removable formic acid.

4.3. Resolution of chemical families

Individual InsPs in tissues and cells are commonly analyzed by TLC, anion exchange cartridges and HPLC after labeling the PtdInsPs with a radionucleotide, such as $[\gamma\text{-}^{32}\text{P}]$ATP and/or myo-$[^3\text{H}]$inositol. The commonly used HPLC methods separate InsPs on anion exchange columns using both isocratic elution and gradients of aqueous mobile phases. The separation of InsPs depends upon the type and size of the column, the composition of the solvent mixture and the gradient of the solvents. Anion exchange cartridge columns such as Dowex formate, AG 1 X-8 formate or DEAE-cellulose columns provide excellent separation of InsPs in biological samples. The InsP, InsP_2, InsP_3 and InsP_4 may be eluted with increasing concentration of ammonium formate in dilute formic acid (Berridge et al., 1983; Ellis et al., 1963).

More recently, the anion exchange chromatography has been adopted to cartridge separation. A simple separation of InsPs (Wreggett and Irvine, 1989; Wreggett et al., 1990) has been achieved on Am-Prep (prepacked anion exchange capsules sold by Amersham (UK), which has the advantages of the earlier Sep-Pak separation (Wreggett and Irvine, 1987).

One of the major disadvantages of ion-exchange chromatography is its inability to separate InsP isomers. Previous studies have shown that InsPs exist in several isomeric forms that differ in their metabolism and biological activity (Rana and Hokin, 1990; Agranoff et al., 1985). Therefore, ion-exchange chromatography may not be suitable for studying InsP metabolism or the possible relationship between InsPs and their biological activity. Anion exchange HPLC has been found to be more suitable for this purpose because of higher resolving capacity.

4.3.1. TLC

In early studies silica gel TLC was used for the separation of InsPs (Hokins-Neaverson and Sadeghian, 1976; Koch-Kallnbach and Diring, 1977), although the methods were not very sensitive and exhibited poor resolution. Emilsson and Sundler (1984) first described the use of polyethyleneimine (PEI)-cellulose plates for the TLC separation of $InsP_1s$, $InsP_2s$ and $InsP_3s$. Ryu et al. (1987) reported that these plates also permitted the separation of $Ins(1,3,4,5)P_4$ from ATP. Spencer et al. (1990) later adapted TLC on PEI-cellulose plates for the separation of higher InsPs. By developing the plates in HCl it was possible to separate $Ins(1,4,5)P_3$ from $Ins(1,3,4,5)P_4$, $InsP_5$ and $InsP_6$. These separations were possible while the InsPs were still in presence of conventional assay reagents, which enabled the assays for InsPs kinases. Excellent separations were obtained for P_i, $InsP_3$, ATP, $InsP_4$ following development with 0.5 M HCl, while $InsP_5$ and $InsP_6$ were separated following development with 1.0 M HCl. The phosphorus-containing spots were located by molybdate spray. The spray was freshly prepared by mixing three parts H_2O/conc., $HCLO_4$/conc. and HCl (100:10:1, by vol.) to be part of ammonium molybate/$HClO_4$ (4%:9%, w/v, in H_2O) as described by Clarke and Dawson (1981). Commercially available InsPs can be used as markers, but phytic acid hydrolysate made by autoclaving a solution of $InsP_6$ could also be used to provide crude markers for constituent families of InsPs.

Hatzack and Rasmussen (1999) have described a simple and inexpensive HPTLC method for the analysis of $InsP_1$ to $InsP_6$ on cellulose precoated plates. Plates were developed in 1-propanol/ 25% ammonia solution/water (5:4:1, by vol.) and substance quantities as low as 100–200 pmol were detected by molybdate staining. Chromatographic mobilities of nucleotides and phosphorylated carbohydrates are also characterized. Charcoal treatment was employed to separate nucleotides from InsPs with similar R_f values prior to HPTLC analysis. Figure 4.1 shows the separation of six different InsPs, four nucleotides, three phosphorylated carbohydrates and Pi on the HPTLC cellulose plate. Hatzack and Rasmussen (1999) demonstrated a practical application of the HPTLC system by analysis of grain extracts from wild type and low phytate mutant barley as well as phytate degradation products resulting from barley phytase activity.

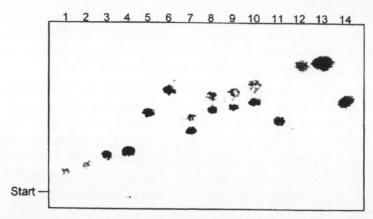

Fig. 4.1. HPTLC analysis of InsPs. Samples were loaded in the following order. 1, InsP$_6$ (2.1 nmol); 2, Ins(1,3,4,5,6)P$_5$ (1.9 nmol); 3, Ins(3,4,5,6)P$_4$ (2.3 nmol); 4, Ins(1,4,5)P$_3$ (2.4 nmol); 5, Ins(5,6)P$_2$ (2.2 nmol); 6, Ins(1)P (2.3 nmol); 7, GTP (0.9 mol); 8, UTP (0.9 nmol); 9, CTP (0.9 nmol); 10, TP (0.9 nmol); 11, Fru(1,6)P$_2$ (1.9 nmol); 12, Fru(6)P (3.1 nmol); 13, Glc(1)P(2.7 nmol); 14, Pi(2.4 nmol). The plate (10 × 10 cm^2) was developed with a 10 ml volume of mobile phase made up of 1-propanol/25% ammonia/water (5:4:1, by vol.) at room temperature (5 h). The phosphates were detected by spraying with acidic molybdate followed by heating exposure to UV light. Reproduced from Hatzack and Rasmussen (1999) with permission of publisher.

The TLC method is simple and rapid, but it may not be suitable for the routine analysis of InsPs in small quantities of tissue or cell samples because of its low sensitivity and poor resolution. The PEI-cellulose TLC method shares with other anion exchange methods the inability to separate positional isomers of the individual InsPs.

4.3.2. Anion exchange cartridge chromatography

The simplest and quickest method for analyzing InsPs of numerous samples has been anion exchange chromatography using small open-topped columns (Downes and Michell, 1981). Ammonium formate elutes InsPs because formate has greater affinity than phosphate for the resin. However, a very high concentration of formate may be needed for the elution of InsP$_3$ and InsP$_4$ which may exhibit some cross-contamination (Ellis et al., 1963). On standard batch-eluted formate-form Dowex columns it is difficult to resolve InsP$_4$s, InsP$_5$s and InsP$_6$s.

Bartlett (1982) has developed an improved ion exchange method that permits the separation of poly-InsPs from P to InsP$_6$ in blood samples. In this method, the red blood cells are extracted with PCA and the PCA extract is neutralized with KOH. The neutralized extract is poured onto a Dowex formate column and the InsPs are eluted with a linear gradient of 0–5 M ammonium formate followed by a linear gradient of 0–1 M HCl. Fractions of 5–10 ml are collected and assayed for InsPs by measuring the phosphate concentration (Lanzetta et al., 1979). This method provided excellent separation of InsP, InsP$_2$, InsP$_3$, InsP$_4$, InsP$_5$ and InsP$_6$. However, the method is not suitable for routine analysis of InsPs since it requires large sample volumes and large amounts of elution reagents.

Heathers et al. (1989) have reported an excellent separation of InsPs on an AG1-X8 resin. The samples for the illustration were obtained from myocytes loaded with [^{3}H]Ins. The myocytes were extracted twice with TCA. The TCA extracts were pooled and washed with water-saturated diethyl ether. The final extract was lyophilized to dryness and the dried sample suspended in 2 ml of Tris buffer (0.01 M, pH 8.5), neutralized with NaHCO$_3$ and applied to the resin column. The mono-, bis-, tris- and tetrakis-phosphates were eluted with increasing concentrations of ammonium sulfate. Similar separations were obtained by DaTorre et al. (1990).

Christensen and Harbak (1990) described an ion exchange method that is suitable for the separation of poly-InsPs in small tissue samples. A carefully prepared anion column allowed separation of each group of InsPs from InsP$_1$ up to InsP$_6$. The ammonium formate buffers used to elute the InsPs can be easily removed by repeated lyophilization.

Maslanski and Busa (1990) have described an effective separation of *myo*-InsPs using Sep-Pak Accell Plus QMA cartridges (Waters, Millipore Corp). The Sep-Pak columns required a smaller volume of eluant and can be run at a higher flow rate than the Dowex columns. For the separation of InsPs Maslanski and Busa (1990) recommend the use of Sep-Pak Accell Plus QMA cartridges (Waters, Millipore Corp., Milford, MA) instead of Dowex columns (Maslanski and Busa, 1990). The Sep-Pak columns require a smaller volume of eluant and can be run at a higher flow rate than the Dowex columns. This greatly reduces the time spent preparing the InsPs for the assay. Additionally, the eluant employed is readily removed via vacuum centrifugation. Wreggett and Irvine (1987) had established conditions whereby InsPs could be separated on Sep-Pak cartridges, but then retracted the method when the manufacturer changed the resin. Maslanski and Busa (1990) have achieved separation of the InsPs with the new Sep-Pak resin using triethylammonium bicarbonate (TEAB) as the eluant, which they describe in detail. Elution of the InsPs was accomplished by sequential addition of 4 ml of 0.1, 0.3, 0.4 M, and 6 ml of 0.5 M TEAB to elute Ins(4)P, Ins(1,4)P$_2$, Ins(1,4,5)P$_3$ and Ins(1,3,4,5)P$_4$, respectively.

Additionally, the eluant employed is readily removed via vacuum centrifugation. Before first use, columns are washed with 10 ml of deionized water, 10 ml of 1.0 M TEAB (flow rate less than 0.5 ml/min) and 10 ml more deionized water. Neutralized extracts are diluted to 5–10 ml with deionized water and applied to the washed Sep-Pak columns. Columns should not be allowed to run dry. After the samples are applied the columns are washed with 4 ml of 0.02 M TEAB. The InsPs are eluted from the column with 4 ml each of 0.1, 0.3, 0.4 M and 6 ml of 0.5 M TEAB to recover the InsP$_1$s, InsP$_2$s, InsP$_3$s and InsP$_4$s, respectively, at a flow rate not exceeding 2 ml/min. Before reuse, the columns are washed with 4 ml of 1.0 M TEAB followed by 10 ml of deionized water. The above procedure is sufficient to elute free *myo*-Ins and InsPs (InsP$_1$s through InsP$_3$s).

Goldschmidt (1990) has described a preparative resolution of myo-Ins bis- and tris-phosphate isomers by anion exchange chromatography.

Singh (1992) has described an application of the Sep-Pak column in the separation of InsPs from synaptosomes. Synaptosomes were loaded with [^{3}H]Ins and then incubated with Ringers buffer alone or a buffer containing carbamylcholine or NE. The stimulated and unstimulated samples were precipitated with TCA and centrifugation. The clear supernatant was collected and neutralized with 10 M KOH. Then the sample was applied to a Dowex-50W column and the eluate was collected. The column was washed with 5 column volumes of water and the eluates were pooled, and lyophilized overnight. The dried residue was redissolved in deionized water and poured into Sep-Pak column. Individual InsPs were eluted by using a linear gradient of TEAB (1 M) from 10 to 90% in 60 min.

Anderson et al. (1992) have described a method for rapid and selective determination of radiolabeled InsPs in cancer cells using solid-phase extraction with Bond Elut strong anion exchange minicolumns. The InsP$_1$, InsP$_2$ and InsP$_3$ are selectively eluted with 0.05, 0.3 and 0.8 M ammonium formate−0.1 M formic aid, respectively. Cancer cells are extracted with 10% PCA which is then neutralized prior to loading samples on to the minicolumns. Recovery is 54.1, 66.6 and 61.3% for InsP$_1$, InsP$_2$, and InsP$_3$ with between-day coefficients of variation of 7.6, 6.8 and 1.9%, respectively.

The ion exchange methods described above provide excellent separation of InsP classes, but do not resolve different InsP isomers.

4.3.3. HPLC analysis

The HPLC systems usually employ two pumps and a gradient controller although isocratic runs may also be employed. A spectrometer or Refraction Index detector is used to monitor the column effluent. A radioactivity flow detector may also be used. Ion exchange HPLC with both strong (SAX) (Irvine et al., 1985; Hansen et al., 1986) and weak (WAX) anion exchange (Binder et al., 1985; Wilson et al., 1985) have been extensively utilized for the separation of the families of InsPs as well as for the resolution of most positional isomers (see below). Anion exchange HPLC, however, does not allow the separation of nucleotides from InsPs, especially ATP from InsP$_3$. Thus, the analysis of [^{32}P]InsPs requires various other chromatographic or enzymatic pre-treatments to remove the nucleotides, 2,3-bisphosphoglycerate and pyrophosphate (Ishii et al., 1986; Tarver et al., 1987; Meek, 1986; Daniel et al., 1987).

McConnell et al. (1991) used HPLC to separate neutralized extracts of T5-1 cells on a 10 cm Partisphere 5 SAX column (Whatman, Clifton, NJ, USA) (Fig. 4.2). Cells were labeled for 24 h with [^{3}H]inositol and quenched with TCA. The inositol phosphates were eluted with a gradient obtained by mixing water (Buffer A) with NH$_4$H$_2$PO$_4$, pH 3.8, with H$_3$PO$_4$ (Buffer B) according to the protocol given in the figure. The degree of phosphorylation of each peak was determined from the elution times of [^{3}H]-labeled standards run separately before and after the samples.

Sasakawa et al. (1990; 1993) used anion exchange HPLC to separate inositol polyphosphates, e.g. Ins(1,4,5) P$_3$ and Ins(1,3,4,5)P$_4$.

Singh and Jiang (1995) have assembled a large table listing various HPLC methods that have been used more or less successfully for the separation of InsPs (Table 4.1). In general short columns provide more rapid and sharper separation of InsPs than large columns shallower gradients provide better separation but longer chromatographic run-times than steep gradients; stepwise gradient increments provide better resolution of P isomers than linear increments; and solvents with pH < 4 provide sharper separation of InsPs. Exposure to low pH may hydrolyze InsPs and produce false values of P$_i$, which is being detected for the identification of the InsPs. Most HPLC procedures use linear or stepwise gradients of ammonium formate, ammonium acetate and/ or ammonium phosphate as the mobile phase. Irvine et al. (1985) identified InsPs by adding a mixture of AMP, ADP and ATP to the extracts that served as markers for InsP, InsP$_2$ and InsP$_3$,

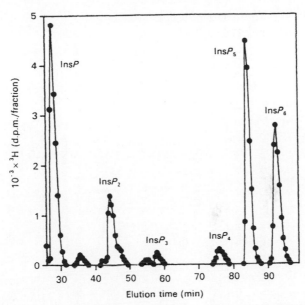

Fig. 4.2. HPLC analysis of InsPs in resting T5-1 cells. Cells were labeled for 24 h with [^{3}H]Ins, quenched with TCA and analyzed by HPLC. Neutralized samples of T5-1 extracts were applied to a 10 cm Partisphere 5 SAX column (Whatman). The InsPs were eluted with a gradient obtained by mixing water (Buffer A) with $NH_4H_2PO_4$, pH 3.8, with H_3PO_4 (Buffer B) as follows: 0–20 min, B increased linearly from 0–0.05 M; 20 min, B increased to 0.2 M; 20–30 min, B = 0.2 M; 30 min, B increased to 0.54 M; 30–60 min B increased linearly to 0.56 M; 60 min, B increased to 0.8 M; 60–75 min, B = 0.8 M; 75 min, B increased to 1.6 M; 75–85 min, B = 1.6 M; 85 min, B increased to 2 M; 85–95 min, B = 2 M. Peaks are identified by reference to standards as shown in the figure. Reproduced, with permission, from McConnell, F. M., Stephens, L. R., and Shears, S. B. (1991) Biochemical Journal 280, 323–329. © The Biochemical Society and the Medical Research Society.

respectively. This was done because InsPs do not absorb UV or visible light, but nucleotides do, and because these nucleotides elute close to the respective InsPs. Several studies have demonstrated direct detection of InsPs by using metal–dye (Mayr, 1988), amperometric (Johnson and LaCourse, 1990) and conductivity detection (Smith et al., 1989).

TABLE 4.1

Separation of chemical classes of InsPs by different HPLC systems

Column	Mobile phase	Detection	Comments	Reference
Partisil SAX (anion exchange)	NH_4–HCOOH, gradient (0–1 M, pH 3.7)	Radioactivity	Separation of InsPs classes/not isomers	Reed and DeBellerache (1988)
Partisil SAX	H_2O/methanol (95:5); 0–95% gradient of 1 M citrate in methanol	Phosphorus	Separation of InsPs classes not isomers	Mathews et al. (1988)
Nucleosil 5/5B	0–350 mM gradient of KH_2PO_4 with 350 mM KCl at pH 2.7	Radioactivity	Separation of InsPs classes, ATP overlaps with $InsP_3$	Brammer and Weaver (1989)
Mono Q HR	A: 0.1 mM $ZnSO_4$, 0.1 mM EDTA, 10 mM Hepes, pH 7.4; B, A with 0.5 M Na_2SO_4; 0–100%B	UV	Rapid separation of InsPs classes/not isomers	Meek (1986)
Mon Q HR or Aminex A27	A: 0.1 mM $ZnSO_4$, 5 mM Hepes, pH 7.4; B, 1 M Na_2SO_4, 5 mM Hepes, 0.1 mM Zn^{2+}	Phosphate	Limited resolution of InsPs classes/not isomers	Meek and Nicoletti (1986)

Modified from Singh and Jiang (1995).

Sulpice et al. (1990) have pointed out that ion-pair chromatoagraphy offers a potential alternative in the separations of InsPs. Ion-pair chromatography is more versatile than anion exchange chromatography, since it is affected by more parameters. The selection of the counterion and its concentration, of the salts, of the organic solvents and of the pH of the elution buffer offers many possibilities to optimise the separation. Sulpice et al. (1990) selected tetrabutylammonium hydrogen sulfate (TBAHS) as the counterion and Sup. RS classic 25 cm × 0.46 cm (Prolabo, Paris, France) and Ultrasphere IP (5 μm) 25 cm × 0.46 cm (Beckman, USA) as the reversed phase columns. The non-end-capped Lichrosorb RP18 and μBondapack C18 also performed well initially but rapidly deteriorated. Optimum separations of the test materials were obtained using KH_2PO_4 gradient. Thus, Solvent I contained 25 mM TBAHS, 40 mM KH_2PO_4/acetonitrile (85:15, v/v) (pH 5.5) and solvent II was 25 mM TBAHS, 150 mM KH_2PO_4/acetonitrle (85:15, v/v) (pH 5.5). Elution was obtained by running 100% I (15 min), followed by a linear gradient of 100% I to 100% II (25 min) and completing the elution with 100% II (30 min). Under these conditions the elution times were 21, 39, 52 and 60 min, respectively, for $Ins(1,4,5)P_3$, $Ins(1,3,4,5)P_4$, $Ins(1,3,4,5,6)P_5$ and $InsP_6$, the last peak being very broad. Some of these separations may also be obtained using isocratic elution conditions (Shayman and BeMent, 1988).

For ion-pair chromatography on HPLC, quaternary ammonium salts such as tetrabutylammonium hydrogen sulfate or dihydrogen phosphate and N-methylimipramine have been used (Sulpice et al., 1989; Shayman and Barcellon, 1990). The eluent is applied to reversed phase HPLC columns and contains acetonitrile as the organic component. The order of elution obtained by this method is about the same as that observed by anion exchange chromatography. The principle relies on the fact that the more charges an InsPs carries, the more lipophilic counterions are bound, therefore leading to longer retention on the reversed phase column. Similar HPLC system involves a micellar mobile phase with an octyldimethylsilyl stationary phase (Brand et al., 1990).

Gradient elution separated GroP, three isomeric $InsP_1$, GroPInsP, three isomeric $InsP_2$, $GroPInsP_2$, three $InsP_3$s and $Ins(1,3,4,5)P_4$.

The advantage of ion-pair chromatography over ion exchange is that it provides good resolution of InsPs from PP_i, 2,3-DPG, ADP and ATP. After a simple charcoal pre-treatment to remove ^{32}P-labeled contaminants co-migrating with InsPs, their quantification by this method was possible, even with incomplete elimination of ADP and ATP.

The mechanism of ion-pair chromatography can be explained by ion-pair formation in the mobile phase, ion exchange at the stationary phase surface or dynamic equilibrium. Arguments in favor of the ion-exchange mechanism have been presented by Shayman and BeMent (1988).

4.3.4. Other

An effective method for separation of InsPs is provided by electrophoresis in pyridine-acetic acid (Dawson and Clarke, 1972), ammonium carbonate (Clarke and Dawson, 1981), or sodium oxalate (Seiffert and Agranoff, 1965). The method has been adapted for cellulose plates and a bench-top electrophoresis apparatus (Whipps et al., 1987). The iontophoresis gives good separation of InsPs and nucleotides and it is therefore suitable for analyzing ^{32}P-labeled tissues (Agranoff et al., 1983; Harrison et al., 1990).

Electrophoresis on Whatman No. 1 paper or cellulose TLC plates provides relatively rapid separation of InsPs in brain extracts (Seiffert and Agranoff, 1965). Electrophoresis effectively separates InsP isomers that ion-exchange columns cannot (Seiffert and Agranoff, 1965). However, the loading capacity of electrophoresis paper is very small. These observations suggest that TLC, paper chromatography or electrophoresis may not be suitable for the routine analysis of InsPs in small quantity of tissue samples. Sekar et al. (1987) have employed electrophoresis for the analysis of Ins [1:2(cyc), 4,5]P_3 and [1,2(cyc), 4]P_2 formed during stimulation of mouse pancreatic minilobules with carbamyl-choline.

Paper chromatography provides better separation of InsPs but exhibits poor sensitivity (Emilsson and Sundler, 1984; Grado and Ballou, 1961; Tomlinson and Ballou, 1962; Brown and Steward, 1966).

Chester et al. (1989) have employed supercritical fluid chromatography with mass spectrometry for the separation, detection and identification of $InsP_3$ and $InsP_6$ derivatives.

4.4. Resolution of positional isomers

It is an essential part of many studies to resolve InsPs into their positional isomers, sometimes by mass, and sometimes by various radiolabeling techniques. Likewise, the study of the intermediary metabolism of the higher InsPs and their metabolic precursors and products has become important in itself. Agranoff (1978) and Agranoff et al. (1985) have offered helpful suggestions to avoid cyclitol confusion.

The chromatographic isolation of the InsPs classes inadvertently leads to some fractionation of positional isomers. Thus, the various $InsP_1s$ are resolved into the Ins(1)P, Ins(2)P, Ins (3)P, Ins (4)P, etc. Likewise, the $InsP_2s$ are resolved into Ins(3,4)P_2 and Ins (4,5)P_2, while the various $InsP_3s$ are resolved into the Ins(1,2,3)P_3, Ins(1,2,4)P_3, and Ins(1,3,5)P_3 derivatives. Meek (1986) and Meek and Nicoletti (1986) provided a detailed description of the separation InsPs on a weak anion exchange resin (Pharmacia Mono Q Hr 5/5). To minimize contamination with iron, steel pump inlet filters in the LKB 2150 pump were removed and steel connecting tubing was replaced with Teflon. The mobile phase gradients were generated from mixtures of a weak eluant, Buffer A (0.1 mM zinc sulfate, 0.1 mM EDTA, 10 mM HEPES, pH 7.4) and a strong eluant, Buffer B (0.5 M sodium sulfate plus the same concentrations as in Buffer A of EDTA, HEPES, and zinc sulfate). The linear gradient used in the experiment was from 5% Buffer B at injection time to 0.25% Buffer B after 25 min with a flow rate of 0.1 ml/min or 5–25% Buffer B in 25 min, then up to 80% Buffer B by 35 min. The column effluent was passed through a variable wavelength detector (260 nm), then a 5 cm × 3 mm i.d. glass column containing phenoxyacetyl-cellulose loaded with 50 μl of alkaline phosphatase. The stream was then mixed with a molybdate reagent pumped at 0.3 /min and monitored at 380 nm to continuously detect inorganic P. Figure 4.3 shows the identification of InsPs in rat brain. A conventional PCA extract of rat brain InsPs was neutralized and treated with charcoal and passed over a NH_2 column. The phosphates were recovered in 300 μl of 1.5 M NaOH and 200 μl of the eluate was chromatographed. Fractions (0.4 min) were collected for measurement of radioactivity (bars at bottom). Absorbance of phosphates at 380 nm (bottom trace) and native absorbance at 260 nm (not shown) were recorded. Also shown (middle trace) is a chromatogram of standards (5 nmol) of $InsP_2$; CABP (the internal standard); $GroP_2$; $Ins(1,4,5)P_3$; and 500 nmol phosphate. The top trace is a mixture of InsPs produced by random phosphorylation. Gradient conditions were 25–125 mM sulfate in 25 min then to 400 mM sulfate by 35 min. Meek (1986) has also provided chromatograms demonstrating the progress of purification of the InsPs prior to the anion exchange HPLC separation.

Dean and Beaven (1989) and Singh and Jiang (1995) have summarized the established analytical methods. The present discussion draws attention to more recent developments, which have been found to be useful or have shown promise. The choice of

Fig. 4.3. Identification of inositol phosphates in rat brain by chromatography on a WAX resin (Pharmacia Mono Q Hr 5/5). The brain was homogenized in 10 ml of $HClO_4$ and centrifuged; 4 ml of the neutralized supernatant was treated with charcoal and passed over a NH_2 column. The phosphates were eluted in 300 μl of 1.5 M NH_4OH; 200 μl of the eluate was chromatographed. Fractions (0.4 min) were collected for measurement of radioactivity (bars at bottom). Absorbance of phosphates at 380 nm (bottom trace) and native absorbance at 260 nm (not shown) were recorded. Also shown (middle trace) is a chromatogram of standards produced by random phosphorylkation. Gradient conditions were 25–125 mM sulfate in 25 min then to 400 mM sulfate by 35 min. The absorbance traces were offset vertically for clarity. Reproduced from Meek (1986) with permission of publisher.

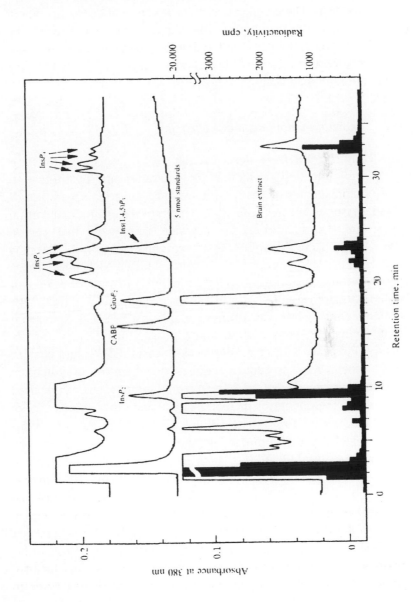

method is important and depends on the availability of equipment and expertise. The simplest and cheapest method to answer a biological question usually provides a good compromise, while the use of the most modern and expensive methodology may not always be necessary. There should be no compromise about using adequate reference standards, however.

4.4.1. Inositol monophosphates (InsP₁s)

Sun et al. (1990) have described an HPLC procedure together with ion conductivity detection for separation and quantification of isomeric species of $InsP_1s$. Figure 4.4 shows the separation of standard isomers of $InsP_1$, $InsP_2$ and $InsP_3$ by a step gradient elution procedure. The conditions for stepwise elution are: InsP region (0–18 min), 165 mM NaOH; $InsP_2$ region (18–36 min), 78.0 mM NaOH; and $InsP_3$ region (36–50 min), 112.5 mM NaOH. These peaks in the figure were obtained following injection of 1.5 μg of each $InsP_1$ isomer and 0.5 μg of each of the other compounds. The peak assignments were as follows: peak 1, Ins(2)P; peak 2, Ins(1)P; peak 3, Ins(4)P; peak 4, Ins(1,4)P₂; peak 5, Ins(2,4)P₂; peak 6, Ins(4,5)P₂; peak 7, Ins(1,3,4)P₃; peak 8, Ins(1,4,5)P₃; and peak 9, Ins(2,4,5)P₃. This procedure requires no chemical modification reactions which is sensitive in the picomolar range, and can be used to measure the quantities of InsP isomers and other anions (Smith and MacQuarrie, 1988). The column (Dionex Ion Pac AS5A 5μ column and guard column) is washed with 10.5 mM NaOH. The separation of Ins(1)P and Ins(4)P is accomplished with an eluant A (150 mM NaOH): eluant B (deionized water) ratio of 7:93, which is also used to pre-equilibrate the column. The conductivity baseline should be approximately 2.5 μS (Maslanski and Busa, 1990). The retention time for Ins(1)P is approximately 13 min and that for Ins(4)P is approximately 14 min. Under normal conditions, brain samples show the presence of both Ins(1)P and Ins(4)P peaks. Occasionally a large unknown peak elutes just prior to

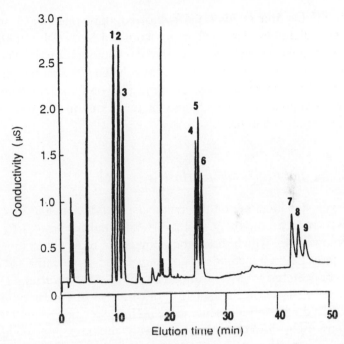

Fig. 4.4. Separation of isomers of inositol mono-, bis-, and trisphosphate standards by a step gradient elution procedure. The conditions for stepwise gradient elution are: InsP region (0–18 min), 1.5 mM NaOH; InsP$_2$ region (18–36 min), 78.0 mM NaOH; and InsP$_3$ region (36–50 min), 112.5 mM NaOH. The peaks in the figure were obtained following injection of 1.5 μg of each inositol monophosphate isomer and 0.5 μg of each of the other compounds. The peak assignments are as follows: peak 1, Ins(2)P; peak 2, Ins(1)P; peak 3, Ins(4)P; peak 4, Ins(1,4)P$_2$; peak 5, Ins(2,4)P$_2$; peak 6, Ins(4,5)P$_2$; peak 7, Ins(1,3,4)P$_3$; peak 8, Ins(1,4,5)P$_3$; and peak 9, Ins(2,4,5)P$_3$. Reproduced from Sun et al. (1990) with permission of the publisher.

the Ins(1)P peak. The order of elution of the InsP$_1$s were InsP(2)P < Ins(1)P < Ins(4)P.

4.4.2. Inositol bisphosphates (InsP$_2$s)

The order of elution of the standard InsP$_2$s in Fig. 4.4 were Ins(1,4)P$_2$ < Ins(2,4)P$_2$ < Ins(4,5)P$_2$. Analysis of brain samples for $InsP_2$ by Sun et al. (1990) showed only one peak, which corresponded to $Ins(1,4)P_2$. The procedure of Sun et al. (1990) for the separation of brain $InsP_2$s required the equilibration of a Dionex Ion Pac AS5A 5μ column with 67.5 mM NaOH (the ratio of eluant A/eluant B is 45:55 v/v, see above). The baseline conductivity should be approximately 5 μS, and the retention time for $Ins(1,4)P_2$ approximately 8.5 min.

4.4.3. Inositol trisphosphates ($InsP_3s$)

The extensive literature on the synthesis, metabolism and function of InsPs is focused largely on $Ins(1,4,5)P_3$ and $Ins(1,3,4,5)P_4$ and the pathways by which these compounds are recycled to inositol. The early HPLC methods were developed especially for the separation of $Ins(1,4,5)P_3$ from $Ins(1,3,5)P_3$ (Irvine et al., 1984; Irvine et al., 1985) or the isolation of different InsPs (Batty et al., 1985; Reed and deBellerache, 1988). However, numerous HPLC methods using different columns solvent mixtures and modes of detection are presently available for the analysis of InsPs. The choice of the HPLC method will depend upon the objectives of the experiment and the questions asked. If one wishes to study changes in total InsPs, then ion-exchange chromatography or simple HPLC methods as described by Reed and deBellerache (1988) or Mathews et al. (1988) should be used. However, if one wants to study InsP metabolism, regulation of Ca^{2+} mobilization by $Ins(1,4,5)P_3$ or the redistribution of Ca^{2+} by $Ins(1,3,4,5)P_4$, then methods that provide effective separation of the InsP isomers will be more suitable. HPLC methods that do not effectively separate ATP from $Ins(1,4,5)P_3$, such as the one described by Brammer and Weaver (1989), are not suitable for experiments that use radiolabeling with $[^{32}P]ATP$.

Isocratic conditions are useful for the separation of the positional isomers of InsPs, e.g. $Ins(1,3,4)P_3$ and $Ins(1,4,5)P_3$ (Wreggett and Irvine, 1987). For the separation of more complex mixtures, a gradient flow technique is necessary. Figure 4.4 also showed the resolution of Ins(1,3,4)P$_3$, Ins(1,4,5)P$_3$ and Ins(2,4,5)P$_3$ in that order as reported by Sun et al. (1990). The procedure described by Sun et al. (1990) for the separation of the InsP$_3$s calls for isocratic washing of the Dionex Ion Pac As5A 5μ column with 112.5 mM NaOH. This is accomplished by using 150 mM NaOH solution in reservoir A (eluant A) and deionized H$_2$O in reservoir B (eluant B) and setting the elution ratio to 75:25. The baseline conductivity at this stage should be approximately 8.5 μS. InsP$_3$s are loaded into the 50 μl sample loop, the conductivity auto-offset and then the standard injected. A standard curve should be constructed with different concentrations of each standard. Substantial peak sizes are observed with 50–100 pmol sample. The InsP$_3$ is eluted isocratically with 112.5 mM NaOH. The retention time for Ins(1,3,4)P$_3$ is approximately 13 min and that for Ins(1,4,5)P$_3$ is approximately 14 min. Brain samples normally show Ins(1,4,5)P$_3$ as the major isomer and only trace amounts of Ins(1,3,4)P$_3$.

Sasakawa et al.(1992) have reported that for the samples from N1E-115 cells, an appropriate gradient is formed by two independent pumps drawing on reservoirs containing water (pump A) and 1 M ammonium phosphate (pH 3.35 with phosphoric acid); (pump B) at a flow rate of 1 ml/min as follows: 0 min, 0% B; 5 min, 0%B; 25 min, 25%B; 45 min, 30% B, 65 min, 65%B; 70 min, 100% B; 75 min, 100% B; 80 min, 0% B. The latter gradient conditions were especially well suited for separation of the isomeric InsP$_3$s.

Adelt et al. (1999) have reported the resolution of a mixture of seven InsP$_3$ regioisomers by anion exchange chromatograph on a Mono Q HR column (Fig. 4.5). A linear gradient of HCl was applied to elute the InsPs in the order: Ins(1,2,6)P$_3$ < Ins(1,5,6)P$_3$ < Ins(1,2,3)P$_3$ < Ins(1,4,6)P$_3$ $\Leftarrow$ Ins(4,5,6)P$_3$ < Ins(1,4,5)P$_3$ < Ins(2,4,5)P$_3$ < Ins(1,2, 4)P$_3$. The peaks were detected photometrically using the metal–dye-detector of Mayr (1988).

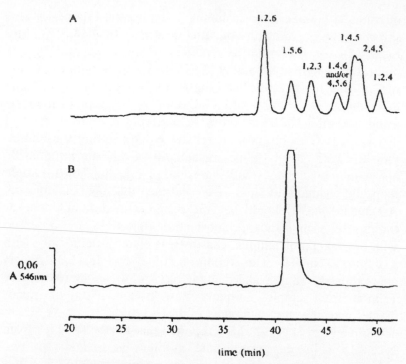

Fig. 4.5. HPLC-MDD (metal–dye detector) chromatograms showing (A) standard mixture of seven InsP₃ regioisomers and (B) Ins(1,5,6)P₃, the product of phosphatase reaction starting from Ins(1,4,5,6)P₄. The compounds were separated by anion-exchange chromatography on a Mono Q HR 10/10 column (Pharmacia). A linear gradient of HCl was applied to elute the InsPs (0 min, 0.2 mM HCl; 70 min 0.5 M HCl; flow rate 1.5 ml/min). Photometric detection at 546 nm was achieved using modified metal–dye reagent (2 M Tris–HCl (pH 9.1), 200 µM 4-(2-pyridylazo)resorcinol (PAR), 30 µM YCl₃, 10% (v/v) MeOH). The isomeric InsP₃s are identified by the ring positions occupied by the phosphate groups. Reproduced from Adelt et al. (1999) with the permission of the publisher.

4.4.4. Inositol tetrakisphosphates (InsP₄s)

Stephens (1990) has pointed out that charcoal-treated acid extracts of ^{32}P-labeled avian erythrocytes still contain substantial quantities of ^{32}P-labeled nucleotide phosphates and [^{32}P]Pi. These contaminating compounds are most effectively removed by applying the sample to a Partisil 10 SAX HPLC column and eluting with a rapidly developing ammonium formate/phosphoric acid gradient. These conditions had been originally designed to separate Ins(1,3,4)P$_3$ and Ins(1,4,5)P$_3$ (Irvine et al., 1985). Under these conditions, Ins(1,3,4,5)P$_4$, Ins(1,3,4,5,6)P$_4$ and Ins(3,4,5,6)P$_4$ elute together but separate from InsP$_5$.

The InsP$_4$ and InsP$_5$ peaks are then collected and desalted. The HPLC purified InsPs are desalted in either ammonium formate/phosphoric acid or ammonium phosphate/phosphoric acid buffers (e.g. Stephens and Downes, 1990; Dean and Moyer, 1987) and by adjusting the acidity of the eluting buffer to pH 6–7 with triethylamine. After diluting with H$_2$O (8–10 fold) the samples are applied to a column containing 100–200 μl of packed AG 1.8 200–400 resin (BioRad) in the formate form. The columns are washed with 6–8 ml of 2 M ammonium formate, 0.2 M formic acid (to elute P$_i$) and then with 2.5 ml of 2 M ammonium formate, 0.1 M formic acid to elute InsP$_4$s or InsP$_5$. The eluants are roughly neutralized with NH$_4$OH and freeze-dried directly. The samples are free of ammonium formate in about 15 h. Since the InsPs stick to glass surfaces they need to be removed by dilute NH$_4$OH.

Better resolution of InsP$_4$s can be obtained on SAX HPLC columns by using phosphate buffers (Dean and Moyer, 1987). Under these conditions, InsP$_4$s elute from a standard analytical scale SAX HPLC column with 1 M Na$_2$H$_2$PO$_4$·NaOH pH 3.8, 25°C in the order Ins(3,4,5,6)P$_4$, Ins(1,3,4,5)P$_4$ and Ins(1,3,4,6)P$_4$. However, Ins(3,4,5,6)P$_5$ is usually poorly resolved from Ins(1,3,4,5)P$_4$. Since the major [^{3}H]InsP$_4$ in extracts from control cells is usually [^{3}H]Ins(3,4,5,6)P$_4$ and the minor one [^{3}H]Ins (1,3,4,5)P$_4$, this is not an ideal system for analyzing the cell-derived extracts.

Stephens (1990) has shown that the Partisphere WAX HPLC columns separate all three species of InsP$_4$ found in the avian erythrocytes. In order of increasing retention time they emerge as follows: Ins(1,3,4,6)P$_4$ < Ins(1,3,4,5)P$_4$ < Ins(3,4,5,6)P$_4$ (Stephens et al., 1988a). For this purpose a new column is eluted isocratically with 0.16 M $(NH_4)_2HPO_4 \cdot H_3PO_4$ (pH 3.2, 22°C) and InsP$_4$s should emerge after 40–80 min. [^{3}H]Ins(1,3,4,5,6)P$_5$ was eluted from a fresh Partisphere WAX HPLC column by approximately 0.4 M $(NH_4)_2HPO_4 \cdot H_3PO_4$ (pH 3.2 at 22°C). The [^{3}H]InsP$_4$ peaks were identified by either spiking the samples with trace quantities of [^{32}P]InsP$_4$ 'standards', which can be detected in the column eluant by Cerenkov counting or by removing small portions from each of the fractions collected and counting them individually for ^{3}H.

Sasakawa et al. (1990; 1992; 1993) have separated isomeric InsP$_4$s using HPLC with Adsorbosphere SAX 5.5 μm column (Alltech Applied Sciences) using a solvent mixture originally described by Balla et al. (1987) with minor modifications. A linear gradient (1% increase/min) of ammonium phosphate (pH 3.35 with phosphoric acid) at a flow rate of 1 ml/min (Sasakawa et al., 1990) yielded the following separation of InsPs from [^{3}H]inositol pre-labeled chromaffin cells: [^{3}H]Ins(1,3,4)P$_3$ < [^{3}H]Ins(1,4,5)P$_3$ < [^{3}H]Ins(1,3,4,6)P$_4$ < [^{3}H]Ins (1,3,4,5)P$_4$ < [^{3}H]Ins(3,4,5,6)P$_4$ < [^{3}H]Ins(1,3,4,5,6)P$_5$ < [^{3}H]InsP$_6$.

Adelt et al. (2001) have resolved five authentic regioisomers of InsP$_4$s essentially completely using HPLC-MDD (metal–dye detector) analysis. Figure 4.6 illustrates the application of the method to the identification of the kinase products after conversion of Ins(1,3,4)P$_3$ and Ins(1,2,4)P$_3$ to InsP$_4$s. The InsP$_4$s were eluted in the following order: Ins(1,2,4,6)P$_4$ < Ins(1,3,4,6)P$_4$ < Ins(1,2,4,5)P$_4$ < Ins(1,3,4,5)P$_4$ < Ins(3,4,5,6)P$_4$. The InsPs were eluted using a linear gradient of HCl as described by Adelt et al. (1999).

4.4.5. Inositol pentakisphosphates (InsP$_5$s)

Two of the six myo-InsP$_5$s possess a plane of symmetry, the remaining four form two pairs of enantiomers (see Chapter 10).

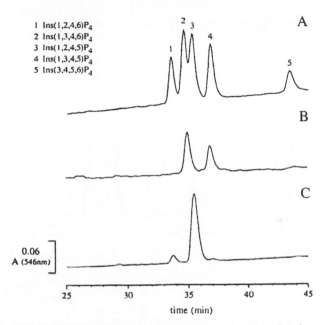

Fig. 4.6. HPLC-MDD (metal–dye detector) chromatograms showing standard mixture of five InsP$_4$ regioisomers (A) and the kinase products after complete conversion of Ins(1,3,4)P$_3$ (B) and Ins(1,2,4)P$_3$ (C). The InsP$_4$ isomers are identified as shown in the figure. The isomeric InsP$_4$s were resolved as described in Fig. 4.5. Reproduced from Adelt et al. (2001) with the permission of the publisher.

Therefore, non-chiral chromatographic techniques would be expected to resolve a maximum of four species of InsP$_5$: Ins(1,2,3,4,6)P$_5$, Ins(1,3,4,5,6)P$_5$ (*meso*-compounds, which can in theory at least, be purified completely by standard chromatographic techniques) and D- and L-Ins(1,2,3,4,5)P$_5$ and D- and L-Ins(1,2,4,5,6)P$_5$ (i.e. the best that non-chiral chromatography could achieve with these isomers is, for example, to obtain an Ins(1,2,4,5,6)P$_5$ peak that did not contain any of the other four non-enantiomeric InsP$_5$s, but could itself contain either D- or L-Ins(1,2,4,5,6)P$_5$).

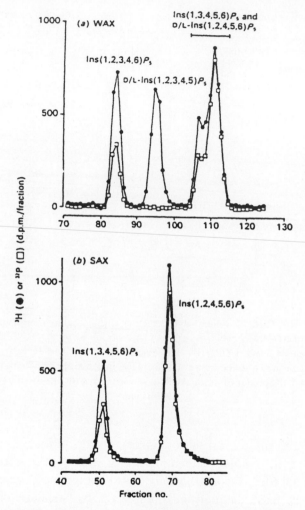

Fig. 4.7. HPLC resolution of isomeric [^{32}P]InsP$_5$s in an acid extract from [^{32}P]P$_i$ prelabeled *Dictyostelium amoebae*. Nc4 amoebae were incubated with [^{32}P]I for 4 h and an acid extract was prepared and mixed with standard [^{3}H]InsPs. The sample was chromatographed on a Partisphere WAX HPLC column. The region in which the [^{3}H]InsP$_5$s were eluted is shown in (a). They were eluted in the order given in the figure. The peak marked with a horizontal bar was pooled, desalted and chromatographed on a Partisphere SAX HPLC column (40% of each fraction

Several published methods have resolved four species of $InsP_5$; These include: moving-paper iontophoresis (Tate, 1968); ion exchange HPLC utilizing an ion exchange column and nitric acid as an eluant (this requires a nitric acid-compatible HPLC-system) (Phillipy and Bland, 1988); anion exchange chromatography on 1 m long Dowex resin columns (the resin is in the chloride form and is eluted with HPLC; run times extended to many hours) (Cosgrove, 1969); and on commercial HPLC columns with HCl as an eluant, which requires an HCL-compatible solvent delivery system (Mayr, 1988). The HCl- and HNO_3-based systems resolve the four $InsP_5$s in one run, but the systems require care to avoid acid-catalyzed migration of the phosphate groups in the InsPs being purified.

Stephens et al. (1991) have described alternative sets of conditions that can fully resolve all four species of $InsP_5$ by using the standard phosphate buffers and anion exchange HPLC columns that are commonly employed to separate $InsP_4$s and $InsP_3$s, although two chromatographic steps are required. Figure 4.7a shows the separation of $InsP_5$s by HPLC on a WAX resin. A mixture of ^{32}P-labeled $Ins(1,2,3,4,6)P_5$, D/L-$Ins(1,2,4,5,6)P_5$, D/L-$Ins(1,2,3,4,5)P_5$ and $Ins(1,3,4,5,6)P_5$ was applied to a Partisphere WAX HPLC column and eluted with a gradient of $0.6-1.0$ M $(NH_4)_2HPO_4 \cdot H_3PO_4$ (pH 3.8, 22°C). The order of elution was $Ins(1,3,4,5,6)P_5$, D/L-$1,2,3,4,5)P_5$ and finally $Ins(1,2,3,4,6)P_5$ and D/L-$Ins(1,2,4,5,6)P_5$, which eluted together. Fractions were collected every 0.5 min and counted for ^{32}P by Cerenkov counting. The peak that eluted last, which contained D/L-$[^{32}P]Ins(1,2,4,5,6)P_5$ and $[^{32}P]Ins(1,2,3,4,6)P_5$ was pooled,

was removed and counted for ^{3}H and ^{32}P radioactivity. The region in which the $[^3H]InsP_5$s were eluted is shown in (b). The $[^3H]$- and $[^{32}P]InsP_5$s were resolved into two peaks, the first-eluted being $[^3H]Ins(1,3,4,5,6)P_5$ and the second D/L-$[^3H]Ins(1,2,4,5,6)P_5$. Peak identification is given in the figure. Reproduced, with permission, from Stephens, L., Hawkins, P. T., Stanley, A. Moore, T., Poyner, D. R., Morris, P. J., Hanley, M. R., Kay, R. R, and Irvine, R. F. (1991). Biochemical Journal 275, 485–499. © The Biochemical Society and the Medical Research Society.

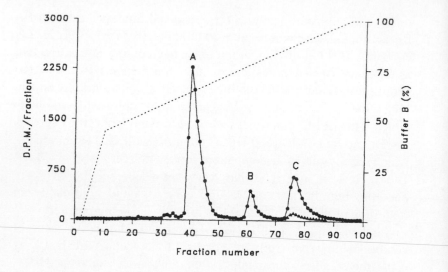

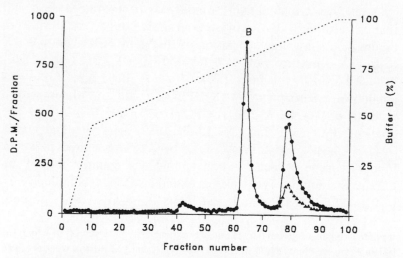

Fig. 4.8. Phosphorylation of InsP$_6$ and PP-InsP$_5$ by liver homogenate. Live homogenates were incubated at a final concentration of 0.9 mg protein/ml for 40 min in 0.5 ml of assay buffer 1 (100 mM KCl, 7 mM MgSO$_4$, 5 M ATP, 10 mM phosphocreatine, 10 mM NaF, 10 mM HEPES (pH 7.2 with KOH), 1 mM Na$_2$EDTA, 1 mM dithiothreitol, and 0.05 mg/ml phosphocreatine kinase) that contained 10 000 dpm [^{3}H]InsP$_6$ (peak A, upper panel) or 5000 dpm

desalted, mixed with [^{3}H]Ins(1,3,4,5,6)P$_5$ and D/L-[^{3}H]Ins(1,2,4,5,6)P$_5$ and applied to Partisphere SAX HPLC column. Figure 4.7b shows the HPLC SAX resolution of the two InsP$_5$s isomers. The Ins(1,3,4,5,6)P$_5$ isomer eluted ahead of the Ins(1,2,4,5,6)P$_5$ isomer.

4.4.6. Inositol hexakis and pyrophosphates (InsP$_6$s, InsP$_7$s and InsP$_8$s)

There is evidence for the occurrence in animal tissues of the more highly phosphorylated InsPs such as InsP$_6$ (Shamsuddin et al., 1988; Shumsuddin, 1999; Grases et al., 2002). The interest in this InsPs has increased substantially following the discovery that it is further phosphorylated to a diphospho InsP$_5$ (PP-InsP$_5$) (Menniti et al., 1993; Stephens et al., 1993), a widely distributed cellular constituent of organisms as phylogenetically diverse as *Dictyostelium discoideum* (Stephens et al., 1993) and rat (Menniti et al., 1993). Figure 4.8 shows the HPLC separation of the phosphorylation products of InsP$_6$ and PP-InsP$_5$ by a rat liver homogenate (Shears et al., 1995). It is seen that InsP$_6$ (Peak A, upper panel) is readily converted to PP-InsP$_5$ (Peak B, upper panel), when NaF is employed to inhibit PP-InsP$_5$ phosphatase (Menniti et al., 1993).

PP-[3H]InsP$_5$ (peak B, lower panel). Reactions were quenched and extracted and spiked with 600 dpm [β^{32}P]PP-InsP$_4$-PP, and analyzed by HPLC as follows. Deproteinized samples were loaded onto a Partisphere 5-μm SAX HPLC column at a flow rate of 1 ml/min using WISP model 712 (Waters), which injected each sample in aliquots interspersed with HPLC buffer (1 mM Na$_2$EDTA) in a ratio of sample: buffer A of about 1:10. The peaks were eluted with a gradient generated from a mixture of buffer A (1 mM Na$_2$EDTA) and Buffer B (Buffer A plus 1.3 M (NH$_4$)$_2$HPO$_4$ (pH 3.8) with H$_3$PO$_4$) at a flow rate of 1 ml/min: 0–2 in, B = 0%; 2–10 min, B increased linearly to 44%; 10–100 min, B increased linearly to 100% and was retained at this level until 101 min. Finally, B was returned to 0%. Circles depict ^{3}H dpm/fraction for the unknown and triangles depict ^{32}P dpm/fraction for the internal standard. Reproduced from Shears et al. (1995) with permission of the publisher.

TABLE 4.2

Separation of positional isomers of InsPs

Mono-1 AE	0.2 mM HCl, 9–18 μM transition metal, and 0.4 mM HCl with 14–28 μM	Metal-dye	Rapid separation of InsP isomers with good resolution	Mayr (1988); Johnson and LaCourse (1990)
Mon Q HR 5/20	0.2 mM HCl/15 μM YCl$_3$ and 0.5 M HCl/22.5 μM YCl$_3$, concave gradient	Metal-dye	Separation of InsP to InsP$_6$ in 60 min	Kerovuo et al. (2000)
ASP 4E	4–37% non-linear gradient of cyanophenol in water	Conductivity	Separation of InsPs, anions and nucleotides	Smith et al. (1989)
Partisil SAX	0–2 M non-linear gradient of NH$_4$COO, pH 3.7	Radioactivity	Separation of InsPs isomers	Prestwich and Bolton (1991)
Dionex AS-7	100 mM NaOH and 100 mM NaOOCCH$_3$	Amperometric	Separation of Ins(1)P and Ins(4)P	Barnaby (1991)
Partisil SAX	0–1.75 M stepwise gradient of (NH$_4$)H$_2$PO$_4$	Radioactivity	Separation of InsP isomers	Sastry et al. (1992)
Partisil SAX	0–100% gradient of 0.5 M (NH$_4$)H$_2$PO$_4$, pH 3.2	Metal-dye, NMR	Separation InsP isomers	Stephens et al. (1991)
Partisil WAX	0–95% of 0.5 M (NH$_4$)H$_2$PO$_4$, pH 3.2	Radioactivity	Separation of four isomers of InsP$_5$	Stephens et al. (1991)
Partisil SAX	0–100% non-linear gradient of (NH$_4$)H$_2$PO$_4$, pH 3.8	Radioactivity	Excellent separation of InsP isomers	Stauderman and Pruss (1990)

Modified from Singh and Jiang (1995).

Approximately 25% of $InsP_6/h/(mg$ protein) was converted to peak C (upper panel). Shears et al. (1995) have suggested that peak C is a *bis*-diphosphoinositol tetrakisphosphate ($PP-InsP_4-PP$).

Table 4.2 lists selected chromatographic systems for the separation of positional isomers of InsPs.

4.5. Determination of chemical structure

The identification of natural or synthetic InsPs relies predominantly classical chemical analysis and magnetic resonance spectroscopy. Other techniques, such as mass spectrometry and X-ray structural analysis are also helpful.

4.5.1. Chemical analysis

In addition to establishing the presence of inositol and phosphate as the only components of the unknown InsPs, the chemical determination of the absolute configuration of InsPs requires their conversion to known molecules in a defined way. The classical chemical procedures for the determination of complete structure of InsPs started with the selective cleavage of the InsP ring structure by periodate oxidation between vicinal hydroxyl groups. After reduction of the resulting aldehyde functions, the phosphates were hydrolyzed, and the open-chain polyols could be identified by comparison. This procedure allows the identification of optical isomers by determining the optical rotation (Grado and Ballou, 1961).

The strategy used for polyol production is shown in Fig. 4.9 (Stephens et al., 1988b). The method is based on previous work by Grado and Ballou (1961), Johnson and Tate (1969) and Irvine et al. (1984). Briefly, a dried sample of a $[^3H]$- or $[^{14}C]$-InsPs is dissolved in 0.1 M-periodic acid (pH adjusted to 2.0 with NaOH; 0.5 ml). The sample is then left in a closed vessel in the dark at 25°C for 36 h. The aldehydes resulting from periodate oxidation, together

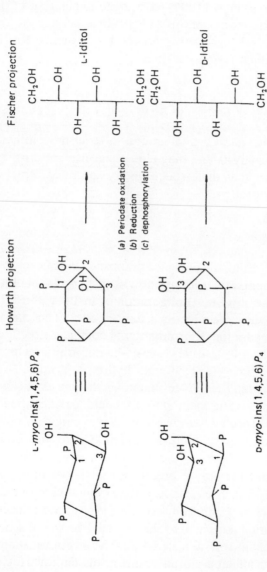

Fig. 4.9. Periodate oxidation, reduction and dephosphorylation of L- and D-Ins(1,4,5,6)P$_4$. The abbreviated conformational diagrams of D- and L-Ins(1,4,5,6)P$_4$ are presented with the numbering system for carbon atoms in the ring indicated. The Haworth projections of these two conformational diagrams are also shown. The polyols derived from the periodate oxidation, reduction and dephosphorylation of these InsPs are drawn in the form of a Fischer projection. L- and D-Ins(1,4,5,6)P$_4$ are streoisomers, but L-Ins(1,4,5,6)P$_4$ is, and could be named (though by convention it is not), D-Ins(3,4,5,6)P$_4$. Reproduced, with permissions, from Stephens, L., Hawkins, P. T., Carter, A. N., Vlahwala, S. B., Morris, A. J., Whetton, A. D. and Downes, C. P. (1988). Biochemical Journal 249, 271–282. © The Biochemical Society and the Medical Research Society.

with excess periodate, are reduced by addition of 1 M-NaBH$_4$ (0.5 ml) and the sample left in the vessel open to air for a further 10–12 h at 25°C. A small amount of appropriate polyols are then added to act as carriers for the ^{3}H-labeled material; these are added as single portions (5 μl) containing 20–25 μg each of inositol, D-glucitol, D-altritol and L-iditol and approximately 2×10^3 dpm of both myo-[^{14}C]-inositol and D-[^{14}C]glucitol. The borohydride is removed by acidification and conversion of the resulting boric acid into a volatile trimethylborane. This is achieved by first passing the sample through a small column (3 ml) of Bio-Rad AG-50W resin (X4; 200–400 mesh; H$^+$ form). The eluate is then freeze-dried, resuspended in methanol (10 ml) and freeze-dried again.

Dephosphorylation is achieved by using alkaline phosphatase under conditions previously designed to lead to the complete dephosphorylation of D-Ins[^{32}P]-(1,4,5)P$_3$ and D-Ins[^{32}P](1,3,4,5)P$_4$. The dried sample is dissolved in 10 mM-ethanolamine (pH 9.5)/1 mM-MgSO$_4$, containing 20 units of bovine intestinal alkaline phosphatase (type P-5521, Sigma; units defined in glycine buffer)/ml in a final volume of 2 ml, and left at 25°C for 10–12 h. Finally the sample was desalted by passing through a column of mixed-bed ion exchange resin (Amberlite MB-3, Sigma; 2 ml). The sample was freeze-dried and then redissolved in a small volume of water.

The polyols were separated on a Brownlee cation-exchange HPLC column (22 cm × 0.46 cm; Polypore-Ca; Anachem, Luton, Beds, UK) and a mobile phase of deionized water. A flow rate of 0.2 ml/min and column temperature of 85–90°C gives a good separation of inositol, altritol, glucitol and iditol. Figure 4.10 shows the resolution obtained for the polyol standard (A) and for polyols derived from InsPs standards (B). Detection of 5–20 μg of each polyol was achieved by monitoring changes in either refractive index or UV absorbance (200 nm). Radioactive polyols were detected by measuring the radioactivity in 0.5 min fractions of the column eluate. Because of peak broadening, it was helpful to include [^{14}C]inositol and [^{14}C]glucitol as internal standards for each run, to

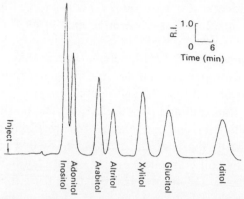

Fig. 4.10. HPLC separation of polyols. A mixture of *myo*-Ins, D-altritol, D-glucitol and -iditol (approx. 25 μg of each, dissolved in 5 μl of water) was resolved by HPLC on a Brownlee 22 cm Polypore-Ca column using deionized water as the mobile phase. A flow rate of 0.2 ml/min and a column temperature of 85–90°C gave a good separation, but it could not distinguish between the two enantiomers of the sugar alcohols. Detection of 5–20 μg of each polyol was achieved by changes in refractive index. Reproduced, with permission, from Stephens, L. R., Hawkins, P. T., and Downes, C. P. (1989). Biochemical Journal 259, 267–282. © The Biochemical Society and the Medical Research Society.

insure a reliable identification of ^{3}H-labeled material. A substitution of the 'Pb^{2+}' for the 'Ca^{2+}' column yielded greater resolution of the polyols when used in an identical manner (Stephens et al., 1988a). Figure 4.11 shows the HPLC separation of polyols derived from InsPs standards D-[^{3}H]Ins(1,4,5)P$_3$, D-[^{3}H]Ins(1,3,4)P$_3$ and D-[^{3}H]Ins-(1,3,4,5)P$_4$ (Stephens et al., 1988b). A sample of each *myo*-[^{3}H]InsP was oxidized with 0.1 M-periodic acid, pH 2.0, for 36 h at 25°C then reduced by addition of NaBH$_4$. An aliquot containing polyol standards, D-[^{14}C]glucitol and *myo*-[^{14}C]Ins in the case of the samples derived from D-[^{3}H]Ins(1,3,4)P$_3$ and D-[^{3}H]Ins(1,3,4,5)P$_4$, and only D-[^{14}C]glucitol in the case of the sample derived from D-[^{3}H]Ins(1,4,5)P$_3$ was added to each sample. Each sample was dephosphorylated by incubation with alkaline phosphatase. The [^{3}H]polyols produced were analyzed by HPLC on a Brownlee-Polypore-Ca^{2+} column as in Fig. 4.10.

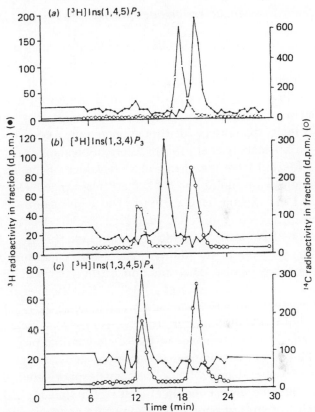

Fig. 4.11. Separation of polyols derived from InsPs standards. A sample of each myo-[³H]InsPs was oxidized with 0.1 M periodic acid, pH 2.0, for 36 h at 25°C, then reduced by addition of NaBH₄. An aliquot containing polyol standards, D-[¹⁴C]glucitol and myo-[¹⁴C]Ins in the case of samples derived from D-[³H]Ins(1,3,4)P₃ and D-[³H]Ins(1,3,4,5)P₄, and only D[¹⁴C]glucitol in the case of the sample derived from D-[³H]Ins(1,4,5)P₃ was added to each sample. Each sample was dephosphorylated by incubation with alkaline phosphatase. The [³H]polyols produced were analyzed by HPLC on a Brownlee Polypore-Ca²⁺ column as described in Fig. 4.10. The flow rate was 0.2 ml/min. Fractions (0.5 min) of the column eluate were collected and their ¹⁴C/³H radioactivities determined by a dual-label liquid scintillation counting. Reproduced, with permission, from Stephens, L., Hawkins, P. T., Carter, A. Chahwala, S. B., Morris, A. J., Whetton, A. D. and Downes, C. P. (1988). Biochemical Journal 249, 271–282. © The Biochemical Society and the Medical Research Society.

4.5.2. Nuclear magnetic resonance spectroscopy (NMR)

MR data for several InsPs including positional isomers of mono-, bis-, tris-, tetrakisphosphates, and cyclic InsP have been reported (Cerdan et al., 1986; Radenberg et al., 1989; Scholz et al., 1990). Due to the low sensitivity of NMR, its application for this purpose is limited to *in vitro* experiments. Since several signals are obtained for each InsP, the correct identification requires complete prior separation (Radenberg et al., 1989). Information about the structure can be extracted from the chemical shift values of the signals in combination with multiplicity of their resonances (Mayr and Dietrich, 1987; Scholz et al., 1990). The combination of NMR experiments detecting different nuclei, e.g. ^{13}C, led to further information (Lindon et al., 1986; Scholz et al., 1990). Selective decoupling as well as two-dimensional NMR experiments later allowed a more precise determination of the chemical structure (Scholz et al., 1990; Johansson et al., 1990). Johansson et al. (1990) showed that 2D ^{1}H-NMR spectroscopy can indicate the number and sites of phosphorylation, and up to three isomers can be analyzed simultaneously. The resonance of the sole equatorial proton (H-2) was shifted furthest downfield (to approximately 47 ppm) when C-2 was phosphorylated; when C-5 is unphosphorylated, the H-5 resonance had the lowest chemical shift (approximately 3.3 ppm). The number of sites of phosphorylation could be determined from the sum of the chemical shifts for the resonances in a given *myo*-InsP. Johansson et al. (1990) constructed an algorithm that can identify any *myo*-InsP on the basis of the chemical shifts.

High resolution NMR has been shown to be an important tool for the identification of isomeric InsPs. Although ^{1}H-NMR and ^{13}C-NMR both have been used for this purpose, the ^{1}H-NMR spectrum is more sensitive and rapid than ^{13}C-NMR spectra (Scholz et al., 1990). For NMR analysis it is important to isolate the compound of interest from the sample matrix and to determine its concentration because approximately $0.1-0.15$ μmol of the compound may be needed to accumulate satisfactory data (Scholz et al., 1990).

Therefore, the samples are separated by HPLC and the fractions containing the individual InsPs are pooled and further purified by ion-exchange column chromatography. The compounds are freeze-dried and the dried residue is dissolved in 1 ml of D_2O (isotopic purity $> 99\%$). The sample is treated with Celex 100 resin and the purified sample (pH 6 or 9) is further freeze-dried (Stephens et al., 1991; Scholz et al., 1990; Radenberg et al., 1989). The dried samples are again dissolved in D_2O and transferred into NMR tubes and dried. The sample is subjected to NMR analysis for either one-dimensional ^1H-NMR or ^{13}C-NMR spectra (Stephens et al., 1991; Mayr, 1990). Figure 4.12 indicates the assignment of multiple isomer structures in a proton NMR spectrum obtained from a mixture of InsPs (Scholz et al., 1990).

Dictyostelium cells contain two diphosphorylated *myo*-InsPs at intracellular concentrations over the range of 0.02–0.30 mM, depending on cell density and growth conditions. In the original studies, these compounds were preliminarily characterized and recognized as diphosphoinositol pentakisphosphate (bis-PP-InsP$_5$) and bisdiphosphoinositol tetrakisphosphate (bis-PP-InsP$_4$) (Stephens et al., 1993; Mayr et al., 1992). Stephens et al. (1993) reported a ^{31}P NMR analysis of compounds A (PP-InsP$_5$) and B (PPInsP$_4$), together with InsP$_6$ for comparison, as shown in Fig. 4.13. The α and β resonances of compounds A and B exhibited characteristic coupling patterns expected for the diphosphate moieties proposed on basis of previous chemical and FAB-MS analyses (see below). A detailed examination of the NMR spectra allowed to suggest that compound A is a PP-InsP$_5$ and B is a bis-PP-InsP$_4$. Recently, their structures have been elucidated by two-dimensional $^1H/^{31}P$ NMR analysis (Laussmann et al., 1996). This investigation showed that PP-InsP$_5$ is either D-4-PP-InsP$_5$ or its enantiomer D-6-PP-InsP$_5$ and bis-PP-InsP$_4$ is either D-4,5-bis-PP-InsP$_4$ or its enantiomer D-5,6-bis-PP-InsP$_4$. A differentiation between enantiomers was not possible by the NMR method used.

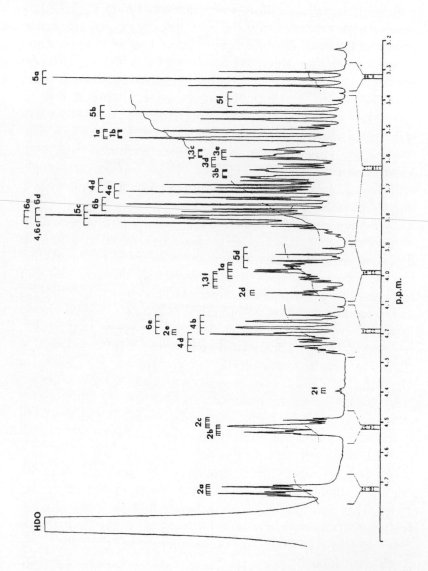

Laussmann et al. (1997) have identified the naturally occurring enantiomers by using defined synthetic PP-InsP$_5$ isomers as substrates for a partially purified PP-InsP$_5$ 5-kinase from *Dictyostelium*. This enzyme specifically phosphorylated the naturally occurring PP-InsP$_5$ and the synthetic D-6-PP-InsP$_5$, leading to D-5,6-bis-PP-InsP$_4$. In contrast, neither D-4-PP-InsP$_5$ nor D-1-PP-InsP$_5$ or D-3-InsP were converted by the enzyme.

Martin et al. (1993) used ^{31}P-NMR to investigate the occurrence of high amounts of two InsPs in PCA extracts of axenic *Entamoeba histolytica*. These components were identified by ^{31}P-NMR, using homonuclear J-resolved and two-dimensional ^{1}H-^{31}P correlative, analyses as *myo*-InsP$_3$ (Ins(2,4,6)P$_3$) and pentakisphospho-*myo*-inositol diphosphate [Ins(1,2,3,4,6)P$_5$(5)P$_2$]. However, a reinvestigation of the structure of InsPs in trophozoites of the parasitic amoeba *Entamoeba histolytica* has shown (Martin et al., 2000) that, rather than being *myo*-inositol derivatives, these compounds belong to a new class of InsPs in which the cyclitol isomer is *neo*-inositol. The structures of *neo*-inositol hexakisphosphate, 2-diphospho-*neo*-InsP$_5$, and 2,5-bisdiphospho-*neo*-InsP$_4$, which are present in *E. histolytica* at concentrations of 0.08–0.36 mM, were solved by two-dimensional ^{31}P-^{1}H NMR spectroscopy. No evidence for the co-existence of their *myo*-inositol counterparts was found. These *neo*-inositol compounds were not substrates of 6-diphospho-InsP$_5$ 5-kinase, an enzyme purified from *D. discoideum* that phosphorylates 6-diphospho-*myo*-InsP$_5$ and more slowly also *myo*-InsP$_6$, specifically on position 5, as indicated by large amounts of the same *neo*-InsP and its diphosphate esters also in another primitive amoeba, *Phreatamoeba balamuthi*. Martin et al. (2000)

Fig. 4.12. Assignment of multiple isomer structures in a proton NMR spectrum obtained from a mixture of InsP$_2$s. The total amount of InsP$_2$s was 15 μmol. Assigned resonances are labeled by the ring-proton number and a letter denoting the identified isomer: a, Ins(1,2)P$_2$ (25% of total); b, Ins(2,4)P$_2$ (25%); c, Ins(2,5)P$_2$ (14%); d, Ins(4,5)P$_2$ (11%); e, Ins(1,6)P$_2$ (6%); f, Ins(1,3)P$_2$ (4%). Reproduced from Scholz et al. (1990) with permission of the publisher.

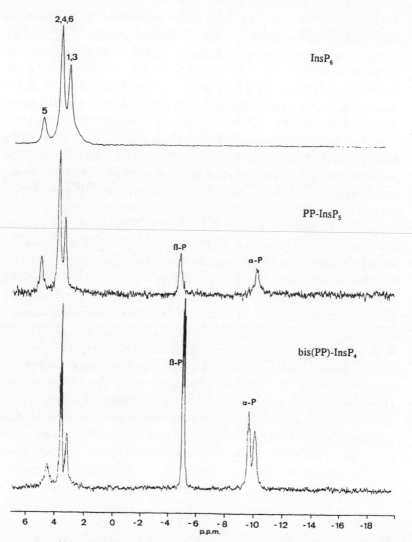

Fig. 4.13. ^{31}P NMR spectra of compounds A and B. ^{31}P NMR spectra of samples of *Dictyostelium*-derived InsP$_6$ (top spectrum: 70 μmol), compound A (middle spectrum: 7.5 μmol), and compound B (bottom spectrum; 8 μmol). The spectra were obtained at pH 9.0. The α and β resonances of compounds A and B exhibit the characteristic coupling patterns expected for the proposed diphosphate

speculate that the occurrence of high concentrations of *neo*-InsPs may be much more general than previously thought.

A new InsPs isolated from *Pyrococcus woes* has been described by Scholz et al. (1992). This compound was obtained by means of anion exchange chromatography, preparative TLC and gel filtration. Alkaline hydrolysis yielded Ins and $InsP_1$ as analyzed by TLC. In accordance with mass spectra, a phosphorus diester structure was determined by ^{1}H-NMR. To establish the stereochemical configuration of D-InsP, all three possible stereoisomers (LL, DD and DL) were prepared synthetically (Van Leeuwen et al., 1994). Comparison of the specific rotation values revealed that the naturally occurring D-InsP had the LL configuration.

The results obtained from ordinary NMR experiments give no information about enantiomeric phosphates (Lindon et al., 1987; Scholz et al., 1990). NMR techniques, when coupled with chirality selective shifting reagents, can also distinguish between enantiomers.

4.5.3. Mass spectrometry

The electron ionization (EI) spectra of all the positional isomers of *myo*-$InsP_1$ and of the *myo*-Ins(1,2-cyc)P have been obtained by GC/MS of the pertrimethylsilyl derivatives (Sherman et al., 1986). The phosphate moiety was found to direct fragmentation to produce fragment ions of useful intensity with specific carbon retention. Only trace-abundance molecular ions were observed, but each of the TMS $InsP_1$ and Ins diphosphates gave $[M-15]^+$ ions (m/z 749 and 901, respectively) of significant intensity. Simple losses from the molecular or related ions of TMS *myo*-$InsP_1$ are shown in Fig. 4.14. The composition of seven of the ions was corroborated by deuterium labeling (Sherman et al., 1986). The formation of ions m/z 318 and 217 with selective

moieties ($^2J_{POP}$ of the Pα-Pβ coupling of about 19 Hz and an unresolved coupling of Pα with the vicinal inositol ring proton with ($^3J_{PCOH}$ of about 10 Hz. Reproduced from Stephens et al. (1993) with the permission of the publisher.

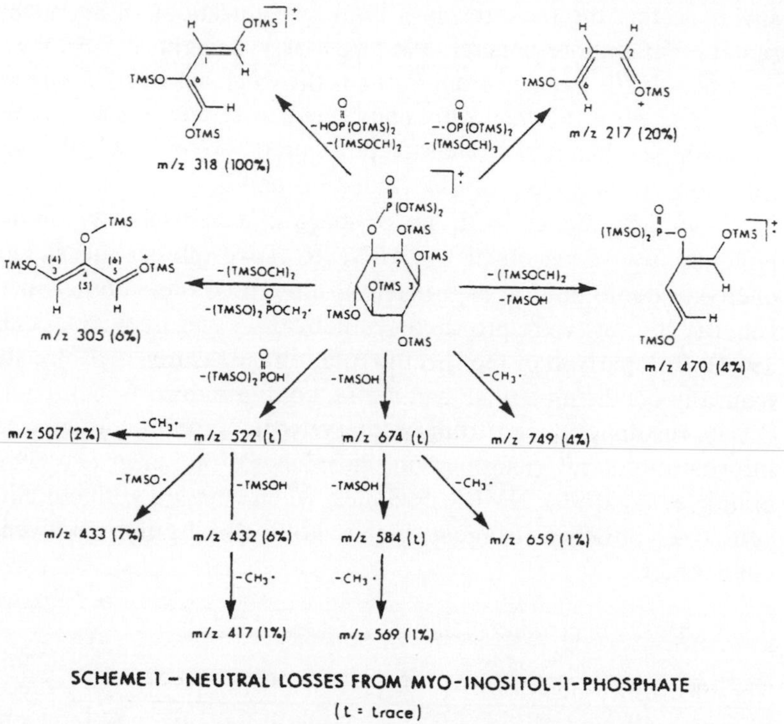

SCHEME 1 – NEUTRAL LOSSES FROM MYO-INOSITOL-1-PHOSPHATE

(t = trace)

Fig. 4.14. Simple neutral losses from the molecular or related ions of trimethylsilyl (TMS) *myo*-Ins(1)P. The composition of seven of the ions in the figure were corroborated by deuterium labeling. Reproduced from Sherman et al. (1986) with the permission of the publisher.

retention of C-1 and adjacent atoms suggested that the phosphate moiety had a direct effect on the fragmentation paths. The stereochemistry of the phosphate in relation to the inositol also had an effect on fragmentation that may be useful in structure analysis. The unusually high intensity of m/z 470 in the spectrum of TMS *myo*-Ins(2)P appeared to be another instance of phosphate retention due to stereochemical factors. However, TMS of *myo*-Ins(2)P was the only example studied. GC/MS of each of the substances was performed by EI at 70 eV on an LKB-9000 using

$4' \times 1/4''$ glass column packed with 0.1% DC-710 silicone on 80–100 mesh etched glass beads. The EI spectrum of TMS *myo*-Ins(1,2-cyc)Pic phosphate was obtained by GC/MS on a Finnigan model 3200 quadrupole at 70 eV using a $5' \times 1/4''$ glass column packed with 3% OV-17 on Chromosorb W-HP (Ohio Valley Chem Co.).

Since InsPs are not suitable for GC/MS even after trimethylsilylation, they were examined by FAB/MS in the sodium and hydrogen form (Sherman et al., 1986). Both the positive and negative ion spectra of the hydrogen form of phytic acid were markedly simpler than those of the sodium form. The negative ion spectrum is shown in Fig. 4.15 along with those of $InsP_5$-, tris- and bis-phosphates. Except for a small amount of sodiated molecular ion, the ion current is mainly due to $[M - H]^-$ (m/z 659). Losses of 80 u (HPO_3) and 98 u (H_3PO_4) produce m/z 579 and 561 (Fig. 4.15A). The positive ion spectrum had essentially the same ions, 2 Da lower in mass. However, a sample of an equivalent size gave a five-fold less-intense spectrum than that of the negative ions. Figure 4.15B shows the negative ion FAB spectrum of 300 nmol of *myo*-Ins(1,2,3,5,6)P_5 in 30 µl of glycerol with $[M - H]^-$ at m/z 579. The only losses sustained are 80 and 98 u giving m/z 499 and 481, supporting the possibility of the unusual HPO_3 loss. Ions at m/z 419 and 401 represented further losses of 80 u (or 98 u from m/z 499 in one case). The four isomers of *myo*-InsP_5$ examined by FAB in glycerol were the Ins(1,2,3,4,6)-, the Ins(1,2,3,5,6)-, the Ins(1,2,4,5,6)- and the Ins(1,3,4,5,6)P_5. All had identical positive ion spectra. The positive and negative ion FAB spectra of *myo*-Ins(1,2,5,6)P_4 showed the tetraphosphate to undergo exactly the same process as the pentaphosphate. In this case, $[MH]^+$ occurs at m/z 501 and $[M - H]^-$ at m/z 499 (spectra not included).

Figure 4.15C shows the negative ion spectrum of 300 nmol of D-*myo*-Ins(1,4,5)P_3- (dihydrogen phosphate) from human red blood cells, while Fig. 4.15D is the spectrum of the same amount of the $InsP_2$ with $[M - H]^-$ at m/z 419 and 339, respectively. The 80 u

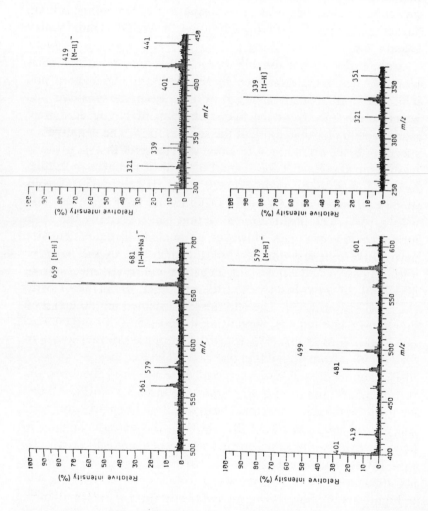

losses are clearly evident in the $InsP_3$ (at m/z 339 and 321) but are at the noise level in the $InsP_2$. Water loss from $[M - H]^-$ is more obvious in these spectra than others in this study and is seen only in the spectra of the free acids. The myo-Ins(2)P, which has a FAB spectrum representative of the $InsP_1$, has, as the free acid, intense $[MH]^+$ and $[M - H]^-$ ions at m/z 261 and 259, respectively (spectra not included here). In the positive ion spectrum, a glycerol adduct peak at m/z 353 is 40% of $[MH]^+$. More recently, Stephens et al. (1993) performed negative ion FAB-MS on a preparation of diphospho-$InsP_5$ and bisdiphospho-$InsP_4$ isolated from *Dictyostelium*. Desalted samples of the mixture of the two were dissolved in 5% aqueous acetic acid, aliquoted into acidified thioglycerol (Dell et al., 1988), and subjected to FAB-MS (using a VG analytical ZAB-HF mass spectrometer fitted with a M-scan FAB gun operated at 10 kV). Spectra were recorded on oscillographic chart paper and manually counted. Samples of $InsP_6$ that had been purified by the same procedure were analyzed in parallel. The negative FAB mass spectra positively identified $InsP_7$ and $InsP_8$ by showing clear quasi-molecular ion clusters centered around m/z 739/761/783/805/827/849 and m/z 841/863/885/907, respectively. These results, however, could have been obtained from a range of phosphate configurations, e.g. the $InsP_8$ could be present as bis-diphospho-$InsP_4$ or a triphospho-$InsP_5$ or a tetraphospho-$InsP_4$, etc.

Fig. 4.15. Negative ion FAB spectra of myo-InsPs. (A) The negative ion FAB spectrum of 200 nmol of myo-InsP$_6$ in glycerol with $[M - H]^-$ at m/z 659. Losses of 80 u (HPO$_3$) and 98 (H$_3$PO$_4$) result in ions m/z 579 and 561; (B) the negative ion FAB spectrum of 300 nmol of myo-Ins(1,2,3,5,6)P$_5$ in 30 μl of glycerol with $[M - H]^-$ at m/z 579. The losses of 80 and 98 u occur to a greater extent than in A; (C) The negative ion FAB spectrum of 300 nmol of D-myo-Ins(1,4,5)P$_3$ from human red blood cells. $[M - H]^-$ is at m/z 419, and the 80 and 98 u losses are again evident; (D) the negative ion FAB spectrum of 300 nmol of D-myo-Ins(1,4)P$_2$ from the same source as the sample in (C). $[M - H]^-$ is at m/z 339. A water-loss is more evident in other spectra of this series. Reproduced from Sherman et al. (1986) with the permission of the publisher.

Therefore, the samples were purified further and the inositol/phosphorus ratios determined. Partial acid hydrolysis of both compounds showed that they were degraded to P_i and a species of InsP with a mobility of $InsP_6$. Furthermore, the stoichiometry of P_i liberated/$InsP_6$ form showed a ratio of 1:1 mol/mol for A and 2:1 for B), confirming that A and B possessed inositol/phosphate ratios of 1:7 and 1:8, respectively. These data independently confirmed the conclusions of the negative FAB-MS, but further suggested that both compounds contain fully phosphorylated *myo*-inositol (Stephens et al., 1993). The positional location of the pyrophosphate bonds was established by NMR (see above).

The GLC method described above for the analysis of *myo*-inositol has been utilized for the indirect analysis of InsPs (Rittenhouse and Sasson, 1985; DaTorre et al., 1990). The method involves separation of InsPs in crude extract, dephosphorylation of individual InsP into *myo*-inositol by using alkaline phosphatase and subsequent analysis of *myo*-inositol by GLC. *myo*-Inositol is detected either by a flame ionization detector (DaTorre et al., 1990) or by mass spectrometry (Rittenhouse and Sasson, 1985). Chromatographic separation of natural and d_6-labeled *myo*-inositol by using selective-ion monitoring GC/MS provides quantitative data. A major disadvantage of the GLC procedures is that InsPs must be extracted in a form that can be dephosphorylated by alkaline phosphatase. Rittenhouse and Sasson (1985) achieved this by eluting InsPs with ammonium formate and then subjecting the InsPs to a second ion-exchange chromatography using LiCl. This method provided poor and variable recovery because the use of ethanol to remove the excess LiCl caused large losses of InsPs. Heathers et al. (1989) and DaTorre et al. (1990) circumvented this problem by using ammonium sulfate for the elution of InsPs from the resin.

DaTorre et al. (1990) and Turk et al. (1986) have developed a GC/MS method that measures TMS derivatized *myo*-Ins by electron-impact and chemical-ionization MS, respectively. For the analysis of InsPs, it is important to separate the different InsPs and dephosphorylate each InsP into *myo*-Ins and measure the concentration of *myo*-Ins (DaTorre et al., 1990). For GC/MS analysis, tissue samples are mixed with d_6-*myo*-Ins as an internal standard and extracted with TCA. The acid is removed by washing the sample with water-saturated diethyl ether and the aqueous phase is lyophylized. The dried sample is dissolved in Tris buffer (pH 8.8, 0.01 M) and poured onto an anion exchange resin. InsP, InsP$_2$, InsP$_3$ and InsP$_4$ are separated by using gradients of ammonium sulfate. Each InsP is dephosphorylated with alkaline phosphatase and *myo*-Ins is derivatized with a reagent containing 25 µl each of pyridine and BSTFA + 0.1% TMCS. The derivatized sample is analyzed by a GC/EI/MS (Myher et al., 1978) or GC/CI/MS (Radenberg et al., 1989).

4.6. Determination of stereochemical structure

Several published methods have been employed for stereochemical classification of InsPs. The stereoselective assignment of D- and L-Ins(1)P has been achieved by use of a gas–liquid chromatography system utilizing chiral stationary phase (Loewus et al., 1982; Leavitt and Sherman, 1982a). This approach has not proven to be versatile, probably because the methods of detection that have been developed for use with GLC systems are not compatible with ^{3}H-labeled samples and they may require micromolar quantities of appropriately derivatized material.

NMR techniques, when coupled with chirally selective shifting reagents, can also distinguish between enantiomers. Again, both the compound of interest and each of the reference standards must be available in milligrams and in pure state. Likewise, enzymes or binding proteins that are stereoselective for either the original inositol phosphate or some derivative of them can be used, provided a source of relevant enantiomers is available along with appropriate detection system.

4.6.1. Optical rotation

In principle, the handedness of a pure inositol phosphate can be established in a number of ways. Grado and Ballou (1961) established the chirality of the PtdIns(4,5)P_2 head group by measuring the direction in which the plane of light from a plane polarized source was rotated when it passes through a solution of iditol derived from PtdIns(4,5)P_2. They compared the fully acetylated derivative of iditol with a standard independently prepared 'L' and 'D'-iditols. The recent synthesis (Taylor et al., 1989) of the D and L species of Ins(1,4,5)P_3 would now allow the comparison to be made with the unmodified Ins(1,4,5)P_3. Irving and Cosgrove (1972) have classified several InsP$_5$s by measuring their specific optical rotations relative to reference compounds. This method, however, requires several milliliters of a 1 mM solution for analysis.

The InsP$_5$s were originally characterized by their specific optical rotation relative to reference compounds (Irving and Cosgrove, 1972). This approach requires sufficient amounts of pure material (several ml of 1 mM solution). The stereoselective assignment of D- and L-Ins(1)P is not generally applicable (Loewus et al., 1982). Since sufficient material is not available, determination of optical rotation is not practical. However, the oxidized, reduced and dephosphorylated samples can be subjected to a sensitive oxidation by a stereoselective yeast-derived polyol dehydrogenase.

4.6.2. Polyol dehydrogenases

At present, analysis of the stereostructure of trace quantities of [³H]InsPs is usually achieved by identifying the ³H-labeled polyols produced by a process of oxidation with periodate, reduction and dephosphorylation of the unknown InsP. The final identification of the stereoisomer is made by the use of stereospecific polyol dehydrogenases (Stephens et al., 1988b; Barker et al., 1988;

Balla et al. 1989). This approach provides great resolving power by converting a wide variety of InsPs into a relatively small family of polyols, many of which can be oxidized by a single dehydrogenase preparation with total D versus L selectivity. Stephens (1990) has tabulated the polyols generated by this process for all-myo-InsP$_1$, myo-InsP$_2$, myo-InsP$_3$ and myo-InsP$_4$, along with the products generated by the polyol dehydrogenases. In this instance all the polyols are generated from myo-inositol, but similar products could be obtained from other inositols with different phosphate arrangements (see discussion on $chiro$-inositol below).

The commercially available, yeast-derived L-iditol dehydrogenase (L-iditol:NAD 2-oxidoreductase, also found in mammalian tissues) is distinguished from a number of other polyol dehydrogenases by its ability to oxidize L- but not D-iditol. According to Stephens (1990) a typical assay is performed by allowing to equilibrate 980 μl of 0.1 M Tris–HCl (ph 8.3, 22°C), 20 mM β-NAD$^+$ with 1.5 units yeast-derived L-iditol dehydrogenase/ml in two 1 ml quartz cuvettes at 22°C in a thermostatically controlled dual-beam spectrophotomer set to measure the difference in absorbance at 340 nm between the two cuvettes. After 10–20 min equilibration, the substrate mixture should be added to one cuvette (20 μl of H$_2$O) and H$_2$O to the reference chamber. The substrate mixture normally contains 1000 dpm of D-[^{14}C]iditol, 200 μM L-iditol, and a [^{3}H]iditol of unknown isomerism. Initially, the basal absorbance at 340 nm changes significantly, but after 15–20 min it stabilizes, at which time the substrate is added to one of the cuvettes and monitoring of the production of NADH in the experimental cuvette and the basal drift in the control is continued. After 90–100 min, the rate of change of absorbance at 340 nm falls to that of the control sample or zero, in case of the dual beam instrument. Once this happens, the contents of assay cuvette are transferred to a microfuge tube and boiled for 3 min. The heat denatured protein is pelleted by centrifugation and the supernatant desalted by mixing with 3 ml of MB3A mixed-bed ion exchange resin and allowed to stand for 2–4 h. The resin is washed and the sample freeze-dried, resuspended in 10 μl of H_2O and applied to a cation exchange HPLC column (Pb^{2+} mode) that separates the reactants from products. In the Pb^{2+} form, the columns offer higher resolution (Stephens et al., 1988b; Stephens et al., 1989a,b). The samples are injected in 10 μl of H_2O and the columns are eluted with H_2O at 0.2 ml/min; thermostatically regulated temperatures of about 25°C have given the best results. Elution times for the polyols have been reported by Stephens et al. (1988b) and Stephens et al. (1989b). Xylulose and fructose elute before inositol, whereas ribulose elutes after arabitol but before glucitol (Stephens, 1990).

The percentage of the total ^3H-labeled sample eluted after 16–17 min can be assumed to be equal to the percentage of L-[^3H]iditol in the original sample (Stephens et al., 1988b; Stephens et al., 1989a). Oxidation of L-iditol to L-sorbose is observed with L-Ins(1,4)P_2, L-Ins(1,4,5)P_3, L-Ins(1,4,6)P_3 and L-Ins(1,4,5,6)P_4, while D-Ins(1,4)P_2, D-Ins(1,4,5)P_3, D-Ins(1,4,6) P_3 and D-Ins(1,4,5,6)P_4 on periodate oxidation and reduction yield D-iditol, which is not oxidized by yeast-derived polyol dehydrogenase (Stephens et al., 1988b). Stephens (1990) has tabulated the non-specific response to the production of non-specific oxidation products of yeast polyol dehydroganases (Table 4.3).

The oxidation of L- but not D-altritol conforms to the McCorkindale-Edson rules (McCorkindale and Edson, 1954), which predict the substrate specificity of L-iditol dehydrogenase, although the reaction proceeds at a substantially lower rate than that against L-iditol. The enzyme oxidizes D-altritol at 1/132nd of the rate of L-atritol, under first order reaction conditions. The product is L-tagatose, which can be resolved from altritol and allulose (another potential product). According to Stephens (1990) the assays are run in a manner identical with that for L-iditol oxidation except that 9.5 units L-iditol dehydrogenase is used and the assays are run for about 2.5 h. For calibration of the

TABLE 4.3

Groups of non-cyclic *myo*-inositol phosphates which can, upon oxidation with periodate, reduction and dephosphorylation, yield *iditols* that can be enantiomerically resolved by a yeast-derived polyol dehydrogenase, along with the oxidation products formed by the enzyme

InsPs yielding polyol	Polyol	Product of oxidation
D-Ins(1,4)P$_2$	D-iditol	Not oxidized
D-Ins(1,4,5)P$_3$	D-iditol	Not oxidized
D-Ins(1,4,6)P$_3$	D-iditol	Not oxidized
D-Ins(1,4,5,6)P$_4$	D-iditol	Not oxidized
L-Ins(1,4)P$_2$	L-iditol	L-sorbose
L-Ins(1,4,5)P$_3$	L-iditol	L-sorbose
L-Ins(1,4,6)P$_3$	L-iditol	L-sorbose
L-Ins(1,4,5,6)P$_4$	L-iditol	L-sorbose

Modified from Stephens (1990).

extent of the reaction accurately $500-1000$ dpm of L-[^{14}C]altritol is included in the assay, along with $200 \, \mu M$ L-altritol and the ^{3}H-altritol of unknown isomerism. The percentage of the L-[^{14}C]altritol oxidized is then used to define the extent of the reaction. Under the above conditions, $86-92\%$ of the initial L-[^{14}C]altritol is oxidized. Oxidation of L-altritol to L-tagatose is observed with D-Ins(1,3,4)P$_3$, D-Ins(1,2,4)P$_3$ and D-Ins(1,2,3,4)P$_4$, while L-Ins(1,3,4)P$_3$, L-Ins(1,2,4)P$_3$ and L-Ins(1,2,3,4)P$_4$ yield D-altritol on periodate oxidation and reduction, and as a consequence are not oxidized by yeast-derived L-iditol dehydrogenase (Table 4.4).

According to the rules of substrate specificity defined by McCorkindale and Edson (1954), D-glucitol (sorbitol) but not

TABLE 4.4

Groups of non-cyclic *myo*-inositol phosphates which can, upon oxidation with periodate, reduction and dephosphorylation, yield *altritols* that can be enantiomerically resolved by a yeast-derived polyol dehydrogenase, along with the oxidation products formed by the enzyme

InsPs yielding polyol	Polyol	Product of oxidation
L-Ins(1,3,4)P$_3$	D-altritol	Not oxidized
L-Ins(1,2,4)P$_3$	D-altritol	Not oxidized
L-Ins(1,2,3,4)P$_4$	D-altritol	Not oxidized
D-Ins(1,3,4)P$_3$	L-altritol	L-tagatose
D-Ins(1,2,4)P$_3$	L-altritol	L-tagatose
D-Ins(1,2,3,4)P$_4$	L-altritol	L-tagatose

Modified from Stephens (1990).

L-glucitol is oxidized by L-iditol dehydrogenase, yielding D-fructose and NADH (Stephens, 1990). Under the assay conditions recommended by Stephens (1990), the Km of L-iditol dehydrogenase for D-glucitol is 25 mM and its Vmax is 330 nmol/min per unit. The oxidation of L-glucitol is undetectable. D-[^{14}C]glucitol is used as an internal calibration standard for each assay, and the product of its oxidation, D-[^{14}C]fructose, is separated from glucitol by the cation-exchange HPLC system described above. The assay can be safely run to completion, without significant oxidation of L-glucitol, by incubating with 3 units polyol dehydrogenase/ml for 2 h. Oxidation of D-glucitol to D-fructose is observed for D-Ins(1)P, L-Ins(2,4,5)P$_3$, D-Ins(1,2,5)P$_3$ and D-Ins(1,2,5,6)P$_4$, while periodate oxidation and reduction of L-Ins(1)P, D-Ins(2,4,5)P$_3$, L-Ins(1,2,5)P$_3$, and L-Ins(1,2,5,6)P$_4$ yield L-glucitol, which is not oxidized by yeast-derived L-iditol dehydrogenase (Stephens, 1990) (Table 4.5).

TABLE 4.5

Groups of non-cyclic *myo*-inositol phosphates which can, upon oxidation with periodate, reduction and dephosphorylation, yield *glucitols* that can be enantiomerically resolved by a yeast-derived polyol dehydrogenase, along with the oxidation products formed by the enzyme

InsPs yielding polyol	Polyol	Product of oxidation
L-Ins(1)P	L-Glucitol	Not oxidized
D-Ins(2,4,5)P$_3$	L-Glucitol	Not oxidized
L-Ins(1,2,5)P$_3$	L-Glucitol	Not oxidized
L-Ins(1,2,5,6)P$_4$	L-Glucitol	Not oxidized
D-Ins(1)P	D-Glucitol	D-Fructose
L-Ins(2,4,5)P$_3$	D-Glucitol	D-Fructose
D-Ins(1,2,5)P$_3$	D-Glucitol	D-Fructose
D-Ins(1,2,5,6)P$_4$	D-Glucitol	D-Fructose

Modified from Stephens (1990).

The oxidation of L-arabitol but not D-arabitol by L-iditol dehydrogenase provides an exception to the McCorkindale-Edson rules (McCorkindale and Edson, 1954) breaking the requirement for the C$_4$ hydroxyl moiety to be L with respect to C-1. L-arabitol can be oxidized at either end by L-iditol dehydrogenase, yielding either L-xylulose or L-ribulose. The kinetics of the reaction is complex. Under the standard conditions of the assay (9.5 units L-iditol dehydrogenase, 100 mM Tris–HCl (pH 8.3, 22C), 20 mM β-NAD$^+$), after 2.5 h roughly only half of the total substrate is converted to product, despite the fact that NADH had ceased to accumulate. Nevertheless, the complete stereo-selectivity of the dehydrogenase preparation enables the chirality of unknown [^{3}H]arabitols to be established. No oxidation is observed for L-Ins(2,4)P$_2$ and D-Ins(1,2,6)P$_3$, which yield D-arabitol on periodate oxidation and reduction, while D-Ins(2,4)P$_2$ and L-Ins(1,2,5)P$_3$ yield L-arabitol which is oxidized to L-ribulose and L-xylulose (Stephens, 1990).

There are six InsP$_5$ isomers, which theoretically divided into four chromatographically distinct groups Ins(1,3,4,5,6)P$_5$, Ins(1,2,3,4,6)P$_5$ (both of which contain a plane of symmetry) and D- and L-Ins(1,2,3,4,5)P$_5$ and D- and L-Ins(1,2,4,5,6)P$_5$ (see Section on nomenclature, Chapter 1).

Stephens et al. (1991) have used standard and high performance anion exchange-chromatographic techniques to purify myo-[^{3}H]InsP$_5$s from various myo-[^{3}H]inositol-prelabeled cells. Slime mold (D. discoideum) contained 8 μM-myo-[^{3}H]Ins (1,3,4,5,6)P$_5$, 16 μM-myo-[^{3}H]Ins(1,2,3,4,6)P$_5$ and 36 μM-D-myo[^{3}H]Ins(1,2,4,5,6)P$_5$ [calculated intracellular concentrations; Stephens and Irvine (1990) germinating mung bean (Phaseolus aureus) seedlings contained both D- and L-myo-[^{3}H]Ins(1,2,4, 5,6)P$_5$ (which was characterized by ^{32}P and two-dimensional proton NMR.) and D- and/or L-myo-[^{3}H]Ins(1,2,3,4,5)P$_5$; HL60 cells contained myo-[^{3}H]Ins(1,3, 4,5,6)P$_5$ (in a 500-fold excess over the other species), myo-[^{3}H]Ins(1,3,4,5,6)P$_5$ and D- and /or L-myo [^{3}H]Ins(1,2,4,5,6)P$_5$; and NG-115-401L-C3 cells contained myo-[3H]Ins(1,3,4,5,6)P$_5$ (in 100-fold excess over the other species), D- and/or L-myo-[^{3}H]Ins (1,2,4,5,6)P$_5$, myo-[^{3}H]Ins(1,2,3, 4,6)P$_5$ and D- and/or L-myo-[^{3}H]Ins(1,2,3,4,5)P$_5$.

Ins(1,3,4,5,6)P$_5$ isomer is the InsP$_5$ found at a high concentration in avian erythrocytes (Johnson and Tate, 1969) and can be readily prepared labeled with either [^{32}P] or [^{3}H] for use as a standard (Stephens et al., 1988a).

D- and/or L-Ins(1,2,3,4,5)P$_5$ was established to be the major InsP$_5$ in germinating mung beans (in terms of both phosphate content and, [^{32}P]Pi- or [^{3}H]Ins-prelabeled cells, [^{32}P] or [^{3}H] content (Stephens et al., 1991).

D- and/or L-[^{3}H]Ins(1,2,4,5,6)P$_5$ was purified by HPLC (by both strong and weak anion exchange HPLC columns) from [^{3}H]Ins-prelabeled amoebae, then partially dephosphorylated with an

TABLE 4.6

Groups of non-cyclic *myo*-inositol phosphates which can, upon oxidation with periodate, reduction and dephosphorylation, yield *arabitols* that can be enantiomerically resolved by a yeast-derived polyol dehydrogenase, along with the oxidation products formed by the enzyme

InsPs yielding polyol	Polyol	Product of oxidation
L-Ins(2,4)P_2	D-Arabitol	Not oxidized
D-Ins(1,2,6)P_3	D-Arabitol	Not oxidized
D-Ins(2,4)P_2	L-Arabitol	L-Ribulose L-Xylulose
L-Ins(1,2,6)P_3	L-Arabitol	L-Ribulose L-Xylulose

Modified from Stephens (1990).

Aspergillus-derived phytase preparations (Irving and Cosgrove, 1972). Two independent preparations of D- and/or L-[^{3}H]Ins(1,2,4,5,6)P_5 were dephosphorylated to different extents (see Table 4.6) and the products were resolved by anion exchange HPLC. The [^{3}H]InsP$_5$ peak yielded [^{3}H]arabitol and was therefore either D- or L-[^{3}H]Ins(1,2,6)P_3; traces of [^{3}H]glucitol were also obtained, but because of the ambiguity in the way [^{3}H]glucitol can be derived from [^{3}H]InsP$_3$s this was not further analyzed. The [^{3}H]InsP$_4$ peak yielded [^{3}H]glucitol, meaning some D- or L-[^{3}H]InsP$_4$ was present.

These incompatibilities largely rule out the use of the two most common techniques for separation of InsPs (namely elution from Dowex columns with ammonium formate and HPLC with phosphate-containing buffers). However, the technique of Spencer et al., (1990), whereby InsPs are eluted with HCl, might be suitable for use in conjunction with this mass assay. The Sep-Pak technique just described can substitute for Dowex (formate) chromatography, and the ion chromatographic separation technique of Sun et al. (1990) may substitute for phosphate-based HPLC techniques.

4.6.3. Inositol phosphate kinases

McConnell et al. (1991) determined the enantiomeric composition of D/L-[^{3}H]Ins(1,2,4,5,6)P$_5$ by incubating it, in duplicate with a 100 000g soluble fraction of *Dictyostelium*, which phosphorylates D-Ins(1,2,4,5,6)P$_5$ but not L-Ins(1,2,4,5,6)P$_5$ (Stephens and Irvine, 1990; Stephens et al., 1991). The D/L-[^{3}H]Ins(1,2,4,5,6)P$_5$ for these experiments was first separated from [^{3}H]Ins(1,3,4,5,6)P$_5$ on a Partisphere 5μ SAX column (see above) and then desalted by using HCL (see above). Some 3000–4000 dpm of ^{3}H was incubated with 250 μl of medium containing 25 mM Hepes (pH 7), 5 mM-MgATP, 1 mM-EGTA, 1 mM MgCl$_2$, 1 mM dithiothreitol, 1 mg of BSA/ml and approximately 600 dpm of a 1:1 racemic mixture of D- and L-[^{32}P]Ins(1,2,4,5,6)P$_5$. After 25 min at 25°C, reactions were quenched with HClO$_4$ and neutralized. The products were analyzed by HPLC using Partisphere 5μ SAX column (see above). Typically, between 30 and 44% of the racemic mixture of D- and L-[^{32}P]Ins(1,2,4,5,6)P$_5$ was phosphorylated to [^{32}P]InsP$_6$. By extrapolating the reaction to the point were 50% of the racemic mixture would have been phosphorylated, the proportion of D-[^{3}H]Ins(1,2,4,5,6)P$_5$ in the original ^{3}H-labeled sample could be estimated.

Laussmann et al. (1997) assayed the enzyme in a reaction mixture containing 30 mM HEPES (pH 6.8), 6 mM MgCl$_2$, 5 mM NaF, 5 mM Na$_2$ATP, 5 mM phosphocreatine, 2 units/ml creatine kinase and 7.7 μM of the particular PP-InsP$_5$ isomer in a final volume of 0.8 ml. The reaction mixture was incubated at 25°C and the reaction was terminated after 2–60 min by diluting the sample with 1.2 ml of ice-cold water. Amounts of PP-InsP$_5$ and bis-PP-InsP$_4$ were immediately determined by the metal–dye-detection (MDD/HPLC) method (Mayr, 1990). Samples were separated on a high-resolution 1 cm × 0.5 cm Source-15 Q column with a 30 min linear gradient of 200–425 mM HCl. Detection at 546 nm was performed by a modified MDD post-column dye reagent [2 M Tris–HCl (pH 8.5)/200 μM 4-(2-pyridylazo)resourcinol/30 μM

$YCl_3/10\%$ (v/v) methanol)]. Amounts of 0.2–19 nmol of PP-InsP$_5$ are detectable using this method. No enzyme activity was observed in zero-time incubations, or after boiling, of enzyme solutions. Figure 4.16 shows the substrate specificity of PP-InsP$_5$ kinase of *Dictyostelium* (Laussmann et al., 1997). All samples were incubated for 10 min at 25°C with pre-purified PP-InsP$_5$ kinase. Only D-6-PP-InsP$_5$ and *D. discoideum* (*D. d.*) PP-InsP$_5$ were almost completely converted into 5,6-bis-PP-InsP$_4$. It should be noted that D-1-PP-InsP$_5$ and D-3-PP-InsP$_5$ have remarkably different retention times in comparison with the natural PP-InsP$_5$. Contamination InsP$_6$ is due to hydrolysis of the particular PP-InsP$_5$ isomers. The structures of the diphosphoinositol phosphates found in other species remain to be characterized.

4.6.4. Inositol phosphate phosphatases

Theoretically, the phosphatases, which have shown stereospecificity could have been used to establish the chirality of the inositol phosphates, but no specific applications appear to have been made. A large family of 5-phosphates has been identified in the soluble and in the particulate fraction of many cell types (see Chapter 7). Some of these enzymes originally identified by their ability to hydrolyze the phosphate at the 5-position of the second messengers Ins(1,4,5)P$_3$ and Ins(1,3,4,5)P$_4$ have been shown to remove the 5-phosphate at the 5-position of InsPs.

Stricker et al. (1999) have demonstrated the experimental feasibility of detecting the specific hydrolysis of Ins(1,3,4,5)P$_4$ by InsP$_4$ 5-phosphatase in pig cerebellar membranes. Ins(1,3,4)P$_3$ formation was demonstrated by HPLC-MDD analysis in relation to a fully resolved set of reference standards: Ins(1,2,6)P$_3$, Ins(3,4,5)P$_3$, Ins(1,3,4)P$_3$, Ins(1,4,5)P$_3$ and Ins(1,3,5)P$_3$, which were eluted in that order using a gradient composed of 50 mM Tris–HCl (pH 8.5; solvent A) and 50 mM Tris–HCl, 0.4 M KCl (pH 8.5; solvent B). The gradient was established as follows:

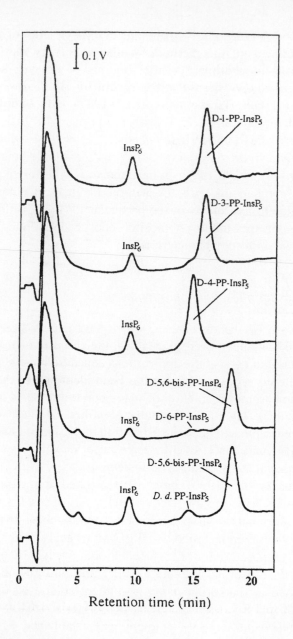

Retention time (min)

0 min, 30% B; 2 min, 40% B; 16 min, 42% B; 20 min, 50% B; 38 min, 60% B; 48 min, 75% B; 50 min, 100% B; 60 in, 100% B (flow rate, 1.5 ml/min). The metal–dye reagent consisted of 2 mM $NH_4OAc/AcOH$ (pH 5.0), 200 μM 4-(2-pyridylazo)resorcinol, 30 μM YCl_3 and 10% (v/v) MeOH (flow rate, 0.75 ml/min).

4.7. Quantification

The InsP content of cells and tissues has been measured by both direct and indirect methods. The InsPs, which make up the metabolic cycle generating and removing the second messengers, have mainly been investigated using radioactive isotopes followed by chromatographic separation. More recently several affinity procedures have been developed for the mass analysis of InsPs. Most methods involve precipitation of the protein and removal of lipids from tissue samples, chromatographic separation of InsPs using TLC, cartridge-column or HPLC, and the detection of individual InsPs by a suitable detector or chemical reaction.

When using radioactive tracers (i.e. [^{3}H]Ins or [^{32}P]P$_i$) to estimate changes in the amounts of InsPs, tissues should be labeled to isotope equilibrium so that changes in the radioactivity of an individual InsP reflect changes in its mass, rather than in its specific radioactivity. [^{32}P]P$_i$ labeling results in labeled nucleotides and other phosphorylated compounds which interfere with an analysis of [^{32}P]P$_i$-containing InsPs (Dangelmaier et al., 1986). Hence the only suitable method for determining the amounts of InsPs present

Fig. 4.16. Substrate specificity of PP-InsP$_5$ kinase from *Dictyostelium*. All samples were incubated for 10 min at 25°C with pre-purified PP-InsP$_5$ kinase. Only D-6-PP-InsP$_5$ and *D. discoideum* (*D. d.*) PP-InsP$_5$ are almost completely converted into 5,6-bis-PP-InsP$_4$. It should be noted that D-1-PP-InsP$_5$ and D-3-PP-InsP$_5$ have remarkably different retention times in comparison with the natural PP-InsP$_5$. Contaminating InsP$_6$ is due to hydrolysis of the particular PP-InsP$_5$ isomers. Reproduced, with permission, from Laussmann, T., Reddy, K. M., Reddy, K. K., Falck, J. R. and Vogel, G. (1997) Biochemical Journal 322, 31–33. © The Biochemical Society and the Medical Research Society.

during receptor stimulation is to measure their masses (Prestwich and Bolton, 1991).

Several reviews describing direct methods for mass measurements of InsP have been published (Dean and Beaven, 1989; Palmer and Wakelam, 1989; Shears, 1989). Some of these methods have been successfully applied to eluates from AG1X8 anion exchange resin (Rittenhouse and Sasson, 1985; Shayman and Kirkwood, 1987; Shayman et al., 1987; Heathers et al., 1989), although this does not allow separation of the individual isomers of InsP$_3$. Stricker et al. (1999) have reported a quantitative analysis of the InsPs in rat brain tissue using HPLC-MDD.

Labeling tissues with [^{3}H]Ins requires long incubation periods which can result in decreased agonist responses, e.g. platelets (Rittenhouse and Sasson, 1985).

4.7.1. Radio-isotope assays

The metabolic changes in PtdIns and InsPs are commonly studied by radiolabeling with [^{3}H]Ins (Prestwich and Bolton, 1991) or [γ^{32}P]ATP and [^{32}P]phosphate (Stephens and Downes, 1990) measuring the radioactivity of the individual InsPs. Tissues can be labeled with ^{32}P either by in vitro incubation of tissue with [γ-^{32}P]ATP or by intracerebral injection of [γ-^{32}P]ATP (Sun and Lin, 1989; King et al., 1987). Singh and Jiang (1995) have reviewed the studies that have demonstrated that the incorporation of [^{32}P]ATP or [^{3}H]Ins into PtdIns (4,5)P$_2$ takes a couple of hours to a couple of days depending upon the type of the radioisotope and tissue samples (Stubbs et al., 1988; Chandrasekhar et al., 1988; Horstman et al., 1988). The incorporation of ^{32}P into PtdIns(4,5)P$_2$ reaches a maximum level within 2 h and then the labeling decreases (Sun and Lin, 1989), while the incorporation of ^{32}P into other phospholipids continues to increase after 2 h (Sun and Lin, 1989). Furthermore, prolonged labeling of tissues with [^{3}H]Ins may affect agonist response (Putney et al., 1989) and the concentration of [^{3}H]Ins at isotopic equilibrium may differ in different tissues (Putney et al., 1989). Despite the above noted shortcomings with respect to isotopic equilibration, the method has been used for radiolabeling of InsPs.

The separation of labeled InsPs has been most successful using the HPLC method of Irvine et al. (1985) or modifications of this method (Batty et al., 1985). The most widely used eluate for analysis of InsPs has been ammonium formate buffered with phosphoric acid (Irvine et al., 1985) and ammonium phosphate buffers (Dean and Moyer, 1987). The measurement of the mass of InPs in eluates of this type has been very difficult, however, especially for the more highly phosphorylated InsPs, due to the presence of high concentrations of phosphate ions. This has led to the development of laborious desalting techniques (Dean and Moyer, 1987; Stephens et al., 1988a), which are not suitable for multiple sample analysis. Portilla and Morrison (1986), using this HPLC method, circumvented the problem by using acid hydrolysis of the InsPs, as dephosphorylation using alkaline phosphatase is inhibited by phosphate. After chromatographic separation, radioactivity is measured by taking small fractions, adding scintillator liquid, and subsequent analysis with liquid scintillation spectrometer.

More recently, the desalting process has been coupled directly to the chromatographic separation in on-line experiments (Foster et al., 1994; Katchintorn et al., 1993). Zhang and Buxton (1998) have provided details for the radio-HPLC method of separation with flow detection developed earlier (Zhang and Buxton, 1991; Zhang et al., 1995). The InsPs and deacylated PtdIns are separated by HPLC on a Whatman Partisil SAX 5 column at a flow rate of 0.5–1.0 ml/min. The molecules of interest are identified by the recorded elution times of radiolabeled standards. Quantification of radioactivity of each InsP is achieved using an on-line liquid scintillation detector to determine HPLC peak height/area as recorded in counts per minute (CPM). Figure 4.17 shows the separation on a Partisil SAX 5 column (Whatman Partisil SAX 5 analytical column, 10 cm, of InsP standard mix (2 μCi): [^{3}H]Ins(1)

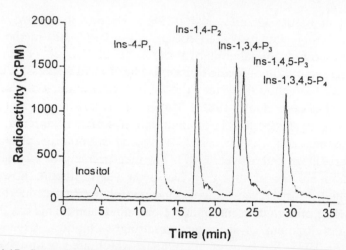

Fig. 4.17. Separation of standard InsPs on a Partisil SAX 5 column using an on-line scintillation flow detector (INUS beta-RAM). Using the following gradient program: $0-2$ min, $100\,H_2O$; $2-25$ min, $100\%\,0.05\,M\,NH_4H_2PO_4$), $25-30$ min, $2.0\,M\,NH_4H_2PO_4$, $30-35$ min, $100\%\,H_2O$. Peaks are identified as shown in the figure. Reproduced from Zhang and Buxton (1998) with the permission of the publisher.

$P < [^3H]Ins(1,4)P_2 < [^3H]Ins(1,4,5)P_3 < [^3H]Ins(1,3,4)P_3 < [^3H]$ $Ins(1,3,4,5)P_4$ (Zhang and Buxton, 1998). The mixture is separated using the following gradient. The column is equilibrated with H_2O at the chosen flow rate (e.g. 0.6 ml/min). The sample is injected in H_2O, and elution is continued for 2 min. This is followed by elution with $0.05\,M\,NH_4H_2PO_4$ for 23 min, and with $2.0\,M\,NH_4H_2PO_4$ for 5 min for a total run of 30 min. The column is returned to starting state by washing with $100\%\,H_2O$ for 5 min. The standard InsPs are well resolved with the retention times ranging from 3.88 min (*myo*-inositol), 12.06 min [Ins(4)P], 17.87 min [Ins(1,4)P_2], 22.43 min [Ins(1,3,4)P_3], 23.40 min [Ins(1,4,5)P_3] and 29.12 min [Ins(1,3,4,5)P_4]. The method can also distinguish Ins(1)P and Ins(4)P (Zhang et al., 1995). The method has been successfully applied to HPLC separation of [^3H]InsPs from vascular tissue extracts (Zhang et al., 1995).

4.7.2. Radio-receptor assays

To circumvent the disadvantages of the radio-labeling method a radioreceptor assay has been proposed for the mass analysis of $Ins(1,4,5)P_3$ and $Ins(1,3,4,5)P_4$. The radioreceptor mass assays are based on the assumption that only the test InsP interacts with the InsP receptor, while all the other InsPs isomers in cell/tissue extracts are much poorer displacers of the radiolabeled test InsP. However, natural InsPs may be present (Vallejo et al., 1987). Several binding assays have been developed specifically for the measurement of the mass of $Ins(1,4,5)P_3$ (Bradford and Rubin, 1986; Challiss et al., 1988; Bredt et al., 1989; Palmer et al., 1989; Nahorski and Potter, 1989; Nunn et al., 1990).

Palmer et al. (1989) and Palmer and Wakelam (1990) reported the first mass assay for $Ins(1,4,5)P_3$ using a crude microsomal fraction of bovine adrenal cortex. The sensitivity of the assay is such that 0.2 pmol of $Ins(1,4,5)P_3$ can be detected. In principle, the $Ins(1,4,5)P_3$ present in a cell extract competes with a fixed quantity of high specific activity [³H]$Ins(1,4,5)P_3$ (although [³²P]$Ins(1,4,5)P_3$ could be used instead) for the $Ins(1,4,5)P_3$-specific binding sites of bovine adrenocortical microsomes. A standard curve using known amounts of unlabeled $Ins(1,4,5)P_3$ is conducted in parallel. Thus, the quantity of $Ins(1,4,5)P_3$ in the cell extract can be calculated.

It provides more accurate information than the radiolabeling method with regard to the mobilization of InsPs (Challiss et al., 1988). This radioreceptor method, however, specifically measures $Ins(1,4,5)P_3$ or $Ins(1,3,4,5)P_4$; therefore, the metabolism of other phospholipids is not studied. Despite the above-mentioned shortcomings, radiolabeling with [³H]Ins has been extensively used for studying receptor-stimulated InsP mobilization. For this purpose the cell culture, homogenate or subcellular fraction, is incubated for 2–4 h at 37°C with [³H]Ins in Ringer's buffer (0.75 μM, 15 Ci/ mmol). After incubation, the samples are washed four times with Ringer's buffer. To study the receptor stimulation of $PtdIns(4,5)P_2$

hydrolysis, the loaded samples were incubated with the agonist and radioactivity in different InsPs is measured at different time intervals.

Subsequent to the original publication of this assay (Challiss et al., 1988), Bredt et al. (1989) reported a similar assay based on the use of a crude microsomal fraction from rat cerebellum. To overcome the higher apparent KD reported for the Ins(1,4,5)P$_3$ receptor in this tissue, a high assay pH was used. According to Challis et al. (1993) alkalinization of the assay buffer decreases the KD value and recommend caution in the use of this procedure (see below).

The Ins(1,4,5)P$_3$ radioreceptor assay developed by Challiss et al. (1988) and Bredt et al. (1989) provide simple, sensitive, and specific methods for the rapid determination of Ins(1,4,5)P$_3$ mass. The major advantage of these methods is that they can quantify endogenous Ins(1,4,5)P$_3$ mass in tissue or cell extracts with little purification. The range and sensitivity of the assays will ultimately depend on the Ins(1,4,5)P$_3$ receptor used. Commercially marketed Ins(1,4,5)P$_3$ radioreceptor assay kits utilize either bovine adrenals (Amersham), or calf cerebellum (DuPont) as a source of the Ins(1,4,5)P$_3$ receptor. Challiss and Nahorski (1993) have described an improved routine for mass determination of Ins(1,4,5)P$_3$ using [^{3}H]Ins(1,4,5)P$_3$. D-Ins(1,4,5)P$_3$ is dissolved in aqueous solution at 1 mM at $-20°$C. Standard curves are constructed by dilution of 40 μM stocks to give final concentrations of 1, 3, 10, 30, and 100 nM. The final volume is 120 μl, and therefore these concentrations correspond to 0.12–12 pmol Ins(1,4,5)P$_3$ per assay. To 30 μl of standard Ins(1,4,5)P$_3$ or unknown is added 30 μl of 100 mM Tris–HCl, 4 mM EDTA, pH 8.0, and 30 μl [^{3}H]Ins(1,4,5)P$_3$ (appropriately diluted to give 6000–8000 dpm/ assay). The assay tubes are maintained at 0–4°C at all times by performing the procedure in an ice-bath. The assay is initiated by addition of 30 μl of the bovine adrenal cortical preparation. Samples are incubated for 30 min on ice with intermittent vortex mixing. The separation of bound and free radioligand was best achieved by vacuum filtration. The wash buffer (25 mM Tris–HCl, 1 mM EDTA, 5 mM NaHCO$_3$, pH 8.0) used for filtration should be ice-cold. Millipore (Bedford, MA) vacuum manifolds are loaded with GF/B filters and wetted with wash buffer. Assay samples are diluted with 3 ml wash buffer and immediately filtered, and the sample is then rapidly washed two times with 3 ml wash buffer (within 5–10 s). The GF/B filter disks are then transferred to vials, and 4 ml of a suitable scintillant is added. Samples are allowed to extract for at least 6 h prior to scintillation counting. A standard displacement curve is prepared by plotting [^{3}H]Ins(1,4,5)P$_3$ (dpm) bound versus [Ins(1,4,5)P$_3$] (pmol/assay). In this instance, in the absence of unlabeled Ins(1,4,5)P$_3$ about 50% binding was obtained, with non-specific binding being less than 4%. Addition of 0.12 pmol unlabeled Ins(1,4,5)P$_3$ resulted in 17.5% displacement of specific binding, whereas 50% displacement of binding occurred at 0.87 pmol/assay.

Willcocks et al. (1988) have shown that myo-Ins (1,4,5)tris-phosphorothioate also binds to specific [^{3}H] Ins(1,4,5)P$_3$ sites in rat cerebellum but is resistant to 5-phosphatase.

More recently, Zhang (1998) has compared two commercial kits for Ins(1,4,5)P$_3$ assay: bovine adrenals (Amersham) and calf cerebellum (DuPont NE). It was found that the kit from DuPont is more economic and easier to perform. Briefly, isolated rat cerebella is minced and suspended in 10 ml of ice-cold buffer A (20 mM Tris–HCl, 20 mM NaCl, 100 mM KCl, 1 mM EDTA, 5 mM DTT, 1 mg/ml BSA, 0.02% sodium azide, pH 7.7). The tissue is homogenized with a Polytron homogenizer at setting 3.5 in two bursts of 15 s each, while the sample is being kept cold at all time. The homogenate is centrifuged at 100 000g for 30 min at 4°C and the supernatant is discarded. The pellet is then rehomogenized in half the original volume of buffer A at setting 3.5 for 15 s, the homogenate centrifuged at 100 000g for 30 min at 4°C, and the supernatant discarded.

The final pellet is resuspended in buffer B (50 mM Tris–HCl, 5 mM EDTA, 5 mM EGTA, 0.02% sodium azide (w/w), pH 8.5) at a protein concentration of 10 mg/ml and saved in 1 ml aliquots at $-70°C$ for subsequent use. For the DuPont kit, the receptor preparation/tracer in a vial was suspended with addition of 2.5 ml of distilled water. Before the assay, the concentrated receptor preparation/tracer was diluted 1:15 (v/v) with assay buffer. The assay tubes were prepared as follows: tubes 1–2 (total counts), 400 μl, receptor/tracer; tubes 3–4 (non-specific binding), 100 μl of blanking solution plus 400 μl receptor/tracer; tubes 5–6 ('0' Standards), 100 μl distilled water and 400 μl of Receptor/tracer; tubes 7–20 (standards) 100 μl standard plus 400 μl receptor/tracer; tubes 21... (samples), 100 μl sample plus 400 receptor/tracer. The tubes were vortex mixed for 3–4 s except the total count tubes, which are directly transferred into counting vials. The tubes were incubated at 2–8°C for 1 h, centrifuged at 4°C for 10 min at 2400g, and the supernatant decanted by inverting the tubes and shaking sharply downward. The tubes were allowed to remain upside down on absorbent paper for 10 s. The pellets were solubilized at room temperature for 10 min. Then the tubes were placed in the counting vials and 5 ml of scintillation cocktail added. The vials were shaken vigorously for 5 s and placed in liquid scintillation counter and read after waiting for 5 min. The amount of Ins(1,4,5)P$_3$ in the sample was determined from the standard curve.

The radioreceptor assay of Ins(1,3,4,5)P$_4$ is performed similarly (Challiss and Nahorski, 1993). Stocks of D-Ins(1,3,4,5)P$_4$ can be stored at millimolar concentrations at $-20°C$. Standard curves are constructed to give 0.3–1000 nM final concentrations (i.e. 0.036–12 Ins(1,3,4,5)P$_4$ per assay for a 120 μl final assay volume). The assay is performed by adding to 30 μl of standard Ins(1,3,4,5)P$_4$ or unknown, 30 μl of 50 mM sodium acetate, 50 mM KH$_2$PO$_4$, 2 mM EDTA, 0.25% bovine serum albumin, pH 5.0 (buffer B), and 30 μl radiolabeled Ins(1,3,4,5)P$_4$ in buffer B. The assay is initiated by the addition of 30 μl of the cerebellar preparation and samples are incubated for 30 min on ice with intermittent vortex mixing. Separation of bound and free radioligand is best achieved by rapid vacuum filtration over GF/B filter disks. It is crucial that the wash buffer (25 mM sodium acetate, 25 mM KH_2PO_4, 5 mM $NaHCO_3$, 1 mM EDTA, pH 5.0) be ice-cold. The sample is diluted with 3 ml wash buffer and immediately filtered, and the sample tube is then rapidly washed two times with 3 ml wash buffer (with 5–10 s). Filter disks are transferred to vials and 4 ml of a suitable scintillant added. Samples are allowed to extract for 6 h prior to scintillation counting. Examples of Ins(1,3,4,5)P_4 displacement curves are given in Fig. 4.18 (Challiss and Nahorski, 1993). The biphasic displacement of [^{32}P]Ins(1,3,4,5)P_4 from preparations of rat and pig cerebellar membranes does not compromise the mass assay, as displacement between 0.036 and 12 pmol Ins(1,3,4,5)P_4 per assay can be adequately modeled using simple curve-fitting programs. The assay can reproducibly detect 0.1 pmol Ins(1,3,4,5)P_4 in a 30 μl sample.

4.7.3. Mass assays of inositol phosphates

Although the radiotracer studies are extremely sensitive, only qualitative information can be obtained concerning InsP$_3$ levels in signaling. In situations where the radiolabel does not have access to in vivo tissues, or if the tissue or cells are not viable for the duration required for radiolabeling, even qualitative assessments of InsP$_3$ signaling are not possible. Therefore, methods for analysis of InsP$_3$ mass have evolved and include gas chromatography (Rittenhouse and Sasson, 1985; Heathers et al., 1988; DaTorre et al., 1990; Leavitt and Sherman, 1982a,b; Rubin et al., 1993), radioimmuno-assay, HPLC postcolumn detection, and ion chromatography/HPLC (Palmer and Wakelam, 1989; Irvine, 1990; Dean and Beaven, 1989). The sensitivity of most of these methods is such, however, that picomole amounts of InsP$_3$ are required, necessitating relatively large quantities of tissue. Many of these methods, however, are not suitable for multi-sample analysis and are not available for routine application in the majority of laboratories.

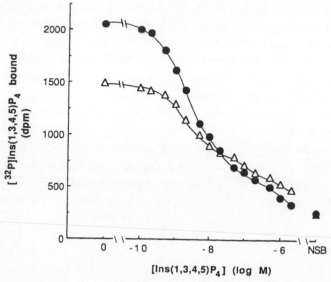

Fig. 4.18. Displacement of [^{32}P]Ins(1,3,4,5)P$_4$ binding from cerebellar membranes by Ins(1,3,4,5)P$_4$. Assays were performed as follows: to 30 μl of standard Ins(1,3,4,5)P$_4$ is added 30 μl of 50 mM sodium acetate, 50 mM H$_2$PO$_4$, 2 mM EDTA, 0.25% bovine serum albumin pH 5.0 (buffer B), and 30 μl radiolabeled Ins(1,3,4,5)P$_4$ in buffer B (10 000 dpm/assay). The assay is initiated by the addition of 30 μl of the cerebellar preparation and homogenization. Samples are incubated for 30 min on ice with intermittent vortex mixing. Separation of bound and free radioligand is best achieved by rapid vacuum filtration over GF/B filter disks. The wash buffer (25 mM sodium acetate, 25 mM KH$_2$PO$_4$, 5 mM NaHCO$_3$, 1 mM EDTA, pH 5.0) should be ice-cold. Filter disks are transferred to vials and 4 ml of suitable scintillant added and the counting performed 6 h later. Nonspecific binding was defined by inclusion of 3 mM 2,3-bisphosphoglycerate or 1 mM InsP$_6$ or 100 μg/ml heparin, all of which gave comparable estimates. Reproduced from Challiss and Nahorski (1993) with the permission of the publisher.

A few methods have been developed which can be applied to eluates from HPLC analysis (Meek, 1986; Portilla and Morrison, 1986; Mayr, 1988; Heathers et al., 1989). A metal–dye reaction has been successfully used for the quantification of pmol amounts of InsPs by HPLC coupled with a post-column reaction chamber and a UV detection (Mayr, 1988; Mayr, 1990; Stephens et al., 1991). The method is based on the observation that yttrium, which is a tervalent transition metal ion, binds with high affinity both to cation-specific dyes such as PAR and to polyanions such as InsPs. The InsPs are separated on anion exchange HPLC column using an HCl gradient. The eluent contains yttrium ions, which bind with high affinity to both polyanions such as InsPs and cation-specific dye 4-(2-pyridylazo) resourcinol (PAR). The PAR is mixed with the eluant for the post-column reaction. The changes in the absorbance of the PAR-yttrium complex are monitored at 546 nm. The HPLC technique separates $InsP_2$ and higher poly-InsP isomers from each other and from interfering phosphorylated substances and the InsPs are quantified by the metal–dye reaction. InsP and $InsP_2$ are detected by this method because GTP, glucose, 1,6-bis-phosphate or sedoheptulose 1,7-bisphosphate interfere with the chromatographic separation of $InsP_2$ and phosphoenolpyruvate; phosphoglycerate and 6-phosphogluconate may interfere with the chromatography of InsP (Mayr, 1988). This method, however, requires a complex and expensive HPLC system and extensive knowledge of HPLC separation techniques. Pure HPLC fractions of InsPs can be assayed for phosphorus (Rouser et al. 1969), but myo-$InsP_6$ may interfere with inorganic P detection (Irving and Cosgrove, 1970).

Kerovuo et al. (2000) have used a modified version of the post-column metal–dye detection HPLC system of Mayr (1990). Acidic elution was performed on a chemically inert HPLC system (10vp-series; Shimadzu, Kyoto, Japan) equipped with Resource Q (1 ml) Mono Q HR 5/20 column configuration (Pharmacia, Uppsala, Sweden). Samples were applied automatically and an LC-10Ai pump delivered solvent A (0.2 mM HCl/15μM YCl_3) and solvent B (0.5 M HCl/22.5 μM YCl_3) at 1 ml/in. The InsPs were separated on a concave gradient: 0 min, 10% B; 10 min, 14.5% B; 23 min, 28% B; 31 min, 44% B; 35 in, 60% B; 40 min, 90% B; 43 min, 100% B; 62 min, 100% B. Detection of InsPs was performed by mixing 2-(4-pyrididylazo) resourcinol reagent [300 μM 2-(4-

pyrdidylazo)resorcinol/1.6 M triethanolamine, made to pH 9.0 with HCl] a 0.55 ml/min into the eluent. Compound detection at 546 nm was performed with a photodiode-array detector.

Prestwich and Bolton (1991) performed their HPLC analysis as previously described by Irvine et al. (1985) and Batty et al. (1985) as modified by Salmon and Bolton (1988). Eluate from the HPLC was collected in 120 0.5 ml fractions using an Amersham 232 fraction collector. The fractions were then divided as follows: 150 μl of each fraction was removed and 3 ml of Hisafe 3 scintillant (LKB) was added. Radioactivity was determined using a Beckman S1701 scintillation counter and, after scaling according to their protein content, the elution profiles of the InsPs containing [³H]Ins were plotted. For mass determination, 300 μl of each fraction was removed and 3 ml of water was added. These samples were either processed immediately or stored at − 20°C ready for desalting. The method described by Prestwich and Bolton (1991) is suitable for multiple sample analysis from HPLC eluates which contain phosphate ions or potentially any other buffer ion contaminant. It comprises several essential steps after HPLC analysis: simple desalting, dephosphorylation with alkaline phosphatase, and oxidation, reduction and measurement using a D-myo-inositol dehydrogenase (IDH); NAD-linked bioluminescence reaction for Ins. This method fulfils all the requirements for anion-radiometric microanalysis and is potentially applicable and specific for any InsP isomer as well as is suitable for multi-sample analysis.

Direct quantification of inositol polyphosphates by GC/MS (Sherman et al., 1986), fast atom bombardment mass spectrometry (Sherman et al., 1986), or thermospray-liquid chromatography/mass spectrometry (Hsu et al., 1990), although promising, has failed to improve on sensitivity or has suffered the limitation of a lack of specificity (Goldman et al., 1990).

Hirvonen et al. (1988) have used GLC with FID for quantification of Ins(1)P in tissue samples. Although the method is simple and sensitive, it does not detect other InsPs and, thus, its application is limited. The above-described procedures have been applied in the analysis of InsP$_3$ levels from two preparations that do not readily lend themselves to mass analysis by other procedures (Heathers et al., 1988). The study reports the basal levels of InsP$_3$ in substrate attached ventricular myocytes and isolated porcine coronary smooth muscle as 235 ± 55 for five preparations (with six determinations/preparation) and 144 ± 37 for four preparations (with eight determinations/preparation) pmol/mg protein, respectively. These InsP$_3$ levels are within the range reported for other tissues analyzed with a variety of quantitative methods (Palmer and Wakelam, 1989).

Heathers et al. (1989) have reported the quantification of mass of Ins in InsPs from biological samples by GLC. The individual InsP fractions were isolated by anion exchange chromatography with a sodium sulfate gradient, followed by purification by HPLC procedures. The individual InsP fractions were subsequently dephosphorylated and desalted. The *myo*-Ins from each fraction was then derivatized to the hexatrimethylsilyl derivative and the *myo*-Ins derivatives were quantified by GLC using the hexatrimethylsilyl derivative of *chiro*-Ins as an internal concentration standard.

The most sensitive (0.2 pmol/injection) GC/MS method (DaTorre et al., 1990) for quantification of InsP$_3$ evolved from procedures designed to measure the dephosphorylated cyclitol inositol (Rittenhouse and Sasson, 1985; Sherman et al., 1977). DaTorre et al. (1990) has shown that the major EI ions produced by the neutral *myo*-Ins are at m/z 305 and 318 and the major ions produced by d$_6$-*myo*-Ins are at m/z 307 and 321. Quantification is performed by determining the peak area ratio for *myo*-Ins and d$_6$-*myo*-Ins. For more sensitivity the GC/MS is programmed to monitor only selected ions, e.g. at m/z 305, 307, 318 and 321, and determining the area under the curve for each peak.

By necessity, therefore, this method requires prior separation of the InsPs species, dephosphorylation of the compounds to free inositol, and then quantification of a derivatized product by GC/MS. Rubin et al. (1993) have described a modification of the basic procedure to obtain femtomole quantification of InsP$_3$ from submilligram quantities of tissue. This procedure utilizes a novel dephosphorylation method, a high molecular weight fluoroalkyl derivative of inositol, and high mass GC/MS analysis. Briefly, 0.3–2 mg of freshly dispersed porcine coronary artery smooth muscle cells are extracted with 1–1.5 ml of ice-cold TCA, the protein is pelleted by centrifugation (4°C, 16 000g × 10 min). All glassware is silanized by coating with 7% dichlorodimethylsilane in toluene. Supernatants from TCA extracts are transferred to 50 ml conical polypropylene tubes and the TCA extracted with water-saturated diethyl ether (4 × 5 ml); residual ether is removed by vaporization at 60°C for 30 min. The InsPs are separated by anion exchange chromatography on AG 1 X8 columns as described by Downes and Michell (1981). This method, however, does not give baseline separation of InsP$_2$, InsP$_3$ and InsP$_4$ and cross-contamination is possible. Separation of InsP$_3$ isomers is also not possible. Nevertheless, anion exchange was used to obtain partially purified material for the development of the high sensitivity method. The InsPs were eluted according to the procedure of Downes and Michell (1981). A sample of [^{3}H]InsP, [^{3}H]InsP$_2$ and [^{3}H]InsP$_3$ was applied to a 1 ml AG 1 X8 column, eluted with increasing concentrations of ammonium formate buffered to pH 5.0 with ammonium hydroxide, and collected as 1 ml fractions. One molar ammonium formate was used to elute InsP$_3$ as there was no InsP$_4$ in the sample. The recovery from the anion exchange column was estimated to be 80–90% of the applied radioactivity. The ammonium formate was removed from the samples by repeated lyophilization (2–3 days). Lyophilization, however, can generate artifacts in chromatographic profiles of InsPs (Woodcock et al., 1993). The samples are resolubilized in 560 µl of distilled water. Ten microliters of d$_6$-myo-inositol (1 pmol/µl) is added to each tube to serve as an internal standard for GC/MS quantification. The tops of the tubes were then heat-sealed and the samples heated for 3 min to 200°C in a sand bath. After cooling, the tubes are opened and the samples evaporated to dryness. The inositols in the samples are esterified with heptafluorobutyric anhydride (HFBA) to produce perheptafluorobutyrylinositol (HFB$_6$-inositol), which is detected with high sensitivity by negative ion chemical ionization (NICI) mass spectrometry. The derivative is formed by adding 20 μl each of anhydrous acetonitrile and HFBA to the tubes containing the dried samples. The tubes are then heat sealed and the derivatization is allowed to proceed at room temperature at least overnight. Tubes should remain sealed until immediately before GC/MS analysis. GLC is performed on a 15 m fused silica capillary column (DB-210, 0.5 μm film thickness, 0.32 mm ID, J&W Scientific, Folsom, CA) in split mode (10:1 to 20:1) using a silanized glass injection port liner and Teflon-faced septa. Chromatographic conditions are as follows: injection port 220°C, He carrier gas, 5 psi, column temp. program 120–220°C at 10°C/min. Quantitative analysis is carried out using selected ion monitoring of m/z 1336, for the endogenous inositol, and m/z 1341 for the internal standard HFB6-d$_6$-myo-inositol. The ions are quantified by comparing the integrated areas originating from the internal standard and the unknown. The ion originating from the internal standard (HFB6-d$_6$-myo-inositol), which corresponds to m/z 1336, is m/z 1341, which arises from the loss of 2HF from the internal standard. A standard curve is prepared, where the intensity ratio of m/z 1336 to that of m/z 1341 is plotted against the total amount of inositol per sample. The identity of the measured inositol is determined by retention time and by ratios of ions in the spectrum of the inositol.

An enzymatic analysis is based on the oxidation of myo-Ins to $scyllo$-inosose and the reduction of NAD- to NADH by myo-Ins dehydrogenase (MIDH). Since the accumulation of NADH inhibits MIDH activity, the reaction is coupled with a scavenge reaction, e.g. malate dehydrogenase (MDH) reaction which reduces oxaloacetate to malate (MacGregor and Matschinsky, 1984), the alcohol dehydrogenase reaction (Shayman et al., 1987) the diphorase reaction with resazurin as the substrate (Singh, 1992; Maslanski and Busa, 1990) or a bioluminescence reaction (Prestwich and Bolton, 1991). The quantification of InsPs by the enzymatic method is a multistep process involving the extraction of tissues with TCA or

PCA, removal of endogenous myo-Ins by passing the extract through a Dowex-50W column, separation of InsP, $InsP_2$, $InsP_3$, and $InsP_4$ by anion exchange chromatography or HPLC, dephosphorylation of each InsP to myo-Ins by alkaline phosphatase and then measuring the concentration of MI in each fraction.

The enzymatic assay has been recently adapted for Sep-Pak eluates where myo-Ins is measured by using diaphorase and resazurin (Singh, 1992) and for HPLC eluates where myo-Ins is measured with a bioluminescence assay (Prestwich and Bolton, 1991). These methods are sensitive and require only small sample size. However, there are some limitations of the enzymatic methods, e.g. the rate of dephosphorylation of $InsP_4$ is slower than that of InsP, $InsP_2$ and $InsP_3$; the $InsP_5$ and $InsP_6$ are very poorly dephosphorylated and the presence of salt interferes with dephosphorylation of InsPs (Singh and Jiang, 1995).

An enzyme-linked determination of InsPs using fluorimetric measurements (MacGregor and Matschinsky, 1986) was applied to eluates from AG1X8 anion exchange columns by Shayman et al. (1987). This technique has also been used by Stephens and Downes (1990) on eluates from HPLC after extensive desalting. Tarver et al. (1987) using an isotopic dilution assay with $Ins(1,4,5)P_3$ 3-kinase, measured the amount of $Ins(1,4,5)P_3$ and deduced the amount of $Ins(1,3,4)P_3$ present after applying the technique of Rittenhouse and Sasson (1985), which measured total $InsP_3$s.

Maslanski and Busa (1990) have described a mass assay for the determination of myo-inositol and inositol phosphates in physiological samples. The assay is based on two coupled enzyme reactions linked by the reduction and reoxidation of NAD. myo-Inositol dehydrogenase (IDH, EC 1.1.1.18), in the presence of NAD, is used to oxidize free myo-inositol to $scyllo$-inosose and NADH (MacGregor and Matschinsky, 1984). The NADH thus formed is then stoichiometrically reoxidized by the enzyme diaphorase. The electron acceptor in this second reaction is the non-fluorescent dye resazurin. Upon reduction, resazurin is converted to the intensely fluorescent compound resorufin, which is then quantified using a fluorometer. The assay is highly sensitive (with a detection limit of about 10 pmol) and is specific for *myo*-inositol.

The *myo*-Ins content of the InsPs resolved on the anion exchange column (see above) is assayed as follows. Briefly, the dried samples are added 45 μl of freshly made alkaline phosphatase solution (e.g. 0.1 M Tris–HCl (pH 9.0), 0.1 mM $ZnCl_2$ at a concentration of 200 units/ml). The samples are incubated for 2 h at 37°C. The alkaline phosphatase is then inactivated by placing the samples in a water-bath at 100°C for 4 min. The tubes are cooled to at least 27°C. Hexose removal is effected by adding 5 μl of hexokinase solution each sample and incubation for 1 h at 37°C. The samples are then placed in water-bath at 100°C for 3 min. Allow tubes to cool to room temperature. For inositol oxidation add 5 μl each of the NAD (0.1 M in deionized water) and IDH (5 units/ml in 10 mM phosphate, 0.02% BSA (pH 6.8) solutions). Incubate at room temperature for 15 min. Decrease pH to approximately 6.5 by adding 5 μl of 0.8 M HCl. The resazurin reduction is effected by adding 10 μl of the resazurin solution (TLC purified reagent at 20 μM concentration) and 5 μl of the diaphorase solution to each sample. Incubate at room temperature in a darkened area for 15 min. For the resorufin measurement, add 920 μl 0.1 M Tris–HCl (pH 9.0) to each sample and transfer to fluorometer cuvette. The fluorescence is measured at 56 nm (excitation) and 585 nm (emission). Besides samples, tissue blanks and internal standards ('spikes') also have to be assayed (Maslanski and Busa (1990)). The calculation of the picomoles in a sample is made by the following equation: pmol in sample = (sample RFU − blank RFU) × (pmol added in spike)/(spike RFU − sample RFU), where RFU stands for relative fluorescence units and blank is equal to the fluorescence of the tissue blank (tissue sample that does not contain IDH) plus the fluorescence of the IDH blank (the tube that contains IDH but no tissue) minus the fluorescence of resazurin (the tube that contains neither tissue not IDH).

While the conditions presented above are sufficient to fully dephosphorylate 97% of Ins(1,4,5)P$_3$, Ins(1,3,4,5)P$_4$ is quite resistant to dephosphorylation by alkaline phosphatase. Shayman et al.

(1987) have observed lower rates of hydrolysis of $InsP_3$ than expected and suggest the inefficiency of alkaline phosphatase in removing vicinal phosphates as the cause. In such instances, acid digestion may be considerate as an alternative. Inclusion of internal standard will allow quantification. Other interferences may result from the presence of *scyllo*- and *epi*-inositol, which are present at 50 and 1500 fold lower level than *myo*-inositol (Sherman et al, 1978).

Smith et al. (1987) performed inositol analyses according to the method of Sherman et al. (1977). Samples were freeze-dried and then hydrolyzed in 6 M HCl at 110°C for 24 h after addition of an internal standard (deuterated *myo*-inositol $C-d_6$ (MSD Isotopes, Montreal, Canada) or *scyllo*-inositol (Calbiochem). The hydrolysates were dried repeatedly from water to remove HCl and reacted with 5% trimethylsilyl chloride, 45% NO-bis(trimethylsilyl)trifluoroacetamide in anhydrous pyridine for at least 24 h at room temperature. Portions were analyzed by GC/MS using electron impact and selected ion monitoring for the characteristic ions [TMSOCH = C(OTMS)-CH = OTMS] + , m/z 305 and [TMSO = CHCH = C(OTMS)CHOTMS] − , m/z 318. Where deuterated *myo*-inositol was used as an internal standard, the ion [TMSOCD = C(OTMS)-CD = OTMS] + , m/z 307 was also monitored. Preparation for the determination of $InsP_1$ isomers was different from that of the other InsPs, because Q-Sepharose does not adequately separate them from P_i; thus not allowing proper desalting with high recovery. Therefore, fractions containing $InsP_1$ isomers were collected, diluted with water and freeze-dried to remove ammonium formate. They were then reconstituted with water and incubated with alkaline phosphatase. After boiling to destroy the alkaline phosphatase and centrifugation, the samples were freeze-dried, reconstituted and assayed in duplicate.

Other InsP isomers were measured by dephosphorylation, oxidation and NADH-coupled bioluminescence assay.

CHAPTER 5

Glycosylphosphatidylinositols

5.1. Introduction

Glycosylphosphatidylinositols (glycosyl PtdInsPs) are a class of eukaryotic glycolipids containing the structural motif Man α1-4GlcNα1-6*myo*-inositol-1-PO$_4$-DG represent an ubiquitous class of glycolipids. This minimal glycosyl PtdIns structure may be enhanced with additional ethanolamine phosphate groups and/or carbohydrate side chains in a species- and tissue-specific manner (Ferguson, 1999). These structures were originally discovered in the form of glycosyl PtdIns anchored cell surface glycoproteins, but more recent work has demonstrated that a significant proportion of the glycosyl PtdIns synthetic output of a cell is unrelated to protein anchoring (Brodbeck, 1998). The discovery and structural characterization of the glycosyl PtdIns anchors has been discussed in comprehensive reviews (Ferguson and Williams, 1988; McConville and Ferguson, 1993; Englund, 1993) while the advances made in the structural characterization of these compounds have been periodically updated (Menon, 1994; Schneider and Ferguson, 1995; McConville and Menon, 2000).

Tiede et al. (1999), McConville and Menon (2000), Kinoshita and Inoue (2000), Nagamune et al. (2000), and Morita et al. (2000a, b,c) have reviewed recent developments in the cell biology and biochemistry of glycosyl PtdIns and have indicated that there is much diversity in the range of structural modifications found in glycosyl PtdIns within and between species and cell types, and in the topological arrangements of their biosynthetic pathway in the endoplasmic reticulum. Consistent with additional functional roles for the glycosyl PtdIns is their wide distribution in the cellular endomembrane system.

5.2. Natural occurrence and isolation

The occurrence of the complex glycosyl PtdIns lipids has been demonstrated by the isolation of PtdIns lipid anchors from cell surface proteins (Ferguson and Williams, 1988; Low, 1989; Thomas et al., 1990; McConville, 1991; McConville and Ferguson, 1993). Complete or partial structures of the glycosyl PtdIns anchors have been obtained from protozoal, yeast, slime mold, fish and mammalian sources. In each case, the C-terminus of the protein is linked via ethanolamine phosphate to a glycan with the conserved backbone sequence $Man\alpha1\text{-}2Man\alpha1\text{-}6Man\alpha1\text{-}4GlcNH_2$, which in turn is linked to the 6-position of the myo-inositol ring of PtdIns (McConville and Ferguson, 1993; McConville, 1996). The tetrasaccharide backbone may be substituted with other sugars in a species-specific and developmental stage specific manner. The lipid moieties of the glycosyl PtdIns anchors can also vary in a species and stage specific manner. These anchors have been found to contain dimyristoylglycerol, lyso-1-O-stearoylglycerol and alkylacylglycerol. Some of these anchors may also contain an additional fatty acid (palmitate) on the inositol ring (Roberts et al., 1988a–c; Field et al., 1991; Treumann et al., 1995). This feature renders the anchor resistant to PtdIns-specific phospholipase C hydrolysis (Roberts et al., 1987). These structures were first discovered covalently linked to eukaryotic cell surface glyco-proteins and recognized to be an important alternative mechanism for anchoring proteins to cell membranes. The glycosyl PtdIns anchored proteins may be found on both apical and basolateral surfaces (McGuire et al., 1999). Over 100 glycosyl PtdIns anchored proteins have been described to date.

The isolation and identification of the glycosyl PtdIns anchors is facilitated by radiolabeling of the molecule, which can be accomplished metabolically, in vivo or in vitro. In parasites metabolic labeling of glycosyl PtdIns may be accomplished with [^{3}H]GlcN (50 μC/ml) and [^{3}H]myristic acid (50 μCi/ml) (Naik et al., 2000).

Alternatively, surface iodinated or surface biotinylated cells may be used for the analysis (Benting et al., 1999). If the protein of interest is an enzyme (e.g. alkaline phosphatase), or if antibodies to the protein are available, then analysis of a labeled sample is not essential as enzymatic activity or immunoreactivity can be followed instead. The PtdIns-anchored macromolecules on the cell surface can be divided in two major groups: cell surface glycoproteins bearing glycosyl PtdIns anchors, and various protein free, glycosylated PtdIns.

5.2.1. Isolation of free glycosyl PtdIns anchors

Free glycosyl PtdIns species with structures resembling the glycosyl PtdIns anchors of proteins have been reported. These compounds were first identified in trypanosomes (Masterson et al., 1989; Menon et al., 1990), but subsequently were found in mammalian cells as more polar mannosylated glycosyl PtdIns species with the same core glycan found in glycosyl PtdIns (Puoti and Conzelmann, 1992, 1993; Hirose et al., 1992; Kamitani et al., 1992; Ueda et al., 1993) and less polar glycosyl PtdIns species devoid of mannose but with GlcNAc or GlcN linked to PtdIns (Stevens and Raetz, 1991; Hirose et al., 1991; Sugiyama et al., 1991; Deeg et al., 1992a,b). Recent results concerning free glycosyl PtdIns have been reviewed by McConville and Menon (2000). Milne et al. (1999) have reported the detection of lipid X, a possible catabolic intermediate of glycosyl PtdIns.

5.2.1.1. Extraction

Free PtdIns glycans may be recovered from metabolically labeled cells by conventional lipid extraction procedures. According to Menon (1994) a differential extraction protocol using chloroform/ methanol (2:1, v/v) and chloroform/methanol/water (10:10:3, by vol.) is particularly useful in fractionating the lipids into two broad categories on the basis of polarity. Individual species in each extract can then be isolated by TLC, anion-exchange chromatography, and/or hydrophobic interaction chromatography.

For the lipid extraction, 1.5 ml of ice-cold chloroform/methanol (Folch et al., 1957) is added to pelleted cells in a glass tube. This should be sufficient to maintain a single phase. Mix by stirring with a long Pasteur pipette; do not vortex or shake. Centrifuge the sample (1500g, 10 min) to pellet the debris, remove the chloroform/ methanol supernatant carefully (using a Pasteur pipette), and repeat the extraction several times. Pool the extracts and leave on ice. Continue the extraction process by stirring the partly delipidated cell debris with 1 ml of ice cold chloroform/methanol/water (10:10:3, by vol.) Centrifuge the sample to pellet the debris and reextract as before and pool the extracts. Remove water-soluble contaminants in both chloroform/methanol and in chloroform/methanol/water extract by separate washing procedures. For the chloroform/ methanol extract, add 0.2 vol. of 4 mM $MgCl_2$, vortex to mix, and separate the phases by centrifugation. Remove the upper phase and reextract the lipid-containing chloroform-rich lower phase with mock upper phase prepared by mixing fresh chloroform/methanol with 4 mM $MgCl_2$ in a separate tube. The chloroform/methanol/ water extract should be processed by drying in a Speedvac and resuspending the residue by vortexing in 500 μl each of *n*-butanol and water and centrifuging to separate the phases. Lipids are quantitatively recovered in the upper, butanol-rich phase (Menon, 1994). Lipids in the chloroform/methanol extract or in chloroform/ methanol/water-derived butanol extract can be resolved by TLC on Silica gel 60 (chloroform/methanol/water, 4:4:1 or 10:10:3, by vol.) (Menon, 1994). Care must be used in using methanol extraction under non-standard conditions, as it may lead to formation of phosphate esters (Brown et al., 1988). Other lipid purification techniques may also be used (Doering et al., 1994; Field and Menon, 1992, 1994).

Free glycolipids of *Plasmodium chabaudi* (labeled with [1-^{3}H]glucosamine) were extracted with chloroform/methanol (1:1, v/v) as described by Ferguson (1993). The organic solvent extracts were pooled and subjected to repeated Folch washing, dried and partitioned between water and water-saturated butan-1-ol (Gerold et al., 1997). The washed extracts were analyzed by TLC on Si-60 plates (Merck, Darmstadt, Germany) using either chloroform/methanol/water (4:4:1, by vol.) or chloroform/methanol/conc. acetic acid/water (25:15:4:2, by vol.). The metabolically labeled glycolipids (TLC purified or within the glycolipid mixture) were identified as glycosyl PtdIns by specific enzymic and chemical treatments including PtdIns-PLC, glycosyl PtdIns-specific phospholipase D and HNO$_2$ deamination, as described previously (Gerold et al., 1997, 1996a,b,c).

Naik et al. (2000) have provided a detailed description of the isolation of free glycosyl PtdIns from *Plasmodium falciparum* as follows. The parasites were radiolabeled in medium containing 5 mM glucose and [^{3}H]GlcN (50 μCi/ml) as described by Gowda et al. (1997). Labeling with [^{3}H] fatty acids (50 μCi/ml) was performed in medium containing 2% serum and 20 mM glucose. All procedures of labeling and isolation of glycosyl PdIns are carried out using acid-washed, siliconized glassware, and sterile water and buffers. According to Naik et al. (2000), the radiolabeled parasites (10 ml) were lyophilized and extracted three times with 50 ml of chloroform/methanol (2:1, v/v) to remove non-glycosylated lipids. The free glycosyl PtdIns were isolated by extracting five times with 50 ml of chloroform/methanol/water (10:10:3, by vol.), drying, and then partitioning between water and water-saturated 1-butanol. The organic layer was dried and the residue was extracted with 80% aqueous 1-propanol. Finally, glycosyl PtdIns were purified by HPLC and HPTLC.

Santos de Macedo et al. (2001) used chloroform/methanol/water (10:10:3, by vol.) for glycolipid extraction as described by Schmidt et al. (1998). The extracted glycolipids were dried in a Speedvac concentrator (Savant Inc.), subjected to repeated 'Folch' washings, and finally, partitioned between water and water-saturated with n- butanol. Washed glycolipid extracts were analyzed by TLC on Silica gel 60 plates using chloroform/methanol/water (4:4:1, by vol.) as the developing solvent. After chromatography, the plates were dried and scanned for radioactivity using Berthod LB 2842 automatic TLC scanner or were analyzed by BAS-1000 Bio-Imaging Analyzer (Fuji Film).

5.2.1.2. Purification

TLC may be effectively employed at various stages of the fractionation of the glycosyl PtdIns. Menon (1994) has listed a large number of solvent systems that have been employed most often for the purification of glycosyl PtdIns and their degradation products. Sevlever et al. (1995) purified the free glycosyl PtdIns by suspending the samples in water-saturated n-butanol and applying the sample to Silica gel 60 (Merck). The TLC plate was developed with chloroform/ethanol/water (10:10:3, by vol.) as described by Sevlever and Rosenberry (1993). The distribution of radioactivity on the plates was determined either by a Bioscan System imaging scanner or by fluorography after spraying the plates with En[3]hance (Dupont).

According to Naik et al. (2000) the glycosyl PtdIns (approximately 10 µg plus 400 000 cpm of [3H]GlcN-labeled glycosyl PtdIns) are chromatographed on a C_4 reversed phase Supelcosil LC-304 HPLC column (4.6 × 250 mm^2, 5 µm particle size; Supelco) using a linear gradient of 20–60% aqueous 1-propanol containing 0.1% TFA over a period of 80 min and held for 30 min at a flow rate of 0.5 ml/min, as originally described by Roberts et al. (1988a). Fractions (1.0 ml) were collected, and elution of glycosyl PtdIns was monitored by measuring radioactivity. Aliquots (0.5 µl) were also assayed by ELISA with Kenyan adult sera. Glycolipids extracted from control erythrocyte membrane debris, total lysate, and ghosts were similarly chromatographed, and fractions were analyzed for reactivity with Kenyan serum.

For HPTLC purification, the glycosyl PtdIns (5 μg) were applied onto 10×10 cm^2 plates as continuous streaks. Parallel spots with [^{3}H]GlcN labeled glycosyl PtdIns (50 000 cpm) were used to monitor glycosyl PtdIns bands by fluorography using En^{3}Hance (Naik et al., 2000). The plates were developed with chloroform/methanol/water (10:10:2.5, by vol). Glycosyl PtdIns from the plates were extracted with chloroform/methanol/water (10:10:3, by vol), dried, dissolved in water-saturated 1-butanol, and washed with water. In separate experiments, HPTLC plates were scraped (0.5 cm-width fractions), and glycosyl PtdIns were extracted and analyzed by Elisa using Kenyan adult sera.

5.2.2. Isolation of glycosyl PtdIns-anchored proteins

In many cases glycosyl PtdIns anchored proteins can be selectively released into the culture medium on treatment of intact cells with bacterial PtdIns-specific phospholipase C and this technique has been used to determine whether or not a protein is glycosyl PtdIns anchored. In other instances, where specific antibodies are available, loss of a particular protein from the cell surface can be monitored by fluorescence-activated cell sorter (FACS) analysis (Low and Kincade, 1985). Another approach commonly used to identify PtdIns anchor proteins involves detergent partitioning with Triton X-114 (Bordier, 1981; Pryde, 1998). Glycosyl PtdIns-anchored proteins will partition into the detergent phase. Treatment with PtdIns specific phospholipase C will remove the hydrophobic component of the glycosyl PtdIns anchor and the resulting hydrophilic proteins will partition into the detergent-poor aqueous medium phase if the sample is again subjected to detergent phase separation. The detergent and aqueous phases can be separately analyzed by SDS-polyacrylamide gel electrophoresis (SDS-PAGE) to assess cleavage and to identify the susceptible polypeptides.

The glycosyl PtdIns-anchored proteins are present in detergent-resistant membrane (DRM) fragments or rafts that can be isolated from cell lysates after extraction (Brown and Rose, 1992; Brown and London, 1998). Cerneus et al. (1993) showed that after treatment of mammalian cells with saponin glycosyl PtdIns-anchored proteins could readily be solubilized by Triton X-100. Since saponin removes cholesterol, it appeared that Triton insolubility of glycosyl PtdIns-anchored proteins requires cholesterol. Similarly, Hanada et al. (1995) found that the Triton X-100 resistance of PLAP depended on the presence of cholesterol, as well as sphingolipids. Schroder et al. (1998) have proposed a mechanism in which the detergent-insolubility is due to ordered domains in the membranes, rather than to a specific interaction between proteins and the sphingolipids or cholesterol. The highly ordered state of the membrane domains was attributed to the saturated fatty chains of the glycosyl PtdIns-anchors and the sphingolipids and cholesterol, and was characterized in physico-chemical terms. In those instances, where the glycosyl PtdIns anchors did not possess two saturated fatty chains in the PtdIns moiety, an extra-saturated chain was found located on the inositol residue, thus allowing the theory to produce two saturated fatty chains for insolubilization. There were no stoichiometric relationships established among the identified components or a complete chemical analysis of the complex presented. Lee et al. (1999) have reported a release of PtdIns-anchored Zn^{2+}-GroPCho choline phosphodiesterase from bovine brain membranes by bee venom PLA_2.

5.2.2.1. Metabolic labeling

Metabolic labeling with [^{3}H]ethanolamine is particularly useful in identifying glycosyl PtdIns structures because metabolic conversion of the precursor is limited and few molecules other than PtdIns glycans are labeled. Although radioactive PtdEtn accounts for a large proportion of the radioactivity incorporated into the lipid fraction, it can be separated from the polar lipids by a simple differential extraction protocol and TLC. [^{3}H]labeled fatty acid samples are essential for complete structural analysis, but labelling of fatty acids is not specific for glycosyl PtdIns-anchored proteins and is therefore not the best choice for initial screening of radio-labeled structures. Identification of glycosyl PtdIns and glycosyl PdIns-anchored proteins via metabolic labeling with [^{3}H]mannose is complicated unless performed in the presence of tunicamycin to eliminate protein labeling due to N-linked glycosylation, as well as the dollichol phosphates.

Other experiments are typically performed with cells metabolically labeled with radioactive amino acids (e.g. ^{35}S-methionine). Detailed protocols for labeling these proteins are described elsewhere (Doering et al., 1994; Rosenberry et al., 1989; Varki, 1991) and numerous examples of glycosyl PtdIns labeling may be found in the literature (Conzelmann et al., 1990; Fatemi et al., 1987; Mayor et al., 1990a,b). Typically, about 10^7 cells are incubated with ^{3}H-labeled precursor (specific activity 120–50 Ci/mol; concentration in the labeling medium, 5–100 µCi/ml) for 1–2 h or overnight. The potential glycosyl PdIns-anchor precursor (glycosyl PtdIns-peptide or glycosyl PtdIns-protein) labeled with [^{3}H]myristic acid was solubilized in 10 µl of methanol and treated with HNO$_2$ (Gerold et al., 1997). The hydrophobic fragment released was subjected to butanol/water phase separation (Gerold et al., 1997), and the organic phase was analyzed by TLC using solvent system A (chloroform/methanol/water 4:4:1, by vol.) along with PtdIns standards derived from fatty acid-labeled P2 (glycolipid A) and P3 (glycolipid C) of T. brucei.

The isolation of the glycosyl PtdIns-anchor starts with the recovery from metabolically labeled cells of the glycosyl PtdIns-anchored protein which must be obtained in a salt and detergent free condition, which can be achieved in several ways. The isolation of the glycosyl PtdIns-anchored protein into a dialyzable detergent solution during the last step of chromatographic purification allows the removal of the detergent by dialysis. For example, proteins retained on ion-exchange columns, lectin affinity columns, or antibody affinity columns can be washed and eluted in detergent solutions containing 0.2% (w/v) n-octylglucopyranoside (nOG) (Schneider and Ferguson, 1995). The glycosyl PtdIns anchored protein can be precipitated with an equal volume of cold acetone (12 h at 0°C). When small amounts of protein require precipitation (< 100 μg/l), carrier protein (defatted bovine serum albumin) can be added to a final concentration of 200 μg/l. Alternatively, the protein can be precipitated by making the solution 5% (w/v) with respect to trichloroacetic acid and incubating for 1 h at 0°C (McConville et al., 1993). Biosynthetic labeling with radioactive fatty acids is the most convenient way of generating labeled material for this purpose, but in some instances exogenous labeling may be necessary.

5.2.2.2. Chemical labeling

The glycosyl PtdIns anchored protein may be labeled chemically. Thus, acetyl cholinesterase (AcChoE) extracted from outdated human erythrocytes with Triton X-100 and purified by affinity chromatography on acridinium resin (Roberts et al., 1987), was radiolabeled with 3-(trifluoromethyl)-3-(m-[^{125}I]iodophenyl) diazirine) [^{125}I]TID (Roberts and Rosenberry, 1986) or radiomethylated with [^{14}C]HCHO in Triton X-100 in slight modification of a previous procedure (Haas et al., 1986). Thus, radio-iodinated photactivatable probes such as ([^{125}I]TID) that react preferentially with hydrophobic structures have been employed with success (Roberts and Rosenberry, 1986; Roberts et al., 1988c). Briefly, samples of AcChoE (100–200 nmol) to which a trace amount of [^{125}I]TID-labeled enzyme had been added as a specific marker for the membrane anchor wee concentrated 10–25-fold in a Speedvac concentrator and reductively methylated by incubation with 10 mM [^{14}C]HCHO (ICN, 1–4 mCi/mmol, by dilution with unlabeled HCHO) and 50 mM NaCNBH$_3$ at 37°C for 15 min. After reductive methylation the samples were dialyzed extensively and repurified by affinity chromatography on acridinium resin (Haas et al., 1986).

Bovine erythrocyte is reductively radio-methylated on free amine groups and treated with PtdIns-PLC to remove the alkylacylglycerol component of the glycosyl PtdIns anchor that tends to reduce recoveries of anchor peptide fragments (Haas et al. (1996). Assay by octyl-Sepharose binding showed that PtdIns-PLC cleavage was essentially complete, as 90% of the PtdIns-PLC-treated enzyme remained unbound and only 4% was bound.

5.2.3. Release of bound glycosyl PtdIns-anchors

Although glycosyl PtdIns anchors have been isolated in the free form, they are usually found covalently bound to proteins, from which they may be released by chemical and enzymatic methods. The protein-bound glycosyl PtdIns anchor may be obtained via chemical and enzymatic cleavage analysis of appropriately radio labeled glycosyl PdIns moiety. The glycosyl PtdIns anchors may be released from the protein in the free form and in various partially degraded forms, which are adequate for the characterization of the PtdIns part of the molecule. The proteins containing the tagged PtdIns anchor are subjected to chemical and enzymic cleavage analysis. Lipid fragments may be generated from a glycosyl PdIns anchored protein in a variety of ways.

The generation and analysis of glycosyl PtdIns lipid fragments involves the following general steps. The radio-labeled glycosyl PtdIns sample is resuspended in a detergent-containing aqueous buffer, reagent (chemical or enzyme) is added, and the sample is incubated for a period of time at ambient temperature or at 37°C. The released lipid fragments are extracted by adding organic solvents, vortexing, and separating the aqueous organic phases by centrifugation. An aliquot of each phase is taken for liquid scintillation counting to assess cleavage efficiency. The remainder of the organic phase is taken for TLC analysis. (For Berthold automatic TLC analyzer, 1000 cpm is sufficient, for auto-radiography, more radioactivity is required.) Nitrous acid treatment of glycosyl PtdIns results in deamination of the glucosamine residue and cleavage of the glucosamine-inositol glycosidic bond to release an inositol-containing phospholipid fragment. Other phospholipid fragments may be generated from glycosyl PtdIns by glycosyl PtdIns-specific phospholipase D found in mammalian serum, irrespective of inositol acylation, to give PtdOH. PtdIns and glycosyl PtdIns specific phospholipases C release a diacylglycerol (diradylglycerol) moiety from glycosyl PtdIns, provided the inositol residue is not acylated (palmitoylated). If the inositol residue is acylated, it is to subject the anchor to acetolysis or chemical dephosphorylation in order to generate diradylglycerols for analysis.

5.2.3.1. Proteolysis

Structural studies of glycosyl PtdIns anchors have utilized several proteolytic cleavage procedures to produce anchor fragments for analysis. Papain digestion of human erythrocyte acetylcholinesterase (E^{hu} AcChoE) produced a C-terminal dipeptide His-Gly linked through ethanolamine to the anchor (Roberts and Rosenberry, 1986; Haas et al., 1986). This proteolysis fragment also contained an additional ethanolamine and a glucosamine with unblocked amino groups (Haas et al., 1986) and 2 fatty acid residues (Roberts and Rosenberry, 1985).

Repurified radio-labeled AcChoE (280 nmol) was dialyzed extensively and digested with activated papain resin (Dutta--Choudhury and Rosenberry, 1984). The anchor-containing proteolysis product was isolated by passage through acridinium resin and gel exclusion chromatography on Sepharose CK-6B and Sephadex LH-60 (Roberts and Rosenberry, 1986).

The glycosyl PtdIns anchors of proteins in neutral detergents, such as Triton X-100 or Nonidet P-40, can be recovered by exhaustive proteolysis (e.g. pronase at 2% (w/w) in 100 mM NH_4HCO_3, 16 h at 37°C) followed by extraction three times with an equal volume of toluene. Repurified radio-methylated AcChoE (100–200 nmol, reduced either with $NaCNBH_3$ or $NaCNBD_3$) was dialyzed against water and reduced in volume in a Speedvac concentrator prior to incubation for 10 h at 50°C in a 3 ml mixture of Pronase (10 mg/l), 0.1 M HEPES (pH 8.0), 15 mM $CaCl_2$, and 1% Triton X-1000. Following the addition of SDS (to 1%) and a second aliquot of Pronase (to 15 mg/l total), digestion was continued for 7 h at 50°C. The digestion mixture was equilibrated in 20 mM sodium phosphate (pH 7) and 0.05% Triton X-100. The anchor-containing peptide was monitored by coelution of ^{14}C and ^{125}I radioactivity and had an elution volume of 0.44 relative to solvent marked $K_2Cr_2O_7$. After vortexing and centrifugation the toluene phases are removed, leaving behind the interface. The glycosyl PtdIns peptides remain in the aqueous phase. If the detergents to be removed are more polar (SDS or nOG) 1-butanol can be used to extract the detergent (Schneider et al., 1990). The glycosyl PtdIns peptides remain in the aqueous phase and can be recovered by rotary evaporation to remove residual solvents and freeze-dried to remove residual ammonium acetate.

More recently, Naik et al. (2000) isolated glycosyl PtdIns that are linked to parasite proteins by digesting the delipidated parasite pellet with pronase (50 U/ml) in 50 ml of 100 mM NH_4HCO_3, 1 mM $CaCl_2$, pH 8, at 37°C for 24 h. The released glycosyl PtdIns (designated as amino acid-linked glycosyl PtdIns) were extracted with water-saturated 1-butanol, washed with water, dried, and purified by HPLC. Control erythrocyte membrane debris, saponin-lysate, and ghosts were similarly extracted and fractionated.

Prior to the cleavage of the anchor proper, the protein carrying the anchor may be subjected to trypsin digestion as described for acetylcholinesterase (AcChoE) by Haas et al. (1996). The peptide fragments were resolved by microbore HPLC. Since reductive radio-methylation (Haas et al., 1986) labels the free ethanolamine and glucosamine residues (Doctor et al., 1990) and the N-terminus of the protein, the tryptic fragment bearing the delipidated anchor should be among the radio-labeled peaks from the digest. To identify this fragment, corresponding peak fractions from a parallel digest of AcChoE radio-methylated at high specific radioactivity are subjected to acid hydrolysis and analyzed by cation-exchange chromatography on an amino acid analyzer. Fractions corresponding to the first major radioactive peak in the microbore HPLC eluate contained radio-methylated glucosamine and ethanolamine, and the major peptide in this peak was denoted Peptide 1 (see below).

5.2.3.2. Nitrous acid deamination

Treatment of anchored proteins with nitrous acid deaminates the anchor glucosamine and cleaves its glycosidic linkage to inositol (Ferguson et al., 1985a). Deamination of trypanosome VSGs released dimyristoyl GroPIns, but deamination of E^{hu}AcChoE generated a novel inositol phospholipid in which an inositol hydroxyl group appeared to be palmitoylated (Roberts et al., 1988b). This palmitoylation was shown to be responsible for the resistance of the E^{hu}AcChoE anchor to PtdIns-specific phospholipase C from S. aureus (Roberts et al., 1988c).

The treatment with nitrous acid is usually applied to purified glycosyl PtdIns (Ferguson, 1992; Menon, 1994). Thus, the HPLC purified glycosyl PtdIns (20 μg) may be treated with HNO_2 in the presence of sodium taurodeoxycholate. The released PtdIns moieties are extracted with water-saturated 1-butanol and washed with water. The aqueous phase is desalted on Bio-Gel P-4 (Vijaykumar et al., 2001).

[^{125}I]TID-labeled E^{hu}AcChoE was treated with nitrous acid, and products extracted into an organic phase were chromatographed by HPLC (Roberts et al., 1988b). Isolated deamination product (2 nmol) was acetylated by adding pyridine/acetic anhydride (1:1) (20 μl) and incubating for 1.5 h at 60°C. Alternatively, the isolated deamination product (2 nmol) was deacylated by the addition of ammonia-saturated methanol and incubation for 2 h at 65°C. Both samples were dried under vacuum and used without further purification.

Both neutral and charged glycosyl PtdIns fragments may be resolved by TLC on silica gel 60 plates developed three times in chloroform/methanol/water (10:10:3, by vol.) with air drying between each development. Specifically, this system has been used to resolve Man1AHM, Man2AHM, and Man3AHM, as well as negatively charged glycans prepared by deaminating mammalians glycosyl PtdIns containing one to three phosphoethanolamine substituents (Kamitani et al., 1992). Other TLC systems for glycosyl PtdIns glycan separation have also been described (Schneider et al., 1993).

5.2.3.3. Hydrofluoric acid

According to Menon (1994) chemical dephosphorylation with aqueous hydrofluoric acid can be used to release diradylglycerols from the glycosyl PtdIns. The dried radioactive fatty acid-labeled glycosyl PtdIns sample in a plastic microcentrifuge tube is cooled in an ice-water mixture, and 50 μl of ice-cold aqueous hydrofluoric acid is added. After incubating the sample in the ice-water mixture for 60 h, the reaction medium is neutralized with saturated LiOH, and the released diradylglycerol is extracted by vortexing with a small amount of toluene or hexane. The diradylglycerols (presumably isomerized) are recovered in the organic extract and analyzed as described above.

These experiments are typically performed with cells metabolically labeled with radioactive amino acids (e.g. ^{35}S-methionine).

According to Vijaykumar et al. (2001) HF cleavage yields both the glycan and the diaradylglycerol moieties, which can be purified by HPLC (Naik et al., 2000). For this purpose, HPLC purified glycosyl PtdIns (10 μg) were treated with 50% aqueous HF (Ferguson, 1992; Menon, 1994), and the reaction mixture was diluted with 20 volumes of ice cold water, lyophilized and portioned between water and water-saturated 1-butanol.

The products of chemical dephosphorylation with aqueous hydrofluoric acid, which releases neutral diacylglycerol fragments, can be analyzed by TLC in a neutral solvent system.

5.3. Determination of chemical structure

The structure of glycosyl PtdIns protein anchor was first elucidated by Ferguson et al. (1988). Since that time, a number of glycosyl

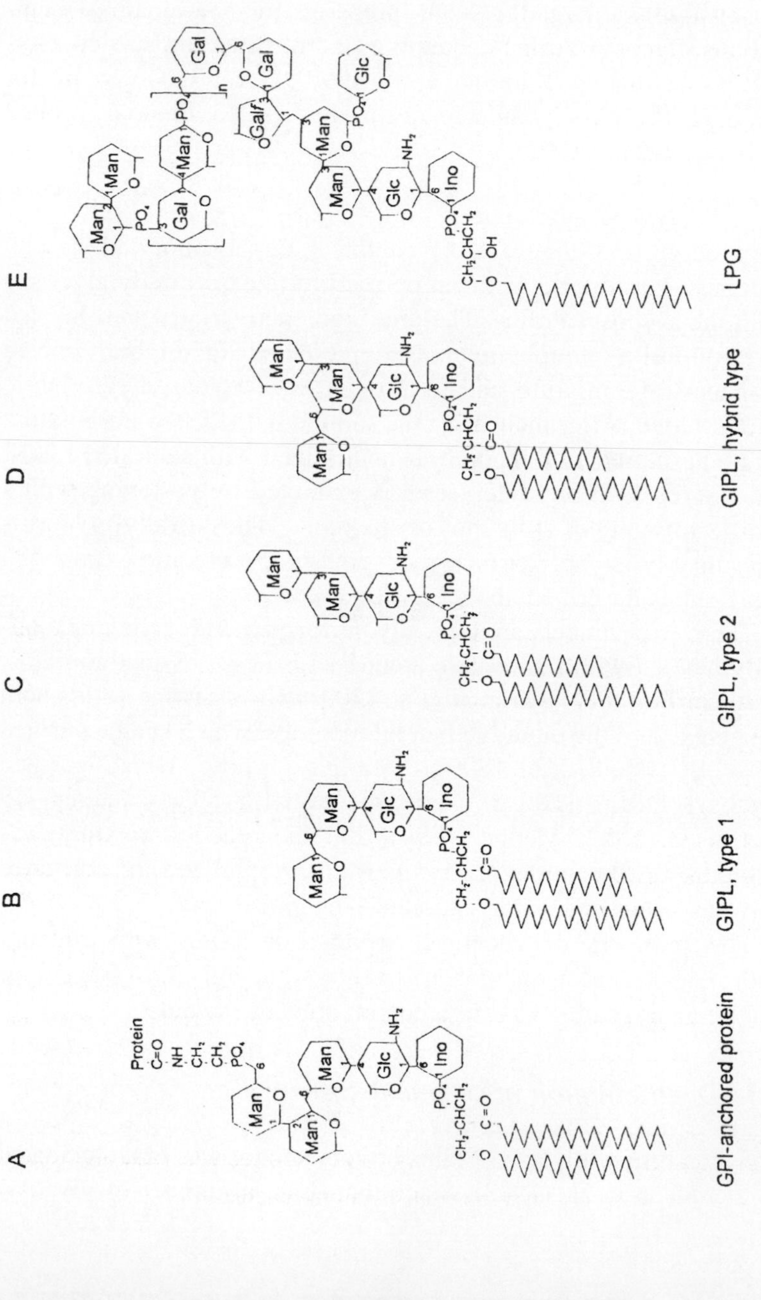
A
B
C
D
E
GPI-anchored protein
GIPL, type 1
GIPL, type 2
GIPL, hybrid type
LPG

PtdIns anchors from different species have been structurally characterized (McConville and Ferguson, 1993). These studies have revealed the glycosyl PtdIns core structure to be completely conserved throughout evolution. Tiede et al. (1999) have summarized graphically the major structures (Fig. 5.1). The carboxyl-terminal amino acid of the glycosyl PtdIns-anchored protein is linked via ethanolamine phosphoate to a tetrasaccharide, which in turn is bound to *myo*-inositol in alpha 1–6 linkage. The tetrasaccharide represents a glycan backbone that can be substituted with additional ethanolamine phosphate groups or with other sugars

Fig. 5.1. Examples of glycosylated PtdIns (GPIs) in eucaryotic cells. The family of GPI-related molecules is defined by the core structure Manα1-4GlcNα1-6*myo*-inositol-1-PO$_4$. (A) GPI membrane anchors, which are found on all eukaryotic cells, contain a conserved core structure of EtN-P-6Manα1-2Manα1-6Manα1-4GlcNα1-6PtdIns. The fatty acid moiety usually contains *sn*-1,2-diacylglycerol or *sn*-1-alkyl-*sn*-2-acylglycerol in mammals (the latter form is shown here). In protozoa, *sn*-1,2-diacylglycerol or *sn*-1-lyso-2-acylglycerol is found. In yeast, the lipid is usually *sn*-1,2-diacylglycerol or a ceramide (see text for details). (B–E) Examples of protein-free glycosylatedPtdIns from *Leishmania* spp. (B) Type-1 glycoinositolphospholipids (GIPLs) have the core structure Manα1-6Manα1-4GlcNα1-6PtdIns and are typical for the *Leishmania* amastigote form, which resides with host macropage endosomes. The PtdIns moiety contains *sn*-1-alkyl-*sn*-2-acylglycerol with predominantly C18:0 alkyl chains at *sn*-1. (C) Type GIPLs share the core Manα1-3Manα1-4GlcNα1-6PtdIns, which can be extensively elongated by galactose and galactofuranose-containing oligosaccharides. These GILs are found on promastigotes, the insect form of *Leishmania*. PtdIns contains *sn*-1-alkyl-*sn*-2-acylglycerol with C$_{18:0}$, C$_{22:0}$, C$_{24:0}$ and C$_{26:0}$ alkyl chains at *sn*-1. (D) Hyrbrid-type GIPLs have a branched core structure of Manα1-6(Manα1-3)Manα1-4GlcNα1-6 PtdIns and are also found on *Leishmania* promastigotes. (E) Lipophosphoglycan (LPG) contains a glycosyl PtdIns anchor of the structure Galα1-6Galα1-3Galfβ1-3(Glcα1-PO$_4$-6) Manα1-3Manα1-4GlcNα1-6(*sn*-1-alkyl-*sn*-2-lyso-glycerol)PtdIns, a partially substituted repeating phospho-disaccharide (Galβ1-4Manα1-PO$_4$-6)$_n$, and a cap of the minimal structure Manα1-2Man that can be elongated by one or two Gal residues. The *sn*-1-fatty acid is preferentially long alkyl chain (C$_{24:0}$, C$_{26:0}$). Reproduced from Tiede et al. (1999) with permission of the publisher.

and acetylneuraminic acid. Substitution with elaborate, branched side chains is also observed. The modifications of the glycan core are specific for distinct species, cell types and developmental stages.

Menon (1994) has provided detailed protocols for the analyses of glycosyl PtdIns structures with special emphasis on the hydrophobic domain, while Schneider and Ferguson (1995) have provided detailed protocols for the microscale analysis of glycosyl PtdIns anchors with special emphasis on the carbohydrate structure. For structure identification, Menon (1994) recommends that the washed extracts containing the free glycosyl PtdIns are purified by TLC on Si-60 plates (Merck, Darmstadt, Germany) using either chloroform/methanol/water (4:4:1, by vol.) or chloroform/methanol/conc. acetic acid/water (25:15:4:2, by vol.).

The general features of the glycosyl PtdIns-anchored proteins may be obtained by mass spectrometry (Roberts et al., 1988c) and one- and two-dimensional ^{1}H-NMR (Ferguson et al., 1988). Various cleavage procedures have been found to produce anchor fragments, which facilitate the determination of the chemical structure of the glycosyl PtdIns by mass spectrometry). Chromatographic analyses are performed with radio labeled glycosyl PtdIns (Menon, 1994; Vijaykumar et al., 2001). Radio-labeled glycolipids (TLC purified or within the glycolipid mixture) are identified as glycosyl PtdIns by specific enzymic and chemical treatments including PtdIns-specific PLC, glycosyl PtdIns-specific PLD and HNO_2 deamination, as described previously (Gerold et al., 1996a–c, 1997).

Hydrophilic fragments are prepared from TLC-purified glycolipids by deamination with HNO_2 as described (Ferguson et al., 1993). Gerold et al. (1997) have analyzed the fragments on a Bio-Gel 4 size-exclusion column.

The hydrophobic fragments generated by HNO_2 deamination are characterized by saponification (Schneider et al., 1993) and TLC analysis (Menon, 1994) using chloroform/methanol/water (4:4:1 or 10:10:3, by vol.). Sensitivity of glycolipids to specific treatments may be assessed by butanol/water phase separation of the reaction mixtures and TLC analysis of the organic phases (Gerold et al., 1997).

The characterization of the glycosyl PtdIns anchors is commonly performed as a series of analyses of specific features of the overall structure, a few examples of which have been discussed below. Menon (1994) has summarized methods for generation of lipid fragments from glycosyl PtdIns anchored proteins and glycosyl PtdIns.

5.3.1. Protein-ethanolamine bridge

Ferguson and Williams (1988) have reviewed the early studies indicating that the carboxy-terminal amino acid of the glycosyl PtdIns-anchored protein is linked via ethanolamine phosphate (Etn-P) to a tetrasaccharide of the sequence Manα1-2Manα1-6Manα1-4GlcN, which in turn is bound to *myo*-inositol in α 1 – 6 linkage. The nature of the protein-ethanolamine bridge was established on the basis that only one ethanolamine per mole was seen, and that this was linked in an amide bond to the α-COOH group of the final amino acid as shown by the fact that NH$_2$-group of ethanolamine in a peptide-glycosyl PtdIns fragment could be dansylated only after removal of the terminal amino acid by Edman sequencing (Holder, 1983; Ferguson and Williams, 1988). The protein-ethanolamine link was demonstrated in a similar way for E[hu]AcChoE (Haas et al., 1986). Fasel et al. (1989) reported the in vitro attachment of glycosyl PtdIns anchor structures to mouse Thy-1 antigen and human DAF. Fraering et al. (2001) have cloned the glycosyl PtdIns transamidase complex from *S. cerevisiae*, while Reid Taylor et al. (1999) have reconstituted the glycosyl PtdIns-anchored protein Thy-1.

Structural studies of glycosyl PdIns anchors have utilized several cleavage procedures to produce anchor fragments for analysis. Papain digestion of E[hu] AcChoE produced a C-terminal dipeptide His-Gly linked through ethanolamine to the anchor (Roberts and Rosenberry, 1986; Haas et al., 1986).

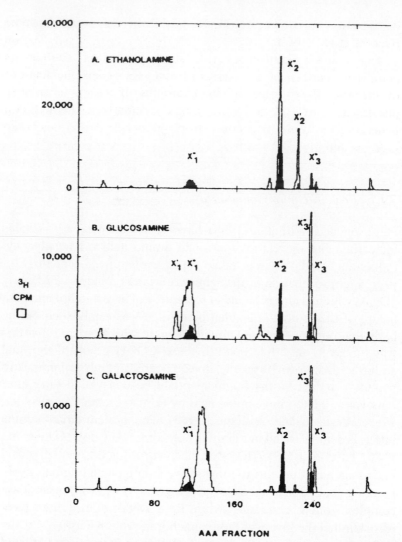

Fig. 5.2. Illustrates the identification of ethanolamine and glucosamine in the hydrophobic domain of red blood cell acetylcholine esterase. Ethanolamine, glucosamine and galactosamine (1 mM) were reductively methylated with 1 mM [^{3}H]HCHO. Aliquots of each labeled standard were mixed with samples of ^{14}C-labeled hydrophobic domain fragment that was produced by papain digestion of

Because of the fatty acyl heterogeneity and the extreme hydrophobicity of this proteolysis product, further purification was achieved by deacylation and reverse phase HPLC (Roberts et al., 1988c). A chromatogram of the base-treated product showed three radioactive peaks, which corresponded to a non-retained fraction (8% of applied ^{14}C) and retained doublet of peaks that contained 20 and 40% of the applied radioactivity. The larger peak was examined by both positive and negative ion FAB-MS. The positive ion spectrum revealed two major ions at m/z 1730.9 and 1752.8, which were assigned to the protonated and sodiated species, respectively. These results indicated the molecular weight of the papain proteolysis product to be 1729.9. Unfortunately, the spectrum did not exhibit fragmentation that could be used to establish a component sequence. Detailed structural information was obtained from a more highly purified product of Pronase digestion of E^{hu} AcChoE that consisted of the glycosyl PtdIns attached only to the C-terminal of glycine residue. A more effective removal of Triton X-100 was obtained by pooling only the heart fractions from the Sephadex LH-60 column and by using a C_3 instead of a C_8 reversed phase HPLC column. Three radioactive

low specific activity ^{14}C-radiomethylated acetylcholoine esterase in Triton X-100 micelles. The mixed samples were dried, hydrolyzed, and analyzed by automated amino acid analyzer. The radioactivity profiles for ^{3}H (open bars) and ^{14}C (closed bars) were determined by dual channel scintillation counting. The ethanolamine hydrolysate showed two major ^{3}H peaks, and these corresponded to the ^{14}C peaks labeled X_2' and X_2''. These two peaks were assigned to N-methylethanolamine and N,N-dimethylethanolamine, respectively. The hydrolysates of ^{3}H-labeled glucosamine and galactosamine were more complex and showed four major ^{3}H peaks. Two of these corresponding to the ^{14}C peaks labeled X_3' and X_3'' were obtained from both hexoamines. The two other ^{3}H peaks from glucosamine corresponded to ^{14}C-labeled X_1' and X_1'', in contrast to the comparable pair from galactosamine. The analysis technique distinguished unequivocally between methylated derivatives of glucosamine and galactosamine. Reproduced from Haas et al. (1986) with permission of the publisher.

HPLC peaks were obtained from the radiomethylated base-hydrolyzed Pronase proteolysis product, with 18% of the ^{14}C not retained and 21 and 44% in a doublet that eluted with mobile phase composition of about 30% 1 propanol. Figure 5.3(A) shows

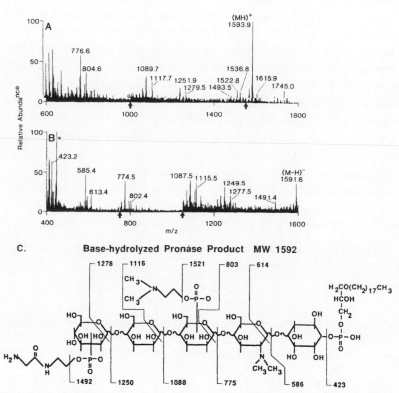

Fig. 5.3. FAB-MS analysis of the base-hydrolyzed pronase proteolysis product of E^{hu} AcChoE. (A) Positive ion FAB mass spectrum. For m/z values between 1000 and 1550 (arrows) the relative abundances were amplified by a factor of 2; (B) Negative ion FAB mass spectrum. For m/z values between 750 and 1050 (arrows) the relative abundances were amplified by a factor 2, and for m/z values greater than 1050, the amplification was by a factor of 10. The triethanolamine matrix gave rise to an ion at m/z 446.3 (*); (C) Proposed structure of the pronase proteolysis product summarizing results from A and B. Reproduced from Roberts et al. (1988c) with permission of the publisher.

a positive ion and Fig. 5.3(B) a negative ion FAB spectrum of the major HPLC peak. In the positive ion mode, the most abundant molecular ion at m/z 1593.9 $(MH)^+$ was accompanied by an associated sodium adduct ion $(M + Na)^+$ that was 22 Da higher at m/z 1615.9. The molecular weight of the Pronase proteolysis product was 137 Da lower in mass than that generated by papain, consistent with amino acid analysis data that indicated only a C-terminal glycine residue and not the penultimate histidine residue found in the papain proteolysis products (Haas et al., 1986). The mass spectrum is interpreted in Fig. 5.3(C) based on the deamination product spectrum (not shown), the anchor component previously identified for $E^{hu}AcChoE$ (Haas et al., 1986; Roberts et al., 1987, 1988c) and previous reports of the anchor structure for trypanosome VSGs (Ferguson et al., 1988; Schmitz et al., 1987).

The bridge between the glycosyl PtdIns moiety and the protein involves the COOH-terminal aspartate in an amide linkage to the aspartyl α-carboxyl group (Holder, 1983). Ferguson et al. (1988) used periodate oxidation to confirm that the ethanolamine is in a phosphodiester linkage to the 6-position of either the terminal or the penultimate 2-O-substituted mannose as suggested by the presence of mannose-6-phosphate in the methanolyzates of the soluble VSG COOH-terminal glycopeptide produced by pronase digestion. Treumann et al. (1995) have succeeded in using mass spectrometry to assess indirectly the nature of the glycosyl PtdIns-peptide in the CD52 antigens extracted from human spleens. Attempts to obtain ESI mass spectra and matrix-assisted laser desorption ionization mass spectra of native CD52-I and CD52-II were unsuccessful due to extensive microheterogeneity of the N-linked oligossaccharides of CD-52. The glycosyl PtdIns-peptide derived from D52-I, however, yielded useful information, when analyzed by positive and negative ion ESI/MS and produced the spectra shown in Fig. 5.4(A) and (B), respectively (Treumann et al., 1995). After transformation, the positive ion spectra, the data indicated the

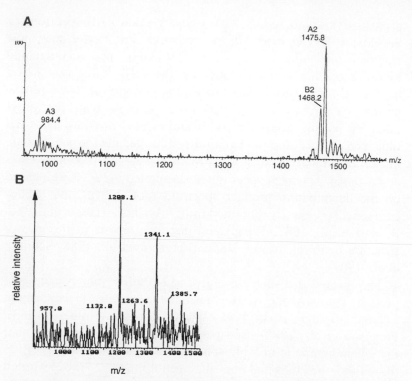

Fig. 5.4. Electrospray mass spectrometric analysis of the glycosyl PtdIns-peptide of CD52-I. *Panel A*: Positive ion spectrum of the glycosyl PtdIns-peptide of CD52-I. The ions at *m/z* 1475.8 (A2) and 984.4 (A3) correspond to the [M + 2H]$^{2+}$ and the [M + ^{3}H]$^{3+}$ pseudomolecular ions of a molecule with a molecular mass of 2951.1 ± 1.6 Da. The peak at *m/z* 1468.2 corresponds to the [M + 2H]$^{2+}$ pseudomolecular ion of a molecule with a molecular mass of 2933.5 ± 0.8 Da; *Panel B*: Negative-ion spectrum of the glycosyl PtdIns peptide of CD52-I after partial alkaline hydrolysis. The ions at *m/z* 2682.2 and 1208.1 correspond to the [M − 2H]$^{2-}$ pseudomolecular ions of the glycosyl PtdIns-peptide minus 1 and 2 stearoyl residues (calculated masses 2682.2 and 2684.2 Da), respectively.

Reproduced from Treumann et al. (1995) with permission of the publisher.

presence of a major molecular species of mass 2951.1 ± 1.6 Da. The theoretical average mass of the CD52-glycosyl PtdIns-peptide shown in Fig. 5.3(A) is 2951.5 Da. The close agreement in the measured and theoretical masses are consistent with the suggested composition of the major CD52-I glycosyl PtdIns-peptide component (i.e. the dodecapeptide sequence, the trimannosyl-glucosaminyl glycan structure, the two ethanol amine phosphate groups, and the distearoylGroPIns moiety). The minor molecular species of 2933.5 ± 0.8 Da (Fig. 5.3(A)) is 17.6 ± 2.4 Da smaller than the major species. This difference could be due to substitution of a Ser residue by Ala within the peptide sequence (16 Da theoretical difference) (Treumann et al., 1995).

Moran et al. (1991) have investigated the glycosyl PtdIns membrane anchor attachment/cleavage site in the decay accelerating factor (DAF). Using [³H]ethanolamine to tag the COOH terminus, a COOH-terminal residue tryptic peptide was isolated and sequenced, thereby identifying Ser-319 as the COOH-terminal residue attached to the glycosyl PtdIns anchor. Moran et al. (1991) analyzed the structural requirements at the cleavage site by replacing Ser-319 with all possible amino acids. A glycosyl PtdIns anchored fusion protein, human growth hormone-DAF, was used for this analysis. It was shown that alanine, aspartate, asparagines, glycine, or serine efficiently supported glycosyl PtdIns attachment while valine and gluta-mate were partially effective. These results support the general rule that the residue at the cleavage/attachment site must be small.

In addition to the bridging ethanolamine, all glycosyl PtdIns-anchors contain additional 1–2 ethanolamine residues. The extra ethanolamine appears to be linked through phosphate to hexose groups on the basis of results from FAB/MS on the glycosyl PtdIns-peptide fragment (Roberts et al., 1987). More recently, the product of glycerol-phosphoryl-ethanolamine has been identified as its TMS derivative by GC/MS after periodate oxidation, sodium borodeuteride reduction, and partial acid hydrolysis with the glycerol coming from carbons 4, 5 and 6 of an oxidized hexose residue (Ferguson et al., 1987).

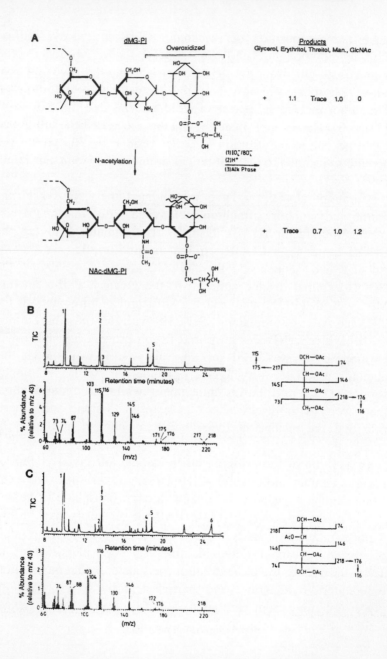

5.3.2. Glucosamine-inositol linkage

The glycosidic nature of the glucosamine-inositol linkage was established by nitrous acid deamination, which selectively cleaves the $GlcNH_2$ glycosidic bond during its rearrangement to 2,5-anhydromannose. When this reaction was applied to glycosyl PtdIns from intact and soluble VSG, the released products were identified as a dimyristyl GroPIns and InsP, respectively (Ferguson et al., 1985; Schmitz et al., 1986). The $GlcNH_2$ linkage has been since shown to be in the α-configuration and linked to the C-6 position of the inositol (Ferguson et al., 1988). For this demonstration deacylated, N-acetylated fraction of the glycosyl PtdIns anchor was used to recover 1,4-dideutero threitol after periodate oxidation, sodium borodeuteride reaction, and acid hydrolysis. This product could only arise from a 1,6-disubstituted inositol ring, where the C-1 position is occupied by the phosphate group of PtdIns and the C-6 position by linkage to the $GlcNH_2$ residue (Fig. 5.5(A)). The periodate-oxidized and NaB^2H_4-reduced

Fig. 5.5. Periodate oxidation of the $Man\alpha1\text{-}4GlcNH_2\alpha1\text{-}6$ myo-inositol core region. Samples (120 nmol) of demyristoylated glycoslyl PtdIns and its N-acetylated form, Nac-demyristoylated glycosyl PtdIns, were periodate oxidized, and NaB^2H_4 reduced, hydrolyzed with 2 M HCl (100°C, 3 h) and treated with bovine alkaline phosphate. Before GC/MS analysis, the products were per-O-acetylated. The GC/MS was performed on Supelcowax 10 column with non-linear temperature programming from 100 to 260°C. Panel A: The periodate oxidation sites are shown by waved lines. The bold lines represent the carbon–carbon bonds resistant to oxidation; Panel B: The GC/MS analysis of the periodate-treated demyristoylated glycosyl PtdIns sample showing the total ion current chromatogram (upper panel) and the mass spectrum of Peak 2, $[1\text{-}^2H]$erythritol (lower panel); Panel C: The GC/MS analysis of the periodate-treated Nac-demyristoylated glycosyl PtdIns sample showing the total ion current chromatogram (upper panel) and the mass spectrum of Peak 3, $[1,4\text{-}di\text{-}^2H]$threitol (lower panel). The other numbered peaks were identified as the acetate derivatives of $[1\text{-}^2H]$glycerol. Peak 1; α- and β-mannose; Peaks 4 and 5; and N-acetylglucosamine, Peak 6. Reproduced from Ferguson et al. (1988) with permission of the publisher.

demyristylated glycosyl PtdIns was hydrolyzed, treated with alkaline phosphatase, and analyzed by GC/MS after peracetylation of the products (Fig. 5.5(B) and (C)). The three predicted major products were found: $[1-^2H]$glycerol, an intact mannose residue (derived from protected 3,6-di-O-substituted mannose branch-point residue), and $[1-^2H]$erythritol (derived from the glucosamine residue). $1-^2H$-NMR data suggested that the glucosamine is linked $\alpha 1-6$ to the myo-inositol residue as evidenced by a NOE linking GlcNH$_2$ H-1 to inositol H-6. The predicted oxidation and reduction product of the inositol ring is therefore $[1,4-di-^2H]$threitol (derived from carbons 1,2,5 and 6) (Ferguson et al., 1988).

5.3.3. Glycan moiety

The glycosyl PtdIns tetrasaccharide (Manα1-2Manα1-6Manα1-4GlcN) represents a glycan backbone that can be substituted with additional Etn-P groups or with other sugars such as mannose (Man), galactose (Gal), N-acetylgalactosamine (GalNAc), N-acetylglucosamine (GlcNAc) and N-acetylneuraminic acid (NANA). Substitution with elaborate, branched side chains is also observed, especially in the protozoon Trypanosoma brucei. The modifications of the glycosyl PtdIns glycan core are often specific for distinct species, cell types, or developmental stages (Tiede et al., 1999).

The structures of the carbohydrate tetramers have been independently determined by a combination of two-dimensional NMR and mass spectrometry (Ferguson et al., 1988) and have been shown to exhibit microheterogeneity except for the three α-mannose residues.

The neutral glycosyl PtdIns glycans can be separated and sequenced by BioGel P4 gel-filtration and/or by high pH anion-exchange chromatography before and after exoglycosidase treatments and partial acetolysis (Ferguson, 1992).

Analysis of neutral glycans by BionGel P-4 gel filtration gives a direct measurement of size (hydrodynamic volume) and, in conjunction with exoglycosidase digestion experiments, provides useful fragments for structural studies (Yamashita et al., 1982). Glycosyl PtdIns-neutral glycan fragments prepared by dephosphorylation, deamination, and reduction elute as follows (size expressed in glucose units): AHM, 1.7; Manα1-4AHM, 2.3; Manα1-6Manα1-4AHM, 3.2; Manα1-2Manα-6Manα1-4AHM, 4.2; Manα1-2Manα1-2Manα1-6Manα1-4AHM, 5.2 (Mayor et al., 1990a; Ferguson, 1992).

The analysis is performed on an HPLC AS6 column (Dionex Corporation, Sunnyvale, CA) using a gradient system. The gradient is began with 100% Solvent A (0.1 M NaOH), 0% Solvent B (0.5 M CH$_3$COONa, 0.1 M NaOH) up to 3 min after sample injection, followed by a linear change to 55% solvent A, 45% solvent B at 33 min, then 0% solvent A, 100% solvent B at 38 min. The flow rate is 1 ml/min. The glucose oligomer standards and carriers are prepared by partial acid hydrolysis of dextran (Yamashita et al., 1982; Ferguson, 1992). Other examples of anion-exchange chromatography with Dionex CarboPac columns are also found in the literature (Guther, 1992; Stahl et al., 1992; Deeg et al., 1992a,b; Roberts et al., 1988b,c; Ferguson, 1992).

Hirose et al. (1992) have outlined fragmentation strategy used for glycan analyses (Hirose et al., 1992). HF, dephosphorylation by hydrogen fluoride; HNO$_2$, nitrous acid hydrolysis; GPI-PLD, hydrolysis by glycosyl PtdIns-specific phospholipase D; PtdIns-PLC, hydrolysis by PtdIns-specific phospholipase C (only if the inositol is not substituted). A palmitoyl chain is present in immature glycosyl PtdIns anchors but is generally removed from mature glycosyl PtdIns-anchored proteins (Doering et al., 1994).

Schneider et al. (1993, 1994) and Schneider and Ferguson (1995) have described detailed methods for the analysis of neutral glycan fractions of glycosyl PtdIns by HPTLC. The neutral glycans are prepared by nitrous acid deamination and NaB^{3}H$_4$ reduction of the glycosyl PtdIns-anchored proteins. The glycosyl PtdIns glycan

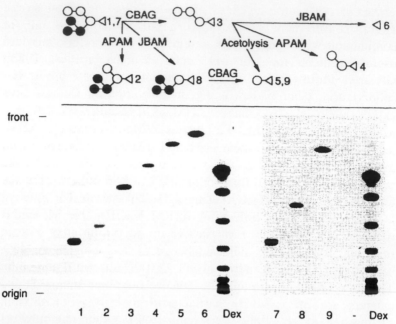

Fig. 5.6. Exoglycosidase sequencing of glycosyl PtdIns neutral glycan by HPTLC. Sequencing of structure Manα1-2Manα1-6[Galα1-6(Galα1-2)Galα1-3]Manα1-4AHM from *T. brucei* VSG (Ferguson et al., 1988) using the combination of exoglycosidases and acetolysis indicated above the fluorograph of the HPTLC plate. The numbers next to the products refer to the lane numbers on the HPTLC fluorograph. The HPTLC plate was developed with solvent system B: first and third developments with 1-propanol/acetone/water (9:6:5, by vol.), second development with 1-propanol/acetone/water (5:4:1, by vol.). Lanes marked Dex contain reduced radioactive dextran oligomer standards. ○, αMan; ●, αGal; ◁, AHM; APAM, *Aspergillus saitoi* α-mannosidase; JBAM, jack bean α-mannosidase; CBAG, coffee bean α-galactosidase. Reproduced from Schneider and Ferguson (1995) with permission of the publisher.

is released by aqueous HF dephosphorylation, re-N-acetylation and desialylation. The feasibility of microsequencing a glycosyl PtdIns anchor glycan using a TLC system is illustrated in Fig. 5.6, using the neutral glycan derived from *T. brucei* variant surface glycoprotein anchor (structure H) (Schneider and Ferguson, 1995). Structure H was sensitive to an α-mannosidase specific for α1-2 linkages, α-mannosidase with broad specificity, as well as to α-galactosidase, indicating a branched structure with two α Man residues, the terminal one being linked α1-2, and three α Gal residues. The product of the α-glactosidase digestion was sensitive to jack bean α-mannosidase, indicating the presence of three αMan residues. The αGal branch was located to the αMan residue at the reducing end of the oligosaccharide, by sequential digestion with α-mannosidase followed by α-galactosidase which yielded the product Man_1AHM. Taken together, the results indicate the following structure: $Man\alpha1-2Man\alpha1-6[\alpha Gal_3] \alpha ManAHM$.

The sequence strategies for an abundant glycosyl PtdIns glycolipid from L. major are shown elsewhere (Schneider et al. (1993). This glycolipid is sensitive to alpha-galactosidase, indicating the presence of a terminal alpha-galactose. The resulting structure was resistant to α-mannosidase, unless an acid-labile residue, most likely Gal, was first removed by mild acid treatment. α-Mannosidase was able to remove two alpha-mannose residues from the acid treated glycan to yield anhydromannitol (AHM). From these results, the following partial structure was deduced: $\alpha Gal_p-Gal_f-\alpha Man-\alpha Man-AHM$.

The 3 hexose residues in the linear core glycan previously observed by FAB/MS (Roberts et al., 1988c) have been subsequently confirmed to be Man in the more abundant AcChoE anchor glycans α and β obtained from radiomethylated $E^{hu}AcChoE$ following complete proteolysis and PtdIns PLC cleavage. The samples α and β were analyzed by ESI/MS. Predominant $(MH_2)^{+2}$ ions with m/z 634.1 and 709.7 were observed for α and β, respectively. The CID/MS spectra derived from these molecular ions are shown in Fig. 5.7. The CID-MS spectra yielded major fragments for rupture of Man glycosidic bonds or phosphate diester linkages. The mass difference between glycans α and β $(1417 - 1266 = 151)$ is accounted for by the additional $EthN(Me)_2$ phosphate linked to Man_2 (Deeg et al., 1992a,b).

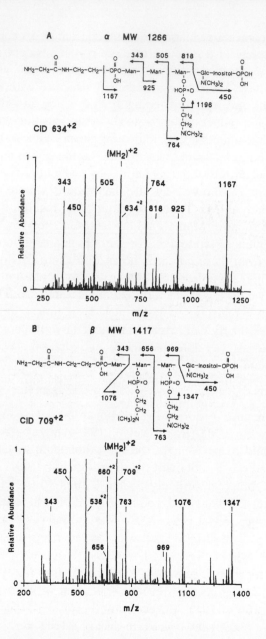

These features are consistent with those previously reported for VSG (Ferguson et al., 1988) and rat Thy-1 (Homans et al., 1988).

Haas et al. (1996) subjected a radiomethylated trypsin digestion product of bovine AcChoE to ESI/MS and examined the resulting m/z 3798 ion in the $+4$ state by CID/MS, as shown in Fig. 5.8. Characteristic glycan fragmentation peaks are marked, and the inferred glycan structure is shown at the top of the figure. Most features are entirely consistent with the human AcChoE anchor structure (Roberts et al., 1988b,c; Deeg et al., 1992a,b). One significant difference involved a glycan substituent: the hexose next to glucosamine bears an additional N-acetylhexosamine residue in the bovine anchor.

Glycans from several other anchors show this same variation in addition of N-acetylhexosamine (Haas et al., 1996). As expected from the susceptibility of the anchor to PtdIns-PLC, however, the inositol ring is not palmitoylated.

Nitrous acid deamination cleaves the glycosidic bond at the 1 position Gerold et al. (1996b,c) have prepared neutral core glycans from TLC purified [^{3}HG]glucosamine labeled glycosyl PtdIns anchor of *T. congolese*, by dephosphorylation, deamination and reduction as described by Mayor et al., (1990a,b). Desalted neutral core glycans were analyzed by TLC on Si-60 HPTLC plates (Merck) using propan-1-ol/acetone/water (9:6:5, by vol.) as solvent

Fig. 5.7. MS/CID/MS analysis of glycans α and β derived from E^{hu} AcChoE. *Panel A*: collision spectrum of glycan α obtained by selecting the doubly charged parent ion at m/z $(634)^{2+}$. All fragments were observed to be singly charged, and they were produced as indicated in the scheme at the *top*; *Panel B*: collision spectrum of glycan β obtained by selecting the doubly charged parent ion at m/z $(709)^{2+}$. The fragment ions at m/z 538 and 660 were observed to be doubly charged. The m/z 538 ion is the doubly charged counterpart of a singly charged m/z 1076 fragment. The m/z 660 ion is consistent with loss of a phosphate from the parent ion, which would represent a decrement of 49 Da for these doubly charged ions. The fragmentation scheme is indicated at the *top*. Reproduced from Deeg et al. (1992a) with permission of the publisher.

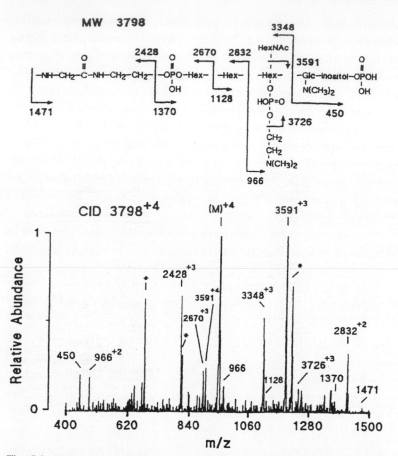

Fig. 5.8. MS of peptide-linked glycosyl PtdIns and its deduced glycan structure. Peptide 1 sample was subjected to ESI/MS, and the collision spectrum (MS/CID/MS) obtained by selecting the m/z 3798 parent ion of +4 charge is shown. Mass values for the fagments are marked (e.g. doubly charged 966^{+2} at m/z 483); $[M]^{+4}$ denotes the quadruply charged 3798^{+4} parent ion at m/z 950. Asterisks mark major peptide peaks resulting from preferential cleavage before Pro_{14}; b_{13}^{+2} at m/z 680, y_{13}^{+3} at m/z 813 and y_{13}^{+2} at m/z 1220. The deduced glycan structure is indicated at the top, together with the molecular masses of fragments from this structure which were found in the spectrum. Reproduced, with permission, from Haas, R., Jackson, B. C., Reinhold, V. R. Foster, J.D. and Rosenberry, T. L. (1996). Biochemical Journal 314, 817–825. © The Biochemical Society and the Medical Research Society,

(Schneider et al., 1993) or by high-pH anion exchange chromatography HPAEC). The HPAEC analysis was accomplished by gradient elution using a Carbopak PA1 column (4 mm × 250 mm). The elution program for HPAEC was 100% buffer A (0.1 M NaOH), 0% buffer B (0.1 M NaOH, 0.25 M sodium acetate) up to 6 min after injection, then increase in buffer B to 30% over 36 min, at a flow rate of 1 ml/min. The isolated core-glycan was analyzed by MALDI-TOF-MS as the native oligosaccharide and by liquid secondary ion mass spectrometry (LSIMS) as the permethylated glycan. Monosaccharide linkage positions were assigned by methylation analysis. MALDI/TOF/MS led to pseudomolecular ions of m/z 1038.9 ($[M + Na]^+$) and 1054.7 ($[M + K]^-$) reflecting an average molecular mass of 1015.9 in agreement with a monosaccharide composition of $Hex_4HexNcAHM-ol$. This assignment was corroborated by LSIMS.

Vijaykumar et al. (2001) used HPTLC to analyze various cleavage products of glycosyl PtdIns anchor of *P. falciparum.* The glycosyl PtdIns were subjected to jack bean α-mannosidase, bee venom phospholipase A_2, and HF treatment and the radioactive products resolved using chloroform/methanol/water (10:10:2.5, by vol.). Similar solvent systems have been used for HPTLC analysis of HF-treated headgroups contained in HPLC peaks derived from complex ethanolamine phosphate in addition to glycosyl PtdIns anchors of mammalian cells (Kamitani et al., 1992).

Glycosyl PtdIns glycans prepared by nitrous acid treatment of non-protein-linked glycosyl PtdIns can be analyzed by conventional anion-exchange chromatography to determine their net negative charge (Mullis et al., 1990). The analysis is conveniently performed with a Mono Q column (HR5/5 bed size 5 × 50 mm²) and a gradient fast protein liquid chromatography (FPLC) system (Pharmacia, Piscataway, NJ). The elution is begun with 100% solvent A (filtered deionized distilled water) up to 5 min after sample injection, then a linear change to 75% solvent A, 25% solvent B (0.5 M ammonium acetate) at 30 min, followed by changes to 50% solvent A, 50% solvent B at 40 min, and 0% solvent A, 100% solvent B at 45 min. The flow rate is 1 ml/min. The column is run with appropriate radioactive standards including [^{3}H]mannose, inositol–PO$_4$–glycerol, and UDP[^{3}H]galactose. [^{3}H]Mannose and other neutral molecules will elute in the void volume (fractions 1–2), inositol–PO$_4$–glycerol elutes in fractions 12-13, and UDP[^{3}H]galactose elutes in fractions 34–35. Glycosyl PtdIns glycans prepared by deaminating non-protein-linked glycosyl PdIns containing one, two or three phosphoethanolamine substituents elute in fractions 12, 20, and 37–38, respectively (Mullis et al., 1990).

5.3.4. PtdIns moiety

The PtdIns moieties are released when free glycosyl PtdIns (20 μg) suspended in 150 μl 0.2 M NaOAc, pH 3.8, are treated with 150 μl 1 M NaNO$_2$ at room temperature for 24 h (Ferguson, 1992). The released PtdIns moieties are extracted with water-saturated 1-butanol. The glycan moieties were recovered by chromatography on a Bio-Gel P-4 column (1 × 90 cm^2) in 100 mM pyridine, 100 M HOAc, pH 5.2 (Naik et al., 2000), glucosamine (Shively and Conrad, 1976) and has been shown to release PtdIns from glycolipid-anchored proteins (Ferguson et al., 1985). According to Menon (1994) the dried glycosyl PtdIns sample is dissolved in 100 μl of buffer (0.2 M sodium acetate, pH 3.7, prepared by mixing 10 ml of 0.2 M sodium acetate with 90 ml of 0.2 acetic acid) plus 2 μl of 10% Nonidet P-40 in water. Then 100 μl of freshly prepared 400 mM solution of sodium nitrite in buffer is added to the glycosyl PtdIns solution. The sample is left at room temperature and the acidity is maintained in a narrow range centered at pH 4, which is critical for the reaction to proceed. After 4 h, 50 μl of a freshly prepared 1 M solution of sodium nitrite in buffer is added and the reaction is allowed to continue at room temperature for an additional 4–8 h. The reaction is terminated by extracting twice with 200 μl of water-saturated *n*-butanol. The extracts are pooled and the PtdIns isolated by TLC using chloroform/methanol/acetic acid/water (25:15:4:2, by vol.). The released PtdIns has been identified as *sn*-1,2-diacylglycerol or *sn*-1-alkyl-2-acylglycerol-3-phospho-*myo*-inositol. However, lysoforms containing only one hydrocarbon chain have been found as well, for instance in the *Trypanosoma* procyclic acidic repetitive protein (PARP) (Field et al., 1991). In the yeast *Saccharomyces cerevisiae* and in *Dictyostelium discoideum*, the glycolipid is often replaced by ceramide (see below). The diradylglycerols are released from the parent PtdIns by hydrolysis with phospholipase C as described in Chapter 2.

The structure of the PtdIns moiety of glycosyl PtdIns carrying a palmitoyl group at the 2-position of the inositol ring cannot be hydrolyzed by PLC (Roberts et al., 1988b). In such instances, the PtdIns structure may be determined while still bound to the glycan moiety. Thus, Roberts et al. (1987) treated E^{hu} AcChoE labeled on the anchor lipid with a trace amount of [^{125}I]TID with nitrous acid and purified the released glycosyl PtdIns by HPLC. A major peak of radioactivity was well separated from unlabeled UV absorbing peaks that represented residual Triton X-100. The labeled peak contained 2 mol of fatty acid/mol of *myo*-inositol. Direct analysis of this purified product by negative ion FAB/MS revealed several clusters of molecular weight-related ions [M − H]⁻ with a predominant group containing two major ions at *m/z* 1135.7 and 1137.7. The adjacent ions and clusters could be attributed to lipid heterogeneity arising from alkyl and acyl chain unsaturation (2 or 4 double bonds) and chain length variations in the number of methylene groups (14–28 methylene units), a feature characteristic of naturally occurring phospholipids. The observed molecular ion heterogeneity was consistent with the alkyl and acyl chain composition determined for the inositol phospholipid (Roberts et al., 1988c). Moreover, fragment ions shown below to contain lipid could be distinguished by their heterogeneity (e.g. *m/z* 735.5)

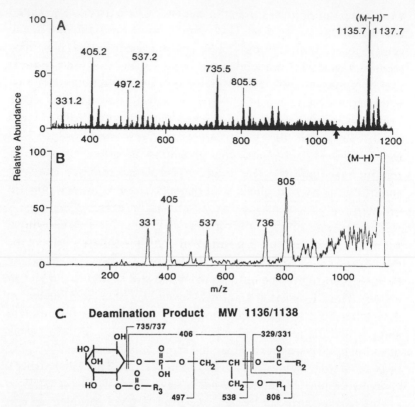

Fig. 5.9. Analysis of the nitrous acid deamination product by negative ion FAB/MS and CID/MIKES. The E^{hu} AcChoE nitrous acid deamination product was purified by normal phase HPLC with isocratic 2-propanol/exane/water (8:6:0.4, by vol.). *Panel A*: negative ion FAB mass spectrum of the deamination product. For *m/z* values less than 1050 (arrow) the relative abundances were amplified by a factor of 2. *Panel B*: MIKE spectrum of the FAB *m/z* 1135.7 and 1137.7 [M − H]⁻ molecular ions. *Panel C*: proposed structure of the deamination product summarizing results from FAB and MIKES analysis. Bond cleavage denoted by vertical lines produced ion fragments in the direction indicated by the *horizontal lines* and *fragment masses*. The observed fragment ions are consistent with $R_1 = C_{18}H_{35}$ and $C_{18}H_{37}$, $R_2 = C_{21}H_{33}$ and $C_{21}H_{35}$ and $R_3 = C_{15}H_{31}$. Assignment of a 1-alkyl-2-acylglycerol constituent is based on the identification of 1-alkylglycerols following methanolysis of anchor fragments (Roberts et al.,

from more homogeneous fragments free of lipid (e.g. m/z 497.2 and 537.2).

To establish the structure of the inositol phospholipid generated from E^{hu} AcChoE by nitrous acid treatment, the most intense molecular ion [M − H]$^-$, m/z 1135.7 (Fig. 5.9(A)) produced by FAB ionization was focused into a helium-filled collision cell, and the daughter ions were studied in a two-sector mass spectrometer (Fig. 5.9(B)). Predominant fragments in the FAB mass spectrum (Fig. 5.9(A)) correspond to those in the FAB mass spectrum (Fig. 5.9(B)) and thus clearly derive from the m/z 1135.7 and 1137.7 molecular ions. To ascertain further the daughter ion composition of these fragments, each ion in the MIKE spectrum was also analyzed by MIKES. The products of this analysis frequently produced single constituents or conjugate groups, and this lineage relationship assisted in composing the final structure. For example, MIKE analysis of the nitrous acid deamination product [MH]$^-$, m/z 1135.7), produces a daughter ion fragment of m/z 736 the difference in mass, 400 Da, corresponds to a palmitoylated inositol group that was eliminated with a hydrogen transfer back to the ruptured bond. When the m/z 735.5 ion generated by FAB was subjected to CID, the MIKE spectrum indicated a daughter ion at m/z 405 that arose by elimination of the acyl group at the 2-position of glycerol. Further study of the m/z 405.2 ion produced by FAB (Fig. 5.2A) by CID-MIKES indicated a daughter ion equal in mass to an ionized dihydrogen phosphate group (m/z 97). Thus, the m/z 405.2 ion can be reconstructed as an 18:0 alkylglycerol phosphate, a component expected on the basis of previous methanolysis and GLC which revealed 1-stearylglycerol (Roberts et al., 1988c). A structure, which summarizes these data is

1988b). Attachment of the palmitoyl group (R_3CO) to the inositol 2-hydroxyl is arbitrary, and acylation of any inositol hydroxyl except that on the 1-position cannot be ruled out by the current data. Reproduced from Roberts et al. (1988c) with permission of the publisher.

presented in Fig. 5.9(C) and corresponds to a plasmanylinositol palmitoylated on the inositol group.

The CD52 antigen extracted from human spleens with organic solvents and purified by immunoaffinity is resolved into two CD52 species, both of which possess glycosyl PtdIns anchors. One of them is sensitive to PtdIns PLC (CD52-I), while the other is resistant (CD52-II). Treumann et al. (1995) deaminated the intact CD52-I and the glycosyl PtdIns-peptide of CD52-II to release their PtdIns moieties which were recovered by solvent extraction. Analysis of these fractions by negative ion ESI/MS revealed major pseudomolecular ions at m/z 866 for CD52-I and m/z 1124 for CD52-II (Fig. 5.10(A) and (B)). The m/z 866 from CD52-I was attributed $[M - 1]^-$ ion of a distearoyl GroPIns, an assignment which is consistent with the compositional data for this molecule (Xia et al., 1993a) and with the ESI mass spectral data described above for the glycosyl PtdIns-peptide. The m/z 1124 ion from CD52-II was attributed to the $[M - 1]^-$ ion of a palmitoylated (stearoylarachidonoyl) GroPIns. This assignment was confirmed by negative ion tandem mass spectrometry. The daughter ion spectrum of m/z 1124 (Fig. 5.10(C)) and a daughter ion spectrum of authentic 1-stearoyl-2-arachidonoyl GroPIns (Fig. 5.10(D)) possess two major fragment ions at m/z 283 and 303 corresponding to the carboxylate ions of stearic and arachidonic acid, respectively. The presence of a palmitoyl group in the parent ion of m/z 1124 can be inferred by the presence of the carboxylate fragment ion at m/z 255. The intensity of the palmitate ion is weaker than those of the stearate and arachidonate ions. This appears to be in agreement with the result of mass spectrometric study (Treumann et al., 1998) on the *T. brucei* procyclic acidic repetitive protein (PARP), that fatty acid residues attached to the inositol ring produce weaker carboxylate fragment ions than those attached to the glycerol backbone. The daughter ion spectrum of the 1-stearoyl-2-arachidonoyl GroPIns standard (Fig. 5.10(D)) shows an intense fragment ion at m/z 241 that corresponds to Ins-1,2-cyclic phosphate (Sherman et al., 1985). This ion is absent from the

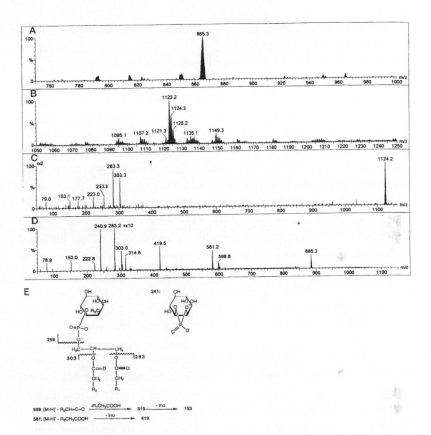

Fig. 5.10. Electrospray mass spectrometric analysis of the PtdIns moieties of CD52-I and CD52-II. *Panel A*: Negative-ion spectrum of the CD52-I PtdIns fraction; *Panel B*: Negative ion spectrum of the CD52-II PtdIns fraction; *Panel C*: Daughter ion spectrum of the m/z 1124 pseudomolecular ion of CD52-II PtdIns; *Panel D*: Daughter ion spectrum of the m/z 885 pseudomolecular ion of bovine liver 1-stearoyl-2-arachidonoyl GroPIns; *Panel E*: Fragmentation scheme for the CID of 1-stearoyl-2-arachidonoyl GroPIns. R_3 is H in the case of bovine liver PtdIns and $CH_3(CH_2)_{14}-CO$ (palmitoyl) in the case of CD52-II PtdIns. R_1 is $CH_3[(CH = CH)_4(CH_2)_9]$ (for the arachidonoyl group) and R_3 is $CH_3-(CH_2)_{15}$ (for the stearoyl) group. Reproduced from Treumann et al. (1995) with permission of the publisher.

corresponding spectrum of the palmitoylated (stearoyl/arachido-noyl GroPIns) from CD52-II (Fig. 5.10(C)). Thus, the collision spectrum suggests that the palmitoyl component of CD52-II is predominantly linked to the 2-position of the inositol ring preventing the formation of this ion. However, the presence of minor ions suggested the presence of several molecular species of acylated PtdIns moieties in the CD52-II.

The glycosylated PtdIns anchor of PARP carrying palmitoyl group on its Ins moiety has been characterized by differential enzymic and chemical degradation (Field et al., 1991). Purification of [^{3}H]myristate-labeled PARP to radiochemical purity and reanalysis of the sensitivity of the fatty acid to release by various treatments confirmed the presence of an acyl-inositol glycosyl PtdIns anchor in PARP. The assignment of the anchor as an acyl-inositol glycosyl PtdIns was consistent with previous studies (Roberts et al., 1988a–c; Mayor et al. 1990b).

It may be noted that GlcN(acyl)PtdIns, in which glucosamine and a fatty acid are linked to inositol groups, has been found to accumulate in HeLa S3 cells as well as in yeast mutants. Sevlever et al. (1995) used octyl-Sepharose and TLC to purify GlcN(acyl)PtdIns following metabolic labeling with [^{3}H]inositol and treatment with venom PLA$_2$. Analysis of GlcN(acyl)PtdIns produced by HF fragmentation showed that palmitate was the acyl group attached to inositol, while stearic and oleic acid were in the glycerolipid. Although GlcN(acyl)PtdIns has been proposed as an early intermediate in the glycosyl PtdIns biosynthetic pathway, in He La D cells it accumulates largely in a compartment that is inert to subsequent mannosylation (Wongkajornsilp et al., 2001).

5.3.5. Diradylglycerol moiety

The hydrolysis of PtdIns glycans is illustrated using the PtdIns-specific phospholipase C from S. aureus or B. thuringiensis (ICN Co). The sample of PtdIns glycan was from bovine erythrocyte acetylcholine esterase. Place the protein (3–12 nmol) in 20 mM phosphate buffer, pH 7, sodium deoxycholate is treated for 90 min at 37°C with the enzyme. Diradylglycerols are extracted with three 1 ml portions of hexane (Low, 1992). The diacylglycerols released from the glycosyl PtdIns-anchored protein or free glycosyl PtdIns by PtdIns-specific PLC can be recovered by TLC using petroleum ether/diethyl ether/acetic acid (80:20:1, by vol.). This solvent system is also suitable for the recovery of the diacylglycerol acetates released by acetolysis (Mayor et al., 1990a). The PtdOH moieties released by treatment of the glycosyl PtdIns-anchor protein or free glycosyl PtdIns by glycosyl PtdIns-specific PLD can be resolved by TLC using the solvent systems made up of chloroform/methanol/water (4:4:1 or 10:10:3, by vol.). Alternatively, chloroform/methanol/acetic acid/water (25:15:4:2, by vol.) or chloroform/methanol/90% formic acid (50:30:7, by vol.) may be used. The nitrous acid deamination products can be best resolved by TLC with chloroform/methanol/water 4:4:1, by vol.) or chloroform/methanol/acetic acid/water (25:15:4:2, by vol.). The nature of the lipid moiety of the glycosyl PtdIns anchors is usually determined by gas chromatography–mass spectrometry (GC/MS) of lipid derivatives from hydrolyzed material (Ferguson, 1992) and/or FAB/MS of the PtdIns or diradylglycerol moieties, released by nitrous acid deamination and PtdIns-specific phospholipase C treatment, respectively (Roberts et al., 1988a; Schmitz et al., 1986). Although the molecular association of the radyl chains in the PtdIns molecules can be determined directly by reversed phase HPLC, FAB-MS/MS or ESI/MS/MS (see below), it is more accurate to asses it by examining the diradylglycerols moieties of PtdIns. This procedure allows the prefractionation of the molecular species into subclasses on the basis of the alkylacyl, alkenylacyl, and diacylglycerol content. Furthermore, a preliminary resolution of the enantiomers and reverse isomers of the rac-1,2-diradylglycerols by chromatography allows the estimation of these subclasses of diradylglycerols, which direct mass spectrometry cannot accomplish.

The hydrolysis of glycosyl PtdIns is illustrated using the PtdIns-specific PLC from *S. aureus* or *Bacillus thuringiensis* (ICN Co). A sample of glycosyl PtdIns from bovine erythrocyte acetylcholine esterase (3–12 nmol) in 20 mM phosphate buffer, pH 7, and sodium deoxycholate is treated for 90 min at 37°C with the enzyme. Diradylglycerols are extracted with three 1 ml portions of hexane (Roberts et al., 1988a).

Butikofer et al. (1992) have released the diradylglycerols from the glycosyl PtdIns moiety of human acetylcholine esterase by a two-step enzyme hydrolysis. First the human erythrocyte enzyme (approximately 50 nmol) was treated with glycosyl PtdIns specific PLD purified from bovine serum (Hoerner, 1992) in 10 mM Tris/HCl pH 7.4, 144 mM NaCl, 0.5 mM $CaCl_2$, 0.01% Triton X-100 overnight at room temperature. Subsequently, the lipids were extracted according to Bligh and Dyer (1959) and the extract was dried under nitrogen. Lipids were resuspended in 300 μl 37.6 mM sodium citrate, pH 4.9, by brief sonication in a bath sonicator, and the PtdOH was dephosphorylated by addition of acidic phosphatase (1 unit; Boeringer, Mannheim, FRG) for 2 h at 37°C. Under these conditions, more than 95% of the PtdOH was dephosphorylated, as tested by commercially available PtdOH standard. Generated diradylglycerols were extracted with toluene, dried under nitrogen, and immediately benzoylated for chromatographic identification. Butikofer et al. (1992) note that the presence of a fatty acid esterified to the inositol residue is sufficient to partition the protein into the detergent phase even after cleavage of the diradylglycerol moiety of the glycosyl PtdIns anchor. In order to determine by phase partitioning in Triton X-114 the extent of hydrolysis of human erythrocyte acetylcholinesterase by glycosyl PtdIns PLD, the inositol linked fatty acid must be removed, e.g. by incubation of the enzyme with 0.8 M hydroxylamine hydrochloride, 0.1 M triethylamine, pH 11.0 (Butikofer et al., 1992).

According to Menon (1994) the release of diradylglycerols from glycosyl PtdIns-anchored proteins or from free glycosyl PtdIns may be effected with PtdIns-specific PLC as follows. The sample is dissolved in 100 μl of buffer (0.1 M Tris–HCl, pH 7.4, 0.1%, w/v, sodium deoxycholate), and PtdIns-PLC is added to a final concentration of 0.1–1 unit/ml). The mixture is vortexed briefly and incubated at 37°C for 1–4 h. The reaction is terminated by adding 5 μl of glacial acetic acid and the reaction products are extracted twice with 100 μl of toluene or hexane. The organic extract is dried in a SpeedVac evaporator and the residue suspended in 15 μl of toluene or hexane for TLC analysis.

Alternatively, acetolysis may be employed as a method of preparation of diradylglycerol acetates from glycosyl PtdIns that cannot be hydrolyzed with phospholipase C because of palmitoylation of the inositol moiety (Roberts et al., 1988b). Acetolysis leads to isomerization of the diradylglycerols, which must be resolved into the sn-1,2- and sn-2,3-isomers before further analysis. According to Menon (1994) the acetolysis of glycosyl PtdIns may be performed as follows. The glycosyl PtdIns sample is dried in a ReactiVial (Pierce Chemical C., Rockford, IL) and 200 μl of a 3:2 (v/v) acetic acid/acetic anhydride mixture is added. The reaction mixture is incubated at 105°C for 3 h. The sample is dried in a SpeedVac evaporator, the residue suspended in chloroform/methanol (2:1, v/v) and applied to TLC plates. The sn-1,2- and sn-1,3-diradylglycerol isomers are resolved by developing the plate with petroleum ether/diethyl ether/acetic acid 80:20:1 (by vol.).

Acetolysis has been found to be useful for releasing diradylglycerols from glycosyl PtdIns that cannot be hydrolyzed with phospholipase C because of palmitoylation of the inositol moiety (Roberts et al., 1988b,c). Acetolysis, however, leads to migration of acyl groups (Privett and Nutter, 1977). The mixture of 1,2- and 1,3-diacylglycerols acetates generated by acetolysis makes detailed analysis of the molecular species more difficult, although the 1,2- and 1,3-forms can be purified before analysis by TLC.

The sn-1,3-diradylglycerol acetates migrate ahead of the sn-1,2-diradylglycerol acetates. However, confusion may arise when both diacetyl and alkylacetylglycerol acetates are present. Diradylglycerol acetate standards should be run along with the unknowns during the TLC purification. The acetates also permit improved HPTLC resolution of the diacyl, alkylacyl and alkenylacylglycerol subclasses (Nakagawa et al., 1985).

In order to remove the products of acyl migration, the diradylglycerols are purified in the free form by TLC on borate-treated silica gel with chloroform/methanol (96:4, v/v) as the developing solvent (Thomas, 1965). The conversion of the diradylglycerols into appropriate derivatives prevents further isomerization and improves their chromatographic separation into subclasses and their detection.

Thus, the diradylglycerols may be converted into TMS or TBDMS ethers by reaction with the corresponding trialkylchlorosilane under conditions that do not lead to isomerization during the reaction (see Chapter 2). Briefly, The diradylglycerols are converted into TMS or TBDMS ethers by reaction with the corresponding trialkylchlorosilane under conditions that do not lead to isomerization during the reaction. Thus, the TMS ethers are prepared (Myher and Kuksis, 1975) by reaction with 100 μl of pyridine/hexamethyldisilazane/trimethylchlorosilane (15:5:2, by vol.) for 30 min at room temperature. The reagents are then evaporated under nitrogen gas and the products dissolved in 2 ml of hexane. After brief centrifugation (1000g), the supernatant is evaporated under nitrogen gas and the sample dissolved in an appropriate volume of hexane. The centrifugation step can be avoided by using pyridine/bis(trimethylsilyl)-trifluoroacetamide/trimethylchlorosilane (50:49:1, by vol.), which yields volatile by-products.

The TBDMS ethers are prepared (Myher et al., 1978) by reacting the diradylglycerols with 150 μl of a solution of 1 M t-butyldimethylchlorosilane and 2.5 M imidazole in dimethylformamide for 20 min at 80°C. After the reaction mixture is dissolved in 5 ml of hexane, the products are washed three times with H_2O and then dried by passage through a small column containing anhydrous sodium sulfate.

The TBDMS ethers of alkenylacyl glycerols ($R_f = 0.66$) and diacylglycerols ($R_f = 0.35$) are well resolved by TLC on Silica Gel

H with benzene (or toluene) as developing solvent (Myher and Kuksis, 1984a,b). BHT has an $R_f = 0.82$ in this system. When present in significant amounts, the alkylacylglycerols migrate to a position just above the diacylglycerols in this solvent system and it is difficult to obtain a pure alkylacylglycerol fraction, when the diacylglycerol fraction is present in larger quantities.

The diradylglycerol TMS or TBDMS ethers are readily resolved into subclasses by normal phase HPLC on a silica column (Supelco, Inc., Bellefonte, PA) with 0.3% (v/v) isopropanol, in hexane (flow rate = 1 ml/min) (Kuksis and Myher, 1986). Fractions corresponding to reference alkenylacyl (3.8–4.8 min), alkylacyl (4.8–6.3 min) and diacyl (6.3–8.5 min) glycerol TMS ethers are collected and evaporated to dryness. The residues are taken up in hexane and saved for analysis of molecular species by polar capillary GLC with flame ionization detection of by GC/MS (see below).

Similar HPLC separations of the subclasses of the diradylglycerols are obtained for the benzoyl (Blank et al., 1987), dinitrobenzoyl (Takamura et al., 1986), naphthylurethane (Rabe et al., 1989) and 1-anthroyl (Ramesha et al., 1989) derivatives.

For preparation of the benzyl esters Blank et al., (1987), used benzoyl chloride. The diradylglycerol benzyl esters are resolved into the alkenylacyl, diacyl and alkylacyl diacylglycerols by normal phase HPLC using a shallow solvent gradient.

For preparation of the naphthylurethanes (Rabe et al., 1989), the diradylglycerols (200 nmol) are dissolved in 100 µl of dimethylformamide. α-Naphthylisocyanate (10 µl) yielding at least 500-fold excess of the reagent is then added along with 1,4-diazabicyclo(2,2,2)octane (0.1 M in dimethylformamide). The stoppered vial is heated at 85°C for 30 min. The excess reagent is destroyed by addition of 300 µl of methanol and the reaction mixture centrifuged to obtain a clear supernatant, which is evaporated to a final volume of 100 µl. The separation of the alkylacyl and diacylglycerol naphthylurethanes (Rabe et al., 1989) is obtained on a Lichrosorb Si 100 column (250 mm × 4 mm) with hexane/tetrahydrofuran (95:5, v/v) as the eluting solvent (flow rate = 1 ml/min). The alkylacyl (12–15 min) and the diacyl (22–27 min) derivatives were clearly resolved from each other but partly contaminated with UV absorbing unknowns, which, however, did not contain significant amounts of fatty acids.

Frequently, the free diradylglycerols released from PtdIns have become isomerized or contaminated with tissue diacylglycerols of unknown origin. These diradylglycerols may contain both regional isomers and racemates. Such racemates are readily resolved into enantiomers by chiral phase HPLC of the dinitrophenylurethane (DNPU) derivatives. The DNPU derivatives are prepared by reacting diradylglycerol samples of less than 1 mg with about 2 mg of 3,5-dinitrophenylisocyanate (Sumimoto Chemical Co., Osaka, Japan) in 4 ml of dry toluene in the presence of dry pyridine (40 μl) at ambient temperature for 1 h. The crude urethane derivatives of diacyl and dialkylglycerols are purified by TLC on silica Gel GF plates (20 cm × 20 cm × 0.25 mm, Analtech Inc., Newark, DE) using hexane/ethylene dichloride/ethanol (40:10:3 and 40:10:1, by vol., respectively). Prior to use, the plates are activated by heating at 110–120°C for 2 h. The DNPU of the diacylglycerols ($R_f = 0.52$) and dialkylglycerols ($R_f = 0.63$) are recovered by eluting the gel scrapings with diethyl ether (Takagi and Itabashi, 1987). DNPU derivatives of synthetic enantiomers of diacyl and alkylacylglycerols are used as reference standards (Takagi et al., 1990). Takagi and Itabashi (1987) have shown, that from chiral phase HPLC columns the dialkylglycerol derivatives emerge well ahead of the diacylglycerol derivatives using the OA-4100 stationary phase and hexane/ethylene dichloride/ethanol (80:20:1, by vol.) as the mobile phase, when analyzed as the dinitrophenylurethane derivatives. The DNPU derivatives of enantiomeric alkylacylglycerols have been prepared and resolved by Takagi et al. (1990).

The resolution of the unknown diradylglycerols into the individual enantiomer subclasses is best performed following a preliminary separation of the diradylglycerols into alkylacyl and diacylglycerol subclasses either at the free diradylglycerol stage or following the preparation of the TMS ethers. The recovered TMS ethers of the diradylglycerol subclasses can be converted into the DNPU derivatives without isomerization. For this purpose, the TMS ethers are reacted directly with the 3,5-dinitrophenylisocyanate or after a prior hydrolysis with H_2O (Itabashi et al., 1990a,b).

Itabashi et al. (1990a,b) have obtained excellent resolution of the enantiomeric diacylglycerols derived from natural triacylglycerols by Grignard degradation on a chiral phase consisting of co-valently bonded (R)-(+)-1-(1-naphthyl)ethanolamine to 300 Å wide pore spherical silica (5 μm particles, YMC-Pack A-KO3, YMC Inc., Kyoto, Japan) using isocratic elution with n-hexane/1,2-dichloro-ethane/ethanol (40:10:1, by vol.). The recovered diradylglycerol enantiomers can be subjected to GLC analysis of fatty acids following transmethylation or to GLC analysis or molecular species following silolysis (see below). The DNPH derivatives of the enantiomeric diradylglycerols may be directly subjected to a resolution of molecular species by reversed phase HPLC with mass spectrometry (see below).

Resolution of enantiomeric diacylglycerols may also be obtained by normal phase HPLC following preparation of the diastereomeric naphthylethylurethane (NEU) derivatives (Christie et al., 1991). For this purpose, the diradylglycerols (1–2 mg) are dissolved in dry toluene (300 μl) and treated with the (R)- or (S)-forms of 1-(1-naphthyl)ethyl isocyanate (10 μl) in the presence of 4-pyrrolidinopyridine (approximately 10 μmol) overnight at 50°C. The products are extracted with hexane/diethyl ether (1:1, v/v) and washed with 2 M HCl and H_2O. The organic layer is taken to dryness under a stream of nitrogen gas and the sample is purified on a short column of Florisil eluted with ether. For optimum resolution, two columns of silica gel (Hypersil, 3 μ, 25 cm × 4.6 mm i.d., HiChrom, Reading, UK) in series were utilized with 0.4 to 0.33% (v/v) 1-propanol (containing 2% water) in isooctane as mobile phase at a flow rate of 1 ml/min. The effluents of the normal phase columns are monitored by UV absorption and can be subjected to on-line mass spectrometry (see below). Alternatively, the NEU derivatives of the diaradylglycerols may be subjected to reversed phase HPLC, which results in an effective resolution of molecular species (see below).

5.3.6. Composition of fatty chains

As a general strategy, it is not recommended to release fatty acids in the free or simple ester form because they may become contaminated with fatty acids from others sources, including solvents and equipment. Nevertheless, the fatty acid and alkyl ether composition of the glycosyl alkylacyl GroPIns is frequently determined by following chemical or enzymic degradation. The total composition of fatty acids can be determined by transmethylation even on very small quantities of material provided sample contamination is avoided. The presence of alkenylethers can be monitored by isolation and identification of the dimethylacetals appearing in the GLC runs of the methyl esters. In contrast, the stable alkylether glycerol moieties must be isolated separately from the transmethylation mixture and reanalyzed by GLC or HPLC as the acetate, TMS or other suitable derivative.

The fatty acid and alkylglycerol composition of alkylacylglycerols TMS ethers isolated by HPLC are determined following a basic methanolysis for 15 min with 0.5 ml $NaOCH_3$ (1 M in methanol/toluene (3:2, v/v), at 20°C. The mixture is then neutralized with 1% (v/v) acetic acid in hexane and extracted by adding 200 µl of chloroform, 0.5 ml of H_2O and 50 µl of 3 M aqueous ammonia. The organic phase is washed with 250 µl of H_2O and then dried by passing it through a small column of anhydrous sodium sulfate. After being dried under nitrogen gas, the sample is acetylated for 0.5 h at 80°C with acetic anhydride/pyridine (1:1, v/v). The reagents are removed by evaporation under nitrogen gas and the products purified by normal phase HPLC on a silica column with 0.8% (v/v) isopropanol in hexane (flow rate = 1 ml/min) (Myher et al., 1989a,b; Kuksis and Myher, 1986). The fractions that correspond to standard fatty acid methyl esters (4.3–7.3 min) and alkylglycerol derivatives (7.3–10.6 min) are collected and evaporated to dryness. The residues are taken up in hexane and examined by GLC on both polar and non-polar capillary columns, as described (Myher et al., 1989a,b; Roberts et al., 1988b).

The ester linked fatty acids can be removed from glycosyl PtdIns preparations by alkaline hydrolysis with methanolic ammonia. According to Menon (1994) the glycosyl PtdIns is suspended in 200 µl of freshly prepared methanol/30% ammonia (1:1, v/v) mixture and incubated at 37°C for 2 h. At the end of the incubation, the reaction mixture is evaporated and the residue resuspended in 200 µl of buffer (0.1 M Tris–HCl, pH 7.4, containing 0.1% sodium deoxycholate). The fatty acids along with any methyl esters are extracted several times by vortexing with 200 µl of toluene or hexane, and centrifuging to separate the aqueous and organic phases.

Alternatively, Menon (1994) recommends to add 100 µl of 50% methanol to the dried sample and to dry again to remove residual ammonia. Then partition the products between 25 µl of 10 mM HCl and 250 µl of diethyl ether. Transfer the ether phase to a clean ReactiVial and reextract the aqueous phase twice with 250 µl of ether. The released fatty acids are quantitatively recovered in the ether phase. The fatty acids are esterified by heating the dried residue with 200 µl of BF$_3$-methanol at 100°C for 2 min. Add 200 µl of 5 M NaCl in 0.5 M acetic acid and extract the fatty acid methyl esters into toluene.

According to Naik et al. (2000), HPLC- and HPTLC-purified, free and amino acid-linked glycosyl PdIns (4–5 µg each), and hydrofluoric acid released inositol-acylated carbohydrate moiety (2–3 µg), and diacylglycerol moiety (1–2 µg) are treated with 8 anhydrous methanolic KOH for 2 h at 120°C. After acidification with cold dilute HCl, the fatty acid methyl esters were extracted with chloroform and analyzed by GLC.

The fatty acid methyl esters are identified by GLC. The deacylated glycosyl PtdIns sample remaining in the aqueous phase may be further treated with PtdIns-specific phospholipase C to give a lyso-alkyl glycerol, which may be identified by GLC or HPLC after appropriate derivatization.

The fatty acid and alkylglycerol composition of the alkylacyl-glycerophospholipids is determined by GLC following chemical degradation. The composition of fatty acids can be determined on very small quantities of material provided sample contamination is avoided.

For this purpose (Roberts et al., 1988a,b) 1 M anhydrous methanolic HCl (100 μl) is added to the dried sample and the reaction mixture is heated at 65°C for 16 h. For analysis of fatty acid methyl esters, samples are extracted with 2,2,4-trimethylpentane. For combined analysis of fatty acid ethyl esters and alkylglycerols, 200 μl of chloroform and 75 μl of H_2O are added to the sample, and, after vortexing, the lower organic phase is removed. Samples for subsequent analysis of TLC are then dried. For GLC analysis, the aqueous phase is reextracted with chloroform (100 μl) and the combined organic extracts are acetylated by incubation with 10 μl of acetic anhydride/pyridine (1:1, v/v) for 30 min at 80°C, dried, and the sample is resuspended in 10 μl of 2,2,4-trimethylpentane.

The release of alkyl- and alk-1-enylglycerols from diradylgly-cerophospholipids can also be achieved by reduction with sodium bis(2-ethoxy-ethoxy)aluminium hydride (Vitride™ or Red-Al™) (Schmid et al., 1975). Compared to LiAlH₄, Vitride™ reagent gives better recoveries of alk-1-enylglycerols and can be purchased as a solution in benzene or toluene (Aldrich Chemical Co.). The resulting alkyl- and alk-1-enylglycerols along with the fatty alcohols are isolated by TLC with neutral lipid solvent systems and are converted to trimethylslyl (TMS) ethers, acetates or isopropylidines for GLC analysis.

Table 5.1 compares the fatty acid composition of the glycosyl PtdIns anchors of acetylcholine esterase of human and bovine erythrocytes (Roberts et al., 1987). The total quantity of fatty acids was 1.4 mol% of bovine compared to 2.0 mol% of human enzyme.

TABLE 5.1

Fatty acid compositions of bovine (E^{bo}) and human (E^{hu}) acetylcholine esterase
(AcChoE)

Fatty acid	Mol/mol of AcChoEase monomer	
	E^{bo}	E^{hu}
14:0	0.03 ± 0.01	0.06 ± 0.02
16:0	0.08 ± 0.02	0.75 ± 0.04
16:1	0.04 ± 0.02	0.09 ± 0.03
18:0	1.13 ± 0.04	0.23 ± 0.03
18:1	0.13 ± 0.02	0.24 ± 0.04
18:2	ND	0.07 ± 0.05
20:4	ND	0.06 ± 0.01
22:4	ND	0.28 ± 0.02
22:5	ND	0.21 ± 0.01
22:6	ND	0.05 ± 0.01
Total	1.41 ± 0.01	2.02 ± 0.07

Reproduced from Roberts et al. (1987) with permission of publisher.

In addition, the fatty acid composition of the two enzymes differ
considerably. Stearate is the most abundant fatty acid in the bovine
enzyme in contrast to palmitate in the human enzyme. Docosahexaenoate and docosapentaenoate are major components in the
human enzyme but are not detected in the bovine enzyme. These
estimates of the fatty chain composition do not make allowance for
the presence of a few percent of alkyl groups estimated
subsequently (Roberts et al. 1988a,b). The fatty acid composition
of the sn-2-position of the glycosyl PtdIns could be determined by
conventional phospholipase A_2 hydrolysis of the PtdIns moiety
released from the anchor by deamination with nitrous acid. The
positional analysis of fatty acids in PtdIns by phospholipaseA_2 is
discussed in Chapter 2.

Table 5.2 compares the alkyl chain composition of alkylacyl-
GroPIns of red blood cells (Lee et al., 1991) and the glycosyl
PtdIns anchor (Roberts et al. 1988a) of acetylcholinesterase of
bovine erythrocytes. The alkyl groups presumably occupied the

TABLE 5.2

Comparison of alkyl chain composition of alkylacyl GroPIns and the glycosyl PtdIns anchor of acetylcholine esterase in bovine erythrocytes

Alkyl chain composition	Inositol Phospholipid (mol%)[a]		Acetylcholine esterase (mol%)[b]
	Exp. 1	Exp. 2	
16:0	13.2	12.7	2.3
17:0	2.0	0.7	3.8
18:0	71.6	70.5	83.7
18:1	13.2	16.3	10.2

[a]Data calculated from molecular species composition assuming that the fraction containing [16:0–18:1 and 18:0–18:2] species and the first listed aliphatic chain on the molecular species corresponds to the alkyl moiety.
[b]Data from Roberts et al. (1988a).
Reproduced from Lee et al. (1991) with permission of publisher.

sn-1-position. On the basis of this resemblance, Lee et al. (1991) have suggested that alkylacylGroPIns may serve as precursors for the glycosyl PtdIns anchor of bovine erythrocyte acetylcholinesterase and other related proteins containing alkylacylGroPIns.

The purified deamination and acetolysis products from human acetylcholinesterase on fatty acid analysis following acidic methanolysis revealed that two fatty acid residues were associated with the deamination product and only one with the alkylglycerol acetolysis product (Roberts et al. 1988b). The other fatty acid residue was primarily palmitate and was present in ester linkage to an inositol hydroxyl as indicated by treatment of the deamination product of deacylated glycosyl PtdIns with base. This linkage was shown to be responsible for the resistance of the glycosyl PtdIns to cleavage by $S. aureus$ PtdIns-specific phospholipase C. Digestion of the acetylcholine esterase with the anchor-specific phospholipase D resulted in release of plasmanic acids (1-alkyllysoGroPIns) from the intact palmitoylated plasmanylinositol (Roberts et al., 1988b).

5.3.7. Inositol moiety

The inositol residue of PtdIns anchors has been identified by GC/MS as *myo*-inositol and is substituted at the 6-position by the glycan moiety (see above). The *myo*-inositol is believed to be of the D form because of the susceptibility of glycosyl PtdIns-tails to cleavage by PtdIns-specific phospholipase C, whose natural substrate is Ptd-D-*myo*-inositol (Parthasarathy and Eisenberg, 1986).

However, both *neo*- and *scyllo*-inositols have also been found in natural PtdIns (see Chapter 3), although no reports of their occurrence in glycosyl PtdIns have thus far appeared. In addition, glycosyl PtdIns anchors of many mammalian proteins, e.g. human acetylcholine esterase (Roberts et al., 1988b), CD52 (Treumann et al., 1995) and protozoal proteins. e.g. PARP (Field et al., 1991), contain a fatty acid attached to the 2-position of the inositol ring. The glycosyl PtdIns anchor of the major *Plasmodium falciparum* mereozoite surface proteins are inositol-myriostoylated (Gerold et al., 1997; Schmidt et al., 1998).

Tiede et al. (1999), McConville and Menon (2000), Kinoshita and Inoue (2000), and Morita et al. (2000a–c) have reviewed the more recent developments in the cell biology and biochemistry of glycosyl PtdIns and have indicated that there is much diversity in the range of structural modifications found in glycosyl PtdIns within and between species and cell types, and in the topological arrangements of their biosynthetic pathway in the endoplasmic reticulum. Consistent with additional functional roles for the glycosyl PtdIns is their wide distribution in the cellular endomembrane system.

5.4. Determination of molecular species

As a prelude to the determination of molecular species of glycosyl PtdIns, a compositional analysis of fatty acids is usually performed.

It provides an estimate of the range of molecular weights to be anticipated during chromatographic and mass spectrometric characterization (see above).

5.4.1. Mass spectrometry

In recent years mass spectrometry has been increasingly employed for the characterization of the structure and the molecular species of glycosyl PtdIns. The analyses have ranged from the mass spectrometric characterization of the intact glycosyl PtdIns anchored protein to determination of the structure of biochemical degradation products.

5.4.1.1. Intact protein

Fini et al. (2000) studied the structure of ecto-5-nucleotidase from bull seminal plasma, containing a glycosyl PtdIns anchor, by MALDI/MS. The analyses of intact protein indicated a mass of 65,568.2 Da for the monomeric form. It also showed a heterogeneous population of glycoforms with the glycosidic moiety accounting for approximately 6000 Da. MALDI/MS analysis showed that Asn53, Asn311, Asn333 and Asn403 were four sites of N-glycosylation. GC/MS analysis provided information on the glycosidic structures linked to the four asparagines. Asn3, sn11 and sn333 were linked to high-mannose saccharide chains, whereas the glycan chains linked to Asn403 contained heterogeneous mixture of oligosaccharides, the high mannose type structure being the most abundant and hybrid or complex type glycans being minor components. By combining enzymatic and/or chemical hydrolysis with GC/MS analysis, detailed characterization of the glycosyl PtdIns anchor was obtained. MALDI/MS analysis indicated that the glycosyl PtdIns core contained EtN(P)Man3GlcNH$_2$-*myo*-inositol-glycerol, principally modified by stearoyl and palmitoyl residues or by stearoyl and myristoyl residues to a lesser extent. Moreover, 1-palmitoylglycerol and

1-stearoyl glycerol outweighed 2-palmitoylglycerol and 2-stearoylglycerol. The combination of chemical and enzymatic digestions of the protein with the mass spectral analysis also yielded a complete pattern of S–S bridges.

Taguchi et al. (1999a) had applied the ES/MS/MS and MALDI/ TOF/MS to analysis of the glycosyl PtdIns anchored C-terminal peptide derived from $5'$-nucleotidase. ES/MS/MS analysis was applied to the core structure (MW = 2743). In the CID spectrum, single-charged ions such as m/z 162 (glucosamine), 286 (mannose-phosphate-ethanolamine), and 447 ([mannose-phosphate-ethanolamine]-glucosamine) were clearly detected as characteristic fragment ions of the glycosyl PtdIns anchored peptide. On MALDI/TOF/MS analysis, heterogeneous peaks of glycosyl PtdIns anchored peptides were detected as single-charged ions in the positive ode. Product ions were obtained by post-source decay of m/z 2905 using curved field reflection of TOF/MS. Most of the expected product ions derived from the glycosyl PtdIns-anchored peptide, containing the core structure and an additional mannose side chain, were successfully obtained.

Taguchi et al. (1999b) in a parallel study have observed that by treatment with PdIns-specific PLC several candidates of glycosyl PtdIns-anchored proteins (e.g. 55, 42, 40 and 30 Da) are obtained from bovine erythrocyte membrane, in addition to the well-known glycosyl PtdIns-anchored acetylcholinesterase. The 42-Da protein was further analyzed by ES/MS after hydrolysis by lysyl endoprotease. By LC/ES/MS, the C-terminal peptides bearing the products of glycosyl PtdIns were effectively detected by combination with CID and multifunctional scanning for these several characteristic fragment ions from the glycosyl PtdIns structure. Existence of microheterogeneity was also observed in the C-terminal glycosyl PtdIns-peptides from the 42-kDa protein. The latter result was confirmed by analysis with TOF/MS. On the basis of MS/MS analysis, this glycosyl PtdIns-anchor structure was found to contain an additional N-acetyl hexosamine.

5.4.1.2. Intact glycosyl PtdIns

The molecular species of diradylglycerols can be resolved, identified and quantified by FAB-MS of the intact original glycosyl PtdIns (Sherman et al., 1985). The mass spectra were obtained on a VG ZAB-HF operated as a double-focusing instrument with 8 keV energy and about 1000 resolution. Mass analyzed ion kinetic spectra (MIKES) were obtained on the same instrument. FAB ionization was performed with xenon atoms using the gun arrangement normally fitted by VG. Samples were applied directly into 2–3 µl of glycerol on the sample holder. It was shown that analysis on a triple sector mass spectrometer with and without collisional activation was necessary for complete composition information especially when the parent ion contains isobaric species. Analysis of the [M − H]⁻ ions for fatty ester composition by means of MIKES was not adequate to resolve fatty acyl daughter ions when the parent ion contains isobaric species. In both instances quantitative analysis of the fatty ester content of individual molecular species was complicated by dissimilar ion yields form fatty acyl-bearing fragments from compositionally different parent ions.

FAB/MS has also been utilized in the identification of the glycosyl Ins backbone and molecular species of the deacylated glycosyl PtdIns from the lipid anchor of human red blood cell acetylcholine esterase (Roberts et al., 1988c).

Although Brennan (1968) reported some evidence for triacylated PtdIns in *Corynebacterium xeroxis*, the FAB/MS data in Figs. 5.2 and 5.3 provide a documentation of fatty acid acylation of inositol and complement the fatty acid composition data (Roberts et al., 1988b) which support this conclusion. FAB/MS data also indicate that treatment with ammonia-saturated methanol preferentially deacylates this palmitoyl group from the plasmanylinositol deamination products. This finding was used to demonstrate that palmitoylation of the inositol is solely responsible for the resistance of the palmitoylated plasmanylinositol to PtdIns specific phospholipase C from *S. aureus* (Roberts et al., 1988c). A structure for the rat brain Thy-1 glycosyl PtdIns anchor has recently been reported by Homans et al. (1988) and illustrates further heterogeneities in anchor structures.

5.4.1.3. PtdIns moiety

MS analyses of the various degradation products of glycosyl PtdIns have been carried out as part of the characterization of their chemical structure. These analyses have also provided some information about the molecular species composition of the mixture. Thus, FAB/MS analyses have been made of the PtdIns resulting from HF-treatment of the glycosyl PtdIns. The products released by both PtdIns specific phospholipases C and D have been extensively characterized by mass spectrometry, as described for PtdIns (Chapter 2) and PtdInsP (Chapter 3). The MS analyses of the diradylglycerols released from glycosylated PtdIns by PtdIns-specific phospholipase C are discussed in detail below.

5.4.2. Chromatography/mass spectrometry

Although less efficient and more complicated, analyses of molecular species of intact PtdIns are possible and necessary when assessing the molecular association of the radyl chains with radio- or stable-isotope labeled polar head groups. These analyses are best performed with reversed or normal phase HPLC with mass spectrometry. The most detailed studies of the molecular species of the diradylglycerol moieties have been obtained following partial degradation of the glycosylated PtdIns. For this purpose the glycosylated PtdIns are subjected to partial degradation by glycosidases and phospholipases.

5.4.2.1. Intact glycosyl PtdIns

Reversed phase HPLC resolution of molecular species of PtdIns requires a prior isolation of pure glycosyl PtdIns. This is usually accomplished by HPLC or HPTLC.

Thus, Naik et al. (2000a) have used HPLC for purification of glycosyl PtdIns. The glycosyl PtdIns (approximately 10 μg plus 400 000 cpm of [³H]GlcN-labeled glycosyl PtdIns) were chromatographed on a C_4 reversed phase Supelcosil LC-304 HPLC column (4.6 × 250 mm², 5 μm particle size; Supelco) using a linear gradient of 20–60% aqueous 1-propanol containing 0.1% TFA over a period of 80 min and held for 30 min at a flow rate of 0.5 l/min (Roberts et al., 1988c). Fractions (1.0 ml) were collected, and elution of glycosyl PtdIns was monitored by measuring radioactivity. The HPLC purified glycosyl PtdIns were further purified by HPTLC. The HPLC purified glycosyl PtdIns (5 μg) were applied to 10 × 10 cm² HPTLC plates as continuous streaks. Parallel spots with [³H]GlcN-labeled glycosyl PtdIns (50 000 cpm) were used for monitoring glycosyl PtdIns bands by fluorography using En³Hance (Naik et al., 2000a,b).

5.4.2.2. Diradyl GroPIns

Reversed phase HPLC of intact PtdIns can be performed without prior derivatization of the PtdIns molecule, although it complicates detection and quantification (Patton et al., 1982; Myher and Kuksis, 1984a,b). Thus, excellent separations of the molecular species of PtdIns from rat liver have been obtained on a 250 mm × 4.6 mm column of 5 μm particle size (Ultrasphere OS, supplier). The column is eluted with 20 mM choline chloride in methanol/water/acetonitrile (90:7::2.5, by vol.) (Patton et al., 1982). The eluate is monitored at 205 nm. The molecular species are resolved in order of increasing partition number. The major species of rat liver PtdIns are eluted in the following order: 16:0/18:1; 18:0/20:4; 18:0/18:2; 1:0/22:5; 18:0/20:3; 18:0/22:4 and 18:0/18:1. Kim and Salem (1987) have reported the resolution of the molecular species of intact PtdIns from bovine liver by a reversed phase LC/MS with thermospray interface. The column was eluted with methanol/hexane/1 M ammonium acetate in H_2O (500:25:25, by vol.) with a flow rate of 1 ml/min of 2 ml/mn. Under the thermospray conditions, the spectrum of PtdIns species is dominated by monoacyl- ($[RO + 74]^+$), and diacylglycerol-type fragments. There is also an ion due to the ammonium adduct of inositol at m/z 198. Similar applications have not been reported for the PtdIns species derived from glycosyl PtdIns anchor preparations.

5.4.2.3. Diradyl GroPOH

Although the molecular association of the radyl chains in the PtdIns molecules can be determined directly by reversed phase HPLC, FAB-MS/MS or ESI/MS/MS (see below), it is more accurate to asses it by examining the diradylglycerol moieties of PtdIns. This procedure allows the prefractionation of the molecular species on basis of the alkylacyl, alkenylacyl, and diacylglycerol content. Furthermore, a preliminary resolution of the enantiomers and isomers of the diradylglycerols by chromatography allows the estimation of these subclasses of diacylglycerols, which direct mass spectrometry cannot accomplish.

The PtdOH is released from PtdIns, PtdInsPs and PtdIns glycans by PtdIns-specific phospholipase D. PtdOH may also occur free in the tissues from which they are isolated along with other phospholipids. Since PtdOH readily becomes associated with salts, great care must be used to avoid contamination with metal salts which leads to adsorption and peak tailing.

Glycosyl PtdIns specific phospholipase D (GPI-PLD) is abundant in mammalian serum, but the source of the circulating enzyme is unknown.

The PtdOH is released from the precipitate carrying the suspected PdIns glycan anchor as follows (Menon, 1995): the protein precipitate or free glycosyl PtdIns is dissolved in 100 μl of buffer and 2 μl of serum added. After a brief vortexing the mixture is incubated at 37°C for 1–4 h. The reaction is terminated by adding 5 μl of glacial acetic aid and extracting the reaction products twice with 200 μl of water-saturated n-butanol. An aliquot of the combined butanol phases is taken for liquid scintillation counting. The remainder of the butanol extract is dried in a SpeedVac evaporator (Savant Instruments) and the residue is resuspended in 15 μl of water-saturated n-butanol for TLC analysis on Silica 60 with an appropriate solvent system (see below). The enzyme is completely inhibited by preincubation of the serum for 30 min on ice with buffer containing 0.5 mM 1,10-o-phenanthroline. EDTA (1 mM) in the assay system also inhibits the enzyme.

When extremely low levels of the PtdIns are suspected, the phospholipase D digestion and the PtdOH isolation may be performed in presence of 25 μg each of myristic acid, dimyristoylglycerol, and dimyristoyl GroP as carriers (Deeg and Vechere, 1997). The samples are spotted on silica gel 60 TLC plates (E. M. Merck, Gibbstown, NJ) and the plates were developed with chloroform/methanol/0.25% KCl (55:45:5, by vol.). The PtdOH spots on the plates were identified by iodine vapors. Radioactive peaks were also recognized that migrated with myristic acid and dimyristoylglycerol.

PtdOH, the expected product from phospholipase D hydrolysis of PtdIns or other sources of PtdIns accounted for 95% of the total radioactive products. The molecular species of PtdOH may be resolved by reverse phase HPLC following methylation as described in Chapter 2. The molecular species of PtdOH can also be readily recognized and quantified by FAB/MS/MS. This method disposes of the analytical column, or reduces its service to that of a convenient injection port. The identification of the species is based on MS/MS fragmentation of the parent molecules. Both free acids and the dimethyl esters are equally well suited for this purpose.

Normal phase HPLC on its own is incapable of resolving any molecular species. However, in combination with mass spectrometry, it can provide a rapid and detailed account of most molecular species.

The molecular species of PtdOH can also be readily recognized and quantified by FAB/MS/MS. This method disposes of the analytical column, or reduces its service to that of a convenient injection port. The identification of the species is based on MS/MS fragmentation of the parent molecules. Both free acids and the dimethyl esters are equally well suited for this purpose.

5.4.2.4. Diradylglycerols

Earlier workers determined the molecular species of the diradylglycerol moieties by GLC analysis of the fatty acids (Ferguson et al., 1985) or FAB/MS analysis of the diradylglycerols (Schmitz et al., 1986; Roberts et al., 1988a–c) released from the glycosyl PtdIns anchors by chemical or enzymatic degradation.

Capillary GLC (Kuksis and Myher, 1990) and reversed phase HPLC (Butikofer et al., 1990; 1992), however, provide highly sensitive and reliable alternative methods for separation and quantification of molecular species within each diradylglycerol subclass following simple derivatization. Non-polar capillary GLC separates diradylglycerols based on molecular weight or carbon number. The TMS ethers of the diradylglycerols are resolved on an 8 m fused silica capillary column coated with a non-polar permanently bonded SE-54 liquid phase. The samples are injected on-column and the oven temperature is programmed from 40 to 150°C at 30°C/min, then to 230°C at 20°C/min and to 340°C at 5°C/min (Myher and Kuksis, 1984a,b). Hydrogen at 6 ψ (1 ψ = 6.9 kPa) head pressure is used as a carrier gas. This separation may be combined with a preliminary resolution of the diacylglycerols by AgNO$_3$-TLC, AgNO$_3$-HPLC or reversed phase HPLC (Myher and Kuksis, 1982).

The molecular species of the diradylglycerols are most extensively resolved by GLC on polar capillary columns, which provide separations based on both carbon and double bond number. The analysis is performed by GLC at 250°C on a 10 m capillary column coated with an SP-2330 polar liquid phase (68% cyanopropyl, 32% phenylsiloxane) supplied by Supelco (Myher and Kuksis, 1982). The temperatures of the injector and detector are maintained at 270 and 300°C, respectively.

More recently (Myher et al., 1989a; Kuksis and Myher, 1990), diradylglycerol TM ethers have been separated on fused silica columns (15 m × 0.32 mm) coated with cross-bonded RTx 2330 (Restek Corp., Port Matilda, PA) or with stabilized SP230 (Supelco) (Myher et al., 1989a). The carrier gas was hydrogen at 3 p.s.i. head pressure and the column temperature isothermal at 250°C. Figure 5.11 compares the separation of the alkylacylglycerol moiety of the glycosyl PtdIns anchor of bovine AcChoE on non-polar and polar capillary columns (Roberts et al., 1988b). The species were made up largely of combinations of C_{18} and $C_{18:1}$ glyceryl ethers and 18:0 and 18:1 fatty acids. The glyceryl ethers and fatty acids were identified by GLC on a polar capillary column following transmethylation and trimethylsilylation. About 96% of the total was alkylacylglycerol, of which sn-1-stearyl-2-stearoyl-glycerol, sn-1-stearyl-2-oleoylglycerol and sn-1-oleyl-2-stearoyl-glycerol accounted for 69, 13 and 10%, respectively. The composition of these diradylglycerols is in marked contrast to the diradylglycerol composition of bovine red blood cell PtdIns, which contains minimal amounts of alkylacylglycerol species. Thus, Lee et al. (1991) reported 4.8–9.5% alkylacylglycerols along with 2.0–2.4% alk-1-enylglycerols of the total cellular diradylglycerols of PtdIns of bovine red cells. Butikofer et al. (1992), however, showed only about one half as much of the alkylacylglycerols and no alkenylacylglycerols among the molecular species of the diradylglycerols derived from bovine red cell PtdIns. Furthermore, the glycosyl PtdIns anchor of bovine red cell acetyl ChoE contained a much higher proportion of the more saturated species than the cellular PtdIns.

Figure 5.12 compares the separation of the alkylacylglycerol moiety of the glycosyl PtdIns anchor of human AcChoE on non-polar and polar capillary columns (Roberts et al., 1988a–c; Kuksis and Myher, 1990). The samples were analyzed as the diradylgly-cerol acetates released from E^{hu}AcChoE by acetolysis. On the non-polar capillary column, the acetates gave two minor peaks for corresponding to C36 and C38, and one major peak corresponding to a C40 diradylglycerol acetate. On the polar capillary column, there was considerable tailing of the peaks possible due to a partial

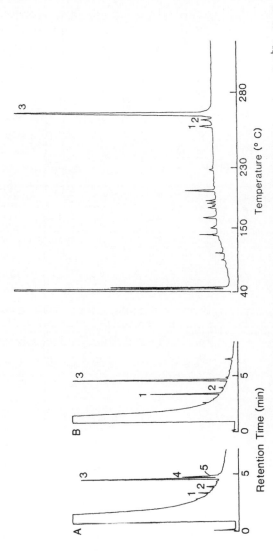

Fig. 5.11. GLC separation of the alkylacylglycerol moieties of the glycosyl PtdIns lipid anchor of E^{bo}AcChoE as trimethylsilyl ethers on (A and B) polar and (C) non-polar capillary columns. Peak identification (A): **1**, 16:0′–18:0 plus 18:0′–16:0; **2**, 17:0′–18:0; **3**, 18:0′–18:0; **4** and **5**, 18:0′–18:1 and 18:1′–18:0. Peak identification (synthetic standards, (B): **1**, 16:0′–18:0; **2**, 17:0′–18:0; **3**, 18:1′–18:0. Peak identification (C): **1**, 34:0; **2**, 35:0; **3**, 36:1. The polar capillary column RTx 2330 (15 m × 0.32 mm ID, Restek Corp., Port Matilda, PA) was operated isothermally at 250°C with hydrogen carrier gas at 3 ψ, while the non-polar capillary column SE-54 (8 m × 0.3 mm ID, Restek) was temperature programmed non-linearly from 40 to 340°C. Sample: 1–2 μl of approximately 0.01% solution of the TMS ethers in hexane. Reproduced from Roberts et al. (1988a) with permission of the publisher.

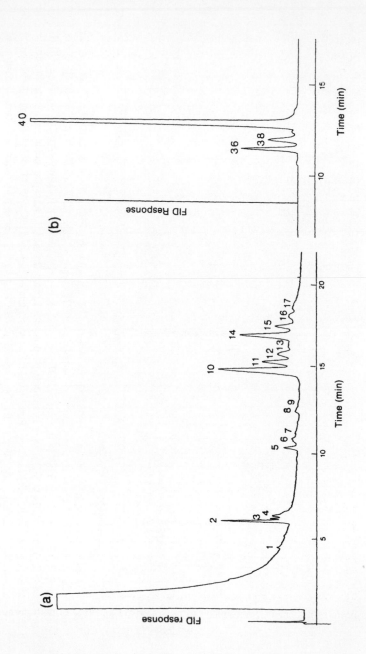

isomerization of the samples during acetolysis. Nevertheless, it was possible to demonstrate that 1-octadecyl-2-docosatetraenoyl-glycerol- (32.4%), 1-octadecyl-2-docosapentaenoyl (docosahexaenoyl)diacylglycerol- (25.2%), 1-octadecenyl-2-docosatetraenoyl glycerol- (14.2%), and 1-octadecenyl-2-docosapentaenoyl (docosahexaenoyl)glycerol- (11%) made up the major components. The identification and quantification of the alkylacylglycerols was obtained by combining fatty acid and glyceryl ether composition data (not shown) with analysis of the intact alkylacylglyceryl acetates on the non-polar and polar capillary columns (Roberts et al., 1988a–c). Again the molecular species composition of the diradylglycerol moieties of the glycosyl PtdIns anchor of the E^{hu}AcChoE differs greatly from that of the molecular species of the diradyl GroPIns of the human red blood cells. Butikofer et al. (1992) has estimated that the alkylacyl content of the PtdIns of human red cells ranges from 1.5 to 3.5% and there was no evidence for the presence of any alkenylacylglycerols. About 96% of the total was alkylacylglycerol, of which sn-1-stearyl-2-stearoylglycerol, sn-1-stearyl-2-oleoylglycerol and sn-1-oleyl-2-stearoylglycerol accounted for 69, 13 and 10%, respectively. The composition of these diradylglycerols is in marked contrast to

Fig. 5.12. GLC separation of the alkylacylglycerol moieties of the glycosyl PtdIns lipid anchor of E^{hu}AcChoE as acetates on (a) polar and (b) non-polar capillary columns. Peak identification (a): **1**, 34:0; **2**, 1,2–18:0′–18:0; **3**, 1,3–18:0′–18:0; **4**, 1,2–18:1′–18:0; **5**, 1,2–18:0′–20:4; **6**, 1,3–18:0′–20:4; **7**, unidentified; **8**, 1,2–17:0′–22:4; **9**, 1,3–17:0′–22:4; **10**, 1,2–18:0′–22:4; **11**, 1,3–18:0′–22:4; **12**, 1,2–18:1′–22:4; **13**, 1,3–18:1′–22:4; **14**, 1,2–18:0′–22:5 plus 18:0′–22:6; **15**, 1,3–18:0′–22:5 plus 18:1′–22:6; **16**, 1,2–18:1′–22:5 plus18:1′–22:6; **17**, 1,3–18:1′–22:5 plus 18:1′–22:6. A prime (′) indicates an alkyl group. Peak identification (b): Peaks 36, 38 and 40, diradylglycerols with a combined alkyl and acyl carbon number of 36, 38 and 40. Instrumentation and analytical conditions as given in Fig. 5.11. Sample: 1 μl of approximately 0.01% diradylglycerol acetates in hexane. Reproduced from Roberts et al. (1988b) with permission of the publisher.

the diradylglycerol composition of bovine red blood cell PtdIns, which contains minimal amounts of alkylacylglycerol species. Thus, Lee et al. (1991) reported 4.8–9.5% alkylacylglycerols along with 2.0–2.4% alk-1-enylglycerols of the total cellular diradylglycerols of PtdIns of bovine red cells. Butikofer et al. (1992), however, showed only about one half as much of the alkylacylglycerols and no alkenylacylglycerols among the molecular species of the diradylglycerols derived from bovine red cell PtdIns. Furthermore, the glycosyl PtdIns anchor of bovine red cell AcChoE contained a much higher proportion of the more saturated species than the cellular PtdIns.

Kuksis and Myher (1990) have provided a detailed discussion of the mass analyses of the molecular species of diradylglycerol moieties of PtdIns by GLC and by GC/MS. While these techniques have been extensively utilized in the analyses of the diradylglycerols of natural glycerolipids and PtdIns, they have found only limited application in the analyses of the molecular species of the diradylglycerol moieties of glycosyl PtdIns anchors. These methods are discussed further in Chapters 2 and 3.

Alternatively, the molecular species of the diradylglycerols can be resolved by reversed phase HPLC. This method is especially convenient if UV absorbing or fluorescent derivatives are prepared. However, HPLC with light scattering can monitor all species of diacylglycerols regardless of the chromogenic properties of the derivative. Butikofer et al. (1992) have presented a complete outline for the isolation, release, derivatization and quantification by reversed phase HPLC of the diradylglycerol moieties of glycosyl PtdIns species from human and bovine erythrocytes and the corresponding AcChoE anchors (Fig. 5.13). Since PtdIns PLC could not be used to release the diradylglycerols from the lipid anchor of human AcChoE, these glycosyl PtdIns were subjected to glycosyl PtdIns PLD to produce PtdOH, which was converted into diradylglycerols by the action of acidic phosphatase. The resulting diradylglycerols were converted into the benzoates, which could be resolved into the diacyl, alkylacyl and alkenylacyl subtypes by

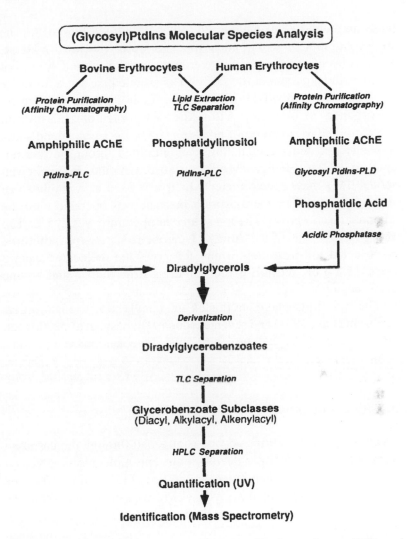

Fig. 5.13. Molecular species analysis of human and bovine erythrocyte PtdIns and AcChoE glycosyl PtdIns by combined HPLC mass spectrometry of diradylglycerobenzoate derivatives. Reproduced from Butikofer et al. (1992) with permission of the publisher.

TLC in benzene/hexane/diethyl ether (50:45:4, by vol.). The molecular species within each subclass were resolved by HPLC with an octadecyl reverse phase column using a solvent system of acetonitrile/isopropanol (80:20, v/v) as mobile phase at a flow rate of 1 ml/min. Individual peaks were quantified by measuring absorbance at 230 nm. After the UV detector, 0.8 ml/min methanol/0.2 M ammonium acetate (10:90, v/v) was added via a T-connector, and the total flow was admitted through a thermospray interface (Finnigan MAT, San Jose, CA) into Finnigan Mat model TSQ mass spectrometer which was used as a single-stage mass separator. The thermospray interface was operated with the filament discharge off. The ion source temperature was 275°C. The $[M + NH_4]^+$ ions of the diradylglycerobenzoates were monitored by selected ion recording. Table 5.3 gives the molecular species composition of AcChoE glycosyl-PtdIns from human and bovine erythrocytes.

The 3,5-dinitrophenylurethanes of diacylglycerol enantiomers (Itabashi et al., 1991) and reverse isomers (Itabashi et al., 2000) can be resolved by HPLC on chiral and reversed phase columns, respectively.

5.5. Quantification

Since the glycosyl PtdIns anchor is attached through the carboxyl terminal to the protein molecule, only one molecule of glycosyl PtdIns is found per membrane bound protein. It follows that the various constituents of the glycosyl PtdIns can be quantitatively related to *myo*-inositol, a typical constituent of all glycosyl PtdIns so far characterized. Any excess of glycosyl PtdIns must therefore occur in the free or unbound form. The glycosyl PtdIns moiety is quantified on the basis of radioactivity following appropriate prelabeling with *myo*-[2-³H]inositol or any other biosynthetic precursor of glycosyl PtdIns. The presence of the label may be quantified in any subfraction of glycosyl PtdIns that

TABLE 5.3

Molecular species composition of acetylcholine esterase (AcChoE) glycosyl-PtdIns from human and bovine erythrocytes (%)

Molecular species	Amount in AcChoE glycosyl-PtdIns	
	Human alkylacyl	Bovine alkylacyl
18:2–22:5	0.9	
16:1–22:4	1.3	
18:2–20:4	0.9	
18:1–22:6	3.3	
18:1–20:5 + 16:0–22:6	0.9	
18:1–22:5	4.8	
18:0–20:5	2.4	
18:0–22:6	10.6	
18:1–22:4	4.2	
16:0–22:4	0.8	
18:0–22:5	33.2	
18:0–20:4	2.5	
18:0–22:4	26.4	98
18:0–20:3	2	
18:0–18:1		23.4
16:0–18:0		1.0
18:0–18:0	3.9	72.9
Other	3.9	2.6
Total	100.0	99.9

Reproduced from Butikofer et al. (1992) with permission of publisher.

can be released in representative manner by enzymatic or chemical methods. The mass measurements of glycosyl PtdIns are usually made by mass spectrometry, but mass measurements of the released lipid moieties can also be obtained by chromatography using variety of detectors.

5.5.1. Mass spectrometry

Originally FAB/MS was used for the analysis of permethylation products of glycosyl PtdIns anchors and their N-acetylation and hydrazinolysis products. Later, electrospray ionization mass spectrometry (ESI-MS) in combination with high performance liquid chromatography pump was used for the characterization and quantification of the various degradation products of glycosyl PtdIns. More recently, ESI-MS analyses are performed following normal phase or reversed phase HPLC resolution of the unknown mixture (LC/ESI/MS) and LC/ESI/MS/MS.

GC/MS is commonly employed for the analysis of the methylation products. Inositol, neutral sugar, phosphosugar, and methylation linkage analyses are performed as described by Ferguson (1992) and Schneider and Ferguson (1995). The analyses can be conveniently performed with a Hewlett–Packard 5890-MSD system using SE-54 column (30 m × 0.25 mm, Alltech). Methylation products are analyzed using SP-2380 column (30 m × 0.25 mm, Supelco). Alkylglycerols and fatty acids in relation to each other and to inositol may be estimated using heptadecanol or 1-hexadecyl glycerol as internal standard. GC/MS quantification of *myo*-inositol in presence of *scyllo*-inositol as internal standard was described in Chapters 3 and 4.

5.5.2. Chromatography

HPTLC conditions for analysis of neutral glycan fractions of glycosyl PtdIns have been described by Schneider et al. (1993) and Treumann et al. (1995) and may be summarized as follows. The chromatographic separations are performed on Silica Gel 60 HPTLC plates using either a solvent system A: 1-propanol/acetone/water (9:6:5, by vol.) for the first and third developments; solvent system B: 1-butanol/ethanol/water (4:3:3, by vol.), three or four developments; or solvent system C: 1-propanol/acetone/water (9:6:4, by vol.) for one development. Labeled glycans were detected by fluorography after the sheets were sprayed with EN^{3}Hance spray (DuPont NEN).

Santos de Macedo et al. (2001) have used Silica Gel 60 TLC plates for the quantitative analysis of glycolipids using chloroform/methanol/water (4:4:1, by vol.) as the solvent system. After chromatography, the plates were dried and scanned for radioactivity using a Berthold LB 2842 automatic TLC scanner or analyzed by a BAS-1000 Bio-Imaging Analyser (Fuji Film).

Both polar and non-polar capillary GLC has been extensively utilized for the quantification of molecular species of diradylglycerols as the TMS or TBDMS ethers using hydrogen flame ionization detector (Kuksis and Myher, 1990). Likewise, polar capillary GLC with flame ionization detector has been routinely employed for the quantification of the fatty acid esters and alkylglycerol ethers derived from glycated PtdIns (Roberts et al., 1988a,b). In a micromethod described by Sevlever et al. (2001), a sample scraped from silica beads was combined with a 17:0 fatty acid internal standard (5 nmol) and hydrolyzed in 100 μl 1 M anhydrous methanolic HCl for 16–20 h at 65°C. Chloroform (200 μl) and water (75 μl) were added, the partitioned aqueous phase was washed with chloroform (100 μl) and the combined organic phases were dried. Samples for fatty acid methyl ester determination were resuspended in 20 μl isooctane. Aliquots were analyzed on a Hewlett–Packard Model 5890 gas chromatograph containing SP 2380 column (15 m × 0.25 mm ID, Supelco) and a flame ionization detector using previously determined response factors (Roberts et al., 1988b).

Both normal and reversed phase HPLC has been extensively utilized for the quantification of diradylglycerols as UV absorbing (Butikofer et al., 1992). The UV absorbing peaks, however, may require confirmation of identity by mass spectrometry, which also may be used for quantification assuming identical response factors (Butikofer et al., 1992).

Ion exchange chromatography is frequently employed for the isolation of various degradation products of glycosyl PtdIns. Measurements of radiolabeled amines in acid hydrolysates by cation-exchange chromatography have been described by

Sevlever et al. (1995) based on previous protocols. Briefly, a [^{14}C]methylated or [^{3}H]inositol-labeled bands scraped from TLC plates and the silicon beads were hydrolyzed in 6N HCl at 100°C for 10 h. The products were chromatographed on a Beckman 119CL amino acid analyzer in pH 2.2 buffer and identified by characteristic elution times (Deeg et al., 1992a,b).

Monosaccharides in glycolipid hydrolysates (4 M HCl for 4 h at 100°C) may be determined by high pH anion exchange chromatography on a Dionex Basic Chromatography System (Dionex Corp) using a CarboPac PA-1 column (4 mm × 250 cm, Bio-LC, Dionex Co., Sunnyvale, CA), and isocratic conditions (10 mM NaOH) (Santos de Macedo et al., 2001). Fractions of 0.3 ml were collected and subjected to liquid scintillation.

5.5.3. Chemical analysis

Conventional chemical analyses are usually limited to determination of phosphorus (Chapter 2), carbohydrate (orcinol staining), while nitrogen content has been estimated from amino acid and aminosugar analyses by an amino acid analyzer.

5.6. Related structures

The structure of glycosyl PtdIns anchors has similarities to some glycolipids of mycobacteria and yeasts and the phytoglycolipids of plants. The extensive methodologies developed for these molecules have been invaluable in studying the structure of glycosyl PtdIns anchors.

5.6.1. Glycosylinositol phosphoceramides

Within the cell membrane glycosphingolipids and cholesterol cluster together in distinct domains or lipid rafts, along with glycosyl PtdIns anchored proteins in the outer leaflet and acylated proteins in the inner leaflet of the bilayer. These lipid rafts are characterized by insolubility in detergents such as Triton X-100 at 4°C. Brown and London (1998), Rietveld and Simons (1998) and Hooper (1999) have reviewed the biophysical and biochemical properties of membrane microdomains that make up the lipid rafts. Denny et al. (2001) have reported the segregation of glycosyl PtdIns-anchored proteins and glycoconjugates into lipid rafts in *Kineto-plastida*.

In glycosyl PtdIns, the released lipid moiety is alkyl/acylglycerol or a diacylglycerol. Ceramide-based lipids have been found in yeast (Conzelmann et al., 1992; Fankhauser et al., 1993) and *Dictyostelium discoideum* (Stadler et al., 1989; Haynes et al et al., 1993). Fankhauser et al. (1993) have presented a detailed scheme of reactions for analysis of the structure of yeast glycosyl PtdIns anchors, including FAB/MS, ending in the identification of the majority of glycosyl PtdIns anchors with a ceramide lipid moiety, while the minor glycosyl PtdIns anchors had lyso- or diacylglycerol as the lipid moiety. Figure 5.14 gives the structural variants found in yeast glycosyl PtdIns anchor (Guillas et al., 2000).

Azzouz et al. (1995, 2001) have determined the structure of the glycosylinositol-phosphoceramide produced in a cell-free system prepared from *Paramecium primaurelia*. The glycan moiety of the polar glycolipid is substituted with an acid labile unit, which was identified as a mannosyl phosphate. The structure of the most polar of the glycosylinositol phosphoceramides was suggested to consist of ethanolamine phosphate-6Man-α1-2-Man-α1-6Man (mannosyl phosphate) α1-4glucosamine-inositolphosphoceramide (Azzouz et al., 1995). The lipid moieties of the *Paramaecium* glycolipids were shown not to be susceptible to mild alkali treatment, indicating the absence of ester-linked fatty acids, but were found to be sensitive to sphingomyelinase treatment, suggesting a ceramide-based lipid. According to Azzouz et al. (1995) the glycosylinositol-phosphoceramides were extracted twice using 1 ml chloroform/methanol (2:1, v/v), followed by chloroform/methanol/water (10:10:3, by vol.) to obtain the more polar glycolipids. The glycolipids recovered in the chloroform/methanol extracts were subjected to repeated

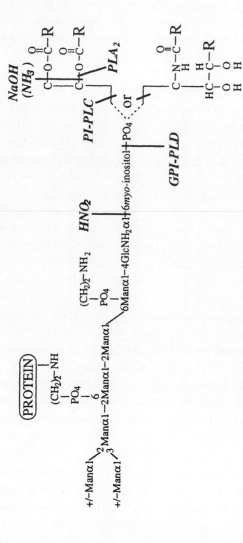

Fig. 5.14. Structural variants of the yeast glycosyl PtdIns anchors. Relevant cleavage procedures are indicated. Anchors can contain an optional fifth mannose residue linked either to α1,2 or α1,3. The residue is added to glycosyl PtdIns proteins in the Golgi (Fankhauser et al., 1993; Sipos et al., 1997). They also bear a second ethanolaminephosphate on the α1,4-linked mannose. GPI-PLD, glycosyl PtdIns-specific phospholipase D; PI-PLC, PtdIns-specific phospholipase C. Reproduced from Guillas et al. (2000) with permission of the publisher.

'Folch' wash. The resulting lower phases were pooled, dried in a speed-Vac and partitioned between water and water-saturated n-butanol. The glycolipids recovered in butanol phases were analyzed by TLC on silica 60 plates (Merck) using either solvent system A (chloroform/methanol/0.25% KCl, 10:10:3, by vol.) or system B (chloroform/methanol/ acetic acid/water, 25:15:4:2, by vol.) or on silica 50 000 HPTLC plates (Merck) with chloroform/methanol (9:1, v/v). For structural analysis, glycolipids were purified by scraping the corresponding areas and extracting for 15 min under sonication with chloroform/methanol/water (10:10:3, by vol.) or chloroform/methanol (2:1, v/v). The methods of enzymatic and chemical analysis of the glycolipids paralleled those just described for glycosyl PtdIns.

Jennemann et al. (1999, 2001) have isolated from the edible mushroom, the basidiomycete *Agaricus*, a novel carbohydrate-homologous series of four glyco-inositol-phosphosphingolipids, designated basidiolipids. The chemical structures of the basidiolipids were elucidated to be: Manβ1-2-inositol1-phospho-ceramide, Galpα-6[Fucpα-2]Galpβ-6 Manβ-2-inositol1-phosphoceramide, Galpα-6Galpα-6[Fucpα-2]Galpβ-6Manpβ-2inositol1-phosphoceramide and Galpα-6Galpα-6[Fucpα-2]Galpβ-6Manpβ-2-inositol 1-phosphoceramide.

All four glycolipids contained a ceramide which was composed of phytosphingosine and predominantly α-hydroxy-behenic and α-hydroxy-lignoceric acid. For extraction of the glycolipids with organic solvents, a prior lyophylization step was necessary, because of the high water content of the mushrooms. However, an extraction procedure using water was also established. The extraction with organic solvents was performed by suspending 1 g of dry powder in 20 ml of methanol, or propanol-2/hexane/water (55:35:10, by vol), or chloroform/methanol/water (60:35:8, by vol.). Each extraction mixture was allowed to stand for 1 h at 50°C. Supernatants were collected after centrifugation at 4000 rpm. for 10 min at room temperature. The extraction was repeated once and the extracts concentrated and resolved by TLC. The four basidio compounds that were detected on TLC as major carbohydrate-positive bands, were separated, using a glass column filled with silica gel Si60 LiChroprep material, washed with 500 ml chloroform/methanol (8:2, v/v). The column was washed with the same solvent mixture followed by chloroform/methanol/water (65:25:4, by vol.), 1 l each. Single basidiolipids were eluted from the column with 3 l of chloroform/methanol/water (60:27:5, by vol.). Fractions of 100 ml were collected, and screened for their content by TLC.

Loureiro y Penha et al. (2001) have characterized novel structures of glycoinositolphosphoceramide from the infective yeast form of *Sporothrix schenckii* by methylation analysis, mass spectrometry and NMR spectroscopy. The lipid portion was characterized as a ceramide composed of C_{18} phytosphingosine N-acylated by either 2-hydroxylignoceric (80%), lignoceric (15%) and 2,3-dihydroxy-lignoceric (5%) acids. The ceramide was linked through a phosphodiester to *myo*-inositol, which is situated on position O-6 by an oligomannose chain. Glycoinositolphosphorylceramide-derived Ins oligomannosides were liberated by ammonolysis and characterized as a novel family of fungal glycoinositolphosphoryl-ceramides as they contained Manpα1-6Ins substructure, which has not been previously characterized unambiguously, and may be acylated with a 2,3-dihydroxylignoceric acid, a feature hitherto undescribed in fungal lipids.

Heise et al. (2002) have reported a molecular analysis of a novel family of complex glycosyl InsP ceramides from *Cryptococcus neoformans*, which show structural differences between encapsulated and acapsular yeast forms.

5.6.2. Lipophosphoglycans and glycoinositolphospholipids

The phytoglycolipids of plants are also similar to glycosyl PtdIns. In their glycan moieties the plant structures again have an α mannose residue linked to the C-2 position of *myo*-inositol and a 'core' region of $GlcNH_2$ or GlcNAcα1-4αGlc UA linked to the C6 position (Ferguson and Williams, 1988; Thompson and Okuyama, 2000).

A variety of glycosyl PtdIns-related structures, such as lipophosphoglycans and glycoinositolphospholipids are expressed by other trypanosomatid parasites (Ferguson, 1999). Likewise, *Plasmodium* (Gerold et al., 1996b,c), *Toxoplasma* (Striepen et al., 1997, 1999), *Trichomonas* (Singh et al., 1994) and *Entamoeba* (Moody-Haupt et al., 2000) also have abundant glycosyl PtdIns-anchored glycoproteins and/or glycoinositolphospholipids.

The parasitic protozoan *Leishmania donovani* expresses a glycosylated PtdIns (Turco et al., 1987; Turco and Descoteaux, 1992). The lipid portion is an unusual 1-alkyl lysoPtdIns that contains just one ether linked fatty alcohol (Orlandi and Turco, 1987). The *myo*-inositol head-group is substituted at the C6 position by Manα1-4αGlcNH$_2$, which is identical to the beginning of the VSG glycosyl PtdIns-glycan (Homans et al., 1988). Structurally, the glycosylinositol phospholipid consists of a polymer of repeating PO$_4$-6Gal(β1-4)Manα1 units (average of 16 units) linked via phosphosaccharide core to a novel lyso-1-O-alkyl PtdIns lipid anchor (Turco, 1990). According to Carver and Turco (1991) [^{3}H]galactose- or [^{14}C]mannose-labeled product was cleaved by PtdIns specific PLC, deaminated by nitrous acid, and degraded into radioactive, low molecular weight fragments upon hydrolysis with mild acid.

5.6.3. Lipoarabinomannans and arabinogalactans

The lipoarabinomannans, which exhibit a large spectrum of immunological activities and emerge as the major antigens of mycobacterial envelopes, possess a structure based on a PtdIns anchor. It has been shown that the integrity of the lipoarabino-mannan structure is crucial for its biological activity and particularly for presentation to CD4/CD8 double negative $\alpha\beta$ cells by CD1 molecules. Nigau et al. (1999) have developed an analytical approach for high resolution ^{31}P-NMR analysis of native, i.e. multiacylated, lipoarabinomannans. The one-dimensional ^{31}P-spectrum of cellular lipoarabinomannans, from

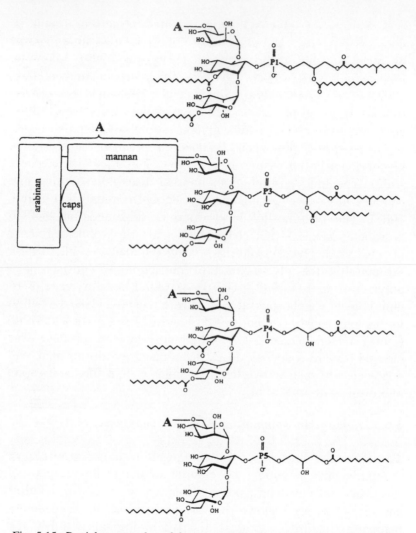

Fig. 5.15. Partial structural models proposed for the mannosylated lipoarabino-mannans (ManLAM) characterized by P1, P3 P4 and P5 phosphates. The PtdIns anchor structure is detailed. The acylation state of position 6 of Man*p* linked to C-2 of *myo*-inositol was not established in this study, but this position is acylated in the fourth model since it is a potential acylation site (Khoo et al., 1995). The nature of the fatty aids on glycerol is in agreement with previous data on the

Mycobacterium bovis Bacillus Calmette-Guerin, exhibited four ^{31}P resonances typifying four types of lipoarabinomannans. Two-dimensional [1]H–[31]P heteronuclear multiple-quantum-correlation/ homonuclear Hartmann–Hahn analysis of the native molecules showed that these four types of lipoarabinomannan differed in the number and localization of fatty acids (from 1 to 4) esterifying the anchor. Besides the three acylation sites described previously (Nigau et al., 1997), i.e. positions 1 and 2 of glycerol and 6 of the mannosyl unit linked to the C-2 of *myo*-inositol, Nigau et al. (1999) showed the existence of a fourth acylation position at the C-3 of *myo*-inositol. Figure 5.15 shows the structures of the native multiacylated lipoarabinomannans, establishing the structure of the intact PtdIns anchor.

Zawadzki et al. (1998) have shown that one of the glycoino-sitolphospholipids of *Leishmania panamensis* contains unusual glycan and lipid moieties. The glycolipids were purified by HPTLC and their structures were determined by GLC/MS, fast-atom bombardment, methylation analysis and chemical and enzymatic sequencing of the glycan headgroups. The major glycosylino-sitol phospholipids contained two glycan core sequences, Manα 1-3Manα 1-4GlcN-PtdIns (type 2 series) or Manα 1-3|Manα 1-2Manα 1-6]Manα 1-4GlcN-PtdIns (hybrid series), which were elaborated with Galα1-2Galβ 1- or Galα 1-2/3Galα 1-2Galβ 1-extensions that were attached to the 3-position of the α 1-3 linked mannose. The PtdIns moiety contained exclusively diacylglycerol with palmitoyl, stearoyl and heptadecanoyl chains.

ManLAM acylglycerol structure (Nigau et al., 1997); only the major forms are represented but other combinations are possible. Man*p* are esterified by palmitic acid in agreement with previous work (Khoo et al., 1995). The nature of *myo*-inositol acyl substituents remains unknown. P1 ManLAMs represent approximately 47%; P3 ManLAMs, 40%; P4 ManLAMs, 6%; and P5 ManLAMs, 7%. A corresponds to the carbohydrate moiety of ManLAMs. Reproduced, with permission, from Nigau, J., Gilleron, M. and Puzo, G. (1999). Biochemical Journal 337, 453–460. © The Biochemical Society and the Medical Research Society.

Non-galactosylated species with the same core structures were also found.

Oxley and Bacic (1999) have determined the structure of the glycosylPtdIns anchor of an arabinogalactan protein from *Pyrus communis*. The glycosyl PtdIns anchor had the minimal core oligosaccharide structure consistent with those found in animals, protozoa, and yeast, but with a partial β(1-4)-galactosyl substitution of the 6-linked Man residue, and had a phosphoceramide lipid composed primarily of phytosphingosine and tetracosanoic acid.

Jones et al. (2000) have determined the structure of a complex glycosyl PtdIns-anchored glucoxyllan from the kinematoplastid protozoan *Leptomonas samueli*.

CHAPTER 6

Biosynthesis of inositol phospholipids

6.1. Introduction

The biosynthesis and metabolism of PtdIns is of enormous interest due to the involvement of PtdIns and its phosphorylated derivatives in intracellular signal transduction. The biosynthesis of PtdIns and PtdIns phosphates in mammalian systems has been extensively discussed in formal reviews (Hokin, 1985; Kent, 1995) and in leading original publications (Carstensen et al., 1995; Lykidis et al., 1997). Eukaryotic PtdIns is synthesized from CDP-diacylglycerol and inositol in a reaction catalyzed by PtdIns synthase, also known as CDP-diacylglycerol:inositol transferase. The PtdIns phosphates are synthesized by successive phosphorylation at C-4, C-5 and C-3 of the inositol ring. The breakdown products of the inositol phosphatides are ubiquitous second messengers downstream of many G protein-coupled receptors and tyrosine kinases involved in mitogenesis, the regulation of calcium mobilization, and protein kinase C activation (Cantley et al., 1991; Noh et al., 1995). Inositol phospholipids are also involved in vesicular movement within cells (DeCamilli et al., 1996), cytoskeletal organization (Hartwig et al., 1995; Janmey, 1994; Pike and Casey, 1996), and stimulation of protein kinase cascades (Franke et al., 1997). Other reviews have dealt with the biosynthesis of glycosyl PtdIns (Stevens, 1995; Ferguson et al., 1999; Tiede et al., 1999). Separate chapters have been included elsewhere in this book on PtdIns kinases (Chapter 7), PtdInsP phosphatases (Chapter 8) and PtdIns phospholipases

(Chapter 9). In addition, certain aspects of the biosynthesis of PtdIns and PtdInsPs have been considered under PtdInsP and InsP signaling (see Chapter 11).

Although a rapid agonist-dependent burst of PtdIns biosynthesis was the first feature of the PtdInsP signaling pathway to be discovered (Hokin, 1985), much of the recent work on these signal transduction pathways has focused on the regulation of the kinases and phospholipases involved in generating PtdInsP derived second messengers. Little is known about the biochemical mechanisms responsible for the control over PtdIns synthesis.

6.2. Phosphatidylinositols

PtdIns may be derived via de novo synthesis, glycerophospholipid head group exchange and transphosphatidylation. In addition, the acyl groups of PtdIns and GPtdIns are subject to acyl exchange. It is not known to what extent the PtdIns phosphates participate in the latter transformations. In yeast, the diacylglycerol moiety of GPtdIns may be exchanged en bloc for a ceramide moiety.

6.2.1. De novo synthesis

The biosynthesis of PtdIns was shown by Agranoff et al. (1958) and Paulus and Kennedy (1960) to proceed via PtdOH, CDP-diacylglycerol and *myo*-inositol (Kent, 1995). Kinnard et al. (1995) have isolated *scyllo*-inositol-containing PtdIns from plant cells. Carstensen et al. (1999) have demonstrated that *scyllo*-inositol can substitute for *myo*-inositol during PtdIns biosynthesis in barley seeds. The overall pathway of the biosynthesis of PtdIns is summarized in Fig. 6.1 based on the recent study in barley seeds by Carstensen et al. (1999). Ferguson et al. (1999, 2000) have claimed the biosynthesis of glycosyl PtdIns as a drug target African sleeping sickness.

There are two enzymes in the PtdIns biosynthetic pathway, CDP-DAG synthetase (CTP:phosphatidate cytidylyltransferase, EC 2.7.7.4.1) and PtdIns synthase (CDP-diacylglycerol:*myo*-inositol 3-phosphatidyltransferase, EC 2.7.8.11). CDP-DAG synthase is found at a branch point in phospholipid metabolism, where phosphatidic acid is partitioned between diacylglycerol and CDP-DAG, the key intermediate in the formation of anionic phospholipids (Kent, 1995).

The enzyme CDP-diacylglycerol synthase responsible for reaction [2], (Figure 6.1) purified from *E. coli* (Icho et al., 1985) had a molecular weight of 27 000, while that from *Drosophila* (Wu et al., 1995) and yeast (Carman and Kelley, 1992; Shen et al., 1996) possessed molecular weights of 49 000 and 52 000, respectively. cDNA sequences for CDP-DAG synthase have been obtained for human (Weeks et al., 1997) and rat brain (Saito et al., 1997). The molecular mass of this rat enzyme was 53 000. At the amino acid sequence, the rat enzyme shared 55.5, 31.7 and 20.9% identity with the already known *Drosophila*, *Saccharomyces cerevisiae* and *E. coli* enzymes, respectively. The rat CDP-DAG synthase preferred 1 stearoyl 2-arachidonoyl GroP as a substrate, and its activity was strongly inhibited by PtdIns(4,5)P$_2$ (Saito et al., 1997). Much of the arachidonoyl GroPIns, however, is believed to be derived by subsequent acyl exchange (see below). A direct biosynthetic origin has not been established for the small amounts of alkylacyl and alkenylacyl GroPIns found in human and bovine erythrocyte PtdIns (Lee et al., 1991; Butikofer et al., 1992).

PtdIns synthase, the enzyme responsible for reaction [3], (Figure 6.1) was first solubilized and purified from rat brain (Rao and Strickland, 1974) and liver (Takenawa and Egawa, 1977) microsomal fraction. It was activated by either Mg^{2+} or Mn^{2+} and phospholipid. The PtdIns synthase from rat liver gave on SDS gel a molecular weight of 60 000. However, PtdIns synthase purified from human placenta (Antonsson, 1994) and rat liver (Monaco et al., 1994) more recently, yielded proteins with estimated masses of 24 and 21 kDa, respectively. Furthermore, Tanaka et al. (1996)

Note: Asterisk (*) indicates presence of radiolabel.

cloned a PtdIns synthase cDNA from rat brain by functional complementation of the yeast pis mutation, which is defective in PtdIns synthase, and obtained a deduced protein composed of 213 amino acids with a calculated molecular mass of 23 613 Da.

The PtdIns synthase purified from yeast gave a molecular weight of 34 000 (Carman and Fischl, 1992), while a molecular mass deduced from the gene for yeast PtdIns synthase had given earlier value of for the enzyme as 24 823 (Nikawa et al., 1987). Antonsson and Klig (1996) have isolated PtdIns synthase from the microsomal cell fraction of *Candida albicans* and have compared it to the PtdIns synthases isolated from *S. cerevisiae* and human placenta. Both similarities and minor differences were observed. Carman and Henry (1999) have reviewed the phospholipid biosynthesis in yeast.

The CDP-DAG synthetase associated with the cytoplasmic side of the endoplasmic reticulum is thought to operate in the PtdIns biosynthetic pathway, whereas the enzyme located on the matrix side of the inner mitochondrial membrane appears to be involved in the synthesis of PtdGro and cardiolipin.

The cellular inositol may be derived from diet or uptake from plasma, de novo biosynthesis, and recycling. Biosynthesis of inositol from glucose can occur in the brain and testes and to a lesser extent in other tissues. The rate-limiting step appears to be the synthesis of Ins(3)P from glucose 6-phosphate (Downes and McPhee, 1990). Ins(3)P is converted to Ins by InsP phosphatase.

Fig. 6.1. Biosynthesis of *myo*- and *scyllo*-PtdIns. Step 1, diacylglycerol kinase; Step 2, CDP-DAG synthase (CTP:phosphatidate cytidylyltransferase, EC 2.7.7.4.1); Step 3, PtdIns synthase (CDP-DAG:*myo*-inositol 3-phosphatidyltransferase, EC 2.7.8.11); The reaction of CDP-DAG with *myo*-inositol to yield *myo*-PtdIns and cytidine monophosphate is catalyzed by CDP-DAG:inositol 3-phosphatidyltransferase (EC 2.7.8.11), also called *myo*-PtdIns synthase.; Step 4, analogous to Step 3, but utilizing *scyllo*-inositol; Step 5, head group exchange; Step 6, *myo*-PtdIns:*myo*-inositol phosphatidyltransferase. Reproduced from Carstensen et al. (1999) with permission of the publisher.

myo-Inositol is transported into live cells by a carrier-mediated process that does not involve active transport.

By analogy with CTP-phosphocholine cytidylyltransferase, the rate-limiting enzyme in PtdCho biosynthesis, cytidylyltransferase has been proposed (Kent, 1995) as a regulatory point in PtdIns biosynthesis. Lykidis et al. (1997, 1998) have challenged this suggestion by showing that an overexpression of the cytidylyltransferase does not result in significantly enhanced incorporation of Ins into PtdIns or result in higher cellular levels of cellular PtdIns. Similarly, cells overexpressing the PtdIns synthase did not increase PtdIns formation. Equally unsatisfactory were explanations advanced on basis of inhibition of cytidylyltransferase 1 by PtdInsPs (Saito et al., 1997) and/or activation of PtdOH phosphatase 1 as pointed out by Lykidis et al. (1997). The data of Lykidis et al. (1997), however, were consistent with DAG kinase as a potential rate controlling enzyme in the production of PtdIns. There are several isoforms of DAG kinase, each of which contains distinct domains postulated to regulate its activity and mediate protein:protein interactions at specific cellular sites (Carman and Fischl, 1992; Goto and Kondo, 1996). Weeks et al. (1997) constructed stable cell lines that overexpressed twice the normal levels of human Cds1. Such cells appeared to amplify cellular signaling systems. These results together with those of Lykidis et al. (1997, 2001) have prompted the conclusion that the CDP-DAG synthase level exerts a determinant role in regulating the rate of PtdIns turnover rather than controlling the rate of de novo PtdIns biosynthesis. It remains to be determined if the other CDP-DAG synthase isoform, CDP-DAG synthase 2, is responsible for regulating the de novo component of PtdIns metabolism. Lykidis et al. (1998; 2001; see also Lykidis and Jackowski, 2001).

Carstensen et al. (1999) have described the isolation of PtdIns synthase from barley seeds. Specifically investigated were the subcellular localization of *scyllo*-PtdIns and the relative rates of biosynthesis and accumulation of [^{32}P]Pi-labeled *scyllo*- and *myo*-PtdIns in the plasma membrane and intracellular membrane pools in the aleurone cells of barley seeds. About 25% of the [^{32}P]-labeled phospholipids were present in the plasma membrane and 75% in intracellular membranes. Incorporation of [^{32}P] into *scyllo*-PtdIns was greater than into *myo*-PtdIns in both the plasma membranes and intracellular membranes, thus suggesting a higher rate of biosynthesis. However, the data do not preclude breakdown of labeled *scyllo*-PtdIns as a contributing factor. Carstensen et al. (1999) conducted in vitro studies to investigate the presence of cytidinediphosphate diacylglycerol (CDP-DAG):*scyllo*-inositol 3-phosphatidyl-transferase (*scyllo*-PtdIns synthase) and to optimize enzymatic activity. The inclusion of non-ionic detergents (Brij 58 and Triton X-100) effected significant enhancement in the biosynthesis of *scyllo*-PtdIns, whereas anionic, cationic, and zwitterionic detergents had little or no effect. This is the first evidence for CDP-DAG:*scyllo*-inositol 3-phosphatidyltransferase activity. Figure 6.2 shows the separation of in vivo [^{32}Pi]-labeled phospholipids (middle lane) alongside in vitro *myo*-[2-^{3}H(N)ino-sitol-labeled (left lane) and *scyllo*-[^{3}H(N)inositol-labeled (right lane) phospholipids from PtdIns-synthase assay (Carstensen et al., 1999). Aliquots of the [^{32}Pi]- and [^{3}H]-labeled phospholipids were mixed with carrier phospholipids and separated on a high-performance TLC plate. The in vivo [^{32}P]-labeled phospholipids (in the middle lane) were localized by autoradiography and silica gel (left and right lanes) corresponding to *myo* and *scyllo*-PtdIns, PtdCho, and PtdIns-P; other regions of the plate were scraped and the [^{3}H] contents determined by scintillation counting. The numbers in the left and right lanes indicate [^{3}H] values above background.

Ostlund et al. (1993) have described the D-*chiro*-inositol metabolism in diabetes mellitus.

6.2.1.1. Synthesis of CDP-DAG via CTP

Carman and Kelley (1992) assayed the CDP-DAG synthetase using cells suspended in 50 mM Tris–HCl, pH 8.0, incubated on ice for

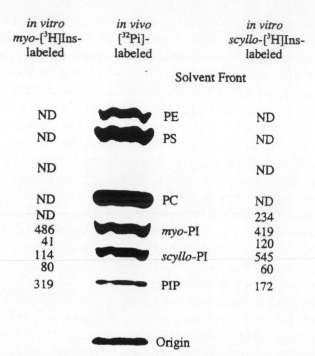

in vitro myo-[³H]Ins-labeled	in vivo [³²Pi]-labeled		in vitro scyllo-[³H]Ins-labeled
		Solvent Front	
ND		PE	ND
ND		PS	ND
ND			ND
ND		PC	ND
ND			234
486		myo-PI	419
41			120
114		scyllo-PI	545
80			60
319		PIP	172
		Origin	

Fig. 6.2. Separation of in vivo [³²P]-labeled phospholipids (middle lane) along-side in vitro myo-[2-³H(N)]inositol-labeled (left lane) and scyllo-[³H(N)]inositol-labeled (right lane) phospholipids from PtdIns-synthase assay. The numbers in the left and right lanes indicate [³H] values above background. PE, PtdEtn; PC, PtdCho; PS, PtdSer; PIP, PtdInsP; Ins, inositol. Reproduced from Carstensen et al. (1999) with permission of the publisher.

30 min, and lysed by sonication in an ice bath for 3 × 30 s. The assays contained 0.69 μM [³H]CTP (spec. act. 14.5 Ci/mmol), 10 mM MgCl₂, 2 mM PtdOH (sonicated suspension in water), 100 mM Tris–HCl, pH 6.5, and 5–25 μg of protein in a final volume of 50 μl. The MgCl₂ was the last compound added to the reaction mixture. The assay mixtures were incubated for 5 min at 37°C, and the assays were terminated by the addition of 180 μl of chloroform/methanol/conc. HCl (1:2:0.02, by vol). Next, 60 μl

of chloroform and 60 µl of 2 M KCl were added, and following vortexing, the phases were separated by centrifugation. The amount of [^{3}H]CDP-DAG formed was determined by counting the radioactivity in 40 µl of the organic phase. CDP-DAG formation was confirmed by TLC of the organic phase on silica gel 60 plates developed with chloroform/methanol/acetic acid/water (50:30:8:4, by vol). The radioactivity co-chromatographed with the CDP-DAG standard. The Heacock et al. (1996) have cloned the CDP-DAG synthase from human neuronal cells.

6.2.1.2. Synthesis of PtdIns via CDP-DAG

Carman and Fischl (1992) assayed the PtdIns synthase using 0.3 µM CDP-DAG (sonicated suspension in water), 3.3 µM myo-[^{3}H]inositol (spec. act. 30 Ci/mmol), 2 mM MnCl$_2$, 50 MgCl$_2$, 100 mM Tris–HCl, pH 8.0, and 5–35 µg of protein in a final volume of 50 µl. The mixtures were incubated for 15 min at 37°C and the assays terminated by the addition of 180 µl of chloroform/methanol/conc. HCl (1:2:0.02 by vol). Next, 60 µl of chloroform and 60 µl of 2 M KCl were added and after vortexing the phases were separated by centrifugation. The amount of PtdIns formed was determined by counting the radioactivity present in 40 µl of the organic phase. The formation of PtdIns was confirmed by TLC of the organic phase on Silica Gel 60 plates developed in chloroform/methanol/ammonium   hydroxide/water/0.25 M   EDTA (45:35:1.5:8.34:0.16 by vol). All radioactivities co-chromatographed with the PtdIns standard.

According to Ghalayini and Eichberg (1993), solubilized PtdIns synthase is assayed in a mixture containing 1.0 mM CDP-DAG (prepared by sonication in 10 mM glycylglycine, pH 8.6), 6.0 mM myo-[^{3}H]inositol [50 000 counts/min (cpm)], 0.6% Triton X-100, 48 mM MgCl$_2$, 0.5 mM DTT, and 50 mM glycylglycine buffer (pH 8.6) and 0.2–0.5 mg protein in a final volume of 0.2 ml. Incubations are conducted at 37°C for 30 min and terminated by the addition of 4.8 ml chloroform/methanol (2:1, v/v). The lower phase is subsequently washed first with 1 ml of 0.01N HCl, and then twice with 2 ml of 1 mM myo-inositol in 0.01N HCl/methanol/chloroform (47:48:3, by vol). An aliquot of the lower phase was dried under nitrogen, and the radioactivity incorporated into PtdIns was quantified by scintillation counting. Ghalayini and Eichberg (1993) assayed the activity of the purified enzyme in a mixture containing 50 mM glycylglycine buffer (pH 8.6), 0.3% Triton X-100, 0.5 mM DTT, 5–10 μg phosphorus/ml asolectin or PtdCho, 1–4 μg enzyme protein, 10 mM MgCl$_2$, 0.5 mM EGTA, 6.0 mM myo-[^{3}H]inositol (30 000–50 000 cpm), and 0.5 mM CDP-DAG in a final volume of 0.2 ml. Incubations were carried out at 37°C for 8–15 min.

Salman and Pagano (1997) have studied the uptake, metabolism, and distribution of a fluorescent analog of CDP-diacylglycerol [cytidine diphosphate-1,2-oleoyl, [N-4(nitrobenzo-2-oxa-1,3-diazole) aminocaproyl]glycerol; CDP-NBD-DAG]. Incubation of CDP-NBD-DAG-treated cells with inhibitors of PtdOH phospho-hydrolase and diacylglycerol kinase resulted in a dramatic increase in the amount of fluorescent PtdIns formed (64% of all CDP-NBD-DAG metabolites). Salman and Pagano (1997) concluded that CDP-NBD-DAG can be used for the de novo synthesis of fluorescent PtdIns, and in combination with ^{32}P-labeling, provides a convenient method of studying PtdIns turnover. Salman et al. (1999) used this method to study PtdIns biosynthesis in mycobacteria, including the effect of structural analogs.

6.2.1.3. Synthesis of PtdIns via CDP-DAG in plants

Carstensen et al. (1999) assayed the PtdIns synthase activity at 30°C in a solution containing 20 mM Tris–HCl pH 8.0), 0.23 M sucrose, 7.5 M KCl, 1 M CDP-DAG, made from egg PtdCho, 5 mM MnCl$_2$, 1 mM myo-[2-^{3}H(N)inositol (40 000–200 000 cpm per assay) or scyllo-[^{3}H(N)]inositol (100 000–200 000 cpm per assay, and 0.03% Brij 58 (polyoxyethene 20 cetyl ether) or 0.2% Triton X-100 in a total volume of 0.4 ml. CDP-DAG was added to the solution containing Tris–HCl, sucrose, and KCl and sonicated for 30–60 s. After the solution appeared homogeneous, inositol and $MnCl_2$ were added. The reaction was initiated by addition of the PtdIns synthase enzyme (150–300 μg protein) and was terminated after 3 h by the addition of 3.3 ml $CHCl_3/CH_3OH/H_2O$ (1:2:0.3, by vol). The phospholipids were extracted by the acidic Bligh-Dyer method.

Justin et al. (2002) have reported PtdIns synthesis and exchange of the Ins head to be catalyzed by a single PtdIns synthase 1 from *Arabidopsis* (see below).

6.2.2. Remodeling

The de novo formed PtdIns is subject to acyl exchange at both *sn*-1- and *sn*-2-positions of the glycerol moiety. In addition, the *myo*-inositol head group may be exchanged for a radiolabelled one, or for an isomeric inositol (e. g. *scyllo*-inositol). There is a need to distinguish between a simple head-group exchange (no release of PtdOH) and a transphosphatidylation.

6.2.2.1. Acyl group exchange

In tissues such as brain and liver, PtdIns has considerably more stearic and arachidonic acids than the phosphatidic acid from which it is synthesized (Holub and Kuksis, 1978). Specific modifications may take place in response to a dietary excess of long chain polyunsaturated fatty acids (Nair et al., 1999). The enrichment is produced by deacylation and reacylation cycles. As discussed elsewhere (see Chapter 9) PtdIns can be directly deacylated via phospholipase A_2 with the release of arachidonic acid. Alternatively, deacylation via phospholipase A_1 and subsequent hydrolysis of the isomerized lyso(2-acyl) PtdIns by lysophospholipase (Irvine et al., 1978) can also provide for the release of the eicosanoid precursor. Baker and Thompson (1972) showed that [³H]arachidonic acid was incorporated in vivo into brain PtdIns by such a cycle. The same authors (1973) described the acylation of 1-acyl-glycero-3-phosphoinositol by a brain microsomal fraction, arachidonoyl CoA being the most effective acylating agent. Holub (1976) and Holub and Piekarski (1979) have shown that the microsomal fraction of rat liver also has the necessary acyltransferases. With the 1-acylglycerol-3-phosphoinositol as acceptor, arachidonoyl CoA is the preferred substrate. With the 2-acyl compound, stearoyl CoA is a better donor than palmitoyl CoA. Figure 6.3 summarizes the acyl exchange of PtdIns based on studies in rat liver by Holub (1976) and Holub and Piekarski (1979).

A preferential introduction of arachidonic acid into 1-acyl-*sn*-glycero-3-phosphoinositol was demonstrated in vitro as follows (Holub, 1976). A Potter–Elvehjem homogenate of rat liver was centrifuged at $800\,g$ for 10 min to remove cellular debris. The assay medium for monitoring PtdIns formation contained 50 mM ATP, 0.13 mM CoA, 4 mM $MgCl_2$, 68.7 mM NaF, 11.5 μM 1-acyl 2-lysoGroPIns[^{3}H] 200 μl of purified homogenate (5.1 mg protein) in a final volume of 1.0 ml. The mixture was incubated for 15–45 min at 37°C in the absence of added exogenous acyl groups. Conventional TLC was used to isolate PtdIns[^{3}H] and the radioactivity determined by scintillation counting. To confirm the positional location of the radioactive fatty acids in the product, 0.8 μmol of carrier lipid was added, and then PtdIns was hydrolyzed with phospholipase A_2 (*Crotalus adamanteus*).

A preferential introduction of stearic acid into 2-arachidonoyl-*sn*-glycero-3-phosphoinositol was demonstrated similarly (Holub and Piekarski, 1979). The in vitro acylation of 32 μM 1-lyso-2-acyl-GroPIns (prepared from rat liver PtdIns) by 16 μM mixed acylCoA (8 μM [^{14}C]palmitoyl-CoA) plus 8 μM [^{3}H]stearoyl-CoA) containing 5000–9000 cpm/nmol was monitored following addition of 20 μg of microsomal (rat liver) protein in 500 μM Tris–HCl buffer (pH 7.4). Incubations were conducted at 37°C in a shaking water bath for 1 or 2 min after which the reaction was terminated and PtdIns formation was determined by differential liquid scintillation counting of appropriate TLC fractions. To

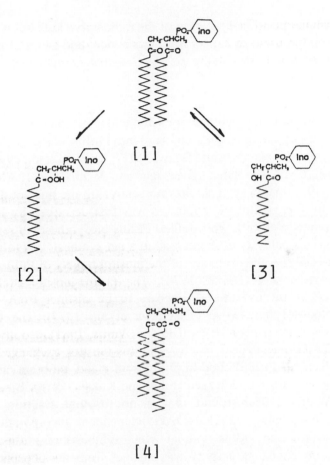

[4]

Fig. 6.3. Fatty acid remodeling during PtdIns biosynthesis in rat liver microsomes. The series of lipid remodeling reactions starts with phospholipase A_2 followed by sn-1-acyl GroPIns:sn-2-acylCoA transferase or phospholipase A_1 followed by sn-2-acylGroPIns:sn-1-acylCoA transferase. The major product of the remodeling of the de novo synthesized sn-1-palmitoyl-sn-2-linoleoylGroPIns [1] is sn-1-stearoyl sn-2-arachidonoyl GroPIns [4]. Drawn from the data of Holub and Piekarski (1979) with permission of the publisher.

confirm the positional location of the radioactive fatty acids in the product, 0.8 μmol of carrier lipid was added, and then PtdIns was hydrolyzed with phospholipase A_2 (*Crotalus adamanteus*).

6.2.2.2. Head-group exchange

In addition to PtdIns synthase, two alternative mechanisms exist for the incorporation of free *myo*-inositol into PtdIns: one is nucleotide independent (Takenawa and Egawa, 1977; McPhee et al., 1991) the other is CMP-stimulated (Bleasdale and Wallis, 1981; McPhee et al., 1991) pathway. The enzyme catalyzing the nucleotide-free pathway for inositol incorporation was first described by Paulus and Kennedy (1960), who studied PtdIns biosynthesis in rat liver tissue. This pathway does not lead to a net synthesis of PtdIns but rather catalyzes the exchange of the inositol moiety of PtdIns with free *myo*-inositol (Holub, 1974). The PtdIns-inositol exchange enzyme of rat liver tissue appears to be distinct from CDP-DAG:inositol transferase (Takenawa et al., 1977). The CMP-stimulated mechanism (Bleasdale and Wallis, 1981) also involves CDP-DAG-independent reaction. PtdIns-inositol exchange reactions have also been found in pig thyroid gland, rat brain, and rat lens tissues (for a review, see Holub and Kuksis, 1978). Bleasdale and Wallis (1981) found that a microsomal fraction from rabbit lung catalyzed CDP-DAG-independent incorporation of [³H]inositol into PtdIns without a net synthesis of PtdIns. The PtdIns-Ins exchange activity was distinct from the phospholipid base exchange enzymes and was specific for inositol. The PtdIns-Ins exchange enzyme differed from the CDP-DAG:Ins transferase on the basis of different degrees of inhibition by either Ca^{2+}, Hg^{2+} or heat. The exchange is especially rapid in avian species. The reaction was dependent on divalent cations, either Mg^{2+} or Mn^{2+} (Irvine, 1998). McPhee et al. (1991) have investigated the effects of analogues with modification of the substituent at the 1-, 2-, 3-, 4-, and 5-positions on incorporation of [³H]inositol into PtdIns during both synthesis and exchange reaction. The 2-deoxy-2-fluoro-*myo*-inositol was a poor substitute for *myo*-inositol in the synthesis and exchange reaction.

Klezovitch et al. (1993) have reinvestigated the nature of the PtdIns synthase reactions by expressing the yeast PtdIns synthase in *E. coli*. A single enzyme was shown to carry out both CDP-DAG-dependent incorporation of inositol into PtdIns and a CDP-DAG-independent exchange reaction between PtdIns and Ins. The exchange reaction and reversal of PtdIns synthase were both stimulated by CMP, but had different optimum pH and substrate requirements. These results suggested that CMP-stimulated exchange and CMP-dependent reverse reactions are distinct processes catalyzed by the same enzyme, PtdIns synthase. Cubitt and Gershengorn (1990) have suggested that CMP may act by enhancing the head group exchange reaction and that this reaction is different from the reversal of PtdIns synthase.

Figure 6.1 also, includes the enzyme that catalyzes the exchange of the head group, *myo*-inositol, or *myo*-PtdIns:*myo*-inositol phosphatidyltransferase, which has been observed in a number of cells including plant cells (Carstensen et al., 1999). The exchange enzyme is believed to be localized in the endoplasmic reticulum and Golgi apparatus. Recent cloning studies in mammalian cells indicate that a single polypeptide exhibits both synthetase and exchange activities (Likydis et al., 1997; Lykidis and Jackowski, 2001), thereby providing evidence that in mammalian cells the CMP-dependent exchange activity is due to PtdIns synthase catalyzing the reverse and forward reactions sequentially. The physiological significance of the exchange reaction is unknown. Analogous to the biosynthesis of *myo*-PtdIns, *scyllo*-PtdIns could be produced by the reaction of CDP-DAG with *scyllo*-inositol (Carstensen et al., 1999). Alternatively, the head group of *myo*-inositol, could be exchanged with *scyllo*-inositol to form *scyllo*-PtdIns.

Another potential pathway for PtdIns formation may be provided by the transphosphatidylation reaction catalyzed by phospholipase D. This enzyme catalyzes a transphosphatidylation reaction in which the phosphatidyl moiety of glycerophospholipids is accepted by primary alcohols, thereby producing a PtdEt instead of PtdOH. Polyols, such as glycerol and sugar alcohols, and alcoholic bases, including choline, ethanolamine and serine also serve with varying efficiencies as phosphatidyl acceptors (Singer et al., 1997). Transphosphatidylation by PLD is often confused with the activity of base exchange enzymes, which provide a biosynthetic route for certain phospholipid-bound head groups. Neither transphosphatidylation nor base exchange produces PtdOH as do hydrolases. It is unclear whether transphosphatidylation or base exchange represents a normal physiological function.

Measurement of PtdEt is the most commonly used method for assessing phospholipase D activity in cells owing to its relative metabolic stability compared to PtdOH. [^{3}H]Myristic acid is commonly used as a pre-label of the cellular pool because it is preferentially incorporated into cellular pool of PtdCho. It is not known whether or not PtdIns(4,5)P$_2$ serves as an in vivo cofactor of the transphosphatidylation activity of phospholipase D, but it is being included in in vitro assays (Siddiqui et al., 1995).

The head-group exchange may be demonstrated as follows (Klezovitch et al., 1993). The CDP-DAG-dependent incorporation of Ins is assayed in a medium containing 0.1 mM myo[2-^{3}H]Ins (50 000 cpm/nmol), 30 mM Tris–HCl buffer (pH 8.0), 3 mM MnCl$_2$, 2 mM DTT, 0.26% (w/v) Triton X-100, 0.3 mM CDP-DAG and 600–800 μg of protein in a total volume of 0.33 ml. The mixture is incubated for 30 min at 30°C. The reaction is terminated by the addition of 5 ml chloroform/methanol/HCl (100:50:1, by vol) solution. The chloroform phase was washed twice with 0.4 ml of a 2% NaCl/1% Ins solution. A 150 μl aliquot of the washed lower phase was then mixed with 10 ml of toluene/PPO (5 g/l) and radioactivity was determined by scintillation counting. Exchange activity was measured without addition of CDP-DAG. Reverse activity was measured by following the release of radioactive free Ins in the upper aqueous phase from [^{3}H]labeled PtdIns added to the reaction mixture (3500 cpm/nmol), or the decrease of the label of radiolabeled PtdIns in the lower organic phase.

Lykidis et al. (1997) have assayed the exchange reaction in COS-7 cells essentially as described by Klezovitch et al. (1993). The assay contained 0.5 μM PtdIns (sonicated suspension in water), 3.3 μM myo-[^{3}H]inositol (spec. act. 30 Ci/mmol), 2 mM MnCl$_2$, 50 mM MgCl$_2$, 100 mM Tris–HCl, pH 8.0, and 5–50 μg of protein in a final volume of 50 μl. The assay mixtures were incubated for 15 min at 37°C, and the assays were terminated by the addition of 180 μl of chloroform/methanol/conc. HCl (1:2:0.02 by vol). Next, 60 μl of chloroform and 60 μl of 2 M KCl were added, and following vortexing, the phases were separated by centrifugation. The amount of PtdIns formed was determined by counting the radioactivity present in 40 μl of the organic phase. The formation of PtdIns was confirmed by TLC of the organic phase on Silica Gel 60 plates developed as described above in the PtdIns-synthase assay.

Justin et al. (2002) have reported that the PtdIns synthase 1 from *Arabidopsis thaliana*, when expressed in *E. coli*, is able to catalyze both de novo synthesis of PtdIns as well as exchange of the polar head group. Exchange was observed in the absence of CMP, but was greatly enhanced in the presence of 4 μM CMP. The enzyme required the presence of free manganese ions. Justin et al. (2002) conducted the exchange reaction by incubating 50–200 μg membrane proteins in 50 mM Tris–HCl pH 8.0, 0.36 mM PtdIns from soybean (Sigma), 0.5 mM myo-Ins containing tritium-labeled myo-Ins for a final activity of 500 Bq/nmol, 2.4 mM Triton X-100, 2.5 mM MnCl$_2$ in the presence or in absence of 4 μM CMP, but without CDP-DAG. The incubation time was 20–30 min at 30°C.

Siddiqui et al. (1995) have assayed the transphosphatidylation reaction catalyzed by phospholipase D as follows. The phospholipase D activity was measured in unlabeled membranes, cytosol, and column fractions, using phospholipid (PtdEtn/PtdInsP$_2$/PtdCho, 16:1.4:1) vesicles (Brown et al., 1993). Dipalmitoyl[2-palmitoyl-9,10-[^{3}H]GroPCho (0.5 μCi) (for [^{3}H]PtdEt formation) or dipalmitoyl[methyl-[^{3}H]cholineGroPCho (0.5 μCi) for [^{3}H]choline release) was included in the phospholipid mixture for each assay. For a 60 μl assay volume, 10 μl of vesicles were added to the assay buffer (50 mM HEPES, (pH 7.2), 3 mM EGTA, 80 mM KCl, 1 mM dithiothreitol, 3 mM MgCl$_2$, and 2 mM CaCl$_2$) containing membranes, cytosol, or column fractions and were incubated at 37°C for 30 min (unless otherwise indicated). For the [^{3}H]choline release assay, the reactions were stopped by the addition of 200 μl of 10% trichloroacetic acid and 100 μl of 1% bovine serum albumin, and the mixtures were processed as described by Brown et al. (1993). For [^{3}H]PtdEt formation, ethanol (1% (v/v), for membranes or 2% (v/v) for cytosol, and the column fractions) was included in the reaction mixtures, and the reaction terminated by the addition of 375 μl of chloroform/methanol/HCl (50:98:2), and the lipids were extracted. Lipids from the incubation mixtures were suspended in chloroform/methanol (2:1) and spotted onto TLC plates. The plates were developed with ethyl acetate/ isooctane/acetic acid/water (50:25:10:50, by vol). Separated phospholipids were detected by exposure to iodine vapor. Spots corresponding to PtdOH and PtdEt were scraped, mixed with ready organic scintillant, and the radioactivity determined.

6.3. Phosphatidylinositol phosphates

The existence of PtdIns 4-kinases (PtdIns4Ks) has been appreciated for some 30 years, and the understanding of these enzymes has been steadily advancing. Seven different phosphorylated forms of PtdIns have been found to occur in nature (see Chapter 3). In mammalian cells, the major phosphorylated forms of PtdIns were the PtdIns(4)P and PtdIns(4,5)P$_2$. However, the discovery of a PtdInsP kinase that phosphorylates the D-3 position of PtdIns, PtdIns(4)P and PtdIns(4,5)P$_2$ has led to the production of three additional lipids: PtdIns(3)P, PtdIns(3,4)P$_2$ and PtdIns(3,4,5)P$_3$. Although these lipids are less abundant than PtdIns(4)P and PtdIns(4,5)P$_2$, the enzyme that produces them, PtdIns 3-kinase

(PtdIns 3K), has been shown to play a major role in cellular regulation (Toker and Cantley, 1997; Vanhaesebroeck et al., 1997a,b). In fact, three families of PtdIns 3Ks with different mechanisms of regulation and different substrate specificity have been described in higher eukaryotes. Two additional phosphoinositides have been discovered recently. PtdIns(5)P was found in mammalian cells (Rameh et al., 1997) and PtdIns(3,5)P$_2$ in yeasts and in mammalian cells (Dove et al., 1997; Whiteford et al., 1997).

There is evidence that PtdIns(3,4)P$_2$ can be produced by phosphorylation of PtdIns(4)P at the D-3 position by a PtdIns 3K and by phosphorylation of PtdIns(3)P at the D-4 position by a 5-phosphatase. The most extensively characterized of these enzymes is the PtdInsP 3-kinase. However, progress has also been made in the characterization of PtdIns 4-kinases and PtdIns(4)P 5-kinases. In addition, new pathways involving PtdIns(5)P 4-kinases, PtdIns(3)P 5-kinases and PtdIns (3)P 4-kinases have recently been described. The pathways for the production of the various phosphoinositides in mammalian cells are summarized in Fig. 6.4 along with the corresponding kinases (Tolias and Cantley, 1999).

It is not known to what extent if any the polyphosphoinositides undergo remodeling via acyl and head group exchange. In an early study (Holub et al., 1970) demonstrated that the mono-, di- and triphosphoinositides of bovine brain have similar compositions of molecular species. More recently, Gaudette et al. (1993) have shown that the fatty acid composition of the 3-phosphorylated PtdInsP$_2$ of human platelets does not undergo significant change during U46619 stimulation. The authors suggested that this validates the use of HPLC analysis of radiolabeled PtdInsPs for the estimation of PtdIns(3,4)P$_2$ mass in agonist-stimulated platelets.

6.3.1. PtdIns monophosphates

A number of kinases responsible for the formation of PtdIns monophosphates have been identified, purified, and cloned. These

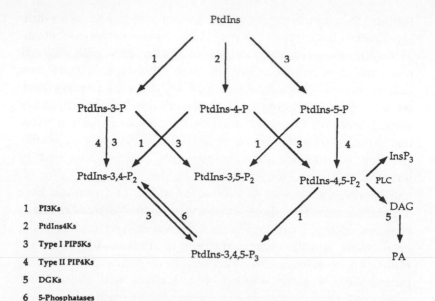

Fig. 6.4. Biosynthetic pathways of all known PtdInsPs in mammalian cells. Step 1, PtdIns 3-kinases; Step 2, PtdIns 4-kinases; Step 3, PtdInsP 5-kinases (Type I); Step 4, PtdInsP 4-kinases (Type II); Step 5, diacylglycerol kinases; Step 6, 5-phosphatases. The biosynthetic pathways depicted are based on in vitro evidence and have not yet been demonstrated in vivo. Reproduced from Tolias and Cantley (1999) with permission of the publisher.

enzymes are ubiquitously expressed and are most abundant in cellular membranes including the Golgi, lysosomes, endoplasmic reticulum (ER), plasma membrane, and a variety of vesicles that include Glut 4-containing vesicles, secretory vesicles, and coated pits (Pike, 1992; Fruman et al., 1998; Tolias and Cantley, 1999).

6.3.1.1. PtdIns(4)P

PtdIns-4Ks convert PtdIns to PtdIns(4)P by phosphorylating the inositol ring at the D-4 position. (Balla, 1998). The PtdIns(4)P thus generated can be further phosphorylated by both PtdIns-3Ks and PtdIns(4)P 5-kinases (PIP4Ks) to yield PtdIns(3,4)P$_2$ and

PtdIns(4,5)P$_2$, respectively. Unlike the functionally related PtdIns 3Ks and PtdIns(4)P 5Ks, the PtdIns 4Ks appear to use only PtdIns as a substrate and cannot phosphorylate either the singly or doubly phosphorylated lipids generated by the other enzymes. The PtdIns 4Ks were classically subdivided into two types (II and III) based on biochemical differences of the partially purified enzymes. Previous workers had obtained two forms of the kinase from microsomes, one having a subunit molecular weight of 45 000 (Buxeda et al., 1991) and the other a subunit molecular weight of 55 000 (Nickels et al., 1994).

The genes encoding the PtdIns 4-kinases were first isolated from S. cerevisiae. PtdIns K1, a 125 kDa PtdIns 4-kinase, was found to be indispensable for growth (Flanagan et al., 1993) STT4, another PtdIns 4-kinase of 200 kDa in size, was found to be dispensable for growth in the presence of osmotic stabilizers (Yoshida et al., 1994). Sequence analysis revealed that PtdIns K1 and STT4 share homology with PtdIns 3-kinase family genes, such as the bovine p110 and Vps 34s. In vitro, the PtdIns 4Kα/STT4 isoforms phosphorylate only PtdIns.

In mammalian cells, at least three different genes encode kinases that phosphorylate the D-4 position of PtdIns cDNA clones of two of these enzymes (PtdIns 4Kα and PtdIns 4Kβ) have been isolated, whereas the cDNA for 55 type II PtdIns 4-kinase has not yet been identified. Wong and Cantley (1994) reported the cloning and characterization of a mammalian PtdIns 4-kinase, named PtdIns 4Kα. The 2.6 kb cDNA encoded a protein of 854 amino acids that was highly homologous to the yeast STT4 enzyme. A bovine gene encoding a 170–200 kDa PtdIns 4-kinase that shares more than 95% identity with the human PtdIns 4Kα in the overlapping region has been cloned by Gehrmann et al. (1996). Likewise, Nakagawa et al. (1996a,b) reported the cloning and characterization of an alternative splice of rat PtdIns 4-Kα gene that generates a protein predicted to be 230 kDa. Subsequently, Myers and Cantley (1997) isolated another human cDNA that encodes a 110 kDa PtdIns 4-kinase (named PtdIns 4-Kβ) that is more homologous to the yeast

PtdIns K1 gene. The PtdIns 4-Kβ is wortmannin-sensitive and is thought to be the same as that involved in the hormone sensitive pools of inositol phospholipids (Nakanishi et al., 1995). Wong et al. (1997) have determined the subcellular locations of these enzymes in CHO-IRS cells and have provided evidence for other yet unidentified isoforms present in specific organelles. PtdIns 4-kinase α was mostly membrane bound and located at the endoplasmic reticulum, whereas PtdIns 4-kinase β was in the cytosol and also present in the Golgi region. Neither of these isoforms accounted for the major type II PtdIns 4-kinase activity detected in the lysosomes and plasma membrane fraction.

The majority of the PtdIns 4-kinase activity detected in most mammalian cells is due to the 55 kDa type II PtdIns 4-kinase, an enzyme that has been purified from a number of tissues but not yet sequenced or cloned. This enzyme (or family of enzymes), unlike PtdIns 4-kinase α and PtdIns 4-kinase β is present at high levels in the plasma membrane (Wong et al., 1997). Wong et al. (1997) have extensively investigated the possibility that the type II 55-kDa enzyme is a proteolytic product of the 97-kDa PtdIns 4-kinase α, since they share similar enzymatic properties. However, it was not possible to detect proteins of 55 kDa resulting from the processing of the 97 kDa PtdIns 4-kinase α. It was concluded that the 55 kDa type II PtdIns 4-kinase appears to be encoded by a distinct but related gene.

A general consensus in regard to the distinction between two structurally different isoforms of the PtdIns 4-kinase has emerged from the cloning and sequencing of PtdIns 4-kinase from yeast, human, rat, and bovine tissues (Pike, 1992; Wong et al., 1997; Fruman et al., 1998; Tolias and Cantley, 1999). The two isoforms differ in size, amino acid sequence homology, and putative location and function within the cell. The smaller type encodes a polypeptide of about 110–125 kDa that is found in the cytosol and is associated with the Golgi apparatus. The second type encodes a larger protein of about 200–230 kDa that is membrane-associated. In most mammalian cells, the levels of PtdIns(4)P do not appreciably change in response to growth factors or osmotic stress, which suggests that if PtdIns 4-kinases are regulated by extracellular signals, with the regulation likely at specific regions of the cell. A distinctive feature of the group of PtdIns 4-kinases is that they contain putative pleckstrin homology (PH) domains, which are poorly conserved protein modules of about 100 amino acids in length that associate with membranes during signal transduction (Fruman et al., 1998).

Stevenson et al. (1998) have cloned and characterized a 205 kDa PtdIns 4-kinase from higher plants. The deduced amino acid sequences of the plant PtdIns 4-kinase clones were 41% identical and 68% similar to rat and human PtdIns 4-Kα lipid kinase unique, PH, and lipid kinase catalytic domains. The recombinant PH domain had the highest affinity for PtdIns(4)P, the product of the reaction, but PtdIns(4,5)P$_2$ and PtdOH were bound. The 3-phosphoinositides were not bound.

Yamakawa and Takenawa (1993) have synthesized the PtdIns(4)P by PtdIns 4-kinase in bovine brain. The enzyme activity from bovine brain was assayed by measuring the formation of [^{32}P]PtdIns(4)P from PtdIns and [γ-^{32}P]ATP. The formed PtdIns(4)P is extracted with chloroform/methanol/conc. HCl (200:100:1, v/v/v). The standard incubation mixture contains 50 mM Tris–HCl (pH 7.4), 20 mM MgCl$_2$, 1 mM EGTA, 0.4% Triton X-100, 1 mM PtdIns, 200 μM [γ-^{32}P]ATP (1.0–0.5 μCi), and 1–100 μg of enzyme protein in a total volume of 50 μl. The reaction is started by the addition of [γ-^{32}P]ATP, continued for 10 min at 30°C, and then stopped by the addition of 2 ml of chloroform/methanol/conc. HCl (200:100:1, v/v/v). Then 0.5 ml of 1N HCl is added, and the mixture is vortexed and separated into two phases by centrifugation at 2000g for 5 min at room temperature. The upper phase is discarded, and the lower phase is washed with 0.5 ml of the synthetic upper phase. After centrifugation, the resultant lower phase is removed and dried under a stream of dry nitrogen.

For the estimation of the activity of crude enzymes, the [^{32}P]PtdIns(4)P product must be separated by TLC to avoid contamination with other [^{32}P]-labeled lipids. The contribution of PtdIns 3-kinase activity is minimal because PtdIns 3-kinase is inhibited in the presence of detergents. PtdIns(4)P is separated from other phospholipids with one-dimensional TLC. For this purpose 10 μg PtdIns(4)P is added as carrier. The TLC plates (Merck and Co., Darmstadt, Germany) are treated with methanol/water (2:3, v/v) containing 1% sodium oxalate and then activated for 1 h at 110°C. The lipids are spotted on the plates and developed with chloroform/methanol/4.3 NH$_4$OH (90:65:20, by vol). The spots are visualized with iodine vapor and then scraped off the plates for measuring the radioactivity.

Stevenson et al. (1998) have synthesized PtdIns(4)P by PtdIns 4-kinase from plants. The PtdIns 4-kinase activity in each fraction from the immunoaffinity column was assayed in duplicate (60 μl/assay). The reaction mixture contained final concentrations of 7.5 mM MgCl$_2$, 1 mM sodium molybdate, 0.5 mg/ml PtdIns, 0.1% (v/v) Triton X-100, 0.9 mM ATP, 30 mM Tris (pH 7.2), and 20 μCi of [γ-^{32}P]ATP (7000 Ci/mmol) in a total volume of 100 μl. Stock PtdIns (5 mg/ml) was solubilized in 1% (v/v) Triton X-100. Reactions were incubated at 25°C for 2 h with intermittent shaking. The reactions were stopped with 1.5 ml of ice-cold CHCl$_3$/methanol (1:2) and kept at 4°C until the lipids were extracted. Lipids were extracted as described previously (Cho et al., 1993. Extracted lipids were vacuum-dried, solubilized on CHCl$_3$/MeOH (2:1), and spotted onto Whatman LK5D silica gel plates that had been completely dried in a microwave oven for 5 min after pre-soaking in 1% (w/v) potassium oxalate for 80 s. The lipids were separated in either a CHCl$_3$/MeOH/NH$_4$OH/H$_2$O (86:76:6:16, v/v/v) solvent system (Cho et al., 1993) or a pyridine-based solvent (Walsh et al., 1991) and quantified with a Bioscan System 500 Imaging Scanner. The plate was subsequently exposed to phosphor screen for two days and visualized by a Storm PhosphorImager (Molecular Dynamics).

Mani et al. (1996) have reported increased formation of PtdIns(4)P in human platelets stimulated by PtdOH.

6.3.1.2. PtdIns(3)P

The D-3 position of PtdInsP is phosphorylated by a family of PtdIns 3-kinases, which transfer the γ-phosphate group of ATP to the D-3 position of the inositol ring. Numerous PtdIns 3-kinases have been isolated from many mammalian and non-mammalian sources and have been identified by various biochemical approaches and various cloning strategies. The structure and function of the PtdIns 3-kinases has been extensively reviewed by Carpenter et al. (1990), Toker and Cantley (1997), Vanhaeseb-roeck et al. (1997a,b, 2001), Fruman et al. (1998), Wymann and Pirola (1998) and Tolias and Cantley (1999). On the basis of their in vitro substrate specificities, the PtdIns 3-kinases have been grouped into Classes I, II and III. Class I PtdIns 3-kinases are heterodimers of approximately 200 kDa, composed of a 110–120 kDa catalytic subunit and a 50–100 kDa adaptor subunit, which are able to phosphorylate in vitro PtdIns, PtdIns(4)P and PtdIns(4,5)P$_2$. The preferred in vivo substrate for Class I PtdIns 3-kinases has been shown to be PtdIns(4,5)P$_2$. Class II PtdIns 3-kinases phosphorylate only PtdIns and PtdIns(4)P, and possess masses ranging from 170 to 210 kDa (Gaidarov et al., 2001). Class III PtdIns 3-kinases are homologues of S. cerevisiae Vps34p (vacuolar protein sorting mutant) and phosphorylate exclusively PtdIns (Odorizzi et al., 2000).

Members of the three different classes of PtdIns 3-kinases can be distinguished by their in vitro substrate specificity, structure, and their mode of regulation (Wymann and Pirola, 1998; Fruman et al., 1998; Vanhaesebroeck et al., 2001). The class I PtdIns 3-kinases are capable of phosphorylating the D-3 position of PtdIns, as well as PtdIns(4)P or PtdIns(4,5)P$_2$ in vitro. Their preferred in vivo substrate is PtdIns(4,5)P$_2$. The class I PtdIns 3-kinases are heterodynamic enzymes with regulatory and catalytic subunits. These enzymes can be further divided into two subfamilies, Class Ia and Class Ib PtdIns 3-kinases, which are subject to regulation. The class II PtdIns 3-kinases have a larger catalytic subunits (170–210 kDa) and contain a carboxy-terminal C2 domain. Based on their in vitro activities, Class II PtdIns 3-kinases are likely to produce PtdIns(3)P in vivo, but they might also contribute to PtdIns(3,4)P$_2$ or PtdIns(3,4,5)P$_3$ production. Mammals have three class II isoforms (PtdIns 3-kinase C2α, β and γ) encoded by three separate genes (Vanhaesebroeck et al., 2001).

Ono et al. (1998) observed increased expression of class II PtdIns 3-kinase during liver regeneration. Soos et al. (2001) report that class II PtdIns 3-kinase is activated by insulin but not by contraction in skeletal muscle. PtdIns 3-kinase is considered a key switch in insulin signaling (Shepherd et al., 1998).

Class III PtdIns 3Ks have been found in other yeasts, plants, *Drosophila*, algae, and mammals (Wymann and Pirola, 1998). All these Class III enzymes are restricted to PtdIns and thus produce PtdIns(3)P as the sole product.

Based on the homologies within the catalytic domain, close relatives of PtdIns 3-kinases have been identified among the PtdIns 4-kinase family (Hunter, 1995; Balla, 1998) and elsewhere (see Wymann and Pirola, 1998).

While higher eukaryotes from *D. discoideum* to *C. elegans* to mammals have enzymes of all three classes of PtdIns 3Ks, *S. cerevisiae* have only the Class II PtdIns 3-kinase. Thus, PtdIns(3)P is found in yeast, but PtdIns(3,4)P$_2$ and PtdIns(3,4,5)P$_3$ have not been detected. Recently, PtdIns(3)P was shown to bind directly to FYVE domains in proteins involved in vesicle trafficking (Burd and Emr, 1998).

Alternatively, PtdIns(3)P may be formed by isomerization of PtdIns(4)P (Chapter 10). In addition, PtdIns(3)P can be derived from PtdIns(3,4)P$_2$ and PtdIns(3,5)P$_2$ by the action of the appropriate phosphatases (see Chapter 8).

Soltoff et al. (1993) have described an in vitro synthesis of PtdIns(3)P by PtdIns 3-kinase in Anti-P-Tyrosine Immunoprecipitates. Since PtdIns 3-kinase associates with the PDGF receptor, middle T/pp60c-src tyrosine kinase, and other receptors that are phosphorylated on tyrosine in response to growth factors, the enzyme is commonly immunoprecipitated using anti-Ptyr antibodies and the assay is performed on the immunoprecipitate. The PtdIns 3-kinase in PC12 cells was assayed in a final volume of 50 μl. To about 30 μl (about 20 μl of protein A-Sepharose beads plus about 10 μl TNE) in a microcentrifuge tube is added 10 μl of a lipid mixture containing PtdIns combined with 10 μl of a [γ-^{32}P]ATP mixture to initiate the assay. PtdIns 3-kinase assay is performed by adding 10 μl containing 20–40 μCi [γ-^{32}P]ATP (in 250 μM ATP, 25 MgCl$_2$, and 5 mM HEPES, 1 mM EGTA, pH 7.5) to the immunoprecipitates for 10 min at room temperature. The reaction is stopped by adding 80 μl of HCl (1 M) and vortexing the sample. Subsequently, 160 μl of methanol/chloroform (1:1, v/v) is added, and the sample is vortexed thoroughly. The sample is centrifuged (about 10 s) to separate the organic and aqueous layers. The lipid-containing chloroform phase is spotted onto oxalate-coated TLC plates (Silica gel 60, 0.2 mm thick adsorbent layer). The 20 × 20 cm TLC plates are cut in half and developed (10 cm vertical) in a mixture of chloroform/methanol/water/7.7N NH$_4$OH (60:47:11.3:2, v/v). The solvent is allowed to migrate nearly to the top, which takes about 30 min. The solvent system resolves PtdIns, PtdInsP, PtdInsP$_2$, but PtdIns(3,4,5)P$_3$ and ATP remain at the origin. Authentic PtdIns(4)P is spotted as a standard and visualized using iodine vapor. [^{32}P]PtdInsP appears just below the non-radiolabeled PtdIns. The [^{32}P]PtdInsP spots are excised and quantified by scintillation counting or by measuring Cerenkov radiation. PtdIns(3)P is not separable from PtdIns(4)P using this technique. The identity of the PtdIns(3)P is confirmed by HPLC of the deacylated phosphatide, as described elsewhere (see Chapter 3). Linassier et al. (1997) have reported cloning and biochemical characterization of PtdIns 3-kinase from *Drosophila*.

Conway et al. (1999) assayed PtdIns 3-kinase in resuspended anti-phosphotyrosine or Grb-2 immunoprecipitates (40 μl), which were combined with 20 μl of PtdIns (3 mg/ml) in incubation buffer containing 1% cholate. To each, 40 μl of [^{32}P]ATP (3 μM Na$_2$ATP, 7.5 mM MgCl$_2$ and 0.25 mCi of [^{32}P]ATP/ml) was added. The reaction was performed at 37°C for 15 min and was terminated by adding 450 μl of chloroform/methanol (1:2, v/v). Organic and aqueous phases were resolved by adding 150 μl of chloroform and 150 μl 0.1 M HCl. Samples were mixed and centrifuged (4200g for 10 min). This was repeated and the lower phase harvested, dried, and [^{32}P]PtdIns(3)P resolved by TLC using chloroform/methanol/ammonia/water (42:60:9:15, by vol) in parallel with non-radioactive standard. Radioactive bands were revealed by autoradiography, and samples corresponding to [^{32}P]PtdIns(3)P scraped from the plate and subjected to Cerenkov counting.

Valverde et al. (1997) and Valverde et al. (1999) have reported an in vitro synthesis of PtdIns(3)P by PtdIns 3-kinase of fetal adipocytes. The PtdIns 3-kinase activity in fetal brown adipocytes was assayed by in vitro phosphorylation of PtdIns. The adipocytes were incubated with IGF-I in the presence and absence of PtdIns 3-kinase inhibitors. At the end of the culture time cells were washed with ice-cold PBS, solubilized in lysis buffer [10 M μNa$_3$VO$_4$/1% Triton X-100 (pH 7.6)] containing leupeptin (10 μg/ml), aprotinin (10 μg/ml) and 1 mM PMSF. Lysates were clarified by centrifugation at 15 000g for 10 min at 4°C, and proteins were immunoprecipitated with a monoclonal antibody anti-Tyr(P) (Py72). The immunoprecipitates were washed successively in PBS containing 1% Triton X-100 and 100 μM Na$_3$VO$_4$ (twice), 100 mM Tris, pH 7.5, in PBS containing 0.5 mM LiCl, 1 mM EDTA and 100 μM Na$_3$VO$_4$ (twice), and in 25 mM Tris pH 7.5) containing 100 mM NaCl and 1 mM EDTA (twice). To each pellet were added 25 μl of 1 mg/ml PtdIns/PtdSer sonicated in 25 mM HEPES (pH 7.5) and 1 mM EDTA. The PtdIns 3-kinase reaction was started by the addition of 100 nM [γ-^{32}P]ATP (10 μCi) and

300 μM ATP in 25 μl of 25 mM HEPES (pH 7.4)/10 M MgCl$_2$/ 0.5 mM ETA. After 15 min at room temperature, the reaction was stopped by the addition of 500 μl of chloroform/methanol (1:2, v/ v) a 1% concentration of HCl plus 125 μl of chloroform/125 μl of HCl (10 mM). The samples were centrifuged, and the lower organic phase was extracted, dried under vacuum, and resuspended in chloroform. Samples were applied to a silica gel TLC plate (Whatman, Clifton, NJ). TLC plates were developed in propanol/ acetic acid (2 M), 65:35, v/v) dried, visualized by autoradiography and quantified by scanning laser densitometry.

Stack et al. (1993) have described the in vitro synthesis of PtdIns(3)P by PtdIns 3-kinase in yeast. Yeast cell lysates were prepared and centrifuged at 2000g for 10 min, and the supernatant used for the assays. Approximately 0.05 OD$_{660}$ equivalents (4 μg of protein) of lysate were assayed for PtdIns 3-kinase activity as described by Stack et al. (1993). Fifty-microliter reactions were performed in 20 mM HEPES, pH 7.5, 10 mM MgCl$_2$, 0.2 mg/l sonicated PtdIns, 60 μM ATP, and 0.2 mCi/ml [γ-^{32}P]ATP and were incubated at 25°C for 5 min. The reaction was terminated by the addition of 80 μl of 1 M HCl, and the lipids were extracted with 160 μl chloroform/methanol (1:1, v/v). The reaction products were dissolved in chloroform and spotted onto Silica Gel 60 TLC plates (Merck), which were then developed in a borate buffer system (Walsh et al., 1991). The position of the products of PtdIns 3-kinase [PtdIns(3)P] and PtdIns 4-kinase [PtdIns(4)P] were located by means of standards. Quantification of the labeled species was performed either by use of a PhosphorImager (Molecular Dynamics) or by scraping the phospholipids from the TLC plates and quantifying by liquid scintillation counting. A similar assay system was employed for the assay of partially purified Vps34p.

6.3.1.3. PtdIns(5)P

Rameh et al. (1997) have demonstrated the existence of PtdIns(5)P in NIH-3T3 fibroblasts, but it is not known whether it is formed by a direct phosphorylation of PtdIns by a PtdIns 5-kinase. This lipid was discovered as an impurity in commercial PtdIns (brain-derived), and was shown to be present in mammalian fibroblasts at approximately the same level as PtdIns(3)P. PtdIns(5)P was overlooked in the past because it is difficult to separate from PtdIns(4)P by conventional methods. Hence, its formation by a PtdIns 5-kinase cannot be absolutely excluded. The type I γ or type I β PtdIns(4)P 5-kinases can phosphorylate PtdIns at the D-5 position in vitro (Tolias et al., 1998a). However, the preferred substrate for these enzymes is PtdIns(4)P. The in vivo significance of this pathway, therefore, remains to be established. Additional homologues of PtdIns (5)P kinases have been cloned but not fully characterized. Since the enzyme that converts PtdIns(5)P to PtdIns(4,5)P_2 (type II PtdIns(5)P 4-kinase; see below) is very abundant and active in most mammalian cells, it is believed that the PtdIns(5)P pathway plays an important role in cellular regulation (Tolias et al., 1998a,b). An alternative possibility is that PtdIns(5)P is a breakdown product of PtdIns(3,5)P_2 by 3-phosphatase (Chapter 7). PtdIns(5)P can be produced in vitro by dephosphorylation of PtdIns(4,5)P_2 (Rameh et al., 1997).

Tolias et al. (1998a,b) have described the in vitro synthesis of PtdIns(5)P by type I α PtdIns 5-kinase from murine cells. Briefly, the PtdInsP 5-kinase was assayed in 50 μl reactions containing 50 mM Tris, pH 7.5, 30 M NaCl, 12 mM MgCl$_2$, 67 μM PtdIns, 133 μM phosphoserine, and 50 μM [γ-^{32}P]ATP (10 μCi/assay). No carrier lipid was used. Reactions were stopped after 1 min by adding 80 μl of 1N HCl and then 160 μl of CHCl$_3$/MeOH (1:1). Lipids were separated by TLC using 1-propanol, 2 M acetic acid (65:35, v/v). Phosphorylated lipids were visualized by autoradiography and quantified using a Bio-Rad Molecular Analyst. The PtdIns used in all experiments was separated from contaminating inositol phosphatides by TLC purification using CHCl$_3$/MeO/H$_2$O/NH$_4$OH (60:47:11:1.6, by vol). After identification using an iodine-stained PtdIns standard, the PtdIns was scraped from the plate and eluted in the same solvent solution. Following lyophilization, PtdIns was resuspended once in MeOH, 0.1 M EDTA (1:0.9, v/v).

The reaction products were identified by HPLC analysis. For this purpose the deacylated lipids were mixed with ^{3}H-labeled standards and analyzed by anion-exchange HPLC using Partisphere SAX column and an ammonium phosphate gradient. The compounds were eluted with 1 M $(NH_4)_2HPO_4$, pH 3.8, and water using the following gradient: 0% 1 M $(NH_4)_2HPO_4$ for 5 ml, 0–1% 1 M $(NH_4)_2HPO_4$ over 5 ml, 1–3% 1 M $(NH_4)_2HPO_4$ over 40 ml, 3–10% 1 M $(NH_4)_2HPO_4$ over 10 ml, 1–13% 1 M $(NH_4)_2HPO_4$ over 25 ml, and 13–65% 1 M $(NH_4)_2HPO_4$ over 25 ml. Eluate from the HPLC column flowed into an on-line continuous flow scintillation detector (Radiomatic Instruments, FL) for isotope detection.

The phosphorylation products were further identified by periodate oxidation (Stephens et al., 1988) using the following modifications. Deacylated lipids were incubated in the dark at 25°C with 10 or 100 mM periodic acid, pH 4.5, for 30 min or 36 h, respectively. The remaining oxidizing reagent was removed by adding 500 mM ethylene glycol and incubating the reaction in the dark for 30 min. Two percent 1,1-dimethylhydrazine, pH 4.5, was then added to a final concentration of 1% and the reaction was allowed to proceed for 4 h at 25°C. The mixture was then purified by ion exchange using Dowex 50W-X8 cation exchange resin (20–50 mesh, acidic form), dried, and applied to an HPLC column. PtdIns(3,4)P_2 and PtdIns(3,4,5)P_3 controls used in this experiment were prepared using recombinant Sf9 cell-expressed PtdIns 3-kinase.

6.3.2. PtdIns bisphosphates

Although some of the PtdIns bisphosphates are synthesized by the same enzymes as the monophosphates, for the present purposes the synthetic reactions have been classified by the starting materials and the final products. By selecting the appropriate starting material, the course of the reaction can be directed to the production of a specific final product at least as part of the overall biosynthetic process (Tolias et al., 1998a,b; Rameh and Cantley, 1999; Anderson et al., 1999).

The PtdIns bisphosphates may be produced by sequential phosphorylation of PtdIns and by dephosphorylation of the appropriate PtdIns trisphosphates. These pathways have been demonstrated in vitro and the relative contribution of each of the possible mechanisms to the total production of a given bisphosphate is not known with certainty.

6.3.2.1. PtdIns(4,5)P$_2$

The PtdIns(4,5)P$_2$ has been shown to be formed by the phosphorylation of the D-5 position of PtdIns(4)P by PtdIns(4)P 5-kinase and by the phosphorylation of the D-4 position of PtdIns(5)P by PtdIns(5)P 4-kinase (Hinchliffe et al., 1998; Toker, 1998). Three different type I PtdIns(4)P 5-kinases (1α, 1β, and 1γ) from mammalian cells have been described (Ishihara et al., 1998; Itoh et al., 1998). Two yeast homologues have also been identified (MSS4 and FAB1). The MSS4-gene product has been shown to have PtdIns(4)P 5-kinase activity (Desrivieres et al., 1998). As indicated above, these enzymes also convert PtdIns to PtdIns(5)P.

In addition, PtdIns(4,5)P$_2$ is produced in human 293 cells via D-3 dephosphorylation of PtdIns(3,4,5)P$_3$ by the PtdIns(3,4,5)P$_3$ 3-phosphatase (Chapter 7). It is not known whether the latter pathway contributes significantly to PtdIns(4,5)P$_2$ levels in non-transformed cells. Under non-equilibrium radiolabeling conditions, the D-5 phosphate has a higher specific activity, suggesting that it is added to the molecule later than the D-4 phosphate, which implies that production of PtdIns(4,5)P$_2$ by PtdIns(4)P phosphorylation predominates (Stephens et al., 1991; Hawkins et al., 1992). The contribution of PtdIns(5)P phosphorylation to PtdIns(4,5)P$_2$ formation in vivo remains in doubt in view of the comparatively small amount of PtdIns(5)P relative to PtdIns(4)P within the cells (Hinchliffe et al., 1998; Rameh et al., 1997).

The PtdIns(4)P 5-kinase and PtdIns(5)P 4-kinases are two closely related families of enzymes. They were originally called type I and type II PtdIns(4)P 5-kinases until it was discovered that the type II enzymes are actually PtdIns(5)P 4-kinases (Rameh et al., 1997). The PtdInsP 4-kinases are about 35% identical to the PtdInsP 5-Ks in their kinase domains but are dissimilar in their amino- and carboxy-terminal extensions.

To date, clones for three different type I PtdIns(4)P 5-kinase family members (PtdInsP 5-kinase I γ, PtdInsP 5-kinase I β and PtdInsP 5-kinase I α) and three different type II PtdIns(5)P 4-kinase family members (PtdInsP 4-kinase γ), PtdInsP 4-kinase β and PtdInsP 4-kinase α? from mammalian cells have been published (Fruman et al., 1998; Itoh et al., 1998). Two yeast homologues have also been identified (MSS4 and FAB1). The MSS4 gene product was recently shown to have PtdIns(4)P 5-kinase activity and to be critical for actin structures in *S. cerevisiae* (Desrivieres et al., 1998). Based on in vitro studies with purified enzymes, the mammalian type I γ and I β PtdInsP 5-kinases are relatively promiscuous with respect to substrate utilization and position of phosphorylation on the inositol ring. Thus, although the preferred substrate is PtdIns(4)P, when provided with PtdIns(3)P, these enzymes produce PtdIns(3,5)P$_2$, PtdIns(3,4)P$_2$, and PtdIns(3,4,5)P$_3$ (Zhang et al., 1997; Tolias et al., 1998a,b).

The most extensively studied reaction catalyzed by PtdIns P 5-kinases is the conversion of PtdIns(4)P to PtdIns(4,5)P$_2$. Using enzyme purified from erythrocytes the K_m for PtdIns(4)P in micelles, liposomes, or native membranes ranges from 1 to 10 μM (Loijens and Anderson, 1996). This reactions stimulated by heparin and spermine and inhibited by the product, PtdIns(4,5)P$_2$. Importantly, PtdInsP 5-kinases are stimulated as much as 50-fold by phosphatidic acid (Moritz et al., 1992; Jenkins et al., 1994). This lipid can be generated from PtdCho by phospholipase D and DAG by DAG-kinases. PtdIns(3,4,5)P$_3$ is also produced when recombinant PtdInsP 5-kinases are mixed with PtdIns(3)P (Zhang et al., 1997). The amount of PtdIns(3,4,5)P$_3$ produced by PtdInsP 5-kinase α approaches as much as 50% of the PtdInsP$_2$ produced in a 10 min reaction (Zhang et al., 1997).

Type I PtdInsP 5-kinase activity in vivo may be influenced by small G proteins (Hinchliffe et al., 1998; Tolias et al., 1998a,b). Tolias et al. (1998a,b) have reported the in vitro synthesis of PtdIns(4,5)P$_2$ by PtdIns(4)P 5-kinase from murine cells. The reaction is carried out as described for PtdIns 5-kinase, except that PtdIns(4)P is the substrate now. The PtdIns(4,5)P$_2$ product needs to be isolated by TLC as a PtdInsP$_2$ before counting. The structure needs to be confirmed by HPLC and periodate oxidation of the deacylated product (Chapter 3).

Itoh et al. (1998) have described an in vitro synthesis of PtdIns(4,5)P$_2$ using II γ PtdIns(5)P 4-kinase. Forty-eight hours after electroporation, the expression vector—transfected COS-7 cells were lysed with lysis buffer (20 mM HEPES, pH 7.2, 50 mM NaCl, 30 mM sodium pyrophosphate, 1% Nonidet P-40, 1 mM EGTA, 25 mM NaF, 0.1 mM sodium vanadate, and 1 mM phenylmethylsulfonyl fluoride). The expressed enzyme was immunoprecipitated with monoclonal anti-Myc antibody and washed three times with lysis buffer and once with reaction buffer (50 mM Tris–HCl, pH 7.5, 10 mM MgCl$_2$, and 1 mM EGTA). The reaction was started by adding 50 μM PtdIns(5)P (purified from bovine spinal chord PtdInsP by a neomycin column), 50 μM ATP, and 10 μCi of [γ-^{32}P]ATP in 50 μl. After incubating for 10 min at room temperature, the lipids were extracted with 1N HCl and chloroform/methanol (2:1, v/v) and spotted on TLC plates. The plates were developed in chloroform/methanol/ammonia/water (14:20:3:5, by vol), and the products were monitored by autoradiography or quantified by image analyzer. The structure of the PtdIns(4,5)P$_2$ product needs to be confirmed by HPLC and periodate oxidation of the deacylated product (Chapter 3).

6.3.2.2. PtdIns(3,4)P$_2$

There are three potential pathways for in vivo production of PtdIns(3,4)P$_2$ based on the in vitro activities that have been characterized (Wymann and Pirola, 1998). The Class I and Class II PtdIns 3-Ks can convert PtdIns(4)P to PtdIns(3,4)P$_2$ in vitro. However, PtdIns(4)P is not the optimal substrate for either of these enzymes. The formation of PtdIns(3,4)P$_2$ from PtdIns(3)P by PtdIns(3)P 4-kinase has also been demonstrated in platelets and red blood cells and plants. The identity of this enzyme has not been definitely established, however, the type I and Type II PtdInsP kinases, both of which have PtdIns(3)P 4-kinase activity in vitro, could be involved, although recent evidence to the contrary has also been obtained (Banfic et al., 1998; Hinchliffe et al., 1998).

As indicated above, the type I PtdInsP 5-kinases and the type II PtdInsP 4-kinases can phosphorylate PtdIns(3)P at the D-4 position to produce PtdIns(3,4)P$_2$ (Fruman et al., 1998). Although PtdIns(3)P is a poor substrate for the type I and type II enzymes that have been characterized, it is clear that there exists an enzymatic activity in platelets that converts PtdIns(3)P to PtdIns(3,4)P$_2$ in response to aggregation of the integrin, GPIIb3a (Banfic et al., 1998). In thrombin-stimulated platelets, however, the bulk of PtdIns(3,4)P$_2$ results from a pathway in which the D-4 phosphate is phosphorylated before the D-3 phosphate (Carter et al., 1994).

PtdIns(3,4)P$_2$ can also be produced in vitro via hydrolysis of PtdIns(3,4,5)P$_3$ by a D-5 specific phosphatases such as SHIP and synaptojanin family members (see Chapter 7).

Serunian et al. (1989) reports the in vitro biosynthesis of PtdIns(3,4)P$_2$ by PtdIns(4)P 3-kinase from transformed cells. Briefly. [^{32}P]PtdIns(3)P, [^{32}P]PtdIns(3,4)P$_2$ and [^{32}P]PtdIns(3,4,5)P$_3$ substrates were prepared from immunoprecipitates of polyoma virus-transformed cells as described by Whitman et al. (1988). PtdIns 3-kinase activity was immunoprecipitated from polyoma-transformed cells using a polyclonal rabbit anti-polyoma middle T-specific antiserum, which was added to cell lysates as described by Whitman et al. (1987). Proteins were collected on Protein A-Sepharose CL-4B (Sigma), and the immunoprecipitate was washed three times with phosphate-buffered saline containing 1% Nonidet P-40, two times with 500 mM LiCl in 100 mM Tris, pH 7.5, and three times in TNE buffer (10 mM Tris, pH 7.5, 100 mM NaCl, 1 mM EDTA, pH 7.5, 1 mM EGTA, pH 7.5). After washing, the immunoprecipitate was mixed with either PtdIns, PtdIns(4)P, or PtdIns(4,5)P_2 (final concentration, 0.2 mg/ml) that had been sonicated in 20 mM HEPES, pH 7.5, 1 mM EDTA, pH 7.5. Enzyme reactions were started by adding [γ-^{32}P]ATP and 10 mM MgCl$_2$ in 20 mM HEPES buffer, pH 7.5. The reaction was incubated at room temperature in final volume of 150 μl for 20 min and was stopped by adding 0.25 ml of 1 M HCl and 0.5 ml of methanol/chloroform (1:1, v/v). After vigorous mixing and centrifuging to separate the phases, the organic phase was removed, washed two times with methanol/0.1 M EDTA, pH 7.5 (1:1, v/v) and stored under nitrogen at $-70°$C until further use.

In all cases, an aliquot of the resultant aqueous and organic phases was removed and counted for radioactivity. The remaining organic phase was washed two times with 0.5 volume of methanol/0.1 M EDTA, pH 7.5 (1:0.9 v/v), dried under nitrogen, and stored at $-70°$C until deacylation. The deacylation of phospholipids was performed using methylamine reagent described by Whitman et al. (1988). Specifically, methylamine reagent (42.8% of 25% methylamine in H$_2$O, 45.7% methanol, 11.4% butanol) was added to the phospholipids and incubated at 53°C for 50 min. After cooling to room temperature, the sample was dried under vacuum, water was added to the dried sample, and the contents were dried again. Samples were resuspended in H$_2$O and extracted two times with an equal volume of 1-butanol/light petroleum ether/ethyl formate (20:4:1, by vol). A Partisphere SAX anion-exchange column (Whatman) was used to analyze the deacylation products by HPLC.

Graziani et al. (1992) have reported the in vitro biosynthesis of PtdIns(3,4)P_2 by type III PtdIns(3)P 4-kinase of bovine brain as follows. PtdIns(3)P 4-kinase was assayed at 37°C for 15 min in a 20 ml reaction mixture containing [^{32}P]PtdIns(3)P (0.5–1 μM), [^{3}H]PtdIns(4)P (0.5–1 μM), 9 μM PtdIns, 8 μM PtdSer, 500 μM ATP and 10 mM MgCl$_2$ in 50 mM HEPES buffer (pH 8)/1 mM EGTA (lipids were dried under nitrogen and sonicated for 10 min in a bath sonicator). The reaction was started by adding enzyme and was terminated with chloroform/methanol/HCl. The lipids were extracted and identified by TLC and ion exchange chromatography. The lipids were deacylated by incubation with methylamine at 54°C for 40 min; the derived GroPIns phosphates were applied to a strong anion-exchange HPLC (Whatman Partisil 10 SAX) column and eluted with a gradient of ammonium formate. The [^{32}P]PtdIns(3)P substrate was prepared by incubating PtdIns with [γ^{32}P]ATP and purified PtdIns 3-kinase.

6.3.2.3. PtdIns(3,5)P$_2$

Whiteford et al. (1997) and Jones et al. (1999) obtained conclusive identification of PtdIns(3,5)P$_2$ in mammalian cells and Dove et al. (1997) in yeast. The bulk of this lipid is made by a pathway in which the D-3 position is phosphorylated prior to the D-5 position. The mammalian type Iα and type Iβ PtdIns(3)P 5-K activity may be responsible for the production of some of the PtdIns(3,5)P$_2$ in cells.

As mentioned above, the mammalian type I γ and type I β PtdInsP 5-Ks have a significant PtdIns(3)P 5-kinase activity and could be responsible for the production of some of the PtdIns(3,5)P$_2$ in cells. Although PtdIns(3,5)P$_2$ is far less abundant than PtdIns(4,5)P$_2$ in mammalian cells (approximately 1%), in yeast PtdIns(3,5)P$_2$ levels can approach that of PtdIns(4,5)P$_2$, under conditions of osmotic stress (Dove et al., 1997). PtdIns(3,5)P$_2$ is produced in yeast by the yeast PtdInsP 5-K homologue, Fab1 (Gary et al., 1998).

In vitro, PtdIns(3,5)P$_2$ can also be generated through phosphorylation of the novel lipid PtdIns(5)P by the Class I$_A$ PtdIns 3-kinase (Rameh et al., 1997). Tolias et al. (1998a,b) have demonstrated that Type I PtdIns(4)P 5-kinases synthesize the novel lipid PtdIns(3,5)P$_2$.

Whiteford et al. (1997) have observed the in vivo synthesis of PtdIns(3,5)P$_2$ via PtdIns(3)P 5-kinase in mouse fibroblasts. For myo-[^{3}H]inositol labeling, subconfluent cultures of mouse fibroblasts in 60 mm dishes were first starved of inositol by incubation in inositol-free Delbecco's modified essential medium (DMEM) containing 20 mM HEPES, pH 7.1, ITS-Plus- and 10% dialyzed calf serum. After 6–8 h, this medium was replaced with 2 ml of inositol-free DMEM containing 20 mM HEPES, pH 7.1, ITS-Plus- and 50 μCi of myo-[2-^{3}H]inositol, and labeling was allowed to progress for 18–24 h. The incubations were stopped by addition of perchloric acid. Perchlorate extraction, deacylation of PtdInsPs, and separation of GroPInsPs were carried out as described elsewhere (Chapter 3). Under these conditions, (2–6) × 10^6 dpm were recovered in the lipid-containing perchlorate precipitates. After deacylation and HPLC, peaks corresponding to GroPIns, GroPIns(3)P, GroIns(4)P, GroIns(3,5)P$_2$ and Gro(4,5)P$_2$ contained 80–90, 0.15–0.18, 5.5–6.0, 0.03–0.04 and 4–7%, respectively, of the total radioactivity recovered in this fraction from untreated Ph-N2 cultures. An identical protocol was utilized for large-scale isolation of GroPIns(3,5)P$_2$ from approximately 1 × 10^8 Ph-N2F5 cells labeled with 1.5 mCi of myo-[2-^{3}H]inositol, with an overall recovery of approximately 60 000 dpm of [^{3}H]GroPIns(3,5)P$_2$.

For [^{32}P]P labeling, nearly confluent cultures in 100 mm-diameter dishes were incubated for 24 h in serum-free DMEM containing 20 mM HEPES, pH 7.1, and ITS-Plus, then incubated for 30 min in phosphate-free DMEM containing 20 mM HEPES, pH 7.1. Cultures were labeled with 3–4 mCi of [^{32}P]P in 1 ml of this medium in an open water bath. The labeling medium was removed after 3–5 min, and 0.72 ml of 1 M HCl and 0.9 ml of methanol were added. Cells were scraped from the dishes, transferred to glass tubes, and mixed vigorously with 1.8 ml of methanol/chloroform (1:1, v/v). After 215 min on ice, 0.9 ml of water and 0.9 ml of chloroform were added and phases were separated by centrifugation at 24°C. The organic phase was reextracted once with 1.8 ml of methanol/1 M HCl (1:1, v/v), then dried and deacylated (Downes et al., 1986). Between 1800 and 5400 cpm of GroPIns(3,5)P$_2$ and 116 000–624 000 cpm of GroPIns(4,5)P2 were recovered by this procedure.

Rameh et al. (1997) have reported an in vitro synthesis of PtdIns(3,5)P$_2$ via PtdIns(5)P and Class I$_A$ PtdInsP 3-kinase. Thus, dipalmitoyl GroPIns(5)P was converted into the corresponding PtdIns(3,5)P$_2$. For the kinase reaction, recombinant murine type I PtdInsP kinase β was used. Deacylated lipids were analyzed by anion-exchange HPLC using a Partisphere SAX column (Chapter 3).

6.3.3. PtdIns trisphosphates

Although PtdIns possesses five free hydroxyl groups only the 3, 4 and 5 positions are known to be phosphorylated in PtdInsPs. Therefore, only one trisphosphate is known, but it may be derived by a variety of phosphorylation sequences. Like the PtdIns bisphosphates, the trisphosphate may be formed by the same enzymes that form the mono- and bisphosphates. The route taken to the final product is determined by the nature of the starting bisphosphate. A reaction started with PtdIns or PtdInsP usually leads to a mixture of products, which must be resolved by chromatography usually following deacylation.

6.3.3.1. PtdIns(3,4,5)P$_3$

PtdIns(3,4,5)P$_3$ is nominally absent in quiescent cells, but appears within seconds to minutes of cell stimulation with a variety of activators (Traynor-Kaplan et al., 1989). The bulk of the PtdIns(3,4,5)P$_3$ is believed to be generated by phosphorylation of PtdIns(4,5)P$_2$ at the 3-position of the inositol ring (Hawkins et al., 1992). This is believed to occur via the Class I PtdIns 3-kinases, which are the only enzymes known to use PtdIns(4,5)P$_2$ as a substrate. The Class I PtdIns 3-kinases are activated by growth factor stimulation of cells, which is mediated in rat by interaction of their SH2 domain with tyrosine-phosphorylated proteins (Toker and Cantley, 1997).

Alternatively, PtdIns(3,4,5)P$_3$ can be synthesized by a pathway independent of PtdIns(4,5)P$_2$. The PtdInsP 5-kinases α and β have been shown to utilize PtdIns(3)P as a substrate to produce PtdIns(3,4,5)P$_3$ by phosphorylating the 4- and 5-positions of the inositol ring in a concerted reaction (Zhang et al., 1997; Tolias et al., 1998a,b). Zhang et al. (1997) identified the PtdInsP$_2$ product as PtdIns(3,4)P$_2$. The relative contribution of this pathway to intracellular levels of PtdIns(3,4,5)P$_3$ is not known. The α and β isoforms convert PtdIns(3,4)P$_2$ to PtdIns(3,4,5)P$_3$ with K_{cat}/K_m ratios that are 3-fold and 100-fold lower, respectively, than those observed using PtdIns(4)P as the substrate (Zhang et al., 1997). Type I PtdInsP 5-Ks also utilize PtdIns(3)P as a substrate in vitro.

Hawkins et al. (1992) have demonstrated in vivo synthesis of PtdIns(3,4,5)P$_3$ via PtdIns(4,5)P$_2$ 3-kinase in Swiss 3T3 cells. Swiss 3T3 cells were grown to confluence over six days in small glass bottles and then serum-starved for 16 h before the experiment. Each bottle was then removed from the incubator, washed with HEPES-buffered DMEM (112.5 mM NaCl, 5.37 mM KCl, 0.81 mM MgSO$_4$, 1.8 mM CaCl$_2$, 25 mM glucose, 1 mM NaHCO$_3$, 25 mM HEPES, pH 7.4, at 37°C, 0.1% fatty acid-free BSA, 1 × DMEM amino acids and vitamins) and then incubated with 0.4 ml HEPES-buffered DMEM containing 0.25 mCi [^{32}P]P for 70 min at 37°C. Each bottle was then washed twice and incubated at 37°C in a total volume of 0.4 ml HEPES—buffered DMEM either with or without PDGF (human recombinant form BB used at a final concentration of 30 ng/ml). Incubations were terminated by addition of 1.5 ml CHCl$_3$/MeOH (1:2, v/v). Lipids were extracted using a Folch distribution of organic and aqueous phases, with 1.0 M HCl, 25 mM EDTA, 5 mM tetrabutylammonium sulfate included in the aqueous phase. Lipids were deacylated with methylamine and the GroPIns esters separated by HPLC using Partisphere SAX column as described (Stephens et al., 1991). The identities of the appropriate radioactive compounds obtained using this strategy has been rigorously characterized (Stephens et al., 1991).

Petitot et al. (2000) demonstrated in vitro synthesis of PtdIns(3,4,5)P$_3$ via Class I PtdIns(4,5)P$_2$ 3-kinase from HT-29 cells. The PtdIns(3,4,5)P$_3$ formation was determined by a modification of the method of Burgering et al. (1991). Briefly, immunoprecipitates (prepared with anti-phosphotyrosine serum and collected on protein A-Sepharose beads) and the supernatant of the immunoprecipitation were resuspended in 80 µl of buffer containing 30 mM HEPES and 300 µM adenosine. To both samples, 50 µl of lipid mixture (synthetic lipid and PtdSer) was added for 20 min. The assay of Class 1 PtdIns 3-kinase activity was performed on the immunoprecipitate using diC16GroPIns(4,5)P$_2$ as substrate in the presence of Mg^{2+}. The reaction was initiated by the addition 20 µCi of [γ-^{32}P]ATP, 100 µM ATP, 100 µM MgCl$_2$ and terminated after 15 min by the addition of 80 µl of 1 M HCl and 200 µl of chloroform/methanol, 1:1 (v/v). After vigorous mixing and centrifugation, the organic layer was removed, dried under N$_2$ and resuspended in chloroform/methanol. Extracted lipids were separated by TLC in developing solvents: isopropyl alcohol/H$_2$O/acetic acid, 65:34:1 (by vol), and then exposed to hyperfilm. Unlabeled standards (1 mg/ml) were revealed by exposure to iodine vapor.

Zhang et al. (1997) reported in vitro synthesis of PtdIns(3,4,5)P$_3$ via PtdIns(3,4)P$_2$ and recombinant human PtdInsP 5-kinase isozymes. The biosynthesis of PtdIns(3,4,5)P$_3$ via PtdInsP 5-kinases was determined in 50 µl reactions containing 50 mM Tris, pH 5, 30 mM NaCl, 12 mM MgCl$_2$, 67 µM lipid substrate [including dipalmitoyl GroPIns(3,4)P$_2$ among others], 133 µM phospho-serine, and 50 µM [γ-^{32}P]ATP (10 µCi/assay). When PtdIns was used as the lipid substrate, no carrier lipid was used, and when [^{32}P]PtdIns(3)P was used as substrate, unlabeled ATP was used in place of [γ-^{32}P]ATP. Reactions were stopped after 10 min by adding 80 μl of 1N HCl and then 160 μl of CHCl$_3$/MeOH (1:1, v/v). Lipids were separated by TLC using 1-propanol/2 M acetic acid (65:35, v/v). Phosphorylated lipids were visualized by autoradiography and quantified using Bio-Rad Molecular Analyst.

The reaction products were analyzed by HPLC following deacylation. The deacylated lipids were mixed with ^{3}H-labeled standards and analyzed by anion-exchange HPLC using Partisphere SAX column (Chapter 3). To separate PtdIns(4)P from PtdIns(5)P and PtdIns(3,4)P$_2$ from PtdIns(3,5)P$_2$, a modified ammonium phosphate gradient was used. The compounds were eluted with 1 M (NH$_4$)$_2$HPO$_4$, pH 3.8, and water using the following gradient: 0% 1 M (NH$_4$)$_2$HPO$_4$ for 5 ml, 0–1% 1 M (NH$_4$)$_2$HPO$_4$ over 5 ml, 1–3% 1 M (NH$_4$)$_2$HPO$_4$ over 40 ml, 3–10% 1 M (NH$_4$)$_2$HPO$_4$ over 10 ml, 10–13% 1 M (NH$_4$)$_2$HPO$_4$ over 25 l, and 13–65% 1 M (NH$_4$)$_2$HPO$_4$ over 25 ml. Eluate from the HPLC column flowed into an on-line continuous flow scintillation detector (Radiomatic Instruments, FL) for isotope detection. The structures were confirmed by periodate oxidation as described elsewhere (Chapter 3).

6.4. Glycosyl phosphatidylinositols

Glycosyl PtdIns (GPtdIns) are synthesized by all eukaryotic cells examined to date, and they are typically found covalently linked to cell-surface glycoproteins. GPtdIns serve an important alternative mechanism for anchoring proteins to cell membranes, and a wide spectrum of functionally diverse proteins rely on PtdIns anchor for membrane association. The PtdIns moiety is synthesized in the endoplasmic reticulum and then transferred to proteins containing a carboxyl-terminal GPtdIns-attachment signal sequence. General information on the structure of GPtdIns and methods of determination was presented in Chapter 4, which may be consulted for analytical details. The present discussion is based on major original papers and recent reviews on biosynthesis of GPtdIns membrane anchors (Stevens, 1995), radiolabeling techniques for studying GPtdIns biosynthesis in cell lysates, subcellular fractions and permeabilized cells (Vidugiriene and Menon, 1994, 1995), intracellular precursors of mammalian GPtdIns-anchored proteins (Hirose et al., 1995) and on the contributions of *Trypanosoma* research to structure, biosynthesis and functions of GPtdIns anchors (Milne et al., 1999; Ferguson et al., 1999), as well as the biosynthesis of GPtdIns in mammals and unicellular microbes (Tiede et al., 1999). Schmidt et al. (1998) reported that two structurally distinct free- and protein-bound glycosyl PtdIns are formed by *P. falciparum* in maturation dependent manner. Schmidt et al. (2002) described the substrate specificity of the *P. falciparum* glycosyl PtdIns bio-synthetic pathway and its inhibition. McConville and Menon (2000) have provided a most thorough summary of recent developments in the biology and biochemistry of glycosyl PtdIns lipids with special emphasis on the compartmentalization of the GPtdIns biosynthetic pathways and the redistribution of GPtdIns from their site of synthesis to other cellular membranes.

GPtdIns membrane anchors are synthesized by a sequential transfer of sugars and phosphocthanolamine to PtdIns. The complete anchor is then transferred en bloc to the nascent GPtdIns-anchored protein. The entire pathway, including transfer to protein, takes place in the rough ER with the initial steps occurring on the cytoplasmic surface of ER membrane and the final steps on the lumenal side of the organelle. The biochemical analysis of the biosynthetic pathway, as well as the partial characterization of its enzymatic machinery, has been made possible by the development of cell-free synthetic systems.

6.4.1. GPtdIns protein anchors

GPtdIns is used to anchor various eukaryotic proteins to the cell-surface membrane. GPtdIns-anchored proteins are found in yeast, mammals, and plants but are especially abundant in protozoan

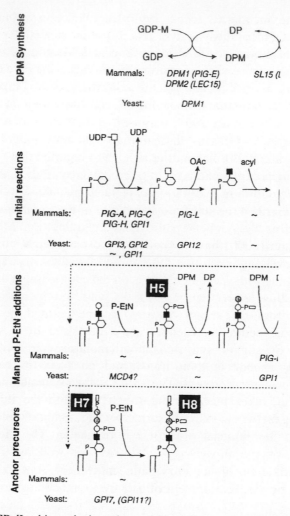

Fig. 6.5. GPtdIns biosynthetic pathway. Mammalian gene products required at each step are listed along with their yeast equivalents. Gene products involved in the assembly and consumption of dolichol-phosphomannose (DPM) are shown in the top panel. The activities of Pig-B, Pig-F, Mcd4, Gpi7, Gpi11, YII031c and SI15/Lec35 are implied and have not been explicitly tested. Molecular species of the PtdIns substrate can vary and further remodeling reactions may occur after assembly of the complete head-group (see text). H5, H7 and H8 indicate P-Etn

parasites (Tiede et al., 1998; Ferguson, 1999; Okazaki and Moss, 1999). The core structure and the biosynthetic steps of GPtdIns are basically conserved in various organisms (McConville and Ferguson, 1993; Englund, 1993; Kinoshita et al., 1997; Kinoshita and Inoue, 2000). Figure 6.5 summarizes the basic steps in the GPtdIns biosynthetic pathway (McConville and Menon, 2000).

Detailed structural analyses of several of these protein anchors has led to the proposal of an evolutionary conserved core structure made up of PtdIns-lipid linked to a linear core glycan consisting of non-acetylated glucosamine and three mannose residues. The structure is attached to the C-terminal amino acid of the protein via an ethanolamine bridge.

Mayor et al. (1990a,b) have described the glycolipid precursors of membrane anchor of *T. brucei*, the lipid structures of the PtdIns-specific PLC resistant and sensitive glycolipids.

6.4.1.1. GlcNAc-PtdIns

The first step of GPtdIns biosynthesis involves the transfer of GlcNAc from UDP-GlcNAc to PtdIns to form *N*-acetylglucosaminyl-PtdIns (GlcNAc-PtdIns) (Doering et al., 1989). The second substrate in the reaction, PtdIns, is supplied by the membranes or cellular fractions, which also contain the enzymatic activity, which catalyzes the reaction. In mammalian cells, there is significant metabolism of UDP-GlcNAc to non-GPtdIns-related products (Stevens, 1993), which decreases the amount of this substrate available for GlcNAc-PtdIns biosynthesis. This competition does not occur in trypanosome membranes. The α1-6 GlcNAc transferase for this step was found to be a rather complex aggregate. The availability of three mammalian cell lines defective in this step allowed identification of three genes involved in this

containing mammalian GPtdIns. DG, diacylglycerol; DP, dolichol phosphate; GDP, guanosine diphosphate; UDP, uridine diphosphate. Graphical symbols are as shown in the insert. Reproduced from McConville and Menon (2000) with permission of the publisher.

step: PIG-A (for PtdIns glycan class A, Miyata et al., 1993); PIG-C (Inoue et al., 1996), and PIG-H (Kamitani et al., 1993). By analogy, three *S. cerevisiae* genes involved in GlcNAc-PtdIns synthesis have been cloned: GPI1 (Leidich and Orlean, 1996), GPI2 (Leidich et al., 1995), and GPI3 (Leidich et al., 1995). Sequence comparisons revealed that human PIG-A and yeast GPI3, as well as human PIG-C and yeast GPI2, encode homologous proteins, while the products of the two remaining genes, human PIG-H and yeast GI1, lacked any significant homology. A further mammalian gene encoding a protein similar to yeast Gpi1p was cloned (Tiede et al., 1998; Watanabe et al., 1998). The enzyme complex consisting of at least four proteins, Pig-a, Pig-c, Pig-h and Gpi1, has been partially purified from ER of transfected cells by affinity chromatography, and has been shown to be capable of in vitro GlcNAc-PtdIns synthesis (Watanabe et al., 1998). Stevens (1993) and Tiede et al. (1999) have discussed plausible functions for some of these proteins, while the roles of some others have remained obscure. Four proteins in mammals and at least three in yeast make up a complex that carries out the first step in the transfer of GlcNAc from UDP-GlcNAc to PtdIns. Kostova et al. (2000) have used photoaffinity labeling with P3-(4-azidoanilido)uridine 5'-trisphosphate to identify gpi3 as the UDP-GlcNAc-binding subunit of the enzyme that catalyzes formation of GlcNAc-PtdIns.

Despite the apparent specificity for variations in structure of the inositol lipid portion of the GPtdIns anchor in many species, the available evidence suggests that this is not determined by the selection of particular inositol phospholipids for incorporation into the GlcNAc-PtdIns. In mammalian cells, alkylacyl GroPIns represents less than 5% of total cellular PtdIns (Stevens, 1993) yet 15–20% of the GlcNAc-PtdIns synthesized by lysates of normal lymphoma cell line contains the *sn*-1-alkyl chain as judged by the sensitivity to alkaline hydrolysis (Stevens, 1993). Furthermore, microsomes prepared from EL4 (Thy-1-f) cells, which are unable to make alkyl lipids because of a defect in ether lipid biosynthesis, produce normal levels of GlcNAc-PtdIns

(Stevens, 1993 review). It appears that mammalian cells can make GPtdIns anchors having only acyl-linked fatty acids without any deleterious consequences. Wongkajjornsilp et al. (2001) have demonstrated that administration of sn-1-alkyl-sn-2-lyso-GlcN-PtdIns in HeLa D cells leads to an accumulation of the unaltered lyso compound, the GlcN-[^{3}H]PtdIns and GlcN(acyl)[^{3}H]PtdIns in a metabolically inert compartment.

The final type of specificity seen in the PtdIns portion of GPtdIns is in the fatty acid composition of the VSG membrane anchor. Detailed analysis has established that trypanosomes synthesize GlcNAc-PtdIns and other GPtdIns precursors using PtdIns with a variety of fatty acids (Doering et al., 1994) and introduce myristate into the anchor after complete assembly of the core carbohydrates. Much of the myristate is apparently derived from a recently discovered specialized fatty acid synthesis system (Paul et al., 2001).

Hirose et al. (1991) reported the following in vitro synthesis of GPtdIns precursors. The demonstration started with a preparation of the lysates of both JY25 and CHO cells. Lysates of JY25 cells equivalent to 4×10^8 cells were incubated with 80 μCi of UDP-[6-^{3}H]GlcNAc (13.3 μCi/ml) for 3 h at 37°C in a buffer consisting of 50 mM HEPES/NaOH, pH 7.4, 25 M KCl, 5 mM MgCl$_2$, 0.1 mM Tos-Lys-CH$_2$Cl, 1 μg/ml leupeptin, 5 M MnCl$_2$, 1 m ATP, 0.5 M DTT and 0.2 μg/ml tunamycin. The lipids were extracted with CHCl$_3$/MeOH/water (10:10:3, by vol). The lipid extract was dried, dissolved in water-saturated butan-1-ol and partitioned with water. The butanol fraction was dried. The glycolipids were analyzed by TLC with the solvent system CHCl$_3$/CH$_3$OH/1 M NH$_4$OH (10:10:3, by vol) followed by image analysis. To quantify the GlcNAc-PtdIns and GlcN-PtdIns generated, the intensities of the spots were measured with an Image Analyzer after exposure for 2−5 days.

Hirose et al. (1991) have also reported the acetylation of [6-^{3}H]GlcN-PtdIns to generate [6-^{3}H]GlcNAc-PtdIns. For this purpose, lysates of JY25 cells equivalent to 4×10^8 cells were incubated with 80 μCi of UDP-[6-^{3}H]GlcNAc (13.3 μCi/ml) for

3 h at 37°C in a buffer containing 50 mM HEPES/NaOH, pH 7.4, 25 M KCl, 5 mM $MgCl_2$, 0.1 mM Tos-Lys-CH_2Cl, 1 µg/ml leupeptin, 5 M $MnCl_2$, 1 m ATP, 0.5 M DTT and 0.2 µg/ml tunamycin. Glycolipids extracted by butan-1-ol and dried were dissolved in 2.8 ml of methanol and then mixed with 2.8 ml of $NaHCO_3$-saturated water. The addition of acetic acid anhydride (280 µl) was followed by incubation on ice for 10 min. Further 280 µl of acetic anhydride was added and the mixture was incubated at room temperature for 50 min. GlcNAc-PtdIns was extracted as described above and a small aliquot was used to measure radioactivity.

Subramanian et al. (2000) have concluded that the intestinal protozoan Giardia lamblia can use exogenously supplied [^{3}H]PtdIns and [^{3}H]Ins to synthesize glycosyl PtdIns of the invariant surface antigen GP49. While PtdIns is directly incorporated into glycosyl PtdIns molecules, free Ins is first converted into PtdIns by head group exchange enzymes, and this newly formed PtdIns participates in glycosyl PtdIns anchor synthesis.

6.4.1.2. GlcN-PtdIns

The second step in GPtdIns biosynthesis is the deacetylation of GlcNc-PtdIns to form GlcN-PtdIns (Doering et al., 1989; Hirose et al., 1991). This reaction occurs spontaneously to varying extent in the parasite, yeast and mammalian cell-free systems. Mammalian cell lines defective in the second step of GPtdIns biosynthesis have become available recently. The first of these mutants, designated IB, deacetylates GlcNAcc-PtdIns with 10% of the efficiency of wild type K562 cells (Hirose et al., 1992a). Milne et al. (1994) achieved partial purification of the N-deacetylase from the blood stream of *Trypanosoma brucei*. The enzyme rapidly deacetylates GlcNAc-PtdIns without steric specificity for the C-2 to C-5 hydroxyl groups of the *myo*-inositol ring. GlcNc-*myo*-Ins-PGro, an analog lacking the fatty acid residues of the natural substrate, is still recognized and turned over by the enzyme. Further truncated substrates, including GlcNAc-*myo*-inositol and GlcNAc are not recognized, indicating that the enzyme is not a general N-deacetylase (Stevens, 1993; Tiede et al., 1999). The products of hydrolysis, GlcN-PtdIns and acetate, do not seem to play any role in regulation of the enzyme activity. In contrast, mammalian deacetylase activity is specifically stimulated by GTP at an optimal concentration of 1 mM. The nucleoside analog guanosine $5''$-O-(thiotrisphosphate), GTPγS, completely suppresses the effect, indicating that hydrolysis of the nucleoside triphosphate is required for stimulation (Stevens, 1993). In contrast to the mammalian enzyme, the trypanosome N-deacetylase is not activated by GTP (Smith et al., 1996). Studies on substrate specificity (regarding the acyl group that can be removed from the substrate) revealed little tolerance for moieties larger than acetyl groups. The efficiency was observed to drop off dramatically in order GlcN-acetyl-PtdIns > GlcNpropionyl-PtdIns > GlcN-butyryl-PtdIns = GlcN-isobutyryl-PtdIns = GlcN-pentanoyl-PtdIns > GlcN-hexanoyl-PtdIns (Sharma et al., 1997). The nature of the active site of the GlcNAc-PtdIns deacetylase remains uncertain (Stevens, 1993; Tiede et al., 1999). Stevens et al. (1999) have developed a system to study the regulation of deacetylation that uses microsomes from cells defective in the first step in glycosyl PtdIns biosynthesis (Stevens and Raetz, 1991) and the second reaction in the pathway. With this mixed-microsome system, the deacetylation of GlcNAc-PtdIns was almost completely dependent on GTP hydrolysis.

Watanabe et al. (1999) have assayed the GlcNAc-PtdIns de-N-acetylase as follows. Briefly, [6-^{3}H]GlcNAc-PtdIns synthesized in vitro and dissolved in 1 μl of 99% (v/v) ethanol was incubated with enzyme samples with or without 1 mM GTP in 150 μl of a buffer consisting of 50 mM HEPES/NaOH, pH 7.4, 25 mM KCl, 5 mM MgCl$_2$, 5 mM MnCl$_2$, 0.5 M DTT, 0.1 mM Tos-Lys-CH$_2$Cl and 1 μg/ml leupeptin at 37°C for 0–120 min. When studying the effect of metal ions and chelating reagent, Watanabe et al. (1999) used 50 mM HEPES/NaOH (pH 7.4)/25 mM KCl/0.5 mM DTT/0.1 mM Tos-Lys-CH$_2$Cl/1 μg/ml leupeptin as a basic buffer.

The deacylated products were extracted and analyzed as described above.

6.4.1.3. GlcN-PtdIns acyl esters

In mammals and in yeast, acylation of the 2-position of the inositol ring represents the third step in the GPtdIns pathway (Costello and Orlean, 1992; Urakaze et al., 1992). Acylation depends on the presence of acyl CoA and labeling studies in yeast and mammalian cell-free systems suggest that acyl-CoA is the immediate acyl donor (Doerrler et al., 1996). Stevens and Zhang (1994), however, have proposed that the acyl group may be transferred from an endogenous phospholipid by CoA-independent transacylation. Acylation of the Ins moiety renders GPtdIns-anchored proteins resistant to PtdIns-specific PLC (Roberts et al., 1988a). However, most GPtdIns-anchored proteins in mammals and other eukaryotes are PtdIns-PLC sensitive because they do not contain acylated inositol. There is evidence that the acyl group originally attached to the Ins moiety may be subsequently removed by deacylation in the ER within 5 min after protein transfer (Chen et al., 1998).

In trypanosomes, inositol acylation does not take place before first mannose residue has been added to GlcN-PtdIns. After Man1-GlcN-PtdIns, acylated and non-acylated forms of virtually every intermediate are found in a dynamic equilibrium (Guther and Ferguson, 1995). The trypanosomal acyltransferase, but not the mammalian enzyme, can be blocked by the action of the active site serine-directed inhibitor phenylmethylsulfonyl fluoride (PMSF, Guther et al., 1994). In contrast, diisopropylfluorophosphate (DFP) selectively inhibits the removal of the inositol acyl chain in trypanosomes, indicating that acylation and deacylation are carried out by two different enzymatic activities (Guther and Ferguson, 1995).

Doerrler and Lehrman (2000) describe the in vitro acylation of GlcN-PtdIns(C_8) as follows. The required palmitoyl-CoA-dependent GlcN-PtdIns acyltransferase activity (AT-2) was derived from CHO-K1 cells by hypotonic swelling, and microsomal membranes were prepared from either swollen cultured cells or tissues by Dounce homogenization in a solution containing 0.5 M NaCl, 20 mM Tris–HCl (pH 7.4) and 100 μg/ml DNase I. As described in the text AT-2 was displaced from microsomes by NaCl, and an alternative homogenization solution without NaCl containing 20 mM Tris–HCl (pH 7.4), 10% sucrose and 1 mM EDTA was found to give improved yields of microsomally associated AT-2.

Labeling with [^{3}H]palmitoyl-CoA was carried out in a 0.1 ml volume containing 100 ng Glc-PtdIns(C8), 1 μCi (0.25 μM) [^{3}H]Palmitoyl-CoA, 0.1 M Tris–HCl (pH 8.0), 1 mM Na$_3$EDTA and a source of enzyme. Detection of activity required that detergent be present at concentrations below its critical micellar concentration. Therefore, samples containing CHAPS were diluted into reaction buffer to 0.2% or less. Following incubation at 37°C for 20 min, the reaction was partitioned between butanol and water. The butanol phase was backwashed once with butanol-saturated water, dried under a stream of nitrogen, and lipids were dissolved in 50 μl chloroform/methanol (2:1) and applied to a pre-activated Whatman Silica Gel 60 TLC plate. The plate was developed once in chloroform/methanol/water (65:25:4 by vol), sprayed with fluor and exposed to X-ray film overnight.

Chen et al. (1998) have described the in vitro deacetylation of GlcNAc-PtdIns to GlcN-PtdIns. Subsequent deacylation takes place in the endoplasmic reticulum within 5 min after protein transfer. Chen et al. (1998) have demonstrated the deacylation as 70/30 for miniPLAP, 40/60 for DAF and CD52 in human erythroleukemia K562 ATCC cells.

6.4.1.4. GlcN-PtdIns mannosides

In mammalian cells and yeast, GlcN-PtdIns must first be acylated on the inositol residue before mannose addition can occur. In other organisms, inositol acylation may occur at a later stage of the biosynthetic pathway or not at all (see below).

After deacetylation of GlcNAc-PtdIns to GlcN-PtdIns, the GPtdIns core is extended by the addition of three mannoses. The three different GPtdIns mannosyltransferases (GPI-MT-I, II, III) all use dolicholphosphate mannose (Dol-P-Man) as the sugar donor (Menon et al., 1990; Ilgoutz et al., 1999b), which in turn is formed from GDP-mannose and dolichol phosphate. The mannoses of the GPtdIns core can be radiolabeled by adding GDP-[^{3}H]mannose to trypanosome membranes. The D-*myo*-inositol configuration as well as free amine group of GlcN-PtdIns were crucial for substrate binding to the enzyme (Smith et al., 1996). Several synthetic analogs, including GlcPtdIns and 2-deoxy-GlcPtdIns, were neither mannosylated nor able to inhibit mannosylation of the natural substrate, suggesting that they do not bind to the active site of GPtdIns-MT-1 (Smith et al., 1996). Azzouz et al. (2001) have obtained data in *Paramecium* that suggest the involvement of classical heterotrimeric G proteins in the regulation of GPtdIns-anchor biosynthesis through dolichol-phosphate-mannose synthesis via the activation of adenylcyclase and protein phosphorylation.

In yeast and mammalian cells, an acyl group must be added to the inositol ring of GlcN-PtdIns before mannoses can be added. Likewise, in the mammalian cell-free system, transfer of the first mannose residue requires inositol acylation of GlcN-PtdIns (Doerrler et al., 1996; Smith et al., 1997b). The first mannose is added in an α1-4 linkage to the GlcN. The second mannose is then linked α1-6 to the first, and the third is linked α1-2 to the second. In mammalian cells, the pathway appears to be more complicated, and the actual substrates for the second and third mannose additions in mammalian cells have not been established conclusively.

No genes for the GPtdIns-MT-I have been cloned so far, nor has a mutant cell line deficient in this enzyme been reported. Likewise, a GPtdIns-MT-II defective mutant cell line remains to be identified. There is evidence, however, for a GPtdIns-MT-III defective mammalian mutant cell line, for which the corresponding gene, PIG-B, was isolated by expression cloning (Takahashi et al., 1996).

Four yeast genes encoding proteins similar to mammalian Pig-b have been identified, one of which is involved in the addition of the third GPtdIns mannose residue (Canivenc-Gansel et al., 1998; Sutterlin et al., 1998) as indicated by the action of GPtdIns inhibitor YW3548. However, it is now known that YW3548 also inhibits the addition of the first PEtn residue, which precedes the next two mannosylation steps.

Doerrler et al. (1996) have described a method of synthesis of GlcN-PtdIns mannosides. Briefly, labeling of GlcN-PtdIns(C$_8$) with GDP-[^{3}H]mannose is carried out as follows. The zwitterionic form of GlcNα-PtdIns(C$_8$) was dispersed in 0.03% (w/v) Triton X-100 to ensure reproducible dissolution and subsequently diluted 4-fold into a 0.1 ml reaction volume containing (final concentrations) 50 mM Tris–HCl (pH 7.4), 5 mM MgCl$_2$, 5 mM MnCl$_2$, 1 mM 5′-AMP, and 0.26 µM GDP-[^{3}H]Mannose. Reactions were initiated by addition of 50 µg of CHO-K1 or S. cerevisiae membrane protein. Reactions were incubated at 37°C for 20 min (CHO K1, trypanosomes) or at 30°C for 40 min (S. cerevisiae) at which time they were chilled on ice and extracted twice with 0.2 ml of water-saturated butanol. Pooled butanol extracts were back washed once with 0.1 ml of butanol-saturated water and then dried under a stream of nitrogen. Lipids were dissolved in 30 µl chloroform/methanol (2:1, v/v) and applied to a pre-activated Whatman Silica Gel 60 plate. TLC plates were developed in chloroform, methanol, 0.25% KCl (55:45:10, by vol), sprayed with fluor and exposed to X-ray film for two to four days. The identification of mannose-P-dolichol (Man-P-Dol) is based on earlier work (DeLuca et al., 1994). The R_f values of various lipids were: Man-P-Dol, 0.82; lipid 1, 0.65; lipid 2, 0.55/0.57; lipid 2′, 0.42; lipid 1′, 0.38; lipid 2″, 0.24.

In yeast, the GPtdIns transferred to protein bears a fourth, α1,2-linked mannose on the α1,2-mannose that receives the PEtn moiety through which GPtdIns become protein linked (Gerold et al., 1999; Sipos et al., 1995; Heise et al., 1996). Grimme et al. (2001) have isolated from temperature-sensitive mutants an Smp-3 protein, which is required for the addition of the side-branching fourth mannose during assembly of yeast GPtdIns.

6.4.1.5. GlcN-PtdIns(Man)₃EtnP

The biosynthesis of the GPtdIns is completed by the addition of PEtn to the 6-hydroxy group of the third mannose residue. The endogenous donor of the PEtn in yeast and trypanosomes is PtdEtn. This was established in yeast by demonstrating that the mutants which could not transfer exogenous [³H]Etn into PtdEtn because of the disruption of Etn phosphotransferase gene did not incorporate this radiolabeled compound into GPtdIns-anchored proteins (Menon et al., 1993). In yeast and in mammalian cells, the first Etn-P is added to the 2-position of the first mannose residue, most probably at the stage of Man1-GlcN-(acylinositol)PtdIns (Homans et al., 1988; Puoti et al., 1991; Hirose et al., 1992b; Puoti and Conzelmann, 1993; Canivenc-Gansel et al., 1998). A further Etn-P residue can be present at the 6-position of the second mannose. The donor for these Etn-P is believed to be the phospholipid PtdEtn (McConville and Menon, 2000). The attachment of Etn-P to position 6 of the terminal mannose leads to completion of the GPtdIns core structure Etn-P-Man3-GlcN-(acylinositol)PtdIns. This step is an essential part of the GPtdIns pathways in all eukaryotes, since it provides the bridge between the GPtdIns glycan and the anchored protein. In the trypanosome cell-free system, Etn-P addition is much more efficient when the precursor inositol ring is acylated.

Canivenc-Gansel et al. (1998) have demonstrated that the first Etn-P is attached to the α1,4-linked mannose of the complete precursor glycophospholipid. The second Etn-P is attached to position 6 of the terminal mannose, which leads to completion of the GPtdIns core structure Etn-P-Man3-GlcN-(acylinositol)PtdIns (Puoti and Conzelmann, 1993). GPtdIns containing a PEtn substituent on the third mannose residue from GlcN are substrates for GPtdIns transamidase, which attaches a GPtdIns molecule to ER-translocated proteins bearing a carboxy-terminal GPtdIns signal sequence. Some protozoan protein anchors may add aminoethyl-phosphonate instead of PEtn (Acosta Serrano et al., 1995).

A family of three proteins (Mcd4, Gpi7, and YII031) that are highly conserved between yeast and mammalian cells, but which are unrelated to Pig-F/Gpi11 (see above), have been implicated in the addition of PEtn to the GPtdIns mannoses in yeast (Benachour et al., 1999; Gaynor et al., 1999). Taron et al. (2000) have speculated that Mcd4, Gpi7 and YII031c transfer PEtn to the first, second and third GPtdIns mannoses, respectively. Vidugiriene et al. (2001) analyzed photo-adducts resulting from UV irradiation of the samples to demonstrate that during the glycosyl PtdIns-anchoring reaction, a small reporter protein becomes glycosyl PtdIns-anchored when the corresponding mRNA is translated in the presence of microsomes, in conjunction with site-specific photo-crosslinking to identify ER membrane components that are proximal to the reporter during its conversion to glycosyl PtdIns-anchored protein.

Mann and Sevlever (2001) have reported that 1,10-phenanthro-line (PTN) inhibits GPtdIns anchoring by preventing P-Etn addition to GPtdIns anchor precursors. PTN acutely inhibited the synthesis of GPtdIns-anchored proteins, but the synthesis was rapidly restored once the inhibitor was washed out.

Smith et al. (1999) described the first synthetic substrate analogues as selective inhibitors of the glycosyl PtdIns biosynthetic pathway of *T. brucei*. The studies of parasite specific inhibition of the glycosyl PtdIns biosynthetic pathway were later extended to stereoisomeric substrate analogues (Smith et al., 2000) and to the synthesis of parasite-specific suicide substrate inhibitors (Smith et al., 2001). In other studies (Screaton et al., 2000) ectopic expression of various members of the human carcinoembryonic antigen family of intracellular adhesion molecules in murine myoblasts showed that the specificity for the differentiation blocking activity of carcinoembryonic antigen resides in its glycosyl PtdIns anchor.

6.4.1.6. Transfer to protein

Upon completion of its biosynthesis, the GPtdIns anchor is attached to the carboxy-terminus of a pre-synthesized protein at luminal face of the ER membrane. Nascent GPtdIns-anchored proteins are required to present certain structural features to the enzyme that effects anchor transfer. These include an amino-terminal hydrophobic signalpeptide, a carboxy-terminal hydrophobic peptide that transiently anchors the protein to the ER membrane, and a suitable triplet of predominantly small amino acids surrounding the GPtdIns cleavage/attachment site (Tiede et al., 1999). These requirements have been elucidated by mutational analysis of GPtdIns-anchored proteins and appear broadly similar among mammals, yeast and protozoa (Udenfriend and Kodukula, 1995). The GPtdIns-to-protein transfer is thought to occur by transamidation, because there is evidence for the formation of an activated carbonyl intermediate, which is the hallmark of a transamidase (Maxwell et al., 1995). Doering and Schekman (1997) have described an in vitro system for the study of the GPtdIns anchor reaction. The GPtdIns anchoring system can be reproduced in a number of differently formatted cell-free systems, some of which take advantage of endogenous protein acceptors as substrates (Doering and Schekman, 1997; Sharma et al., 1999; Reid-Taylor et al., 1999). Sharma et al. (2000) have shown that GPtdIns anchoring in trypanosome membranes requires the participation of a soluble protein GPI18, which is lost when the membranes are washed at a high pH. The GPtdIns anchoring could be restored by adding the isolated transamidase component.

Vidugiriene et al. (2001) have studied the GPtdIns-anchoring reaction by using a small reporter protein that becomes GPtdIns-anchored when the corresponding mRNA is translated in the presence of microsomes, in conjunction with site-specific photo-crosslinking to identify ER membrane components that are proximal to the reported during its conversion to a GPtdIns-anchored protein. Several ER proteins involved in GPtdIns-anchor attachment were isolated and identified.

The mechanism of glycosyl PtdIns-cell interaction in signaling is poorly understood. Previous studies have postulated a receptor-mediated binding and insertion (see, Almeida et al., 2000). Vijaykumar et al. (2001) have shown that induction of TNF-α by *P. falciparum* glycosyl PtdIns in macrophages is mediated by the recognition of the distal fourth mannose residue. This is critical but not sufficient because for effective cell signaling interaction by the acylglycosyl moiety of the glycosyl PtdIns is also required.

6.4.1.7. Remodeling

The lipid moieties of protein-linked GPtdIns are highly variable and include diacylglycerols, alkylacylglycerols, and ceramides. The lipid moieties may be acquired as a result of an initial selection of a subpool of cellular PtdIns (Lee et al., 1991) or by remodeling of GPtdIns after they have been bound to a protein. Remodeling of ester linked fatty acids in the PtdIns moiety has been best characterized in *T. brucei*, but also occurs in other parasites (Ralton and McConville, 1998), yeast (Sipos et al., 1997) and probably also in animal cells (Butikofer et al., 1992).

There is evidence that the glycosyl PtdIns anchored proteins associate with membrane microdomains (rafts) (Rietveld and Simons, 1998) and that this process involves sterols (Rietveld et al., 1999).

The GPtdIns anchor of VSG has exclusively myristic acid attached in ester linkages *sn*-1 and *sn*-2-positions of the glycerol backbone, although initially GPtdIns is synthesized by trypanosomes with stearic acid in the *sn*-1-position and a mixture of fatty acids including 18:0, 18:1, 20:4 and 22:6 in the *sn*-2-position (Doering et al., 1994). The process of remodeling was investigated by Masterson et al. (1990) using trypanosomal membrane cell-free system. It was shown that the remodeling occurs by the sequential replacement of first the *sn*-2-fatty acid and then the *sn*-1-fatty acid with myristic acid. The fatty acid donor for both of these reacylation reactions is myristoyl CoA. The glycolipid, which is normally the substrate for fatty acid remodeling is the Etn-P

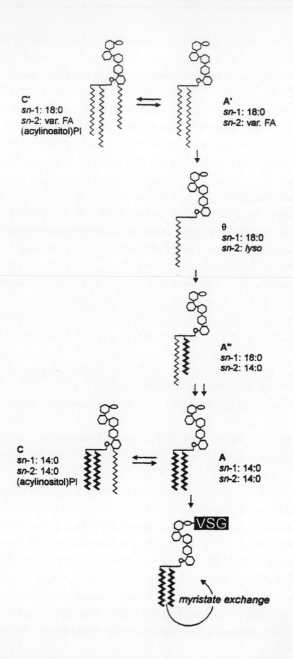

containing GPtdIns core (Masterson et al., 1990). The two myristoyl remodeling reactions have different biochemical properties and probably occur in different subcellular compartments (Buxbaum et al., 1996). Inositol acylated precursors are remodeled beyond the replacement of the *sn*-2-fatty acid (Guther and Ferguson, 1995). Tiede et al. (1999) have reviewed the biosynthesis of GPtdIns in unicellular microbes and have summarized the fatty acid remodeling in *T. brucei* (Fig. 6.6). The mechanism of introduction of the *sn*-1-alkyl group has not been determined. Fatty acid remodeling in GPtdIns of mammalian systems has not been well investigated (Tiede et al., 1999) although there is evidence for differences in chain length composition as well as substitution of alkyl for acyl groups in the final products (Roberts et al., 1988b; Kuksis and Myher, 1990).

A second type of GPtdIns lipid remodeling involves the exchange of a glycerolipid (diacylglycerol) with ceramide to generate an anchor with inositol phosphoceramide lipid. In the yeast, most but not all GPtdIns proteins are resistant to mild alkaline hydrolysis because of the presence of ceramide backbone instead of diacylglycerol (Conzelmann et al., 1992). However, as in

Fig. 6.6. Fatty acid remodeling during glycosyl PtdIns biosynthesis in the *T. brucei* blood stream form. Step 1, Removal of *sn*-2-acyl group from glycolipid A′ (Etn-P-Man$_3$-GlcN-PtdIns), which is the deacylated form of glycolipid C′ (Etn-P-Man3-GlcN-(acylinositol)PtdIns), the major product of GPtdIns biosynthesis in this organism; Step 2, Myristoylation of *sn*-1-acyl-2-lyso-glycolipid A′ (also termed glycolipid θ) to give glycolipid A″; Step 3, Removal of *sn*-1-acyl group from glycolipid A″ to form 1-lyso-2-myristoyl glycolipid A; Step 4, Myristoylation of 1-lyso-2-myristoyl glycolipid A″ to form glycolipid A. The GPtdIns anchor of both newly synthesized and mature VSG is subject to myristate exchange, an additional reaction sequence that replaces the *sn*-1- and *sn*-2-myristate by new myristate residues. The myristate exchange process takes place in a compartment different from that where fatty acid remodeling takes place. Myristoyl CoA is required but divalent cations are not, indicating that the enzymes involved are different from those participating in fatty acid remodeling. Reproduced from Tiede et al. (1999) with permission of the publisher.

mammals and trypanosomatids, the yeast GPtdIns anchor is initially synthesized with an alkali-labile diacylglycerol backbone.

The remodeling from diacylglycerol to ceramide involves en bloc exchange of the complete glycerolipid for ceramide (Sipos et al., 1994; Reggiori et al., 1997). Lipid remodeling occurs both in the ER and in the Golgi apparatus (Sipos et al., 1997). Yeast lipid remodeling occurs subsequent to protein attachment (Sipos et al., 1994; Reggiori et al., 1997). Sipos et al. (1997) compared the PtdIns moieties isolated from myo-[2-^{3}H]inositol-labeled protein anchors and from GPtdIns intermediates. There was no evidence for the presence of long chain fatty acids in any intermediate of GPtdIns biosynthesis. However, GPtdIns-anchored proteins contain either the PtdIns moiety characteristic of the precursor lipids or a version with a long chain fatty acid in the sn-2-position of glycerol. The introduction of long chain fatty acids into sn-2-position occurs in the ER and is independent of the sn-2-specific acyltransferase SLCI. Analysis of ceramide anchors revealed the presence of two types of ceramide, one added in the ER and another more polar molecule, which is found only on proteins, which have reached the mid Golgi. Both ceramide and fatty acid remodeling result in the incorporation of very long acyl chains (C26:0 and C26:0-OH) into the lipid moieties (Sutterlin et al., 1997; Regiori and Conzelmann, 1998). The presence of a ceramide in a protein anchor however does not necessarily indicate ceramide remodeling, as early GPtdIns precursors in *Paramecium* already contain ceramide (Azzouz et al., 1998).

The addition of an extra fatty acyl chain to the inositol headgroup of GPtdIns may be considered as third type of remodeling. It occurs in all eukaryotes, although the timing, extent and possible function of this modification varies among the species. It was noted above that this modification was essential for subsequent mannosylation steps in mammalian cells, because synthetic GPtdIns substrates having a 2-hydroxyl group on the inositol ring are neither acylated nor used as substrates by the mannosyltransferases (Smith et al., 1997a,b). In animal cells, mature GPtdIns precursors with acylated inositol head groups are added to proteins and subsequently rapidly deacylated while the protein is still in the ER (Chen et al., 1998). It has been proposed that this lipid modification occurs to facilitate the transbilayer flipping of precursors between the cytosolic side of the ER, where the fatty acid remodeling enzyme may reside, and the lumenal face of ER, where the transfer to protein occurs (Guther and Ferguson, 1995).

Reggiori et al. (1997) have described the en bloc exchange of DAG for CER in the Golgi apparatus. Figure 6.7 outlines the labeling procedure and shows the distribution of Golgi remodelase and GPtdIns proteins before and after labeling. Briefly, *S. cerevisiae* cells were labeled with [^{3}H]dihydrosphingosine (DHS). Exponentially growing cells (2.5 OD_{600}) were harvested and suspended in 250 µl of SDC medium and labeled. Thermosensitive strains were pre-incubated and labeled at 24 or 37°C in a shaking water bath by adding 25 µCi of [^{3}H]Ins or 25 µCi of [4,5-^{3}H]DHS. After 40 min, the samples were diluted with 750 µl of fresh SDC medium and were incubated further for 80 min. Labeling was terminated by adding NaF and NaN$_3$ (10 mM final concentrations). To enhance the labeling with [^{3}H]DHS, myriocin was frequently added at a final concentration of 40 µg/ml, Chx (Fluka Chemie AG) was used at 200 µg/ml in order to prevent protein synthesis. At the end of incubation, the cells were washed and disrupted by vortexing with glass beads. A protein pellet was isolated and delipidated with chloroform/methanol/water (10:10:3 by vol) and was dried in a Speed Vac evaporator.

For phospholipase treatment, the delipidated proteins were dissolved in 50 µl of PtdIns-PLC Buffer I (20 mM Tris–HCl, pH 7.5, 0.2 mM EDTA, 0.1% Triton-X-100) and boiled for 5 min. The sample was centrifuged at 10 000g for 5 min and the supernatant divided into two 25 µl aliquots. Then 0.01 U of PtdIns-PLC was added to one aliquot and the tubes were incubated at 37°C for 5 h. The reaction was stopped by adding 25 µl of twice concentrated reducing sample buffer and boiling for 5 min. The same procedure was used for PtdIns-PLD treatment, but proteins

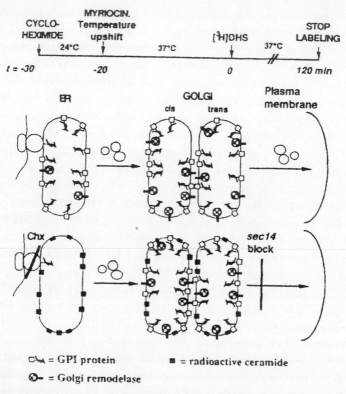

Fig. 6.7. GPtdIns anchor remodeling in the Golgi of *S. cerevisiae*. The top diagram outlines the labeling procedure. The diagram in the middle shows the distribution of Golgi remodelase and GPtdIns proteins before the start of the pre-incubation; the bottom diagram shows the presumed situation during labeling with [³H]DHS. Protein labeling can only occur if GPtdIns proteins, labeled Cer and remodelase are present concomitantly in the same compartment. The strategy was to deplete the Golgi of GPtdIns proteins by pre-incubation in the presence of cycloheximide (Chx) and to see if [³H]DHS was still incorporated into GPtdIns proteins. Incorporation of [³H]DHS under these conditions was only efficient in the presence of myriocin, which blocks the export of the GPtdIns-anchored proteins. Symbols are as explained in the figure. Reproduced from Reggiori et al.

(1997) with permission of the publisher.

were dissolved in PtdIns-PLD Buffer (50 mM Tris–HCl, pH 4.5, 10 mM NaCl, 2.6 mM CaCl$_2$, 0.018% Triton X-100), 5 U of GPtdIns-PLD from bovine serum were added to one of two aliquots and samples were incubated at 37°C overnight. Lipids were desalted and partitioned between *n*-butanol and water followed by a back extraction of the butanol phase with water. The desalted lipids were analyzed by ascending TLC using 0.2 mm thick silica gel plates with Solvent system 1 [chloroform/methanol/2 M NH$_4$OH (40:10:1, by vol)] or Solvent system 2 [chloroform/methanol/0.2 5% KCl (55:45:10, by vol)]. Radioactivity was detected and quantified by two-dimensional radioscanning and fluorography as described above.

Specific release of inositol-linked acyl groups was performed using methanolic NH$_3$ in 200 μl for 2 h at 30°C in an air-heated incubator; controls were incubated only with methanol. Mild base treatment complete deacylation of glycerophospholipids was done using 8N NH$_3$ in 300 μl methanol/water (1:1, v/v) (Chapter 4). GPtdIns-PLD and GPtdIns-PLC treatments were performed as described elsewhere (Chapter 7). For PLA$_2$ treatment, lipids were resolubilized in 30 μl buffer (25 mM Tris–HCl, pH 7.5, 2 mM CaCl$_2$, 0.1% sodium deoxycholate) and treated with 2 U of enzyme (bee venom) for 5 h at 37°C. For TLC, lipids were desalted by butanol extraction if significant salt concentrations had to be removed, lipids were dried and resolubilized in chloroform/methanol (1:1, v/v) for application. Pre-equilibration tanks with the following solvent systems were used for ascending TLC: Solvent system 1: [chloroform/methanol/0.25% KCl (55:45:10, by vol)]; Solvent system 2 [chloroform/methanol/water (10:10:3, by vol)]; Solvent system 3 [chloroform/methanol/0.25% KCl (55:45:5, by vol)]. In most cases, radioactivity on TLC plates was monitored and quantified by one- and two-dimensional radio-scanning using a Berthold radioscanner.

Dickson et al. (1997) have reported that the synthesis of mannose-(inositol-P)$_2$-ceramide, the major sphingolipid in *S. cerevisiae*, requires the IPT1 (YDRO72c) gene, while Fischl et al. (1999) have described the inositol-phosphoryl ceramide synthase from yeast. Lester and Dixon (1993) have reviewed the biochemistry of sphingolipids with inositol-phosphate-containing head groups.

6.4.2. Protein-free GPtdIns

There is increasing evidence that a large proportion of the biosynthesis of GPtdIns is directed towards formation of non-protein linked species. In mammals, GPtdIns biosynthetic intermediates have been shown to escape the site of their synthesis, the endoplasmic reticulum (ER), and to populate other cellular membranes (van't Hof et al., 1995; Menon et al., 1997). In contrast to the parasite GPtdIns-related glycolipids, which are specialized structures synthesized along distinct biochemical pathways, these molecules originate from a pool of excess GPtdIns precursors that have not been attached to protein. The functional relevance of these structures is unclear (Tiede et al., 1999; McConville and Menon, 2000).

In the protozoal parasites, the protein-free GPtdIns frequently differ from the protein-conjugated counterparts in structural modifications, which may be a result of separate pathways of biosynthesis (Ralton and McConville, 1998; McConville and Menon, 2000). These modifications include the attachment to the core mannose of the highly immunogenic $Glc\alpha1$-$4Gal$-$NAc\beta1$-4 side chain in *Toxoplasma gondii* (Striepen et al., 1997) and the attachment to the core GlcN of a Man_1-phosphate in the ciliate *Paramecium primaurelia* (Azzouz et al., 1998). Some of the compositional differences between protein-bound and free GPtdIns could have arisen as a result of preferential utilization of certain molecular species of Man1GlcN-PtdIns for protein binding. As noted above, lipid remodeling reactions may lead to further changes in the final lipid composition of both free GPtdIns and mature protein-linked GPtdIns.

The GPtdIns are subject to complex subcellular distribution and intracellular trafficking, which may involve further biochemical transformations. These aspects of the GPtdIns metabolism are discussed in detail by McConville and Menon (2000). Pal et al. (2002) have demonstrated the presence of GPtdIns anchor specific endosomal pathway in the protozoan pathogen *T. brucei*. There exists a highly complex developmentally regulated endocytic network, which is vital for nutrient uptake and evasion of the immune response.

6.4.2.1. Protozoal GPtdIns-related structures
In several protozoal parasites, the major cellular glycoconjugates share part of the GPtdIns core sequence (Manα1-6GlcNα1-6-*myo*-inositol) and are therefore regarded as GPtdIns-related structures. Several other GPtdIns-related glycolipids are known in different protozoa, most of them expressed during the insect stage. *Trypanosoma cruzi*, for instance, forms various lipopeptidophosphoglycans (LPGs) as well as GIPLs (de Lederkremer et al., 1991; Carreira et al., 1996). In a variation on the general consensus given above, LPPGs contain ceramide instead of PtdIns and an aminoethylphosphonic acid group attached to the 6-position of the GLcN residue. Like GPtdIns protein anchors, these protein-free GPtdIns are synthesized by the sequential addition of sugar residues to PtdIns. Given their structural relationship, these molecules were initially thought to have common precursors. However, recent studies in *Leishmania mexicana* demonstrate that the pathways for the various types of GIPLs are separate from the LPG and protein anchor pathways (Ralton and McConville, 1998). In particular, GIPLs are assembled on a pool of 1-*O*-(C18:0)alkyl-2-stearoyl-GroPIns, whereas LPG and protein anchor synthesis starts on 1-*O*-(C24:1/C26:0)alkyl-2-stearoyl-GroPIns. These different PtdIns species make up 20 and 1% of the total cellular PtdIns pool, respectively. The remarkable specificity of this selection could be accomplished either by the use of different enzymes for the initial biosynthetic steps or by efficient

compartmentalization of the different GPtdIns pathways. Further, the LPG and protein GPtdIns biosynthetic pathways might also be physically separated, although these are utilizing the same PtdIns species. Thus, it has been shown that cytoplasmic expression of the trypanosome phospholipase C inhibits protein anchor, but not LPG synthesis in *Leishmania major* (Mensa-Wilmot et al., 1994).

6.4.2.2. Lipophosphoglycans

Examples of these glycolipids include the *Leishmania* low-molecular weight GPtdIns and lipophosphoglycans (LPGs). The GPtdIns anchor of *L. major* LPG is an unusual branched hexasaccharide of GlcN, Man, Glc-P, Gal, and galactofuranose (Galf) residues attached to *sn*-1-alkyl-, 1-1-*sn*-2-lyso-PtdIns (Chapter 4). The phosphoglycan moiety is a linear chain of repeated (-6Galβ1-4Manα1-PO_4)-units in an extended helical conformation with the galactose residue being substituted by monosaccharides or complex oligosaccharide side chains (Ilg et al., 1992). The terminal phosphoglycan unit is capped by a neutral oligosaccharide (McConville et al., 1990). Three distinct lineages of the low molecular weight GPtdIns lipids are expressed in *Leishmania* species.

The pathway leading to the putative LPG precursor Gal-Gal-(Glc-P)Man-Man-GlcN-(1-alkyl-2-lyso)PtdIns is not completely understood. However, biosynthesis starts from an 1-alkyl-2-acyl-PtdIns (see above), and involves sequential addition of nucleotide or dolicholphosphate-activated monosaccharides as well as a phospholipase A_2-like activity that removes the *sn*-2-fatty acid, most probably at the stage of Man1-GlcN-PtdIns (McConville et al., 1993; Smith et al., 1997a).

The repetitive phosphoglycan chain is built up on the preformed LPG precursor by the sequential addition of Man-1-P and Gal from GDP-Man and UDP-Gal, respectively (Carver and Turco, 1991). There are distinct Man-P transferase activities for the initiation and elongation phases of phosphoglycan synthesis (Brown et al., 1996).

6.4.2.3. Glycoinositolphospholipids (GIPLs)

The early steps in the assembly of *Leishmania* GIPLs are the same as those required for protein GPtdIns anchor biosynthesis, except that they utilize a different PtdIns pool containing exclusively C_{18} alkyl chains (see above). At least up to the formation of the first mature glycan head groups, all biosynthetic reactions take place on the cytoplasmic face of microsomal vesicles. Dol-P-Man is utilized as a mannose donor in these apparently cytoplasmically oriented reactions (Ralton and McConville, 1998). The flux of intermediates through the GIPL pathway appears very high and is independent of the flux through the LPG and protein anchor pathways, further confirming that these pathways are independently regulated. At some stage past the Man_2-GlcN-PtdIns intermediate, GIPLs undergo a series of fatty acid remodeling reactions in which the *sn*-2-fatty acid (primarily stearate) is replaced by myristate (Ralton and McConville, 1998).

6.4.2.4. Glucosyl PtdIns glycolipids

T. gondii is an obligate intracellular parasite of the phylum apicomplexa and a common and often life-threatening opportunistic infection associated with AIDS. Striepen et al. (1997) have identified a family of parasite-specific glycosyl PtdIns containing a novel glucosylated side chain, which is highly immunogenic in humans. In contrast to trypanosomes in *T. gondii* side chain modification takes place before addition to protein in the endoplasmic reticulum. Striepen et al. (1992) have identified this antigen to be a family of protein-free GPtdIns glycolipids and have recently elucidated the structures of these GPtdIns. Two types of core glycans were identified, glycan A modified by GalNAc linked β1-4 to the core mannose adjacent to the non-acetylated glucosamine and glycan B containing a novel Glcα1-4GalNAc side branch. (Striepen et al., 1997). Subsequent immunological analysis revealed that only glucosylated GPtdIns containing glycan B were recognized by sera from infected humans suggesting that the unique glucose modification is required for immunogenicity

(Striepen et al., 1997). Striepen et al. (1999) have unidentified UDP-glucose as the direct donor for side chain modification in *T. gondii* using carbohydrate analogues. Detailed analysis of glycolipids synthesized in vitro in the presence of UDP and GDP derivatives of D-glucose and 2-deoxy-D-glucose rule out an involvement of dolichol phosphate-glucose and demonstrated direct transfer of glucose from UDP-glucose.

CHAPTER 7

Phosphatidylinositol and inositol phosphate kinases

7.1. Introduction

PtdIns is unique among phospholipids in that its head group can be phosphorylated at multiple free hydroxyl groups. Several phosphorylated PtdIns and InsPs have been identified in eukaryotic cells ranging from yeast to mammals. PtdInsPs and InsPs are involved in the regulation of diverse cellular functions, including cell proliferation, survival, cytoskeletal organization, vesicle trafficking, glucose transport and platelet function. The enzymes that phosphorylate PtdIns and its derivatives are termed PtdIns kinases. Many of the kinases have been purified to homogeneity and studies of their substrate specificity have led to the categorization of PtdIns kinases into different classes and subclasses. Frequently, the same lipid kinases also phosphorylate the water soluble forms (InsPs) of the PtdInsPs. The genes of several of the enzymes have been cloned and the sequence homology observed has provided support for their further detailed classification. The present chapter discusses the individual kinases, their substrate specificity, and the assay systems. The specific cellular functions of individual phosphatides have been considered in Chapter 11.

7.2. Phosphatidylinositol kinases

On the basis of their positional specificity the PtdIns kinases have been divided into three general families: PtdIns 3-kinases (PtdIns 3Ks), PtdIns 4-kinases (PtdIns 4Ks) and PtdIns 5-kinases (PtdInsP 5Ks). The PtdIns 5Ks are also referred to as the PtdInsP kinases (PIPKINS), Fig. 7.1.

On the basis of their in vitro substrate specificities, the PtdIns 3-kinases have been grouped into Classes I, II and III. Class I PtdIns 3Ks are heterodimers of approximately 200 kDa, composed of a 110–120 kDa catalytic subunit and a 50–100 kDa adaptor

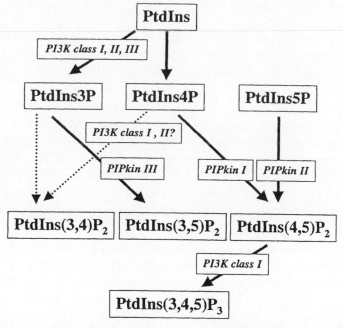

Fig. 7.1. The PtdIns kinases. The solid arrows indicate the major routes of 3-phosphorylation of PtdIns, PtdInsP and PtdInsP$_2$. The dashed arrows indicate less well-defined synthetic routes. Reproduced from Vanhaesebroeck et al. (2001) with permission, from the Annual Review of Biochemistry Volume 70 © 2001 by Annual Reviews www.annualreviews.org.

subunit, which are able to phosphorylate in vitro PtdIns, PtdIns(4)P and PtdIns(4,5)P_2. The preferred in vivo substrate for Class I PtdIns 3Ks has been shown to be PtdIns(4,5)P_2. Class II PtdIns 3Ks phosphorylate only PtdIns and PtdIns(4)P, and possess masses ranging from 170 to 210 kDa. Class III PtdIns 3Ks are homologues of *Saccharomyces cerevisiae* Vps34p (vacuolar protein sorting mutant) and phosphorylate exclusively PtdIns. All PtdIns 3Ks catalytic subunits share a homologous region that consists of a catalytic core domain (HR1) linked to a PtdIns kinase (PIK) homology domain (HR2) and a C2 domain (HR3).

The structure and function of the PtdIns 3-kinases have been extensively reviewed by Carpenter and Cantley (1990), Toker and Cantley (1997), Vanhaesebroeck et al. (1997a,b), Fruman et al. (1998), Wymann and Pirola (1998), Tolias and Cantley (1999) and Vanhaesebroeck et al. (2001). Vanhaesebroeck et al. (2001) have summarized the modular structure of PtdIns 3-kinases in a graphical form (Fig. 7.2).

The PtdIns 4Ks have been divided into Type II and III subclasses (Endemann et al., 1987; Whitman et al., 1987). A Type I activity was later shown to be a PtdIns 3K. On the basis of homologies within the catalytic domain some PtdIns 4Ks are close relatives of PtdIns 3Ks (Hunter, 1995; Balla, 1998; Wymann and Pirola, 1998). Type II PtdIns4K was characterized as a 55 kDa protein that could be renatured from a sodium dodecyl sulfate gel (Walker et al., 1988; Ling et al., 1989) and could be inhibited by the monoclonal antibody 4C5G (Endemann et al., 1991). The 55 enzyme has been implicated in many signaling pathways through its association with a variety of molecules and subcellular compartments, including epidermal growth factor (EGF) receptor (Thompson et al., 1985), chromaffin granules (Wiedemann et al., 1996), CD4-p56[lck] (Prasad et al., 1993), integrins (Berditchevski et al., 1997), and PKCs. Type I PtdInsP 5Ks exhibit little homology to PtdIns 3Ks or PtdIns 4Ks. The PtdInsP 5Ks are also divided into subclasses.

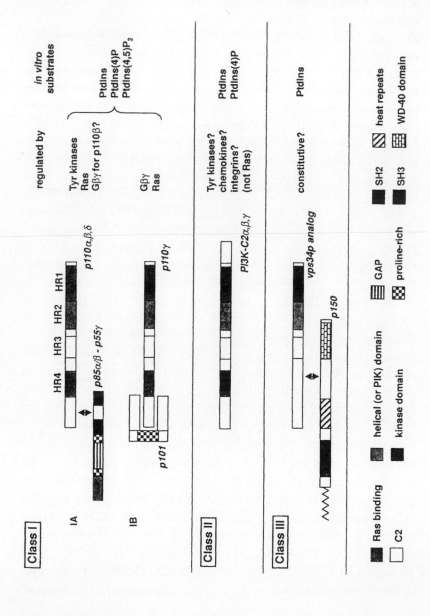

7.2.1. PtdIns 3-kinase

PtdIns(3)P is present in cells under all conditions and serves as a precursor for the polyphosphorylated 3-phosphate-containing PtdIns (Cunningham and Majerus, 1991). The D3 position of PtdIns is phosphorylated by PtdIns 3Ks, which transfer the γ-phosphate group of ATP to the D3 position of the inositol ring. Members of the three different classes of PtdIns 3-Ks can be distinguished by their in vitro substrate specificity, structure, and their mode of regulation (Wymann et al., 1996; Wymann and Pirola, 1998; Fruman et al., 1998).

Based on their in vitro activities, Class II PtdIns 3Ks are likely to produce PtdIns(3)P in vivo, but they might also contribute to PtdIns(3,4)P$_2$ or PdIns(3,4,5)P$_3$ production. While higher eukarytes from *Dictyostelium discoideum* to *C. elegans* to mammals have enzymes of all three classes of PtdIns 3Ks, *S. cerevisiae* have only the Class II PtdIns 3K. Thus, PtdIns(3)P is found in yeast, but PtdIns(3,4)P$_2$ and PtdIns(3,4,5)P$_3$ have not been detected.

The class II PtdIns 3-kinase C2α is activated by clathrin (Gaidarov et al., 2001). Turner et al. (1998) have reported that certain chemotactic peptides activate both the class I 85/p110 PtdIns 3-kinase and class II PtdIns 3-kinase 2α.

PtdIns 3-kinase is activated by growth factors (Carter and Downes, 1992), by association with insulin receptor substrate-1 during insulin stimulation (Backer et al., 1992; Cengel et al., 1998b), okadaic acid (Cengel et al., 1998a), and by interaction with Ras (Rodriguez-Viciana et al., 1996).

Fig. 7.2. Modular structure of the PtdIns 3-kinases. In p150, the kinase domain represents a protein kinase domain, and the zigzag line represents N-terminal myristoylation. Reproduced from Vanhaesebroeck et al. (2001) with permission, from the Annual Review of Biochemistry, Volume 70 © 2001 by Annual Reviews www.annualreviews.org. PtdIns 3-kinase is activated by growth factors (Carter and Downes, 1992), by association with insulin receptor substrate-1 during insulin stimulation (Backer et al., 1992; Cengel et al., 1998b), okadaic acid (Cengel et al., 1998a), and by interaction with Ras (Rodriguez-Viciana et al., 1996).

Class III PtdIns 3Ks have been found in other yeasts, plants, *Drosophila*, algae, and mammals (Wymann and Pirola, 1998). All these Class III enzymes are restricted to PtdIns and thus produce PtdIns(3)P as the sole product. This lipid modification of Vps15p targets Vps34p to the membrane in yeast, and a similar role is anticipated for p150 in mammals. Class III PtdIns 3Ks are the homologues of the yeast vesicular-protein–sorting protein Vps34p (Prior and Clague, 1999; Odorizzi et al., 2000). These PtdIns 3Ks can use only PtdIns as a substrate in vitro, and are likely to be responsible for the generation of most of the PtdIns(3)P in cells. A single class III PtdIns 3K catalytic subunit has been identified in all eukaryotic species.

The progress in the studies on the role of PtdIns 3-kinases in cellular processes has been facilitated by the availability of a range of inhibitors which have greatly simplified experimental investigations in the area. The most widely used of these are two specific and cell permeable inhibitors of PtdIns 3-kinase, namely wortmannin (Arcaro and Wymann, 1993; Ui et al., 1995) and LY294002 (Vlahos et al., 1994). Wortmannin has an IC_{50} in the low-nanomolar range for mammalian class Ia (Ui et al., 1995), class Ib (Stoyanova et al., 1997) and class 3 (Volinia et al., 1995) PtdIns 3-kinases, although the IC_{50} for class II PtdIns 3-kinases is greater (Virbasius et al., 1996; Domin and Waterfield, 1997). Wortmannin's specificity exists over a concentration range of two orders of magnitude, although at concentrations greater than 100 nM, wortmannin will inhibit some forms of PtdIns 4Ks (Nakanishi et al., 1995) and also phospholipase A_2. LY294002 is another highly specific inhibitor of class I PtdIns 3Ks, while class II PtdIns 3Ks are relatively resistant to this inhibitor (Domin and Waterfield, 1997). PtdIns 3-kinase is also selectively inhibited by PtdOH and related lipids (Lauener et al., 1995).

7.2.1.1. Assay of immunoprecipitates

Soltoff et al. (1993) described an in vitro synthesis of PtdIns(3)P by PtdIns 3-kinase in Anti-P-tyrosine immunoprecipitates. Since

PtdIns 3-kinase associates with the PDGF receptor, middle T/pp60c-src tyrosine kinase, and other receptors that are phosphorylated on tyrosine in response to growth factors, the enzyme is commonly immunoprecipitated using anti-Ptyr antibodies and the assay is performed on the immunoprecipitate. The PtdIns 3-kinase in PC12 cells was assayed in a final volume of 50 µl. To about 30 µl (about 20 µl of protein A-Sepharose beads plus about 10 µl TNE) in a microcentrifuge tube is added 10 µl of a lipid mixture containing PtdIns combined with 10 µl of a [γ-^{32}P]ATP mixture to initiate the assay. PtdIns 3-kinase assay is performed by adding 10 µl containing 20–40 µCi [γ-^{32}P]ATP (in 250 µM ATP, 25 mM $MgCl_2$, and 5 mM HEPES, 1 mM EGTA, pH 7.5) to the immunoprecipitates for 10 min at room temperature. The reaction is stopped by adding 80 µl of HCl (1 M) and vortexing the sample. Subsequently, 160 µl of methanol/chloroform (1:1, v/v) is added, and the sample is vortexed thoroughly. The sample is centrifuged (about 10 s) to separate the organic and aqueous layers. The lipid-containing chloroform phase is spotted onto oxalate-coated TLC plates (Silica gel 60, 0.2 mm thick adsorbent layer). The 20 × 20 cm^2 TLC plates are cut into half and developed (10 cm vertical) in a mixture of chloroform/methanol/water/7.7N NH_4OH (60:47:11.3:2, by vol.). The solvent is allowed to migrate nearly to the top, which takes about 30 min. The solvent system resolves PtdIns, PtdInsP, and PtdInsP$_2$, but PtdIns(3,4,5)P$_3$ and ATP remain at the origin. Authentic PtdIns(4)P was spotted as a standard and visualized using iodine vapor. [^{32}P]PtdInsP appears just below the non-radiolabeled PtdIns. The [^{32}P]PtdInsP spots are excised and quantified by scintillation counting or by measuring Cerenkov radiation. PtdIns(3)P is not separable from PtdIns(4)P using this technique. The identity of the PtdIns(3)P is confirmed by HPLC of the deacylated phosphatide, as described elsewhere (see Chapter 3).

Valverde et al. (1999) assayed the PtdIns 3-kinase activity in fetal brown adipocytes by in vitro phosphorylation of PtdIns as described by Valverde et al. (1997). At the end of the culture time cells were washed with ice-cold PBS, solubilized in lysis buffer

[10 mM Tris/HCl/5 mM EDTA/50 mM NaCl/30 mM sodium pyrophosphate/50 mM NaF/100 μM Na$_3$VO$_4$/1% Triton X-100 (pH 7.6)] containing leupeptin (10 μg/ml), aprotinin (10 μg/ml) and 1 mM PMSF. Lysates were clarified by centrifugation at 15 000g for 10 min at 4°C, and proteins were immunoprecipitated with a monoclonal antibody anti-Tyr(P) (Py72). The immunoprecipitates were washed successively in PBS containing 1% Triton X-100 and 100 μM Na$_3$VO$_4$ (twice), 100 mM Tris (pH 7.5), in PBS containing 0.5 mM LiCl, 1 mM EDTA and 100 μM Na$_3$VO$_4$ (twice), and in 25 mM Tris (pH 7.5) containing 100 mM NaCl and 1 mM EDTA (twice). To each pellet were added 25 μl of 1 mg/ml PtdIns/dSer sonicated in 25 mM HEPES (pH 7.5) and 1 mM EDTA. The PtdIns 3-kinase reaction was started by the addition of 100 nM [γ ^{32}P]ATP (10 μCi) and 300 μM ATP in 25 μl of 25 mM HEPES (pH 7.4)/10 M MgCl$_2$/0.5 mM EDTA. After 15 min at room temperature the reaction was stopped by the addition of 500 μl of chloroform/methanol (1:2, v/v), a 1% concentration of HCl; plus 125 μl of chloroform/125 μl of HCl (10 mM). The samples were centrifuged, and the lower organic phase was extracted, dried under vacuum, and resuspended in chloroform. Samples were applied to a silica gel TLC plate (Whatman, Clifton, NJ). TLC plates were developed in propanol/acetic acid (2M; 65:35, v/v) dried, visualized by autoradiography and quantified by scanning laser densitometry.

Conway et al. (1999) assayed PtdIns 3-kinase in resuspended anti-phosphotyrosine or Grb-2 immunoprecipitates (40 μl), which were each combined with 20 μl of PtdIns (3 mg/ml) in incubation buffer containing 1% cholate. To each, 40 l of [^{32}P]ATP (3 μM Na$_2$ATP, 7.5 mM MgCl$_2$ and 0.25 mCi of [^{32}P]ATP/ml) was added. The reaction was performed at 37°C for 15 min and was terminated by adding 450 μl of chloroform/methanol (1:2, v/v). Organic and aqueous phases were resolved by adding 150 μl of chloroform and 150 μl 0.1 M HCl. Samples were mixed and centrifuged (4200g for 10 min). This was repeated and the lower phase harvested, dried, and [^{32}P]PtdIns(3)P resolved by TLC using chloroform/methanol/ammonia/water (42:60:9:15, by vol.) in parallel with non-radioactive standards. Radioactive bands were revealed by autoradiography, and samples corresponding to $[^{32}]$PtdIns(3)P scraped from the plate and subjected to Cerenkov counting.

7.2.1.2. Assay of cell lysates

Wurmser and Emr (1998) have described an assay for PtdIns 3-kinase based on earlier description. *S. cerevisiae* spheroplast lysates were prepared and separated into P100 and S100 fractions by a 30 min 100 000*g* spin as previously described (Schu et al., 1993; Stack et al., 1995). Because active Vps34p resides in the P100 fraction (Schu et al., 1993), the activity found in this fraction was assayed. PtdIns 3-kinase reactions were carried out by incubating a 2 µg P100 extract in 50 µl 20 mM HEPES pH 7.5, 10 mM MgCl$_2$, 0.2 mg/ml sonicated PtdIns, 60 µM ATP and 0.2 mCi/ml $[\gamma-^{32}P]$ATP at 25°C for 5 min. Reactions were stopped with the addition of 240 µl 1 M HCl/chloroform/methanol (1:1:1, by vol.) followed by vortexing. The organic phase was analyzed by TLC as described by Schu et al. (1993) and Stack et al. (1995). Total cell lipids were deacylated and analyzed by HPLC as described by Schu et al. (1993) and Stack et al. (1995).

Pahan et al. (1999) assayed PtdIns 3K as follows: after stimulation in serum-free DMEM/F-12 cells were lysed with ice-cold lysis buffer containing 1% v/v Nonidet P-40, 100 mM NaCl, 20 mM Tris (pH 7.4), 10 mM iodoacetamide, 10 mM NaF, 1 mM sodium orthovanadate, 1 mM phenylmethylsulfonyl chloride, 1 µg/ml leupeptin, 1 µg/ml antipain, 1 µg/ml aprotinin, and 1 µg/ml pepstatin A. Lysates were incubated at 4°C for 15 min followed by centrifugation at 13 000*g* for 15 min. The supernatant was precleared with protein G-Sepharose beads (Amersham Pharmacia Biotech) for 1 h at 4°C followed by the addition of 1 µg/ml p85α monoclonal antibody. After 2-h incubation at 4°C, protein G-Sepharose beads were added, and the resulting mixture was further incubated for 1 h at 4°C. The immunoprecipitates were washed twice with lysis buffer, once with phosphate-buffered saline, once with 0.5 M LiCl and 100 mM Tris (pH 7.6), once with water, and once in kinase buffer (5 mM $MgCl_2$, 0.25 M EDTA, 20 M HEPES (pH.4)). PtdIns 3-kinase activity was determined as described (Ward et al., 1992) using a lipid mixture of 100 μl of 0.1 mg/ml PtdIns and 0.1 mg/ml PtdSer dispersed by sonication in 20 mM HEPES (pH 7.0) and in 1 mM EDTA. The reaction was started by the addition of 20 μCi of [γ-^{32}P]ATP (3000 Ci/mmol; NN Life Science Products) and 100 μM ATP, and terminated after 15 min by the addition of 80 μl of 1N HCl and 200 μl of chloroform/methanol 1:1 (v/v). Phospholipids were separated by TLC and visualized by exposure to iodine vapor and autoradiography (Ward et al., 1992).

7.2.2. PtdIns 4-kinases

The PtdIns 4Ks convert PtdIns to PtdIns(4)P by phosphorylating the inositol ring at the D4 position. The PtdIns(4)P thus generated can be further phosphorylated by both PtdIns 3Ks and PtdIns(4)P 5-kinases (PIP4Ks) to yield PtdIns(3,4)P_2 and PtdIns(4,5)P_2, respectively (see below). In this way, the PtdIns 4Ks play a central role in signaling by feeding into multiple pathways. These enzymes are ubiquitously expressed and are most abundant in cellular membranes including the Golgi, lysosomes, endoplasmic reticulum (E), plasma membrane, and a variety of vesicles that include Glut 4-containing vesicles, secretory vesicles, and coated pits (Pike, 1992). Unlike the functionally related Ptdns 3Ks and PtdInsP 5Ks, these enzymes appear to use only PtdIns as a substrate and cannot phosphorylate either the singly or doubly phosphorylated lipids generated by the other enzymes.

A clear distinction between two structurally different isoforms of the PtdIns 4-kinase has emerged from the cloning and sequencing of PtdIns 4-kinase from yeast (Flanagan et al., 1993; Garcia-Bustos et al., 1994; Yoshida et al., 1994a,b), human (Suzuki et al., 1997; Wong and Cantley, 1994; Meyers and Cantley, 1997), rat (Gehrmann et al., 1996; Gehrmann and Heilmeyer, 1998; Verghese et al., 1999), and bovine (Gehrmann et al., 1996; Balla et al., 1997) tissues. The two isoforms differ in size, amino acid sequence homology, and putative location and function within the cell. The smaller type, II enzyme, encodes a polypeptide of about 110–125 kDa that is found in the cytosol and is associated with the Golgi apparatus (Flanagan et al., 1993; Wong et al., 1997). The second type (III enzyme, encodes a larger protein of about 200–230 kDa that is membrane-associated (Yoshida et al., 1994a, b; Nakagawa et al., 1996; Wong et al., 1997). A distinctive feature of the group of higher molecular mass PtdIns 4-kinases is that they contain putative PH domains. Antibodies raised against the expressed lipid kinase unique, PH, and catalytic domains identified a polypeptide of 205 kDa in *Arabidopsis* microsomes and an F-actin-enriched from carrot cells. The 205 kDa immunoaffinity-purified *Arabidopsis* protein had PtdIns 4-kinase activity. Stevenson et al. (1998) expressed the PH domain to characterize its lipid binding properties. The recombinant PH domain selectively binds to PtdIns(4)P, PtdIns(4,5)P_2 and phosphatidic acid and does not bind to the 3-phosphorylated PtdInsPs. The PH domain had the highest affinity for PtdIns(4)P, the product of the reaction.

Type III PtdIns4Ks were defined as membrane-bound enzymes whose lipid kinase activity is unaffected by high concentrations of adenosine and is maximally active in non-ionic detergent (Endemann et al., 1987). The enzyme, purified from bovine and rat brain, had an apparent molecular weight of 220 kDa by sucrose gradients and gel filtration (Endemann et al., 1987; Li et al., 1989), and it was resistant to the monoclonal antibody 4C5G (Endemann et al., 1991). The prototype PdIns4Ks were first cloned from yeast and designated PK1 (Flanagan et al., 1993) and TT4 (Yoshida et al., 1994a,b). PtdIns 4Kα is homologous to yeast PtdIns 4K, the STT4 gene product that was identified as a staurosporine- and temperature-sensitive mutant in yeast (Yoshida et al., 1992; Carman et al., 1996; Han et al., 2002).

In vitro, the PtdIns 4Kα/Stt4 isoforms phosphorylate only PtdIns. Although PtdIns4Kα is relatively resistant to wortmannin, recent data suggest that the p220 PtdIns 4Kα protein can be completely inhibited at a high concentration of wortmannin (10 μM) (Nakagawa et al., 1996). By contrast, the yeast tt4 is >90% inhibited by 10 nM wortmannin in vitro (Cutler et al., 1997). Recent studies have localized the mammalian PtdIns 4Ks (PtdIns 4Kα and PtdIns 4Kβ) to distinct intracellular membranes (ER and Golgi, respectively), where they may regulate the local levels of PtdIns(4)P (Nakagawa et al., 1996; Wong et al., 1997).

Sequence analysis revealed that PtdIns K1 and STT4 share homology with PtdIns 3-kinase family genes, such as the bovine p110 and Vps 34s. In vitro, the PtdIns 4Kα/STT4 isoforms phosphorylate only PtdIns.

In mammalian cells, at least three different genes encode kinases that phosphorylate the D4 position of PdIns cDNA clones of two of these enzymes (PtdIns 4Kα and PtdIns 4Kβ) have been isolated, whereas the cDNA for the 55 type II PtdIns 4-kinase has not yet been identified. Wong and Cantley (1994) reported the cloning and characterization of a mammalian PtdIns 4-kinase, named PtdIns 4Kα. The 2.6 kb cDNA encoded a protein of 854 amino acids that was highly homologous to the yeast STT4 enzyme. A bovine gene encoding a 170–200 kDa PtdIns 4-kinase that shares more than 95% identity with the human PtdIns 4Kα in the overlapping region has been cloned by Gehrmann et al. (1996). Likewise, Nakagawa et al. (1996) reported the cloning and characterization of an alternative splice of rat PtdIns 4-Kα gene that generates a protein predicted to be 230 kDa. Nakagawa et al. (1996a) also cloned a 92 kDa PtdIns 4-kinase from rat brain, while Downing et al. (1996) characterized a soluble adrenal PtdIns 4-kinase. Subsequently, Meyers and Cantley (1997) isolated another human cDNA that encodes a 110 kDa PtdIns 4-kinase (named PtdIns 4-Kβ) that is more homologous to the yeast PtdIns K1 gene. The PtdIns 4-Kβ is wortmannin-sensitive and may be the same as that involved in the hormone sensitive pools of inositol phospholipids (Nakanishi et al., 1995). Wong et al. (1997) have determined the subcellular locations of these enzymes in CHO-IRS cells and have provided evidence for other yet unidentified isoforms present in specific organelles. PtdIns 4-Kα was mostly membrane-bound and located at the endoplasmic reticulum, whereas PtdIns 4-Kβ was in the cytosol and also present in the Golgi region. Neither of these isoforms accounted for the major type II PtdIns 4-kinase activity detected in the lysosomes and plasma membrane faction.

The majority of the PtdIns 4K activity detected in most mammalian cells is due to the 55 kDa type II PtdIns 4-K, an enzyme that has been purified from a number of tissues but not yet sequenced or cloned. This enzyme (or family of enzymes), unlike PtdIns 4Kα and PtdIns 4Kβ is present at high levels in the plasma membrane (Wong et al., 1997). Wong et al. (1997) have extensively investigated the possibility that the type II 55 kDa enzyme is a proteolytic product of the 97 kDa PtdIns 4-Kα, since they share similar enzymatic properties. However, it was not possible to detect proteins of 55 kDa resulting from the processing of the 97 kDa PtdIns 4-Kα. It was concluded that the 55 kDa type II PtdIns 4-kinase appears to be encoded by a distinct but related gene. Tolias and Carpenter (2000) have summarized the structural features of the PtdInsP 4-kinase family members in a graphical form (Fig. 7.3).

Xue et al. (1999) have described a plant 126 kDa PtdIns 4-kinase with a novel repeat structure while Barylko et al. (2001) have described a novel class of PtdIns 4-kinases, which are highly conserved evolutionarily, but unrelated to previously characterized PtdIns kinases. The novel PtdIns 4-kinases, which are widely expressed in cells, only phosphorylate PtdIns, are potently inhibited by adenosine, but are insensitive to wortmannin or phenylarsine oxide. Although they lack an obvious transmembrane domain, they are strongly attached to membranes by palmitoylation. Moritz et al. (1992) have claimed that PtdOH is a specific activator of PtdIns 4-phosphate kinase.

Waugh et al. (1998) reported the synthesis of PtdIns(4)P in immunoisolated caveolae-like vesicles and in low buoyant density non-caveolar membranes.

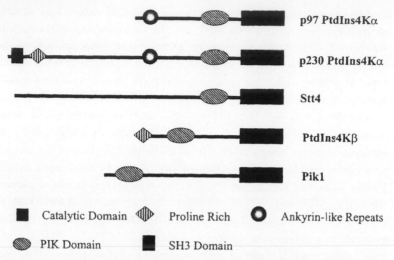

Fig. 7.3. Structural features of the PtdIns 4-kinase family members. The catalytic domain, PtdIns kinase domain, Src homology 3, ankyrin-like repeats, and proline-rich regions of the PtdIns 4-kinases are shown. Reproduced from Tolias and Carpenter (2000) with permission of the publisher.

7.2.2.1. Assays in micelles

Yamakawa and Takenawa (1993) have assayed PtdIns 4-kinase in bovine brain by measuring formation of $[^{32}P]PtdIns(4)P$ from PtdIns and $[\gamma-^{32}P]ATP$. The standard incubation mixture contained 50 mM Tris–HCl (pH 7.4), 20 mM $MgCl_2$, 1 mM EGTA, 0.4% Triton X-100, 1 mM PtdIns, 200 μM $[\gamma-^{32}P]ATP$ (1.0–0.5 μCi), and 1–100 μg of enzyme protein in a total volume of 50 μl. The reaction was started by the addition of $[\gamma-^{32}P]ATP$, continued for 10 min at 30°C, and then stopped by the addition of 2 ml of chloroform/methanol/conc. HCl (200:100:1, by vol.). Then 0.5 ml of 1N HCl was added, and the mixture was vortexed and separated into two phases by centrifugation at 2000g for 5 min at room temperature. The upper phase was discarded, and the lower phase was washed with 0.5 ml of the synthetic upper phase. After centrifugation, the resultant lower phase was removed and dried under a stream of dry nitrogen.

PtdIns(4)P was separated from other phospholipids with one-dimensional TLC. For this purpose 10 μg PtdIns(4)P was added as carrier. The TLC plates (Merck and Co., Darmstadt, Germany) were treated with methanol/water (2:3, v/v) containing 1% sodium oxalate and the plates activated for 1 h at 110°C. The lipids were spotted on the plates and the plates developed with chloroform/methanol/4.3 NH$_4$OH (90:65:20, by vol.). The spots were visualized with iodine vapor and then scraped off the plates for measuring the radioactivity.

For the estimation of the activity of crude enzymes, the [^{32}P]PtdIns(4)P product must be separated by TLC to avoid contamination with other [^{32}P]-labeled lipids. The contribution of PtdIns 3-kinase activity is minimal because PtdIns 3-kinase is inhibited in the presence of detergents.

Barylko et al. (2001) measured the novel PtdIns 4-kinase activity by phosphorylation of PtdIns micelles using [γ-^{32}P]ATP (10 mCi/mmol) as phosphate donor. Typically, kinase (in 20 mM Tris, pH 7.5, 10% glycerol, 0.1 M NaCl, 1% Triton X-100, 1 mM dithiothreitol, and protease inhibitors) was preincubated for 10 min with PtdIns prepared by sonication with 50 mM Tris, pH 7.5, 1 mM EGTA, 0.4% Triton X-100, and 0.5 mg/ml bovine serum albumin. Reactions were initiated by adding ATP and MgCl$_2$ at final concentrations of 0.2 and 15 mM, respectively, and carried out at room temperature for 15 min. Phospholipids were extracted with chloroform/methanol 1:1 (v/v) and separated by TLC in n-propyl alcohol/H$_2$O/NH$_4$OH (65:20:15, by vol.) solvent system. Radioactive spots were detected by autoradiography, scraped, and radioactivity measured by scintillation counting.

7.2.2.2. Assays of recombinant proteins

Suer et al. (2001) expressed human PtdIns 4K, isoform PtdIns 4K92, as His6 tagged protein in Sf9 cells reaching a level of 5% of cellular protein. The isolated enzyme produced PtdIns(4)P as product. The enzyme activity was assayed at 37°C in an optimized test as described by Varsanyi et al. (1989) and Gehrmann et al. (1996). The phospholipids produced were extracted with chloroform and washed with 1 M HCl/methanol (1:1, v/v). The chloroform phase was transferred to scintillation vials or spotted onto HPTLC plates NH₂ (Merck, Germany). Plates were developed in 1-propyl acetate/2-propanol/ethanol/6% aqueous ammonia (3:9:3:9, by vol.) (Hegewald, 1996; Mayer et al., 2000) and visualized by phosphorImager.

7.2.2.3. Plant tissue assays

Stevenson et al. (1998) assayed the synthesis of PtdIns(4)P by PtdIns 4-kinase in plants. The PtdIns 4-kinase activity in each fraction from an immunoaffinity column was assayed in duplicate (60 μl/assay). The reaction mixture contained final concentrations of 7.5 mM MgCl₂, 1 mM sodium molybdate, 0.5 mg/ml PtdIns, 0.1% (v/v) Triton X-100, 0.9 mM ATP, 30 mM Tris (pH 7.2), and 20 μCi of [γ-^{32}P]ATP (7000 Ci/mmol) in a total volume of 100 μl. Stock PtdIns (5 mg/ml) was solubilized in 1% (v/v) Triton X-100. The reactions were incubated at 25°C for 2 h with intermittent shaking. The reactions were stopped with 1.5 ml of ice-cold CHCl₃/methanol (1:2, v/v) and kept at 4°C until the lipids were extracted. Lipids were extracted as described previously (Cho et al., 1993). Extracted lipids were vacuum-dried, solubilized on CHCl₃/MeOH (2:1, v/v), and spotted onto Whatman LK5D silica gel plates that had been completely dried in a microwave oven for 5 min after presoaking in 1% (w/v) potassium oxalate for 80 s. The lipids were separated in either a CHCl₃/MeOH/NH₄OH/H₂O (86:76:6:16, by vol.) solvent system (Cho et al., 1993) or a pyridine-based solvent (Walsh et al., 1991) and quantified with a Bioscan System 500 Imaging Scanner. The plate was subsequently exposed to phosphor screen for 2 days and visualized by a Storm PhosphorImager (Molecular Dynamics).

7.2.3. PtdIns 5-kinases

Rameh et al. (1997) have demonstrated the existence of PtdIns(5)P in NIH-3T3 fibroblasts, but it was not established whether or not it is formed by a direct phosphorylation of PtdIns by a PtdIns 5-kinase. This lipid was discovered as an impurity in commercial PtdIns (brain-derived), and was shown to be present in mammalian fibroblasts at approximately the same level as PtdIns(3)P. PtdIns(5)P had been overlooked in the past because it was difficult to separate it from PtdIns(4)P by conventional methods. Hence its formation by a PtdIns 5-kinase cannot be absolutely excluded. It is now known that the type I γ or type I β PtdIns(4)P 5-kinases can phosphorylate PtdIns at the D5 position in vitro (Tolias et al., 1998). However, the preferred substrate for these enzymes is PtdIns(4)P. The in vivo significance of this pathway, therefore, remains to be established. Additional homologues of PtdIns (5)P kinases have been cloned but not fully characterized. Since the enzyme that converts PtdIns(5)P to PtdIns(4,5)P$_2$ (type II PtdIns(5)P 4-kinase; see below) is very abundant and active in most mammalian cells, it is believed that the PtdIns(5)P pathway plays an important role in cellular regulation (Tolias et al., 1998). An alternative possibility is that PtdIns(5)P is a breakdown product of PtdIns(3,5)P$_2$ by 3-phosphatase (Chapter 7). PtdIns(5)P can be produced in vitro by dephosphorylation of PtdIns(4,5)P$_2$ (Rameh et al., 1997).

Shisheva et al. (1999) have recently isolated a novel mammalian PtdIns kinase, p235, of undetermined substrate specificity. Sbrissa et al. (1999) have now shown that the mouse p235 (renamed PIKfyve) displays a strong in vitro activity for PtdIns(5)P and PtdIns(3,5)P$_2$ generation. The enzymatic activity was observed only if PIKfyve sequence was intact. Truncation of the N-terminus, C-terminus, or middle region of the molecule yielded kinase-deficient mutants. Morris et al. (2000) have

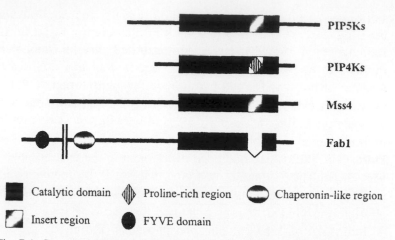

Fig. 7.4. Structural features of the PdIns(4)P 5-kinases and PtdIns(5)P 4-kinases. These enzymes share significant homology in their catalytic domains despite their different enzymatic activities. Various family members also contain proline-rich regions, a kinase insert region, a chaperonin-like region, and a FYVE finger domain. Reproduced from Tolias and Carpenter (2000) with permission of the publisher.

reported that thrombin stimulation causes an increase in PtdIns 5-phosphate revealed by mass assay.

Subsequently, Sbrissa et al. (2000) have shown that PIKfyve is a large protein of 2052 amino acids, which displays an intrinsic, wortmannin-resistant lipid kinase activity and, besides itself, can phosphorylate exogenous proteins in a substrate specific manner (dual-specificity enzyme). Tolias and Carpenter (2000) have summarized the structural features of the PtdIns(4)P 5-kinases and PtdIns(5)P 4-kinases in graphical form (Fig. 7.4).

7.2.3.1. Assays of cellular proteins

Tolias et al. (1998) assayed the in vitro synthesis of PtdIns(5)P by type I α PtdIns 5-kinase from murine cells as follows. The PtdInsP 5-kinase was assayed in 50 μl reactions containing 50 mM Tris, pH 7.5, 30 M NaCl, 12 mM MgCl$_2$, 67 μM PtdIns, 133 μM phosphoserine, and 50 μM [γ-^{32}P]ATP (10 μCi/assay). No carrier lipid was used. Reactions were stopped after 1 min by adding 80 μl of 1N HCl and then 160 μl of CHCl$_3$/MeOH (1:1, v/v). Lipids were separated by TLC using 1-propanol/2 M acetic acid (65:35, v/v). Phosphorylated lipids were visualized by autoradiography and quantified using a Bio-Rad Molecular Analyst. The PtdInsP used in all experiments was separated from contaminating PtdInsPs by TLC purification using CHl$_3$/MeO/H$_2$O/NH$_4$OH (60:47:11:1.6, by vol.). After identification using an iodine-stained PtdIns standard, the PtdIns was scraped from the plate and eluted in the same solvent solution. Following lyophilization, the PtdInsP was resuspended once in MeOH/0.1 M EDTA (1:0.9, v/v).

Tolias et al. (1998) identified the reaction products by HPLC analysis. For this purpose the deacylated lipids were mixed with ^{3}H-labeled standards and analyzed by anion-exchange HPLC using Partisphere SAX column and an ammonium phosphate gradient. The compounds were eluted with 1 M (NH$_4$)$_2$HPO$_4$, pH 3.8, and water using the following gradient: 0% 1 M (NH$_4$)$_2$HPO$_4$ for 5 ml, 0–1% 1 M (NH$_4$)$_2$HPO$_4$ over 5 ml, 1–3% 1 M (NH$_4$)$_2$HPO$_4$ over 40 ml, 3–10% 1 M (NH$_4$)$_2$HPO$_4$ over 10 ml, 1–13% 1 M (NH$_4$)$_2$HPO$_4$ over 25 ml, and 13–65% 1 M (NH$_4$)$_2$HPO$_4$ over 25 ml. Eluate from the HPLC column flowed into an on-line continuous flow scintillation detector (Radiomatic Instruments, FL) for isotope detection.

Tolias et al. (1998) identified the phosphorylation products further by periodate oxidation (Stephens et al., 1988) using the following modifications. Deacylated lipids were incubated in the dark at 25°C with 10 or 100 mM periodic aid, pH 4.5, for 30 min or 36 h, respectively. The remaining oxidizing reagent was removed by adding 500 mM ethylene glycol and incubating the reaction in the dark for 30 min 2% 1,1-dimethylhydrazine, pH 4.5, was then added to a final concentration of 1% and the reaction was allowed to proceed for 4 h at 25°C. The mixture was then purified by ion exchange using Dowex 50W-X8 cation exchange resin (20–50 mesh, acidic form), dried, and applied to an HPLC column.

PtdIns(3,4)P_2 and PtdIns(3,4,5)P_3 controls used in this experiment were prepared using recombinant Sf9 cell-expressed PtdIns 3-kinase.

7.2.3.2. Assays of recombinant protein

Sbrissa et al. (1999) assayed the PtdIns 5-kinase activity in the immunoprecipitates of P235 from 3T3-L1 adipocytes or COS-7 cells. Briefly, after immunoprecipitation the beads were washed with RIPA buffer; twice with 50 mM HEPES, pH 7.4, 1 mM EDTA, 150 mM NaCl; thrice with 100 mM Tris–HCl, pH 7.5, 500 mM LiCl; twice with 10 mM Tris–HCl, pH 7.5, 100 mM NaCl, 1 mM EDTA; and twice with 'assay buffer' (50 mM Tris–HCl, pH 7.5, 1 mM EGTA, containing typically 10 mM MgCl$_2$). The kinase reaction, carried out in the assay buffer (50 μl final volume) for 15 min at 37°C, contained 50 μM ATP (12.5 μCi) (or 15 μM ATP, 30 μCi for HPLC analysis) and 100 μM lipid substrate [PtdIns, PtdIns(3)P, PtdIns(4)P, PtdIns(5)P, PtdIns(3,4)P_2, or PtdIns(4,5)P_2] sonicated individually in 'lipid buffer' (20 mM HEPES, pH 7.5, 1 mM EDTA) by 2 × 30 s intervals in batch sonicator. When lipid substrates and lipid carriers were provided as a mixture they were sonicated together in a 1:1 molar ratio in the lipid buffer. The reaction was stopped with 200 μl of 1N HCl, and the lipids were extracted with 160 μl of chloroform/methanol (1:1, v/v). The lower layer was collected and washed twice with 100 μl of 1:1 (v/v) methanol, 1N HCl. Aliquots of the resulting organic layer were applied on pre-eluted (1.2% potassium oxalate) and activated (110°C, 16 h) TLC plates (Whatman, PE SIL G, 250 μm). The lipids were separated by chromatography at room temperature with either chloroform/methanol/water/30% ammonia (90:90:20:7, by vol.) or with n-propanol/2 M acetic acid (65:35, v/v), as the solvent systems (Tolias et al., 1998; Jenkins et al., 1994). The radioactive products were detected by autoradiography. Lipid standards run in parallel were detected under UV light following spraying with 0.001% primulin.

Following TLC separation of ^{32}P-labeled PtdIns products Sbrissa et al. (1999), the radiolabeled spots on the plate corresponding to that indicated on the autoradiogram were recovered and deacylated at 54°C for 50 min with methylamine reagent as described by Serunian et al. (1991). The deacylating reagent and the lower aqueous phase following the butanol/petroleum ether/ethyl formate extraction step were removed at 45–50°C by blowing nitrogen, and drying was effected in the same manner by the addition of methanol. The products were dissolved in 400 μl of water and filtered prior to injection of 100–200 μl on a Waters HPLC system (Millipore Corp). Analyses were performed on a Whatman 235 mm × 4.60 mm column packed with 5μ Partisphere SAX (H$_3$PO$_4$) and eluted at a flow rate of 1.0 ml/min under a slightly modified protocol (Rameh et al., 1997; Tolias et al., 1998). For recovery of PtdInsP, elution was conducted with water for 5 min, and then water was diluted discontinuously and linearly with 1 M ammonium phosphate buffer, pH 3.8, to 99% over 5 min, to 97% over 40 min, to 90% over 20 min, and finally with 90% for 20 min. For the higher phosphorylated PtdIns, the gradient was as above up to 70 min and was linearly increased thereafter to 13% buffer over 25 min and to 65% buffer over the next 25 min [^{3}H]GroPIns(4)P and [^{3}H]GroPIns(4,5)P$_2$ (deacylated from [^{3}H]PtdIns(4)P and [^{3}H]PtdIns(4,5)P$_2$, respectively, as described above and stored in water at − 20°C or [^{3}H]GroPIns(5)P were co-injected as internal HPLC standards. Deacylated [^{32}P]PtdIns were prepared to be used as internal standard. Fractions were collected every 0.25 min and their radioactivity was analyzed simultaneously for ^{3}H- and ^{32}P-labeled standards and products with 2.0 ml of ScintiVerse liquid scintillation mixture.

7.3. Phosphatidylinositol monophosphate kinases

The PtdInsP kinases are also known as PIPKINS (Hinchcliffe et al., 1998). PtdInsP kinases phosphorylate PtdIns(3)P, PtdIns(4)P and

PtdIns(5)P to the corresponding PtdInsP$_2$ and PtdInsP$_3$ derivatives. These enzymes are widely distributed but their physiological role in cellular signaling and regulation is poorly understood (Anderson et al., 1999).

7.3.1. PtdInsP 3-kinases

PtdInsP 3-kinases phosphorylate the D3 position of PtdIns(4)P, PtdIns (5)P, and PtdIns(4,5)P$_2$. Each of the PtdInsP 3Ks is made up of several isozymes.

Class I PtdInsP 3-kinases possess a 110 kDa catalytic subunit that associates with a 85 kDa regulatory subunit (p85) (Rani et al., 1999). Class I PtdInsP 3Ks are further subdivided into Class IA and Class IB enzyme, which signal downstream of tyrosine kinases and heterotrimeric G-protein-coupled receptors, respectively (Vanhaesebroeck et al., 2001). In vitro, these PtdInsP 3-kinases utilize PtdIns(4)P and PtdIns(4,5)P$_2$. The catalytic subunit of Class IA PtdIns 3Ks exists in a complex with an adapter protein that has two Src homology 2 (SH2) domains. Many mammals have three class IA p110 isoforms, p110α, β and γ, which are encoded by three separate genes (Vanhaesebroeck et al., 1997a,b).

Class II PtdIns 3Ks are approximately 170 kDa, and their defining feature is a C-terminal C2 domain, which can bind in vitro to phospholipids in a Ca^{2+} independent manner (MacDougall et al., 1995; Araco et al., 1998). In contrast to class I PtdIns 3Ks, which appear to be mainly cytosolic, class II PtdIns 3Ks are predominantly associated with the membrane fraction of cells (Arcaro et al., 1998; Prior and Clague, 1999; Domin et al., 1997).

In vitro, class II PtdIns 3K enzymes have a lipid substrate specificity that is clearly distinct from that of class I and class III enzymes. Mammals have three class II isoforms (PtdIns 3K-C2α, β and γ, endcoded by three separate genes). PtdInsP-C2α, and β are ubiquitous, in contrast to PtdIns 3K-C2γ, which is mainly found in liver. Although class II PtdIns 3Ks can use PtdInsP, PtdIns(4)P and

PtdIns(4,5)P$_2$ as substrates, they prefer PtdIns to PtdIns(4)P and PtdIns(4,5)P$_2$ (see above).

7.3.1.1. PtdIns(4)P 3-kinase

Class II PtdInsP 3-kinase generates PtdIns(3,4)P$_2$ by phosphorylation of PtdIns(4)P (Zhang et al., 1998; Arcaro et al., 2000), although the enzyme shows strong preference for phosphorylation of PtdIns over PtdIns(4)P or PtdIns(4,5)P$_2$ (Zhang et al., 1998).

Serunian et al. (1989) have described an in vitro synthesis of PtdIns(3,4)P$_2$ by PtdIns(4)P 3-kinase in transformed cells. [^{32}P]PtdIns(3)P, [^{32}P]PtdIns(3,4)P$_2$ and [^{32}P]PtdIns(3,4,5)P$_3$ substrates were prepared from immunoprecipitates of polyoma virus-transformed cells as described by Whitman et al. (1988). Proteins were collected on Protein A-Sepharose CL-4B (Sigma), and the immunoprecipitate was washed three times with phosphate-buffered saline containing 1% Nonidet P-40, two times with 500 mM LiCl in 100 mM Tris, pH 7.5, and three times in TNE buffer (10 mM Tris, pH 7.5, 100 mM NaCl, 1 mM EDTA, pH 7.5, 1 mM EGTA, pH 7.5). After washing, the immunoprecipitate was mixed with either PtdIns, PtdIns(4)P, or PtdIns(4,5)P$_2$ (final concentration, 0.2 mg/ml) that had been sonicated in 20 mM HEPES, pH 7.5, 1 mM EDTA, pH 7.5. Enzyme reactions were started by adding [γ-^{32}P]ATP and 10 mM MgCl$_2$ in 20 mM HEPES buffer, pH 7.5. The reaction was incubated at room temperature in final volume of 150 μl for 20 min and was stopped by adding 0.25 ml of 1 M HCl and 0.5 ml of methanol/chloroform (1:1, v/v). After vigorous mixing and centrifuging to separate the phases, the organic phase was removed, washed two times with methanol/0.1 M EDTA, pH 7.5 (1:1, v/v) and stored under nitrogen at $-70°$C until further use.

In all cases, an aliquot of the resultant aqueous and organic phases was removed and counted for radioactivity. The remaining organic phase was washed two times with 0.5 volume of methanol/0.1 M EDTA, pH 7.5 (1:0.9, v/v), dried under nitrogen, and stored at $-70°$C until deacylation. The deacylation of phospholipids was done using the methylamine reagent described by Whitman et al. (1988). Briefly, methylamine reagent (42.8% of 25% methylamine in H_2O, 45.7% methanol, 11.4% butanol) was added to the phospholipids and incubated at 53°C for 50 min. After cooling to room temperature, the sample was dried under vacuum, water was added to the dried sample, and the contents were dried again. Samples were resuspended in H_2O and extracted two times with an equal volume of 1-butanol/light petroleum ether/ethyl formate (20:4:1, by volume). A Partisphere SAX anion-exchange column (Whatman) was used to analyze the deacylation products by HPLC.

More recently, Arcaro et al. (2000) have used PtdIns(4)P as a substrate for PtdIns 3K-C2α and PtdIns 3K-C2β in the presence of Mg^{2+}. The PtdIns(4)P phosphorylation by the 3K-isozymes was performed in a total volume of 50 μl containing 20 mM HEPES (pH 7.4), 100 mM NaCl, 0.1 mM EGTA, 0.1 mM EDTA, and 200 mM PtdInsP. After preincubation of sonicated lipid with samples for 10 min, reactions were initiated by the addition of divalent cation (Ca^{2+} or Mg^{2+}, 6 mM) and 100 μM ATP (0.2 μCi of [γ-^{32}P]ATP). Reaction mixtures were incubated at 30°C for 20 min, and reactions were terminated with acidified chloroform/methanol (1:1, v/v). The extracted lipid products were fractionated by TLC. To separate PtdIns(3)P and PtdIns(4)P, a borate solvent system was used (Walsh et al., 1991). Phosphorylated lipids were visualized by either autoradiography or PhosphorImager analysis.

7.3.1.2. PtdIns(5)P 3-kinase

In vitro, PtdIns(3,5)P$_2$ has been generated through phosphorylation of PtdIns(5)P by Class IAPtdIns 3-kinase (Rameh et al., 1997). The PtdIns 3-kinase converted PtdIns(5)P to a molecule whose deacylation product migrated ahead of deacylated PtdIns(3,4)P$_2$ at the position expected for deacylated PtdIns(3,5)P$_2$ (Whiteford et al., 1997). Dowler et al. (2000) have demonstrated that centaurin-β2, an Arf-GAP, shows a weak binding specificity for PtdIns(3,5)P$_2$. Dipalmitoyl GroPIns(5)P was converted to the corresponding PtdIns(3,5)P$_2$. For the kinase reaction, recombinant murine type I PtdInsP kinase β was used. Deacylated lipids were analyzed by anion-exchange HPLC using a Partisphere SAX column as described by Serunian et al. (1991).

7.3.2. PtdInsP 4-kinases

It is known that the type I PtdIns 3K and type III PtdIns 4K enzymes have related sequences and belong to the PtdIns 3K family (Fruman et al., 1998). However, the fraction containing the type II PtdIns 4K (PtdIns 4KII) has so far not been characterized at the molecular level. Minogue et al. (2001) have partially purified the human type II isoform from plasma membrane rafts of human A431 epidermoid carcinoma cells and obtained peptide mass and sequence data. The results allowed the conclusion that the purified protein type IIα and a second human isoform, type IIβ belong to a novel, third family of PtdIns kinases.

7.3.2.1. PtdIns(3)P 4-kinase

The formation of PtdIns(3,4)P$_2$ from PtdIns(3)P by Ptdns(3)P 4-kinase has been demonstrated in platelets and red blood cells (Yamamoto et al., 1990; Banfic et al., 1998) and plants (Brearley and Hanke, 1993; Munnik et al., 1998) and yeast (Carman et al., 1992). The identity of this enzyme has not been definitely established, however, the type I and type II PtdInsP kinases, both of which have PtdIns(3)P 4-kinase activity in vitro, could be involved, although recent evidence to the contrary has also been obtained (Banfic et al., 1998; Hinchcliffe et al., 1998).

As indicated above, the type I PtdInsP 5Ks and the type II PtdInsP 4Ks can phosphorylate PtdIns(3)P at the D4 position to produce PtdIns(3,4)P$_2$ (Fruman et al., 1998). Although PtdIns(3)P is a poor substrate for the type I and type II enzymes that have been characterized, it is clear that there exists an enzymatic activity in platelets that converts PtdIns(3)P to PtdIns(3,4)P$_2$ in response to aggregation of the integrin, GPIIb3a (Banfic et al., 1998). In thrombin-stimulated platelets, however, the bulk of PtdIns(3,4)P_2 results from a pathway in which the D4 phosphate is phosphorylated before the D3 phosphate (Carter et al., 1994).

Graziani et al. (1992) have purified a PtdIns(3)P 4-kinase from human erythrocytes and have described an in vitro synthesis of PtdIns(3,4)P_2 by type III PtdIns(3)P 4-kinase of the erythrocytes. PtdIns(3)P 4-kinase was assayed at 37°C for 15 min in a 20 ml reaction mixture containing [^{32}P]PtdIns(3)P (0.5–1 μM), [^{3}H]PtdIns(4)P (0.5–1 μM), 9 μM PtdIns, 8 μM PtdSer, 500 μM ATP and 10 mM MgCl$_2$, in 50 mM HEPES buffer (pH 8)/1 mM EGTA (lipids were dried under nitrogen and sonicated for 10 min in a bath sonicator). The reaction was started by adding enzyme and was terminated with chloroform/methanol/HCl. The lipids were extracted and identified by TLC and ion exchange chromatography. The lipids were deacylated by incubation with methylamine at 54°C for 40 min; the derived GroPIns phosphates were applied to a strong anion-exchange HPLC (Whatman Partisil 10 SAX) column and eluted with a gradient of ammonium formate. The [^{32}P]PtdIns(3)P substrate was prepared by incubating PtdIns with [γ-^{32}P]ATP and purified PtdIns 3-kinase.

7.3.2.2. PtdIns(5)P 4-kinase

Based on the ability to phosphorylate commercial PtdIns(4)P to produce PtdIns(4,5)P_2, a second form of PtdInsP-kinase was purified (Ling et al., 1989; Bazenet et al., 1990) and cloned (Boronenkov and Anderson, 1995; Divecha et al., 1995). This enzyme was thought to be a PtdIns(4)P 5-kinase and was named type II α PtdInsP 5K. However, by gene cloning, this PtdInsP-kinase has been shown to be a PtdIns(5)P 4-kinase (PtdInsP 4Kα). A second highly related gene (78% identical) has also been cloned (Castellino et al., 1997). Previously termed type II β PtdInsP 5K, this enzyme has since been designated PtdInsP 4Kβ. The PtdInsP 4Ks are about 35% identical to the PtdInsP 5Ks in their kinase domains but are dissimilar in their amino- and carboxy terminal extensions. Through the use of pure preparations of PtdIns(4)P and PtdIns(5)P, Rameh et al. (1997) found that PtdInsP 4Kα and β can phosphorylate only the latter. This conclusion is supported by the observation that PtdInsP 4Ks do not synthesize PdIns(3,4,5)P$_3$ when PtdIns(3,4)P$_2$ is presented as a substrate (Zhang et al., 1997; Toker, 1998).

Little is known about the regulation of PtdInsP 4K activity. In vitro activity is inhibited by PtdIns(4,5)P$_2$ and heparin. PtdInsP 4Kβ associates directly with the p55 subunit of the tumor necrosis factor (TNF) receptor, and indirect evidence suggests that its activity increases following TNF treatment of cells (Castellino et al., 1997). Itoh et al. (1998) have identified a novel rat PtdIns(5)P 4-kinase, PtdInsP kinase IIγ (PtdInsP KIIγ). PtdInsP KIIγ was found to have PtdIns(5)P 4-kinase activity as demonstrated for other type II kinases such as PtdInsP KIIα. It is expressed predominantly in kidney. Phosphorylation is induced by treatment with mitogens such as serum and epidermal growth factor. PtdInsP KIIγ comprises 420 amino acids with a molecular mass of 47 048 Da, showing a greater homology to the type IIα and IIβ isomers of PtdInsP kinase than to the type-I isoforms.

Itoh et al. (1998) have reported the in vitro synthesis of PtdIns(4,5)P$_2$ by II γ PtdIns(5)P 4-kinase. Briefly, 48 h after electroporation, the expression vector-transfected COS-7 cells were lysed with lysis buffer (20 mM HEPES, pH 7.2, 50 mM NaCl, 30 mM sodium pyrophosphate, 1% Nonidet P-40, 1 mM EGTA, 25 mM NaF, 0.1 mM sodium vanadate, and 1 mM phenylmethyl-sulfonyl fluoride). The expressed enzyme was immunoprecipitated with monoclonal anti-Myc antibody and washed three times with lysis buffer and once with reaction buffer (50 mM Tris–HCl, pH 7.5, 10 mM MgCl$_2$, and 1 mM EGTA). The reaction was started by adding 50 μM PtdIns(5)P (purified from bovine spinal chord PtdInsP by a neomycin column), 50 μM ATP, and 10 μCi of [γ-^{32}P]ATP in 50 μl. After incubating for 10 min at room temperature, the lipids were extracted with 1N HCl and chloroform/methanol (2:1, v/v) and spotted on TLC plates. The plates were developed in chloroform/methanol/ammonia/water

(14:20:3:5, by vol.), and the products were monitored by autoradiography or quantified by image analyzer. The structure of the PtdIns(4,5)P$_2$ product was confirmed by HPLC and periodate oxidation of the deacylated product (Chapter 3).

7.3.3. PtdInsP 5-kinases

The PtdIns(4)P 5-kinase and PtdIns(5)P 4-kinases are two closely related families of enzymes. They were originally called type I and type II PtdIns(4)P 5-kinases until it was discovered that the type II enzymes are actually PtdIns(5)P 4-kinases (Rameh et al., 1997). The PtdInsP 4Ks are about 35% identical to the PtdInsP 5Ks in their kinase domains but are dissimilar in their amino- and carboxy terminal extensions.

The kinase domains of PtdInsP 5Ks are related to two proteins in *S. cerevisiae*: Mss4 and Fab1. Genetic evidence places MSS4 downstream of the yeast PtdIns 4-kinase STT4, consistent with Mss4 acting as a PtdIns(4)P kinase (Yoshida et al., 1994a,b).

7.3.3.1. PtdIns(3)P 5-kinase
Whiteford et al. (1997) obtained conclusive identification of PtdIns(3,5)P$_2$ in mammalian cells and Dove et al. (1997) in yeast. The bulk of this lipid is made by a pathway in which the D3 position is phosphorylated prior to the D5 position.

As mentioned above, the mammalian type I γ and 1β PtdInsP 5Ks have a significant PtdIns(3)P 5-kinase activity and could be responsible for production of some of the PtdIns(3,5)P$_2$ in cells. Although PtdIns(3,5)P$_2$ is far less abundant than PtdIns(4,5)P$_2$ in mammalian cells (approximately 1%), in yeast PtdIns(3,5)P$_2$ levels can approach those of PtdIns(4,5)P$_2$, under conditions of osmotic stress (Dove et al., 1997). Hyperosmotic stress induces rapid synthesis of PtdIns(3,5)P$_2$ in plant cells (Meijer et al., 1999). PtdIns(3,5)P$_2$ is produced in yeast by the yeast PtdInsP 5K homologue, Fab1 (Gary et al., 1998). The stress-activated PtdIns(3)P

5-kinase Fablp is essential for vacuole function in *S. cerevisiae* (Cooke et al., 1998). McEwen et al., (1999) have shown that *S pombe* and murine Fablp homologues are PtdIns(3)P 5-kinases.

Whiteford et al. (1997) reported in vivo synthesis of PtdIns(3,5)P$_2$ via PtdIns(3)P 5-kinase in mouse fibroblasts. For *myo*-[^{3}H]inositol labeling, confluent cultures of mouse fibroblasts in 60 mm dishes were first starved of inositol by incubation in inositol-free Dulbecco's modified essential medium (DMEM) containing 20 mM HEPES, pH 7.1, ITS-Plus- and 10% dialysed calf serum. After 6–8 h, this medium was replaced with 2 ml of inositol-free DMEM containing 20 mM HEPES, pH 7.1, ITS-Plus- and 50 μCi of *myo*-[2-^{3}H]inositol, and labeling was allowed to progress for 18–24 h. The incubations were stopped by addition of perchloric acid. Perchlorate extraction, and deacylation and separation of inositol phospholipids were carried out as described in Chapter 3. Under these conditions, $2-6 \times 10^6$ dpm were recovered in the lipid-containing perchlorate precipitates. After deacylation and HPLC, peaks corresponding to GroPIns, GroPIns(3)P, GroPIns(4)P, GroPIns(3,5)P$_2$ and GroPIns(4,5)P$_2$ contained 80–90, 0.15–0.18, 5.5–6.0, 0.03–0.04 and 4–7%, respectively, of the total radioactivity recovered in this fraction from untreated Ph-N$_2$ cultures. An identical protocol was utilized for large-scale isolation of GroPIns(3,5)P$_2$ from approx. 1×10^8 Ph-N$_2$F5 cells labeled with 1.5 mCi of *myo*-[2-^{3}H]inositol, with an overall recovery of approx. 60 000 dpm of [^{3}H]GroPIns(3,5)P$_2$.

For [^{32}P]P labeling, nearly confluent cultures in 100 mm-diameter dishes were incubated for 24 h in serum-free DMEM containing 20 mM HEPES, pH 7.1, and ITS-Plus, then incubated for 30 min in phosphate-free DMEM containing 20 mM HEPES, pH 7.1. Cultures were labeled with 3–4 mCi of [^{32}P]P in 1 ml of this medium in an open water bath. The labeling medium was removed after 3–5 min and 0.72 ml of 1 M HCl and 0.9 ml of methanol were added. Cells were scraped from the dishes, transferred to glass tubes, and mixed vigorously with 1.8 ml of methanol/chloroform (1:1, v/v). After 215 min on ice, 0.9 ml of water and 0.9 ml of chloroform were added and phases were separated by centrifugation at 24°C. The organic phase was reextracted once with 1.8 ml of methanol/1 M HCl (1:1, v/v), then dried and deacylated (Downes et al., 1986). Between 1800 and 5400 cpm of GroPIns(3,5)P$_2$ and 116 000–624 000 cpm of GroPIns(4,5)P$_2$ were recovered by this procedure.

7.3.3.2. PtdIns(4)P 5-kinase
Phosphorylation of PtdIns(4)P by PtdIns(4)P 5-kinase to form PtdIns(4,5)P$_2$ constitutes the second step in the biosynthesis of PtdIns(3,4,5)P$_3$. Detailed studies of PtdIns 4-kinase and PtdIns 5-kinase activities have been made in plant membranes (Sommarin and Sandelius, 1988; Sandelius and Sommarin, 1990; Gross et al., 1992; Cho et al., 1993) and cytoskeletal (Tan and Boss, 1992; Xu et al., 1992) and soluble (Okpodu et al., 1995) fractions. Recently, a putative PtdIns(4)P 5-kinase was cloned from *Arabidopsis* (Satterlee and Sussman, 1997), but the genes encoding PtdIns 4-kinases in plants have remained unknown.

Genes encoding the two distinct families of PtdInsP kinases have now been cloned and found to contain significant sequence similarity (Bazenet and Anderson, 1992; Boronenkov and Anderson, 1995; Ishihara et al., 1996; Loijens and Anderson, 1966). As noted above, recent studies have revealed that the two families of enzymes selectively phosphorylate the 5th and 4th positions on the inositol ring (Rameh et al., 1997; Zhang et al., 1997).

The best studied reaction catalyzed by PtdInsP 5Ks is the conversion of PtdIns(4)P to PtdIns(4,5)P$_2$. According to Jones et al., (2000a,b) type I PtdIns (4)P 5-kinase directly interacts with ADP-ribo-sylation factor 1 and is responsible for PtdIns (4,5)P$_2$ synthesis in the Golgi. Using enzyme purified from erythrocytes the Km for PtdIns(4)P in micelles, liposomes, or native membranes ranges from 1 to 10 μM (Loijens et al., 1996). This reaction is stimulated by heparin and spermine and inhibited by the product, PtdIns(4,5)P$_2$. Importantly, PtdInsP 5Ks are stimulated as much as 50-fold by phosphatidic acid (Moritz et al., 1992; Jenkins et al.,

1994). Chong et al. (1994) have shown that the small GTP-binding protein ρ regulates a PtdIns (4)P 4-kinase in mammalian cells.

PtdInsP 5Ks have been shown to phosphorylate other PtdIns in vitro (Tolias et al., 1998). The α and β isoforms convert PtdIns(3,4)P$_2$ to PtdIns(3,4,5)P$_3$ with Kcat/Km ratios that are 3- and 100-fold lower, respectively, than those observed using PtdIns(4)P as the substrate (Zhang et al., 1997). Type I PtdInsP 5Ks also utilize PtdIns(3)P as a substrate in vitro (Zhang et al., 1997). Zhang et al. (1997) identified the PtdIns-bisphosphate product as PtdIns(3,4)P$_2$. In contrast, work in other laboratories has shown that the bisphosphorylated product includes PtdIns(3,5)P$_2$, a lipid that was recently demonstrated to exist in vivo (Whiteford et al., 1997; Dove et al., 1997).

Under non-equilibrium radiolabeling conditions, the D5 phosphate has a higher specific activity, suggesting that it is added to the molecule later that the D4 phosphate, which implies that production of PtdIns(4,5)P$_2$ by PtdIns(4)P phosphorylation predominates (Stephens et al., 1991; Hawkins et al., 1992). The contribution of PtdIns(5)P phosphorylation to PtdIns(4,5)P$_2$ formation in vivo remains in doubt in view of the comparatively small amount of PtdIns(5)P relative to PtdIns(4)P within the cells (Hinchcliffe et al., 1998; Rameh et al., 1997).

To date, clones for three different type I PtdIns(4)P 5-kinase family members (PtdInsP 5K I γ, PtdInsP 5K I β and PtdInsP 5-K I α) and three different type II PtdIns(5)P 4-kinase family members (PtdInsP 4-K γ, PtdInsP 4-K β and PtdInsP 4-Kα) from mammalian cells have been published (Fruman et al., 1998; Itoh et al., 1998, 2000). Two yeast homologues have also been identified (MSS4 and FAB1). The MSS4 gene product was recently shown to have PtdIns(4)P 5-kinase activity and to be critical for actin structures in *S. cerevisiae* (Desrivieres et al., 1998). Based on in vitro studies with purified enzymes, the mammalian type I γ and I β PtdInsP 5-Ks are relatively promiscuous with respect to substrate utilization and position of phosphorylation on the inositol ring. Thus, although the preferred substrate is PtdIns(4)P, when provided with PtdIns(3)P, these enzymes produce PtdIns(3,5)P$_2$, PtdIns(3,4)P$_2$, and PtdIns(3,4,5)P$_3$ (Zhang et al., 1997; Tolias et al., 1998). Type I PtdInsP 5-K activity in vivo may be influenced by small G proteins (Hinchcliffe et al., 1998; Tolias et al., 1998).

Tolias et al. (1998) report an in vitro synthesis of PtdIns(4,5)P$_2$ by PtdIns(4)P 5-kinase, which parallels the above described routine for PtdInsP 4-kinase, except that PtdIns(4)P is the starting material now. The PtdIns(4,5)P$_2$ product was isolated by TLC as a PtdInsP$_2$ before counting. The structure of the product was confirmed by HPLC and periodate oxidation of the deacylated product (Chapter 3).

7.4. Phosphatidylinositol bisphosphate kinases

7.4.1. PtdIns (4,5)P$_2$ 3-kinase

Phosphorylation of PtdIns(4,5)P$_2$ via PtdInsP 3-kinase completes the biosynthesis of PtdIns(3,4,5)P$_3$. This is the major pathway as demonstrated by studies with radioactive tracers (Soriski et al. (1992); Hawkins et al., 1992; Carter et al., 1994) and selective inhibitors. Remarkably, PtdIns(3,4,5)P$_3$ is also produced when recombinant PtdInsP 5Ks are mixed with PtdIns(3)P (Zhang et al., 1997). The amount of PtdIns(3,4,5)P$_3$ produced by PtdInsP 5Kα is significant, approaching 50% as much as the PtdInsP$_2$ produced in a 10 min reaction (Zhang et al., 1997). The synthesis of PtdIns(4,5)P$_2$ is subject to regulation by small GTPases, including Arf. The mechanisms by which Arfs increase PtdIns(4,5)P$_2$ values include an activation of PtdIns 4-kinase and PtdIns(4)P 5-kinase. Vanhaesebroeck et al. (2001) have summarized the roles of small GTPases in PtdIns(4,5)P$_2$ and PtdInsP$_3$ synthesis (Fig. 7.5).

Hawkins et al. (1992) reported an in vivo assay of PtdIns(3,4,5)P$_3$ synthesis via PtdIns(4,5)P$_2$ 3-kinase. Swiss 3T3 cells were grown to confluence over 6 days in small glass bottles

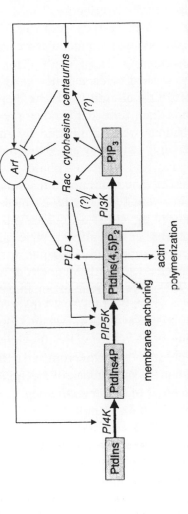

Fig. 7.5. The roles of small GTPases in PtdIns(4,5)P_2 and PtcInsP$_3$ synthesis, and possible feedback loops. Reproduced from Vanhaesebroeck et al. (2001) with permission from the Annual Review of Biochemistry, Volume 70 © 2001 by Annual Reviews www.annualreviews.org.

and then serum-starved for 16 h before the experiment. Each bottle was then removed from the incubator, washed with HEPES-buffered DMEM (112.5 mM NaCl, 5.37 mM KCl, 0.81 mM MgSO$_4$, 1.8 mM CaCl$_2$, 25 mM glucose, 1 mM NaHCO$_3$, 25 mM HEPES, pH 7.4, at 37°C, 0.1% fatty acid-free BSA, 1 × DMEM amino acids and vitamins) and then incubated with 0.4 ml HEPES-buffered DMEM containing 0.25 mCi [^{32}P]P for 70 min at 37°C. Each bottle was then washed twice and incubated at 37°C in a total volume of 0.4 ml HEPES-buffered DMEM either with or without PDGF (human recombinant form BB used at a final concentration of 30 ng/ml). Incubations were terminated by addition of 1.5 ml CHCl$_3$/MeOH (1:2, v/v). Lipids were extracted using a Folch distribution of organic and aqueous phases, with 1.0 M HCl, 25 mM EDTA, 5 mM tetrabutylammonium sulfate included in the aqueous phase. Lipids were deacylated with methylamine and the GroPIns esters separated by HPLC using Partisphere SAX column as described (Stephens et al., 1991). The identities of the appropriate radioactive compounds obtained using this strategy have been rigorously characterized (Stephens et al., 1991).

Petitot et al. (2000) have described an in vitro synthesis of PtdIns(3,4,5)P$_3$ via Class I PtdIns(4,5)P$_2$ 3-kinase from HT-29 cells. PtdIns(3,4,5)P$_3$ formation was determined by a modification of the method of Burgering et al. (1991). Briefly, immunoprecipitates (prepared with anti-phosphotyrosine serum and collected on protein A-Sepharose beads) and the supernatant of the immunoprecipitation were resuspended in 80 μl of buffer containing 30 mM HEPES and 300 μM adenosine. To both samples, 50 μl of lipid mixture (synthetic lipid and PtdSer) was added for 20 min. The assay of Class 1 PtdIns 3-kinase activity was performed on the immunoprecipitate using diC$_{16}$GroPIns(4,5)P$_2$ as substrate in the presence of Mg^{2+}. The reaction was initiated by the addition 20 μCi of [γ-^{32}P]ATP, 100 μM ATP, 100 μM MgCl$_2$ and terminated after 15 min by the addition of 80 μl of 1 M HCl and 200 μl of chloroform/methanol, 1:1 (v/v). After vigorous mixing and centrifugation, the organic layer was removed, dried under N$_2$ and resuspended in chloroform/methanol. Extracted lipids were separated by TLC in the developing solvent: isopropyl alcohol/H_2O/acetic acid (65:34:1, by vol), and then exposed to hyperfilm. Unlabeled standards (1 mg/ml) were revealed by exposure to iodine vapor. Lipids were identified by comparison with unlabeled standards.

7.4.2. PtdIns(3,5)P_2 4-kinase

This enzyme may not have been discovered as yet. Presumably, PtdIns(3,5)P_2 does not accumulate to serve as a substrate in PtdIns(3,4,5)P_3 synthesis, but a synthetic standard could have been assayed.

7.4.3. PtdIns(3,4)P_2 5-kinase

PtdIns(3,4,5)P_3 is nominally absent in quiescent cells, but appears within seconds to minutes of cell stimulation with a variety of activators (Traynor-Kaplan et al., 1988). The bulk of the PtdIns(3,4,5)P_3 is believed to be generated by phosphorylation of PtdIs(4,5)P_2 at the 3-position of the inositol ring (Hawkins et al., 1992). This is believed to occur via the Class I PtdIns 3-kinases, which are the only enzymes known to use PtdIns(4,5)P_2 as a substrate. The Class I PtdIns 3-kinases are activated by growth factor stimulation of cells, which is mediated in rat by interaction of their SH2 domain with tyrosine-phosphorylated proteins (Toker and Cantley, 1997).

Alternatively, PdIns(3,4,5)P_3 can be synthesized by a pathway independent of PtdIns(4,5)P_2. The PtdInsP 5-kinases α and β have been shown to utilize PtdIns(3)P as a substrate to produce PtdIns(3,4,5)P_3 by phosphorylating the 4- and 5-positions of the inositol ring in a concerted reaction (Zhang et al., 1997; Tolias et al., 1998). Zhang et al. (1997) identified the PtdInsP$_2$ product as PtdIns(3,4)P_2. The relative contribution of this pathway to intracellular levels of PtdIns(3,4,5)P_3 is not known. The α and β isoforms convert PtdIns(3,4)P_2 to PtdIns(3,4,5)P_3 with K_{cat}/K_m

ratios that are 3-fold and 100-fold lower, respectively, than those observed using PtdIns(4)P as the substrate (Zhang et al., 1997). Type I PtdInsP 5-Ks also utilize PtdIns(3)P as a substrate in vitro.

Vanhaesebroeck et al. (2001) have discussed the role of small GTPases in PtdIns(4,5)P_2 and PtdInsP$_3$ synthesis along with possible feedback loops, which have been graphically summarized in Fig. 7.5.

Zhang et al. (1997) have described an in vitro synthesis of PtdIns(3,4,5)P_3 via PtdIns(3,4)P_2 by recombinant human PtdInsP 5-kinase isozymes as follows. The biosynthesis of PtdIns(3,4,5)P_3 via PtdInsP 5-kinases was performed in 50 μl reactions containing 50 mM Tris, pH 7.5, 30 mM NaCl, 12 mM MgCl$_2$, 67 μM lipid substrate (including dipalmitoyl GroPIns(3,4)P_2 among others), 133 μM phosphoserine, and 50 μM [γ-^{32}P]ATP (10 μCi/assay). When PtdIns was used as the lipid substrate, no carrier lipid was used, and when [^{32}P]PtdIns(3)P was used as substrate, unlabeled ATP was used in place of [γ-^{32}P]ATP. Reactions were stopped after 10 min by adding 80 μl of 1N HCl and then 160 μl of CHCl$_3$/ MeOH (1:1, v/v). Lipids were separated by TLC using 1-propanol, 2 M acetic acid (65:35, v/v). Phosphorylated lipids were visualized by autoradiography and quantified using Bio-Rad Molecular Analyst.

The reaction products were analyzed by HPLC following deacylation. The deacylated lipids were mixed with ^{3}H-labeled standards and analyzed by anion-exchange HPLC using Partisphere SAX column (Chapter 3). To separate PtdIns(4)P from PtdIns(5)P and PtdIns(3,4)P_2 from PtdIns(3,5)P_2, a modified ammonium phosphate gradient was used. The compounds were eluted with 1 M (NH$_4$)$_2$HPO$_4$, pH 3.8, and water using the following gradient: 0% 1 M (NH$_4$)$_2$HPO$_4$ for 5 ml, 0–1% 1 M (NH$_4$)$_2$HPO$_4$ over 5 ml, 1–3% 1 M (NH$_4$)$_2$HPO$_4$ over 40 ml, 3–10% 1 M (NH$_4$)$_2$HPO$_4$ over 10 ml, 10–13% 1 M (NH$_4$)$_2$HPO$_4$ over 25 ml, and 13–65% 1 M (NH$_4$)$_2$HPO$_4$ over 25 ml. Eluate from the HPLC column flowed into an on-line continuous flow scintillation detector (Radiomatic Instruments, FL) for isotope detection.

The structures of the products were confirmed by periodate oxidation as described above (see also Chapter 3).

7.5. Inositol monophosphate kinases

Early studies showed that all tissues contained GroPIns, InsP, Ins(1,4)P$_2$ and Ins(1,4,5)P$_3$, which were identified by using anion-exchange and high resolution anion-exchange chromatography, high voltage electrophoresis and paper chromatography. Berridge et al. (1983) developed a simple anion-exchange chromatographic method for separating these inositol phosphates for quantitative analysis. Of the six possible InsP derivatives, only four isomers, Ins(1)P, Ins(2)P, Ins(3)P, and Ins(4)P have been detected in plants and animals. InsPs are subject to phosphorylation by InsP kinases, which have been best characterized from plants, where they are involved in phytic acid formation. Ins kinases that yield other isomers of InsP have not yet been identified, although they are known to be produced by the action of phytases and phospholipase C.

Subsequent studies using similar chromatographic methods led to the demonstration of natural occurrence of isomeric tetrakis-, pentakis-, and hexakis-phosphates and their diphosphorylated derivatives (Shears, 1998a,b). Some of these phosphates are generated by phospholipase C hydrolysis of inositol phosphatides (see Chapter 6), while others are generated by the action of phosphatases (see Chapter 8) on InsPs, and the action of kinases on InsPs, as discussed below.

7.5.1. InsP 1-kinases

7.5.1.1. Ins(3)P 1-kinase
The conversion of glucose-6-phosphate to myo-Ins(3)P is the only biosynthetic route to myo-Ins. It is widely accepted that the myo-

InsP synthase from plants, mammals, and yeast all function in the same manner (Murthy, 1996). Ins(3)P is also produced by an Ins kinase, an ATP- and Mg^{2+}-dependent enzyme (Murthy, 1996).

Early work suggested that phytic acid synthesis involved sequential phosphorylation of inositol by one or more kinases, while later work showed that an alternative pathway involving sequential phosphorylation, either partially or completely, of an inositol derivative such as X-P-Ins, which would then be hydrolyzed to phytic acid (Chakrabarti and Biswas, 1981; Igaue et al., 1982). In any event, Ins(1,3)P$_2$ was recognized as an intermediate using suspension cultures of rice cells and [31]P$_i$ and [^{3}H]Ins as precursors.

7.5.2. InsP 6-kinases

7.5.2.1. Ins(3)P 6-kinases
Studies by Stephens and Irvine (1990) indicate that, in *Dyctiostelium* the biosynthesis of phytic acid proceeds by the formation of Ins(3,6)P$_2$ as intermediate.

7.5.2.2. Ins(2)P 6-kinases
The presence of Ins(2)P and its conversion to phytic acid by both rice cells and germinating mung bean is interesting. Although no evidence for an Ins kinase that phosphorylates at the 2-position is available, Ins(2)P is an intermediate in the phytase catalyzed hydrolysis of phytic acid. However, Murthy (1990) points out that ^{31}P shifts dispersed over a narrow range, result in significant overlap of resonances and additional details of structural assignment may be necessary.

7.6. Inositol bisphosphate kinases

InsP$_2$ have been demonstrated to serve as intermediates in the biosynthesis phytic acid via one or more kinases, the exact intermediate depending on the nature of the postulated pathway (see above).

7.6.1. InsP₂ 4-kinases

7.6.1.1. Ins(3,6)P₂ 4-kinases

According to Stephens and Irvine (1990), Ins(3,6)P₂ becomes phosphorylated by a 4-kinase to Ins(3,4,6)P₃ as an intermediate in phytic acid biosynthesis in *Dictyostelium*.

7.6.2. InsP₂ 5-kinases

7.6.2.1. Ins(1,3)P₂ 5-kinase

According to Igaue et al. (1980, 1982), Ins(1,3)P₂ is apparently converted by a 5-kinase to Ins(1,3,5)P₃ as an intermediate step in the formation of phytic acid in rice cell culture suspensions.

7.6.2.2. Ins(2,6)P₂ 5-kinase

Igaue et al. (1980, 1982) also recognized the formation of Ins(2,5,6)P₃ by the same or another 5-kinase as an another intermediate during phytic acid formation in rice cell suspensions.

7.6.3. InsP₂ 6/3-kinase

Ins(1,4)P₂ is generated by phospholipase C hydrolysis of PtdIns(4)P (Chapter 9). Ins(1,4)P₂ is subject to hydrolysis by InsP₂ phosphatases (Chapter 8). It is not known whether or not Ins(1,4)P₂ is subject to phosphorylation by a kinase. Ongusaha et al. (1998) have shown that the dual-specificity Ins(1,4,5)P₃ 6/3 kinase does not recognize Ins(1,4)P₂ or InsP₃ isomers other than Ins(1,4,5)P₃ and Ins(1,4,6)P₃. All identified substrates have at least one equatorial pair of *para*-phosphate groups [Ins(1,3,4,5,6)P₅ has two pairs] and most inositol phosphates that did not affect its kinase activity lack this feature (Ongusaha et al., 1998).

Donie and Reiser (1989) have described a binding assay for the determination of Ins(1,4,5)P$_3$ using a binding protein preparation from beef liver. Ins(1,4,5)P$_3$ was determined in a reaction mixture that contained (final concentrations) EDTA (1 mM), Tris–HCl (25 mM), pH 9.0, and bovine serum albumin (1 mg/ml). The binding reaction was initiated by adding the binding protein.

7.7. Inositol trisphosphate kinases

According to Igaue et al. (1980, 1982), InsP$_3$s serve as intermediates in phytic acid biosynthesis in rice cell suspensions.

7.7.1. InsP$_3$ 1-kinases

7.7.1.1. Ins(3,4,6)P$_3$ 1-kinase
According to Stephens and Irvine (1990), *Dictyostelium* contains 1-kinase, which converts Ins(3,4,6)P$_3$ to Ins(1,3,4,6)P$_4$.

7.7.2. InsP$_3$ 3-kinases

7.7.2.1. Ins(1,4,5)P$_3$ 3-kinase
Ins(1,4,5)P$_3$ is generated by the receptor mechanism (phospholipase C) from PtdIns(4,5)P$_2$. Ins(1,4,5)P$_3$ 3-kinase catalyzes the ATP-dependent phosphorylation of InsP$_3$ to Ins(1,3,4,5)P$_4$ (Irvine et al., 1986). Morris et al. (1988) have obtained a purified preparation of this enzyme, which has allowed to investigate the substrate specificity and potential control mechanisms. It was shown that Ins(2,4,5)P$_3$ was not phosphorylated, while Ins(4,5)P$_2$ and GroPIns(4,5)P$_2$ were phosphorylated much more slowly than Ins(1,4,5)P$_3$. CTP, GTP and adensine 5'[γ-thio]trisphosphate were unable to substitute for ATP. Takazawa et al. (1988) showed the presence of a Ca^{2+}/CaM-sensitive InsP$_3$ 3-kinase in rat and bovine brain. Maximal stimulation at 1 μM InsP$_3$ substrate level was two-to-three-fold. A cDNA clone encoding rat brain Ca^{2+}/CaM-sensitive InsP$_3$ 3-kinase has been isolated

(Takazawa et al., 1990b) and expressed in *E. coli*. This enzyme is now referred to as InsP$_3$ 3-kinase A. Using the rat cDNA as a probe, Takazawa et al. (1990b) obtained and expressed human InsP$_3$ 3-kinase A from human hippocampus cDNA library (which had 93% sequence identity with rat brain InsP$_3$ 3-kinase A amino acid sequence (Takazawa et al., 1991a) and a novel InsP$_3$ 3-kinase isozyme, referred to as InsP$_3$ 3-kinase B (Takazawa et al., 1991b). The expressed InsP$_3$ 3-kinase B was more sensitive to Ca^{2+}/CaM (i.e. stimulation factor of 7–10-fold) compared with expressed InsP$_3$ 3-kinase A (two- to three-fold stimulated). Communi et al. (1994) have isolated from human platelets an InsP$_3$ 3-kinase isozyme of a molecular weight of 70 kDa, which is distinct from previously reported InsP$_3$ 3-kinase A (50 kDa) and InsP$_3$ 3-kinase B (63 kDa) as well as in the stimulation by Ca^{2+}/calmodulin, which was 17-fold at saturating calmodulin and 10 μM free Ca^{2+}.

Shears (1998a,b) has reviewed the Ins(1,4,5)P$_3$/Ins(1,3,4,5)P$_4$ cycle and points out that type A (51 kDa) and type B (74 kDa) have been the best characterized 3-kinases. The most intensely studied process is the stimulation of 3-kinase activity by Ca^{2+}/calmodulin, which increases the Vmax of the kinases (see Chapter 9), Shears (1998a,b) also points out that the tissue-specific differences in the levels of expression of the two isoforms have not been established, yet may be accompanied by distinctive mechanisms of regulating Ins(1,4,5)P$_3$ metabolism.

Morris et al. (1988) assayed the phosphorylation of Ins(1,4,5)P$_3$ by 3-kinase as follows. The samples for Ins(1,4,5)P$_3$ kinase activity were incubated at 30°C in a buffer containing 70 mM potassium glutamate, 10 mM HEPES, pH 7.2, 11 mM MgSO$_4$, 10 mM ATP, 30 mM NaCl, [^{3}H]Ins(1,4,5)P$_3$ (10 μM; spec. act. 5 dpm/pmol) and bovine serum albumin (10 mg/ml). Incubations were terminated by the addition of an equal volume of ice-cold HClO$_3$ (10%, v/v), centrifuged (2000g), and the acid was removed from the supernatant by the method of Sharpes and McCarl (1982). EDTA was added to a final concentration of 20 mM, and the InsP$_4$ produced was separated from Ins(1,4,5)P$_3$ by anion-exchange chromatography on Bio-Rad AGI (200–400 mesh; formate form) with ammonium formate/formic acid eluents as described by Downes et al. (1986).

Communi et al. (1994, 1995) have assayed the high-molecular mass human platelet Ins(1,4,5)P_3 3-kinase isozyme as follows. InsP$_3$ 3-kinase assay was performed at 37°C for 8 min in 50 µl reaction mixture containing 84 mM HEPES/NaOH (pH 7.4), 5 mM ATP, 1 mg/ml BSA, 1 mM EGTA, 20 mM MgCl$_2$, 7.5 mM 2,3-bisphosphoglycerate, 12 mM 2-mercaptoethanol, 50 µg/ml Pefabloc, 5 µM leupeptin, 50 mM benzamidine, 10 µg/ml calpain inhibitors I and II, 1500 cpm of [^{3}H]InsP$_3$, 5 µM InsP$_3$ and 5–15 µl of enzyme. In assays performed in the presence of Ca^{2+}/CaM, 0.1 µM CaM was added as CaCl$_2$ to adjust the free Ca^{2+} concentration to 10 µM. The addition of calpain inhibitors during the purification procedure was critical for limiting proteolysis and stabilizing native enzyme (Takazawa et al., 1990a).

Donie and Reiser (1989) have described a quantitive binding assay based on a membrane preparation from porcine cerebellum which displays high-affinity binding sites for [^{3}H]Ins(1,3,4,5)P_4. The receptor site was specific for InsP$_4$, since Ins(1,3,4,5,6)P_5 and Ins(1,4,5,6)P_4 displaced binding of InsP$_4$ with EC50 values of 0.2 and 0.3 µM, respectively. Ins(1,4,5)P_3 and other InsPs were less effective. Using this InsP$_4$ receptor, an assay for measuring tissue content of InsP$_4$ was developed. The detection limit of the assay was 0.1 pmol.

Donie and Reiser (1989) prepared the binding protein for Ins(1,3,4,5)P_4 from freshly collected pig cerebellum. The tissue was homogenized in 10 vols buffer containing (mM) Tris–HCl, EDTA (Berridge and Irvine, 1984), and mercaptoethanol (Berridge and Irvine, 1984), pH 7.7, at 4°C. The homogenate was centrifuged for 30 min at 35 000g, the pellet resuspended and homogenized again, centrifuged (30 min, 35 000g), washed with buffer and centrifuged. The final pellet, resuspended in buffer at a concentration of 10–20 mg/ml was homogenized using a Potter-Elvejhem and kept at $-20°C$. A similar procedure was applied to obtain the binding protein for Ins(1,4,5)P$_3$ from fresh beef liver.

For radio-ligand-binding studies (Donie and Reiser, 1989), binding protein from cerebellum (0.3–1.0 mg per test tube) was added to a solution containing (final concentrations in mM), EDTA (1), bovine serum albumin (2.5 mg/ml), sodium acetate (25) and potassium phosphate (25), pH 5.0, [³H]Ins(1,3,4,5)P$_4$ (approx. 10 000 dpm = 84 fmol) as tracer to give a final concentration of 0.22 nM and, where indicated, unlabeled Ins(1,3,4,5)P$_4$, tissue sample or other reagents. Incubations were carried out on ice. A solution of sucrose (5%), sodium acetate (25 mM), pH 5.0 was carefully injected into the reaction vial to obtain a sucrose density gradient with the reaction mixture (400 μl) forming the upper phase. After 20 min the reaction vials were centrifuged (10 000g) in a cooled Heraeus biofuge A for 3 min, thus separating the membranes from unbound ligand. Finally, the pellet containing the bound label was quantitatively transferred to 5 ml scintillation vials by centrifugation for 5 min at 1500g and 3.5 ml scintillation liquid were added.

7.7.2.2. Ins(2,5,6)P$_3$ 3-kinase

Igaue et al. (1980, 1982) also noted the conversion of Ins(2,5,6)P$_3$ to Ins(2,3,5,6)P$_4$ in rice cell suspensions and assumed that this transformation took place via 3-kinase. Ins(2,3,5,6)P$_4$ was produced as the other one of the two isomers of InsP$_4$.

7.7.3. InsP$_3$ 6/3 kinases

7.7.3.1. Ins(1,4,5)P$_3$ 6/3-kinase

Ongusaha et al. (1998) have shown that the concentration of InsP$_6$ in S. pombe is regulated by stress, and that the Ins(1,4,5)P$_3$ 6-kinase of this yeast is a multi-functional kinase that phosphorylates Ins(1,4,5)P$_3$ to both Ins(1,4,5,6)P$_4$ and Ins(1,3,4,5)P$_4$, and also converts InsP$_4$, via Ins(1,3,4,5,6)P$_5$, into InsP$_6$. Addition of a single phosphate to Ins(1,4,5)P_3 could yield Ins(1,2,4,5)P_4, Ins(1,3,4,5)P_4 or Ins(1,4,5,6)P_4. It was confirmed that the initial kinase products were Ins(1,3,4,5)P_4 and Ins(1,4,5,6)P_4 by hydrolysis of a mixed [^{3}H]P_4 fraction made up of the purified kinase with ammonia and analyzed the liberated InsPs. Ins(1/3)P, Ins(4/6)P and Ins(5)P species were formed, but no Ins(2)P was detected, so the kinase could not have yielded Ins(1,2,4,5)P_4. Periodate oxidation, reduction and dephosphorylation converted the major [^{3}H]P_4 to [^{3}H]iditol, confirming that it was Ins(1,4,5,6)P_4. The identity of the minor InsP$_4$ product was thereby confirmed as Ins(1,3,4,5)P_4 (Ongusaha et al., 1998). Thus, the purified kinase phosphorylated Ins(1,4,5)P_3 on the 6- or the 3-position. The phosphorylation of the 6-position proceeded faster than that of the 3-position.

The dual-specificity Ins(1,4,5)P_3 6/3 kinase does not recognize Ins(1,4)P_2 or InsP$_3$ isomers other than Ins(1,4,5)P_3 and Ins(1,4,6)P_3. All identified substrates have at least one equatorial pair of *para*-phosphate groups [Ins(1,3,4,5,6)P_5 has two pairs] and most InsPs that did not affect its kinase activity lack this feature (Ongusaha et al., 1998). Ongusaha et al. (1998) have prepared energy-minimized views of its substrates and have made predictions about the relative kinase activity.

Ongusaha et al. (1998) assayed the Ins(1,4,5)P_3 6/3-kinase using [^{3}H]Ins(1,4,5)P_3 [1–5 nM; either 20 000 dpm (for analysis of products by Dowex mini-column) or 10 000 dpm (for analysis on HPLC anion-exchange) or other [^{3}H]labeled inositol polyphosphates as substrate, under first-order conditions, with 5 mM ATP as phosphate donor. ^{3}H-labeled assays were stopped with 10% (w/v) perchloric acid, left for 10 min on ice and centrifuged (15 000g, 5 min). After neutralizing with 20 mM HEPES/KOH (pH 7.2) and standing for 30 min on ice, precipitated salt was removed by centrifugation.

Alternatively, Ongusaha et al. (1998) conducted the kinase assays with unlabeled inositol phosphates and [γ-^{32}P]ATP (0.5 μM in 0.1 ml assays containing approximately 5×10 dpm [γ-^{32}P]ATP, purified by HPLC immediately before use). Yeast cells fractions or purified kinase preparations were incubated at pH 7.5 and 28°C in 20 mM HEPES/1 mM EGTA containing ATP, 10 mM creatine phosphate and 10 units/ml creatine kinase. The assays with $[\gamma^{-32}P]$ATP lacked creatine phosphate and creatine kinase, and incubations (for 2 h at 28°C) were terminated with 1 ml ice-cold water and deproteinized on an NAP10 desalting column (Pharmacia). The eluates from the $[^{32}P]$ATP assays were spiked with $[^3H]$Ins(1,3,4,5)P$_4$ and $[^3H]$Ins(1,4,5,6)P$_4$ and analyzed by HPLC on a Partisil 10 SAX column, using the pH 3.8 $(NH_4)_2PO_4$ gradient described below. This gradient separates the InsP$_4$ isomers from closely eluting contaminants in the $[\gamma^{-32}P]$ATP better than the pH 4.4 gradient used in other experiments.

Ongusaha et al. (1998) separated the inositol phosphate classes on a 6×25 mm AG-1 (x8, 200–400 mesh, formate) columns. Samples were diluted 10-fold before loading. They were eluted with 0.18 M ammonium formate (AF)/0.1 M formic acid (FA) (for InsPs species); 0.4 M AF/0.1 M FA (InsP$_2$s); 0.8 M AF/0.1 M FA (Ins P$_3$s); 1.2 M AF/0.1 M FA (InsP$_4$s) and 2 MAF/0.1 M FA (InsP$_5$s + InsP$_6$). Each fraction was mixed with 3 ml of 5 M AF and 10 ml of scintillant radioactivity was measured.

Before individual InsPs species were separated by anion-exchange HPLC, samples were neutralized with 20 mM HEPES (pH 7.4). A 235 × 4.6 mm Partisphere 5-SAX column was eluted with a phosphate gradient at 1 ml/min. Buffer A was water and Buffer B 1.25 M diammonium orthophosphate, pH 4.4 (or pH 3.8). The gradient was: 0–15 min, 0% B; 46 min, 6% B; 50 min, 15% B; 90 min, 22%; B, 95 min, 38% B; 150 min, 46% B; 180–189 min, 100% B; 190–200 min, 0% B. The eluate was either collected (0.25 ml/fraction) or continuously mixed with scintillant (3 ml/min) and fed into an on-line scintillation detector (0.1 min averaging periods).

Eckmann et al. (1997) have reported data that show the infection of intestinal epithelial cells with *Salmonella*, but not other invasive bacteria, induces Ins(1,4,5,6)P$_4$ production. This InsP isomer has a function in epithelial cells by antagonizing signaling through PtdIns 3-kinase pathway.

For the identification of enantiomeric Ins(1,4,5,6)P$_4$, Eckmann et al. (1997) fractionated cell extracts by HPLC using Adsorbosphere SAX column. The [^{3}H]Ins(3,4,5,6)P$_4$/[^{3}H]Ins(1,4,5,6)P$_4$ peak was separated from all other [^{3}H]InsP$_4$ isomers, desalted, and then incubated with partially purified Ins(1,4,5,6)P$_4$ 3-kinase with an internal standard of [^{32}P]Ins(1,4,5,6)P$_4$ (Vajanaphanich et al., 1994). The enzyme preparation did not phosphorylate Ins(3,4,5,6)P$_4$. Equal amounts of ^{3}H and ^{32}P radioactivity were added to the reaction, and the ratio of [^{3}H]Ins(3,4,5,6)P$_4$/[^{3}H]Ins(1,4,5,6)P$_4$ in the original peak was determined by comparing the relative amounts of ^{3}H- and ^{32}P-labeled Ins(1,3,4,5,6)P$_5$ formed.

7.7.4. InsP$_3$ 6/5 kinase

7.7.4.1. Ins(1,3,4)P$_3$ 6/5-kinase

Ins(1,3,4)P$_3$ is at a branch point of InsP metabolism. It is dephosphorylated by specific phosphatases to either Ins(3,4)P$_2$ or Ins(1,3)P$_2$ (see Chapter 8). Hansen et al. (1988) and Abdullah et al. (1992) have shown that it is phosphorylated to Ins(1,3,4,5)P$_4$ by Ins(1,3,4)P$_3$ 5/6 kinase. Ins(1,3,4,6)P$_4$ is the first intermediate in the pathway leading to the higher InsPs including other InsP$_4$, InsP$_5$, InsP$_6$, and PP-InsP forms of these (Menniti et al., 1993a,b). All of these InsPs are ubiquitously found in tissues.

Ins(1,3,4)P$_3$ 5/6 kinase has been partially purified from rat liver (Hansen et al., 1988; Abdullah et al., 1992), porcine brain and bovine testes (Hughes et al., 1994) and in each case were reported to phosphorylate Ins(1,3,4)P$_3$ on either the 5 or 6 position yielding a mixture of two products, although the 5 and 6 positions of *myo*-inositol are on opposite faces of the ring. Abdullah et al. (1992), Hughes et al. (1994), Wilson and Majerus (1996) and Wilson et al. (2001) have described a mammalian kinase that phosphorylates Ins(1,3,4)P$_3$ to Ins(1,3,4,6)P$_4$ and Ins(1,3,4,5)P$_4$ in a 6:1 ratio. Specifically, Wilson and Majerus (1996) showed that a purified calf brain and recombinant proteins produced both $Ins(1,3,4,6)P_4$ and $Ins(1,3,4,5)P_4$ as products in a ratio of 2.3–5:1. This finding suggested that a single kinase phosphorylated inositol in both the D5 and D6 positions.

A related *Arabidopsis* $Ins(1,3,4)P_3$ kinase makes $Ins(1,3,4,6)P_4$ and $Ins(1,3,4,5)P_4$ in a 1:3 ratio, in contrast with the human enzyme, which gives a product ratio of 3:1 (Wilson and Majerus, 1997). According to Wilson and Majerus (1997) there was no evidence of phosphorylation of $Ins(1,4,5)P_3$, $Ins(1,4)P_2$, $Ins(1)P$, $Ins(2)P$, $Ins(3)P$ or $cIns(1,2)P$ by the recombinant *Arabidopsis* $Ins(1,3,4)P_3$ 5/6-kinase.

Prior to this time, Shears (1989a,b) had demonstrated $Ins(1,3,4)P_3$ phosphorylation in liver. Two $InsP_4$ products and an $InsP_5$ were detected. The $InsP_4$ isomers were unequivocally identified as $Ins(1,3,4,5)P_4$ and $Ins(1,3,4,6)P_4$ by HPLC of InsPs, periodate oxidation, alkaline hydrolysis, and stereospecific polyol dehydrogenase.

Wilson and Majerus (1996) assayed the $Ins(1,3,4)P_3$ 5/6-kinase activity using a procedure modified from Hansen et al. (1988).

Wilson and Majerus (1996) identified the products of $Ins(1,3,4)P_3$. The assay was conducted at 37°C in 25 μl containing 20 mM HEPES, pH 7.2, 5 mM ATP, 6 mM $MgCl_2$, 10 mM $LiCl_3$, 100 mM KCl, 10 mM phosphocreatine, 10 units phosphocreatine kinase, 1 mM DTT, and 0.05 pmol $[^3H]Ins(1,3,4)P_3$. Reactions were stopped by the addition of 1 ml of water, and the samples were loaded onto a 400 μl Dowex-formate column equilibrated in water. Each column was washed with 8 ml of 0.8 M ammonium formate, pH 3.5 to elute the substrate. The products were eluted with 2 ml of 1.6 M ammonium formate and was counted in a liquid scintillation counter. Activity was expressed as a first-order rate constant using the equation $[S] = [S]_0 e^{-kt}$, as described previously (Caldwell et al., 1991).

Wilson and Majerus (1996) identified the products of Ins(1,3,4)P$_3$ phosphorylation by 5/6-kinase as follows. The samples were applied to a 250 × 4.6 mm Adsorbosphere SAX column (Alltech/Applied Science) and eluted with a linear gradient of 0–1.0 M ammonium phosphate, pH 3.5, over 200 min at a flow rate of 0.6 ml/min, using a modification of the method of Hughes et al. (1989). Fractions (480 μl) were collected, and radioactivity was measured using a liquid scintillation counter. An internal standard of [^{32}P]Ins(1,3,4,5)P$_4$ was prepared by phosphorylating [^{32}P]Ins(1,4,5)P$_3$ with recombinant Ins(1,4,5)P$_3$ 3-kinase from *E. coli* as described by Wilson and Majerus (1996). The minor product co-chromatographed with the internal standard of [^{32}P]Ins(1,3,4,5)P$_4$ and the major product eluted just before this in a position previously shown for Ins(1,3,4,6)P$_4$ (Shears, 1989a,b).

Field et al. (2000) have shown that amoebae have an Ins(1,3,4)P$_3$ 5/6-kinase, which also has a Ins(1,4,5)P$_3$ 3-kinase activity.

7.7.5. Ins(1,3,5)P$_3$ 6-kinase

According to Igaue et al. (1980, 1982) Ins(1,3,5)P$_3$ becomes phosphorylated presumably by a 6-kinase to yield Ins(1,3,5,6)P$_4$ as an intermediate in the biosynthesis of phytic acid in rice cell suspensions. Ins(1,3,5,6)P$_4$ was produced as one of the two isomers of InsP$_4$.

7.8. Inositol tetrakisphosphate kinases

InsP$_4$s serve as intermediates in the biosynthesis of phytic acid in rice cell culture suspensions. The InsP$_4$s are formed by one or more kinases.

7.8.1. InsP$_4$ 1-kinases

7.8.1.1. Ins(3,4,5,6)P$_4$ 1-kinase

Ins(3,4,5,6)P$_4$ has been suggested to provide an intracellular signal that regulates calcium-dependent chloride conductance. Tan et al. (1997) have studied the inactivation of $Ins(3,4,5,6)P_4$ signal by the 1-kinase that phosphorylates it. The enzyme was purified from rat liver 1600-fold with a 1% yield. The native molecular mass was found to be 46 kDa by gel filtration. $Ins(1,3,4)P_3$ was shown to be a potent, specific, and competitive inhibitor of 1-kinase.

Tan et al. (1997) assayed $Ins(3,4,5,6)P_4$ 1-kinase as follows. During purification, $10-40$ μl samples were incubated at 37°C for $10-30$ min in a final volume of 100 μl containing 4000 dpm of $[^3H]Ins(3,4,5,6)P_4$ (adjusted to a concentration of 5 μM with non-radioactive substrate), 20 mM HEPES (pH 7.2), 6 mM $MgSO_4$, 0.4 mg/ml saponin, 100 mM KCl, 0.3 mg/ml bovine serum albumin, 2 μM $InsP_6$, 5 mM ATP, 10 mM phosphocreatine, 2.5 Sigma units of phosphocreatine kinase. The reaction mixture was quenched with 1 ml of ice-cold medium containing 1 mg/ml $InsP_6$, 0.2 M ammonium formate, 0.1 M formic acid. The quenched reactions were diluted to 10 ml with water and chromatographed on Bio-Rad gravity-fed columns using AG 1-X8 ion-exchange resin. The final preparations of enzyme were assayed using a quench medium that did not contain $InsP_6$. In some experiments, incubations were quenched with 40 μl of 2 M perchloric acid plus 1 mg/ml $InsP_6$, neutralized with Freon/octylamine, and chromatographed on HPLC using Partisphere SAX column (Vajanaphanich et al., 1994).

Yang et al. (1999) and Yang and Shears (2000) have shown that $Ins(1,3,4)P_3$ inhibited $Ins(3,4,5,6)P_4$ 1-kinase activity that was either in lysates of AR4-2J pancreatoma cells or in a preparation purified 22 500-fold from bovine aorta. The 1-kinase was also inhibited relatively potently by the non-physiological trisphosphates, D/L $Ins(1,2,4)P_3$, while metabolites of $Ins(1,3,4)P_3$ such as $Ins(1,3)P_2$ and $Ins(3,4)P_2$ were > 100-fold weaker inhibitors of the 1-kinase compared to $Ins(1,3,4)P_3$-mediated inhibition of the 1-kinase.

7.8.2. $InsP_4$ 2/4 kinases

7.8.2.1. $Ins(1,3,5,6)P_4$ 2/4-kinase

According to Igaue et al. (1980, 1982) the $Ins(1,3,5,6)P_4$ is converted to $Ins(1,2,3,5,6)P_5$ and $Ins(1,3,4,5,6)P_5$ by a 2/4-kinase in rice cell suspensions. Either one or two separate kinases may be involved in this phosphorylation. To complete the conversion to InsP6, Biswas et al. (1978) and Chakrabarti and Biswas (1981) proposed the action of a phytic acid-adenosine diphosphate phosphotransferase that was identified in germinating mung beans (Biswas et al., 1978).

7.8.3. $InsP_4$ 3-kinases

Stephens et al. (1988) demonstrated formation of Ins $(1,3,4,5,6)P_5$ from Ins $(1,3,4,5,6)P_4$ by 3-kinase from chicken erythrocytes. Craxton et al. (1994) showed that Ins $(1,4,5,6)P_4$ is phosphorylated directly by 3-kinase, and that Ins $(3,4,5,6)P_4$ is not an obligatory intermediate, as previously claimed.

7.8.4. $InsP_4$ 4-kinases

7.8.4.1. $Ins(2,3,5,6)P_4$ 4-kinase

This kinase is believed to catalyze the phosphorylation of $Ins(2,3,5,6)P_4$ to $Ins(2,3,4,5,6)P_5$ as a semifinal step in the alternative route of biosynthesis of phytic acid in rice cell culture suspensions (Igaue et al., 1980, 1982).

7.8.5. $InsP_4$ 5-kinases

7.8.5.1. $Ins(1,3,4,6)P_4$ 5-kinase

Stephens and Irvine (1990) have proposed the phosphorylation of $Ins(1,3,4,6)P_4$ to $Ins(1,3,4,5,6)P_5$ by a 5-kinase as the penultimate step in the biosynthesis of phytic acid in rice cell cultures.

7.9. Inositol pentakisphosphate kinases

There are several kinases that convert $InsP_5$ to $InsP_6$ (phytic acid). Phytic acid is the most abundant inositol phosphate in nature and presumably has a variety of functions, which have remained largely unknown. Stephens and Irvine (1990) have reported that $InsP_6$ synthesis in the cellular slime mold *Dictyostelium* is catalyzed by a series of soluble ATP-dependent kinases independently of the mechanism of both PtdIns and $Ins(1,4,5)P_3$. The intermediates between *myo*-inositol and $InsP_6$ are $Ins(3)P$, $Ins(3,6)P_2$, $Ins(3,4,6)P_3$, $Ins(1,3,4,6)P_4$ and $Ins(1,3,4,5,6)P_5$. The 3- and 5-phosphates of $InsP_6$ take part in apparently futile cycles in which $Ins(1,2,4,5,6)P_5$ and $Ins(1,2,3,4,6)P_5$ are rapidly formed by dephosphorylation of $InsP_6$, followed by rephosphorylation to yield again $InsP_6$.

7.9.1. $InsP_5$ 1-kinases

7.9.1.1. $Ins(2,3,4,5,6)P_5$ 1-kinase
Likewise, there is a possibility that $Ins(2,3,4,5,6)P_5$ may be phosphorylated to $InsP_6$ (phytic acid) via 1-kinase (Igaue et al., 1980, 1982; Stephens et al., 1991). Stevens et al. (1991) assayed $InsP_5$ kinases as follows: Assays were $200-300$ μl total volume and contained 0.1 M KCl, 10 mM phosphocreatine, 1 mM dithiothreitol, 5 units of creatine kinase (Sigma)/ml, 5 mM ATP, 6 mM $MgCl_2$, 2 mM EGTA, 25 or 50 mM HEPES, 0.3 mM $CaCl_2$ (pH 7.0; 25°C for *Dictyostelium*-derived and mung bean derived kinases; 37°C for rat brain-derived kinases). It should be noted that D-$Ins(2,3,4,5,6)P_5$ and D-$Ins(1,2,3,5,6)P_5$ are referred to by their alternative names, L-$Ins(1,2,4,5,6)P_5$ and L-$Ins(1,2,3,4,5)P_5$, respectively. Assays were quenched with 4.5% $HClO_4$ and spiked with phytate hydrolysate (10 μl of a solution containing 1.2 mg of total phosphorus/ml) and $500-1000$ dpm of $^{32}PInsP_6$. The acid-precipitated protein was pelleted by centrifugation, and the

supernatants were removed and mixed with 2 M KOH/0.1 m Mes/ 25 mM EDTA (200 μl). The final pH of the samples was 6.3–6.8/ KClO$_4$ was sedimented by centrifugation and the supernatants were applied to an anion-exchange HPLC column or to batch-eluted columns (Chapter 4).

7.9.2. InsP$_5$ 2-kinases

7.9.2.1. Ins(1,3,4,5,6)P$_5$ 2-kinase

The possibility that Ins(1,3,4,5,6)P$_5$ is converted to InsP$_6$ (phytic acid) by a 2-kinase was expressed by Chakrabarti and Biswas (1981), but it remained to be demonstrated experimentally. On the basis of the specific activities of [^{3}H]Ins-labeled intact *Dyctioste-lium*, Stephens and Irvine (1990) suggested that the main precursor in the de novo synthesis of InsP$_6$ both in vivo and in vitro is Ins(1,3,4,5,6)P$_5$. Other InsP$_5$ were either unlabeled or insufficiently labeled to serve as precursors of InsP$_6$.

Stephens and Irvine (1990) used anion-exchange HPLC to assay of InsP$_5$ 2-kinase and to separate the ^{3}H-labeled products formed in a *Dictyostelium* homogenate in the presence of [^{3}H]Ins and an ATP-generating system. Aliquots (150 μl) of the homogenate were incubated in 200 μl assays containing (final composition) 100 mM KCl, 5 mM ATP, 6 mM MgCl$_2$, 25 mM HEPES, 1 mM EGTA, 10 mM creatine phosphate, 5 units/ml creatine phosphokinase, 1 mM DTT, 1 mg/ml BSA (pH 7.0, 25°C) and 100 μCi [^{3}H]Ins (Amersham). After 50 min at 25°C, the assay was quenched with 600 μl 3.16% (v/v) perchloric acid and centrifuged at 15 000 rpm (2 min), the supernatant was removed and added immediately to 200 μl 2 M KOH, 0.1 M MES and 50 μl 0.1 M EDTA (pH 6.0, 25°C). The resulting solution was confirmed to be pH 6.0 ^{32}P- and ^{14}C-labeled standards were then added. The InsP$_4$s, in order of increasing retention time, were Ins(1,3,4,5)P$_4$ and Ins(1,3,4,6)P$_4$. The sample was filtered, applied to a Partisphere SAX anion-exchange column and eluted with a Na$_2$PO$_4$ gradient (pH 3.75) (Stephens, 1990). Fractions were collected every 30 s and, except for those containing $Ins(1,2,4,5,6)P_5$ and $Ins(1,2,3,4,6)P_5$ were pooled, desalted and reapplied to a Partisphere WAX HPLC column, eluted with 0.35 M $(NH_4)_2HPO_4$ (pH 3.2 with H_3PO_4) and the resolved isomers finally quantified by liquid scintillation counting. $[^3H]Ins$-labeled peaks were identified as follows: first, $[^3H]Ins(3)P$ co-migrated with D/L-$Ins(1)P$ but not $Ins(2)P$, $Ins(4)P$ or $Ins(5)P$ during anion-exchange chromatography (Stephens et al., 1988), yielded $[^3H]glucitol$ upon partial periodate-oxidation, reduction and dephosphorylation (Stephens, 1990) and served as a substrate for a soluble $Ins(3)$ kinase activity isolated from *Dictyostelium* homogenates; second, $[^3H]Ins(3,6)P_2$ co-migrated with $Ins(1,4)P_2$ (and not $Ins(4,5)P_2$), lost all of its ^3H on complete periodate oxidation and yielded $[^3H]L$-iditol and $[^3H]D$-altritol upon partial periodate-oxidation, reduction and dephosphorylation (Stephens, 1990); third, $[^3H]Ins(3,4,6)P_3$ eluted from a Partisphere SAX anion-exchange HPLC column significantly later than $Ins(1,4,5)P_3$ and yielded $[^3H]L$-iditol on periodate-oxidation, reduction and dephosphorylation; fourth, $[^3H]Ins(1,3,4,6)P_4$ co-migrated with $[^{32}P]Ins(1,3,4,6)P_4$ but not with $Ins(1,3,4,5)P_4$, $Ins(1,2,4,5)P_4$ or $Ins(2,3,4,6)P_4$ ($[^3H]Ins(2,3,4,6)P_4$ eluted substantially later than $Ins(1,3,4,6)P_4$ from standard Partisphere SAX-anion-exchange HPLC columns); fifth, $[^3H]Ins(1,3,4,5,6)P_5$ co-migrated with $[^{32}P]Ins(1,3,4,5,6)P_5$ on both Partisphere SAX and WAX anion-exchange HPLC columns, but not with $D/L[^{32}P]Ins$ $(1,2,4,5,6)P_5$, $D/L[^{32}P]Ins(1,2,3,4,5)P_5$ or $[^{32}P]Ins(1,2,3,4,5,6)P_5$; sixth, $[^3H]Ins(1,2,3,4,6)P_5$ ran appropriately relative to internal standards in the HPLC protocol described above; seventh, $[^3H]Ins(1,2,4,5,6)P_5$ ran appropriately relative to internal standards in the HPLC protocol described above and when enzymatically dephosphorylated yielded $[^3H]Ins(1,2,6)P_3$ and $[^3H]Ins(1,2,5,6)P_4$ (as opposed to $[^3H]Ins(2,3,4)P_3$ or $[^3H]Ins(2,3,4,5)P_4$).

Stephens and Irvine (1990) have shown that the rat brain also contains $InsP_5$ hydroxy-kinase activities capable of phosphorylating $Ins(1,2,4,5,6)P_5$, $Ins(1,2,3,4,6)P_5$ and $Ins(1,3,4,5,6)P_5$, although the rates of the reactions are two to three orders of magnitude lower than in *Dictyostelium*.

According to Stephens and Irvine (1990) 2-kinase is one of the kinases that completes the phosphorylation of Ins(1,3,4,5,6)P_5 to InsP$_6$ or phytic acid in *Dictyostelium*. Stephens et al. (1991) and Ji et al. (1989) have published studies with animal cells, which also contained an Ins(1,3,4,5,6)P_5 2-kinase and this observation has been confirmed by results in AR42J homogenates (Shears, 1998a,b). In cell extracts, the 2-kinase activity is extremely slow, which may help explain why others have not reproduced these results (Rudolf et al., 1997). There are many other InsP$_5$ kinase activities: Ins(2,3,4,5,6)P_5 1-kinase (Stephens et al., 1991), Ins(1,2,4,5,6)P_5 3-kinase (Craxton et al., 1994; Stephens et al., 1991; Rudolf et al., 1997) and D/L-Ins(1,2,3,4,6)P_5 4/6 kinase (Stephens et al., 1991). Shears (1998a,b) points out that theoretically, any of these InsP$_5$ isomers could serve as a precursor of InsP$_6$, except that there is no evidence in animal systems for the existence of the necessary InsP$_4$ kinases that could synthesize these alternative InsP$_5$ isomers. Only Ins(1,3,4,5,6)P_5 is an established metabolite that could serve as a precursor of InsP$_6$. All the others are products of multiple InsPs phosphatases.

Ongusaha et al. (1998) have shown that purified *S. pombe* kinase converts Ins(1,3,4,5,6)P_5 to InsP$_6$ via the 2-kinase activity.

Ives et al. (2000) have shown that purified recombinant scIPK1 kinase (*S. cerevisiae* InsP$_5$ K1) activity is highly selective for InsP$_5$ substrate and exhibits apparent Km values 644 nM and 62.8 μM for InsP$_5$ and ATP, respectively. Cells lacking scIPK1 are deficient in InsP$_6$ production and exhibit lethality in combination with a gle1 mutant allele. Ives et al. (2000) have obtained data that suggest that the mechanism for InsP$_6$ production is conserved across species.

The enzymes responsible for the production of InsP$_6$ in vertebrate cells are not known. Based on limited conserved sequence motifs among five yeast Ipk1 proteins from different fungal species, Verbsky et al. (2002) have identified a human genomic DNA sequence on chromosome 9 that encodes human Ins(1,3,4,5,6)P_5 2-kinase. Recombinant human enzyme was produced in Sf21 cells, purified, and shown to catalyze the synthesis of $InsP_6$ in vitro.

7.9.3. $InsP_5$ 3-kinases

7.9.3.1. $Ins(1,2,4,5,6)P_5$ 3-kinase

Stephens et al. (1991) detected multiple ATP-dependent myo-$InsP_5$ kinase activities in slime mold, rat brain and in mung-bean seedling homogenates, including Ins $(1,2,4,5,6)P_5$ 3-kinase.

7.9.4. $InsP_5$ 5/6 kinases

7.9.4.1. D/L-$Ins(1,2,3,4,6)P_5$ 5/6-kinase (Stephens et al., 1991)

Stephens et al. (1991) also demonstrated the phosphorylation of D/L-Ins $(1,2,3,4,5)P_5$, $Ins(1,2,3,4,6)P_5$ and $Ins(1,3,4,5,6)P_5$ to $InsP_6$.

The substrates and products were resolved by HPLC anion exchange techniques (Stephens, 1990)

7.9.5. $InsP_5$ pyrophosphokinase

7.9.5.1. $Ins(1,3,4,5,6)P_5$ pyrophosphokinase

Mammalian cells phosphorylate $Ins(1,3,4,5,6)P_5$ to PP-$InsP_4$ (Menniti et al., 1993a,b) (Fig. 1.3, Chapter 1).

7.10. Inositol hexakisphosphate kinases

Interest in $InsP_6$ kinases has increased following the discovery that $InsP_6$ is further phosphorylated to a diphosphoinositol pentakisphosphate (PP-$InsP_5$) (Menniti et al., 1993a,b; Stephens et al., 1993), a widely distributed cellular constituent of organisms as phylogenetically diverse as *D. discoideum* (Stephens et al., 1993) and rat (Menniti et al., 1993a,b). Shears et al. (1995) have elucidated the pathway of synthesis and metabolism of PP-$InsP_4$-PP by HPLC using [3H]- and [32P]-labeled substrates. Metabolites were identified by using two purified phosphatases in a structurally diagnostic manner, e.g. tobacco 'pyrophosphatase' (Shinshi et al., 1976) and rat hepatic multiple inositol polyphosphate phosphatase (Craxton et al., 1995). In liver homogenates, $InsP_6$ was phosphorylated first to a $PP-InsP_5$ and then to $PP-InsP_4-PP$. These kinase reactions were reversed by phosphatases, establishing two coupled substrate cycles.

The high energy diphosphoinositol polyphosphates ($PPInsP_5$ and $[PP]_2InsP_4$) in mammalian cells are considered candidate molecular switches (Safrany et al., 1999; Voglemaier et al., 1996). The molecules are rapidly metabolized through substrate cycles that are regulated by cAMP, cGMP and Ca^{2+}. Saiardi et al. (2000) have now identified these compounds in yeast.

7.10.1. InsP₆ 5-kinase

The $InsP_6$ kinase, which in mammals adds a phosphate at the 5-position (Albert et al., 1997) and the $PP-InsP_5$ kinase (the positional specificity is unknown at this point), have both been purified and studied (Voglmaier et al., 1996). Saiardi et al. (1999) have reported the cloning of two mammalian $InsP_6$ kinases and a yeast $InsP_6$ kinase. The authors also show that ArgRIII, is an inositol-polyphosphate kinase that can convert $InsP_3$ to $InsP_4$, $InsP_5$ and $InsP_6$. The newly identified family of highly conserved inositol-polyphosphate kinases contains a unique consensus sequence.

It has been proposed that the high-energy pyrophosphates might participate in protein phosphorylation. Saiardi et al. (1999) have purified $InsP_6$ kinase and $PP-InsP_5$ kinase, both of which display ATP synthase activity, transferring phosphate to ADP.

Shears et al. (1995) investigated the InsPs metabolism using trace amounts of radiolabeled substrate at 37°C in 0.5–1 ml aliquots of assay buffer 1 (100 mM KCl, 7 mM $MgSO_4$, 5 mM ATP, 10 mM phosphocreatine, 10 mM NaF, 10 mM HEPES (pH 7.2 with KOH), 1 mM Na_2EDTA, 1 mM dithiothreitol, and

0.05 mg/ml phosphocreatine kinase). The reactions were initiated by the addition of aliquots of appropriately diluted tissue and quenched with 4 ml of ice-cold medium comprising 0.1 M Tris (pH 7.7 with HCl at 25°C), 10 mM triethylamine, 2 mM Na_2EDTA. Protein was subsequently removed by a deproteonizing cartridge, which was primed with 10 ml of methanol, followed by 10 ml of quench buffer. The quenched reaction was added to the column, which was then washed with 2 × 4 ml aliquots of quench medium. The flow rate was 3 ml/min. Recoveries of InsPs exceeded 95% in the flow through from the column, which was analyzed by HPLC either immediately or after storage at $-20°C$.

Shears et al. (1995) loaded the deproteinized samples onto Partisphere 5 μm SAX HPLC column at a flow rate of 1 ml/min using WISP model 712 (Waters Associates), which injected each sample in aliquots interspersed with HPLC Buffer A (1 M Na_2EDTA) in a ratio of sample to Buffer A of approximately 1:10. When required, fractions of 2.5 ml were collected during this loading procedure. A gradient was then generated from a mixture of Buffer A (1 mM Na_2EDTA, disodium salt) and Buffer B (Buffer A plus 1.3 M $(NH_4)_2HPO_4$ (pH 3.8) with H_3PO_4) at a flow rate of 1 ml/min (unless otherwise indicated): 0–2 min, B = 0%, 2–10 min, B increased linearly to 44%; 10–100 min, B increased linearly to 100% and was then retained at this level until 101 min (for older columns) or 110 min (new columns). Finally, B then returned to 0%. A slightly different gradient was used to resolve inositol polyphosphates in extracts of intact cells: 0–10 min, B = 0%; 10–30 min, B increased linearly to 40%; 30–130 min, B increased linearly to 75%; 130–131 min, B increased linearly to 100%; 131–157 min, B = 100%; 157 min, B returned to 0%. Radioactivity was quantified by liquid scintillation spectrometry after mixing with 4 volumes of Monoflow 4 scintillant.

Schell et al. (1999) have cloned a cDNA which has the ability to stimulate inorganic phosphate uptake in *Xenopus* oocytes (phosphate uptake stimulator, PiUS), which shows significant similarity to inositol 1,4,5-trisphosphate 3-kinase. However, the expressed PiUS protein showed no detectable activity against Ins(1,4,5)P_3, nor the Ins(1,3,4,5)P_4 or Ins(3,4,5,6)P_4-isomers of InsP_4, whereas it was very active in converting InsP_6 to InsP_7.

Saiardi et al. (1999) have reported that the ARGIII gene of *S. cerevisiae* encodes a transcriptional regulator that also has inositol-polyphosphate multikinase (IPMK) activity. Saiardi et al. (2000) have investigated how InsPs regulate gene expression by disrupting the ARGIII gene. This mutation impaired nuclear mRNA export, slowed cell growth, increased cellular InsP_3 170-fold and decreased InsP_6 100-fold, indicating reduced phosphorylation of InsP_3 to InsP_6. Levels of PP-InsPs were decreased much less dramatically than was InsP_6. Low levels of InsP_6 and considerable quantities of Ins(1,3,4,5)P_4, were synthesized by an ipmk-independent route.

Morrison et al. (2001) have provided a detail description of the determination of PP-IP$_5$ formation using InsP_6 as a substrate in the presence of ATP. The InsP_6 kinase activity (Voglemaier et al., 1996) was determined in whole cell extracts of 107 NIH-OVCAR-3 cells by Dounce homogenization on ice in 100 ml KCl, 20 mM NaCl, 1 mM EGTA, 20 mM HEPES, pH 7.4, and adjusted to 1.5 mg of protein/ml. Using 0.15-ml extracts, kinase assays were run in 0.5 ml of 100 mM KCl, 25 mM HEPES, pH 7.2, 5 mM Na$_2$ATP, 6 mM MgSO$_4$, 10 mM phosphocreatine, 20 units creatine phosphokinase, 0.1 mg saponin using [^{3}H]P_6 as substrate (Perkin Elmer Life Sciences). Reactions were run at 37°C for 20 min, and were terminated on ice and then quenched with 0.25 ml of 6% v/v perchloric acid and 0.5 mg/ml InsP_6 (Calbiochem) followed by extraction with 1:1 Freon/octylamine and concentration by Speedvac. Products were separated using polyethyleneimine cellulose TLC (PEI-LC; Meck). The reaction mixture was spotted onto PEI-TLC plates and developed in 1.1 M KH$_2$PO$_4$, 0.8 M K$_2$PO$_4$ and 2.3 M HCl. Lanes were divided into 1 cm fractions, and the PEI cellulose matrix was scraped from the TLC plates, shaken with 0.5 ml of 16 M HCl, mixed with 0.5 ml of H$_2$O, 3 ml of scintillant, and counted. Approximately 80% of applied ^{3}H was recovered by this method. $InsP_6$ and $PP-InsP_5$ migrated with an R_f value of 0.75 and 0.45, respectively, and comigrated with standard preparations. $[^3H]PP-InsP_5$ standard was prepared by incubation of 20 ng of recombinant $InsP_6$ K2 and $[^3H]InsP_6$ in a kinase reaction as above (60 min incubation), followed by HPLC purification on a 4.6 × 125 mm Partisphere strong anion-exchange column (Whatman). Gradient elution utilized Buffer A (1 M Na_2EDTA) and Buffer B (Buffer A and 1.3 M $(NH_4)_2HPO_4$).

7.10.2. PP-InsP4 kinase

Shears et al. (1995) assayed the PP-InsPs for the presence of pyrophosphate esters by tobacco 'pyrophosphatase' using a modification of an earlier procedure (Menniti et al., 1993a,b). Trace amounts of ^3H-labeled substrate were incubated at 37°C with 100–300 μl of medium containing 30 mM sodium acetate buffer (pH 5.5), 10 mM Na_2EDTA, 2 M dithiothreitol, 0.01% (v/v) Triton X-100, 0.4 mg/ml bovine serum albumin. Immediately prior to use, the stock solution of pyrophosphatase was diluted 1:1 with 50 mM Na_2EDTA (pH 7.0) and added to the incubation buffer to a final concentration of 150–170 Sigma units/ml. After 16–22 h, incubations were quenched with 4 volumes of ice-cold water, followed by 2 volumes of 0.2 ml of 2 M perchloric acid plus 1 mg/ml $InsP_6$. Samples were neutralized with freon-octylamine. There was no hydrolysis of substrate in control samples to which enzyme was not added.

Saiardi et al. (2000) have resolved the inositol polyphosphates on two complementary HPLC systems (Glennon and Shears, 1993; Balla et al., 1991). In one system, a Partisphere SAX HPLC column (Krackler Scientific, NC, USA) was eluted with a gradient generated by mixing Buffer A (1 mM Na_2EDTA and buffer B (buffer A plus 1.3 M $(NH_4)_2HPO_4$, pH 3.85, with HPO_4) as follows: 0–5 min, 0% B; 5–10 min, 0–50% B; 10–60 min, 50–100% B; 60–70 min, 100% B. (ii) A SynChropak Q100 SAX HPLC column (Thompson Instrument, VA, USA) was eluted with a gradient generated by mixing buffer A (1 mM Na_2EDTA) with Buffer B (Buffer A plus 2 M $(NH_4)H_2PO_4$, pH 3.35, with H_3PO_4): 0–5 min, 0% B; 5–120 min, 0–65% B. InsPs were identified by their co-elution with the following authentic standards: [3H]Ins(1,4,5)P$_3$, [3H]Ins (1,3,4,5)P$_4$ and InsP$_6$. [3H]Ins(1,3,4,5,6) P$_5$ (Stephens and Downes, 1990), [3H]Ins(1,2,4,5,6)P$_5$ (Stephens et al., 1991), [3H[PP-InsP$_5$ and [3H][PP]$_2$-InsP$_4$ (Safrany et al., 1999) and 14C-labeled Ins(1,3,4,5)P$_4$, Ins(1,3,4,6)P$_4$, D/L-Ins(1,4,5,6)P$_4$ and Ins(1,3,4,5,6) P$_5$ isolated from [14C]inositol-labeled, parotid glands (Hughes et al., 1989).

7.10.3. bis-PP-InsP$_4$

Huang et al. (1998) have purified a PP-InsP$_5$ kinase from rat brain which uses PP-InsP$_5$ as a substrate to form bis-PP-InsP$_4$. The purified protein is a 56 kDa monomer. The enzyme activity was assayed in 10 μl of reaction mixture containing 20 mM HEPES (pH 6.8), 1 mM DTT, 6 mM $MgCl_2$, 5 mM Na_2ATP, 10 mM phosphocreatine, 40 units/ml creatine phosphokinase, 5 mM NaF, 5 μM PP-InsP$_5$, and 10–20 nM [3H]PP-InsP$_5$ and incubated at 37°C for 10–30 min, under zero-order kinetics. Reaction was stopped by addition of 1 μl of 1 M HC1. The reaction products were separated using polyethyleneimine-cellulose TLC using InsP$_6$, PP-InsP$_5$, bis-PP-InsP$_4$ and ATP as standards (Spencer et al., 1990).

CHAPTER 8

Phosphatidylinositol phosphate and inositol phosphate phosphatases

8.1. Introduction

In addition to a host of inositol and PtdIns kinases that form many signaling molecules, there are many phosphatases that degrade the signaling molecules that act in the system. In contrast to the kinases, the PtdInsP phosphatases are less well characterized, and fall into a number of phosphatase classes with little similarity to each other. The PtdInsP phosphatases identified from a number of cellular sources fall into four major categories, hydrolyzing the 1-, 3-, 4- or 5-phosphorylated InsPs or PtdInsPs (Woscholski and Parker, 2000). The PtdInsP 5-phosphatases consist of a number of well-characterized protein families (Mitchell et al., 1996; Woscholski and Parker, 1997; Erneux et al., 1998). The type I 5-phosphatases do not hydrolyze PtdInsPs, while type II 5-phosphatases do. The type II 5-phosphatases share with type I phosphatases a conserved catalytic domain, but additionally have an extended region N-terminal to the catalytic domain, termed type II domain. The type II phosphatases are characterized by further regulatory domains each particular to subsets of enzymes, e.g. N-terminal SH2 (Src homology 2) domain (SH2 domain-containing PtdInsP 5-phosphatases; SHIPs) or a C-terminal GAP (GTPase-activating protein) domain (GAP domain-containing PtdInsP 5-phosphatases; GIPs) (Woscholski and Parker, 2000). Another subset of type II 5-phosphatases is characterized by the presence of an N-terminal Sac domain. The Sac domain-containing PtdInsP phosphatases (SCIPs) are discussed further under specific phosphatases (see below).

The complexity of the system is indicated by the finding that there are six inositol phospholipids and more than 20 soluble InsPs that have been found in mammalian cells (Majerus et al., 1999). The yeast *S. cerevisiae* has at least six species of acid and alkaline phosphatases with different cellular localizations, as well as inorganic phosphate transporters. This phosphatase system has special advantages for studying gene regulation due to the simplicity of phenotype determination in genetic analysis (Oshima, 1997). During the last 10–15 years a large number of phosphatases have been discovered, purified and cloned. In general, the inositol phosphatases tend to inhibit or terminate signaling reactions, although there are exceptions. In many cases the same enzymes hydrolyze phosphate from both water-soluble InsPs and the corresponding lipids with the same arrangement of phosphate groups. The present chapter summarizes the biochemical activity of the phosphatases described to date along with their substrates and products with emphasis on methods of enzyme assay and product identification. The degradation of the water-soluble and lipid-soluble substrates is discussed under separate subtitles.

In view of the overwhelming number of publications in the area, it has not been possible to refer to all of them. An attempt, however, has been made to cover most of the analytical methodologies employed for the assay of the phosphatases, including TLC, HPLC on reversed phase and anion exchange columns, using mainly [^{3}H]Ins and [^{32}P]Pi labeled substrates. GLC and GC/MS have been used less frequently on conventional and chiral phase capillary columns to measure the mass of inositol, inositol phosphate, and acylglycerol moieties of PtdIns and PtdInsPs. Recently highly sensitive and specific methods have been developed for the mass measurement of inositol and inositol derivatives by enzymatic assays, receptor binding and isotope dilution, which are likely to replace some of the lengthy and hazardous procedures of the past, e.g. Van der Kaay et al. (1997, 1998), Rivera et al. (1998) and Chengalvala et al. (1999).

8.2. PtdIns phosphate phosphatases

The general structure and biosynthesis of the PtdInsPs were discussed in Chapters 2 and 6, respectively. The present chapter discusses the PtdInsP phosphatases, which have gained much attention in the recent years because of their ability to counteract and diversify PdIns 3-kinase signaling. The known PtdInsP-phosphatases have been divided into three structurally different families, the 3- and 4-phosphatases from the Cx_5R families and the type II 5-phosphatases. The Cx_5R family members share a limited consensus sequence ($CxxxxxR$) in their catalytic core region (Maehama and Dixon, 1999). The type II 5-phosphatases have two more extensive signature motifs in the catalytic core domain (Majerus et al., 1999). The modular structures of some of the PtdIns P phosphatases are shown in Fig. 8.1 (Vanhaesebroeck et al., 2001). Sac-domain phosphatases show negligible activity towards soluble inositol polyphosphates (Guo et al., 1999). Krystal (2000) has reviewed lipid phosphatases in the immune system, while Nandurkar and Huysmans (2002) have reviewed the myotubularin-related gene family proteins that exhibit lipid phosphatase activity in relation to human disease.

8.2.1. PtdIns monophosphate phosphatases

8.2.1.1. PtdIns (3)P 3-phosphatase
Lips and Majerus (1989) have characterized a phosphatase activity from the soluble fraction of NIH 3T3 cells, which hydrolyzes the monoester phosphate in the D-3 position of the inositol ring of PtdIns(3)P. This phosphatase is specific as it does not hydrolyze monoester phosphates such as PtdIns(4)P, the diester phosphate, PtdIns(4,5)P_2, and has little or no activity on Ins(1,3)P_2. Further

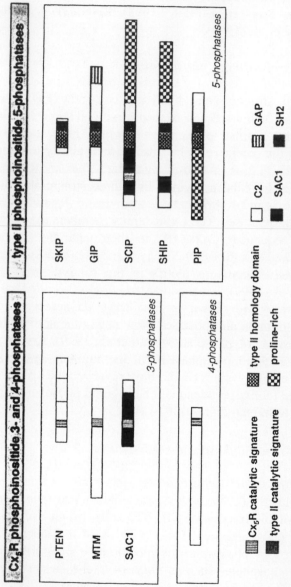

Fig. 8.1. Modular structure of the PtdInsP phosphatases. Reproduced from Vanhaesebroeck et al. (2001) with permission, from the Annual Review of Biochemistry, Volume 70 (C) 2001 by Annual Reviews www.annualreviews.org.

studies are required to determine if this phosphomonoesterase hydrolyzes the other D-3 phosphate-containing PtdIns. The enzyme does not require added metal ions for activity since it is maximally active in the presence of EDTA and is inhibited by Ca^{2+}, Mg^{2+}, and Zn^{2+}. It is also inhibited by the phosphatase inhibitor VO_4^{3-}.

Caldwell et al. (1991) have demonstrated that the PtdIns(3)P 3-phosphatase can hydrolyze also Ins(1,3)P_2 and that there are two identical forms of the enzyme. Caldwell et al. (1991) have isolated the two forms of the enzyme, designated Type I and Type II 3-phosphatases, from rat brain. Type I 3-phosphatase consisted of a protein doublet that migrated at a relative Mr of 65 000 on SDS/ PAGE. The Type II enzyme consisted of equal amounts of a Mr = 65 000 and a Mr 78 000 band upon SDS/PAGE, indicating that the enzyme was a heterodimer. The Type II 3-phosphatase catalyzed the hydrolysis of Ins(1,3)P_2 with a catalytic efficiency of one-nineteenth of that measured for the Type I enzyme, whereas PtdIns(3)P was hydrolyzed by the Type II 3-phosphatase at three times the rate measured for the Type I 3-phosphatase. The Mr = 65 000 subunits of the two forms of 3-phosphatase appear to be the same based on co-migration on SDS/PAGE and peptide maps generated with protease. Caldwell et al. (1991) have proposed that the differing relative specificities of the Type I and Type II 3-phosphatases for Ins(1,3)P_2 and PtdIns(3)P are due to the presence of the Mr = 78 000 subunit of the Type II enzyme. Recently, Nandurkar et al. (2001) have cloned and characterized the cDNA encoding the human 3-phosphatase adapter subunit (3-PAP). Sequence alignment showed that 3-PAP shares significant sequence similarity with the protein and lipid 3-phosphatase myotubularin, and with several other members of the myotubularin gene family including SET-binding factor. However, unlike myotubularin gene family, 3-PAP did not contain a consensus HCX(5)R catalytic motif.

Taylor et al. (2000a,b) have shown that myotubularin, a protein of tyrosine phosphatase mutated in myotubular myopathy, dephosphorylates the lipid second messenger, PtdIns(3)P.

Recombinant human myotubularin specifically dephosphorylated PtdIns(3) in vitro. Overexpression of a catalytically inactive substrate-trapping myotubularin mutant (C3755) in human 293 cells increased PtdIns(3)P levels relative to that of cells over-expressing the wild-type enzyme, demonstrating that PtdIns(3)P is a substrate for myotubularin in vivo.

Maehama et al. (2001) have reviewed the phosphatase properties of myotubularin. Myotubularin displays the greatest similarity to the dual specificity PTPs and is active towards artificial protein and peptide substrates phosphorylated on tyrosyl and seryl/threonine residues. In addition to the phosphatase domain, myotubularin-related proteins possess several motifs known to mediate protein–protein interactions and lipid binding.

Taylor et al. (2000a,b) have reported that recombinant myotubularin dephosphorylates the lipid second messenger, PtdIns(3)P. The activity of myotubularin towards PtdIns(3)P is from 20- to 500 000-fold greater than that observed with other PtdInsPs, soluble InsPs, or protein substrates. Overexpression and deletion gene experiments have suggested that PtdIns(3)P is a physiological substrate for myotubularin. It is not known how myotubularin achieves its high degree of specificity for PtdIns(3). It could be predicted that interactions between the active site Asp residues and phosphoryl groups at the D4 and D5 positions of the inositol ring would prohibit substrate access to the active site cleft.

Two other human myotubularin-related proteins, Sbf1 and K1AA0371, are also of interest because they contain unique lipid-binding domains. KIAA0371 contains an N-terminal myotubularin-like catalytic domain, as well as a C-terminal FYVE domain. FYVE domains are modified Zn^{2+}-finger motifs that specifically bind PtdIns(3)P and serve to target proteins to specific sites within cells. Although the role of FYVE domains in the function of myotubularin-related proteins has yet to be determined, Maehama et al. (2001) point out that it is possible that they serve as targeting motifs to direct the lipid phosphatase domains to specific subcellular environments where PtdIns(3) is abundant. Zhao et al. (2000) have reported a novel dual specificity phosphatase that contains a FYVE domain at the C-terminus (FYVE-DSP1). Its sequence alignment of the catalytic phosphatase domain closely resembled that of myotubularin.

Walker et al. (2001a) have observed that the myotubularin-related protein 3 (MTMR3) shows extensive homology to myotubularin, including the catalytic domain, and additionally possess a C-terminal extension encompassing a FYVE domain. Thus MTMR3 is an inositol lipid 3-phosphatase, with unique substrate specificity. It is able to hydrolyze PtdIns(3)P and PtdIns(3,5)P$_2$, both in vitro and when heterologously expressed in *Saccharomyces cerevisiae*, and thus provide a clearly defined route for the cellular production of PtdIns(5)P. MTMR3 failed to hydrolyze PtdIns(4)P, PtdIns(5)P, PtdIns(3,4)P$_2$, PtdIns(4,5)P$_2$ and PtdIns(3,4,5)P$_3$ (Fig. 8.2).

Walker et al. (2001b) have observed that PTEN and newly discovered human homologue of PTEN (TPIPα) show activity against PtdIns(3)P (see below).

Radioisotope assays. According to Lips and Majerus (1989) PtdIns(3)P 3-phosphatase is assayed as follows. [^{32}P]PtdIns (3) P in chloroform/methanol is dried under a stream of nitrogen and solubilized in the assay buffer. Initial assays are performed in 100 mM KCl, 20 mM HEPES, pH 6.8, 10 mM EDTA, 1 mM EGTA, 0.6% (w/v) octyl glucoside, and 1 mg/ml bovine serum albumin (BSA). Generally, 1000 cpm of [^{32}P]PtdIns(3)P $(3-10 \times 10^{-15} \text{ mol})$ is used in a 20 µl assay (PtdIns(3)P = 0.2–0.5 nM). This amount of substrate also contained approximately 50 pmol of PtdIns that functioned as a carrier lipid. Samples of Mono Q fractions (1–5 µl) were assayed in a total volume of 20 µl for times ranging from 5 to 20 min at 37°C. The assays were initiated by the addition of substrate and terminated by addition of 500 µl of 10% (w/v) trichloroacetic acid followed by 50 µl of 20% (v/v) Triton X-100. The mixture is centrifuged for 2 min, and the supernatant and pellet, representing water-soluble and lipid radioactivity, respectively, are counted in a liquid scintillation

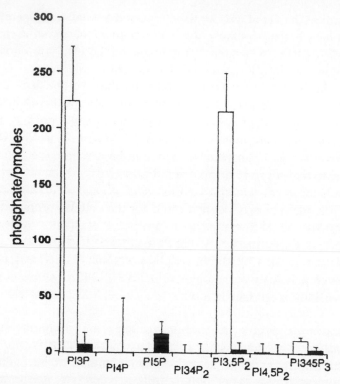

Fig. 8.2. Phosphoinositide phosphatase activity of MTMR3. Full length MTMR3 (open bars) and MTMR3 (1–488) (closed bars) were contemporaneously prepared as GST-tagged fusion proteins from Sf9 cells. Equimolar quantities of enzyme were incubated with potential PtdInsP substrates for 10 min at room temperature. Liberation of free phosphate was determined by the Malachite Green assay and is expressed as OD units. Significant activity is found for the full length protein with PtdIns(3)P and PtdIns(3,5)P₂. Reproduced, with permission, from Hughes, W. E., Cooke, F. T. and Parker, P. (2000) Biochemical Journal 350, 337–352. (c) the Biochemical society and the Medical Research Society.

counter with Scintiverse I. These conditions typically resulted in breakdown of 10–50% of the substrate. The lipid products resulting from PtdIns 3-phosphatase treatment of $[^{32}P]$PtdIns(3)P, Ptd$[^{3}H]$Ins(4)P, and Ptd$[^{3}H]$Ins(4,5)P₂, 2000 cpm of $[^{32}P]$Ptd Ins(3)P and 40 000 cpm each of Ptd$[^{3}H]$Ins(4)P and Ptd$[^{3}H]$Ins

(4,5)P_2 were treated in separate reactions with 6 µl of PtdIns 3-phosphatase from Mono Q fraction 7 in a volume of 20 µl for 2.5 h at 37°C. The products of each reaction were fractionated into aqueous and organic phases and the organic phases resolved by TLC (Lips and Majerus, 1989). Silica Gel 60 thin-layer plates (20 × 20 cm) were dipped in a solution of 1% potassium oxalate and allowed to air dry. A standard solution containing 10 nmol each of PtdIns, PtdIns(4)P, and PtdIns(4,5)P_2 was added to the sample, and this mixture was applied and chromatographed in $CHCl_3$/CH_3OH/4 N NH_4OH (9:7:2, by vol.). Standards were visualized by staining with iodine vapor. TLC plates containing [^3H] were sprayed with ENHANCE prior to autoradiography. Caldwell et al. (1991) assayed the PtdIns(3)P hydrolysis as follows: [^{32}P]PtdIns(3)P in chloroform/methanol and PtdIns (carrier lipid) were dried under a stream of nitrogen, suspended in 500 mM KCl, 100 mM MES, pH 6.5, 12.5 mM EDTA, and 1.5% (w/v) octyl glucoside; and sonicated for 30 s in a bath sonicator. Reaction mixtures contained 1 nM [^{32}P]PtdIns(3)P, 20 µM PtdIns, 100 mM KCl, 20 mM MES, pH 6.5, 2.5 mM EDTA, 0.2 mg of cytochrome C/ml, and 0.3% (w/v) octyl glucoside in 20 µl. Reactions were started by addition of enzyme, incubated at 37°C for 3–10 min, and terminated by addition of 500 µl of 10% (w/v) trichloroacetic acid followed by 50 µl of 20% (v/v) Triton X-100.

Mass assays. Walsh et al. (1991) have described a TLC separation of PtdInsP in the presence of boric acid. Silica gel 60 plates (5 × 20 cm) are immersed face up for 10 s with gentle swirling in *trans*-1,2-diaminocyclohexane-N,N,N',N'-tetraacetic acid (CDTA) solution. The plates are allowed to air-dry by standing up to 1 h and then baked at 100°C for 10 min. The TLC developing solution is prepared by stirring together methanol (75 ml), $CHCl_3$ (60 ml), pyridine (45 ml), and boric acid (12 g) until the boric acid is dissolved. Water (7.5 ml), 88% (v/v) formic acid (3.5 ml), BHT (0.375 g), and technical grade ethoxyquin (75 µl) are then added. PdIns(4)P migrated with a Rf of 0.46, whereas the Rf of PtdIns (3)P was 0.51. TLC bands were detected with phosphomolybdate stain. The products were eluted from the silica plates by extraction twice with 2 ml of chloroform/methanol/pyridine/acetic acid/water (1:2:1:1:1, by vol.) and were assayed for radioactivity when using radioactive substrates.

Walker et al. (2001a) used the malachite green method of Maehama et al. (2000) for assaying inorganic phosphorus in the 3-phosphatase digests. Malachite green solution (80 μl) was added to 20 μl of the reaction mixture and color was allowed to develop for 20 min at room temperature. Absorbance at 620 nm was measured, and phosphate release was quantified by comparison to inorganic phosphate standards.

8.2.1.2. PtdIns(4)P 4-phosphatase

Rass Hope and Pike (1994) have isolated a new membrane-bound PtdInsP phosphatase from rat brain, which utilizes PtdIns(4)P in addition to PtdIns(3)P and PtdIns(4,5)P$_2$ as substrates. In case of PtdIns(4,5)P$_2$ the substrate is doubly dephosphorylated to yield PtdIns. The enzyme does not hydrolyze the water-soluble InsPs.

Guo et al. (1999) have characterized the yeast 5-phosphatases Inp52p and Inp53p and mammalian synaptojanin and have observed a second phosphatase activity associated with these proteins. This was attributed to the Sac domain, which was demonstrated to exhibit a broader-specificity phosphatase activity capable of hydrolyzing phosphate from PtdIns(3)P, PtdIns(4)P and PtdIns(3,5)P$_2$. Hughes et al. (2000a,b) have since demonstrated that the Sac domain from Sac1p itself exhibits phosphatase activity directed against phosphate from the same lipids (Fig. 8.3). Schorr et al. (2001) have reported that PtdIns phosphatase Sac1p controls trafficking of the yeast Chs3p chitin synthase. Sac1p acts as an antagonist of the PtdIns 4-kinase Pik1p in Golgi trafficking.

Radioisotope assays. According to Rass Hope and Pike (1994), the PtdIns(4)P 4-phosphatase is assayed by measuring the release of [^{32}P]PO$_4^{3+}$ from [^{32}P]PtdIns(4)P or PtdIns (3)P. Briefly, the

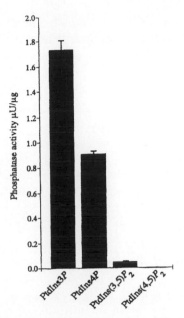

Fig. 8.3. Phosphatidylinositol phosphate specificity of Sac-domain phosphatases. Glutathione S-transferase-tagged Sac1p was expressed and purified from yeast and tested for phosphatase activity against ^{32}P-labeled PtdIns(3)P, PdIns(4)P, PtdIns(3,5)P$_2$, PtdIns(4,5)P$_2$ and PtdIns(3,4)P$_2$. The protein shows activity principally against monophosphorylated PtdIns, and no activity could be detected for PtdIns(4,5)P$_2$ or PtdIns(3,4)P$_2$. Reproduced, with permission, from Hughes, W. E., Cooke, F. T., and Parker, P. (2000) Biochemical Journal 350, 337–352. © The Biochemical Society and the Medical Research Society.

reactions contained 50 mM HEPES, pH 7.2, 0.1% Triton X-100, 1 mM EDTA, and 2 μM [^{32}P]PtdIns(4)P or 5–10 nM [^{32}P]PtdIns (3)P (20 000–40 000 cpm/assay) plus enzyme in a final volume of 100 μl. The reactions were started by addition of the labeled lipid and were stopped by the addition of 0.5 ml of methanol/ conc. HCl (10:1). A 0.5 ml aliquot of water and 1 ml chloroform were then added to each tube, the tubes vortexed and the phases separated by centrifugation. A 0. 5 ml aliquot of the aqueous phase was added to 5 ml of ECONOSAFE scintillation fluid and counted for ^{32}P in a liquid scintillation counter. The assay was specific for PtdInsPs phosphatase activity. When [^{32}P]PtdIns(4)P and [^{3}H]PtdIns(4)P labeled at the 2-position of the inositol ring were used independently as substrates, radiolabel was released into the aqueous phase only in assays containing [^{32}P]PtdIns(4)P substrate. These data show that the assay detects lipid phosphatase activity but not phospholipase C activity.

Mass assays. Palmer (1990) determined PtdIns(4)P 4-phosphatase activity by measuring the release of inorganic phosphate in 0.15 ml reaction mixtures containing 50 mM HEPES (pH 7.2), 5 mM dithiothreitol, 0.2% (w/v) Triton X-100, 1 mM CTAB, and 1 M substrate. The cation-independent PtdIns(4)P 4-phosphatase was measured in the presence of 5 mM EDTA. The reaction was stopped by adding 0.35 ml of 5% (w/v) SDS containing 50 mM EDTA (pH 4.0) and the released inorganic phosphate was determined by an automated phosphorus method (Hegyvary et al., 1979).

8.2.1.3. PtdIns(5)P 5-phosphatase
Neither the soluble nor membrane-bound PtdIns(5)P phosphatase activity appears to have been assayed or commented upon (Guo et al., 1999; Hughes et al., 2000a,b). However, there is evidence for the formation of genuine PdIns(5)P in mammalian tissues (Chung et al., 1997).

8.2.2. PtdIns bisphosphate phosphatases

8.2.2.1. PtdInsP$_2$ 3-phosphatases
PTEN is capable of dephosphorylating the D-3-phosphorylated PtdInsPs, PtdIns(3)P, PtdIns(3,4)P$_2$, and the soluble InsP, Ins(1,3,4)P$_3$ (Maehama et al., 2001). Walker et al. (2001a) have observed that the MTMR3 shows extensive homology to myotubularin, including the catalytic domain, and additionally possess a C-terminal extension that includes a FYVE domain. Thus MTMR3 is an inositol lipid 3-phosphatase, with unique substrate

specificity. It is able to hydrolyze PtdIns(3)P and PtdIns(3,5)P$_2$, both in vitro and when heterologously expressed in *S. cerevisiae*, and thus provide a clearly defined route for the cellular production of PtdIns(5)P. MTMR3 failed to hydrolyze PdIns(4)P, PtdIns(5)P, PtdIns(3,4)P$_2$, PtdIns(4,5)P$_2$ and PtdIns(3,4,5)P$_3$.

Walker et al. (2001b) have shown that a newly discovered human homologue of PTEN (TPIPα) shows activity against PtdIns(3,4)P$_2$ and PtdIns(3,4)P$_2$ (see below).

Radioisotope assays. According to Walker et al. (2001a) yeast strains were radiolabeled (10 μCi/ml [^{3}H]*myo*-inositol) in 5 ml of inositol-free synthetic media lacking uracil and methionine for four to six divisions (to a density of $2 \times 10^6 - 4 \times 10^6$ cells/ml) at 30°C. Cells were immediately killed by the addition of 2 vol. of ice-cold methanol/11.5 M HCl (100:1, v/v). The lipids were deacylated and resolved by an anion exchange HPLC on a 25 cm Partisphere 5 μm SAX column, as described elsewhere (Stephens et al., 1989; Dove et al., 1997; Hughes et al., 2000a,b). An exception was the inclusion of 5 mM tetrabutylammonium hydrogen sulfate in the acid used to split the Folch extract into two phases. The glycerophosphoinositols were resolved on the HPLC gradient of Stephens et al. (1989).

Mass assays. For this purpose, Walker et al. (2001) suspended the PtdInsPs in assay buffer (20 mM HEPES, 100 mM KCl, 1 μM ZnCl$_2$ and 2 mM DDT [pH 7.2]) at 2500 pmol per 50 μl per experimental point and shaken. The protein to be tested was added for various time periods and the reaction was stopped by the addition of Malachite green solution. The reaction was measured by the change in absorption at 650 nm as in Section 8.3.1.

8.2.2.2. PtdInsP$_2$ 4-phosphatases

Bansal et al. (1990) first described an InsPs 4-phosphatase that releases the phosphate from the 4-position of both Ins(3,4)P$_2$ and Ins(1,3,4)P$_3$. Norris and Majerus (1994) and Norris et al. (1995, 1997) showed that the InsPs 4-phosphatase that hydrolyzes the phosphate from the 4-position of both Ins(3,4)P$_2$ and Ins(1,3,4)P$_3$ also rapidly metabolizes PtdIns(3,4)P$_2$ by a 4-phosphatase preparation isolated from using an affinity elution purification method. Vyas et al. (2000) have demonstrated an inositol polyphosphate 4-phosphatase type I activity in megakaryocytes from transcription factor GATA-1 lacking mice. This enzyme hydrolyzes PtdIns(3,4)P$_2$ and also has lesser activity against soluble analogues of this lipid, Ins(3,4)P$_2$ and Ins(1,3,4)P$_3$. Reintroduction of 4-phosphatase into both primary GATA-1$^-$ and wild-type megakaryocytes significantly retarded cell growth.

Radioisotope assays. Using the general conditions of Bansal et al. (1990), Norris and Majerus (1994) determined the first order rate constant for [^{32}P]Ins(3,4)P$_2$ by carrying out the reaction in 20 μl of an assay buffer made up of 50 mM Mops, 10 mM EDTA, 200 mM NaCl, and 0.3% n-octyl glucoside, and the reaction was stopped by the addition of 30 μl of chloroform/methanol (1:1, v/v), and the fraction of [^{32}P]PtdIns(3,4)P$_2$ converted to [^{32}P]PtdIns(3)P was determined by TLC.

In the general PtdInsPs phosphatase assay described by Rass Hope and Pike (1994), a measurement is made of the release of [^{32}P]3-4 from [^{32}P]PtdIns (4)P or [^{32}P]PtdIns(3)P. Briefly, the reactions contained 50 mM HEPES, pH 7.2, 0.1% Triton X-100, 1 mM EDTA, and 2 μM [^{32}P]PtdIns(4)P or 5-10 nm [^{32}P]PtdIns(3)P (20 000-40 000 cpm/assay) plus enzyme in a final volume of 100 μl. The reactions were started by the addition of the labeled lipid and were stopped by the addition of 0.5 ml of methanol/concentrated HCl (10:1). A 0.5 ml aliquot of water and 1 ml chloroform were then added to each tube. After vortexing the phases were separated and aliquots counted. For experiments examining the utilization of [^{3}H]PtdIns(4,5)P$_2$ as a substrate, a similar assay was used except that the phosphatase activity was assessed by measuring the generation of [^{3}H]PtdIns from [^{3}H]PtdIns(4,5)P$_2$. No detectable [^{3}H]PtdInsP was formed in these assays. The reactions were stopped as usual, and the lipids were extracted in CHCl$_3$/methanol. The organic phase was evaporated to dryness and analyzed by TLC (Pike and Eakes, 1987). The TLC plates were sprayed with ENHANCE and exposed to X-ray film for 3 days (Rass Hope and Pike, 1994).

Mass assays. According to Norris and Majerus (1994) the TLC separations are performed on Silica Gel 60 plates treated with a solution of 1% potassium oxalate in 50% ethanol. The plates are activated for at least 30 min at oven temperature of 90°C prior to use. TLC plates are developed using a solvent mixture of chloroform/acetone/methanol/glacial acetic acid/water (80:30: 26:24:1, by vol.) as described by Traynor-Kaplan et al. (1988). Phospholipid standards (PtdIns, PtdIns(4)P, and PtdIns(4,5)P$_2$) are stained with iodine. Radiolabeled phospholipids are detected by autoradiography using Hyperfilm-MP (Amsterdam). Radio-labeled phospholipids are scraped from the plates, mixed with 10 ml of Aquassure scintillation mixture and counted in a liquid scintillation counter. Alternatively, the fractions were analyzed for phosphorus by the method of Ames and Dubin (1960).

Palmer (1990) was one of the first to distinguish between specific 4- and 5-phosphatase activities when examining PtdIns(4,5)P$_2$ hydrolysis. The activities and subcellular distributions of the hydrolases were compared in the developing chick central nervous system. PtdIns(4,5)P$_2$ was found to be largely in the soluble fraction in embryonic and myelinated brain. Furthermore, the specific activity of the 4-phosphatase increased coincidentally with myelination.

Palmer (1990) assayed PtdIns(4,5)P$_2$ 4-phosphatase activity by measuring the release of inorganic phosphate in 0.15 ml reaction mixture containing 50 mM HEPES (pH 7.2), 5 mM dithiothreitol, 0.2% (w/v) Triton X-100, 1 mM CTAB, and 1 mM substrate. Assays for the Mg^{2+}-dependent PtdIns(4,5)P$_2$ also contained 0.5 mM MgCl$_2$ and 1 mM EGTA.The reaction was stopped by adding 0.35 ml of 5% (w/v) SDS containing 50 mM EDTA (pH 4.0) and the released inorganic phosphate was determined by an automated method (Hegyvary et al., 1979).

8.2.2.3. PtdInsP₂ 5-phosphatases

The 5-inositol phosphatase from skeletal muscle and kidney possesses a catalytic core and type II domain (Ijuin et al., 2000) and hydrolyzes most PtdInsPs but seems to prefer PtdIns(4,5)P$_2$. The 5-phosphatase II and OCRL gene product of the GAP-containing inositol 5-phosphatases dephosphorylate PtdIns(4,5)P$_2$ and PdInsP$_3$, as well as their corresponding soluble InsPs (Jefferson and Majerus, 1995; Jackson et al., 1995; Suchy et al., 1995). However, recent work has shown that the SHIPs can also dephosphorylate PtdIns(4,5)P$_2$ under certain conditions (Kisseleva et al., 2000).

The proline-rich inositol 5-phosphatases are characterized by an N-terminal proline-rich domain. The first discovered member was pharbin, a phosphatase that upon transfection in fibroblasts induces a dendritic-like outgrowth similar to neurite growth. Pharbin is found in the Golgi and hydrolyzes PtdIns(3,4)P$_2$ (Kong et al., 2000), a second member of proline-rich inositol polyphosphate 5-phosphatase (PIPP). Both enzymes have been shown to hydrolyze PtdIns(4,5)P$_2$ and PtdInsP$_3$, as well as the corresponding soluble inositol phosphates (Mochizuki and Takenawa, 1999).

PtdIns(4,5)P$_2$ 5-phosphatases were first purified from a protozoan and from human erythrocyte cytosol. Matzaris et al. (1994) demonstrated that the 75-kDa Ins(1,4,5)P$_3$ 5-phosphatase represents the major PtdIns(4,5)P$_2$ 5-phosphatase in human platelets. The platelet PtdIns(4,5)P$_2$ 5-phosphatase activity was magnesium but not calcium dependent. The elution profile of platelet cytosolic PtdIns(4,5)P$_2$ 5-phosphatase matched that of the 75-kDa Ins(1,4,5)P$_3$ 5-phosphatase. Purified 75-kDa Ins(1,4,5)P$_3$ 5-phosphatase hydrolyzed PdIns(4,5)P$_2$ forming PtdIns(4)P. In contrast, purified membrane-associated 43 kDa Ins(1,4,5)P$_3$ 5-phosphatase did not hydrolyze PtdIns(4,5)P$_2$.

Palmer (1994) purified two PtdIns(4,5)P$_2$ 5-phosphatases from bovine brain cytosol that have similar reaction characteristics including InsP$_3$ phosphatase activity. Both enzymes were larger (155 and 115 kDa by SDS-PAGE) than the protozoa and erythrocyte PtdIns(4,5)P$_2$ phosphatases. A monoclonal antibody raised to the purified preparations recognized both forms of the phosphatases indicating that the two enzymes were related. The determination of the precise relationship between the bovine brain PtdIns(4,5)P$_2$ 5-phosphatase and the 75 kDa Ins(1,4,5)P$_3$ 5-phosphatase awaits comparison of the amino acid sequences of the platelet and bovine brain enzymes. Palmer et al. (1994) have suggested that the high molecular weight PtdIns(4,5)P$_2$ phosphatases may be precursors of lower molecular weight soluble Type II InsPs 5-phosphatases shown to account for the PtdIns(4,5)P$_2$ phosphatase activity in platelets (Matzaris et al., 1994).

Zhang et al. (1995) have reported that the oculocerebrorenal (OCRL) syndrome protein, encoded in Lowe syndrome locus, is a PtdIns(4,5)P$_2$ 5-phosphatase. OCRL hydrolyzes the other 4-phosphatase substrates, although less well than the other isozymes. PtdInsPs 5-phosphatases have also been described which may play an important role in regulating the functions of PtdIns(4,5)P$_2$. Synaptojanin-1 represents one member of a family of Sac domain-containing inositide phosphatases (Woscholski and Parker, 1997) that also include the yeast proteins Inp51p, Inp52p and Inp53p (Stolz et al., 1998a,b). Synaptojanin 2 is a recently discovered 5-phosphatase and its catalytic motifs are identical to those of synaptojanin-1 (Nemoto et al., 1997).

PtdIns(4,5)P$_2$ reorganizes actin filaments by modulating the functions of a variety of actin-regulatory proteins. Until recently, it was thought that bound PtdIns(4,5)P$_2$ is hydrolyzed only by tyrosine-phosphorylated PLCγ after the activation of tyrosine kinases. Sakisaka et al. (1997) have shown that bound PtdIns(4,5)P$_2$ is hydrolyzed by synaptojanin. The 5-phosphatase activity of a native 150 kDa protein purified from bovine brain was not inhibited by profilin, cofflin, or α-actinin, which markedly inhibit the activity of PLC $\delta 1$ activity. The data suggest that the 150 kDa protein (synaptojanin) hydrolyzes PtdIns(4,5)P$_2$ bound to actin regulatory proteins, resulting in the rearrangement of actin filaments downstream of tyrosine kinase and Ash/Grb2 (Sakisaka et al., 1997).

The Sac phosphatases do not hydrolyze either PtdIns(3,4)P$_2$ or PtdIns(4,5)P$_2$, which contain adjacent phosphate groups, but retain reduced activity towards PtdIns(3,4)P$_2$, where the two phosphates are separated by a hydroxylated carbon. The Sac domain is approximately 400 residues in length and proteins containing this domain show approximately 35% identity with other Sac domains throughout this region (Guo et al., 1999).

Kisseleva et al. (2000) have demonstrated that PtdInsPs 5-phosphatase IV, a member of a large family of inositol polyphosphate phosphatases, is a lipid-specific inositol-phospha-tase in vivo. Only the lipid substrates, PtdIns(4,5)P$_2$ and PtdIns(3,4,5)P$_3$ are utilized by this enzyme in vitro (Kisseleva et al., 2000). The enzyme has an exceptionally high affinity for PtdIns(3,4,5)P$_3$, making it a potential candidate for regulation of the PtdIns(3,4,5)P$_3$ level in cells.

Rass Hope and Pike (1994) have isolated and purified from rat brain a PtdInsPs phosphatase, which can catalyze the removal of both phosphates from PtdIns(4,5)P$_2$.

Radioisotope assays. Matzaris et al. (1994) assayed PtdIns(4,5)P$_2$ 5-phosphatase by suspending [^{3}H]PtdIns(4,5)P$_2$ in 20 mM Tris, pH 7.2, containing 150 mM NaCl and 2 M cetyltriethylammonium bromide with vigorous agitation and brief sonication on ice. The assays were performed in a reaction volume of 50 μl containing 20 mM Tris, pH 7.2, 150 mM NaCl, 200 μg/ml BSA, 3 M MgCl$_2$, 2 mM cetyltriethylammonium bromide, and 250 μM [^{3}H]PtdIns(4,5)P$_2$ (3500 cpm/nmol). Reactions were run for 30 min at 37°C and were terminated by the addition of 100 μl of 1 M HCl. Lipids were extracted by the addition of 500 μl of 2 M

KCl and 200 μl of chloroform/methanol (1:1, v/v) and separated by TLC in chloroform/methanol/acetic acid/H_2O (43:38:5:7, by vol.). Chromatographed lipids were visualized by iodine staining and compared to the migration of known standard PtdIns, PtdIns(4)P, and PtdIns(4,5)P_2. Products of the reaction, [^{3}H]PtdIns(4)P and [^{3}H]PtdIns(4,5)P_2, were excised from the TLC plate and quantified by liquid scintillation counting.

The reaction products derived from PtdIns(4,5)P_2 5-phosphatase were assayed further as follows: Silica gel was scraped from the TLC plate and the phospholipid products of the PtdIns(4,5)P_2 5-phosphatase assayed were deacylated with methylamine (Sun and Lin, 1990). Deacylated phospholipids were analyzed by HPLC SAX chromatography (Waters) and compared to deacylated [^{3}H]PtdIns(4)P and [^{3}H]PtdIns(4,5)P_2 standards (Sun and Lin, 1990). The authenticity of radiolabeled phospholipids was confirmed by HPLC analysis of both deacylated and deglycerated lipid products, as previously described (Divecha et al., 1991; Stephens et al., 1991).

Alternatively, the PtdIns(4,5)P_2 5-phosphatase activity was determined using [^{3}H]PtdIns(4,5)P_2. The assay was performed using the same conditions as those for the PtdIns(3,4,5)P_3 5-phosphatase activity determination (see below). The lipids were extracted with $CHCl_3$, dried under vacuum, and separated by TLC using n-propyl alcohol, 2N acetic acid (2:1) (Kodaki et al., 1994). The corresponding spots for PtdIns(4)P and PtdIns(4,5)P_2 were identified using [^{32}P]-labeled standards, subsequently scraping off and counting by liquid scintillation (Woscholski et al., 1995).

Kisseleva et al. (2000) assayed the hydrolysis of PtdIns(4,5)P_2 using the novel 5-phosphatase by suspending [^{3}H]PtdIns(4,5)P_2 (3000 cpm/assay) in buffer containing 50 mM Tris-HCl, pH 7.5, 3 mM $MgCl_2$, 150 mM NaCl, and 2 mM CTAB followed by brief sonication. Alternatively, the radio-labeled [^{3}H]PtdIns(4,5)P_2 was resuspended in buffer containing 50 mM Tris-HCl, pH 7.5, 3 mM $MgCl_2$, 50 mM NaCl, and 0.03 n-octyl β-glucopyranoside. Reactions were started by addition of the enzyme and carried out at 37°C. Extraction of the lipids and analysis of the reaction products were as previously described (Norris and Majerus, 1994; Matzaris et al., 1994). The proof of the structure of the product was established by deacylation and ion exchange chromatography (Lips and Majerus, 1989). GroPIns derivatives were separated by HPLC on a Partisphere SAX column (Whatman) with the gradient of 0–1.25 M NaH_2PO_4, pH 4.5. The gradient consisted of a 0–5 min linear rise to 29% for pump B, a 5–35 min linear rise to 60% B, and a 35–45 min linear rise to 100% B followed by a 5-min wash with 100% B. Radiolabeled deacylated phospholipids were detected by a β-RAM Flow-Through System (In/US Systems, Inc, Tampa, Florida).

For experiments examining the utilization of [³H]PtdIns(4,5)P$_2$ as a substrate, an assay similar to that just described for the 4-phosphatase, was used except that phosphatase activity was assessed by measuring the generation of [³H]PtdIns from [³H]PtdIns(4,5)P$_2$ (Kisseleva et al., 2000). No detectable [³H]PtdInsP was formed in these assays. The reactions were stopped by adding the organic solvents, and the lipids extracted into $CHCl_3$/methanol. The organic phase was evaporated to dryness and analyzed by TLC (Pike and Eakes, 1987). The TLC plates were sprayed with ENHANCE and exposed to X-ray film for 3 days. The bands corresponding to PtdIns, PtdInsP, and PtdInsP$_2$ were scraped and counted for ³H in Econo-Safe (RPI).

Hughes et al. (2000a,b) have pointed out that the HPLC protocol commonly employed for the assay of the 5-phosphatase products as the deacylation products GroPIns(4)P and Gro-PIns(5)P cannot be efficiently separated (Rameh et al., 1997). Such a separation could be performed, however, by HPLC on a Partisphere 5 μm SAX column (250 × 40 mm) with a flow rate of 1 ml/min using the following gradient separation: 0 min, 0% solvent B; 5 min, 0% solvent B; 15 min, 2% solvent B; and 100 min 2% solvent B (solvent A = H_2O and solvent B = 1.25 M $(NH_4)_2HPO_4$ (pH 3.8).

Mass assays. The mass assays of PtdInsP$_2$ and PtdInsPs are difficult to execute and it is necessary to determine the recovery of each PtdInsPs. Palmer et al. (1994) determined Mg^{2+}-dependent PtdIns(4,5)P$_2$ 5-phosphatase activity by measuring the release of inorganic phosphate from PtdIns(4,5)P$_2$ in optimized reaction mixtures containing non-ionic and cationic detergents as described earlier (Palmer, 1990). Specifically, PtdIns(4,5)P$_2$ 5-phosphatase activity was determined (Palmer et al., 1994) by measuring the release of inorganic phosphate in 0.15 ml reaction mixtures containing 50 mM HEPES (pH 7.2), 5 mM dithiothreitol, 0.2% (w/v) Triton X-100, 1 mM CTAB, and 1 mM substrate. The assays for Mg^{2+}-dependent PtdIns(4,5)P$_2$ 5-phosphatase also contained 0.5 mM MgCl$_2$ and 1 mM EGTA. The reaction was stopped by adding 0.35 ml of 5% (w/v) SDS containing 50 mM EDTA (pH 4.0) and the released inorganic phosphate was determined by an automated method (Hegyvary et al., 1979).

Stein et al. (1990) have described a method for acetylation of PtdIns(4)P and PtdIns(4,5)P$_2$ with [^{3}H]acetic anhydride and for separation of the products from each other and from unchanged starting material. The separations are performed on oxalate impregnated TLC plates with chloroform/methanol/4 M ammonia (60:40:10, by vol.) as the developing solvent. The spots were detected and quantified by either autoradiography of TLC plates or by a Berthold automatic TLC-linear analyzer. Total radioactivity (^{32}P or ^{14}C + ^{3}H) was measured.

8.2.3. PtdIns trisphosphate phosphatases

Phosphatases that terminate the signal-transduction process and regulate the levels of PtdInsP are essential for signaling and proper cell function. Distinct forms and isoforms of various PtdInsPs phosphatases have been characterized in mammalian cells biochemically and by molecular cloning. Many of the InsPs phosphatases discussed in the earlier sections of this chapter also act on the PtdInsPs. All are magnesium-dependent phosphomonoesterases.

8.2.3.1. PtdInsP₃ 3-phosphatases

PtdIns(3,4,5)P$_3$ is a key molecule involved in cell growth signaling. The tumor suppressor PTEN/MMAC1 (phosphatase and tensin homologue or mutated in multiple advanced cancers) has been implicated in a large number of tumors and is conserved from humans to worms (Maehama and Dixon, 1999). Characterization of PTEN protein has shown that it is a phosphatase that acts on proteins and on 3-phosphorylated PtdIns, including PtdIns $(3,4,5)$P$_3$, and can therefore modulate signal transduction pathways that involve lipid second messengers (Maehama and Dixon, 1998; Maehama et al., 2001; Leslie and Downes, 2002). Mammalian PTEN-proteins are highly similar, consisting of an N-terminal phosphatase domain and a C-terminal domain, which may function as a phospholipid-targeting motif (Fig. 8.4) (Maehama et al., 2001). Enzymatic characterization of recombinant human PTEN has revealed that protein is a poor catalyst towards artificial substrates, with the activity towards p-nitrophenylphosphate being approximately 1000-fold lower than the other tyrosine specific or dual specificity protein tyrosine phosphatases. Other comparisons of activity have shown that although Km of PTEN activity towards PtdInsP$_3$ and Ins(1,3,4,5)P$_4$ differs only two-fold, dephosphorylation of PtdInsP$_3$ proceeds at a 400-fold greater rate than that of Ins(1,3,4,5)P$_4$ (Downes et al., 2001). The crystal structure of human PTEN has revealed that the C-terminal region is similar to the C2 domains of PLCδ$_1$, cytoplasmic PLA$_2$, and protein kinase Cδ (Lee et al., 1999). The C2 domain of PTEN lacks the CBR1 and CBR2 loops, but retains CBR3, and can bind to phospholipid vesicles in a Ca^{2+}-independent manner in vitro. Leslie et al. (2001) have shown that deletion of the C-terminal PDZ-binding sequence retained full PTEN activity against soluble substrates. The PDZ-binding sequence was dispensable for the efficient down-regulation

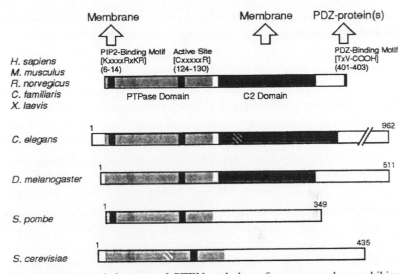

Fig. 8.4. Structural features of PTEN orthologs from mammals, amphibian, worm, fly, fission yeast and budding yeast are illustrated. The PTP domain, active site residues, PIP2-binding motif and C2 domain, as well as the C-terminal PDZ-binding motif are shown as indicated. Reproduced from Maehama et al. (2001) with permission, from the Annual Review of Biochemistry, Volume 70 © 2001 by Annual Reviews www.annualreviews.org.

of cellular PtdIns(3,4,5)P$_3$ levels and a number of PtdIns 3-kinase dependent signaling activities, including PKB and p70S6K.

Walker et al. (2001a) have identified and characterized two human homologues of PTEN, which differ with respect to their subcellular localization and lipid phosphatase activities. The previously cloned, but uncharacterized, transmembrane phosphatase with tensin (TPTE) homology is localized to the plasma membrane, but lacks detectable PtdInsP 3-phosphatase activity. TPIP (TPTE and PTEN homologous inositol lipid phosphatase) is a novel phosphatase that occurs in several differentially spliced forms of which two, TPIPα and TPIPβ, appear to be functionally distinct. TPIPα displays similar PtdInsP 3-phosphatase activity compared with PTEN against PtdIns(3,4,5)P$_3$, PtdIns(3,5)P$_2$, PtdIns(3,4)P$_2$ and PtdIns(3)P, has N-terminal transmembrane domains and appears to be localized on the endoplasmic reticulum. TPIPβ lacks detectable phosphates activity and is cytosolic.

Rohrschneider et al. (2000) have recently reviewed the structure, function and biology of SHIP proteins, which are components of many growth factor receptor signaling pathways and represent SH2-containing Ins 5-phosphatase (SHIP). Its enzymatic activity removes the phosphate group from the $5'$-position of both PtdIns(3,4,5)P$_3$ and Ins(1,3,4,5)P$_4$ (Damen et al., 1996). The $3'$-position of the inositol phospholipid must be phosphorylated before SHIP can dephosphorylate the $5'$-position, implying that SHIP acts sequentially with PtdIns 3-kinase in a PtdIns pathway. Although SHIP acts on the same substrate as PTEN, it is not a tumor suppressor as it yields a different product, PtdIns(4,5)P$_2$.

Radioisotope assays. Van der Kaay et al. (1997) analyzed the recovery of PtdIns (3,4,5)P$_3$ by spiking tracer [^{32}P]PtdIns(3,4,5)P$_3$ into total cell lipid extracts or lipids from Folch fraction I from bovine brain. A single back-extraction of the upper phase of such extracts with synthetic lower phase was sufficient to recover 98% of the spiked radioactivity. The recovery of the radiolabeled Ins(1,3,4,5)P$_4$ was then followed during an anion exchange HPLC analysis of the products. The recovery of Ins(1,3, 4,5)P$_4$ through this procedure, which is part of the analytical routine, was dependent on the relative amounts of lipid and KOH. Thus, hydrolysis of 0.3 mg of astrocytoma cell lipids with 0.1N and 0.5 M KOH yielded 17.0 and 52% Ins(1,3,4,5)P$_4$, respectively. All subsequent experiments used 50 μl of 1 M KOH to hydrolyze sample containing not more than 0.6 mg of lipid; the volume of 1 M KOH was increased correspondingly for samples containing greater amounts of lipid.

According to Maehama and Dixon (1998) the PtdIns(3,4,5)P$_3$ phosphatase assay is performed at 37°C in a buffer (20 μl) consisting of 100 mM Tris-HCl (pH 8), 10 mM DTT, [^{32}P]Ptd Ins(3,4,5)P$_3$ and 1 μg of purified PTEN. The reaction is terminated by the addition of 0.47 ml of CH$_3$OH/CHCl$_3$/6% HClO$_4$ (30:15:2, v/v/v). The phospholipids were extracted and separated on a TLC plate. The $[^{32}P]PtdIns(3,4,5)P_3$ was prepared by phosphorylation of $PtdIns(4,5)P_2$ by PtdIns 3-kinase using $[\gamma^{32}P]ATP$ as phosphate donor. For identification of the dephosphorylation site, dephosphorylation of $PtdIns(3,4,5)P_3$ by PTEN was carried out in a buffer (20 μl) consisting of 100 mM Tris-HCl (pH 8, 10 mM DTT, 0.1 mg/ml of $PtdIns(3,4,5)P_3$, 0.15 mg/ml of PtdSer and 1 μg of purified PTEN. The reaction was terminated by the addition of 0.47 ml of $CH_3OH/CHCl_3/6\%HClO_4$ (30:15:2, vv/v). The phospholipids were extracted, dried, and then used for the PtdIns 3-kinase-catalyzed reaction. Guo et al. (1999) have shown that the SAC1-like domains of yeast SAC1 have intrinsic enzymatic activity that defines a new class of polyphosphoinositide phosphatases. Purified recombinant SAC1-like domains convert yeast lipids PtdIns(3)P, PtdIns(4)P and $PtdIns(3,5)P_2$ to PtdIns. Whereas $PtdIns(4,5)P_2$ is not a substrate. Of significant interest is the regulation of $PtdIns(3,5)P_2$. The data of Guo et al. (1999) demonstrate that the PtdInsPs phosphatase activities of Inp52, Inp53p, synaptojanin, and Sac1p are biologically relevant regulators of $PtdIns(3,5)P_2$.

Taylor et al. (2000a,b) conducted the phosphatase assays at 30°C in a buffer consisting of 50 mM sodium acetate, 25 mM bis–Tris, 25 mM Tris (pH 6.0), and 2 mM DTT. Radiolabeled protein substrates and p-nitrophenylphosphate were prepared and assayed as described (Taylor et al., 1997). PtdInsPs substrates used for activity measurements were synthetic D($+$)-sn-1,2-di-O-octanoyl derivatives obtained from Echelon Research Laboratories (Salt Lake City).

HEK293 cells radiolabeled with $[^3H]Ins$ as described by Taylor et al. (2000a,b) were resuspended and washed twice with PBS. Total cellular lipids, including 10 μmol (each) of unlabeled PtdIns(3)P and PtdIns(4)P added as carrier lipids, were extracted by using acidified chloroform/methanol, and were deacylated as described by Auger et al. (1989). Total cellular protein and radioactivity varied less than 5% between samples. After deacylation, greater than 98% of the total radioactivity from the lipid extract was present in the aqueous phase. Radiolabeled GroPInsPs were separated by using a Partisphere 5-SAX column (Whatman) as described by Auger et al. (1989) with the following modifications. A discontinuous gradient from 10 mM (pump A) to 1.0 M $(NH_4)_2HPO_4$ pH 3.8) (pump B) was established as follows: 0% B for 10 min; 0–12% B over 30 min; 12% B to 100% B over 5 min: 100% B for 5 min; 100% B to 0% B over 1 min; 0% B for 24 min. Radioactivities eluted from the column were quantified by a continuous-flow in-line scintillation detector. The elution positions of GroPIns(3)P, GroP(4), and GroPIns(4,5)P_2 were determined using authentic radiolabeled standards. PtdIns(3) was prepared by immunoprecipitated PtdIns 3-kinase, $\gamma[^{32}P]$ATP, and PtdIns as previously described (Stack et al., 1995).

Mass assays. Van der Kaay et al. (1997) have described a novel, rapid and highly sensitive mass assay for PtdIns $(3,4,5)P_3$ and its application to measure insulin-stimulated PtdIns$(3,4,5)P_3$ production in skeletal muscle in vivo. The method is based on simple, reproducible isotope dilution assay which detects PtdIns$(3,4,5)P_3$ at subpicomole sensitivity, suitable for measurements of both basal and stimulated levels of PtdIns$(3,4,5)P_3$ obtained from samples containing approximately 1 mg of cellular protein. Total lipid extracts, containing PtdIns$(3,4,5)P_3$, are first subjected to alkaline hydrolysis which results in the release of the polar head group Ins$(1,3,4,5)P_4$. The latter is measured by its ability to displace $[^{32}P]$Ins$(1,3,4,5)P_4$ from a highly specific binding protein present in cerebellar membrane preparations. This approach, which avoids tedious chromatographic procedures, is based on a previously reported assay for PtdIns$(4,5)P_2$, which utilized alkaline hydrolysis coupled to the radioligand displacement-based measurement of Ins$(1,4,5)P_3$ (Chilvers et al., 1991). The Ins$(1,3,4,5)P_4$-binding protein was obtained from sheep cerebellum. Briefly, cerebella were homogenized in ice-cold buffer (20 mM NaHCO$_3$, pH 8.0, 1 mM dithiothreitol, 2 mM EDTA) and centrifuged for 10 min at 5000g. The pellet was washed twice and resuspended in homogenization buffer at a final protein concentration of 10–20 mg/ml. Ins(1,3,4,5)P_4 concentration was determined as described by Challis and Nahorski (1990). Assays of 320 μl comprised 80 μl of assay buffer (0.1 M NaAc, 0.1 M KH$_2$PO$_4$, pH 5.0, 4 mM EDTA, 80 μl of 3 × 10^5 dpm of [^{32}P]Ins(1,3,4,5)P_4), 80 μl of sample, and 80 μl of binding protein. Both standard Ins(1,3,4,5)P_4 and samples were in 0.5 M KOH/acetic acid (pH 5.0) and were assayed directly. Samples were diluted in 0.5 M KOH/acetic acid (pH 5.0) to allow measurements in the most sensitive range of the displacement curves. After addition of binding protein, samples were incubated on ice for 30 min and subsequently subjected to rapid filtration using GF/C filters. Filters were washed twice with 5 ml of ice cold buffer (25 mM NaAc, 25 mM KH$_2$PO$_4$, pH 5.1, 1 mM EDTA, and 5 mM NaHCO$_3$). Radioactivity was determined after the filters were extracted for 12 h in 4 ml of scintillant.

Van der Kaay et al. (1997) have shown that this assay solely detects PtdIns(3,4,5)P_3 and does not suffer from interference by other compounds generated after alkaline hydrolysis of total cellular lipids. Measurements on a wide range of cells, including rat-1 fibroblasts, 1321N1 astrocytoma cells, HEK 293 cells, and rat adipocytes, showed wortmannin-sensitive increased levels of PtdIns(3,4,5)P_3 upon stimulation with appropriate agonists. The enhanced utility of this procedure was further demonstrated by measurements of PtdIns(3,4,5)P_3 levels in tissue derived from whole animals. Specifically, Van der Kaay et al. (1997) have also shown that simulation with insulin increases PtdIns(3,4,5)P_3 levels in rat skeletal muscle in vivo with a time course which parallels the activation of protein kinase B in the same samples.

8.2.3.2. PtdInsP$_3$ 4-phosphatases

Although Ins(1,3,4)P_3 4-phosphatase hydrolyzes the 4-position of PtdIns(3,4)P_2, the 4-postion of PtdInsP$_3$ is not known to be hydrolyzed by a 4-phosphatase (Norris et al. 1997).

Radioisotope assays. The 4-phosphatase activity was determined using both soluble and lipid substrates as described by Norris and Majcrus (1994), except that 0.1% Triton X-100 was substituted for the octylglucoside in the lipid assay buffer.

Mass assays. Mass assay for phosphorus by one of the above described methods for quantification based on phosphorus determination by one of the micromethods (Ames and Dubin, 1960; Hegyvary et al., 1979).

8.2.3.3. PtdInsP₃ 5-phosphatases

Jackson et al. (1995) have demonstrated that at least two enzymes are capable of hydrolyzing the 5-position phosphate from PtdIns(3,4,5)P$_3$, the 75 kDa 5-phosphatase and a previously unidentified 5-phosphatase, which forms a complex with the p85/p110 form of PtdIns 3-kinase. This enzyme is immunologically and chromatographically distinct from the platelet 43 kDa and 75 kDa 5-phosphatase and is unique in that it removes the 5-position phosphate from PtdIns(3,4,5)P$_3$, but does not metabolize PtdIns(4,5)P$_2$, Ins(1,4,5)P$_3$ or Ins(1,3,4,5)P$_4$.

Woscholski et al. (1995) have characterized a PtdIns(3,4,5)P$_3$ 5-phosphatase activity from rat brain tissue. A cytosolic PtdIns(3,4,5)P$_3$ 5-phosphatase activity was purified using ion exchange, affinity, and size exclusion chromatography. The enzyme was a magnesium dependent 5-phosphatase that was able to hydrolyze PtdIns(4,5)P$_2$ and PtdIns(3,4,5)P$_3$. Woscholski et al. (1997) have subsequently demonstrated that this protein is encoded by a particular splice variant of the recently described synaptojanin gene. It is further shown that a recombinant synaptojanin protein has intrinsic PtdIns(3,4,5)P$_3$ 5-phosphatase activity. These studies suggest a unifying concept in which inositol phosphate and inositol phospholipid second messengers are metabolized by a common set of cellular phosphatases (Jackson et al., 1995).

Damen et al. (1996) have isolated a 145-kDa tyrosine-phosphorylated protein that becomes associated with Shc (an SH-

2-containing adopter protein) in response to multiple cytokines from murine hemopoetic cell line B6SUtA1. Cell lysates immunoprecipitated with antiserum to this protein exhibited both PtdIns(3,4,5)P$_3$ and Ins(1,3,4,5)P$_4$ 5-phosphatase activity. To determine whether the 145 kDa protein was indeed a 5-phosphatase, lysates from B6SutA1 cells were immunoprecipitated with anti-15-mer, anti-Shc, or normal rabbit serum, and the immunoprecipitates were tested with various 5-phosphate substrates (Zhang et al., 1995; Jefferson and Majerus, 1995). It was shown that, anti-15-mer, but not normal rabbit serum, immunoprecipitates hydrolyzed [^{3}H]Ins(1,3,4,5)P$_4$ to [^{3}H]Ins(1,3,4)P$_3$. The product of the reaction was found to be [^{3}H]Ins(1,3,4)P$_3$ by incubation with recombinant InsPs 4-phosphatase (Norris et al., 1995) or InsPs 1-phosphatase (York et al., 1994), followed by separation of the bisphosphate product on Dowex/formate (Zhang et al., 1995; Jefferson and Majerus, 1995). In the presence of 3 mM EDTA, no hydrolysis of [^{3}H]Ins(1,3,4,5)P$_4$ was observed, suggesting that this 5-phosphatase is Mg^{2+}-dependent. Of the 5-phosphatases cloned to date (Zhang et al., 1995; Jefferson and Majerus, 1995), the 145 kDa protein is the first to possess an SH2 domain and to be tyrosine phosphorylated.

Synaptojanin is a major GrB2-binding protein of the brain, which contains two domains, which are linked to inositol phosphate metabolism (thus the name from the Roman god with two faces, Janus). The central region of synaptojanin defines it as a member of the inositol-5-phosphate family, which includes the product of the gene that is defective in the OCRL syndrome of Lowe (Attree et al., 1992; Zhang et al., 1995; Ross et al., 1991; Laxminarayan et al., 1994; Jackson et al., 1995). Its amino terminal domain is homologous with the yeast protein Sac1 (Rsd1), which is genetically implicated in phospholipid metabolism and in the function of the cytoskeleton. McPherson et al. (1996) have obtained evidence, which suggests a link between PtdIns metabolism and synaptic vesicle recycling. To prove that synaptojanin is a 5-phosphatase, as suggested by its primary sequence, McPherson et al. (1996) tested the enzyme activity of the purified protein. Anti-synpatojanin immunoprecipitates obtained from adult brain extracts converted the $[^3H]Ins(1,4,5)P_3$, but not $[^3H]Ins(1,3,4)P_3$, to $[^3H]InsP_2$. These results were confirmed by analysis of the purified protein. Synaptojanin used both $Ins(1,4,5)P_3$ and $Ins(1,3,4,5)P_4$ as substrates and also dephosphorylated $PtdIns(4,5)P_2$ to PtdInsP, as determined by TLC. Synaptojanin is now recognized as a nerve terminal protein of relative molecular mass 145 kDa, which appears to participate with dynamin in synaptic vesicle recycling (McPherson et al., 1994a,b).

Nemoto et al. (1997) have identified and characterized a novel form of synaptojanin, synaptojanin 2, which has broader tissue distribution. Synaptojanin 2 cDNA from rat brain library encodes a protein of 1248 amino acids with a predicted Mr of 138 268. The two synaptojanin isoforms share 57.2 and 53.8% amino acid identity in their Sac I and phosphatase domains, respectively.

The recently purified 145 kDa phosphoprotein and SH2-containing 5-phosphatase SHIP has been shown to associate with Shc by multiple cytokines including the interleukin 2, interleukin 3 and GM-CSF. Cell lysates immunoprecipitated with antiserum to this protein exhibit both $Ins(1,3,4,5)P_4$ and $PtdIns(3,4,5)P_3$ hydrolyzing activity (Damen et al., 1996). Pesesse et al. (1997) have reported the identification of cDNAs for a new SH2-domain-containing protein showing homology to the Ins 5-phosphatase SHIP and therefore referred to as SHIP2. SHIP2 differs at both N- and C-terminal ends with the sequence of INPPL-1. The translated sequence of SHIP2 encodes a 1258 amino acid protein with a predicted molecular mass of 142 kDa. Pesesse et al. (2001) have since shown that the SH2 domain containing Ins 5-phosphatase SHIP2 is recruited to the EGF receptor where it dephosphorylates $PtdIns(3,4,5)P3$ in EGF stimulated COS-7 cells.

Kisseleva et al. (2000) have reported the cDNA cloning and characterization of a novel human InsPs 5-phosphatase that has substrate specificity unlike any previously described members of this large family. This enzyme hydrolyzes only lipid substrates,

PtdIns(3,4,5)P_3 and Ptdns(4,5)P_2. The isolated cDNA comprises 3110 base pairs and predicts a protein product of 644 amino acids and Mr = 70 023. Kisseleva et al. (2000) have designated this 5-phosphatase as Type IV. It is a highly basic protein (pI = 8.8) and has the greatest affinity towards PtdIns(3,4,5)P_3 of all known 5-phosphatases. The activity of 5-phosphatase Type IV is sensitive to the presence of detergents in the in vitro assay. Thus, the enzyme hydrolyzes lipid substrates in the absence of detergents or in the presence of n-octyl β-glucopyranoside or Triton X-100, but not in the presence of cetyltriethylammonium bromide, the detergent that has been used in other studies of the hydrolysis of PtdIns(4,5)P_2.

Kisseleva et al. (2000) also showed that SHIP, a 5-phosphatase previously characterized as hydrolyzing only substrates with D-3 phosphates, also readily hydrolyzed PtdIns(4,5)P_2 in the presence of n-octyl β-glucopyranoside but not cetyltriethylammoniumbromide (Fig. 8.5).

Radio assays. According to Jackson et al. (1995), the PtdIns(3,4,5)P_3 5-phosphatase activity is best determined by a radioisotope assay. For this purpose, the necessary PtdIns([^{32}P]3,4,5)P_3 is generated by incubating partially purified PtdIns 3-kinase [30 nmol PtdIns(3)P formed/mg/min] with PtdIns(4,5)P_2 (50 μm) for 5 min at room temperature in the presence of PtdSer (75 μM) and [^{32}P]-labeled ATP (1 μCi/nmol). The assays were stopped with 100 μl 1N HCl and extracted with 200 μl chloroform/methanol (1:1) and 500 μl 2 M KCl. PtdIns([^{32}P]3,4,5)P_3 (10 pmol) was dried under nitrogen and finally resuspended in 20 mM HEPES, pH 7.2, 5 mM MgCl$_2$, 1 mM EDTA. The suspended lipid was sonicated (100 W) for 6 min on ice, prior to the addition of purified enzymes or cell extracts. Assays were performed in the presence or absence of 10 mM EDTA for various times at room temperature. Hydrolysis of PtdIns([^{32}P]3,4,5)P_3 to PtdIns([^{32}P]3,4)P_2 was determined by TLC analysis of the lipid products (Jackson et al., 1994). Individual lipids were scraped from the TLC plate and quantified by liquid

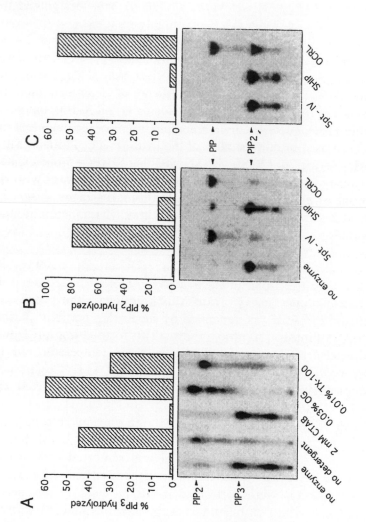

scintillation counting. PtdIns($[^{32}P]$3,4,5)P$_3$ 5-phosphatase activity was expressed as pmol PtdIns($[^{32}P]$3,4)P$_2$ formed/mg/min.

According to Woscholski et al. (1995), the $[^{32}P]$PtdIns(3,4,5)P$_3$/PtdIns(4,5)P$_2$ mixture was dried under vacuum and then dissolved in 2% cholate in 0.2 M Tris, pH 7.4, by sonication. The phosphatase activity was determined in the presence of 5 mM MgCl$_2$, 0.5 mM EGTA, 50 mM Tris, pH 7.4, and 0.5% cholate using between 3 and 30 μM $[^{32}P]$PtdIns(3,4,5)P$_3$ (containing 3–20-fold excess of PtdIns(4,5)P$_2$ in order to block detection of PtdIns(4,5)P$_2$ 5-phosphatases, which also served as a carrier lipid). The assay was stopped after 5 min of incubation, and the lipids were subsequently extracted with chloroform and dried down under vacuum. The extracted inositol lipids were then separated by TLC using n-propyl alcohol/2N acetic acid (1:1, v/v) (Kodaki et al., 1994). The spots corresponding to $[^{32}P]$PtdIns(3,4,5)P$_3$ and $[^{32}P]$PtdIns(3,4)P$_2$ were scraped off and counted. The calculated ratio of the $[^{32}P]$PtdIns(3,45)P$_2$/$[^{32}P]$PtdIns(3,4,5)P$_3$ was used to assess the 5-phosphatase activity in order to correct for any losses of lipid due to processing.

Kisseleva et al. (2000) assayed the novel InsPs 5-phosphatase by determining the hydrolysis of PtdIns(3,4,5)P$_3$. For this purpose PtdIns(3$[^{32}P]$4,5)P$_3$ was synthesized by incubation with PtdIns 3-kinase (Norris and Majerus, 1994).Thus, 0.5 mg of PtdIns(4,5)P$_2$ and 0.5 mg of PtdSer were dried under nitrogen and resuspended in 1 ml containing 20 mM HEPES, pH 7.6, 2 mM EDTA, and 5 mM

Fig. 8.5. Effect of detergents on enzymatic activity of inositol polyphosphate 5-phosphatases. (A) Effect of detergent on hydrolysis of PtdIns(3,4,5)P$_3$ by PtdIns 5-phosphatase type IV. (His-tagged PtdIns 5-phosphatase type IV (3.3 ng) was incubated for 20 min at 37°C with PtdIns(3$[^{32}P]$,4,5)P$_3$, 400 cpm/assay, in the presence of indicated detergents in the assay. (B and C) Hydrolysis of PtdIns(4,5)P$_2$ by 3.3 ng of PtdIns 5-phosphatase type IV 40 ng of OCRL, and 3 ng of SHIP in the presence of 0.03% n-octyl β-glucopyranoside (B) or 2 mM CTB (C) in the assay. PIP$_3$, position of PtdIns(3,4,5)P$_3$; PIP$_2$, position of PtdIns(4,5)P$_2$ and PIP, PtdInsP standards are marked with arrowheads.

Reproduced from Kisseleva et al. (2000) with permission of the publisher.

$MgCl_2$. After brief sonication, the PtdIns 3-kinase immunoprecipitate was added followed by $[\gamma^{32}P]ATP$ (2 mCi, 6000 Ci/mmol). After a 1 h incubation, the lipids were extracted twice with chloroform/methanol (1/1, v/v). The lipids were separated on TLC, and radio-labeled $PtdIns(3,4,5)P_3$ was scraped from the plate and extracted from silica gel with chloroform/methanol. For the phosphatase assay, radiolabeled lipids were dried under N_2 and resuspended in buffer containing 50 mM Tris-HCl, pH 7.5, 3 mM

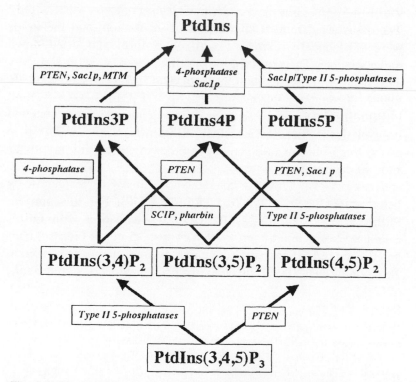

Fig. 8.6. The PtdInsP phosphatases. The arrows show the currently identified major flows of 3-phosphorylated PtdInsP metabolism. The appropriate substrates and products of the 3-, 4- and 5-phosphatases are indicated. Reproduced from Vanhaesebroeck et al. (2001) with permission, from the Annual Review of Biochemistry, volume 70 © 2001 by Annual Reviews www.annualreviews.org.

$MgCl_2$, and 0.03% n-octyl β-glucopyranoside and briefly sonicated on ice. The activity was assessed as described previously (Norris and Majerus, 1994; Matzaris et al., 1994). Palmer (1987) has described a lysoPtdIns lysophosphatase.

Proof of product. Kisseleva et al. (2000) deacylated the inositol lipids as described by Lips and Majerus (1989) and separated GroPIns derivatives on a PartiSphere SAX column with a gradient of 0–1.25 M NaH_2PO_4, pH 4.5. The gradient consisted of a 0–5 min linear rise to 29% for pump B, a 5–35 min linear rise to 60% B, and a 35–45 min linear rise to 100% B. Radio-labeled deacylated phospholipids were detected by a beta-RAM Flow Through System (In/US System, Inc., Tampa, FL).

Vanhaesebroeck et al. (2001) have recently summarized the major flows of intermediates in 3-phosphorylated PtdInsP metabolism including the pathways catalyzed by the PtdInsP 3-, 4-, and 5-phosphatases (Fig. 8.6).

8.3. Inositol phosphate phosphatases

Inositol monophosphate phosphatases were the first to be discovered and exhibit the least positional and stereochemical specificity (Eisenberg, 1967). Subsequently, a great variety of other InsP phosphatases have been discovered (Murray and Greenberg, 2000) as well as InsP pyrophosphate phosphatases (Majerus et al., 1999; Murray and Greenberg, 2000; Irvine and Schell, 2001). Shears (2001) has outlined several rigorous criteria for assessing the metabolic status of $InsP_6$, which may be indirectly applicable to the assessment of the metabolic status of other InsPs.

8.3.1. Inositol monophosphate phosphatases

8.3.1.1. Ins(1)P phosphatases
myo-Ins(1)P phosphatase was first studied in yeast as a component of the pathway by which D-glucose-6-phosphate is

converted to myo-Ins via the intermediate L-myo-Ins(1)P. Eisenberg (1967) found that myo-Ins(1)P phosphatase from several rat organs hydrolyzed D- and L-Ins(1)P, but was inactive with myo-Ins(2)P. The testis enzyme was found to hydrolyze (−)-chiro-Ins(3)P and α-glycerophosphate, while other phosphate esters were not substrates. myo-Ins(1)P phosphatase has been partially purified from bovine brain (Hallcher and Sherman, (1980). The enzyme has a molecular weight of about 58 000. Both L-myo-Ins(1)P and D-myo-Ins(1)P are hydrolyzed by the enzyme as well as (−)chiro-Ins(3)P and 2′-AMP. myo-Ins(2)P and PtdInsP$_3$ are not substrates. The enzyme is completely dependent on Mg^{2+}, while calcium and manganese ions are competitive inhibitors of Mg^{2+}. LiCl inhibits the hydrolysis of both L- and D-myo-Ins(1)P to the extent of 50% at a concentration of 0.8 mM. Ins(1)P phosphatase is frequently employed as part of the assay systems for InsPs 1-phosphatase.

Ackermann et al. (1987) have examined several InsP- and InsPs as potential substrates of Ins(1)P- phosphatase along with the products formed by the action of the enzyme. The Ins(1)P, Ins(4)P and Ins(5)P, which occur in the cerebral cortex of LiCl treated rats in the relative proportions of 10:1:0.2, were hydrolyzed at comparable rates by the myo-Ins(1)P-phosphatase from bovine brain. The hydrolysis of Ins(4)P was inhibited by Li$^+$ to a greater degree than the hydrolysis of Ins(1)P and Ins(5)P. D-Ins(1,4,5)P$_3$ and D-Ins(1,4)P$_2$ were neither substrates nor inhibitors of Ins(1)P-phosphatase. The enzyme hydrolyzed Ins(1,4)P$_2$ but released Ins(4)P at a greater rate than Ins(1)P. The following other InsPs were neither substrates nor inhibitors of Ins(1)P-phosphatase: Ins(1,6)P$_2$, Ins(1,2)P$_2$, Ins(1,2,5,6)P$_4$, Ins(1,2,4,5,6)P$_5$ and Ins(1,3,4,5,6)P$_5$. Also, myo-Ins(1,2)-cycP was neither substrate nor inhibitor for Ins(1)P-phosphatase.

Diehl et al. (1990) used oligonucleotide probes based on partial amino acid sequence data for the bovine brain enzyme and obtained several overlapping cDNA clones. All contained an open reading frame encoding a 277-amino acid protein. The protein expressed in a bacterium reacted with an anti-inositol monophosphatase monoclonal antibody and gave Li^+-sensitive hydrolysis of Ins(1)P. Parthasarathy et al. (1993) purified to homogeneity the *myo*-Ins(P) phosphatase from rat testes. It showed an apparent native molecular weight of 58 000 as determined by gel filtration chromatography and was composed of two identical subunits of molecular weight of 29 000 as determined by SDS-PAGE. Li^+ ions inhibited this phosphatase non-competitively. The enzyme acted on both the D- and L-forms of *myo*-Ins(1)P. This specific phosphatase was not inhibited by free *myo*-Ins or by *scyllo*-, *allo*-, and D-*chiro*-Ins isomers at 5 mM concentrations. In addition to *myo*-Ins(1)P, the *myo*-Ins(3)P, *myo*-Ins(4)P and *myo*-Ins(6)P (which are equatorial) all acted as substrates, but not *myo*-Ins(2)P, which is equatorial. According to Gee et al. (1988) the D-*myo*-Ins(4)P and D-*myo*-Ins(6)P, are also substrates. The phosphatase was notably heat stable (80°C for 15 min). The brain Ins(1)P phosphatase has been subsequently identified as galactose 1-phosphatase (Parthasarathy et al., 1997). Hydrolysis of both Gal(1)P and Ins(1)P was similarly sensitive to both the monovalent and divalent cations lending support that the enzymatic activity is mediated by a single enzyme (Table 8.1).

Lopez et al. (1999) have shown that the yeast InsP monophosphatases are lithium and sodium sensitive enzymes encoded by a non-essential gene pair. These enzymes participate in the inositol cycle of calcium signaling and in inositol biosynthesis. Two open reading frames with homology to animal and plant InsP are present in the yeast genome. The two recombinant purified proteins were shown to catalyze Ins(1)P hydrolysis sensitive to lithium and sodium. Lopez-Coronado et al. (1999) have reported a mammalian lithium-sensitive enzyme with $3'$-phosphoadenosine $5'$-phosphate and InsPs 1-phosphatase activity.

Chen et al. (1999) have recently overexpressed in *Escherichia coli* an extremely heat-stable tetrameric form of Ins(1)P phosphatase from the hyperthermophilic bacterium *Thermotoga maritime*. In addition to its quaternary structure, this enzyme

TABLE 8.1

Specific IMPase activities (micromoles of phosphate liberated/min/mg protein) from different sources. The final enzyme preparations were electrophoretically homogeneous as evidenced by a single 29 kDa protein on SDS-PAGE analysis

Enzyme source	D-1-*myo*-InsP	D-3-*myo*-InsP	D-Galactose-1P
Bovine brain	6.1 ± 0.10	5.9 ± 0.12	6.0 ± 0.14
Human brain	8.1 ± 0.19	8.1 ± 0.18	8.1 ± 0.11
Rat brain	7.6 ± 0.08	7.7 ± 0.13	7.8 ± 0.08
Bovine brain (heated)	5.7 ± 0.16	5.6 ± 0.14	5.7 ± 0.08
Human brain (heated)	7.7 ± 0.32	8.1 ± 0.34	7.7 ± 0.41
Rat brain (heated)	7.2 ± 0.15	7.2 ± 0.16	7.3 ± 0.62

A final concentration of 0.7 mM of substrates was used for the determination of enzyme activity. Each value represents mean ± SD of triplicate determinations. Values in the lower half of the table were obtained after heating at 80°C for 15 min. A small decrease of specific activity was observed by non-specific heat inactivation. Reproduced from Parthasarathy et al. (1997) with permission of the publisher.

displayed a 20-fold higher rate of hydrolysis of D-Ins(1)P than of the L-isomer. The subunit mass of the enzyme was 29 kDa, which is similar to that of the subunit mass of the usual dimeric Ins(1) phosphatases.

$T.\ maritime$ is a eubacterium, but its Ins(1)P phosphatase is more similar to Ins(1)P phosphatase from another hyperthermophile, $Methanococcus\ jannaschii$, the other known bacterial or mammalian Ins(1)P phosphatases with respect to substrate specificity, Li^+ inhibition, inhibition by high Mg^{2+} concentrations, metal ion activation, heat stability, and activation energy. In parallel studies, Stec et al. (2000) have shown that the protein product of $M.\ jannaschii$ gene MJ0109, which had been tentatively identified as an inositol monophosphatase, has both the monophosphatase and the fructose-1,6-bisphosphatase activities. Stec et al. (2000) note that by this finding they have located 'the missing' archaeal fructose-1,6-bisphosphatase.

Nigou and Besra (2002) have recently characterized the inositol monophosphatase activity in $Mycobacterium\ smegmatis$. The enzyme hydrolyzed Ins(1)P to free Ins, but other mono-, di- or poly InsPs were not tested.

Radioisotope assays. Although the hydrolysis of the InsP monophosphatase activity can be readily followed by the liberation of either [^{3}H]Ins or [^{32}P]Pi from appropriately labeled substrates, there has been usually sufficient material for mass analysis based on colorimetry or mass spectrometry.

Nigau and Besra (2002) have described the following radio-active assay for conversion of [^{3}H]Ins(1)P into [^{3}H]Ins. The reaction mixture contained, in a total volume of 100 μl, 0.2 mM [^{3}H]Ins(1)P (3.7 × 10^3 cpm/nmol), 6 mM MgCl$_2$ and 25 μg of protein in 50 mM Tris/HCl, pH 7.5. The reaction mixture was incubated at 37°C for 20 min. The reaction was terminated by the addition of 0.9 ml of water and cooled on ice. Ins and Ins(1)P were resolved by passage of the reaction mixture through a 1 ml SAX (strong ion exchange) column. Free Ins was eluted with 3 ml of water, while residual Ins(1)P remained bound to the SAX column. The total cpm of radiolabeled material eluted from the column was measured by scintillation counting.

Mass assays. For routine work assaying large numbers of chromatographic fractions for phosphorus the calorimetric Malachite method described by Eisenberg and Parthasarathy (1987) has proven useful. Three tenths of a milliliter of the reaction mixture consisting of 0.05 ml of 4.2 mM HCl, pH 7.8, 0.05 ml of 18 mM MgCl$_2$, 0.05 ml of 4.2 mM L-*myo*-inositol 1-phosphate or D-*myo*-inositol 1-phosphate, 0.1 ml of water, and 0.05 ml of enzyme was incubated at 37°C for 10–15 min. Triton X-100 at a final concentration of 0.5% was included in the assay mixture after the heated step to reduce the aggregation of the enzyme. The reaction was stopped by the addition of 0.05 ml of 20% trichloroacetic acid. The suspension was centrifuged and 0.1 ml of supernatant was used to estimate the liberated inorganic phosphate by the method utilizing Malachite Green reagent.

The activity of inositol monophosphatase can be also determined by a continuous fluorescence assay using a foreign substrate, the fluorescent compound 4-methylumbelliferyl phosphate (Gore et al., 1992). The hydrolysis of the phosphate group from this compound is readily detected by a resultant large red shift in the emission spectrum from 390–450 nm.

Ackermann et al. (1987) determined the activity of the Ins(1)P-phosphatase with the various substrates by incubations in presence of 50 mM Tris/HCl, pH 7.8, and 3 mM Mg^{2+}. Incubations in all cases were for 30 min. Ins(1)P, Ins(4)P, Ins(5)P and myo-inositol were measured by GC/MS of their TMS derivatives as described by Sherman et al. (1986). The phosphate moiety was found by direct fragmentation to produce ions of useful intensity with specific carbon retention. Separation of the enantiomers of Ins(1)P was carried out on pentakis(trimethylsilyl)-myo-inositol dimethylphosphate as described by Leavitt and Sherman (1982), but with a fused-silica column coated with Chirasil-Val (Alltech Associates, Waukegan, IL, USA). The Chirasil-Val column is suitable for GC/MS also. The composition of each substrate (i.e. the inositol/phosphate content) was determined by FAB/MS (Sherman et al., 1986). The inositol phosphates, particularly the InsPs, had been difficult to study by chromatography because of their extreme polarity and fast atom bombardment/mass spectrometry (FAB/MS) offered a new approach to their analysis. The high sensitivity of FAB/MS allowed the measurement of the lesser InsPs to a level of about 10 nmol. Despite this success, most workers in the area have preferred other methods, such as TLC and ion exchange chromatography for their work (see below).

8.3.1.2. Ins(3)P phosphatases

It was already noted (Section 8.2.1) that the three InsP, i.e. Ins(1)P, Ins(3)P and Ins(4)P, were converted to Ins by Ins(1)P-phosphatase (Ackermann et al., 1987).

Van Dijken et al. (1996) investigated Dictyostelium discoideum inositol monophosphatase activity. Partial purification of the proteins in the soluble cell fraction using anion-change chromatography revealed the presence of at least three enzyme activities capable of degrading inositol monophosphate isomers. One activity was similar to the monophosphatase found in all mammalian cells, as it degraded Ins(4)P, Ins(1)P and to a lesser extent Ins(3)P, was dependent on $MgCl_2$ and inhibited by LiCl in a uncompetitive manner. The second enzyme activity was specific for Ins(4)P, not dependent on $MgCl_2$ and not inhibited by LiCl. The third monophosphatase activity degraded especially Ins(3)P, but also Ins(4)P and Ins(1)P, and was inhibited by increasing concentrations of $MgCl_2$, whereas LiCl had no effect.

Radioisotope assays. Van Dijken et al. (1996) assayed the Ins monophosphatase activities, including the 4-phosphatase activity, using a 10 μl incubation containing about 2000 dpm of $[^3H]InsP$ and various concentrations of $MgCl_2$, LiCl and unlabeled InsP with 10 μl enzyme fraction obtained after Mono Q separation. Reactions were stopped by the addition of 500 μl of chloroform/methanol/fuming HCl (20:40:1, by vol.). Phase separation was induced by the addition of 200 μl of water. After brief centrifugation (2 min, 14 000g), the aqueous phase was applied to 0.25 ml Dowex columns that were eluted with 6 ml of water of 6 ml of 150 mM ammonium formate/5 mM sodium tetraborate to elute inositol and inositol monophosphates, respectively. After addition of scintillation cocktail, radioactivity in the samples was determined in a liquid scintillation counter.

Mass assays. Ackermann et al. (1987) performed incubations for the assessment of $InsP_2$ phosphatase activity in either a synthetic buffer made 1 mM in EGTA or in a 'physiological' buffer containing NaCl (10 mM), potassium glutamate (70 mM), KCl (30 mM), $MgSO_4$ (4 mM) and EGTA (1 mM) in 10 mM HEPES adjusted to pH 7.2, as used by Storey et al. (1984). The released *myo*-Ins was measured by GC/MS of its TMS derivative.

Van Dijken et al. (1996) determined the activity of the InsP monophosphatases, including the 4-phosphatase, by incubating fractions obtained after anion exchange chromatography in a mixture containing 30 mM p-nitrophenylphosphate and various concentrations of LiCl and $MgCl_2$. The reaction was followed in a 96-well microtitre plate by measuring A_{415} with a Biorad microplate reader. Reactions were linear with time for at least 3 h.

Chen and Roberts (1999) determined the activity of the tetrameric monophosphatase from the hyperthermophilic bacteria by measuring colorimetrically the released inorganic phosphate by the micromethod of Itaya and Ui (1966). To monitor the InsP phosphatase activity of column fractions, each reaction mixture was adjusted to contain approximately 1 to 2 μl of 10 mM D-Ins(1)P (in 50 mM Tris buffer, pH 8.0), approximately 1 to 2 μl of 200 mM $MgCl_2$ (in 50 mM Tris buffer, pH 8.0), and 2 to 10 μl of a fraction. The amount of the enzyme used was adjusted to give an absorbance at 600 nm of approximately 0.3–0.4 within approximately 1–2 min. After incubation at 95°C for 2 min, the mixtures were quickly chilled on ice. All the substrates were quite stable (i.e. no detectable Pi was released) for at least 5 min at 100°C. The liberated Pi was measured by performing a colorimetric phosphate assay with ammonium molybdate Malachite Green reagent (Itaya and Ui, 1966). The released Pi was quantified by reference to standard curves and the specific activity of the enzyme was calculated by normalizing data to protein concentrations.

8.3.2. Inositol bisphosphate phosphatases

The $InsP_2$ phosphates are dephosphorylated in a series of reactions, the primary function of which would appear to be the recovery of Ins (Downes, 1988). These pathways are catalyzed by a minimum of four distinct enzymes, two of which are Mg^{2+}-dependent and potently inhibited by the anti-manic drug. Li^+. The other two are metal-ion independent enzymes (Majerus et al., 1988).

8.3.2.1. $InsP_2$ 1-phosphatase

According to Ackermann et al. (1987) L-Ins(1,4)P_2 is a substrate for brain Ins(1)P phosphatase. Li^+, but not Ins(2)P, inhibited the hydrolysis of L-Ins(1,4)P_2. D-Ins(1,4)P_2 or D-Ins(1,4,5)P_3 were neither substrates nor inhibitors of the enzyme. The following other InsPs were neither substrates nor inhibitors of Ins(1)P phosphatase: Ins(1,6)P_2, Ins(1,2)P_2, Ins(1,2,5,6)P_4, Ins(1,2,4,5,6)P_5, Ins(1,3,4,5,6)P_5 and phytic acid.

Inhorn and Majerus (1987) have also described the properties of an enzyme which cleaves the 1-phosphate from Ins(1,4)P_2 yielding Ins(4)P. Inhorn and Majerus (1988) have purified the enzyme to homogeneity from calf brain. The enzyme hydrolyzes 50.3 μmol of Ins(1,4)P_2/min/mg protein. It has a mass of 44 000 Da and appears to be monomeric. Calcium inhibits hydrolysis of Ins(1,4)P_2 with about 40% inhibition occurring at 1 μM free Ca^{2+}. The enzyme also hydrolyzes Ins(1,3,4)P_3.

Radioisotope assays. The [^{3}H]inositol assay of Ins(1,4)P_2 1-phosphatase is performed as described by Inhorn and Majerus (1987). Enzymatic hydrolysis of Ins(1,4)P_2 is assayed by one of two methods. Assay 1 is performed with [^{3}H]Ins(1,4)P_2 and InsPs 1-phosphatase incubated in 50 mM HEPES, pH 7.5, 3 mM $MgCl_2$/100 mM KCl, 0.5 mM EGTA and an excess of partially purified Ins(1)P monophosphatase (Hallcher and Sherman, 1980). The reaction mixture is loaded onto a 0.4 ml AG 1-X8 formate column and [^{3}H]inositol, the product of this coupled enzyme reaction, is eluted with water.

Assay 2 was performed as above except that inositol mon-ophosphate phosphatase was omitted. The product, [^{3}H]Ins(4)P, was eluted from the Dowex-formate column with 0.05 M NH_4-HCOOH containing 0.1 M HCOOH. Briefly, Assay 1 involved incubation of [^{3}H]Ins(1,4)P_2 and InsPs 1-phosphatase with 50 mM HEPES (pH 7.5), 3 mM $MgCl_2$, 100 mM KCl, 0.5 mM EGTA, and an excess of partially purified InsP phosphatase (Hallcher and Sherman (1980)). The reaction mixture was loaded onto a 0.4 ml

Dowex-formate column, and [^{3}H]Ins, the product of this coupled enzyme reaction, was eluted with water (Inhorn and Majerus, 1988). Assay 2 was performed as above except that inositol monophosphate phosphatase was not added. The product, [^{3}H]Ins(4)P, was eluted with 0.05 M NH$_4$COOH containing 0.1 M HCOOH.

Mass assays. Ackermann et al. (1987) determined the concentrations and purities of the inositol phosphates after alkaline phosphatase digestion by measurement of inositol by GLC (see above) and Pi by colorimetry (see Section 8.2.1.2). Incubations with Ins(1)P phosphates contained 50 mM Tris-HCl, pH 7.8, and 3 mM Mg^{2+}, with changes or additions as required. Incubations for the assessment of InsP$_2$ phosphatase activity were carried out either in the latter made 1 mM in EGTA or in a 'physiological' buffer, containing NaCl (10 mM), potassium glutamate (70 mM), KCl (30 mM), MgSO$_4$ (4 mM) and EGTA (1 mM) in 10 mM HEPES adjusted to pH 7.2, as used by Storey et al. (1984). All incubations were terminated after 30 min.

Sun et al. (1990) have described the separation and quantification of isomers of InsPs by ion chromatography. A gradient elution with 16.5–112.5 mM NaOH is used to obtain an excellent sequential elution of InsP, InsP$_2$ and InsP$_3$ with nearly complete baseline resolution for the common isomers of each phosphoinositol subclass.

8.3.2.2. InsP$_2$ 3-phosphatases

Caldwell et al. (1991) have demonstrated that the InsPs 3-phosphatase can hydrolyze Ins(1,3)P$_2$ and PtdIns(3)P. The substrate specificity of the 3-phosphatase appears to be regulated by a 78 kDa non-catalytic subunit, which when present increases the rate of hydrolysis of PtdIns(3)P three-fold. The 75-kDa Ins(1,4,5)P$_3$ 5-phosphatase demonstrates similar function to the 3-phosphatase, as it can metabolize both a water-soluble inositol phosphate and a phosphoinositide.

Inhorn et al. (1987) have reported the presence in crude extracts of brain of an enzyme that utilizes Ins(3,4)P$_2$ as substrate forming

Ins(3)P. The enzyme is magnesium-independent and is not inhibited by Li^+. Bansal et al. (1987) have shown that three different $InsP_2$ formed from calf brain supernatant are each further metabolized by a separate enzyme. The three inositol monophosphates, i.e. Ins(1)P, Ins(3)P and Ins(4)P, are converted to Ins by InsP 1-phosphatase.

A recent addition to the InsP 3-phosphatase group is the product of the tumor suppressor gene PTEN, which was originally identified from a screen directed at a human genetic locus deleted in a variety of human cancers (Maehama and Dixon, 1999). Although it possesses some tyrosine protein phosphatase activity, PTEN has a higher turnover number for dephosphorylation of the 3-phosphate group of phosphatidylinositol phosphates and soluble inositol polyphosphates (Maehama and Dixon, 1998). Specifically, PTEN hydrolyzed $Ins(1,3,4,5)P_4$ to produce $Ins(1,4,5)P_3$.

Radioisotope assays. Caldwell et al. (1991) assessed the hydrolysis of $Ins(1,3)P_2$ by incubating $[^3H]Ins(1,3)P_2$ (1300–2800 cpm/pmol, 2000 cpm) with the enzyme preparation in 50 mM MES, pH 6.5, 5 mM EDTA, 0.2 mg of BSA/ml, and 8 mM 2-mercaptoethanol for 10 min at 37°C in 25 μl. Reactions were stopped by addition of 1 ml of water and applied to a 0.5 ml Dowex A G1-X8 column previously equilibrated with 40 mM ammonium formate. The product, $[^3H]Ins(1)P$ was separated from $[^3H]Ins(1.3)P_2$ by sequential elution in two 2.5 ml fractions with 0.2 M ammonium formate, 0.1 M formic acid and then counted in a liquid scintillation counter.

According to Bansal et al. (1987, 1990) $Ins(1,3)P_2$ 3-phosphatase activity is measured by incubating calf brain supernatant in a 25 μl reaction mixture containing $[^3H]Ins(1,3)P_2$ (14 μM, 1200 cpm/nmol), 50 mM MES (pH 6.5), and 5 mM EDTA for 10 min at 37°C. The product $[^3H]Ins(1)P$ was separated from $[^3H]Ins(1,3)P_2$ by the elution of the former from 0.50 ml column with 8 ml of 0.1 M formic acid, 0.2 M ammonium formate, followed by elution of the latter with 10 ml of 1 M formic acid,

0.4 M ammonium formate. The product was counted in a scintillation counter.

Mass assays. The GC/MS systems described by Ackermann et al. (1987) are obviously applicable here, even though the researchers actually conducting the above experiments did not utilize them.

A mass assay for the 3-phosphatase may be performed as described by Itaya and Ui (1966), except that after phase separation the removed water phase was treated with ammonium molybdate, and subsequently the phosphomolybdate complex was extracted with ethyl acetate. The extracted phosphomolybdate complex represented the orthophosphate in the reaction mix, whereas the remaining water phase contained the water-soluble InsPs, the potential products of any phospholipase activity, if present. HPLC analysis was performed as described by Kodaki et al. (1994).

Inhorn et al. (1987) first described that in crude extracts of calf brain, Ins(1,3,4)P$_3$ is first converted to Ins(3,4)P$_2$, then the Ins(3,4)P$_2$ is further converted to Ins(4)P. Actually, the release of Ins(4)P from the Ins(3,4)P$_2$ substrate was not specifically demonstrated.

The released Ins monophosphates could have been detected and identified by GC/MS as described by Ackermann et al. (1987).

8.3.2.3. InsP$_2$ 4-phosphatase

Ackermann et al. (1987) have also investigated the action of rat brain InsP$_2$ 4-phosphatase on Ins(1,4)P$_2$, which releases the 4-phosphate group. The 1-phosphate moiety of Ins(1,4)P$_2$ is actually removed more rapidly than the 4-phosphate, and both the Li$^+$ sensitivity and the activity of the bisphosphatase are less than for the monophosphatase.

Inhorn et al. (1987) reported that in crude extracts of calf brain, Ins(1,3,4)P$_3$ is first converted to Ins(3,4)P$_2$, then the Ins(3,4)P$_2$ is further converted to Ins(3)P. It was found that the same enzyme also converted Ins(1,4)P$_2$ to Ins(4)P. in vitro, the PtdInsP 4-phosphatases catalyze the hydrolysis of the D-4 position of the soluble analogues of these lipids, $Ins(3,4)P_2$ and $Ins(1,3,4)P_4$ (Bansal et al., 1990; Norris et al., 1995; 1997). Vyas et al. (2000) have demonstrated an inositol polyphosphate 4-phosphatase type I activity in megakaryocytes from transcription factor GTA-1 lacking mice. This enzyme hydrolyzes $PtdIns(3,4)P_2$ and also has lesser activity against soluble analogues of this lipid, $Ins(3,4)P2$ and $Ins(1,3,4)P_3$. Reintroduction of 4-phosphatase into both primary GATA-1$^-$ and wild-type megakaryocytes significantly retarded cell growth.

Radioisotope assays. Assays for 4-phosphatase could be performed as described for $Ins(3,4)P_2$ 4-phosphatase activity as described by Bansal et al. (1987). It could be measured by incubating calf brain supernatant with 50 mM MES, pH 6.5, 5 mM EDTA and $[^3H]Ins(3,4)P_2$ (0.3 μM, 500 cpm) in 50 μl of incubation mixture at 37°C. The product $[^3H]Ins(1)P$ is separated from $[^3H]Ins(1,4)P_2$ on Dowex formate. Separation of $Ins(1)P$ and $Ins(1,4)P_2$ is achieved by elution of the former from 0.5 ml columns with 8 ml of 0.1 M formic acid, 0.2 M ammonium formate, followed by elution of the latter with 10 ml of 0.1 M formic acid, 0.4 M ammonium formate.

Mass analysis. The mass analyses of the substrates and products of the $InsP_2$ 4-phosphatase were assayed as described by Ackermann et al. (1987) for $Ins(1,4)P_2$ 1-phosphatase (see above). $[^3H]Ins(3,4)P_2$ was prepared by incubation of $[^3H]Ins(1,3,4)P_3$ with InsPs 1-phosphatase and isolated by Dowex-formate chromatography using a linear gradient (200 ml) of 0.05–1 M NH_4COOH (pH 3.45). $[^3H]Ins(3,4)P_2$ (0.03 nmol) was incubated 20 min at 37°C with 25 μg of crude calf brain supernatant, 50 mM MES (pH 6.5), and 3 mM $MgCl_2$ in 0.1 ml, and the products were separated by Dowex-formate chromatography, in relation to elution positions of standards (Inhorn et al., 1987).

According to Bansal et al. (1987, 1990) $Ins(3,4)P_2$ 4-phosphatase activity was measured by incubating calf brain supernatant with 50 mM MES (pH 6.5), 5 mM EDTA, and [^{3}H]Ins(3,4)P$_2$ (5–100 μM, 500–1000 cpm/nmol) in 50 μl of incubation mixture at 37°C. The reaction was stopped with 1 ml of cold water and applied to a 0.4 ml Dowex formate column. The product [^{3}H]Ins(3)P was separated from[^{3}H]Ins(3,4)P$_2$ on Dowex formate. Separation of Ins(3)P and Ins(3,4)P$_2$ was achieved by elution of the former from 0.50 ml columns with 8 ml of 0.1 M formic acid, 0.2 M ammonium formate, followed by elution of the latter with 10 ml of 1 M formic acid, 0.4 M ammonium formate. The product was counted in a scintillation counter.

The Ins(4)P product of the action of the 3-phosphatase can be detected by GC/MS as described by Ackermann et al. (1987) although such an assay was not performed by Inhorn et al. (1987).

8.3.2.4. InsP$_2$ 5-phosphatases

InsPs 5-phosphatases hydrolyze the D-5 position of InsPs and corresponding phospholipids. Eight InsPs 5-phosphatases and several splicing variants have been isolated and classified into three groups according to their substrate specificity (Ijuin et al., 2000). Type I 5-phosphatases hydrolyze only water-soluble substrates, Ins(1,4,5)P$_3$ and Ins(1,3,4,5)P$_4$, but apparently does not hydrolyze Ins(4,5)P$_2$ to Ins(4)P. The 5-phosphatases, which convert PtdIns(4,5)P$_2$ to PtdIns(4)P (see below) apparently do not hydrolyze Ins(4,5)P$_2$, although there is no evidence that this transformation would have been tested (Bansal et al., 1990; Zhang et al., 1995).

8.3.3. Inositol trisphosphate phosphatases

Enzymes that terminate the signal-transduction processes and regulate the levels of soluble inositol phosphate messengers are essential for proper cell function. Positionally specific phosphatases have been discovered, isoforms identified, and characterized in mammalian cells and have been characterized biochemically and by molecular cloning.

8.3.3.1. InsP$_3$ 1-phosphatase

InsPs 1-phosphatase hydrolyzes the 1-phosphate ester from two substrates of the PtdIns signaling pathway, Ins(1,3,4)P$_3$ and Ins(1,4)P$_2$ (Inhorn et al., 1987; Inhorn and Majerus, 1987). Other 1-phosphate-containing InsPs are not substrates (Inhorn et al., 1987). InsPs 1-phosphatase is non-competitively inhibited by Li$^+$ ions, a drug used to treat patients with psychiatric disorders. The enzyme converts Ins(1,3,4)P$_3$ to Ins(3,4)P$_2$. Another enzyme, which hydrolyzes Ins(1,3,4)P$_3$ is InsPs 4-phosphatase, yielding Ins(1,3)P$_2$ (Bansal et al., 1987). It is Mg^{2+} independent and Li$^+$-insensitive. York and Majerus (1990) have described an improved method of preparation of the enzyme from bovine brain. Bovine InsPs 1-phosphatase is a monomeric protein with a molecular mass of 44 000 Da (York et al., 1994). The low abundance of InsPs 1-phosphatase in tissues has precluded structural studies requiring large quantities of enzyme. York et al. (1994) has used recombinant *Baculovirus* harboring the cDNA of bovine InsPs 1-phosphatase to infect *Spodoptera frugiperda* (Sf9) insect cells. Recombinant protein was purified to homogeneity. The enzyme produced in Sf9 cells was similar to the native purified protein as determined by immunoblotting catalytic properties, and inhibition by Li$^+$ ions. Van Lookeren Campagne et al. (1988) have found that Ins(1,4,5)P$_3$ may be dephosphorylated by 1 phosphatase to yield Ins(4,5)P$_2$.

Majerus et al. (1999) have reviewed the role of InsPs 1-phosphatase in diverse cellular signaling reactions. In contrast to the large 5-phosphatase gene family (see below), there is a single gene that encodes InsPs 1-phosphatase. It is a member of the magnesium dependent, lithium-inhibited gene family that is characterized by a common structural fold (York et al., 1995). The authors point out that 1-phosphatase is a bona fide target for the therapeutic action of lithium, and other inhibitors of this enzyme might be used in the treatment of this disease.

Radioisotope assays. The purified recombinant InsPs 1-phosphatase was assayed utilizing [^{3}H]Ins(1,3,4)P$_3$ as substrate. After incubation at 37°C reaction mixtures were loaded onto 0.2 ml Dowex-formate column, equilibrated with 0.425 M NH_4COOH, 0.1 M HCOOH and the product [^3H]Ins(3,4)P_2 was eluted with 20 column volumes of buffer A. Radioactivity was measured by liquid scintillation counting. Enzyme reactions utilizing [^3H]Ins(1,4)P_2 as a substrate were loaded onto 0.2 ml Dowex-formate columns equilibrated with 0.05 M NH_4COOH, 0.2 M HCOOH and the product [^3H]Ins(4)P was eluted with the same buffer (York et al., 1994).

The radio assay of Ins(1,3,4)P_3 1-phosphatase is performed as described by Inhorn and Majerus (1987). In each assay, the enzyme was diluted appropriately so that less than 15% of the substrate was utilized. Inhorn and Majerus (1988) estimated the relative rates of degradation of Ins(1,3,4)P_3 via Mg^{2+}-dependent InsPs 1-phosphatase and by Mg^{2+}-independent InsPs 4-phosphatase. The relative activity of these two pathways in crude tissue homogenates was estimated by comparing the rates of hydrolysis in the presence and absence of EDTA.

Inhorn and Majerus (1988) have reported an experiment which confirms that Ins(3,4)P_2 is in fact, the major product of Ins(1,3,4)P_3 degradation in rat liver homogenates. For this demonstration crude rat liver homogenate supernatant (50 µg), 50 mM HEPES, pH 7.2, and 3 mM $MgCl_2$ in 50 µl total volume was incubated with [^3H]Ins(1,3,[^{32}P]4)P_3 (1.7 nmol) for 20 min at 37°C. The reaction mixture was chromatographed on an HPLC strong anion exchange column as described (Bansal et al., 1987). Elution positions of standards are indicated: Ins, inositol; Ins-1-P, Ins(1)P; Ins-3-P, Ins(3)P; Pi, inorganic phosphate. Ins(1,4)P_2 comigrates with Insi(1,3)P_2 in this chromatographic system.

Mass assays. Mass assays of Ins(1,3,4)P_3 have been reported by Mayr (1990) using HPLC with post-column metal-dye detection (MDD) system. MDD is a post-column derivatization method that does not involve any covalent modification of the substances to be detected; it is simply a kind of 'on-line complexometry'.

The reactions take into account a 1:1 stoichiometry of the dye and the detector metal in the complex, as in the case of yttrium. The basal absorbance level is usually high and the negative absorbance peak observed follows precisely the formation of metal-anion complex.

8.3.3.2. $InsP_3$ 3-phosphatases

Bansal et al. (1987) observed that InsPs 3-phosphatase degraded $Ins(1,3)P_2$ to $Ins(1)P$ in the absence of Mg^{2+} and was not inhibited by lithium. The 3-phosphatase did not act upon $Ins(1,3,4)P_3$, $Ins(3,4)P_2$, or $Ins(3)P$. However, a 3-phosphatase enzyme activity that converts $Ins(1,3)P_2$ to $Ins(1)P$ was also recognized. This activity was also Mg^{2+}-independent and was not inhibited by Li^+.

Recent studies suggest a role for InsPs phosphatase in the virulence of *Salmonella* infections. SopB, a protein secreted by *Salmonella dublin*, has been shown to have a sequence similarity to the 4-phosphatase active site motif (Norris et al., 1998). SopB encodes a phosphatase that hydrolyzes many but not all InsPs including $Ins(1,3,4,5,6)P_5$ yielding the product $Ins(1,4,5,6)P_4$ (Eckmann et al., 1997).

A recent addition to the 3-phosphatase group is the product of the tumor suppressor gene PTEN homologue, which functions as phospholipid phosphatase (Maehama and Dixon, 1999). Kinetic parameters for PTEN determined using either $PtdInsP_3$ or its soluble headgroup analog, $Ins(1,3,4,5)P_4$ have provided a clear indication as to why PTEN is such an inefficient catalyst with other PP substrates. The Km of PTEN towards $PtdIns(3,4,5)P_3$ or $Ins(1,3,4,5)P_4$ is at least 250-fold lower than that of pNPP. Furthermore, although the Km of PTEN towards $PtdInsP_3$ and $Ins(1,2,3,4,5)P_4$ differs only two-fold, dephosphorylation of $PtdInsP_3$ proceeds to a 400-fold greater rate than that of $Ins(1,3,4,5)P_4$, confirming that the lipid is its preferred substrate. PTEN is also capable to hydrolyze the soluble $Ins(1,3,4)P_3$ (Maehama et al., 2001).

Radioisotope assays. Norris and Majerus (1998) assayed the recombinant SopB (InsPs 3-phosphatase) (0.6 µg) by incubating with Ins(1[^{32}PO$_4$],3,4,5,6)P$_5$ in 50 mM Mops (pH 7.0), 10 mM EDTA, and 1 mM DTT for 30 min at 37°C. The SopB hydrolysis products of Ins1[^{32}P],3,4,5,6)P$_5$ were determined by HPLC on Adsorbosphere SAX columns as described by Wilson and Majerus (1996). Excellent separations were obtained for Ins(1,4)P$_2$, Ins(1,3,4)P$_3$, Ins(1,4,5)P$_3$, Ins(1,3,4,6)P4, Ins(1,3,4,5) P$_4$, Ins(1,4,5,6)P$_4$, and Ins(1,3,4,5,6)P$_5$. The peaks were collected and the radioactivity counted.

Mass assays. Stephens (1990) has described SAX and WAX HPLC columns for the separation of isomeric InsP$_5$, while Spencer et al. (1990) have described polyethyleneimine–cellulose TLC for the same purpose. The PEI-cellulose plates were developed on 0.5 M HCl and sprayed for phosphate to quantify the product present in an extract, appropriate areas of the TLC plate were cut out, extracted and the radioactivity or inorganic phosphate determined. See also Mayr (1990).

8.3.3.3. InsP$_3$ 4-phosphatases

Bansal et al. (1987) demonstrated in a calf brain supernatant that Ins(1,3,4)P$_3$ is first converted by an InsPs 4-phosphatase to Ins(1,3)P$_2$, which is further hydrolyzed to Ins(1)P by an InsPs 3-phosphatase. The enzymes do not require Mg^{2+} and are not inhibited by lithium ions. A similar 4-phosphatase was found to degrade Ins(3,4)P$_2$ to Ins(3)P. Three different InsP$_2$ formed from calf brain supernatant were each further metabolized by a separate enzyme. The three inositol monophosphates, i.e. Ins(1)P, Ins(3)P and Ins(4)P, were converted to inositol by InsP phosphatase (Ackermann et al., 1987).

In preliminary studies, Bansal et al. (1987) observed that in the presence of 5 mM EDTA 4-phosphatase did not utilize Ins(1,3,4,5)P$_4$, Ins(1,4,5)P$_3$, Ins(1,4)P$_2$, or Ins(4)P as substrates.

Bansal et al. (1990) have isolated and characterized InsPs 4-phosphatase. This enzyme specifically hydrolyzes the 4-phos-phate from both Ins(2,3,4)P$_3$ and Ins(3,4)P$_2$. Norris and Majerus

(1994) have shown that the enzyme also possesses $PtdIns(3,4)P_2$ hydrolytic activity.

Norris et al. (1995) have reported the isolation of cDNAs encoding rat and human brain 4-phosphatases that are 97% identical at the amino acid level. Recombinant rat protein expressed in *E. coli* catalyzed the hydrolysis of all three substrates of the 4-phosphatase: $Ins(3,4)P_2$, $Ins(1,3,4)P_3$ and $PtdIns(3,4)P_2$. The rat 4-phosphatase is one amino acid longer than the human 4-phosphatase as a result of a serine inserted at position 489.

In a subsequent study, Norris et al. (1997) have reported the cloning of cDNA that encodes an isozyme of 4-phosphatase with 37% amino acid identity to the 4-phosphatase characterized originally. Furthermore, alternatively spliced cDNA of both isozymes of 4-phosphatases were identified that appear to encode C-terminal transmembrane domains. Norris et al. (1997) have designated these 4-phosphatase isozymes as type I and type II with the hydrophilic and hydrophobic spliced forms as α and β, respectively. A conserved motif between 4-phosphatase I and II is the sequence CKSAKDRT that contains the Cys-Xaa5-Arg active site consensus sequence identified for other Mg^{2+}-independent phosphatases. Munday et al. (1999) have recently reported that InsPs 4-phosphatase forms a complex with PtdIns 3-kinase in human platelet cytosol and that the complex serves to localize the 4-phosphatase to sites of $PtdIns(3,4)P_2$ production.

Majerus et al. (1999) have reviewed the role for InsP phosphatases in diverse cellular signaling reactions discovered recently. It is now obvious that InsPs 4-phosphatases are a family of enzymes that participate in the regulation of PtdIns 3-kinase signaling. These magnesium-independent phosphatases catalyze the hydrolysis of the 4-position phosphate of the second messenger, $PtdIns(3,4)P_2$, yielding the product $PtdIns(3)P$. 4-Phosphatase cDNA clones have been isolated that are derived from two 4-phosphatase genes.

Of the few 4-phosphatases identified from mammalian sources, there are no significant homologues in the *S. cerevisiae* genome to either rat or human type I or type II polyphosphate 4-phosphatases (Norris et al., 1997). This is taken to indicate that metabolism of 4-phosphorylated PdInsPs is via PLC or that there exists a distinct class of monoesterases that will act on these lipids (Hughes et al., 2000a,b).

Radioisotope assays. According to Bansal et al. (1987, 1990) $Ins(1,3,4)P_3$ 4-phosphatase activity was measured by incubating $[^3H]Ins(1,3,4)P_3$ (40 μM, 1300 cpm) with 50 mM MES (pH 6.5), 5 mM EDTA, and crude calf brain supernatant in a final volume of 25 μl at 37°C. The reaction was stopped by dilution to 1 ml with cold water and poured on to 0.4 ml Dowex formate column. Separation of $[^3H]Ins(1,3)P_2$ was obtained from $[^3H]Ins(1,3,4)P_3$ using Dowex formate column and eluting with 8 ml of 0.4 M ammonium formate, 0.1 M formic acid, and radioactivity measured by scintillation counting. Separation of $Ins(3,4)P_2$ from $Ins(1,3,4)P_3$ was achieved by elution of the former with 10 ml of 0.1 M formic acid, 0.4 M ammonium formate followed by elution with 10 ml of 0.1 M formic acid, 0.8 M ammonium formate.

The $[^3H]$inositol assay of $Ins(1,3,4)P_3$ 4-phosphatase can also be performed as reported by Norris and Majerus (1994).

Mass assays. As noted above, mass assays of $Ins(1,3,4)P_3$ may be performed as reported by Mayr (1990) using HPLC with post-column MDD system. Using a strongly acidic elution system with an appropriate elution protocol, an adequate separation was obtained for $Ins(1,3,5)P_3$, $Ins(1,3,5)P_3$ and $Ins(1,4,5)P_3$ in order of increasing retention time.

8.3.3.4. InsP₃ 5-phosphatases

The InsPs 5-phosphatase enzymes have been extensively studied and a variety have been described. Of the 5-phosphatases, type I enzymes dephosphorylate inositol phosphates (Verjans et al., 1992, 1994a,b; Zhang et al., 1995), whereas type II enzymes hydrolyze inositol phosphates as well as PtdInsPs (Jefferson and Majerus,

1995; Kavanaugh et al., 1996). Type II 5-phosphatases form a growing family and may be subdivided into five groups based on their domain structure (Vanhaesebroeck et al., 2001).

The 5-specific dephosphorylation of $Ins(1,4,5)P_3$ and $Ins(1,3,4,5)P_4$ by InsPs 5-phosphatase inactivates these intracellular messenger molecules and is essential for the provision of free inositol for inositol lipid resynthesis (see Chapter 6). Downes et al. (1982) first described an $Ins(1,4,5)P_3$ 5-phosphatase in human erythrocyte membranes, where its activity was Mg^{2+} dependent and inhibited by 2,3-bisglycerophosphate. The product of dephosphorylation was $Ins(1,4)P_2$. The same enzyme specifically removes the 5-phosphate from $Ins(1,3,4,5)P_4$ to form $Ins(1,3,4)P_3$, the most abundant $InsP_3$ isomer in many cells (Batty et al., 1985). Hansen et al. (1987) showed that the soluble fraction of rat brain separated by an ion exchange chromatography containing two different $InsP_3$ 5-phosphatases. The enzymes were designated as Type I and Type II as they were eluted from the column. Type II enzyme represented 80–90% of the total activity. Two forms of $InsP_3$ 5-phosphatase have also been described for bovine brain: 50–70% Type I and 30–50% Type II (Erneux et al., 1989). Further studies by other groups (Majerus, 1991; Mitchell et al., 1996; Drayer et al., 1996) have led to the recognition that this reaction terminates the calcium signaling because the substrate stimulates cellular calcium mobilization whereas the product does not. Subsequent work has provided evidence for a large family of 5-phosphatase enzymes, which also remove the 5-position phosphate from PtdInsPs (Mitchell et al., 1996; Drayer et al., 1996).

InsPs 5-phosphatases have been identified in both the soluble and particulate fractions of multiple cell types (reviewed by Majerus, 1992; Shears, 1992; Verjans et al., 1994a,b). The soluble enzymes have been purified from the cytosolic fraction of human platelets (Connolly et al., 1985; Mitchell et al., 1989). The Type I enzyme has a molecular mass of 40–45 kDa and is phosphorylated and activated by protein kinase C (Connolly et al., 1986). Type II enzyme has a molecular mass of 75 kDa and the cDNA encoding this enzyme shares sequence homology with the gene that is defective in Lowe's OCRL syndrome, a rare X-linked congenital disorder characterized by growth defects and mental retardation (Ross et al., 1991; Attree et al., 1992). A 43 kDa membrane-associated 5-phosphatase which appears to be identical to the Type I soluble enzyme has been purified and cloned by several groups (Erneux et al., 1989; Laxminarayan et al., 1992; 1993; De Smedt et al., 1994; Hodgkin et al., 1994; Verjans et al., 1994a,b). This enzyme is functionally distinct from the Type II 75 kDa 5-phosphatase, however, in that it is unable to remove the 5-position phosphate from PtdIns(4,5)P_2 (Matzaris et al., 1994, 1998).

5-Phosphatases are also found in plants, *Caenorhabditis elegans* and *Drosophila*, and there are four genes that encode 5-phosphatase enzymes (Majerus, 1996; Stolz et al., 1998a,b).

There are four known substrates for 5-phosphatases: Ins(1,4,5)P_3, Ins(1,3,4,5)P_4 and the lipids PtdIns(4,5)P_2 and PtdIns(3,4,5)P_3. Majerus et al. (1999) and Erneux et al. (1998) have classified the various 5-phosphatases into four groups. They are defined by two signature motifs: (F/I)WXGDDXN(F/Y)R and (R/N)XP(S/A)(W/Y)(C/T)DR(I/V)(L/I). These motifs are separated by 60–75 amino acids except for 5-phosphatase I, where the motifs are 103 amino acids apart. Group I 5-phosphatases hydrolyze only the water-soluble substrates Ins(1,4,5)P_3 and Ins(1,3,4,5)P_4. The platelet type I enzyme is a representative of this group, which includes members from dog thyroid and placenta. These enzymes are membrane-associated presumably through isoprenylation (Jefferson and Majerus, 1995; De Smedt et al., 1996). 5-Phosphatase I in platelets is in a stoichiometric complex with pleckstrin that regulates its activity (Auethavekiat et al., 1997). When platelets are stimulated with thrombin, pleckstrin is rapidly phosphorylated on serine and threonine residues by protein kinase C, and this activates the 5-phosphatase.

Group II 5-phosphatases hydrolyze all four 5-phosphate substrates although with varying degree of efficiency. Platelet type II 5-phosphatase belongs to this group. It is 51% identical over a span of 744 amino acids with OCRL-1 (oculocerebrorenal dystrophy; Lowe syndrome), which is a 5-phosphatase (Zhang et al., 1995). Both of these enzymes are membrane associated without membrane spanning domains. 5-Phosphatase II is isoprenylated (Jefferson and Majerus, 1995) and is bound to mitochondria and plasma membrane. OCRL-1 is not modified by any lipid moiety and is found on the surface of lysosomes, as reviewed by Majerus et al. (1999). Synaptojanin (McPherson et al., 1994a,b; 1996) and synaptojanin 2 (Nemoto et al., 1997; Seet et al., 1998) are related group of II enzymes that participate in synaptic vesicle trafficking. These enzymes occur in several alternatively spliced forms of uncertain significance (Seet et al., 1998). The two synaptojanins are over 50% identical in the yeast protein *Sac1* and 5-phosphatase domains (Sha et al., 1998).

Group III 5-phosphatases hydrolyze only substrates with a 3-position phosphate group, i.e. Ins(1,3,4,5)P_4 and PtdIns(3,4,5)P_3. There are two such enzyme designated as SHIP and SHIP2 (Damen et al., 1996; Pesesse et al., 1997). They are 50% identical in amino acid sequence but have very different tissue distributions. They contain the N-terminal SH2 domain and phosphotyrosine binding domains and upon activation of receptors form complexes with cellular signaling proteins. The SHIP's C-terminus is essential for its hydrolysis of PtdInsP_3 and inhibition of mast cell degranulation (Damen et al., 2001).

Group IV 5-phosphatases remain poorly characterized and are on activity that only hydrolyzes PtdIns(3,4,5)P_3 (Jackson et al., 1995) and form complexes with PtdIns 3-kinase. Kisseleva et al. (2000, 2002) have recently isolated a cDNA encoding the Group IV enzyme.

Laxminarayan et al. (1993) have identified and characterized a membrane-associated InsPs 5-phosphatase from the particulate fraction of human placenta. The enzyme was purified to apparent chromatographic homogeneity and had a molecular mass of 43 kDa as determined by SDS-PAGE. The enzyme hydrolyzes Ins(1,4,5)P_3

to Ins(1,4)P$_2$ as well as Ins(1,3,4,5)P$_4$ to Ins(1,3,4)P$_3$. The enzyme requires magnesium for activity and is inhibited by calcium concentrations higher than 100 μM. Polyclonal antibodies developed against the membrane-associated enzyme immunoprecipitated the purified membrane-associated placental 5-phosphatase and the platelet Type I cytosolic enzyme, but not the 75 kDa platelet Type II 5-phosphatase.

Hodgkin et al. (1994) have purified membrane-associated Ins(1,4,5)P$_3$/Ins(1,3,4,5)P$_4$ 5-phosphatases from bovine testis and human erythrocytes by chromatography on several media, including a novel 2,3-bisphosphoglycerate affinity column. The enzymes had apparent molecular masses of 42 kDa (testis) and 70 kDa (erythrocyte), as determined by SDS/PAGE, and affinities for Ins(1,4,5)P$_3$ of 14 and 22 μM, respectively. The two enzymes hydrolyze both Ins(1,4,5)P$_3$ and Ins(1,3,45)P$_4$ and are therefore Type I Ins(1,4,5)P$_3$ 5-phosphatases (Hansen et al., 1987). Hodgkin et al. (1994) have proposed that these be termed Type Ia [typified by the testis enzyme, 40 kDa, a higher affinity for Ins(1,4,5)P$_3$] and Type Ib [typified by the erythrocyte enzyme 60 kDa, lower affinity for Ins(1,4,5)P$_3$].

De Smedt et al. (1994) have reported the first cloning, expression and production of recombinant Type I InsP$_3$ 5-phosphatase from human brain and have established its localization in cerebellar Purkinje cell bodies. The protein sequence showed a putative C-terminal isoprenylation site (CVVQ).

Jefferson and Majerus (1995) have isolated cDNA clones encoding Type II InsPs 5-phosphatase resulting in a combined cDNA of 3076 nucleotides encoding a protein of 942 amino acids. The 5-phosphatase was of Type II and hydrolyzed both Ins(1,4,5)P$_3$ to Ins(1,4)P$_2$ and PtdIns(4,5)P$_2$ to PtdIns(4)P both in vitro and in vivo.

Auethavekiat et al. (1997) have reported that InsPs 5-phosphatase Type I in platelets occurs in complex with platelet pleckstrin. This enzyme hydrolyzes the 5-phosphate from Ins(1,4,5)P$_3$ and Ins(1,3,4,5)P$_4$ and thus serves as a calcium signal-terminating enzyme. Thus, plextrin terminates calcium signaling in platelets when it is phosphorylated by binding to an activating 5-phosphatase Type I.

According to Jefferson et al. (1997) an InsPs 5-phosphatase (SIP-110) that binds the SH3 domains of the adapter protein GRB2 is produced in Sf9 cells. The enzyme hydrolyzes Ins(1,3,4,5)P_4 to Ins(1,3,4)P_3. It does not hydrolyze Ins(1,4,5)P_3 that is a substrate for previously described 5-phosphatases nor does it hydrolyze PtdIns(4,5)P_2. However, SIP-110 did hydrolyze PtdIns(3,4,5)P_3 to PtdIns(3,4)P_2 as did recombinant forms of two other 5-phosphatases designated as inositol polyphosphate 5-phosphatase II, and OCRL (the protein that is mutated in OCRL syndrome). As a result, the InsPs 5-phosphatase enzyme family now is represented by at least nine distinct genes and includes enzymes that fall into four subfamilies based on their activities toward various 5-phosphatase substrates.

Stricker et al. (1997, 1999) have identified a membrane associated Ins(1,3,4,5)P_4 5-phosphatase, which yields Ins(1,3,4)P_3 as part of a mechanism for translocation between membranes and cytosol of 42^{IP4}, a specific inositol 1,3,4,5-tetrakisphosphate/phosphatidylinositol 3,4,5-trisphosphate-receptor protein from brain.

Ijuin et al. (2000) have identified a cDNA encoding a novel InsPs 5-phosphatase. It contains two highly conserved catalytic motifs for 5-phosphatase with a molecular mass of 51 kDa, which is ubiquitously expressed and especially abundant in skeletal muscle, heart, and kidney. Ijuin et al. (2000) designated this 5-phosphatase as skeletal muscle, and kidney enriched inositol phosphatase (SKIP). SKIP is a simple 5-phosphatase with no other motifs. Baculovirus-expressed recombinant SKIP protein exhibited 5-phosphatase activities toward Ins(1,4,5)P_3, Ins(1,3,4,5)P_4, PtdIns(4,5)P_2 and PtdIns(3,4,5)P_3 but had six-fold greater substrate specificity for PtdIns(4,5)P_2 than for Ins(1,4,5)P_3.

Kong et al. (2000) have cloned and characterized a novel 5-phosphatase, which demonstrates a restricted substrate specificity and tissue expression. The 3.9 kb cDNA predicts a 72 kDa protein with an N-terminal proline rich domain, a central 5-phosphatase domain, and C-terminal CAAX motif. Immunoprecipitated recombinant 72 kDa 5-phosphatase hydrolyzed PtdIns(3,4,5)P_3 and PtdIns(3,4,5)P_2, forming PtdIns(3,4)P_2 and PtdIns(3)P, respectively. Kong et al. (2000) proposed that the novel 5-phosphatase hydrolyzes PtdIns(3,4,5)P_3 and PtdIns(3,5)P_2 on the cytoplasmic Golgi membrane and thereby regulate Golgi-vesicular trafficking.

Erneux et al. (1998) have reviewed the diversity and possible functions of the InsPs 5-phosphatases. Distinct forms of InsPs and PtdInsPs 5-phosphatase selectively remove the phosphate from the 5-position of the inositol ring from both water-soluble and lipid substrates, e.g. Ins(1,4,5)P_3 and Ins(1,3,4,5)P_4, PtdIns(4,5)P_2 or PtdIns(3,4,5)P_3. In mammalian cells, this family contains a series of distinct genes and splice variants. All InsPs 5-phosphatases share a 5-phosphatase domain and various protein modules responsible for specific cell localization or recruitment (H2 domain, proline rich sequences, prenylation sites, etc.)

SHIP-2 is a PtdIs(3,4,5)P_3 5-phosphatase that contains an NH_2-terminal SH2 domain, a central 5-phosphatase domain, and a COOH-terminal proline-rich domain. SHIP2 hydrolyses the 5-position phosphate from PtdIns(3,4,5)P_3 and PtdIns(4,5)P_2, and in some, but not all, studies it has been shown to hydrolyze the soluble inositol phosphate Ins(1,3,4,5)P_4 (Pesesse et al., 1997; Wisniewski et al., 1999; Taylor et al., 2000a,b). SHIP-2 bears significant sequence identity with the 5-phosphatase, SHIP-1, except in the proline-rich domain. Unlike SHIP-1, which has a restricted hematopoietic expression, SHIP-2 is widely expressed and distributed. Bruyns et al. (1999) have reported that SHIP-1 and SHIP-2 are co expressed in human lymphocytes while Dyson et al. (2001) have shown that SHIP-2 binds filamin and regulates submembranous actin.

Tsujishita et al. (2001) have reported the enzymatic characterization of the inositol 5-phosphatase catalytic (IPP5C) domain of synaptojanin. The structure of the IPP5C domain revealed how a conserved Dnase-like catalytic core has been expanded to yield a diverse family of inositol polyphosphate signaling enzymes.

Variations in the surface surrounding the active site contribute to specificity for lipid versus soluble inositol polyphosphates. Differences in the $\alpha7$-$\beta9$ loop appeared to be primarily responsible for differences in preferences for 3-phosphorylated versus other substrates. These results provide potential explanation for the observation that 5-phosphatases are competent to bind soluble and lipid inositol derivatives phosphorylated at combinations of the 3-,4- and 5 positions, yet they dephosphorylate only the 5-position.

Radioisotope assays. Erneux et al. (1989) assayed the InsPs 5-phosphatase activity at 37°C using 1–30 μM [^{3}H]Ins(1,4,5)P$_3$ and/or Ins(1,4,5)[5-^{32}P]P$_3$ or 0.1–4 μM [^{3}H]Ins(1,3,4,5)P$_4$ in 50 mM HEPES/NaOH pH 7.4, 5 mM 2-mercaptoethanol, 1 mg/ml BSA, 2 mM MgCl$_2$, and the appropriate enzyme dilution in a final volume of 0.1 ml. The assay was initiated by adding the substrate, stopped after 5–10 min by addition of 1 ml ice-cold 0.2 M ammonium formate/0.1 M formic acid and the resulting solution was immediately applied to Dowex columns [^{32}P]Pi, the product of Ins(1,4,5)[^{32}P]P$_3$ 5-phosphatase activity was separated on 0.2 ml Dowex columns. [^{3}H]Ins(1,3,4)P$_3$, the product of [^{3}H]Ins(1,4,5)P$_3$ 5-phosphatase was separated on 0.6 ml Dowex columns. The specificity of dephosphorylation was checked using a mixture of Ins(1,4,5)[5-^{32}P]P$_3$ and [^{3}H]Ins(1,4,5)P$_3$ as substrate and determining the ratio of the two radioactive labels.

Mitchell et al. (1989) performed the hydrolysis of Ins(1,[^{32}P]4,[^{32}P]5)P$_3$ and [^{3}H]Ins(1,3,4,5)P$_4$ (10 μM) with the 75 kDa InsPs 5-phosphatase from human platelets as described by Connolly et al. (1985). Ins(1,[^{32}P]4,[^{32}P]5)P$_3$ was isolated from erythrocyte ghosts as described by Downes et al., 1982).

A radioisotope/ratio assay for Ins(1,4,5)P$_3$ 5-phosphatase has been described also by Erneux et al. (1989) who measured the dephosphorylation of Ins(1,4,5)P$_3$ by bovine brain 5-phosphatase at 37°C using 1–30 μM [^{3}H]Ins(1,4,5)P$_3$ and/or Ins(1,4,5)[5-^{32}P]P$_3$ in 50 mM HEPES/NaOH (pH 7.4), 5 mM MgCl$_2$, and the appropriate enzyme dilution in a final volume of 0.1 ml. The assay was initiated by adding the substrate, stopped after 5–10 min by addition of 1 ml ice-cold 0.2 M ammonium formate/0.1 M formic acid and chromatographed the resulting solution immediately on Dowex columns. [^{32}P]Pi, the product of Ins(1,4,5)[5-^{32}P]P$_3$ 5-phosphatase was separated on 0.2 ml Dowex columns (Erneux et al., 1986). The ratio of the [^{32}P] label recovered with respect to the [^{3}H] label in the InsP$_2$ column fraction was determined. As the Ins(1,4,5)[^{32}P]P$_3$ was labeled in the 5-position, a low ^{32}P/^{3}H ratio in the InsP$_2$ fraction indicates essentially 5-phosphatase activity, whereas a high ratio in the InsP$_2$ fraction indicates 1- and/or 4-phosphatase activity. ^{32}P/^{3}H ratios in the InsP$_2$ fraction were 10–16% of the original ratio in the Ins(1,4,5)P$_3$ substrate suggesting 5-phosphatase activity.

Erneux et al. (1989) assayed the dephosphorylation of Ins(1,3,4,5)P$_4$ by bovine brain 5-phosphatase at 37°C using 0.1–4 μM [^{3}H]Ins(1,3,4,5)P$_4$ in 50 mM HEPES/NaOH (pH 7.4), 5 mM MgCl$_2$, and the appropriate enzyme dilution in a final volume of 0.1 ml. The assay was initiated by adding the substrate, stopped after 5–10 min by addition of 1 ml ice-cold 0.2 M ammonium formate/0.1 M formic acid and the resulting solution was immediately applied to Dowex columns. [^{3}H]Ins(1,3,4)P$_3$, the product of [^{3}H]Ins(1,2,4,5)P$_4$ 5-phosphatase, was separated on 0.6 ml Dowex column (Takazawa et al., 1988).

A 'physiological medium' made up of 10 mM NaCl 70 mM K$^+$-glutamate, 30 mM KCl, 4 mM MgSO$_4$, 10 mM HEPES and 1 mM EGTA, pH 7.2 for the assay of Ins(1,4,5)P$_3$ 5-phosphatase activity gave more than half of that in the ionic strength medium used previously (Downes et al., 1982), both for erythrocyte membranes and liver cell fractions. Incubations were in a final volume of 1 ml, containing 50 μl of the homogenate or cell fraction, and were continued for an appropriate period of time (1 min to 4 h). The substrate concentrations were in the range 40–300 nM for Ins(1,4,5)P$_3$ and Ins(1,4)P$_2$, and 8–24 μM for Ins(1)P. The incubations were stopped by adding 100 μl of 12 M perchloric acid containing 1 mM Pi and released ^{32}P-Pi was estimated as described previously (Downes et al., 1982).

Erneux et al. (1993) have given a general method for the assay of Ins(1,4,5)P$_3$ 5-phosphatase activity. The enzyme preparation is incubated in 50 mM HEPES (pH 7.4), 2 mM MgCl$_2$, 48 mM mercaptoethanol, 1 mg/ml BSA, 30 μM cold InsP$_3$, and [^{3}H]InsP$_3$ (specific activity 17 Ci/mmol) in a final volume of 0.1 ml. InsP$_3$ is provided in 85% purity by Sigma. (Crude soluble enzymatic preparations of bovine brain are diluted 50-fold in 50 mM HEPES (pH 7.4), 2 mM MgCl$_2$, 48 mM mercaptoethanol, and 1 mg/ml BSA, 10 μl of this dilution is used for each assay per tube.) Each sample is incubated at 37°C for 8 min, stopped by the addition of 1 ml of 0.1 M HCOOH, 0.4 M NH$_4$COOH, and then loaded on a 0.2 ml column of AG 1-X8 resin (200–400 mesh), formate form (Erneux et al., 1986). The InsPs were eluted in 4 ml of 0.1 M HCOOH, 0.4 M NH$_4$COOH; the remaining [^{3}H]InsP$_3$ was eluted in 5 ml of 0.1 M HCOOH, 0.8 M NH$_4$COOH. Instagel (10 ml) was added to estimate radioactivity.

Matzaris et al. (1994, 1998) determined the Ins(1,4,5)P$_3$ 5-phosphatase activity by measuring the release of [^{32}PO$_4$] by hydrolysis of Ins(1,[^{32}P]4,[^{32}P]5)P$_3$ as described by Connolly et al. (1985). Ins(1,[^{32}P]4,[^{32}P]5)P$_3$ was isolated from erythrocyte ghosts as described by Downes et al. (1982). Ijuin et al. (2000) measured the activity of the new InsPs 5-phosphatase as described by Connolly et al. (1985), using [^{3}H]Ins(1,3,4)P$_3$ and [^{3}H]Ins(1,3,4,5)P$_4$ as substrates and purified Baculovirus-expressed recombinant proteins as enzymes. InsPs were separated by HPLC using Partisphere SAX column (Whatman) according to the method of Zhang and Buxton (1998). Collected fractions were diluted 20:1 with distilled water and quantified by liquid scintillation counting. Phosphatase activity for Ins(1,3,4)P$_3$ was measured as follows: [^{3}H]Ins(1,3,4,5)P$_4$ was incubated with recombinant partial SHIP protein at 37°C for 15 min; after that, recombinant SKIP protein was added and further incubated at 37°C for 15 min. The reaction products were isolated and separated on SAX columns as above. Fractions were collected and assayed for radioactivity by scintillation counting.

Mass assays. Meek (1986) reported an HPLC system with inorganic phosphate analysis for the quantification of the mass of inositol bis-, tris- and tetrakis-phosphates. The method employed involved anion-exchange HPLC with on-line alkaline phosphatase hydrolysis of the phosphate esters. Detection of the inorganic phosphate formed was performed by on-line mixing with the molybdate reagent and monitoring the effluent at 380 nm to continuously detect inorganic phosphate. The content of $Ins(1,4,5)P_3$ in brain and salivary gland from rats killed by decapitation was found to be 10–60 times higher than that from rats killed by focused microwave irradiation to block post-mortem metabolism. Mayr (1990) has reported a mass assay for $InsP_3$ based on post-column metal dye detection of phosphate as noted above.

8.3.4. Inositol tetrakisphosphate phosphatases

8.3.4.1. InsP$_4$ 3-phosphatases
Nogimori et al. (1991) have purified an $Ins(1,3,4,5)P_4$ 3-phosphatase activity from rat liver and evaluated its substrate specificity. A self-activated $Ins(1,3,4,5)P_4$ 3-phosphatase has been reported to exist at the inner surface of the human erythrocyte membrane (Estrada-Garcia et al., 1991). The identity of the latter enzyme is unknown, but it exhibits several similarities to the ER-based Minpp1 (Chi et al., 2000). The two enzymes have epitopes in common, and both can cleave the 3-position from $Ins(1,3,4,5)P_4$ in vitro (Craxton et al., 1995; Estrada-Garcia et al., 1991). Chi et al. (2000) reports that both the ER and the erythrocyte enzyme are encoded by the Minpp1 gene.

8.3.4.2. InsP$_4$ 4-phosphatases
Bansal et al. (1987; 1990) observed that the Ins polyphosphate 4-phosphatase does not utilize $Ins(1,3,4,5)P_4$. Bansal et al. (1990) reported less than 0.01 μmol/min/μg hydrolysis of $Ins(1,3,4,5)P_4$.

8.3.4.3. InsP$_4$ 5-phosphatases

Stricker et al. (1999) used the HPLC-MDD method of Mayr (1988) to separate and quantify the Ins(1,3,4)P$_3$ formed as a result of the enzymatic degradation of Ins(1,3,4,5)P$_4$. The inositol phosphates from the brain tissue were extracted as follows. Brain cerebellum and cortex from 12-week old rats, dissected directly after decapitation, were frozen in liquid nitrogen and stored at $-80°C$. The inositol phosphates were extracted from the samples by homogenization of 1 g of frozen tissue in 2 ml of 2 M PCA supplemented with 10 mM EDTA and 1 mM NaF. To remove nucleotides, 400 μl of a charcoal suspension (20%, w/v, in 100 mM NaCl, 50 mM sodium acetate, pH 4.0) were added to the neutralized supernatants. After 15 min on ice, the samples were centrifuged and the sedimented charcoal washed with 1 ml washing solution (100 mM NaCl, 5 mM NaF, 1 mM EDTA) and recentrifuged. The pooled supernatants were lyophilized and dissolved in water. The compounds were separated by anion exchange chromatography on Mono Q HR 10/10 (Pharmacia) column. A linear gradient of HCl was applied to distinguish between InsP$_5$ and InsP$_3$ isomers (0 min, 0.2 mM HCl; 70 min, 0.5 M HCl; flow rate, 1.5 ml/min; gradient I). Photometric detection at 546 nm was achieved employing a post-column derivatization method [metal-dye reagent, 2 M Tris/HCl (pH 9.1), 200 μM 4-(2-pyridylazo)resorcinol, 30 μM YCl$_3$, 10% (v/v) MeOH; flow rate: 0.75 ml/min]. The identity of the InsPs, formed as a result of the enzymatic degradation of Ins(1,3,4,5)P$_4$, was determined by the use of a modified, slightly alkaline gradient elution system (Freund et al., 1992).

Adelt et al. (2001) have exploited the regiospecificity of a partially purified 5-phosphatase and phytase (1-phosphatase) from *Dictyostelium* to convert Ins(1,2,4,5)P$_4$ and Ins(1,2,3,6)P$_4$ to the corresponding Ins(1,2,4)P$_3$ and Ins(2,3,6)P$_3$.

Mass assays. Luzzi et al. (2000) have employed capillary electrophoresis combined with a biological detector cell to quantify InsP$_3$ in small regions of a *Xenopus* oocyte. Thus far, only experimental parameters that influence the sensitivity, accuracy, and reliability of InsP$_3$ detector cell coupled to capillary electrophoresis have been identified and characterized.

Meek (1986) has described an HPLC method for the resolution of Ins(1,3,4,5)P$_4$ in tissues, but quantification had to await synthesis of a standard. The method employed anion-exchange separation with on-line enzymatic hydrolysis of the phosphate esters and detection of the inorganic phosphate formed by on-line molybdate color reaction. The mass determination method of Mayr (1990) includes the Ins(1,3,4,5)P$_4$ as one of the components quantified by metal-dye complexing. Ins(1,3,4,5)P$_4$ has been quantified also by the enzymatic assay method of Maslanski and Busa (1990), and by modifications of this method (Singh, 1992), and Ashizawa et al. (2000). Stephens (1990) has described the purification and separation of isomeric InsP$_4$ on SAX and WAX HPLC columns.

8.3.5. Inositol pentakisphosphate phosphatases

8.3.5.1. InsP$_5$ 1-phosphatases
Ho et al. (2002) have recently shown that the synthesis of Ins(3,4,5,6)P4 in T84 cells takes place by Ins(1,3,4,5,6)P$_5$ 1-phosphatase activity which had previously been characterized as Ins(3,4,5,6)P$_4$ 1-kinase. Rationalization of this phenomenon with a ligand binding model unveiled that Ins(1,3,4)P$_3$ was not simply an alternative kinase substrate, but also an activator of Ins(1,3,4,5,6)P$_5$ 1-phosphatase. Stable overexpression of the enzyme in epithelial monolayers verified its physiological role in elevating Ins(3,4,5,6)P$_4$ levels and inhibiting secretion.

Mass assays. Ho et al. (2002) assayed the inositol phosphate phosphatase activity in buffer containing 100 mM KCl, 20 mM HEPES (pH 7.2), 5 mM ADP, 6 mM MgSO$_4$ and 0.3 mg/ml BSA. Assays were acid quenched, neutralized, and analyzed by HPLC using a Synchropak Q100 column (Caffrey et al., 2001). One milliliter fractions were collected for 70 min, followed by 0.5 ml fractions. Other reactions were heat inactivated (95°C, 3 min) and analyzed by a metal dye detection, HPLC method (Mayr, 1988; Adelt et al., 2001).

8.3.5.2. InsP$_5$ 3-phosphatases

Nogimori et al. (1991) purified an Ins(1,3,4,5)P$_4$ 3-phosphatase from rat liver, which also dephosphorylated Ins(1,3,4,5,6)P$_5$ to Ins(1,4,5,6)P$_4$. Norris et al. (1998) have reported that SopB, a protein secreted by *Salmonella dublin*, has sequence homology to mammalian InsPs 4-phosphatases and that recombinant SopB has InsPs phosphatase activity in vitro. SopB encodes a phosphatase that hydrolyzes many but not all InsPs including Ins(1,3,4,5,6)P$_5$ yielding the product Ins(1,4,5,6)P$_4$. Norris et al. (1998) have expressed recombinant glutathione S-transferase-SopB and found that it hydrolyzed several inositol phosphates. The enzyme specificity was tested with a synthetic 1[^{32}PO$_4$],3,4,5,6)P$_5$ and several InsP$_4$ isomers were found to be formed, including Ins(1,3,4,6)P$_4$, Ins(1,3,4,5)P$_4$, as well as Ins(1,4,5,6)P$_4$. Table 8.2 compares the phosphatase activity of recombinant SopB on soluble

TABLE 8.2

Phosphatase activity of recombinant SopB

Substrate	nmol/min/mg protein
PtdIns(3)P	47
PtdIns(3,4)P$_2$	51
PtdIns(3,4,5)P$_3$	68
Ins(1,2,4)P$_3$	9.1
Ins(1,4,5)P$_3$	13
Ins(1,3,4,5)P$_4$	24
Ins(1,3,4,5,6)P$_5$	4.5

Assays were performed with 10 μM substrate in 50 mM Mops (pH 7.0), 10 mM EDTA, and 1 mM DTT. Reproduced from Norris et al. (1998) with permission of the publisher.

and lipid inositol phosphates. The enzyme has significantly higher activity with the lipid phosphates as already pointed out.

The most likely candidate enzyme for controlling the size of the metabolic pools of $InsP_5$ and $InsP_6$ is the multiple inositol polyphosphate phosphatase (MIPP) identified in the rat and chick and the histidine phosphatase of the endoplasmic reticulum (HiPER) on the basis of in vitro activity (Craxton et al., 1997). Craxton et al. (1997) have described the isolation of a 2.3 kb cDNA clone of a rat hepatic form of MIPP. The predicted amino acid sequence of MIPP included 60% identity with the catalytic domain of a fungal $InsP_6$ phosphatase (phytase A). The similarity included conservation of the RHGXRXP signature of the histidine acid phosphatase family. Craxton et al. (1997) expressed in *E. coli* a histidine-tagged, truncated form of MIPP and determined the enzymic specificity of the recombinant protein. $Ins(1,3,4,5,6)P_5$ was hydrolyzed, first to $Ins(1,4,5,6)P_4$ and then to $Ins(1,4,5)P_3$ by consecutive 3- and 6-phosphatase activities. $InsP_6$ was catabolized without specificity towards a particular phosphate group. In contrast, MIPP only removed the β-phosphate from the 5-diphosphate group of PP-$InsP_5$. Previously, cruder preparations of the enzyme had been shown to attack also the $(PP)_2$-$InsP_4$ (Shears et al., 1995).

Chi et al. (2000) has reported biochemical analyses that demonstrate that $InsP_5$ and $InsP_6$ are in vivo substrates for ER based MIPP (renamed Minpp1). Specifically, it was shown that these InsPs in Minpp1-deficient embryonic fibroblasts were 30–45% higher than in wild-type cells. This increase was reversed by reintroducing exogenous Minpp1 into ER. Among the phosphatases that hydrolyze InsPs, Minpp1 is unique by virtue of possessing an active site histidine, cleavage of the 3-phosphate from multiple InsPs, and compartmentalization in the lumen of the ER (Chi et al., 2000). In vitro, Minpp1 hydrolyzes $Ins(1,3,4,5)P_4$, $InsP_5$, and $InsP_6$. Kinetic experiments have indicated that $InsP_5$ and $InsP_6$ are the most important substrates for Mnpp1. Although it was shown that the enzyme favored $InsP_5$ and $InsP_6$ as substrates over other InsPs, the exact nature of the products of hydrolysis was not established in these studies.

In contrast to a previous conclusion that InsPs are not physiologically relevant substrates for PTEN (Maehama and Dixon, 1998), Caffrey et al. (2001) have demonstrated that PTEN is an active $Ins(1,3,4,5,6)P_5$ 3-phosphatase when expressed and purified from bacteria or HEK cells. Kinetic data indicated that $Ins(1,3,4,5,6)P_5$ and $PtdIns(3,4,5)P_3$ competed for PTEN in vivo. Transient transfection of HEK cells with PTEN decreased $Ins(1,3,4,5,6)P_5$ levels.

8.3.5.3. $Ins(1,3,4,5,6)P_5$ 6-phosphatases

Craxton et al. (1997) expressed in *E. coli* a histidine-tagged, truncated form of MIPP and determined the enzymic specificity of the recombinant protein. $Ins(1,3,4,5,6)P_5$ was hydrolyzed, first to $Ins(1,4,5,6)P_4$ and then to $Ins(1,4,5)P_3$ by consecutive 3- and 6-phosphatase activities.

Radioisotope assay. Chi et al. (2000) assayed aliquots of column-purified enzyme fractions with approximately 10 000 dpm of either $[^3H]Ins(1,3,4,5)P_4$ or $[^3H]Ins(1,3,4,5,6)P_5$ for 60–120 min at 37°C in a final volume of 100 μl of buffer containing 50 mM KCl, 50 mM HEPES (pH 7.0 with KOH), 1 mM EDTA, 4 mM CHAPS, and 0.5 mg of BSA/ml. Assays were quenched with 0.9 ml of 0.5 mM EDTA–0.1 M formic acid–0.1 ammonium formate, and the degree of InsPs hydrolysis was determined after chromatography on gravity-fed anion-exchange columns.

8.3.6. Inositol hexakisphosphate phosphatases

Phytases catalyze the hydrolysis of phytate (myo-InsP$_6$), thereby releasing inorganic phosphate (Wodzinski and Ullah, 1996). Based on the position of the first phosphate hydrolyzed, three classes of acid phytases have been recognized, all of which yield $Ins(2)P$ (Loewus and Murthy, 2000). In recent years, these enzymes have become of interest for biotechnological applications, specifically for improving dietary phytate-phosphorus utilization by swine and poultry. Engelen et al. (1994) have described a simple and rapid method for determining the enzymatic activity of microbial phytase. The method is based on the determination of inorganic P released on hydrolysis of sodium phytate at pH 5.5.

8.3.6.1. InsP₆ 6-phytase

Van der Kaay and Van Haastert (1995) have demonstrated that *Paramecium* phytase removes the phosphates of $InsP_6$ in the sequence $6 > 5 > 4 > 1$. Two phytases have been purified from *E. coli* (Greiner et al., 1993). Greiner et al. (2000) have demonstrated that enzyme P2 of *E. coli* dephosphorylates *myo*-$InsP_6$ in a stereospecific manner by sequential removal of phosphate groups via D-Ins(1,2,3,4,5)P_5, D-Ins(2,3,4,5)P_4, D-Ins(2,4,5)P_3, Ins(2,5)P_2 to finally produce Ins(2)P (notation 6/1/3/4/5).

Rodriguez et al. (1999) have isolated an *E. coli* strain from pig colon and have cloned, sequenced and expressed the acid phosphatase gene (appA2). Lim et al. (2000) have determined the crystal structures of *E. coli* phytase and its complex with phytate.

8.3.6.2. InsP₆ 3-phytase

Several phytases have been cloned and characterized, including fungal phytases from *Aspergillus niger* (Piddington et al., 1993) and mammalian phytase (Craxton et al. (1997). These enzymes do not show any apparent sequence similarity to each other. They do share, however, a highly conserved sequence motif, RHGXRXP, which is found at the active site of acid phosphatases (Ullah et al., 1991). Furthermore, they contain a remote C-terminal His-Asp motif (HD-motif) that is likely to take part in the catalysis.

Kostrewa et al. (1997) have reported the X-ray crystal structure of a 3-phytase from *A. niger*. The structure consists of a large α/β-domain, which shows structural similarity to a high molecular weight acid phosphatase in rats, and a smaller α-domain. It was proposed that the conserved histidine (His 59) was involved in nucleophilic attack at the 3-phosphate. Mullaney et al. (2000) have characterized the native phytase activity of *A. fumigatus* isolates to further delineate the role the primary structure has in determining activity levels.

8.3.6.3. *InsP₆ 5-phytase*

An unusual constitutive alkaline phytase has been isolated from lily pollen and seeds (Barietos et al., 1994). It initiates hydrolysis by first removing the 5-phosphate of phytic acid. Subsequent hydrolytic steps remove phosphate from the 4 and 6 position to yield $Ins(1,2,3)P_3$ as the final product. This final product, $Ins(1,2,3)P_3$ has been shown to inhibit iron-catalyzed free radical formation by chelating iron. Other phytases yielding *myo*-inositol trisphosphate as the end products are the alkaline phytase $[Ins(1,2,3)P_3]$ (Hara et al., 1985), the phytase from *Typha latifolia* pollen, and the rat hepatic MIPP $[Ins(1,4,5)P_3]$ (Craxton et al., 1997).

A reaction mechanism of *E. coli* phytase has been suggested (Ostanin and Van Etten, 1993) based on positive charge of the guanido group of the aginine residue in the tripeptide RHG interacting directly with the phosphate group in the substrate making it more susceptible to nucleophilic attack, and the histidine residue serving as a nucleophile in the formation of a covalent phosphohistidine intermediate, while the aspartic residue from the C-terminal HD motif protonates the substrates leaving group.

Minpp1 genes conserved as homologous sequences have been identified in mammals, chicks, fruit flies, and plants. Minpp1 enzymes also share distant homology with yeast phytases ($InsP_6$ phosphatases) (Romano et al., 1998; Caffrey et al., 1999; Chi et al., 2000). $InsP_6$ was catabolized without specificity towards a particular phosphate group.

Mass assays. Mullaney et al. (2000) assayed the extracellular phytase activity of *A. fumigatus* isolates immediately after purification in 1.0 ml vol. at designated temperatures in the appropriate buffer. The buffer used for pH 1.0–2.5 was 50 mM glycine HCl; pH 3.0–6.0 was 50 mM sodium acetate; and pH 7.0–9.0 was 50 mM imidazole. The liberated inorganic ortho-phosphates were quantified spectrophotometrically by a modified method from Heinonen and Lahti (1981), using a freshly prepared acetone ammonium molybdate reagent consisting of acetone, 5.0N sulfuric acid, and 10 M ammonium molybdate (2:1:1, by vol.). Adding 2.0 ml of the reagent solution per assay tube terminated phytase assay. After 30 s 0.1 ml of 1.0 M citric acid was added to each tube. Absorbance was read at 355 nm.

Greiner et al. (2000) used HPLC for the separation of myo-InsP$_6$ degradation products.

8.3.7. Inositol pyrophosphate phosphatases

Menniti et al. (1993) have discovered that pancreatoma cell homogenates, incubated with 5 mM fluoride and 5 mM ATP, convert both InsP$_5$ and InsP$_6$ to more polar products. The novel products were determined to be inositol pyrophosphates because of their relatively specific hydrolysis by tobacco pyrophosphatase and alkaline phosphatase. The pyrophosphates were metabolized rapidly by cell homogenates back to their InsP$_5$ and InsP$_6$ precursors. In addition, in mammalian cells there is another diphosphate derivative of InsP$_6$, which Shears et al. (1995) have identified as bis-diphosphoinositol tetrakisphosphate (PP-InsP$_4$-PP). Such material has been isolated from *Dictyostelium* (Stephens et al., 1993).

8.3.7.1. (PP)$_2$InsP$_4$ pyrophosphatase

Safrany et al. (1998) have demonstrated a specific receptor-dependent regulation of the turnover of bis-diphosphoinositol tetrakisphosphate ([PP]$_2$-InsP$_4$). [PP]$_2$-InsP$_4$ is dephosphorylated polyphosphate, the synthesis and metabolism of which is regulated by two coupled kinase/phosphatase substrate cycles. Interest in this group of compounds has in part arisen from the substantial free energy change associated with hydrolysis of the β-phosphate in the diphosphate groups (Stephens et al., 1993; Laussmann et al., 1997). The β-phosphatases that attack PP-InsP$_4$ and PPInsP$_5$, are inhibited by F$^-$, which acts as a metabolic trap that causes the levels of both PP-InsP$_4$ and PPInsP$_5$ to increase (Menniti et al., 1993). By this means it has been estimated that, every hour, 30–50% of the entire cellular pools of InsP$_5$ and InsP$_6$ cycle through the diphosphorylated polyphosphates (Menniti et al., 1993; Albert et al., 1997).

In contrast, MIPP only removed the β-phosphate from the 5-diphosphate group of PP-InsP$_5$ (Chi et al., 2000). Previously, cruder preparations of the enzyme had been shown to attack also the (PP)$_2$-InsP$_4$ (Shears et al., 1995).

Safrany et al. (1998) have purified a rat hepatic diphosphoinositol polyphosphate phosphohydrolase (DIPP) that cleaves a β-phosphate from the diphosphate groups in PP-InsP$_5$ and [PP]$_2$InsP$_4$. Interestingly, microsequencing of the phosphohydrolase revealed a 'MutT' domain, which in other instances guards cellular integrity by dephosphorylating 8-oxo-dGTP, which causes AT and CG transversion mutations (Bessmann et al., 1996). DIPP activity is magnesium dependent with 1–2 mM required for maximal activity. The apparent molecular mass was 18 kDa. DIPP was very specific for diphosphoinositol polyphosphates; InsP6, Ins(1,3,4,5,6)P$_5$, Ins(1,2,4,5,6)P$_5$ and Ins(1,3,4,5)P$_4$ were not hydrolyzed.

Shears (1998) has discussed the known mechanisms by which PP-InsP$_5$ and [PP]$_2$-InsP$_4$ can be dephosphorylated back to InsP$_6$. These mechanisms include the 'reverse' actions of the InsP$_6$ and PP-InsP$_5$ kinases and the phosphohydrolase which attacks the β-position of the diphosphate group. Shears (1998) points out that while the 5β-phosphate is cleaved from PP-InsP$_5$, it is the other diphosphate group in [PP]$_2$-InsP$_4$, which is hydrolyzed. Both reactions appear to be performed by the same enzyme, DIPP. The enzyme is inhibited by fluoride (Menniti et al., 1993).

8.3.7.2. PPInsP₅ pyrophosphatases

Caffrey et al. (2000) have discovered molecular and catalytic diversity among human diphosphoinositol polyphosphate phosphohydrolases (hDIPP1). Following earlier identification of a prototype hDIPP1, new 21-kDa human isoforms, hDIPP2α and hDPP2β homologues in rat and mouse have been identified. The rank order for catalytic is hDIPP1 > hDIPP2α > hDIPP2β. The 76% identity between hDIPP1 and hDIPP2s includes conservation of an emerging signature sequence, a Nudt (MutT) motif with a GX(2)GX(6)G carboxy extension.

Caffrey and Shears (2001) have provided a genetic rationale for the microheterogeneity of human DIPP type 2 and have thus expanded the repertoire of molecular mechanisms regulating InsP pyrophosphate metabolism and function. 'Intron boundary skidding' by spliceosomes was suggested as a mechanism for yielding both hDIPP2α and hDIPP2β mRNAs.

Radioisotope assays. According to Safrany and Shears (1998) DIPP activity against PP-InsP₅ and [PP]₂-InsP₄ is determined routinely at 37°C in 500 μl of medium containing 50 mM KCl, 50 mM HEPES (pH 7.2 with KOH), 4 mM CHAPS, 0.05 mg/ml BSA, 1 mM EDTA, 2 mM MgSO₄. The reactions are quenched by the addition of ice-cold 8% perchloric acid (500 μl) containing 1 mg/ml InsP₆, and then neutralized as previously described (Safrany and Shears, 1998). In some experiments, release of Pi was recorded using an assay (Hoenig et al., 1989) that does not detect PPi.

In other experiments, the metabolism of [³H]labeled substrates were assayed by HPLC (Safrany and Shears, 1998), or ion exchange chromatography to measure [³²P]Pi release from either [5β-³²P]PP-InsP₅ or 5-PP-[5-β-³²P]PP-InsP₄. The second diphosphate group is only tentatively assigned to the 6-carbon. The samples, which had been stored at −20°C prior to being loaded onto 4.6 × 125 mm Partisphere 5 μm SAX HPLC columns, were eluted at 1 ml/min by the following gradient generated by mixing Buffer A (1 M Na_2-EDTA) and Buffer B (buffer A plus 1.3 M $(NH_4)_2HPO_4$, pH 3.85 with H_3PO_4; total [Pi] = 2.6 M): 0–10 min 0% B; 10–25 min, 0–35% B; 25–105 min, 35–100% B; 105–115 min, 100% B. Fractions of 1 ml were collected and radioactivity counted by scintillation (Shears et al., 1995).

Mass assays. The analyses of DIPP have been technically challenging. Specialist HPLC techniques (Mayr, 1990) have been utilized following calibration with chemically synthesized standards (Albert et al., 1997). In addition, two-dimensional NMR and stereospecific enzyme assays have also been used (Laussmann et al., 1997).

8.4. Other phosphatases and phytases

Phosphomonoesterases are a diverse group of enzymes that encompass a wide range of structures and reaction mechanisms. These enzymes have been classified as alkaline phosphatases, purple acid phosphatases (PAPs), low molecular weight acid phosphatases and protein phosphatases. The alkaline phosphatases have also been referred to as serine phosphates, while the low molecular weight acid phosphates have been referred to as cysteine phosphatases (Ostanin et al., 1994).

8.4.1. Alkaline phosphatases

Most fungal phytases hitherto characterized hydrolyze phytic acid to *myo*-inositol monophosphate (Wyss et al., 1999). Alkaline phosphatases contain two Zn^{2+} ions and one Mg^{2+} ion per enzyme subunit. The two Zn^{2+} ions form a binuclear center bridged by the product phosphate, whereas the Mg^{2+} ion is not directly in contact with the phosphate. A base-labile, acid stable phosphoserine intermediate has a critical role in alkaline phosphatases (Coleman, 1992).

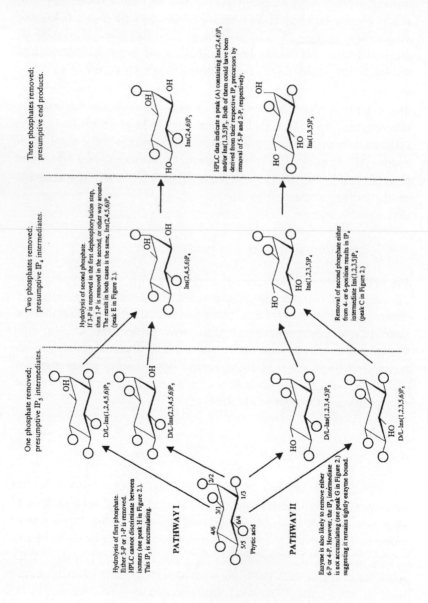

One phosphate removed; presumptive IP₅ intermediates.

Two phosphates removed; presumptive IP₄ intermediates.

Three phosphates removed; presumptive end products.

Hydrolysis of second phosphate.
If 3-P is removed in the first dephosphorylation step, then 1-P is removed in the second, or other way around. The result in both cases is the same, Ins(2,4,5,6)P₄ (peak E in Figure 2).

HPLC data indicate a peak (A) containing Ins(2,4,6)P₃ and/or Ins(1,3,5)P₃. Both of them could have been derived from their respective IP₄ precursors by removal of 5-P and 2-P, respectively.

D/L-Ins(1,2,4,5,6)P₅

D/L-Ins(2,3,4,5,6)P₅

Ins(2,4,5,6)P₄

Ins(2,4,6)P₃

Ins(1,3,5)P₃

PATHWAY I

Hydrolysis of first phosphate. Either 3-P or 1-P is removed. HPLC cannot discriminate between isomers (see peak H in Figure 2). This IP₅ is accumulating.

Phytic acid

PATHWAY II

Enzyme is also likely to remove either 6-P or 4-P. However, the IP₅ intermediate is not accumulating (see peak G in Figure 2), suggesting it remains tightly enzyme bound.

D/L-Ins(1,2,3,4,5)P₅

D/L-Ins(1,2,3,5,6)P₅

Removal of second phosphate either from 4- or 6-position results in IP₄ intermediate Ins(1,2,3,5)P₄ (peak C in Figure 2)

Ins(1,2,3,5)P₄

Inositol phosphates can be hydrolyzed by alkaline phosphatase, and the liberated inorganic phosphate measured by a calorimetric method using Malachite Green reagent (Eisenberg and Parthasarathy, 1987). The hydrolysis may be performed as follows: Combine 50 μl each of sodium-carbonate bicarbonate buffer (0.6 M, pH 9.7), magnesium chloride solution (18 mM), zinc acetate solution (6 μM), alkaline phosphatase (calf intestinal alkaline phosphatase is prepared by diluting the enzyme sample at least 100-fold with 20 mM sodium carbonate–bicarbonate buffer with 2 mM magnesium chloride), and the phosphate fraction (100 μl) in a total volume of 0.3 ml, and incubated at 37°C for 15 min. The reaction is stopped by the addition of 1.0 ml of 3N HCl. The liberated inorganic phosphate is measured by the Malachite Green reagent. Potassium dihydrogen phosphate is used to calibrate the standard curve.

8.4.2. Acid phosphatases

Acid phosphatases lack the metal-ion cofactors. In general, acid phosphatases hydrolyze monophosphate esters in a two-step mechanism. This involves a covalent phosphoryl-enzyme and a non-covalent enzyme-inorganic phosphate complex. There is good evidence for base-stable, acid-labile phosphohistidine as the phosphorylated intermediate (Ostanin et al., 1992).

The PAPs comprise a family of binuclear metal-containing hydrolases, members of which have been isolated from plants, mammals and fungi. The biological roles of PAPs are not well

Fig. 8.7. Scheme of the hydrolysis pathway(s) of *Bacillus* phytase (PhyC). Carbon atoms in the inositol ring are numbered for the D,L-configuration (D/L, respectively). Presumptive reaction intermediates and end products are referred to the corresponding assigned peaks on the basis of chromatographic retention time. Reproduced, with permission, from Kerovuo, H., Rouvinen, J. and Hatzack, F. (2000). Biochemical Journal 352, 623–628. © The Biochemical Society and the Medical Research Society.

defined in plants or animals. Kidney bean PAP exhibited phosphatase activity on a broad range of substrates, including polyphosphate as the preferred substrate, but lacked any phytase activity (Cashikar et al., 1997).

Hegeman and Grabau (2001) have recently purified a phytase from cotyledons of germinated soybeans. The soybean phytase was unrelated to previously characterized microbial or maize phytases, which were classified as histidine acid phosphatases. The soybean sequence exhibited a higher degree of similarity to PAPs, a class of metallophosphoesterases.

8.4.3. *Bacillus* phytases

Highly thermostable phytases have been isolated and cloned from *Bacillus* species. These enzymes do not align with any other phytases or phosphatases in the sequence data bank, nor do they contain the conserved active-site motif RHGXRXP. Unlike other phytases, these enzymes are dependent on Ca^{2+} ions for stability and activity (Kerovuo et al., 1998). Kerovuo et al. (2000) have described the hydrolysis of $InsP_6$ by *Bacillus* phytase (PhyC), which releases only three phosphates from phytic aid. The enzyme seems to prefer the hydrolysis of every second phosphate to that of adjacent ones. The authors suggest that the enzyme likely has two alternative pathways for the hydrolysis of phytic acid, resulting in two different *myo*-inositol end products, $Ins(2,4,6)P_3$ and $Ins(1,3,5)P_3$ as illustrated in Fig. 8.7. Ha et al. (2000) have obtained the crystal structure of a thermostable phytase from *B. amyloliquefaciens* in partially and fully calcium-loaded states. In the Ca^{2+}-loaded state, the enzyme is thermostable but inactive while in the fully Ca^{2+}-loaded state, the enzyme is thermostable and active.

CHAPTER 9

Phosphatidylinositol phospholipases

9.1. Introduction

The hydrolysis of PtdIns and PtdIns phosphates by phospholipases is an important mechanism for the production of many lipid derived signal transduction molecules. The formation of these lipid mediators is dependent on the intrinsic catalytic activity of the enzyme and the qualities of the phospholipid interface that allow the process of interfacial binding and activation as a crucial first step in overall reaction. Some of the phospholipases are specific for PtdIns, while others attack PtdIns along with other diradylglycerophospholipids. The PtdInsPs and GPtdIns are not at all attacked or attacked at much lower rate by the nonspecific phospholipases.

The assay strategies and methods for phospholipases have been reviewed by Reynolds et al. (1991). The simplest techniques for following PtdIns degradation by various enzymes is the use of radiolabeled phospholipids. Ortho-[^{32}P]phosphate is used to label cellular ATP pools, which are in rapid equilibrium with the monoester phosphates of the inositol phospholipids. The degradation of inositol phospholipids may also be followed by [^{3}H]inositol- and [^{14}C]fatty acid-labeling. The phospholipids are usually extracted with chloroform/methanol (Folch et al., 1957; Bligh and Dyer, 1959) and resolved by TLC, HPTLC or HPLC and assayed for radioactivity. The phospholipid mass is determined by phosphorus analyses, or by combinations of HPLC with light scattering or mass spectrometry. Depending on whether the assay is intended to characterize the enzyme or the substrate, different reaction conditions and the substrate and enzyme concentrations may be required.

In the early studies of substrate specificities, relative velocities for the reaction of PLA_2 and other lipolytic enzymes on a variety of different substrate vesicles composed of pure phospholipid classes were reported but no effort was made to normalize for the amount of enzyme bound to the interface. In order to minimize the kinetic complexities of interfacial catalysis arising from a reversible association of the phospholipase with the interface, Jain and Gelb (1991) have developed a special method for studying the action of PLA_2 on substrate vesicles in the 'scooting' mode in which all the enzyme is tightly bound to the interface. In such a case, reversible binding is no longer a part of the catalytic turnover within the interface. The most important use of the scooting assay is in the determination of the absolute substrate specificities and the interfacial rate constants of lipolytic enzymes (Gelb et al., 1995). Roberts (1996) has described other structural and functional motifs for working with phospholipases at an interface. Continuous fluorescence assays for PLA_2, PLC and PLD have proven valuable for kinetic studies (Wilton et al., 1990; Huang et al., 1994; Cho et al., 1999).

For the purposes of the present chapter, the discussion has been arranged to cover separately the PtdIns-nonspecific and PtdIns-specific lipases. Furthermore, the activation of the phospholipases by $PtdInsP_2$ and $PtdInsP_3$ has been covered separately. The coverage of the individual phospholipases includes isolation, substrate specificity, and updated assay procedures for PtdIns, its phosphates and glycans, along with references to elaborate more on general reviews, when available, found elsewhere.

9.2. PtdIns-nonspecific phospholipases

The PtdIns nonspecific PLA_1, PLA_2, PLC and PLD attack the PtdIns along with other glycerophospholipids, although frequently at considerably slower rate. Some of these phospholipases show partial specificity for glycerophospholipids other than PtdIns. However, the phospholipases occur in various isomeric forms and not all isomers have been assayed for their activity with PtdIns. The nonspecific enzymes of snake venom and bacterial or plant origin are available in high purity and in sufficient amounts for the quantitative degradation of PtdIns provided it is first isolated in pure state. Recently, several phospholipases of animal tissues have also been purified and cloned and shown to be specific for Cho, Etn and Gro glycerophospholipids, and PtdOH. The phospholipases lacking the phospholipid class specificity can nevertheless exhibit stereo, regio and fatty acid specificity, as shown below.

The phospholipases A_1 (PLA$_1$) and A_2 (PLA$_2$) attack the fatty acid ester bonds at the *sn*-1 and *sn*-2-positions, respectively, of the glycerol molecule, while phospholipases C (PLC) and D (PLD) attack the phosphodiester bonds on either the left or the right hand side of the phosphorus atom, respectively, as shown in Fig. 9.1.

9.2.1. Phospholipase A₁

PLA$_1$ attacks the *sn*-1-position of glycerophospholipids, including PtdIns, and releases a free fatty acid along with the corresponding

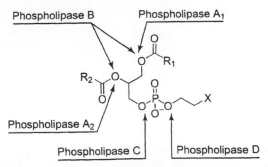

Fig. 9.1. Reactions catalyzed by phospholipases A_1, A_2, B, C and D on glycerophospholipid molecules. X denotes the nitrogenous base (choline, ethanolamine) or other (serine, inositol, glycerol, aliphatic alcohol or water) polar head group. Reproduced with permission from Ravandi and Kuksis (2000).

2-lysoglycerophospholipid (Pete et al., 1994). Enzymes with PLA_1 activity have been found to occur in certain molds and animal tissues with maximum activity in the acidic pH region. In contrast, the enzyme identified in snake venoms and bee venom has shown a maximum activity in the alkaline region. The PLA_1 of venoms, like PLA_2 from the same source, is extremely stable to heat, and is activated by both Ca^{2+} and Mg^{2+} ions. These properties contrast with those of PLA_1 from animal tissues, which are less stable to heat and are inactivated by Ca^{2+} and Mg^{2+}. *Escherichia coli* contains two kinds of PLA_1. One, termed detergent resistant PLA_1, is bound to the outer membrane and retains its activity in the presence of detergent or organic solvent. The other, named detergent-sensitive PLA_1, is present in the cytosol fraction. Both cytosolic and membrane-bound activities have been identified in mammalian tissues (Higgs and Glomset, 1996). However, it has not been clearly distinguished from lipase and lysophospholipase activity (Pete et al., 1994). Thus, a study of PtdIns turnover using a rat brain fraction (Hirasawa et al., 1981) or bovine arterial cells (Martin and Wysolmerski, 1987) showed that most PtdIns deacylation occurred by the actions of PLA_1 and $lysoPLA_2$. However, addition of Ca^{2+} caused a further contribution to PtdIns deacylation by PLC and diacylglycerol lipase. Some PLA_1s also attack triacylglycerols, thus complicating the assays (Slotboom et al., 1970).

9.2.1.1. Substrate specificity

In a few instances genuine PLA_1 activity appears to have been identified and purified. The molecular weight of the detergent resistant PLA_1 of *E. coli* has been estimated to be 21 000, 28 000 or 30 809 kDa depending on the methodology used for the determination. This PLA_1 catalyzes the hydrolysis of the glycerophospholipids of *E. coli* in the following order of decreasing activity: PtdEt > PtdGro > $(Ptd)_2Gro$ > PtdCho (Nakagawa et al., 1991). The chain length and degree of unsaturation of the acyl groups of glycerophospholipids have little effect on the activity of the enzyme. The detergent resistant PLA_1 of $E.$ $coli$ possesses both PLA_1 and PLA_2 activities. Furthermore, the enzyme also catalyzes the hydrolysis of 1-acyl- and 2-acyl-lysoglycerophospholipids.

Intracellular PLA_1 has been partially purified from rat liver (Dawson et al., 1983) and heart (Nalbone and Hostetler, 1985), and from rat kidney lysosomes (Hostetler et al., 1991; Hostetler and Gardner, 1991), and from bovine brain (Ueda et al., 1993). Pete et al. (1994) have succeeded in the purification of PLA_1 from a soluble fraction of bovine brain. The purified PLA_1 eluted from the Sephacryl S-300HR column in a volume corresponding to a molecular mass of 365 kDa and migrated as two bands (Mr = 112 000 and 95 000) when separated by polyacrylamide gel electrophoresis in sodium dodecyl sulfate. The bovine brain enzyme catalyzed the specific hydrolysis of acyl groups from the sn-1-position of a broad range of phospholipid substrates, including hydrolysis of the anionic phospholipids 1-stearoyl-2-[1-^{14}C]arachidonoyl sn-GroPIns and 1,2-diolcoylglyccro-sn-glycero-3-phospho-[3-^{14}C]serine to some extent. Hydrolysis of neutral lipids could not be detected under any conditions.

Higgs and Glomset (1994) have identified a cytosolic PLA_1 activity in bovine brain and testis that preferentially hydrolyzes PtdOH substrates. Using a Triton X-100 mixed micelle system they showed that the cytosolic PLA_1 preferentially hydrolyzed PtdOH by 4-fold over PtdIns, 5-fold over PtdSer, 7.5-fold over PtdEtn and 10-fold over PtdCho. Higgs and Glomset (1996) have also purified a PtdOH-preferring PLA_1 from bovine testis. Although this enzyme was not specifically assayed for its activity on PtdIns, it was shown that the PLA_1 isolated by Pete et al. (1994) and which hydrolyzes PtdIns, preferred PtdOH once the inhibitory $MgCl_2$ was excluded from the incubation medium. Higgs et al. (1998) have since cloned a PtdOH-preferring PLA_1 from bovine testis. More recently, the PtdOH-preferring cytosolic PLA_1 has been shown to interact with membrane lipids, which affected both, the activity and the specificity of the enzyme (Lin et al., 2000).

Gassama-Diagne et al. (1991) have discussed the determination of PLA$_1$ activity of guinea pig pancreatic lipase. The authors had purified two cationic lipases displaying PLA$_1$ activity from guinea pig pancreas, which lacks the classic secretory PLA$_2$ enzyme. The following order of hydrolytic activity was established using radiochemical assays with pure substrates under optimal conditions of measurement: dioleoylglycerol > 1(3)-monooleoylglycerol > trioleoylglycerol > PtdCho = PtdIns > lysoPtdCho > lysoPtdEtn = PtdGro. A striking feature which differentiates guinea pig pancreatic PLA$_1$ from classical PLA$_2$ is the absolute lack of stereospecificity.

A nonstereospecific PLA$_1$ activity has been demonstrated for the PLB purified from *Penicillium notatum* (Saito et al., 1991). The enzyme hydrolyzes the primary ester bonds of both *sn*-1,2- and *sn*-2,3-glycerophospholipids. The enzyme hydrolyzes the 2-acyl esters bond of *sn*-1,2-diacyl GroPCho, but not that of *sn*-2,3-diacyl GroPCho. The PLB of *P. notatum* hydrolyzes phospholipids in the following order: PtdSer > PtdIns > PtdOH > PtdEtn > Ptd$_2$Gro (Saito and Kates, 1974).

A Ca^{2+}-independent PLB-like activity, which hydrolyzes both *sn*-1- and *sn*-2-acyl groups of glycerophospholipids, has been isolated from the intestinal brush border membranes (Pind and Kuksis, 1987, 1988, 1989, 1991; Diagne et al., 1987). The enzyme has been shown (Gassama-Diagne et al., 1992) to hydrolyze triacyl-, diacyl- and monoacylglycerols in addition to the glycerophospholipids. The enzyme has shown activity with retinyl esters (Rigtrup et al., 1994). Subsequent work (Tojo et al., 1998; Takemori et al., 1998) has led to the identification and cloning of the functional domain of this enzyme and the determination of its tissue distribution. The purified enzyme exhibited broad substrate specificity including esterase, PLA$_2$, lysophospholipase, and lipase activities. SDS-gel electrophoretic and reverse phase HPLC analyses demonstrated that a single enzyme catalyzes these activities. It preferred hydrolysis of the *sn*-2-position of diacylglycerophospholipids and diacylglycerols without strict stereospecificity (Tojo et al., 1998).

A Ca^{2+}-independent PLA, which releases various fatty acids from sn-1- and sn-2-positions of glycerophospholipids has been partially purified from the rat brain soluble fraction (Yoshida et al., 1998). The enzyme showed an approximate molecular mass of 300 kDa on gel filtration column chromatography. This enzyme produced a radiolabeled lysoPtdCho when 1-palmitoyl-2-[1-^{14}C]-oleoyl-sn-glycero-3-phosphocholine was used as substrate. By using a series of synthetic PtdCho, the enzyme cleaved oleic, linoleic, and arachidonic acids like phospholipase A_2 and released palmitic and stearic acids like phospholipase A_1. PtdCho, PtdEtn, PtdIns and PtdOH were hydrolyzed with almost equal efficiencies by this enzyme, while PtdSer was an inefficient substrate. Although the enzyme isolated in this study was not pure, the PLA_1 and PLA_2 activities were eluted from several columns in a concurrent fashion suggesting that the enzyme possessed both catalytic activities as an intrinsic property (Yoshida et al., 1998).

9.2.1.2. Methods of assay

The enzyme may be assayed by loss of substrate or appearance of product. Both changes are best measured using appropriate radioactive substrates. When sufficient substrate is available, mass assays may be employed. PLA_1 does not attack the alkyl or alkenyl ether linkages in the sn-1-position of the glycerophospholipids, but it may release the sn-1-acyl groups at different rates depending on their degree of unsaturation and chain length. The fatty acid specificity is best determined by analyzing the changes in the substrate composition, unless a full spectrum of radiolabeled fatty acids is available for release in the test substrate.

Experiments with unilamellar liposomes appear promising for this purpose (Zhang et al., 1991). Jain and Gelb (1991) have found that PLA_2 (and probably PLA_1) hydrolyzes vesicles of anionic phospholipids such as dimyristoylGroPMe in the scooting mode.

Hydrolysis of the vesicles in the scooting mode is monitored with a pH-stat equipped with a high speed mechanical stirrer and a water-jacketed thermostated vessel maintained at 21°C. It is also possible to adopt other methods of product analysis, e.g. radioactivity released can be measured by scintillation counting.

PLA$_1$ activity may be assayed by TLC as follows (Pete et al., 1994). The standard assay measures the site-specific deacylation of lipids and employs a radioactive substrate (10 μM, in the absence of PtdSer) in Tris–HCl (100 mM, pH 7.5), 325 μM CHAPS, 3 mM MgCl$_2$, 1 mM EDTA, and 70% (v/v) glycerol. The reaction is initiated by the addition (5 μl) of substrate (10 μM) to enzyme in assay buffer (final volume, 50 μl). The reaction mixture is incubated at 37°C for 10–60 min and terminated by the addition of 0.5 ml of methanol/chloroform (1:1, v/v) followed by 0.15 ml of H$_2$O. The mixture is vortexed and centrifuged. The bottom layer is collected, mixed with 5 μg each of the appropriate reaction product (e.g. lysophospholipid and free fatty acid), and dried under a stream of N$_2$. The dried sample is redissolved in 35 μl of chloroform/methanol (65:35, v/v) and spotted on a silica gel LK6D plate Whatman, Hilsboro, OR). The TLC plate is developed in chloroform/methanol/H$_2$O (65:35:5, by vol.) until the solvent front is 10 cm from the top of the plate. The plate is dried and rechromatographed in hexane/diethyl ether/formic acid 90:60:4 (by vol.) until the solvent front is 2 cm from the top of the plate. The lipids are visualized using a solution containing Coomassie Brilliant Blue R-250 (0.03%, w/v), 100 mM NaCl, and 30% (v/v) methanol. The R_f values obtained for lysoPtdCho, lysoPtdEtn, PtdCho, PtdEtn, and free fatty acid were 0.08, 0.16, 0.22, 0.36, and 0.76, respectively. The appropriate zones are scraped from the plate, mixed with 0.5 ml ethanol/conc. HCl (100/1, v/v) and then 7.5 ml of scintillation fluid is added, and the radioactivity counted. Alternatively, the lipid extract is taken to dryness and redissolved in chloroform and applied to the TLC plate (Silica gel G, Merck and Co.), which is developed with petroleum ether/

diethyl ether/acetic acid 60:40:1 (by vol.) for the separation of the fatty acid and phospholipids. Lipids are stained with iodine, and the fatty acid spot is scraped off into a vial. Scintillation fluid is added for radioactive counting.

A more convenient and faster method than TLC for separation of fatty acid is the method developed by Dole and Meinertz (1960) based on liquid–liquid partition.

For the liquid–liquid separation of liberated fatty acid from phospholipid, the reaction is terminated by the addition of 3 ml of Dole's reagent (2-propanol/heptane/1N H_2SO_4, 20:5:2 by vol.). The tubes are shaken vigorously with a vortex mixer for 15 s. Distilled water (1.6 ml) and then heptane (1.8 ml) are added. Mixing of the heptane and water phases with vortex mixer for 10 s is essential for the quantitative extraction of fatty acid. The emulsion is clearly separated by centrifugation at 3000g for 10 min. A portion (1.5 ml) of the upper phase is removed to another tube containing 30 mg of silica gel (Wako gel, Wako Pure Chemical, Osaka, Japan). After shaking for 20 s, the silica gel is precipitated by centrifugation at 100g for 10 min. Heptane (0.8 ml) is removed for radioactivity counting in 7 ml of ACS II scintillation fluid (Amersham, Buckinghamshire, UK).

LysoPLA I and II have been recognized to be present in a macrophage-like cell line. Wang et al. (1997) have described the stereospecificity and catalytic triad of lysoPLA$_1$. These enzymes hydrolyze the fatty acid ester bond of lysophospholipids, liberating water-soluble glycerophosphocholine (ethanolamine) and fatty acids (Fig. 9.2). These enzymes are generally assayed using the Dole assay (Zhang et al., 1991), followed by separation of the reactants and products by TLC. The lipid components are separated by elution with chloroform/methanol/acetic acid/water (25:15:4:2, by vol.). The lipids are visualized with I_2 vapor, and the zones corresponding to fatty acid and lysoPtdCho are scraped directly into scintillation vials and counted with 6 ml of scintillation fluid.

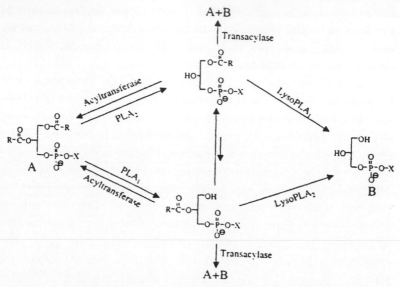

Fig. 9.2. Regiospecificity and metabolic interconversions involving lysophospholipases. X denotes a polar head group as in Fig. 9.1. Reproduced in modified form from Wang et al. (1997) with permission of the publisher.

9.2.2. Phospholipase A_2

PLA$_2$ catalyzes the hydrolysis of the acyl ester bond at the sn-2-position of 1,2-diradyl-sn-glycerophospholipids. Traditionally, the sPLA$_2$s have been subdivided according to their primary structure into two main groups: type I and type II (Dennis, 1994; Murakami et al., 1997). Snake venom PLA$_2$ from *Elapidae* and *Hydrophidae* are called type I PLA$_2$, while the others from *Crotalidae* and *Viperidae* are type II PLA$_2$. Pancreatic PLA$_2$ has structural similarities to snake venom type I PLA$_2$s and therefore is now termed mammalian type I PLA$_2$.

PLA$_2$ enzymes have been found in many mammalian cell types. In the 1980s, an extracellular (secretory) PLA$_2$ (sPLA$_2$) was isolated from a wide variety of mammalian tissues and inflammation sites. It is now termed mammalian type II PLA$_2$ and has structural characteristics similar to those of snake venom type II PLA$_2$s. Evidence is accumulating that type II PLA$_2$ is induced by inflammatory stimuli and plays crucial roles in cellular responses at sites of inflammation (Pruzanski and Vadas, 1991).

Based on their cellular localization, they are divided into two general groups: secreted and intracellular (Dennis, 1994, 1997; Six and Dennis, 2000). The secreted PLA$_2$s have low molecular masses, generally between 14 and 16 kDa. They require millimolar calcium concentrations for catalytic activity, and possess 5–7 disulfide bonds. Tishfield (1997) has reassessed the low molecular weight PLA$_2$s gene family in mammals. The intracellular PLA$_2$s are further divided into Groups IV and VI based on the Ca^{2+} requirements needed for basal catalytic activity. These enzymes have larger molecular masses of about 80–85 kDa and are insensitive to thio-reducing agents such as TT. According to biochemical properties and structural features, the PLA$_2$ super-family can be subdivided into three main types, i.e. the Ca^{2+}-dependent secretory enzymes (sPLA$_2$), the Ca^{2+}-dependent cytosolic enzymes (cPLA$_2$) and Ca^{2+}-independent cytosolic enzymes iPLA$_2$ (Balsinde et al., 1997). Balboa et al. (1996) have demonstrated that the sPLA$_2$ of group V is actively involved in arachidonic acid signaling in macrophages.

The work of Bingham et al. (1999) has shown that the group IIA PLA$_2$ is present in the secretory granules of mouse bone marrow-derived mast cells, while the group V PLA$_2$ is associated with various membranous organelles including the Golgi apparatus, nuclear envelope and plasma membrane. The spatial segregation of group IIA PLA$_2$ and group V PLA$_2$ implies that these enzymes are not functionally redundant.

Other PLA$_2$s have been identified that do not depend on Ca^{2+} for activity (Dennis, 1994; Ackerman and Dennis, 1995). Only a handful of such Ca^{2+}-independent PLA$_2$s has been purified and characterized. The Ca^{2+}-independent PLA$_2$ (iPLA$_2$) include enzymes isolated from brush border membranes (Pind and Kuksis, 1987; Diagne et al., 1987), macrophages (Ackerman and Dennis, 1995), rat brain (Yoshida et al., 1998), and include PAF-acetylhydrolase (Stafforini et al., 1991). Takemori et al. (1998) have reported the cloning, expression and tissue distribution of the rat intestinal PLB (lipase).

Mizenina et al. (2001) have isolated a novel group IIA phospholipase A_2 (srPLA$_2$), which interacts with v-Src oncoprotein from RSV-transformed hamster cells. This protein contains all the conserved functional residues typical for group IIA PLA$_2$ enzymes and, in addition, eight amino acids at its C terminus. In vitro, srPLA$_2$ protein interacts with v-SrcHM as well as with v-SrcLM oncoproteins. The 17-kDa precursor version of srPLA$_2$ was identified by co-immunoprecipitation with v-Src protein; it was found to be phosphorylated on a tyrosine. Mizenina et al. (2001) assayed the novel srPLA$_2$ directly on the Sepharose beads as follows. The beads were resuspended in the reaction mixture containing 140 mM NaCl, 30 mM KCl, 20 mM ATP, and 20 mM HEPES, pH 7.5. Lipid lipase assays were initiated by the addition of mixture of 10 μM PtdCho and 0.5 μC/ml [^{14}C]PtdCho (Amersham). Reaction mixtures were incubated at room temperature for 40 min, then lipids were extracted by a chloroform/methanol mixture (1:1, v/v) and subsequently dotted onto silica gel plates. [^{14}C]Arachidonic acid was separated from [^{14}C]PtdCho by HPTLC in chloroform/methanol/11% ammonium hydroxide/water (65:35:6:3, by vol.). The solvent was run up to 2 cm from the top of the plate. Plates were air dried, fixed, and autoradiographed to detect labeled lipids. Lipids were located by autoradiography or visualized with iodine vapor.

PLA$_2$ activity can be regulated via heterotrimeric G-proteins, phosphorylation (by receptor-tyrosine kinases, MAP kinase and PKC) and Ca^{2+}. Some PLA$_2$s, such as those that contain a CalB domain, can be activated by translocation to their substrate in the membrane (reviewed by Exton, 1994; Roberts, 1996; Serhan et al., 1996). The G-protein activator mastoparan activates PLA$_2$ in zucchini and other studies have implicated G-proteins in auxin signaling and plant defense responses (Munnik et al., 1998).

PLA_2 is not stimulated by G-proteins in all cell types. The lysoPtdIns generated by PtdIns-specific PLA_2 can be further deacylated by lysophospholipase which may yield GroPIns (4)P from PtdIns (4)P. GroPIns (4)P is a biologically active-water soluble molecule, whose cellular concentrations are reported to increase in Ras-transformed cells (Falasca et al., 1997).

9.2.2.1. Substrate specificity

Phospholipids containing most of the naturally occurring polar head groups, including Cho, Etn, Ins, Ser and Gro as well as other molecular classes such as Ptd_2Gro and PtdOH, are hydrolyzed to some degree by the extracellular PLA_2s. The snake venom PLA_2s are completely stereospecific for *sn*-1,2-diacylglycerophospholipids, including PtdCho, PtdEtn, PtdIns, PtdGro, PtdSer, PtdOH and Ptd_2Gro (Holub and Kuksis, 1978) but not the PtdIns phosphates. The substrate specificity of human nonpancreatic $sPLA_2$ was examined by Bayburt et al. (1993) using the double-radiolabeled approach described by Ghomashchi et al. (1991). The synthesis and fluorometric analysis of new substrates of 14- and 85 kDa PLA_2s has since been described by Bayburt et al. (1995).

Of particular interest is the release of the arachidonic acid from the *sn*-2-position of the phospholipids. It is crucial in this type of experiment to take into account the sizes of different endogenous pools of arachidonate and the rates of incorporation of exogenously provided arachidonic acid in these pools. Chilton and Murphy (1986) have studied the remodeling of arachidonate-containing glycerophospholipids within the human neutrophil. Chilton (1991) has shown that [^{3}H]arachidonic acid (<0.1 μM) will reach a constant radiospecific activity in all phospholipid molecular species of the mast cell after 24–36 h in culture.

Cytosolic PLA_2 ($cPLA_2$, or type IV PLA_2) is apparently one of the most important PLA_2 isozymes involved in regulating the lipid mediator generation resulting from cell activation (Dennis, 1994; Kudo et al., 1993; Murakami et al., 1995, 1997). Only one isomer of this PLA_2 family has been identified so far. $cPLA_2$ has a molecular mass of 85 kDa, an apparent requirement for submicromolar Ca^{2+} concentrations for PLA_2 activity and exhibits preferential hydrolysis of phospholipids bearing arachidonic acid. When recombinant $cPLA_2$ was incubated with membrane vesicles prepared from mammalian cells, arachidonic acid was released in marked preference to other fatty acids, even though oleic acid was much more abundant in the vesicles (Clark et al., 1991; Diez et al., 1994). More detailed analyses using natural and synthetic membranes demonstrated that $cPLA_2$ prefers polyunsaturated fatty acids, especially those with three *cis*-double bonds between carbons 5 and 6, 8 and 9, and 11 and 12 (Diez et al., 1992; Hanel et al., 1993). Thus, linolenate or eicosapentaenoate are also good substrates for $cPLA_2$, but these acids make up only a small proportion in relation to arachidonate. Hanel et al. (1993) reported that the order of preference of *sn*-2-fatty acids is 20:4 > 18:3 > 18:2 > 18:1 > 16:0, and that of the 20-carbon acyl chains is 20:4 > 20:3 > 20:2 > 20:1 > 20:0, and there is a preference for positional isomers with double bonds closest to the *sn*-2 ester. Diacyl- and alkenylacyl GroPEtn with docosahexaenoic acid are poor substrates for $cPLA_2$ (Shikano et al., 1994). Caramelo et al. (2000) mapped the catalytic pockets of PLA_2 and PLC using radioiodinatable PtdChos derivatized with bulky end groups of the sn-2-fatty chain as a molecular ruler.

There was some preference of head group specificity in that the order of preference was found to be PtdCho > PtdIns > PtdEtn by Diez et al. (1992) and PtdCho = PtdIns > PtdEtn > PtdOH = PtdSer by Hanel et al. (1993), although the difference between the rates of hydrolysis of best and worst substrates was only 3.5-fold. In contrast, studies using natural membranes demonstrated no selectivity among PtdCho, PtdIns and PtdEtn, and the rate of hydrolysis was based on the content of arachidonic acid in each phospholipid subclass. Rabbit platelet $cPLA_2$ hydrolyzed PtdIns only when a submillimolar Ca^{2+} concentration was present, compared with PtdCho and PtdEtn that were hydrolyzed in the presence of submicromolar Ca^{2+} concentrations (Kim et al., 1991).

Cupillard et al. (1997) have reported the cloning, chromosomal mapping, and recombinant expression of a novel $sPLA_2$. Based on its structural properties, this $sPLA_2$ appears as a first member of a new group of mammalian $sPLA_2$s, called group X, according to the PLA_2 nomenclature defined by Dennis (1997). A cDNA coding for the novel $sPLA_2$ was isolated from human fetal lung. The mature $sPLA_2$ protein with molecular weight of 13.6 kDa, is acidic (pI 5.3), and is made up of 123 amino acids. Key structural features of the $sPLA_2$ included a long prepropeptide ending with an arginine doublet, a total of 16 cysteines located at positions that are characteristic of both group I and group II $sPLA_2$s, a C-terminal extension typical of group II $sPLA_2$s, and the absence of elapid and pancreatic loops that are characteristic of group I $sPLA_2$s. It was maximally active at physiological pH and with 10 mM Ca^{2+}. Group X $sPLA_2$ was found to prefer PtdEtn and PtdCho liposomes to those of PtdSer.

A similar cloning and characterization of group X $sPLA_2$ from mouse and human sources has been reported by Ishizaki et al. (1999). Among the phospholipid classes examined (no PtdIns) L-α-1-palmitoyl-2-linoleoyl-GroPEtn was hydrolyzed most efficiently. The results also indicated the absence of a preference for the arachidonic acid-containing phospholipids. However, in a subsequent paper (Hanasaki et al., 1999) the same authors reported that a purified group X $sPLA_2$ induced the release of arachidonic acid from PtdCho more efficiently than other human $sPLA_2$ groups and elicited a prompt and marked release of arachidonic acid from human monocytic THP-1 cells compared with group IB $sPLA_2$ and group IIA $sPLA_2$ with concomitant production of prostaglandin E_2. Experiments using purified phospholipids used as substrates showed that type I PLA_2 hydrolyzes both PtdCho and PtdEtn at almost identical rates, whereas type II PLA_2 hydrolyzes PtdGro, PtdEtn, and PtdSer in preference to PtdCho and PtdIns (Murakami et al., 1999).

Bezzine et al. (2000) have recently demonstrated that exogenously added human group X sPLA$_2$ but not the group IB, IIA and V enzymes efficiently release arachidonic acid from adherent mammalian cells. Bezzine et al. (2000) give an LC/MS method for detecting released lysophospholipids. The PAF-acetylhydrolase is the most completely characterized of the Ca^{2+}-independent PLA$_2$s, it inactivates PAF and degrades oxidized phospholipids (Stafforini et al., 1991; Roberts, 1996). Interestingly, lipid ozonization products have been reported to activate phospholipase A$_2$, C and D (Kafoury et al., 1998).

While glycosylPtdIns anchors can be hydrolyzed by specific PLCs and PLDs, an equivalent mechanism has yet to be established for PLA$_2$. It has been demonstrated, however, that 10% of the PtdIns 4-K in carrot plasma membranes appeared to be released by PLA$_2$ treatment (Gross et al., 1992).

9.2.2.2. Methods of assay

PLA$_2$ activity also may be assayed by measuring the loss of radioactivity from an appropriately labeled substrate, or by measuring the appearance of radiolabeled product from an appropriately labeled substrate. In addition, mass assays may be used when substrate levels permit it. In addition to the release of the fatty acids from the sn-2-position, PLA$_2$ may yield variable amounts of 1-alkyl and 1-alkenyl 2-lysoGroPCho or GroPEtn, which may be determined to further characterize the enzyme activity and substrate specificity. Reynolds et al. (1992) developed a spectrophotometric assay for microtiter plate reader using short chain PtdCho micelles to examine the high molecular weight human synovial fluid PLA$_2$.

Bayburt et al. (1993) have used the scooting method of Ghomashchi et al. (1991) to determine the substrate specificity of human nonpancreatic secreted PLA$_2$ using the double-radiolabel approach. Specifically, vesicles containing radiolabeled lipids, including PtdIns, were prepared by dissolving 5.4 mg of oleoyl-palmitoyl GroPCho, 0.6 mg of dioleoyl GroPOH, approximately

3.6 μCi of ^{3}H-substrate, and approximately $1-2$ μCi of ^{14}C substrate in CHCl$_3$/MeOH (1:1). The hydrolysis of these mixed-lipid vesicles in the scooting mode was followed by the pH-stat method. From the amounts of competing substrates in the vesicle at the beginning of the reaction and the amounts of products formed from each of the two substrates after partial hydrolysis, the ratio of kcat/Km values for the two competing substrates was obtained. The data showed that phospholipids containing most of the naturally occurring polar head groups (including Ins) are all hydrolyzed with similar relative kcat/Km values. Thus, the previously claimed (Franken et al., 1992) preference of sPLA$_2$ for PtdEtn over PtdCho is incorrect and sPLA$_2$ cannot be classified as a PtdEtn-specific enzyme.

To date, cPLA$_2$-IV, iPLA$_2$-VI, sPLA$_2$-II, and sPLA$_2$-V are the major PLA$_2$ types that have been well characterized in mammalian tissues (Dennis, 1994, 1997; Chen and Dennis, 1998). It is difficult to demonstrate specificity of function for a single PLA$_2$ isoform in vivo because most of the PLA$_2$ inhibitors currently available are not isoform-specific. However, Yang et al. (1999) have been able to combine the use of the available inhibitors along with the exact working conditions to distinguish among the four major cellular PLA$_2$.

According to Yang et al. (1999), appropriate amounts of cold and ^{14}C-labeled phospholipid (100 000 cpm per assay) are dried under a stream of N$_2$ gas. The dried lipid is lyophilized for 1 h to remove the remaining organic solvent. The lipid is then suspended in an adequate amount of assay buffer to give a 10-fold lipid suspension after vigorous mixing. For mixed micelles, an appropriate amount of Triton X-100 (to give the final Triton X-100 concentration of 400 μM) is added to the 10-fold lipid suspension and the solution is vortexed. Small unilamellar vesicles are prepared by bath-sonication of the 10-fold lipid suspension using an 80-W Branson sonifier. The dried lipid is suspended in 1 ml of 100 mM HEPES, pH 7.5, in 70×10 mm^2 thick-wall test tube.

Routinely, 50 μl of the sample is added to 450 μl of substrate to start the reaction at 40°C for 30 min. The reaction is stopped by the addition of 2.5 ml of the Dole reagent (2-propanol:heptane:0.5 M H_2SO_4, 400:100:20, by vol.) followed by vortexing. The product mixture is subsequently processed according to the modified Dole assay (Conde-Frieboes et al., 1996). Briefly, 200 mg of silicic acid (200–425 mesh, Fisher Sci.) is added to the mixture. The fatty acids are extracted with 1.5 ml of heptane and 1.5 ml of H_2O. One milliliter of the upper layer is loaded onto a Pasteur pipette containing silicic acid (2.0–2.5 cm in length, 200–425 mesh); the eluent is collected in a scintillation vial. After the sample passes through the column, 1 ml of diethyl ether is added to elute the remaining free fatty acid. The radioactivity is counted after mixing with a scintillation fluid (5 ml).

iPLA$_2$-VI assay. The iPLA$_2$-VI assay utilizes 100 μM dipalmitoyl GroPCho (containing 100 000 cpm of 1-palmitoyl-2-[1-^{14}C]palmitoyl-sn-glycerol-3-phosphorylcholine) in 400 μM Triton X-100 mixed micelles in 100 mM HEPES, pH 7.5, 5 mM EDTA, 2 M TT, and 1 M ATP. The presence of 1 mM ATP is essential to stabilize the activity of iPLA$_2$ in the crude mixture.

cPLA$_2$-IV assay. The cPLA$_2$-IV assay utilizes 100 μM palmitoylarachidonoyl GroPCho/PtdIns-P$_2$ (97/3) (containing 100 000 cpm of 1-palmitoyl-2-[1-^{14}C]arachidonoyl-sn-glycerol-3-phosphorylcholine) in Triton X-100 (400 μM) mixed micelles in 100 mM HEPES, pH 7.5, 80 μM Ca^{2+}, 2 mM DTT, and 0.1 mg/ml BSA. A higher concentration of bovine serum albumin (BSA) can adversely affect the activity of cPLA$_2$ in this assay.

sPLA$_2$-V assay. The sPLA$_2$-V assay contained 100 μM DPPC/POPS (3/1) in the form of small unilamellar vesicles (containing 100 000 cpm of 1-palmitoyl-2-[1-^{14}C]palmitoyl-sn-glycerol-3-phosphorylcholine) in 100 mM HEPES, pH 7.5, with 5 mM Ca^{2+} and 1 mg/ml BSA.

sPLA$_2$-IIA assay. The sPLA$_2$-II assay contains 100 μM palmitoyllinoleoyl GroPEtn/palmitoyloleoyl GroPSer (1/1) in the form of single unilamellar vesicles (SUVs) (containing

100 000 cpm of 1-palmitoyl-2-[1-^{14}C]linoleoyl-sn-glycerol-3-phosphorylethanolamine) in 100 mM HEPES, pH 7.5, with 1 mM Ca^{2+} and 1 mg/ml BSA.

Yang et al. (1999) have developed special sets of equations to determine the absolute amounts of each enzyme in any given sample. A limitation of this assay system is seen when one form is more than two orders of magnitude more abundant that other types.

Clark et al. (1991) have described an assay for arachidonate release by cPLA$_2$ from natural membranes, which is still frequently employed in the current literature.

According to Clark et al. (1991), membrane vesicles are prepared by suspending U937 cells in Tris–HCl buffer (25 mM at pH 7.4) containing 136 mM NaCl and 0.5 mM EGTA and lysing by N$_2$ cavitation (600 psi 3x). The resulting lysate is centrifuged (8500g, 30 min) at 4°C followed by a high speed centrifugation (105 000g, 60 min) of the supernatant to pellet the membranes. This resulting pellet was washed once with the Tris–HCl buffer containing 0.25 mg/ml BSA and resuspended in water for use as substrate. The fatty acids in the sn-2-position were determined by gas chromatography. cPLA$_2$ from the CHO cell line expressing cPLA$_2$ was purified to a specific activity of 10 U/mg. Recombinant cPLA$_2$ (10 μg) was incubated at 37°C with the U937 membranes (360 nmol of phospholipid) in HEPES buffer (50 mM, pH 7.4) containing 0.975 mM Ca^{2+}, 1 mM EGTA (free Ca^{2+} concentration = 1 μM), and 0.25 mg/ml BSA. The reaction was stopped at each time point by adding 360 μl of methanol, and the released fatty acids were quantified by GLC.

According to Cupillard et al. (1997) the activity of X sPLA$_2$ is best assayed on mixed phospholipid liposomes prepared by co-sonication of PtdEtn, PtdCho, or PtdSer, with D-α-dipalmitoyl-GroPCho, 75:25%). D-α-PtdCho is a nonhydrolyzable phospholipid that is used for the formation of liposomes with the various hydrolyzable phospholipids. The phospholipids are dissolved in chloroform, dried under a stream of nitrogen, and resuspended at

0.33 mM in 100 mM Tris–HCl buffer, pH 7.4. The lipid suspension is then sonicated twice for 2 min with 20 kHz MSE tip probe at 100 W. Incubations are carried out for 15 min at 37°C in a total volume of 500 μl containing 200 nmol of hydrolyzable phospholipids resuspended in 100 mM Tris–HCl, pH 7.4, 10 mM $CaCl_2$, and 0.1% fatty acid free BSA. Group X $sPLA_2$ activity is measured by adding 20 μl of a 72-h COS cell supernatant, and incubation is carried out for 15 min at 37°C. Released fatty acids are extracted by a modification of Dole's procedure, methylated with diazomethane, and quantified by GC/MS measurements. Control incubations in the absence of added $sPLA_2$ must be carried out in parallel and used to calculate specific hydrolysis. Group X $sPLA_2$ activity on erythrocyte membrane phospholipids (white ghosts) is determined similarly. The assay is performed in a total volume of 500 μl containing erythrocyte membranes (representing 200 nmol of hydrolyzable phospholipids) resuspended in 100 mM Tris–HCl, pH 7.4, 10 mM $CaCl_2$, and 0.1% fatty acid-free BSA. Incubations are carried out for 15 min at 37°C with translational shaking. Twenty microliter of a 72-h COS cell supernatant are used to measure group X $sPLA_2$ activity. The released fatty acids are extracted by a modification of the Dole' procedure (Tsujita et al., 1994), methylated and quantified by GC/MS.

Finally, Williams et al. (2000) have assayed the calcium-independent PLA_2 activity in nuclei prepared from isolated rat hearts by measuring the release of [^{14}C]arachidonic acid from 1-hexadecanoyl-2-[1-^{14}C]eicosatetra-5′,8′,11′,14′-enoyl-sn-GroPEtn as substrate (Ma et al., 1999). Lipids from isolated nuclei were extracted by the method of Bligh and Dyer (1959) in the presence of internal standards. Chloroform extracts from 200 μg of nuclear protein were used for analysis of individual phospholipid molecular species, utilizing mass spectrometry (Williams et al., 2000) (see Chapter 2). Fatty acid remodeling by PLA_2 during the biosynthesis of glycosyl PtdIns membrane anchors in *Trypanosoma* has been reported by Masterson et al. (1990).

9.2.3. Phospholipase C

PLC catalyzes the hydrolysis of a phosphodiester bond in phospholipids to provide a DAG and phosphorylated head group as products. Nonspecific PLC was first identified in bacteria. Crude preparations of PLC from *Bacillus cereus* (such as type III and some batches of type V) have activity against PtdIns and can be used to hydrolyze that phospholipid as well as PtdCho, PtdEtn, and PtdSer (Kuksis and Myher, 1990). Purer preparations (such as type XIII) have no PLC activity with PtdIns, but prefer PtdCho. The PtdCho preferring PLC from *B. cereus* (PLC$_{Bc}$) has now been purified and shown to be a 28.5 kDa monomeric enzyme containing three zinc ions in its active site (Hough et al., 1989). Mammalian cells also contain PtdCho preferring PLC (Wolf and Gross, 1985; Clark et al., 1986), where it is subject to induction by agonists. These observations have been confirmed by a more recent work and the PtdCho-PLC from mammalian sources has been purified despite low concentrations in the tissues. Since Clark et al. (1986) have demonstrated immunological similarity between PLC$_{Bc}$ and its mammalian counterpart, this bacterial enzyme has been used as a model for the poorly characterized mammalian PtdCho-PLCs (Hergenrother and Martin, 1997; Martin et al., 2000a,b).

9.2.3.1. Substrate specificity

Early studies with crude enzyme preparations (Holub and Kuksis, 1978) showed reduced activity for PtdEtn, PtdSer and PtdIns, when compared to PtdCho. Subsequent studies with purified preparations confirmed the significantly lower activity for PtdEtn and PtdSer along with a loss of activity towards PtdIns (Ries et al., 1992). However, the length of the acyl side chains in these substrates varied from C_{12} to C_{18}; thus direct comparisons of different classes as substrates for PLC$_{Bc}$ was not possible. Furthermore, these assays were conducted with the pH-stat method or by extraction of the residual mixture followed by analysis of the water-soluble products. More recently, Hergenrother and Martin (1997) have determined the kinetic parameters for PLC_{Bc} on different phospholipid classes using a chromogenic assay based on the quantitation of inorganic phosphate (Fig. 9.3). The assay compared the PLC_{Bc} hydrolysis of 1,2-dihexanoyl-*sn*-glycero-3-phosphocholine, 1,2-dihexanoyl-*sn*-glycero-3-phosphoethanolamine and 1,2-dihexanoyl-*sn*-glycero-3-phospho-L-serine. It was found that these compounds are substrates for the enzyme with their V_{max} being in order of PtdCho > PtdEtn > PtdSer. Earlier, Lewis et al. (1990) had determined susceptibility of the synthetic short-chain PtdCho with a total of 14 carbons in the acyl chains (ranging from 1-lauroyl-2-acetylGroPCho to 1-hexanoyl-2-octanoyl-GroPCho) to PLC_{Bc}. The results showed that with the exception of monomyristoylGroPCho, each of the asymmetric short-chain PtdChos exhibited high reactivity, comparable to that of the symmetrical diheptanoylGroPCho.

Martin et al. (2000a) have elucidated the roles that a number of amino acid residues play as zinc ligands and in binding and catalysis using three soluble substrates, 1,2-dihexanoyl-*sn*-glycero-3-phosphocholine (C_6GroPCho), 1,2-dihexanoyl-*sn*-glycero-3-phosphoethanolamine (C_6GroPEtn), and 1,2-dihexanoyl-*sn*-glycero-3-phospho-L-serine (C_6GroPSer) at concentrations below their corresponding critical micelle concentration. An X-ray structure of PLC_{Bc} complexed with a phosphonate inhibitor related to PtdCho revealed that the three amino acids residues Glu4, Tyr56, and Phe66 comprise the choline binding pocket. Martin et al. (2000a,b) constructed by PCR mutagenesis, a series of site-specific mutants for Glu4, Tyr56 and Phe66. It was shown that replacement of Phe66 with nonaromatic residue dramatically decreased kcat (about 200-fold) and reduced PLC_{Bc} activity toward C_6GroPCho, C_6GroPEtn and C_6GroPSer, whereas changes to Glu4 and Tyr56 typically led to much more modest losses in catalytic efficiency. The presence of an aromatic residue at position 56 seemed to confer some substrate specificity for C_6GroPCho and C_6PGroPEtn, which bear a positive charge on the head group, relative to C_6GrPSer, which has no net charge on the head group.

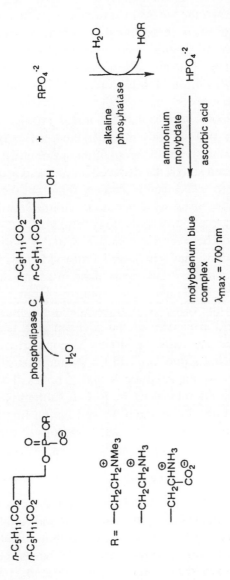

Fig. 9.3. Reaction scheme for PLC_{Bc} assay. The phosphorylated head group produced from the PLC_{Bc}-catalyzed cleavage of phospholipids is converted to a blue chromogene through the action of alkaline phosphatase, ammonium molybdate, and ascorbic acid. Abbreviations are as explained in figure. Reproduced from Hergenrother and Martin (1997) with permission of the publisher.

Phospholipase C of *B. cereus* hydrolyzes the *sn*-1,2-enantiomers many times faster than the *sn*-2,3-enantiomers and this difference in the hydrolysis rate has been exploited (Myher and Kuksis, 1984) for the preparation of enantiomeric diacylglycerols from *rac*-PtdCho (Chapter 10). The nonspecific PLCs do not hydrolyze PtdIns(4)P and PtdIns(4,5)P$_2$ (Holub et al., 1970) and probably other PtdIns phosphates.

Furuya et al. (2000) have isolated a novel PtdIns-PLC from *Trypanosoma cruzi* that is lipid modified and activated during trypomastigote to amastigote differentiation. In contrast to other PtdIns-PLCs described so far, the deduced amino acid sequence of the new enzyme revealed some unique features such as an N-myristoylation consensus sequence at its amino-terminal end, lack of a apparent pleckstrin homology domain and a highly charged linker region between the catalytic X and Y domains.

Munnik et al. (1998) have reviewed PtdIns-PLC signaling in plants in comparison to the earlier work done in animal, yeast and slime mold systems. Most animal mechanisms seem to exist in plants, even though the evidence is often scattered and fragmentary. The amino acid sequences of the soybean and *Arabidopsis* PLCs have provided the basis for structural comparisons to the mammalian, yeast, and slime mold PLCs (Munnik et al., 1998). The plant PLCs contain the catalytic X and Y boxes. The overall structure is most closely related to the PLC-δ subfamily but they are smaller and lack the PH domain as well as the first EF-hand lobe. Sequence identities with the X- and Y-regions of other PLC-δs ranged from 25 to 70%. A single amino acid substitution in the pleckstrin homology domain of PLC deltal has been shown to enhance the rate of hydrolysis (Bromann, 1997).

While mammalian PLCs are mainly activated by heterotrimeric G-proteins, tyrosine kinases and Ca^{2+} (Williams and Katan, 1996), only the presence of the heterotrimeric G-proteins in plants has been well documented. The tyrosine kinases have been reported but none has been linked to PLC signaling (Munnik et al., 1998).

In plants, PLC seems to be involved in osmotic regulation.

9.2.3.2. Methods of assay

The PLC activity is usually determined by measuring the loss of radioactive substrate or by the formation of radioactive product. Where appropriate, the loss of the substrate can be compared to the formation of the product. The release of a radioactive polar head group, such as choline phosphate into the water phase of a lipid extract may be compared to the appearance of the diradylglycerol in the organic phase. PLC may also be assayed spectrophotometrically using thioester or p-nitrophenyl-linked substrates. The released diacylglycerols may be resolved into molecular species and/or fatty acid composition determined.

Hergenrother and Martin (1997) and Martin et al. (2000a,b) assayed the recombinant PLC$_{Bc}$ mutants using a sensitive method based on the quantification of inorganic phosphate. The water-soluble phospholipid substrates for the assay were analyzed at concentrations below their CMC so that they were monomeric. All assays were performed in a constant-temperature water bath at 37°C. A 50 μl solution of the phospholipid substrate, which was made up in water three times the desired final concentration, was first added to the wells of a 96-well plate. Different rows on the plate represented different substrate concentrations, and the columns represented different time points. Fifty-microliter of 0.1 M 3,3-dimethylglutaric acid (DMG) (pH 7.3) was added to each well, and the plate was preincubated for 10 min at 37°C. A solution of PLC$_{Bc}$ in 1.0 mM DMG and 0.1 mM ZnSO$_4$ containing 0.167 mg/ml serum albumin (pH 7.3) was also preincubated. Assays were typically conducted with PLC$_{Bc}$ (50 μl total volume), but other trials were made with final concentrations in the well being varying from 2.5 to 15 nM. After incubating for 10 min, 50 μl of the enzyme solution was added to the wells using a multichannel pipettor so that the substrate concentrations received the enzyme at exactly the same time. After 20, 40, 60 and 100 s, the reactions were quenched by the addition of 50 μl of 2 M Tris containing 0.4% SDS. Tris is a known PLC$_{Bc}$ inhibitor (Hergenrother et al., 1995). A blank is also run in which the Tris solution was added to the substrate before the addition of PLC_{Bc}. After quenching with Tris, 25 μl of alkaline phosphatase solution in H_2O at a concentration of 60 units/ml was added to all the wells, and the plates were covered and incubated at 37°C for 1 h. After incubation with alkaline phosphatase, the P_i liberated was converted to the molybdenum blue complex by the addition of freshly prepared reagent solutions: Solution A, 2% ammonium molybdate in H_2O; solution B, 10.5% ascorbic acid in 37.7% aqueous TCA; and solution C, 2% sodium metaarsenite, 2% trisodium citrate in 2% acetic acid in H_2O. Solutions A and B were combined in a 1:1.5 ratio and thoroughly mixed immediately before use. A 50 μl aliquot of the A:B mixture was then added to each of the wells, and the contents of the wells were mixed by pipetting up and down with a multichannel pipettor. Two minutes after the addition of the A:B solution, 50 μl of solution C was added to all wells, and the solutions were mixed thoroughly with a multichannel pipettor as before. The blue color was allowed to develop for 20 min and then the absorbencies of the wells were read by a microplate reader at 700 nm. This blue color was found to be stable for at least 2 h.

The phosphorylated head group produced by the PLC_{Bc}-catalyzed hydrolysis of phospholipids is treated with alkaline phosphatase to liberate inorganic phosphate (P_i), which then forms a complex with ammonium molybdate. This complex is reduced to molybdenum blue state with ascorbic acid to give a blue solution with λ_{max} at 700 nm. Each of the three substrates was assayed at concentrations below their respective cmc values. Circular dichroism experiments were performed at enzyme concentrations of 20 μg/ml, on a Jasdco 6000 as described (Martin and Hergenrother, 1998).

The measurement of the released diacylglycerols can be performed using the diacylglycerolkinase kit from LIPIDEX. The lipid extracts of the cells are used as substrate for *E. coli* diacylglycerol kinase in the presence of 0.5 mM [^{32}P]ATP (0.1 μCi/nmol). Radioactive products are separated on silica gel G layers developed with chloroform/acetone/methanol/acetic acid/water (10:4:3:2:1, by vol.), and the region corresponding to an authentic PtdOH standard is scraped and quantified by liquid scintillation counting.

PtdCho-PLC activity in rabbit aortic smooth muscle cells was measured by Nakahata et al. (2000) by monitoring the release of [³H]PCho and [³H]Cho in cells prelabeled with [³H]Cho. Endogenous PLC activity may be assayed as follows (Kobayashi et al., 2000): Cells (1.5×10^5 per well) in a six-well plate were cultured in DMEM supplemented with 5% FCS for 3 days and then exposed to differentiation medium for 2 days. The cells were labeled with DMEM containing [³H]choline chloride (2 μCi/ml) for 18–24 h. The cells were then washed twice and preincubated in incubating medium containing 118 mM NaCl, 4.7 mM KCl, 1.8 mM $CaCl_2$, 1.2 mM $MgSO_4$, 1.2 mM KH_2PO_4, 10 mM glucose, and 20 mM HEPES (pH 7.4) at 37°C for 10 min. The reaction was initiated by the addition of 9,11-epithio-11,12-methano-TXA2 (STA2). When carbonodithioic acid ester (D-609) was used, it was added 10 min before the addition of STA2. After incubation for 1–5 min, the incubation medium containing the metabolites of [³H]PtdCho were collected in a tube, followed by centrifugation at 1000g for 5 min. [³H] radioactivity of supernatant (0.5 ml) was measured with a liquid scintillation counter.

Endogenous PtdCho-PLC activity may also be assayed following prelabeling of the cells with radioactive 1-O-alkylGroPCho (Xu et al., 1993). Thus, cells in tissue culture can be prelabeled to contain over 80% of 1-O-[³H]octadecyl-sn-glycero-3-phosphocholine in the intracellular diradylGroPCho (in the form of alkylGroPCho). Subsequent stimulation of cells (leukocytes, granulocytes and hepatocytes) labeled with this compound induces the formation of 1-O-[³H]octadecyl-2-acylglycerol consistent with the hydrolysis of PtdCho by PLC.

The molecular species of the released diacylglycerols are determined by reversed phase HPLC or high temperature GLC, while the fatty acids are determined by conventional GLC (Chapter 2), after appropriate derivatization. The chromatographic resolution of certain reverse isomers of the asymmetric diradylglycerols has been achieved using either conventional or chiral chromatographic systems (Chapter 2).

9.2.4. Phospholipase D

PLD is involved in diverse physiological processes such as membrane trafficking, mitogenesis, inflammation, and secretion (Exton, 1997, 1999; Frohman et al., 1999; Liscovitch et al., 1999). It is now known to occur in bacteria, fungi, plants, and animals. It is widely distributed in mammalian cells, where it is regulated by a variety of hormones, growth factors, and other extracellular signals. Its major substrate is PtdCho. PLD also catalyzes a phosphatidyl transfer reaction in which a primary alcohol acts as nucleophilic acceptor in the place of water (Kobayashi and Kanfer, 1987; Yang et al., 1967). The resulting PtdEt represents a specific assay for PLD.

PLD activity was initially identified in cabbage leaves. A PtdCho-PLD has been purified from many sources (Exton, 1997) and has recently been cloned from yeast, bacteria, plant, and mammalian sources (Morris et al., 1997). PtdCho PLDs have been purified and characterized from pig lung (Okamura and Yamashita, 1994) and cloned and characterized from rat brain (Park et al., 1997), while Park et al. (2000) have characterized a cardiac PLD_2. Earlier, Horwitz and Davis (1993) had determined the substrate specificity of PtdCho PLD of rat brain microsomes using radioactive butanol as PtdOH acceptor. A PLD activity has been reported in *Dictyostelium discoideum* (Cubitt et al., 1993) and yeasts (Ella et al., 1995). These enzymes exhibit significant similarities among their nucleotide sequences. The enzymes from *Saccharomyces*, *Ricinus* and *Streptomyces* have several sequences that are conserved in the human enzyme. A PLD of 92 kDa has recently been isolated and cloned from castor bean endosperm (Wang et al., 1993, 1994), and homologous enzymes have been identified in rice and maize (Ueki et al., 1995).

Pappan et al. (1997a) have demonstrated that the two types of PLD activities may be selectively assayed in plant extracts, the PtdInsPs-independent enzyme being assayed at mM Ca^{2+} concentrations, and the PtdIns requiring μM Ca^{2+}. Pappan et al. (1997a,b) have demonstrated that while the former activity comes from PLDα, the latter activity can result from PLDβ, PLDγ or both.

Complete PLD cDNA sequences are now known for 10 plant species (Wang, 2000). The cloning and expression of PLD cDNAs have demonstrated unambiguously that the hydrolytic and transphosphatidylation reactions are catalyzed by the same protein (Wang et al., 1994). All the cloned PLDs including plant PLDα, β and γ, mammalian PLD1a, b and PLD 2, and yeast PLD1, possess transphosphatidylation activity (Wissing et al., 1996). The characterization of the cloned PLDs has also led to the revelation of new regulatory and catalytic properties. Thus, it has been observed that PLDα is an acidic phospholipase, whereas PLDβ and PLDγ are neutral (Pappan and Wang, 1999).

The activities of PLDα, β and γ all require Ca^{2+} (Qin et al., 1997). However, the requirement varies among the enzymes. The PLDα requires millimolar levels of Ca^{2+}, while PLDγ and β require only micromolar concentrations of Ca^{2+}, being inhibited by higher concentrations. PLDα was active at micromolar Ca^{2+} concentrations when assayed at pH 4.5–5.5, in the presence of PtdInsPs. Sequence analysis of plant PLD suggests that the C2 domain and the catalytic region could be involved in Ca^{2+} binding (Ueki et al., 1995).

PLD activity in mammalian cells is low and is transiently stimulated upon activation by G-protein-coupled and receptor-tyrosine kinase cell surface receptors. Two mammalian PLD enzymes have been cloned and their intracellular regulators identified as ARF and Rho proteins, PKCα as well as the lipid, PtdInsP$_2$. Exton (1997, 1998, 1999, 2000) has repeatedly reviewed the field and recent reviews by Cockcroft (2001), Xie et al. (2002a,b) and Du et al. (2002a) have also appeared on the expression, regulation and signaling roles of mammalian PLD1 and PLD2. The PLD isoforms are directly regulated by classic PKC isozymes and members of the families of the two small GTPases ARF and Rho (Cai and Exton, 2001). The PLD2 isoforms show high basal activity and respond minimally to protein kinase C or the small GTPases. PLD2 requires PtdIns(4,5)P$_2$ as cofactor (Colley et al., 1997; Kodaki and Yamashita, 1997). A novel membrane-bound PLD from *Streptoverticilium cinnamoneum* has been identified by Ogino et al. (2001). The enzyme possesses only hydrolytic activity and differs from an extracellular enzyme cloned earlier from the same bacterial strain (Ogino et al., 1999).

9.2.4.1. Substrate specificity
Depending on the source and assay, the cytosolic PLD can hydrolyze PtdEtn, PtdIns, and PtdCho and require Ca^{2+} for activity (Balsinde et al., 1989; Siddiqi et al., 1995; Liscovitch et al., 2000). Martin et al. (2000a) have determined the substrate specificity for PLD from *Streptomyces chromofuscus*. 1,2-Dihexanoyl GroPCho, GroPEtn, GroPSer, GroPOH and an unnatural phospholipid bearing a neohexyl head group were examined as substrates. The results indicate that the catalytic efficiency for PLD on the substrates is PtdCho ≫ PtdSer ≥ PtdEtn > PtdGro ≫ PtdDB.

A unique property of PLD is that it catalyzes a transphosphatidylation reaction in which the phosphatide moiety of glycerophospholipids is accepted by primary alcohols, thereby producing phosphatidylalcohol instead of PtdOH. By analogy, the formation of PtdOH can be viewed as a transphosphatidylation reaction in which water serves as the phosphatidyl acceptor. Inclusion of a primary alcohol such as ethanolamine in the reaction mixture resulted in the transphosphatidylation of plasmenylcholine to plasmenylethanolamine. Figure 9.4 illustrates schematically the PLD-catalyzed reactions (Liscovitch et al., 2000). Although PtdEt are formed only by PLDs, not all PLDs can catalyze this reaction.

Virto et al. (2000) have reported the hydrolysis and transphosphatidylation of lysoPtdCho with a partially purified preparation

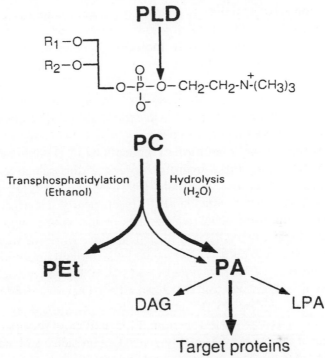

Fig. 9.4. PLD-catalyzed reactions. A phosphatidyl-enzyme intermediate is believed to form transiently which normally is hydrolyzed by water, generating PtdO. Primary short-chain alcohols (e.g. ethanol) can substitute for water in a competing, transphosphatidylation reaction. PC, phosphatidylcholine; PEt, phosphatidylethanol; PA, phosphatidic acid; DG, diacylglycerol; LPA, lysophosphatidic acid. Reproduced, with permission, from Liscovitch, M., Czarny, M., Fiucci, G. and Tang, X. (2000). Biochemical Journal 345, 401–415. © The Biochemical Society and the Medical Research Society.

of PLD from Savoy cabbage. These reactions were 20 times slower than the hydrolysis of PtdCho in a micellar system. The enzyme had an absolute requirement for Ca^{2+}, with an optimum at about 40 mM $CaCl_2$. The enzyme was inactivated by the detergent CTAB.

The conventional PLDs in plant tissues have shown activity towards most glycerophospholipid substrates, including PtdCho, PtdEtn, PtdGro, although the hydrolytic activity towards PtdIns,

PtdSer and Ptd$_2$Gro has been low. In addition to head group preference, the PLDs also show preference for the more unsaturated and medium-chain molecular species of PtdCho (for a review, see Wang, 2000).

An exception is provided by the four PLD isozymes from *Catharathus roseus* which hydrolyze PtdIns but not PtdCho, PtdEtn, Ptd$_2$Gro or PtdSer and also exhibit no transphosphatidylation activity (Wissing et al., 1996).

In mammals, the major form of intracellular PLD prefers PtdCho as substrate, but PtdEtn-hydrolyzing and PtdIns-selective PLDs have also been observed (Billah, 1993). Nakamura et al. (1996) have observed that mammalian PLD reacts nearly selectively with PtdCho in the form of mixed micelles or membranes with other phospholipids, especially PtdEtn. Dipalmitoyl GroPEtn was inert, while sphingomyelin was 30% as active as unsaturated PtdEtn.

Banno et al. (1996) have described two different types of PLD activity in a neural cell line PC12 and HL60. Oleate-dependent and GTP γS-dependent PLD activities were examined by using exogenous [^{3}H]PtdCho as substrate. PLD activity of the membrane fraction of PC12 cells was highly dependent on oleate and independent of GTP γS. This activity profile was in sharp contrast to that observed in HL60 cells showing the profound GTP γS-induced activation of PLD. The oleate-dependent PLD activity of PC12 membrane was inhibited by high concentrations of Ca^{2+} and Mg^{2+}.

9.2.4.2. Methods of assay

Martin et al. (2000a,b) assayed the activity of PLD from *S. chromofuscus* with a variety of synthetic substrates utilizing an assay based on the quantification of inorganic phosphate (see Methods of assay of PLC; Hergenrother and Martin, 1997). Martin et al. (2000a,b) did not assay PtdIns.

An electrochemical immunoassay for PLD detection has also been described. Becker et al. (1997) have reviewed the problems and have proposed a flow injection analysis for the detection and characterization of PLD. The choline liberated by PLD action on liposome preparations made from PtdCho is converted by choline oxidase to betaine and hydrogen peroxide, which is detected by chemiluminescence.

PLD assay is performed using a special experimental set-up (Becker et al., 1997). The substrate solution (0.1 mg/ml PtdCho liposomes in 20 mM sodium borate, pH 8.0, and 10 mM $CaCl_2$, stored at 25°C) is mixed with an equal volume of PLD solution (0.1 U/ml PLD in H_2O, stored at 0°C) in the mixing coil and stopped in one of the six parallel tube reactors between the six-way valve and the confluence point. Each tube reactor has a volume of 75 µl. After 30 min at 25°C, an equal volume of choline oxidase (1 U/ml choline oxidase in 10 mM sodium borate, pH 8.0) was added. After 1 min at 25°C, the luminol reagent (20 µM luminol and 20 µM $CaCl_2$ in 0.1 M $NaHCO_3$, pH 10.0) was added to detect hydrogen peroxide. Since chemiluminescence is immediately generated and rapidly quenched, the luminol reagent is added just before the detector cell. Following calibration of the method with different choline concentrations, the correlation between this chemiluminescence signal and the PLD activity was linear in the range between 1 and 100 mU/ml PLD.

Accumulation of phosphatidylalcohol is the most common method for assessing PLD activity in cells owing to its unique position as a metabolic signature for PLD activity and its relative metabolic stability compared to PtdOH.

PLD is commonly determined on the basis of its transphosphatidylation activity, which leads to the production of [3H]PtdEt when cells containing [3H]PtdCho are incubated in the presence of ethanol (Morris et al., 1997). Macrophages are washed once with DMEM containing 0.1% BSA and then incubated for 3 h with this same medium containing 1 µCi of [3H]myristate/ml to label cell PtdCho. The radioactive medium is then aspirated and the cells are washed twice with nonradioactive DMEM containing 0.1% BSA. The macrophages are incubated for further 2.5 h in BSA- or serum-free DMEM. No intermediate washes are carried out along the procedure to prevent the burst of sphingolipids and diacylglycerol that occurs rapidly after changing the medium (Smith et al., 1997). Ethanol (1% final concentration) is added 5 min prior to the addition of agonists. The macrophages are incubated for varying times, then washed once with ice-cold Ca^{2+}-free PBS and extracted with chloroform/methanol as follows. Cells are scraped into 0.5 ml of methanol, and wells are washed further with 0.5 ml of methanol. The two aliquots are combined and mixed with 0.5 ml of chloroform. Lipids are extracted by separating phases with a further 0.5 ml chloroform and 0.9 ml of 2 M KCl and 0.2 M H_3PO_4. Chloroform phases are blown down under N_2 and lipids are separated by TLC using Silica Gel 60 coated glass plates. TLC plates are developed for 50% of their heights with chloroform/methanol/acetic acid 9:1:1 (by vol.) and then dried. The plates were then redeveloped for their full height with petroleum ether/diethyl ether/acetic acid 60:40:1 (by vol.). The position of lipids is identified after staining with I_2 vapor by comparison with authentic standards. Radioactive lipids are quantified after scraping from the plates by liquid scintillation counting.

Alternatively, PtdIns(P)$_2$-independent PLD activity was assayed by a modified procedure of Wang et al. (1993). The reaction mixture contained 100 mM MES (pH 7.0), 1 mM SDS, 1% ethanol, 2.5 µg cabbage PLD and 0.2 mM dipalmitoyl-[2-palmitoyl-9,10-^{3}H]PtdCho with varying concentrations of CaCl$_2$ at the mM level. To assay PLD activity with PtdEtn as substrate, PtdCho was replaced by unlabeled, plus 1-palmitoyl-2-arachidonoyl (arachidonoyl-1-^{14}C]PtdEtn in the assay system. The final concentration of PtdEtn was 0.4 mM. The assays were conducted at 37°C for 20 min in a total assay volume of 60 µl.

Pappan et al. (1997a,b) have subsequently assayed the high Ca^{2+}-dependent PLD activity under further modified conditions. This reaction mixture contained 100 mM MES (pH 6.5), 25 mM CaCl$_2$, 0.5 mM SDS, 1% (v/v) ethanol, 5–15 µg of protein, and 2 mM PtdCho (egg yolk) containing dipalmitoylGro-3-P[methyl-^{3}H]Cho. The assay volume was reduced to 100 µl and then 100 μl of 2 M KCl was added to the 2:1 (v/v) chloroform/ methanol extraction.

A new spectrophotometric assay for the transphosphatidylation activity of PLD is described by Hagishita et al. (1999). This assay measures p-nitrophenol liberated by PLD reaction of Ptd-p-nitrophenol and ethanol in an aqueous–organic emulsion system. The reaction mixture comprises 200 μl of benzene containing 4 μmol Ptd-p-nitrophenol and 150 μl of sodium acetate buffer (2 μmol, pH 5.5) containing 200 μmol ethanol and 1 mg BSA in a 1.5 ml sample tube. The mixture is sonicated for 15 min and incubated for 5 min at 37°C. The enzyme solution (50 μl) is added to the reaction mixture, and it is incubated for 10 min at 37°C. The reaction is terminated by the addition of 100 μl of 1N HCl, followed by the addition of 150 μl of 1N NaOH. The lipids are then extracted with 400 μl of chloroform/methanol (3:1, v/v). After centrifuging for 10 min at 4°C, the aqueous phase of 20 μl was taken and brought to 200 μl of Tris–HCl buffer (0.1 M, pH 8.0) in a 96-well microtiter plate. The amount of p-nitrophenol liberated was estimated from the absorbance at 405 nm with a microplate reader.

Aurich et al. (1999) have described an emulsion system consisting of the buffer and a nonpolar organic solvent. The assay is based on the transphosphatidylation of PtdCho with 1-butanol in dichloromethane/buffer with subsequent densitometric quantification of the products after their separation by HPTLC and staining with $CuSO_4/H_3PO_4$ reagent. The method was applied to the determination of PLD in *Streptomyces*.

9.3. PtdIns and GPtdIns-specific phospholipases

Although some of the nonspecific phospholipases also attack PtdIns and PtdIns phosphates, other lipases are known to be specific for PtdIns, PtdIns phosphates, and PtdIns glycans. Since the latter enzymes have been shown to be involved in various signaling and metabolic control pathways, they have been gathered here for special discussion.

A highly efficient method for purification and assay of bee venom phospholipase A_2 has been described (Ameratunga et al., 1995). The functional assay of bee venom PLA_2 was based on its propensity to cause hemolysis of guinea pig red blood cells.

9.3.1. Phospholipase A_1 and A_2

An EDTA-insensitive PLA_1 activity, which hydrolyzes PtdIns was first demonstrated in rat brain cytosol fractions (Hirasawa et al., 1981), but the specificity for phospholipids was not fully defined. A PLA_1 enzyme, which has shown substrate preference for PtdIns was later studied in rat brain by Ueda et al. (1993). Gray and Strickland (1982) have purified and characterized a PLA_2 activity from the microsomal pellet of bovine brain acting on PtdIns. The purified enzyme gave a single band on SDS-gel electrophoresis. Molecular weight estimations gave values of 18 300 by SDS-gel electrophoresis and 18 521 by amino acid analysis. The purified enzyme, with a pH optimum at 7.4, was heat stable to 70°C and was activated by Ca^{2+} (5 mM). The microsomal location and preference for PtdIns shown by the enzyme lend support to the view that it may function to form lysoPtdIns in a deacylation–reacylation cycle for altering the distribution of the fatty acyl groups in PtdIns.

Thompson (1969) found that bee venom PLA_2 readily hydrolyzed PtdIns P_3 and PtdIns P_2, which are resistant to hydrolysis by PLA_2s of snake venom (Van Deenen and DeHaas, 1964). The preferential hydrolysis of PtdIns phosphates by bee venom PLA_2 is not fully understood, although it is known that bee venom PLA_2 is bound more strongly to anionic surfaces than other low molecular weight PLA_2s. Using charge reversal mutagenesis, Ghomashchi et al. (1998) have obtained evidence that the preferential binding of bee venom PLA_2 to anionic vesicles versus

PtdCho vesicles is mainly due to hydrogen-bonding and hydrophobic interactions rather than electrostatic binding, which is believed to be involved in the anionic surface binding of snake venom PLA_2s.

9.3.1.1. Substrate specificity

The specificity for phospholipid classes of the EDTA-insensitive PLA_1 of rat brain was determined by deacylation of [^{3}H]glycerol-labeled PtdIns, PtdEtn and PtdCho and measuring the formation of corresponding lysophospholipids (Ueda et al., 1993). The enzyme was highly selective for PtdIns; the hydrolysis rates for PtdEtn and PtdCho being below 5% of that for PtdIns. No accumulation of radioactive water-soluble products was observed, indicating that the enzyme did not have phospholipase B activity.

Gascard et al. (1993) have used bee venom PLA_2 to characterize the structural and functional PtdIns domains in human erythrocyte membranes. For this purpose, Gascard et al. (1993) measured the accessibility of PtdInsPs to the bee venom PLA_2 in native erythrocyte membranes, or after treatments designed to remove peripheral proteins and cytoplasmic domains of integral proteins. The results indicated that the fractions of PtdInsPs sensitive to PLA_2 represented the rapidly turned over and PLC-sensitive metabolic pools. It was observed that 85–95% of PtdOH, PtdCho, and PtdEtn was hydrolyzed at a PLA_2 concentration of 4 IU/μmol of phospholipid. In contrast, PtdIns(P)$_2$ and to a smaller extent PtdInsP were resistant to the hydrolytic action of the enzyme, requiring between 16 and 24 IU/μmol of phospholipid to reach the same level of hydrolysis. It was proposed that the rapidly metabolized pool of PtdIns(P)$_2$ and PtdInsP, involved in the regulation of major cellular functions, would be maintained in its functional state through interactions with integral proteins.

9.3.1.2. Methods of assay

The assays again are based on the disappearance of substrate or the appearance of product or both. As mentioned above, the appearance of the products may be measured as the total amount of fatty acid or lysophospholipid released, or the release of specific molecular species, depending on the needs of the experiment and the availability of equipment.

Ueda et al. (1993) determined the specific PtdIns hydrolyzing PLA_1 activity in column fractions as follows: The enzyme activity was measured in a reaction mixture (100 μl) containing 50 mM Tris–HCl (pH 8.0), 1 mM EDTA, 100 μM [2-^{3}H]glycerol-labeled PtdIns (approx. 3000 dpm/nmol), and usually 20–40 μl of an enzyme solution. The mixture is incubated at 37°C for 15 min and the activity was determined as the amount of accumulated [2-^{3}H]lysoPtdIns essentially as described by Darnell and Saltiel (1991). For characterization of the partially purified PLA_1, the assay is performed in 100 mM Tris–HCl (pH 7.0), 2 mM EDTA in the presence of 0.002% Triton X-100 and 100–200 μg/ml of the pooled Ultrahydrogel chromatographic fraction in a total volume of 100 μl. The reaction mixtures were incubated at 37°C for 5 or 10 min. After incubation, the reaction is terminated by adding 100 μl n-butanol. The products extracted in the butanol phase are applied to TLC plates (Silica Gel 60, Merck), and developed with $CHCl_3/CH_3OH/CH_3COOH/H_2O$ (25:15:4:2, by vol.). The lysoPt-dIns regions are scraped off into vials and radioactivities are counted in a toluene/Triton X-100/water scintillator cocktail.

Ueda et al. (1993) used 1-stearoyl-2-[^{14}C]arachidonoyl species of PtdIns as substrate to determine the positional specificity of PLA_1. Only 2-[^{14}C]arachidonoyl lysoGroPIns was detected without significant formation of acylglycerols and free arachidonate. In comparison to other molecular species of PtdIns, the arachidonoyl-containing ones were the poorer substrate for this enzyme.

According to Thompson (1969) 1 μmol PtdIns(P)$_3$ is incubated with 0.2 ml of 0.1% aqueous solution of bee venom (Sigma Chemical Co.) in 1.8 ml of 0.01 M Tris–HCl buffer (pH 7.2), 3.0 ml ether is then added, and the stoppered tubes are kept at room temperature. There was a rapid hydrolysis of PtdIns(P)$_3$ until 50% of the fatty acids were liberated in the free form, as determined by titration. Prolonged reaction up to 20 h, gave no further net release of free fatty acids. The bee venom enzyme selectively removed unsaturated fatty acids from PtdIns(P)$_3$.

Gascard et al. (1993) have described the preparation of SUV containing PtdInsP and PtdIns(P)$_2$ along with the conditions of hydrolysis with bee venom PLA$_2$. SUV were prepared from pure phospholipids (PtdCho, PtdCho/PtdEtn, PtdCho/PtdInsP, or PtdCho/PtdIns(P)$_2$). CH$_3$Cl/CH$_3$OH solutions of PtdCho and PtdEtn were evaporated under a stream of nitrogen at 4°C. After addition of medium B (150 mM NaCl/1.5 mM HEPES-NaOH, pH 7.4) and, when required, PtdInsP or PtdIns(P)$_2$ (sodium salt) in aqueous solution, the mixture was vortexed for 2 min and subjected to probe sonication (4 cycles of 7 min, 50 W at 4°C). The sonicates were centrifuged (30 000g, 60 min, 4°C) to remove undispersed lipids and multilamellar liposomes. SUV, recovered in the supernatants, were incubated for 5 min, at 25°C, in medium B containing 500 μM CaCl$_2$ (2–4 μmol of phospholipid/ml) with or without different concentrations of PLA$_2$. The reaction was stopped by the addition of 1 volume of medium B containing 1 mM EGTA. Phospholipid extraction, separation, and assay was carried out as described previously (Gascard et al., 1991).

9.3.2. Phospholipase C

The mammalian PtdIns-specific PLCs have been extensively investigated and reviewed in detail (Cockcroft and Thomas, 1992; Williams and Katan, 1996; Williams, 1999; Cussac et al., 2002). PtdIns-PLCs are soluble enzymes that catalyze the cleavage of membrane PtdIns(4,5)P$_2$ releasing two second messengers, lipid-soluble DAG and a water-soluble myo-Ins(1,4,5)P$_3$ (Exton, 1996; Rhee, 2001). These enzymes have a multidomain structure containing, besides the N-terminal PH domain, four elongation factor (EF) hands, a catalytic X–Y domain and a C2 domain (Rhee, 2001). There are three known families of mammalian PLCs (β, γ and δ), each with a distinct set of protein regulators. PLC-β isoenzymes are activated by αq and βγ subunits of heterotrimeric GTP binding proteins (G-proteins). PLC-γ enzymes are regulated by receptor-tyrosine kinases. Protein regulators of PLC-γ enzymes are unknown now, although RhoGAP and transglutaminase have been implicated (Rhee, 2001). For effective catalysis, PLCs must associate with membranes, which involves the PH domain. The PH domain of PLC-γ binds specifically to membranes containing PtdIns(3,4,5)P$_3$ (James and Downes, 1997) and the POH domain of PLC-δ binds strongly to membranes containing PtdIns(4,5)P$_2$ (Garcia et al., 1995; Lemmon et al., 1995). The PH domain of PLC-β binds strongly and nonspecifically to membranes which helps to promote its lateral association with protein subunits and it has been found that the isolated PH domain of PLC-β$_2$ binds to G-βγ on membranes with similar strong affinity (Wang et al., 2000). PLC-β$_2$ is stimulated by Rho GTPases (Illenberger et al., 2000).

These messengers then promote the activation of protein kinase C and the release of Ca^{2+} from intracellular stores, respectively. Eleven distinct isoforms of PtdIns-specific phospholipase C (PtdIns-PLC), which are grouped into four subfamilies (β, γ, δ and ε), have been identified in mammals. An isoform originally termed PLC-α was subsequently shown to be a proteolytic fragment of PLC-δ1. The molecular size of the PLC-δ isozymes is about 85 kDa, while those of β- and γ-type enzymes are in the range of 120–155 kDa, and that of PLC-ε is 230–260 kDa. All PtdIns-PLC isoforms contain X and Y domains, which form the catalytic core, as well as various combinations of regulatory domains that are common to many other signaling proteins. These regulatory domains serve to target PLC isozymes to the vicinity of their substrate or activators through protein–protein or protein–lipid interactions. These domains include the pleckstrin homology (PH) domain, the COOH-terminal region including the C2 domain of PLC-β, the PH domain and Src homology of PLC-γ, the PH domain and C2 domain of PLC-δ, and the Ras binding domain of PLC-ε. These groups also differ in the mechanisms by which the isozymes are activated in response to ligand interaction with various receptors. Rhee (2001) has presented a detailed review of the regulation of PtdIns-specific PLC.

Intracellular PtdIns-PLC are prevalent in mammalian cells (Rhee et al., 1989; Rhee and Bae, 1997) and are involved in PtdIns signal transduction pathways producing the two second messenger molecules, DAG, that are responsible for the activation of protein kinase C, and Ins(1,4,5)P_3, which is involved in intracellular calcium mobilization (Berridge, 1987). The first mammalian PLC purified to homogeneity was from rat liver. It had a molecular mass of 68 kDa. Subsequently, numerous PLC activities have been resolved chromatographically from a variety of tissues and shown to differ in molecular mass, isoelectric point and pH optima, and calcium dependency, indicating the existence of isozymes (Shukla, 1982; Low et al., 1984) and their characterization with respect to size, behaviour, localization, activities, and antigenicity (Rhee et al., 1989; Crooke and Bennett, 1989).

These enzymes have been isolated from both, prokaryotes and eukaryotes (Griffith and Ryan, 1999). Of the better characterized PtdIns-PLCs, the bacterial enzymes are secreted while those of other organisms are intracellular enzymes. The PtdIns-PLCs of about 35 kDa are produced by a variety of aerobic and anaerobic Gram-positive bacteria (Katan, 1998). Rhee (2001) has summarized the domain organization of the four types of PLC isozymes. The X and Y catalytic domains as well as the PH, EF-hand, C2, SH2, SH3, RasGEF, and RA domains are indicated (Fig. 9.5).

Several laboratories have reported evidence for the existence of membrane-bound PLC by demonstrating that endogenous or exogenous substrates ([^{3}H]-PtdIns) are hydrolyzed in membrane preparations isolated from platelets (Banno and Nozawa, 1987). Banno et al. (1988) have extended the previous work and resolved two forms (mPLC-I and mPLC-II) of membrane-bound PLC from human platelets, and achieved purification of the major enzyme (mPLC-II) to homogeneity, and the two enzymes were biochemically characterized. Both enzymes hydrolyzed PtdInsP$_2$ at low

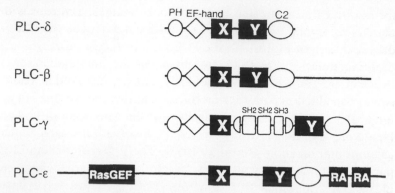

Fig. 9.5. Domain organization of the four types of phospholipase C isozymes. Reproduced from Rhee (2001) with permission of the publisher.

Ca^{2+} concentration (0.1 – 10 μM) and exhibited higher V_{max} for PtdInsP$_2$ than for PtdIns. PtdIns-P$_2$-hydrolyzing activities of both enzymes were enhanced by various detergents and lipids, such as deoxycholate, cholate, PtdEtn and PtdCho. GTP-binding proteins had no significant effect on the PLC-II activity.

9.3.2.1. Substrate specificity

Hendrickson and Hendrickson (1999) have determined the binding of PtdIns-specific PLC (*B. cereus* and mammalian PLC-δ1) to phospholipid interfaces by fluorescence resonance energy transfer. Dansyl-labeled lipid probes were used as acceptors with intrinsic tryptophan of the enzyme as the donor. The bacterial PLC bind with high-affinity to PtdIns interfaces and have slightly lower activity to PtdCho interfaces, while the mammalian PLC showed significant binding only to PtdIns interfaces.

Three PLC isozymes of 150, 145 and 85 kDa were purified to homogeneity from bovine brain (Ryu et al., 1986a, 1987a,b). The three enzymes were specific for PtdIns and the PtdIns polyphosphates (Ryu et al., 1987a,b) and did not hydrolyze other phospholipids. Rhee et al. (1989) studied the catalytic properties of the three isozymes by using small unilamellar vesicles prepared from either PtdIns or PtdIns(4,5)P_2 as substrates. Hydrolysis of both PtdIns and PtdIns(4,5)P_2 by the three enzymes was Ca^{2+}-dependent. However, at low Ca^{2+} concentration, PtdInsP$_2$ is the preferred substrate for all the three enzymes.

Homma et al. (1988) have isolated and purified to homogeneity two different forms of PtdIns specific PLC from crude extract of rat brain. Both enzymes had Mr of 85 000. One of the enzymes (PLC-II) hydrolyzed PtdIns(4,5)P_2 as well as PtdIns(4)P and PtdIns, although at slower rate, while the other enzyme (PLC-III) hydrolyzed PtdIns(4,5)P_2 preferentially with little activity towards PtdIns(4)P and PtdIns.

It is possible, however, that some of these comparisons are not entirely valid, as the assay conditions may not have been the same. Thus, the experimental outcome may vary with the form of substrate presentation (micelles versus liposomes) and the nature and concentration of the divalent ions (Ca^{2+} versus Mg^{2+}), which may stimulate or inhibit the activity. In general higher concentrations of Ca^{2+} favor the following order: PtdIns > PtdIns(4)P > PtdIns(4,5)P_2 (summarized from Munnik et al., 1998).

Determination of the three-dimensional structures both of a PLC-δ1 lacking the PH domain and of the PH domain of this isoenzyme have revealed a four-module organization of the enzyme consisting of the PH domain, the EF-hand domain, the catalytic domain (comprising tightly associated X and Y domains), and the C2 domain (Katan, 1998). The PH domain of PLC-δ1 binds to Ins(1,4,5)P_3 and to PtdIns(4,5)P_2 (Yagisawa et al., 1998). The binding of PLC-δ1 to its lipid substrate and, consequently, its catalytic activity are dependent on the concentration of PtdIns(4,5)P_2, and the catalytic activity of this isozyme is inhibited by Ins(1,4,5)P_3. The crystal structure of the PH domain of rat PLC-δ1 complexed with Ins(1,4,5)P_3 revealed that the 4- and 5-phosphoryl groups of Ins(1,4,5)P_3 interact with the side chains of Lys32 and Lys57 and with those of Lys30, Arg40, and Lys57, respectively, thus explaining the mechanism by which these phosphoryl groups contribute to high-affinity binding (Ferguson et al., 1995).

Although PLC-β and PLC-γ isozymes possess a PH domain at their NH_2-termini, the basic amino acids located inside the inositol phosphate binding pocket of PLC-δ1 are not well conserved in these proteins. As a result, the NH_2-terminal PH domains of PLC-γ isozymes bind to PtdIns(3,4,5)P_3 but not to PtdIns(4,5)P_2 (Razzini et al., 2000), which suggests the possibility of recruitment of these isozymes by PtdIns(3,4,5)P_3. The PH domain of PLC-β1 has been shown to specifically bind PtdIns(3)P, and this interaction appears to be responsible for the membrane recruitment of this isozyme in cell in which PtdIns 3-kinase is activated (Razzini et al., 2000). All four types of PLC require Ca^{2+} for catalytic function. Recent evidence indicates that the EF-hand domain of PLC-δ1 binds Ca^{2+} and that the bound Ca^{2+} is necessary for the efficient interaction of the PH domain with PtdIns(4,5)P_2 (Yamamoto et al., 1999). Ca^{2+} ions bound to the C2 domain of PLC-δ1 have been recently shown to enhance the enzyme activity by promoting the formation of an enzyme–PtdSer–Ca^{2+} ternary complex, therefore by increasing the affinity of the enzyme for substrate vesicles (Lomasney et al., 1996, 1999).

The known selectivity (PtdInsP$_2$ > PtdInsP > PtdIns) of the eukaryotic enzymes for substrate was illuminated by the structure of PLC-δ1 (Essen et al., 1996), which showed interaction of active site residues with all the functional groups of InsP$_3$ except the 6-hydroxyl moiety while the interaction with phosphates at the 4 and 5-positions explains a preference for PtdInsP$_2$, the lack of space to accommodate a phosphate at the 3-position accounts for the lack of action of these enzymes on PtdInsP$_3$ (Serunian et al., 1989). The differential association of the pleckstrin homology domains of PLC-β1, PLC-β2 and PLC-δ1 with lipid bilayers and the βγ subunits of heterotrimeric G-proteins has been determined by Wang et al. (1999). Like the intact enzymes, the PH domains of PLC-β1 and PLC-β2 bind to membrane surfaces composed of zwitterionic PtdCho with moderate affinity. Inclusion of the anionic PtdSer or PtdIns(4,5)P_2 along with G-$\beta\gamma$ subunits had little effect on their membrane affinity. In contrast, binding of PLC-$\delta 1$ or its PH domain was highly dependent on PtdIns(4,5)P_2.

Hydrolysis of inositol phospholipids by mammalian PLC enzymes produces cyclic inositol phosphates as minor products. These enzymes catalyze the conversion of PtdIns to InsP with overall retention of the stereochemical configuration at the phosphorus atom (Bruzik et al., 1992). Zhou et al. (1997a) have used water-soluble cycInsP as a substrate to study the phospho-diesterase activity of PtdIns-PLC and to investigate the possibility of interfacial activation/surface allosteric activation towards a water-soluble, monomeric substrate. Zhou et al. (1997a) have performed a detailed NMR analysis of the conformation of the short-chain PtdIns molecules in both monomer and micellar states. Coupling constant analysis and NOE data address the difference between D- and L-Ins isomers as well as the similarity of 2-methoxy-PtdIns to diC$_7$ GroPIns. QUANTA computational modeling suggests that the inositol ring is nearly parallel to the chain packing direction, leaving the phosphate ester accessible to attack by PtdIns-specific PLC enzymes. DiC$_7$ GroPIns-2-O-methylinositol is a good inhibitor of PtdIns-specific PLC because it blocks the initial phosphotransferase step in PtdIns hydrolysis. Introduction of the methyl group on the C-2 hydroxyl group lowered the CMC of the derivative compared to diC$_7$ GroPIns. Zhou et al. (1997a) have discussed these results in light of PtdIns-specific PLC and nonspecific PLC activities.

Zhou et al. (1997b) have reported a detailed NMR analysis of the conformation of the short-chain PtdIns molecules in both monomer and micellar states. Coupling constant analysis and NOE data address the difference between D- and L-Ins isomers as well as the similarity of 2-methoxy-PtdIns to diC$_7$GroPIns. In the PtdIns-PLC catalytic reaction, the first step is intramolecular and requires a hydroxyl group at position 2 of the inositol ring in order to form the cyclic product [D-myo-Ins(1,2)cycP]. The hydrogen bond between hydroxyl group at position 2 with the pro-R oxygen may orient the hydroxyl group for an 'in line' attack. Replacing the hydroxyl group at position 2 with a methoxy group generated a good inhibitor. This indicated that a free hydroxyl group at position 2 is not important for ligand binding to the protein in contrast to the hydroxyl group at position 3 (Bruzik et al., 1994).

Bruzik et al. (1994) have subjected both diastereomers of *chiro*-Ins phosphatides to cleavage by PtdIns-PLC from *Bacillus thuringiensis*. The reaction of 1-L-*chiro*-PtdIns produced *chiro*-Ins(1,2)cycP, however, at the rate of 10^{-3} of that attained with the natural substrate, PtdIns. On the other hand, 1-D-*chiro*-PtdIns was found to be resistant to PtdIns-PLC. These results suggest that the natural *chiro*-Ins derivatives should have the 1-L-configuration if they are produced by PtdIns-PLC, which is in contrast to the 1D-configuration reported by others. Bruzik et al. (1994) therefore have isolated *chiro*-inositol from the total bovine liver lipid and determined its absolute configuration. The obtained *chiro*-inositol was found to be exclusively of the 1-L-configuration, with the enantiomeric purity exceeding 99%. Nevertheless, the results imply that 1-L-*chiro*-PtdIns can be metabolized by enzymes known to accept *myo*-Ins (Johnson et al., 1993).

The bacterial PtdIns-PLC is a small, water-soluble enzyme that cleaves the natural membrane lipids such as PtdIns, lysoPtdIns and GPtdIns. Griffith and Ryan (1999) have reviewed the crystal structure, NMR and enzymatic mechanism of this enzyme. The bacterial PtdIns-PLC consisting of a single domain folded as a $(\beta\alpha)_8$ barrel (TIM barrel), are calcium-independent, and interact weakly with membranes. Sequence similarity among PtdIns-PLCs from different bacterial species is extensive, and includes the residues involved in catalysis. Bacterial PtdIns-PLCs are structurally similar to the catalytic domain of mammalian PtdIns-PLCs.

The PtdIns-PLCs from *B. cereus* and *B. thuringiensis*, which have been most extensively investigated, have been shown to catalyze the hydrolysis of the phospholipid in two steps. In the first, DAG is produced along with a D-*myo*-Ins(1,2)cycP, while in the second (Volwerk et al., 1990), the water-soluble intermediate is hydrolyzed to yield acyclic D-*myo*-Ins(1)P (Fig. 9.6). The second reaction proceeds at a rate of 10^3 times slower than the first leading to an accumulation of the intermediate cyclic phosphate. The bacterial PtdIns-PLCs are stereospecific for D-*myo*-Ins (Lewis et al., 1993). The L-enantiomer is neither a substrate nor an inhibitor. Experiments with 2-methoxy substituted *myo*-inositol head group show that a free axial 2-OH group is absolutely necessary for the PtdIns molecule to be a substrate. The second reaction is regiospecific since only the 1-phosphate is produced. Furthermore, the bacterial PtdIns-PLCs do not attack PtdIns with phosphorylated 4- or 5-positions (Volwerk et al., 1989). This is in contrast to the eukaryotic PtdIns-PLCs, which show the highest catalytic activity with PtdIns phosphorylated at the 4- and 5-hydroxyl groups (Ellis et al., 1993, 1998). The strict requirements for the head group configuration do not extend to the stereochemistry of the *sn*-2-carbon, which can accommodate various natural variations in the

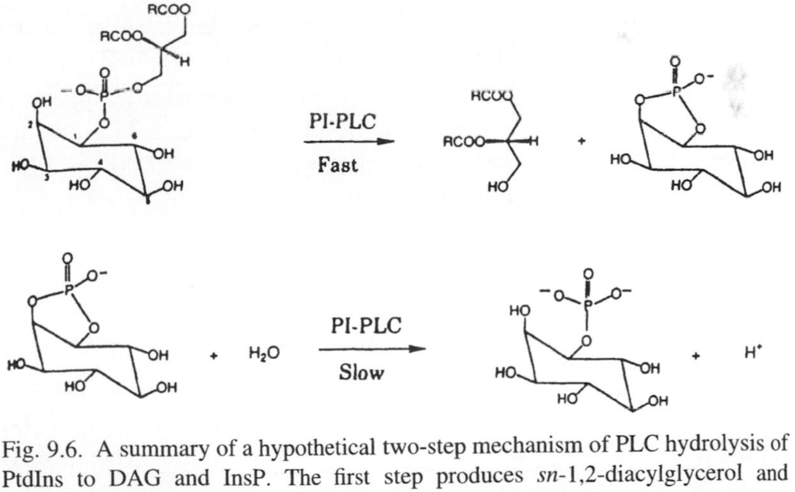

Fig. 9.6. A summary of a hypothetical two-step mechanism of PLC hydrolysis of PtdIns to DAG and InsP. The first step produces *sn*-1,2-diacylglycerol and D-Ins(1:2*cyc*)P with the enzyme acting as phosphotransferase. In the second step, the enzyme acts as a regio and stereospecific cyclic phosphodiesterase producing D-Ins(1)P. Reproduced from Volwerk et al. (1990) with permission of the publisher.

lipid portion of the substrate molecule (Bruzik et al., 1992). A recent X-ray structure of PtdIns-PLC from *B. cereus* suggests that PtdIns hydrolysis may be mediated by general acid and base catalysis using two conserved histidines (Heinz et al., 1995). Interactions of active site residues in the *B. cereus* enzyme with the 4- and 5-hydroxyl groups explain the specificity of this enzyme for PtdIns.

The bacterial PLCs have the ability to cleave the GPtdIns moieties of membrane protein anchors (Low and Saltiel, 1988; Ikezawa, 1991). These PtdIns-PLCs, which can be isolated in milligram quantities, are important tools in the study of membrane proteins that are attached to the membrane via a PtdIns-PLC sensitive GPtdIns-anchor (Ferguson and Williams, 1988). A GPtdIns specific PLC releases the DAG moiety from the GPtdIns anchors of proteins. The enzyme has been purified from *Staphylococcus aureus* (Low, 1981), *Trypanosoma brucei* (Hereld et al., 1986) and other sources (Ferguson, 1999). Hereld et al. (1986) appear to have been the first to isolate a PLC specific for GPtdIns. The purified enzyme released the dimyristoylglycerol moiety of the glycolipid in the surface coat of *T. brucei*. There was no apparent cleavage of other myristate-containing lipids of trypanosomes or 1-stearoyl-2-arachidonoyl GroPIns. The purified enzyme had an estimated molecular weight of 37 000 on SDS-PAGE gels, with a molecular mass of 43 kDa estimated for the native enzyme. The purified enzyme did not require Ca^{2+}. Bacterial PtdIns-PLC cleaves GPtdIns anchors by the same mechanism as PtdIns substrates, yielding DAG and a glycosylated polypeptide terminating in Ins(1,2)cycP. Palmitoylation of the 2-OH group of *myo*-inositol renders the anchor molecule resistant to hydrolysis by bacterial PtdIns-PLC (Treumann et al., 1995), presumably because this modification prevents the formation of Ins(1,2)cycP. GPtdIns-PLC activity is not found for any of the eukaryotic PtdIns-PLCs (Griffith and Ryan, 1999). It is unique to the bacterial PtdIns-PLCs and a G PtdIns-PLC of *T. brucei*.

Flick and Thorner (1993) have described the identification and isolation of a yeast homologue (PLC1) of the mammalian PtdIns-PLC-δ subtype and provided a direct demonstration that its product (designated PilcIp) is a bona fide PtdIns-PLC. PLC activity was measured as the release of water-soluble radioactivity from $[2-^3H]$Ins-labeled PtdIns and from $[2-^3H]$Ins-labeled $PtdInsP_2$ by the procedures described by Hofmann and Marjerus (1982), except that the reaction volume was reduced to 0.1 ml.

McDonald and Mamrack (1988) have reported that $AlCl_3$ has a differential effect of the purified PLC activity from bovine heart cytosol. It stimulated PtdIns hydrolysis while inhibiting $PtdIns(4,5)P_2$ hydrolysis in a concentration dependent manner from 1 to 100 μM. The effect of $AlCl_3$ was attributed to the free ion, Al^{3+}. Complexation of Al^{3+} with phosphate, citrate, fluoride or hydroxide could block the stimulatory effect on PtdIns hydrolysis. Only fluoride or hydroxide could partially reverse the inhibition of $PtdInsP_2$ hydrolysis by $AlCl_3$.

Razzini et al. (2000) have demonstrated that a cooperative mechanism involving PtdIns(3)P and the Gβγ subunit regulates the plasma membrane localization and activation of PLC-β1-PH.

Farenza et al. (2000) have demonstrated a role for nuclear PLC-β1 in cell cycle control. PLC-$β_1$ exists as two polypeptides of 150 and 140 kDa generated from a single gene by alternative RNA splicing. Both contained in the COOH-terminal tail a cluster of lysine residues responsible for nuclear localization. The mammalian PLC-β1 is activated in insulin-like growth factor 1 mediated mitogenesis and undergoes down-regulation during murine erythroleukemia differentiation.

Broad et al. (2001) obtained evidence for the requirement of PLC and PtdInsPs for the activation of capacitative calcium entry. The results challenge the previously suggested roles of $InsP_3$ and $InsP_3$ receptor in this mechanism, at least in the lacrimal acinar cells.

Lopez et al. (2001) have recently reported that PLC-ε is a novel bifunctional enzyme that is regulated by the heterotrimeric G-protein Gα12 and activates the small G-protein

Ras/mitogen-activated protein kinase signaling pathway. Schmidt et al. (2001) have shown that G(s)-coupled receptors can stimulate PLC-ε, apparently via formation of cyclic AMP and activation of the Ras-related GTPase Rap2B. More recently, Evellin et al. (2002) have reported stimulation of PLC-ε by the M3 muscarinic acetylcholine receptor mediated by cyclic AMP and GTPase Rap2B. Song et al. (2001) have reported that the novel human PLC-ε is regulated through membrane targeting by Ras. A Ras-associating domain of PLC-ε specifically binds to the GTP-bound forms of Ha-Ras and Rap1A. Using a liposome-based reconstitution assay, it was shown that the PtdIns(4,5)P_2-hydrolyzing activity of PLC-ε is stimulated in vitro by Ha-Ras in a GTP-dependent manner.

Mitchell et al. (2001) have reported that hydrolysis of PtdIns and PtdInsP$_2$ are independently regulated in pancreatic islets and that PLC-γ1 selectively mediates the breakdown of PtdIns. The activation mechanism of PLC-γ involves tyrosine phosphorylation (but not PLC-γ directly) and PtdIns-3-kinase. These findings point to a novel bifurcation of signaling pathways downstream of muscarinic receptors and suggests that hydrolysis of PtdIns and PtdInsP$_2$ might serve different physiological ends.

Lehto and Sharom (2002) have investigated the host membrane properties affecting the GPtdIns-anchor cleavage using purified porcine lymphocyte ecto-5'-nucleotidase reconstituted into lipid bilayer vesicles and cleaved with PtdIns-specific PLC from *B. thuringiensis*. The PLC activity was highly dependent on the chain length and unsaturation of the constituent phospholipids. The activation energy of GPtdIns-anchor cleavage was substantially higher in gel phase bilayers compared to those in the liquid crystalline phase. Addition of cholesterol simultaneously abolished the phase transition and the large difference in cleavage rates observed above and below the phase transition temperature. Lipid fluidity and packing appeared to be the most important modulators of PLC hydrolysis of GPtdIns anchors.

Hondal et al. (1998) have attempted to rigorously evaluate the mechanism of PtdIns-PLC from *B. thuringiensis* by examining the functional and structural roles of His-32 and His-82, along with the two nearby residues Asp-274 and Asp-33, using site directed mutagenesis. Twelve mutants were constructed, which showed variable decrease in specific activity towards PtdIns and (2R)-1,2-dipalmitoyloxypropane-3-(thiophospho-1D-*myo*-inositol) as substrates. The obtained results suggested a mechanism in which His-32 functions as a general base to abstract the proton from 2-OH and facilitates the attack of the deprotonated 2-oxygen on the phosphorus atom. Its base function was augmented by the carboxylate group of Asp-274, which forms a diad with His-32. His-82 functioned as a general acid with assistance from Asp-33, facilitating the departure of the leaving group by protonation of the O3 oxygen. The complete mechanism also included activation of the phosphate group towards nucleophilic attack by a hydrogen bond between Arg-69 and a nonbridged oxygen atom.

Mensa-Wilmot et al. (1995) have reported the purification and use of recombinant glycosyl PtdIns PLC, while Armah and Mensa-Wilmot (1999) have described a covalent lipid modification of GPtdIns-specific PLC, which is an integral membrane constituent in the kinetoplastid *T. brucei*. Myristic acid was detected on cysteine residue(s) (i.e. thiomyristoylation). Reversible thioacylation was proposed as a novel mechanism for regulating the activity of PLC. More recently, Armah and Mensa-Wilmot (2000) have demonstrated that oligomerization of GPtdIns-PLC is associated with high enzyme activity both in vivo and in vitro.

Sawai et al. (2000) have shown that ISC1 (YER019w), which has homology to bacterial neutral sphingomyelinase, encodes inositol phosphosphingolipid-PLC in *S. cerevisiae*. Deletion of ISC1 eliminated the endogenous PLC activity. Labeling of yeast cells with [^{3}H]dihydrosphingosine showed that inositol phosphosphingolipids were increased in the deletion mutant cells.

9.3.2.2. Methods of assay

A frequently used assay for PLC is based on the measurement of the release of water-soluble radioactivity from [2-^{3}H]inositol-labeled

PtdIns and from [2-^{3}H]inositol-labeled PtdInsP$_2$ (Hofmann and Marjerus, 1982; Flick and Thorner, 1993; Furuya et al., 2000). Briefly, stock solutions containing either PtdIns or PtdInsP$_2$ in organic solvent were dried under a stream of nitrogen and suspended just prior to use in reaction buffer by sonication for 5 min in an ultrasonic water bath. Reaction mixtures containing 250 or 500 μM PtdIns (or PtdInsP$_2$), 12 000–20 000 cpm of [^{3}H]PtdIns or [^{3}H]PtdInsP$_2$, 1 mg of sodium deoxycholate per ml, 0.5 mg of BSA per ml, 100 mM NaCl, 50 mM HEPES-HCl (pH 7.0), 2 mM EGTA, and 0–3 mM CaCl$_2$. Reactions were initiated by the addition of the enzyme source. Free Ca^{2+} concentrations in the range of 0.5 μM to 1 mM were achieved by using CaCl$_2$–EGTA buffer. The reactions were carried out in 1.5 ml screw cap polypropylene microcentrifuge tubes for 10 min at 37°C and were terminated by the addition of 0.5 ml of chloroform/methanol/HCl (100:100:0.6, by vol.) followed by 0.15 ml of 1N HCl–5 mM EGTA. Samples were subjected to vigorous vortex mixing for 5 s and spun in a microcentrifuge for 30 s to separate the organic and aqueous phases. A 0.25 ml aliquot of the aqueous phase was removed and dissolved in 20 ml of a liquid scintillation fluid and the radioactivity counted.

Homma and Emori (1997), Shibatohge et al. (1998) and Song et al. (2001) have employed a more complex buffer for the incubations. Thus, a sample containing fusion protein was assayed for PLC activity by incubating 50 μl reaction mixtures containing 50 mM 2-(N-morpholino)ethanesulfonic (MES) acid, pH 6.8, 10 μM Ca^{2+}/EGTA, 100 mM NaCl, 0.2 mg/ml BSA 0.1 mM dithiothreitol, 90 μM [^{3}H]PtdInsP$_2$ (20 000 cpm), and 80 μM PtdEtn for 30 min at 30°C. [^{3}H]InsP$_3$ produced was extracted and quantified by liquid scintillation counting.

The isolated radiolabeled products may be resolved by TLC or HPLC and the radioactivity of any resolved fractions assayed. HPTLC analysis of [^{32}P]-labeled PtdInsPs cannot be used for direct chemical estimation of the levels of these lipids, although it can be used to determine the relative alterations of activation of PLC.

An analysis of the radioactive deacylation products may provide more precise comparisons between different inositol lipids, which label to different specific activities with ^{32}P. For this purpose, the lipids are analyzed after chemical removal of the fatty acid chains and then resolved as GroPInsP species on high performance liquid anion-exchange chromatography (Stephens et al., 1989). Thus, cells prelabeled to isotopic steady state with [^{3}H]Ins (1–10 μCi/ml) and exposed to the desired stimulant can be assessed for changes in the PtdInsPs. A comparison of the basal pattern of labeled phospholipids [PtdOH, PtdIns, PtdIns(4)P, PtdIns(4,5)P$_2$] is made to the pattern of labeled phospholipids following a brief period of stimulation (Hanley et al., 1991).

According to Hanley et al. (1991) the activity of PtdIns-specific PLC is assayed by HPTLC using [^{32}P]prelabeled PtdInsPs in a selected cell culture in response to a stimulant. At the end of the incubation, the cells are quenched and extracted with 3.75 volumes of chloroform/methanol/HCl (40:80:1, by vol.). The lipids are separated from the labeled water-soluble products by the addition of 1.25 volumes each of chloroform and 0.1 M HCl. The mixture is split into two phases by brief centrifugation in a bench top centrifuge. The lower, organic phase contains the labeled lipids, and aliquots are taken for direct counting, to assess later yields, and for drying under nitrogen stream. The dried sample is redissolved in 0.1 ml chloroform/methanol/water (75:25:2, by vol.), and an aliquot is taken for the determination of total radioactivity. The remaining redissolved sample is applied to a strip on a 10 × 10 cm HPTLC plate, which has been impregnated with oxalate by prechromatography in 1% potassium oxalate, followed by activation at 110°C for 10 min (Downes et al., 1983). The HPTLC plates are developed in chloroform/methanol/acetone/acetic acid/water (40:15:13:12:8, by vol.), or another suitable solvent system (Tysnes et al., 1985), for approximately 60 min, dried, and then exposed to X-ray film for 10–16 h, depending on the radioactivity. The mobilities of standard unlabeled inositol lipids are determined by running a separate lane with 10 μg each of inositol phospholipids, which can be visualized after development of the autoradiogram by one of several standard techniques, such as iodine vapor staining. The radioactivity in the TLC bands is determined by scraping the silica gel and counting in a Cerenkov counter in water.

According to Stephens et al. (1989), the prelabeled cells are quenched with 2 ml ice-cold 20% TCA, and the lipid containing precipitate is centrifuged and washed with 1 ml of 5% TCA/2 mM EDTA. The washed pellet is extracted by mixing with 1 ml of chloroform/methanol/10 mM EDTA, 0.1 mM inositol, 0.1 M HCl (5:9:4:4, by vol.). Two phases are obtained by adding 0.37 ml each of chloroform and 10 mM EDTA, 0.1 mM inositol, 0.1 M HCl, and the upper phase is removed after brief centrifugation. The lower phase and interfacial material are washed with 1.33 ml of synthetic upper phase, and the resulting upper phase is again discarded. The presence of EDTA during the lipid extraction is essential to remove divalent cations (e.g. Mg^{2+}, Ca^{2+}), which would interfere with the subsequent deacylation of PtdInsPs. The lower phase and particulate material are dried in vacuo and taken up in 2 ml methylamine deacylation reagent [57.66 ml methylamine in water (25–30%), 61.757 ml methanol, 1541 ml n-butanol] which is heated in 2 ml water and 2.4 ml n-butanol/light petroleum/ethyl formate (20:40:1, by vol.), and mixed well. The phases are split by a brief centrifugation, and the upper phase is removed. After taking aliquots for counting, the remainder is fractionated on HPLC (see Chapter 3). Cells are labeled with [^{3}H]Ins to isotopic steady state and quenched with 20% TCA as before. However, for analysis of the labeled water-soluble products, the extract is centrifuged and the supernatant is taken for analysis. The pellet is reextracted with 5% TCA/2 mM EDTA which is combined with the previous supernatant. The acid is removed from the combined extracts by the addition of 9 l of 1,1,2-trichlorotrifluoroethane (Freon)/tri-octylamine (1:1, v/v) and vigorous mixing (Hawkins et al., 1986). The upper aqueous phase is removed and analyzed as follows. Samples are fractionated by HPLC on a Partisil SAX column

(Jones Chromatography, 10 m, 25 cm) using a flow rate of 1.25 ml/ min. The stepwise gradient elution uses HPLC-grade water (reservoir A) and 3.5 M ammonium formate/H_3PO_4, pH 3.7 (reservoir B): A alone, 5 min; 0–20% B, 10 min; 20–50% B, 45 min; 50–100% B, 20 min; 100% B, 6 min.

Membrane-associated PLC activity is determined by measuring the amount of DAG or phosphorylated head group produced from added exogenous substrate.

The reaction is carried out in tubes containing membrane protein (50 μg); 20 mM Tris maleate; pH 7.4; 100 nM free calcium (adjusted with Ca^{2+}/EGTA buffer); and in the presence or absence of GTP 100 μM, GTPγS 10 μM and BK 10^{-7} M (Portilla et al., 1988). The reaction was initiated by the addition of 50 μl of the plasma membranes to the labeled liposomes in a final volume of 100 μl. The reaction was carried out at 37°C for 10 and 60 s of incubation and terminated by the addition of ice-cold $CHCl_3$, MeOH, H_2O (1:2:1, by vol.). The radioactive phosphorylated base is isolated by HPLC using a mobile phase buffer of 20 mM ethanolamine, pH 9.5. The fractions were counted by liquid scintillation spectrometry.

Lomasney et al. (1996) have determined the hydrolysis of PtdIns in phospholipid vesicles as follows: Hydrolysis of 30 μM PtdIns was carried out in PtdIns/PtdSer/PtdCho (molar ratio of 1:5:5) small unilamellar vesicles (SUV) incorporated with varying amounts of PtdIns-P_2. Single bilayer vesicles were prepared as described previously (Hofmann and Marjerus, 1982) with slight modification.

According to Lomasney et al. (1996), a stock solution of lipids, 30 nmol of [^{3}H]-labeled PtdIns (4 000 000 cpm), 150 nmol of PtdSer, 150 nmol of PtdCho, and the indicated amount of PtdInsP₂ (0–7.5 nmol) are added and dried under a stream of nitrogen. The dried lipids are resuspended in 0.95 ml of 50 mM HEPES, pH 7.0, 100 mM NaCl, and 1 mM EGTA. The mixture is vortexed and followed by two cycles of ultrasonication for 15 s with 45 s cooling intervals. Samples were then centrifuged at 120 000g for 60 min, and the clear supernatant was carefully removed. Fifty-microliter of BSA was added to give a final concentration of 200 μg/ml. The final concentration of PtdIns, PtdSer, and PtdCho was 30, 150, and 150 μM, respectively, plus the indicated concentration of PtdInsP$_2$. Lipids are used within 24 h. To assay the PtdIns-PLC activity, 50 μl of SUV containing 30 μM PtdIns (20 000 cpm) was preincubated at 37°C for 5 min with 0.01–10 ng of enzyme. The reaction is initiated by adding 2.5 μl of 60 mM CaCl$_2$ and incubated at 37°C for another 1–15 min. The reaction was terminated by adding 0.2 ml of chloroform/methanol/HCl (100:100:0.6, by vol.), followed by 0.06 ml of 1N HCl containing 5 mM EGTA. The aqueous and organic phases were separated by centrifugation, and a 100 μl portion of the upper aqueous phase was counted by liquid scintillation.

A modification of this general routine has been described (Kim et al., 1998; Lopez et al., 2001). In this case, the substrate was provided as mixed phospholipid vesicles containing PtdIns(4,5)P$_2$ (28 mM) and PtdEtn (280 mM) in a ratio of 1:10 with 25 000–35 000 cpm of [^{3}H]PtdInsP$_2$/assay. PLC assays were performed at 37°C for varying time periods in a mixture (70 ml) containing 50 mM HEPES, pH 7.3, 3 mM EGTA, 0.2 mM EDTA, 1 mM MgCl$_2$, 20 mM NaCl, 30 mM KCl, 4 mM dithiothreitol, 0.1 mg/ml acetylated bovine albumin, 1.6 mM sodium deoxycholate, and 1 mM CaCl$_2$. The reactions were started by the addition of the transfected cell membranes and terminated by adding 350 μl of chloroform/methanol/concentrated HCl (500:500:3, by vol.). The samples were vortexed, and 100 μl of 1 M HCl containing 5 mM EGTA was added. The phases were separated by centrifugation and assayed for radioactivity by liquid scintillation counting.

Other variations to this standard assay have been described, including the use of small unilamellar vesicles. According to Irvine et al. (1984), PtdInsP$_2$, PtdEtn (1:10, molar ratio), and [^{3}H]PtdInsP$_2$ are mixed in chloroform, evaporated under nitrogen gas stream, and then dispersed into the assay buffer by vigorous vortexing and sonicated for 2 min, to yield 2 nmol of [^{3}H]PtdInsP$_2$ (25 000 dpm) with 20 nmol of PtdEtn/assay. To examine the effects of various lipids on PLC activity, each lipid (2 mM) was added to the assay mixture which contained 6 μM Ca^{2+}, 20 mM Tris maleate buffer, pH 6.0, 80 μM KCl, [^{3}H]PtdInsP$_2$ (25 000 dpm; 0.2 M), and 0.001% sodium cholate.

The water-soluble reaction products from incubation of the PLC-I and mPLC-I with PtdIns and PtdInsP$_2$ are analyzed by chromatography on a column of AG 1xX8 according to Downes and Mitchell (1981). More than 90% of the phosphorylated inositol product derived from PtdIns and PtdInsP$_2$ hydrolysis corresponded to the carrier inositol monophosphate and Ins(1,4,5)P$_3$, respectively. Simple variations of this procedure were used to assay other enzymes and substrates. The analysis is illustrated by a combination of chromatograms shown in Fig. 9.7 (Hanley et al., 1991).

As with deacylated lipids, inositol phosphates may be provisionally identified using radiolabeled internal standards, many of which are commercially available. However, the formal problems of Ins-P identification are now potentially quite elaborate, given the recognized complexity of Ins-P species (Shears, 1989). The predicted major product of PtdIns(4,5)P$_2$ breakdown, Ins(1,4,5)P$_3$, can occur in an HPLC fraction with as many as five other InsP$_3$ species. However, on activation of intracellular PLC, the only species of InsP$_3$ in the Ins(1,4,5)P$_3$ HPLC fraction which exhibits increase in levels is authentic Ins(1,4,5)P$_3$. This conclusion has been based in part on conversion of the labeled InsPs to characteristic noncyclic polyols by sequential periodate oxidation, borohydride reduction, and dephosphorylation, which has been described in detail (Stephens et al., 1988). The Ins(1,4,5)P$_3$ can also be assayed by binding to its intracellular receptor (Bredt et al., 1989).

According to Tan and Roberts (1996) hydrolysis of PtdIns by PtdIns-PLC may be effectively determined as follows: PtdIns (20 mg, final concentration 5.65 mM) was dissolved in 4 l of D$_2$O/ H$_2$O (1:1, v/v) containing detergent (sodium deoxycholate or

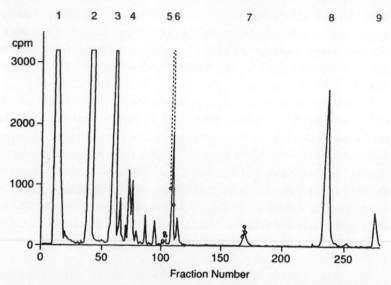

Fig. 9.7. Representative HPLC analysis of [³H]inositol-labeled acid-soluble metabolites. Samples were prepared from [³H]inositol-labeled (20 (μCi/ml, 24 h) NG115-401L neuroblastoma x glyoma cells. The samples were fractionated on a Partisil SAX column (Jones Chromatography, 10 μm, 25 cm) using a flow rate of 1.25 ml/min. The stepwise gradient elution used HPLC grade water (reservoir A) and 3.5 M ammonium formate/phosphoric acid, pH 3.7 (reservoir B): A alone, 5 min; 0–20% B, 10 min; 20–50% B, 45 min; 50–100% B, 20 min; 100% B, 6 min. The solid line indicates the basal levels of [³H]inositol-labeled compounds, and the dotted line illustrates the stimulated changes in the inositol tris- and tetrakis-phosphate fractions following bradykinin stimulation (15 s, maximal dose). The positions of the standards are as follows: 1, Ins; 2, GroPIns; 3, InsP; 4, InsP₂; 5, Ins(1,3,4)P₃; 6, Ins(1,4,5)P₃; 7, InsP₄; 8, InsP₅; 9, InsP₆ (phytic acid).

Reproduced from Hanley et al. (1991) with permission of the publisher.

Triton X-100, final concentration 12.5 mM) and buffer. The buffers used were 50 mM Tris/acetate (pH 5.5, 7.5, or 8.5) and 50 mM sodium borate/HCl (pH 7.5). The reaction was initiated by the addition of 1–10 μl of PtdIns-PLC (0.3–30 ng) in 0.1% BSA (pH 7.5). The mixture was then quickly transferred to an NMR tube and placed in the probe of the instrument (Tan and Roberts, 1996).

The PtdIns-PLC-catalyzed hydrolysis of Ins(1,2)cycP was studied by ^{31}P NMR essentially as described above for PtdIns. Samples contained 2–3 mg of Ins(1,2)cycP and the buffers described above. The reaction was initiated by the addition of 20–30 μg of PtdIns-PLC.

PtdIns specific PLCs from several bacterial species release GPtdIns-anchored proteins from cell membranes by cleaving the DAG moiety of PtdIns. Figure 9.8 illustrates the degradation pathway (Brodbeck, 1998).

Tsujioka et al. (1999) have described the determination of PtdIns-PLC cleavage of PtdIns in He-La cells by quantitative determination of diacylglycerol as follows. He-La cells (5×10^6 cells per dish) were homogenized in PBS and separated into two portions: one was used directly for lipid extraction and the other was incubated at 37°C for 1 h with 0.5 units of bacterial PtdIns-PLC (Funakoshi, Tokyo). The cells were extracted by adding chloroform/methanol (1:2, v/v) to bring the aqueous volume to 0.8 ml. After mixing, 1 ml of chloroform and 1 ml of 1 M NaCl were added to effect phase separation. After centrifugation at 5000g for 2 min, the lower phase was analyzed with DAG kinase assay (Preiss et al., 1986; see Chapter 2).

The HPLC routine of Auger et al. (1990) has also been extensively used for the isolation of fractions corresponding to the retention times of PtdInsP and PtdinsP$_2$ standards.

9.3.3. Phospholipase D

PtdIns-PLD activities have been found that attack PtdIns preferentially and that act on PtdIns glycans. Progress in understanding the regulation and function of PLD activity has been slow. Cockcroft et al. (1984) obtained the first evidence of PtdIns hydrolysis to PtdOH in intact human neutrophils suggesting a signal-activated PLD in mammalian cells. Subsequently, PtdIns-specific PLD activities were identified in human neutrophils,

A: Protein - CO—NH—CH$_2$—CH$_2$—CH$_2$—O—P—O—6 Man α1-2Man α1-6 Man α1-6 Man α1-4GlcN α1-6 Pdtlns

B: P$_{1-3}$ − (αGal) $_{2-4}$ - βGlcNAc - Hex-GlcN - Pdtlns

C:

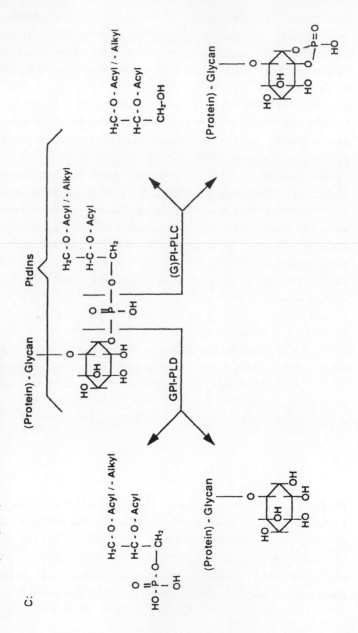

bovine lung, canine kidney cells and in B lymphocytes (Exton, 1997, 1998). These enzymes are located in cytosol, stimulated by Ca^{2+}, and have been reported to utilize PtdEtn and PtdCho, or PtdEtn and PtdCho also. This is in contrast to other PLDs, which are activated by oleate and are highly specific for PtdCho as substrate (see above).

A PLD specific for glycosyl PtdIns (GPtdIns-PLD) was discovered in mammalian serum simultaneously by Davitz et al. (1989), Low and Prasand (1988) and Cordoso de Almeida et al. (1988). It has been purified and its basal biochemical parameters have been established by several groups (Davitz et al., 1989; Huang et al., 1990; Hoerner and Brodbeck, 1992). Determination of its amino acid sequence suggested the presence of lipid and Ca^{2+} binding domains, phosphorylation sites and N-glycosylation sites. Deeg and Davitz (1994, 1995) have prepared an early review on the secreted PtdIns-PLD that specifically cleaves the PtdIns-glycan anchor of numerous plasma membrane proteins, while Brodbeck (1998) has recently reviewed the role of GPtdIns-anchor hydrolysis in signal transduction.

GPtdIns-PLD is abundant in mammalian serum, but the source of the circulating enzyme is unknown. Deeg and Verchere (1997) showed that β TC3 cells secrete GPtdIns-PLD in response to insulin secretagogues and suggest that GPtdIns-PLD may be secreted via the regulated pathway in these cells. Rhode et al. (1999) have shown that the GPtdIns-PLD activity in serum is positively correlated with

Fig. 9.8. Structures of glycosyl PtdIns and cleavage specificity of glycosyl PtdIns-Hydrolyzing phospholipases. (A) Core structure of glycosyl PtdIns membrane protein anchor. (B) Model structure of glycosyl PtdIns phospholipids participating in signal transduction. Several forms exist differing in the number of phosphates (1–3) attached to one galactose residue in positions 2–4. C, Cleavage of glycosyl PtdIns by glycosyl PtdIns-PLC and glycosyl PtdIns-PLD and respective products formed. PtdIns represent both diacyl- and alkylacyl-Gro moieties. Reproduced from Brodbeck (1998) with permission of the publisher.

inflammatory markers and counts of monocytes and stab cells and negatively correlated with polymorphonuclear neutrophils and lymphocytes in severe diseases. The study suggested that liver is not the only source of enzyme. O'Brien et al. (1999) used human GPtdIns-PLD specific antibodies to demonstrate cell-associated GPtdIns-PLD in 95% of atherosclerotic segments, primarily on a subset of macrophages, and extracellular GPtdIns-PLD in 30% of atherosclerotic segments localized with apoA-I. GPtdIns-PLD was not detected in nonatherosclerotic segments. It was suggested that oxidative processes may regulate GPtdIns-PLD expression and that GPtdIns-PLD may play a role in inflammation and pathogenesis of atherosclerosis.

Civenni et al. (1999) have reported the in vitro phosphorylation of purified bovine serum GPtdIns-PLD by cAMP-dependent protein kinase (PKA) and by tyrosine kinase. Phosphorylation by PKA occurred in the 110 kDa native form of GPtdIns-PLD as well as in multiple proteolytic degradation products and caused significant decrease in enzyme activity. Dephosphorylation by AP completely restored GPtdIns-PLD activity. The site of phosphorylation by PKA was assigned to Thr-286. Tyrosine phosphorylation was only observed in a proteolytically processed fragment of GPtdIns-PLD but not in the 110 kDa native form and had no effect on GPtdIns-PLD activity. The physiological function of GPtdIns-PLD in serum is not clear because the enzyme does not liberate PtdOH from the anchor moiety of glycosyl PtdIns proteins under physiological conditions (Deng et al., 1996). Tsujioka et al. (1999) have obtained evidence suggesting that GPtdIns-PLD is involved in intracellular cleavage of the GPtdIns-anchor, which is a potential source of diacylglycerol production to activate PKCα.

LeBoeuf et al. (1998) have published a map position and discussed genetic regulation for GPtdIns-PLD1. There was a single structural gene (Gpld1) mapping to mouse Chromosome 13, with liver showing the greatest abundance of GPtdIns-PLD mRNA. Lee et al. (1999) have investigated the regulation of brain GPtdIns-PLD by natural amphiphiles. Monoacylglycerols enhanced significantly the conversion of the amphiphilic form of Zn^{2+}-GroPCho-cholinephosphodiesterase into hydrophilic form. The enhancing effect depends on the size of the acyl group (C_8-C_{18}). Presence of double bond in acyl chain further enhanced the conversion by GPtdIns-PLD. LysoPtdCho and PtdOH inhibited GPtdIns-PLD activity. It was suggested that GPtdIns-PLD activity in an in vivo system may be regulated by natural amphiphiles. Recently, Du and Law (2001) have shown that the level of cellular GPtdIns-PLD expression is tightly regulated in murine macrophages, and overexpression of GPtdIns-PLD is cytotoxic to cells. Du et al. (2002a,b) have confirmed this observation in Chinese hamster ovary cells. However, overexpression of GPtdIns-PLD could be tolerated by Chinese hamster ovary cell mutants with aberrant GPtdIns biosynthesis.

9.3.3.1. Substrate specificity
The PtdInsPs activation of plant PLD is a property shared by PLDs cloned from mammals and yeast (Hammond et al., 1995; Colley et al., 1997; Waksman et al., 1996). However, the mechanism by which PtdInsPs affect PLD activities has not been elucidated. Qin et al. (1997) have provided evidence indicating that plant PLDs directly bind PtdInsP$_2$ and that this binding is required for the activities of PLDγ and PLDβ, but not for PLDα. This binding ability correlates with the degree of conservation of a basic PtdInsP$_2$-binding motif located near the putative catalytic site. Qin et al. (1997) have reported the cloning and molecular analysis of a new *Arabidopsis* PLD, PLDγ, whose activity is regulated by PtdInsPs. A feature unique to PLDγ is the presence of a putative myristoylation site, MGXXXS, near its N terminus. There is evidence that the PtdIns-PLDγ isolated from plants by Qin et al. (1997) prefers PtdIns(3)P (see below).

Ching et al. (1999) have now isolated three PtdIns-PLD isozymes from rat brain, tentatively designated as PtdIns-PLDα, PtdIns-PLDβ, and PtdIns-PLDγ, and have shown that they differ in their substrate preferences. Thus, PtdIns-PLDγ preferentially utilized PtdIns(3)P as substrate, followed by, in sequence, PtdIns(3,4,5)P$_3$, PtdIns(4)P, PtdIns(3,4)P$_2$ and PtdIns(4,5)P$_2$. None of these enzymes reacted with PtdCho, PtdSer or PtdEtn. All three PtdIns-PLDs were Ca^{2+}-dependent. Furthermore, all these enzymes required sodium deoxycholate for optimal activities, while other detergents including Triton X-100 and Nonidet P-40 were inhibitory. The discovery of these enzymes is of great importance in view of the established function of PtdIns 3-kinase in signal transduction and intracellular membrane traffic (Rameh and Cantley, 1999). Liscovitch et al. (2000) have reviewed the area and have suggested that one function of the newly discovered PtdIns-PLDs might be to turn off PtdInsP$_3$-mediated signaling.

Rhode et al. (2000) have reported the modulation of activity of GPtdIns-PLD by naturally occurring amphiphiles. The study was performed with an enzyme purified 6000-fold from human serum and used pure alkaline phosphatase containing one anchor moiety per molecule as substrate. The enzyme was stimulated by n-butanol, but in contrast to other PLDs this activation was not produced by transphosphatidylation reaction. The enzyme was found to be inhibited by PtdSer, PtdIns, PtdGro, PtdOH, gangliosides, cholesteryl esters and sphingomyelins. PtdEtn and various monoacylglycerols were found to be activators. Rhode et al. (2000) have discussed the presence of high serum concentrations of strongly inhibiting compounds and compartmentalization effects as possible reasons for the observed inactivity of the native enzyme.

In contrast to serum GPtdIns-PLD, intracellular GPtdIns-PLDs hydrolyze glycosyl PtdIns-proteins in the natural environments without detergent addition. Thus, endogenous GPtdIns-PLD activities have been demonstrated in a variety of cell cultures (Brunner et al., 1994; Metz et al., 1994; Lierheimer et al., 1997; Tsujioka et al., 1998, 1999; Wilhelm et al., 1999).

9.3.3.2. Methods of assay
Rhode et al. (2000) determined the glycosyl PtdIns-PLD activity as described by Rhode et al. (1995) with slight modifications. Briefly,

Fraction II of electrophoretically homogeneous GPtdIns alkaline phosphatase (AP) from calf intestine, bearing one anchor moiety per active dimer (Bublitz et al., 1993; Gerghardt et al., 2000) was used as substrate. Totally 30 IU/ml GPtdIns-AP, corresponding to 0.271 nM of hydrophobic GPtdIns-anchor moiety, were incubated with glycosyl PtdIns-PLD containing samples at 37°C in 10 μl final volume in 100 mM triethanolamine/HCl, pH 7.0, containing 1 mM MgCl$_2$, 40 μM CaCl$_2$ (buffer C), and 0.025% Triton X-100. The reaction was stopped by adding 100 μl of 1.0 M diethanolamine/ HCl, pH 9.8. Separation of hydrophobic GPtdIns-AP and hydrophilic AP and determinations of both AP activities were performed in microtiter plates at ambient temperature according to Rhode et al. (1995). GPLD activities were obtained from measurements of liberated hydrophilic AP activities, taking into account that GPtdIns-AP Fraction II is a dimer with 118 049 kDa molecular mass and 936.6 IU/mg specific activity, and that it bears one anchor moiety per dimer.

Tsujioka et al. (1998) have described an in vitro assay for glycosyl PtdIns-PLD partially purified from mammalian CHO or insect H5 cells by affinity chromatography through concanavalin A-Sepharose column. A substrate for GPtdIns-PLD was prepared from COS-1 cells pulse-labeled with [^{35}S]methionine for 30 min and chased for 4 h. A post-nuclear fraction obtained from the cells was separated by centrifugation at 105 000g for 1 h into a pellet (membranes) and a supernatant (cytosol). The membranes were resuspended in 20 mM Tris–HCl (pH 7.4) containing 0.1 mM CaCl$_2$. Equal amounts of the membrane suspension were incubated at 37°C for 2 h with the indicated GPtdIns-PLD preparation with or without 1% Triton X-100. The samples were recentrifuged at 105 000g for 1 h, and the resulting supernatants were subjected to immunoprecipitation with antihuman placental AP antibodies and analyzed by SDS-PAGE. Quantitative analysis of released placental alkaline phosphatase (PLAP) was performed by NIH Image software. GPtdIns-anchor cleavage activity of GPtdIns-PLD was expressed as percentage of that exhibited by PtdIns-PLC.

The cleavage by PtdIns-PLC was assayed by quantifying the release of DAG as described elsewhere (see above).

Qin et al. (1997) assayed the PtdInsP$_2$-independent PLD using a reaction mixture of 100 mM MES (pH 6.5), 0.5 mM SDS, 1% (v/v) ethanol, 30 μg of protein, and 0.4 mM PtdCho (egg yolk) containing dipalmitoylglycero-3-phospho[methyl-^{3}H]choline. CaCl$_2$ was added to reactions as noted below. The substrate preparation, reaction conditions, and separation of products were performed as described previously (Wang et al., 1994). Release of [^{3}H]choline into the aqueous phase was quantified by scintillation counting. Control assays used 30 μg of protein from lysed bacteria harboring the pBluescript SK (−)plasmid without PLD cDNA insert.

9.4. PtdIns(4,5)P$_2$ activated phospholipases

Some of the PtdInsPs serve as both substrates and activators of the PtdIns-specific phospholipases. Current widespread interest in the PtdIns(4,5)P$_2$ activation of the phospholipases stems from the putative role of this activation in the formation of signaling molecules (Chapter 11).

9.4.1. Phospholipase A$_2$

Cytosolic PLA$_2$ (cPLA$_2$) has unique structural and regulatory properties within the PLA$_2$ superfamily (Dennis, 1994, 1997). Much interest in the cytosolic PLA$_2$ stems from its putative role as a key enzyme involved in the inflammatory response, a provider of arachidonic acid, the precursor for prostaglandins and leukotrienes (Goetzl et al., 1995; Kramer and Sharp, 1995). It was noted early (Leslie and Channon, 1990) that several anionic phospholipids including PtdIns(4,5)P$_2$ activated cPLA$_2$. This stimulatory effect was hypothesized to result from an enhancement of the partitioning of $cPLA_2$ into lipid membranes caused in general by all anionic lipids. Mosior et al. (1998) have demonstrated that $cPLA_2$ binds in a 1:1 stoichiometry with high-affinity and specificity to $PtdIns(4,5)P_2$ in lipid vesicles, and the effect is quite distinct from that of other anionic lipids. In addition, an apparent functional similarity between $cPLA_2$ and phospholipase $C\delta_1$ ($PLC-\delta_1$) has allowed these workers to propose the location of a pleckstrin homology (PH) domain in $cPLA_2$.

The group IV $cPLA_2$ exhibits a potent and specific increase in affinity for lipid surfaces containing $PtdIns(4,5)P_2$ at physiologically relevant concentrations. Specifically, the presence of 1 mol% $PtdIns(4,5)P_2$ in PtdCho vesicles results in a 20-fold increase in the binding affinity of $cPLA_2$. This increased affinity is accompanied by an increase in substrate hydrolysis of a similar magnitude. The binding studies and kinetic analysis indicate that $PtdIns(4,5)P_2$ binds to $cPLA_2$ in a 1:1 stoichiometry (Mosior et al., 1998).

Activation of group IV $cPLA_2$ by agonists has been correlated with the direct phosphorylation of the enzyme by members of the mitogen-activated protein kinase (MAPK) cascade (Leslie, 1997). Balboa et al. (2000) have demonstrated that phosphorylation of cytosolic group IV PLA_2, while necessary, is not sufficient for arachidonic acid release in macrophages.

9.4.1.1. Affinity assay

The $cPLA_2$ has been shown to possess a pleckstrin homology domain through which the enzyme is thought to strongly interact with $PtdIns(4,5)P_2$. This interaction may help facilitate enzyme activation (Ackerman and Dennis, 1995). Mosior et al. (1998) have demonstrated that $cPLA_2$ binds in a 1:1 stoichiometry with high-affinity and specificity to $PtdIns(4,5)P_2$ in lipid vesicles, and the effect is quite distinct from that of other anionic lipids. The group IV $cPLA_2$ exhibits a potent and specific increase in affinity for lipid surfaces containing $PtdIns(4,5)P_2$ at physiologically relevant concentrations.

Mosior et al. (1998) quantified the membrane affinity using the apparent membrane association constant K_{app}, defined as the reciprocal of the total concentration required for half-maximal association of the protein with the membrane. Determination was performed in the standard binding solution (100 mM KCl, 20 mM PIPES, pH 6.8) at 2 μM free Ca^{2+}. The membrane consisted of the large unilamellar vesicles composed of 1-palmitoyl-2-[1-^{14}C]-arachidonoyl GroPCho with variable mol% of PtdIns(4,5)P$_2$. The suspension was allowed to equilibrate for 10 min and was then centrifuged for 15 min at 25°C at 150 000g.

9.4.1.2. Catalytic assay

A substrate assay with cPLA$_2$-IV would normally be performed as described for other PLA$_2$s (see above), except that PtdIns-P$_2$ would be included as an activator at several levels of concentration. cPLA$_2$-IV possesses a preference for phospholipids containing arachidonic acid (Dennis, 1994; Leslie, 1997; Clark et al., 1995) and a product assay would be most appropriate. According to Yang et al. (1999), the product assay is best performed with 100 μM 16:0/20:4 GroPCho/PtdInsP$_2$ (97/3) (containing 100 000 cpm of 1-palmitoyl-2-[1-^{14}C]arachidonoyl-sn-GroPCho) in Triton X-100 (400 μM) mixed micelles in 100 mM HEPES, pH 7.5, 80 μM Ca^{2+}, 2 mM DTT, and 0.1 mg/ml BSA. Higher concentration of BSA can adversely affect the activity of cPLA$_2$ in this assay. A control assay with 1-palmitoyl-2-[1-^{14}C]linoleoyl-sn-GroPCho would be required in order to claim preferential release of arachidonate.

Mosior et al. (1998) quantified the cPLA$_2$ catalytic activity towards 1-palmitoyl-2-[1-^{14}C]arachidonoyl GroPCho using a modified Dole assay, as previously described by Zhang et al. (1991). The structurally similar PtdIns(3,4,5)P$_3$ activated cPLA$_2$ to roughly 60% the level of PtdIns(4,5)P$_2$. PtdIns(3,4)P$_2$ activated cPLA$_2$ to 63% of that of PtdIns(4,5)P$_2$, but PtdIns(3)P only 32% of that of PtdIns(4)P demonstrating the stereoselectivity of cPLA$_2$ for PtdIns(4,5)P$_2$.

9.4.2. Phospholipase C

Most studies have focused on the regulation of PLC activity by various protein modulators. Several lines of evidence suggest that the substrate, $PtdInsP_2$, may also be a key modulator of PLC activity. Initial studies with PLC-δ1 indicated that the relative concentration of $PtdInsP_2$ can affect hydrolytic rates through interaction with allosteric site in the molecule (Cifuentes et al., 1993). Cifuentes et al. (1993) found that high-affinity $PtdInsP_2$ binding involves a stereospecific recognition of the polar head group by PLC-δ1; the most important feature is the presence of vicinal phosphates at the 4 and 5-positions of the inositol ring. Such specificity by the anchoring site could provide an explanation for the observation that PLC-δ1 hydrolyzes $PtdInsP_2$ at significantly faster rates than PtdIns under the same conditions. Furthermore, loss of high-affinity $PtdInsP_2$ binding by the catalytically active fragments supports the model that PLC-δ1 has at least two sites for substrate binding, a high-affinity membrane-anchoring site and a low affinity catalytic site. Sternfeld et al. (2000) have reported that fMLP induced arachidonic acid release in db-cAMP-differentiated HL-60 cells is independent of $PtdIns(4,5)P_2$-PLC activation and $cPLA_2$ activation.

The PtdIns-PLC from *Trypanosoma cruzi* is active on inositophosphoceramide (Salto et al., 2002).

9.4.2.1. Affinity assays

Tall et al. (1997) tested for the presence of high-affinity $PtdIns(4,5)P_2$ and $PtdIns(3,4,5)P_3$ binding sites in four PLC isozymes (δ1, β1, β2, and β3) by probing these proteins with analogs of InsPs, D-Ins(1,4,5)P_3, D-Ins(1,3,4,5)P_4 and $InsP_6$, and $PtdIns(4,5)P_2$ and $PtdIns(3,4,5)P_3$, which contain a photoactivatable benzoyldihydrocinnamide moiety. Only PLC-δ1 was specifically radiolabeled. More than 90% of the label was found in tryptic and chymotryptic fragments, which reacted with antisera against the pleckstrin homology domain, whereas 5% was recovered in

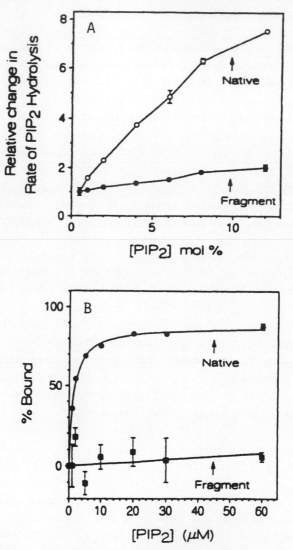

Fig. 9.9. Effects of PtdInsP$_2$ concentration on binding of PLC-δ_1 to PtdCho/PtdInsP$_2$ vesicles and rate of substrate hydrolysis. (A) The effects of PtdInsP$_2$ mole percentage on the rates of substrate hydrolysis catalyzed by PLC-δ_1 or the 77-kDa fragment in dodecyl maltoside/PtdInsP$_2$ mixed micelles. The rate of

fragments that encompassed the catalytic core. The catalytic activity of PLC-$\delta 1$ was inhibited by the product PtdIns(1,4,5)P$_3$, whereas no inhibition of PLC-$\beta 1$, PLC-$\beta 2$, or PLC-$\beta 3$ activity was observed. These results suggest that the PH domain is the sole high-affinity PtdIns(4,5)P$_2$ binding site of PLC-$\delta 1$ and that a similar site is not present in PLC-$\beta 1$, PLC-$\beta 2$, or PLC-$\beta 3$.

Tall et al. (1997) assayed PLC catalytic activity using LUVs composed of PtdEtn/PtdCho/[^{3}H]PtdIns-P$_2$ (15:4:1) and prepared by extrusion through polycarbonate filters as previously described by Cifuentes et al. (1993) to yield a final PtdIns(4,5)P$_2$ concentration of 60 μM. The reactions, which were initiated by addition of substrate vesicles, were carried out in buffer containing 150 mM NaCl, 20 mM HEPES (pH 7.5), 2.5 mM DTT, 0.1% gelatin, 120 μM CaCl$_2$, and 4 μg/ml leupeptin and pepstatin, in the presence or absence of 30 μM Ins(1,4,5)P$_3$.

9.4.2.2. Catalytic assay

Cifuentes et al. (1993) measured the rates of PtdInsP$_2$ hydrolysis catalyzed by native PLC-$\delta 1$ or a purified 77-kDa fragment as a function of PtdInsP$_2$ mole fraction in dodecyl maltoside mixed micelles at a constant total concentration of PtdInsP$_2$ (30 μM). The change in mole percentage of PtdInsP$_2$ was achieved by decreasing the detergent concentration. The concentration of free calcium was buffered to 10 μM with EGTA. Figure 9.9 demonstrates the effect of PtdInsP$_2$ mole percentage on the rate of substrate hydrolysis

PtdInsP$_2$ hydrolysis catalyzed by native PLC-δ_1 (open circles) or the purified 77-kDa fragment (closed circles) were measured as a function of PtdInsP$_2$ mole fraction in dodecyl maltoside mixed micelles at a constant total concentration of PtdInsP$_2$ (30 μM). (B) Binding of native PLC-δ_1 and the 77-kDa fragment to PtdCho/PtdInsP$_2$ (98:2, mol/mol) bilayer vesicles. Native enzyme (filled circles) or the purified 77-kDa fragment (filled squares) was incubated with sucrose-loaded PtdCho/PtdInsP$_2$ LUV in buffer. PLC bound to vesicles was separated from free PLC by sedimentation of the sucrose-loaded LUV. Reproduced from Cifuentes et al. (1993) with permission of the publisher.

catalyzed by PLC-δ1 or the 77-kDa fragment in dodecyl maltoside/ PtdInsP$_2$ mixed micelles.

9.4.3. Phospholipase D

Multiple PLD activities exist in mammalian tissues. These can be subdivided into two classes (Table 9.1). The first of these are stimulated by free fatty acids whereas the second class is inhibited by these agents but dependent on PtdInsPs (Brown et al., 1995; Chalifour and Kanfer, 1982). PLD1 and PLD2 belong to the latter classes of enzymes and may account for all the PtdInsP-dependent PLD activity described (Hammond et al., 1995; Colley et al., 1995). Fatty acid stimulated PLD activity has been reported in a number of tissues and cell types with brain and lung being the most extensively investigated, but no corresponding cDNA sequences have been reported (Massenburg et al., 1994). Laine et al. (2000) have detected an oleate-dependent PLD activity that co-localized with PLD1 enzyme in crude plasma membrane and microsomal fractions from rat pancreas. Xie et al. (2001) have recently reported a requirement for covalent palmitoylation of rat brain PLD1. The palmitoylation is required for membrane association.

The first to be identified with exogenous substrate was in a detergent extract of a rat brain particulate fraction (Saito and Kanfer, 1975). The first purified mammalian PtdCho-PLD has a molecular mass of 124 kDa (Hammond et al., 1995). A PtdCho-PLD purified from pig lung has a molecular mass of 190 kDa and pH optimum of 6.6. Sodium oleate has been established as a stimulator of PtdCho-PLD activity (Kobayashi and Kanfer, 1987).

cDNAs for two mammalian PLDs have been cloned (Hammond et al., 1995, 1997; Jenco et al., 1998; Colley et al., 1997). PLD1 has low basal activity and can be activated by small GTP-binding proteins such as ARF and Rho and by PKCα and β (Hammond

TABLE 9.1

Phospholipase A_2

Group	Sources	Location	Size (kDa)	Ca^{2+}-requirement	Di-sulfides	Molecular characteristics
IA	Cobras, kraits	Secreted	13–15	mM	7	His-Asp pair
IB	Porcine/human pancreas	Secreted	13–15	mM	7	His-Asp pair, elapid loop
IIA	Rattlesnakes, human synovial fluid/platelets	Secreted	13–15	mM	7	His-Asp pair, carboxyl extension
IIB	Gaboon viper	Secreted	13–15	MM	6	His-Asp pair, carboxyl extension
IIC	Rat/mouse testis	Secreted	15	MM	8	His-Asp pair, carboxyl extension
III	Bees, lizards	Secreted	16–18	MM	5	His-Asp pair
IVA	Rat kidney, human platelets	Cytosolic	85	$<\mu M$		Ser-228; Ar-200; Asp-549; Ser 505; Cal B, PH domain
IVB	Human brain	Cytosolic	100	$<\mu M$		N-terminal extension, Ser-228
IVC	Human heart/skeletal muscle	Cytosolic	65	None		Prenylated; Ser-228; lacks CalB domain; Ser-505/727
V	Human/rat/mouse heart/lung, P338D₁ macrophages	Secreted	14	M	6	His-Asp pair, no elapid loop, no carboxyl extension
VI	CHO cells, P338D₁ macrophages	Cytosolic	80–85	None		GXSXG consensus, ankyrin repeats, 340-kDa complex
VIIA	Human plasma	Secreted	45	None		GXSXG consensus sequence, Ser-273, Asp-296, His-351
VIIB	Bovine brain	Cytosolic	42	None		Myristoylated at N terminus
VIII	Bovine brain	Cytosolic	29	None		Ser-47
IX	Marine snail	Secreted	14	MM	6	His-Asp pair
X	Human leukocytes	Secreted	14	mM	7	His-sp pair

et al., 1995). The PLD isoform has a high basal activity and is relatively insensitive to activation by ARF and Rho (Exton, 1999; Frohman et al., 1999; Jenco et al., 1998). Human PLD1 and PLD2 are also dependent on PtdInsP$_2$ or PtdIns(3,4,5)P$_3$ for activity (Exton, 1999; Frohman et al., 1999; Bandoh et al., 1999). It is possible that PtdInsP$_2$ may be a physiological activator for PLD (Exton, 1999; Frohman et al., 1999; Liscovitch et al., 1999), and the concentration of this lipid is likely to be limiting in cells for the expression of PLD activity.

Thus far, two isoforms of PLD have been described, PLD1 and PLD2, both of which exist as two splice variants (Colley et al., 1997; Hammond et al., 1997; Park et al., 1997). The genes have been cloned and shown to be about 50% identical. The PLD1 isoforms are directly regulated by classic protein kinase C (PKC) isozymes and members of the families of the two small GTPases ARF and Rho (Exton, 1999; Frohman et al., 1999). The PLD2 isoforms show high basal activity and respond minimally to PKC or the small GTPases. Other isoforms of comparable properties have been since cloned from other mammalian tissues. Petitt et al. (2001) have observed that PLD$_{1b}$ and PLD$_{2a}$ generate structurally identical PtdOH species in mammalian cells.

Singh et al. (2001) have demonstrated that ceramides inhibit PLD activity in several mammalian cell types. These effects have been related to preventing activation by ARF1, RhoA, and protein kinase Cα and β and therefore indicate that PLD1 is inhibited. Singh et al. (2001) have subsequently investigated in Sf9 cells the effects of ceramides in inhibiting both PLD1 and PLD2 and the interaction with another activator, PtdIns(4,5)P$_2$. Partially purified PLD2 exhibited an absolute requirement for PtdInsP$_2$ when the activity was measured using Triton X-100 micelles. It was concluded that ceramides reversibly inhibit mammalian PLD2 as well as PLD1 activities. Oh et al. (2000) have reported regulation of PLD$_2$ by H$_2$O$_2$ in PC12 cells, while Kam and Exton (2002) have reported homologous and heterologous dimerization of PLD$_1$ and

PLD_2 which does not appear to affect the enzymatic activity of the isozymes.

In animals, three different PLD activities can be distinguished (Liskovitch, 1996; Roberts, 1996): membrane-bound (mPLD), cytosolic (cPLD) and secreted (sPLD). The mPLDs seem to be specific for PtdCho and do not depend on Ca^{2+} for activity, although it can stimulate it (Okamura and Yamashita, 1994). The substrate specificity of cPLDs is less strict for they can hydrolyze PtdEtn, PtdCho, and PtdIns, but only if Ca^{2+} is present (Huang et al., 1992; Wang et al., 1991). The sPLDs are specific for GPtdIns anchors and cannot hydrolyze PtdCho at all (Scallon et al., 1991).

The enzyme is enriched in plasma membranes from many tissues but is also present in high activity in Golgi and nuclei, as well as in cytosol (Exton, 1997). Another PtdCho-PLD from pig brain has a molecular mass of 95 kDa and is markedly stimulated by proteins ARF and RhoA. The enzyme is stimulated by $PtdIns-P_2$ (see below).

PLD has recently been identified as a group of enzymes with diverse functions, termed the PLD superfamily (Ponting and Kerr, 1997). The hallmark of the PLD superfamily is the presence of conserved HxKxxxx motifs in the enzyme amino acid sequence (Hammond et al., 1995; Koonin, 1996). It is believed that duplicated HKD motifs are key structural components of the active site (except in Nuc which has a single HKD motif) and that members of the superfamily therefore share similar catalytic mechanisms (Zhao et al., 1997; Koonin, 1996). Li and Fleming (1999) have proposed that AlF_4^- is a competitive inhibitor of purified cabbage PLD, with a mechanism of action based on its phosphate-mimicking property. Specifically, it is proposed that AlF_4^-, and other phosphate analogs, inhibit plant PLD by competing with a substrate phosphate group for a substrate binding site, thereby preventing the formation of an enzyme-phosphatidyl intermediate. This may be a conserved feature of PLD superfamily of enzymes.

Gomez-Munoz et al. (1999) have shown that lysoPtdCho stimulates the activity of nonspecific PLD in murine peritoneal macrophages by a mechanism involving both protein kinase C activation and tyrosine phosphorylation, and that at least part of this effect is mediated by PAF receptor activation. Furthermore, acetylation of lysoPtdCho substantially increased its potency in activating PLD, suggesting that a cellular metabolite of lysoPtdCho such as 1-acyl-2-acetyl GroPCho might be responsible for at least par of the effect of lysoPtdCho on PLD.

The mechanisms controlling inositol phospholipid dependent PLD activity have been subject to intense investigation. It is obvious that intact cells contain PLD activities that are controlled by extracellular ligands that act through specific receptors and are subject to lipid and protein regulators (Exton, 1997; Liscovitch et al., 2000). PtdInsP$_2$ was first identified as an essential cofactor for HL-60 cell ARF activated PLD (Brown et al., 1995). PLD2 activity is also strongly dependent on PtdInsP$_2$ (Hammond et al., 1995, 1997). PtdIns(3,4,5)P$_3$ can also activate the enzymes with equal potency but reduced efficiency (see below). Other acidic phospholipids and soluble InsPs neither stimulate activity of PLD nor block activation by PtdInsP$_2$ (Hammond et al., 1997; Jenco et al., 1998). The physiological role of PtdInsPs in PLD regulation is not clear. One possibility is that binding to PtdInsPs anchors the PLD enzymes to the membrane surface placing them in close proximity with their substrates (Sciorra et al., 1999). Studies using antibody that blocks PtdIns(4)P 5-kinase demonstrated that synthesis of PtdInsP$_2$ is required for stimulation of PLD by nonhydrolyzable guanine nucleotides in U937 cell membranes (Pertile et al., 1995). Likewise, the inositol phospholipid-dependent PLD activity is inhibited in vitro by hydrolysis of the PtdInsP$_2$ by the PtdIns(4,5)P$_2$ 5-phosphatase, synaptojanin (Chung et al., 1997). Lee et al. (2000) have shown that the PtdIns P$_2$ PLD is also inhibited by amphiphysins I and II.

The complex regulation of mammalian PLD is illustrated through the early identification of four unique factors that modulate its activity. The first of these factors was identified during the development of an exogenous substrate assay to measure guanine nucleotide-dependent PLD activity associated with membranes from HL60 cells (Brown et al., 1993a,b). This activity displayed dependence on the inclusion of PtdIns(4,5)P$_2$ in phospholipid vesicles containing radiolabeled PtdCho as a substrate. The necessity for this regulatory lipid was subsequently shown for PLD activity derived from rat brain membranes in a study that also demonstrated the similar efficacy of PtdIns(3,4,5)P$_3$ (Liscovitch et al., 1994). Brown et al. (1995) have reported partial purification and characterization of Arf-sensitive PLD from porcine brain. Extraction of porcine brain membranes with detergent and subsequent chromatography with SP-Sepharose revealed a large peak of Arf-sensitive PLD activity. PLD that was highly enriched, retained a requirement for PtdIns(4,5)P$_2$ for efficient expression of activity. The uniqueness of PtdInsP$_2$ and PtdInsP$_3$ in these exogenous substrate assays was demonstrated by the near or total ineffectiveness of PtdSer, PtdIns(4)P, PtdIns and PtdOH. PtdInsP$_2$ was also found to be essential for the activity of the SPO14 enzyme from yeast (Rose et al., 1995) and the recombinant human PLD (Hammond et al., 1995). A requirement of PtdInsP$_2$ for PLD activity in vivo was suggested in experiments with permeabilized U937 cells (Pertile et al., 1995). Lopez et al. (1998) have cloned and characterized a human PLD2, which was stimulated by ARF-1 about two-fold compared twenty-fold of human PLD1 activity, when expressed in insect cells.

Min et al. (1998) have since shown that rPLD1 activity is stimulated significantly by both PtdIns(4,5)P$_2$ and PtdIns(3,4,5)P$_3$, while PtdIns(4)P and PtdIns(3,4)P$_2$ are largely ineffective. The activation of PLD by PtdInsP$_2$ and PtdInsP$_3$ was dependent on the molar fraction of PtdInsP$_2$ or PtdInsP$_3$ in the vesicles. Furthermore, oleate had a strong inhibitory effect on the enzyme, whereas other fatty acids (palmitate, stearate, arachidonate) were ineffective. The rPLD1 enzyme was shown to be specific for PtdCho, when compared to PtdEtn and PtdIns under standard assay conditions (PtdEtn/PtdIns(4,5)P$_2$/PtdCho, 16:1.4:1 by weight, containing 0.5 μCi of [choline-methyl-^{3}H]PtdCho) (Bae et al., 1998a,b).

Ching et al. (1999) have discovered three distinct PtdIns-specific PLD isozymes from rat brain, tentatively designated as PtdIns-PLDα, PtdIns-PLDβ, and PtdIns-PLDγ. These enzymes convert [^{3}H]PtdIns(3,4,5)P$_3$ into a novel InsP, D-myo-[^{3}H]Ins(3,4,5)P$_3$ and PtdOH. In contrast to PtdCho-PLDs, these isozymes are predominantly associated with the cytosol. PtdIns-PLDα and PtdIns-PLDβ display a high degree of substrate specificity for PtdIns(3,4,5)P$_3$, with a relative potency of PtdIns(3,4,5)P$_3$ ≫ PtdIns(3)P or PtdIns(4)P > PtdIns(4,5)P$_2$. In contrast, PtdIns-PLDγ preferentially utilizes PtdIns(3)P as substrate, followed by, in sequence, PtdIns(3,4,5)P$_3$, PtdIns(4)P, PtdIns(3,4)P$_2$ and PtdIns(4,5)P$_2$. None of these enzymes react with PtdCho, PtdSer or PtdEtn. All the three PtdIns-PLDs are Ca^{2+}-dependent. Furthermore, all the three enzymes require sodium deoxycholate for optimal activities, while other detergents examined including Triton X-100 and Nonidet P-40 are inhibitory. Interestingly, PtdIns(3,4,5)P$_3$ was necessary for the PtdIns(4,5)P$_2$ stimulation of the basal enzyme activity. This stimulating effect (up to 2.5-fold) occurred only within a narrow range of PtdIns(4,5)P$_2$/PdtIns(3,4,5)P$_3$ molar ratios between 1:1 and 2:1. Excess amounts of PtdIns(4,5)P$_2$ either inhibited or had no effect on the PtdIns(3,4,5)P$_3$-metabolizing activity. Ching et al. (1999) have suggested that the inhibition was due to competition for enzyme binding by the PtdIns(4,5)P$_2$, which was a poor substrate, thereby counteracting its stimulating effect. It remains to be investigated, whether this stimulatory effect is mediated by direct enzyme activation or by affecting the PtdIns(3,4,5)P$_3$ packaging in lipid vesicles. Hodgkin et al. (1999) have reported the regulation of PLD activity in the detergent insoluble fraction of HL-60 cells by protein kinase C and small G proteins, while Hodgkin et al. (2000) have reported that the regulation and localization is dependent upon PtdIns(4,5)P$_2$-specific PH domain.

Banno et al. (1999) have shown in in vitro experiments that gelsolin inhibits recombinant PLD1 and PLD2 activities but not the oleate-dependent PLD, which requires PtdInsP$_2$ as a cofactor.

Likewise, sphingosine 1-phosphate, phorbol myristate acetate or Ca^{2+} ionophore A23187 induced PLD activation could not be modified by gelsolin. In contrast, gelsolin overexpression suppressed bradykinin induced activation of PLD. It was concluded that gelsolin modulates bradykinin-mediated PLD activation by suppression of PLC and PKC activities.

Complete PLD cDNA sequences are now known for 10 plant species (Wang, 2000). The cloning and expression of PLD cDNAs have demonstrated unambiguously that the hydrolytic and transphosphatidylation reactions are catalyzed by the same protein (Wang et al., 1994). All the cloned PLDs including plant PLDα, β and γ, mammalian PLD1a, b and PLD 2, and yeast PLD1, possess transphosphatidylation activity (Wissing et al., 1996). The characterization of the cloned PLDs has also led to the revelation of new regulatory and catalytic properties. Thus, it has been observed that PLDα is an acidic phospholipase, whereas PLDβ and PLDγ are neutral (Pappan and Wang, 1999).

Pappan et al. (1997b) have isolated from *Arabidopsis* a novel PLD activity, which is dependent on $PtdIns(4,5)P_2$ and nanomolar concentrations of Ca^{2+}. The enzyme activity of this $PtdInsP_2$-dependent enzyme has been designated as PLDβ in order to distinguish it from PLDα, the earlier characterized $PtdInsP_2$-independent PLD that requires millimolar Ca^{2+} for optimal activity. Sequence analysis reveals that PLDβ is evolutionary divergent from PLDα and that its N-terminus contains a regulatory Ca^{2+}-dependent phospholipid-binding domain that is found in a number of signal transducing and membrane trafficking proteins.

Both in vitro and in vivo systems are used to measure the activity of PLD.

9.4.3.1. In vitro assay
For in vitro assays, exogenous PtdCho labeled with either radioisotopes or fluorophores are commercially available. Some investigators have used endogenous substrates for in vitro reaction

by prelabeling cells and then preparing membranes that contain labeled phospholipids. The reaction products that are generated can be separated by a variety of chromatographic procedures.

Xie and Meier (2002) have provided a brief protocol for a PLD assay with membrane preparations, which utilizes a fluorescent substrate, BODIPY-PtdCho (BPC). BPC is a 1-alkyl-2-acyl GroPCho analog (Meier and Gibbs, 1999) and serves as a substrate for plant, yeast and mammalian PLD enzymes. The assay takes advantage of the unique property of PLDs to catalyze a transphosphatidylation reaction, in which a Ptd alcohol is produced as a stable product, when a source of PLD is incubated in the presence of alcohol (1–5% v/v); BPC is converted to BODIPY-Ptd alcohol. Thus, cells grown in 60 or 100 mm dishes are incubated with or without agonist, washed with phosphate buffered-saline (PBS), resuspended in ice-cold lysis buffer [20 M HEPES (pH 7.5), 80 mM β-glycerophosphate, 10 mM EGTA, 2 mM EDTA, 5 mM dithiothreitol (DTT)], sonicated, and sedimented at 100 000g for 20 min at 4°C. The membrane pellet is resuspended in lysis buffer before use in PLD assays. An aliquot of 1 mM BPC (Molecular Probes, Eugene, OR) in ethanol is dried under nitrogen and briefly sonicated in 500 μM octylglucoside, 400 mM NaCl, 66 mM HEPES (pH 7.0) before use to make a 250 μM solution. The reaction mixture (12.5 μl) contains 6 μl of membrane suspension (1–10 μg of protein) in lysis buffer, 5 μl; of 250 μM BPC, and 1.5 μl of 9% (v/v) butanol. The reaction is initiated by the addition of membrane protein and incubated for 60 min at 30°C. A 5 μl aliquot of the reaction mixture is loaded on a TLC plate (60 Å, 20 × 20 cm^2, without fluorescent indicator; EM Science, Gibbstown, NJ) and developed with chloroform/methanol/water/acetic acid (45:45:10:2, by vol.). Developed plates can be photographed on Polaroid-type film under UV light, imaged under UV light with digital camera system, or imaged and quantified with a STORM system (Molecular dynamics, Sunnyvale, CA). PLD activity is represented by the formation of BODIPY-Ptd butanol (B-PBt) and PA (Horwitz and Davis, 1993). Separate spots may be seen on the TLC plate in ascending order: lyso-B-PtdCho < B-PtdCho < B-PtdOH < lyso-B-PBt (BODIPY-lysoPtd butanol) < B-PBt (BODIPY-Ptd butanol) < B-MG (BODIPY-monoacylglycerol) < B-DG (BODIPY-diacylglycerol).

Alternatively, PLD may be assayed in an in vitro system that Frohman et al. (2000) have adapted from Brown et al. (1993). The assay consists of adding PLD and activators to a reaction mixture of lipid vesicles containing tritiated PtdCho (where the label is on the choline moiety) and buffer for a fixed period of time. At the end of the assay, the unhydrolyzed labeled lipid is pelleted using BSA to absorb the lipid and TCA used to precipitate it. The labeled choline released by PLD remains soluble in this setting, and scintillation determination of the supernatant accordingly indicates the amount of PLD activity that was present during the assay period. As an example, Frohman et al. (2000) utilized an assay mixture consisting of 50 µl lipids (tritiated PtdCho plus premade lipid mixture), 20 µl 5 × assay buffer, 10 µl PLD source, 1 µl RhoA (1–2 µg/ml, GTPγS preloaded), and 19 µl H_2O. Total reaction volume was 100 µl. The mixture is assembled on ice and incubated for 30 min at 37°C. After transferring to ice, 100 µl BSA (10 mg/ml) and 200 µl 10% TCA (in that order) are added to stop the reaction. The reaction mixture is centrifuged for 5 min at maximum speed and the supernatant transferred to scintillation vials, and counted. The lipid mixture is premade in the molar ratio of PtdEtn/PtdP$_2$/PtdCho (16:1.4:1) and 100 µM total lipid is used in each assay sample.

Hammond et al. (1995) have described in vitro assays of PLD using a protocol not unlike that of Brown et al. (1993) for release of choline. For analysis of phospholipid products, the vesicles were of standard composition except that approximately 10–20 × 10^3 dpm of [^{32}P]PtdCho, [^{32}P]PtdEtn, or [^{32}P]PtdIns (specific activity, 10 000 dpm/nmol) was substituted for the [^{3}H]PtdCho. Incubations were exactly as described by Brown et al. (1993a,b), except that transphosphatidylation assays contained 2% (v/v) EtOH. The assays were terminated by adding

222 μl of chloroform/methanol (1:1, v/v). After mixing and centrifugation, the lower phases were removed, dried under vacuum, and analyzed by TLC on oxalate-impregnated Whatman 60A silica gel plates in a solvent system of chloroform/methanol/ acetic acid (13:3:1, by vol.). Products were visualized by autoradiography and identified by their mobility relative to authentic standards.

The in vitro system is used to assess the interaction and direct activation of PLD by its regulators, which have been well characterized as important components in the signaling pathways mediated by G-protein-coupled receptors.

PtdIns-P_2-dependent PLD activity may be assayed in a cell-free system (Li and Fleming, 1999) containing phospholipid vesicles (PtdEtn/PtdnsP$_2$/PtdCho molar ratio 16:1.4:1) with a final concentration of 8.7(μM PtdCho prepared according to Brown et al. (1993a,b). Ten-microliter lipid vesicles was added to the assay system (50 mM HEPES, 80 mM KCl, 3 mM MgCl$_2$, 3 mM DTT, pH 7.0) which contained 2.5 μg cabbage PLD. The free calcium concentration was EGTA-buffered to 7.8 μM Ca^{2+} unless otherwise indicated. PLD activity is assayed either by the transphosphatidylation reaction in the presence of 1% ethanol, with dipalmitoyl-[2-palmitoyl-9,10-^{3}H$_2$]GroPCho as substrate as described previously (Li and Fleming, 1999), or radiolabeled choline-release with dipalmitoylGroP[methyl-^{3}H]Cho as substrate. The assays were conducted at 37°C for 20 min in a total assay volume of 60 μl. For the transphosphatidylation assay, ^{3}H-labeled PtdEtn was separated by TLC and quantified by scintillation counting as described previously (Li and Fleming, 1999). For the choline-release assay, the reaction is stopped by adding 360 μl chloroform/methanol (2:1, v/v). After vigorous vortexing and centrifugation at 6000g for 5 min radiolabeled choline was measured in 50 μl aliquots of the aqueous phase by scintillation counting.

Qin et al. (1997) have assayed the plant PtdInsP$_2$-dependent PLD activity using conditions described previously (Pappan et al., 1997a,b) with some modifications. The basic assay mixture contains 100 mM MES (pH 7.0), 0.5 mM Mg_2Cl, 80 mM KCl, 0.4 mM lipid vesicles, and 30 μg of Bluescript SK-expressed protein in a total volume of 100 μl. $CaCl_2$ is added to the reaction at the concentrations noted below. The lipid vesicles contained 35 nmol of PtdEtn, 3 nmol of $PtdInsP_2$, and 2 nmol of PtdCho. PLD-mediated hydrolysis of PtdCho was measured using either 1-palmitoyl-2-[1-^{14}C]oleoylGroPCho or dipalmitoyl-GroP[methyl-^{3}H]Cho as substrate. The reaction was initiated by the addition of substrate and incubation at 30°C for 30 min in a shaking water bath. Cho, PtdOH, and PtdEt produced in the PLD reaction were separated and quantified as described previously (Pappan et al., 1997a,b). In the phospholipid activation experiments, $PtdIns-P_2$ was replaced with 3 mmol of PtdEtn, PtdOH, PtdGro, P, PtdIns or PtdInsP.

Ching et al. (1999) performed the catalytic activity and substrate specificity study by determining the radioactive InsPs liberated by the enzymatic hydrolysis of [^{3}H]PtdIns(3,4,5)P₃, [^{3}H]PtdIns(3,4)P₂, [^{3}H]PtdIns(3)P and [^{3}H]PtdIns(4)P by HPLC. The enzyme incubation was effected by exposing the radioactive phospholipids (0.8 μg; total radioactivity, 0.2 μC) in PtdEtn (40 μg) and PtdSer (5 μg) suspended in 1 ml of 20 mM HEPES, pH 7, containing 120 mM KCl, 10 mM NaCl, 2 mM EGTA, and 0.8 mM sodium deoxycholate to the enzyme. The suspension was sonicated for 5 min and the reaction was initiated by adding 50 μl of the phospholipid mixture and incubated at 37°C for 30 min. The reaction was terminated by adding 100 μl of $HClO_4/CHCl_3$ (1:0.33, v/v) followed by 100 μl of 10 g/ml BSA. The mixture was centrifuged at 13 000g for 10 min. The supernatant was collected and was extracted immediately with 300 μl of tri-n-octylamine/Freon (1:1, v/v) twice to remove $HClO_4$. The neutralized solution was transferred to a new vial, and the radioactive Ins-P was analyzed by HPLC on an Adsorbosphere Sax column equilibrated with water. The InsPs were eluted with a linear gradient of 0–0.9 M $NH_4H_2PO_4$ in 60 min at a flow rate of 1 ml/min. Fractions were collected for every 1 ml and their radioactivity was measured by liquid scintillation. Synthetic [^{3}H]Ins(1,3,4,5)P$_4$, [^{3}H]Ins(3,4,5)P$_3$, [^{3}H]Ins(1,4,5)P$_3$, [^{3}H]Ins(4,5)P$_2$, [^{3}H]Ins(3,4)P$_3$, [^{3}H]Ins(4)P and [^{3}H]Ins(3)P were used as standards. The relative retention times were 60, 50, 48, 43, 41, 31 and 29 min, respectively.

9.4.3.2. In vivo assay

In this PLD assay, PLD substrates are labeled with radioisotope in intact mammalian cells. Meier and Gibbs (1999) have provided a detailed description of this procedure. Briefly, cells growing in six-well plates are incubated with ^{3}H-labeled fatty acid (e.g. palmitic acid) for 10–24 h. The labeled cells are then incubated with 0.5% (v/v) ethanol and the desired experimental agents in Dubelco's modified Eagle's medium supplemented with 10 mM HEPES (pH 7.5) at 37°C in a cell culture incubator for the desired time. Cellular lipids are extracted by adding 1 ml of methanol/HCl (50:2, v/v), followed by 0.5 ml of chloroform and 0.28 ml of 1 M NaCl. Dried lipid samples and PtdEt standard are loaded on TLC plates (60 Å, 20 × 20 cm^2, 250 μm thickness, 19 channels, with preabsorbent strip, without fluorescent indicator; Whatman, Clifton, NJ) and then separated with a solvent consisting of the upper phase of ethyl acetate/trimethylpentane/acetic acid/water (90:50:20:100, by vol.). Spots containing PtdEt, as well as the remainder of each sample lane, are scraped and quantified by liquid scintillation spectrometry.

Frohman et al. (2000) also describe an in vivo assay, which they have adapted from Wakelam et al. (1997) for determination of PLD activity in COS-7, HEK-293 cells. Briefly, the cells (COS-7, HEK-293) are plated to density of about 10–20% confluence in DMEM containing Pen-Strep, glutamine, 10% FCS in 35 mm TC dishes. On the second day, the cells are transfected using Lipofectamine PLUS in 1 ml Opti-mem media with 1 μg plasmid per dish and after 4 h, replace the transfection medium with 1 ml of media containing 10% FCS containing [^{3}H]palmitate label. After 24 h, the labeling medium is replaced with 2 ml of warm, fresh Opti-mem media for 1–2 h, following which the media is replaced with fresh Opti-mem media containing 0.3% butanol and any combinations of stimulators (100 nM final concentration). At the end of the treatment (15–30 min), the stimulating medium is replaced with 300 μl ice-cold methanol and the dishes are placed on ice. The cells are scraped off using a rubber policeman and the plates are washed and rescraped with a second 300 μl of cold methanol, which is combined with the first methanol extract. The methanol extract is mixed with 0.5 ml of chloroform, vortexed well, and after 25 min at room temperature diluted with 0.4 ml water. The mixture is vortexed and lower phase collected and dried by speed-vac centrifugation and resuspended in 2 × 25 μl of chloroform/methanol (19:1, v/v) containing 50 μg of authentic PtdBt (dipalmitate; Avanti Polar Lipids, Alibaster, AL) for spotting on TLC plates. The plates are developed in the upper phase of a mixture of ethyl acetate/isooctane/acetic acid/water (110:50:20:100, by vol.; R_f of PtdBt = 0.4). The location of the TLC spots is obtained by iodine staining. The radioactivity is determined by scintillation counting of the powder.

In addition, PtdIns(4,5)P$_2$ stimulates in a dose-dependent manner the newly discovered PtdIns specific PLDs isozymes (Ching et al., 1999). The enhancement in the enzyme activity was noted only when the molar ratio of PtdIns(4,5)P$_2$ to PtdIns(3,4,5)P$_3$ was between 1:1 and 2:1(see below). The relative affinity of the isomeric PtdIns-PLD for different PtdInsPs, however, does not appear to have been determined (awaiting more recent results).

Divecha et al. (2000) assayed PLD in vivo as follows: Transfected cells were left for 24 h and labeled overnight with [^{32}P]Pi (10 μCi) in phosphate-free Dubelcco's modified Eagle's medium (DMEM). The cells were washed with RPMI-salts buffer and incubated in this buffer containing butan-1-ol (0.3% v/v) for 30 min. This medium was aspirated, and 0.45 ml of 2.4 M HCl added. The cells were maintained and scraped on ice, removed to clean Eppendorf tube and the dishes washed with 0.5 ml of methanol, which was pooled with HCl. Half-a-milliliter of chloroform containing 5 μg of Folch lipid extract was added, together with 0.25 ml of water. The two phases were mixed vigorously, centrifuged and the lower phase was removed, and was washed once with theoretical upper phase (chloroform/methanol/ 1M HCl, 15:245:235, 0.7 ml), with the lower phase being removed to a clean Eppendorf tube. After further washing and centrifugation and combining the first and second lower phases, the samples were dried and kept at − 20°C. Samples were analyzed for PtdBut formation and for PtdIns/PytdInsP/PtdInsP$_2$ labeling by TLC. Either using an acidic (chloroform/methanol/acetone/glacial acetic acid/water 240:78:72:70:42, by vol.) or an alkaline solvent (chloroform/methanol/ammonia (28%)/water 45:35:2:8, by vol.).

9.5. PtdIns(3,4,5)P$_3$ activated phospholipases

Like PtdInsP$_2$, PtdInsP$_3$ can serve both as substrate and as an activator of the phospholipases. This activation is believed to be important in lipid signaling.

9.5.1. Phospholipase A$_2$

Mosior et al. (1998) have reported that Group IV cPLA$_2$, which binds with high-affinity and specificity to PtdIns(4,5)P$_2$ resulting in dramatic increase in activity, is also activated by the structurally similar PtdIns(3,4,5)P$_3$ although to a lesser extent (roughly 60% of the level of PtdIns(4,5)P$_2$). Furthermore, PtdIns(3,4)P$_2$ activated cPLA$_2$ to 63% of the activity of PtdIns(4,5)P$_2$, but PtdIns(3)P only 32% that of PtdIns(4)P. Also the soluble head groups of both PtdIns(4,5)P$_2$ and PtdIns(3,4,5)P$_3$ were investigated on cPLA$_2$ activity and its PtdIns(4,5)P$_2$ activation. Soluble Ins(1,4,5)P$_3$ and Ins(1,3,4,5)P$_4$ were added at a 10- and 5-fold molar excess, respectively, relative to PtdIns(4,5)P$_2$ and shown not to have any effect on the PtdIns(4,5)P$_2$ activation of cPLA$_2$. Additionally, the head groups showed no appreciable activation of cPLA$_2$. The significance of the limited activation of cPLA$_2$ by PtdIns(3,4,5)P$_3$ therefore remains to be established. The activation of cPLA$_2$ by PtdIns(3,4,5)P$_3$ was assayed as described above for PtdIns(4,5)P$_2$ activation (Mosior et al., 1998) Table 9.2 shows that cPLA$_2$ selectively releases arachidonic acid from sn-2-position of phospholipids (Clark et al., 1991).

9.5.2. Phospholipase C

Signal transduction across cell membranes often involves the activation of both PtdIns-specific PLC and PtdIns 3-kinase. PtdIns(4,5)P$_2$, a substrate for both enzymes, is converted to PtdIns(3,4,5)P$_3$ by the action of PtdIns 3-kinase. Other characterized mechanisms for the activation of PLC isozymes include the

TABLE 9.2

Fatty acid selectivity of cPLA$_2$

Fatty Acids	Composition of Fatty Acids Liberated by cPLA$_2$ (%)[a]			Composition of fatty acids in sn-2-position (%)[b]
	Time (min)			
	2	5	15	
Arachidonic	83.6 ± 0.9	77 ± 0.8	70.4 ± 2	19 ± 2.0
Linoleic	1.2 ± 0.6	4.0 ± 0.8	3.8 ± 0.2	6.8 ± 1.0
Oleic	16 ± 0.4	17.7 ± 0.4	19.9 ± 0.8	69 ± 5.0
Palmitic	ND	1.3 ± 1.3	6.0 ± 2.5	3.0 ± 3.0

[a]cPLA$_2$ was incubated with U937 membranes in pH buffer containing 1 μM free Ca^{2+}, and the fatty acids released at each time point were analyzed by gas chromatography.
[b]No free palmitic acid could be detected after 2 min incubation. The values shown are mean ± SE ($n = 5$). Reproduced from Clark et al. (1991) with permission of the publisher.

phosphorylation of PLC-γ isoforms by protein-tyrosine kinases and the interaction of PLC-β isozymes with G-proteins (Rhee and Bae, 1997). PLC-γ1 is a widely expressed and abundant enzyme, whereas PLC-γ2 is abundant in cells of hematopoietic origin.

Bae et al. (1998a,b) have shown that PtdIns(3,4,5)P$_3$ activates purified PLC-γ isozymes by interacting with their Src homology 2 domains. The sigmoidal response to PtdIns(3,4,5)P$_3$ also suggests that PLC-γ activation would be minimal until the lipid concentration exceeds a certain threshold. More than 60 different receptors are known to stimulate PLC (Rhee and Bae, 1997).

Scharenberg et al. (1998) have directly demonstrated that PtdIns(3,4,5)P$_3$ interacting with the PH domain acts as an upstream activation signal for Tec kinases, resulting in Tec kinase-dependent PLC-γ tyrosine phosphorylation and InsP$_3$ production. More specifically, the results of Scharenberg et al. (1998) suggest that PtdIns(3,4,5)P$_3$ initiates PLC-γ2 dependent InsP$_3$ production at least in part through its ability to interact with and activate Tec kinases. In a more recent study, Pasquet et al. (2000) have demonstrated that PtdIns(3,4,5)P$_3$ regulates Ca^{2+} entry via tyrosine kinase Btk in platelets and megakaryocytes without increasing PLC activity. The mechanism through which PtdIns(3,4,5)P$_3$ and Btk regulate Ca^{2+} entry in platelets/megakaryocytes remains to be clarified.

Bae et al. (1998a,b) measured the activities of PLC-β1, PLC-γ1, PLC-γ2, and PLC-δ1 using a mixed micellar substrate containing [^{3}H]PtdIns(4,5)P$_2$, PtdEtn, and PtdSer in a molar ratio of 1:3:3 together with various amounts of synthetic dipalmitoyl-GroPIns(3,4,5)P$_3$ or dipalmitoyl-GroPIns(3,4)P$_2$ in 0.1% deoxycholate. The final assay mixture (100 μl) contained 10 μM [^{3}H]PtdIns(4,5)P$_2$ (26 000 cpm), 50 mM HEPES-NaOH (pH 7.0), 10 mM NaCl, 120 mM KCl, 2 mM EGTA, 0.05% deoxycholate, BSA (5 μg/ml), 1 μM free Ca^{2+}, and the indicated concentrations of PtdIns(3,4,5)P$_3$ or PtdIns(3,4)P$_2$. After incubation for 10 min at 30°C, the reactions were terminated by the addition of 200 μl of

10% (w/v) TCA and 100 μl of 10% (w/v) BSA, followed by centrifugation. The amount of PLC isozymes (4–7 ng) was adjusted to give similar basal activity.

The PtdIns(3,4)P$_2$ and PtdIns(3,4,5)P$_3$ are not substrates of any known PLC (Serunian et al., 1989) and are normally absent from resting cells. However, they appear within seconds to minutes of stimulation of cells with various growth factors or other cellular activators. In contrast, the concentration of PtdIns(3)P does not change substantially in response to cell stimulation.

The assay for DAG produced is performed according to Divecha et al. (1991) using DAG kinase enzyme purified from rat brain. DAG was extracted from the tissue preparation, dissolved in 20 μl of CHAPS (9.2 mg/ml), and sonicated at room temperature for 15 s. After the addition of 80 μl of reaction buffer (50 mM Tris acetate, pH 7.4, 80 mM KCl, 10 mM magnesium acetate, 2 mM GTA), the assay was started by the addition of 20 μl of DAG kinase enzyme followed by 80 μl of reaction buffer containing 5 μM ATP, and 1 μCi of [γ-^{32}P]ATP. Incubation was for 1 h at room temperature; then PtdOH was extracted, chromatographed, and autoradiographed, and its radioactivity was counted in a liquid scintillation system. Standard curves are obtained as reported by Divecha et al. (1991), using 1,2-dioleoyl-3-palmitoyl-sn-glycerol as substrate.

Glycosyl PtdIns-specific PLD from mammalian serum is a 115 kDa glycoprotein consisting of 816 amino acids. Stadelmann et al. (1997) found that C-terminal deletions of only two to five amino acids reduced enzymatic activity by 70% compared to wild type protein. Stambuk and Curdoso de Almeida (1996) have described an assay for glycosyl PtdIns-anchor degrading phospholipases.

9.5.3. Phospholipase D

PtdInsP cofactor was initially used to distinguish special PLDs from the conventional PLDs. More recent work has shown that this requirement varies with the assay conditions (Pappan et al., 1997a,b). PLDα activity is known to be independent of PtdInsPs when Ca^{2+} is present at or near optimal stimulating levels (Wang, 2000). Earlier, Liskowitch et al. (1994) and Pertile et al. (1995) had shown that PtdCho-PLDs were strongly stimulated by PtdIns(4,5)P$_2$ and PtdIns(3,4,5)P$_3$ with an equal potency (see above). PLDβ, PLDγ and PLDα are known to bind PtdInsP$_2$ in the above order of binding affinity. Two types of PtdInsP$_2$-binding sites are present on PLD. One is provided by the PtdInsPs binding motif rich in basic residues, (R/K)xxxx(R/K)x(R/K)(R/K), near the second HKD motif (Qin et al., 1997). The differences in PtdInsP$_2$ requirements, primary structures, and dose dependence suggests that PLDα interacts with inositol phospholipids differently from PLDβ and PLDγ. The requirement of PtdInsPs for activity and the presence of PtdInsP-binding sites on PLD suggest that PLD activation is interconnected with the metabolism and signaling of PtdInsPs.

Pappan et al. (1997a,b) assayed the PtdnsP$_2$-dependent PLD activity by using either 1-palmitoyl-2-oleoyl-[oleoyl-1-^{14}C]glycero-3-P-Cho or dipalmitoylglycero-3-P-[methyl-^{3}H]Cho as substrates. The acyl labeled PtdCho was used for assaying transphosphatidylation activity whereas the choline-labeled PtdCho was used in all other studies. In both cases, 2.5 μCi of radiolabeled PtdCho was mixed with 3.5 μmol of PtdEtn, 0.3 μmol PtdInsP$_2$, and 0.2 μmol of unlabeled PtdCho in chloroform, and the solvent evaporated under a stream of N$_2$. In the phospholipid specificity experiments, PtdInsP$_2$ was replaced with 0.3 μmol of Ptdtn, PtdGro, PtdSer, PtdIns, or PtdInsP. The phospholipid substrate was dispersed in 1 ml H$_2$O by sonication at room temperature. Previously reported conditions were adopted to yield an enzyme assay mixture that contained 100 mM MES (pH 7.0), 5 μM CaCl$_2$, 2 mM MgCl$_2$, 80 mM KCl, 0.4 mM lipid vesicles, and 5–25 μg of expressed protein in a total volume of 100 l (Brown et al., 1993a,b). In the Ca^{2+} dependence experiments, the concentrations of free Ca^{2+} and Mg^{2+} in the reaction mixture were determined using Ca^{2+}/Mg^{2+}-EGTA buffers at pH 7.5.

The reaction was initiated by the addition of substrate and incubation at 30°C for 30 min in a shaking water bath. When choline-labeled PtdCho was used, the reaction was stopped by the addition of 1 ml of 2:1 (v/v) chloroform/methanol and 100 μl of 2 M KCl. After vortexing and centrifugation at 12 000g for 5 min, a 200 μl aliquot of the aqueous phase was mixed with 3 ml of scintillation fluid, and the release of [^{3}H]choline was counted.

When acyl-labeled PtdCho was used, the reaction mixture included ethanol to a final concentration of 0.5% (v/v) for assaying the transphosphatidylation activity of PLD. The reaction was stopped by adding 375 μl of 1:2 (v/v) chloroform/methanol. Additionally, 100 μl of chloroform and 100 μl of 2 M KCl were added and the sample was vortexed. The chloroform and aqueous phases were separated, the aqueous phase removed and chloroform phase dried. TLC was conducted as described previously using 65:35:5 (by vol.) chloroform/ethanol/ammonia as the developing solvent (Wang et al., 1993). Lipids separated on the TLC plates were visualized by exposure to iodine vapor.

Ching et al. (1999) assayed the PtdIns-PLD activity in all enzyme preparations by monitoring the liberation of [^{3}H]-labeled PtdIns head group from [1-^{3}H]PtdIns(3,4,5)P$_3$ into the medium. [1-^{3}H]PtdIns(3,4,5)P$_3$ (0.8 μg; total radioactivity, 0.2 μCi), PtdIns (40 μg), and PtdSer (5 μg) were suspended in 1 ml of 20 M HEPES, pH 7, containing 120 mM KCl, 10 mM NaCl, 2 mM EGTA, and 0.8 mM sodium deoxycholate. Various amounts of CaCl$_2$ were added to the mixture before assays, and the free Ca^{2+} concentration was calculated by a computer program. The suspension was sonicated in a water bath-type sonicator for 5 min, and mixed vigorously with a vortex mixer before assays. Various enzyme preparations (10 μl) were incubated with 40 μl of the aforementioned HEPES buffer. The reaction was initiated by adding 50 μl of the phospholipid mixture, incubated at 37°C for 30 min, and stopped by adding 200 μl of 10% TCA and 150 μl of 10 mg/ml BSA. The mixture was centrifuged at 15 000g for 5 min, and the radioactivity in the supernatant was measured by liquid scintillation. The composition of the control was identical to that mentioned above except that the enzyme preparation was replaced by an equal amount of distilled water.

In order to assess the substrate specificity of the PtdIns-specific PLD, Ching et al. (1999) used HPLC to identify and quantify the radioactive InsPs liberated by the enzymatic hydrolysis of $[^3H]PdtIns(3,4,5)P_3$, $[^3H]PtdIns(3,4)P_2$, $[^3H]PtdIns(4,5)P_2$, $[^3H]PtdIns(3)P$, and $[^3H]PtdIns(4)P$. The enzyme incubation, under the same conditions as described above for $[^3H]PtdIns(3,4,5)P_3$, was terminated by adding 100 μl of $HClO_4$/$CHCl_3$ (1:0.33, v/v) followed by 100 μl of 10 mg/ml BSA. The mixture was centrifuged at 13 000g for 10 min. The supernatant was collected and was extracted immediately with 300 μl of tri-n-octylamine/Freon (1:1, v/v) twice to remove $HClO_4$. The neutralized solution was transferred to a new vial and the radioactive InsP was analyzed by HPLC on an Adsorbosphere Sax column (5 μm; 4.6 × 200 mm^2) equilibrated with H_2O. The InsP was eluted with a linear gradient of 0–0.9 M NH_4PO_4 in 60 min at a flow rate of 1 ml/min. Fractions were collected for every 1 ml, and their radioactivity was measured by liquid scintillation. Synthetic $[^3H]Ins(1,3,4,5)P_4$, $[^3H]Ins(3,4,5)P_3$, $[^3H]Ins(1,4,5)P_3$, $[^3H]Ins(4,5)P_2$, $[^3H]Ins(3,4)P_2$, $[^3H]Ins(4)P$ and $[^3H]Ins(3)P$ were used as standards. The respective retention times were 60, 50, 48, 43, 41, 31, and 29 min, respectively.

The cells are washed three times with physiological saline to remove excess radioactivity, and then cell stimulants are added. At the end of the incubation, cells are quenched and extracted with chloroform/methanol. The lipids are separated from the labeled water-soluble products by the addition of chloroform and dilute HCl. Both lipid and aqueous phases may be counted by scintillation, or be analyzed by choosing an appropriate chromatographic method.

CHAPTER 10

Preparation of standards

10.1. Introduction

Most InsPs and PtdInsPs have now been synthesized by chemical or combined chemical and enzymatic procedures. These procedures, however, require skilful staff and are not likely to be readily employed in a biochemical laboratory, although they might provide an access to these compounds for commercial distribution. Furthermore, metabolic studies require radiolabeled compounds, which must be prepared in the biochemical laboratory not equipped to handle involved chemical transformations. Biochemical preparations of phosphatide standards therefore must rely on low yielding metabolic transformations. In cells labeled with [^{3}H]Ins, for example, PtdIns(3)P is labeled to < 1% the extent of PtdIns(4)P (Lips et al., 1989). In addition to small abundance of the metabolites, the study is made difficult by the isomeric nature of the metabolites, the overall complexity of the structures involved, and the general lack of commercial standards. It must be pointed out that conventional methods have frequently failed to separate isomeric PtdInsPs and InsPs. Thus, Auger et al. (1990) point out that the initial studies with PtdIns(3)P, PtdIns(3,4)P$_2$ and PtdIns(3,4,5)P$_3$ were badly handicapped by the inability to resolve the novel isomers from the well-known and more abundant PtdInsPs. The PtdIns(3)P and PtdIns(4)P co-migrate nearly all one- and two-dimensional TLC systems. Likewise, PtdIns(3,4)P$_2$ and PtdIns(4,5)P$_2$ also co-migrate in conventional TLC systems.

Since that time, adequate methods for the separation and detection of these isomers have became available (Auger et al., 1990).

As a result, it has been necessary to develop more efficient methods for extraction and chromatographic resolution of the radioactive markers prepared. In the following the preparation of [^{3}H] and [^{14}C]Ins labeled [^{32}P]P$_i$ labeled PtdIns is described as a practical aid to the analytical and metabolic studies with PtdInsPs. In many instances, the preparation of the PtdInsPs also serves as a route to the preparation of individual InsPs, which is also discussed in detail. Some non-radioactive and selected radioactive reference PtdIns, PtdInsP and PtdInsP$_2$ as well as the GroPIns, GroPInsP and GroPInsP$_2$ have become commercially available in recent years.

The inositol head group of PtdIns has five free hydroxyl groups, but only three of them become phosphorylated in cells, in different combinations. The 2 and 6-positions have not been found to be phosphorylated, although such phosphorylations have been accomplished chemically.

10.2. Phosphatidylinositol phosphates

The preparation of PtdInsPs is usually limited to the isolation of a specific PtdInsP class. The composition of the component fatty acids or molecular species is ignored or assumed to represent the general composition, e.g. mainly *sn*-1-stearoyl/*sn*-2-arachidonoyl GroPIns. This, assumption, however, may not be always justified, especially when dealing with PtdIns of plant and microbial origin, which are known to be rich in other molecular species. Furthermore, enzymatic transformations may lead to selective utilization of different molecular species and the final product may not represent the composition of the original substrate. There exist, however, chromatographic (Kuksis and Myher, 1990; Wiley et al., 1992; Zhu and Eichberg, 1993) and mass spectrometric (Cronholm et al., 1992; Gunnarsson et al., 1997; Schiller et al., 1999; Hsu and Turk, 2000) methodologies that permit the characterization of the molecular species provided sufficient mass is available. The preparation of the individual PtdInsPs includes verification of purity and characterization of the chemical identity, which is usually accomplished by highly selective chromatographic methods that may involve deacylation and deglyceration of the product. The stereochemistry of the product may be assayed by enzymatic means. As a result, some of the preparations have been amplified to include the preparation of the deacylation products, while the characteristics of the deglyceration products may be found described under the preparation of inositol phosphates (see below).

10.2.1. PtdIns monophosphates

10.2.1.1. PtdIns(4)P

Low (1990) has modified the original procedure of Hendrickson and Ballou (1964) for the purification of PtdIns(4)P and PtdIns(4,5)P_2 from natural sources. Briefly, bovine brain Folch fraction I (1 g) is dissolved in 12 ml of chloroform. The PtdInsPs are precipitated by mixing with 22 ml of methanol followed by centrifugation at 1000g for 5 min. The supernatant is discarded, and the pellet redissolved in 12 ml of chloroform. This precipitation with methanol is repeated a total of six times. The final methanol precipitate is dissolved in 15 ml of chloroform, 15 ml of methanol, 5 ml of 1 M HCl. Additional 1 M HCl (8.5 ml) is added to give two phases. After centrifugation the upper phase is discarded and, to the lower phase, 15 ml of methanol and 5 ml of 2 M NaCl are added and the sample mixed until it becomes clear. The phases are separated by the addition of 2.5 ml of 2 M NaCl and 6 ml of H_2O. The upper phase is removed and the lower phase washed once more with methanol and 2 M NaCl as above. All the steps are done at 0–4°C. The lower phase is transferred to a clean tube, evaporated to dryness under nitrogen gas, the lipids redissolved in 2 ml of chloroform and stored at −20°C until required. One gram of starting material yields 150–250 μmol of precipitated phospholipid (as determined by organic phosphorus).

The chromatographic purification step is best performed (Low, 1990) by preparative HPLC on an amino column. The brain phospholipids (prepared as described above) dissolved in 5 ml of solvent A (chloroform/methanol/H_2O, 20:9:1, by vol.) are applied at a flow rate of 2.5/min to a 10 mm × 250 mm amino-NP column (5 μm spherical silica with n-propylamine bonded phase (IBM Instruments, Danbury, CT) with a 4.5 mm × 50 mm guard column (same packing material) equilibrated with solvent A. The column is eluted at a flow rate of 2.5 ml/min with 25 ml of solvent A, a 125 ml gradient of 100% solvent A to 100% solvent B (solvent A containing 0.6 M ammonium acetate), and 100 ml of 100% solvent B; 5 ml fractions being collected. The fractions are assayed for organic phosphorus. Average recovery of organic phosphorus is approximately 80%. The PtdIns(4,5)P_2 is eluted last and is preceded by PtdIns(4)P, as determined by TLC. The HPLC procedure is superior in resolution, speed and overall convenience to chromatography on a DEAE-cellulose column, which also has been used for this purpose.

The concentration of the fractions is critical and Low (1990) has proposed an appropriate routine for this purpose. The volume of the pooled fractions is adjusted to 18 ml with solvent A. To the 18 ml portion (in a 50 ml glass tube), methanol (6.6 ml) and 2 M NaCl (10.2 ml) are added to separate the phases. The lower phase is washed with 12 ml of methanol and 10.8 ml of 2 M NaCl. The above steps are done at 0–4°C. The lower phase is removed, evaporated to dryness under nitrogen gas in a clean tube and dissolved in 5 ml of chloroform, 0.7 ml methanol, 0.05 ml of H_2O. The purified PtdInsPs (in contrast to crude starting material) do not dissolve readily in chloroform or chloroform/methanol mixtures. Addition of small amount of water is therefore essential to dissolve the purified lipids at the 2–3 mM level. The composition of the final solvent given here was chosen to minimize the amount of methanol and H_2O so that the solvent can be rapidly removed by evaporation for preparation of substrates. The procedure produces approximately 7–15 μmol of PtdIns(4)P.

Low (1990) has utilized platelets as a convenient source of PtdIns(4)[^{32}P]P. Platelets prepared from 60 ml of fresh human blood were incubated in 3 ml of 140 mM NaCl, 5 mM KCl, 0.05 mM $CaCl_2$, 0.1 mM $MgCl_2$, 16.5 mM glucose, 0.1 mg bovine serum albumin/ml, 15 mM HEPES (pH 7.4), containing 2–6 mCi of [^{32}P]P_i (carrier-free, ICN Radiochemicals, Irvine, CA) for 2 h at 37°C with occasional gentle shaking. Prostaglandin E_1 (PGE$_1$) (2.8 μM) is also added, since it presents activation and increases the yield of Ins[^{32}P]Ps. The lipids are extracted with 11.25 m of chloroform/methanol/HCl (50:100:1:3, by vol.) for 30 min at 25°C. Chloroform (3.75 ml) and 0.1 M HCl (3.75 ml) are then added to separate the phases. The lower phase is washed twice with 5 ml of methanol, 4.5 ml of 2 M NaCl, 0.5 ml of 100 mM EDTA–NaOH (pH 7.4) and then evaporated to dryness in a clean tube with approximately 1.5 μmol of acid-washed, methanol-precipitated bovine brain PtdInsPs (material prepared as above) to act as a carrier. The extract is finally redissolved in 0.5 ml of solvent A (chloroform/methanol/H_2O, 20:9:1, by vol.) and applied at a flow rate of 1 ml/min to a 4.5 mm × 250 mm amino-Np column (same packing as column mentioned above) equilibrated in solvent A. The column is eluted with 5 ml of solvent A, a 25 ml linear gradient from 100% solvent A to 100% solvent B, and 20 ml of solvent B. One milliliter of fractions is collected and radioactivity is determined by liquid scintillation counting. Average recovery is approximately 70%. Peak fractions are pooled, the volume adjusted to 6 ml with solvent A and methanol (2.2 ml) and 3.4 ml of 1 M NaCl added to separate the phases. The extraction is completed as described for the isolation of the unlabeled PtdIns(4)P using proportionally smaller solvent volumes. Approximately 2–6 μCi of [^{32}P]PtdIns(4)P and 1–3 μCi of [^{32}P]PtdIns(4,5)P_2 are obtained by this procedure.

Letcher et al. (1990) have provided a detailed description for preparation of Ptd [^{14}C] Ins(4)P from PtdIns(4)P using human erythrocyte PtdIns 4-kinase. Briefly, PtdIns (74 kBq (2 μCi)) radiolabeled with ^{14}C in the inositol ring, specific activity >8.1 GBq/mol, is dried down and resuspended in 50 μl of 4.8% (w/v) Triton X-100. The incubation is carried out in 400 μl giving a final concentration of 15 mM 2-mercaptoethanol, 0.1 mM PMSF, 1 mg bovine serum albumin/ml, 5 mM ATP, 10 mM creatine phosphate, 5 units creatine phosphokinase/ml, 80 mM KCl, 6 mM MgCl$_2$, 50 mM HEPES (pH 7.0) and 2 mM EGTA. This mixture is made up in 325 μl to which is added the [^{14}C]PtdIns in 4.8% Triton X-100 and 25 μl packed ghosts. Incubation is continued for 90 min at 37°C. The reaction is stopped by adding 3.75 vol. (1.5 ml) chloroform/methanol/12 M HCl (200:400:1, by vol.) and 200 nmol of unlabeled PtdIns(4)P as a carrier. After the mixture has stood for 10 min, 0.5 ml of chloroform, and then 0.5 ml of 0.1 M HCl are added to form two phases. After being shaken, the tube is spun in a bench top centrifuge, the upper phase and interface are removed by aspiration and the lower phase is washed with (1 mM EDTA, 0.1 M HCl)/methanol (0.9:1, by vol.). The lower phase is evaporated to dryness and loaded on to a TLC plate (Kieselgel 60F254 20 × 20 cm obtained from Merck) that has been sprayed with 1% (w/v) potassium oxalate and held at 100°C for 60 min. External standards of PtdIns and PtdInsP are also spotted on the plate. The plate is developed in chloroform/methanol/4 M ammonium hydroxide (9:7:2, by vol.) for 60 min and is allowed to dry. The sample is covered with a glass plate while the PtdIns and PtdIns standard lanes are sprayed with a phosphate detecting spray (Vaskovsky and Kostetsky, 1968).

Letcher et al. (1990) have provided a detailed description of the preparation of GroP[^{14}C]Ins(4)P by deacylation of [^{14}C] PtdIns(4)P. Thus, the area of the TLC lane of the sample corresponding to PtdInsP is scraped off and transferred to a suitable vessel for deacylation. Deacylation of the [^{14}C]PtdIns(4)P is carried out by adding 3 ml of monomethylamine reagent

(Clarke and Dawson, 1981) directly to the TLC scrapings. The presence of silicic acid affected neither the deacylation reaction nor the subsequent recovery of [^{14}C]GroPIns(4)P. The deacylation step is carried out for 60 min at 53°C, after which the monomethylamine reagent is removed by drying in a rotary evaporator, starting at room temperature and gradually raising the temperature to 50°C. The residue is dissolved in 2 ml of H_2O and 2.4 ml of n-butanol/ petroleum ether (b.p. 40–60°C/ethyl formate, 20:4:1, by vol.) is added. After mixing, the two phases are allowed to separate. The upper phase is removed by aspiration and the lower phases are further washed with 1.5 ml of the n-butanol/petroleum ether/ethyl formate mixture. The final lower phase is then dried in vacuo.

10.2.1.2. PtdIns(3)P

Walsh et al. (1991) have reported a method for synthesis of PtdIns(3)P by chemical isomerization of PtdIns(4)P. The method employs carbodiimide to promote migration of phosphate from the 4-position of inositol via a cyclic phosphodiester intermediate. The PtdIns(3)P and PtdIns(4)P are resolved by TLC. Briefly, PdIns(4)P (5 mg, 5.0 μmol) is dissolved in 2.5 ml of $CHCl_3$/methanol (3:2, v/v). Addition of 5 μl of 1.0 M HCl is required to achieve solution. Five microcuries (1 Ci = 37 GBq) of [^{3}H]PtdIns(4)P and 5 μl of ethoxyquin (purified grade) are then added. This mixture is passed over a 0.5 ml column of pyridinium Dowex 50W-X8 that was equilibrated with CH_2Cl_2/methanol/pyridine (5:4:1, by vol.) and eluted with 4 ml of the same solvent. The solvents are removed on a rotary evaporator. Traces of solvents are removed by evaporation of an additional 5 ml pyridine. The syrupy residue is then dissolved in 2.5 ml of pyridine containing 5.0 mM diisopropylethylamine. The reaction was initiated by addition of 12.5 μmol of N,N'-diisopropylcarbodiimide (62.5 μl of a freshly prepared 200 mM solution in tetrahydrofuran). The reaction vessel is capped with a rubber septum, purged with argon, and stirred magnetically in the dark at room temperature for 2 days. At the end of this period, the pyridine is removed by rotary evaporation, and the residue is dissolved in 5 ml of 0.1 M HCl in tetrahydrofuran/water (9:1, v/v). This solution is allowed to stand at room temperature for 1 h to hydrolyze the cyclic phosphates and any phosphorylurea. The solution is then concentrated to 50 µl by rotary evaporation and dissolved in 1.0 ml of CHCl$_3$. The CHCl$_3$ solution is extracted twice with 1.0 ml of 1.0 M HCl in methanol/water 1:1 (v/v). The volume of the final CHCl$_3$ solution is 1.1 ml.

For purification, 60 µl of the product PtdInsP are streaked onto 5 × 20 cm trans-1,2-diaminocyclohexane-$N,N,N'N'$-tetraacetic acid (CDTA)-treated silica plates. Standards of ^{32}P-labeled PtdIns(3)P and PtdIns(4)P are alternatively spotted at 1 cm intervals. The TLC plates are developed in borate solution made up of 70 ml methanol/60 ml chloroform/45 ml pyridine/12 g boric acid plus 7.5 ml water, 3 ml 88% formic acid, 0.375 BHT and 75 µl ethoxyquin. The migration of the PtdIns[^{32}P](3)P and PtdIns[^{32}P](4)P is determined by autoradiography at −70°C. The silica gel containing the [^{3}H]PtdIns(3)P is scraped into a glass centrifuge tube and pulverized with a glass rod. The product is eluted from the silica by extraction twice with 2 ml of CHCl$_3$/methanol/pyridine/acetic acid/water, 1:2:1:1:1 (by vol.). The extracts are combined, and 1 ml of CHCl$_3$ and 3 ml of H$_2$O were added. This mixture is vortexed, and the lower chloroform phase is transferred to a new tube and the upper phase is extracted with an additional 1 ml of CHCl$_3$. The two CHCl$_3$ phases are combined and extracted twice with 2 ml of 1 M HCl in methanol/water, 1:1 (v/v). TLC scrapings are washed with CHCl$_3$ to remove the antioxidants prior to elution of the [^{3}H]PtdIns(3)P.

Whitman et al. (1988) has described the preparation of PtdIns(3)P by type I PtdIns kinase, which specifically phosphorylates the D-3 position. The preparation requires an initial isolation of type I PtdIns kinase, which can be accomplished by immunoprecipitation of middle T/pp60^{c-src} from middle T-transformed NIH 3T3 fibroblasts as described by Whitman et al. (1985). Briefly, immunoprecipitates are washed and assayed in 0.2 mg/ml sonicated detergent-free PtdIns (Avanti) with 10 µCi [^{32}P]ATP,

0.02 mM ATP, 10 mM $MgCl_2$, 0.1 M NaCl, 20 mM Tris, pH 7.6 in a final volume of 50 μl for 5 min. After extraction, reaction products are separated by TLC in $CH_3Cl/MeOH/2.5$ M NH_4OH (9:7:2, by vol.) and the plates are visualized by autoradiography. The type 1 kinase product [PtdIns(3)P] migrates slightly more slowly than the reference [PtdIns(4)P], which is also the product of type II PtdIns kinase.

Since reversed phase HPLC of the type I and type II PtdInsPs showed similar composition of fatty acids, the slight difference in the TLC migration rate was attributed to different sites of phosphorylation of the inositol ring, which was demonstrated by anion exchange chromatography of the deacylated $Gro[^{32}P]PIns$ and confirmed by periodate oxidation of the derived $InsP_2s$ (see Chapter 4).

The glycerol moiety is removed from the $[^{14}C]GroPIns(4)P$ by a modification of the mild periodate oxidation method of Brown and Stewart (1966) as described for $[5-^{32}P]GroPIns(4,5)P_2$. After passing through acid Dowex and having been neutralized with potassium hydroxide, the $[^{14}C]Ins(1,4)P_2$ is dissolved in 2 ml of H_2O ready for HPLC. The column used as before, is a 25 cm × 4.6 mm Partisil 10 SAX column (Technicol Ltd.). The flow rate is 1.5 ml/min and the column is equilibrated with 0.1 M NaH_2PO_4 adjusted to pH 3.8 with H_3PO_4. After loading, the eluting buffer is increased to 0.24 M NaH_2PO_4 (pH 3.8) and a 15 min isocratic run is carried out. A parallel run is carried out with an ADP standard (20 mg). As $Ins(1,4)P_2$ elutes shortly after ADP, its expected elution time can be estimated and fractions are collected every 15 s in the sample run from 1 min before the demonstrated elution time of ADP. The ADP is located with an ultraviolet detector at 254 nm.

A portion (1%) of each fraction is added to scintillation fluid for counting. The relevant fractions are pooled, diluted 10-fold with H_2O and loaded on to an Amprep SAX 100 mg cartridge (Amersham International). The phosphate is eluted with 0.15 M ammonium formate, 0.015 M formic acid and the $[^{14}C]Ins(1,4)P_2$

is eluted in 10 ml of 1 M triethylamine bicarbonate (freshly prepared by bubbling CO_2 through 14% (v/v) triethylamine). A typical yield is $15-18$ kBq of $[^{14}C]Ins(1,4)P_2$ from 74 Bq of $[^{14}C]PtdIns$.

Letcher et al. (1990) note that if the sole use of the $[^{14}C]Ins(1,4)P_2$ and $[^{32}P]Ins(1,4,5)P_3$ is as internal markers for HPLC, neither of them need be purified by HPLC as described; a single run of each with some $[^{3}H]Ins(1,4)P_2$ or $[^{3}H]Ins(1,4,5)P_3$ from commercial sources will establish whether the major radioactive compound is the desired one, and if it is, then each can be used in impure form as an internal marker.

Auger et al. (1990) have described methods for the preparation of $PtdIns(3)P$, $PtdIns(3,4)P_2$ and $PtdIns(3,4,5)P_3$ from the corresponding $PtdIns$, $PtdIns(4)P$, and $PtdIns(4,5)P_2$ and an immune complex containing PtdIns 3-kinase. The synthesis is performed in vitro with PtdIns 3-kinase immunoprecipitated from transformed cells in the exponential growth phase. Plates of cells (usually 10 cm in diameter) are washed two times in ice-cold phosphate buffered saline (PBS) and lysed in a standard lysis buffer (1% NP-40 (Sigma), 137 mM NaCl, 1 mM $MgCl_2$, 1 mM $CaCl_2$, 150 μM vanadate, 1 μg leupeptin/ml, 1 μg aprotinin/ml). After incubation on a rocker platform at 4°C for $15-20$ min, the lysate is cleared by centrifugation at 12 000g for 5 min at 4°C. Immunoprecipitation with anti-phosphotyrosine antibody or anti-growth factor receptor antibody is done at 4°C on a rocker platform for at least $2-3$ h. Immune complexes are collected on Protein A-Sepharose Cl 4B that has been pre-washed in 10 mM Tris-HCl (pH 7.5),% bovine serum albumin, and stored in PBS containing 0.02% (w/v) sodium azide. Immune complexes are washed three times with 1 ml of 1% NP-40 in PBS, two times with 0.5 M LiCl in 100 mM Tris-HCl (pH 7.5), and finally two times in TNE (10 mM Tris-HCl (pH 7.5) 100 mM NaCl, 1 mM EDTA). PtdIns kinase assays are performed directly on the beads. Routinely, all three phospholipid substrates, namely, PtdIns, PtdIns(4)P and $PtdIns(4,5)P_2$ are used. The final concentration of the PtdIns, PtdInsP and $PtdInsP_2$ substrates is

0.03 mg/ml each, in carrier of PtdSer at a final concentration of 0.1 mg/ml. Alternatively, a crude PI mixture from bovine brain can be obtained from Sigma and used at a final concentration 0.2 mg/ml. The mixture contains all four phospholipids and thus eliminates the need to mix individual lipids. The phospholipids stored in chloroform are placed in a 1.5 ml microfuge tube, dried under a stream of nitrogen as, and resuspended in 10 mM HEPES (pH 7.5), 1 mM EGTA, prior to sonication.

The PtdIns 3-kinase reaction is usually done in a total volume of 50 μl and is initiated by the addition of 10–50 μCi of $[\gamma\text{-}^{32}\text{P}]\text{ATP}$ (Dupont NEN, 3000 mCi/mmol) in a carrier of 50 μM unlabeled ATP, 10 mM Mg^{2+} and 20 mM HEPES (pH 7.5) to the washed immune complexes that have been pre-incubated for 5 min with the phospholipid substrates. The enzyme reaction is incubated at room temperature for 5–10 min. After stopping the reaction with 80 μl of 1 M HCl and extracting the lipids with 160 μl of chloroform/ methanol (1:1, v/v), the ^{32}P-labeled phospholipid products in the bottom organic phase are collected after a brief centrifugation and stored at $-70°C$ until further use.

The samples are often analyzed and resolved on TLC plates (MCB Reagents, Merck, Silica Gel 60, 0.2 mm thickness) that have been pre-coated with 1% (w/v) potassium oxalate and baked at 100°C for 30 min to 1 h immediately before use. Unlabeled phospholipid standards are run in parallel to monitor the lipid migration and are visualized by exposure to iodine vapor. In order to separate the highly phosphorylated $PdIns(3,4,5)P_3$ from the radioactivity remaining near the origin and any $[\gamma\text{-}^{32}\text{P}]\text{ATP}$ and $[^{32}\text{P}]P_i$ carried over with the organic phase during lipid extraction, an acidic solvent system of n-propanol/2.0 M acetic acid (13:7, v/v) is used instead of the more commonly used $CHCl_3/MeOH/2.5$ M NH_4OH (9:7:2, by vol.) solvent system. To achieve maximum resolution of each of the phospholipids from each other and from the material close to the origin, the solvent is allowed to migrate nearly to the top of a 20 cm TLC plate, a process that is routinely accomplished in 5 to 6 h.

10.2.1.3. PtdIns(5)P

This lipid has been reported as an impurity in commercial PtdIns(4)P (brain derived), and has been shown to be present in mammalian fibroblasts at approximately the same level as PtdIns(3)P (Rameh et al., 1997). Tolias et al. (1998a) have shown that type Iα or type Iβ PtdIns(4)P 5-kinases can phosphorylate PtdIns(3)P and PtdIns at the D-5-position in vitro. However, the preferred substrate for these enzymes is PtdIns(4)P. Briefly, Tolias et al. (1998a) isolated the PtdInsP 5-kinases from 293 E1A cells. Cells were rinsed with PBS and lysed in 600 μl of lysis buffer (50 mM Tris, pH 7.5, 50 mM NaCl, 5 mM MgCl$_2$, 1% Nonidet P-40, 10% glycerol, 1 mM dithiothreitol, 4 μg/ml each leupeptin and pepstatin, and 200 μM AEBSF. The clarified lysates were incubated with 1.5 μg of an anti-HA antibody (12CA5 from Boehringer Mannheim) and protein A-Sepharose beads for 3 h at 4°C. The beads were then washed twice with lysis buffer and twice with TNM (50 mM Tris, pH 7.5, 50 mM NaCl, and 5 mM MgCl$_2$). The PtdIns(5)P synthesis was performed in 50 μl reactions containing 50 mM Tris, pH 7.5, 30 mM NaCl, 12 mM MgCl$_2$, 67 μM PtdIns, 133 μM phosphoserine, and 50 μM [γ-^{32}P]ATP (10 μCi/assay). Reactions were stopped after 10 min by adding 80 μl of 1N HCl and then 160 μl of CHCl$_3$/CH$_3$OH (1:1). Lipids were separated by TLC using 1-propanol/2 M acetic acid (63:35, v/v). Phosphorylated lipids were visualized by autoradiography and quantified by a Bio-Rad Molecular Analyst. The PtdIns used in the experiments was separated from contaminating PtdInsPs by TLC purification using CHCl$_3$/MeOH/H$_2$O/NH$_4$OH (60:47:11:1.6, by vol.). After identification using an iodine-stained PtdIns standard, the PtdIns was scraped from the TLC plate and eluted in the same solvent solution. Following lyophilization, PtdIns was resuspended in chloroform and then extracted once with MeOH/1N HCl (1:1, v/v) and once with MeOH, 0.1 M EDTA (1:0.9, v/v). The reaction was stopped with 1 mM EDTA and 50 μl of 3N HCl, and the lipids extracted with 200 μl of CHCl$_3$/CH$_3$OH (1:1, v/v).

Tolias et al. (1998) identified the PtdIns(5)P following de-acylation and mixing with ^{3}H-labeled standards and analyzed by anion-exchange HPLC using Partisphere SAX column (Whatman) as described by Serunian et al. (1991). To separate PtdIns(4)P from PtdIns(5)P and PtdIns(3,4)P$_2$ from PtdIns(3,5)P$_2$, a modified ammonium phosphate gradient was used. The compounds were eluted with 1 M (NH$_4$)$_2$HPO$_4$, H 3.8, and water using the following gradient: 0% 1 M (NH$_4$)$_2$HPO$_4$ for 5 ml, 0–1% 1 M (NH$_4$)$_2$HPO$_4$ over 5 ml, 1–3% 1 M (NH$_4$)$_2$HPO$_4$ over 40 ml, 3–10% 1 M (NH$_4$)$_2$HPO$_4$ over 10 ml, 10–13% 1 M (NH$_4$)$_2$HPO$_4$ over 25 ml, and 13–65% 1 M (NH$_4$)$_2$HPO$_4$ over 25 ml. Eluate from the HPLC column flowed into an on-line continuous flow scintillation detector (Radiomatic Instruments, FL) for isotope detection. The polyol structure of the PtdIns(5)P was confirmed by periodate oxidation. Periodate oxidation was performed as described by Stephens et al. (1988b) with the following modifications. Deacylated lipids were incubated in the dark at 25°C with 10 or 100 mM periodic acid, pH 4.5, for 30 min or 36 h, respectively. The remaining oxidizing agent was removed by adding 500 mM ethylene glycol and incubating the reaction in the dark for 30 min. Two percent 1,1-dimethylhydrazine, pH 4.5, was then added to a final concentration of 1% and the reaction was allowed to proceed for 4 h at 25°C. The mixture was then purified by ion exchange using Dowex 50W-X8 cation-exchange resin (20–50 mesh, acid form) dried, and applied to and HPLC anion exchange column.

10.2.2. PtdIns bis-phosphates

10.2.2.1. PtdIns(4,5)P$_2$
Low (1990) has modified the original procedure of Hendrickson and Ballou (1964) for the purification of Ins(4,5)P$_2$ from natural sources. Bovine brain Folch fraction I (1 g) is dissolved in 12 ml of chloroform. The PtdInsPs are precipitated by mixing with 22 ml of

methanol followed by centrifugation at 1000g for 5 min. The supernatant is discarded, and the pellet redissolved in 12 ml of chloroform. This precipitation with methanol is repeated a total of six times. The final methanol precipitate is dissolved in 15 ml of chloroform, 15 ml of methanol, 5 ml of 1 M HCl. Then 1 M HCl (8.5 ml) is added to give two phases. After centrifugation the upper phase is discarded and, to the lower phase, 15 ml of methanol and 5 ml of 2 M NaCl are added and the sample mixed until it becomes clear. The phases are separated by the addition of 2.5 ml of 2 M NaCl and 6 ml of H_2O. The upper phase is removed and the lower phase washed once more with methanol and 2 M NaCl as above. All the steps are done at $0-4°C$. The lower phase is transferred to a clean tube, evaporated to dryness under nitrogen gas, the lipids redissolved in 2 ml of chloroform and stored at $-20°C$ until required. One gram of starting material yields $150-250$ μmol of precipitated PtdInsPs (as determined by organic phosphorus).

The PtdInsPs extract of bovine brain is purified by chromatography on DEAE cellulose. The DEAE-cellulose (100 g) (Sigma) is washed on a filter with 500 ml of 1 M NaOH, H_2O to neutrality, 500 ml of 10% (v/v) acetic acid and finally H_2O to neutrality. The bulk of the H_2O is removed by stirring the DEAE-cellulose for about 5 min with two volumes of methanol and then allowing the mixture to stand for approximately 30 min. The methanol is decanted and the treatment repeated twice with methanol and then twice with two volumes of solvent A (chloroform/methanol/H_2O, 20:9:1, by vol.). The DEAE is finally suspended in solvent A and packed into a glass column with solvent resistant fittings (2×38 cm). The brain phospholipids (prepared as described above) are dissolved in 50 ml of solvent A and applied to the DEAE-cellulose column. The column is then eluted at a flow rate of approximately 100 ml/h with a liter linear gradient from 100% solvent A to 100% solvent B (solvent A containing 0.6 M ammonium acetate) and $13-15$ ml fractions collected. The fractions are assayed for organic phosphorus and the peak fractions are pooled and filtered. Average recovery of organic phosphorus is approximately 90%. The PtdIns(4,5)P_2 is eluted last and is preceded by PtdIns(4)P, as determined by TLC.

[^{32}P]PtdIns(4,5)P_2 may be prepared by phosphorylation of PtdIns(4)P. According to Letcher et al. (1990), PtdIns(4)P (200 nmol) is dissolved in chloroform in a thin-walled glass tube and dried at room temperature in a current of nitrogen gas. H_2O (0.5 ml) is added and the tube is placed in a bath sonicator at room temperature for 2 min, or until all the lipid has been displaced from the tube wall. The sample is stored on ice, and to it are added the other components of the incubation mixture (see below) giving a final volume of 1.3 ml. The incubation mixture is 40 mM Tris-acetate, pH 7.4, 15 mM magnesium acetate, 40 μM ATP, 1.85 MBq (50 μCi) adenosine 5'-[γ-^{32}P]trisphosphate, triethylammonium salts in aqueous solution (Amersham), and 0.5 ml of an enzyme preparation from rat brain. The incubation is started by adding the enzyme and is continued for 4 h at room temperature. The reaction is stopped by adding 5 ml of chloroform/methanol (1:2 v/v), 20 μl of 12 M HCl. After vortexing, the tube is stored on ice for 10 min, after which 1.67 ml of chloroform and then 1.67 ml of 1 M HCl are added to complete the extraction mixture and to form two phases. The lower phase is washed six times with theoretical upper phase chloroform/methanol/1 M HCl (3:48:47, by vol.), with mixing and centrifugation after each wash. The final upper phase is removed by aspiration and the lower phase dried thoroughly to remove all the HCl. The [^{32}P]PtdIns(4,5)P_2 is dissolved in a small amount of chloroform.

The [5-^{32}P]GroPIns(4,5)P_2 may be prepared by deacylation. Letcher et al. (1990) transferred the purified [^{32}P]PdIns(4,5)P_2 to a vessel suitable for monomethylamine treatment, and evaporated the chloroform. Monomethylamine reagent (3 ml) [prepared as described by Clarke and Dawson (1981)] is added, and the vessel is securely stoppered and incubated at 53°C for 60 min in a water bath. The vessel is cooled on ice and the sample is dried using a rotary evaporator, starting with the vessel immersed in cold H_2O and finally heating at 50°C to remove all the monomethylamine.

The procedure of gradually heating the flask prevents 'bumping' and subsequent loss of the sample. Alternatively, the monomethylamine reagent may be removed by a Vac-Fuge. The [5-^{32}P]GroPtdIns(4,5)-P$_2$ sample thus obtained is dissolved in 2 ml of water and 2.4 ml of n-butanol/petroleum ether (40–60°C)/ethyl formate (20:41, by vol.) is added (Clarke and Dawson, 1981). The mixture is thoroughly shaken and is allowed to stand to form two clear phases. The upper phase containing n-methyl fatty acid amides is discarded and the lower phase is washed a second time with 1.5 ml of the same n-butanol/petroleum ether (40–60°C)/ethyl formate mixture before being taken to dryness.

10.2.2.2. PtdIns(3,5)P$_2$

The existence of PtdIns(3,5)P$_2$ in mammalian cells was demonstrated some 10 years ago (Auger et al., 1989), the conclusive identification of this lipid in mammalian cells (Whiteford et al., 1997) and in yeast (Dove et al., 1997) was obtained only recently. Tolias et al. (1998) found that the type I PtdInsP 5-kinase synthesizes the novel phospholipid PtdIns(3,5)P$_2$. The synthesis conditions were identical to those just discussed for the synthesis of PtdIns(5)P. The intact PtdIns(3,5)P$_2$ is partially resolved by TLC from PtdIns(3,4)P$_2$, which migrates slightly more slowly (Tolias et al., 1998). To confirm the presence of either or both PtdInsP$_2$s, a separation is effected using the deacylated derivatives of PtdIns(3,4)P$_2$ and PtdIns(3,5)P$_2$. An ammonium phosphate gradient modified from Serunian et al. (1991) is used (Tolias et al., 1998). The compounds are eluted with 1 M (NH$_4$)$_2$HPO$_4$, pH 3.8, and water using the following gradient: 0% 1 M (NH$_4$)$_2$HPO$_4$ for 5 ml, 0–1% 1 M (NH$_4$)$_2$HPO$_4$ over 5 ml, 1–3% 1 M (NH$_4$)$_2$HPO$_4$ over 40 ml, 3–10% 1 M (NH$_4$)$_2$HPO$_4$ over 10 ml, 10–13% 1 M (NH$_4$)$_2$HPO$_4$ over 25 ml and 13–65% 1 M (NH$_4$)$_2$HPO$_4$ over 25 ml. Eluate from the HPLC column flowed into an on-line continuous flow scintillation detector (Radiomatic Instruments, FL) for isotope detection. The GroPIns(3,5)P$_2$ migrates ahead of the GroPIns(3,4)P$_2$ and is clearly separated from it.

PtdIns(3,5)P_2 is produced in larger proportions in yeast under osmotic stress (Dove et al., 1997).

10.2.2.3. PtdIns(3,4)P_2

Auger et al. (1990) have described methods for the preparation of PtdIns(3)P, PtdIns(3,4)P_2 and PtdIns(3,4,5)P_3 from the corresponding PtdIns, PtdIns(4)P, and PtdIns(4,5)P_2 and an immune complex containing PtdIns 3-kinase. The synthesis is performed in vitro with PtdIns 3-kinase immunoprecipitated from transformed cells in the exponential growth phase. Plates of cells (usually 10 cm in diameter) are washed two times in ice-cold PBS and lysed in a standard lysis buffer (1% NP-40 (Sigma), 137 mM NaCl, 1 mM MgCl$_2$, 1 mM CaCl$_2$, 150 μM vanadate, 1 μg leupeptin/ml, 1 μg aprotinin/ml). After incubation on a rocker platform at 4°C for 15–20 min, the lysate is cleared by centrifugation at 12 000g for 5 min at 4°C. Immunoprecipitation with anti-phosphotyrosine antibody or anti-growth factor receptor antibody is done at 4°C on a rocker platform for at least 2–3 h. Immune complexes are collected on Protein A-Sepharose Cl 4B that has been pre-washed in 10 mM Tris-HCl (pH 7.5),% bovine serum albumin, and stored in PBS containing 0.02% (w/v) sodium azide. Immune complexes are washed three times with 1 ml of 1% NP-40 in PBS, two times with 0.5 M LiCl in 100 mM Tris–HCl (pH 7.5), and finally two times in TNE (10 mM Tris–HCl (pH 7.5) 100 mM NaCl, 1 mM EDTA). PtdInsP kinase assays are performed directly on the beads. Routinely, all three phospholipid substrates, namely, PtdIns, PtdIns(4)P and PtdIns(4,5)P_2 are used. The final concentration of the PtdIns, PtdInsP and PtdInsP$_2$ substrates is 0.03 mg/ml each, in carrier of PtdSer at a final concentration of 0.1 mg/ml. Alternatively, a crude PtdInsPs mixture from bovine brain can be obtained from Sigma and used at a final concentration of 0.2 mg/ml. The mixture contains all four phospholipids and thus eliminates the need to mix individual lipids. The phospholipids stored in chloroform are placed in a 1.5 ml microfuge tube, dried under a stream of nitrogen as, and resuspended in 10 mM HEPES (pH 7.5), 1 mM EGTA, prior to sonication. After sonication, the lipid suspension, which is initially cloudy, becomes clear.

The PtdIns 3-kinase reaction is usually done in a total volume of 50 μl and is initiated by the addition of 10–50 μCi of [γ^{32}-P]ATP (Dupont NEN, 3000 mCi/mmol) in a carrier of 50 μM unlabeled ATP, 10 mM Mg^{2+} and 20 mM HEPES (pH 7.5) to the washed immune complexes that have been pre-incubated for 5 min with the phospholipid substrates. The enzyme reaction is incubated at room temperature for 5–10 min. After stopping the reaction with 80 μl of 1 M HCl and extracting the lipids with 160 μl of chloroform/methanol (1:1, v/v), the ^{32}P-labeled phospholipid products in the bottom organic phase are collected after a brief centrifugation and stored at $-70°C$ until further use.

The samples are often analyzed and resolved on TLC plates (MCB Reagents, Merck, Silica Gel 60, 0.2 mm thickness) that have been pre-coated with 1% (w/v) potassium oxalate and baked at 100°C for 30 min to 1 h immediately before use. Unlabeled phospholipid standards are run in parallel to monitor the lipid migration and are visualized by exposure to iodine vapor. In order to separate the highly phosphorylated PtdIns(3,4,5)P_3 from the radioactivity remaining near the origin and any [γ-^{32}P]ATP and [^{32}P]phosphate carried over with the organic phase during lipid extraction, an acidic solvent system of n-propanol/2.0 M acetic acid (13:7, v/v) is used instead of the more commonly employed $CHCl_3$/MeOH/2.5 M NH_4OH (9:7:2, by vol.) solvent system. To achieve maximum resolution of each of the phospholipids from each other and from the material close to the origin, the solvent is allowed to migrate nearly to the top of a 20 cm TLC plate, a process that is routinely accomplished in 5–6 h.

Tolias et al. (1998) found that the type I PtdInsP 5-kinase synthesizes the novel phospholipid PtdIns(3,5)P_2. The synthesis conditions are identical to those just discussed for the synthesis of PtdIns(5)P. The intact PtdIns(3,4)P_2 is partially resolved by TLC from PtdIns(3,5)P_2, which migrates slightly ahead (Tolias et al., 1998) when 1-propanol/2 M acetic acid (65:32, v/v) as solvent.

To confirm the presence of either or both $PtdInsP_2s$, a separation is effected using the deacylated derivatives of $PtdIns(3,4)P_2$ and $PtdIns(3,5)P_2$. An ammonium phosphate gradient modified from Serunian et al. (1991) is used (Tolias et al., 1998). The compounds are eluted with 1 M $(NH_4)_2HPO_4$, pH 3.8, and water using the following gradient: 0% 1 M $(NH_4)_2HPO_4$ for 5 ml, 0–1% 1 M $(NH_4)_2HPO_4$ over 5 ml, 1–3% 1 M $(NH_4)_2HPO_4$ over 40 ml, 3–10% 1 M $(NH_4)_2HPO_4$ over 10 ml, 10–13% 1 M $(NH_4)_2HPO_4$ over 25 ml and 13–65% 1 M $(NH_4)_2HPO_4$ over 25 min. Eluate from the HPLC column flowed into an on-line continuous flow scintillation detector (Radiomatic Instruments, FL) for isotope detection. The $GroPIns(3,4)P_2$ is eluted after $GroPIns(3,5)P_2$ and clearly separated from it.

10.2.3. PtdIns trisphosphates

10.2.3.1. PtdIns(3,4,5)P₃

The bulk of $PtdIns(3,4,5)P_3$ in mammalian cells is likely produced by class I PtdIns 3-kinase. This lipid is nominally absent in quiescent cells, but appears within seconds to minutes of cell stimulation with a variety of activators (Auger et al., 1989; Traynor-Kaplan et al., 1989).

Auger et al. (1990) have described methods for the preparation of $PtdIns(3)P$, $PtdIns(3,4)P_2$ and $PtdIns(3,4,5)P_3$ from the corresponding PtdIns, PtdIns(4)P, and $PtdIns(4,5)P_2$ and an immune complex containing PtdIns 3-kinase. The synthesis is performed in vitro with PtdIns 3-kinase immunoprecipitated from transformed cells in the exponential growth phase. Plates of cells (usually 10 cm in diameter) are washed two times in ice-cold PBS and lysed in a standard lysis buffer (1% NP-40 (Sigma), 137 mM NaCl, 1 mM $MgCl_2$, 1 mM $CaCl_2$, 150 μM vanadate, 1 μg leupeptin/ml, 1 μg aprotinin/ml). After incubation on a rocker platform at 4°C for 15–20 min, the lysate is cleared by centrifugation at 12 000g for 5 min at 4°C. Immunoprecipitation with anti-phosphotyrosine antibody or anti-growth factor receptor antibody is done at 4°C on a rocker platform for at least 2–3 h. Immune complexes are collected on Protein A-Sepharose Cl 4B that has been pre-washed in 10 mM Tris-HCl (pH 7.5),% bovine serum albumin, and stored in PBS containing 0.02% (w/v) sodium azide. Immune complexes are washed three times with 1 ml of 1% NP-40 in PBS, two times with 0.5 M LiCl in 100 mM Tris-HCl (pH 7.5), and finally two times in TNE (10 mM Tris-HCl (pH 7.5) 100 mM NaCl, 1 mM EDTA). PtdInsP kinase assays are performed directly on the beads. Routinely, all three phospholipid substrates, namely, PtdIns, PtdIns(4)P and PtdIns(4,5)P_2 are used. The final concentration of the PtdIns, PtdInsP and PtdInsP$_2$ substrates is 0.03 mg/ml each, in carrier of PtdSer at a final concentration of 0.1 mg/ml. Alternatively, a crude PtdIns mixture from bovine brain can be obtained from Sigma and used at a final concentration of 0.2 mg/ml. The mixture contains all four phospholipids and thus eliminates the need to mix individual lipids. The phospholipids stored in chloroform are placed in a 1.5 ml microfuge tube, dried under a stream of nitrogen gas, and resuspended in 10 mM HEPES (pH 7.5), 1 mM EGTA, prior to sonication. After sonication, the lipid suspension, which is initially cloudy, becomes clear.

The PtdIns 3-kinase reaction is usually done in a total volume of 50 μl and is initiated by the addition of 10–50 μCi of [γ^{32}-P]ATP (Dupont NEN, 3000 mCi/mmol) in a carrier of 50 μM unlabeled ATP, 10 mM Mg^{2+} and 20 mM HEPES (pH 7.5) to the washed immune complexes that have been pre-incubated for 5 min with the phospholipid substrates. The enzyme reaction is incubated at room temperature for 5–10 min. After stopping the reaction with 80 μl of 1 M HCl and extracting the lipids with 160 μl of chloroform/methanol (1:1, v/v), the ^{32}P-labeled phospholipid products in the bottom organic phase are collected after a brief centrifugation and stored at − 70°C until further use.

However, PtdIns(3,4,5)P$_3$ is difficult to purify from biological sources but several groups have synthesized it by various synthetic routes and tested its in vitro actions on purified enzymes.

Jiang et al. (1998) have prepared membrane-permeant PtdIns(3,4,5)P$_3$, while Niggli (2000) has demonstrated that this synthetic product can serve as an activator of human neutrophil migration. Carter et al. (1994) have demonstrated that PtdIns(3,4,5)P$_3$ is synthesized from PtdIns(4,5)P$_2$ in thrombin-stimulated platelets. PtdIns(3,4,5)P$_3$ and PtdIns(3,4)P$_2$ were synthesized by briefly (10 min) incubating platelets with high activities of [^{32}P]P$_i$, followed by 20 or 30 s exposure to thrombin. Incubations were terminated with chloroform/methanol (1:2, v/v), followed by 0.75 ml of 3.1 M HCl and 1 ml of chloroform. Vigorously mixed samples were centrifuged, and the upper phases were extracted twice with 1.5 ml of chloroform. The pooled lower phases were evaporated to dryness and the extracted lipids were resolved by TLC and the PtdInsPs detected by autoradiography (Carter and Downes, 1992). Carter et al. (1994) have provided a detailed flow chart outlining the analysis of the PtdIns(3,4,5)P$_3$ product.

10.3. Inositol phosphates

Inositol phosphates can be isolated from natural sources (Irvine et al., 1985) as well as prepared by semisynthetic procedures from the corresponding phospholipids (Bird et al., 1989) or enzymatic hydrolysis of phytic acid (Grado and Ballou, 1961). The fully synthetic routes leading to the formation of InsPs have been recently reviewed (Schultz et al., 1996).

Tritium-radiolabeled standards are commercially available in 1–2 µCi quantities for Ins(1,3,4,5)P$_4$, Ins(1,4,5)P$_3$, Ins(1,3,4)P$_3$, Ins(1,4)P$_2$, Ins(4)P, and Ins(1)P. Unlabeled standards of several mono-, bis-, tris, tetrakis- and pentakis-phosphates have recently become available from Sigma. They are extremely costly with a few micrograms costing several hundred dollars. Laboratory preparation including the necessary structural characterization may therefore represent a bargain. Other standards must be prepared in the laboratory.

The preparations of InsPs described here are largely limited to natural compounds. In few instances reference is made to preparation of membrane-permeable derivatives. However, their preparation may require more expertise than commonly available in a biochemical laboratory. The preparation of various other InsPs and other inositol derivatives synthesized as experimental inhibitors of various metabolic and degradative enzymes has not been covered here. This subject has been discussed by Schultz et al. (1996).

10.3.1. Inositol monophosphates (InsP)

Sigma-Aldrich (Catalogue 2000) now lists Ins(2)P, DL-*meso*-Ins(1,2-cyc)P and the radiolabeled ones from DuPont/NEN.

10.3.1.1. Ins(1)P

Ins(1)P synthase catalyzes the irreversible conversion of (D)-glucose-6-phosphate to (L)-*myo*-Ins(1)P (Eisenberg and Parthasarathy, 1987). Chen et al. (2000) have recently demonstrated that the enzyme is a class II aldolase. The enzyme activity, however, does not provide a practical means of synthesizing Ins(1)P. Early studies (Pizer and Ballou, 1959) had demonstrated that *myo*-InsP are susceptible to acid-catalyzed migration and that acid hydrolysis cannot be used for the preparation of InsP of definite positional distribution. In contrast, base hydrolysis of PtdIns yields optically active L-*myo*-Ins(1)P (Ballou and Pizer, 1960), which was eventually proven to have the same linkage as in the original phospholipid (Grado and Ballou, 1961). Efficient chemical methods for the synthesis of *myo*-Ins(1)P preparation have been described by Kieley et al. (1974) and Gero et al. (1972).

10.3.1.2. Ins(2)P

Ins(2)P is not known to occur naturally, but it may appear in synthetic or isomerized preparations of Ins(1)P. Since the alkaline hydrolysis proceeds via a cyclic 1,2-phosphate ester, a small amount of *myo*-Ins(2)P is also formed, which can be chromatographically resolved from the *myo*-Ins(1)P (Grado and Ballou, 1961). Ins(2)P may be isolated from the dephosphorylation mixture of phytic acid (see Chapter 4).

10.3.1.3. Ins(3)P

Ins(3)P should be available from phospholipase D hydrolysis of PtdIns(3)P (see above). Ins(3)P also should be available from enzymatic hydrolysis of phytic acid (see Chapter 4).

10.3.1.4. Ins(4)P

Ins(4)P should be available from phospholipase D hydrolysis of PtdIns(4)P. Alternatively, it should be an intermediate in phytic acid hydrolysis (Chapter 4).

10.3.1.5. Ins(5)P

Walker et al. (2001) prepared authentic [^{32}P]GroPIns(5)P from radiolabeled yeast that had been stressed with 0.9 M NaCl for 10 min. Total [^{32}P]radiolabeled lipids were extracted, deacylated, and resolved by HPLC, and the [^{32}P]GroPIns(3,5)P$_2$ peak was collected and desalted. [^{32}P]GroPIns(3,5)P$_2$ (500,000 dpm) was then treated with 0.7 ml of erythrocyte ghost membranes for 4 h at 37°C in presence of 5 mM EDTA and 1 mM EGTA (pH 7.5). The reaction was stopped with PCA, protein and perchlorate precipitated, and supernatant collected by centrifugation. The supernatant was found to contain a 1:1:1 mixture of [^{32}P]GroPIns(5)P, [^{32}P]GroPIns(3,5)P$_2$ and [^{32}P]PO$_4$.

10.3.2. Ins bisphosphates

Chemical synthesis of InsP$_2$ has been attempted and the resulting asymmetrically substituted *myo*-InsP$_2$ have been resolved into optical antipodes (Shvets et al., 1973). These procedures cannot yet be recommended for the preparation of reference standards. Therefore, semisynthetic routes for preparation of both radioactive and non-radioactive standards must be chosen. Sigma-Aldrich now offers the following $InsP_2$ reference compounds: $Ins(1,4)P_2$; $Ins(2,4)P_2$; $Ins(4,5)P_2$; and $Ins(5,6)P_2$.

10.3.2.1. $Ins(1,2)P_2$

D,L-Ins$[^{32}P](1,2)P_2$ can be obtained by boiling an aliquot of [3-phosphate-^{32}P]Ins$(1,3)P_2$ in 50 μl of 1.0 M HCl for 8 min (Stephens et al., 1989a). Under these conditions, phosphate migration across *cis*-related hydroxy groups in *myo*-Ins (i.e. the 1 and 2 and/or 2 and 3 positions) occurs without significant migration between *trans*-related neighboring hydroxy groups (Pizer and Ballou, 1959). The sample is then cooled, neutralized precisely with 2 M KOH and 0.2 M HEPES, diluted with equal volume of water and desalted as described elsewhere (Chapter 4). Stephens et al. (1989a) could resolve two Ins$[^{32}P]P_2$ species by a Partisphere SAX column. One possessed the retention time expected for $Ins(1,3)P_2$ and the second peak, comprising 50% of the recovered ^{32}P, was eluted after $[^3H]Ins(1,4)P_2$ and was presumed to be Ins$[^{32}P](1,2)P_2$. This was confirmed by processing in parallel a ^{3}H-labeled sample, which yielded a bisphosphate that gave $[^3H]$erythritol upon periodate oxidation, reduction and dephosphorylation. Unlabeled and ^{14}C-labeled D/L-Ins$(1,2)P_2$s can be prepared as described by Stephens et al. (1988a,b).

10.3.2.2. $Ins(1,3)P_2$ and $Ins(3,4)P_2$

These are readily prepared from commercially available $Ins(1,3,4)P_3$ by incubation with a high-speed supernatant prepared from homogenized rat brain as shown by Batty et al.(1989). For this purpose, $[^3H]Ins(1,3,4)P_3$ was prepared from authentic $[^3H]Ins(1,3,4,5)P_4$ by incubation with human erythrocyte membranes as described by Batty et al. (1985). $[^3H]Ins(1,3)P_2$ and $[^3H]Ins(3,4)P_2$ were isolated by HPLC. Identification of $[^3H]InsPs$ was based on co-elution with internal ^{14}C or ^{32}P-labeled InsPs standards. For routine analysis of samples by HPLC, fractions (0.5–1.0 min) were collected only across the chromatographic windows appropriate for Ins, $InsP_1(s)$, $InsP_2(s)$, $InsP_3(s)$ and $InsP_4$. To overcome problems associated with variations in retention times, samples were routinely spiked with $50-100$ mmol each of adenosine and guanosine mono-, di-, tri- and tetra-phosphates before injection. HPLC analyses of [^{3}H]InsPs were performed by a modification of the method of Dean and Moyer (1987). Separation was achieved with a Partisil ($10 \mu M$) SAX analytical column packed with Whatman pellicular anion-exchange resin and eluted with gradients comprising water and $(NH_4)H_2PO_4$, adjusted to pH 3.7 with H_3PO_4. After sample injection (2 ml), free [^{3}H]Ins was eluted by washing the column for $5-15$ min with water. [^{3}H]InsPs were then separated by applying three consecutive gradients at a flow rate of 1 ml/min. GroPIns and $InsP_1(s)$ were resolved by applying a linear gradient of $0-60$ mM $(NH_4)H_2PO_4$ over 30 min. GroPInsP and $InsP_2(s)$ were then separated by isocratic elution at 190 mM $(NH_4)H_2PO_4$ for 15 min, followed by a linear increase in eluent concentration to 300 mM over a further 15 min. $GroPInsP_2$ and $InsP_3(s)$ were separated by isocratic elution for 35 min at 500 mM $(NH_4)H_2PO_4$. $Ins(1,3,4,5)P_4$ was then displaced from the column by a 15 min wash at 1.4 M $(NH_4)H_2PO_4$.

Stephens et al. (1989a) obtained [^{3}H,3-phosphate-^{32}P]Ins(1,3)P_2 from a mixture of [^{3}H]Ins(1,3,4)P_3 and [3-phosphate-^{32}P]-Ins(1,3,4)P_3 (prepared as described below by incubating the trisphosphates with rat brain cytosol in the presence of EDTA for 10 min (the final assay volume was 1 ml and contained 40 mM HEPES, 2 mM EDTA, 80 mM KCl, 250 μl of rat brain cytosol fraction and 0.7 nmol of substrate). The incubation was terminated with $HClO_4$, neutralized, applied to Partisphere SAX HPLC column and eluted as described by Stephens et al. (1988a). Approximately 60% of the starting substrate was recovered as [^{3}H,3-phosphate-^{32}P]Ins(1,3)P_2.

Stephens et al. (1989a) prepared [3-phosphate-^{32}P]Ins(1,3)P_2 and [3-phosphate-^{32}P]Ins(3,4)P_2 by incubating [3-phosphate-^{32}P]Ins(1,3,4,5)P_4 with rat brain homogenate (300 μl total assay volume, containing 50 mM HEPES, 2 mM EGTA,

1 mM $MgCl_2$, 10 mM LiCl and 40 μl of rat brain homogenate, pH 7.0 at 37°C) for 15 min. The products of this reaction, which have been established to be $Ins(1,3)P_2$ and $Ins(3,4)P_2$ (Stephens et al., 1988c), were obtained in 22 and 23% yields and were resolved by HPLC on a Partisphere SAX column, neutralized and desalted as described by Stephens et al. (1988b).

10.3.2.3. Ins(1,4)P₂

$Ins(1,4)P_2$ can be prepared by phospholipase C hydrolysis of $PtdIns(4)P$ as described by Irvine et al. (1979). Radiolabeled $Ins(1,4)P_2$ may be prepared from the corresponding radiolabeled $PtdIns(4)P$. Thus, $[^{14}C]PtdIns(1,4)P_2$ is prepared first by using human erythrocyte PtdIns 4-kinase to catalyze the phosphorylation of ^{14}C-labeled PtdIns to $[^{14}C]PtdIns(4)P$ (Letcher et al., 1990). The product is isolated as described above and deacylated with 3 ml of monomethylamine (Clarke and Dawson, 1981). The glycerol is removed from the $[^{14}C]GroPIns(4)P$ by a modification of the mild periodate oxidation method of Brown and Stewart (1966) exactly as described above for $[5-^{32}P]GroPIns(4,5)P_2$. After passing through acid Dowex and having been neutralized with potassium hydroxide, the $[^{14}C]Ins(1,4)P_2$ is dissolved in 2 ml of H_2O ready for HPLC. The column used as before, is a 25 cm × 4.6 mm Partisil 10 SAX column (Technicol Ltd.). The flow rate is 1.5 ml/min and the column is equilibrated with 0.1 M NaH_2PO_4 adjusted to pH 3.8 with H_3PO_4. After loading, the eluting buffer is increased to 0.24 M NaH_2PO_4 (pH 3.8) and a 15 min isocratic run is carried out. A parallel run is carried out with an ADP standard (20 mg). As $Ins(1,4)P_2$ elutes shortly after ADP, its expected elution time can be estimated and fractions are collected every 15 s in the sample run from 1 min before the demonstrated elution time of ADP. The ADP is located with a UV detector at 254 nm.

A portion (1%) of each fraction is added to scintillation fluid for counting. The relevant fractions are pooled, diluted 10-fold with H_2O and loaded on to an Amprep SAX 100 mg cartridge (Amersham International). The phosphate is eluted with 0.15 M ammonium formate, 0.015 M formic acid and the $[^{14}C]Ins(1,4)P_2$ is eluted in 10 ml of 1 M triethylamine bicarbonate (freshly prepared by bubbling CO_2 through 14% (v/v) triethylamine). A typical yield is 15–18 kBq of $[^{14}C]Ins(1,4)P_2$ from 74 Bq of $[^{14}C]PtdIns$. Letcher et al. (1990) note that if the sole use of the $[^{14}C]Ins(1,4)P_2$ and $[^{32}P]Ins(1,4,5)P_3$ is as internal markers for HPLC, neither of them need be purified by HPLC as described; a single run of each with some $[^3H]Ins(1,4)P_2$ or $[^3H]Ins(1,4,5)P_3$ from commercial sources will establish whether the major radioactive compound is the desired one, and if it is, then each can be used in impure form as an internal marker.

Stephens et al. (1989a) prepared $Ins[^{32}P](1,4)P_2$ from $[^{32}P]P_i$-prelabeled human erythrocytes and purified it by HPLC as described by Downes et al. (1986).

Batty et al. (1989) prepared $[^3H]Ins(1,4)P_2$ from authentic $[^3H]Ins(1,4,5)P_3$ by incubation with human erythrocyte membranes as described by Batty et al. (1985). $[^3H]Ins(1,4,5)P_3$ was prepared as described by Downes et al. (1982).

10.3.2.4. Ins(1,5)P₂
Theoretically, $Ins(1,5)P_2$ may be prepared by phospholipase C hydrolysis of $PtdIns(5)P$ (Chapter 9).

10.3.2.5. Ins(3,5)P₂
Theoretically, $Ins(3,5)P_2$ could be prepared by phospholipase D hydrolysis of $PtdIns(3,5)P_2$ (Bird, 1998).

10.3.2.6. Ins(4,5)-P₂
Stephens et al. (1989a) prepared $Ins[^{32}P](4,5)P_2$ by boiling $GroPIns[^{32}P](4,5)P_2$ with 2 M KOH as described by Grado and Ballou (1961). After 1 h the reaction mixture is cooled and neutralized with $HClO_4$. The $KClO_4$ formed is pelleted by centrifugation and the ^{32}P-labeled reaction products are resolved on a Partisphere SAX column. The fractions containing the major $Ins[^{32}P]P_2$ product (approximately 50% of the recovered

radioactivity) are pooled and desalted. The identity of this peak was confirmed by treating an aliquot of $GroP[^3H]Ins[^{32}P](4,5)P_2$ in an identical manner. The major product had a $^3H/^{32}P$ ratio identical with that of the starting material.

Batty et al. (1989) prepared $[^3H]Ins(4,5)P_2$ in mixture with $[^3H]Ins(1,4,5)P_3$ and $[^3H]Ins(2,4,5)P_3$ by alkaline hydrolysis of authentic $[^3H]PtdIns(4,5)P_2$ (see above). The products were resolved and identified by HPLC in relation to reference standards (Dean and Moyer, 1987).

10.3.3. InsP₃

Not all of the theoretically possible $InsP_3$ have been shown to occur in nature and of those that have been identified and/or isolated, only a few have been prepared as reference compounds. Purely chemical methods for the synthesis of $Ins(1,4,5)P_3$ have been reported by numerous laboratories (including Aquilo et al., 1992; Salamonczyk and Pietrusiewicz, 1991; Liu and Chen, 1989), while chemo-enzymic synthesis has been reported by Ling and Okazaki (1994). Again these compounds have been unlabeled and chemical methods for the introduction labeled carbon, hydrogen or phosphate moieties have not been described. Sigma-Aldrich (Catalogue 2000) now offers for sale the following myo-$InsP_3$ isomers: $Ins(1,3,4)P_3$; $Ins(1,4,5)P_3$; $Ins(1,5,6)P_3$, $Ins(2,4,5)P_3$. Certain radiolabeled isomers are available from NEN-DuPont (Catalogue 2000).

10.3.3.1. Ins(1,3,4)P₃

Irvine et al. (1986a,b) obtained this by incubation of 3 μmol of $Ins(1,3,4,5)P_4$ with two additions (at zero time and 30 min) of 8 ml of human red cell membranes (Downes et al., 1982) at pH 7.5 and 4 mM magnesium acetate in 8 ml initial volume (Irvine et al., 1984). This treatment specifically removes the 5-phosphate (Batty et al., 1985) and the $InsP_3$ resulting was purified and desalted as for the other InsPs. The $InsP_3$ was at least 99% pure by electrophoresis.

From earlier data on the $InsP_3$ isomer formed (Batty et al., 1985) it was deduced that it must be $>95\%$ the D-1,3,4-isomer. Further purification was carried out by HPLC using $[^3H]Ins(1,4,5)P_3$ as marker to give a $>99\%$ pure sample.

Batty et al. (1989) prepared $[^3H]Ins(1,3,4)P_3$ from authentic $[^3H]Ins(1,3,4,5)P_4$ by incubation with human erythrocyte membranes as described by Batty et al. (1985). The reaction product was isolated and identified by HPLC using a modification of the method described by Dean and Moyer (1987).

Stephens et al. (1989a) has prepared $[^3H]Ins(1,3,4)P_3$ by using the method of Stephens et al. (1988a). Stephens et al. (1988a) incubated authentic $[^3H]Ins(1,3,4,5)P_4$ with human erythrocyte ghosts (1 h at 37°C) in a medium containing 2 ml of packed ghosts and 5 ml of a buffer containing 2 mM $MgCl_2$, 50 mM HEPES, 2 mM EGTA and 1 mg of BSA/ml, pH 7.0. $[^3H]InsP_3$ was purified from the assay mixture by chromatography on Partisil 10-SAX anion exchange HPLC column. The product was characterized by periodate oxidation, reduction, dephosphorylation and identification of the polyol as $[^3H]$altritol.

Stephens et al. (1989a) prepared similarly the [3-phosphate-^{32}P]Ins(1,3,4)P_3 by incubating 0.15 nmol of [3-phosphate-^{32}P]Ins(1,3,4,5)P_4 with 200 μl of packed human erythrocyte ghosts [prepared as described previously (Stephens et al., 1988a] in 50 mM HEPES/2 mM EGTA/2 mM $MgCl_2$/BSA (1 mg/ml), pH 7.0 (final volume 1 ml) at 37°C for 60 min. The reaction was quenched with $HClO_4$, neutralized with tri-n-octylamine and the products resolved on a Partisil 10-SAX HPLC column as described previously (Stephens et al., 1988c). Of the starting $Ins[^{32}P]P_4$, 87% was recovered as [3-phosphate-^{32}P]Ins(1,3,4)P_3.

Batty et al. (1985) prepared D-$[^3H]Ins(1,3,4)P_3$ by 5-phosphatase-specific dephosphorylation of D-$[^3H]Ins(1,3,4,5)P_4$, using the $InsP_4$ 5-phosphatase present in human erythrocyte membranes.

Theoretically, $Ins(1,3,4)P_3$ could be prepared by deacylation and deglyceration of $PtdIns(3,4)P_2$.

10.3.3.2. Ins(1,3,5)P₃

Theoretically, Ins(1,3,5)P₃ could be prepared by deacylation and deglyceration of PtdIns(3,5)P₂.

10.3.3.3. Ins(1,3,6)P₃

According to Stephens et al. (1989b) myo-[³H]Ins(1,3,6)P₃ made up 3.3% of the total myo-[³H]InsP₃ recovered from myo-[³H]Ins prelabeled avian erythrocytes. It was isolated along with [³H]Ins(1,3,4)P₃. It gave d-altritol on oxidation with periodate, reduction and dephosphorylation.

10.3.3.4. Ins(1,4,5)P₃

Irvine et al. (1986a,b) have provided a detailed description of the preparation of Ins(1,4,5)P₃ from PtdIns(4,5)P₂ using a modification of the method of Brown and Stewart (1966). The PtdIns(4,5)P₂ (Folch, 1949) was acid washed (0.1 M HCl) twice to remove divalent cations, which interfere with deacylation (Clarke and Dawson, 1981). The extract was dried down thoroughly, suspended in chloroform and redried. To 60 mg P of the lipid was added 300 ml of Clarke and Dawson's (1981) deacylation reagent. This gives a clean removal of the fatty acids with no cleavage of the GroInsP diester linkage. After 60 min at 59°C, the mixture was cooled on ice and then processed exactly as described by Clarke and Dawson (1981) with all volumes increased 100-fold. The resulting deacylated phospholipid preparation was thoroughly dried and resuspended in 100 ml of water and the pH checked to be <6.5. An equal volume of 0.03 M sodium periodate was added and the A_{254} was monitored; in general 15 min was sufficient to complete the rapid oxidation which is due to removal of the sn-1-carbon of the glycerol moiety (Brown and Stewart, 1966). The reaction was quenched with 10 ml of 3% ethylene glycol, and after 20 min, 50 ml of Brown and Stewart's (1966) dimethylhydrazine reagent was added. After 4 h under N₂, 60 ml of washed Dowex-W (H⁺ form) was added and the solution was filtered with Celite 545 prewashed with 0.1 M formic acid.

To the resulting filtrate was added 10 ml of Dowex 1 × 8–400 resin in the formate form to bind all the Ins(1,4,5)P$_3$. After filtration, the Dowex was washed with 20 ml of 0.1 M formic acid/0.4 ammonium formate, and then suspended in 40 ml of 0.1 M formic acid/1.2 M ammonium formate and filtered to remove the InsP$_3$. After six-fold dilution with water, the solution was loaded onto a 2 ml Dowex formate column and after washing with 30 ml of 0.1 M formic acid/0.4 M ammonium formate the InsP$_3$ was eluted with 30 ml of 0.1 M formic acid/0.8 M ammonium formate. The 0.8 M solution was diluted five-fold, loaded onto 1 ml Dowex chloride column, and eluted with 4 ml of 1 M LiCl which was then removed by drying and ethanol washes as described by Burgess et al. (1984b). The Ins(1,4,5)P$_3$ obtained was >99% pure. From each 60 mg P batch of PtdIns(4,5)P$_2$, 7–10 mg P Ins(1,4,5)P$_3$ (approximately 40 mg of InsP$_3$) (Irvine et al., 1986a,b). Liu and Chen (1989) prepared optically active D-myo-Ins(1,4,5) P3.

Letcher et al. (1990) prepared [^{32}P]Ins(1,4,5)P$_3$ from ^{32}P-labeled PtdIns(4,5)P$_2$. For this purpose, [^{32}P]PtdIns(4,5)P$_2$ is prepared first using human erythrocyte PtdIns 4-kinase to catalyze the phosphorylation of ^{14}C-labeled PtdIns to [^{14}C]PtdIns(4)P (see above under Preparation of PtdInsPs), and subsequently deacylated and the glycerol is removed by oxidation with mild periodate. According to Letcher et al. (1990) the glycerol moiety is removed from the [5-^{32}P]GroPIns(4,5)P$_2$ by a modification of the mild periodate oxidation method of Brown and Stewart (1966). To the dry [5-^{32}P]GroPIns(4,5)P$_2$ 1 ml of 50 mM sodium periodate is added. After mixing to dissolve the [5-^{32}P]GroPIns(4,5)P$_2$, the tube is placed in the dark for 30 min at room temperature. If autoxidation is allowed to continue for longer, the periodate will start to oxidize the inositol ring. Ethylene glycol (150 μl of 10% (v/v)) is added to stop the reaction and, after 15 min, 0.5 ml of 1% 1,1-dimethylhydrazine/ formic acid (pH 4) (Brown and Stewart, 1966) is added, and the mixture is allowed to stand for a further 60 min. The mixture is passed through 30 ml of a cation exchange resin (BioRad AG 50 W × 8 200–400 in the acid form) in a 1.6 cm diameter glass column.

The column is washed with 30 ml of H_2O, the pooled eluates being adjusted to pH 6 with dilute KOH and evaporated to dryness on a rotary evaporator. Finally, the $[^{32}P]Ins(1,4,5)P_3$ is dissolved in a small volume of H_2O for application to HPLC. Letcher et al. (1990) recommend that 2 ml of H_2O is used and the purification is performed in two runs. The column used is a 25 cm × 4.6 mm Partisil 10 SAX HPLC column supplied by Technicol Ltd. (Cheshire, UK). A flow-rate of 1.5 ml/min is used and the column is equilibrated with 0.24 M NaH_2PO_4 (adjusted to pH 3.8 with H_3PO_4). The 1 ml sample is injected on to the column and the eluting buffer is immediately increased to 0.55 M NaH_2PO_4 (pH 3.8, adjusted as before). An isocratic run in this buffer is continued for 20 min. A 20 μg standard of cold ATP is included in the sample injection and this is located with an ultraviolet detector at 254 nm, and would otherwise be eluted 10 min after injection. Fractions are collected at 15 s intervals from the appearance of the ATP peak, and collection is carried out for 10 min. The fractions are collected in 6 ml plastic vials and are counted by Cerenkov radiation: the $[^{32}P]Ins(1,4,5)P_3$ usually elutes at fractions 20–25. The relevant fractions are pooled, diluted 10 times with H_2O and loaded on to an Amprep SAX 100 mg cartridge (Amersham International, p.l.c) prewashed with 3 M ammonium formate, 0.1 M formic acid followed by 10 ml H_2O. The loaded Amprep is washed with 20 ml of 0.15 M ammonium formate, 0.015 M formic acid to remove the phosphate, and the $[^{32}P]Ins(1,4,5)P_3$ is eluted with 10 ml of 1 M triethylamine bicarbonate (freshly prepared by bubbling CO_2 through a 14% (w/v) triethylamine solution). The triethylamine bicarbonate is removed using a rotary evaporator and the residual $[^{32}P]Ins(1,4,5)P_3$ is dissolved in a small volume of water. A typical yield is 0.1 MBq $[^{32}P]Ins(1,4,5)P_3$ starting with 1.85 MBq of $[^{32}P]ATP$; i.e. 5.4% incorporation. The $[^{32}P]Ins(1,4,5)P_3$ can be converted to $[^{32}P]Ins(1,3,4,5)P_4$ if required by the method of Irvine et al. (1986).

Batty et al. (1985) prepared D-$[^{14}C]Ins(1,4,5)P_3$ by incubating primary cultured macrophages with $[^{14}C]Ins$ (20 μCi/ml;

Amersham) in a 3 cm diameter culturing dish for 24 h. A phospholipid extract was prepared and the product isolated by HPLC using a Partisil 10-SAX column and a gradient of ammonium formate (Batty et al., 1985). After desalting (Downes et al., 1986) the sample was deacylated with methylamine (Clarke and Dawson, 1981) and deglycerated by periodate oxidation and treatment with an aq. 1% 1:1 dimethylhydrazine. The mixture of the $[^{14}C]$InsPs produced by this procedure was passed through 5 ml of Bio-Rad AG 50W cation exchange resin, neutralized, freeze-dried and applied to an HPLC column (Partisil 10-SAX) in 1 ml of 10 mM EDTA, pH 7.0. The fractions containing D-$[^{14}C]$ Ins(1,4,5)P$_3$ amounting to 6% of the total $[^{14}C]$InsP preparation, were pooled, neutralized, desalted, yielding a sample of radio-chemically pure D-$[^{14}C]$Ins(1,4,5)P$_3$.

Stephens et al. (1989b) have reported that avian erythrocytes prelabeled for 24–48 h with myo-$[^{3}H]$Ins contain 36% myo-$[^{3}H]$Ins(1,4,5)P$_3$, which they isolated and identified along with other $[^{3}H]$InsP$_3$s.

Leung and Bittman (1997) have reported a convenient chemical synthesis of D-myo-Ins(1,4,5)P$_3$ and L-myo-Ins(1,4,5)P$_3$. Crystallization of the diastereomers of (+/−)-1-O-[(+)-menthoxycarbonyl]-6-O-benzyl-2,3:4,5-di-O-isopropylidene-myo-inositol diastereomers from methanol gives one diastereomer. Alkaline hydrolysis gives the useful inositol derivative (−)-6-O-benzyl-2,3:4,5-di-O-isopropylidene-myo-inositol. Likewise, crystallization of the diastereomers of (+/−)-3-O-[(−)menthoxycarbonyl]-4-O-benzyl-1,2:5,6-di-O-isopropylidene-myo-inositol from methanol gave a pure compound which could be hydrolyzed to give (+)-4-O-benzyl-1,2:5,6-di-O-isopropylidene-myo-inositol, a precursor to D-myo-inositol 3,5,6-trisphosphate [Ins(3,5,6)P3, (+)].

Ling and Okazaki (1994) have described a new practical route for the synthesis of D-myo-inositol 1,4,5-trisphosphate. Enzyme-catalyzed esterification of racemic 2,3-O-cyclohexylidene-myo-inositol (DL-1) proceeded exclusively in 1,4-dioxane to give optically pure L-1-O-acetyl-2,3-O-cyclohexylidene-myo-inositol (L-2) and D-2,3-O-cyclohexylidene-myo-inositol (D-1).

10.3.3.5. Ins(1,4,6)P₃

Stephens et al. (1989b) reported that avian erythrocytes prelabeled for 24–48 h with myo-[^{3}H]Ins contain 4.4% myo-[^{3}H]Ins(1,4,6)P₃, which they isolated and identified along with other [^{3}H]InsP₃s. It was isolated along with [^{3}H]Ins(1,4,5)P₃. It was resistant to human erythrocyte ghosts, yielding D-[^{3}H]iditol.

10.3.3.6. Ins(3,4,5)P₃

Stephens et al. (1989b) have reported that avian erythrocytes prelabeled for 24–48 h with myo-[^{3}H]Ins contain 13% myo-[^{3}H]Ins(3,4,5)P₃, which they isolated and identified. Human erythrocyte ghosts and 40 kDa protein in rat brain cytosol removed the 6-phosphate from [^{3}H]Ins(3,4,5,6)P₄ to yield [^{3}H]Ins(3,4,5)P₃.

Falck and Abdeli (1993) reported chemical synthesis of D-myo-Ins(1,5,6)P3 but neglected to include values for their optical activity. Adelt et al. (1999) have reported total syntheses of both optical antipodes of the enantiomeric pair D-myo-Ins(3,4,5)P₃ and D-myo-Ins(1,5,6)P₃. The ring system characteristic of myo-inositol was constructed de novo from p-benzoquinone. Subsequent transformations under stereo-controlled conditions led to enantio-pure C₂-symmetrical 1,4-(di-O-benzyldiphospho)conduritol B derivatives. These intermediates were used to prepare Ins(3,4,5,6)P₄ and Ins(1,4,5,6)P₄ in three steps. When subjected to hydrolysis by a partially purified InsP₅/InsP₄ phosphatase from *Dictyostelium discoideum*, these enantiomers could be converted to Ins(3,4,5)P₃ and Ins(1,5,6)P₃ on a preparative scale. The phosphatase possesses high regiospecificity and low stereospeci-ficity, allowing mirror positions of enantiomerically pure substrates to be converted with nearly equal rates.

10.3.3.7. Ins(3,4,6)P₃

According to Stephens et al. (1989b) myo-[^{3}H]Ins(3,4,6)P₃ makes up 9.7% of the total myo-[^{3}H]InsPs extracted. It was isolated together with myo-[^{3}H]Ins(3,5,6)P$_3$, which was resistant to erythrocyte ghosts, yielding L-[^{3}H]-iditol.

10.3.3.8. Ins(4,5,6)P$_3$

Stephens et al. (1989b) prepared Ins[^{32}P](4,5,6)P$_3$ by incubation of Ins[^{32}P](3,4,5,6)P$_4$ (Stephens et al., 1988c) in a solution containing 15 units of alkaline phosphatase/ml (units as defined by Sigma; type P5521/0.1 mM ZnCl$_2$/0.1% (w/v) BSA/10 mM ethanolamine (pH 9.5, 12°C)). Under these conditions optimum yields of Ins[^{32}P](4,5,6)P$_3$ were obtained after 10 min incubation. The reaction was quenched with HClO$_4$ (4%, v/v, final), neutralized with tri-n-octylamine/Freon (1:1, v/v; Sharpes and McCarl, 1982) and filtered before application to a Partisphere 5-SAX HPLC column (Stephens et al. 1988c).

10.3.3.9. Ins(1,2cyc4,5)P$_3$

Irvine et al. (1986a) prepared it by using non-radiolabeled red blood cells exactly as in Irvine et al. (1985), spiking the preparation either with ^{32}P-labeled Ins(1,4,5)P$_3$ purified by HPLC, or with [^{3}H]Ins(1,4,5)P$_3$ (Amersham). Fractions corresponding to cyclic InsP$_3$ (Irvine et al., 1985, 1986a) were collected, and after five-fold dilution and adjustment to a final concentration of 0.1 M formic acid and 0.15 M ammonium formate, poured down a 0.5 ml Dowex formate column. Inorganic phosphorus was removed by 10 ml of 0.1 M formic acid/0.4 M ammonium formate, and the cyclic InsP$_3$ was eluted with 0.1 M formic acid/0.8 M ammonium formate and desalted as for Ins(1,4,5)P$_3$ above. Cyclic InsP$_3$ and non-cyclic InsP$_3$ were separated by electrophoresis (Irvine et al., 1985, 1986). The samples were >90% pure.

10.3.3.10. Ins(1,2,4)P$_3$

Adelt et al. (2001) have reported the use of regiospecific phosphohydrolases from *Dictyostelium* as tools for the chemoenzymatic synthesis of the enantiomers D-myo-Ins(1,2,4)P$_3$ and D-myo-Ins(2,3,6)P$_3$ as non-physiological, potential analogues of biologically active D-*myo*-Ins(1,3,4)P$_3$. Starting with enantiopure dibromocyclohexenediol, several C$_2$ symmetrical building blocks were synthesized, which gave access to D-*myo*-Ins(1,2,4,5)P$_4$ and D-*myo*-Ins(1,2,3,6)P$_4$.

10.3.4. InsP$_4$

All the known natural InsP$_4$s believed to be part of the InsPs metabolism, including Ins(1,3,4,5)P$_4$ (Watanabe et al., 1990; Baudin et al., 1988) and Ins(1,3,4,6)P$_4$ (Watanabe et al., 1989) have been chemically synthesized. Likewise, the syntheses of Ins(3,4,5,6)P$_4$ and Ins(1,4,5,6)P$_4$ by conventional chemical (Pietrusiewicz et al., 1992) and chemo-enzymatic (Ling and Okazaki, 1994) procedures have been reported. Sigma-Aldrich (Catalogue 2000) now offer for sale the following reference standards: Ins(1,2,5,6)P$_4$; Ins(1,3,4,6)P$_4$ and Ins(3,4,5,6)P$_4$. [^{3}H]Ins(1,3,4,5)P$_4$ is also commercially available.

Some of the radiolabeled derivatives of InsP$_4$ required for metabolic studies, however, may have to be prepared in the biochemical laboratory. The high polarity of the InsP$_4$s presents special problems due to overlap with nucleotide polyphosphates, which are commonly removed by treatment with charcoal. A charcoal treated acid extract of ^{32}P-labeled avian erythrocytes, however, may still contain substantial quantities of ^{32}P-labeled nucleotide phosphates and [^{32}P]P$_i$. The contaminating compounds are most effectively removed by applying the sample to a Partisil 10 SAX HPLC column and eluting with a rapid developing ammonium formate/phosphoric acid gradient, i.e. conditions originally designed to separate Ins(1,3,4)P$_3$ (Irvine et al., 1987).

10.3.4.1. D/L [^{3}H]-or [^{32}P]Ins(1,2,3,4)P$_4$

According to Stephens (1990) D/L [^{3}H]-or [^{32}P]Ins(1,2,3,4)P$_4$ (in a roughly equal mixture of the two enantiomers) and [^{3}H]- or [^{32}P]Ins(1,2,5,6)P$_4$ can be obtained from mung bean seeds germinated in the dark with solutions of [^{3}H]Ins or [^{32}]P$_i$ and extracted with acid (see below for description of preparation of [^{32}P]InsP$_5$s from mung beans).

D/L-[^{3}H]Ins(1,2,4,6)P$_4$ has been prepared from [^{3}H]Ins(1,3,4,6)P$_4$ by controlled acid-catalyzed phosphate migration at 80°C for 100°C for 8 min with 1 M HCl (Stephens et al., 1988a,b). Ins(1,2,4,6)P$_4$ is eluted from a Partisphere SAX column substantially later than Ins(1,3,4,6)P$_4$. Both Ins(1,2,3,4)P$_4$ and Ins(1,2,5,6)P$_4$ are eluted from a Partisphere SAX column after Ins(1,3,4,5)P$_4$; Ins(1,2,5,6)P$_4$ is retained longer (Stephens, 1990).

10.3.4.2. Ins(1,2,4,5)P$_4$

Chung et al. (1998) have synthesized D- and L-myo-inositol 1,2,4,5-tetrakisphosphate. D- and L-myo-Ins(1,2,4,5)P$_4$, which are analogues of D-myo-Ins(1,4,5)P$_3$, a calcium mobilizing second messenger, were obtained via resolution of the camphanate ester of a myo-inositol derivative.

10.3.4.3. Ins(1,3,4,5)P$_4$

Irvine et al. (1986b) prepared it by phosphorylation of Ins(1,4,5)P$_3$ by a rat brain supernatant under incubation conditions similar to those described by Irvine et al. (1986a), except that a pH of 9 was used, as at this pH there is negligible InsP$_3$ or InsP$_4$ phosphatase activity. Each batch contained 20 μmol of Ins(1,4,5)P$_3$ and the final volume of the reaction, containing 0.1 M Tris/maleate, pH 9.0, 20 mM Mg^{2+} and 10 mM ATP was 40 ml. After quenching of the reaction, removal of TCA and neutralization (Irvine et al., 1986a), the solution (50 ml) was mixed with 40 ml of 0.2 M glycine/NaOH, pH 8.6, containing 0.1 M magnesium acetate and poured down a 1.7 cm × 20 cm column of phenyl boronate (Amincon)(pre-equilibrated with the glycine/Mg^{2+} buffer), which holds back ATP. The ATP free eluate was mixed with one vol. of 0.2 M formic acid/0.4 M ammonium formate and 0.5 volume of water, and the InsPs were removed by adding 5 ml of Dowex formate. The column was eluted as described for Ins(1,4,5)P$_3$ isolation (Irvine et al., 1986a), except that the final separation of InsP$_3$ and InsP$_4$ was achieved on a 2 cm × 0.8 cm Dowex formate column (Batty et al., 1985). Ins(1,3,4,5)P$_4$ was obtained in 50–56 μmol yield from 20 μmol of Ins(1,4,5)P$_3$ and was found to be 99% free from P$_i$, adenine nucleotide or InsP$_3$ contamination when examined by electrophoresis. The purity of the preparation could be further increased by HPLC separation (Batty et al., 1985; Irvine et al., 1986b), with the InsP$_4$ fractions then desalted as for Ins(1,2cyc4,5)P$_3$ (see above).

Stephens et al. (1989b) prepared [3-phosphate-^{32}P]Ins(1,3,4,5)P$_4$ by incubating 2.0 nmol of Ins(1,4,5)P$_3$ (Amersham International) with 57.2 μCi of [γ-^{32}P]ATP (tetra-ethylammonium stabilized salt, which had been purified by anon exchange HPLC as described previously (Stephens et al., 1988b) before being used as a substrate, and 200 μl of rat brain cytosol in 50 mM HEPES/2 m EGTA/15 mM 2-mercaptoethanol/0.5 mM MgNa$_2$ATP/2 M MgCl$_2$ (pH 7.0, 37°C) in a total volume of 1 ml for 2.5 min. The reaction was terminated with HClO$_4$, treated with charcoal, neutralized, and purified by HPLC as described previously (Stephens et al., 1988b). The fractions (containing Ins[^{32}P]P$_4$ species were neutralized with triethylamine and desalted as described above. The yield was 0.76 mol of [3-phos-phate-^{32}P]Ins(1,3,4,5)P$_4$] or 38% of the starting material. In order to assess the homogeneity of the [3-phosphate-^{32}P]Ins(1,3,4,5)P$_4$ preparation, an aliquot was mixed with [^{3}H]Ins(1,3,4,6]P$_4$ and applied to a Partisphere WAX anion exchange column and eluted as described by (Stephens et al., 1988c). The ^{32}P-labeled compounds were eluted as a single peak with baseline separation from the [^{3}H]Ins(1,3,4,6)P$_4$.

Total synthesis of D-myo-Ins(1,3,4,5)P$_4$ has been reported by Falck and Abdali (1993), who also prepared Ins(3,4,5)P$_3$ by chemical synthesis. Hirata et al. (1991) synthesized the Ins(1,3,4,5)P$_4$ analogues, 2-O-(4-aminobenzoyl)myo-inositol (1,3,4,5)tetrakisphosphate and 2-O-(4-aminocyclohexanecarbonyl)-*myo*-inositol(1,3,4,5)tetrakisphosphate, as follows. A solution of 6-*O*-benzyl-1,3,4,5-tetrakis-(1′,5′-dihydrobenzodioxaphosphe-lynyl)2-*O*-(4′-nitrobenzoyl)-*myo*-inositol (271 mg) and 5% palla-dium/carbon (20 mg) in 10 ml of methanol/water (4:1, v/v) was stirred under H_2 gas for 24 h. After filtration of the mixture to remove Pd/C, the filtrate was evaporated and chromatographed on an Avicel column to afford 2-*O*-(4-aminobenzoyl)*myo*-inositol (1,3,4,5)tetrakisphosphate (79.9 mg, 50%). A mixture of the product (14.9 mg) and ruthenium oxide (21.5 mg) in water (3.75 ml) and H_2 gas [8 Mpa, 80 atm) were charged in a 50 ml autoclave and heated at 60°C for 2 h. The catalyst was filtered off and the filtrate was evaporated. The residue was chromatographed on a cellulose column using *n*-propanol/conc. aq. NH_4/water (5:4:1, by vol.) to give 2-*O*-4-aminocyclohexanecarbonyl)-*myo*-inositol(1,3,4,5)tetrakisphosphate (14.3 mg, 96.8%; Rf 0.24).

10.3.4.4.   Ins(1,3,4,6)P$_4$

$[^3H]$Ins(1,3,4,6)P$_4$ has been prepared from $[^3H]$Ins(1,3,4)P$_3$, which is commercially available, by several groups (Shears et al., 1987; Balla et al., 1987; Stephens et al., 1988a,b) using purified Ins(1,3,4)P$_3$ 6-OH kinase preparations (Stephens et al., 1988a,b; Hansen et al., 1987). However, reasonable yields (greater than 40% of starting material) can be obtained using crude cytosol preparation from rat brain. In the latter case the product must be purified by HPLC because 5–10% of the total $[^3H]$InsP$_4$ recovered is $[^3H]$Ins(1,3,4,5)P$_4$ due to the presence of Ins(1,3,4)P$_3$ 5-OH kinase activity in the rat brain (Shears, 1989).

According to Stephens (1990) the reaction is run in 400 μl of buffer and this volume can be loaded on a typical analytical-scale anion exchange HPLC column without adversely affecting column performance. The buffer contains 5 mM ATP, 6 mM $MgCl_2$, 10 mM creatine phosphate, 1 mM dithiothreitol, 100 mM KCl, 50 mM HEPES (pH 7.0, 37°C), 2 mM EGTA, 5 units creatine phosphokinase/ml, and 80 μl of a 10 000*g* rat brain supernatant (prepared from one 250 g male rat brain homogenized in 5 ml of

0.25 M sucrose, 50 mM HEPES (pH 7.0, 4°C), 2 mM EGTA, 0.1 mM phenylmethyl sulphonylfluoride, 1 mM dithiothreitol). Incubation is at 37°C for 20 min.

However, when rat brain cytosol is used as source of Ins(1,3,4)P$_3$ 6-kinase activity, the product needs to be purified on HPLC anion column, because 5–10% of the total [^{3}H]InsP$_4$ recovered would be [^{3}H]Ins(1,3,4,5)P$_4$ as a consequence of an Ins(1,3,4)P$_3$ 5-OH kinase activity found in rat brain cytosol (Stephens, 1990).

By this means, Ins(1,3,4,5)P$_4$, Ins(1,3,4)P$_3$ and Ins(3,4,5,6)P$_4$ are eluted together but separated from InsP$_5$. The InsP$_4$ and InsP$_5$ peaks are then collected and desalted (see below and Stephens and Downes, 1990). Better resolution of InsP$_4$s can be obtained on SAX HPLC columns by using phosphate buffers (Dean and Moyer, 1987). Under these conditions, InsP$_4$s is eluted from a standard analytical-scale SAX HPLC column with 1 M NaH$_2$PO$_4$·NaOH (pH 3.8, 25°C) in the order Ins(3,4,5,6)P$_4$, Ins(1,3,4,5)P$_4$ and Ins(1,2,4,6)P$_4$ (Stephens, 1990), which can then be detected in the column eluant by Cerenkov counting; or by removing small portions of each of the fractions collected and counting them individually for ^{3}H.

Ongusaha et al. (1998) have purified a soluble Mg^{2+}-dependent kinase of molecular mass of about 41 kDa that converts Ins(1,4,5)P$_3$ into Ins(1,4,5,6)P$_4$ and Ins(1,3,4,5)P$_4$ in a constant ratio of about 5:1 and thence to Ins(1,3,4,5,6)P$_5$ and also InsP$_6$. [^{3}H]Ins(1,4,5,6)P$_4$ was made by phosphorylating [^{3}H]Ins(1,4,5)P$_3$ with partially purified *Schizosaccharomyces pombe* Ins(1,4,5)P$_3$ kinase. The experiments used [^{3}H]Ins(1,4,5)P$_3$ [1–5 nM; either 20 000 dpm or 100 000 dpm as substrate], under first order conditions, with 5 mM ATP as phosphate donor. Yeast cell fractions or purified kinase preparations were incubated at pH 7.5 and 28°C in 20 mM HEPES/1 mM EGTA containing ATP, 10 mM creatine phosphate and 10 units/ml creatine kinase. ^{3}H-labeled assays were stopped with 10% PCA, left for 10 min on ice and centrifuged. After deproteinization and desalting the eluates from the [^{32}P]ATP assays were spiked with [^{3}H]Ins(1,3,4,5)P$_4$ and [^{3}H]Ins(1,4,5,6)P$_4$

and analyzed by HPLC on a Partisil 10 SAX column, using a pH 3.8 $(NH_4)_2PO_4$ gradient. (Ongusaha et al., 1998).

However, $Ins(3,4,5,6)P_4$ is usually poorly resolved from $Ins(1,3,4,5)P_4$ and, as the major [^{3}H]$InsP_4$ in extracts from control cells is usually [^{3}H]$Ins(3,4,5,6)P_4$ and the minor one [^{3}H]$Ins(1,3,4,5)P_4$, this is not an ideal system for analyzing cell derived extracts; WAX (described below) is more suitable. The preparations of Partisil 10 SAX-purified [^{32}P]$InsP_4$ standards, or alternatively, the crude, neutralized ^{3}H-labeled cell extracts, are then resolved on a Partisphere WAX HPLC column. These columns separate all three species of $InsP_4$ found in the avian erythrocytes in order of increasing retention time: $Ins(1,3,4,6)P_4$, $Ins(1,3,4,5)P_4$ and $Ins(3,4,5,6)P_4$ (Stephens et al., 1988a). A new column is eluted isocratically with 0.16 M $(NH_4)_2HPO_4 \cdot H_3PO_4$ (pH 3.2, 22°C) and $InsP_4$s should emerge after 40–80 min.

10.3.4.5. *Ins(1,4,5,6)P₄*

The major $InsP_4$ in chicken erythrocytes has been identified as either D- or L-$Ins(1,4,5,6)P_4$ (Johnson and Tate, 1969), as indicated by the formation of iditol upon periodate oxidation, reduction and dephosphorylation. Stephens et al. (1988b) have demonstrated that both avian and mammalian cells contain L-$Ins(1,4,5,6)P_4$, which may be an intermediate in the synthesis or breakdown of $InsP_5$, which is also synthesized in both chick erythrocytes and macrophages.

Ongusaha et al. (1998) prepared [^{3}H]$Ins(1,4,5,6)P_4$ by phosphorylating [^{3}H]$Ins(1,4,5)P_3$ with partially purified *S. pombe* $Ins(1,4,5)P_3$ kinase. The product was purified by HPLC, and peak fractions were pooled, neutralized with 20 mM HEPES/KOH (pH 7.2), diluted 10-fold, and loaded on to 200 µl AG-1(formate form) resin columns. These were washed with 10 ml of 0.3 M ammonium formate/0.1 M formic acid and 10 ml water. Triethylammonium hydrogen carbonate (TEAB) (10 ml of 1 M) then eluted Ins(1,4,5,6)P$_4$ and 20 ml of 1 M TEAB eluted Ins(1,3,4,5,6)P$_5$; TEAB was removed by lyophilization.

10.3.4.6. Ins(3,4,5,6)P$_4$

^{3}H- and ^{32}P-labeled Ins(3,4,5,6)P$_4$ have so far been successfully prepared only from appropriately radiolabeled intact cells. A very convenient cell type for this purpose is the chick erythrocyte, as it is easily prepared, robust and readily incorporates both [^{3}H]Ins and [^{32}P]P$_i$ into InsPs. Avian erythrocytes also contain Ins(1,3,4,5)P$_4$, Ins(1,3,4,6)P$_4$ and Ins(1,3,4,5,6)P$_5$ (see below), which also become rapidly labeled by exogenously supplied [^{32}P]P$_i$ and [^{3}H]Ins. Therefore, the same avian erythrocyte preparations can be used to obtain [^{3}H]Ins or [^{32}P]P$_i$-labeled Ins(1,3,4,6)P$_4$, Ins(1,3,4,5)P$_4$, and Ins(1,3,4,5,6)P$_5$, provided effective anion exchange columns are available to ensure complete separation of the InsPs produced.

According to Stephens (1990) approximately 0.3 ml of packed erythrocytes is all that can be effectively chromatographed on an analytical-scale Partisphere WAX HPC column. The incubations can then be done in a total volume of 0.8–3.0 ml of medium containing 0.3 ml of packed cells. The cell suspensions are usually incubated in a single 1 ml well of a typical 12 well tissue-culture plate, with the lid held in place by a strip of Parafilm (to reduce water loss through evaporation), and the entire plate is attached to the bottom of a large, flat-bottomed, glass trough (which is then covered over with Parafilm) using a sheet of Plasticene. The entire assembly is then placed in a shaking, thermostatically controlled (38–40°C) water bath. After 20–36 h the cells are removed, placed in 10 ml of ice-cold isotonic saline in a 10 ml pointed-tipped polypropylene test tube, and pelleted by centrifugation. Well over 90% of the cells should be recovered. The saline supernatant should be rapidly aspirated and the cells lysed by adding 1 ml of ice-cold H$_2$O, vortexing for 5 s, adding 50 μl of 70% v/v PCA, and vortexing for 20–30 s. The cell debris produced during the above procedure is pelleted by centrifugation for 5 min at 4°C. The acid soluble supernatants from [^{3}H]Ins labeled extracts are then neutralized immediately. ^{32}P-labeled extracts are incubated for 15 min, on ice, with charcoal to remove nucleotides, then neutralized.

If a charcoal extraction, using Norit GSX or Darco-G60 is to be incorporated into the purification procedure, to remove the majority of [^{32}P]ATP from cell extracts or assays, it should be thoroughly acid-washed (five times with ice-cold 5% PCA). It should then be mixed only with InsPs extracts under acidic conditions, e.g. in presence of 4% ice-cold PCA (Stephens and Downes, 1990) or at high salt concentrations (Mayr, 1988). The charcoal is added as a 25 mg/ml slurry in 4% PCA with an air displacement pipette (100 μl is used with an extract from 0.3 ml of packed cells). The charcoal is recovered after 15 min (on ice) by adding 40 μl of 10% (w/v in H$_2$O) BSA, vortexing, and then centrifuging the samples.

Stephens (1990) recommends quenching all assays and cell preparations with ice-cold PCA (final concentrations range from 2 to 5%) because PCA has the advantage that a number of techniques can be employed that simultaneously remove the acid and neutralize the extracts, hence minimizing the ionic strength of the final sample. Two neutralization strategies, which are employed most often consist of KOH and tri-n-octylamine/Freon (1:1, v/v) (Sharpes and McCarl, 1982). The former takes advantage of the relatively low solubility of KClO$_4$. The PCA precipitated suspension is vortexed and then a pellet is obtained by centrifugation. The supernatant above the PCA-insoluble material is removed and added directly to a previously established fixed volume of 2 M KOH, 0.1 M MES buffer (pKa around 6) and 10–20 mM EDTA, such that the final pH is 6.0. The advantage of this procedure is that absolutely no problems are encountered with non-specific losses of InsPs despite an absence of carrier, low ionic strength conditions or the presence of divalent cations (so long as sufficient EDTA has been added). If tri-n-octylamine/Freon is to be used as the neutralizing agent, the ice-cold supernatant from the centrifugation of the PCA quenched samples is added to 1.1 vol. of tri-n-octylamine/Freon (freshly prepared at room temperature) into which an appropriate portion of Na_2EDTA (e.g. 50 μl of 0.1 M) solution has been placed. The sample is then very well mixed, and centrifuged at the maximum setting of a bench-top centrifuge or in a microfuge for 5–10 min (again at room temperature). A three-layered system should result: an upper aqueous phase, that should be pH 5–6 (depending on the pH of the original sample); a viscous middle layer of tri-n-octylamine perchlorate; and a lower phase containing the Freon and any excess tri-n-octylamine. The disadvantages are that some organic residues invariably contaminate the upper phase and these may occasionally interfere with subsequent assays, e.g. Ins(1,4,5)P$_3$ binding assays (Palmer et al., 1989).

Recently, Adelt et al. (1999) have reported total syntheses of both optical antipodes of the enantiomeric pair D-myo-Ins(3,4,5,6)P$_4$ and D-myo-Ins(1,4,5,6)P$_4$. The ring system characteristic of myo-inositol was constructed de $novo$ from p-benzoquinone. Subsequent transformations under sterically-controlled conditions led to enantiopure C_2-symmetrical 1,4-(di-O-benzyldiphospho)conduritol B derivatives. These intermediates were used to prepare Ins(3,4,5,6)P$_4$ and Ins(1,4,5,6)P$_4$ in three steps, one of which involved flash dihydroxylation. The Ins(3,4,5,6)P$_4$ and Ins(1,4,5,6)P$_4$ have been synthesized previously (Pietrusiewicz et al., 1992; Ozaki et al., 1994; Romer et al., 1995, 1996) but there remains confusion in the assignment of the absolute configuration (Shears, 1996) as is apparent from the opposite signs of the optical rotation reported for the same compounds.

10.3.5. InsP$_5$s

There are six possible myo-InsP$_5$s: two pairs of enantiomers (D- and L-Ins(1,2,4,5,6)P$_5$ and D- and L-Ins(1,2,3,4,5)P$_5$, and two isomers in which the phosphates do not disturb the plane of symmetry in the parent myo-Ins moiety Ins(1,2,3,4,6)P$_5$ and Ins(1,3,4,5,6)P$_5$.

There are six myo-Ins$_5$s, two of which possess a plane of symmetry; the remaining four form two pairs of enantiomers (see above). Consequently, non-chiral chromatographic techniques would be expected to resolve a maximum of four species of InsP$_5$ Ins(1,2,3,4,6)P$_5$, Ins(1,3,4,5,6)P$_5$ ($meso$-compounds which can, in theory at least, be purified completely by standard chromatographic techniques) and D- and L-Ins(1,2,3,4,5)P$_5$ and D- and L-Ins(1,2,4,5,6)P$_5$ [i.e. the best that non-chiral chromatography could achieve with these isomers is, for example, to obtain an Ins(1,2,4,5,6)P$_5$ that did not contain any of the other four non-enantiomeric InsP$_5$s, but could itself contain either D- or L-Ins(1,2,4,5,6)P$_5$].

Sigma-Aldrich lists Ins(1,3,4,5,6)P$_5$ as the only commercially available InsP$_5$ reference standard. However, other companies, such as NEN-DuPont, may provide additional standards in both ^{3}H and ^{32}P-form.

[3]H]InsP$_6$ or ^{32}P]InsP$_6$ can be dephosphorylated under alkaline conditions to yield a mixture of all the InsP$_5$s. Heating a solution of InsP$_6$ (in 50 mM Na$_2$HPO$_4$·NaOH, pH 10.5, 25°C) at 120°C in an autoclave for 3 h results in a fairly equal mixture of InsP$_5$ (in about 15–20% yield from the InsP$_6$ (Phillippy et al., 1987); however, it must be emphasized that the Ins(1,2,3,4,5)P$_5$ and Ins(1,2,4,5,6)P$_5$ species will be racemic mixtures. The four chromatographically distinct types of InsP$_5$ can all be purified from the hydrolysate by anion exchange HPLC (see below) after the pH of the sample has been titrated back to pH 6–7 (with formic acid), and a 50 µl sample of 0.1 M EDTA·NaOH (pH 7.0; 25°C) has been added.

Several published methods have resolved four species of InsP$_5$. These include: moving paper electrophoresis (Tate, 1968); ion exchange HPLC utilizing an ion exchange column and nitric acid as an eluant [this requires a nitric acid-compatible HPLC system (Phillippy and Bland, 1988)]; anion exchange chromatography on 1 m long Dowex resin columns [the resin is in the chloride form and is eluted with HCl; run times extend to many hours (Cosgrove, 1969)] on commercial HPLC columns with HCl as an eluant [this requires an HCl-compatible solvent delivery system etc. (Mayr, 1988)]. All of the HCl- or HNO_3-based systems require considerable care to avoid acid-catalyzed migration of the phosphate groups in the InsPs being purified. All of these systems have the significant advantage of resolving the $InsP_5$s in one run. Mayr (1988) has developed an alternative set of conditions that can fully resolve all four species of $InsP_5$ by using the standard phosphate buffers and anion exchange HPLC columns that are commonly employed to separate $InsP_4$s and $InsP_3$s. The disadvantage of the procedure based on phosphate-buffered eluants is that two chromatographic steps are required; however, the resolution of the four $InsP_5$ species is complete even with well used columns and no phosphate migration problems occurs.

On standard SAX HLC column (e.g. Partisil 10 SAX or Partisphere 5 SAX), all $InsP_5$s can be effectively separated by eluting with a gradient of $0.6-1.0$ M $(NH_4)_2HPO_4 \cdot H_3PO_4$ (pH 3.8, 22°C) (Mayr, 1990). The order of elution is $Ins(1,3,4,5,6)P_5$, D/L-$Ins(1,2,3,4,5)P_5$ and finally $Ins(1,2,3,4,6)P_5$ and D/L-$Ins(1,2,4,5,6)P_5$, which elute almost precisely together. On a standard WAX HPLC column (e.g. Partisphere WAX) eluted with a gradient of $0.3-0.4$ M $(NH_4)_2HPO_4 \cdot H_3PO_4$ (pH 3.2 at 22°C); these ionic strengths are appropriate for a new column, but the concentration of phosphate buffer needs to be reduced to maintain resolution as the column ages. The order of elution is $Ins(1,2,3,4,6)P_5$, D/L-$Ins(1,2,3,4,5)P_5$, and finally $Ins(1,3,4,5,6)P_5$ and D/L-$Ins(1,2,4,5,6)P_5$ eluting very close together. Because of column instability and aging, it is desirable to apply unknown samples (mixed with appropriate standards) to a SAX HPLC column. If any material elutes in the $Ins(1,2,3,4,6)P_5/(1,2,4,5,6)P_5$ region then a portion of the same sample is applied to a WAX column. Alternatively, the sample applied to the SAX column could be desalted—from the scintillation fluid in which it was counted it could be desalted, if necessary (Stephens and Downes, 1990).

10.3.5.1. Ins(1,2,3,4,5)P_5 and Ins(1,2,4,5,6)P_5

D,L-Ins(1,2,3,4,5)P_5 is one of the principal types of InsP$_5$ that can be isolated chromatographically from germinating mung bean (*Phaseolus aureus*) seedlings (Stephens et al., 1991). Typically, one mung bean is germinated in 0.75 ml of sterile H$_2$O containing [^{3}H]Ins or [^{32}P]Ins, in the dark, for 50–60 h at 25°C. The resulting seedling is homogenized in 2 ml of ice-cold 5% PCA using a 'studded' glass-on-glass hand homogenizer. The acid-insoluble debris is pelleted in a microfuge and the supernatant neutralized with tri-*n*-octylamine/Freon as described above. Two InsP$_5$s can be isolated from mung bean seedlings: D/L-Ins(1,2,3,4,5)P_5 (proportion of enantiomers undetermined) and D/L-Ins(1,2,4,5,6)P_5 (largely the L-isomer). The D/L-Ins(1,2,4,5,6)P_5, which is retained longer during HPLC on both SAX and WAX, has been completely characterized by ^{31}P and two-dimensional ^{1}H NMR.

Ins(1,2,3,4,5)P_5, with either little or no L-Ins(1,2,3,4,5)P_5 in it, was determined to be a minor product of a fungal phytase (Irving and Cosgrove, 1972) acting on InsP$_6$. Its enantiomeric configuration was determined by measuring the specific optical rotation of the product and comparing it to a characterized Ins(1,2,3,4,5)P_5 preparation. Stephens (1990) utilizes a commercially available preparation of the enzyme activity with ^{3}H- or ^{32}P-labeled InsP$_6$ (see below) to prepare ^{3}H- or ^{32}P-labeled Ins(1,2,3,4,5)P_5. Thus, [^{3}H]- or [^{32}P]InsP$_6$ (at concentration of about 3 μM, i.e. in the first-order range), 20 mM CH$_3$COOH·NaOH (pH 5.0, 37°C), 1 mg BSA/ml, 0.0008 units *Aspergillus phytase* units defined by Sigma. Optimal yields of InsP$_5$s [the major isomer, Ins(1,2,4,5,6)P_5] are obtained after 60 min at 37°C. The InsP$_5$ products are resolved by anion exchange HPLC.

According to Tomlinson and Ballou (1962), D-Ins(1,2,3,5,6)P_5 [= L-Ins(1,2,3,4,5)P_5] is the major product of a crude preparation of wheat-bran phytase. It has been more recently resolved both by differing properties (e.g. inhibitors, pH) and by chromatography. The F$_1$ component of this enzyme preparation is largely responsible for the specific removal of the 4-phosphate group (Lim and Tate, 1973).

This pioneering work has been exploited to prepare ^{3}H- and ^{32}P-labeled Ins(1,2,4,5,6)P$_5$ from ^{3}H- or ^{32}P-labeled InsP$_6$. Thus, Stephens (1990) proceeds as follows: a commercially available, crude wheat-bran phytase (Sigma) is incubated in a buffer containing 20 mM CH$_3$COONa (pH 5.0, 37°C), 1 mg BSA/ml, 3 µM InsP$_6$, and 0.00072 units of wheat-bran phytase. After 60 min, approximately 10% of the InsP$_6$ phosphorus would have been released yielding the optimal amounts of InsP$_5$. The assay is quenched with PCA and processed with KOH/MES, as described above. The [^{3}H]- or [^{32}P]InsP$_5$ is purified by anion exchange HPLC and desalted.

Stephens et al. (1991) have described a preparation of high-specific radioactivity D/L-[^{3}H]Ins(1,2,4,5,6)P$_5$, D/L-[^{3}H]Ins(1,2,3,4,5)P$_5$ or [^{3}H]Ins(1,2,3,4,6)P$_5$ by phosphorylation of [^{3}H]Ins with polyphosphoric acid as described by Cosgrove (1980), except that high-specific radioactivity [^{3}H]Ins (80–120 Ci/mmol; Amersham) was included in the reaction mixture without additional carrier and the quantities of the reagents employed were substantially decreased (Hawkins et al., 1990). After the reaction was complete, the sample was neutralized, diluted and the [^{3}H]InsPs were isolated by ion-exchange chromatography on an AG-1 (formate form) resin column and desalted by freeze-drying as described by Hawkins et al. (1990). [^{3}H]InsP$_5$ isomers and [^{3}H]InsP$_6$ were purified from the mixture by anion-exchange HPLC. The samples (20 000 dpm) or the isomeric [^{3}H]InsP$_5$s were mixed with 5000 dpm of Ins[^{32}P]P$_6$ and applied to 12 cm × 0.6 cm columns of Bio-Rad AG 1 × 8 200–400 resin (chloride form) and then eluted with sequential 10 ml batches of 0.65 M HCl or 0.63 M HCl followed by two 10 ml batches of 0.65 M HCl.

10.3.5.2. D-Ins(1,2,3,5,6)P$_5$

Stephens et al. (1991) prepared both ^{3}H- and ^{32}P-labeled D-Ins(1,2,3,5,6)P$_5$ from [^{3}H]- or [^{32}P]InsP6 using commercially available wheat-bran phytase. Assays contained [^{3}H]- or [^{32}P]InsP$_6$ (final concentrations in the 10 nm–10 µM range), 0.0006 units of phytase/ml, 1 mg of BSA/ml and 20 mM acetic acid (pH 5.0 with

NaOH, 37°C). After 60 min at 37°C the assays were quenched and the reaction products resolved by HPLC as described for the products of the fungal phytase attack on $InsP_6$ above. $8-15\%$ of the starting material (for $[^3H]InsP_6$) was typically recovered as $InsP_5$.

10.3.5.3. D-Ins(1,2,4,5,6)P₅

D-Ins(1,2,4,5,6)P_5 is the major product of a fungal phytase (Irving and Cosgrove, 1972) and can be prepared with a commercially available crude preparation of *Aspergillus ficuum* (Sigma). Optimal yields of $InsP_5$ were obtained when approximately $10-15\%$ of $InsP_6$ phosphorus had been released as P_i. Incubation of $InsP_6$ with 0.0008 units of *A. ficuum*/ml in 20 mM acetic acid (pH 5.0 with NaOH, 37°C)/1 mg BSA/ml resulted in 10% of the starting $InsP_6$ being dephosphorylated in 60 min at 37°C. D/L-Ins(1,2,4,5,6)P_5 is usually the major product formed during the early stages of the alkaline hydrolysis of $InsP_6$ and so this technique provides the best source of racemic mixtures of the two enantiomers. McConnell et al. (1991) have detected substantial amounts of three $[^3H]InsP_6$ isomers in $[^3H]$inositol-labeled human lymphoblastoic (T5-1) cells. Their structures were determined by HPLC (Phillippy and Bland, 1988) and by utilizing a stereo-specific D-Ins(1,2,4,5,6)P_5 3-kinase from *D. discoideum* (Stephens and Irvine, 1990). The structures were: D-Ins(1,2,4,5,6)P_5, L-Ins(1,2,4,5,6)P_5 and Ins(1,3,4,5,6)P_5.

10.3.5.4. Ins(1,3,4,5,6)P₅

Ins(1,3,4,5,6)P_5 is the only isomer of $InsP_5$ that has been described in animal cells. Johnson and Tate (1969) originally characterized the $InsP_5$ extracted from avian erythrocytes as Ins(1,3,4,5.6)P_5 and these cells still remain the best source of this $InsP_5$ isomer either for labeling with ^3H- or ^{32}P or for larger scale preparations. Stephens et al. (1991) prepared Ins$[^{32}P]$(1,3,4,5,6)P_5 from $[^{32}P]P_i$-labeled avian erythrocytes precisely as described by Stephens et al. 1988b). Stephens et al. (1988b) have demonstrated that Ins(1,3,4,5,6)P_5 is a product of phosphorylation of L-Ins(1,4,5,6)P_4 by a widely distributed cytosolic InsPs kinase.

McConnell et al. (1991) have demonstrated that multiple isomers of InsP$_5$ are formed in vivo and in vitro by a transformed lymphocyte cell line. In incubations with permeabilized T5-1 cells, both 1,3,4,6- and 3,4,5,6- isomers of InsP$_4$ were phosphorylated solely to Ins(1,3,4,5,6)P_4.

Ongusaha et al. (1998) prepared [^{3}H]Ins(1,3,4,5,6)P_5 by incubating [^{3}H]Ins(1,3,4)P_3 with rat liver cytosol (Shears, 1989). The product was purified by HPLC, and peak fractions were pooled, neutralized with 20 mM HEPES/KOH (pH 7.2), diluted 10-fold, and loaded on to 200 µl AG-1(formate form) resin columns. These were washed with 10 ml of 0.3 M ammonium formate/0.1 M formic acid and 10 ml water. TEAB (10 ml of 1 M) then eluted Ins(1,4,5,6)P_4 and 20 ml of 1 M TEAB eluted Ins(1,3,4,5,6)P_5; TEAB was removed by lyophilization.

10.3.5.5. Ins(1,2,3,4,6)P$_5$

Stephens et al. (1991) have shown that Ins(1,2,3,4,6)P_5 and Ins(1,3,4,5,6)P_5 are obtained pure, but D/L-Ins(1,2,4,5,6)P_5 and D/L-Ins(1,2,3,4,5)P_5 as racemic mixtures by alkaline hydrolysis of InsP$_6$ as described by Phillippy et al. (1987). Essentially, HPLC pure InsP$_6$ (either ^{3}H- or ^{32}P-labelled) is dissolved in 0.5 ml of 50 mM Na$_2$HPO$_4$ (pH 10.5 with NaOH) and then autoclaved (121°C) for 3 h. After autoclaving, the sample was cooled, diluted to 2.5 ml and its pH adjusted to 7.0 with formic acid. InsP$_5$ isomers were resolved from this mixture by anion exchange HPLC, first on a Partisphere SAX column, and then on a Partisphere WAX column.

10.3.6. InsP$_6$

10.3.6.1. InsP$_6$

InsP$_6$ (phytic acid) is commercially available. InsP$_6$ can be prepared by chemical phosphorylation as described below (Stephens et al., 1990; Hawkins et al., 1990).

10.3.6.2. [³H]InsP₆ and [³²P]InsP₆

[³H]InsP₆ is now commercially available, but it can be easily prepared at high-specific radioactivity and yield by direct phosphorylation of inositol. According to Stephens (1990) [³H]Ins is dried down into a conical glass vial, then dissolved in 10 μl of H_3PO_4 by heating to 60°C for 15 min; 0.18 g of polyphosphoric acid is added and the mixture placed under vacuum and heated to 150°C. After 6 h, 0.5 ml H_2O is added then the mixture is heated to 100°C for 3 h (no vacuum) and cooled. The reaction mixture is washed out with H_2O, diluted to 100 ml, the pH adjusted to 6.5 with NaOH and the sample applied to a small desalting column. After the sample has been freeze-dried the [³H]InsP₆ is purified by HLC. Typically, 20% of the starting [³H]Ins will be recovered as [³H]InsP₆.

Hawkins et al. (1990) have provided a similar description of the preparation of [³H]InsP₆ by chemical phosphorylation and purification of the product by HPLC on Partisil 10 SAX columns.

[³²P]InsP₆ can be easily prepared (although not very high specific activity) from [³²P]Pᵢ-labeled germinating mung beans. A bean is labeled and extracted as described for the preparation of [³²P]InsP₅s (Stephens, 1990), and the [³²P]InsP₆ is purified on an anion exchange HPLC column. The [³²P]InsP₆ can be readily identified as the most polar ³²P-labeled metabolite eluting from the column in significant quantities (a typical yield from one bean germinated with 0.5 mCi of [³²P]Pᵢ for 50–60 h would be 15 μCi of [³²P]InsP₆).

10.4. Inositol pyrophosphates

Dictyostelium cells contain two diphosphorylated *myo*-InsPs, which were preliminarily characterized as diphosphoinositol pentakisphosphate (PP-InsP₅) and bisdiphosphoinositol tetrakis-phosphate (bis-PP-InsP₄) (Mayr et al., 1992; Stephens et al., 1993).

10.4.1. PP-InsP$_5$

PP-Ins-P$_5$ was additionally isolated from *D. discoideum* by Laussmann et al. (1996). Cells were grown in maltose medium, harvested by centrifugation, washed with phosphate buffer, denatured with 2 M HClO$_4$ containing EDTA, centrifuged and the supernatant immediately neutralized with 4 M KOH and the precipitated KClO$_4$ removed by filtration. The resulting extract was treated with charcoal, filtered and diluted with distilled water to a final conductivity of 3–4 mS/cm and applied to an 2.5 cm anion exchange column. All InsPs, including most of the InsP6 were eluted with 200 ml of 250 mM HCl. Diphospho-*myo*-InsPs containing some InsP6 were eluted with 200 ml of 550 mM HCl. The appropriate fractions were collected immediately neutralized on ice with 4 M LiOH and freeze-dried. The dried samples were desalted by dissolving LiCl in 100% ethanol (20 ml/g) and the insoluble InsPs were collected from ethanolic suspension by centrifugation. InsP$_6$, PP-InsP$_5$ and bis-PP-InsP$_4$ were separated on a high resolution 1.6 cm × 10 cm anion exchange column (Resource Q; Pharmacia) and eluted isocratically with 375 mM HCl. The fractions containing PP-InsP$_5$ and bis-PP-InsP$_4$ were combined, neutralized with LiOH, and freeze-dried. LiCl was removed with ethanol as described above. The yield and purity of the isolated compounds was examined by the metal-dye-detection HPLC method (Mayr, 1990).

Laussmann et al. (1996) showed by ^{1}H/^{31}P NMR that PP-InsP$_5$ is either D-4-PP-InsP$_5$ or its enantiomer D-6-PP-InsP$_5$, and that bis-PP-InsP$_4$ is either D-4,5-bis-PP-InsP$_4$ or its enantiomer D-5,6-bis-PP-InsP$_4$. Laussmann et al. (1997) have subsequently identified the naturally occurring enantiomers using synthetic PP-InsP$_5$ isomers as substrates for partially purified PP-InsP$_5$ 5-kinase from *Dictyostelium*. This enzyme specifically phosphorylates the naturally occurring PP-InsP$_5$ and the synthetic D-6-PP-InsP$_5$, leading to D-5,6-bis-PP-InsP$_4$. In contrast, neither D-4-PP-InsP$_5$, D-1-PP-InsP$_5$ nor D-3-PP-InsP$_5$ were converted by the enzyme.

10.4.2. PP-[³H]InsP₄-PP

PP-[³H]InsP₄-PP, PP-[³H]InsP₅ are available from DuPont NEN. Shears et al. (1995) prepared PP-[³H]InsP₄ as described by Menniti et al. (1993) by incubation with AR4-2J cells with [³H]Ins in the presence of 10 mM fluoride. Other preparations of PP-[³H]InsP₅ and PP-[³H]InsP₄-PP were obtained by phosphorylation of [³H]InsP₆ and PP-[³H]InsP₅, respectively, using a liver homogenate. The InsPs were isolated by loading the deproteinized samples onto Partisphere 5 μm SAX HPLC columns (Shears et al., 1995).

10.5. Structural analogues

To assist in the elucidation of the mechanisms of action and structural requirements within the PtdInsP moieties that are necessary for recognition and binding of proteins and activation of receptors, various structural analogues and derivatives with increased water solubility have been prepared. Other derivatives have been prepared to facilitate their detection following membrane binding or cellular uptake.

10.5.1. Deoxy-myo-InsPs

Several laboratories have prepared the 2-deoxy derivatives of InsPs and have shown that some of them possess biological activity (Wilcox et al., 1994; Rudolf et al., 1998). Starting with common inositol phosphates (Rudolf et al., 1998), various forms of Barton-McCombie deoxygenation and classical protection/deprotection procedures yielded the desired precursors rac-1-O-butyryl-2-deoxy-myo-inositol ent-3-O-butyryl-2-deoxy-myo-inositol (ent-12), and rac-2-O-butyryl-1-deoxy (rac-19), respectively. Phosphorylation and subsequent deprotection yielded rac-3. Compared to the 1,2-di-O-butyryl-myo-inositol 3,4,5,6-tetrakis-phosphate octakis (acetoxymethyl) ester, the membrane-permeant

Ins(3,4,5,6)P$_4$, the 2-deoxy derivative (rac-5) exhibited a slightly weaker response, while the enantiomerically pure 2-deoxy-Ins(1,4,5,6)P$_4$ (ent-5) and 1-deoxy (rac-6) were inactive. The effect was stereoselective, and the 1-hydroxy is apparently essential for binding and inhibitory effect of Ins(3,4,5,6)P$_4$ on the secretion, whereas the 2-hydroxyl group plays a less important role. Wilcox et al. (1994) demonstrated that modification of the 2-position of Ins(1,4,5)P$_3$, even with an anionic group, did not affect Ins(1,4,5)P$_3$ binding interaction or Ca^{2+} release, suggesting that the 2-hydroxyl of Ins(1,4,5)P$_3$ fails to interact significantly with the binding site of its receptor. A modification remote from the crucial vicinal 4,5-bisphosphate can affect Ca^{2+} release. Horne and Potter (2001) have reported the synthesis of 2-deoxy-myo-Ins(1,3,4,5)P$_4$ from D-glucose.

Qiao et al. (1998) have reported that the 3-deoxy-D-myo-inositol 1-phosphate, 1-phosphonate, and ether analogues are inhibitors of PtdIns 3-kinase signaling and cell growth.

Horne and Potter (2001) have prepared both enantiomers of 6-deoxy-Ins(1,3,4,5)P$_4$. The synthesis of 6-deoxy-Ins(1,3,4,5)P$_4$ in racemic and chiral form was obtained by resolution of partially blocked myo-inositol derivatives using chiral auxiliary (S)-(+)-O-acetylmandelic acid. The racemic tetrakisphosphate was synthesized from DL-1,2-O-isopropylidine-myo-Ins in eight steps. Deoxygenation at C6 was achieved following the Barton-McCrombie procedure. Absolute configuration was confirmed by synthesis of the known D-6-deoxy-myo-inositol. Preliminary data indicate that the D-isomer binds to GAP^{IB4BP}, interacts potently with Ins(1,4,5)P$_3$/Ins(1,3,4,5)P$_4$ 5-phosphatase and facilitates Ins(1,4,5)P$_3$-mediated activation of the store operated Ca^{2+} operated Ca^{2+} current I$_{CRAC}$, but to a lesser extent than Ins(1,3,4,5)P$_4$ itself.

10.5.2. Membrane permeators

Early attempts to increase the solubility of the phosphatides in aqueous media involved the replacement of the long chain fatty acids of the diacylglycerol moiety by short and medium chain fatty acids. Thus, Reddy et al. (1995) showed that di-C_8-GroPInsP$_3$ was more soluble and tractable than PtdInsP$_3$ with the more physiological C_{18} and C_{20} fatty acid composition. Others have increased the aqueous solubility of the phosphatides by utilizing the lysoPtdIns, relying on an endogenous reacylation of the lyso compounds for restoration of full biological activity. Filthuth and Eibl (1992) synthesized enantiomerically pure lsoPtdIns and its alkyl analogue. Starting from *myo*-inositol, penta-*O*-acetyl-*myo*-inositol was made in five steps. Then enantiomeric purification was done by diastereomeric salts separation method, and the purity of each enantiomer was spectroscopically measured (^{19}F-NMR). The phosphodiester was made via phosphoramidites.

Jiang et al. (1998) have stereospecifically synthesized di-C_8-GroPInsP$_3$/AM and di-C_{12}-GroPInsP$_3$/AM, the heptakis(acetoxymethyl)esters of dioctanoyl- and dilauroyl GroPIns(3,4,5)P$_3$, in 14 steps from *myo*-inositol. The desirable feature of AM esters is that they are readily hydrolyzed by intracellular esterases, which presumably regenerate PdInsP$_3$ inside the cells. The ability of these uncharged lipophilic derivatives to deliver PtdIns(3,4,5)P$_3$ across cell membranes was demonstrated on 3T3-L1 adipocytes and T$_{84}$ colon carcinoma monolayers.

The synthesis of the AM esters of inositol polyphosphates has been previously reported (Vajanaphanich et al., 1994; Li et al., 1997, 1998).

10.5.3. Fluorescent derivatives

Chen et al. (1996) have synthesized photoactivatable 1,2-diacyl-*sn*-glycerol derivatives of 1-L-phosphatidyl-D-*myo*-inositol 4,5-bisphosphate (PtdInsP$_2$) and 3,4,5-trisphosphate (PtdInsP$_3$). The ability to detect specific binding of inositol lipids to their protein targets by changes in the fluorescence of NDP- and pyrene-modified lipids provides a more direct measurement of the state of lipid than has previously been possible from studies of changes in protein structure or function.

Tuominen et al. (1999) use the method of Chen et al. (1996) to prepare I-palmitoyl-2-(N-nitrobenz-2-oxa-1,3-diazo)aminocaproyl-GroPIns(4,5)P_2 [NBD-PtdIns(4,5)P_2]. The NBD-PtdIns(4,5)P_2 was prepared from the corresponding 2-(6-aminohexanoyl) derivative of PtdIns(3,4,5)P_3 as described for the NDP-PtdIns(4,5)P_2. The fluorescent derivatives of PtdIns(4,5)P_2 were used to test the effects of the PtdIns(4,5)P_2-regulated proteins gelsolin, tau, cofilin, and profiling on labeled PtdIns(4,5)P_2 that was either in micellar form or mixed with PtdCho in bilayer vesicles. Gelsolin increased the fluorescence of 7-nitrobenz-2-oxa-1,3-diazole (NBD)- or pyrene-labeled PtdIns(4,5)P_2 and NBD-PtdIns(3,4,5)P_3, reaching a ratio of 1:1 at saturation. Ozaki et al. (2000) have prepared a shorter chain sn-1-O-(6-aminohexanoyl)-sn-2-O-hexanoyl PtdIns(4,5)P_2 derivative as described by Chen et al. (1966). Fluorescent derivatives of polyamines were synthesized from the amine and either the isocyanate or succinimidyl ester derivatives of the fluorescent reagent in 1.0 M TEAB (aqueous triethylammonium bicarbonate, pH 7.5) (Ozaki et al., 2000). The polyamine carrier allowed an efficient translocation of biologically active PtdInsPs, InsPs, and their fluorescent derivatives into living cells in a physiologically relevant context.

10.5.4. Affinity reagents

One way to identify PtdInsP-binding proteins is to use this lipid in an immobilized form and to isolate proteins that bind to it. Manifava et al. (2001) have prepared and employed a 3-(ω-amino-acyl)-functionalized GroPIns(4,5)P_2 coupled via the amino function to an Affi-Gel 10 solid support. Details of the synthesis are to be reported separately (see Manifava et al., 2001).

CHAPTER 11

InsPs and PtdInsPs as Signaling Molecules

11.1. Introduction

It has been known for several decades that agonist activation of membrane receptors stimulates PtdIns turnover (Hokin, 1985), but only within the last decade a coherent picture of the physiological function of PtdInsPs has begun to emerge. By 1990, it was established that the cellular Ins metabolite Ins(1,4,5)P$_3$ functions as a second messenger for Ca^{2+} mobilization from intracellular stores in response to stimulation by a variety of hormones, neurotransmitters and growth factors (Downes and Carter, 1990; Mitchell, 1989). Ins(1,4,5)P$_3$ is cleaved from its phospholipid precursor, PtdIns(4,5)P$_2$ by PtdIns specific PLC. This process simultaneously generates 1,2-diacylglycerol, which is the second messenger for activation of protein kinase C (Nishizuka, 1984), emphasizing the central importance of PLC activity in coordinating cellular responses to hormonal stimulation. However, in recent years it has become apparent that the PtdInsP metabolism is extremely complex, and that stimuli as diverse as growth factors and light activate molecular programs lead to the production of Ins polyphosphate messenger molecules. Over 30 lipid and water-soluble InsP molecules have been identified in eukaryotic cells, many of which have not yet been assigned a function (Irvine and Schell, 2001).

While many of these compounds are direct or indirect products of PtdInsP hydrolysis, the metabolic origin of some of them appears totally independent. Thus, the $InsP_5$ and $InsP_6$ are derived via InsPs kinases. In addition, several InsPs pyrophosphates have also been isolated and their independent biosynthesis recognized. A large number of InsP phosphatases have been identified and cloned using sequence comparisons derived from structural studies. Subsequent gene disruption showed that InsP metabolism played a critical role in a specific signaling pathway.

Recent work has revealed signaling roles for PtdInsPs that do not involve their hydrolysis (Cantley, 2002). As pointed out in earlier chapters (Chapters 6 and 7) PtdIns 3-kinase generates $PtdIns(3)P$, $PtdIns (3,4)P_2$ and $PtdIns(3,4,5)P_3$ by phosphorylating PtdIns, PtdIns (4)P and $PtdIns(4,5)P_2$, respectively. Since PtdInsPs with a monoester phosphate in the 3-position are resistant to cleavage by a variety of purified PLCs (Lips et al., 1989; Serunian et al., 1989; Lips and Majerus, 1989), it is probable that the phospholipid products of PtdIns 3-kinase themselves, rather than the InsPs derived from them, are functionally significant. There is specific evidence that $PtdIns(3)P$, $PtdIns(3,4,5)P_3$, and $PtdIns(4,5)P_2$ function as site-specific signals on membranes that recruit and/or activate proteins for assembly of spatially localized functional complexes (Smith and Scott, 2002; Czech, 2003). A large number of PtdInsP binding proteins have been identified as the potential effectors for PtdInsPs signals. Additional signaling pathways involve receptor-stimulated $PtdIns(3,4,5)P_3$ synthesis from $PtdIns(4,5)P_2$ by PtdIns 3-kinases (Stephens et al., 1993; Toker and Cantley, 1997) and $PtdIns(3,5)P_2$ synthesis by a stress-activated $PtdIns(3)P$ 5-kinase (Whiteford et al., 1997; Dove et al., 1997). The methods of chemical genetics now allow signal transduction pathways to be probed in a domain specific manner (Specht and Shokat, 2002).

The present chapter reviews the evidence for the signaling roles of the InsPs and PtdInsPs as obtained by analytical, metabolic and molecular analyses. Regulation of the activity of the kinases and phosphatases, as well as protein and receptor binding are also considered. A detailed discussion of the signaling pathways themselves, however, is beyond the scope of the present chapter and the book.

11.2. InsPs as cellular signals

A signaling role for InsPs was established in the 1980s, when it became clear that PLC-mediated hydrolysis of $PtdIns(4,5)P_2$ generates the intracellular signals $Ins(1,4,5)P_3$ and diacylglycerol for regulating Ca^{2+} mobilization and protein phosphorylation mechanisms, respectively (Berridge and Irvine, 1984). The ubiquitous occurrence of InsPn $>$ 1, including $InsP_5$ and $InsP_6$, in eukaryotic cells, and the complex pathways concerned with their synthesis and degradation, suggests that these compounds also are functionally significant, although their cellular roles remain obscure at present (Shears, 1998; Irvine and Schell, 2001).

11.2.1. Evidence for signaling

The following sections consider the analytical, metabolic and molecular evidence for the signaling role of the various InsPs.

11.2.1.1. Analytical

Direct measurement of the InsPs released from PtdInsPs or synthesized via kinases in response to appropriate stimulation continue to provide primary evidence for a signaling role of these molecules. Examples of such measurements are given in Chapter 7 under the assays for specific products of InsP kinases. Detection of the InsPs in vivo, however, has been hampered by the fact that these esters are present in relatively low abundance in almost all cells that have been investigated. This problem has been overcome to some extent by incorporating high levels of radioactivity into the total cellular PtdInsPs prior to stimulation. The original studies relied on *ortho* $[^{32}P]P_i$ labeling, but in recent years the most popular strategy has involved selective radiolabeling using myo-$[^3H]Ins$ to achieve head group labeling of PtdInsPs and InsPs.

The ability of HPLC techniques to separate a complex mixture of InsPs into individual isomers now plays a central role in monitoring changes in the radiolabeling and/or mass of the different InsPs on agonist stimulation. The original routines were described by Irvine et al. (1986), Hawkins et al. (1986), and Batty et al. (1989), while Stephens et al. (1990), Auger et al. (1990) and Bird et al. (1992) have presented improved versions of the silica based organic resin anion exchange columns or MonoQ columns.

The need for accurate and rapid quantification of the cellular concentrations of important intermediates such as Ins(1,4,5)P$_3$ and Ins(1,3,4,5)P$_4$ has been met by the radio-receptor assays. The first description of a mass assay for Ins(1,4,5)P$_3$ employed a crude microsomal fraction of bovine adrenal cortex (Palmer et al., 1988; Challis et al., 1988; Palmer and Wakelam, 1989). In principle, the Ins(1,4,5)P$_3$ present in a cell extract competes with a fixed quantity of high specific activity [^{32}P]Ins(1,4,5)P$_3$ or [^{3}H]Ins(1,4,5)P$_3$ for the Ins(1,4,5)P$_3$-specific binding sites of bovine adrenocortical microsomes. A mass assay for Ins(1,3,4,5)P$_4$ has been described based on a fraction of the porcine (Donie and Reiser, 1989) and rat (Challis and Nahorski, 1993) cerebellum. Both reports present Ins(1,3,4,5)P$_4$ mass determinations employing the binding site as the basis for mass measurements. An acidic assay condition prevented possible interference from the high density of Ins(1,4,5)P$_3$ receptor sites present (Challis et al., 1991). These methods can be performed without extensive isomer resolution prior to assay or the need for specialized instrumentation. The Ins(1,4,5)P$_3$ radio-receptor assay proposed by Zhang (1998) was given in Chapter 4. These methods are not suitable to metabolic profiling of the InsPs cycles or to establishing precursor product relationships.

Furthermore, it must be recognized that all in vivo mass measurements are subject to error, as the mass determined is likely to reflect a sum total of the metabolic activities contributing to the level of a particular metabolite measured. The relation of the level of a specific metabolite to a particular stimulus and cellular response must therefore rely on other evidence including metabolic measurements (Berridge, 1993).

11.2.1.2. Metabolic

The strength and duration of a signal depends on its rate of metabolism (Shears, 1998). Thus, the factor that regulates the Ca^{2+}-mobilizing Ins(1,4,5)P_3 signal is its rate of phosphorylation. A complication arises from the fact that one of the products is Ins(1,3,4,5)P_4, which also influences the overall Ca^{2+} response by promoting Ca^{2+} entry into the cells (Irvine, 1992a). This evidence that Ins(1,3,4,5)P_4 is an intracellular signal extends the importance of Ins(1,4,5)P_3 3-kinase beyond inactivation of Ins(1,4,5)P_3. The signaling activities of both Ins(1,4,5)P_3 and Ins(1,3,4,5)P_4 are terminated by hydrolysis of their 5-phosphate groups, forming Ins(1,4)P_2 and Ins(1,3,4)P_3, respectively. Furthermore, the 3-kinase initiates a metabolic pathway that yields Ins(1,3,4)P_3, which in turn sets in motion a chain of signaling events that ultimately leads to down-regulation of Ca^{2+}-activated Cl^- channels (Shears, 1998).

Ins(1,3,4)P_3 sits at a branch-point of InsP metabolism. One of its fates is to be dephosphorylated to replenish the pool of Ins. Two classes of phosphatases are responsible, a 4-phosphatase that forms Ins(1,3)P_2 and 1-phosphatase that yields Ins(3,4)P_2 (Wilson and Majerus, 1996). In addition, Ins(1,3,4)P_3 serves as a precursor for the higher InsPs. It is phosphorylated first to Ins(1,3,4,6)P_4 and then to Ins(1,3,4,5,6)P_5. Ins(1,3,4)P_3 is the dephosphorylated metabolite of Ins(1,3,4,5)P_4 and is itself rephosphorylated by two Ins(1,3,4)P_3 kinase activities to give two InsP$_4$ products and an InsP$_5$ product. Wilson and Majerus (1996) have purified the Ins(1,3,4)P_3 5/6 kinase from calf brain using chromatography on heparin agarose and affinity elution with InsP$_6$, and have demonstrated that the D-5 and D-6 positions are phosphorylated by a single kinase. Alternatively, $Ins(1,3,4)P_3$ is dephosphorylated by specific phosphatases to either $Ins(3,4)P_2$ or $Ins(1,3)P_2$.

$Ins(1,3,4,6)P_4$ is further phosphorylated to $Ins(1,3,4,5,6)P_5$. In addition to the interconversion of $Ins(1,3,4,5,6)P_5$ to $Ins(3,4,5,6)P_4$, there appears to exist a separate $Ins(1,3,4,5,6)P_5$ 3-phosphatase/ $Ins(1,4,5,6)P_4$ 3-kinase substrate cycle. The 3-phosphatase activity is catalyzed by a multiple InsP phosphatase (MIPP) that is concentrated inside endoplasmic reticulum, although there is no evidence that InsPs can enter the ER. MIPP remains inactive until the ER membranes are permeabilized by detergents. The presence of MIPP and multiple $InsP_5$ kinases in cells suggests that they contain a complex network of substrate cycles that interconvert $InsP_5$ and $InsP_6$. Thus, a large series of kinases and phosphatases interconvert the six known PtdInsPs and the more than twenty InsPs that exist in eukaryotic cells. The levels of the InsPs appear to be maintained and regulated via distinct metabolic cycles (Shears, 1998).

The biosynthesis of $InsP_6$ has been extensively studied in the slime mold, *Dictyostelium* and in the duckweed *Spirodela*. In both instances $Ins(1,3,4,5,6)P_5$ was identified as the immediate precursor of $InsP_6$. Two major diphosphoinositol phosphates have been identified from *D. discoideum* as 6-PP-InsP_5 and $5,6\text{-bis-PP-}$ $InsP_4$ (Laussmann et al., 1997, 1998).

$InsP_6$, which was thought to be the metabolic endpoint of *myo*-inositol metabolism, has been recently shown in mammals to be converted into inositol pyrophosphates, which have been identified as PP-InsP_5 ($InsP_7$) and bis-PP-InsP_4 ($InsP_8$) (Menniti et al., 1993a, b; Stephens et al., 1993). Voglmaier et al. (1996) have shown that $InsP_7$ and $InsP_8$ are synthesized by separate $InsP_6$ and $InsP_7$ kinases. $InsP_7$ and $InsP_8$ apparently turn over much more rapidly in animal cells than does $InsP_6$. Menniti et al. (1993a,b) have calculated that as much as 20% of the cells's $InsP_6$ might be cycling through $InsP_7$ every hour.

The metabolic evidence for InsPs signaling is further discussed under Regulation of InsP kinases (see below).

11.2.1.3. Molecular

It is now well recognized that if an InsP is to serve as a messenger, there must be a protein present in the cell that binds it (Irvine and Cullen, 1996). The early recognition that $Ins(1,4,5)P_3$ exerted its second messenger action by interaction with the endoplasmic reticulum was based on the demonstration of saturable binding sites in permeabilized cells using $[^{32}P]Ins(1,4,5)P_3$. This led to the development of a competitive binding assay, which allowed estimation of $Ins(1,4,5)P_3$ mass in cell extracts using saponin-permeabilized neutrophils as a source of cellular $Ins(1,4,5)P_3$ recognition sites (Bradford and Rubin, 1986). Detailed characterization of the cellular $Ins(1,4,5)P_3$ binding sites defined the pharmacology of ligand–receptor interaction and demonstrated the stereospecificity and positional specificity of $Ins(1,4,5)P_3$ receptor for $Ins(1,4,5)P_3$ (Willcocks et al., 1987; Nahorski and Potter, 1989).

An $Ins(1,4,5)P_3$ binding protein was purified from cerebellum (Worley et al., 1987), and characterized as an $Ins(1,4,5)P_3$-gated Ca^{2+} channel present in the endoplasmic reticulum (Ferris et al., 1989). Enyedi et al. (1989) were able to distinguish $Ins(1,4,5)P_3$- and $Ins(1,3,4,5)P_4$- selective binding sites in membranes prepared from bovine parathyroid gland. The $Ins(1,3,4,5)P_4$-binding site exhibited an acidic PH optimum and a high degree of selectivity for $Ins(1,3,4,5)P_4$ over $Ins(1,4,5)P_3$, $Ins(1,3,4)P_3$ and $InsP_5$.

Hirata and Kanematsu (1993) have described the purification of two $Ins(1,4,5)P_3$-binding proteins from rat brain cytosolic fractions using $Ins(1,4,5)P_3$ affinity matrices developed in parallel. The molecular masses, as estimated by SDS-PAGE, were 130 and 85-kDa, respectively. Partial amino acid sequence determinations following proteolysis and reversed phase HPLC revealed that the 85 kDa protein is the δ-isozyme of PLC, and the 130 kDa protein a newly isolated $Ins(1,4,5)P_3$-binding protein. Biological function and structural features of the 130 kDa protein have also been investigated (Kanematsu et al., 1996).

It is now realized that the signaling units are multimeric and that many of the proteins that participate in signal transduction contain protein–protein interaction domains that promote the assembly of large macromolecular structures. This arrangement generally promotes fast response time, enhances specificity, and minimizes cross-talk from other signaling pathways (Tsunoda et al., 1997). Shears (1998) has reviewed additional contributions to signaling specificity that result from combining into heterotetramers of functionally distinct subtypes of InsPs receptors. Thus, three major classes of the tetrameric receptor of Ins(1,4,5)P$_3$ are currently recognized. The receptor subunits differ in their properties and lead to differences in receptor activity upon assembling into hetero-tetramers. There are other differences between the receptor subtypes that may be functionally significant, and thereby impact upon the versatility of InsPs signaling (see below).

11.2.2. InsP-protein interaction

Since InsPs are water soluble, it has been assumed that they would exert any signaling effects through an initial contact with cytosolic proteins rather than plasma or subcellular membranes. There is good evidence now that InsPs interact with both proteins in the cytosol as well as with membrane bound receptors, although the differentiation between the two may be largely operational.

Several coat proteins have been isolated in an effort to identify InsPs-binding proteins. Clathrin adaptor proteins AP-2 and AP-3 as well as coatomer protein interact with InsPs and PtdInsPs (Fleischer et al., 1994; Norris et al., 1995). Several InsPs-binding proteins such as centaurin-α and synaptotagmin II were purified based on InsPs-binding (Fukuda et al., 1994; Cullen et al., 1995a; Theibert et al., 1997; Mikoshiba et al., 1999). However, PtdInsPs are probably the natural ligands for these proteins (Hammonds-Odie et al., 1996; Schiavo et al., 1996; Mehrotra et al., 1997). The InsPs-binding domain of synaptotagmin consists of the highly basic sequence (GKRLKKKKTTVKKK in the C2B domain) (Fukuda et al., 1994; 1996; Sutton et al., 1995). This portion of the C2 domain was disordered (Sutton et al., 1995) but adopted an ordered structure upon binding InsPs. Mehrotra et al. (2000) have described the binding kinetics and ligand specificity for the interactions of the C2B domain of synaptotagmin II with InsPs and PtdInsPs.

11.2.2.1. Ins(1,4,5)P₃ receptors

$Ins(1,4,5)P_3$ is now recognized as a second messenger that controls many cellular processes by generating internal calcium signals (Berridge, 1993). Taylor (1998) has provided a detailed description of the $Ins(1,4,5)P_3$ receptor, while Shears (1998) has described the variability of the receptor properties that contribute to the versatility of signaling by $Ins(1,4,5)P_3$. Three types of $Ins(1,4,5)P_3$ receptors, namely, types I, II and III, have been defined. They have similar sizes (2670–2749 amino acids) and the same basic structure (reviewed by Joseph, 1996). For the type I $Ins(1,4,5)P_3$ receptor, which is the predominant type in neuronal cells (Joseph, 1996), three domains have been defined: an $Ins(1,4,5)P_3$-binding domain within the N-terminal 650 amino acids, a transmembrane or channel-forming domain. Several lines of evidence indicate that a conformational change occurs upon $Ins(1,4,5)P_3$ binding and that this is responsible for channel opening (Joseph, 1996).

According to Taylor (1998), $Ins(1,4,5)P_3$ first binds to residues within the N-terminal domain of a single subunit of the receptor causing a large conformational change that both exposes a Ca^{2+} binding site and initiates the inactivation process that leads within approximately 1 s to the receptor both binding $Ins(1,4,5)P_3$ with increased affinity and becoming less capable of opening to its fully active state. As the cytosolic Ca^{2+} concentration increases, it can bind to a second site which is accessible whether or not the receptor has $Ins(1,4,5)P_3$. The inhibition of channel opening that follows Ca^{2+} binding to this site is relatively slow.

The primary effect of Ins(1,4,5)P$_3$ is to trigger calcium release from the endoplasmic reticulum, thus raising cytoplasmic free Ca^{2+} concentration (Berridge, 1993; Clapham, 1995). This is achieved by interaction of Ins(1,4,5)P$_3$ with Ins(1,4,5)P$_3$ receptors, proteins that form tetrameric complexes in the endoplasmic reticulum membrane and that act as Ca^{2+} channels (Joseph, 1996; Yoshida and Imai, 1997).

It has recently been found that Ins(1,4,5)P$_3$ receptors are also subject to down-regulation upon stimulation of PtdIns-specific PLC-linked cell surface receptors (Wojcikiewicz and Nahorski, 1991; Wojcikiewicz, 1995; Bokkala and Joseph, 1997), providing a novel locus of adaptation. Activation of certain PLC-linked cell surface receptors is known to cause an acceleration of the proteolysis of Ins(1,4,5)P$_3$ receptors and thus lead to Ins(1,4,5)P$_3$ receptor down-regulation. Zhu et al. (1999) have shown that the down-regulation is mediated by Ins(1,4,5)P$_3$ binding. A mutant Ins(1,4,5)P$_3$ receptor was resistant to down-regulation, whereas wild-type Ins(1,4,5)P$_3$ receptor was down-regulated. Ins(1,4,5)P$_3$ binding induces a substantial but undefined conformation change in the ligand-binding domain of type I formational change in the ligand-binding domain of the type I receptor (Mignery and Sudhof, 1993), which appears to be the primary consequence of Ins(1,4,5)P$_3$ binding (Zhu et al., 1999). It can be envisaged that such conformational change might expose regions of the receptor that either are cleavage sites for proteases or are sites that facilitate ubiquitin conjugation (Bokkala and Joseph, 1997). It has also been shown that Ins(1,4,5)P$_3$ receptor down-regulation can be induced by receptor-independent activation of PLC (Honda et al., 1995; Quick et al., 1996).

Ins(1,4,5)P$_3$ receptor may be purified as described by Supattapone et al. (1988) with the following modifications: CHAPS (1%) was used as the detergent to allow dialysis for reconstitution, and the receptor was purified using only two affinity column chromatography steps. Partially purified receptor was obtained by fractionating CHAPS solubilized proteins on Heparin agarose as described by Supattapone et al. (1988). The purified receptor was obtained by fractionating the proteins, which elute from heparin agarose on concanavalin sepharose. The reconstitution strategy was based on a previously described procedure used for the nicotinic acetylcholine receptor. The molecular and physiological properties of the Ins(1,4,5)P$_3$ receptor closely resemble the calcium-mobilizing ryanodine receptors of muscle (MacLennan, 2000). Munger et al. (2000) have characterized a Ins(1,4,5)P$_3$ receptor in lobster olfactory receptor neurons.

11.2.2.2. Ins(1,3,4,5)P$_4$ receptors

The early work on the high-affinity Ins(1,3,4,5)P$_4$ receptor from cerebellum has been discussed by Reiser (1993). Theibert et al. (1992) had used an Ins(1,3,4,5)P$_4$ affinity probe on solubilized rat brain to identify proteins of 182 and 84 kDa as specific Ins(1,3,4,5)P$_4$-binding proteins. Koppler et al. (1996) have isolated a 74 kDa protein from rat liver nuclei which, binds Ins(1,3,4,5)P$_4$ with high affinity and a certain degree of isomeric specificity. Therefore, the criterion of Ins(1,3,4,5)P$_4$ being a second messenger appears to be fulfilled by demonstration of occurrence of Ins(1,3,4,5)P$_4$ binding proteins. In fact many intracellular proteins bind radio-labeled Ins(1,3,4,5)P$_4$. To serve as a receptor, the binding protein should greatly favor Ins(1,3,4,5,)P$_4$ over all other InsPs in the cytosol. Fukuda and Mikoshiba (1996) have described the determination of the Ins(1,3,4,5)P$_4$ binding domain in mouse Gap1m.

Cullen et al. (1997) have isolated an Ins(1,3,4,5)P$_4$-binding protein (GAP1^{IP4BP}) which demonstrates isomeric specificity coupled with an inability to accommodate substitution on the 1-phosphate and have identified it as a GTPase-activating protein. This protein was originally purified from detergent solubilized pig platelet membranes. It is a high-affinity Ins(1,3,4,5)P$_4$-binding protein, and a member of the GAP1 family. Although GAP1^{IP4BP} is predominantly membrane localized, it does not contain a membrane spanning domain (Cullen, 1998). In vitro, GAP^{IP4BP} stimulates the GTPase activity of all members of the Ras family, but not members of the Rab, Rac and Rho family (Cullen ct al., 1995). The true physiological targets for either $GAP1^{IP4BP}$ or a related protein $GAP1^m$ are unknown (Cullen, 1998). Cullen (1998, 2002) has recently reviewed the evidence for $Ins(1,3,4,5)P_4$ signaling, including the regulation of its production, the properties of its binding proteins, distinguishing $Ins(1,3,4,5)P_4$ receptors from $Ins(1,4,5)P_3$ receptors, along with its physiological target and function.

In case of $Ins(1,3,4,5)P_4$, Bottomley et al. (1998a,b) have shown that expressed full-length $GAP1^{IP4BIP}$ binds $Ins(1,3,4,5)P_4$ with an affinity and specificity similar to that of the original purified protein. The binding activity is dependent on a functional PH/Btk domain. Furthermore, a fundamental distinction between $GAP1^{IOB4BP}$ and its homologue GAP1m was recognized in the fact that both proteins function as Ras GAPs but only $GAP1^{IP4BP}$ displays Rap GAP activity.

11.2.2.3. $Ins(3,4,5,6)P_4$ receptors

The first direct evidence that receptor-dependent increases in $Ins(3,4,5,6)P_4$ had a physiological function came from studies into the control of salt and fluid secretion by T84 colonic epithelial cells (Vajanaphanich et al., 1994). In this study, a cell-permeant bioactivatable analogue of $Ins(3,4,5,6)P_4$ was shown to uncouple Cl^- secretion from stimulation by Ca^{2+}. $Ins(3,4,5,6)P_4$ has now been established as a cellular signal that inhibits the conductance of Ca^{2+}-activated Cl^- channels in the plasma membrane, thereby making an important contribution to the complex homeostatic control of salt and fluid secretion from epithelial cells (Ho et al., 2000). The general importance of $Ins(3,4,5,6)P_4$ necessitates that the cell exerts close control over its synthesis and metabolism. Current understanding of this regulatory process is based on the paradigm that the pathway leading to $Ins(3,4,5,6)P_4$ synthesis terminates within a metabolic cul-de-sac. The levels of $Ins(3,4,5,6)P_4$ are dynamically regulated by the competing activities of a receptor-regulated $Ins(1,3,4,5,6)P_5$ 1-phosphatase/ $Ins(3,4,5,6)P_4$ 1-kinase metabolic cycle (Shears, 1998). Yang and Shears (2000) have recently described a human cDNA encoding 1-kinase activity that inactivates $Ins(3,4,5,6)P_4$, an inhibitor of chloride-channel conductance that regulates salt and fluid secretion, as well as membrane excitability. In addition, it was discovered that this enzyme had an alternative positional specificity (5/6-kinase activity) towards a different substrate, $Ins(1,3,4)P_3$. Kinetic data from a recombinant enzyme indicated that $Ins(1,3,4)P_3$ and $Ins(3,4,5,6)P_4$ actively competed for phosphorylation in vivo. The new $Ins(3,4,5,6)P_4$ 1-kinase has only one amino-acid difference from the protein previously shown to be $Ins(1,3,4)P_3$ 5/6 kinase (Wilson and Majerus, 1996). The new enzyme also phosphorylates $Ins(1,3,4)P_3$ to form $Ins(1,3,4,6)P_4$, thus potentially prolonging the half-life of a cellular signal that mediates receptor-activated Ca^{2+} entry (Irvine, 1992a). The discovery that this enzyme has substrate-dependent positional specificity towards $Ins(1,3,4)P_3$ and $Ins(3,4,5,6)P_4$ is of general enzymological significance, and these findings may be pertinent to characterizing the structure and evolution of this catalytic motif. These observations make them of particular interest to signal transduction and ion transport in higher organisms, where the two substrates perform critical regulatory functions. Thus, the prevailing level of $Ins(3,4,5,6)P_4$ in the cell is believed to set the degree of ionic conductance through Ca^{2+}-activated Cl^- channels in the plasma membrane (Ho et al., 2000), while $Ins(1,3,4)P_3$, by competing for phosphorylation, can act to reduce the cellular levels of $Ins(3,4,5,6)P_4$. At the same time, increases in $Ins(1,3,4)P_3$ levels inhibit the 1-kinase and elevate $Ins(3,4,5,6)P_4$ levels (Yang et al., 1999). Ho et al. (2001) have further demonstrated that only the CaMKII-dependent activation process is inhibited by $Ins(3,4,5,6)P_4$ in CFPAC-1 cells, direct channel activation by Ca^{2+} not being inhibited by $Ins(3,4,5,6)P_4$.

11.2.2.4. InsP$_6$ receptors
The cellular functions of InsP$_6$ remain enigmatic, despite the fact that InsP$_6$ is the most abundant InsPs in nature. This is in part due to an apparent lack of response of InsP$_6$ to extracellular signals, except for the response of *S. pombe* to hypertonic shock, which leads to rapid increase in the levels of InsP$_6$ (Ongusaha et al., 1998). A role for InsP$_6$ in cell signaling has been suggested by recent work in two other systems. Three genes, whose products constitute a path linking Ins(1,4,5)P$_3$ to InsP$_6$ have been identified as members of a group that functionally complement lesions in mRNA export from the nucleus in yeast. These genes, respectively encode PLC, an Ins(1,4,5)P$_3$ 3-/Ins(1,3,4,5)P$_4$ 6-kinase and Ins(1,3,4,5,6)P$_5$ 2-kinase (York et al., 1999; Odom et al., 2000). In insulin-secreting pancreatic β-cells, it was shown (Larsson et al., 1997) that 10 μM InsP$_6$ in the patch pipette increased an L-type Ca^{2+} current, an effect which may reflect the ability of InsP$_6$ to inhibit protein phosphatase, types PP1, PP2A, and PP3, with subsequent activation of the L-type Ca^{2+}-channels.

Lemtiri-Chlieh et al. (2000) have recognized a function for InsP$_6$ in plants as a signal generated in guard cells in response to the stress hormone, (RS)-2-*cis*,4-*trans*-abscisic acid (ABA), along with an electrophysiological target (see below).

11.2.3. Regulation of InsP kinases

There is evidence for the existence of elaborate metabolic pathways maintaining controlled levels of the signals by regenerating them from starting materials or recycling them. Shears (1998) has pointed out the versatility of InsPs as cellular signals. Three specific metabolic cycles were recognized to provide supporting mechanisms for the generation and maintenance of the second messenger levels. The following discussion summarizes and updates the regulation of the enzymes associated with these product cycles, and other cycles, which may represent separate metabolic pools. The pathways of metabolism of InsPs have been summarized in Chapter 7.

11.2.3.1. Ins(1,4,5)P₃/Ins(1,3,4,5)P₄ cycle

$Ins(1,4,5)P_3$ is formed by receptor-activated hydrolysis of $PtdIns(4,5)P_2$, which releases Ca^{2+} from specialized intracellular stores (Streb et al., 1983). The rapid cytosolic conversion of $Ins(1,4,5)P_3$ to $Ins(1,3,4,5)P_4$ following receptor stimulation has been demonstrated in a wide range of mammalian cells (Irvine, 1992a,b). Virtually every animal cell uses up ATP to phosphorylate $Ins(1,4,5)P_3$ to $Ins(1,3,4,5)P_4$ in a carefully controlled reaction catalyzed by a specific InsP₃ 3-kinase (Batty et al., 1985). Although $Ins(1,3,4,5)P_4$ could simply be a metabolic intermediate, there is evidence that $Ins(1,3,4,5)P_4$ has its own distinct second messenger function (Irvine et al., 1999; Irvine and Schell, 2001).

Two isoforms of 3-kinase have been characterized: type A (51 kDa) and type B (74 kDa). The amino acid sequences of 3-kinases A and B are relatively conserved within the C-terminal catalytic domain; their N-terminal regulatory domains are more diverse (Chapter 7). The most intensively studied process is the stimulation of 3-kinase activity by Ca^{2+}/calmodulin (CaM), which increases the V_{max} of the kinases (Shears, 1998). The effect of Ca^{2+} upon the type B enzyme (20-fold) is larger in the case of the type A enzyme (about three-fold). Phosphorylation of the 3-kinase type A by CaM KII brings about a 25-fold increase in the affinity of the kinase for calmodulin, and in addition improves to about 10-fold the degree to which Ca^{2+}/CaM activates this 3-kinase. The type A 3-kinase is activated by protein kinase A. However, protein kinase A attenuates the ability of Ca^{2+}/calmodulin to stimulate type B activity.

$Ins(1,4,5)P_3$ 3-kinase A and B are 68% homologous in the C-terminal catalytic region while the N-terminal regulatory region shows little sequence homology. Expression is tissue specific, with the A isoform detected in brain and testes, whereas the B isoform has been found in several tissues including lung, thymus, testes, brain and heart. While the A isoform is largely soluble, the B isoform is now known to be present in substantial amounts attached to the outer face of the endoplasmic reticulum and the ER to Golgi intermediate compartment. Both enzymes can be activated between 2- to 20-fold by Ca^{2+}/CaM (Sims and Allbritton, 1998) and in vitro via phosphorylation by protein kinase C and cyclic AMP-dependent protein kinase. There is evidence for a high degree of regulation, which suggests that $Ins(1,4,5)P_3$ 3-kinase catalyzes a vitally important reaction: however, it does not distinguish whether it produces directly a second messenger or simply degrades one.

In mammals, where $Ins(1,4,5)P_3$ 3-kinase has been most comprehensively studied, there are a least three isoforms, all of which are regulated by CaM and two of which are regulated by Ca^{2+}/CaM-dependent protein kinase II (CaMKII). $Ins(1,4,5)P_3$ 3-kinase is the most active InsP kinase detectable in mammals, having prominent role in rapidly metabolizing the pools of $Ins(1,4,5)P_3$ that is generated when PLC-coupled receptors are activated. $Ins(1,2,3,4,5)P_4$ is hydrolyzed by the same 5-phosphatase that hydrolyzes $Ins(1,4,5)P_3$, but the enzyme has a 10-fold higher affinity and 100-fold lower V_{max} for $Ins(1,3,4,5)P_4$ than it does for $Ins(1,4,5)P_3$ (Connelly et al., 1987). $Ins(1,3,4,5)P_4$ is believed to protect $Ins(1,4,5)P_3$ against hydrolysis (Hermosura et al., 2000; Irvine, 2001). $Ins(1,3,4,5)P_4$ can also activate Ca^{2+} channels in the plasma membrane (Luckhoff and Clapham, 1992) and it can interact with $Ins(1,4,5)P_3$ receptors on the endoplasmic reticulum membrane, although only at high concentrations. Most $PtdIns(3,4,5)P_3$ receptors have been shown to bind $Ins(1,3,4,5)P_4$ in vitro (Lockyer et al., 1999), which would be anticipated on basis of similarity between the head groups. Other receptors, however, can discriminate between $Ins(1,4,5)P_3$ and $Ins(1,3,4,5)P_4$ (Cozier et al., 2000).

Type I and type II 5-phosphatases (Chapter 8) are directed to the plasma membrane by isoprenylation of the carboxy terminus. In polarized epithelial cells, 5-phosphatase activity may be subsequently redistributed by vesicle trafficking to specific domains of the plasma membrane. There is little evidence that the type I and II Ins(1,4,5)P₃/Ins(1,3,4,5)P₄ 5-phosphatase activities are regulated by cross-talk from other cellular signals. It has been shown that one of the products of 5-phosphatase activity, Ins (1,3,4)P₃, is a potent inhibitor of the Ins(3,4,5,6)P₄ 1-kinase. By inhibiting this kinase, and altering the equilibrium of the 1-kinase/1-phosphatase cycle, Ins(1,3,4)P₃ provides a link between activation of PLC and increases in levels of Ins(3,4,5,6)P₄, an inhibitor of Ca^{2+}-regulated Cl$^-$ secretion (Shears, 1998; Yang et al., 1999).

11.2.3.2. Ins(1,3,4,5,6)P₅/Ins(3,4,5,6)P₄ cycle

Ins(1,3,4)P₃ serves as a precursor for the higher InsPs. It is phosphorylated first to Ins(1,3,4,6)P₄ and then to Ins(1,3,4,5,6)P₅. In vitro, the activity of the Ins(1,3,4)P₃ 6-kinase is restricted by product inhibition and by its nearly complete inhibition by physiologically relevant levels of Ins(3,4,5,6)P₄ (see Chapter 7).

Ins(3,4,5,6)P₄ binds to its own intracellular receptor to exert a function that is entirely different from that of Ins(1,4,5)P₃ (Cullen, 1998). The Ins(1,3,4,5)P₄ binding protein GAP1[IPA4BP] fulfills all the criteria set out for a putative Ins(1,3,4,5)P₄ receptor, namely isomerically specific InsP₄-binding coupled with an inability to accommodate substitution on the 1-phosphate (Cullen et al., 1997). Originally purified from detergent solubilized pig platelet membranes (Cullen et al., 1995a,b), this high affinity Ins(1,3,4,5)P₄-binding protein, is a member of the GAP1 family. Mutational analysis has established that only those constructs containing the PH/Btk domain of GAP1[IPA4BP] will bind Ins(1,3,4,5)P₄. A C2A/C2B-deletion mutant, GAP1[m] will bind GAP1[m] with a similar affinity and specificity to the corresponding mutant within GAP1[IP4BP]. Whether or not Ins(1,3,4,5)P₄ can control Ca^{2+} entry (or mobilization) by interacting with specific receptors has been a controversy for a number of years (Irvine, 1992a,b). This subject has been reviewed by Cullen (1998), who has suggested a simple model summarizing a possible role of GAP1[IP4BP] in regulating

Ca^{2+} entry into cells. It remains unclear whether the Ca^{2+} entry channel activated by $Ins(1,4,5)P_3$ and $Ins(1,3,4,5)P_4$ is a separate one from that controlled by Ca^{2+} store depletion, or whether it is a subset of the same channels (Irvine, 1993).

Shears (1998) has discussed the regulation of the 1-kinase/1-phosphatase cycle and has pointed out that $Ins(1,3,4)P_3$ is a potent inhibitor of the $Ins(3,4,5,6)P_4$ 1-kinase. By inhibiting the 1-kinase, and altering the equilibrium of the putative 1-kinase/1-phosphatase cycle, $Ins(1,3,4)P_3$ is believed to be well suited to link the activation of PLC to increases in levels of $Ins(3,4,5,6)P_4$ (Tan et al., 1997). Yang and Shears (1999) found that $Ins(3,4,5,6)P_4$ 1-kinase is not affected by either Ca^{2+}, protein kinase A, or protein kinase C. The evidence for the receptor-dependent increases in $Ins(3,4,5,6)P_4$ is discussed under regulation of Ca^{2+} levels (uncoupling of Cl^- secretion) (see below).

In vitro, the activity of the $Ins(1,3,4,)P_3$ 6-kinase is constrained by product inhibition (Abdullah et al., 1992) and by its near complete inhibition by physiologically relevant levels of $Ins(3,4,5,6)P_4$ (Hughes et al., 1989; Hildebrant and Shuttleworth, 1992). However, upon receptor activation, $Ins(1,3,4,6)P_4$ is readily formed (Hughes et al., 1989). Control over the levels of 6-kinase could also be mediated by regulation of its rate of proteolysis (Shears, 1998).

In epithelial cells, $Ins(3,4,5,6)P_4$ seems to be a physiologically important inhibitor of Ca^{2+}-regulated Cl^- channels. The activation, however, is transient even though Ca^{2+} level remains elevated. The time course of inhibition of Cl^- secretion matched that of the production $Ins(3,4,5,6)P_4$ best (Vajanaphanich et al., 1994). This observation was confirmed by using cell permeable analogues of natural $Ins(3,4,5,6)P_4$. The molecular mechanism of $Ins(3,4,5,6)P_4$ action was shown by inhibition of a Cl^- channel (Xie et al. 1996; Ismailov et al., 1996). In human epithelial cells, the probable target of $Ins(3,4,5,6)P_4$ is a novel 1 pS Cl- channel, which is activated by Ca^{2+} or by CaMKII (Ho et al., 2001). The mechanism by which levels of $Ins(3,4,5,6)P_4$ increase when $Ins(1,4,5)P_3$-generating agonists are used has been clarified by Yang et al. (1999), who have shown that Ins(3,4,5,6)P_4 is removed by a dual-specificity kinase, whose catalysis of an Ins(3,4,5,6)P_4 1-kinase reaction is inhibited by its co-substrate, Ins(1,3,4)P_3.

11.2.3.3. Ins(1,3,4,5,6)P_5/Ins(1,4,5,6)P_4 cycle

Ins(1,3,4,5,6)P_5 is interconverted with Ins(3,4,5,6)P_4 as shown above as well as participates in a separate Ins(1,3,4,5,6)P_5 3-phosphatase/Ins(1,4,5,6)P_4 3-kinase cycle (Oliver et al., 1992; Craxton et al., 1994). In order to ascertain the activity of this cycle in intact AR4-2J cells, Oliver et al. (1992) used actinomycin A to reduce cellular ATP and hereby inhibit Ins(1,4,5,6)P_4 3-kinase activity. In such experiments, levels of Ins(1,4,5,6)P_4 increased four-fold within 1 h.

Furthermore, Ins(1,4,5,6)P_4 has a high affinity for the plextrin homology domain of a protein, p130, that has considerable similarity to PLC but does not hydrolyze PtdInsPs. It has not been possible to determine how the Ins(1,3,4,5,6)P_5 3-phosphatase/Ins(1,4,5,6)P_4 3-kinase cycle is regulated in vivo, because the 3-phosphatase activity is catalyzed by a MIPP kinase that is concentrated inside endoplasmic reticulum (Ali et al., 1993). When exogenous InsPs were added to purified endoplasmic reticulum vesicles, MIPP activity remained undetectable until the membranes were permeabilized with detergents (Ali et al., 1993).

Ins(1,4,5,6)P_4 is the main InsP$_4$ formed when cell homogenates are incubated with Ins(1,3,4,5,6)P_5 in vitro, and the one that accumulates in cells transformed with src (Mattingly et al., 1991). Majerus et al. (1999) have observed that a *Salmonella* virulence protein, SopB, has some sequence similarity to the Ins polyphosphate 4-phosphatases. Among a number of reactions that it catalyzes is one that generates Ins(1,4,5,6)P_4 from Ins(1,3,4,5,6)P_5 (Norris et al., 1998). In addition, transfection of mammalian cells with SopB increases Cl$^-$ transport (Feng et al., 2001), although recent studies with *Salmonella* virulence factors (Zhou et al., 2001) suggest that the effect may involve complex changes in cell architecture and function.

$Ins(1,3,4,5,6)P_5$ is believed to serve as metabolic hub in higher InsP metabolism, but it is not known whether or not it serves any specific physiological function in mammalian cells. In several animal species with nucleated erythrocytes, $Ins(1,3,4,5,6)P_5$ is believed to decrease the affinity of hemoglobin for O_2, while increasing the cooperativity for O_2 binding, which is a function fulfilled by ATP or 2,3-biphosphoglycerate in most animals. However, this generally accepted function of $Ins(1,3,4,5,6)P_5$, to modulate hemoglobin-O_2 interaction, now appears to require reexamination (Irvine and Schell, 2001).

11.2.3.4. $InsP_6/InsP_5$ cycle

The cellular functions of $InsP_6$ have remained largely unknown, despite the fact that $InsP_6$ is the most abundant InsPs in nature. This is in part due to an apparent lack of response of $InsP_6$ to extracellular signals (Shears, 1998). An exception is provided by the response of S. pombe to hypertonic shock, which leads to rapid increases in the levels of $InsP_6$ (Ongusaha et al., 1998). A role for $InsP_6$ in cell signaling also has been suggested by recent work in two other systems. Firstly, three genes, whose products constitute a path linking $Ins(1,4,5)P_3$ to $InsP_6$ have been identified as members of a group that functionally complement lesions in mRNA export from the nucleus in yeast; the genes, respectively encode PLC, an $Ins(1,4,5)P_3$ 3-/$Ins(1,3,4,5)P_4$ 6-kinase, and $Ins(1,3,4,5,6)P_5$ 2-kinase (York et al., 1999; Odom et al., 2000). Secondly, in insulin-secreting pancreatic β-cells (Larsson et al., 1997), it was shown that 10 μM $InsP_6$ in the patch pipette increased an L-type Ca^{2+} current; the effect reflecting the ability of $InsP_6$ to inhibit protein phosphatase, types PP1, PP2A, and PP3, with subsequent activation of the L-type Ca^{2+}-channels.

Recently, Lemtiri-Chlieh et al. (2000) have recognized a function for $InsP_6$ in plants as a signal generated in guard cells in response to the stress hormone, (RS)-2-*cis*,4-*trans*-abscissic acid

(ABA), along with an electrophysiological target. Specifically, evidence is presented that $InsP_6$ regulates inward K^+ current, which is the best characterized electrophysiological target of ABA in guard cells (Blatt and Grabov, 1997). Lemtiri-Chlieh et al. (2000) have shown that $InsP_6$ in the patch pipette is a potent inhibitor of guard cell plasma lemma inward K^+ current, and that $InsP_6$-dependent inhibition of inward K^+ current is manifested in a calcium-dependent manner. Other experiments showed that the potent effects of $InsP_6$ are not a consequence of its metabolism to $Ins(1,4,5)P_3$.

In *Dictyostelium*, the synthesis of $InsP_6$ proceeds via $Ins(3)P$, $Ins(3,6)P_2$, $Ins(3,4,6)P_3$, $Ins(1,3,4,6)P_4$, and $Ins(1,3,4,5,6)P_5$. It is not known whether these steps are catalyzed by one or multiple kinases (Murthy, 1996).

The originally proposed storage function of $InsP_6$ in plants remains the most likely, but new possibilities, including the activation of K^+ channels in GUARD CELLS has also been suggested (Lemtiri-Chlieh et al., 2000). In recent years, several more suggestions for $InsP_6$ functions have been made, including protein phosphatase inhibition (Larsson et al., 1997), and the activation of protein kinase (Efanov et al., 1997). $InsP_6$ has also been shown to interact in vitro with several intracellular proteins, and these all seem to fall into the general area of secretion or vesicular recycling. Shears (2001) has reviewed this area and raised several concerns about possible artifacts because of the strong interaction with positively charged groups on proteins or low molecular weight compounds. Recently, Hilton et al. (2001) have purified an $InsP_6$-stimulated protein kinase that specifically phosphorylates pacsin/syndapin I, a protein involved in synaptic vesicle recycling. Furthermore, York et al. (1999) have suggested that $InsP_6$ is involved in mRNA export from the nucleus, while a nuclear function for $InsP_6$, has been proposed by Hanakahi et al. (2000), who showed that DNA end-joining was dependent on the addition of a constituent of a cell fraction identified as $InsP_6$. The target for $InsP_6$ was shown to be the DNA-dependent protein kinase, and a possible binding domain for $InsP_6$ was identified (Hanakahi et al., 2000).

11.2.3.5. Pyrophosphate cycle

Stephens et al. (1993b) and Menniti et al. (1993a,b) have described the discovery in *Dictyostelium* of $InsP_7$ and $InsP_8$. In mammalian cells, the only $InsP_7$ identified has the pyrophosphate in the 5-position (Albert et al., 1997). In *D. discoideum* the principal $InsP_8$ has pyrophosphates in the 5- and 6-positions (Laussmann et al., 1997). It is thought that all proteins that bind $InsP_6$ also bind $InsP_7$ and $InsP_8$. (Ye et al., 1995). The $InsP_7$ and $InsP_8$ are involved in the assembly of multimolecular structures or membrane fusion events and may provide a source of local energy (Irvine and Schell, 2001). The enzymes that phosphorylate $InsP_6$ to $InsP_7$ have been cloned and are related to some other InsP kinases (Saiardi et al., 1999; 2000a,b; Schell et al., 1999).

There is evidence that $[PP]_2$-$InsP_4$ levels respond to a cAMP-dependent signaling pathway (Safrany and Shears, 1998). Thus, in response to activation of adenylate cyclase by β-adrenergic agonists, cellular levels of $[PP]_2$-$InsP_4$ steadily declined by up to 70% over a 30 min period. According to Safrany and Shears (1998), the cyclic nucleotides (cAMP and cGMP) appear to regulate $[PP]_2$-$InsP_4$ levels independently of the activation of both protein kinase A and protein kinase G. Voglmaier et al. (1997) have shown that the PP-$InsP_5$ and $[PP]2$-$InsP_4$ can be dephosphorylated in vitro back to $InsP_6$ by reverse reactions of the $InsP_6$ and PP-$InsP_5$ kinases. In addition, PP-$InsP_5$ and $[PP]_2$-$InsP_4$ can be dephosphorylated by a phosphohydrolase attack upon the β-phosphates of the diphosphate groups (Safrany et al., 1999). F^- blocks the dephosphorylation of PP-$InsP_5$ and $[PP]_2$-$InsP_4$ in intact cells by diphosphoinositol phosphate phosphohydrolase (DIPP) (Safrany et al., 1999).

11.2.4. Specific biological effects

The discovery of Ins(1,4,5)P$_3$ as a second messenger led to a rapid identification of many other InsPs along with their synthetic pathways and suggestions of possible physiological functions. Recent studies point to specific involvement of InsPs in diverse aspects of cell biology, which may be summarized under the general topics of ion channel physiology, membrane dynamics and nuclear signaling (Irvine and Schell, 2001).

11.2.4.1. Ion channel physiology

Receptor-induced changes in cytoplasmic and nuclear Ca^{2+} levels as a result of release of membrane-bound Ca^{2+} stores through InsP$_3$ or entry of extracellular Ca^{2+} through plasma membrane ion channels constitute an important signaling mechanism (Berridge and Irvine, 1989). The discovery of calmodulin as a major Ca^{2+}-sensing protein in the cells, and identification of protein targets for Ca^{2+}/CaM complexes including a family of Ca^{2+}/CaM activated protein kinases, has provided an explanation how Ca^{2+} release is translated into molecular consequences. Furthermore, the development of cell-permeant Ca^{2+} sensor fluorophores has afforded detailed spatio-temporal picture of signaling events occurring in the cell.

Experimental evidence, ranging from the first observations of Ins(1,4,5)P$_3$-stimulated Ca^{2+} mobilization in permeabilized cell preparations to the recent purification, cDNA sequencing and functional reconstitution of the Ins(1,4,5)P$_3$ receptor, provide unequivocal evidence that Ins(1,4,5)P$_3$ is the second messenger responsible for hormone-stimulated intracellular Ca^{2+} release (Shears, 1998). Taylor (1998) has discussed the mechanisms of Ca^{2+} regulation of Ins(1,4,5)P$_3$ receptors, including the number of possible Ca^{2+} bindings sites on Ins(1,4,5)P$_3$ receptors, along with the roles of accessory proteins. A range of practical problems, the likely need for accessory proteins, the importance of the rate of change of Ca^{2+} concentration and the existence of receptor subtypes have apparently confused the issue (see above).

Receptor-evoked Ca^{2+} signals include Ca^{2+} release from internal stores and subsequent Ca^{2+} influx across the plasma membrane (Putney and Mckay, 1999). The two activities linked in that Ca^{2+} release from the stores are obligatory for activation of Ca^{2+} influx. This gating behavior has led to the formulation of the capacitative Ca^{2+} entry hypothesis (Putney, 1990; Putney et al., 2001). The hypothesis was originally proposed to account for a gating behavior with the identification of the Ca^{2+} release activated Ca^{2+} current (I_{crac}). Experimental evidence has now been presented in support of two mechanisms to account for the capacitative calcium entry phenomenon (Putney, 1999), One is a secretion mechanism, which proposes that store depletion triggers a delivery of conducting units containing CCE channels to the plasma membrane. The other is based on the conformational-coupling hypothesis (Irvine, 1991; Berridge, 1995) in which $InsP_3$-activated Ca^{2+} release channels ($InsP_3$ receptors) in the store membrane couple to and gate Ca^{2+} influx channels in the plasma membrane. A major unresolved problem is the role played by $Ins(1,4,5)P_3$ in this process. Previous studies have clearly demonstrated a requirement for $Ins(1,4,5)P_3$ to gate these channels (Kiselyov et al., 1999; Zubov et al., 1999).

Recently, Boulay et al. (1999) have provided compelling evidence for the involvement of a member of the TRP family of channel proteins in this signaling pathway, and for signaling activation of the TRP channels through interaction with $Ins(1,4,5)P_3$ receptors. Mammalian TRP proteins are homologues of the *Drosophila* photoreceptor mutants TRP and TRPL. There are at least seven mammalian genes, designated TRP1 through TRP7 (Okada et al., 1999). A 289-amino acid sequence just down stream from the $Ins(1,4,5)P_3$ binding domain of the $Ins(1,4,5)P_3$ receptor, designated F2r (positions 638–926), strongly bound TRP3 and within this sequence, a number of small interacting sequences were identified (Boulay et al., 1999). As pointed out by Boulay et al.

(1999), the simplest mechanism by which signaling might occur from depleted calcium stores would involve a calcium-dependent conformational change in the $Ins(1,4,5)P_3$ receptor. Berridge et al. (2000) have recently commented upon the versatility and universatility of calcium signaling.

$Ins(1,3,4,5)P_4$, a product of the specific phosphorylation of $Ins(1,4,5)P_3$, has been proposed as a candidate for regulating Ca^{2+} homeostasis within the cell interior (Irvine et al., 1988). Hermosura et al. (2000) have shown that $Ins(1,3,4,5)P_4$ facilitates store-operated calcium influx by inhibition of $InsP_3$ 5-phosphatase and therefore increase Ca^{2+} entry near the plasma membrane, where the 5-phosphatase is predominantly localized. Irvine and Schell (2001) have reviewed the evidence that $Ins(1,3,4,5)P_4$ can activate Ca^{2+} channels in the plasma membrane of endothelial cells and neurons. Zhu et al. (2000) have obtained evidence that $InsP_4$ is a frequency regulator in calcium oscillations in HeLa cells. The effects of $Ins(3,4,5,6)P_4$ on Cl^- secretion is of a special interest for patients with cystic fibrosis in whom the epithelial cyclic MP-regulated Cl^- channel is compromised (Irvine and Schell, 2001).

There is evidence that in epithelial cells, $Ins(3,4,5,6)P_4$ is a physiologically important inhibitor of Ca^{2+}-regulated Cl^- channels (Kachintorn et al., 1993). This demonstration has been confirmed by subsequent work with cell-permeable $Ins(3,4,5,6)P_4$ (Vajanaphanich et al., 1994).

In order to specifically study the actions of $Ins(3,4,5,6)P_4$, Carew et al. (2000) delivered it into the interior of the cells in the polarized monolayer, by using the cell permeant and bioactive analogue, $Bt_2Ins(3,4,5,6)P_4/AM$ (Vajanaphanich et al., 1994). The study showed that $Ins(3,4,5,6)P_4$ inhibits the Ca^{2+} activated Cl^- conductance (CaCC) in the apical membrane, in a physiologically relevant context of a polarized monolayer. The obtained results provided the first unequivocal demonstration that $Ins(3,4,5,6)P_4$ is an active constituent of a second messenger system designated to carefully regulate the participation of CaCC in the important process of fluid secretion.

Irvine and Schell (2001) have pointed out four specific consequences resulting from generation of $Ins(1,3,4,5)P_4$. First, $Ins(1,3,4,5)P_4$ inhibits the 5-dephosphorylation of $Ins(1,4,5)P_3$ (Hermosura et al., 2000), thus enhancing the mobilization of Ca^{2+} by $Ins(1,4,5)P_3$. Second, $Ins(1,3,4,5)P_4$ has a direct action on Ca^{2+} channels in the plasma membrane (Luckhoff and Clapham, 1992) and, third, it might have a complex action on the Ca^{2+} stores mobilized by $Ins(1,4,5)P_3$ (Smith et al., 2000). Fourth, $Ins(1,3,4)P_3$ derived from $Ins(1,3,4,5)P_4$ inhibits $Ins(3,4,5,6)P_4$ 1-kinase (Yang et al., 1999; Yang and Shears, 2000), which results in an increase in $Ins(3,4,5,6)P_4$ and thus a decrease of chloride efflux (Ho et al., 2000).

Ching et al. (2001) have presented evidence that PtdIns 3-kinase plays a concerted role with PLC γ in initiating antigen-mediated Ca^{2+} signaling in mast cells via $PtdIns(3,4,5)P_3$-sensitive Ca^{2+} entry pathway. This study refutes the second messenger role of D-myo-$Ins(1,3,4,5)P_4$ in regulating FcϵRI-mediated Ca^{2+} response. Ching et al. (2001) suggest that $PtdIns(3,4,5)P_3$ directly stimulates a Ca^{2+} transport system in plasma membranes. They present pharmacological and biochemical evidence that PtdIns3K plays a concerted role with PLC γ in the regulation of Ca^{2+} influx in RBL-2H3 mast cells via $PtdIns(3,4,5)P_3$-sensitive Ca^{2+} entry pathway. This unique Ca^{2+} influx mechanism was also demonstrated in platelets and Jurkat T cells (Ching et al., 2001a,b). Although several research groups have isolated an $Ins(1,3,4,5)P_4$-binding protein, GAP1^{IP4BP}, from platelet plasma membranes (Cullen et al., 1995; 1997), the study of Ching et al. (2001a,b) refutes the second messenger role of $Ins(1,3,4,5)P_4$ in regulating FcϵRI-mediated Ca^{2+} inflow in mast cells. This finding is consistent with earlier data in mouse lacrimal acinar cells (Bird and Putney, 1996) and Jurkat cells (Hsu et al., 2000) that have ruled out the involvement of $Ins(1,3,4,5)P_4$ in receptor-activated Ca^{2+} influx.

According to the hypothesis of Ching et al. (2001a,b) the Ca^{2+} influx is regulated by two discrete Ca^{2+} entry mechanisms on plasma membranes: (1) the capacitative Ca^{2+} entry that is secondary to the depletion of the intracellular Ca^{2+} store by

Ins(1,4,5)P$_3$, and (2) the PtdIns(3,4,5)P$_3$-sensitive Ca^{2+} entry that is independent of the filling state of internal Ca^{2+} stores.

Bony et al. (2001) have proposed a specific role for PtdIns 3-kinase γ in the regulation of Ca^{2+} oscillations in cardiac cells. PtdIns 3-kinase γ activation was shown to be a crucial step in the purinergic regulation of cardiac cell spontaneous Ca^{2+} spiking. Other data suggested that Btk tyrosine kinase Tec works in concert with a Src family kinase and PtdIns 3-kinase γ to fully activate PLC γ in ATP-stimulated cardiac cells. This cluster of kinases was believed to provide the cardiomyocyte with a tight regulation of InsP$_3$ generation and thus cardiac autonomic activity.

An interesting new possibility for InsP$_6$ in plants has been suggested in the activation of K$^+$ channels in GUARD CELLS (Lemtiri-Chlieh et al., 2000). Application of submicromolar concentrations of InsP$_6$ specifically inhibited the same inward-rectifying K$^+$ channel that is an intrinsic part of the response of these cells to abscissic acid.

Finally, Camina et al. (1999) have proposed an intracellular mechanism of action for a new and extremely potent lipid activator of the Ca^{2+} mobilization, which is active in the vitreous body and is independent of Ins(1,4,5)P$_3$. Initial work had suggested that this molecule is a high molecular weight phospholipid with sn-1- and sn-2-acyl groups and a complex polar head, including at least one terminal phosphate group and possibly a short peptide chain. Based on further experimental data, a signaling pathway involving a PtdCho-specific PLC coupled to DAG kinase was suggested. Ins(1,4,5)P$_3$ and Ca^{2+}/calmodulin-dependent factors have been shown by Valhmu and Raia (2002 to mediate transduction of compression-induced signals in articular chondrocytes.

11.2.4.2. Membrane dynamics

Within a short time of the discovery that Ins(1,4,5)P3 was a Ca^{2+}-mobilizing second messenger, it was found that its immediate dephosphorylation products Ins(1,4)P$_2$ and InsP were increasing in stimulated cells along with Ins(1,3,4,5)P$_4$ and its catabolic product

Ins(1,3,4)P_3 (Irvine et al., 1984; Batty et al., 1985). Ins(1,3,4,5)P_4 is hydrolyzed by the same 5-phosphatase that hydrolyses Ins(1,4,5)P_3, but the enzyme has a 10-fold higher affinity than it does for Ions(1,4,5)P_3. Thus, Ins(1,3,4,5)P_4 can protect Ins(1,4,5)P_3 against hydrolysis, and therefore increase its effectiveness in promoting the Ca^{2+} entry (Hermosura et al., 2000). Other consequences of Ins(1,3,4,5)P_4 generation may be summarized as follows: inhibition of the 5-dephosphorylation of Ins(1,4,5)P_3, direct action of Ins(1,3,4,5)P_4 on Ca^{2+} channels in the plasma membrane, and complex action on the Ca^{2+} stores mobilized by Ins(1,4,5)P_3 (Irvine and Schell, 2001). Ins(1,3,4)P_3 derived from Ins(1,3,4,5)P_4 inhibits Ins(3,4,5,6)P_4 1-kinase (Yang and Shears, 2000). Triple InsP$_3$ receptor knockout Dt40 chicken B-cells were used by Ma et al. (2001) to assess the role of InsP$_3$ receptors in store-operated channel activation. Venkatachalan et al. (2002) have recently reviewed the cellular and molecular basis of store-operated calcium entry.

Ins(3,4,5,6)P_4 appears to be a physiologically important inhibitor of Ca^{2+}-regulated Cl^- channels. Kachintorn et al. (1993) found that the inhibition of Cl^- secretion was paralleled by the production of various InsP$_4$s, of which Ins(3,4,5,6)P_4 correlated best (Vajanaphanich et al. (1994). In human epithelial cells, the target of Ins(3,4,5,6)P_4 may be a novel I ps Cl^- channel, which is activated by Ca^{2+} or by CaMKII (Ho et al., 2001). Ho et al. (2002) have since shown that Ins(3,4,5,6)P_4 is synthesized by Ins(1,3,4,5,6)P_5 1-phosphatase activity by an enzyme previously characterized as an Ins(3,4,5,6)P_4 1-kinase. Ho et al. (2002) point out that it is exceptional for a single enzyme to catalyze two opposing signaling reactions (1-kinase/1-phosphatase) under physiological conditions.

The evidence that Ins(1,4,5,6)P_4 has a physiological function which is indirect and largely conjectural. Ins(1,4,5,6)P_4 levels have been observed to increase markedly when cells are infected with *Salmonella dublin* (Eckmann et al., 1997), which leads to inhibition of EGF receptor stimulated PtdIns 3-kinase signaling possibly because the high levels of Ins(1,4,5,6)P_4 antagonize interactions between PtdIns(3,4,5)P_3 and its targets (Majerus et al., 1999).

Ins(1,3,4,5,6)P_5 is the InsP$_5$ isomer that predominates greatly in most mammalian cells. It serves as a metabolic hub in higher InsP metabolism, but it is not known whether or not it is involved in any other physiological function. In several animal species with nucleated erythrocytes, Ins(1,3,4,5,6)P_5 is believed to decrease the affinity of haemoglobin for O_2, while increasing the cooperativity for O_2 binding, although this generally accepted function may require reexamination (Irvine and Schell, 2001).

In recent years, it has been shown that InsP$_6$ interacts in vitro with several intracellular proteins, and that these seem to fall into the general area of secretion or vesicular recycling. They include synaptotagmins (Fukuda et al., 1994; Mehrotra et al., 2000), the vesicle adapter proteins AP-2 (Voglmaier et al., 1992), AP-180 (Ye et al., 1995) and arrestin (Gaidarov et al., 1999). Another interesting function for InsP$_6$ in plants and animals appears to lie in its antioxidant properties (Hawkins et al., 1993). Shears (2001) has recently reviewed this area and raised several concerns about possible artefact formation by the extremely high charge of InsP$_6$. Shears (2001) has suggested several criteria, which should be considered when testing the physiological relevance of InsP$_6$-dependent phenomena, including presence of Ca^{2+}/Mg^{2+} cations at a physiologically relevant ionic strength, effect not imitated by ion chelators, such as EDTA/EGTA, effect less potent than that of InsP$_6$, effect not imitated by other conformers, appropriate ligand protein ratio, and the purity of InsP$_6$.

Hilton et al. (2001) have purified an InsP$_6$-stimulated protein kinase that specifically phosphorylates pacsin/syndapin I, a protein involved in synaptic vesicle recycling. This study appears to have met most of the criteria advanced by Shears (2001). The kinase is almost completely dependent on InsP$_6$, with high specificity for this InsP over many other InsPs except InsP$_7$, which was equipotent. Hilton et al. (2001) used the 5-PP-InsP$_5$ isomer and an isomer (2-PP-InsP$_5$) that does not occur naturally to determine the ability of inositol pyrophosphates to stimulate phosphorylation of pacsin/syndapin I. It was found that both isomers of PP-InsP$_5$ stimulated phosphorylation of GST-pascin/syndapin I.

11.2.4.3. Nuclear signaling

It has been known for some time that the conventional enzymes and inositol derivatives involved in the PtdIns cycle-kinases, phospholipases, InsP kinases, PtdInsPs and InsPs are all found in the nucleus and that their activities and abundance seem to be linked to cell function (Cocco et al., 1998, 2001; York et al., 1998; Maraldi et al., 1999, 2000; D'Santos et al., 1998). York et al. (1999) and Odom et al. (2000) have screened yeast genes that are involved in mRNA transport out of the nucleus, and isolated four (York et al. 1999). Three of these turned out to encode proteins that could constitute as previously suggested pathway to synthesize InsP$_6$: yeast's only PtdIns-specific C; an Ins(1,3,4,5,6)P$_5$ 2-kinase; and a third gene that was subsequently identified as an enzyme that can phosphorylate Ins(1,4,5)P$_3$ to Ins(1,4,5,6)P$_4$ and then phosphorylate that to Ins(1,3,4,5,6)P$_5$ (InsP multikinase). The possibility that InsP$_6$ is the active component, rather than InsP$_7$ or InsP$_8$ (see above), is supported by the observation that deletion of InsP$_6$ kinase from yeast has no effect on mRNA transport (Saiardi et al., 2000a). Van der Kaay et al. (1995) had earlier shown that in the nucleus InsP$_6$ is synthesized from Ins(1,4,5)P$_3$.

Recent work by York et al. (1999, 2001) and Odom et al. (2000) have reported that inositol signaling in the nucleus regulates gene expression and Chi and Crabtree (2000) have commented upon the implied new level of control over this critical process. There is evidence that a nuclear Ins(1,4,5)P$_3$ kinase activity may be involved in transcriptional regulation (Odom et al., 2000). It was found that InsP$_4$ has a unique role in the nucleus in regulating transcription, and that a specific inositol kinase is a component of the yeast transcription complex. Further comments upon this observation have been provided by Irvine and Schell (2001), Shears (2001) and York et al. (2001).

Another nuclear function for $InsP_6$ arises from the observation of Hanakahi et al. (2000), that $InsP_6$ is necessary for DNA end-joining in vitro. The target for $InsP_6$ was shown to be the DNA-dependent protein kinase, and a possible binding domain for $InsP_6$ was identified (Hanakahi et al., 2000). The requirement for $InsP_6$ appeared to be specific since $Ins(1,3,4,5)P_4$ and $Ins(1,3,4,5,6)P_5$ were less active and $InsS_6$ was inactive. However, $InsP_7$ and $InsP_8$ as well as other InsPs substituted in the 1-, 2- and 3-positions were not tested. Shears (2001) has commented upon several of these findings and has pointed out that stoichiometry, specificity, and the affinity of ligand binding are important factors that remain to be determined.

11.3. PtdInsPs as cellular signals

The signaling functions of PtdInsPs, originally, were attributed to the breakdown products of $PtdIns(4,5)P_2$ that were generated by the activity of PtdIns-specific PLC. The resulting diacylglycerol activated protein kinase C and the $Ins(1,4,5)P_3$ opened Ca^{2+} channels in the endoplasmic reticulum. It is now known that cells use PtdInsPs directly for regulatory functions. They form membrane-binding sites for soluble proteins, recruit cytosolic proteins to membranes, stabilize protein complexes on membranes, and activate membrane proteins. PtdInsPs are involved in signaling, cytoskeleton-membrane interactions, and in membrane-vesicle budding and fusion. The general biological effects of PtdInsPs have been discussed in several recent reviews (Martin, 1998; Hinchcliffe et al., 1998; Vanhaesebroeck et al., 1997, 2001; Hunter, 2000). The specific role of the lipid products of PtdIns 3-kinase in cell function has been reviewed by Rameh and Cantley 1999. Specificity of each PtdInsPs is based on its structure, its location and the timing of its synthesis, modification and hydrolysis (Irvine, 1998). PtdInsPs are products of and substrates for various kinases, phosphatases, and for PLC and PLD, which may rapidly alter the levels of particular PtdInsPs in specific regions of the membrane. The regulated activity of these enzymes provides the basis for an efficient spatial and temporal regulation of vesicular transport (Sprong et al., 2001).

11.3.1. Evidence for signaling

Evidence for a signaling role of intact PtdInsPs comes from detailed analytical determinations, metabolic transformations, and the demonstration of molecular association between specific PtdInsPs and proteins, and the identification of membrane microdomain complexes. The discovery of PtdIns 3-kinases, and the family of $3'$-phosphorylated PtdInsPs they generate led to the elucidation of a specific phospholipid-based signaling system in which proteins are recruited *via* PtdInsP$_3$-binding pH domains to the membrane where they are activated (e.g. the PKB/Akt protein-serine kinase) (Kapeller and Cantley, 1994).

11.3.1.1. Analytical
Precise analytical determinations have been essential for establishing the nature of the PtdInsPs involved in the cellular processes and for detection of changes in the levels of the PtdInsPs in response to cellular activation. There have been many obstacles to establish the full outline of membrane-based signaling processes. Many of the PtdInsPs are in low abundance in cells but at high local concentrations in membrane domains that are not readily detected by conventional biochemical approaches. Thus, it has been helpful to use methods that allow for detection of PtdInsP concentrations localized in membrane domains, such as immunochemistry with PtdInsP antibodies (Voorhout et al., 1992; Tran et al., 1993).

The rise in PtdInsP$_3$ levels correlates with activation of downstream responses. The levels of PtdIns(3,4,5)P$_3$ are very low in unstimulated cells and are increased within seconds of agonist stimulation in a variety of cell types (Stephens et al., 1991;

Hawkins et al., 1992; Conricode, 1995; Tsakiridis et al., 1995; Nave et al., 1996). The rise in PtdIns(3,4,5)P$_3$ levels precedes the activation of signaling molecules thought to lie downstream of PtdIns 3-kinase (Cross et al., 1997; Van der Kaay et al., 1997). Pretreatment with the specific PtdIns 3-kinase inhibitors wortmannin and LY294002 blocks the agonist-stimulated accumulation of PtdIns(3,4,5)P$_3$ in parallel with their inhibitory effects on activation of downstream response (Van der Kaay et al., 1997; Vlahos et al., 1994; Yeh et al., 1995; Arcaro et al., 2000). These findings are consistent with PtdIns(3,4,5)P$_3$ acting as an early upstream element in signal-transduction cascades. Batty and Downes (1996) had earlier analyzed the PtdIns 3-kinase response to insulin and had observed a complex control of PtdInsPs concentrations in 132N1 cells. Insulin did not influence InsP$_3$ concentrations but stimulated accumulation of PtdIns(3,4,5)P$_3$ and PtdIns(3,4)P$_2$.

Furthermore, the rise in PtdInsP$_3$ levels is transient. The duration of the rise in PtdIns(3,4,5)P$_3$ levels depends on cell type and the agonist used to stimulate the cells, and can range from a matter of seconds up to several hours. Hormone stimulation of PLC causes a rapid (within seconds) loss of PtdIns(4,5)P$_2$, PtdIns(4)P, but slower loss of PtdIns, together with a correspondingly rapid (within seconds) formation of Ins(1,4,5)P$_3$ and Ins(1,4)P$_2$ and possibly Ins(1,3,4,5)P$_4$, but delayed rise in InsP (Conricode, 1995; Cross et al., 1997; Van der Kaay et al., 1997). The transient increase in PtdIns(3,4,5)P$_3$ is consistent with a role as an agonist-regulated molecular on/off switch. Cellular PtdIns(3,4,5)P$_3$ levels are regulated by the balance between the rate of PtdIns(3,4,5)P$_3$ production, which relies on continued stimulation, and PtdInsP$_3$ degradation.

A complication in monitoring changes in the PtdInsPs alone is the ability of cells to resynthesize PtdIns rapidly, and therefore PtdInsP and PtdInsP$_2$. Biosynthetic assays (See Chapter 7), however, are not readily applicable to measurements of PtdIns(3,4,5)P$_3$ in extracts of animal tissues. Moreover, they can be misleading since the association of PtdIns 3-kinases in molecular complexes is not necessarily correlated with the enzyme's activity state. Direct measurement of PtdIns(3,4,5)P$_3$ would therefore be desirable since its concentration may be subject to additional control mechanisms such as activation or inhibition of the phosphatases responsible for PtdIns(3,4,5)P$_3$ metabolism. Watt et al. (2002) have recently developed a highly selective method for mapping the PtdIns(4,5)P$_2$ distribution within cells at high resolution, and their data provide direct evidence for the presence of PtdIns(4,5)P$_2$ at key functional locations.

11.3.1.2. Metabolic

For some cellular processes, PtdInsPs may be constitutively produced co-factors, the levels of which may not significantly change. In such instances, evidence of participation of specific PtdInsPs may be obtained by metabolic means, such as introduction of PtdInsP phosphatases (Zhang, 1998) and phospholipases (Rhee and Bae, 1997) of defined specificity or of PtdInsP-specific antibodies (Fukumi et al., 1988) and PtdInsPs-binding peptides (Hartwig et al., 1995) into cells, which inhibit cellular responses mediated by the appropriate PtdInsP signal. Cell-permeant lipid kinase inhibitors with high specificities have also been utilized. Furthermore, over expression wild type or constitutively active PtdInsP kinases of defined substrate specificity (Shibasaki et al., 1997) or introduction of membrane permeable PtdInsP itself (Franke et al., 1997a,b) mimics the effects of cellular activation or enhances a PtdInsP-dependent process.

The first committed step in the biosynthesis of PtdIns(4,5)P$_2$ is the phosphorylation of PtdIns by PtdIns 4-kinase to form PtdIns(4)P. PtdIns (4)P is phosphorylated by PtdIns(4)P 5-kinase to form PtdIns (4,5)P$_2$ (Chapter 7).

Termination or modification of signals mediated by PtdInsPs have classically been attributed to the action of cytoplasmic phospholipases and phosphatases. For example, in mammalian cells PtdIns(4,5)P$_2$ is cleaved to distinct second messengers by PLC in response to tyrosine kinase and G-protein-coupled receptor activation (Rhee, 1991; Steinweiss and Smrcka, 1992). PtdIns(4,5)P$_2$ turnover is also mediated by Type II 5-phosphatases like synaptojanin and OCRL (Attree et al., 1992; McPherson et al., 1996).

D-3 PtdInsPs represent a separate pathway of PtdIns metabolism. These are produced by PtdIns 3-kinases and include PtdIns(3)P, PtdIns(3,4)P$_2$, and PtdIns(3,4,5)P$_3$. The latter two compounds are formed transiently in cells in response to agonists and growth factors (Chapter 6). Degradation of the PtdInsPs leads to termination of the signal. The mechanism for the degradation of PtdInsP$_3$ is specific and is also regulated. Thus, cytoplasmic phosphatases may carry out the turnover of PtdIns(3)Ps as proteins exhibiting PtdIns(3,4,5)P$_3$, PtdIns(4,5)P$_2$ 5-phosphatase activity and PtdIns(3)P 3-phosphatase activity have been identified and purified (Woscholski and Parker, 1997). Unlike its precursor PtdIns(4,5)P$_2$, PtdIns(3,4,5)P$_3$ is not subject to degradation by any known phospholipases.

11.3.1.3. Molecular

The most compelling evidence for the role of PtdInsPs in cellular signaling comes from the binding of the phosphatides to specific domains of enzyme proteins and membrane receptors (Cullen et al., 2001). This view has been accepted only after some controversy. Bottomley et al. (1998a) have discussed the landmark discoveries and recent advances describing specific protein-phospholipid interactions, which play key roles in many signal transduction pathways. Bottomley et al. (1998a,b) have summarized in a tabular form the early data describing the interactions between PH domains and PtdInsPs and InsPs.

Many mammalian cell surface receptors signal through class I PtdIns 3-kinases, which transmit signals by virtue of the specific interactions of their products with PH domain-containing proteins. PH domains termed pleckstrin homology from the sequences initially identified in the platelet protein kinase C substrate pleckstrin, comprise approximately 120 amino acid collinear regions identified by sequence comparison in nearly 100 proteins (reviewed by Lemmon and Ferguson, 2000). PH domains are present in cytoskeletal components (spectrin, (α-actinin), guanine nucleotide exchange proteins of GTPase-regulating proteins and GTPases (dynamin), PtdInsP-regulated protein kinase (Akt/PKB, PDK1), other protein kinases (BTK, βARK), and PLC.

Dynamin is a GTPase required for membrane internalization during synaptic vesicle recycling and receptor mediated endocytosis (see below). Barylko et al. (1998) have demonstrated a synergistic activation of dynamin GTPase by Grb2 and PtdIns(4,5)P$_2$. PtdIns(4)P was a weak activator and PtdIns(3,4,5)P$_3$ did not activate GTPase at all.

The large majority of pH domains do not specifically bind D3-PdInsPs (Honda et al., 1999; Kavran et al., 1998) but those of the kinases Akt, PDK1 and Btk, and of Arf GTP exchange factors ARNO and Grp1 do, and have received much recent attention (James et al., 1996; Venkateswarlu et al., 1998). Thus, ARF6 exchange factor ARNO has been shown to participate in a rapid recruitment to the plasma membrane (t1/2 about 30 s), following agonist activation of PtdIns 3-kinase, via a specific interaction of its pH domain with the lipid products of PtdIns 3-kinase activation, PtdIns(3,4,5)P$_3$. In contrast the pH domain of PLC δ1, which specifically binds PtdIns(4,5)P$_2$, constitutively targets this protein to the plasma membrane, which leads to the activation of the enzyme (Stauffer et al., 1998).

Stephens et al. (1998) have described experiments aimed at characterizing PtdIns(3,4,5)P$_3$- and PtdIns(3,4)P$_2$-sensitive protein kinases B (PKB) and the mechanisms by which they are regulated. It was found that [^{32}P]PtdIns(3,4,5)P$_3$ binding protein or proteins co-purify with PKB kinase activity and four distinct forms of PKB kinase can be resolved. All four activities phosphorylate and activate phosphorylation of myelin basic protein (MBP) by PKB in the presence of the biological stereoisomers of PtdIns(3,4,5)P$_3$ or PtdIns (3,4)P$_2$ (for example, (1-stearoyl 2-arachidonoyl)-*sn*-phosphoryl-D-

myo-inositol(3,4,5)P$_3$). PKB kinases B, C, and D gave very similar results.

Misra and Hurley (1999) and Hurley et al. (2002) have shown that PtdIns(3)P, which is involved in endosomal traffic direction (see below), interacts with the small Zn^{2+}-containing Fab1-YOTP-Vac1-EEAI (FYVE) domain, rather than with the PH domain. The monomeric FYVE domain of Vps27p is about half the size of the pH domain. In this instance an effective binding is obtained by interaction with tandem FYVE domains (Gillooly et al., 2000; Mao et al., 2000) or additional regions, such as Rab5 interaction domain of EEA1 as suggested earlier (Burd and Emr, 1998).

PtdIns(4,5)P$_2$, which is involved in regulation of the cytoskeleton, plasma membrane channels and transporters, as well as in membrane trafficking (see below) is bound by a large number of binding proteins, including proteins with PH domains, 4.1-ezrin-radixin-moesin (FERM) domains. The FERM domain containing proteins ezrin, radixin and moesin have become of interest in regulating the cytoskeleton by linking actin filaments to adhesion proteins (Hurley and Meyer, 2001). The activity of these proteins is regulated by PtdInsP$_2$, although the precise mechanism is not known. The location of the PtdInsP kinases responsible for controlling the synthesis of PtdInsP$_2$ is also highly regulated. The PtdInsP kinases contain an active site loop that corresponds structurally to the activation loop of the protein kinases. A 20 amino acid loop is responsible for nearly all of the PtdInsP specificity observed in vitro (Kunz et al., 2000). Kunz et al. (2002) have subsequently reported that stereospecific substrate recognition of PtdInsP kinases may be swapped by changing a single amino acid residue.

11.3.2. PtdInsPs-protein interactions

It is difficult to make a clear distinction between protein and receptor binding as well as between receptor and membrane binding (Fruman et al, 1999; Holz et al., 2000). The more general term of PtdInsP protein interaction has therefore been adopted for the following discussion. Furthermore, it is often difficult to appreciate the in vivo significance of the PtdInsPs-protein binding data obtained in vitro. This is due to a lack of consistency in the methodology, and in particular to differences in the method of lipid presentation and in the type of lipid side chain in PtdInsPs used in these studies. The identification of protein constituents that exhibit stereoselective interactions with PtdInsPs would be a further important step toward defining proteins with effector roles in PtdInsP-dependent signaling pathways.

Sequence analysis of the genome has revealed protein domains with sequence homology of potential interest to lipid based membrane targeting (Hurley and Meyer, 2001). The epsin amino-terminal homology (ETNH) domain (Kay et al., 1999; Itoh et al., 2001) and Vps27p, Hrs and STAM (VHS) domain (Lohi and Lehto, 1998) are two such examples. The domains were discovered by sequence analysis and have received considerable attention, but their function has not been established. It remains to be seen whether ENTH-domain-mediated and VHS-domain-mediated localization represents an important new class of lipid based membrane targeting mechanisms or protein–protein interactions is more important (Wendland et al., 1999; Hyman et al., 2000).

11.3.2.1. PtdInsP₃ binding

In many cases binding of PtdIns(3,4,5)P$_3$ to target proteins via PH domains, is subject to a great deal of variability in the relative preference of PH domains for other PtdInsPs (Rameh et al., 1997a, b; Salim et al., 1996). This suggests that a distinct subclass of PH domains exists with the ability to bind either PtdIns(3,4,5)P$_3$ or PtdIns(3,4)P$_2$. Proteins containing PtdIns(3,4,5)P$_3$-binding PH domains include a range of molecules which affect the GTP loading status of small GTP proteins, including a molecule having homology with ARF GTPase-activating protein (GAP) (Tanaka et al., 1997; Stricker et al., 1997, 2000; Hammonds-Odie et al.,

1996), cytohesin and GRP-1, both of which show homology with ARF GEF (Klarlund et al., 1997, 2000), the Ras-GEF Sos (Rameh et al., 1997a,b) and the rho GEF TIAM-1 (Rameh et al., 1997a,b). PtdIns(3,4,5)P$_3$-binding PH domains are also found in a range of kinases, including BTK (Salim et al., 1996; Kojima et al., 1997), PKB and PDK-1 (Stokoe et al., 1997; Alessi and Downes, 1998). PtdIns(3,4,5)P$_3$ also binds to some isoforms of PKC (Toker et al., 1994; Nakanishi et al., 1993; Palmer et al., 1995), although these do not contain PH domains.

A hexapeptide from neurogranin (WAAKIQASFRGHMARKK) interacts with high affinity and specificity with PtdInsPs preferring PtdIns(3,4,5)P$_3$ over other phospholipids, and acquires structure upon binding to PtdInsPs (Lu and Chen, 1997).

The studies of Baraldi et al. (1999) have revised the early ideas of PtdInsP$_3$ binding to the PH domains based on the head group structure of PtdInsP$_2$ to the PH domain of PLC-δ by showing that in the PtdInsP$_3$ binding to the PH domain from Btk, the inositol ring was flipped nearly 180° about the axis between the 1 and 4 positions. This led to exchange of positions between the 3-phosphate and 5-phosphate groups of PtdInsP$_3$ as compared with the 3-hydroxyl and 5-phosphate of PtdInsP$_2$ bound to PLC-δ. The Btk PH domain, however, contains a large basic insertion in its β1-β2 loop that forms extensive interactions with the 5-phosphate. There is evidence that at least two major subsets of PH domains specifically recognize PtdInsP$_3$ using residues from different parts of the structure.

The B-cell specific Btk, has an N-terminal PH domain that binds selectively to PtdIns(3,4,5)P$_3$ (Salim et al., 1996; Kojima et al., 1997). Btk activation requires the prior stimulation of PtdInsP 3-kinase to generate PtdIns(3,4,5)P$_3$ and thereby to recruit Btk to the plasma membrane. These interactions are important since mutations that inactivate the Btk PH or SH2 domains cause B0-cell immunodeficiency (X-linked agammaglobulinemia), as does loss of Btk kinase activity (Smith et al., 2001).

The exact buffer conditions are also likely to have profound impact on the pattern of binding observed as, for example, it has been demonstrated that increasing the concentration of calcium can shift the synaptotagmin C2 domain binding preference from $PtdIns(3,4,5)P_3$ to $PtdIns(4,5)P_2$ (Shiavo et al., 1996).

Another problem with much of the data regarding binding of $PtdIns(3,4,5)P_3$ to targets is that the selectivity of this interaction is often weak. If these interactions are to represent a mechanism by which PtdIns 3-kinase may selectively activate a downstream signaling pathway, then there must be a high degree of specificity for binding $PtdIns(3,4,5)P_3$ or $PtdIns(3,4)P_2$. Particularly important is the specificity relative to $PtdIns(4,5)P_2$, the major substrate for PtdIns 3-kinase, which must necessarily be present wherever $PtdIns(3,4,5)P_3$ is produced, and which is present at more than 100 times the level of $PtdIns(3,4,5)P_3$, even in stimulated cells. Specificity must also be demonstrated relative to InsPs, which are present at relatively high concentrations in the cytosol and have structures which might mimic the head groups of $PtdIns(3,4,5)P_3$ or $PtdIns(3,4)P_2$.

The most appropriate form of $Ptdns(3,4,5)P_3$ to use in these binding experiments would appear to be sn-1-stearoyl-2-arachi-donoyl $GroPIns(3,4,5)P_3$, as this is the form in which it is found in cell membranes. The importance of this is highlighted by the studies of (Alessi and Downes, 1998; Alessi, 2001), who have performed one of the few direct comparisons of the efficacy of different forms of $PtdIns(3,4,5)P_3$. As described below, they found that PKB was a much better substrate for D3-PtdInsPs-dependent PK-1 (PDK-1) when incubated with sn-1-stearoyl-2-arachid-onoyl $GroPIns(3,4,5)P_3$ than with dipalmitoyl-$GroPIns(3,4,5)P_3$. Therefore, physiologically relevant evaluations of protein-$PtdIns(3,4,5)P_3$ interactions will be aided by the availability of synthetically produced sn-1-stearoyl-2-arachidonoyl Gro-$PIns(3,4,5)P_3$, with the availability of the D and L enantiomers also being important to facilitate appropriate control experiments (Stokoe et al., 1997; Alessi and Downes, 1998).

The PH domain of Grp1, a PtdIns 3-kinase activated exchange factor for Arf Gtases, selectively binds PtdIns(3,4,5)P$_3$ with high affinity. Lietzke et al. (2000) have determined the structure of the Grp1 PH domain in the unliganded form and bound to Ins(1,3,4,5)P$_4$. The structures revealed a novel mode of PtdInsP$_3$ recognition in which a 20-residue insertion relative to the canonical PH fold accounts for the ability of the Grp1 PH domain to selectively bind PtdIns(3,4,5)P$_3$, but not PtdIns(3,4)P$_2$ or PtdIns(4,5)P$_2$. The absence of this insertion region provides a simple explanation for the promiscuous binding of PtdIns 3-kinase products typical of other PH domains including that of PKB.

11.3.2.2. PtdInsP$_2$ binding

Studies on gelsolin (Yu et al., 1992) led to the assignment of PtdInsP$_2$ binding to two separate regions near the N terminus (CKSGLKYKKGGVASGF and KHVVPNEVVVQRLFQVK-GRR). Peptides corresponding to these sequences exhibit PtdIns(4,5)P$_2$ binding similar to that of gelsolin (Janmey et al., 1992). The first of these lysine/arginine-rich sequences was suggested to constitute a motif (K/RXXXXKXK/RK/R) that was present in other PtdInsP-binding cytoskeletal proteins such as gCap39, villin, cofilin, and profilin (Yu et al., 1992). The binding was thought to involve the hydrophobic amino acids that are present in high percentage of the binding site that interacts with the diacylglycerol moiety, because a simple electrostratic interaction between lysine/arginine and PtdInsPs could not fully account for the PtdIns-binding properties of gelsolin, which does not bind deacylated/deglycerinated PtdInsPs (Janmey, 1994).

The PtdInsP-binding sites for several other actin-associated proteins are not homologous to those of gelsolin. It was suggested that a short peptide sequence (FSMDLRTKST) in profiling was responsible for PtdInsP binding (Sohn et al., 1995). A 12-residue sequence (WAPECAPLKSKM) of cofilin was reported to bind PtdIns(4,5)P$_2$ (Yonezawa et al., 1991). A linear sequence in α-actinin (TAPYRNVNIQNFHLSWK) accounted for PtdIns(4,5)P$_2$ binding, which was eliminated by mutagenesis of the two arginine and lysine residues (Fukami et al., 1996).

Harlan et al., (1994, 1995) originally reported that the PH domains of pleckstrin, ras-GAP, β-ARK, and T cell kinase interacted with PtdIns(4,5)P$_2$. Subsequent studies by a number of laboratories confirmed that PH domains from a large number of proteins bind PtdInsPs but with different affinities and specificities. A study of six PH domains (Rameh et al., 1997a) found that four exhibit a selectivity for PtdIns(3,4,5)P$_3$ over other PtdInsPs, whereas two others bound PtdIns(3,4,5)P$_3$ and PtdIns(4,5)P$_2$ with similar affinities. The PH domain of Akt/PKB has a higher affinity and specificity for PtdIns(3,4)P$_2$ over PtdIns(4,5)P$_2$ and PtdIns(3,4,5)P$_3$ (Franke et al., 1997a,b; Klippel et al., 1997; Frech et al., 1997), whereas the PH domain of a kinase that phosphorylates PKB/Akt, PDK1, prefers PtdIns(3,4,5)P$_3$ over PtdIns(3,4)P$_2$ (Stokoe et al., 1997).

Dowler et al. (2000) have searched expressed sequences tag databases for novel proteins containing PH domains possessing a putative PtdIns(3,4,5)P$_3$ binding motif (PPBM) and have found that many of the PH domains did not bind PtdIns(3,4,5)P$_3$, but instead possessed unexpected and novel PtdIns-binding specificities in vitro. These included proteins possessing PH domains that interact specifically with PtdIns(3,4)P$_2$ (tandem PH-domain-containing protein-1), PtdIns(4)P (PtdIns(4)P adaptor protein-1), PtdIns(3)P (PtdIns(3)P-binding PH domain protein-1) and PtdIns(3,5)P$_2$ (centaurin-β$_2$).

TUBBY proteins feature a characteristic TUBBY domain of about 260 amino acids at the COOH-terminus that forms a unique helix-filled barrel structure, which binds avidly to double stranded DNA. TUBBY localizes to the plasma membrane by binding PtdIns(4,5)P$_2$ through its carboxyl terminal TUBBY domain. X-ray crystallography reveals the atomic-level basis of this interaction and implicates TUBBY domains as PtdInsP binding factors. Santagata et al. (2001) have shown that TUBBY functions in signal transduction from heterotrimeric GTP-binding protein (G protein)-coupled receptors. Of the lipids tested in the nitrocellulose phospholipid binding assay, TUBBY COOH-terminal domain protein bound avidly only to $PtdIns(3,4)P_2$, $PdIns(4,5)P_2$, and $PtdIns(3,4,5)P_3$. No appreciable binding was observed for $PtdIns(3,5)P_2$ or PtdInsP. This suggested that TUBBY bound preferentially to $PtdInsP_2$s with adjacent ring phosphate groups. To define the atomic-level basis of the interaction of TUBBY with PtdInsPs, Boggon et al. (1999) infused preformed TUBBY COOH-terminal domain crystals with soluble lipid head-group analogs and determined cocrystal structures of the complexes formed. A 1.95 A resolution was obtained of the crystal structure of the complex formed with L-α-glycerophospho-D-*myo*-inositol-4,5-bisphosphate (GPMI-P_2), an analog of the head group from $PtdIns(4,5)P_2$. The amino acids that interacted with GPMI-P_2 were mostly in β-strands forming a pocket located at the one end of the putative DNA binding groove. As a central feature of the binding interface, K330 coordinated between the 4- and 5-position phosphates of GPMI-P_2, thus providing a structural basis for the preference for ligands that are phosphorylated on adjacent positions of the inositol ring. A similar binding mechanism has been observed for GPMI-P_2 binding to the clathrin adaptor protein CALM-N (Ford et al., 2001) and $Ins(1,4,5)P_3$ binding to the PLC-δ_1 PH domain (Ferguson et al., 2000). In both the PLC-δ_1 and CALM-N structures, two lysisne side chains intervene peripherally between the phosphorylated ring positions of GPMI-P_2, whereas in TUBBY K330 coordinates directly between them. In the TUBBY structure, the 4-position phosphate is additionally stabilized by a salt bridge to the side chain of R332. Oxygen atoms of both the 4- and 5-position phosphodiester bonds hydrogen bond to the side-chain NH_2 group of N310. Santagata et al. (2001) and Cantley (2001), who reviewed this work, suggest that the pattern of residue conservation among TUBBY like proteins indicates that

PtdIns(4,5)P$_2$ binding is likely to be a function common to all members of this protein family. Dysfunction of the TUBBY protein results in maturity-onset obesity in mice.

Yue et al. (2002) have shown by immunoprecipitation (anti-PtdInsP$_2$ antibody) and Western blot analysis that both G protein α subunit (Giα-3) and PtdIns(4,5)P$_2$ bind β and γ epithelial Na$^+$ channels but not α Na$^+$ channels.

Zheng et al. (2002) have recently shown that plant PLDβ is stimulated by different PtdInsPs among which PtdIns(4,5)P$_2$ is most effective. A new PtdIns(4,5)P$_2$ binding region was identified in the catalytic domain of PLDβ.

Immunological studies with PtdIns(4,5)P$_2$ specific antibodies conducted with permeable PC12 cells indicate that high concentrations of PtdIns(4,5)P$_2$ are synthesized on secretory granule membranes during ATP-dependent priming (Loyet et al., 1998). A small number of proteins, including dynamin, CapZ and CAPS have been shown to exhibit a specificity for binding D4 PtdInsPs over D5 PtdInsPs.

11.3.2.3. PtdInsP$_1$ binding

Recent work has identified proteins which bind PtdIns(3)P that may act downstream of type III PtdIns 3-kinase involved in membrane trafficking, such as AP-2 and EEA1 (Rappoport et al., 1997; Patki et al., 1997).

Cheever et al. (2001) have shown that the vesicle trafficking yeast Vam7p membrane association requires a functional PX domain and PtdIns(3)P generation in vivo, and that the Vam7p PX domain binds to PtdIns(3)P in vitro. The PX domain was initially identified as a conserved motif of approximately 130 residues within the p40^{phox} and p47^{phox} subunits of the neutrophil NDPH oxidase superoxide-generating complex. PX domains have since been found in a wide variety of proteins involved in cell signaling, vesicle trafficking, and control of yeast bud emergence and polarity. Some 57 human and 15 yeast proteins that contain PX domains have been identified

(Simonsen and Stenmark, 2001; Wishart et al., 2001; Ellson et al., 2002). Further evidence for the role of the PX domain as a PtdIns(3)P directed membrane-targeting module has been provided by the analysis of the human sorting nexin, SNX3 (Xu et al., 2001).

Stevenson et al. (1998) have provided biochemical and molecular evidence for a large molecular mass PtdIns 4-kinase in both carrot (DcPtdIns 4-kinase α) and *Arabidopsis* (AtPtdIns 4-kinase α), and have demonstrated the affinity of a PtdIns 4-kinase PH domain for specific PtdInsPs. Antibodies raised against the expressed lipid kinase unique, PH, and catalytic domains identified a polypeptide of 205 kDa in *Arabidopsis* microsomes and an F-actin-enriched from carrot cells. The 205 kDa immunoaffinity-purified β-protein had PtdIns 4-kinase activity. Stevenson et al. (1998) expressed PH domain to characterize lipid binding properties. The recombinant PH domain selectively binds to PtdIns(4)P, PtdIns(4,5)P_2 and PtdOH and does not bind to the PtdIns(3)Ps. The PH domain had the highest affinity for PtdIns(4)P, the product of the reaction.

Snyder et al. (2001) have shown that a family of guanine nucleotide exchange factors (intersection, Dbs and Tiam1), which contain a Dbi homology (DH) domain adjacent to a pleckstrin homology (PH) domain, selectively bind lipid vesicles only when PdInsPs are present. Snyder et al. (2001) demonstrate that PtdInsP binding localizes to the PH domains with no appreciable binding attributable to the DH domains. While the DH/PH fragments of intersection and Dbs bind various PtdInsPs, Tiam1 binds preferentially PtdIns(3)P with low micromolar affinity.

The recent identification of the FYVE finger as a protein domain that binds specifically to PtdIns(3)P provides a number of potential effectors for PtdIns(3)P (Stenmark et al., 1996; Patki et al., 1998; Gaullier et al., 1998, 1999; Gillooly et al., 2001). The FYVE finger is a double-zinc binding domain conserved in more than 30 proteins from yeast to mammals (Stenmark and Aasland, 1999; Shisheva et al., 2001). It is named after the first letter of the four proteins it contains: Fab1p, YOTB, Vac1p and EEA1. The interaction of FYVE fingers with PtdIns(3)P serves to recruit cytosolic FYVE finger proteins to plasma membranes and to enrich membrane microdomains in these proteins, and to ultimately modulate their activity.

11.3.3. Regulation of PtdInsP kinases

PtdIns kinases are a family of enzymes involved in a multiplicity of cellular functions including cell proliferation and transformation, lymphocyte activation, G protein signaling, DNA repair, intracellular vesicle trafficking, and inhibition of programmed cell death (Carpenter and Cantley, 1996a,b; Toker and Cantley, 1997; Vanhaesebroeck et al., 1997a,b; Domin and Waterfield, 1997).

Biochemical evidence for membrane domains enriched in receptor-regulated pools of PtdIns(4,5)P$_2$ and enzymes has been reported (Pike and Casey, 1996; Hope and Pike, 1996). Studies employing fluorescent PtdIns(4,5)P$_2$ in liposomes revealed that PtdInsP-binding peptides can stabilize membrane microdomains in which both protein and phospholipid constituents are segregated (Glaser et al., 1996). Much attention has received the possibility that PtdInsPs specifically interact with molecules in downstream signal-transduction pathways.

Direct evidence for a signaling role of PdIns(3,4,5)P$_3$ has come from studies showing that addition of membrane-permeant forms of PtdIns(3,4,5)P$_3$ to cells can directly modulate downstream events (Jiang et al., 1998). Two mechanisms have been suggested. One is by directly affecting the properties of cellular membranes. For example, it has been proposed that localized production of PtdInsP products could cause membrane curvature and thus contribute to vesicle budding (Schu et al., 1993). This may be particularly relevant in the case of PtdIns(3,4,5)P$_3$, owing to its extremely high charge density and its ability to be produced at distinct cellular locations adjacent to activated signaling complexes. However, the biophysical effects of PtdIns(3,4,5)P$_3$ on vesicle budding and fusion events have not been formally tested.

Wurmser et al. (1999) have summarized recent studies elucidating the downstream effectors and degradation mechanism of PtdIns(3)P as well as the role of PtdIns(3)P in the biosynthesis of PtdIns(3,5)P$_2$. The specific regulation of PtdIns 3-kinases by cell-surface receptors and the function of 3-phosphorylated lipids has been reviewed by Stephens et al. (2000).

11.3.3.1. PtdIns 3-kinases

PtdIns 3-kinase is a dual specificity enzyme (Dhand et al., 1994). It has a regulatory 85 kDa adapter subunit whose SH2 domains bind phosphotyrosine in specific recognition motifs, and a catalytic 110 kDa subunit. The downstream effects of PtdIns kinases (e.g. p110 PtdIns 3-kinase) are carried out by phosphorylated derivatives of PtdIns, which serve as second messengers that recruit effector proteins to specific subcellular localizations and/or influence their activity (Toker and Cantley, 1997; Vanhaesebroeck et al, 1997; 2001).

Enzymes of the class Ia subfamily are regulated by protein-tyrosine (Tyr) kinase signaling pathways. They have regulatory subunits with src-homology-2 (SH2) domains that can interact with phosphotyrosine (pTyr) residues of growth factor receptors and adopter proteins. The catalytic subunits of class Ia PtdIns 3-kinases have a domain that interacts with the activated form of Ras (Rodriguez-Viciana et al., 1994, 1996).

The class Ib PtdIns 3-kinases is regulated by the β/γ subunits of heterodimeric G proteins (Stephens et al., 1994; Stoyanov et al., 1995; Stephens et al., 1997). The catalytic subunit (p110γ) is very similar to that of class Ia enzymes and includes a predicted *Ras*-binding domain, but lacks the region required for interaction with p85 subunits. Instead, this enzyme has a regulatory subunit (p101) that is unrelated to p85 and which appears to assist in mediating activation of β/γ subunits (Stephens et al., 1997). Thus, both class

Ia and class Ib PtdIns 3-kinases are regulated by extracellular stimuli, with the former responding to factors that activate protein-Tyr kinases and the latter responding to factors that activate heterodynamic G proteins. Interestingly, there is evidence that a class Ia PtdIns 3-kinase can be activated in vitro by a combination of Tyr peptides and β/γ subunits (Okada et al., 1996; Tang and Downes, 1997), raising the possibility that this enzyme requires activation of both protein-Tyr kinases and heterotrimeric G proteins for maximal stimulation. Relatively little is known about the regulation of these enzymes in vivo, although there is evidence that they are phosphorylated on Tyr in response to growth factor stimulation (Molz et al., 1996).

The best characterized type of PtdIns 3-kinase is the heterodimeric enzyme that consists of a 100 kDa catalytic subunit (p110α or p110β) and a 85 kDa regulatory subunit (p85α or p85β), and that is utilized for signaling by activated growth factor, cytokine, and antigen receptors (Chapter 7). Several additional modes of PtdIns 3-kinase regulation have been demonstrated, and it is likely that they act in concert to control the production of PtdIns(3)Ps in response to a variety of stimuli.

Each of these proteins contains an N-terminal region that interacts with regulatory subunits, a domain that binds to the small G protein ras, a 'PIK domain' homologous to a region found in other PtdIns kinases, and a C-terminal catalytic domain. Together these gene products are termed class I_A PtdIns 3-kinases.

It has been demonstrated that PtdIns(3,4)P$_2$ and PtdInsP$_3$ levels in the cell can be increased (Rodriguez-Viciana et al., 1996; Rodriguez-Viciana et al., 1994), and various downstream pathways can be activated (Rodriguez-Viciana et al., 1996; Frevert and Kahn, 1997; Hu et al., 1995; Didichenko et al., 1996) by overexpression of wild-type or constitutively active forms of the class 1a PtdIns 3-kinase catalytic subunit. Stephens et al. (2002) have recently discussed the roles of PtdInsP$_3$ kinases in leukocyte chemotaxis and phagocytosis.

11.3.3.2. PtdIns-4-kinases

PtdIns 4-kinases play a central role in signaling by feeding into multiple pathways. A distinctive feature of the group of higher molecular mass PtdIns 4-kinases is that they contain putative PH domains. PtdIns 4-kinase activity is principally membrane associated and found in plasma membrane, nuclear membrane, lysosomes, Golgi, endoplasmic reticulum and constitutive secretory vesicles (Pike, 1992; Olsson et al., 1995) depending on the tissue.

The cDNAs for PtdIns 4-kinases encode hydrophilic proteins that lack transmembrane domains. The basis for the membrane association and targeting of PtdIns 4-kinase isoforms to specific membranes is poorly understood. Possible mechanisms for localization involve mediation through a PH domain that is present in the α isoform (Wong and Cantley, 1994; Nakagawa et al., 1996a) and association with cytoplasmic domains of transmembrane and receptor proteins (Pertile and Cantley, 1995; Berditchevski et al., 1997).

Hemler et al. (1996) have shown that PtdIns 4-kinase may associate with transmembrane 4 superfamily (TM4SF, tetrapsain) proteins, but critical specificity issues were not addressed. Subsequently, Yauch and Hemler (2000) demonstrated that at least five different TM4SF proteins can associate with a similar or identical 55 kDa type II PtdIns 4-kinase. Furthermore, these associations are specific, since no evidence for other PtdIns kinases (PtdIns 3-kinase and PtdIns(4)P 5-kinase) associating with TM4SF proteins were found and many other TM4SF proteins did not associate with PtdIns 4-kinase. These results suggest that a specific subset of TM4SF proteins may recruit PtdIns 4-kinase to specific membrane locations, and thereby influence PtdIns-dependent signaling. More specifically, Yauch and Hemler (2000) have shown that TM4SF-PtdIns-4-kinase complexes are highly specific. Only a single type of PtdIns kinase could associate with TM4F proteins under the conditions utilized and only a subset of TM4SF proteins could associate with this kinase. The TM4SF-PtdIns 4-kinase appeared to occur only at distinct cellular locations.

Little is known about the regulation of PtdInsP 4-kinase activity. In vitro this activity is inhibited by $PtdIns(4,5)P_2$ and heparin. PtdInsP 4-kinase β associates directly with the p55 subunit of the tumor necrosis factor (TNF) receptor, and indirect evidence suggests that its activity increases following TNF treatment of cells (Castellino et al., 1997). The type II β PtdInsP 4-kinase was shown to associate with the cytosolic domain of the p55 TNF receptor (Castellino et al., 1997), raising the possibility that TNF regulates the PtdIns(5)P 4-kinase pathway for $PtdIns(4,5)P_2$ synthesis.

The common D-4 PtdInsPs are primarily synthesized by the sequential phosphorylation of PtdIns by PtdIns 4-kinase and PtdIns(4)P 5-kinase. A PtdInsP 4-kinase activity that converts PtdIns(3)P to $PtdIns(3,4)P_2$ is stimulated by an integrin-mediated pathway in platelets (Banfic et al., 1998). An alternative pathway for the synthesis of $PtdIns(4,5)P_3$ is suggested by the finding that the type II isoform of PtdIns(4)P 5-kinase is a PtdIns 5-kinase (Rameh et al., 1997b).

Novel dual substrate specificities for the PtdInsP kinases have also been reported (Zhang et al., 1997), which would allow sequential phosphorylation of PtdIns(3)P to $PtdIns(3,4,5)P_3$.

11.3.3.3. PtdIns 5-kinases

The existence of the PtdIns 5-kinase was suggested by the finding of PtdIns(5)P as a minor lipid present in mammalian cells. The type I PtdIns(4)P 5-kinase also phosphorylates PtdIns(3)P (Rameh et al., 1997b) to yield $PtdIns(3,5)P_2$ (Whiteford et al., 1997; Dove et al., 1997).

PtdIns(4)P 5-kinase activity is principally cytosolic and is enriched in neural synaptosomes (Stubbs et al., 1988). Position 5 of PtdIns is targeted by two subclasses of enzymes: PtdIns 5-kinase (Fruman et al., 1998; Anderson et al., 1999; Toker, 1998) (or type I PtdInsP kinases) and PtdIns kinases FYVE (Sbrissa et al., 1999).

While the two subclasses share substantial homology in their catalytic region, they display different substrate specificity. Thus, type I PtdIns 5-kinases (α, (β and γ; Ishihara et al., 1996, 1998) show preferences for PtdIns already phosphorylated at D-4 (Zhang et al., 1997; Tolias et al., 1998), while PtdIns kinase FYVE prefers PtdIns and 3′-phosphorylated PtdIns (Sbrissa et al., 1999). Sbrissa et al. (2000; 2002) have obtained results that indicate that PtdInsP kinase FYVE is a dual specificity kinase, which can generate and relay protein phosphorylation signals to regulate the formation of its lipid products, and possibly other events, in the context of living cells. The PtdInsP kinase FYVE possessed an intrinsic protein kinase activity inseparable from its lipid kinase activity and, besides itself, phosphorylated exogenous proteins in a substrate specific manner. A decrease of 70% in the lipid product formation was associated with PtdIns kinase FYVE autophosphorylation, which was reversed upon treatment with phosphatases (Sbrissa et al., 2000).

The central importance of PtdIns(4)P 5-kinase for the synthesis of PtdIns(4,5)P$_2$ and PtdIns(3,4,5)P$_3$ in specific membrane compartments implies that regulatory and targeting mechanisms can operate on this enzyme, but only preliminary accounts of this have been reported. The type II isoform contains a proline-rich domain that may be an SH-3 binding site and thus able to mediate the coupling of type I PtdIns 3-kinases for channeled synthesis of PtdIns(3,4,5)P$_3$ (Boronenkov and Anderson, 1995; Zhang et al., 1997). A splicing isoform of the type II enzyme associates with a TNF receptor (Castellino et al., 1997). The PtdIns(4)P 5-kinase (type I) is uniquely stimulated by PtdOH (Jenkins et al., 1994), which is believed to be an important element of a positive feedback circuit with PLD that generates increased PtdIns(4,5)P$_2$ and may participate in membrane budding or fusion events (Liscovitch and Cantley, 1995) (see below). Non-hydrolyzable guanine nucleotides stimulate PtdIns(4)P phosphorylation, suggesting a G protein regulation of 5-kinase activity (Smith and Chang, 1989).

Honda et al. (1999) have found that the GTP-γ-S-dependent activator of PtdIns(4)P 5-kinase α is the small G protein ADP-ribosylation factor (ARF) and that the activation strictly requires PtdOH, the product of PLD. In vivo, ARF6 and PLD are co-localized in ruffling membranes formed upon AIF4 and EGF stimulation and is blocked by dominant-negative ARF6. Honda et al. (1999) have concluded that PtdIns(4)P 5-kinase α is activated by agonist stimulation and that it triggers recruitment of a diverse but interactive set of signaling molecules into sites of active cytoskeletal and membrane rearrangement.

The signaling phospholipid, PtdIns(4,5)P$_2$, is primarily synthesized by PtdIns(4)P 5-kinase, which has been reported to be regulated by RhoA and Rac1. The regulation of type I PtdInsP 5-kinases appears to be complex. These enzymes can associate with the carboxy-terminal region of the GTP-binding protein, *Rac1* by a mechanism independent of the nucleotide-bound state (Tolias et al., 1995, 1998). A DAG kinase activity is also found in the *rac*/PtdInsP 5-kinase complex (Tolias et al., 1998.). This interaction could be important for regulation since the product of the DAG kinase, PtdOH, is an activator of type I PtdInsP 5-kinase. Jones et al. (2000) have provided direct evidence for the production of saturated/monounsaturated PtdOH (through two different routes, PLD and DAG kinase) and the in vivo activation of type Iα PtdIns(4)P 5-kinase by this lipid second messenger.

11.3.4. Specific biological effects

The specific biological effects of PtdInsPs have been considered in several recent reviews. Hunter (2000) and Vanhaesebroeck et al. (1997, 2001) have focused on the overall biological processes in which PtdIns containing 3-phosphate groups have been implicated. More detailed reviews of the molecular and biological characteristics of each individual downstream target of PtdIns 3-kinases have been referred to below under the topics of membrane and vascular trafficking, cell growth and differentiation, cytoskeletal organization, apoptosis and DNA synthesis.

Munnik et al. (1998) have reviewed the phospholipid signaling in plants, while Dohlman and Thorner (2001) have reviewed recent advances made in yeast in identifying molecules responsible for regulating the action of the components of the G-protein-initiated signal transduction, using primarily genetic methods.

The following discussion is limited to those aspects of the various biological phenomena, where the effect of the lipid products of PtdIns 3-kinase activity have been specifically examined.

11.3.4.1. Membrane and vesicular trafficking

Detailed reviews of vesicle trafficking and cell growth have been recently presented by Lemmon (1999), Krasilnikov (2000), Cullen et al. (2001), Martin (2001) and Czech (2003).

A great deal of evidence has accumulated to suggest that the products of PtdIns 3-kinase activity play a central role in a diverse range of cellular responses, including membrane and vesicle trafficking (Holman and Kassuga, 1997; De Camilli et al., 1996; Shepherd et al., 1996; Kurzchalia and Parton, 1999). The immediate downstream effector of PtdIns(3)P remains to be identified.

According to De Camilli et al. (1996), vesicle-mediated delivery within the cell entails the formation and packaging of cargo into transport intermediates (vesicles), a process requiring the activity of coat proteins, and the docking/fusion of such transport intermediates with the appropriate target organelle, which depends upon SNARE proteins and Rab GTPases, as well as the recycling of transport components (e.g. receptors and SNARES (Burd and Emr, 1998)). The lipid composition of the vesicular transport intermediates is also critical. In particular, PtdInsPs, which can be modified at specific sites of the inositol ring, either singly or in combination, represent versatile molecules through which the cell generates distinct second messengers. Wurmser et al. (1999) and

Simonsen et al. (2001) have reviewed the requirement for PtdIns(3)P in vesicular transport in yeast and mammalian systems.

According to Stack et al. (1995) PtdIns(3)P could alter the structure of the bilayer to facilitate budding, recruit coat proteins or adaptors to the site of budding, or function as a vesicle membrane component essential for subsequent docking or fusion reactions. In yeasts, PtdIns(3,5)P$_2$ synthesis is dependent upon the Vps34p PtdIns 3-kinase (Dove et al., 1997; 1999), which could indicate that the role of PtdIns(3)P in vacuolar sorting is to serve as a precursor of PtdIns(3,5)P$_2$, which links stress responses to membrane trafficking events. A mechanism by which PtdIns(3)P could regulate the sorting of cargo in the endosomal/lysosomal pathway has been suggested by the observation that PtdIns(3)P enhances the affinity of the adaptor protein AP-2 complex for binding tyrosine-based signals present on membrane receptor cytoplasmic tails (Rappoport et al., 1997; Kirchhausen et al., 1997).

Heraud et al. (1998) have presented data that suggest the lipid products of PtdIns 3-kinase are required but not sufficient for ADP-induced spreading of adherent platelets and that PtdIns(4,5)$_2$ could be a downstream messenger of this signaling pathway. These authors pretreated human platelets with specific PtdIns 3-kinase inhibitors LY294002 and wortmannin and found that platelets adhered to the fibrinogen matrix and extended pseudopodia but did not spread.

In the investigation of the role of the products of PtdIns 3-kinases in membrane trafficking in mammalian cells, extensive use has been made of wortmannin, an irreversible inhibitor of some but not all PtdIns 3-kinases (Carpenter and Cantley, 1996a). Wortmannin and LY290042 treatment of mammalian cells results in a limited number of membrane trafficking defects that are restricted to the late Golgi-lysosomal-endosomal pathway (Reaves et al., 1996; Shpetner et al., 1996). However, studies relying exclusively on wortmannin and LY294002 are difficult to interpret and other methods must be employed to assess the role of PtdIns

3-kinase in membrane trafficking reactions, e.g. dominant-negative mutants (Haruta et al., 1995).

Neurotransmitters are stored in synaptic vesicles and, after the nerve has received the appropriate stimulus, they are released by a Ca^{2+}-stimulated fusion of the vesicles with the presynaptic plasma membrane of the nerve cell. It has been shown that PtdInsPs play a crucial role in this context (Cremona et al., 1999). The fusion of exocytic vesicles with the presynaptic plasma membrane is preceded by an ATP-dependent priming step, which requires both PtdIns 4-kinase and a PtdIns 5-kinase activities (Martin, 1998), indicating that $PtdIns(4,5)P_2$ is needed. It has been suggested that this $PtdIns(4,5)P_2$ is likely to modify the activity of synaptic vesicle membrane and/or plasma membrane proteins. One protein that is a good candidate for modification by $PtdInsP_2$ is synaptotagmin, an abundant integral membrane protein of synaptic vesicles. Synaptotagmin has low affinity for $PtdInsP_2$ at low Ca^{2+} concentrations, but an increased affinity at high Ca^{2+} concentrations, such as those found in the terminal of a stimulated nerve cell. This protein is thought (Martin, 1998; Stenmark, 2000) to act as a Ca^{2+}-sensor that triggers exocytosis by interacting with membranes and with the SNARE complexes that are involved in membrane fusion. Synaptotagmin is also required for the recycling of synaptic vesicles (Jergensen et al., 1999). The study of Cremona et al. (1999) reveals that there is in fact a need for $PtdInsP_2$ turnover, which involves the participation of synaptojanin, a protein that is abundant in nerve terminals. Synaptojanin is a 5-phosphatase that converts $PtdInsP_2$ into PtdInsP. Cremona et al. (1999) generated synaptojanin 'knock out' mice and that in contrast to nerve terminals from normal mice, the synaptojanin-deficient nerve terminals showed accumulation of clathrin-coated vesicles. This observation was confirmed in a cell-free assay that reconstitutes coated vesicle formation from crude preparations of brain membranes in the presence of cytosol. These studies demonstrate the role for a lipid cycle in the synaptic vesicle cycle, in which the conversion of PtdIns(4)P to $PtdIns(4,5)P_2$ is required for exocytosis and endocytosis, and the conversion of

PtdIns(4,5)P_2 to PtdIns (4)P is essential for clathrin-coat shedding and hence synaptic vesicle recycling, as illustrated by Stenmark (2000).

CAPS co-purifies with dense-core vesicles and plasma membrane from brain tissue (Berwin et al., 1998). A model in which CAPS on vesicles and plasma membrane undergoes a conformational change in response to PtdIns(4,5)P_2 synthesis on the vesicle has been proposed by Loyet et al. (1998).

There is evidence that other PtdInsP phosphatases also play a role in protein trafficking. Sac1p PtdInsP phosphatase is an integral membrane protein localized at the endoplasmic reticulum and at the Golgi (Hughes et al., 2000). Schorr et al. (2001) have shown that Sac1 homology domains exhibit 3- and 4-phosphatase activity in vitro and were also found, in addition to rat and yeast Sac1p, in yeast Inp/Sjl proteins (Stolz et al., 1998) and mammalian synaptojanins (Kochendorfer et al., 1999). Schorr et al. (2001) have reported that Sac1p has a specific role in secretion and acts as an antagonist of the PtdIns 4-kinase Pik1p in Golgi trafficking. Elimination of Sac1p leads to excessive forward transport of chitin synthetases and this causes specific cell wall defects. Sac1p is critically required for the termination of this signal.

A potentially important role for PtdInsPs in membrane trafficking has been recognized in coat protein recruitment mediated by the ADP ribosylation factors (ARFs) as indicated by the dependence of ARF regulators and effectors in the presence of PtdInsPs in the membrane (Paris et al., 1997; Klarlund et al., 1997, 2000). ARF itself reacts with PtdInsPs (Randazzo, 1997), which promote its nucleotide exchange (Paris et al., 1997) and its interaction with ARF and GTPase-activating factor (GAP) (Randazzo, 1997). ARF GAP also contains PH domains and exhibit PtdInsP-regulated activity (Tanaka et al., 1997).

There is evidence that PLD functions as an effector for ARF in coat protein recruitment (Ktistakis et al., 1996; Roth and Sternweis, 1997). PLD is an additional potential effector for PtdInsPs in membrane budding reactions because PLD activity and its stimulation of ARF stringently require PtdInsPs (Brown et al., 1993; Liscovitch et al., 1994). Either D4 or D3 PtdInsPs function equally effectively as cofactors for PLD (Hammond et al., 1997). Liscovitch and Cantley (1995) have proposed a basis for an autocatalytic cycle for ARF-dependent activation of PLD, which suggests that production of PtdOH by PLD could result in the enhanced production of PtdIns(4,5)P$_2$ by activation of the type I PtdIns(4)P 5-kinase (Jenkins et al., 1994). The requirement for PtdInsPs in all aspects of the ARF activation/deactivation cycle and for PLD activation suggests a critical role for PtdInsPs in Golgi budding and membrane coat recruitment, although there is no evidence for this role in vivo at the present time (Jost et al., 1998; Kurzchalia and Parton, 1999).

Earlier, Hay and Martin (1993) had suggested that PtdInsPs are essential for exocytic fusion reactions on the basis of the discovery that PtdIns transfer protein is required for the reconstitution of Ca^{2+}-dependent priming step in neurotransmitted secretion in permeable PC12 cells. The priming process is a reversible step, and the reversal of priming appears to be catalyzed by a PtdInsP 5-phosphatase (Martin et al., 1997). The essential role of intact PtdIns(4,5)P$_2$, rather than a derived metabolite, was suggested by the failure of all potential metabolites tested to significantly alter exocytosis in the absence or presence of Ca^{2+} (Hay and Martin, 1995; Martin et al., 1997) and the ability of PtdIns(4,5)P$_2$ antibody to inhibit exocytosis from ASTP-primed cells (Hay et al., 1995).

There is evidence for a link between early endosome fusion and PtdIns 3-kinase activity, which suggests that lipid kinases or their D3 PtdInsPs products function upstream of a fusion mechanism that involves activation of the small GTPase Rab5 (Li et al., 1995). Evidence for a direct effect of D3 PtdInsPs on a potential effector for D3 PtdInsP regulation of Rab5 function was provided by the identification of EEA1 (early endosomal antigen) as a PtdIns(3)P-binding protein that resides in part on endosomes (Patki et al., 1997). This work has provided a model for the mechanism of action of D3 PtdInsPs in membrane fusion acting as a signal that recruits EEA1 to endosomes (Horiuchi et al., 1997).

Genetic and biochemical evidences indicate that synaptotagmin I, an abundant secretory vesicle protein, is an essential component of the Ca^{2+}-sensing mechanism in the regulated exocytosis of synaptic vesicles in nerve cells (Sudhof, 1995). Synaptotagmin I interacts with PtdInsPs via its membrane distal C2B domain, and Ca^{2+} increases binding to $PtdIns(4,5)P_2$ but decreases binding to $PtdIns(3,4,5)P_3$ (Schiavo et al., 1996). $PtdIns(4,5)P_2$ formed on secretory vesicles during priming in neuroendocrine cells may also play a role in the endocytic retrieval of the vesicle membrane (Martin, 1998).

Wurmser and Emr (1998) have found that interruption of endosome-to-vacuole transport leads to elevation in PtdIns(3)P levels by disrupting the turnover of PtdIns(3)P. The authors found that the majority of the total pool of PtdIns(3)P, which has been synthesized, but not PtdIns(4)P, requires transport to the vacuole in order to be turned over. Lipids in the cytoplasmic leaflet of the membrane are transferred to the lumen of the vacuole through the invagination and budding of vesicles into the lumen of the vacuole or prevacuolar endosomes, which then fuse with the vacuole. Therefore, it was concluded that PtdIns(3)P is not only a cargo of the endosome, but may also regulate this late step in the vacuolar protein transport pathway.

Phagosomal maturation is a fundamental biological process governed by vesicular and intracellular membrane and protein trafficking. The findings of Fratti et al. (2001) have defined the generation of PtdIns(3)P and EEA1 recruitment as important regulatory events in phagosomal maturation and critical molecular targets affected by *Mycobacterium tuberculosis*.

Maroun et al. (1999) have demonstrated a conserved PtdIns binding site within the pleckstrin homology of the docking protein, which is required for epithelial morphogenesis.

Activation of ADP-ribosylation factor (Arf) 6 has been implicated in regulated exocytosis and recycling of plasma membrane. Brown et al. (2001) have demonstrated that PtdInsP 5-kinase activity and PtdIns(4,5)P$_2$ turnover controlled by activation and inactivation of Arf6 is critical for trafficking through the Arf6 PM-endosomal recycling pathway. Kinuta et al. (2002) have developed an assay to monitor vesicle formation from liposomes incubated in the presence of ATP, GTP and dynamin. By using a cell free system, they have demonstrated that PtdIns(4,5)P$_2$ plays a critical role in the early steps of vesicle formation, possibly in the recruitment of coats and fission factors to membranes.

11.3.4.2. Cell growth and differentiation
A great deal of evidence has accumulated to suggest that the products of PtdIns 3-kinase activity play a central role in cell growth and differentiation (Carpenter and Cantley, 1996a,b; Toker and Cantley, 1997). Of special interest are the recent reviews on the action of insulin (Alessi and Downes, 1998; Saltiel and Pessin, 2002; Bryant et al., 2002). Insulin and IGF-1 activate signaling pathways important in the differentiation processes of two major insulin target tissues, namely muscle and adipose tissue, and PtdIns 3-kinase activity is necessary for these effects. Insulin plays a key role in regulating a wide range of cellular processes. Shepherd et al. (1998) have reviewed the evidence that has been accumulated establishing the PtdIns 3-kinase plays a pivotal role in signal-transduction pathways linking insulin with many of its specific cellular responses.

A number of findings in different experimental systems indicate that PtdIns 3-kinase and consequent generation of its lipid products plays a critical role in growth factor signaling to cell growth and proliferation. First, studies using PDGF (platelet derived growth factor)-receptor mutants unable to bind PtdIns 3-kinase in parallel with add-back mutants suggested that PtdIns 3-kinase was a downstream mediator of the PDGF-receptor mitogenic signal (Valius and Kazlauskas, 1993). Similar mechanisms appeared to be involved in the insulin/IGF-1 (insulin-like growth factor 1) signaling system as the mitogenic effects of insulin and IGF-1 are attenuated by inhibitors of PtdIns 3-kinase (Cheatham et al., 1994).

Wortmannin also blocks differentiation of 3T3-L1 cells into adipocytes, and this process is stimulated by overexpression of constitutively active PKB (protein kinase B), suggesting that a kinase cascade is downstream of PtdIns 3-kinase in this process (Magun et al., 1996). IGF-1 stimulated myoblast differentiation is blocked by PtdIns 3-kinase inhibitors and dominant negative forms of p85 (Kaliman et al., 1999).

Insulin stimulates the recruitment of PtdIns 3-kinase from the cytosol to a low-density fraction in the cell. A more detailed evaluation of the components of the low-density fractions revealed that insulin is recruiting PtdIns 3-kinase to IRS-1 (insulin receptor substrate 1), which is associated with a low density cytoskeletal structure that is distinct from, but co-sediments with, microsomal membranes upon subcellular fractionation (Clark et al., 1998). The elements downstream of PtdIns 3-kinase in the insulin-stimulating signaling pathways regulating GLUT4 (glucose transporter 4) translocation are poorly understood (Okada et al., 1994; Shepherd et al., 1998; Bryant et al., 2002; Somwar et al., 2001).

These findings have subsequently been confirmed by a number of groups using both wortmannin and LY295002 in a range of insulin-sensitive tissues, including isolated rat adipocytes, 3T3-Li adipocytes, L6 myotubes and isolated muscle (for a review see Shepherd et al., 1998). The effects of wortmannin on insulin-stimulated glucose transport can be overcome by addition of membrane-permeable forms of PtdInsP$_3$ (Jiang et al., 1998), providing further evidence that PtdIns 3-kinase is involved. Specific evidence for the involvement of class 1a PtdIns 3-kinase in this process was provided by studies with dominant negative forms of the p85 and p55^{PIK} adapter/regulatory subunits which block insulin stimulation of glucose transport and glucose-transporter translocation (Mothe et al., 1997).

It is now apparent that the activation of class 1a PtdIns 3-kinase is necessary and in some cases sufficient to elicit many of insulin's effects on glucose and lipid metabolism. The lipid products of PtdIns 3-kinase act as both membrane anchors and allosteric regulators, serving to localize and activate downstream enzymes and their protein substrates. One of the major ways these lipid products of PdIns 3-kinase act in insulin signaling is by binding to PH domains of PDK and PKB and in the process regulating the phosphorylation of PKB and PDK. Using mechanisms such as this, PtdIns3-kinase is able to act as a molecular switch to regulate the activity of serine/threonine-specific kinase cascades important in mediating insulin's effects on endpoint responses.

There is evidence that other actions of insulin on glucose metabolism involve PtdIns 3-kinase activity, including gene expression (Shepherd et al., 1998). Thus, both insulin and PDGF cause a similar large recruitment of PtdIns 3-kinase into tyrosine-phosphorylated signaling complexes in 33-L1 adipocytes, but insulin stimulation of glucose transport and glycogen synthase greatly exceeds that by PDGF in these same cells (Nave et al., 1996; Brady et al., 1998). The explanation for the apparent discrepancy between activation of PtdIns 3-kinase and production of PtdInsP$_3$ may be that different growth factors activate different subsets of downstream pathways, some of which may potentiate and some of which may antagonize, a particular endpoint response. One potential source of differential signaling is PLC, as it has been demonstrated that activation of PLC by thrombin blocks insulin-stimulated PtdInsP$_3$ production (Batty and Downes, 1996). Further, stimulation of PtdInsP$_3$ production of PDGF is greatly enhanced by deletion of the sites on the PDFG receptor that allow binding of PLC (Kinghoffer et al., 1996). As PtdIns(4,5)P$_2$ is the major substrate for both class 1a PtdIns 3-kinases and PLC, they could potentially compete for substrate. Therefore, the explanation as to why PDFG has very little effect on PtdInsP$_3$ production in comparison with insulin in 3T3-L1 adipocytes (Conricode, 1995) may lie with the fact that PDGF activates PLCγ, while insulin does not.

Arcaro et al. (2000) have reported that class II PtdIns 3-kinases are downstream targets of activated polypeptide growth factor receptors.

Despite the remarkable progress in dissecting the signaling pathways that are crucial for the metabolic effects of insulin, the molecular basis for the specificity of its cellular actions is not fully understood. Saltiel and Pessin (2002) have summarized recent evidence which suggests that signaling molecules and pathways are localized to discrete compartments in cells by specific protein and lipid interactions. Thus, the insulin receptor, which is a tyrosine kinase, catalyzes the phosphorylation of several intracellular substrates, including the insulin receptor substrate proteins. Each of these substrates recruits a distinct subset of signaling proteins containing SH2 domains, which interact specifically with sequences surrounding the phosphotyrosine residue. Furthermore, each of these substrates may be confined to distinct locations in the cell by specific sequences that direct interactions with other proteins and lipids.

Saltiel and Pessin (2002) suggest that insulin must generate at least two independent and separate signals to stimulate glucose transport. The specific compartmentalization of Glut4 provides a mechanism by which insulin can stimulate robust translocation of Glut4 to the plasma membrane, while only mildly stimulating the translocation of other recycling proteins.

The mechanism of PtdIns 3-kinase activation via IRS is distinct from that of other tyrosine kinase receptors such as PDGF and epidermal growth factor, which recruit PtdIns 3-kinase directly to their receptors. This suggests that the IRS proteins uniquely activate PtdIns 3-kinase through the recruitment of the enzyme to specific sites at or distal from the plasma membrane. Saltiel and Pessin (2002) conclude that the localized generation of PtdIns(3,4,5)P_3 could define the specificity of this signaling pathway. A potential indication of the identity of the PtdIns 3-kinase independent arm of insulin action has emerged from the idea that signal initiation might be segregated into discrete compartments in the plasma membrane.

One candidate for such compartments is caveolae—small invagi-
nations of the plasma membrane containing caveolin that are a
subset of lipid raft domains. Insulin stimulates the tyrosine
phosphorylation of caveolin, the major structural protein of
caveolae. The phosphorylation of another insulin receptor substrate, the proto-oncogene cCbl, led to the identification of
Cbl-associated protein (CAP), which contained three C-terminal
SH3 domains. Once phosphorylated, Cbl could recruit the SH2-
containing adapter protein CrkII to lipid rafts, along with FEF C3G
(Chiang et al., 2001). The C3G then appears to activate the Rho
family GTPase TC10, which is a small GTP-indin protein produced
in fat and muscle and can be actively activated by insulin in CAP-
dependent by PtdIns 3-kinase independent manner. Watson et al.
(2001) have reported that insulin stimulation of GLUT4 translocation in adipocytes requires the spatial separation and distinct
compartmentalization of the PtdIns 3-kinase and TC10 signal
pathways. Interestingly, Thomsen et al. (2002) have reported that
caveolae are highly immobile plasma membrane microdomains,
which are not involved in constitutive endocytic trafficking.

Upon dissociation of insulin, both the receptor and its substrates
undergo a rapid dephosphorylation, implicating protein tyrosine
phosphatases in signal termination. However, insulin action can
also be temporally controlled by lipid posphatases that dephosphorylate PtdIns(3,4,5)P$_3$. Vollenweider et al. (1999) have shown
that microinjection of the PtdInsP$_3$ phosphatase SHIP2, which
specifically removes the 5-phosphate from PtdIns(3,4,5)P$_3$,
and blocks insulin action. There are other factors that attenuate
insulin signaling by reducing PtdIns(3,4,5)P$_3$ levels (Clement et al.,
2001; Lizcano and Alessi, 2002; Jiang et al., 2002).

The potential role of PtdIns 3-kinase in signaling to cell growth
and proliferation has sparked interest in its role in cell
transformation. This has been heightened by the recent identification of transforming retrovirally encoded PtdIns 3-kinase
isolated from chicken haemangiosarcomas (Chang et al., 1997).

11.3.4.3. Cytoskeletal organization and cell motility

A great deal of evidence has accumulated to suggest that the products of PtdIns 3-kinase activity play a central role in cytoskeletal organization and cell motility (Toker and Cantley, 1997; Carpenter, 1996; Tolias et al., 2000; Takenawa and Itoh, 2001). Recently, Caroni (2001) has proposed the regulation of actin through modulation of PtdIns(4,5)P$_2$ rafts. According to Denker and Barber (2002) ion transport proteins anchor and regulate the cytoskeleton.

The rearrangement of actin to form lamellapodia and subsequent membrane ruffles is the most easily observable effect of insulin on the cytoskeleton in most cell types. There is strong evidence that activation of PtdIns 3-kinase is necessary and sufficient for lamellapodia formation, as this process is blocked by wortmannin and dominant negative forms of p85 (Wennstrom et al., 1994; Nobes and Hall, 1995), and it has subsequently been established that activated PtdIns 3-kinase can mimic growth-factor effects on lamellapodia formation (Reif et al., 1996).

Constitutively active PtdIns 3-kinase activates Rac-dependent lamellapodia and Rho-dependent stress fiber assembly in fibro-blasts (Reif et al., 1996). The PH domain of some Db1 family members exhibits selectivity for binding D3 PtdInsPs (Rameh et al., 1997a). This indicates that D3 PtdInsP synthesis could promote recruitment of exchange factors to specific sites on the membrane for activation of Rho family GTPases (Carpenter et al., 1997).

Wennstrom et al. (1994) have shown that activation of PtdIns 3-kinase is required for PDGF-stimulated membrane ruffling, since wortmannin inhibited it in fibroblasts. Martys et al. (1996) had demonstrated the presence of wormannin-sensitive enzyme in three distinct steps of the endocytic cycle in Chinese hamster ovary cells: internalization, transit from early endosomes to the recycling and degradative compartments, and transit from the recycling compartment back to the cell surface. The wortmannin-sensitive enzymes critical for endocytosis and recycling are from those involved in sorting newly synthesized lysosomal enzymes.

Calmodulin is a ubiquitous Ca^{2+}-dependent effector protein which regulates multiple processes in eukaryotic cells, including cytoskeletal organization, vesicular trafficking, and mitogenesis. PtdIns 3-kinase participates in events downstream of the receptors for insulin and other growth factors. Joyal et al. (1997) have demonstrated by coimmunoprecipitation and affinity chromatography that Ca^{2+}/CaM associates with *src* homology 2 domains in the 85 kDa regulatory subunit of PtdIns 3-kinase, thereby significantly enhancing PtdIns 3-kinase activity in vitro and in intact cells. Furthermore, CGS9343B, a CaM antagonist, inhibited basal and Ca^{2+}-stimulated phosphorylation of PtdIns in intact cells. These data demonstrate a novel mechanism for modulating PtdIns 3-kinase and provide a direct link between components of two fundamental signaling pathways.

PtdInsPs have been implicated in the regulation of actin cytoskeleton assembly at several levels. Extensive connections between PtdnsPs and the Rho family of GTP-binding proteins, which mediate extracellular signal regulation of cytoskeletal rearrangements, have been identified (Hall, 1998).

Mutations in FAB1 cause enlargement of the vacuole, consistent with Fab1 and perhaps PtdIns(3,5)P_2 playing a role in vesicle trafficking. The mechanism of regulation of Fab1 is not yet clear. Interestingly, Fab1 contains a FYVE domain that binds PtdIns(3)P, allowing this enzyme to bind to membranes that contain this PtdIns(3,5)P_2 lipid (Burd and Emr, 1998).

Tsakiridis et al. (1996) have reported that insulin activates a p21-activated kinase in muscle cells via PtdIns 3-kinase. The authors describe a renaturable kinase of 65 kDa (PK65) that becomes rapidly activated by insulin in differentiated L6 muscle cells (myotubes) and can phosphorylate histones immobilized in polyacrylamide gels. Insulin activation of PK65 was abolished by the tyrosine kinase inhibitor erbstatin and by the PtdIns 3-kinase inhibitor wortmannin, but was unaffected by inhibitors of protein kinase C or of the activation of $p70^{S6K}$.

The identification of numerous Rho/Rac-binding proteins (Tapon and Hall, 1997) has opened up many new options for downstream signaling to the cytoskeleton. One mechanism for the Rho-dependent induction of stress fiber formation involves activation of a Rho kinase that phosphorylates both myosin light chain (Amano et al., 1996) and myosin phosphatase (Kimura et al., 1996), which leads to increased binding of myosin and the bundling of actin filaments. Rac activation causes an increased synthesis of PtdIns(4,5)P$_2$, which may be mediated by Rac stimulation of a type I PtdIns(4)P 5-kinase (Hartwig et al., 1995). There are other potential downstream effectors among the PtdInsP-binding proteins that regulate actin assembly (Janmey, 1994; Janmey et al., 1999). PtdInsPs induce conformational changes in vinculin that allow vinculin to mediate the cross-linking of talin to actin, which plays a role in focal adhesion assembly (Gilmore and Burridge, 1996).

Other efforts to assess physiological roles for PtdInsPs in regulating actin assembly, overexpression of the type I PtdIns(4)P 5-kinase in COS-7 cells has been found to cause a dramatic increase in the assembly of short-chain actin filaments (Shibasaki et al., 1997). Conversely, overexpression of a type II 5-phosphatase in COS-7 cells reduced the number of actin stress fibers (Sakisaka et al., 1997). These studies designed to alter cellular levels of PtdInsPs in cells provide important link between in vitro studies of PtdInsP-binding by cytoskeletal proteins and the regulation of actin assembly in vivo.

Schell et al. (2001) have reported that Ins(1,4,5)P$_3$ 3-kinase associates with F-actin and dendritic spines via its N terminus.

The Nck adapter protein consists of three Src homology (SH) 3 domains and one SH3 domain and has been postulated to link changes in tyrosine phosphorylation to actin assembly. Rohatgi et al. (2001) have observed that Nck and PtdIns(4,5)P$_2$ synergistically activate actin polymerization through the N-WASP-Arp2/3 pathway. It is known that Wiskott-Aldrich syndrome protein (WAP) and its relative neural WASP (N-WASP) regulate the nucleation of actin filaments through their interaction with the Arp2/3 complex and are regulated in turn by binding to GTP-bound Cdc42 and PtdIns(4,5)P$_2$. The Nck Src homology (SH) 2/3 adaptor binds via its SH3 domains to a proline-rich region on WASP and N-WASP and has been implicated in recruitment of these proteins to sites of trysine phosphorylation. Rohatgi et al. (2001) show that Nck SH3 domains dramatically stimulate the rate of nucleation of actin filaments by purified N-WASP in the presence of Arp2/3 in vitro. All three Nck H3 domains were required for maximal activation. Nck-stimulated actin nucleation by N-WASP-Apr2/3 complexes were further stimulated by PtdIns(4,5)P$_2$, but not by GTP-Cdc42, suggesting that Nck and Cdc activate N-WASP by redundant mechanisms.

Raucher et al. (2000) have recently shown that PtdIns(4,5)P$_2$-mediates control of plasma membrane interaction with the underlying cytoskeleton. By the use of optical tweezers tether force measurements, Raucher et al. (2000) have shown that plasma membrane PtdIns(4,5)P$_2$ acts as a second messenger that regulates the adhesion energy between the cytoskeleton and the plasma membrane. Receptor stimuli could either increase or decrease local PtdIns(4,5)P$_2$ concentrations by activating PtdIns(4)P 5-kinase or by activating PLC, PtdIns 3-kinase and PtdIns 5-phosphatases, respectively. Local changes in PtdIns(4,5)P$_2$ concentration could regulate cortical plasma membrane-cytoskeletal structure by directly altering interactions between PtdIns(4,5)P$_2$ and cytoskeletal anchoring proteins and/or by regulation of actin polymerization via gelsolin and other enzymes that increase or decrease actin polymerization.

Ijuin et al. (2000) have identified a novel InsP 5-phosphatase, which is ubiquitously expressed and especially abundant in skeletal muscle, heart, and kidney. It has been designated as SKIP (representing the first letters from Skeletal muscle and Kidney enriched Inositol Phosphatase). The enzyme exhibited greatest activity towards PtdIns(4,5)P$_2$ but also hydrolyzed PtdIns(3,4,5)P$_3$.

The results suggested that SKIP plays a negative role in regulating the actin cytoskeleton through hydrolyzing PtdIns(4,5)P$_2$.

11.3.4.4. Apoptosis

Apoptotic cell death is a fundamental process of normal development and tissue homeostasis of multicellular organisms (Jacobson et al., 1997). Suppression of apoptotic mechanisms causes autoimmune diseases and is a hallmark of cancer (Hanahan and Weinberg, 2000). Shi (2001) have presented a structural view of mitochondria mediated apoptosis. Sears and Nevins (2002) have reviewed the signaling networks that link cell proliferation and cell fate.

Apoptosis is characterized by typical structural changes including cell shrinkage, membrane blebbing, chromatin condensation, and nuclear DNA fragmentation. In the following the effects of both D3 and D4 phosphorylated PtdIns on the process of apoptosis are briefly reviewed. Galectic et al. (1999) and Tang et al. (2000) have claimed that stimulation of PKB activity protects cells from apoptosis by phosphorylation and inactivation of the pro-apoptotic protein BAD. The crucial role of lipid second messengers in PKB activation was demonstrated through the use of the PtdIns 3K-specific inhibitors wortmannin and LY294002. Membrane attachment of PLB is mediated by its pleckstrin homology domain to PtdIns(3,4,5)P$_3$ or PtdIns(3,4)P$_2$ with high affinity. The D3 phosphorylated PtdInsPs, PtdIns(3,4,5)P$_3$ and PtdIns(3,4)P$_2$ have been implicated in the promotion of cell survival and in prevention of apoptosis. These phospholipids stimulate the phosphorylation of Akt/PKB, a serine/threonine kinase that inactivates multiple components of the apoptotic machinery (Downward, 1998; Krasilnikov, 2000; Vanhaesebroeck et al., 2001). PKB/Akt has emerged as a critical downstream target for PtdIns 3-kinase, being regulated by a combination of direct binding of PtdInsPs to its pleckstrin homology domain and phosphorylation by upstream kinases such as PDK1 that are also controlled by these lipids. Cellular substrates for PKB/Akt include the metabolic regulators

GSK3 and PFK-2, and the Bcl-2 related apoptosis regulatory protein, BAD (Bcl-2/Bcl-Xl-antagonist, causing cell death). Vanhaesebroeck et al. (2001) point out that the absence of functional PtdIns 3-kinase does not necessarily result in cell death. In many different cell lines, PtdIns 3-kinase inhibitors appear to have little effect on cell survival even under prolonged periods of incubation, which indicates that such cells have PtdIns 3-kinase-independent survival pathways and/or that the role of PtdIns 3-kinase in apoptosis may not be as critical or universal as initially thought. Although PtdIns 3-kinase is thought to protect cells from apoptosis via its downstream target Akt/PKB, no general explanation has been provided as to how Akt/PKB delays cell death. Akt/PKB would be expected to mediate survival by phosphorylation and inhibition of proteins that are involved in programmed cell death. The proposed substrates include the Bcl-2 family member BAD and human caspase-9. Caspase-9 is a protease crucial in the initiation of apoptosis (Shi, 2002). All caspases are produced in cells as catalytically inactive zymogens and must undergo proteolytic activation during apoptosis.

Gelsolin, an actin regulatory protein, can inhibit apoptosis induced by various agents, including anti-Fas antibody, ceramide, and dexamthasone (Ohtsu et al., 1997). Gelsolin is a substrate for caspase-3 and the N-terminal cleavage product has been shown to accelerate morphological changes associated with apoptosis (Kothakota et al., 1997). Mitiochondria play a prominent role in cell death as a central organelle involved in the signal transduction and amplification of the apoptotic response. Koya et al. (2000) have demonstrated that gelsolin inhibits apoptosis by blocking potential loss of mitochondrial membrane and release of cytochrome C, which activate the caspases. Koya et al. (2000) have shown that the protective function of gelsolin, which was not due to simple calcium sequestration, was inhibited by PtdIns(4)P and PtdIns(4,5)P$_2$ binding. It was concluded that gelsolin acts on an early step in the apoptotic signaling at the level of mitochondria.

The D4 phosphorylated PtdInsPs may also contribute to cell survival because PtdIns(4,5)P$_2$ is a substrate for PtdInsP 3-kinase. Mejillano et al. (2001) have recently reported that PtdIns(4,5)P$_2$ is a direct inhibitor of initiator caspases 8 and 9, and their common effector caspase 3. Specifically, PtdInsP$_2$ inhibited procaspase 9 processing in cell extracts and in a reconstituted procaspase 9/Apaf1 apoptosome system. Overexpression of PtdIns 5-kinase Iα, which synthesizes PtdInsP$_2$, suppressed apoptosis, whereas a kinase-deficient mutant did not. Mejillano et al. (2000) further substantiated the anti-apoptotic role for PtdInsP$_2$ by the finding that PtdInsP 5-kinase Iα was cleaved by caspase 3 during apoptosis, and cleavage inactivated PtdInsP 5-kinase Iα in vitro. On the basis of further experiments, it was concluded that PtdInsP$_2$ is a direct regulator of apical and effector caspases in the death receptor and mitochondrial pathways, and that PtdInsP 5-kinase Iα inactivation contributes to the progression of apoptosis.

It had been reported earlier (Azuma et al., 2000) that PtdInsP$_2$ complexed with gelsolin inhibits caspase 3 and caspase 9, but not caspase 8, and that PtdInsP$_2$ alone does not inhibit caspases, while Koya et al. (2000) had shown that gelsolin inhibits the loss Δψ$_m$ and cytochrome c release from mitochondria resulting in the lack of activation of caspase-3, -8, and -9 in Jurkat cells treated with staurosporine, thapsigargin, and protoporphyrin IX. Gelsolin thus acts on an early step in the apoptotic signal at the level of mitochondria. Barber et al. (2001) have obtained evidence showing that insulin is a survival factor for retinal neurons by activating the PtdIns 3-kinase/Akt pathway and by reducing caspase-3 activation. Recently, Acehan et al. (2002) have described a three-dimensional structure of an Apaf-1 -cytochrome C complex (apopsome) along with the implications for assembly, procaspase-9 binding and activation.

PtdIns(3,4,5)P$_3$ serves an essential signaling role in mediating the effects of a wide range of extracellular stimuli on cell proliferation, cell survival, and metabolism (Toker and Cantley, 1997). Recent studies have elucidated an important effector pathway that involves D-3 PtdInsPs-mediated membrane recruitment and activation of several protein kinases. The PKB/Akt kinase is activated downstream of receptor-regulated PtdIns 3-kinase in part by direct interaction of PtdIns(3,4)P_2 with its PH domain (Franke et al., 1997a,b; Klippel et al., 1997a,b; Frech et al., 1997; Cohen et al., 1997). Recruitment of PKB/Akt to the membrane requires its PH domain (Andjelkovic et al., 1997), and possible dimerization of PKB/Akt at the membrane has been suggested (Franke et al., 1997b). PKB/Akt undergoes phosphorylation and activation by additional PtdInsPs-dependent protein kinases such as PDK1 (Cohen et al., 1997; Stephens et al., 1998; Alessi, 2001). The phosphorylation of PKB/Akt by PDK1 requires direct PtdInsPs binding to the PH domain that binds PtdIns(3,4,5)P_3; however, its kinase activity may be constitutive, and PtdInsPs binding may serve to recruit PDK1 to a membrane site near its membrane bound substrate (Alessi, 2001). This system represents the clearest example of PtdInsP signaling where the identity of the lipid(s) involved, the basis of its regulated synthesis by a defined kinase, and the role of physiologically relevant PtdInsP-binding effector are partially defined.

Several substrates of PKB/Akt that act downstream in signaling pathways regulating metabolism (Cohen et al., 1997) and cell survival (del Peso et al., 1997) have been identified. Expression of constitutively active PtdIns 3-kinase (Martin et al., 1996; Frevert and Kahn, 1997) or a PKB/Akt (Kohn et al., 1996) enhances Glut4 translocation to the plasma membrane of 3T3-Li adipocytes, suggesting that the PKB/Akt-catalyzed phosphorylation of an unidentified protein.

Ptasznik et al. (1997) have shown that the blockade of PtdIns 3-kinase activity in human fetal undifferentiated cells induces morphological and functional endocrine differentiation. This was associated with an increase in mRNA levels of insulin, glucagon, and somatostatin, as well as an increase in the insulin protein content and secretion in response to secretagogues. The activity of PtdIns 3-kinase activity was inversely correlated with the hepatocyte growth factor/scatter induced down-regulation or nicotinamide-induced upregulation of islet-specific gene expression, giving support to the role of PtdIns 3-kinase, as a negative regulator of endocrine differentiation. The precise molecular mechanisms that link the inhibition of the PtdIns 3-kinase to the induction of endocrine differentiation remain to be elucidated. Only a few potential biochemical targets of PtdInsPs have been found in mammalian cells (for a review see Carpenter and Cantley, 1996a,b). Thus, the exact role and immediate downstream molecular targets of PtdIns(3,4)P_2 and PtdIns(3,4,5)P_3 have not been identified. This is true for all known cellular functions of PtdIns 3-kinase in various systems.

Mithieux et al. (1998) have reported that glucose-6-phosphatase (Glc6Pase) is inhibited in the presence of the lipid products of PtdIns 3-kinase. In order of efficiency in untreated microsomes was: PtdIns(3,4,5)P_3 > PtdIns(3,4)P_2 = PtdIns(4,5)P_2 > PtdIns(3)P = PtdIns(4)P > PtdIns. The mechanism of Glc6Pase inhibition by PtdIns(4,5)P_2, PtdIns(3,4)P_2, and PtdIns(3,4,5)P_3 is competitive in both untreated and detergent treated microsomes. The phenomenon was thought to be of special importance with regard to the insulin's inhibition of hepatic glucose production. Glc6Pase is a crucial enzyme of glucose homeostasis since it catalyzes the ultimate biochemical reaction of both glycogenolysis and gluconeogenesis.

The normal EGF receptor is capable of initiating a variety of signaling cascades upon ligand activation. One such effector whose importance in tumorigenesis is becoming increasingly apparent is PtdIns 3-kinase. PtdIns 3-kinase was first shown to be important in transformation by the observations that it associates with polyome virus middle T protein upon phosphorylation of c-Src, and that mutants of middle T which fail to recruit PtdIns 3-kinase activity are impaired in their tumorigenic activity (Hunter, 1997). The potential role of PtdIns 3-kinase in signaling to cell growth and proliferation has sparked interest in its role in cell transformation. This has been heightened by the recent identification of transforming retrovirally encoded PtdIns 3-kinase isolated from chicken haemangiosarcomas (Chang et al., 1997).

Berggren et al. (1993) have found that the antitumor ether lipid analogue, 1-O-octadecyl-2-O-methyl-rac-glycero-3-phosphocholine (ET-18-OCH$_3$) is an inhibitor of Swiss mouse 3T3 fibroblast and bovine brain PtdIns 3-kinases. The concentration of ET-18-OCH$_3$, causing 50% inhibition (IC50) was 35 μM. The inhibition of PtdIns 3-kinase by ET-18-OCH3 was non-competitive with ATP. The results of the study suggested that inhibition of PtdIns 3-kinase might contribute to the antiproliferative activity of the antitumor ether lipid analogues.

Abreu et al. (2001) have recently investigated the inter-relationship between the Fas death receptor and PtdIns 3-kinase signaling pathways in human intestinal epithelial cells. It was found that PtdIns 3-kinase-dependent pathways opposed Fas-induced apoptosis and limited chloride secretion in the epithelial cells.

Tsuruta et al. (2002) have recently reported that the PtdIns 3-kinase/Akt pathway suppresses Bax translocation to mitochondria. Bax is a proapoptotic member of the Bcl-2 family, localized largely in the cytoplasm, which redistributes to mitochondria in response to apoptotic stimuli, where it induces cytochrome c release.

The tumor suppressor p53 can induce growth arrest and cell death via apoptosis in response to a number of cellular stresses. Pyrzynska et al. (2002) have recently shown that tumor suppressor p53 mediates apoptotic cell death in rat glioma cells triggered by cyclosporin A. The cyclosporin treatment resulted in increased levels of the p53 tumor suppressor, its nuclear accumulation, and transcriptional activation of p53-dependent genes. The increase of p53 correlated with the elevation of p21^{wafl} and Bax protein expression. The increased level of Bax was accompanied by its increased association with mitochondria. The latest advances in the mechanisms of caspase activation and inhibition during apoptosis have been reviewed by Shi (2002). Marte and Downward (1997) have discussed PKB/Akt in regards to the role of PtdIns 3-kinase in cell survival and beyond. Saito et al. (2001) have reported that the interaction between the Btk PH domain and PtdIns(3,4,5)P$_3$ directly regulates Btk.

Kisseleva et al. (2002) has recently reported that PtdInsP 5-phosphatase IV inhibits Akt/PKB phosphorylation and leads to apoptotic cell death. Expression of PtdInsP 5-phosphatase IV in a tetracycline dependent manner was obtained in human 293 cells. Cell proliferation ceased immediately upon overexpression of the enzyme and cellular levels of PtdIns(4,5)P$_2$ and PtdIns(3,4,5)P$_3$ are reduced with corresponding increases in the products, PtdIns(4)P and PtdIns(3,4)P$_2$. These changes are followed by decreased phosphorylation of Akt and an increase in apoptotic cell death. The results indicate that PtdIns(3,4,5)P$_3$ is the PtdIns 3-kinase product that is responsible for activation of Akt in cells and that elevated levels of PtdIns(3,4)P$_2$ are unable to activate Akt. It is concluded that PtdInsP 5-phosphatase IV is an inhibitor of PtdIns 3K/Akt pathway.

According to Morrison et al. (2001) InsP$_6$ kinase 2 mediates growth suppressive and apoptotic effects of interferon-beta in ovarian carcinoma cells.

11.3.4.5. DNA synthesis
There is evidence that activation of PtdIns 3-kinase is involved in the control of DNA synthesis (reviewed in Cocco et al., 1998; D'Santos et al., 1998; Maraldi, 2000 and Vanhaesebroeck et al., 2001). The activation of PtdIns 3-kinase was first shown to correlate with, and then to be required for, the ability of various mitogens and oncoproteins to stimulate DNA synthesis. Also, treating certain cells with such inhibitors of PtdIns 3-kinase as wortmannin and LY294002 or injecting neutralizing antibodies blocks DNA synthesis. Jones et al. (1999) have conducted detailed analysis of PtdInsP$_3$ and PtdIns(3,4)P$_2$ production following treatment of quiescent mammalian HepG2 hepatocarcinoma cells with PDGF and have observed two waves of production of D3-phosphorylated PtdIns. The first wave is acute and appears within

minutes of PDGF treatment as the cells proceed from the resting state (G0) to enter the cell cycle (G1). The second wave is much broader and appears 3–7 h post-stimulation (mid- to late G1). Addition of PtdIns 3-kinase inhibitors showed that the second but not the first wave of PtdIns 3-kinase activity was required for DNA synthesis and entry into the S phase (8–12 h post-stimulation). Addition of synthetic PtdInsP$_3$ and PtdIns(3,4)P$_2$ to PDGF-treated cells that were unable to activate PtdIns 3-kinase rescued the DNA synthesis. Vanhaesebroeck et al (2001) have discussed possible mechanisms for this effect and have concluded that PtdIns 3-kinase is likely to promote cell entry into S phase by increasing cyclin D transcription and translation by preventing its degradation and by reducing levels of CDKI, p27^{kip1} and that these effects could be mediated at the molecular level by Akt/PKB and p70-S6 kinase, since both Akt/PKB and p70-S6K are established PtdIns 3-kinase targets. Vanhaesebroeck et al. (2001) point out that the ability of PtdIns 3-kinase dependent p70-S6K activation to promote both translation and S phase entry also provides a link between the ability of PtdIns 3-kinase to stimulate both growth and S phase entry, but that it remains to be established whether or not the ability of PtdIns 3-kinase to stimulate S phase entry is dependent on its ability to stimulate protein translation and growth.

Another nuclear function for InsP$_6$ arises from the observation of Hanakahi et al. (2000), that InsP$_6$ is necessary for DNA end-joining in vitro. The target for InsP$_6$ was shown to be the DNA-dependent protein kinase, and a possible binding domain for InsP$_6$ was identified as DNA-PK$_{cs}$ rather than Ku, a protein which binds to DNA ends. The requirement for InsP$_6$ appeared to be specific since Ins(1,3,4,5)P$_4$ and Ins(1,3,4,5,6)P$_5$ were less active and InsS$_6$ was inactive. However, InsP$_7$ and InsP$_8$ as well as other InsPs substituted in the 1-, 2- and 3-positions were not tested. Luo et al. (2002) have described the requirement of Ins pyrophosphates for DNA hyper recombination in protein kinase C1 mutant yeast.

Melo and Toczyski (2002) have reviewed recent work on the DNA-damage and S-phase checkpoints. Central to all DNA-damage-induced checkpoint responses is a pair of large protein kinases with homology to PtdIns 3-kinases: the ATM (Ataxia telangiectasia mutated) and ATR (Ataxia and Rad-related) kinases. Either ATM or ATR is required for each of the DNA-damage-responsive checkpoints in yeast and mammals. In view of the homology, the study of Hanahaki et al. (see above) was of special interest. However, Ma and Lieber (2002) have since shown that InsP$_6$ associates with Ku. The binding of DNA ends and InsP$_6$ to Ku were independent of each other. Ma and Lieber (2002) have speculated about the potential significance of the observation and have concluded that the elucidation of the function of Ku-InsP$_6$ interaction requires further studies.

In other studies, Phillips-Mason et al. (2000) have presented evidence that PtdIns 3-kinase is required for (-thrombin-stimulated DNA synthesis in Chinese hamster embryonic fibroblasts.

References

Chapter 1

Agranoff, B. W. (1978). Cyclitol confusion. Trends Biochem. Sci. *3*, N283–N285.

Agranoff, B. W. and Fisher, S. K. (1991). Phosphoinositides and their stimulated breakdown. In: Inositol Phosphates and Derivatives, ACS Symposium Series. American Chemical Society, Washington, DC, pp. 20–32.

Anonymous (1974). IUPAC–IUB Commission on Biochemical Nomenclature (CBN). Nomenclature of cyclitols, recommendations 1973. Pure Appl. Chem. *37*, 285–297.

Anonymous (1976a). IUPAC Commission on the Nomenclature of Organic Chemistry and IUPAC–IUB Commission on Biochemical Nomenclature, 1976, Nomenclature of cyclitols. Biochem. J. *153*, 23–31.

Anonymous (1976b). IUPAC–IUB Enzyme Nomenclature Recommendations 1975. Supplement I: corrections and additions. Biochim. Biophys. Acta *429*, 1–2.

Anonymous (1977). The Nomenclature of Lipids. Recommendations (1976) IUPAC–IUB Commission on Biochemical Nomenclature. Lipids *12*, 455–468.

Anonymous (1989). Numbering of atoms in *myo*-inositol. Recommendations 1988. Biochem. J. *258*, 1–2.

Azzouz, N., Striepen, B., Gerold, P., Capdeville, Y. and Schwarz, R. T. (1995). Glycosylinositol-phosphoceramide in the free-living protozoan *Paramaecium primaurelia*: modification of core glycans by mannosyl phosphate. EMBO J. *14*, 4422–4433.

Azzouz, N., Striepen, B., Gerold, P., Capdeville, Y. and Schwarz, R. T. (1998). Glycosylinositol-phosphoceramide in the free-living protozoan *Paramaecium primaurelia*: modification of core glycans by mannosyl phosphate. EMBO J. *14*, 4422–4433.

Berridge, M. J. (1993). Inositol trisphosphate and calcium signaling. Nature *361*, 315–325.

Berridge, M. J. and Irvine, R. F. (1989). Inositol phosphates and cell signaling. Inositol trisphosphate and diacylglycerol: Two interactive second messengers. Nature *341*, 197–205.

Billington, D. C. (1993). The Inositol Phosphates. VCH Verlagsgesellschaft GmbH, Weinheim, pp. 1–8.

Brockerhoff, H. and Ballou, C. E. (1961). The structure of the phosphoinositide complex of beef brain. J. Biol. Chem. *236*, 1907–1911.

Brodbeck, U. (1998). Signaling properties of glycosylphosphatidylinositols and their regulated release from membranes in the turnover of glycosylphosphatidylinositol-anchored proteins. Biol. Chem. *379*, 1041–1044.

Brown, D. M. and Stewart, J. C. (1966). The structure of triphosphoinositide from rat brain. Biochim. Biophys. Acta *125*, 413–421.

Bruzik, K. S., Hakeem, A. A. and Tsai, M-D. (1994). Are D- and L-*chiro*-phosphoinositides substrates of phosphatidylinositol-specific phospholipase C? Biochemistry *33*, 8367–8374.

Butikofer, P., Kuypers, F. A., Shackleton, C., Brodbeck, U. and Stieger, S. (1990). Molecular species analysis of the glycosylphosphatidylinositol anchor of *Torpedo marmorata* acetylcholinesterase. J. Biol. Chem. *265*, 18983–18987.

Carew, M. A., Yang, X., Schultz, C. and Shears, S. B. (2000). *myo*-Inositol 3,4,5, 6-tetrakisphosphate inhibits an apical calcium-activated chloride conductance in polarized monolayers of a cystic fibrosis cell line. J. Biol. Chem. *275*, 26906–26913.

Conzelman, A., Puoti, A., Lester, R. L. and Desponds, C. (1992). Two different types of lipid moieties are present in glycerophosphoinositol-anchored membrane proteins of *Saccharomyces cerevisiae*. EMBO J. *11*, 457–466.

Cosgrove, D. J. (1980). Inositol Phosphates, Their Chemistry, Biochemistry and Physiology. Elsevier, Amsterdam.

Cote, G. G. and Crain, R. C. (1993). Biochemistry of phosphoinositides. Annu. Rev. Plant Physiol. Plant Mol. Biol. *44*, 333–356.

Creba, J. A., Downes, C. P., Hawkins, P. T., Brewster, G., Mitchell, R. H. and Kirk, C. J. (1983). Rapid breakdown of phosphatidylinositol-4-phosphate and phospatidylinositol 4,5-bisphosphate in rat hepatocytes stimulated by vasopressin and other calcium-mobilizing hormones. Biochem. J. *212*, 733–747.

Dawson, R. M. C. and Dittmer, J. C. (1961). Evidence for the structure of brain triphosphoinositide from hydrolytic degradation studies. Biochem. J. *81*, 540–545.

Denny, P. W., Field, M. C. and Smith, D. F. (2001). GPI-anchored proteins and glycoconjugates segregate into lipid rafts in *Kinetoplastida*. FEBS Lett. *491*, 148–153.

Divecha, N. and Irvine, R. F. (1995). Phospholipid signalling. Review. Cell *80*, 269–278.

Drobak, B. K. (1992). The plant phosphoinositide system. Biochem. J. *288*, 697–712.

Drobak, B. K. (1993). Plant phosphoinositides and intracellular signaling. Plant Physiol. *102*, 705–709.

Duckworth, B. C. and Cantley, L. C. (1996). PI 3-kinase and receptor-linked signal transduction. In: Handbook of Lipid Research, Lipid Second Messengers (Bell, R. M., eds.)Vol. 8. Plenum Press, New York, pp. 125–175.

Eisenberg, F., Jr. (1967). 1-Myoinositol 1-phosphate as product of cyclization of glucose 6-phosphate and substrate for a specific phosphatase in rat testis. J. Biol. Chem. *242*, 1375–1382.

Englund, P. T. (1993). The structure and biosynthesis of glycosyl phosphatidyl-inositol protein anchors. Annu. Rev. Biochem. *62*, 121–138.

Fankhauser, C., Homans, S. W., Thomas-Oates, J. E., McConville, M. J., Desponds, C., Conzelmann, A. and Ferguson, M. A. J. (1993). Structures of glycosyl-phosphatidyl-inositol membrane anchors from *Saccharomyces cerevisiae*. J. Biol. Chem. *268*, 26365–26374.

Farquhar, J. W., Insull, W., Jr., Rosen, P., Stoffel, W. and Ahrens, E. H., Jr. (1959). Nutr. Rev. *17*(Suppl.), 1–30.

Field, M. C. and Menon, A. K. (1993). Glycolipid anchoring of cell surface proteins. In: Lipid Modification of Proteins (Schlesinger, M. J., ed.). CRC Press, Boca Raton, USA, pp. 83–134.

Fisher, S. K. and Agranoff, B. W. (1987). Receptor activation and inositol lipid hydrolysis in neutral tissues. J. Neurochem. *48*, 999–1100.

Folch, J. (1949). Complete fractionation of brain cephalin: isolation from it of phosphatidylserine, phosphatidylethanolamine and diphosphoinositide. J. Biol. Chem. *177*, 497–504.

Fruman, D. A., Meyers, R. E. and Cantley, L. C. (1998). Phosphoinositide kinases. Annu. Rev. Biochem. *67*, 481–507.

Gascard, P., Tran, D., Sauvage, M., Sulpice, J. C., Fukami, K., Takenawa, T., Claret, and Giraud, F. (1991). Asymmetric distribution of phosphoinositides and phosphatidic acid in the human erythrocyte membrane. Biochim. Biophys. Acta *1069*, 27–36.

Hawthorne, J. N. (1960). The inositol phospholipids. J. Lipid Res. *1*, 255–280.

Hawthorne, J. N. (1982). Inositol phospholipids. In: Phospholipids, New Comprehensive Biochemistry (Hawthorne, J. N. and Ansell G. B., eds.), Vol. 4. Elsevier Biomedical, Amsterdam, pp. 263–278.

Hawthorne, J. N. and Kemp, P. (1964). The brain phosphoinositides. Adv. Lipid Res. *2*, 127–166.

Haynes, P. A., Gooley, A. A., Ferguson, M. A. J., Redmond, J. W. and Williams, K. L. (1993). Post-translational modification of the *Dictyostelium discoideum* glycoprotein PsA. Glycosylphosphatidylinositol membrane anchor and composition of O-linked oligosaccharides. Eur. J. Biochem. *216*, 729–737.

Hetherington, A. M. and Drobak, B. K. (1992). Inositol-containing lipids in higher plants. Prog. Lipid Res. *31*, 53–63.

Hinchcliffe, K. (2000). Intracellular signaling: Is PIP(2) a messenger too? Curr. Biol. *10*, R104–R105.

Hinchcliffe, K. A., Ciruela, A. and Irvine, R. F. (1998). PIPkins1, their substrates and their products: new functions for old enzymes. Biochim. Biophys. Acta *1436*, 87–104.

Hoffmann-Ostenhof, O. and Pittner, F. (1982). The biosynthesis of *myo*-inositol and its isomers. Can. J. Chem. *60*, 1863–1871.

Hokin, L. E. (1985). Receptors and phosphoinositide-generated second messengers. Annu. Rev. Biochem. *54*, 205–235.

Hokin, L. E. (1996). History of phosphoinositide reseach. Subcell. Biochem. *26*, 1–41.

Holub, B. J., Kuksis, A. and Thompson, W. (1970). Molecular species of mono-, di- and triphosphoinositides of bovine brain. J. Lipid Res. *11*, 558–564.

Hooper, N. M. (1999). Detergent-insoluble glycosphingolipid/cholesterol-rich membrane domains, lipid rafts and caveolae (Review). Mol. Membr. Biol. *16*, 145–156.

Irvine, R. F., Letcher, A. J., Lander, D. J., Heslop, J. P. and Berridge, M. J. (1987). Inositol(3,4)bisphosphate and inositol(1,3)bisphosphate in GH4 cells— evidence for complex breakdown of inositol(1,3,4)trisphosphate. Biochem. Biophys. Res. Commun. *143*, 353–359.

Irvine, R. F. and Schell, M. J. (2001). Back in the water: the return of the inositol phosphates. Nat. Rev. *2*, 327–338.

Isaacs, R. E. and Harkness, D. R. (1980). Erythrocyte organic phosphates and hemoglobin function in birds, reptiles and fishes. Am. Zool. *20*, 115–129.

Ives, E. B., Nichols, J., Wente, S. R. and York, J. D. (2000). Biochemical and functional characterization of inositol 1,3,4,5,6-pentakisphosphate 2-kinases. J. Biol. Chem. *275*, 36575–36583.

Jefferson, A. B. and Majerus, P. W. (1996). Mutation of the conserved domains of two inositol polyphosphate 5-phosphatases. Biochemistry *35*, 7890–7894.

Jennemann, R., Bauer, B. L., Bertalanffy, H., Geyer, R., Gschwind, M., Selmer, T. and Wiegandt, H. (1999). Novel glycoinositolphosphospingolipids, basidioli-pids, from *Agaricus*. Eur. J. Biochem. *259*, 331–338.

Kinnard, R. L., Narisimhan, B., Pliska-Matyshak, G. and Murthy, P. P. N. (1995). Characterization of *scyllo*-inositol-containing phosphatidylinositol in plant cells. Biochem. Biophys. Res. Commun. *210*, 549–555.

Kuksis, A. and Myher, J. J. (1990). Mass analysis of molecular species of diradylglycerols. In: Methods in Inositide Research (Irvine, R. F., ed.). Raven Press Ltd, New York, pp. 187–216.

Larner, J., Huang, L. C., Schwartz, C. F. W., Oswald, A. S., Shen, T.-Y., Kinter, M., Tang, G. and Zeller, K. (1988). Rat liver insulin mediator which stimulates pyruvate dehydrogenase phosphatase contains galactosamine and D-chiro-inositol. Biochem. Biophys. Res. Commun. *151*, 1416–1426.

Lee, T.-C., Malone, B., Buell, A. B. and Blank, M. L. (1991). Occurrence of ether-containing inositol phospholipids in bovine erythrocytes. Biochem. Biophys. Res. Commun. *175*, 673–678.

Liscovitch, M. and Cantley, L. C. (1994). Lipid second messengers. Cell *77*, 329–334.

Loureiro y Penha,, Tadeschini, A. R., Lopes-Bezera, L. M., Wait, R., Jones, C., Mattos, K. A., Heise, N., Mendonca-Previato, L. and Previato, J. O. (2001). Characterization of novel structures of mannosylinositolphosphorylceramides from the yeast forms of *Sporothrix schenckii*. Eur. J. Biochem. *268*, 4243–4250.

Majerus, P. W., Connolly, T. M., Bansal, V. S., Inhorn, R. C., Ross, T. S. and Lips, D. L. (1988). Inositol phosphates: Synthesis and degradation. J. Biol. Chem. *263*, 3051–3054.

Majerus, P. W., Kisseleva, M. V. and Norris, F. A. (1999). The role of phosphatases in inositol signaling reactions. J. Biol. Chem. *274*, 10669–10672.

Martin, T. F. J. (1998). Phosphoinositide lipids as signaling molecules: common themes for signal transduction, cytoskeletal regulation, and membrane trafficking. Annu. Rev. Cell. Dev. Biol. *14*, 231–264.

Martin, J.-B., Bakker-Grunwald, T. and Klein, G. (1993). ^{31}P-NMR analysis of *Entamoeba histolytica*. Occurrence of high amounts of two inositol phosphates. Eur. J. Biochem. *214*, 711–718.

Martin, J.-B., Laussmann, T., Bakker-Grunwald, T., Vogel, G. and Klein, G. (2000). *neo*-Inositol polyphosphates in the amoeba *Entamoeba histolytica*. J. Biol. Chem. *275*, 10134–10140.

Mato, J. M., Kelly, K. L., Abler, A., Jarett, L., Corkey, B. E., Cashel, J. A. and Zopf, D. (1987). Partial structure of an insulin-sensitive glycophospholipid. Biochem. Biophys. Res. Commun. *146*, 764–770.

Mayr, G. W. (1989). Inositol phosphates: structural components, regulators and signal transducers of the cell—a review. In: Topics in Biochemistry. Mannheim-Boehringer, Mannheim, pp. 1–17.

McConville, M. J. and Ferguson, M. A. (1993). The structure, biosynthesis and function of glycosylated phosphatidylinositols in the parasitic protozoa and higher eukaryotes. Biochem. J. *194*, 305–324.

Medof, M. E., Nagarajan, and Tykocinski, M. L. (1996). Cell-surface engineering with GPI-anchored proteins. Review. FASEB J. *10*, 574–586.

Menniti, F. S., Miller, R. N., Putney, J. W., Jr. and Shears, S. B. (1993). Turnover of inositol polyphosphate pyrophosphates in pancreatoma cells. J. Biol. Chem. *268*, 3850–3856.

Michell, R. H. (1975). Inositol phospholipids and cell surface receptor function. Biochim. Biophys. Acta *415*, 81–147.

Munday, D., Norris, F. A., Caldwell, K. K., Brown, S., Majerus, P. W. and Mitchell, C. A. (1999). The inositol polyphsphate 4-phosphatase forms a complex with phosphatidylinositol 3-kinase in human platelet cytosol. Proc. Natl Acad. Sci. USA *96*, 3640–3645.

Murthy, P. P. N. (1996). Inositol phosphates and their metabolism in plants. Subcell. Biochem. *26*, 227–255.

Nishizuka, Y. (1984). The role of protein kinase C in cell surface signal transduction and tumour promotion. Nature *308*, 693–698.

Nishizuka, Y. (1995). Protein kinase C and lipid signaling for sustained cellular responses. FASEB J. *9*, 484–496.

Norris, F. A. and Majerus, P. W. (1994). Hydrolysis of phosphatidyinositol 3,4-bisphosphate by inositol polyphosphate 4-phosphatase isolated by affinity elution chromatography. J. Biol. Chem. *269*, 8716–8720.

Ostlund, R. E., Jr., McGill, J. B., Herskowitz, I., Kipnis, D. M., Santago, J. V. and Sherman, W. R. (1993). D-*chiro*-Inositol metabolism in diabetes mellitus. Proc. Natl Acad. Sci. USA *90*, 9988–9992.

Pak, Y. and Larner, J. (1992). Identification and characterization of *chiro*-inositol-containing phospholipids from bovine liver. Biochem. Biophys. Res. Commun. *184*, 1042–1047.

Parthasarathy, R. and Eisenberg, F., Jr. (1986). The inositol phospholipids: a stereochemical view of biological activity. Biochem. J. *235*, 313–322.

Parthasarathy, R. and Eisenberg, F., Jr. (1991). Biochemistry, stereochemistry, and nomenclature of the inositol phosphates. In: Inositol Phosphates and Derivatives: Synthesis, Biochemistry, and Therapeutic Potential. ACS Symposium Series 463 (Reiz, A. B., ed.). American Chemical Society, Washington, DC, pp. 1–19.

Posternak, S. (1921). Synthesis of inosite hexaphosphoric aid acid. J. Biol. Chem. *46*, 453–457.

Prieschl, E. E. and Baumruker, T. (2000). Sphingolipids: second messengers, mediators and raft constituents in signaling. Immunol. Today *21*, 555–560.

Ptasznik, A., Beattie, G. M., Mally, M. I., Cirulli, V., Lopez, A. and Hayek, A. (1997). Phosphatidylinositol 3-kinase is a negative regulator of cellular differentiation. J. Cell Biol. *137*, 1127–1136.

Rameh, L. E. and Cantley, L. C. (1999). The role of phosphoinositide 3-kinase lipid products in cell function. J. Biol. Chem. *274*, 8347–8350.

Reggiori, F., Canivenc-Gansel, E. and Conzelmann, A. (1997). Lipid remodeling leads to the introduction and exchange of defined ceramides on GPI proteins in the ER and Golgi of *Saccharomyces cerevisiae*. EMBO J. *16*, 3506–3518.

Reitz, A. B. (ed.) (1991). Inositol Phosphates and Derivatives: Synthesis, Biochemistry and Therapeutic Potential, ACS Symposium Series 463, American Chemical Society, Washington, DC, p. 236.

Rincon, M. and Boss, W. F. (1990). Second messenger role of phosphoinositides. In: Inositol Metabolism in Plants (Morre, D. J., Boss, W. F. and Loewus F. A., eds.). Wiley-Liss, New York, pp. 173–200.

Rittenhouse, S. E. (1996). Phosphoinositide 3-kinase activation and platelet function. Blood *88*, 4401–4414.

Roberts, W. L., Myher, J. J., Kuksis, A., Low, M. G. and Rosenberry, T. L. (1988). Lipid analysis of the glycoinositol phospholipid membrane anchor of human erythrocyte acetylcholinesterase. J. Biol. Chem. *263*, 18766–18775.

Safrany, S. T., Caffrey, J. J., Yang, X. and Shears, S. B. (1999). Diphosphoinositol polyphosphates: the final frontier for inositide research? Biol. Chem. *380*, 945–951.

Saiardi, A., Caffrey, J. J., Snyder, S. II. and Shears, S. B. (2000). The inositol hexakisphosphate kinase family. Catalytic flexibility and function in yeast vacuole biogenesis. J. Biol. Chem. *275*, 24686–24692.

Schultz, C., Burmeister, A. and Stadler, C. (1966). Synthesis, separation, and identification of different inositol phosphates. Subcell. Biochem. *26*, 371–413.

Shears, S. B., Storey, D. J., Morris, A. B., Cubitt, A. B., Parry, J. B., Mitchell, R. H. and Kirk, C. J. (1987). Dephosphorylation of myo-inositol 1,4,5-trisphosphate and myo-inositol 1,3,4-trisphosphate. Biochem. J. *242*, 393–402.

Singer, W. D., Brown, H. A. and Sternweis, P. C. (1997). Regulation of eukaryotic phosphatidylinositol-specific phospholipase C and phospholipase D. Annu. Rev. Biochem. *66*, 475–509.

Stadler, J., Keenan, T. W., Bauer, G. and Gerisch, G. (1989). The contact site A glycoprotein of *Dictyostelium discoideum* carries a phospholipid anchor of a novel type. EMBO J. *8*, 371–377.

Stephens, L. R. (1990). Preparation and separation of inositol tetrakisphosphates and inositol pentakisphosphates and the establishment of enantiomeric configurations by the use of L-iditol dehydrogenase. In: Methods in Inositide Research (Irvine, R. F., ed.). Raven Press, New York, pp. 9–30, .

Stephens, L., Hawkins, P. T., Carter, N., Chahwala, S. B., Morris, A. J., Whetton, A. D. and Downes, C. P. (1988). *l-myo*-Inositol 1,4,5,6-tetrakisphosphate is present in both mammalian and avian cells. Biochem. J. *249*, 271–282.

Stephens, L., Radenberg, T., Thiel, U., Vogel, G., Khoo, K.-H., Dell, A., Jackson, T. R., Hawkins, P. T. and Mayr, G. W. (1993). The detection, purification, structural characterization, and metabolism of diphosphoinositol pentakisphosphate(s) and bisphosphatidylinositol tetrakisphosphate(s). J. Biol. Chem. 268, 4009–4015.

Streb, H., Irvine, R. F., Berridge, M. J. and Schulz, I. (1983). Release of Ca^{2+} from a nonmitochondrial intracellular store in pancreatic acinar cells by inositol-1,4, 5-trisphosphae. Nature 306, 67–69.

Toker, A. and Cantley, L. C. (1997). Signalling through the lipid products of phosphoinositide-3-OH kinase. Nature 387, 673–676.

Tolias, K. T. and Carpenter, C. L. (2000). Enzymes involved in the synthesis of PtdIns(4,5)P$_2$ and their regulation: PtdIns kinases and PtdInsP kinases. In: Biology of Phosphoinositides (Cockcroft, S., ed.). Oxford University Press, Oxford, UK, pp. 109–130.

Traynor-Kaplan, A. E., Harris, A. L., Thompson, B. L., Taylor, P. and Sklar, L. A. (1988). An inositol tetrakisphosphate-containing phospholipid in activated neutrophils. Nature 334, 353–356.

Vajanaphanich, M., Schultz, C., Rudolf, M. T., Wasserman, M., Enyedi, P., Craxton, A., Shears, S. B., Tsien, R. Y., Barrett, K. E. and Traynor-Kaplan, A. (1994). Long-term uncoupling of chloride secretion from intracellular calcium levels by Ins(3,4,5,6)P$_4$. Nature 371, 711–714.

Vallejo, M., Jackson, T., Lightman, S. and Hanley, M. R. (1988). Occurrence and extracellular actions of inositol pentakis- and hexakis-phosphate in mammalian brain. Nature 330, 565–658.

Vanhaesebroeck, B., Leevers, S. J., Ahmadi, K., Timms, J., Katso, R., Driscoll, P. C., Woscholski, R., Parker, P. J. and Waterfield, M. D. (2001). Synthesis and function of 3-phosphorylated inositol lipids. Annu. Rev. Biochem. 70, 535–602.

Vanhaesebroeck, B., Leevers, S. J., Panayotou, G. and Waterfield, M. D. (1997). Phosphoinositide 3-kinases: A conserved family of signal transducers. Trends Biochem. Sci. 22, 267–272.

Varela-Nieto, I., Leon, Y. and Caro, H. N. (1996). Cell signaling by inositol phosphoglycans from different species. Comp. Biochem. Physiol. 115B, 223–241.

Vijaykumar, M., Naik, R. S. and Gowda, D. C. (2001). Plasmodium falciparum glycosylphosphatidylinositol-induced TNF-α secretion by macrophages is mediated without membrane insertion or endocytosis. J. Biol. Chem. 276, 6909–6912.

Whitman, M., Downes, C. P., Keeler, M., Keller, T. and Cantley, L. (1988). Type I phosphatidylinositol kinase makes a novel inositol phospholipid, phosphatidylinositol-3-phosphate. Nature 332, 644–646.

Woscholski, R., Waterfield, M. D. and Parker, P. J. (1995). Purification and biochemical characterization of a mammalian phosphatidylinositol 3,4,5-trisphosphate 5-phosphatase. J. Biol. Chem. *270*, 31001–31007.

Chapter 2

Agren, J., Kuksis, A. (2002). Analysis of diastereomeric diacylglycerol naphthylethylurethanes by normal phase HPLC with on-line electrospray mass spectrometry. J. Chromatogr (submitted).

Ames, B. N. (1966). Assay of inorganic phosphate, total phosphate and phosphatases. Methods Enzymol. *8*, 115–118.

Ansell, G. B. and Hawthorne, J. N. (1964). Phospholipids. Elsevier Publishing Co, Amsterdam, The Netherlands.

Ashizawa, N., Yoshida, M. and Aotsuka, T. (2000). An enzymatic assay for myo-inositol in tissue samples. J. Biochem. Biophys. Methods *44*, 89–94.

Bartlett, G. R. (1959). Phosphorus assay in column chromatography. J. Biol. Chem. *234*, 466–468.

Blank, M. L., Robinson, M., Fitzgerald, V. and Snyder, F. (1984). Novel quantitative method for determination of molecular species of phospholipids and diglycerides. J. Chromatogr. *298*, 473–482.

Bligh, E. G. and Dyer, W. J. (1959). A rapid method of total lipid extraction and purification. Can. J. Biochem. Physiol. *37*, 911–917.

Brugger, B., Erben, G., Sandhoff, R., Wieland, F.T. and Lehmann, W.D. (1997). Quantitative analysis of biological membrane lipids at the low picomole level by nano-electrospray ionization tandem mass spectrometry. Proc. Natl. Acad. Sci. USA *94*, 2339–2344. [published erratum appears in Proc. Natl. Acad. Sci. USA *96*, 10943, 1999].

Bruzik, K. S., Hakeem, A. A. and Tsai, M.-D. (1994). Are D- and L-*chiro*-phosphoinositides substrates for phosphatidylinositol-specific phospholipase C? Biochemistry *33*, 8367–8374.

Bruzik, K. S. and Tsai, M.-D. (1991). Phospholipase stereospecificity at phosphorus. Methods Enzymol. *197*, 258–265.

Butikofer, P., Kuypers, F. A., Schackleton, C., Brodbeck, U. and Stieger, S. (1990). Molecular species analysis of the glycosylphosphatidylinositol anchor of Torpedo marmorata acetylcholinesterase. J. Biol. Cem. *265*, 18983–18987.

Butikofer, P., Zollinger, M. and Brodbeck, U. (1992). Alkylacyl glycerophosphoinositol in human and bovine erythrocytes. Molecular species composition and comparison with glycosyl-inositolphospholipid anchors of erythrocyte acetylcholinesterases. Eur. J. Biochem. *208*, 677–683.

Carrapiso, A. I. and Garcia, C. (2000). Development in lipid analysis: some new extraction techniques and in-situ tansesterification. Lipids *35*, 1167–1177.

Carstensen, S., Pliska-Matyshak, G., Bhuvarahamurthy, N., Robbins, K. M. and Murthy, P. P. N. (1999). Biosynthesis and localization of phosphatidyl-*scyllo*-inositol in barley aleurone cells. Lipids *34*, 67–73.

Casu, M., Anderson, G. T., Choi, G. and Gibbons, W. A. (1991). NMR Lipid profiles of cells, tissues and body fluids. I-1D and 2D proton NMR of lipids from rat liver. Magn. Reson. Chem. *29*, 594–602.

Cho, M.H., Chen, Q., Okpodu, C.M. and Boss, W.F. (1992). Separation and quantitation of [^{3}H]inositol phospholipids using thin-layer chromatography and a computerized ^{3}H imaging scanner. LC-GC *10*, 464–468.

Christiansen, K. (1975). Lipid extraction procedure for in vitro studies of glyceride synthesis with labeled fatty acids. Anal. Biochem. *66*, 93–99.

Cronholm, T., Viesta-Rains, M. and Sjovall, J. (1992). Decreased content of arachidonoyl species of phosphatidylinositol phosphates in pancreas of rats fed on an ethanol-containing diet. Biochem. J. *287*, 925–928.

DaTorre, S. D., Corr, P. B. and Creer, M. H. (1990). A sensitive method for the quantification of the mass of inositol phosphates using gas chromatography-mass spectrometry. J. Lipid Res. *31*, 1925–1934.

Deeg, M. and Verchere, C. B. (1997). Regulation of glycosylphosphatidylinositol-specific phospholipase D secretion from βTC3 cells. Endocrinology *138*, 819–826.

DeLong, C. J., Baker, P. R. S., Samuel, M., Cui, Z. and Thomas, M. J. (2001). Molecular species composition of rat liver phospholipids by ESI-MS/MS: the effect of chromatography. J. Lipid Res. *42*, 1959–1968.

Dittmer, J. C. and Wells, M. A. (1969). Quantitative and qualitative analysis of lipids and lipid components. Methods Enzymol. *14*, 482–530.

Dobson, G. and Deighton, N. (2001). Analysis of phospholipid molecular species by liquid chromatography–atmospheric pressure chemical ionisation mass spectrometry of diacylglycerol nicotinates. Chem. Phys. Lipids *111*, 1–17.

Dobson, G., Itabashi, Y., Christie, W. W. and Robertson, G. W. (1998). Liquid chromatography with particle-beam electron-impact mass spectrometry of diacylglycerol nicotinates. Chem. Phys. Lipids *97*, 27–39.

Downes, C. P. and Michell, R. H. (1981). The polyphosphoinositide phospho-diesterase of erythrocyte membranes. Biochem. J. *198*, 133–140.

Dugan, L. L., Demediuk, P., Pendley, C. III, and Horrocks, L. L. (1986). Separation of phospholipids by high performance liquid chromatography: all major classes, including ethanolamine and choline plasmalogens, and most minor classes, including lysophosphatidylethanolamine. J. Chromatogr. *378*, 317–327.

Eder, K., Reichlmayer-Lais, A. M. and Kirchgessner, M. (1993). Studies on the extraction of phospholipids from erythrocyte membranes in the rat. Clin. Chim. Acta *219*, 93–104.

Folch, J., Lee, M. and Sloane-Stanley, S. M. (1957). A simple method for the isolation and purification of total lipids from animal tissues. J. Biol. Chem. *226*, 497.

Fridriksson, E. K., Shipkova, P. A., Sheets, E. D., Holowka, D., Baird, B. and McLafferty, F. W. (1999). Quantitative analysis of phospholipids in functionally important membrane domains from RBL-2H3 mast cells using tandem high-resolution mass spectrometry. Biochemistry *38*, 8056–8063.

Gassama-Diagne, A., Fauvel, J. and Chap, H. (1991). Phospholipase 1 activity of guinea pig pancreatic lipase. Methods Enzymol. *197*, 316–325.

Han, X. and Gross, W. R. (1994). Electrospray ionization mass spectroscopic analysis of human erythrocyte plasma membrane phospholipids. Proc. Natl. Acad. Sci. USA *91*, 10635–10639.

Han, X., Gubitosi-Klug, R. A., Collins, B. J. and Gross, R. W. (1996). Alterations in individual molecular species of human platelet phospholipids during thrombin stimulation: Electrospray ionization mass spectrometry-facilitated identification of the boundary conditions for the magnitude and selectivity of thrombin-induced platelet phospholipid hydrolysis. Biochemistry *35*, 5822–5832.

Hansbro, P. M., Byard, S. J., Bushby, R. J., Turnbull, P. J. H., Boden, N., Saunders, M. R., Novelli, R. and Reid, D. G. (1992). The conformational behaviour of phosphatidylinositol in model membranes: ²H-NMR studies. Biochim. Biophys. Acta *1112*, 187–196.

Hara, A. and Radin, N. S. (1978). Lipid extraction of tissue with a low-toxicity solvent. Anal. Biochem. *90*, 420–426.

Haroldsen, P. E. and Murphy, R. C. (1987). Analysis of phospholipid molecular species in rat lung as dinrobenzoate diglycerides by electron capture negative chemical ionization mass spectrometry. Biomed. Environ. Mass Spectrom. *14*, 573–578.

Hawthorne, J. N. (1960). The inositol phospholipids. J. Lipid Res. *1*, 255–280.

Hawthorne, J. N. (1982). Inositol phospholipids. In: Phospholipids, New Comparative Biochemistry (Hawthorne, J. N. and Ansell G. B., eds.), Vol. 4. Elsevier Biomedical, Amsterdam, pp. 263–278.

Heathers, G. P., Juehne, T., Rubin, L. J., Corr, P. B. and Evers, A. S. (1989). Anion exchange chromatographic separation of inositol phosphates and their quantification by gas chromatography. Anal. Biochem. *176*, 109–116.

Hendrickson, H. S., Giles, A. N. and Vos, S. E. (1997). Activity of phosphatidylinositol-specific phospholipase C from Bacillus cereus with thiophosphate analogs of dimyristoyl phosphatidylinositol. Chem. Phys. Lipids *89*, 45–53.

Holbrook, P. G., Pannell, L. K., Murata, Y. and Daly, J. W. (1992). Molecular species analysis of a product of phospholipase D activation.

Phosphatidylethanol is formed from phosphatidylcholine in phorbol ester- and bradikinin-stimulated PC12 cells. J. Biol. Chem. 267, 16834–16840.

Holub, B. J. and Kuksis, A. (1978). Metabolism of molecular species of diacylglycerophospholipids. Adv. Lipid Res. 16, 1–125.

Holub, B. J., Kuksis, A. and Thompson, W. (1970). Molecular species of mono-, di- and triphosphoinositides of bovine brain. J. Lipid Res. 11, 558–564.

Hondal, R. J., Riddle, S. R., Kravcuk, A. V., Zhao, Z., Liao, H., Bruzik, K. S. and Tsai, M. D. (1997). Phosphatidylinositol-specific phospholipase C: kinetic and stereochemical evidence for an interaction between arginine-69 and the phosphate group of phosphatidylinositol. Biochemistry 36, 6633–6642.

Hsu, F.-F. and Turk, J. (2000). Characterization of phosphatidylinositol, phosphatidylinositol-4-phosphate, and phosphatidylinositol-4,5-bisphosphate by electrospray ionization tandem mass spectrometry: a mechanistic study. J. Am. Soc. Mass Spectrom. 11, 986–999.

Hvattum, E., Rosjo, C., Gjon, T., Rosenlund, G. and Ruyter, B. (2000). Effect of soybean oil and fish oil on individual molecular species of Atlantic salmon head kidney phospholipids determined by normal-phase liquid chromatography coupled to negative ion electrospray tandem mass spectrometry. J. Chromatogr. B748, 137–149.

Irvine, R. F. (1990). In: Methods in Inositide Research (Irvine, R. F., ed.). Raven Press, pp. 1–219.

Itabashi, Y. and Kuksis, A. (2001). Search for reverse isomers of natural glycerophospholipids by reversed-phase high-performance liquid chromatography using sn-1,2-diradylglycerol 3,5-dinitrophenylurethane derivatives. In: Abstracts, 24th World Congress and Exhibition. International Society for Fat Research, Berlin, Germany, 16–20 September, p. 57.

Itabashi, Y., Kuksis, A., Marai, L. and Takagi, T. (1990). HPLC resolution of diacylglycerol moieties of natural triacylglycerols on a chiral phase consisting of bonded (R)-(+)-1-(1-naphthyl)ethylamine. J. Lipid Res. 31, 1711–1717.

Itabashi, Y., Myher, J. J. and Kuksis, A. (2000). High performance liquid chromatographic resolution of reverse isomers of 1,2-diacyl-rac-glycerols as 3, 5-dinitrophenylurethanes. J. Chromatogr. A893, 261–279.

Janero, D. R. and Burghardt, C. (1990). Solid-phase extraction on silica cartridges as an aid to platelet-activating factor enrichment and analysis. J. Chromatogr. 526, 11–24.

Jensen, N. J., Tomer, K. B. and Gross, M. L. (1987). FAB MS/MS for phosphatidylinositol,-glycerol,-ethanolamine and other complex phospholipids. Lipids 22, 480–489.

Jett, M. and Alwing, C. R. (1983). Selective cytotoxicity of tumor cells induced by liposomes containing plant phosphatidylinositol. Biochem. Biophys. Res. Commun. *114*, 863–869.

Jolles, J., Zwiers, H., Dekker, A., Wirtz, K. W. A. and Gispen, W. H. (1981). Corticotropin-(1-24)-tetracosapeptide affects protein phosphorylation ad polyphosphoinositide metabolism in rat brain. Biochem. J. *194*, 283–291.

Karlsson, A. A., Michelsen, P., Larsen, A. and Odham, G. (1996). Normal phase liquid chromatography class separation and species determination of phospholipids utilizing electrospray mass spectrometry/tandem mass spectrometry. Rapid Commun. Mass Spectrom. *10*, 775–780.

Kates, M. (1972). Techniques of Lipidology. Isolation, analysis and identification of lipids. In: Laboratory Techniques in Biochemistry and Molecular Biology (Work, T. S. and Work E., eds.), Vol. 3. North-Holland/American Elsevier, Amsterdam, pp. 150.

Khaselev, N. and Murphy, R. C. (1999). Susceptibility of plasmenyl glycerophosphoethanolamine lipids containing arachidonate to oxidative degradation. Free Radical Biol. Med. *26*, 275–284.

Kim, H. Y. and Salem, N., Jr. (1987). Application of thermospray high performance liquid chromatography/mass spectrometry for the determination of phospholipids and related compounds. Anal. Chem. *59*, 722–726.

Kinnard, R. L., Narasimhan, B., Pliska-Matyshak, G. and Murthy, P. P. N. (1995). Characterization of *scyllo*-inositol-containing phosphatidylinositol in plant cells. Biochem. Biophys. Res. Commun. *210*, 549–555.

Koivusalo, M., Haimi, P., Heikinheimo, L., Kostiainen, R. and Somerharju, P. (2001). Quantitative determination of phospholipid compositions by ESI-MS: effects of acyl chain length, unsaturation, and lipid concentration on instrument response. J. Lipid Res. *42*, 663–672.

Kolarovic, L. and Fournier, N. C. (1986). A comparison of extraction methods for the isolation of phospholipids from biological sources. Anal. Biochem. *156*, 244–250.

Kuksis, A., Marai, L., Myher, J. J. and Itabashi, Y. (1989). Molecular speciation of natural glycerolipids and glycerophospholipids by liquid chromatography with mass spectrometry. In: Proceedings 15th Scandinavian symposium on lipids (Shukla, V. K. S. and Holmer G., eds.). Lipidforum, Rebild Bakker, Denmark, pp. 336–370.

Kuksis, A., Marai, L., Myher, J. J., Itabashi, Y. and Pind, S. (1991). Applications of GC/MS, LC/MS and FAB/MS to determination of molecular species of glycerolipids. In: AOCS monograph on lipid and lipoprotein analysis (Perkins, E., ed.). AOCS Press, Champaign.

Kuksis, A. and Myher, J. J. (1990). Mass analysis of molecular species of diradylglycerols. In: Methods in Inositide Research (Irvine, R. F., ed.). Raven Press, New York, pp. 187–216.

Kurvinen, J.-P., Aaltonen, J. J., Kuksis, A., and Kallio, H. (2002). Software algorithm for automatic interpretation of mass spectra: application to spectra of glycerophospholipids. Rapid Commun. Mass Spectrom (Submitted).

Kurvinen, J.-P., Kuksis, A., Sinclair, A. J., Abedin, L. and Kallio, H. (2000). The effect of low α-linolenic acid diet on glycerophospholipid molecular species in guinea pig brain. Lipids 35, 1001–1009.

Kuypers, F. A., Butikofer, P. and Shackleton, C. H. L. (1991). Application of liquid-chromatography-thermospray mass spectrometry in the analysis of glycerophospholipid molecular species. J. Chromatogr. Biomed. Appl. 562, 191–206.

Laakso, P. and Christie, W. W. (1990). Chromatographic resolution of chiral diacylglycerol derivatives: Potential in the stereospecific analysis of triacyl-sn-glycerols. Lipids 25, 349–353.

Lanzetta, P. A., Alvarez, L. J., Reinach, P. S. and Candia, O. A. (1979). An improved assay for nanomole amounts of inorganic phosphate. Anal. Biochem. 100, 95–97.

Larner, J., Huang, L. C., Schwartz, C. F. W., Oswald, A. S., Shen, T.-Y., Kinter, M., Tang, G. and Zeller, K. (1988). Rat liver insulin mediator which stimulates pyruvate dehydrogenase phosphatase contains galactosamine and D-chiro-inositol. Biochem. Biophys. Res. Commun. 151, 1416–1426.

Leavitt, A. L. and Sherman, W. R. (1982). Direct gas-chromatographic resolution of DL-myo-inositol 1-phosphate and other sugar enantiomers as simple derivatives on a chiral capillary column. Carbohydr. Res. 103, 203–212.

Lee, T.-C., Malone, B., Buell, A. B. and Blank, M. L. (1991). Occurrence of ether-containing inositol phospholipids in bovine erythrocytes. Biochem. Biophys. Res. Commun. 175, 667–673.

Leondaritis, G. and Galanopoulou, D. (2000). Characterization of inositol phospholipids and identification of a mastoparan-induced polyphosphoinositide response in Tetrahymena pyriformis. Lipids 35, 525–532.

Leray, C. and Pelletier, X. (1987). Thin-layer chromatography of human platelet phospholipids with fatty acid analysis. J. Chromatogr. 420, 411–416.

Lewis, K. A., Garigapati, V. R., Zhou, C. and Roberts, M. F. (1993). Substrate requirements of bacterial phosphatidylinositol-specific phospholipase C. Biochemistry 32, 8836–8841.

Li, L. and Fleming, N. (1999). Aluminum fluoride inhibition of cabbage phospholipase D by a phosphate-mimicking mechanism. FEBS Lett 461, 1–5.

Li, C., McClory, A., Wong, E. and Yergey, J. A. (1999). Mass spectrometric analysis of arachidonyl-containing phospholipids in human U937 cells. J. Mass Spectrom. 34, 521–536.

Luthra, M. G. and Sheltawy, A. (1972a). The fractionation of phosphatidylinositol into molecular species by thin-layer chromatography on silver nitrate-impregnated silica gel. Biochem. J. *126*, 1231–1239.

Luthra, M. G. and Sheltawy, A. (1972b). The distribution of molecular species of phosphatidylinositol in ox brain and its subcellular fractions. Biochem. J. *128*, 587–595.

Luthra, M. G. and Sheltawy, A. (1976). The metabolic turnover of molecular species of phosphatidylinositol and its precursor phosphatidic acid in guinea-pig cerebral hemispheres. J. Neurochem. *27*, 1503–1511.

MacGregor, L. C. and Matschinsky, F. M. (1984). An enzymatic fluorimetric assay for *myo*-inositol. Anal. Biochem. *141*, 382–389.

Mahadevappa, V. G. and Holub, B. J. (1987). Chromatographic analysis of phosphoinositides and their breakdown products in activated blood platelets/neutrophils. Chromatography of lipids in biomedical research and clinical diagnosis J. Chromatogr. Libr., (Kuksis, A., eds.), Vol. 37. Elsevier, Amsterdam, pp. 225–265.

Maslanski, J. A. and Busa, W. B. (1990). A sensitive and specific mass assay for *myo*-inositol and inositol phosphates. In: Methods in Inositide Research (Irvine, R. F., ed.). Raven Press, New York, pp. 113–126.

Mitchell, K. T., Ferrell, Jr. and Huestis, W. H. (1986). Separation of phosphoinositides and other phospholipids by two-dimensional thin-layer chromatography. Anal. Biochem. *158*, 47–453.

Murphy, R. C. and Harrison, K. A. (1994). Fast atom bombardment mass spectrometry of phospholipids. Mass Spectrom. Rev. *13*, 57–75.

Myher, J. J. and Kuksis, A. (1975). Improved resolution of natural diacylglycerols by gas–liquid chromatography on polar siloxanes. J. Chromatogr. Sci. *13*, 138–145.

Myher, J. J. and Kuksis, A. (1982). Resolution of diacylglycerol moieties of natural glycerophospholipids by gas chromatography on polar capillary columns. Can. J. Biochem. *60*, 638–650.

Myher, J. J. and Kuksis, A. (1984a). Molecular species of plant phosphatidyl-inositol with selective cytotoxicity towards tumor cells. Biochem. Biophys. Acta *795*, 85–90.

Myher, J. J. and Kuksis, A. (1984b). Resolution of diacylglycerol moieties of natural glycerophospholipids by gas–liquid chromatography on polar capillary columns. Can. J. Biochem. Cell Biol. *62*, 352–362.

Myher, J. J. and Kuksis, A. (1989). Relative gas–liquid retention factors of trimethylsilyl ethers of diradylglycerols on polar capillary columns. J. Chromatogr. *471*, 187–204.

Myher, J. J., Kuksis, A., Marai, L. and Yeung, S. K. F. (1978). Microdetermination of molecular species of oligo- and polyunsaturated diacylglycerols by gas chromatography-mass spectrometry of their tert-butyldimethylsilyl ethers. Anal. Chem. *50*, 557–561.

Myher, J. J., Kuksis, A. and Pind, S. (1989a). Molecular species of glycerophospholipids and sphingomyelins of human erythrocytes: Improved method of analysis. Lipids *24*, 396–407.

Myher, J. J., Kuksis, A. and Pind, S. (1989b). Molecular species of glycerophospholipids and sphingomyelins of human plasma: comparison to red blood cells. Lipids *24*, 408–418.

Nair, S. S. D., Leitch, J. and Garg, M. L. (1999). Specific modifications of phosphatidylinositol and non-esterified fatty acid fractions in cultured porcine cardiomyocytes supplemented with n-3 polyunsaturated fatty acids. Lipids *34*, 697–704.

Nakagawa, Y., Fujishima, K. and Waku, K. (1986a). Separation of dimethylphosphatidates of alkylglycerophosphocholine and their molecular species by high performance liquid chromatography. Anal. Biochem. *157*, 172–178.

Nakagawa, Y. and Waku, K. (1986). Improved procedure for the separation of the molecular species of dimethylphosphatidate by high performance liquid chromatography. J. Chromatogr. *381*, 225–231.

Nakagawa, Y., Setaka, M. and Nojima, S. (1991). Detergent resistant phospholipase A₁ from *Escherichia coli* membranes. Methods Enzymol. *197*, 309–316.

Narasimhan, B., Plisa-Matyshak, G., Kinnard, R., Carstense, S., Ritter, M. A., von Weyman, L. and Murthy, P. P. N. (1997). Novel phosphoinositides in barley aleurone cells. Plant Physiol. *113*, 1385–1393.

Nasuhoglu, C., Feng, S., Mao, J., Yamamoto, M., Yin, H. L., Earnest, S., Barylko, B., Albanesi, J. P. and Hilgemann, D. W. (2002). Non-radioactive analysis of phosphoinositides and other anionic phospholipids by anion-exchange high-performance liquid chromatography with suppressed conductivity detection. Anal. Biochem. *301*, 243–254.

Norris, F. W. and Darbre, A. (1956). The microbial assay of inositol with a strain of Schizosaccharomyces pombe. Analyst *81*, 394–400.

Ostlund, R. E., Jr., McGill, J. B., Herskowitz, I., Kipnis, D., Santiago, J. V. and Sherman, W. R. (1993). D-*chiro*-Inositol metabolism in diabetes mellitus. Proc. Natl. Acad. Sci. USA *90*, 9988–9992.

Pak, Y. and Larner, J. (1992). Identification and characterization of *chiro*-inositol-containing phospholipids from bovine liver. Biochem. Biophys. Res. Commun. *184*, 1042–1047.

Patton, G. M., Fasulo, J. M. and Robins, S. J. (1982). Separation of phospholipids and individual molecular species of phospholipids by high performance liquid chromatography. J. Lipid Res. *23*, 190–196.

Petkovic, M., Schiller, J., Muller, M., Bnad, S., Reichl, S., Arnold, K. and Arnhold, J. (2000). Detection of individual phospholipids in lipid mixtures by matrix assisted laser desorption/ionization time-of-flight mass spectrometer: phosphatidylcholine prevents the detection of further species. Anal. Biochem. *268*, 202–215.

Pind, S., Kuksis, A., Myher, J. J. and Marai, L. (1984). Resolution and quantification of diacylglycerol moieties of neutral glycophospholipids by reversed-phase liquid chromatography with direct liquid inlet mass spectrometry. Can. J. Biochem. Cell Biol. *62*, 301–309.

Rabe, H., Reichmann, G., Nakagawa, Y., et al. (1989). Separation of alkylacyl and diacyl glycerophospholipids and their molecular species as naphthylurethanes by high performance liquid chromatography. J. Chromatogr. Biomed. Appl. *493*, 353–360.

Ravandi, A., Kuksis, A., Marai, L. and Myher, J. J. (1995). Preparation and characterization of glucosylated aminoglycerophospholipids. Lipids *30*, 885–891.

Rouser, G., Fleischer, S. and Yamamoto, A. (1970). Two dimensional thin layer chromatographic separation of polar lipids and determination of phospholipids by phosphorus analysis of spots. Lipids *5*, 494–496.

Rouser, G., Kritchevsky, G., Yamamoto, A., Simon, G., Galli, C. and Bauman, A. J. (1969). Diethylaminoethyl and triethylaminoethyl cellulose column chromatographic procedures for phospholipids, glycolipids and pigments. Methods Enzymol. *14*, 272–317.

Rubin, L. J., Hsu, F.-F. and Sherman, W. R. (1993). Measurement of inositol trisphosphate by gas chromatography/mass spectrometry: femtomole sensitivity provided by negative-ion chemical ionization mass spectrometry in submilligram quantities of tissue. In: Lipid Metabolism in Signaling Systems, (Fain, J. N., eds.), Vol. 18. Methods in Neurosciences, Academic Press, pp. 201–212.

Schacht, J. (1981). Extraction and purification of polyphosphoinositides. Methods Enzymol. *72*, 626–631.

Schiller, J., Arnhold, J., Benard, S., Muller, M., Reichl, S. and Arnold, K. (1999). Lipid analysis by matrix-assisted laser desorption and ionization mass spectrometer: A methodological approach. Anal. Biochem. *267*, 46–57.

Shaikh, N. A. (1984). Assessment of various techniques for the quantitative extraction of lysophospholipids from myocardial tissues. Anal. Biochem. *216*, 313–321.

Shayman, J. A., Morrison, A. R. and Lowry, O. H. (1987). Enzymatic fluorometric assay for *myo*-inositol trisphosphate. Anal. Biochem. *162*, 562–568.

Sherman, W. R., Ackermann, K. E., Bateman, R. H., Green, B. N. and Lewis, I. (1985). Mass-analyzed ion kinetic energy spectra and B1E-B2 triple sector mass spectrometric analysis of phosphoinositides by fast atom bombardment. Biomed. Mass Spectrom. *12*, 409–413.

Sherman, W. R., Packman, P. M., Laid, M. H. and Boshans, R. L. (1977). Measurement of myo-inositol in single cells and defined areas of the nervous system by selected ion monitoring. Anal. Biochem. 78, 119–131.

Shibata, T., Uzawa, J., Sudiura, Y., Hayashi, K. and Takizawa, T. (1984). Chem. Phys. Lipids 34, 107–113.

Singer, W. D., Brown, H. A. and Sternweis, P. C. (1997). Regulation of eukaryotic phosphatidylinositol-specific phospholipase C and phospholipase D. Annu. Rev. Biochem. 66, 475–509.

Singh, A. K. (1992). Quantitative analysis of inositol lipids and inositol phosphates in synaptosomes and microvessels by column chromatography: comparison of the mass analysis and the radiolabelling methods. J. Chromatogr. 581, 1–10.

Singh, A. K. and Jiang, Y. (1995). Quantitative chromatographic analysis of inositol phospholipids and related compounds. J. Chromatogr. B 671, 255–280.

Skipski, V. P., Peterson, R. F. and Barclay, M. (1964). Quantitative analysis of phospholipids by thin-layer chromatography. Biochem. J. 90, 374–378.

Slotboom, A. J., De Haas, G. H., Bonsen, P. P. M., Burbach-Westerhuis, G. J. and Van Deenen, L. L. M. (1970). Hydrolysis of phosphoglycerides by purified lipase preparations. I. Substrate-, positional- and stereospecificity. Chem. Phys. Lipids 4, 15–29.

Stull, J. T. and Buss, J. E. (1977). Phosphorylation of cardiac troponin by cyclic adenosine 3′5′-monophosphate-dependent protein kinase. J. Biol. Chem. 252, 851–857.

Sun, G. Y., Huang, H. M., Kelleher, J. A., Stubbs, E. B. and Sun, A. Y. (1988). Marker enzymes, phospholipids and acyl group composition of a somal plasma membrane fraction isolated from rat cerebral cortex: a comparison with microsomes and synaptic plasma membranes. Neurochem. Int. 12, 69–77.

Sun, G. Y. and Lin, T.-N. (1990). Separation of phosphoinositides and other phospholipids by high-performance thin-layer chromatography. In: Methods in Inositide Research (Irvine, R. F., ed.). Raven Press, New York, pp. 153–158.

Takamura, H., Narita, H., Urade, R. and Kito, M. (1986). Quantitative analysis of polyenoic phospholipid molecular species by high performance liquid chromatography. Lipids 21, 356–361.

Takamura, H., Tanaka, K., Matsuura, T. and Kito, M. (1989). Ether phospholipid molecular species in human platelets. J. Biochem. 105, 168–172.

Tomlinson, R. V. and Ballou, C. E. (1962). Myo-inositol polyphosphate intermediates in the dephosphorylation of phytic acid by phytase. Biochemistry 1, 166–171.

Touchstone, J. C., Chen, J. C. and Beaver, K. M. (1980). Improved separation of phospholipids in thin-layer chromatography. Lipids 15, 61–62.

Uran, S., Larsen, A., Jacobsen, P. B. and Skotland, T. (2001). Analysis of phospholipid species in human blood using normal-phase liquid chromatography coupled with electrospray ionization ion-trap tandem mass spectrometry. J. Chromatogr. *B758*, 265–275.

Vallejo, M., Jackson, T., Lightman, S. and Hanley, M. R. (1987). Occurrence and extracellular actions of inositol pentakis- and hexakisphosphate in mammalian brain. Nature *330*, 656–658.

Volwerk, J. J., Shashidhar, M. S., Kuppe, A. and Griffith, O. (1990). Phosphatidylinositol-specific phospholipase C from Bacillus cereus combines intrinsic phospho ferase and cyclic phosphodiesterase activities: a ^{31}P NMR study. Biochemistry *29*, 8056–8062.

Wiley, M. G., Przetakiewicz, M., Takahashi, M. and Lowenstein, J. M. (1992). An extended method for separating and quantitating molecular species of phospholipids. Lipids *27*, 295–301.

Wilson, R. and Sargent, J. R. (1993). Lipid and fatty acid composition of brain tissue from adrenoleucodystrophy patients. J. Neurochem. *61*, 290–297.

Wu, L. N. Y., Genge, B. R., Kang, M. W., Arsenault, A. L. and Wuthier, R. E. (2002). Changes in phospholipid extractability and composition accompany mineralization of chicken growth plate cartilage matrix vesicles. J. Biol. Chem. *277*, 5126–5133.

Wuthier, R. E. (1966). Purification of lipids from non-lipid contaminants on Sephadex bead columns. J. Lipid Res. *7*, 558–561.

Wuthier, R. E. (1968). Lipids of mineralizing epiphyseal tissues in the bovine fetus. J. Lipid Res. *9*, 68–78.

Zhang, W., Asztalos, B., Roheim, P. S. and Wong, L. (1998). Characterization of phospholipids in pre-α HDL: selective phospholipid efflux with apolipoprotein A-I. J. Lipid Res. *39*, 1601–1607.

Zhou, C., Wu, Y. and Roberts, M. F. (1997a). Activation of phosphatidylinositol-specific phospholipase C toward inositol 1,2-(cyclic)-phosphate. Biochemistry *36*, 347–355.

Zhou, C., Garigapati, V. and Roberts, M. F. (1997b). Short-chain phosphatidylinositol conformation and its relevance to phosphatidylinositol-specific phospholipase C. Biochemistry *36*, 15925–15931.

Zhu, X. and Eichberg, J. (1993). Molecular species composition of glycerophospholipids in rat sciatic nerve and its alteration in streptozotocin-induced diabetes. Biochem. Biophys. Acta *1168*, 1–12

Chapter 3

Abdel-Latif, A. A. (1986). Calcium-mobilizing receptors, polyphosphoinositides, and the generation of second messengers. Pharmacol. Rev. *38*, 227–272.

Abdel-Latif, A. A., Aktar, R. A. and Hawthorne, J. N. (1977). Acetylcholine increases the breakdown of triphosphoinositide of rabbit iris muscle prelabeled with [^{32}P]phosphate. Biochem. J. *162*, 61–73.

Akhtar, R. A., Taft, C. and Abdel-Latif, A. A. (1983). Effects of ACTH on polyphosphoinositide metabolism and protein phosphorylation in rabbit iris subcellular fractions. J. Neurochem. *41*, 1460–1468.

Ansell, G. B. and Hawthorne, J. N. (1964). Chemical Structures. In: Phospholipids: Chemistry, Metabolism and Function (Ansell, G. B. and Hawthorne J. N., eds.), Vol. 3. BBA Library, Elsevier Publishing Co, pp. 11–39.

Auger, K. R., Serunian, L. A. and Cartleg, L. C. (1990). Separation of novel phosphoinositides. In: Methods in Inositide Research (Irvine, R. F., ed.). Raven Press, New York, pp. 159–166.

Auger, K. R., Serunian, L. A., Soltoff, S. P., Libby, P. and Cantley, L. C. (1988). PDFG-dependent tyrosine phosphorylation stimulates production of novel polyphosphoinositides in intact cells. Cell *57*, 167–175.

Baker, R. R. and Thompson, W. (1972). Positional distribution and turnover of fatty acids in phosphatidic acid, phosphoinositides, phosphatidylcholine and phosphatidylethanolamine in rat brain in vivo. Biochim. Biophys. Acta *270*, 489–503.

Baker, R. R. and Thompson, W. (1973). Selective acylation of 1-acylglycerophosphoinositol by rat brain microsomes. J. Biol. Chem. *248*, 7060–7065.

Ballou, C. E. and Lee, Y. C. (1966). Cyclitols and Phosphoinositides (Kindl, H., ed.) Vol. 2. Pergamon, Oxford, pp. 41.

Ballou, C. E. and Pizer, L. I. (1960). The absolute configuration of the *myo*-inositol-1-phosphates and a confirmation of the bornesitol configuration. J. Am. Chem. Soc. *82*, 3333–3335.

Berridge, M. (1993). Inositol trisphosphates and calcium signalling. Nature *361*, 121–121.

Berrie, C. P., Dragani, L. K., van der Kaay, J., Iurisci, C., Brancaccio, A., Rotilio, D. and Corda, D. (2002). Maintenance of PtdIns(4,5)P$_2$ pools under limiting inositol conditions, as assessed by liquid chromatography-tandem mass spectrometry and PtdIns(4,5)P$_2$ mass evaluation in Ras-transformed cells. Eur. J. Cancer *38*, 2463–2475.

Berry, N. and Nishizuka, Y. (1990). Protein kinase C and T cell activation. Eur. J. Biochem. *189*, 205–214.

Bottomley, M. J., Salim, K. and Panayotou, G. (1998). Phospholipid-binding protein domains. Biochim. Biophys. Acta *1436*, 165–183.

Brockerhoff, H. and Ballou, C. E. (1961). The structure of the phosphoinositide complex of the beef brain. J. Biol. Chem. *236*, 1907–1911.

Brown, D. M. and Stewart, J. C. (1966). The structure of triphosphoinositide from rat brain. Biochim. Biophys. Acta *125*, 413–421.

Butikofer, P., Kuypers, F. A., Shackleton, C., Brodbeck, U. and Stieger, S. (1990). Molecular species analysis of the glycosylphosphatidylinositol anchor of *Torpedo marmorata* acetylcholinesterase. J. Biol. Chem. *265*, 18983–18987.

Butikofer, P., Zollinger, M. and Brodbeck, U. (1992). Alkylacylglycerophosphoinositol in human and bovine erythrocytes. Molecular species composition and comparison with glycosyl-inositolphospholipid anchors of erythrocyte acetylcholinesterases. Eur. J. Biochem. *208*, 677–683.

Carter, A. N., Huang, R., Sorisky, A., Downes, C. P. and Rittenhouse, S. E. (1994). Phosphatidylinositol 3,4,5-trisphosphate is formed from phosphatidylinositol 4,5-bisphosphate in thrombin-stimulated platelets. Biochem. J. *301*, 415–420.

Challis, R. A. J., Batty, I. H. and Nahorski, S. R. (1988). Mass measurements of inositol(1,4,5)trisphosphate in rat cerebral cortex slices using a radioreceptor assay: effects of neurotransmitters and depolarization. Biochem. Biophys. Res. Commun. *157*, 684–691.

Chilvers, E. R., Batty, I. H., Challiss, R. A. J., Barnes, P. J. and Nahorski, S. R. (1991). Determination of mass changes in phosphatidylinositol 4,5-bisphosphate and evidence for agonist-stimulated metabolism of inositol 1,4,5-trisphosphate in airway smooth muscle. Biochem. J. *275*, 373–379.

Christensen, S. (1986). Removal of haem from lipids extracted from intact erythrocytes with particular reference to polyphosphoinositides. Biochem. J. *233*, 921–924.

Clark, N. G. and Dawson, R. M. C. (1981). Alkaline O–N transacylation new method for the quantitative deacylation of phospholipids. Biochem. J. *195*, 301–306.

Cronholm, T., Viestam-Rains, M. and Sjovall, J. (1992). Decreased content of arachidonoyl species of phosphatidylinositol phosphates in pancreas of rats fed on an ethanol-containing diet. Biochem. J. *287*, 925–928.

Dawson, R. M. C. and Dittmer, J. C. (1961). Evidence for the structure of brain triphosphoinositide from hydrolytic degradation studies. Biochem. J. *81*, 540–545.

Divecha, N., Banfic, H. and Irvine, R. F. (1991). The polyphosphoinositide cycle exists in the nuclei of Swiss 3T3 cells under the control of a receptor (for IGF-I) in the plasma membrane, and stimulation of the cycle increases nuclear diacylglycerol and apparently induces translocation of protein kinase C to the nucleus. EMBO J. *10*, 3207–3214.

Divecha, N. and Irvine, R. F. (1990). Mass measurement of phosphatidylinositol 4-phosphate and sn-1,2-diacylglycerols. In: Methods in Inositide Research (Irvine, R. F., ed.). Raven Press, New York, pp. 179–185.

Donie, F. and Reiser, G. A. (1989). A novel, specific binding protein assay for quantification of intracellular inositol 1,3,4.,5-tetrakisphosphate (InsP$_4$) using a high affinity InsP$_4$ receptor from cerebellum. FEBS Lett. 254, 155–158.

Dugan, L. L., Demediuk, P., Pendley, C. E. and Horrocks, L. A. (1986). Separation of phospholipids by high performance liquid chromatography: all major classes, including ethanolamine and choline plasmalogens, and most minor classes, including lysophosphatidylethanolamine. J. Chromatogr. 378, 317–317.

Ferguson, K. M., Kavran, J. M., Sankaran, V. G., Fournier, E., Isakoff, S. J., Skolnik, E. Y. and Lemmon, M. A. (2000). Structural basis for discrimination of 3-phosphoinositides by pleckstrin homology domains. Mol. Cell 6, 373–384.

Ferrell, J. E. and Huestis, W. H. (1984). Phosphoinositide metabolism and the morphology of human erythrocytes. J. Cell Biol. 98, 1992–1998.

Fewster, M. E., Burns, B. J. and Mead, J. F. (1969). Quantitative densitometric thin-layer chromatography of lipids using copper acetate reagent. J. Chromatogr. 43, 120–126.

Fine, J. B. and Sprecher, H. (1982). Unidimensional thin-layer chromatography of phospholipids on boric acid-impregnated plates. J. Lipid Res. 23, 660–663.

Folch, J. (1949). Complete fractionation of brain cephalin: isolation from it of phosphatidylserine, phosphatidylethanolamine and diphosphoinositide. J. Biol. Chem. 177, 497–505.

Folch, J., Lee, M. and Sloane-Stanley, S. M. (1957). A simple method for the isolation and purification of total lipids from animal tissues. J. Biol. Chem. 226, 497–509.

Gabev, E., Kasianowicz, J., Abbott, T. and McLaughlin, S. (1989). Binding of neomycin to phosphatidylinositol 4,5-bisphosphate (PIP$_2$). Biochim. Biophys. Acta 979, 105–112.

Gatelli, R., Stanfil, R. E., Kabra, P. M., Farina, F. A. and Martin, L. J. (1978). Simultaneous determination of phosphatidylglycerol and the lecithin/sphingomyelin ratio in amniotic fluid. Clin. Chem. 24, 1144–1146.

Gaudette, D. C., Aukema, H. M., Jolly, C. A., Chapkin, R. S. and Holub, B. J. (1993). Mass and fatty acid composition of the 3-phosphorylated phosphatidylinositol bisphosphate isomer in stimulated human platelets. J. Biol. Chem. 268, 13773–13776.

Geurts Van Kessel, W. S. M., Hax, W. M. A., Demel, R. A. and de Gier, J. (1977). High performance liquid chromatographic separation and direct ultraviolet detection of phospholipids. Biochim. Biophys. Acta 486, 524–530.

Gonzalez-Sastre, F. and Folch-Pi, J. (1968). Thin-layer chromatography of the phosphoinositides. J. Lipid Res. *9*, 532–533.

Grado, C. and Ballou, C. E. (1961). Myo-inositol phosphates obtained by alkaline hydrolysis of beef brain phosphoinositide. J. Biol. Chem. *236*, 54–60.

Grado, C. and Ballou, C. E. (1961). *Myo*-inositol phosphates obtained by alkaline hydrolysis of beef brain phospholipids. J. Biol. Chem. *236*, 54–60.

Gray, A., Van der Kaay, J. and Downes, C. P. (1999). The pleckstrin homology domains of protein kinase B and GRP1 (general receptor for phosphoinositides-1) are sensitive and selective probes for the cellular detection of phosphatidylinositol 3,4-bisphosphate and/or phosphatidylinositol 3,4,5-trisphosphate in vivo. Biochem. J. *344*, 929–936.

Gunnarsson, T., Ekblad, L., Karlsson, A., Michelsen, P., Odham, G. and Jergil, B. (1997). Separation of polyphosphoinositides using normal-phase high performance liquid chromatography and evaporative light scattering detection or electrospray mass spectrometry. Anal. Biochem. *254*, 293–296.

Harvey, D. J. (1995). Matrix-assisted laser desorption/ionization mass spectrometry of phospholipids. J. Mass Spectrom. *30*, 1333–1346.

Hawkins, P. T., Mitchell, R. H. and Kirk, C. J. (1984). Analysis of the metabolic turnover of the individual phosphate groups of phosphatidylinositol-4-phosphate and phosphatidyl-4,5-bisphosphate. Biochem. J. *218*, 785–793.

Hawthorne, J. N. and Kemp, P. (1964). The brain phosphoinositides. Adv. Lipid Res. *2*, 127–166.

Hegewald, H. (1996). One dimensional thin-layer chromatography of all known D-3 and D-4 isomers of phosphoinositides. Anal. Biochem. *242*, 152–155.

Hegewald, H., Muller, E., Klinger, R., Wetzker, R. and Frunder, H. (1987). Influence of Ca^{2+} and Mg^{2+} on the turnover of the phosphomonoester group of phosphatidylinositol 4-phosphate in human erythrocyte membranes. Biochem. J. *244*, 183–190.

Hendrickson, H. S. and Ballou, C. E. (1964). Ion exchange chromatography of intact brain phosphoinositides on diethylaminoethyl cellulose by gradient salt elution in a mixed solvent system. J. Biol. Chem. *239*, 1369–1373.

Hirvonen, M. R., Lihtamo, H. and Savolainen, K. (1988). A gas chromatographic method for the determination of inositol monophosphates in rat brain. Neurochem. Res. *13*, 957–962.

Holub, B. J. and Kuksis, A. (1978). Metabolism of molecular species of diacylglycerophospholipids. Adv. Lipid Res. *16*, 1–100.

Holub, B. J. and Kuksis, A. (1978). Metabolism of molecular species of diacylglycerophospholipids. In: Advances in Lipid Research (Paoletti, R. and Kritchevsky D., eds.) Vol. 16. Academic Press, New York, pp. 1–125.

Holub, B. J., Kuksis, A. and Thompson, W. (1970). Molecular species of mono-, di-, and triphosphoinositides of bovine brain. J. Lipid Res. *11*, 558–570.

Holub, B. J. (1986). Metabolism and function of *myo*-inositol and inositol phospholipids. Ann. Rev. Nutr. *6*, 563–597.

Hsu, F. F. and Turk, J. (2000). Characterization of phosphatidylinositol, phosphatidylinositol-4-phosphate, and phosphatidylinositol-4,5-bisphosphate by electrospray ionization tandem mass spectrometry: a mechanistic study. J. Am. Soc. Mass Spectrom. *11*, 136–144.

Ikezawa, H., Yamanegi, M., Taguchi, R., Miyashita, T. and Ohyabu, T. (1976). Studies on phosphatidylinositol phosphodiesterase (phospholipase C type) of *Bacillus cereus*. I. Purification, properties and phosphatase-releasing activity. Biochim. Biophys. Acta *450*, 154–164.

Irvine, R. F. (1986). The structure, metabolism and analysis of inositol lipids and inositol phosphates. In: Phosphoinositides and Receptor Mechanisms (Putney, J. W., Jr., ed.). Alan R. Ross, New York, pp. 89–107.

Irvine, R. F. (1990). Introduction, and survey of other methods. In: Methods in Inositide Research (Irvine, R. F., ed.). Raven Press, New York, pp. 1–7.

Itabashi, Y., Myher, J. J. and Kuksis, A. (2000). High performance liquid chromatographic resolution of reverse isomers of 1,2-diacyl-rac-glycerols as 3, 5-dinitrophenylurethanes. J. Chromatogr. A *893*, 261–279.

Jackson, T. R., Stevens, L. R. and Hawkins, P. T. (1992). Receptor specificity of growth factor-stimulated synthesis of 3-phosphorylated inositol lipids in Swiss 3T3 cells. J. Biol. Chem. *267*, 16627–16636.

Jensen, N. J., Tomer, K. B. and Gross, M. L. (1987). FAB MS/MS for phosphatidyl-inositol, -glycerol, -ethanolamine and other complex phospholipids. Lipids *22*, 480–489.

Jolles, J., Zwiers, H., Dekker, A., Wirtz, K. W. A. and Gispen, W. H. (1981). Corticotropin-(−24)-tetracosapeptide affects protein phosphorylation and polyphosphoinositide metabolism in rat brain. Biochem. J. *194*, 283–291.

Jones, D. R., Gonzalez Garcia, A., Diez, E., Martinez-A, C., Carrera, A. C. and Merida, I. (1999). The identification of phosphatidylinositol 3,5-bisphosphate in T-lymphocytes and its regulation by interleukin-2. J. Biol. Chem. *274*, 18407–18413.

Katan, M. B. and Parker, P. J. (1987). Purification of phosphoinositide-specific phospholipase C from a particulate fraction of bovine brain. Eur. J. Biochem. *168*, 413–418.

Kayganich-Harrison, K. A. and Murphy, R. C. (1994). Fast-atom bombardment tandem mass spectrometry of [^{13}C]arachidonic acid labeled phospholipid molecular species. J. Am. Soc. Mass Spectrom. *5*, 144–150.

King, C. E., Stephens, L. R., Hawkins, P. T., Guy, G. R. and Mitchell, R. M. (1987). Multiple metabolic pools of phosphoinositides and phosphatidate in human erythrocytes incubated in a medium that permits rapid transmembrane exchange of phosphate. Biochem. J. *244*, 209–217.

Klyashchitskii, B. A., Sokolov, S. D. and Shvets, V. I. (1969). Russ. Chem. Rev (Engl. Transl.) *38*, 345–345.

Kuksis, A. and Myher, J. J. (1990). Mass analysis of molecular species of diradylglycerols. In: Methods in Inositide Research (Irvine, R. F., ed.). Raven Press, New York, pp. 187–216.

Kuksis, A. and Myher, J. J. (1995). Application of tandem MS for the analysis of long chain carboxylic acids/title. J. Chromatogr. B *671*, 35–70.

Lee, T.-C., Malone, B., Buell, A. B. and Blank, M. L. (1991). Occurrence of ether-containing inositol phospholipids in bovine erythrocytes. Biochem. Biophys. Res. Commun. *175*, 73–678.

Letcher, A. J., Stephens, L. R. and Irvine, R. F. (1990). Preparation of ^{32}P-labeled inositol 1,4,5-trisphosphate and ^{14}C-labeled inositol 1,4-bisphosphate. In: Methods in Inositide Research (Irvine, R. F., ed.). Raven Press, New York, pp. 31–37.

Low, M. G. (1990). Purification of phosphatidylinositol 4-phosphate and phosphatidylinositol 4,5-bisphosphate by column chromatograpy. In: Methods in Inositide Research (Irvine, R. F., ed.). Raven Press, New York, pp. 145–151.

Low, M. G. and Fincan, J. B. (1977). Modification of erythrocyte membranes by a purified phosphatidylinositol specific phospholipase C (*Staphylococcus aureus*). Biochem. J. *162*, 235–240.

Maroun, C. R., Moscatello, D. K., Naujokas, M. A., Holdago Madruga, M. Wong, A. J. and Park, M. (1999). A conserved inositol phospholipid binding site within the pleckstrin homology domain of the Gab1 docking protein is required for epithelial morphogenesis. J. Biol. Chem. *274*, 31719–31726.

Marto, J. A., White, F. M., Seldomridge, S. and Marshall, A. (1995). Structural characterization of phospholipids by matrix-assisted laser desorption Fourier transform ion cyclotron resonance mass spectrometry. Anal. Chem. *67*, 3979–3984.

Matuoka, K., Fukam, N., Nakamshi, O., Kawai, S. and Takenawa, T. (1988). Mitogenesis in response to PDGF and bombesin abolished by microinjection of an antibody to PtdIns(4,5)P$_2$. Science *239*, 640–643.

Mayr, G. W. (1990). Mass determination of inositol phosphates by high-performance liquid chromatography with postcolumn complexometry (Metal-dye detection). In: Methods in Inositide Research (Irvine, R. F., ed.). Raven Press, New York, pp. 83–108.

McCorkindale, J. and Eson, N. L. (1954). Polyol dehydrogenase. 1. The specificity of rat liver polyol dehydrogenase. Biochem. J. *57*, 518–523.

Michelsen, P., Jergil, B. and Odham, G. (1995). Quantification of polyphosphoinositides using selected ion monitoring electrospray mass spectrometry. Rapid Commun. Mass Spectrom. 9, 1109–1114.

Mitchell, K. T., Ferrell, J. E., Jr. and Huestis, W. H. (1986). Separation of phosphoinositides and other phospholipids by two-dimensional thin-layer chromatography. Anal. Biochem. 158, 447–453.

Mueller, M., Schiller, J., Petkovic, M., Oehrl, W., Heinze, R., Wetzker, R., Arnold, K. and Arnhold, J. (2001). Limits for the detection of (poly)phosphoinositides by matrix-assisted laser desorption and ionization time-of-flight mass spectrometry (MALDI-TOF MS). Chem. Phys. Lipids 110, 151–164.

Muenster, H., Stein, J. and Budzikiewicz, H. (1986). Structure analysis of underivatized phospholipids by negative ion fast atom bombardment mass spectrometry. Biomed. Environ. Mass Spectrom. 13, 423–423.

Munnik, T., Irvine, R. F. and Musgrave, A. (1994). Rapid turnover of phosphatidylinositol 3-phosphate in the green alga Chlamidomonas eugametos: signs of a phosphatidylinositide 3-kinase signalling pathway in lower plants? Biochem. J. 298, 269–273.

Murphy, R. C. and Harrison, K. A. (1994). Fast atom bombardment mass spectrometry of phospholipids. Mass Spectrom. Rev. 13, 57–75.

Nakamura, T., Hatori, Y., Yamada, K., Ikeda, M. and Yuzuriha, T. (1989). Anal. Biochem. 179, 127–127.

Nijjar, M. S. and Hawthorne, J. N. (1977). Purification and properties of polyphosphoinositide phosphomonoesterase from rat brain. Biochim. Biophys. Acta 480, 390–402.

Nasuhoglu, C., Feng, S., Mao, J., Yamamoto, M., Yin, H. L., Earnest, S., Barylko, B., Albanese, J. P. and Hilgemann, D. W. (2002). Nonradioactive analysis of phosphatidylinositides and other anionic phospholipids by anion-exchange high-performance liquid chromatography with suppressed conductivity detection. Anal. Biochem. 301, 243–254.

Okada, T., Sakuma, L., Fukui, Y., Hazeki, O. and Ui, M. (1994). Blockage of chemotactic peptide-induced stimulation of neutrophils by Wortmannin as a result of selective inhibition of phosphatidylinositol 3-kinase. J. Biol. Chem. 269, 3563–3567.

Palmer, F. B. St. C. (1981). Chromatography of acidic phospholipids on immobilized neomycin. J. Lipid Res. 22, 1296–1300.

Palmer, S. and Wakelam, M. J. O. (1990). Mass measurement of inositol 1,4,5-trisphosphate using a specific binding assay. In: Methods in Inositide Research (Irvine, R. F., ed.). Raven Press, New York, pp. 127–134.

Petkovic, M., Schiller, J., Mueller, M., Bernard, S., Reichl, S., Arnold, K. and Arnhold, J. (2001). Detection of individual phospholipids in lipid mixtures by matrix-assisted laser desorption/ionization time-of-flight mass spectrometry:

phosphatidylcholine prevents the detection of further species. Anal. Biochem. *289*, 202–216.

Pignataro, O. M. and Ascoli, M. (1990). Epidermal growth factor increases labeling of phosphatidylinositol 3,4-bisphosphates in MA-10 Leydig tumor cells. J. Biol. Chem. *265*, 1718–1723.

Prestwich, G. D., Chaudhary, A., Chen, J., Feng, L. B., Mehrotra, and Peng, J. (1998). Probing phosphoinositide polyphosphate binding to proteins. Advances in Phosphoinositides, (Bruzik, K.,, eds.), Vol. 718. American Chemical Society, Washington, DC, pp. 24–37.

Ptasznik, A., Prossnitz, E. R., Yoshikawa, D., Smrcka, A., Traynor-Kaplan, A. E. and Bokoch, G. M. (1996). A tyrosine kinase signaling pathway accounts for the majority of phosphatidylinositol 3,4,5-trisphosphate formation in che-moattractant-stimulated human neutrophils. J. Biol. Chem. *271*, 25204–25207.

Ptasznik, A., Beattie, G. M., Mally, M. I., Cirulli, V., Lopez, A. and Hayek, A. (1997). Phosphatidylinositol 3-kinase is a negative regulator of cellular differentiation. J. Cell Biol. *137*, 1127–1136.

Rivera, J., Lozano, M. L., Gonzalez-Canejero, R., Corral, J., De Arriba, F. and Vicente, V. (1998). A radioreceptor assay for a measurement of inositol (1,4, 5)trisphosphate using saponin-permeabilized outdated human platelets. Anal. Biochem. *256*, 117–121.

Rouser, G., Kritchevsky, G., Yamamoto, A., Simon, G., Galli, C. and Bauman, A. J. (1969). Diethylaminoethyl and triethylaminoethyl cellulose column chromatography procedures for phospholipids, glycolipids and pigments. Meth. Enzymol. *14*, 272–317.

Ryu, S. H., Cho, K. S., Lee, K. Y., uh, P. G. and Rhee, S. G. (1986). Two forms of phosphatidylinositol-specific phospholipase C from bovine brain. Biochem. Biophys. Res. Commun. *141*, 137–144.

Schacht, J. (1978). Purification of polyphosphoinositides on immobilized neomycin. J. Lipid Res. *19*, 1063–1067.

Schacht, J. (1981). Extraction and purification of polyphosphoinositides. Meth. Enzymol. *72*, 626–631.

Schiller, J. and Arnold, K. (2000). Mass spectrometry in structural biology. In: Encyclopedia of Analytical Chemistry (Myers, R. A., ed.). Wiley, Chichester, pp. 559–585.

Schiller, J., Arnhold, J., Bernard, M., Mueller, M., Reichl, S. and Arnold, K. (1999). Lipid analysis by matrix-assisted laser desorption and ionization mass spectrometry: a methodological approach. Anal. Biochem. *267*, 46–56.

Schiller, J., Arnhold, J., Glander, H-J. and Arnold, K. (2000). Lipid analysis of human spermatozoa and seminal plasma by MALDI-TOF mass spec-troscopy—effects of freezing and thawing. Chem. Phys. Lipids *106*, 145–156.

Serunian, L. A., Auger, K. R. and Cantley, L. C. (1991). Identification and quantification of polyphosphoinositides produced in response to platelet-derived growth factor stimulation. Meth. Enzymol. 198, 78–87.

Shaikh, N. A. (1986). Phospholipid analysis. The Heart and Cardiovascular System, (Fozzard, H. A. S., Haber, E., Jennings, R. B. and Katz A., eds.), Vol. 1. Raven Press, New York, pp. 289–302.

Shaikh, N. A. (1994). Assessment of various techniques for the quantitative extraction of lysophospholipids from myocardial tissues. Anal. Biochem. 216, 313–321.

Sherman, W. R., Ackerman, K. E., Bateman, R. H., Green, B. N. and Lewis, I. (1985). Mass-analyzed ion kinetic energy spectra and B1E-B2 triple sector mass spectrometric analysis of phosphoinositides by fast atom bombardment. Biomed. Mass Spectrom. 12, 409–413.

Shum, T. Y. P., Gra, N. C. C. and Strickland, K. P. (1979). The deacylation of phosphatidylinositol by rat brain preparations. Can. J. Biochem. 57, 1359–1367.

Singh, A. K. (1992). Quantitative analysis of inositol phosphates in synaptosomes and microvessels by column chromatography: comparison of the mass analysis and the radiolabelling methods. J. Chromatogr. 581, 1–10.

Singh, A. K. and Jiang, Y. (1995). Quantitative chromatographic analysis of inositol phospholipids and related compounds. J. Chromatogr. B 671, 255–280.

Slotboom, A. J., De Haas, G. H., Bonen, P. P. M., Burbach-Westerhuis, G. J. and Van Deenen, L. L. M. (1970). Hydrolysis of phosphoglycerides by purified lipase preparations. 1. Substrate-, positional- and stereo-specificity. Chem. Phys. Lipids 4, 15–29.

Stein, J. M., Smith, G. and Luzio, J. P. (1990). Quantification of polyphosphoinositides by acetylation with [^{3}H]acetic anhydride. In: Methods in Iniositide Research (Irvine, R. F., ed.). Raven Press, New York, pp. 167–177.

Stein, J. M., Smith, G. A. and Luzio, J. P. (1991). An acetylation method for the quantification of membrane lipids, including phospholipids, polyphosphoinositides and cholesterol. Biochem. J. 274, 375–379.

Stephens, L. R. (1990). Preparation and separation of inositol tetrakisphosphates and inositol pentakisphosphates and the establishment of enantiomeric configurations by the use of L-iditol dehydrogenase. In: Methods in Inositide Research (Irvine, R. F., ed.). Raven Press, New York, pp. 9–30.

Stephens, L., Hawkins, P. T. and Downes, C. P. (1989). Metabolic and structural evidence for the existence of a third species of polyphosphoinositide in cells: D-phosphatidyl-myo-inositol 3-phosphate. Biochem. J. 259, 267–276.

Sun, G. Y. and Lin, T.-N. (1990). Separation of phosphoinositides and other phospholipids by high-performance thin-layer chromatography. In: Methods in Inositide Research (Irvine, R. F., ed.). Raven Press, New York, pp. 153–158.

Thompson, W. (1969). Positional distribution of fatty acids in brain polyphosphoinositides. Biochim. Biophs. Acta *187*, 1153–1158.

Toner, M., Vaio, G., McLaughlin, A. and Mclaughlin, S. (1988). Adsorption of cations to phosphatidylinositol 4,5-bisphosphate. Biochemistry *27*, 7435–7443.

Tomlinson, R. V. and Ballou, C. E. (1961). Complete characterization of *myo*-inositol polyphosphates from beef brain phosphoinositides. J. Biol. Chem. *236*, 1902–1906.

Traynor-Kaplan, A. E., Harris, A. L., Thompson, B. L., Taylor, P. and Sklar, L. A. (1988). An inositol tetrakisphosphate-containing phospholipid in activated neutrophils. Nature *334*, 353–356.

Traynor-Kaplan, A. E., Thompson, B. L., Harris, A. L., Taylor, P., Omann, G. M. and Sklar, L. A. (1989). Transient increase in phosphatidylinositol 3,4-bisphosphate and phosphatidylinositol trisphosphate during activation of human neutrophils. J. Biol. Chem. *264*, 15668–15673.

Van der Kaay, J., Batty, I. H., Cross, D. A. E., Watt, P. W. and Downe, C. P. (1997). A novel, rapid, and highly sensitive mass assay for phosphatidylinositol 3,4,5-trisphosphate (PtdIns (3,4,5)P$_3$) and its application to measure insulin-stimulated PtdIns (3,4,5)P$_3$ production in rat skeletal muscle in vivo. J. Biol. Chem. *272*, 5477–5481.

Van der Kaay, J., Cullen, P. J. and Downes, C. P. (1998). Phosphatidylinositol(3,4,5)trisphosphate [PtdIns(3,4,5)P$_3$] mass measurement using a radioligand assay. Methods Mol. Biol. *105*, 109–125.

Van der Kaay, J., Beck, M., Gray, A. and Downs, C. P. (1999). Distinct phosphatidylinositol 3-kinase lipid products accumulate upon oxidative and osmotic stress and lead to different cellular responses. J. Biol. Chem. *274*, 35963–35968.

Varticovski, L., Harrison-Findik, D., Keeler, M. and Susa, M. (1994). Role of PI 3-kinase in mitogenesis. Rev. Biochim. Biophys. Acta *1226*, 1–11.

Vickers, J. D. (1995). Extraction of polyphosphoinositides from platelets: comparison of a two-step procedure with a common single-step extraction procedure. Anal. Biochem. *224*, 449–451.

Walsh, J. P., Caldwell, K. K. and Majerus, P. W. (1991). Formation of phosphatidylinositol 3-phosphate by isomeratization from phosphatidylinositol 4-phosphate. Proc. Natl Acad. Sci. USA *88*, 9184–9187.

Wang, T., Pentyala, S., Rebecchi, M. J. and Scarlata, S. (1999). Differential association of the pleckstrin homology domains of phospholipases C-β1, C-β2, and C-δ1 with lipid bilayers and the βγ subunits of heterotrimericG proteins. Biochemistry *38*, 1517–1524.

Whitman, M., Downes, C. P., Keeler, M., Keller, T. and Cantley, L. (1988). Type I phosphatidylinositol kinase makes a novel inositol phospholipid, phosphatidylinositol 3-phosphate. Nature *332*, 644–646.

Wu, L. N. Y., Genge, B. R., Kang, M. W., Arsenault, A. L. and Wuthier, R. E. (2002). Changes in phospholipid extractability and composition accompany mineralization of chicken growth plate catilage matrix vesicles. J. Biol. Chem. *277*, 5126–5133.

Yamada, K., Abe, S., Katayama, and Sato, T. (1988). Sensitive high performance liquid chromatographic method for the determination of phosphatidic acid. J. Chromatogr. *424*, 367.

Zhu, X. and Eichberg, J. (1993). Molecular species composition of glycerophospholipids in rat sciatic nerve and its alteration in streptozotocin-induced diabetes. Biochim. Biophys. Acta *1168*, 1–12.

Chapter 4

Adelt, S., Plettenburg, O., Stricker, R., Reiser, G., Altenbach, H.-J. and Vogel, G. (1999). Enzyme-assisted total synthesis of the optical antipodes D-*myo*-inositol 3,4,5-trisphosphate and D-*myo*-inositol 1,5,6-risphosphate: Aspects of their structure-activity relationship to biologically active inositol phosphates. J. Med. Chem. *42*, 1262–1273.

Adelt, S., Plettenburg, O., Dallmann, G., Ritter, F. P., Shears, S. B., Altenbach, H.-J. and Vogel, (2001). Regiospecific phosphohydrolases from *Dictyostelium* as tools for the chemoenzymatic synthesis of the enantiomers D-*myo*-inositol 1, 2,4-trisphosphate and D-*myo*-inositol 2,3,6-trisphosphate: non-physiological, potential analogues of biologically active D-*myo*-inositol 1,3,4-trisphosphate. Bioorg. Med. Chem. Lett. *11*, 2705–2708.

Agranoff, B. W. (1978). Cyclitol confusion. Trends Biochem. Sci. *3*, N283–N285.

Agranoff, B. W., Murthy, P. and Seguin, E. B. (1983). Thrombin-induced phosphodiesteratic cleavage of phosphatidylinositol bisphosphate in human platelets. J. Bio. Chem. *258*, 2076–2078.

Agranoff, B. W., Eisenberg, F., Hauser, G., Hawthorne, N. S. and Mitchell, R. H. (1985). In: Inositol and Phosphoinositides (Blersdale, J. E., Eichberg, J. and Hauser G., eds.). Humana Press, Clifton, pp. xxi.

Anderson, L., Cummings, J. and Smyth, J. F. (1992). Rapid and selective isolation of radiolabelled inositol phosphates from cancer cells using solid-phase extraction. J. Chromatogr. *574*, 150–155.

Balla, T., Guillemette, G., Baukall, A. J. and Catt, K. J. (1987). Metabolism of inositol 1,3,4-trisphosphate to a new tetrakisphosphate isomer in angiotensin-stimulated adrenal glomerulosa cells. J. Biol. Chem. 262, 9952–9955.

Balla, T., Hunyady, L., Baukal, A. J. and Catt, K. J. (1989). Structures and metabolism of inositol tetrakisphosphates and inositol pentakisphosphates in bovine adrenal glomerulosa cells. J. Biol. Chem. 264, 9386–9390.

Barker, C. J., Morris, A. J., Kirk, C. J. and Mitchell, R. H. (1988). Insitol tetrakisphosphates in WRK-1 cells. Biochem. Soc. Trans. 16, 984–985.

Barnaby, R. J. (1991). Mass assay for inositol 1-phosphate I rat brain by high-performance liquid chromatography and pulsed amperometric detection. Anal. Biochem. 199, 75–80.

Bartlett, G. R. (1982). Isolation of and assay of red-cell inositol polyphosphates. Anal. Biochem. 124, 425–431.

Batty, I. R., Nahososki, S. R. and Irvine, R. F. (1985). Rapid formation of inositol 1,3,4,5-tetrakisphosphate following muscarinic receptor stimulation of rat cerebral cortical slices. Biochem. J. 232, 211–215.

Berridge, M. J., Dawson, R. M., Downes, C. P., Heslop, J. P. and Irvine, R. F. (1983). Changes in the level of inositol phosphates after agonist-dependent hydrolysis of membrane phosphoinositides. Biochem. J. 212, 473–482.

Binder, H., Weber, P. C. and Siess, W. (1985). Separation of inositol phosphates and glycerophosphoinositol phosphates by high-performance liquid chromatography. Anal. Biochem. 148, 220–227.

Bird, I. M., Sadler, I. H., Williams, B. C. and Walker, S. W. (1989). The preparation of myo-inositol 1,4-bisphosphate and D-myo-inositol 1,4,5-trisphosphate in milligram quantities from a readily available starting material. Mol. Cell. Endocrinol. 66, 215–229.

Bradford, G. P. and Rubin, R. P. (1986). Quantitative changes in inositol 1,4,5-trisphosphate in chemo attractant-stimulated neutrophils. J. Biol. Chem. 261, 15644–15647.

Brammer, M. and Weaver, K. (1989). Kinetic analysis of A23187-mediated polyphosphoinositide breakdown in rat cortical synaptosomes suggests that inositol trisphosphate does not arise primarily by degradation of inositol trisphosphatide. J. Neurochem. 53, 399–407.

Brand, C., Hoffman, T. and Bonvini, E. (1990). High performance liquid chromatographic separation of inositol phosphate isomers employing a reversed phase column and a micellar mobile phase. J. Chromatogr. 529, 65–80.

Bredt, D. S., Mourey, R. J. and Snyder, S. H. (1989). A simple, sensitive, and specific radioreceptor assay for inositol trisphosphate receptor binding in brain: regulation by pH and calcium. J. Biol. Chem. 262, 12132–12136.

Brown, D. M. and Steward, J. C. (1966). The structure of triphosphoinositides from rat brain. Biochim. Biophys. Acta 125, 413–421.

Cerdan, S., Hansen, C. A., Johanson, R., Inubushi, T. and Williamson, J. R. (1986). Nuclear magnetic resonance spectroscopic analysis of myo-inositol phosphates including inositol 1,3,4,5-tetrakisphosphate. J. Biol. Chem. 261, 14676–14680.

Challiss, R. A. J., Batty, I. H. and Nahorski, S. R. (1988). Mass measurements of inositol (1,4,5) trisphosphate in rat cerebral cortex slices using a radioreceptor assay, effects of neurotransmitters and depolarization. Biochem. Biophys. Res. Commun. 157, 684–691.

Challiss, R. A. J. and Nahorski, S. R. (1993). Measurement of inositol 1,4,5-trisphosphate, inositol 1,3,4,5-tetrakisphosphate, and phopshatidylinositol 4,5-bisphosphate in brain. In: Lipid Metabolism in Signaling Systems: Methods in Neurosciences (Fain, J. N., ed.)Vol. 18. Academic Press, New York, pp. 224–244.

Chandrasekhar, R., Huang, H. M. and Sun, G. Y. (1988). Alterations in rat brain polyphosphoinositide metabolism due to acute ethanol administration. J. Pharmacol. Exp. Ther. 245, 120–123.

Chester, T. L., Pinkston, J. D., Innis, D. P. and Bowling, D. J. (1989). Separation, detection and identification of inositol trisphosphate and phytic acid derivatives by supercritical fluid chromatography and SFC-mass spectrometry. J. Microcolumn Sep. 1, 182–189.

Christensen, S. and Harbak, H. (1990). Serial separation of inositol phosphates including pentakis and hexakisphosphates on small anion-exchange column. J. Chromatogr. 533, 201–206.

Clarke, N. and Dawson, R. M. C. (1981). Alkaline O–N transacylation. A new method for the quantitative deacylation of phospholipids. Biochem. J. 195, 301–306.

Cosgrove, D. J. (1969). Ion-exchange chromatography of inositol polyphosphates. Ann. NY Acad. Sci. 165, 677–686.

Dangelmaier, C. A., Daniel, J. L. and Smith, J. B. (1986). Determination of basal and stimulated levels of inositol trisphosphate in [^{32}P]orthophosphate-labeled platelets. Anal. Biochem. 154, 414–419.

Daniel, J. L., Dangelmaier, C. A. and Smith, J. B. (1987). Formation and metabolism of inositol 1,4,5-trisphosphate in human platelets. Biochem. J. 246, 109–114.

DaTorre, S. D., Corr, P. B. and Creer, M. H. (1990). A sensitive method for the quantification of the mass of inositol phosphates using gas chromatography-mass spectrometry. J. Lipid Res. 31, 1925–1934.

Dawson, A. P., Lea, E. J. and Irvine, R. F. (2003). Kinetic model of the inositol trisphosphate receptor that shows both steady-state and quantal patterns of Ca^{2+} release from intracellular stores. Biochem. J. 370, 621–629.

Dawson, R. M. C. and Clarke, N. G. (1972). D-*myo*-inositol 1:2 cyclic phosphate 2-phosphohydrolase. Biochem. J. *127*, 113–118.

Dean, N. M. and Beaven, M. A. (1989). Methods for the analysis of inositol phosphates. Anal. Biochem. *183*, 199–209.

Dean, N. M. and Moyer, J. D. (1987). Separation of multiple isomers of inositol phosphates formed in GH₃ cells. Biochem. J. *242*, 361–366.

Dell, A., Rogers, M. E., Thomas-Oates, J. E., Huckerby, T. N., Sanderson, P. N. and Nieduszynski, I. A. (1988). Fast-atom bombardment mass spectrometric strategies for sequencing sulphated oligosaccharides. Carbohydrate Res. *179*, 7–19.

Downes, C. P. and MacPhee, C. H. (1990). *myo*-Inositol metabolites as cellular signals. Eur. J. Biochem. *193*, 1–18.

Downes, C. P. and Michell, R. H. (1981). The polyphosphoinositide phosphodiesterase of erythrocyte membranes. Biochem. J. *198*, 133–140.

Ellis, R. B., Galliard, T. and Hawthorne, J. N. (1963). Phosphoinositides. 5. The inositol lipids of ox brain. Biochem. J. *88*, 125–131.

Emilsson, A. and Sundler, R. (1984). Differential activation of phosphatidylinositol deacylation and a pathway via diphosphoinositide in macrophages responding to zymogen and ionophore A23187. J. Biol. Chem. *259*, 3111–3116.

Folch, J., Lee, M. and Stanley, S. M. (1957). A simple method for the isolation and purification of total lipid from animal tissues. J. Biol. Chem. *226*, 497–509.

Foster, P. S., Hogan, S. P., Hansbro, P. M., O'Brien, R., Potter, B. V. L., Ozaki, S. and Denborough, M. A. (1994). The metabolism of D-*myo*-inositol trisphosphate and D-*myo*-inositol 1,3,4,5-tetrakisphosphate by porcine skeletal muscle. Eur. J. Biochem. *222*, 955–964.

Goldman, H. D., Hsu, F.-F. and Sherman, W. R. (1990). Studies on the permethylation/dephosphorylation of iositol polyphosphates: An approach to a more sensitive assay. Biomed. Environ. Mass Spectrom. *19*, 771–776.

Goldschmidt, B. (1990). Preparative separation of *myo*-inositol bis- and trisphosphate isomers by anion exchange chromatography. Carbohydrate Res. *208*, 105–110.

Grado, C. and Ballou, C. E. (1961). *Myo*-inositol phosphates obtained by alkaline hydrolysis of beef brain phospholipids. J. Biol. Chem. *236*, 54–60.

Grases, F., Simonet, B. M., Vucenik, I., Perelo, J., Prieto, R. M. and Shamsuddin, A. M. (2002). Effects of exogenous inositol hexakisphosphate (InsP₆) on the levels of InsP₆ and of inositol trisphosphate (InsP₃) in malignant cells, tissues and biological fluids. Life Sci. *71*, 1535–1546.

Hansen, C. A., Mah, S. and Williamson, J. R. (1986). Formation and metabolism of inositol 1,3,4,5-tetrakisphosphate in liver. J. Biol. Chem. *261*, 8100–8103.

Harrison, R., Rodan, E., Lander, D. and Irvine, R. F. (1990). Ram spermatozoa produce inositol 1,4,5-trisphosphate but not inositol 1,3,4,5-tetrakisphosphate during Ca^{2+}/ionophore induced acrosome reaction. Cellular Signaling 2, 273–284.

Hatzack, F. and Rasmussen, S. K. (1999). High performance thin-layer chromatography method for inositol phosphates analysis. J. Chromatogr. B 736, 221–229.

Heathers, G. P., Corr, P. B. and Rubin, L. J. (1988). Transient accumulation of inositol (1,3,4,5)-tetrakisphosphate in response to α_1-adrenergic stimulation in adult cardiac myocytes. Biochem. Biophys. Res. Commun. 156, 485–492.

Heathers, G. P., Juehne, J., Rubin, L. J., Corr, P. B. and Evers, A. S. (1989). Anion exchange chromatographic separation of inositol phosphates and their quantification by gas chromatography. Anal. Chem. 176, 109–116.

Hirvonen, M. R., Lihtamo, H. and Savolainen, K. (1988). A gas chromatographic method for the determination of inositol monophosphates in rat brain. Neurochem. Res. 13, 957–962.

Hokins-Neaverson, M. and Sadeghian, K. (1976). Separation of [^{3}H]inositol monophosphates and [^{3}H]inositol on silica gel glass-fiber sheets. J. Chromatogr. 120, 502–505.

Horstman, D. A., Takemura, H. and Putney, J. W., Jr. (1988). Formation and metabolism of [^{3}H]inositol phosphates in AR42J pancreatoma cells. J. Biol. Chem. 263, 15297–15303.

Hsu, F.-F., Goldman, H. D. and Sherman, W. R. (1990). Thermospray liquid chromatographic/mass spectrometric studies with inositol phosphates. Biomed. Environ. Mass Spectrom. 19, 597–600.

Irvine, R. F. (1990). In: Methods in Inositide Research (Irvine, R. F., ed.). Raven, New York, pp. 1–7.

Irvine, R. F., Anggard, E. E., Letcher, A. J. and Downes, C. P. (1985). Metabolism of inositol 1,4,5-trisphosphate and inositol 1,3,4-trisphosphate in rat parotid glands. Biochem. J. 229, 505–511.

Irvine, R. F., Letcher, A. J., Lander, D. J. and Downes, C. P. (1984). Inositol trisphosphates on carbachol-stimulated rat parotid glands. Biochem. J. 223, 237–243.

Irvine, R. F., McNulty, T. J. and Schell, M. J. (1999). Inositol 1,3,4,5-tetrakisphosphate as a second messenger—a special role in neurons? Chem. Phys. Lpids 98, 49–57.

Irving, G. C. J. and Cosgrove, D. J. (1970). Interference by myo-inositol hexaphosphate in inorganic phosphate determinations. Anal. Biochem. 36, 381–388.

Irving, G. C. J. and Cosgrove, (1972). Inositol phosphates phosphatases of microbiological origin: the inositol pentaphosphate products of Aspergillus ficuum phytases. J. Bacteriol. 112, 434–438.

Ishii, H., Connolly, T. M., Bross, T. E. and Majerus, P. W. (1986). Inositol cyclic trisphosphate [inositol 1,2-(cyclic)-4,5-trisphosphate] is formed upon thrombin stimulation of human platelets. Proc. Natl. Acad. Sci. USA *83*, 6397–6401.

Johansson, C., Koerdel, J. and Drakenberg, T. (1990). Analysis of *myo*-inositol phosphates by 2D ^{1}H-n.m.r. spectroscopy. Carbohydrate Res. *207*, 177–183.

Johnson, D. C. and LaCourse, W. R. (1990). Lipid chromatography with pulsed electrochemical detection at gold and platinum electrodes. Anal. Chem. *62*, 589A–597A.

Johnson, L. F. and Tate, M. E. (1969). The structure of phytic acids. Can. J. Chem. *47*, 63–73.

Joseph, S. K. and Williamson, J. R. (1989). Inositol polyphosphates and intracellular calcium release. Arch. Biochem. Biophys. *273*, 1–15.

Katchintorn, U., Vajanaphanich, M., Barrett, K. E. and Traynor-Kaplan, A. E. (1993). Elevation of inositol tetrakisphosphate parallels inhibition of Ca^{2+}-dependent Cl^- secretion in C_{84} cells. Am. J. Physiol. *264*, C671–C676.

Kerovuo, J., Rouvinen, J. and Hatzack, F. (2000). Analysis of *myo*-inositol hexakisphosphate hydrolysis by *Bacillus* phytase: indication of a novel reaction mechanism. Biochem. J. *352*, 623–628.

King, C. E., Stephens, L. R., Hawkins, P. T., Guy, G. R. and Mitchell, R. H. (1987). Multiple metabolic pools of phosphoinositides and phosphatidate in human erythrocytes incubated in a medium that permits rapid transmembrane exchange of phosphate. Biochem. J. *244*, 209–217.

Koch-Kallnbach, and Diring, H. (1977). Isolation and separation of inositol 1-phosphate, cyclic inositol 1,2-phosphate, and glycerol phosphoinositol from tissue culture cells labeled with [^{3}H]inositol. Z. Hoppe-Seyler, Physiol. Chem. *358*, 367–375.

Lanzetta, P. A., Alvarez, L. J., Reinach, P. S. and Candia, O. A. (1979). An improved assay for nanomole amounts of inorganic phosphate. Anal. Biochem. *100*, 95–97.

Laussmann, T., Eujen, R., Weisshuhn, M., Thiel, U. and Vogel, G. (1996). Structures of diphospho-*myo*-inositol pentakisphosphate and bisdiphospho-*myo*-inositol tetrakisphosphate from *Dictyostelium* resolved by NMR analysis. Biochem. J. *315*, 715–720.

Laussmann, T., Reddy, K. M., Reddy, K. K., Falck, J. R. and Vogel, G. (1997). Diphospho-*myo*-inositol phosphates from *Dictyostelium* identified as D-6-diphospho-*myo*-inositol pentakisphosphate and D-5,6-bisdiphospho-*myo*-inositol tetrakisphosphate. Biochem. J. *322*, 31–33.

Leavitt, A. L. and Sherman, W. R. (1982a). Resolution of DL-*myo*-inositol 1-phosphate and other sugar enantiomers by gas chromatography. Meth. Enzymol. *89*, 3–9.

Leavitt, A. L. and Sherman, W. R. (1982b). Determination of inositol phosphates by gas chromatography. Meth. Enzymol. *89*, 9–18.

Lindon, J. C., Baker, D. J., Farrant, R. D. and Williams, J. M. (1986). ¹H, 13C and ³¹P n.m.r spectra and molecular conformation of *myo*-inositol 1,4,5-trisphosphate. Biochem. J. *233*, 275–277.

Lindon, J. C., Baker, D. J., Williams, J. M. and Irvine, R. F. (1987). Conformation of the identities of inositol 1,3,4-trisphosphate and inositol 1,3,4,5-tetrakisphosphate by the use of one-dimensional and two-dimensional n.m.r. spectroscopy. Biochem. J. *244*, 591–595.

Loewus, M. W., Sasaki, K., Leavitt, A. L., Munsell, L., Sherman, W. R. and Loewus, F. A. (1982). Enantiomeric form of *myo*-inositol-1-phosphate produced by *myo*-inositol phosphate synthase and *myo*-inositol kinase in higher plants. Plant Physiol. *70*, 1661–1663.

MacGregor, L. C. and Matschinsky, F. M. (1984). An enzymatic fluorimetric assay for *myo*-inositol. Anal. Biochem. *141*, 382–389.

MacGregor, L. C. and Matschinsky, F. M. (1986). Altered retinal metabolism in diabetes. I. Microanalysis of lipid, glucose, sorbitol, and *myo*-inositol in the choroids and in the individual layers of the rabbit retina. J. Biol. Chem. *261*, 4046–4051.

Martin, J. B., Bakker-Grunwald, T. and Klein, G. (1993). ³¹P-NMR analysis of *Entamoeba histolytica*, Occurrence of high amounts of two inositol phosphates. Eur. J. Biochem. *214*, 711–718.

Martin, J. B., Laussmann, T., Bakker-Grunwald, T., Vogel, G. and Klein, G. (2000). *neo*-Inositol polyphosphates in the amoeba *Entamoeba histolytica*. J. Biol. Chem. *275*, 10134–10140.

Maslanski, J. A. and Busa, W. B. (1990). A sensitive and specific mass assay for *myo*-inositol and inositol phosphates. In: Methods in Inositide Research (Irvine, R. F., ed.). Raven Press, New York, pp. 113–126.

Mathews, W. R., Guido, D. M. and Huff, R. M. (1988). Anion exchange high performance liquid chromatographic analysis of inositol phosphates. Anal. Biochem. *168*, 63–70.

Mayr, G. W. (1988). A novel-metal–dye detection system permits picomolar-range HPLC analysis of inositol polyphosphates from non-radioactively labelled cell or tissue specimens. Biochem. J. *254*, 585–591.

Mayr, G. W. (1990). Mass determination of inositol phosphates by high-performance liquid chromatography with postcolumn complexometry (metal–dye detection). In: Methods in Inositide Research (Irvine, R. F., ed.). Raven Press, New York, pp. 83–108.

Mayr, G. W. and Dietrich, W. (1987). The only inositol tetrakisphosphate detectable in avian erythrocytes is the isomer lacking phosphate at position 3: a NMR study. FEBS Lett. *213*, 278–282.

Mayr, G. W., Radenberg, T., Thiel, U., Vogel, and Stephens, L. R. (1992). Carbohydr. Res. *234*, 247–262.

McConnell, F. M., Stephens, L. R. and Shears, S. B. (1991). Multiple isomers of inositol pentakisphophate in Epstein-Bar virus-transformed (T5-1) B-lymphocytes, Identification of insitol 1,3,4,5,6-pentakisphosphate, D. inositol 1,2,4,5, 6-pentakisphosphate. Biochem. J. *280*, 323–329.

McCorkindale, J. and Edson, N. L. (1954). Polyol dehydrogenases. I. The specificity of rat liver polyol dehydrogenases. Biochem. J. *57*, 518–523.

Meek, J. L. (1986). Inositol bis-, tris-, and tetrakis(phosphate)s: Analysis in tissues by HPLC. Proc. Natl. Acad. Sci. USA *83*, 4162–4166.

Meek, J. L. and Nicoletti, F. (1986). Detection of inositol trisphosphate and other organic phosphates by high performance liquid chromatography using an enzyme-loaded post column reactor. J. Chromatogr. *351*, 303–311.

Menniti, F. S., Miller, R. N., Putney, J. W., Jr. and Shears, S. B. (1993). Turnover of inositol polyphosphate pyrophosphates in pancreatoma cells. J. Biol. Chem. *268*, 3850–3856.

Murphy, C. T., Bullock, A. J., Lindley, C. J., Mills, S. J., Riley, A. M., Potter, B. V. and Westwick, J. (1996). Enantiomers of *myo*-inositol-1,3,4-trisphosphate and *myo*-inositol-1,4,6-trisphosphate: stereospecific recognition by cerebellar and platelet *myo*-inositol-1,4,5-trisphosphate receptors. Mol. Pharmacol. *50*, 1223–1230.

Myher, J. J., Kuksis, A., Marai, L. and Yeung, S. K. F. (1978). Microdetermination of molecular species of oligo- and polyunsaturated diacylglycerols by gas chromatography-mass spectrometry of their tert-butyl dimethylsilyl ethers. Anal. Chem. *50*, 557–561.

Nahorski, S. R. and Potter, B. V. L. (1989). Molecular recognition of inositol polyphosphates by intracellular receptors and metabolic enzymes. Trends Pharmacol. Sci. *10*, 13–144.

Nakamura, T., Hatori, Y., Yamada, K., Ikeda, M. and Yuzuriha, T. (1989). A high-performance liquid chromatographic method for the determination of polyphosphoinositides in brain. Anal. Biochem. *179*, 127–130.

Nunn, D. L., Potter, B. V. L. and Taylor, C. W. (1990). Molecular target sizes of inositol 1,4,5-trisphosphate receptors in liver and cerebellum. Biochem. J. *265*, 393–398.

Palmer, S., Hughes, K. T., Lee, D. Y. and Wakelam, M. J. O. (1989). Development of a novel Ins(1,4,5)P$_3$-specific binding assay: its use to determine the intracellular concentrations of Ins(1,4,5)P$_3$ in unstimulated and vasopresion-stimulated rat hepatocytes. Cellular Signalling *1*, 147–156.

Palmer, S. and Wakelam, M. J. O. (1989). Mass measurement of inositol phosphates. Biochim. Biophys. Acta *1014*, 239–246.

Palmer, S. and Wakelam, M. J. O. (1990). Mass measurement of inositol 1,4,
5-trisphosphte using a specific binding assay. In: Methods in Inosite Research
(Irvine, R. F., ed.). Raven Press, New York, pp. 127–134.

Phillipy, B. and Bland, J. M. (1988). Gradient ion chromatography of inositol
phosphates. Anal. Biochem. 175, 162–166.

Portilla, D. and Morrison, A. R. (1986). Bradykinin-induced changes in
inositol trisphosphate mass MDCK cells. Biochem. Biophys. Res. Commun.
14, 644–649.

Prestwich, S. A. and Bolton, T. B. (1991). Measurement of picomole amounts of
any inositol phosphate isomer separable by high performance liquid
chromatography by means of a bioluminescence assay. Biochem. J. 274,
663–672.

Putney, J. W., Jr, Hughes, A. R., Horstman, D. A. and Takemura, H. (1989).
Inositol phosphate metabolism and cellular signal transduction. Adv. Exp.
Med. Biol. 255, 37–48. (Substitute for Horstman et al (1989).

Rana, R. S. and Hokin, L. E. (1990). Role of phosphoinositides in transmembrane
signalling. Physiol. Rev. 70, 115–164.

Radenberg, T., Scholz, P., Bergmann, G. and Mayr, G. W. (1989). The
quantitative spectrum of inositol phosphate metabolites in avian erythrocytes
analyzed by proton n.m.r. and h.p.l.c. with direct isomer detection. Biochem. J.
264, 323–333.

Reed, L. J. and deBellerache, J. (1988). Increased polyphosphoinositide
responsiveness in the cerebral cortex induced by cholinergic denervation.
J. Neurochem. 50, 1566–1571.

Rittenhouse, S. E. and Sasson, J. P. (1985). Mass changes in myo-inositol
trisphosphate in human platelets stimulated by thrombin. J. Biol. Chem. 260,
8657–8660.

Rouser, G., Kritchevsky, G., Yamamoto, A., Simon, G., Galli, C. and Bauman,
A. J. (1969). Diethylaminoethyl and triethylaminoethyl cellulose column
chromatographic procedures for phospholipids, glycolipids, and pigments.
Meth. Enzymol. 14, 272–272.

Rubin, L. J., Hsu, F.-F. and Sherman, W. R. (1993). Measurement of inositol
trisphosphate by gas chromatography/mass spectrometry: femtomole sensitivity provided by negative-ion chemical ionization mass spectrometry in
submilligram quantities of tissue. In: Lipid Metabolism in Signaling Systems,
Methods in Neurosciences (Fain, J. N., eds.), Vol. 18. Academic Press, New
York, pp. 201–212.

Ryu, S. H., Lee, S. Y., Lee, K. Y. and Rhee, S. G. (1987). Catalytic properties of
inositol trisphosphate kinase: activation by Ca^{2+} and calmodulin. Fed. Am.
Soc. Xp. Biol. 1, 389–393.

Salmon, D. M. and Bolton, T. B. (1988). Early events in inositol phosphate metabolism in longitudinal smooth muscle from guinea-pig intestine stimulated with carbachol. Biochem. J. *254*, 553–557.

Sasakawa, N., Nakaki, T. and Kato, R. (1990a). Rapid increase in inositol pentakisphosphate accumulation by nicotine in cultured adrenal chromaffin cells. FEBS Lett. *261*, 378–380.

Sasakawa, N., Nakaki, T. and Kato, R. (1990b). Stimulus-responsive and rapid formation of inositol pentakisphosphate in cultured adrenal chromaffin cells. J. Biol. Chem. *265*, 17700–17705.

Sasakawa, N., Nakaki, T. and Kato, R. (1993). Characterization of inositol phosphates by high-performance liquid chromatography. In: Lipid Metabolism in Signaling Systems, Methods in Neurosciences (Fain, J. N., eds.), Vol. 18. Academic Press, New York, pp. 213–223.

Sasakawa, N., Nakaki, T., Kashima, R., Kanba, S. and Kato, R. (1992). Stimulus-induced accumulation of inositol tetrakis-, pentakis-, and hexakisphosphate in N1E-115 neuroblastoma cells. J. Neurochem. *58*, 2116–2123.

Sastry, P. S., Dixon, J. F. and Hokin, L. E. (1992). Agonist-stimulated inositol polyphosphate formation in cerebellum. J. Neurochem. *58*, 1079–1086.

Scholz, P., Bergmann, G. and Mayr, G. W. (1990). Nuclear magnetic resonance spectroscopy of *myo*-inositol phosphates. In: Methods in Inositide Research (Irvine, R. F., ed.). Raven Press, New York, pp. 65–82.

Scholz, S., Sonnenbichler, J., Schafer, W. and Hensel, R. (1992). Di-*myo*-inositol-1,1'-phosphate: a new inositol phosphate isolated from *Pyrococcus woesi*. FEBS Lett. *306*, 239–242.

Seiffert, U. B. and Agranoff, B. W. (1965). Isolation and separation of inositol phosphates from hydrolysis of rat tissues. Biochim. Biophys. Acta *98*, 574–581.

Sekar, M., Dixon, J. F. and Hokin, L. E. (1987). The formation of inositol 1:2-cyclic 4,5-trisphosphate and 1:2-cyclic-4-bisphosphate on stimulation of mouse pancreatic minilobules with carbamyl-choline. J. Biol. Chem. *262*, 340–344.

Shamsuddin, A. M., Elsayed, A. M. and Ullah, A. (1988). Suppression of large intestinal cancer in F344 rats by inositol hexaphosphate. Carcinogenesis *9*, 577–580.

Shamsuddin, A. M. (1999). Metabolism and cellular functions of IP6: a review. Anticancer Res. *19*, 3733–3736.

Sharpes, E. S. and McCarl, R. L. (1982). A high performance liquid chromatographic method to measure ^{32}P incorporation into phosphorylated metabolites in cultured cells. Anal. Biochem. *124*, 421–424.

Shayman, J. A. and Kirkwood, M. T. (1987). Bradykinin-stimulated changes in inositol phosphate mass in renal papillary collecting tubule cells. Biochem. Biophys. Res. Commun. *145*, 1112–1119.

Shayman, J. A. and Barcellon, F. S. (1990). Ion-pair chromatography of inositol polyphosphates with N-methylimipramine. J. Chromatogr. 528, 143–152.

Shayman, J. A. and BeMent, D. M. (1988). The separation of myo-inositol phosphates by ion pair chromatography. Biochem. Biophys. Res. Commun. 151, 114–122.

Shayman, J. A., Morrison, A. S. R. and Lowry, O. H. (1987). Enzymatic fluorometric assay of myo-inositol trisphosphate. Anal. Biochem. 162, 562–568.

Shears, S. B. (1989). The pathways of myo-inositol 1,3,4-trisphosphate phosphorylation in liver. J. Biol. Chem. 264, 19879–19886.

Shears, S. B., Ali, N., Craxton, A. and Bembenek, M. E. (1995). Synthesis and metabolism of bis-diphosphoinositol tetrakisphosphate in vitro and in vivo. J. Biol. Chem. 270, 10489–10497.

Sherman, W. R., Ackerman, K. E., Berger, R. G., Gish, B. G. and Zinbo, M. (1986). Analysis of inositol mono- and polyphosphates by gas chromatography/mass spectrometry and fast atom bombardment. Biomed. Environ. Mass Spectrom. 13, 333–341.

Sherman, W. R., Packman, M. H., Laird, M. H. and Boshans, R. L. (1977). Measurement of myo-inositol in single cells and defined areas of the nervous system by selected ion monitoring. Anal. Biochem. 78, 119–131.

Sherman, W. R., Simpson, P. C. and Goodwin, S. L. (1978). scyllo-Inositol and myo-inositol levels in tissues of the skate Raja erinacea. Comp. Biochem. Physiol. B 59, 201–202.

Singh, A. K. (1992). Quantitative analysis of inositol lipids and inositol phosphates in synaptosomes and microvessels by column chromatography: comparison of the mass analysis and the radiolabeling methods. J. Chromatogr. 581, 1–10.

Singh, A. K. and Jiang, Y. (1995). Quantitative chromatographic analysis of inositol phospholipids and related compounds. J. Chromatogr. B 671, 255–280.

Smith, R., Braun, P. E., Ferguson, M. A. J., Low, M. G. and Sherman, W. R. (1987). Direct measurement of inositol in bovine myelin basic protein. Biochem. J. 248, 285–288.

Smith, R. E. and MacQuarrie, R. A. (1988). Determination of inositol phosphates and other biologically important anions by ion chromatography. Anal. Biochem. 170, 308–315.

Smith, R. E., MacQuarrie, R. A. and Jope, R. S. (1989). Determination of inositol phosphates and other anions in rat brain. J. Chromatogr. Sci. 27, 491–495.

Spencer, C. E. L., Stephens, L. R. and Irvine, R. F. (1990). Separation of higher inositol phosphates by polyethylenimine-celluose thin-layer chromatography and by Dowex chloride column chromatography. In: Methods in Inositide Research (Irvine, R. F., ed.). Raven Press, New York, pp. 39–43.

Stauderman, K. A. and Pruss, R. M. (1990). Different patterns of agonist-stimulated increases of ^{3}H-inositol phosphate isomers and cytosolic Ca^{2+} in bovine adrenal chromaffin cells: comparison of the effects of histamine and angiotensin II. J. Neurochem. *54*, 946–953.

Stephens, L. R. (1990). Preparation and separation of inositol tetrakisphosphates and inositol pentakisphosphates and the establishment of enantiomeric configurations by the use of L-iditol dehydrogenase. In: Methods in Inositide Research (Irvine, R. F., ed.). Raven Press, New York, pp. 9–30.

Stephens, L. R. and Downes, C. P. (1990). Precursor-product relationships amongst inositol polyphosphates. Biochem. J. *265*, 435–452.

Stephens, L. R., Hawkins, P. T., Barker, C. J. and Downes, C. P. (1988a). Synthesis of *myo*-inositol 1,3,4,5,6-pentakisphosphate from inositol phosphates generated by receptor activation. Biochem. J. *253*, 721–733.

Stephens, L. R., Hawkins, P. T., Carter, A. N., Chahwala, S. B., Morris, A. J., Whetton, A. D. and Downes, C. P. (1988b). L-*myo*-inositol 1,4,5,6-tetrakisphosphate is present in both mammalian and avian cells. Biochem. J. *249*, 271–282.

Stephens, L. R., Hawkins, P. T. and Downes, C. P. (1989b). Metabolic and structural evidence for the existence of a third species of polyphosphoinositide in cells: D-phosphatidyl-*myo*-inositol 3-phosphate. Biochem. J. *259*, 267–276.

Stephens, L. R., Hawkins, P. T. and Downes, C. P. (1989a). Analysis of *myo*-[^{3}H]inositol trisphosphates found in *myo*-[^{3}H]inositol prelabeled avian erythrocytes. Biochem. J. *262*, 727–737.

Stephens, L. R., Hawkins, P. T., Morris, A. J. and Downes, C. P. (1988c). L-*myo*-inoitol 1,4,5,6-tetrakisphosphate (3-hydroxyl) kinase. Biochem. J. *249*, 283–292.

Stephens, L. R., Hawkins, P. T., Stanley, A. F., Moore, T., Poyner, D. R., Morris, P. J., Hanley, M. R., Kay, R. R. and Irvine, R. F. (1991). *myo*-Inositol pentakisphosphates. Biochem. J. *275*, 485–499.

Stephens, L. R. and Irvine, R. F. (1990). Stepwise phosphorylation of *myo*-inositol leading to *myo*-inositol hexakisphosphate in *Dictyostelium*. Nature (London) *346*, 580–583.

Stephens, L., Radenberg, T., Thiel, U., Vogel, G., Khoo, K.-H., Dell, A., Jackson, T. R., Hawkins, P. T. and Mayr, G. W. (1993). The detection, purification, structural characterization, and metabolism of diphosphoinositol pentakisphosphate(s) and bisdiphosphoinositol tetrakisphosphate(s). J. Biol. Chem. *268*, 4009–4015.

Stricker, R., Adelt, S., Voge, G. and Reiser, R. (1999). Translocation between membranes and cytosol of p42^{IP4}, a specific inositol 1,3,4,5-tetrakisphosphate/phosphatidyl inositol 3,4,5-trisphosphate-receptor protein from brain, is induced by inositol 1,3,4,5-tetrakisphosphate and regulated by a membrane-associated 5-phosphate. Eur. J. Biochem. *265*, 815–824.

Stubbs, E. B., Jr., Kelleher, J. A. and Sun, G. Y. (1988). Phosphatidylinositol kinase, phosphatidylinositol 4-phosphate kinase and diacylglycerol kinase activities in rat brain subcellular fractions. Biochim. Biophys. Acta *958*, 247–254.

Sulpice, J. C., Bachelot, C., Gascard, P. and Giraud, F. (1990). Ion-pair chromatography method for the separation of inositol phosphates after labeling of cells with [^{32}P]phosphate. In: Methods in Inositide Research (Irvine, R. F., ed.). Raven Press, New York, pp. 45–63.

Sulpice, J. C., Gascard, P., Journet, E., Rendu, F., Renard, D., Poggioli, J. and Giraud, F. (1989). The separation of [^{32}P]inositol phosphates by ion-pair chromatography: optimization of the method and biological applications. Anal. Biochem. *179*, 90–97.

Sun, G. Y. and Lin, T. N. (1989). Time course for labeling of brain membrane phosphoinositides and other phospholipids after intracerebral injection of [^{32}P]ATP. Evaluation by an improved HPTLC procedure. Life Sci. *44*, 689–696.

Sun, G. Y., Lin, T.-N., Prekumar, N., Carter, S. and MacQuarrie, R. A. (1990). Separation and quantification of isomers of inositol phosphates by ion chromatography. In: Methods in Inositide Research (Irvine, R. F., ed.). Raven Press, New York, pp. 135–143.

Sylvia, V., Curtin, G., Norman, J., Stec, J. and Busbee, D. (1988). Activation of a low specific activity form of DNA polymerase α by inositol-1,4-bisphosphate. Cell *54*, 651–658.

Tarver, A. P., King, W. G. and Rittenhouse, S. E. (1987). Inositol 1,4,5-trisphosphate and inositol 1,2-cyclic 4,5-trisphosphate are minor components of total mass of inositol trisphosphate in thrombin-stimulated platelets. J. Biol. Chem. *262*, 17268–17271.

Tate, M. E. (1968). Separation of *myo*-inositol pentakisphosphates by moving paper electrophoresis. Anal. Biochem. *23*, 141–149.

Taylor, C. W., Berridge, M. J., Cooke, A. M. and Potter, B. V. L. (1989). Inositol 1,4,5-trisphosphorothioate, a stable analogue of inositol trisphosphate which mobilizes intracellular calcium. Biochem. J. *254*, 645–650.

Taylor, C. W. (2002). Controlling calcium entry. Cell *111*, 767–769.

Tomlinson, R. V. and Ballou, C. E. (1962). *Myo*-inositol polyphosphate intermediates in the dephosphorylation of phytic acid by phytase. Biochemistry *1*, 166–171.

Turk, J., Wolf, B. A. and McDaniel, M. L. (1986). Quantitation of *myo*-inositol as its hexakis(trifluoroacyl) derivative with negative ion chemical ionization mass spectrometry. Biomed. Environ. Mass Spectrom. *13*, 237–244.

Vallejo, M., Jackson, T., Lightman, S. and Hanley, M. R. (1987). Occurrence and extracellular actions of inositol pentakis- and hexakisphosphate in mammalian brain. Nature *330*, 656–658.

Van Leeuwen, S. H., van der Marel, G. A., Hensel, R. and van Boom, J. H. (1994). Synthesis of L,L-di-*myo*-inositol-1,1'-phosphate: a novel inositol phosphate from *Pyrococcus woesei*. Recl. Trav. Cim. Pays-Bas Belg. *113*, 335–336.

Whipps, D. E., Armston, A., Pryor, H. and Halestrap, A. P. (1987). Effects of glucagons and Ca^{2+} on the metabolism of phosphatidylinositol 4-phosphate and phosphatidylinositol 4,5-bisphosphate in isolated rat hepatocytes and plasma membranes. Biochem. J. *241*, 835–845.

Willcocks, A. L., Potter, B. V. L., Cooke, A. M. and Nihorski, S. R. (1988). *Myo*-Inositol (1,4,5) trisphosphorothioate binds specific [^{3}H]inositol(1,4,5)trisphosphate sites in rat cerebellum and is resistant to 5-phosphatase. Eur. J. Pharmacol *155*, 181–183.

Wilson, D. B., Bross, T. E., Sherman, W. R., Berger, R. A. and Majerus, P. W. (1985). Inositol cyclic phosphates are produced by cleavage of phosphatidyl-inositols (polyphosphoinositides) with purified sheep seminal vesicle phospholipase C enzymes. Proc. Natl. Acad. Sci. USA *82*, 4013–4017.

Woodcock, E. A., Anderson, K. E. and Land, S. L. (1993). Lyophilization can generate artifacts in chromatographic profiles of inositol phosphates. J. Chromatogr. *619*, 121–126.

Wreggett, K. A. and Irvine, R. F. (1987). A rapid separation method for inositol phosphates and their isomers. Biochem. J. *245*, 655–660.

Wreggett, K. A. and Irvine, R. F. (1989). Automated isocratic high-performance liquid chromatography of inositol phosphate isomers. Meth. Enzymol. *191*, 707–718.

Wregget, K. A., Lander, D. J. and Irvine, R. F. (1990). Two-stage analysis of radiolabeled inositol phosphate isomers. Meth. Enzymol. *191*, 707–718.

Zhang, L. (1998). Inositol 1,4,5-trisposphate mass assay. In: Methods in Molecular Biology, Phospholipid Signaling Protocols (Bird, I. M., eds.), Vol. 105. Humana Press, Torowa, pp. 77–87.

Zhang, L., Bradley, M. E. and Buxton, I. L. O. (1995). Inositol polyphosphate binding sites and their likely role in calcium regulation in smooth muscle. Int. J. Biochem. Cell Biol. *27*, 1231–1248.

Zhang, L. and Buxton, I. L. O. (1991). Muscarinic receptors in canine colonic circular smooth muscle. II. Signal transduction pathways coupled to the muscarinic receptors. Mol. Pharamacol. *40*, 952–959.

Zhang, L. and Buxton, I. L. O. (1998). Meaaurement of phosphoinositols and phosphoinositides using radio high-performance liquid chromatography flow detection. In: Methods in Molecular Biology, Phospholipid Signaling Protocols (Bird, I. M.,, eds.), Vol. 105. Humana Press, Torowa, pp. 47–63.

Chapter 5

Azzouz, N., Gerold, P., Kedees, M. H., Shams-Eldin, H., Werner, R., Capdeville, Y. and Schwarz, R. T. (2001). Regulation of *Paramecium primaurelia* glycosylphosphatidyl-inositol biosynthesis via dolichol phosphate mannose synthesis. Biochimie *83*, 801–809.

Azzouz, N., Striepen, B., Gerold, P., Capdeville, Y. and Schwarz, R. T. (1995). Glycosylinositol-phosphoceramide in the free-living protozoan Paramecium primaurelia: modification of core glycans by mannosyl phosphate. EMBO J. *14*, 4422–4433.

Benghezal, M., Lipke, P. N. and Conzelmann, A. (1995). Identification of six complementation classes involved in the biosynthesis of glycosylphosphatidylinositol anchors in *Saccharomyces cerevisiae*. J. Cell Biol. *130*, 1333–1344.

Benting, J. H., Rietveld, A. G. and Simmons, K. J. (1999). N-Glycans mediate the apical sorting of GPI-anchored, raft-associated protein in Madin-Darby canine kidney cells. J. Cell Biol. *146*, 314–320.

Blank, M. L. m., Cress, E. A. and Snyder, F. (1987). Separation and quantitation of phospholipid subclasses as their diradylglycerobenzoate derivatives by normal phase high performance liquid chromatography. J. Chromatogr. *392*, 421–425.

Bligh, E. G. and Dyer, W. J. (1959). A rapid method of total lipid extraction and purification. Can. J. Biochem. Physiol. *37*, 911–917.

Bordier, C. (1981). Phase separation of integral membrane proteins in Triton X-114 solution. J. Biol. Chem. *256*, 1604–1607.

Brennan, P. J. (1968). Phosphoinositides of *Corynebacterium xerosis*. Biochem. J. *109*, 158–160.

Brodbeck, U. (1998). Signalling properties of glycosylphosphatidylinositols and their regulated release from membranes in the turnover of glycosylphosphatidylinositol-anchored proteins. Biol. Chem. *379*, 1041–1044.

Brown, D. A. and London, E. (1998). Functions of lipid rafts in biological membranes. Annu. Rev. Cell Dev. Biol. *14*, 111–136.

Brown, D. A. and Rose, J. K. (1992). Sorting of GPI-anchored proteins to glycolipid-enriched membrane subdomains during transport to the apical cell surface. Cell *68*, 533–544.

Brown, J. E., Rudnick, M., Letcher, A. J. and Irvine, R. F. (1988). Formation of methylphosphorylinositol phosphates by extractions that employ methanol. Biochem. J. *253*, 703–710.

Butikofer, P., Kuypers, F. A., Shackleton, C., Brodbeck, U. and Stieger, S. (1990). Molecular species analysis of the glycosylphosphatidylinositol anchor of Torpedo marmorata acetylcholinesterase. J. Biol. Chem. *265*, 18983–18987.

Butikofer, P., Zollinger, M. and Brodbeck, U. (1992). Alkylacyl glycerophosphoinositol in human and bovine erythrocytes. Eur. J. Biochem. 208, 677–683.

Carver, M. A. and Turco, S. J. (1991). Cell-free biosynthesis of lipophosphoglycan from Leishmania donovani. J. Biol. Chem. 266, 10974–10981.

Cerneus, D. P., Ueffin, E., Poshuma, G., Strous, G. J. and Van der Ende, A. (1993). Detergent insolubility of alkaline phosphatase during biosynthetic transport and endocytoasis. Role of cholesterol. J. Biol. Chem. 268, 3150–3155.

Christie, W. W., Nikolova-Damyanova, B., Laakso, P. and Herslof, B. (1991). Stereospecific analysis of triacyl-sn-glycerols via resolution of diastereomic diacylglycerol derivatives by high-performance liquid chromatography on silica. J. Am. Oil Chem. Soc. 68, 695–701.

Conzelmann, A., Fankhauser, C. and Desponds, C. (1990). Myo-Inositol gets incorporated into numerous membrane glycoproteins of Saccharomyces cerevisiae; incorporation is dependent on phosphomannomutase (SEC53). EMBO J. 9, 653–661.

Conzelmann, A., Puoti, A., Lester, R. L. and Desponds, C. (1992). Two different types of lipid moieties are present in glycophosphoinositol-anchored membrane proteins of Saccharomyces cerevisiae. EMBO J. 11, 457–466.

Deeg, M. A., Hamphrey, D. R., Yang, S. H., Ferguson, T. R., Reinhold, V. N. and Rosenberry, T. L. (1992a). Glycan components in the glycoinositol phospholipid anchor of human erythrocyte acetylcholinesterase. J. Biol. Chem. 267, 18573–18580.

Deeg, M. A., Murray, N. R. and Rosenberry, T. L. (1992b). Identification of glycoinositol phospholipids in rat liver by reductive radiomethylation of amines but not in H4IIE hepatoma cells of isolated hepatocytes by biosynthetic labeling with glucosamine. J. Biol. Chem. 267, 18581–18588.

Deeg, M. A. and Vechere, C. B. (1997). Regulation of glycosylphosphatidylinositol-specific phospholipase D secretion from β-TC3 cells. Endocrinlogy 138, 819–826.

Denny, P. W., Field, M. C. and Smith, D. F. (2001). GPI-anchored proteins and glycoconjugates segregate into lipid rafts in Kinetoplastida. FEBS Lett. 491, 148–153.

Doctor, B. P., Chapman, T. C., Christner, C. E., Dea, C. D., De La Hoz, D. M., Gentry, M. K., Ogert, R. A., Rush, R. S., Smyth, K. K. and Wolfe, A. D. (1990). Complete amino acid sequence of fetal bovine serum acetylcholinesterase and its comparison in various regions with other cholinesterases. FEBS Lett. 266, 123–127.

Doering, T. L., Masterson, W. J., Hart, G. W. and Englund, P. T. (1990). Biosynthesis of glycosyl phosphatidylinositol membrane anchors. J. Biol. Chem. 265, 611–614.

Doering, T. L., Pessin, M. S., Hart, G. W., Raben, D. M. and Englund, P. T. (1994). The fatty acids in unremodeled trypanosome glycosyl-phosphatidyl-inositols. Biochem. J. *299*, 741–746.

Dutta-Choudhury, T. A. and Rosenberry, T. L. (1984). Human erythrocyte acetylcholinesterase is an amphipathic protein whose short membrane-binding domain is removed by papain digestion. J. Biol. Chem. *259*, 5653–5660.

Englund, P. T. (1993). The structure and biosynthesis of glycosyl phosphatidyl-inositol protein anchors. Annu. Rev. Biochem. *62*, 121–138.

Fankhauser, C., Homans, S. W., Thomas-Oates, J. E., McConville, M. J., Desponds, C., Conzelmann, A. and Ferguson, M. A. (1993). Structures of glycosylphosphatidyl inositol membrane anchors from *Saccharomyces cerevisiae*. J. Biol. Chem. *268*, 26365–26374.

Fasel, N., Rousseaux, M., Schaerer, E., Medof, M. E., Tykocinski, M. L. and Bron, C. (1989). *In vitro* attachment of glycosyl-inositolphospholipid anchor structures to mouse Thy-1 antigen and human decay-accelerating factor. Proc. Natl. Acad. Sci. USA *86*, 6858–6862.

Fatemi, S. H., Haas, R., Jentoft, N., Rosenberry, T. L. and Tartakoff, A. M. (1987). The glycophospholipid anchor of Thy-1. Biosynthetic labeling experiments with wild-type and class E Thy-1 negative lymphomas. J. Biol. Chem. *262*, 4728–4732.

Ferguson, M. A. J. (1992). In: Lipid Modification of Proteins: A Practical Approach (Hooper, N. M. and Turner A. J., eds.). IRL Press, Oxford, UK, pp. 191.

Ferguson, M. A. J. (1993). In: Glycobiology: A Practical Approach (Fukuda, A. and Kobata A., eds.). IRL Press, Oxford, pp. 349–384.

Ferguson, M. A. J. (1999). The structure, biosynthesis and functions of glycosylphosphatidylinositol anchors, and the contributions of trypanosome research. J. Cell Sci. *112*, 2799–2809.

Ferguson, M. A. J., Homans, S. W., Dwek, R. A. and Rademacher, T. W. (1988). Glycosyl-phosphatidylinositol moiety that anchors *Trypanosoma brucei* variant surface glycoprotein to the membrane. Science *239*, 753–759.

Ferguson, M. A. J., Low, M. G. and Cross, G. A. M. (1985). Glycosyl-*sn*-1,2-dimyristoylphosphatidylinositol is covalently linked to *Trypanosoma brucei* variant surface glycoprotein. J. Biol. Chem. *260*, 14547–14555.

Ferguson, M. A., Murray, P., Rutherford, H. and McConville, M. J. (1993). A simple purification of procyclic acidic repetitive protein and demonstration of a sialylated glycosyl-phosphatidylinositol membrane anchor. Biochem. J. *291*, 51–52.

Ferguson, M. A. J. and Williams, A. F. (1988). Cell-surface anchoring of proteins via glycosyl-phosphatidylinositol structures. Annu. Rev. Biochem. *57*, 285–320.

Field, M. C. and Menon, A. K. (1992). In: Lipid Modification of Proteins: A Practical Approach (Hooper, N. M. and Turner A. J., eds.). IRL Press, Oxford, pp. 155.

Field, C. and Menon, A. K. (1994). In: Lipid Modification of Proteins (Schlesinger, M. J., ed.). CRC Press, Boca Raton, pp. 83–134.

Field, M. C., Menon, A. K. and Cross, G. A. M. (1991). A glycosylphosphatidylinositol protein anchor from procyclic stage Trypanosoma brucei lipid structure and biosynthesis. EMBO J. 10, 2731–2739.

Fini, C., Amoresano, A., Andolfo, A., D'auria, S., Floridi, A., Paolini, S. and Pucci, P. (2000). Mass spectrometry study of ecto-5′-nucleotidase from bull seminal plasma. Eur. J. Biochem. 267, 4978–4987.

Folch, J., Lees, M. and Sloane-Stanley, G. H. (1957). A simple method for the isolation and purification of total lipids from animal tissues. J. Biol. Chem. 226, 497–509.

Fraering, P., Imhof, I., Meyer, U., Strub, J. M., van Dorrselaer, A., Vionnet, C. and Conzelmann, A. (2001). The GPI transamidase complex of Saccharomyces cerevisiae contains Gaalp,Gpi8p, and Gpi16p. Mol. Biol. Cell 12, 3295–3306.

Gerold, P., Dieckmann-Schuppert, A. and Schwarz, R. T. (1994). Glycosylphosphatidylinositols synthesized by asexual erythrocytic stages of the malarial parasite, Plasmodium falciparum. Candidates for plasmodial glycosylpho sphatidylinositol membrane anchor precursors and pathogenicity factors. J. Biol. Chem. 269, 2597–2606.

Gerold, P., Eckert, V. and Schwarz, R. T. (1996a). GPI anchors: an overview. Trends Glycosci. Glycotech. 8, 265–277.

Gerold, P., Schofield, L., Blackman, M. J., Holder, A. A. and Schwarz, R. T. (1996b). Structural analysis of the glycosyl-phosphatidylinositol membrane anchor of the merozoite surface proteins-1 and -2 of Plasmodium falciparum. Mol. Biochem. Parasitol. 75, 131–143.

Gerold, P., Striepen, B., Reitter, B., Geyer, H., Geyer, R., Reinwald, E., Risse, H.-J. and Schwarz, R. T. (1996c). Glycosyl-phosphatidylinositols of Trypanosoma congolese: two common precursors but a new protein-anchor. J. Mol. Biol. 261, 181–194.

Gerold, P., Vivas, L., Ogun, S. A., Azzouz, N., Brown, K. N., Holder, A. A. and Schwarz, R. T. (1997). Glcosylphosphatidylinositols of Plasmodium chabaudi chabaudi: a basis for the study of malarial glycolipid toxins in a rodent model. Biochem. J. 328, 905–911.

Gowda, D. C., Gupta, P. and Davidson, E. A. (1997). Glycosylphosphatidylinositol anchors represent the major carbohydrate modification in proteins of intraerythrocytic stage Plasmodium falciparum. J. Biol. Chem. 272, 6428–6439.

Guillas, I., Pfeferli, M. and Conzelman, A. (2000). Analysis of ceramides present in glycosylphosphatidylinositol anchored proteins of *Saccharomyces cerevisiae*. Methods Enzymol. *312*, 506–515.

Guther, M. L., Cardoso de Almeida, M. L., Yoshida, N. and Ferguson, M. A. J. (1992). Structural studies on the glycosylphosphatidylinositol membrane anchor of *Trypanosoma cruzi* 1G7-antigen. The structure of the glycan core. J. Biol. Chem. *267*, 6820–6828.

Haas, R., Brandt, P. T., Knight, J. and Rosenberry, T. L. (1986). Identification of amine components in a glycolipid membrane-binding domain at the C-terminus of human erythrocyte acetylcholinesterase. Biochemistry *25*, 3098–3105.

Haas, R., Jackson, B. C., Reinhold, V. R., Foster, J. D. and Rosenberry, T. L. (1996). Glycosylinositol phospholipid anchor and protein C-terminus of bovine erythrocyte aceylcholinesterase: analysis by mass spectrometry and by protein and DNA sequencing. Biochem. J. *314*, 817–825.

Hanada, K., Nishijima, M., Akamatsu, Y. and Pagano, R. E. (1995). Both sphingolipids and cholesterol participate in the detergent insolubility of alkaline phosphatase, a glycosylphosphatidylinositol-anchored protein, in mammalian membranes. J. Biol. Chem. *270*, 6254–6260.

Haynes, P. A., Gooley, A. A., Ferguson, M. A., Redmond, J. W. and Williams, K. L. (1993). Post-translational modifications of the *Dictyostelium discoideum* glycoprotein PsA. Glycosylphosphatidylinositol membrane anchor and composition of O-linked oligosaccharides. Eur. J. Biochem. *216*, 729–737.

Heise, N., Gutierrez, A. L., Mattos, K. A., Jones, C., Wait, R., Previato, J. O. and Mendonca-Previato, L. (2002). Molecular analysis of a novel family of complex glycoinositolphosphoryl ceramides from *Cryptococcus neoformans*: structural differences between encapsulated and acapsular yeast forms. Glycobiology *12*, 409–420.

Hirose, S., Prince, G. M., Sevlever, D., Ravi, L., Rosenberry, T. L., Ueda, E. and Medof, M. E. (1992). Characterization of putative glycoinositol phospholipid anchor precursors in mammalian cells. Localization of phosphoethanolamine. J. Biol. Chem. *267*, 16968–16974.

Hirose, S., Ravi, L., Haza, S. V. and Medof, M. E. (1991). Assembly and deacylation of N-acetylglucosaminyl-plasmanylinositol in normal and affected paroxysmal nocturnal hemoglobinuria cells. PNAS USA *88*, 3762–3766.

Hoerner, M. C. and Brodbeck, U. (1992). Phosphatidylinositol-glycan-specific phospholipase D is an amphiphilic glycoprotein that in serum is associated with high density lipoproteins. Eur. J. Biochem. *206*, 747–757.

Holder, A. A. (1983). Carbohydrate is linked through ethanolamine to the C-terminal amino acid of *Trypanosoma brucei* variant surface glycoprotein. Biochem. J. *209*, 261–262.

Homans, S., Ferguson, M. A. J., Dwek, R. A., Rademacher, T. W., Anand, R. and Williams, A. F. (1988). Complete structure of the glycosyl phosphatidylinositol membrane anchor of rat brain Thy-1 glycoprotein. Nature *333*, 269–272.

Hooper, N. M. (1999). Detergent-insoluble glcosylsphingolipid/cholesterol-rich membrane domains, lipid rafts and caveolae (Review). Mol. Membr. Biol. *16*, 145–156.

Itabashi, Y., Kuksis, A., Marai, L. and Takagi, T. (1990a). HPLC resolution of diacylglycerol moieties of natural triacylglycerols on a chiral phase consisting of bonded (RD)-(+)-1-(1-naphthyl)ethylamine. J. Lipid Res. *31*, 1711–1717.

Itabashi, Y., Kuksis, A. and Myher, J. J. (1990b). Determination of molecular species of enantiomeric diacylglycerols by chiral phase high performance liquid chromatography and polar capillary gas–liquid chromatography. J. Lipid Res. *31*, 2119–2126.

Itabashi, Y., Marai, L. and Kuksis, A. (1991). Identification of natural diacylglycerols as the 3,5-dinitrophenylurethanes by chiral phase liquid chromatography with mass spectrometry. Lipids *26*, 951–956.

Itabashi, Y., Myher, J. J. and Kuksis, A. (2000). High-performance liquid chromatographic resolution of reverse isomers of 1,2-diacyl-rac-glycerols as 3, 5-dinitrophenylurethanes. J. Chromatogr. A *893*, 261–279.

Jennemann, R., Bauer, B. L., Bertalanffy, H., Geyer, R., Gschwind, R. M., Selmer, T. and Wiegandt, H. (1999). Novel glycoinositolphosphosphingolipids, basidiolipids, from *Agaricus*. Eur. J. Biochem. *259*, 331–338.

Jennemann, R., Geyer, R., Sandhoff, R., Gschwind, R. M., Levery, S. B., Grone, H.-J. and Wiegandt, H. (2001). Glycoinositolphospholipids (Basidiolipids) of higher mushrooms. Eur. J. Biochem. *268*, 1190–1205.

Jones, C., Wait, R., Previato, J. O. and Mendonca-Previato, L. (2000). The structure of a complex glycosylphosphatidylinositol-anchored glucoxyllan from the kinetoplastid protozoan *Leptomonas samueli*. Eur. J. Biochem. *267*, 5387–5396.

Kamitani, S., Akiyama, Y. and Ito, K. (1992). Identification and characterization of an *Escherichia coli* gene required for the formation of correctly folded alkaline phosphatase, a periplasmic enzyme. EMBO J. *11*, 57–62.

Kamitani, T., Menon, A. K., Hallaq, Y., Warren, C. D. and Yeh, E. T. (1992). Complexity of ethanolamine phosphate addition in the biosynthesis of glycosylphosphatidylinositol anchors in mammalian cells. J. Biol. Chem. *267*, 24611–24619.

Khoo, K. H., Dell, A., Morris, H. R., Brennan, P. J. and Chatterjee, D. (1995). Structural definition of acetylated phosphatidylinositol mannosides from *Mycobacterium tuberculosis* definition of a common anchor for lipomannan and lipoarabinomannan. Glycobiology *5*, 117–127.

Kim, H. Y. and Salem, N., Jr. (1987). Application of thermospray high performance liquid chromatography/mass spectrometry for the determination of phospholipids and related compounds. Anal. Chem. *59*, 722–726.

Kinoshita, T. and Inoue, N. (2000). Dissecting and manipulating the pathway for glycosylphosphatidylinositol-anchor biosynthesis. Curr. Opin. Chem. Biol. *4*, 632–638.

Kuksis, A. and Myher, J. J. (1986). Lipids and their constituents. In: Profility of Body Fluids (Deyl, Z. and Sweeley C. C. Eds.) J. Chromatogr. Biomed. Appl. *379*, 57–90.

Kuksis, A. and Myher, J. J. (1990). Mass analysis of molecular species of diradylglycerols. In: Methods in Inositide Research (Irvine, R. F., ed.). Raven Press, New York, pp. 187–216.

Lee, T.-C., Malone, B., Buell, A. B. and Blank, M. L. (1991). Occurrence of ether-containing inositol phospholipids in bovine erythrocytes. Biochem. Biophys. Res. Commun. *175*, 673–678.

Lee, J.-Y., Kim, M. R. and Sok, D.-E. (1999). Release of GPI-anchored Zn^{2+}-glycero-phosphocholine cholinephosphodiesterase as an amphiphilic form from bovine brain membranes by bee venom phospholipase A_2. Neurochem. Res. *24*, 1043–1050.

Loureiro y Penha, C. V., Rodeschini, A. R., Penha, C. V. L., Todeschini, A. R., Lopes-Bezera, L. M., Wait, R., Jones, C., Mattos, K. A., Heise, N., Mendonca-Previato, L. and Previato, J. O. (2001). Characterization of novel structures of mannosylinositolphosphorylceramides from the yeast forms of *Sporothrix schenckii*. Eur. J. Biochem. *268*, 4243–4250.

Low, M. G. (1989). Glycosyl-phosphatidylinositol: a versatile anchor for cell surface proteins. Fed. Am. Soc. Exp. Biol. *3*, 1600–1608.

Low, M. G. (1992). In: Lipid Modification of Proteins: A Practical Approach (Hooper, N. M. and Turner A. J., eds.). IRL Press, Oxford, UK, pp. 117.

Low, M. G. and Kincade, P. W. (1985). Phosphatidylinositol is the membrane-anchoring domain of the Thy-1 glycoprotein. Nature *318*, 62–64.

Masterson, W. J., Doering, T. L., Hart, G. W. and Englund, P. T. (1989). A novel pathway for glycan assembly: biosynthesis of the glycosyl-phosphatidyl-inositol anchor of the trypanosome variant surface glycoprotein. Cell *56*, 793–800.

Mayor, S., Menon, A. K. and Cross, G. A. M. (1990a). Glycolipid precursors from the membrane anchor of *Trypanosoma brucei* variant surface glycoproteins. I. Glycan structure of the phosphatidylinositol-specific phospholipase C sensitive and resistant glycolipids. J. Biol. Chem. *265*, 6164–6173.

Mayor, S., Menon, A. K. and Cross, G. A. M. (1990b). Glycolipid precursors for the membrane anchor of *Trypanosoma brucei* variant surface glycoproteins II. Lipid structures of the phosphatidylinositol-specific phospholipase C sensitive and resistant glycolipids. J. Biol. Chem. *265*, 6174–6181.

McConville, M. J. (1991). Glycosylated-phosphatidylinositols as virulence factors in *Leischmania*. Cell Biol. Int. Rep. *15*, 779–798.

McConville, M. J. (1996). Glycosylphosphatidylinositols and the surface architecture of parasitic protozoa. In: Molecular Biology of Parasitic Protozoa (Smith, D. F. and Parsons M., eds.). Oxford University Press, Oxford, UK, pp. 205–228.

McConville, M. J., Collidge, T. A., Ferguso, M. A. and Schneider, P. (1993). The glycoinositol phospholipids of *Leishmania mexicana* promastigotes. Evidence for the presence of three distinct pathways of glycolipid biosynthesis. J. Biol. Chem. *268*, 15595–15604.

McConville, M. J. and Ferguson, M. A. (1993). The structure, biosynthesis and function of glycosylated phosphatidylinositols in the parasitic protozoa and higher eukaryotes. Biochem. J. *294*, 305–324.

McConville, M. J. and Menon, A. K. (2000). Recent developments in the cell biology and biochemistry of glycosylphosphatidylinositol lipids (Review). Mol. Memb. Biol. *17*, 1–16.

McGuire, G. B., Becker, R. and Skidgel, R. A. (1999). Carboxypeptidase M, a glycosyl phosphatidylinositol-anchored protein is localized on both the apical and basolateral domains of polarized Madin-Darby canine kidney cells. J. Biol. Chem. *274*, 31632–31640.

Menon, A. K. (1994). Glycosylphosphatidylinositol anchors. Methods Enzymol. *230*, 418–442.

Menon, A. K. (1995). Flippases. Trends in Cell Biol. *5*, 355–360.

Meyer, U., Fraering, P., Bosson, R., Imhof, I., Benghezal, M., Vionnet, C. and Conzelmann, A. (2002). The glycosylphosphatidylinositol (GPI) signal sequence of human placental alkaline phosphatase is not recognized by human Gpi8p in the context of the yeast GPI anchoring machinery. Mol. Microbiol. *46*, 745–748.

Milne, K. G., Ferguson, M. A. and Englund, P. T. (1999). A novel glycosylphosphatidylinositol in African trypanosomes. A possible catabolic intermediate. J. Biol. Chem. *274*, 1465–1471.

Moody-Haupt, S., Patterson, J. H., Mirelman, D. and McConville, M. J. (2000). The major surface antigens of *Entamoeba histolytica* trophozotes are GPI anchored proteophosphoglycans. J. Mol. Biol. *297*, 409–420.

Moran, P., Raab, H., Kohr, W. J. and Caras, I. W. (1991). Glycophospholipid membrane anchor attachment. J. Biol. Chem. *266*, 1250–1257.

Morita, Y. S., Acosta-Serrano, A., Buxbaum, L. U. and Englund, P. T. (2000a). Glycosyl phosphatidylinositol myristoylation in African trypanosomes. New intermediates in the pathway for fatty acid remodeling. J. Bol. Chem. *275*, 14147–14154.

Morita, Y. S., Acosta-Serrano, A. and Englund, P. T. (2000b). The biosynthesis of GPI anchors. In: Oligosaccharides in Chemistry and Biology — A Comprehensive Handbook, (Ernst, P., Sinay, P. and Hart G., eds.). Wiley/ VCH, Weinheim, Germany, pp. 417–433.

Morita, Y. S., Paul, K. S. and Englund, P. T. (2000c). Specialized fatty acid synthesis in African trypanosomes: myristate for GP anchors. Science 288, 140–143.

Mullis, K. R., Haltinwanger, R. S., Hart, G. W., Marchase, R. B. and Engler, J. A. (1990). Relative accessibility of –acetylglucosamine in trimers of the adenovirus types 2 and 5 fiber proteins. J. Virol. 64, 5317–5323.

Myher, J. J. and Kuksis, A. (1975). Improved resolution of natural diacylglycerols by gas–liquid chromatography on polar siloxanes. J. Chromatogr. Sci. 13, 138–145.

Myher, J. J. and Kuksis, A. (1982). Resolution of diacylglycerol moieties of natural glycerophospholipids by as-liquid chromatography on polar capillary columns. Can. J. Biochem. 60, 638–650.

Myher, J. J. and Kuksis, A. (1984a). Molecular species of plant phosphatidyl-inositol with selective cytotoxicity towards tumor cells. Biochim. Biophys. Acta 795, 85–90.

Myher, J. J. and Kuksis, A. (1984b). Determination of plasma total lipid profiles by capillary gas–liquid chromatography. J. Biochem. Biophys. Methods 10, 13–23.

Myher, J. J., Kuksis, A., Marai, L. and Yeung, S. K. F. (1978). Microdetermina-tion of molecular species of oligo- and polyunsaturated diacylglycerols by gas chromatography-mass spectrometry of their tert-butyl dimethylsilyl ethers. Anal. Chem. 50, 557–561.

Myher, J. J., Kuksis, A. and Pind, S. (1989a). Molecular species of glycerophos-pholipids and sphingomyelins of human erythrocytes: improved method of analysis. Lipids 24, 396–407.

Myher, J. J., Kuksis, A. and Pind, S. (1989b). Molecular species of glycerophos-pholipids and sphingomyelins of human plasma: comparison to red blood cells. Lipids 24, 408–418.

Nagamune, K., Nozaki, T., Maeda, Y., Ohishi, K., Fukuma, T., Hara, T., Schwarz, R. T., Sutterlin, C., Brun, R., Riezman, H. and Kinoshita, T. (2000). Critical roles of glycosylphosphatidylinositol for Trypanosoma brucei. Proc. Natl. Acad. Sci. USA 97, 10336–10341.

Naik, R. S., Branch, O. H., Woods, A. S., Vijaykumar, M., Perkins, D. J., Nahlen, B. L., Lal, A. A., Cotter, R. J., Costello, C. E., Ockenhouse, C. F., Davidson, E. A. and Gowda, D. C. (2000b). Glycosylphosphatidylinositol anchors of Plasmodium falciparum: Molecular characterization and naturally elicited antibody response that may provide immunity to malaria pathogenesis. J. Exp. Med. 192, 1563–1575.

Naik, R. S., Davidson, E. A. and Gowda, D. C. (2000a). Develomental stage-specific biosynthesis of glycosylphosphatidylinositol anchors in intraerythro-cytic *Plasmodium falciparum* and its inhibition in a novel manner by mannosamine. J. Biol. Chem. *275*, 24506–24511.

Nakagawa, Y., Sugiura, T. and Waku, K. (1985). The molecular species composition of diacyl, alkylacyl and alkenylacylglycerophospholipids in rabbit alveolar macrophages. High amounts of 1-O-hexadecyl-2-arachidonoyl molecular species in alkylacylglycerophosphocholine. Biochim. Biophys. Acta *833*, 323–329.

Nigau, J., Gilleron, M., Cahuzac, B., Bounery, J. D., Herold, M., Thurnher, M. and Puzo, G. (1997). The phosphatidyl-*myo*-inositol anchor of the lipoarabino-mannans from *Mycobacterium bovis* Bacillus Calmette Guering. J. Biol. Chem. *272*, 23094–23103.

Nigau, J., Gilleron, M. and Puzo, G. (1999). Lipoarabinomannans: characteriz-ation of the multiacylated forms of the phosphatidyl-*myo*-inositol by NMR spectroscopy. Biochem. J. *337*, 453–460.

Orlandi, P. A., Jr. and Turco, S. J. (1987). Structure of the lipid moiety of the *Leishmania donovani* lipophosphoglycan. J. Biol. Chem. *262*, 10384–10391.

Oxley, D. and Bacic, A. (1999). Structure of the glycosylphosphatidylinositol anchor of an arabinogalactan protein from *Pyrus communis* suspension-cultured cells. Proc. Natl. Acad. Sci. USA *96*, 14246–14251.

Parthasarathy, R. and Eisenberg, F., Jr. (1986). The inositol phospholipids: a stereochemical view of biological activity. Biochem. J. *235*, 313–322.

Patton, G. M., Fasulo, J. M. and Robins, S. J. (1982). Separation of phospholipids and individual molecular species of phospholipids by high performance liquid chromatography. J. Lipid Res. *23*, 190–196.

Privett, O. S. and Nutter, L. J. (1977). Determination of the structure of lecithins via the formation of acetylated 1,2-diglycerides. Lipids *12*, 149–154.

Pryde, J. G. (1998). Partitioning of proteins in Triton X-114. Meth. Mol. Biol. *88*, 23–33.

Puoti, A. and Conzelmann, A. (1992). Structural characterization of free glycolipids which are potential precursors for glycophosphatidylinositol anchors in mouse thymoma cell lines. J. Biol. Chem. *267*, 22673–22680.

Puoti, A. and Conzelmann, A. (1993). Characterization of abnormal free glycophosphatidylinositols accumulating in mutant lymphoma cells of classes B, E, F and H. J. Biol. Chem. *268*, 7215–7224.

Rabe, H., Reichmann, G., Nakagawa, Y., et al. (1989). Separation of alkylacyl- and diacyl glycerophospholipids and their molecular species as naphthylur-ethanes by HPLC. J. Chromator. Biomed. Applic., 353–360.

Ramesha, C. S., Pickett, W. C. and Murthy, D. V. K. (1989). Sensitive method for the analysis of phospholipid subclasses and molecular species as 1-anthroyl derivatives of their diglycerides. J. Chromatogr. Biomed. Applic. *491*, 37–48.

Rietveld, A. and Simons, K. (1998). The differential miscibility of lipids as the basis for the formation of functional membrane rafts. Biochim. Biophys. Acta *1376*, 467–479.

Roberts, W. L., Ki, B. H. and Rosenberry, T. L. (1987). Differences in the glycolipid membrane anchors of bovine and human erythrocyte acetylcholinesterases. Proc. Natl. Acad. Sci. USA *84*, 7817–7821.

Roberts, W. L., Myher, J. J., Kuksis, A. and Rosenberry, T. L. (1988a). Alkylacylglycerol molecular species in the glycosylinositol phospholipid membrane anchor of bovine erythrocyte acetylcholinesterase. Biochem. Biophys. Res. Commun. *150*, 271–277.

Roberts, W. L., Myher, J. J., Kuksis, A., Low, M. G. and Rosenberry, T. L. (1988b). Lipid analysis of the glycoinositol phospholipid membrane anchor of human erythrocyte acetycholinesterase. J. Biol. Chem. *263*, 18766–18775.

Roberts, W. L. and Rosenberry, T. L. (1985). Identification of covalently attached fatty acids in the hydrophobic membrane-domain of human erythrocyte acetylcholineesterase. Biochem. Biophys. Res. Commun. *133*, 621–627.

Roberts, W. L. and Rosenberry, T. L. (1986). Selective radiolabeling and isolation of the hydrophobic membrane-binding domain of human erythrocyte acetylcholinesterase. Biochemistry *25*, 3091–3098.

Roberts, W. L., Santikarn, S., Reinhold, V. N. and Rosenberry, T. L. (1988c). Structural characterization of the glycoinositol phospholipid membrane anchor of human erythrocyte acetylcholinesterase by fast atom bombardment mass spectrometry. J. Biol Chem. *263*, 18776–18784.

Rosenberry, T. L., Toutant, J. P., Haas, R. and Roberts, W. L. (1989). Methods Cell Biol. *32*, 231.

Santos de Macedo, C., Gerold, P., Jung, N., Azzouz, N., Kimmel, J. and Schwarz, R. T. (2001). Inhibition of glycosyl-phosphatidylinositol biosynthesis in *Plasmodium falciparum* by C-2 substituted mannose analogues. Eur. J. Biochem. *268*, 6221–6228.

Schmid, H. H. O., Bandi, P. C. and Kwei, L. S. (1975). Analysis and quantitation of ether lipids by chromatographic methods. J. Chromatogr. Sci. *13*, 478–486.

Schmidt, A., Schwarz, R. T. and Gerold, P. (1998). *Plasmodium falciparum*: asexual erythrocytic stage synthesizes two structurally distinct free and protein-bound glycosylphosphatidylinositols in a maturation-dependent manner. Exp. Parsitol. *88*, 95–102.

Schmitz, B., Klein, R., Duncan, I. A., Egge, H., Gunawan, J. and Peter-Katalanic, J. (1987). MS and NMR analysis of the cross-reacting determinant glycan from

Trypanosoma brucei MITat 1.6 variant specific glycoprotein. Biochem. Biophys. Res. Commun. *146*, 1055–1063.

Schmitz, B., Klein, A., Egge, H. and Peter-Katalinic, J. (1986). A study of the membrane attachment site of the membrane-form variant surface glycoprotein from *Trypanosoma brucei* using lipid vesicles as a model of the plasma membrane. Mol. Biochem. Parasitol. *20*, 191–197.

Schneider, P. and Ferguson, M. A. J. (1995). Microscale analysis of glycosylphosphatidylinositol structures. Methods Enzymol. *250*, 614–629.

Schneider, P., Ferguson, M. A. J., McConville, M. J., Mehlert, A., Homans, S. W. and Bordier, C. (1990). Structure of the glycosyl-phosphatidylinositol membrane anchor of the leishmania major promastigote surface protease. J. Biol. Chem. *265*, 16955–16964.

Schneider, P., Ralton, J. E., McConville, M. J. and Ferguson, M. A. J. (1993). Analysis of the neutral glycan fractions of glycosyl-phosphatidylinositols by thin-layer chromatography. Analyt. Biochem. *210*, 106–112.

Schneider, P., Schnur, L. F., Jaffe, C. L., Ferguson, M. A. and McConville, M. J. (1994). Glycoinositol-phospholipid profiles of four serotypically distinct Old World *Leishmania* strains. Biochem J. *304*, 603–609.

Schroder, R. J., Ahmed, S. N., Zhu, Y., London, E. and Brown, D. A. (1998). Cholesterol and sphingolipid enhance the Triton X-100 insolubility of glycosylphosphatidylinositol anchored proteins by promoting the formation of detergent insoluble ordered membrane domains. J. Biol. Chem. *273*, 1150–1157.

Sevlever, D., Huphry, D. R. and Rosenberry, T. L. (1995). Compositional analysis of glucosaminyl(acyl)phosphatidylinositol accumulated in HeLa S3 cells. Eur. J. Biochem. *233*, 384–394.

Sevlever, D., Mann, K. J. and Medof, M. E. (2001). Differential effect of 1,10-phenanthroline on mammalian, yeast, and parasite glycosylphosphatidylinositol anchor synthesis. Biochem. Biophys. Res. Commun. *288*, 1112–1118.

Sevlever, D., Pickett, S., Mann, K. J., Sambamurti, K., Medof, M. E. and Rosenberry, T. L. (1999). Glycosylphosphatidylinositol-anchor intermediates associate with triton-insoluble membranes in subcellular compartments that include the endoplasmic reticulum. Biochem. J. *343*, 627–635.

Sevlever, D. and Rosenberry, T. L. (1993). Mannosamine inhibits the synthesis of putative glycoinositol phospholipid anchor precursors in mammalian cells without incorporating into an accumulated intermediate. J. Biol. Chem. *268*, 10398–10945.

Sherman, W. R., Ackermann, K. E., Bateman, R. H., Gree, B. N. and Lewis, I. (1985). Mass analyzed ion kinetic energy spectra and B_1E-B_2 triple sector mass spectrometric analysis of phosphoinositides by fast atom bombardment. Biomed. Mass Spectrom. *12*, 409–413.

Shively and Conrad (1976).

Singh, B. N., Beach, D. H., Lindmark, D. G. and Costello, C. E. (1994). Identification of the lipid moiety and further characterization of the novel lipophosphoglycan-like glycoconjugates of *Trichomonas vaginalis* and *Trichomonas foetus*. Arch. Biochem. Biophys. *309*, 273–280.

Sipos, G., Puoti, A. and Conzelmann, A. (1995). Biosynthesis of the side chain of yeast glycosylphosphatidylinositol anchors is operated by novel mannosyl-transferases located in the endoplasmic reticulum and the Golgi apparatus. J. Biol. Chem. *270*, 19709–19715.

Sipos, G., Reggiori, F., Vionnet, C. and Conzelmann, A. (1997). Alternative lipid remodeling pathways for glycosylphosphatidylinositol membrane anchors in *Saccharomyces cerevisiae*. EMBO J. *16*, 3494–3505.

Stadler, J., Keenan, T. W., Bauer, G. and Gerisch, G. (1989). The contact site A glycoprotein of *Dicytostelium discoideum* carries a phospholipid anchor of a novel type. EMBO J. *8*, 371–377.

Stahl, N., Baldwin, M. A., Hecker, R., Pan, K. M., Burlingame, A. L. and Prusiner, S. B. (1992). Glycosylinositol phospholipid anchors of the scrapie and cellular prion proteins contain sialic acid. Biochemistry *31*, 5043–5053.

Stevens, V. L. and Raetz, C. R. H. (1991). Defective glycosyl phosphatidylinositol biosynthesis in extracts of three Thy-1 negative lymphoma cell mutants. J. Biol. Chem. *266*, 10039–10042.

Striepen, B., Zinecker, C. F., Damm, J. B., Melgers, P. A., Gerwig, G. J., Koolen, M., Vliegenthart, J. F., Dubremetz, J. F. and Schwarz, R. T. (1997). Molecular structure of the low molecular weight antigen of *Toxoplasma gondii*: a glucose α1-4N-acetylgalactosamine makes free glycosyl-phosphatidylinositols highly immunogenic. J. Mol. Biol. *266*, 797–813.

Sugiyama, E., DeGasperi, R., Urakaze, M., Chang, H. M., Thomas, L. J., Hyman, R., Warren, C. D. and Yeh, E. T. (1991). Identification of defects in glycosylphosphatidyl inositol anchor biosynthesis in the Thy-1 expression mutants. J. Biol. Chem. *266*, 12119–12122.

Taguchi, R., Hamakawa, N., Maekawa, N. and Ikezawa, H. (1999a). Application of electrospray ionization MS/MS and matrix-assisted laser desorption/ ionization-time of flight mass spectrometry to structural analysis of the glycosyl-phosphatidylinositol-anchored protein. J. Biochem (Tokyo) *126*, 421–429.

Taguchi, R., Hamakawa, N., Maekawa, N. and Ikezawa, H. (1999b). Identification of a new glycosylphosphatidylinositol-anchored 42-kDa protein and its C-terminal peptides from bovine erythrocytes by gas chromatography-, time-of-flight, and electrospray-ionization-mass spectrometry. Arch. Biochem. Biophys. *363*, 60–67.

Takagi, T. and Itabashi, Y. (1987). Rapid separation of diacyl- and dialkylglycerol enantiomers by high performance liquid chromatography on a chiral stationary phase. Lipids 22, 596–600.

Takagi, T., Okamoto, J., Ando, Y. and Itabashi, Y. (1990). Separation of the enantiomers of 1-alkyl-2-acyl-rac-glycerol and of 1-alkyl-3-acyl-rac-glycerol by high performance liquid chromatography on a chiral column. Lipids 25, 108–110.

Takamura, H., Narita, H., Urade, R. and Kito, M. (1986). Quantitative analysis of polyenoic phospholipid molecular species by high performance liquid chromatography. Lipids 21, 356–361.

Thomas, J. R., Dwek, R. A. and Rademacher, T. W. (1990). Structure, biosynthesis, and function of glycosylphosphatidylinositols. Biochemistry 29, 5413–5422.

Thomas, A. E., III, Scharoun, J. E. and Ralston, H. (1965). Quantitative estimation of isomeric monoglycerides by thin-layer chromatography. J. Am. Oil Chem. Soc. 42, 789–792.

Thompson, G. A. and Okuyama, H. (2000). Lipid-linked proteins in plants. Progr. Lipid Res. 39, 19–39.

Tiede, A., Bastisch, I., Schubert, J., Orlean, P. and Schmidt, R. E. (1999). Biosynthesis of glycosylphosphatidylinositols in mammals and unicellular microbes. Biol. Chem. 380, 503–523.

Tiede, A., Nischan, C., Schubert, J. and Schmidt, R. E. (2000). Characterization of the enzymatic complex for the first step in glycosylphosphatidylinositol biosynthesis. Int. J. Biochem. Cell Biol. 32, 339–350.

Treumann, A., Lifely, M. R., Schneider, P. and Ferguson, M. A. J. (1995). Primary structure of CD52. J. Biol. Chem. 270, 6088–6099.

Treumann, A., Guther, M. L. S., Schneider, P. and Ferguson, M. A. J. (1998). Analysis of the carbohydrate and lipid components of glycosylphosphatidy-linositol structures. In: Methods in Molecular Biology, (Hounsell, E., ed.), Vol. 76. Humana Press, pp. 213–235.

Turco, S. J. and Descoteaux, A. (1992). The lipophosphoglycan of leishmania parasites. Annu. Rev. Microbiol. 46, 65–94.

Turco, S. J., Hull, S. R., Orlandi, P. A., Shepherd, S. D., Homans, S. W., Dwek, R. A. and Rademacher, T. W. (1987). Structure of the major carbohydrate fragment of the Leishmania donovani lipophosphoglycan. Biochemistry 26, 6233–6238.

Ueda, E., Sevlever, D., Prince, G. M., Rosenberry, T. L., Hirose, S. and Medof, M. E. (1993). A candidate mammalian glycosylinositolphospholipid precursors containing three phosphoethanolamines. J. Bol. Chem. 268, 9998–10002.

Vijaykumar, M., Naik, R. S. and Gowda, D. C. (2001). *Plasmaodium falciparum* glycosylphosphatidylinositol-induced TNF-α secretion by macrophages is mediated without membrane insertion or endocytosis. J. Biol. Chem. *276*, 6909–6912.

White, I. J., Souabni, A. and Hooper, N. M. (2000). Comparison of the glycosylphosphatidylinositol cleavage/attachment site between mammalian cells and parasitic protozoa. J. Cell Sci. *113*, 721–727.

Wongkajornsilp, A., Sevlever, D. and Rosenberry, T. L. (2001). Metabolism of exogenous *sn*-1-alkyl-*sn*-2-lyso-glucosaminyl-phosphatidyinositol in HeLa D cells. Accumulation of glucosaminyl(acyl)phosphatidylinositol in a metabolically inert compartment. Biochem. J. *359*, 305–313.

Xia, M.-Q., Hale, G., Lifey, M. R., Ferguson, M. A. J., Campbell, D., Packman, L. and Waldmann, H. (1993). Structure of the CAMPATH-1 antigen, a glycosyl phosphatidylinositol-anchored glycoprotein which is an exceptionally good target for complement lysis. Biochem. J. *291*, 633–640.

Yamashita, K., Mizouochi, T. and Kobata, A. (1982). Analysis of oligosaccharides by gel filtration. Methods Enzymol. *83*, 105–126.

Zawadzki, J., Scholz, C., Currie, G., Coombs, G. H. and McConville, M. J. (1998). The glycoinositolphospholipids from *Leishmania panamensis* contain unusual glycan and lipid moieties. J. Mol. Biol. *282*, 229–287.

Chapter 6

Almeida, I. C., Camargo, M. M., Procopio, D. O., Silva, L. S., Mehlert, A., Travassos, L. R., Gazzinelli, R. T. and Ferguson, M. A. (2000). Highly purified glycosylphosphatidylinositols from *Trypanosoma cruzi* are potent proinflammatory agents. EMBO J. *19*, 1476–1485.

Anderson, R. A., Boronenkov, I. V., Doughman, S. D., Kunz, J. and Loijens, J. C. (1999). Phosphatidylinositol phosphate kinases, a multifaceted family of signaling enzymes. J. Biol. Chem. *274*, 9907–9910.

Antonsson, B. E. (1994). Purification and characterization of phosphatidylinositol synthase from human placenta. Biochem. J. *297*, 517–522.

Antonsson, B. E. and Klig, L. S. (1996). *Candida albicans* phosphatidylinositol synthase has common features with both *Saccharomyces cerevisiae* and mammalian phosphatidylinositol synthases. Yeast *12*, 449–456.

Azzouz, N., Gerold, P., Kedees, M. H., Shams-Eldin, H., Werner, R., Capdeville, Y. and Schwarz, R. T. (2001). Regulation of *Paramecium primaurelia* glycosylphosphatidylinositol biosynthesis via dolichol phosphate mannose synthesis. Biochemie *83*, 801–809.

Azzouz, N., Striepen, B., Gerold, P., Capdeville, Y. and Schwarz, R. T. (1995). Glycosylinositol-phosphoceramide in the free-living protozoan *Paramecium primaurelia*: modification of core glycans by mannosyl phosphate. EMBO J. *14*, 4422–4433.

Azzouz, N., Striepen, B., Gerold, P., Capdeville, Y. and Schwartz, R. T. (1998). Glycosylinositol-phosphoceramide in the free living protozoan *Paramecium primaurelia*: modification of core glycans by mannosyl phosphate. EMBO J. *14*, 4422–4433.

Balla, T. (1998). Phosphatidylinositol 4-kinases. Biochim. Biophys. Acta *1436*, 69–85.

Banfic, H., Tang, X., Batty, I. H., Downes, C. P., Chen, C. and Rittenhouse, S. E. (1998). A novel integrin-activated pathway forms PKB/Akt-stimulatory phosphatidylinositol 3,4-bisphosphate via phosphatidylinositol 3-phosphate in platelets. J. Biol. Chem. *273*, 13–16.

Baumann, N. A., Vidugiriene, J., Machamer, C. E. and Menon, A. K. (2000). Cell surface display and intracellular trafficking of free glycosylphosphatidylinositols in mammalian cells. J. Biol. Chem. *275*, 7378–7389.

Benachour, A., Sipos, G., Flury, I., Regiori, F., Canivenc-Gansel, E., Vionnet, C., Conzelmann, A. and Benghezal, M. (1999). Deletion of GP17, a yeast gene required for addition of a side chain to the glycosylphosphatidylinositol (GPI) core structure, affects GPI protein transport, remodeling, and cell wall integrity. J. Biol. Chem. *274*, 15251–15261.

Bleasdale, J. E. and Wallis, P. (1981). Phosphatidylinositol inositol exchange in rabbit lung. Biochim. Biophys. Acta *664*, 428–440.

Brown, H. A., Gutowski, S., Moomaw, C. R., Slaughter, C. and Sternweis, P. C. (1993). ADP-ribosylation factor, a small GTP-dependent regulatory protein, stimulates phospholipase D activity. Cell *75*, 1137–1144.

Brown, G. M., Millar, A. R., Masterson, C., Brinacombe, J. S., Nikolaev, A. V. and Ferguson, M. A. (1996). Synthetic phosphooligosaccharide fragments of lipophosphoglycan as acceptors for *Leishmania major* α-D-mannosylphosphate transferase. Eur. J. Biochem. *242*, 410–416.

Burd, C. G. and Emr, S. D. (1998). Phosphatidylinositol(3)-phosphate signalling mediated by specific binding to RING FYVE domains. Mol. Cell 2, 157–162.

Burgering, B. M. T., Medema, R. H., Maassen, J. A., van de Wetering, M. L., van der Eb, A. J., McCormick, F. and Bos, J. L. (1991). Insulin stimulation of gene expression mediated by p21ras activation. EMBO J. *10*, 1103–1109.

Butikofer, P., Zollinger, M. and Brodbeck, U. (1992). Alkykacyl glycerophosphoinositol in human and bovine erythrocytes: molecular species composition and comparison with glycosyl-inositol-phospholipid anchors of erythrocyte acetylcholinesterases. Eur. J. Biochem. *208*, 677–683.

Buxbaum, L. U., Milne, K. G., Werbovetz, K. A. and Englund, P. T. (1996). Myristate exchange on the *Trypanosoma brucei* variant surface glycoprotein. Proc. Natl Acad. Sci. USA *93*, 1178–1183.

Buxeda, R. J., Nickels, J. T., Jr., Belunis, C. J. and Carman, G. M. (1991). Phosphatidylinositol 4-kinase from *Saccharomyces cerevisiae*. Kinetic analysis using Triton X-100/phosphatidylinositol-mixed micelles. J. Biol. Chem. *266*, 13859–13865.

Canivenc-Gansel, E., Imhof, I., Reggiori, F., Burda, P., Conzelmann, A. and Benachour, A. (1998). GPI anchor biosynthesis in yeast: phosphoethanolamine is attached to the α 1,4-linked mannose of the complete precursor glycophospholipid. Glycobiology *8*, 761–770.

Cantley, L. C., Auger, K. R., Carpenter, C., Duckworth, B., Graziani, A., Kapeller, R. and Soltoff, S. (1991). Oncogenes and signal transduction. Cell *64*, 281–302.

Carman, G. M. and Fischl, A. S. (1992). Phosphatidylinositol synthase from yeast. Methods Enzymol. *209*, 305–312.

Carman, G. M. and Henry, S. A. (1999). Phospholipid biosynthesis in the yeast *Saccharomyces cerevisiae* and interrelationship with other metabolic processes. Progr. Lipid Res. *38*, 361–399.

Carman, G. M. and Kelley, M. J. (1992). CDP-diacyglycerol synthase from yeast. Methods Enzymol. *209*, 242–247.

Carpenter, C. L. and Cantley, L. C. (1990). Phosphoinositide kinases. Biochemistry *29*, 11147–11156.

Carpenter, C., Duckworth, B., Auger, K., Cohen, B., Schaffhausen, B. and Cantley, L. (1990). Purification and characterization of phosphoinositide 3-kinase from rat liver. J. Biol. Chem. *265*, 19704–19711.

Carreira, J. C., Jones, C., Wait, R., Previato, J. O. and Mendonca-Previato, L. (1996). Structural variation in the glycoinositolphospholipids of different strains of *Trypanosoma cruzi*. Glycoconjugate J. *13*, 955–966.

Carstensen, S., Pliska-Matyshak, G., Bhuvarahamurthy, N., Robbins, K. M. and Murthy, P. P. N. (1999). Biosynthesis and localization of phosphatidyl-*scyllo*-inositol in barley aleurone cells. Lipids *34*, 67–73.

Carter, A. N., Huang, R., Sorisky, A., Downes, C. P. and Rittenhouse, S. E. (1994). Phosphatidylinositol 3,4,5-trisphosphate is formed from phosphatidylinositol 4,5-bisphosphate in thrombin-stimulated platelets. Biochem. J. *301*, 415–420.

Carver, M. A. and Turco, S. J. (1991). Cell-free biosynthesis of lipophosphoglycan from *Leishmania donovani*. Characterization of microsomal galactosyltransferase and mannosyltransferase activities. J. Biol. Chem. *266*, 10974–10981.

Chen, R., Walter, E. I., Parker, G., Lapurga, J. P., Millan, J. L., Ikehara, Y., Udenfriend, S. and Medof, M. E. (1998). Mammalian glycophosphatidylinositol anchor transfer to proteins and post transfer deacylation. Proc. Natl Acad. Sci. USA *95*, 9512–9517.

Cho, M. H., Shears, S. B. and Boss, W. F. (1993). Changes in phosphatidylinositol metabolism in response to hyperosmotic stress in Daucus carota L. cells grown in suspension culture. Plant Physiol (Bethesda) *103*, 637–647.

Conway, A. M., Rakhit, S., Pyne, S. and Pyne, N. J. (1999). Platelet-derived-growth-factor stimulation of the p42/44 mitogen-activated protein kinase pathway in airway smooth muscle: role of pertussis-toxin-sensitive G-proteins, c-rc tyrosine kinases and phosphoinositide 3-kinase. Biochem. J. *337*, 171–177.

Conzelmann, A., Puoti, A., Lester, R. L. and Desponds, C. (1992). Two different types of lipid moieties are present in glycophosphoinositol-anchored membrane proteins of *Saccharomyces cerevisiae*. EMBO J. *11*, 457–466.

Costello, L. C. and Orlean, P. (1992). Inositol acylation of a potential glycosyl phosphoinositol anchor precursor from yeast requires acyl coenzyme A. J. Biol. Chem. *267*, 8599–8603.

Craxton, A., Erneux, C. and Shears, S. B. (1994). Inositol 1,4,5,6-tetrakisphosphateis phosphorylated in rat liver by a 3-kinase that is distinct fro inositol 1,4,5-trisphopshate 3-kinase. J. Biol. Chem. *269*, 4337–4342.

Cubitt, A. B. and Gershengorn, N. C. (1990). CMP activates reversal of phosphatidylinositol synthase and base exchange by distinct mechanisms in rat pituitary GH3 cells. Biochem. J. *257*, 639–644. Biochem. J. *272*, 813–816.

DeCamilli, P., Emr, D., McPherson, P. S. and Novick, P. (1996). Phosphoinositides as regulators in membrane traffic. Science *271*, 1533–1539.

de Lederkremer, R. M., Lima, C., Ramirez, M. I., Ferguson, M. A., Homans, S. W. and Thomas Oates, J. (1991). Complete structure of the glycan of lipopeptidophosphoglycan from *Trypanosoma cruzi* epimastigotes. J. Biol. Chem. *266*, 23670–23675.

DeLuca, A. W., Rush, J. S., Lehrman, M. A. and Waechter, C. J. (1994). Mannolipid donor specificity of glycosylphosphatidylinositol mannosyltrans-ferase-I (GPIMT-I) determined with an assay system utilizing mutant CHO-K1 cells. Glycobiology *4*, 909–916.

Desrivieres, S., Cooke, F. T., Parker, P. J. and Hall, M. N. (1998). MSS4, a phosphatidylinositol-4-phosphate 5-kinase required for organization of the actin cytoskeleton in *Saccharomyces cerevisiae*. J. Biol. Chem. *273*, 15787–15793.

Dickson, R. C., Nagiec, E. E., Wells, G., Nagiec, M. M. and Lester, R. L. (1997). Synthesis of mannose-(inositol-P)$_2$-ceramide, the major sphingolipid in *Saccharomyces cerevisiae*, requires the IPT1 (YDR072c) gene. J. Biol. Chem. *272*, 29620–29625.

Doering, T. L. and Schekman, R. (1997). Glycosyl-phosphatidylinositol anchor attachment in a yeast *in vitro* system. Biochem. J. *328*, 669–675.

Doering, T. L., Masterson, W. J., Englund, P. T. and Hart, G. W. (1989). Biosynthesis of the glycosylphosphatidylinositol membrane anchor of the trypanosome variant surface glycoprotein: origin of the non-acetylated glucosamine. J. Biol. Chem. *264*, 11168–11173.

Doering, T. L., Pessin, M. S., Hart, G. W., Raben, D. M. and Englund, P. T. (1994). The fatty acids in unremodelled trypanosome glycosyl-phosphatidylinositols. Biochem. J. *299*, 741–746.

Doerrler, W. T. and Lehrman, M. A. (2000). A water-soluble analogue of glucosaminylphosphatidylinositol distinguishes two activities that palmitoylate inositol on GPI anchors. Biochem. Biophys. Res. Commun. *267*, 296–299.

Doerrler, W. T., Ye, J., Falck, J. R. and Lehrman, M. A. (1996). Acylation of glucoaminyl phosphatidylinositol revisited. Palmitoyl-CoA dependent palmitoylation of the inositol residue of a synthetic dioctanoyl glucosaminyl phosphatidylinositol by hamster membranes permits efficient mannosylation of the glucosamine residue. J. Biol. Chem. *271*, 27031–27038.

Dove, S. K., Cooke, F. T., Douglas, M. R., Sayers, L. G., Parker, P. J. and Michell, R. H. (1997). Osmotic stress activates phosphatidylinositol-3,5-bisphosphate synthesis. Nature *390*, 187–192.

Downes, C. P. and Macphee, C. H. (1990). *myo*-Inositol metabolites as cellular signals. Eur. J. Biochem. *193*, 1–18.

Englund, P. T. (1993). The structure and biosynthesis of glycosyl phosphatidylinositol protein anchors. Annu. Rev. Biochem. *62*, 121–138.

Ferguson, M. A. (1999). The structure, biosynthesis and functions of glycosylphosphatidylinositol anchors, and the contributions of trypanosome research. J. Cell Sci. *112*, 299–1809.

Ferguson, M. A. (2000). Glycosylphosphatidylinositol biosynthesis validated as a drug target for African sleeping sickness. Proc. Natl Acad. Sci. *97*, 10673–10675.

Ferguson, M. A., Brimacombe, J. S., Brown, J. R., Crossman, A., Dix, A., Field, R. A., Guther, M. L., Milne, K. G., Sharma, D. K. and Smith, T. K. (1999). The GPI biosynthetic pathway as a therapeutic target for African sleeping sickness. Biochim. Biophys. Acta *1455*, 327–340.

Fischl, A. S., Liu, Y., Browdy, A. and Cremesti, A. (1999). Inositolphosphoryl ceramide synthase from yeast. Methods Enzymol. *311*, 123–130.

Flanagan, C. A., Schnieders, E. A., Emerick, A. W., Kunisawa, R., Admon, A. and Thorner, J. (1993). Phosphatidylinositol 4-kinase gene structure and requirement for yeast cell viability. Science *262*, 1444–1448.

Franke, T. F., Kaplan, D. R., Cantley, L. C. and Toker, A. (1997). Direct regulation of the Akt proto-oncogene product by phosphatidylinositol-3,4-bisphosphate. Science *275*, 665–668.

Fruman, D. A., Meyers, R. E. and Cantley, L. C. (1998). Phosphoinositide kinases. Annu. Rev. Biochem. 67, 481–507.

Gaidarov, I., Smith, M. E. K., Domin, J. and Keen, J. H. (2001). The class II phosphoinositide 3-kinase C2alpha is activated by clathrin and regulates clathrin-mediated membrane trafficking. Mol. Cell 7, 443–449.

Gary, J. D., Wurmser, A. E., Bonangelino, C. J., Weisman, L. S. and Emr, S. D. (1998). Fab1p is essential for PtdIns(3)P 5-kinase activity and the maintenance of vacuolar size and membrane homeostasis. J. Cell Biol. 143, 65–79.

Gaudette, D. C., Aukema, H. M., Jolly, C. A., Chapkin, R. S. and Holub, B. J. (1993). Mass and fatty acid composition of the 3-phosphorylated phosphatidylinositol bisphosphate isomer in stimulated human platelets. J. Biol. Chem. 268, 13773–13776.

Gaynor, E. C., Mondesert, G., Grimme, S. J., Reed, S. I., Orlean, P. and Emr, S. D. (1999). MCD4 encodes a conserved endoplasmic reticulum membrane protein essential for glycosylphosphatidylinositol anchor synthesis in yeast. Mol. Biol. Cell 10, 627–648.

Gehrmann, T., Vereb, G., Schmidt, M., Klix, D., Meyer, H. E., Varsanyi, M. and Heilmeyer, L. M. (1996). Identification of a 200 kDa polypeptide as type 3 phosphatidylinositol 4-kinase from bovine brain by partial protein and cDNA sequencing. Biochim. Biophys. Acta 1311, 53–63.

Gerold, P., Dieckmann-Schuppert, A. and Schwarz, R. T. (1994). Glycosylphosphatidylinositols synthesized by asexual erythrocytic stages of the malarial parasite, Plasmodium falciparum. Candidates for plasmodial glycosylphosphatidylinositol membrane anchor precursors and pathogenicity factors. J. Biol. Chem. 269, 2597–2606.

Gerold, P., Jung, N., Azzouz, N., Freiberg, N., Kobe, S. and Schwarz, R. T. (1999). Biosynthesis of glycosylphosphatidylinositols of Plasmodium falciparum in a cell-free incubation system: inositol acylation is needed for mannosylation of glycosylphosphatidylinositols. Biochem. J. 344, 731–738.

Ghalayini, A. and Eichberg, J. (1993). Purification of phosphatidylinositol synthase from brain. In: Lipid Metabolism in Signaling Systems, Methods in Neurosciences, (Fain, J. N., ed.). Academic Press Inc., New York, pp. 85–92.

Goto, K. and Kondo, H. (1996). Heterogeneity of diacylglycerol kinase in terms of molecular structure, biochemical characteristics and gene expression localization in the brain. J. Lipid Mediat. Cell Signal. 14, 251–257.

Graziani, A., Ling, L. E., Endemann, G., Carpenter, C. L. and Cantley, L. C. (1992). Purification and characterization of human erythrocyte phosphatidylinositol 4-kinase. Biochem. J. 284, 39–45.

Grimme, S. J., Westfall, B. A., Wiedman, J. M., Taron, C. H. and Orlean, P. (2001). The essential Smp3 protein is required for addition of the side-branching fourth mannose during assembly of yeast glycosylphosphatidyl-inositols. J. Biol. Chem. 276, 27731–27739.

Guther, M. L. and Ferguson, M. A. (1995). The role of inositol acylation and inositol deacylation in GPI biosynthesis in Trypanosoma brucei. EMBO J. 14, 3080–3093.

Guther, M. L., Masterson, W. J. and Ferguson, M. A. (1994). The effects of phenylmethylsulfonyl fluoride on inositol-acylation and fatty acid remodeling in African trypanosomes. J. Biol. Chem. 269, 18694–18701.

Hawkins, P. T., Jackson, T. R. and Stephens, L. R. (1992). Platelet-derived growth factor stimulates synthesis of PtdIns(3,4,5)P$_3$ by activating a PtdIns(4,5)P$_2$ 3-OH kinase. Nature 358, 157–159.

Heacock, A. M., Uhler, M. D. and Agronoff, B. W. (1996). Cloning of CDP-diacylglycerol synthase from a human neuronal cell line. J. Neurochem. 67, 2200–2203.

Heise, N., Raper, J., Buxbaum, L. U., Peranovich, T. M. S. and Cardoso de Almeida, M. L. (1996). Identification of complete precursors for glycosylphos-phatidylinositol protein anchors of Trypanosoma cruzi. J. Biol. Chem. 271, 16877–16887.

Hinchliffe, K. A., Ciruela, A. and Irvine, R. F. (1998). PIPkins, their substrates and their products: new functions for old enzymes. Biochim. Biophys. Acta 1436, 87–104.

Hirose, S., Ravi, L., Hazra, S. V. and Medoff, M. E. (1991). Assembly and deacetylation of N-acetylglucosaminyl-plasmanylinositol in normal and affected paroxysmal nocturnal hemoglobinuria cells. Proc. Natl Acad. Sci. USA 88, 3762–3766.

Hirose, S., Mohney, R. P., Mutka, S. C., Ravi, L., Singleton, D. R., Perry, G., Tartakoff, A. M. and Medof, M. E. (1992a). Derivation and characterization of glycoinositol-phospholipid anchor defective human K562 cell clones. J. Biol. Chem. 267, 5272–5278.

Hirose, S., Prince, G. M., Sevlever, D., Ravi, L., Rosenberry, T. L., Ueda, E. and Medof, M. E. (1992b). Characterization of putative glycoinositol phospholipid anchor precursors in mammalian cells. Localization of phosphoethanolamine. J. Biol. Chem. 267, 16968–16974.

Hirose, S., Knez, J. J. and Medof, M. E. (1995). Mammalian glycosylphos-phatidylinositol-anchored proteins and intracellular precursors. Methods Enzymol. 250, 582–614.

Hokin, L. E. (1985). Receptors and phosphoinositide-generated second messen-gers. Annu. Rev. Biochem. 54, 205–235.

Holub, B. J., Kuksis, A. and Thompson, W. (1970). Molecular species of mono-, di-, and triphosphoinositides of bovine brain. J. Lipid Res. *11*, 558–564.

Holub, B. J. (1994). The Mn^{2+}-activated incorporation of inositol into molecular species of phosphatidylinositol in rat liver microsomes. Biochim. Biophys. Acta *369*, 111–122.

Holub, B. J. (1976). Specific formation of arachidonoyl phosphatidylinositol from 1-acyl-*sn*-glycero-3-phosphorylinositol in rat liver. Lipids *11*, 1–5.

Holub, B. J. and Kuksis, A. (1978). Metabolism of molecular species of diacylglycerophospholipids. Adv. Lipid Res. *16*, 1–125.

Holub, B. J. and Piekarski, J. (1979). The formation of phosphatidylinositol by acylation of 2-acyl-*sn*-glycero-3-phosphorylinositol in rat liver microsomes. Lipids *14*, 529–532.

Homans, S. W., Ferguson, M. A., Dwek, R. A., Rademacher, T. W., Anand, R. and Williams, A. F. (1988). Complete structure of the glycosyl phosphatidylinositol membrane anchor of rat brain Thy-1 glycoprotein. Nature *333*, 269–272.

Hunter, T. (1995). Protein kinases and phosphatases: the yin and yang of protein phosphorylation and signaling. Cell *80*, 225–236.

Icho, T., Sparrow, C. P. and Raetz, C. R. H. (1985). Molecular cloning and sequencing of the gene for CDP diglyceride synthetase of *Escherichia coli*. J. Biol. Chem. *260*, 12078–12083.

Ilg, T., Etges, R., Overath, P., McConville, M. J., Thomas Oates, J., Thomas, J., Homans, S. W. and Ferguson, M. A. (1992). Structure of *Leishmania mexicana* lipophosphoglycan. J. Biol. Chem. *267*, 6834–6840.

Ilgoutz, S. C., Mullin, K. A., Southwell, B. R. and McConville, M. J. (1999a). Glycosylphosphatidylinositol biosynthetic enzymes are localized to a stable tubular subcompartment of the endoplasmic reticulum in *Leishmania mexicana*. EMBO J. *18*, 3643–3654.

Ilgoutz, S. C., Zawadzki, J. L., Ralton, J. E. and McConville, M. J. (1999b). Evidence that free GPI glycolipids are essential for growth of *Leishmania mexicana*. EMBO J. *18*, 2746–2755.

Inoue, N., Watanabe, R., Takeda, J. and Kinoshita, T. (1996). *PIGC*, one of the three genes involved in the first step of glycophosphatidylinositol biosynthesis is a homologue of *Saccharomyces cerevisiae GP12*. Biochem. Biophys. Res. Commun. *226*, 193–199.

Irvine, R. F., Hemington, N. and Dawson, R. M. C. (1978). The hydrolysis of phosphatidylinositol by lysosomal enzymes of rat liver and brain. Biochem. J. *176*, 475–484.

Irvine, R. F. (1998). Manganese-stimulated phosphatidylinositol headgroup exchange in rat liver microsomes. Biochim. Biophys. Acta *1393*, 292–298.

Ishihara, H., Shibasaki, Y., Kizuki, N., Wada, T., Yazaki, Y., Asano, T. and Oka, Y. (1998). Type I phosphatidylinositol 4-phosphate 5-kinase. J. Biol. Chem. 273, 8741–8748.

Itoh, T., Ijuin, T. and Takenawa, T. (1998). A novel phosphatidylinositol-5-phosphate 4-kinase (phosphatidylinositol-phosphate kinase IIγ) is phosphorylated in the endoplasmic reticulum in response to mitogenic signals. J. Biol. Chem. 273, 20292–20299.

Janmey, P. A. (1994). Phosphoinositides and calcium as regulators of cellular actin assembly and disassembly. Annu. Rev. Physiol. 56, 169–191.

Jenkins, G. H., Fisette, P. L. and Anderson, R. A. (1994). Type I phosphatidylinositol 4-phosphate 5-kinase isoforms are specifically stimulated by phosphatidic acid. J. Biol. Chem. 269, 11547–11554.

Jones, D. R., Gonzalez-Garcia, A., Diez, E., Martinez-AC, Carrera, A. C. and Meridas, I. (1999). The identification of phosphatidylinositol 3,5-bisphosphate in T-lymphocytes and its regulation by interleukin-2. J. Biol. Chem. 274, 18407–18413.

Justin, A.-M., Kader, J.-C. and Collin, S. (2002). Phosphatidylinositol synthesis and exchange of the inositol head are catalyzed by the single phosphatidylinositol synthase 1 from Arabidopsis. Eur. J. Biochem. 269, 2347–2352.

Kamitani, T., Chang, H. M., Rollins, C., Waneck, G. L. and Yeh, E. T. (1993). Correction of the class H defect in glycosylphosphatidylinositol anchor biosynthesis in Ltk-cells by a human cDNA clone. J. Biol. Chem. 268, 20733–20736.

Kent, C. (1995). Eukaryotic phospholipid biosynthesis. Annu. Rev. Biochem. 64, 315–343.

Kinnard, R. L., Narasimhan, B., Pliska-Matyshak, G. and Murthy, P. P. N. (1995). Characterization of scyllo-inositol-containing phosphatidylinositol in plant cells. Biochem. Biophys. Res. Commun. 210, 549–555.

Kinoshita, T. and Inoue, N. (2000). Dissecting and manipulating the pathway for glycosylphosphatidylinositol-anchor biosynthesis. Curr. Opin. Chem. Biol. 4, 632–638.

Kinoshita, T., Ohnishi, K. and Tanaka, J. (1997). GPI-anchor synthesis in mammalian cells: genes, their products, and a deficiency. J. Biochem (Tokyo) 122, 251–257.

Klezovitch, O., Brandenburger, Y., Geindre, M. and Deshusses, J. (1993). Characterization of reactions catalysed by yeast phosphatidylinositol synthase. FEBS Lett. 320, 256–260.

Kostova, Z., Rancour, D. M., Menon, A. K. and Orlean, P. (2000). Photolabeling with P32-(4-azidoanilido)uridine 5′-trisphosphate identifies Gpi3p as the

UDP-GlcNAc-binding subunit of the enzyme that catalyses formation of GlcNAc-phosphatidylinositol, the first glycolipid intermediate in glycosylphosphatidylinositol synthesis. Biochem. J. *350*, 815–822.

Kuksis, A. and Myher, J. J. (1990). Mass analysis of molecular species of diradylglycerols. In: Methods in Inositide Research (Robin, R. F., ed.). Raven Press, New York, pp. 187–216.

Lee, T.-C., Buell, A. B. and Blank, M. L. (1991). Occurrence of ether-containing inositol phospholipids in bovine erythrocytes. Biochem. Biophys. Res. Commun. *175*, 673–678.

Leidich, S. D. and Orlean, P. (1996). Gpi1, a *Saccharomyces cerevisiae* protein that participates in the first step in glycosylphosphatidylinositol anchor synthesis. J. Biol. Chem. *271*, 27829–27837.

Leidich, S. D., Kostova, Z., Latek, R. R., Costello, L. C., Drapp, D. A., Gray, W., Fassler, J. S. and Orlean, P. (1995). Temperature-sensitive yeast GPI anchoring mutants *gpi2* and *gpi3* are defective in the synthesis of *N*-acetylglucosaminyl phosphatidylinositol. Cloning of the *GPI2* gene. J. Biol. Chem. *270*, 13029–13035.

Lester, R. and Dixon, R. C. (1993). Sphingolipids with inositol-phosphate-containing head groups. Adv. Lipid Res. *26*, 253–274.

Linassier, C., MacDougall, L. K., Domin, J. and Waterfield, M. D. (1997). Molecular cloning and biochemical characterization of a *Drosophila* phosphatidylinositol-specific phosphoinositide 3-kinase. Biochem. J. *321*, 849–856.

Loijens, J. C. and Anderson, R. A. (1996). Type I phosphatidylinositol-4-phosphate 5-kinases are distinct members of this novel lipid kinase family. J. Biol. Chem. *271*, 32937–32943.

Lykidis, A., Jackson, P. and Jackowski, S. (2001). Lipid activation of CTP:phosphocholine cytidyltransferase α: characterization and identification of a second activation domain. Biochemistry *40*, 494–503.

Lykidis, A., Jackson, P. D., Rock, C. O. and Jackowski, S. (1997). The role of CDP-diacylglycerol synthetase and phosphatidylinositol synthetase activity levels in the regulation of cellular phosphatidylinositol content. J. Biol. Chem. *272*, 33402–33409.

Lykidis, A., Murti, K. G. and Jackowski, S. (1998). Cloning and characterization of a second human CTP:phosphocholine cytidyltransferase. J. Biol. Chem. *273*, 14022–14029.

Lykidis, A. and Jackowski, S. (2001). Regulation of mammalian cell membrane biosynthesis. Progr. Nucleic Acid Res. Mol. Biol. *65*, 361–393.

Mani, I., Gaudette, D. C. and Holub, B. J. (1996). Increased formation of phosphatidylinositol-4-phosphate in human platelets stimulated with lysophosphatidic acid. Lipids *31*, 1265–1268.

Mann, K. J. and Sevlever, D. (2001). 1,10-Phenanthroline inhibits glycosylphosphatidylinositol anchoring by preventing phosphoethanolamine addition to glycosylphosphatidylinositol anchor precursors. Biochemistry 40, 1205–1213.

Masterson, W. J., Raper, J., Doering, T. L., Hart, G. W. and Englund, P. T. (1990). Fatty acid remodeling: a novel reaction sequence in the biosynthesis of trypanosome glycosyl phosphatidylinositol membrane anchors. Cell 62, 73–80.

Maxwell, S. E., Ramalingam, S., Gerber, L. D., Brink, L. and Udenfriend, S. (1995). An active carbonyl formed during glycosylphosphatidylinositol addition to a protein is evidence of catalysis by a transamidase. J. Biol. Chem. 270, 19576–19582.

Mayor, S., Menon, A. K., Ctross, G. A. M., Ferguson, M. A. J., Dwek, R. A. and Rademacher, T. W. (1990a). Glycolipid precursors for the membrane anchor of Trypanosoma brucei variant surface glycoproteins. I. Glycan structure of the phosphatidylinositol specific phospholipase C resistant and sensitive glycolipids. J. Biol. Chem. 265, 6164–6173.

Mayor, S., Menon, A. K. and Cross, G. A. M. (1990b). Glycolipid precursors for the membrane anchor of Trypanosoma brucei variant surface glycoproteins. II. Lipid structures of the phosphatidylinositol-specific phospholipase C resistant and sensitive glycolipids. J. Biol. Chem. 265, 6174–6181.

McConville, M. J. and Ferguson, M. A. (1993). The structure, biosynthesis and function of glycosylated phosphatidylinositols in the parasitic protozoa and higher eukaryotes. Biochem. J. 294, 305–324.

McConville, M. J. and Menon, A. K. (2000). Recent developments in the cell biology and biochemistry of glycosylphosphatidylinositols (review). Mol. Membr. Biol. 17, 1–16.

McConville, M. J., Collidge, T. A., Ferguson, M. A. and Schneider, P. (1993). The glycoinositol phospholipids of Leishmania mexicana promastigotes. Evidence for the presence of three distinct pathways of glycolipid biosynthesis. J. Biol. Chem. 268, 15595–15604.

McPhee, F., Lowe, G., Vaziri, C. and Downes, C. P. (1991). Phoshatidylinositol synthase and phosphatidylinositol/inositol exchange reactions in turkey erythrocyte membranes. Biochem. J. 275, 187–192.

Menon, A., Baumann, N. A., van't Hof, W. and Vidugiriene, J. (1997). Glycosylphosphatidylinositols: biosynthesis and intracellular transport. Biochem. Soc. Trans. 25, 861–865.

Menon, A. K., Mayor, S. and Schwarz, R. T. (1990). Biosynthesis of glycosylphosphatidylinositol lipids in Trypanosoma brucei: involvement of mannosylphosphoryldolichol as the mannose donor. EMBO J. 9, 4249–4258.

Menon, A. K., Baumann, N. A., van't Hof, W. and Vidugiriene, J. (1997). Glycosylphosphatidylinositols: biosynthesis and intracellular transport. Biochem. Soc. Trans. 25, 861–865.

Menon, A. K., Eppinger, M., Mayor, S. and Schwartz, R. T. (1993). Phosphatidylethanolamine is the donor of the terminal phosphoethanolamine group in trypanosome glycosylphosphatidylinositols. EMBO J. *12*, 1907–1914.

Mensa-Wilmot, K., LeBowitz, J. H., Chang, K. P., al-Qahtani, A., McGwire, B. S., Tucker, S. and Morris, J. C. (1994). A glycosylphosphatidylinositol (GPI)-negative phenotype produced in *Leishmania major* by GPI phospholipase C from *Trypanosoma brucei*: topography of two GPI pathways. J. Cell Biol. *124*, 935–947.

Milne, K. G., Ferguson, M. A. and Englund, P. T. (1999). A novel glycosylphosphatidylinositol in African trypanosomes. A possible catabolic intermediate. J. Biol. Chem. *274*, 1465–1471.

Milne, K. G., Field, R. A., Masterson, W. J., Cottaz, S., Brimacombe, J. S. and Ferguson, M. A. (1994). Partial purification and characterization of the *N*-acetylglucosaminyl-phosphatidylinositol de-*N*-acetylase of glycosylphosphatidylinositol anchor biosynthesis in African trypanosomes. J. Biol. Chem. *269*, 16403–16408.

Miyata, T., Takeda, J., Lida, Y., Yamada, N., Inoue, N., Takahashi, M., Maeda, K., Kitani, T. and Kinoshita, T. (1993). The cloning of *PGI-A*, a component in the early step of GPI-anchor biosynthesis. Science *259*, 1318–1320.

Monaco, M. E., Feldman, M. and Kleinberg, D. I. (1994). Identification of rat liver phosphatidylinositol synthase as a 21 kDa protein. Biochem. J. *304*, 301–305.

Moritz, A., De Graan, P. N. E., Gispen, W. H. and Wirtz, K. W. A. (1992). Phosphatidic acid is a specific activator of phosphatidylinositol 4-phosphate kinase. J. Biol. Chem. *267*, 7207–7210.

Myers, R. and Cantley, L. C. (1997). Cloning and characterization of a wortmannin-sensitive human phosphatidylinositol 4-kinase. J. Biol. Chem. *272*, 4384–4390.

Nair, S. S. D., Leitch, J. and Garg, M. L. (1999). Specific modifications of phosphatidylinositol and non-esterified fatty acid fractions in cultured porcine cardiomyocytes supplemented with n-3 polyunsaturated fatty acids. Lipids *34*, 697–704.

Nakagawa, T., Goto, K. and Kondo, H. (1996a). Cloning, expression, and localization of 230-kDa phosphatidylinositol 4-kinase. J. Biol. Chem. *271*, 12088–12094.

Nakagawa, T., Goto, K. and Kondo, H. (1996b). Cloning and characterization of a 92 kDa soluble phosphatidylinositol 4-kinase. Biochem. J. *320*, 643–649.

Nakanishi, S., Catt, K. J. and Balla, T. (1995). A wortmannin-sensitive phosphatidylinositol 4-kinase that regulates hormone-sensitive pools of inositol phospholipids. Proc. Natl Acad. Sci. USA *92*, 5317–5321.

Narasimhan, B., Pliska-Matyshak, G., Kinnard, R., Carstensen, S., Ritter, M. A., Von Weymarn, L. and Murthy, (1997). Plant Physiol. *113*, 1385–1393.

Nickels, J. T., Jr., Buxeda, R. J. and Carman, G. M. (1994). Regulation of phosphatidylinositol 4-kinase from the yeast *Saccharomyces cerevisiae* by CDP-diacylglycerol. J. Biol. Chem. *269*, 11018–11024.

Nikawa, J., Kodaki, T. and Yamashita, S. (1987). Primary structure and description of the phosphatidylinositol synthase gene of *Saccharomyces cerevisiae*. J. Biol. Chem. *262*, 4876–4881.

Noh, D.-Y., Shin, S. H. and Rhee, S. G. (1995). Phosphoinositide-specific phospholipase C and mitogenic signalling. Biochim. Biophys. Acta *1242*, 99–114.

Odorizzi, G., Babst, M. and Emr, S. D. (2000). Phosphoinositide signaling and the regulation of membrane trafficking in yeast. Trends Biochem. Sci. *25*, 229–235.

Okazaki, I. J. and Moss, J. (1999). Characterization of glycosylphosphatidylino-sitol-anchored, secreted, and intracellular vertebrate mono-ADP-ribosyltrans-ferases. Annu. Rev. Nutr. *19*, 485–509.

Ono, F., Nakagawa, T., Saito, S., Owada, Y., Sakagami, H., Goto, K., Suzuki, M., Matsumoto, S. and Kondo, H. (1998). A novel class II phosphoinositide 3-kinase predominantly expressed in the liver and its enhanced expression during liver regeneration. J. Biol. Chem. *273*, 7731–7736.

Ostlund, R. E., Jr., McGill, J. B., Herskowitz, I., Kipnis, D. M., Santiago, J. V. and Sherman, W. R. (1993). D-*chiro*-inositol metabolism in diabetes mellitus. Proc. Natl Acad. Sci. USA *90*, 9988–9992.

Pal, A., Hall, B. S., Nebeth, D. N., Field, H. I. and Field, M. C. (2002). Differential endocytic functions of *Trypanosoma brucei* Rab5 isoforms reveal a glycosylphosphatidylinositol-specific endosomal pathway. J. Biol. Chem. *277*, 9529–9539.

Paul, K. S., Jiang, D., Morita, Y. S. and Englund, P. T. (2001). Fatty acid synthesis in African trypanosomes: a solution to the myristate mystery. Trends Parasitol. *17*, 381–387.

Paulus, H. and Kennedy, E. P. (1960). The enzymatic synthesis of inositol monophosphate. J. Biol. Chem. *235*, 1303–1311.

Petitot, A., Ogier-Denis, E., Blommart, E. F. C., Meijer, A. J. and Codogno, P. (2000). Distinct classes of phosphatidylinositol 3'-kinases are involved in signaling pathways that control macroautophagy in HT-29 cells. J. Biol. Chem. *275*, 992–998.

Pike, L. J. (1992). Phosphatidylinositol 4-kinases and the role of polyphos-phoinositides in cellular regulation. Endocr. Rev. *13*, 692–706.

Pike, L. J. and Casey, L. (1996). Localization and turnover of phosphatidylinositol 4,5-bisphosphate in caveolin-enriched membrane domains. J. Biol. Chem. *271*, 26453–26456.

Puoti, A. and Conzelmann, A. (1993). Characterization of abnormal free glycophosphatidylinositols accumulating in mutant lymphoma cells of classes B, E, F, and H. J. Biol. Chem. *268*, 7215–7224.

Puoti, A., Desponds, C., Fanhauser, C. and Conzelmann, A. (1991). Characterization of glycophospholipid intermediate in the biosynthesis of glycophosphatidylinositol anchors accumulating in the Thy-1-negative lymphoma line SIA-b. J. Biol. Chem. *266*, 21051–21059.

Ralton, J. E. and McConville, M. J. (1998). Delineation of three pathways of glycosylphosphatidylinositol biosynthesis in *Leishmania mexicana*. Precursors from different pathways are assembled on distinct pools of phosphatidylinositol and undergo fatty acid remodeling. J. Biol. Chem. *273*, 4245–4257.

Rameh, L. E. and Cantley, L. C. (1999). The role of phosphoinositide 3-kinase lipid products in cell function. J. Biol. Chem. *274*, 8347–8350.

Rameh, L. E., Tolias, K. F., Duckworth, B. C. and Cantley, L. C. (1997). A new pathway for synthesis of phosphatidylinositol-4,5-bisphosphate. Nature *390*, 192–196.

Rao, R. H. and Strickland, K. P. (1974). Phosphatidylinositol synthase assay. Bochim. Biophys. Acta *348*, 306–314.

Reggiori, F., Canivenc-Gansel, E. and Conzelmann, A. (1997). Lipid remodeling leads to the introduction and exchange of defined ceramides on GFI proteins in the ER and Golgi of *Saccharomyces cerevisiae*. EMBO J. *16*, 3506 3518.

Reggiori, F. and Conzelmann, A. (1998). Biosynthesis of inositol phosphoceramides and remodeling of glycosylphosphatidylinositol anchors in *Saccharomyces cerevisiae* are mediated by different enzymes. J. Biol. Chem. *273*, 30550–30559.

Reid-Taylor, K. L., Chu, J. W. and Sharom, F. J. (1999). Reconstitution of the glycosylphosphatidylinositol-anchored protein Thy-1: interaction with membrane phospholipids and galactosylceramide. Biochem. Cell. Biol. *77*, 189–200.

Roberts, W. L., Myher, J. J., Kuksis, A., Low, M. G. and Rosenberry, T. L. (1988a). Lipid analysis of the glycoinositol phospholipid membrane anchor of human erythrocyte acetylcholinesterase. Palmitoylation of inositol results in resistance to phosphatidylinositol-specific phospholipase C. J. Biol. Chem. *263*, 18766–18775.

Roberts, W. L., Myher, J. J., Kuksis, A. and Rosenberry, T. L. (1988b). Alkylacylglycerol molecular species in the glycosylinositol phospholipid membrane anchor of bovine erythrocyte acetylcholinesterase. Biochem. Biophys. Res. Commun. *150*, 271–277.

Rietveld, A., Neutz, S., Simons, K. and Eaton, S. (1999). Association of sterol- and glycosylphosphatidylinositol-linked proteins with *Drosophila* raft lipid microdomains. J. Biol. Chem. *274*, 12049–12054.

Rietveld, A. and Simons, K. (1998). The differential miscibility of lipids as the basis for the formation of functional membrane rafts. Biochim. Biophys. Acta *1376*, 467–479.

Saito, S., Goto, K., Tonosaki, A. and Kondo, H. (1997). Gene cloning and characterization of CDP-diacylglycerol synthase from rat brain. J. Biol. Chem. *272*, 9503–9509.

Salman, M. and Pagano, R. E. (1997). Use of a fluorescent analog of CDP-DAG in human skin fibroblasts: characterization of metabolism, distribution, and application to studies on phosphatidylinositol turnover. J. Lipid Res. *38*, 482–490.

Salman, M., Lonsdale, J. T., Besra, G. S. and Brennan, P. J. (1999). Phosphatidylinositol synthesis in mycobacteria. Biochim. Biophys. Acta *1436*, 437–450.

Schmidt, A., Schwartz, R. T. and Gerold, P. (1998). *Plasmodium falciparum*: asexual erythrocytic stage synthesize two structurally distinct free and protein-bound glycosylphosphatidylinositols in maturation-dependent manner. Exp. Parasitol. *88*, 95–102.

Screaton, R. A., Demate, L., Draber, P. and Stanners, C. P. (2000). The specificity for the differentiation blocking activity of carcinoembryonic antigen resides in its glycophosphatidylinositol anchor. J. Cell Biol. *150*, 613–625.

Serunian, L. A., Haber, M. T., Fukui, T., Kim, J. W., Rhee, S. G., Lowenstein, J. M. and Cantley, L. C. (1989). Polyphosphoinositides produced by phosphatidylinositol 3-kinase are poor substrates for phospholipases C from rat liver and bovine brain. J. Biol. Chem. *264*, 17809–17815.

Sharma, D. K., Hilley, J. D., Bangs, J. D., Coombs, G. H., Mottra, J. C. and Menon, A. K. (2000). Soluble GP18 restores glycosylphosphatidylinositol anchoring in a trypanosome cell-free system depleted of lumenal endoplasmic reticulum proteins. Biochem. J. *351*, 717–722.

Sharma, D. K., Smith, T. K., Crossman, A., Brimacombe, J. S. and Ferguson, M. A. (1997). Substrate specificity of the *N*-acetylglucosaminyl-phosphatidylinositol de-*N*-acetylase of glycosylphosphatidylinositol membrane anchor biosynthesis in African trypanosomes and human cells. Biochem. J. *328*, 171–177.

Sharma, D. K., Smith, T. K., Weller, C. T., Crossman, A., Brimacombe, J. S. and Ferguson, M. A. (1999). Differences between the trypanosomal and human GlcNAc-PI de-N-acetylases of glycosylphosphatidylinositol membrane anchor biosynthesis. Glycobiology *9*, 415–422.

Sharma, D. K., Vidugiriene, J., Bangs, J. D. and Menon, A. (1999). A cell-free assay for glycosylphosphatidylinositol anchoring in African trypanosomes. Demonstration of a transamidation reaction mechanism. J. Biol. Chem. *274*, 16479–16486.

Shen, H., Heacock, P. N., Clancey, C. J. and Dowhan, W. (1996). The COS1 gene encoding CDP-diacylglycerol synthase in *Sacharomyces cerevisiae* is essential for cell growth. J. Biol. Chem. *271*, 789–795.

Shepherd, P. R., Withers, D. J. and Siddle, K. (1998). Phosphoinositide 3-kinase: the key switch mechanism in insulin signaling. Biochem. J. *333*, 471–490.

Siddiqui, R. A., Smith, J. L., Ross, A. H., Qiu, R. G., Symons, M. and Exton, J. H. (1995). Regulation of phospholipase D in HL60 cells. Evidence for cytosolic phospholipase D. J. Biol. Chem. *270*, 8466–8473.

Singer, W. D., Brown, H. A. and Sternweiss, P. C. (1997). Regulaton of eukaryotic phosphatidylinositol-specific phspholipase C and phospholipase D. Annu. Rev. Biochem. *66*, 475–509.

Sipos, G., Puotu, A. and Conzelmann, A. (1994). Glycosylphosphatidylinositol membrane anchors in *Saccharomyces cerevisiae*: absence of ceramides from complete precursor glycolipids. EMBO J. *13*, 2789–2796.

Sipos, G., Puoti, A. and Conzelmann, A. (1995). Biosynthesis of the side chain of yeast glycosylphosphatidylinositol anchors is operated by novel mannosyl-transferases located in the endoplasmic reticulum and the Golgi apparatus. J. Biol. Chem. *270*, 19709–19715.

Sipos, G., Reggiori, F., Vionnet, C. and Conzelmann, A. (1997). Alternative lipid remodelling pathways for glycosylphosphatidylinositol membrane anchors in *Saccharomyces cerevisiae*. EMBO. J. *16*, 3495–3505.

Smith, T. K., Cottaz, S., Brimacombe, J. S. and Ferguson, A. J. (1996). Substrate specificity of the dolichol phosphate mannose: glucosaminyl phosphatidyl-inositol α1-4-mannosyltransferase of the glycosylphosphatidylinositol biosyn-thetic pathway of African trypanosomes. J. Biol. Chem. *271*, 6476–6482.

Smith, T. K., Crossman, A., Borissow, C. N., Paterson, M. J., Dix, A., Brinacombe, J. S. and Ferguson, M. A. J. (2001) Specificity of GlcNAc-PI de-N-acetylase of GPI biosynthesis and synthesis of parasite-specific suicide substrate inhibitors. EMBO J. *20*, 3322–3332.

Smith, T. K., Gerold, P., Crossman, A., Paterson, M. J., Borissow, C. N., Brimacombe, J. S., Ferguson, M. A. and Schwarz, R. T. (2002). Substrate specificity of the *Plasmodium falciparum* glycosylphosphatidylinositol biosynthetic pathway and inhibition by species-specific suicide substrates. Biochemistry *41*, 12395–12406.

Smith, T. K., Milne, F. C., Sharma, D. K., Crossman, A., Brimacombe, J. S. and Ferguson, M. A. (1997a). Early steps in glycosylphosphatidylinositol biosynthesis in *Leishmania major*. Biochem. J. *326*, 393–400.

Smith, T. K., Sharma, D. K., Crossman, A., Dix, A., Brimacombe, J. S. and Ferguson, M. A. (1997b). Parasite and mammalian GPI biosynthetic pathways can be distinguished using synthetic substrate analogues. EMBO J. *16*, 6667–6675.

Smith, T. K., Paterson, M. J., Crossman, A., Brinacombe, J. S. and Ferguson, M. J. (2000). Parasite specific inhibition of the glycosylphosphatidylinositol biosynthetic pathway by stereoisomeric substrate analogues. Biochemistry *39*, 11801–11807.

Smith, T. K., Sharma, D. K., Crossman, A., Brinacombe, J. S. and Ferguson, M. A. J. (1999) Selective inhibitors of the glycosylphosphatidylinositol biosynthetic pathway of Trypanosoma brucei. EMBO J. *18*, 5922–5930.

Soltoff, S. P., Kaplan, D. R. and Cantley, L. C. (1993). Phosphatidylinositol 3-kinase. In: Lipid Metabolism in Signaling Systems, Methods in Neurosciences (Fain, J. N., ed.). Academic Press Inc., New York, pp. 100–112.

Stack, J. H., Herman, P. K., Schu, P. V. and Emr, S. D. (1993). A membrane-associated complex containing the Vps15 protein kinase and the Vps34 PI 3-kinase is essential for protein sorting to the yeast lysome-like vacuole. EMBO J. *12*, 2195–2204.

Stephens, L. R., Hawkins, P. T., Carter, N., Chahwala, S. B., Morris, A. J., Whetton, A. D. and Downes, P. C. (1988). L-*myo*-Inositol 1,4,5,6-tetrakisphosphate is present in both mammalian and avian cells. Biochem. J. *249*, 271–282.

Stephens, L. R., Hughes, K. T. and Irvine, R. F. (1991). Pathway of phosphatidylinositol (3,4,5)-trisphosphate synthesis in activated neutrophils. Nature *251*, 33–39.

Stevens, V. L. (1993). Regulation of glycosylphosphatidylinositol biosynthesis by GTP. Stimulation of N-acetylglucosaminephosphatidylinositol deacetylation. J. Biol. Chem. *268*, 9718–9724.

Stevens, V. L. (1995). Biosynthesis of glycosylphosphatidylinositol membrane anchors. Biochem. J. *310*, 361–370.

Stevens, V. L. and Raetz, C. R. (1991). Defective glycosyl phosphatidylinositol biosynthesis in extracts of three Thy-1 negative lymphoma cell mutants. J. Biol. Chem. *266*, 10039–10042.

Stevens, V. L. and Zhang, H. (1994). Coenzyme A dependence of glycosyl phosphatidylinositol biosynthesis in a mammalian cell-free system. J. Biol. Chem. *269*, 31397–31403.

Stevens, V. L., Zhang, H. and Kristyanne, E. S. (1999). Stimulation of glycosyl phosphatidylinositol biosynthesis in mammalian cell-free systems by GTP hydrolysis: evidence for the involvement of membrane fusion. Biochem. J. *341*, 577–584.

Stevenson, J. M., Perera, I. Y. and Boss, W. F. (1998). A phosphatidylinositol 4-kinase pleckstrin homology domain that binds phosphatidylinositol 4-monophosphate. J. Biol. Chem. *273*, 22761–22767.

Striepen, B., Dubremetz, J.-F. and Schwarz, R. T. (1999). Glucosylation of glycosyl phosphatidylinositol membrane anchors: identification of uridine diphosphate-glucose as the direct donor for side chain modification in *Toxoplasma gondii* using carbohydrate analogues. Biochemistry *38*, 1478–1487.

Striepen, B., Tomavo, S., Dubremetz, J. F. and Schwarz, R. T. (1992). Identification and characterization of glycosylinositolphospholipids in *Toxoplasma gondii*. Biochem. Soc. Trans. *20*, 296–301.

Striepen, B., Zinecker, C. F., Damm, J. B. L., Melgers, P. A. T., Gerwig, G. J., Koolen, M., Vliegenthart, J. F. G., Dubremetz, J. F. and Schwarz, R. T. (1997). Molecular structure of the 'low molecular weight antigen' of *Toxoplasma gondii*: a glucose α1-4 *N* acetylgalactosamine makes free glycosylphosphatidylinositols highly immunogenic. J. Mol. Biol. *266*, 797–813.

Subramanian, A. B., Navarro, S., Carrasco, R. A., Marti, M. and Das, S. (2000). Role of exogenous inositol and phosphatidylinositol in glycosylphosphatidylinositol anchor synthesis of GP49 by *Giardia lamblia*. Biochim. Biophys. Acta *1483*, 69–80.

Sutterlin, C., Escribano, M. V., Gerold, P., Maeda, Y., Mazon, M. J., Kinoshita, T., Schwarz, R. T. and Riezman, H. (1998). *Saccharomyces cerevisiae GPI10*, the functional homologue of human PIG-B, is required for glycosylphosphatidylinositol-anchor synthesis. Biochem. J. *332*, 153–159.

Sutterlin, C., Horvath, A., Gerold, P., Schwarz, R. T., Wang, Y., Dreyfuss, M. and Riezman, H. (1997). Identification of a species-specific inhibitor of glycosylphosphatidylinositol synthesis. EMBO J. *16*, 6374–6383.

Takahashi, M., Inoue, N., Ohishi, K., Maeda, Y., Nakamura, N., Endo, Y., Fugita, T., Takeda, J. and Kinoshita, T. (1996). PIG-B, a membrane protein of the endoplasmic reticulum with a large lumenal domain, is involved in transferring the third mannose of the GPI anchor. EMBO J. *15*, 4254–4261.

Takenawa, T. and Egawa, K. (1977). CDP-diglyceride:inositol transferase from rat liver. J. Biol. Chem. *252*, 5419–5423.

Tanaka, S., Nikawa, J., Imai, H., Yamashita, Y. and Hosaka, K. (1996). Molecular cloning of rat phosphatidylinositol synthase cDNA by functional complementation of the yeast *Saccharomyces cerevisiae* pis mutation. FEBS Lett. *393*, 892–898.

Takenawa, T., Saito, M., Nagai, Y. and Egawa, K. (1977). Solubilization of the enzyme catalyzing CDP-diglyceride-independent incorporation of *myo*-inositol into phosphatidyl inositol and its comparison to CDP-diglyceride:inositol transferase. Arch. Biochem. Biophys. *182*, 244–250.

Tanaka, T., Takimoto, T., Morishige, J.-I., Kikuta, Y., Sugiura, T. and Satouchi, K. (1999). Non-methylene-interrupted polyunsaturated fatty acids: effective substitute for arachidonate of phosphatidylinositol. Biochem. Biophys. Res. Commun. *264*, 683–688.

Taron, C. H., Wiedman, J. M., Grimme, S. J. and Orlean, P. (2000). Glycosylphosphatidylinositol biosynthesis defects in Gpi11p- and Gpi13p-deficient yeast suggest a branched pathway and implicate pgi13p in phosphoethanolamine transfer to the third mannose. Mol. Biol. Cell *11*, 1611–1630.

Tiede, A., Bastisch, I., Schubert, J., Orlean, P. and Schmidt, R. E. (1999). Biosynthesis of glycosylphosphatidylinositols in mammals and unicellular microbes. Biol. Chem. *380*, 503–523.

Tiede, A., Schubert, J., Nischan, C., Jensen, I., Westfall, B., Taron, C. H., Orlean, P. and Schmidt, R. E. (1998). Human and mouse Gpi1 homologues restore glycosyl phosphatidylinositol membrane anchor biosynthesis in yeast mutants. Biochem. J. *334*, 69–616.

Toker, A. (1998). The synthesis and cellular roles of phosphatidylinositol 4,5-bisphosphate. Curr. Opin. Cell Biol. *10*, 254–261.

Toker, A. and Cantley, L. C. (1997). Signalling through the lipid products of phosphoinositide-3-OH-kinase. Nature (London) *387*, 673–676.

Tolias, K. F. and Cantley, L. C. (1999). Pathways for phosphoinositide synthesis. Rev. Chem. Phys. Lipids *98*, 69–77.

Tolias, K. F., Couvillon, A. D., Cantley, L. C. and Carpenter, C. L. (1998a). Characterization of a Rac1- and RhoGD1-associated lipid kinase signaling complex. Mol. Cell. Biol. *18*, 762–770.

Tolias, K. F., Rameh, L. E., Ishihara, H., Shibasaki, Y., Chewn, J., Prestwich, G. D., Cantley, L. C. and Carpenter, C. L. (1998b). Type I phosphatidyl-inositol-4-phosphate 5-kinases synthesize the novel lipids phosphatidylinositol 3,5-bisphosphate and phosphatidylinositol-5-phosphate. J. Biol. Chem. *273*, 18040–18046.

Traynor-Kaplan, A. E., Thompson, B. L., Harris, A. L., Taylor, P., Omann, G. M. and Sklar, L. A. (1989). Transient increase in phosphatidylinositol 3,4-bisphosphate and phosphatidylinositol trisphosphate during activation of human neutrophils. J. Biol. Chem. *264*, 15668–15673.

Udenfriend, S. and Kodukula, K. (1995). How glycosylphosphatidylinositol-anchored membrane proteins are made. Annu. Rev. Biochem. *64*, 563–591.

Urakaze, M., Kamitani, T., DeGasperi, R., Sugiyama, E., Chang, H. M., Warren, C. D. and Yeh, E. T. (1992). Identification of a missing link in glycosylphos-phatidylinositol anchor biosynthesis in mammalian cells. J. Biol. Chem. *267*, 6459–6462.

Valverde, A. M., Lorenzo, M., Navarro, P. and Benito, M. (1997). Phosphatidyl-inositol 3-kinase is a requirement for insulin-like growth factor I—induced differentiation but not for mitogenesis in fetal brown adipocytes. Mol. Endocrinol. *11*, 595–607.

Valverde, A. M., Navarro, P., Teruel, T., Conejo, R., Benito, M. and Lorenzo, M. (1999). Insulin and insulin-like growth factor I up-regulate GLUT4 gene expression in fetal brown adipocytes, in a phosphoinositide 3-kinase-dependent manner. Biochem. J. *33*, 397–405.

Vanhaesebroeck, B., Leewers, S. J., Panayotou, G. and Waterfield, M. (1997a). Phosphoinositide 3-kinases: a conserved family of signal transducers. Trends Biochem. Sci. *22*, 267–272.

Vanhaesebroeck, B., Leevers, S. J., Ahmadi, K., Timms, J., Katso, R., Driscoll, R. C., Woscholski, R., Parker, P. J. and Waterfield, M. D. (2001). Synthesis and function of 3-phosphorylated inositol lipids. Annu. Rev. Biochem. *70*, 535–602.

Vanhaesebroeck, B., Welham, M. J., Kpani, K., Stein, R., Warne, P. H., Zvelebil, M. J., Higashi, K., Volinia, S., Downward, J., Waterfield, M. D. (1997b). p110δ, a novel phosphatidylinositol 3-kinase in leukocytes. Proc. Natl Acad. Sci., USA *94*, 4330–4335.

van't Hof, W., Rodriguez-Boulan, E. and Menon, A. K. (1995). Nonpolarized distribution of glycosylphosphatidylinositols in plasma membrane of polarized Madin-Darby canine kidney cells. J. Biol. Chem. *270*, 24150–24155.

Vidugiriene, J. and Menon, A. K. (1994). The GPI anchor of cell surface proteins is synthesized on the cytoplasmic face of the endoplasmic reticulum. J. Cell Biol. *127*, 333–341.

Vidugiriene, J. and Menon, A. K. (1995). Biosynthesis of glycosylphosphatidylinositol anchors. Methods Enzymol. *250*, 513 535.

Vidugiriene, J., Vainauskas, S., Johnson, A. E. and Menon, A. K. (2001). Endoplasmic reticulum proteins involved in glycosylphosphatidylinositol-anchor attachment: photocrosslinking studies in a cell-free system. Eur. J. Biochem. *268*, 2290–2300.

Vijaykumar, M., Naik, R. S. and Gawda, D. C. (2001). *Plasmodium falciparum* glycosylphosphatidylinositol-induced TNFα secretion by macrophages is mediated without membrane insertion or endocytosis. J. Biol. Chem. *276*, 6909–6912.

Walsh, J. P., Caldwell, K. K. and Majerus, P. W. (1991). Formation of phosphatidylinositol 3-phosphate by isomerization from phosphatidylinositol 4-phosphate. Proc. Natl Acad. Sci. USA *88*, 9184–9187.

Watanabe, R., Inoue, N., Westfall, B., Taron, C. H., Orlean, P., Takeda, J. and Kinoshita, T. (1998). The first step of glycosylphosphatidylinositol biosynthesis is mediated by a complex of PIG-A, PIG-H, PIG-C and GPI1. EMBO J. *17*, 877–885.

Watanabe, R., Ohishi, K., Maeda, Y., Nakamura, N. and Kinoshita, T. (1999). Mammalian and PIG-L and its yeast homologue Gpi12p are *N*-acetylglucosaminylphosphatidylinositol de-N-acetylases essential in glucosylphosphatidylinositol biosynthesis. Biochem. J. *339*, 185–192.

Weeks, R., Dowhan, W., Shen, H., Balantac, N., Meengs, B., Nudelman, E. and Leung, D. W. (1997). Isolation and expression of an isoform of human CDP-diacylglycerol synthase cDNA. DNA Cell Biol. *16*, 281–289.

Whiteford, C. C., Best, C., Kazlauskas, A. and Ulug, E. T. (1996). D-3 phosphoinositide metabolism in cells treated with platelet-derived growth factor. Biochem. J. *319*, 851–860.

Whiteford, C. C., Brearley, C. A. and Ulug, E. T. (1997). Phosphatidylinositol 3,5-bisphosphate defines a novel PI 3-kinase pathway in resting mouse fibroblasts. Biochem. J. *323*, 597–601.

Whitman, M., Downes, C. P., Keeler, T., Keller, T. and Cantley, L. (1988). Type I phosphatidyl kinase makes a novel inositol phospholipid, phosphatidyl-inositol-3-phosphate. Nature *332*, 644–646.

Whitman, M., Kaplan, D., Roberts, T. and Cantley, L. (1987). Evidence for two distinct phosphatidylinositol kinases in fibroblasts. Biochem. J. *247*, 165–174.

Wong, K. and Cantley, L. C. (1994). Cloning and characterization of a human phosphatidylinositol 4-kinase. J. Biol. Chem. *269*, 28878–28884.

Wong, K., Meyers, R. and Cantley, L. C. (1997). Subcellular locations of phosphatidylinositol 4-kinase isoforms. J. Biol. Chem. *272*, 13236–13241.

Wongkajjornsilp, A., Sevlever, D. and Rosenberry, T. L. (2001). Metabolism of exogenous *sn*-1-alkyl-*sn*-2-lyso-glucosaminyl-phosphatidylinositol in HeLa D cells: accumulation of glucosaminyl(acyl)phosphatidylinositol in a metaboli-cally inert compartment. Biochem. J. *359*, 305–313.

Wu, L., Niemeyer, B., Colley, N., Socolich, M. and Zuker, C. S. (1995). Regulation of PLC-mediated signaling in vivo by CDP-diacylglycerol synthase. Nature *373*, 216–222.

Wymann, M. P. and Pirola, L. (1998). Structure and function of phosphoinositide 3-kinases. Biochim. Biophys. Acta *1436*, 127–150.

Yamakawa, A. and Takenava, T. (1993). Phosphatidylinositol 4-kinase from bovine brain. In: Lipid Metabolism in Signaling Systems, Methods in Neurosciences (Fain, J. N., ed.). Academic Press Inc., New York, pp. 93–99.

Yoshida, S., Ohya, Y., Goebl, M., Nakano, A. and Anraku, Y. (1994). A novel gene, STT4, encodes a phosphatidylinositol 4-kinase in the PKC1 protein kinase pathway of *Saccharomyces cerevisiae*. J. Biol. Chem. *269*, 1166–1171.

Zhang, X., Loijens, J. C., Boronenkov, I. V., Parker, G. J., Norris, F. A., Chen, J., Thum, O., Prestwich, G. D., Majerus, P. W. and Anderson, R. A. (1997). Phosphatidylinositol-4-phosphate 5-kinase isozymes catalyze the synthesis of 3-phosphate-containing phosphatidylinositol signaling molecules. J. Biol. Chem. *272*, 17756–17761.

Chapter 7

Abdullah, M., Hughes, P. J., Craxton, A., Gigg, R., Desai, Y., Marecek, J. F., Prestwich, G. D. and Shears, S. B. (1992). Purification and characterization of inositol-1,3,4-trisphosphate 5/6-kinase from rat liver using an inositol hexakisphosphate affinity column. J. Biol. Chem. *267*, 22340–22345.

Albert, C., Safrany, S. T., Bembenek, M. E., Reddy, K. M., Reddy, K. K., Falck, J. R., Broker, M., Shears, S. B. and Mayr, G. W. (1997). Biological variability in the structures of diphosphoinositol polyphosphates in *Dictyostelium discoideum* ad mammalian cells. Biochem. J. *327*, 553–560.

Anderson, R. A., Boronenkov, I. V., Doughman, S. D., Kunz, J. and Loijens, J. C. (1999). Phosphatidylinositol phosphate kinases, a multifaceted family of signaling enzymes. J. Biol. Chem. *274*, 9907–9910.

Arcaro, A., Volinia, S., Zvelebil, M. J., Stein, R., Watton, S. J., Layton, M. J., Gout, I., Ahmadi, K., Downward, J. and Waterfield, M. D. (1998). Human PI3-kinase C2beta – the role of calcium and the C2 domain in enzyme activity. J. Biol. Chem. *273*, 33082–33091.

Arcaro, A. and Wymann, M. P. (1993). Wortmannin is a potent phosphatidyl-inositol 3-kinase inhibitor: the role of phosphatidylinositol 3,4,5-trisphosphate in neutrophil responses. Biochem. J. *296*, 297–301.

Arcaro, A., Zvelebil, M. J., Wallasch, C., Ullrich, A., Waterfield, M. D. and Domin, J. (2000). Class II phosphoinositide 3-kinases are downstream targets of activated polypeptide growth factor receptors. Mol. Cell. Biol. *20*, 3817–3830.

Backer, J. M., Myers, M. G. Jr., Schoelson, S. E., Chin, D. J., Sun, X. J., Miralpeix, M., Hu, P., Margolis, B., Skolnik, E. Y., Schlessinger, J. and White, M. F. (1992). Phosphatidylinositol 3′-kinase is activated by association with IRS-1 during insulin stimulation. EMBO J *11*, 3469–3479.

Balla, T. (1998). Phosphatidylinositol 4-kinases. Biochim. Biophys. Acta *1436*, 69–85.

Balla, T., Downing, G. J., Jaffe, H., Kim, S., Zolyomi, A. and Catt, K. J. (1997). Isolation and molecular cloning of wortmannin-sensitive bovine type III phosphatidylinositol 4-kinases. J. Biol. Chem. *272*, 18358–18366.

Balla, T., Sim, S. S., Iida, T., Choi, K. Y., Catt, K. J. and Rhee, S. G. (1991). Agonist-induced calcium signaling is impaired in fibroblasts overproducing inositol 1,3,4,5-tetrakisphosphate. J. Biol. Chem. *266*, 24719–24726.

Banfic, H., Tang, X., Batty, I. H., Downes, C. P., Chen, C. and Rittenhouse, S. E. (1998). A novel integrin-activated pathway forms PKB/Akt-stimulatory phosphatidylinositol 3,4-bisphosphate via phsphatidylinositol 3-phosphate in platelets. J. Biol. Chem. *273*, 13–16.

Barylko, B., Gerber, S. H., Binns, D. D., Grichine, N., Khvotchev, M., Sudhof, T. C. and Albanesi, J. P. (2001). A novel family of phosphatidylinositol 4-kinases conserved from yeast to humans. J. Biol. Chem. 276, 7705–7708.

Bazenet, C. E. and Anderson, R. A. (1992). Phosphatidylinositol-4-phosphate 5-kinase from human erythrocytes. Meth. Enzyol. 209, 189–202.

Bazenet, C. E., Ruano, A. R., Brockman, J. L. and Anderson, R. A. (1990). The human erythrocyte contains two forms of phosphatidylinositol 4-phosphate 5-kinase which are differentially active towards membranes. J. Biol. Chem. 265, 18012–18022.

Berditchevski, F., Tolias, K. F., Wong, K., Carpenter, C. L. and Hemler, E. (1997). A novel link between integrins, transmembrane-4 superfamily proteins (CD63 and CD81), and phosphatidylinositol 4-kinase. J. Biol. Chem. 272, 2595–2598.

Berridge, M. J., Dawson, R. M. C., Downes, C. P., Heslop, J. P. and Irvine, R. F. (1983). Changes in the levels of inositol phosphates after agonist-dependent hydrolysis of membrane phosphoinositides. Biochem. J. 212, 473–482.

Berridge, M. J. and Irvine, R. F. (1984). Inositol trisphosphate, a novel second messenger in cellular signal transduction. Nature 312, 315–321.

Biswas, S., Maity, I. B., Chakrabarti, S. and Biswas, B. B. (1978). Purification and characterization of myo-inositol hexaphosphate-adenosine diphosphate phosphotransferase from Phaseolus aureus. Arch. Biochem. Biophys. 185, 557–566.

Boronenkov, I. V. and Anderson, R. A. (1995). The sequence of phosphatidyl-inositol 4-phosphate 5-kinase defines a novel family of lipid kinases. J. Biol. Chem. 270, 2881–2884.

Brearley, C. A. and Hanke, D. E. (1993). Pathway of synthesis of 3,4- and 4, 5-phosphorylated phosphatidylinositols in the duckweed Spirodela polyrhiza L. Biochem. J. 290, 145–150.

Caldwell, K. K., Lips, D. L., Bansal, V. S. and Majerus, P. W. (1991). Isolation and characterization of two 3-phosphatases that hydrolyse both phosphatidyl-inositol 3-phosphate and inositol 1,3-bisphosphate. J. Biol. Chem. 266, 18378–18386.

Carman, G. M., Belunis, C. J. and Nichels, J. T., Jr (1992). Phosphatidylinositol 4-kinase from yeast. Meth. Enzymol. 209, 183–189.

Carman, G. M., Buxeda, R. J. and Nickels, J. T. Jr. (1996). Adv. Lipobiol. 1, 367–385.

Carpenter, C. L. and Cantley, L. C. (1996). Phosphoinositide 3-kinase and the regulation of cell growth. Biochim. Biophys. Acta Rev. Cancer 1288, 11–16.

Carter, A. N. and Downes, C. P. (1992). Phosphatidylinositol 3-kinase is activated by nerve growth factor and epidermal growth factor in PC12 cells. J. Biol. Chem. 267, 14563–14567.

Carter, A. N., Huang, R., Sorisky, A., Downes, C. P. and Rittenhouse, S. E. (1994). Phosphatidylinositol 3,4,5-trisphosphate is formed from phosphatidylinositol 4,5-bisphosphate in thrombin-stimulated platelets. Biochem. J. *301*, 415–420.

Castellino, A. M., Parker, G. J., Boronenkov, I. V., Anderson, R. A. and Chao, M. V. (1997). A novel interaction between the juxtamembranes region of the p55 tumor necrosis factor receptor and phosphatidylinositol-4-phosphate 5-kinase. J. Biol. Chem. *272*, 5861–5870.

Cengel, K. A., Godbout, J. P. and Freund, G. G. (1998a). Phosphatidylinositol 3-kinase that is activated by okadaic acid. Biochem. Biophys. Res. Commun. *242*, 513–517.

Cengel, K. A., Kason, R. E. and Freund, G. G. (1998b). Phosphatidylinositol 3′-kinase associates with an insulin receptor substrate-1 serine kinase distinct from its intrinsic serine kinase. Biochem. J. *335*, 397–404.

Chakrabarti, S. and Biswas, B. B. (1981). Two forms of phosphoinositol kinase from germinating mung bean seeds. Phytochemistry *20*, 1815–1817.

Chong, L. D., Traynor-Kaplan, A., Bokoch, G. M. and Schwartz, M. A. (1994). The small GTP-binding protein Rho regulates a phosphatidylinositol 4-phosphate 5-kinase in mammalian cells. Cell *79*, 507–513.

Communi, D., Vanweyenbrug, V. and Erneux, C. (1994). Purification and biochemical properties of a high-molecular mass inositol 1,4,5-trisphosphate 3-kinase isozyme in human platelets. Biochem. J. *298*, 669–673.

Communi, D., Vanweyenbrug, V. and Erneux, C. (1995). Molecular study and regulation of d-*myo*-inositol 1,4,5-trisphosphate 3-kinase. Cell. Signaling *7*, 643–650.

Conway, A. M., Rakhit, S., Pyne, S. and Pyne, N. J. (1999). Platelet-derived-growth-factor stimulation of the p42/44 mitogen-activated protein kinase pathway in airway smooth muscle: role of pertussis-toxin-sensitive G-proteins, c-rc tyrosinekinases and phosphoinositide 3-kinase. Biochem. J. *337*, 171–177.

Cooke, F. T., Dove, S. K., McEwen, R. K., Painter, G., Holmes, A. B., Hall, M. N., Michell, R. H. and Parker, P. J. (1998). The stress-activated phosphatidyl-inositol 3-phosphate 5-kinase Fab1p is essential for vacuole function in *S. cerevisiae*. Curr. Biol. *8*, 1219–1222.

Craxton, A., Ali, N. and Shears, S. B. (1995). Comparison of the activities of a multiple inositol polyphosphate phosphatase obtained from several sources: a search for heterogeneity in this enzyme. Biochem. J. *305*, 491–498.

Craxton, A., Emeaux, C. and Shears, S. B. (1994). Inositol 1,4,5,6-tetrakisphosphate is phosphorylated in rat liver by a 3-kinase that is distinct from inositol 1,4,5-trisphosphate 3-kinase. J. Biol. Chem. *269*, 4337–4342.

Cunningham, T. W. and Majerus, P. W. (1991). Pathway for the formation of D-3 phosphate containing inositol phospholipids in PDGF stimulated NIH 3T3 fibroblasts. Biochem. Biophys. Res. Commun. *175*, 568–576.

Cutler, N. S., Heitman, J. and Cardenas, M. E. (1997). STT4 is an essential phosphatidylinositol 4-kinase that is a target of wortmannin in *Saccharomyces cerevisiae*. J. Biol. Chem. *272*, 27671–27677.

Desrivieres, S., Cooke, F. T., Parker, P. J. and Hall, M. N. (1998). MSS4, a phosphatidylinositol-4-phosphate 5-kinase required for organization if the action cytoskelton in *Saccharomyces cerevisiae*. J. Biol. Chem. *273*, 15787–15793.

Divecha, N., Truong, O., Hsuan, J. J., Hinchliffe, K. A. and Irvine, R. F. (1995). The cloning and sequence of the C isoform of PtdIns4P 5-kinase. Biochem. J. *309*, 715–719.

Domin, J., Pages, F., Volina, S., Rittenhouse, S. E., Zvelebil, M. J., Stein, R. C. and Waterfield, M. D. (1997). Cloning of a human phosphoinositide 3-kinase with a C2 domain that displays reduced sensitivity to the inhibitory wortmannin. Biochem. J. *326*, 139–147.

Domin, J. and Waterfield, M. D. (1997). Using structure to define the function of phpsphoinositide 3-kinase family members. FEBS Lett. *410*, 91–95.

Donie, F. and Reiser, G. (1989). A novel, specific binding protein assay for quantitation of intracellular inositol 1,3,4,5-tetrakisphosphate (insP$_4$) using a high affinity InsP$_4$ receptor from cerebellum. FEBS Lett. *254*, 155–158.

Dove, S. K., Cooke, F. T., Douglas, M. R., Sayers, L. G., Parker, P. J. and Michell, R. H. (1997). Osmotic stress activates phosphatidylinositol-3,5-bisphosphate synthesis. Nature *390*, 187–192.

Dowler, S., Currie, R. A., Cambell, D. G., Deak, M., Kular, G., Downes, C. P. and Alessi, D. R. (2000). Identification of pleckstrin-homology-domain-containing proteins with novel phosphoinositide-binding specificities. Biochem. J. *351*, 19–31.

Downes, C. P., Hawkins, P. T. and Irvine, R. F. (1986). Inositol 1,3,4,5-tetrakisphosphate and not phosphatidylinositol 3,4-bisphosphate is the probable precursor of inositol 1,3,4-trisphosphate in agonist-stimulated parotid gland. Biochem. J. *238*, 501–506.

Downing, G. J., Kim, S., Nakanishi, S., Catt, S. J. and Balla, T. (1996). Characterization of a soluble adrenal phosphatidylinositol 4-kinase reveals wortmannin sensitivity of type II phosphatidylinositol kinases. Biochemistry *35*, 3587–3594.

Eckmann, L., Rudolf, M. T., Ptasznik, A., Schultz, C., Jiang, T., Wolfson, N., Tsien, R., Fierer, J., Shears, S. B., Kagnoff, M. F. and Traynor-Kaplan, A. E. (1997). D-*myo*-Inositol 1,4,5,6-tetrakisphosphate produced I human intestinal epithelial cells in response to *Salmonella* invasion inhibits phosphoinisitide 3-kinase signalling pathways. Proc. Natl. Acad. Sci. USA *94*, 14456–14460.

Endemann, G., Dunn, S. N. and Cantley, L. C. (1987). Bovine brain contains two types of phosphatidylinositol kinase. Biochemistry 26, 6845–6852.

Endemann, G., Graziani, A. and Cantley, L. C. (1991). A monoclonal antibody distinguishes two types of phosphatidylinositol 4 kinase. Biochem. J. 273, 63–66.

Field, J., Wilson, M. P., Mai, Z., Majerus, P. W. and Samuleson, J. (2000). An Enamoeba histolytica inositol 1,3,4-trisphosphate 5/6-kinase has a novel 3-kinase activity. Mol. Biochem. Parasitol. 108, 119–123.

Flanagan, C. A., Schnieders, E. A., Emerick, A. W., Kunisawa, R., Admon, A. and Thorner, J. (1993). Phosphatidylinositol 4-kinase gene structure and requirement for yeast cell viability. Science 262, 1444–1448.

Fruman, D. A., Meyers, R. E. and Cantley, L. C. (1998). Phosphoinositide kinases. Annu. Rev. Biochem. 67, 481–507.

Gaidarov, I., Smith, M. E., Domin, J. and Keen, J. H. (2001). The Class II phosphoinositide 3-kinase C2α is activated by clathrin and regulates clathrin-mediated membrane trafficking. Mol. Cell 7, 443–449.

Garcia-Bustos, J. F., Marini, F., Stevenson, I., Frei, C. and Hall, M. N. (1994). PIK1, an phosphatidylinositol 4-kinase associated with the yeast nucleus. EMBO J 13, 2352–2361.

Gary, J. D., Wurmser, A. E., Bonangelino, C. J., Weisman, L. S. and Emr, S. D. (1998). Fab1p is essential for PtdIns(3)P kinase activity and the maintenance of vacuolar size and membrane homeostasis. J. Cell Biol. 143, 65–79.

Gehrmann, T. and Heilmeyer, L. M. Jr. (1998). Phosphatidylinositol 4-kinases. Eur. J. Biochem. 253, 357–370.

Gehrmann, T., Vereb, G., Schmidt, M., Klix, D., Meyer, H. E., Varsanyi, M. and Heilmeyer, L. M. G. Jr. (1996). Identification of a 200 kDa polypeptide as type 3 phosphatidylinositol 4-kinase from bovine bran by partial protein and cDNA sequencing. Biochim. Biophys. Acta 1311, 53–63.

Glennon, M. C. and Shears, S. B. (1993). Turnover of inositol pentakisphosphates, inositol hexakisphosphate and diphosphoinositol polyphosphates in primary cultured hepatocytes. Biochem. J. 293, 583–590.

Graziani, A., Ling, L. E., Endemann, G., Carpenter, C. L. and Cantley, L. C. (1992). Purification and characterization of human erythrocyte phosphatidylinositol 4-kinase. Biochem. J. 284, 39–45.

Gross, W., Yan, W. and Boss, W. F. (1992). Release of carrot plasma membrane-associated phosphatidylinositol kinase by phospholipase A_2 and activation by a 70 kDa protein. Biochim. Biophys. Acta 1134, 73–80.

Han, G. S., Audhya, A., Markley, D. J., Emr, S. D. and Carman, G. M. (2002). The Saccharomyces cerevsiae LSB6 gene encodes phosphatidylinositol 4-kinase activity. Biol. Chem. 277, 47709–47718.

Hansen, C. A., Vom Dahl, S., Huddell, B. and Williamson, J. R. (1988). Characterization of inositol 1,3,4-trisphosphate phosphorylation in rat liver. FEBS Lett. *236*, 53–56.

Hawkins, P. T., Jackson, T. R. and Stephens, L. R. (1992). Platelet-derived growth factor stimulkaes synthesis of PtdIns(3,4,5)P$_3$ by activating a PtdIns(4,5)P$_2$ 3-OH kinase. Nature *358*, 157–159.

Hegewald, H. (1996). One-dimensional thin-layer chromatography of all known D-3 and D-4 isomers of phosphoinositides. Anal. Biochem. *242*, 152–155.

Hinchcliffe, K. A., Ciruela, A. and Irvine, R. F. (1998). PIPkins, their substrates and their products: new functions for old enzymes. Biochim. Biophys. Acta *1436*, 87–104.

Huang, C. F., Voglmaier, S. M., Bembenek, M. E., Saiardi, A. and Snyder, S. H. (1998). Identification and purification of diphosphoinositol pentakisphosphate kinase, which synthesizes the inositol pyrophosphate bis(diphospho)inositol tetrakisphosphate. Biochemistry *37*, 14998–15004.

Hughes, P. J., Hughes, A. R., Putney, J. W. Jr. and Shears, S. B. (1989). The regulation of the phosphorylation of inositol(1,3,4)-trisphosphate in cell free preparations and its relevance to the formation of inositol(1,3,4,6)-tetraki-sphosphate in agonist-stimulated rat parotid acinar cells. J. Biol. Chem. *264*, 19871–19878.

Hughes, P. J., Kirk, C. J. and Michell, R. H. (1994). Inhibition of porcine brain inositol 1,3,4-trisphosphate kinase by inositol polyphosphates, other polyol phosphates, polyanions and polycations. Biochim. Biophys. Acta *1223*, 57–70.

Hunter, T. (1995). Protein kinases and phosphatases: the yin and yang of protein phosphorylation and signaling. Cell *80*, 225–236.

Igaue, I., Miyaauchi, S. and Saito, K. (1982). Formation of *myo*-inositol phosphates in a rice cell suspension culture. In: Proceedings of the 5th International Congress of Plant Tissue and Cell Culture (Fujwara, A., ed.). Mauruzen, Tokyo, pp. 265–266.

Igaue, I., Shimizu, M. and Miyaauchi, S. (1980). Formation of a series of *myo*-inositol phosphates during growth of rice plant cells in suspension culture. Plant Cell Physiol. *21*, 351–356.

Irvine, R. F., Letcher, A. J., Heslop, J. P. and Berride, M. J. (1986). The inositol tris/tetrakisphosphate pathway – demonstration of Ins(1,4,5)P$_3$ 3-kinase activity in animal tissues. Nature *320*, 631–634.

Ishihara, H., Shibasaki, Y., Kizuki, N., Katagiri, H., Yazaki, Y., et al. (1996). Cloning of cDNAs encoding two isoforms of 68 kDa type I phosphatidyl-inositol-4-phosphate 5-kinase. J. Biol. Chem. *271*, 23611–23614.

Itoh, T., Ishihara, H., Shibasaki, Y., Oka, Y. and Takenawa, T. (2000). Autophosphorylation of Type I phosphatidylinositol phosphate kinase regulates its lipid kinase activity. J. Biol. Chem. *275*, 18389–18394.

Ives, E. B., Nichols, J., Wente, S. R. and York, J. D. (2000). Biochemical and functional characterization of inositol 1,3,4,5,6-pentakisphosphate 2-kinases. J. Biol. Chem. *275*, 3675–3683.

Jenkins, G. H., Fisette, P. L. and Anderson, R. A. (1994). Type I phosphatidylinositol 4-phosphate 5-kinase isoforms are specifically stimulated by phosphatidic acid. J. Biol. Chem. *269*, 11547–11554.

Jones, D. H., Morris, J. B., Morgan, C. P., Kondo, H., Irvine, R. F. and Cockcroft, S. (2000a). Type I phosphatidylinositol 4-phosphate 5-kinase directly interacts with ADP-ribosylation factor 1 and is responsible for phosphatidylinositol 4,5-bisphosphate synthesis in the Golgi compartment. J. Biol. Chem. *275*, 13962–13966.

Jones, D. R., Sanjuan, M. A. and Merida, I. (2000b). Type Iα phosphatidylinositol 4-phosphate 5-kinase is a putative target for increased intracellular phosphatidic acid. FEBS Lett. *476*, 160–165.

Ji, H., Sandbeg, K., Baukal, A. J. and Catt, K. J. (1989). Metabolism of inositol pentakisphosphate to inositol hexakisphosphate in *Xenopus laevis* oocytes. J. Biol. Chem. *264*, 20185–20188.

Lauener, R., Shen, Y., Duronio, V. and Salari, H. (1995). Selective inhibition of phosphatidylinositol 3-kinase by phsphatidic acid and related lipids. Biochem. Biophys. Res. Commun. *215*, 8–14.

Li, Y. S., Porte, F. D., Hoffman, R. M. and Deuel, T. F. (1989). Separation and identification of two phosphatidylinositol 4-kinase activities in bovine uterus. Biochem. Biophys. Res. Commun. *160*, 202–209.

Ling, L. E., Schulz, J. T. and Cantley, L. C. (1989). Characterization and purification of membrane-associated phosphatidylinositol 4-phosphate kinase from human red blood cells. J. Biol. Chem. *264*, 5080–5088.

Loijens, J. C. and Anderson, R. A. (1996). Type I phosphatidylinositol-4-phosphate 5-kinases are distinct members of this novel lipid kinase family. J. Biol. Chem. *271*, 32937–32943.

Loijens, J. C., Boronenkov, I. V., Parker, G. J. and Anderson, R. A. (1996). Phosphatidylinositol 4,5-bisphosphate binding to the plekstrin homology domain of phospholipase C δ1 enhances enzyme activity. Adv. Enzyme Regul. *36*, 115–150.

Mayer, A., Scheglmann, D., Dove, S., Glatz, A., Wicker, W. and Haas, A. (2000). Phosphatidylinositol 4,5-bisphosphate regulates two steps of homotypic vacuole fusion. Mol. Biol. Cell *11*, 807–817.

MacDougall, L. K., Domain, J. and Waterfeld, M. D. (1995). A family of phosphoinositide 3-kinases in Drosophila identifies a new mediator of signal transduction. Curr. Biol. *5*, 1404–1415.

McEwen, R. K., Dove, S. K., Cooke, F. T., Painter, G. F., Holmes, A. B., Shisheva, A., Ohya, Y., Parker, P. J. and Michell, R. H. (1999). Complementation analysis in PtdInsP kinase-deficient yeast mutants demonstrates that *Schizosaccharomyces pombe* and murine Fab1p homologues are phosphatidylinositol 3-phosphate 5-kinases. J. Biol. Chem. *274*, 33905–33912.

Meijer, H. J. G., Divecha, N., Van den Ende, H., Musgrave, A. and Munnik, T. (1999). Hyperosmotic stress induces rapid synthesis of phosphatidylinositol 3, 5-bisphosphate in plant cells. Planta *208*, 294–298.

Menniti, F. S., Miller, R. N., Putney, J. W. Jr. and Shears, S. B. (1993a). Turnover of inositol polyphosphate pyrophosphates in pacreatoma cell. J. Biol. Chem. *268*, 3850–3856.

Menniti, F. S., Oliver, K. G., Putney, J. W. Jr. and Shears, S. B. (1993b). Inositol phosphates and cell signaling: New views of InsP$_5$ and InsP$_6$. Trends Biochem. Sci. *18*, 53–56.

Meyers, R. and Cantley, L. C. (1997). Cloning and characterization of a wortmannin-sensitive human phosphatidylinositol 4-kinase. J. Biol. Chem. *272*, 4384–4390.

Minogue, S., Anderson, J. S., Waugh, M. G., dos Santos, M., Corless, S., Cramer, R. and Hsuan, J. J. (2001). Cloning of a human type II phosphatidylinositol 4-kinase reveals a novel lipid kinase family. J. Biol. Chem. *276*, 16635–16640.

Moritz, A., De Graan, P. N. E., Gispen, W. H. and Wirtz, K. W. A. (1992). Phosphatidic acid is a specific activator of phosphatidylinositol 4-phosphate kinase. J. Biol. Chem. *267*, 7207–7210.

Moritz, A., Westerman, J., De Graan, P. N. E. and Wirtz, K. W. A. (1992). Phosphatidylinositol 4-kinase and phosphatidylinositol-4-phosphate 5-kinase from bovine brain membranes. Meth. Enzymol. *209*, 202–211.

Morris, J. B., Hinchcliffer, K. A., Ciruela, A., Letcher, A. J. and Irvine, R. F. (2000). Thrombin stimulation causes an increase in phosphatidylinositol 5-phosphate revealed by mass assay. FEBS Lett. *475*, 57–60.

Morris, A. J., Muray, K. J., England, P. J., Downes, C. P. and Michell, R. H. (1988). Partial purification and some properties of rat brain inositol 1,4,5-trisphosphate 3-kinase. Biochem. J. *251*, 157–163.

Morrison, B. H., Bauer, J. A., Kalvakolanu, D. V. and Lindner, D. J. (2001). Inositol hexakisphosphate kinase 2 mediates growth suppressive and apoptotic effects of interferon-β in ovarian carcinoma cells. J. Biol. Chem. *276*, 24965–24970.

Munnik, T., Irvine, R. F. and Musgrave, A. (1998). Phospholipid signaling in plants. Biochim. Biophys. Acta *1389*, 222–272.

Murthy, P. P. N. (1996). Inositol phosphates and their metabolism in plants. In: Subcellular Biochemistry*myo*-Inositol Phosphates, Phosphoinoisitides, and Signal Transduction (Biswas, B. B. and Biswas S., eds.), Vol. 26. Plenum Press, New York, pp. 227–255.

Nakagawa, T., Goto, K. and Kondo, H. (1996a). Cloning and characterization of a 92 kDa soluble phosphatidylinositol 4-kinase. Biochem. J. *320*, 643–649.

Nakagawa, T., Goto, K. and Kondo, H. (1996b). Cloning, expression, and localization of 230 kDa phosphatidylinositol 4-kinase. J. Biol. Chem. *271*, 12088–12094.

Nakanishi, S., Catt, K. J. and Balla, T. (1995). A wortmannin-sensitive phosphatidylinositol 4-kinase that regulates hormone sensitive pools of inositol phospholipids. Proc. Natl. Acad. Sci. USA *92*, 5317–5321.

Odorizzi, G., Babst, M. and Emr, S. D. (2000). Phosphoinositide signaling and the regulation of membrane trafficking in yeast. Trends Biochem. Sci. *25*, 229–235.

Okpodu, C. M., Gross, W., Burkhart, W. and Boss, W. F. (1995). Plant Physiol (Bethesda) *107*, 491–500.

Ongusaha, P. P., Hughes, P. J., Davey, J. and Mitchell, R. H. (1998). Inositol hexakisphosphate in *Schizosaccharomyces pombe*: synthesis from Ins(1,4,5)P$_3$ and osmotic regulation. Biochem. J. *335*, 671–679.

Pahan, K., Raymond, J. R. and Singh, I. (1999). Inhibition of phosphatidylinositol 3-kinase induces nitric-oxide synthase in lipopolysaccharide- or cytokine-stimulated C$_6$ glial cells. J. Biol. Chem. *274*, 7528–7536.

Petitot, A., Ogier-Denis, E., Blommaart, E. F. C., Meijer, A. J. and Codogno, P (2000). Distinct classes of phosphatidylinositol 3'-kinases are involved in signaling pathways that control macroautophagy in HT-29 cells. J. Biol. Chem. *275*, 992–998.

Pike, L. J. (1992). Phosphatidylinositol 4-kinases and the role of polyphosphoinositides in cellular regulation. Endocr. Rev. *13*, 692–706.

Prasad, K. V. S., Kapeller, R., Janssen, O., Repke, H., Duke-Cohan, J. S., et al. (1993). Phosphatidylinositol (PI) 3-kinase and PI 4-kinase binding to the CD4-p56[lck] SH3 domain binds to PI 3-kinase but not PI 4-kinase. Mol. Cell. Biol. *13*, 7708–7717.

Prior, I. A. and Clague, M. J. (1999). Localization of a class II phosphatidylinositol 3-kinase, PI3K2α, to clathrin-coated vesicles. Mol. Cell Biol. Res. Commun. *1*(2), 162–166.

Rameh, L. E., Arvidsson, A., Carraway, K. L., III, Couvillion, A. D., Rathbun, G., Crompton, A., et al. (1997). A comparative analysis of the phosphoinositide binding specificities of plekstrin homology domains. J. Biol. Chem. *272*, 22059–22066.

Rameh, L. E., Chen, C. S. and Cantley, L. C. (1995). Phosphatidylinositol (3,4,5)P$_3$ interact with SH2 domain and modulates PI 3-kinase association with tyrosine-phosphorylated proteins. Cell *83*, 821–830.

Rameh, L. E., Tolias, K. F., Duckworth, B. C. and Cantley, L. C. (1997). A new pathway for synthesis of phosphatidylinositol-4,5-bisphosphate. Nature *390*, 192–196.

Rani, M. R. S., Leaman, D. W., Han, Y., Leung, S., Croze, E., Fish, E., Wolan, A. and Ransohoff, R. M. (1999). Catalyticaly active TYK2 is essential for interferon-beta-mediated phosphorylation of TT3 and interferon-alpha receptor-1 (IFNAR-1) but not for activation of phosphoinositol 3-kinase. J. Biol. Chem. *274*, 32507–32511.

Rittenhouse, S. E. (1996). Phosphoinositide 3-kinase activation and platelet function. Blood *88*, 4401–4414.

Rodriguez-Viciana, P., Warne, P. H., Vanhaesebroeck, B., Waterfeld, M. D. and Downward, J. (1996). Activation of phosphoinositide 3-kinase by interaction with Ras and by point mutation. EMBO J. *15*, 2442–2451.

Rudolph, M. T., Kaiser, T., Guse, A. H., Mayr, G. W. and Schultz, C. (1997). Liebigs Ann. *9*, 1861–1869.

Safrany, S. T., Ingram, S. W., Cartwright, J. L., Falck, J. R., McLennan, A. G., Barnes, L. D. and Shears, S. B. (1999). The diadenosine hexaphosphate hydrolases from *Schizosaccharomyces pombe* and *Saccharomces cerevisiae* are homologues of the human diphosphoinositol polyphosphate phosphohydrolase. Overlaping substrate specificities in a MutT-type protein. J. Biol. Chem. *274*, 21735–21740.

Saiardi, A., Caffrey, J. J., Snyder, S. H. and Shears, S. B. (2000). Inositol polyphosphate multikinase (ArgRIII) determines nuclear mRNA export in *Saccharomyces cerevisiae*. FEBS Lett. *468*, 28–32.

Saiardi, A., Erdjument-Bromage, H., Snowman, A. M., Tempst, P. and Snyder, S. H. (1999). Synthesis of diphosphoinositol pentakisphosphate by a newly identified family of higher inositol polyphosphate kinases. Curr. Biol. *9*, 1323–1326.

Sandelius, A. S. and Sommarin, M. (1990). In: Inositol Metabolism in Plants (Morre, D. J., Boss, W. F. and Loewus F. A., eds.). Wiley-Liss, New York, pp. 139–161.

Satterlee, J. S. and Sussman, M. R. (1997). Plant Physiol (Bethesda) *115*, 864–864.

Sbrissa, D., Ikonomov, O. C., Deeb, R. and Shisheva, A. (2002). Phosphatidylinositol 5-phosphate biosynthesis is linked to PIKfyve and is involved in osmotic response pathway in mammalian cells. J. Biol. Chem. *277*, 47276–47284.

Sbrissa, D., Ikonomov, O. C. and Shisheva, A. (1999). PIKfyve, a mammalian ortholog of yeast Fab1p lipid kinase, synthesizes 5-phosphoinositides. J. Biol. Chem. *274*, 21589–21597.

Sbrissa, D., Ikonomov, O. C. and Shisheva, A. (2000). PIKfyve lipid kinase is a protein kinase: downregulation of 5′phosphoinositide product formation by autophosphorylation. Biochemistry *39*, 15980–15989.

Sbrissa, D., Ikonomov, O. C. and Shisheva, A. (2002). Phosphatidylinositol 3-phosphate-interacting domains in PIKfyve. Binding specificity and role in PIKfyve. Endomembrane localization. J. Biol. Chem. *277*, 6073–6079.

Schell, M. J., Letcher, A. J., Brearley, C. A., Biber, J., Murer, H. and Irvine, R. F. (1999). PiUS (Pi uptake stimulator) is an inositol hexakisphosphate kinase. FEBS Lett. *61*, 169–172.

Schu, P. V., Takegawa, K., Fry, M. J., Stack, J. H., Waterfield, M. D. and Emr, S. (1993). Phosphatidylinositol 3-kinase encoded by yeast VPS34 gene essential for protein sorting. Science *260*, 88–91.

Serunian, L. A., Auger, K. R. and Cantley, L. C. (1991). Identification and quantification of polyphosphoinositides produced in response to platelet-derived growth factor stimulation. Meth. Enzymol. *198*, 78–87.

Serunian, L. A., Haber, M. T., Fukui, T., Kim, J. W., Rhee, S. G., Lowenstein, J. M. and Cantley, L. C. (1989). Polyphosphoinositides produced by phosphatidylinositol 3-kinase are poor substrates for phospholipases C from rat liver and bovine brain. J. Biol. Chem. *264*, 17809–17815.

Sharps, E. S. and McCarl, R. L. (1982). A high-performance liquid chromato-graphic method to measure ^{32}P incorporation into phosphorylated metabolites in culture cells. Anal. Biochem. *12*, 421–424.

Shears, S. B. (1998a). The versatility of inositol phosphates as cellular signas. Biochim. Biophys. Acta *1436*, 49–67.

Shears, S. B. (1998b). The versatility of inositol phosphates as cellular signals. Biochim. Biophys. Acta *1436*, 49–67.

Shears, S. B. (1989a). Metabolism of the inositol phosphates produced upon receptor activation. Biochem. J. *260*, 313–324.

Shears, S. B. (1989b). The pathway of *myo*-inositol 1,3,4-trisphosphate phosphorylation in liver. Identification of *myo*-inositol 1,3,4-trisposphate 6-kinase, *myo*-inositol 1,3,4-trisphosphate 5-kinase, and *myo*-inositol 1,3,4,6-tetrakisphosphate 5-kinase. J. Biol. Chem. *264*, 19879–19897.

Shears, S. B., Ali, N., Craxton, A. and Bembenek, M. (1995). Synthesis and metabolism of bis-diphosphoinositol tetrakisphosphate *in vitro* and *in vivo*. J. Biol. Chem. *270*, 10489–10497.

Shinshi, H., Miwa, M., Kato, K., Noguchi, M., Matsushima, T. and Sugimura, T. (1976). A novel phosphodiesterase from cultured tobacco cells. Biochemistry *15*, 2185–2190.

Shisheva, A., Sbrissa, D. and Ikonomov, O. (1999). Cloning, characterization, and expression of a novel Zn^{2+}-binding FYVE finger-containing phosphoinositide kinase in insulin-sensitive cells. Mol. Cell. Biol. *19*, 623–634.

Soltoff, S. P., McMillan, M. K., Talamo, B. R. and Cantley, L. C. (1993). Blockade of ATP binding site of P2 purinoreceptors in rat parotid acinar cells by isothiocyanate compounds. Biochem. Pharmacol. 45, 1936–1940.

Sommarin, M. and Sandelius, A. S. (1988). Phosphatidylinositol and phosphatidylinositol phosphate kinases in plant plasma membranes. Biochim. Biophys. Acta 958, 26–278.

Soriski, A., King, W. G. and Rittenhouse, S. E. (1992). Accummulation of PtdIns (3,4)P$_2$ and PtdIns(3,4,5)P$_3$ in thrombin-stimulated platelets. Biochem. J. 286, 581–584.

Spencer, C. E. L., Stephens, L. R. and Irvine, R. F. (1990). Separation of higher inositol phosphates by polyethyleneimine-cellulose thin-layer chromatography and by Dowex chloride column chromatography. In: Methods in Inositide Research (Irvine, R. F., ed.). Raven Press, New York, pp. 39–43.

Stephens, L. R. (1990). Preparation and separation of inositol tetrakisphosphates and inositol pentakisphosphates and the establishment of enantiomeric configurations by the use of L-iditol dehydrogenase. In: Methods in Inositide Research (Irvine, R. F., ed.). Raven Press, New York, pp. 9–30.

Stephens, L. R. and Downes, C. P. (1990). Product-precursor relationships amongst inositol polyphosphates. Incorporation of [^{32}P] into myo-inositol 1,3,4,6-tetrakisphosphate, myo-inositol 1,3,4,5-tetrakisphosphate, myo-inositol 3,4,5,6-tetrakisphosphate and myo-inositol 1,3,4,5,6-pentakisphosphate in intact avian erythrocytes. Biochem. J. 265, 435–452.

Stephens, L. R., Hawkins, P. T., Morris, A. J. and Downes, P. C. (1988). L-myo-Inositol 1,4,5,6-tetrakisphosphate (3-hydroxy)kinase. Biochem. J. 249, 283–292.

Stephens, L. R., Hawkins, P. T., Sanley, A. F., Moore, T., Poyner, D. R., Morris, P. J., Hanley, M. R., Kay, R. R. and Irvine, R. F. (1991). myo-Inositol pentakisphosphates. Structure, biological occurrence and phosphorylation to myo-inositol hexakisphosphate. Biochem. J. 275, 485–499.

Stephens, L. R. and Irvine, R. F. (1990). Stepwise phosphorylation of myo-inositol leading to myo-inositol hexakisphosphate in Dictyostelium. Nature 346, 580–583.

Stephens, L. R., McGregor, A. and Hawkins, P. (2000). Phosphoinoasitide 3-kinases: regulation by cell-surface receptors and function of 3-phosphorylated lipids. In: Biology of Phosphoinositides, (Cockcroft, S., ed.). Oxford Univ. Press, Oxford, UK, pp. 32–108.

Stephens, L. R., Radenberg, T., Thiel, U., Vogel, G., Khoo, K.-H., Dell, A., Jackson, T. R., Hawkins, O. T. and Mayr, G. W. (1993). The detection, purification, structural characterization, and metabolism of diphosphoinositol pentakisphosphate(s) and bisdiphosphoinositol tetrakisphosphate(s). J. Biol. Chem. 268, 4009–4015.

Stevens, V. L. and Raetz, C. R. (1991). Defective glycosyl phosphatidylinositol biosynthesis in extracts of three Thy-1 negative lymphoma cell mutants. J. Biol. Chem. *266*, 10039–10042.

Stevenson, J. M., Perera, I. Y. and Boss, W. F. (1998). A phosphatidylinsitol 4-kinase pleckstrin homology domain that binds phosphatidylinositol 4-monophosphate. J. Biol. Chem. *273*, 22761–22767.

Stoyanova, S., Bulgarelli-Leva, G., Kirsch, C., Hanck, T., Klinger, R., Wetzker, R. and Wymann, M. P. (1997). Lipid kinase and protein kinase activities of G-protein-coupled phosphoinositide 3-kinase gamma: structure-activity analysis and interactions with wortmannin. Biochem. J. *324*, 489–495.

Suer, S., Sickmann, A., Meyer, H. E., Herberg, F. W. and Heilmeyer, L. M. G. Jr. (2001). Human phosphatidylinositol 4-kinase isoform P14K92. Expression of the recombinant enzyme and determination of multiple phosphorylation sites. Eur. J. Biochem. *268*, 2099–2106.

Suzuki, K., Hirano, H., Okutomi, K., Suzuki, M., Kuga, Y., Fujiwara, T., Kanemoto, N., Isono, K. and Horie, M. (1997). Identification and characterization of an novel human phosphatidylinositol 4-kinase. DNA Res. *31*, 273–280.

Takazawa, K., Lemos, M., Delvaux, A., Lejeune, C., Dumont, J. E. and Erneux, C. (1990). Rat brain inositol 1,4,5-trisphosphate 3-kinase. Ca2 + -sensitivity, purification and antibody production. Biochem. *268*, 213–217.

Takazawa, K., Passareiro, H., Dumont, J. E. and Erneaux, C. (1988). Ca^{2+}/ calmodulin-sensitive inositol 1,4,5-trisphosphate 3-kinase in rat and bovine brain tissues. Biochem. Biophys. Res. Commun. *153*, 632–641.

Takazawa, K., Perret, J., Dumont, J. E. and Erneux, C. (1990a). Human brain inositol 1,4,5-trisphosphate 3-kinase cDNA sequence. Nucleic Acids Res. *18*, 7141.

Takazawa, K., Vandekerckhove, J., Dumont, J. E. and Erneux, C. (1990b). Cloning and expression in *Escherichia coli* of a rat brain cDNA encoding a Ca^{2+}/calmodulin-sensitive inositol 1,4,5-trisphosphate 3-kinase. Biochem. J. *272*, 107–112.

Takazawa, K., Perret, J., Dumont, J. E. and Erneux, C. (1991). Molecular cloning and expression of a new putative inositol 1,4,5-trisphosphate 3-kinase isozyme. Biochem. J. *278*, 883–886.

Takazawa, K., Perret, J., Dumont, J. E. and Erneux, C. (1991). Molecular cloning and expression of a human brain inositol 1,4,5-trisphosphate 3-kinase. Biochem. Biophys. Res. Commun. *174*, 529–535.

Tan, Z. and Boss, W. F. (1992). Plant Physiol (Bethesda) *100*, 2116–2120.

Tan, Z., Bruzil, K. S. and Shears, S. B. (1997). Properties of the inositol 3,4,5,6-tetrakisphosphate 1-kinase purified from rat liver. Regulation of enzyme activity by inositol 1,3,4-trisphosphate. J. Biol. Chem. *272*, 2285–2290.

Thompson, D. M., Cochet, C., Chambaz, E. M. and Gill, G. N. (1985). Separation and characterization of a phosphatidylinositol kinase activity that co-purifies with the epithelial growth factor receptor. J. Biol. Chem. 260, 8824–8830.

Toker, A. (1998). The synthesis and cellular roles of phosphatidylinositol 4,5-bisphosphate. Curr. Opin. Cell Biol. 10, 254–261.

Toker, A. and Cantley, L. C. (1997). Signalling through the lipid products of phosphoinositide-3-OH-kinase. Nature (London) 387, 673–676.

Tolias, K. F. and Cantley, L. C. (1999). Pathways for phosphoinositide synthesis. Chem. Phys. Lipids 98, 69–77.

Tolias, K. F. and Carpenter, C. L. (2000). Enzymes involved in the synthesis of PtdIns(4,5)P$_2$ and their regulation: PtdIns kinases and PtdInsP kinases. In: Biology of Phosphoinositides (Cockcroft, S., ed.). Oxford Univ. Press, Oxford, UK, pp. 109–130.

Tolias, K. F., Rameh, L. E., Ishihara, H., Shibasaki, Y., Chen, J., Prestwich, G. D., Cantley, L. C. and Carpeneter, C. L. (1998). J. Biol. Chem. 273, 18040–18046.

Traynor-Kaplan, A. E., Harris, A. L., Thompson, B. L., Taylor, P. and Sklar, L. A. (1988). An inositol tetrakisphosphate-containing phospholipid in activated neutrophils. Nature (London) 334, 353–356.

Traynor-Kaplan, A. E., Thompson, B. L., Harris, A. L., Taylor, P., Omann, G. M. and Sklar, L. A. (1989). Transient increase in phosphatidyinositol 3,4-bisphosphate and phosphatidylinositol trisphosphate during activation of human neutrophils. J. Biol. Chem. 264, 15668–15673.

Turner, S. J., Domin, J., Waterfield, M. D., Ward, S. G. and Westwick, J. (1998). The CC chemokine monocyte chemotactic peptide-1 activates both the class I 85/p110 phosphatidylinositol 3-kinase and the class II PtdIns 3-kinase-2α. J. Biol. Chem. 273, 25987–25995.

Ui, M., Okada, T., Hazeki, K. and Hazeki, O. (1995). Wortmannin as a unique probe for an intracellular signalling protein, phosphoinositide 3-kinase. Trends Biol. Sci. 20, 303–307.

Vajanaphanich, M., Schultz, C., Rudolf, M. T., Wasserman, M., Enyedi, P., Craxton, A., Shears, S. B., Tsien, R. Y., Barrett, K. E. and Traynor-Kaplan, A. (1994). Long-term uncoupling of chloride secretion from interacellular calcium levels by Ins(3,4,5,6)P$_4$. Nature 371, 711–714.

Valverde, A. M., Lorenzo, M., Navarro, P. and Benito, M. (1997). Phosphatidyl-inositol 3-kinase is a requirement for insulin-like growth factor I—induced differentiation but not for mitogenesis in fetal brown adipocytes. Mol. Endocrinol. 11, 595–607.

Valverde, A. M., Navarro, P., Teruel, T., Cnejo, R., Benito, M. and Lorenzo, M. (1999). Insulin and insulin-like growth factor I up-regulate GLUT4 gene expression in fetal brown adipocytes, in phosphoinositide 3-kinase-dependent manner. Biochem. J. *337*, 397–405.

Vanhaesebroeck, B., Leevers, S. J., Ahmadi, K., Timms, J., Katso, R., Driscoll, P. C., Woscholski, R., Parker, P. J. and Waterfield, M. D. (2001). Synthesis and function of 3-phosphorylated inositol lipids. Annu. Rev. Biochem. *70*, 535–602.

Vanhaesebroeck, B., Leewers, S. J., Panayotou, G. and Waterfield, M. (1997a). Phosphoinositide 3-kinases: a conserved family of signal transducers. Trends Biochem Sci. *22*, 267–272.

Vanhaesebroeck, B., Welham, M. J., Kpani, K., Stein, R., Warne, P. H., et al. (1997b). p110δ, a novel phosphatidylinositol 3-kinase in leukocytes. Proc. Natl. Acad. Sci. USA *94*, 4330–4335.

Varsanyi, M., Messer, M. and Brandt, N. R. (1989). Intracellular localization of inositol-phospholipid-metabolizing enzymes in rabbit fast-twitch skeletal muscle. Can D-myo-inoitol 1,4,5-trisphosphate play a role in excitation–contraction coupling? Eur. J. Biochem. *179*, 473–479.

Verbsky, J. W., Wilson, M. P., Kisseleva, M. V., Majerus, P. W. and Wente, S. R. (2002). The synthesis of inositol hexakisphosphate: Characterization of human inositol 1,3,4,5,6-pentakisphosphate 2-kinase. J. Biol. Chem.

Verghese, M., Fernandis, A. Z. and Subrahmanyam, G. (1999). Purification and characterization of a type II phosphatidylinositol 4-kinase from rat spleen and comparison with a PtdIns 4 kinase from lymphocytes. Indian J. Biochem. Biophys. *36*, 1–9.

Virbasius, J. V., Guilherme, A. and Czech, M. P. (1996). Mouse p170 is a novel phosphatidylinositol 3-kinase containing a C2 domain. J. Biol. Chem. *271*, 13304–13307.

Vlahos, C. J., Matter, W. F., Hui, K. Y. and Brown, R. F. (1994). Specific inhibitor of phosphatidylinositol 3-kinase, 2-(4-morpholinoyl)-8-phenyl-4H-1-benzo-pyran-4-ne (LY294002). J. Biol. Chem. *269*, 5241–5248.

Voglmaier, S. M., Bembenek, M. E., Kaplin, A. I., Dorman, G., Olszewski, J. D., Prestwich, G. D. and Snyder, S. H. (1996). Purified inositol hexakisphosphate kinase is an ATP synthase: diphosphoinositol pentakisphosphate is a high energy phosphate donor. Proc. Natl. Acad. Sci. USA *93*, 4305–4310.

Volinia, S., Dhand, R., Vanhaesebroeck, B., MacDougall, L. K., Stein, R., Zvelebil, M., Domin, J., Panareto, C. and Waterfield, M. D. (1995). A human phosphatidylinsitol 3-kinase complex related to the yeast Vps34p-Vsp15p protein sorting system. EMBO J. *14*, 3339–3348.

Walker, D. H., Dougherty, N. and Pike, L. J. (1988). Purification and characterization of a phosphatidylinositol kinase from A431 cells. Biochemistry *27*, 6504–6511.

Walsh, J. P., Caldwell, K. K. and Majerus, P. W. (1991). Formation of phosphaidylinositol 3-phosphate by isomerization from phosphatidylinositol 4-phosphate. Proc. Natl. Acad. Sci. USA *88*, 9184–9187.

Ward, S. G., Reif, K., Ley, S., Fry, M. J., Waterfield, M. D. and Cantrell, D. A. (1992). Regulation of phosphoinositide kinases in T cells. J. Biol. Chem. *267*, 23862–23869.

Waugh, M. G., Lawson, D., Tan, S. K. and Hsuan, J. J. (1998). Phosphatidylinositol 4-phosphate synthesis in immunoisolated caveolae-like vesicles and low buoyant density non-caveolar membranes. J. Biol. Chem. *273*, 17115–17121.

Whiteford, C. C., Brearley, C. A. and Ulug, E. T. (1997). Phosphatidylinositol 3,5-bisphosphate defines a novel PI 3-kinase pathway in resting mouse fibroblasts. Biochem. J. *323*, 597–601.

Whitman, M., Downes, C. P., Keeler, M., Keller, T. and Cantley, L. (1988). Type I phosphatidylinositol kinase makes a novel inositol phospholipid, phosphatidylinositol-3-phosphate. Nature *332*, 644–646.

Whitman, M., Kaplan, D., Roberts, T. and Cantley, L. (1987). Evidence for two distinct phosphatidylinositol kinases in fibroblasts. Implications for cellular regulation. Biochem. J. *247*, 165–174.

Wiedemann, C., Schafer, T. and Burger, M. M. (1996). Chromaffin granule-associated phosphatidylinositol kinase activity is required for stimulated secretion. EMBO J. *15*, 2094–2101.

Wilson, M. P. and Majerus, P. W. (1996). Isolation of inositol 1,3,4-trisphosphate 5/6 kinase, cDNA cloning, and expression of the recombinant enzyme. J. Biol. Chem. *271*, 11904–11910.

Wilson, M. P. and Majerus, P. W. (1997). Characterization of a cDNA encoding Arabidopsis thaliana inositol 1,3,4-trisphosphate 5/6-kinase. Biochem. Biophys. Res. Commun. *232*, 68–681.

Wilson, M. P., Sun, Y., Cao, L. and Majerus, P. W. (2001). Inositol 1,3,4-trisphosphate 5/6 kinase is a protein kinase that phosphorylates the transcription factors c-Jun and ATF-2. J. Biol. Chem. *276*, 40998–41004.

Wong, K. and Cantley, L. C. (1994). Cloning and characterization of a human phosphatidylinositol 4-kinase. J. Biol. Chem. *269*, 28878–28884.

Wong, K., Meyers, R. and Cantley, L. C. (1997). Subcellular locations of phosphatidylinsotiol 4-kinase isoforms. J. Biol. Chem. *272*, 13236–13241.

Wurmser, A. E. and Emr, S. (1998). Phosphoinositide signalling and turnover: PtdIns(3)P, a regulator of membrane traffic is transported to the vacuole and degraded by a process that requires lumenal vacuolar hydrolase activities. EMBO J *17*, 493–4942.

Wymann, M. P., Bulgerelli-Leva, G., Zvelebil, M. J., Pirola, L., Vanhaesebroeck, B., Waterfield, M. D. and Panayotou, G. (1996). Wortmannin inactivates phosphoinositide 3-kinase by covalent modification of Lys-802, a residue involved in the phosphate transfer reaction. Mol. Cell. Biol. *16*, 1722–1733.

Wymann, M. P. and Pirola, L. (1998). Structure and function of phosphoinositide 3-kinases. Biochim. Bophys. Acta *1436*, 127–150.

Xu, P., Lloyd, C. W., Staiger, C. J. and Drobak, B. K. (1992). Plant Cell *4*, 941–951.

Xue, H. W., Pical, C., Brearley, C., Elge, S. and Muller-Rober, B. (1999). A plant 126 kDa phosphatidylinositol 4-kinase with a novel repeat structure. Cloning and functional expression in baculovirus-infected insect cells. J. Biol. Chem. *274*, 5738–5745.

Yamamoto, K., Graziani, A., Carpenter, C., Cantley, L. C. and Lapetina, E. G. (1990). A novel pathway for the formation of phosphatidylinositol 3, 4-bisphosphate. Phosphorylation of phosphatidyl 3-monophosphate by phosphatidylinositol-3-monophosphate 4-kinase. J. Biol. Chem. *265*, 22086–22089.

Yang, X., Rudolph, M., Carew, M. A., Yoshida, M., Nerreter, V., Riley, A. M., Chung, S. K., Bruzik, K. S., Potter, B. V., Schultz, C. and Shears, S. B. (1999). Inositol 1,3,4-trisphosphate as in vivo as a specific regulator of cellular signaling by inositol 3,4,5,6-tetrakisphosphate. J. Biol. Chem. *274*, 18973–18980.

Yang, X. and Shears, S. B. (2000). Multitasking in signal transduction by a promiscuous human Ins(3,4,5,6)P₄ 1-kinase/Ins(1,3,4)P₃ 5/6-kinase. Biochem. J. *351*, 551–555.

Yoshida, S., Ikeda, E., Uno, I. and Mitsuzawa, H. (1992). Characterization of a staurosporine- and temperature-sensitive mutant, sst1, of *Saccharomyces cerevisiae*: STT1 is allelic to PKC1. Mol. Gen. Genet. *231*, 337–344.

Yoshida, S., Ohya, Y., Goebl, M., Nakano, A. and Anraku, Y. (1994a). A novel gene, STT4, encodes a phosphatidylinositol 4-kinase in the PKC1 protein kinase pathway of Saccharomyces cerevisiae. J. Biol. Chem. *269*, 1166–1171.

Yoshida, S., Ohya, Y., Nakano, A. and Anraku, Y. (1994b). Genetic interactions among genes involved in the STT4-PKC1 pathway of *Saccharomyces cerevisiae*. Mol. Gen. Genet. *242*, 631–640.

Zhang, J., Banfic, H., Straforini, F., Tosi, L., Volinia, S. and Rittenhouse, S. R. (1998). A type II phosphoinositide 3-kinase is stimulated via activated integrin in platelets. J. Biol. Chem. *273*, 14081–14084.

Zhang, X., Loijens, J. C., Boronenkov, I. V., Parker, G. J., Norris, F. A., Chen, J., Thum, O., Prestwich, G. D., Majerus, P. W. and Anderson, R. A. (1997). Phosphatidylinositol-4-phosphate 5-kinase isozymes catalyze the synthesis of 3-phsphate-containing phosphatidylinositol signalling molecules. J. Biol. Chem. *272*, 17756–17761.

Zhang, J., Vanhaesebroeck, B. and Rittenhouse, S. E. (2002). Human platelets contain p110δ phosphoinositide 3-kinase. Biochem. Biophys. Res. Commun. *296*, 178–181.

Chapter 8

Ackermann, K. E., Gish, B. G., Honchar, M. P. and Sherman, W. R. (1987). Evidence that inositol 1-phosphate in brain of lithium-treated rats results mainly from phosphatdylinositol metabolism. Biochem. J. *242*, 517–524.

Adelt, S., Plettenburg, O., Dallmann, G., Ritter, F. P., Shears, S. B., Altenbach, H. J. and Vogel, G. (2001). Regiospecific phosphohydrolases from *Dictyostelium* as tools for the chemoenzymatic synthesis of the enantiomers of D-*myo*-inositol 1,2,4-trisphosphate and D-*myo*-inositol 2,3,6-trisphosphate: non-physiological, potential analogues of biologically active D-*myo*-inositol 1,3,4-trisphosphate. Bioorg. Med. Chem. Lett. *11*, 2705–2708.

Albert, C., Safrany, S. T., Bembenek, M. E., Redd, K. M., Reddy, K. K., Falck, J. R., Broker, M., Shears, S. B. and Mayr, G. (1997). Biological variability in the structures of diphosphoinositol polyphosphates in *Dictyostelium discoideum* and mammalian cells. Biochem. J. *327*, 553–560.

Ames, B. N. and Dubin, D. T. (1960). Role of polyamines in the neutralization of bacteriophage deoxyribonucleic acid. J. Biol. Chem. *235*, 769–775.

Ashizawa, N., Yoshida, M. and Aotsuka, T. (2000). An enzymatic assay for *myo*-inositol. J. Biochem. Biophys. Methods *44*, 89–94.

Attree, O., Olivos, I., Okabe, I., Bailey, C. L., Nelson, D., Lewis, R. A., McInnes, R. R. and Nussbaum, R. L. (1992). The Lowe's oculocerebrorenalsyndrome gene encodes a protein highly homologous to inositol polyphosphate-5-phosphatase. Nature *358*, 239–242.

Auethavekiat, V., Abrams, C. S. and Majerus, P. W. (1997). Phosphorylation of platelet pleckstrin activates inositol polyphosphate 5-phosphatase I. J. Biol. Chem. *272*, 1786–1790.

Auger, K. R., Serunian, L. A., Soltoff, S. P., Libby, P. and Cantley, L. (1989). PDGF-dependent tyrosine phosphorylation stimulates production of novel phosphoinositides in intact cells. Cell *57*, 167–175.

Bansal, V. S., Caldwell, K. K. and Majerus, P. W. (1990). The isolation and characterization of inositol polyphosphate 4-phosphatase. J. Biol. Chem. *265*, 1806–1811.

Bansal, V. S., Inhorn, R. C. and Majerus, P. W. (1987). The metabolism of inositol 1,3,4-trisphosphate to inositol 1,3-bisphosphate. J. Biol. Chem. *262*, 9444–9447.

Barietos, L., Scott, J. J. and Murthy, P. P. N. (1994). Specificity of hydrolysis of phytic acid by alkaline phytase from lily pollen. Plant Physiol. *106*, 1489–1495.

Batty, I. H., Nahorski, S. R. and Irvine, R. F. (1985). Rapid formation of inositol 1,3,4,5-tetrakisphosphate following muscarinic receptor stimulation of rat cerebral cortical slices. Biochem. J. *232*, 211–215.

Bessman, M. J., Frick, D. N. and O'Handley, S. F. (1996). The MuT proteins or "Nudix" hydrolases, a family of versatile, widely distributed, "housecleaning" enzymes. J. Biol Chem. *271*, 25059–25062.

Bruyns, C., Pesesse, X., Moreau, C., Blero, D. and Erneux, C. (1999). The two SH2-domain-containing inositol 5-phosphatases SHIP1 ad SHIP2 are coexpressed in human lymphocytes. Biol. Chem. *380*, 969–974.

Caffrey, J. J., Darden, T., Wenk, M. R. and Shears, S. B. (2001). Expanding coincident signaling by PTEN through its inositol 1,3,4,5,6-pentakisphosphate 3-phosphatase activity. FEBS Lett. *499*, 6–10.

Caffrey, J. J., Hidaka, K., Matsuda, M., Hirata, M. and Shears, S. B. (1999). The human and rat forms of multiple inositol polyphosphate phosphatase: functional homology with a histidine acid phosphatase up-regulated during endochondral ossification. FEBS Lett. *442*, 99–104.

Caffrey, J. J., Safrany, S. T., Yang, X. and Shears, S. B. (2000). Discovery of molecular and catalytic diversity among human diphosphoinositol-polyphosphate phosphohydrolases. An expanding Nudt family. J. Biol. Chem. *275*, 12730–12736.

Caffrey, J. J. and Shears, S. B. (2001). Genetic rationale for microheterogeneity of human diphosphoinositol polyphosphate phosphohydrolase type 2. Gene *269*, 53–60.

Caldwell, K. K., Lips, D. L., Bansal, V. S. and Majerus, P. W. (1991). Isolation and characterization of two 3-phosphatases that hydrolyze both phosphatidylinositol 3-phosphate and inositol 1,3-bisphosphate. J. Biol. Chem. *266*, 18378–18386.

Cashikar, A. G., Kumaresan, R. and Rao, N. M. (1997). Biochemical characterization and subxcellular localization of the red kidney bean purple acid phosphatase. Plant Physiol. *114*, 907–915.

Challis, R. A. J. and Nahorski, S. R. (1990). Neurotransmitter and depolarization-stimulated accumulation of inositol 1,3,4,5-tetrakisphosphate mass in rat cerebral cortex slices. J. Neurochem. *54*, 2138–2141.

Chen, L. and Roberts, M. F. (1999). Characterization of a tetrameric inositol monophosphatase from the hyperthermophilic bacterium *Thermotoga maritima*. Appl. Environ. Microbiol. *65*, 4559–4567.

Chengalvala, M., Kostek, B. and Frail, D. E. (1999). A multi-well filtration assay for quantitation of inositol phosphates in biological samples. J. Biochem. Biophys. Methods *38*, 163–170.

Chi, H., Yang, X., Kingsley, P. D., O'Keefe, R. J., Puzas, E., Rosier, R. N., Shears, S. B. and Reynolds, P. R. (2000). Targeted deletion of Minpp1 provides new insight into the activity of multiple inositol polyphosphate phosphatase in vivo. Mol. Cell Biol. *20*, 6496–6507.

Chilvers, E. R., Batty, I. H., Challis, R. A. J., Barnes, P. J. and Nahorski, S. R. (1991). Determination of mass changes in phosphatidylinositol 4, 5-bisphosphate and evidence for agonist-stimulated metabolism of inositol 1,4, 5-trisphosphate in airway smooth muscle. Biochem. J. 275, 373–379.

Coleman, J. E. (1992). Structure and mechanism of alkaline phosphatase. Annu. Rev. Biophys. Biomol. Struct. 21, 441–483.

Connolly, T. M., Bross, T. E. and Majerus, P. W. (1985). Isolation of a phosphomonoesterase from human platelets that specifically hydrolyzes the 5-phosphate of inositol 1,4,5-trisphosphate. J. Biol. Chem. 260, 7868–7874.

Connolly, T. M., Lawing, W. J. and Majerus, P. W. (1986). Protein kinase C phosphorylates human platelet inositol trisphosphate 5'-phosphomonoesterase, increasing the phosphatase activity. Cell 46, 951–958.

Craxton, A., Ali, N. and Shears, S. B. (1995). Comparison of the activities of a multiple inositol polyphosphate phosphatase obtained from several sources: a search for heterogeneity in this enzyme. Biochem. J. 305, 491–498.

Craxton, A., Caffrey, J. J., Burkhart, W., Saffrany, S. T. and Shears, S. B. (1997). Molecular cloning and expression of a rat hepatic multiple inositol polyphosphate phosphatase. Biochem. J. 328, 75–81.

Damen, J. E., Liu, L., Rosten, P., Humphries, R. K., Jefferson, A. B., Majerus, P. W. and Krystal, G. (1996). The 145-kDa protein induced to associate with Shc by multiple cytokines is an inositol tetraphosphate and phosphatidylinositol 3,4,5-trisphosphate 5-phosphatase. Proc. Natl Acad. Sci. USA 93, 1689–1693.

Damen, J. E., Ware, M. D., Kalesnikoff, J., Hughes, M. R. and Krystal, G. (2001). SHIP's C-terminus is essential for its hydrolysis of PIP3 and inhibition of mast cell degranalation. Blood 97, 1343–1351.

De Smedt, F., Boom, A., Pesesse, X., Schiffmann, S. N. and Erneux, C. (1996). Post-translational modification of human brain type I inositol-1,4,5-trisphosphate 5-phosphatase by farnesylation. J. Biol. Chem. 271, 10419–10424.

De Smedt, F., Verjans, B., Mailleux, P. and Erneux, C. (1994). Cloning and expression of human type I inositol 1,4,5-trisphosphate 5-phosphatase. FEBS Lett. 347, 69–72.

Diehl, R. E., Whiting, P., Potter, J., Gee, N., Ragan, C. I., Linemeyer, D., Schoepfer, R., Bennett, C. and Dixon, R. A. (1990). Cloning and expression of bovine brain inositol monophosphatase. J. Biol. Chem. 265, 5946–5949.

Divecha, N., Banfic, H. and Irvine, R. F. (1991). The polyphosphoinositide cycle exists in the nuclei of Swiss 3T3 cells under the control of a receptor (for IGF-I) in the plasma membrane, and stimulation of the cycle increases nuclear diacylglycerol and apparently induces translocation of protein kinase C to the nucleus. EMBO J. 10, 3207–3214.

Dove, S. K., Cooke, F. T., Douglas, M. R., Sayers, L. G., Parker, P. J. and Michell, R. H. (1997). Osmotic stress activates phosphatidylinositol-3,5-bisphosphate synthesis. Nature *390*, 187–192.

Downes, C. P. (1988). Inositol phosphates: a family of signal molecules? Trends Neurosci. *11*, 336–338.

Downes, C. P., Bennett, D., McConnachie, G., Leslie, N. R., Pass, I., MacPhee, C., Patel, L. and Gray, A. (2001). Antagonism of PI 3-kinase-dependent signaling pathways by the tumour suppressor protein, PTEN. Biochem. Soc. Trans. *29*, 846–851.

Downes, C. P., Mussat, M. C. and Michell, R. H. (1982). The inositol trisphosphate phosphomonoesterase of the human erythrocyte membrane. Biochem. J. *203*, 169–177.

Drayer, A. L., Pesesse, X., De Smedt, F., Communi, D., Moreau, C. and Erneux, C. (1996). The family of inositol and phosphatidylinositol polyphosphate 5-phosphatases. Biochem. Soc. Trans. *24*, 1001–1005.

Dyson, J. M., O'Malley, C. J., Becanovic, J., Munday, A. D., Berndt, M. C., Coghill, I. D., Nandurkar, H. H., Ooms, L. M. and Mitchell, C. A. (2001). The S2-containing inositol polyphosphate 5-phosphatase, SHIP-2, binds filamin and regulates submembranous actin. J. Cell Biol. *155*, 1065–1079.

Eckmann, L., Rudloff, M. T., Ptasznik, A., Schultz, C., Jiang, T., Wolfson, N., Tsien, R., Fierer, J., Shears, S. B., Kagnoff, M. F. and Traynor-Kaplan, A. (1997). D-*myo*-Inositol 1,4,5,6-tetrakisphosphate produced in human intestinal epithelial cells in response to Salmonella invasion inhibits phosphoinositide 3-kinase signaling pathways. Proc. Natl Sci. USA *94*, 14456–15153.

Eisenberg, F. Jr. (1967). D-*myo*-Inositol 1-phosphate as a product of cyclization of glucose 6-phosphate and substrate for a specific phosphatase in rat testis. J. Biol. Chem. *242*, 1375–1382.

Eisenberg, F. Jr. and Parthasarathy, R. (1987). Measurement of biosynthesis of *myo*-inositol from glucose-6-phosphate. Methods Enzymol. *141*, 127–143.

Engelen, A. J., van der Heeft, F. C., Randsdorp, P. H. and Smit, E. L. (1994). Simple and rapid determination of phytase activity. J. AOAC Int. *77*, 760–764.

Erneux, C., Delvaux, A., Moreau, C. and Dumont, J. E. (1986). Characterization of D-*myo*-inositol 1,4,5-trisphosphate phosphates in rat brain. Biochem. Biophys. Res. Commun. *134*, 351–358.

Erneux, C., Govaerts, C., Communi, D. and Pesesse, X. (1998). The diversity and possible functions of the inositol polyphosphate 5-phosphatases. Biochim. Biophys. Acta *1436*, 185–199.

Erneux, C., Lemos, M., Verjans, B., Vanderhaeghen, P., Delvaux, A. and Dumont, J. E. (1989). Soluble and particulate Ins(1,4,5)P$_3$/Ins(1,3,4,5)P$_4$ 5-phosphatase in bovine brain. Eur. J. Biochem. *181*, 317–322.

Erneux, C., Moreau, C., Vandermeers, A. and Takazawa, K. C. (1993). Interaction of calmodulin with a putative calmodulin-binding domain of inositol 1,4,5-trisphosphate 3-kinase. Effects of synthetic peptides and site-directed mutagenesis of Tr165. Eur. J. Biochem. *14*, 497–501.

Erneux, C., Takazawa, K. and Verjans, B. (1993). Inositol 1,4,5-trisphosphate phosphatase and kinase from brain. In: Lipid Metabolism in Signaling Systems (Fain, J. N., ed.). Academic press, New York, pp. 312–319.

Estrada-Garcia, T., Craxton, A., Kirk, C. J. and Michell, R. H. (1991). A salt-activated inositl 1,3,4,5-tetrakisphosphate 3-phosphatase at the inner surface of the human erythrocyte membrane. Proc. Royal Soc. Lond B Biol. Sci. *244*(1309), 63–68.

Freund, W-D., Mayer, G. W., Tietz, C. and Schultz, J. E. (1992). Metabolism of inositol phosphates in the protozoan Paramecium. Characterization of a novel inositol-hexakisphosphate-dephosphorylating enzyme. Eur. J. Biochem. *207*, 359–367.

Gee, N. S., Reid, G. G., Jackson, R. G., Barnaby, R. J. and Ragan, C. I. (1988). Purification and properties of inositol 1,4-bisphosphatase from bovine brain. Biochem. J. *249*, 777–782.

Gore, M. G., Greasley, P. J. and Ragan, C. I. (1992). Bovine inositol mono-phosphatase: development of a continuous fluorescence assay of enzyme activity. J. Biochem. Biophys. Methods *25*, 55–60.

Greiner, R., Konietzny, U. and Jany, K. D. (1993). Purification and characterization of two phytases from *Escherichia coli*. Arch. Biochem. Biophys. *303*, 107–113.

Greiner, R., Carlsson, N. and Alminger, M. L. (2000). Stereospecificity of *myo*-inositol hexakisphosphate dephosphorylation by a phytate-degrading enzyme of *Escherichia coli*. J. Biotechnol. *84*, 53–62.

Guo, S., Stolz, L. E., Lemrow, S. M. and York, J. D. (1999). *SAC1*-like domains of yeast SAC1, INP52, and INP53 and of human synaptojanin encode polyphos-phoinositide phosphatases. J. Biol. Chem. *274*, 12990–12995.

Ha, N-C., Oh, B-C., Shin, S. S., Kim, H-J., Oh, T-K., Kim, Y-O., Choi, K. Y. and Oh, B-H. (2000). Crystal structures of a novel, thermostable phytase in partially and fully calcium-loaded states. Nat. Struct. Biol. *7*, 147–153.

Hallcher, L. M. and Sherman, W. R. (1980). The effects of lithium ion and other agents on the activity of *myo*-inositol-1-phosphatase from bovine brain. J. Biol. Chem. *255*, 10896–10901.

Hansen, C. A., Johanson, R. A., Williamson, M. T. and Williamson, J. R. (1987). Purification and characterization of two types of soluble inositol phosphate 5-phosphomonoesterases from rat retina. J. Biol. Chem. *262*, 17319–17326.

Hara, A., Ebina, S., Kondo, A. and Funaguma, T. (1985). A new type of phytase from pollen of *Typha latifolia*. Agric. Biol. Chem. *49*, 3539–3544.

Hegeman, C. E. and Grabau, E. A. (2001). A novel phytase with sequence similarity to purple acid phosphatases is expressed in cotyledons of germinating soybean seedlings. Plant Physiol. *126*, 1598–1608.

Hegyvary, C., Kang, K. and Bandi, Z. (1979). Automated assay of phosphohydrolases by measuring the released phosphate without deproteinization. Anal. Biochem. *94*, 397–401.

Heinonen, J. K. and Lahti, R. J. (1981). A new and convenient colorimetric determination of inorganic orthophosphate and its application to the assay of inorganic pyrophosphatase. Anal. Biochem. *113*, 313–317.

Hildebrant, J-P. and Shuttleworth, T. J. (1992). Calcium-sensitivity of inositol 1,4,5-trisphosphate metabolism in exocrine cells from the avian salt gland. Biochem. J. *282*, 703–710.

Ho, M. W., Yang, Carew, M. A., Zhang, T., Hua, L., Kwon, Y. U., Chung, S. K., Adelt, S., Vogel, G., Riley, A. M., Potter, B. V. and Shears, S. B. (2002). Regulation of ins(3,4,5,6)P$_4$ signaling by a reversible kinase/phosphatase. Curr. Biol. *12*, 477–482.

Hodgkin, M., Craxton, A., Parry, J. B., Hughes, P. J., Potter, B. V. L., Michell, R. H. and Kirk, C. J. (1994). Bovine testis and human erythrocytes contain different subtypes of membrane-associated Ins(1,4,5)P$_3$/Ins(1,3,4,5)P$_4$ 5-phosphomonoesterases. Biochem. J. *297*, 637–645.

Hoenig, M., Lee, R. J. and Ferguson, D. C. (1989). A microtiter plate assay for inorganic phosphate. J. Biochim. Biophys. Methods *19*, 249–252.

Hughes, W. E., Cooke, F. T. and Parker, P. J. (2000a). Sac phosphatase domain proteins. Biochem. J. *350*, 337–352.

Hughes, W. E., Woscholski, R., Cooke, F. T., Patrick, R. S., Dove, S. K., McDonald, N. Q. and Parker, P. J. (2000b). SAC1 encodes a regulated lipid phosphoinositide phosphatase, defects in which can be suppressed by the homologous Inp52p and Inp53p phosphatases. J. Biol. Chem. *275*, 801–808.

Ijuin, T., Mochizuki, Y., Fukami, K., Funaki, M., Asano, T. and Takenawa, T. (2000). Identification and characterization of a novel inositol polyphosphate 5-phosphatase. J. Biol. Chem. *275*, 10870–10875.

Inhorn, R. C., Bansal, V. S. and Majerus, P. W. (1987). Pathway for inositol 1,3,4-trisphosphate and 1,4-bisphosphate metabolism. Proc. Natl Acad. Sci. USA *84*, 2170–2174.

Inhorn, R. C. and Majerus, P. W. (1987). Inositol polyphosphate 1-phosphatase from calf brain. J. Biol. Chem. *262*, 15946–15952.

Inhorn, R. C. and Majerus, P. W. (1988). Properties of inositol polyphosphate 1-phosphatase. J. Biol. Chem. *263*, 14559–14565.

Irvine, R. F. and Schell, M. J. (2001). Back in the water: the return of the inositol phosphates. Nat. Rev. Mol. Cell Biol. *2*, 327–338.

Itaya, K. and Ui, M. (1966). A new micromethod for the colorimetric determination of inorganic phosphate. Clin. Chim. Acta *14*, 361–366.

Jackson, S. P., Schoenwaelder, S. M., Matzaris, M., Brown, S. and Mitchell, C. A. (1995). Phosphatidylinositol 3,4,5-trisphosphate is a substrate for the 75 kDa inositol polyphosphate 5-phosphatase and a novel 5-phosphatase which forms a complex with the p85/p110 form of phosphoinositide 3-kinase. EMBO J. *14*, 4490–4500.

Jackson, S. P., Schoenwaelder, S. M., Yuan, Y., Rabinowitz, I., Salem, H. H. and Mitchell, C. A. (1994). Adhesion receptor activation of phosphatidylinositol 3-kinase. J. Biol. Chem. *269*, 27093–27099.

Jefferson, A. B., Auethavekiat, V., Pot, D. A., Williams, L. T. and Majerus, P. W. (1997). Signaling inositol polyphosphate 5-phosphatase. Characterization of activity and effect of GRB2 association. J. Biol. Chem. *272*, 5983–5988.

Jefferson, A. B. and Majerus, P. W. (1995). Properties of type II inositol polyphosphate 5-phosphatase. J. Biol. Chem. *270*, 9370–9377.

Jefferson, A. B. and Majerus, P. W. (1996). Mutation of the conserved domains of two inositol polyphosphate 5-phosphatases. Biochemistry *35*, 7890–7894.

Kavanaugh, W. M., Pot, D. A., Chin, S. M., Deuter-Reinhard, M., Jefferson, A. B., Norris, F. A., Masiarz, F. R., Cousens, L. S., Majerus, P. W. and Williams, L. T. (1996). Multiple forms of an inositol polyphosphate 5-phosphatase from signaling complexes Shc and Grb 2. Curr. Biol. *6*, 438–445.

Kerovuo, J., Lauraeus, M., Nurminen, P., Kalkkinen, N. and Apajalahti, J. (1998). Isolation, characterization, molecular gene cloning, and sequencing of a novel phytase from *Bacillus subtilis*. Appl. Environ. Microbiol. *64*, 2079–2085.

Kerovuo, J., Rouvinen, J. and Hatzack, F. (2000). Analysis of *myo*-inositol hexakisphosphate hydrolysis by *Bacillus* phytase: indication of a novel reaction mechanism. Biochem. J. *352*, 623–628.

Kisseleva, M. V., Cao, L. and Majerus, P. W. (2002). Phosphoinositide-specific inositol polyphosphate 5-phosphatase IV inhibits Akt/PKB phosphorylation and leads to apoptotic cell death. J. Biol. Chem. *277*, 6266–6272.

Kisseleva, M. V., Wilson, M. P. and Majerus, P. W. (2000). The isolation and characterization of a cDNA encoding phospholipid-specific inositol polyphosphate 5-phosphatase. J. Biol. Chem. *275*, 20110–20116.

Kodaki, T., Woscholski, R., Emr, S., Waterfield, M. D., Nurse, P. and Parker, P. J. (1994). Mammalian phosphatidylinositol 3′-kinase induces a lethal phenotype expression in *Schizosaccharomyces pombe*; comparison with the VPS34 gene product. Eur. J. Biochem. *219*, 775–780.

Kong, A. M., Speed, C. J., O'Malley, C. J., Layton, M. J., Meehan, T., Loveland, K. L., Cheema, S., Ooms, L. M. and Mitchell, C. A. (2000). Cloning and characterization of a 72-kDa inositol-polyphoshate 5-phosphatase localized to the Golgi network. J. Biol. Chem. *275*, 24052–24064.

Kostrewa, D., Gruninger-Leitch, F., D'Arcy, A., Broger, C., Mitchell, D. and Van Loon, A. P. G. M. (1997). Crystal structure of phytase from *Aspergillus ficuum* at 2.5 A resolution. Nature Struct. Biol. *4*, 185–190.

Kostrewa, D., Wyss, M., D'Arc, A. and van Loon, A. P. G. M. (1999). Crystal structure of *Aspergillus niger* pH 2.5 acid phosphatase at 2.4 A resolution. J. Mol. Biol. *288*, 965–974.

Krystal, G. (2000). Lipid phosphatases in the immune system. Semin. Immunol. *12*, 397–403.

Laussmann, T., Reddy, K. M., Reddy, K., Falck, J. R. and Vogel, G. (1997). Diphospho-*myo*-inositol phosphate from *Dictyostelium* identified as D-6-diphospho-*myo*-inositol pentakisphosphate and D-5,6-bisdiphospho-*myo*-inositol tetrakisphosphate. Biochem. J. *322*, 31–33.

Laxminarayan, K. M., Matzaris, M., Speed, C. J. and Mitchell, C. A. (1993). Purification and characterization of a 43-kDa membrane-associated inositol polyphosphate 5-phosphatase from human placenta. J. Biol. Chem. *268*, 4968–4974.

Laxminarayan, K. M., Chan, B. K., Tetaz, T., Bird, P. I. and Mitchell, C. A. (1994). Characterization of a cDNA encoding the 43-kDa membrane-associated inositol polyphosphate 5 phosphatase. J Biol. Chem. *269*, 17305–17310.

Leavitt, A. L. and Sherman, W. R. (1982). Determination of inositol phosphates by gas chromatography. Methods Enzymol. *89*, 9–18.

Lee, J. O., Yang, H., Georgescu, M. M., Di Cristofano, A., Maehama, T., Shi, Y., Dixon, J. E., Pandofi, P. and Pavletich, N. P. (1999). Crystal structure of the PTEN tumor suppressor: implications for its phosphoinositide phosphatase activity and membrane association. Cell *99*, 232–334.

Leslie, N. R., Bennett, D., Gray, A., Pass, I., Hoang-Xuan, K. and Downes, C. P. (2001). Targeting mutants of PTEN reveal distinct subsets of tomour suppressor functions. Biochem. J. *357*, 427–435.

Leslie, N. R. and Downes, C. P. (2002). PTEN: The down side of PI 3-kinase signalling. Cell Signal. *14*, 285–295.

Lim, D., Golovan, S., Forsberg, C. W. and Jia, Z. (2000). Crystal structures of *Escherichia coli* phytase and its complex with phytate. Nat. Struct. Biol. *7*, 108–113.

Lips, D. L. and Majerus, P. W. (1989). The discovery of a 3-phosphomonoesterase that hydrolyzes phosphatidylinositol 3-phosphate in NIH 3T3 cells. J. Biol. Chem. *264*, 19911–19915.

Loewus, F. A. and Murthy, P. P. N. (2000). *myo*-Inositol metabolism in plants. Plant Sci. *150*, 1–19.

Lopez, F., Leube, M., Gil-Mascarell, R., Navarro-Avino, J. P. and Serrano, R. (1999). The yeast inositol monophosphatase is a lithium- and sodium-sensitive enzyme encoded by a non-essential gene pair. Mol. Microbiol. *31*, 1255–1264.

Lopez-Coronado, J. M., Belles, J. M., Lesage, F., Serrano, R. and Rodriguez, P. L. (1999). A novel mammalian lithium-sensitive enzyme with a dual enzymatic activity, 3′-phosphoadenosine 5′-phosphate phosphates and inositol-polyphosphate 1-phosphatase. J. Biol. Chem. *274*, 16034–16039.

Luzzi, V., Murazina, D. and Allbritton, N. L. (2000). Characterization of a biological detector cell for quantitation of inositol 1,4,5-trisphosphate. Anal. Biochem. *277*, 221–227.

Maehama, T. and Dixon, J. E. (1998). The tumor suppressor, PTEN/MMAC1, dephosphorylates the lipid second messenger, phosphatidylinositol 3,4,5-trisphosphate. J. Biol. Chem. *273*, 13375–13378.

Maehama, T. and Dixon, J. E. (1999). PTEN: a tumour suppressor that functions as a phospholipid phosphatase. Trends Cell Biol. *9*, 125–128.

Maehama, T., Taylor, G. S. and Dixon, J. E. (2001). PTEN and myotubularin: Novel phosphoinositide phosphatases. Annu. Rev. Biochem. *70*, 247–279.

Maehama, T., Taylor, G. S., Slama, J. T. and Dixon, J. E. (2000). A sensitive assay for phosphoinositide phosphatases. Anal. Biochem. *279*, 248–250.

Majerus, P. W. (1991). The George M. Kober Lecture: molecular mechanisms of intracellular signal transduction. Trans. Assoc. Am. Physicians *104*, clxviii–clxxx.

Majerus, P. W. (1992). Inositol phosphate biochemistry. Annu. Rev. Biochem. *61*, 225–250.

Majerus, P. W. (1996). Inositols do it all. Genes Dev. *10*, 1051–1053.

Majerus, P. W., Connolly, T. M., Bansal, V. S., Inhorn, R. C., Ross, T. S. and Lips, D. L. (1988). Inositol phosphates: synthesis and degradation. J. Biol. Chem. *263*, 3051–3054.

Majerus, P. W., Kisseleva, M. V. and Norris, F. A. (1999). The role of phosphatases in inositol signaling reactions. J. Biol. Chem. *274*, 10669–10672.

Maslanski, J. A. and Busa, W. B. (1990). A sensitive and specific mass assay for *myo*-inositol and inositol phosphates. In: Methods in Inositide Research (Irvine, R. V., ed.). Raven Press Ltd, New York, NY, pp. 113–126.

Matzaris, M., Jackson, S. P., Lazminarayan, K. M., Speed, C. J. and Mitchell, C. A. (1994). Identification and characterization of the phosphatidylinositol 4, 5-bisphosphate 5-phosphatase in human platelets. J. Biol. Chem. *269*, 3397–3402.

Matzaris, M., O'Malley, C. J., Badger, A., Speed, C. J., Bird, P. I. and Mitchell, C. A. (1998). J. Biol. Chem. *273*, 8256–8267.

Mayr, G. W. (1988). A novel metal-dye detection system permits picomolar range h.p.l.c. analysis of inositol polyphosphates from non-radioactively labeled cell or tissue specimens. Biochem. J. *254*, 585–591.

Mayr, G. W. (1990). Mass determination of inositol phosphates by high-performance liquid chromatography with post column complexometry metal-dye detection. In: Methods in Inositide Research (Irvine, R. F., ed.). Raven Press, New York, pp. 83–108.

McPherson, P. S., Czernik, A. J., Chilcote, T. J., Onori, F. Benfanati, F., Greengard, P., Schlessinger, J. and De Camilli, P. (1994a). Interaction of *Grb2* via its *Src* homology 3 domains with synaptic proteins including synapsin I. Proc. Natl Acad. Sci. USA *91*, 6486–6490.

McPherson, P. S., Garcia, E. P., Slepnev, V. I., David, C., Zhang, X., Grabs, D., Sossin, W. S., Bauerfeind, R., Nemoto, Y. and De Camilli, P. (1996). A presynaptic inositol-5-phosphatase. Nature *379*, 353–357.

McPherson, P. S., Takei, K., Schmid, S. L. and DeCamilli, P. (1994b). p145, a major Grb2-binding protein in brain, is co-localized with dynamin in nerve terminals where it undergoes activity-dependent dephosphorylation. J. Biol. Chem. *269*, 30132–30139.

Meek, J. L. (1986). Inositol bis-, tris-, and tetrakis(phosphate)s: analysis in tissues by HPLC. Proc. Natl Acad. Sci. USA *83*, 4162–4166.

Menniti, F. S., Miller, R. N., Putney, J. W. Jr. and Shears, S. B. (1993). Turnover of inositol polyphosphate pyrophosphates in pancreatoma cells. J. Biol. Chem. *268*, 3850–3856.

Mitchell, C. A., Brown, S., Campbell, J. K., Munday, A. D. and Speed, C. J. (1996). Regulation of second messengers by the inositol polyphosphate 5-phosphatases. Biochem. Soc. Trans. *25*, 994–1000.

Mitchell, C. A., Connolly, T. M. and Majerus, P. W. (1989). Identification and isolation of a 75-kDa inositol polyphosphate 5-phosphatase from human platelets. J. Biol. Chem. *264*, 8873–8877.

Mochizuki, Y. and Takenawa, T. (1999). Novel inositol polyphosphate 5-phosphatase localizes at membrane ruffles. J. Biol. Chem. *274*, 36790–36795.

Mullaney, E. J., Daly, C. B., Sethumadhavan, K., Rodriguez, E., Lei, X. G. and Ullah, A. H. J. (2000). Phytase activity in *Aspergillus fumigatus* isolates. Biochem. Biophys. Res. Commun. *275*, 759–763.

Munday, A. D., Norris, F. A., Caldwell, K. K., Brown, S., Majerus, P. W. and Mitchell, C. A. (1999). The inositol polyphosphate 4-phosphatase forms a complex with phosphatidylinositol 3-kinase in human platelet cytosol. Proc. Natl Acad. Sci. USA *96*, 3640–3645.

Murray, M. and Greenberg, M. L. (2000). Expression of yeast INM1 encoding inositol monophosphatase is regulated by inositol, carbon source and growth stage and is decreased by lithium and valporate. Mol. Microbiol. *36*, 651–661.

Nandurkar, H. H. and Huysmans, R. (2002). The myotubularin family: novel phosphoinositide regulators. IUBMB Life *53*, 37–43.

Nandurkar, H. H., Caldwell, K. K., Whisstock, J. C., Layton, M. J., Gaudet, E. A., Norris, F. A., Majerus, P. W. and Mitchell, C. A. (2001). Characterization of an adapter subunit to a phosphatidylinositol(3)P 3-phosphatase: Identification of a myotubularin-related protein lacking catalytic activity. Proc. Natl Acad. Sci. USA *98*, 9499–9504.

Nemoto, Y., Arribas, M., Haffner, C. and DiCamilli, P. (1997). Synaptojanin 2, a novel synaptojanin isoform with a distinct targeting domain and expression pattern. J. Biol. Chem. *272*, 30817–30821.

Nigou, J. and Besra, G. S. (2002). Characterization and regulation of inositol monophosphatase activity in *Mycobacterium smegmatis*. Biochem. J. *361*, 385–390.

Nigou, J. and Besra, G. S. (2002). Cytidine diphosphate-diacylglcerol synthesis in *Mycobacterium smegmatis*. Bioce. J. *367*, 157–162.

Nogimori, K., Hughes, P. J., Glennon, M. C., Hodgson, M. E., Putney, J. W. Jr. and Shears, S. B. (1991). Purification of an inositol (1,3,4,5)-tetrakisphosphate 3-phosphatase activity from rat liver and the evaluation of its substrate specificity. J. Biol. Chem. *266*, 16499–16506.

Norris, F. A., Atkins, R. C. and Majerus, P. W. (1997). The cDNA cloning and characterization of inositol polyphosphate 4-phosphatase type II. J. Biol. Chem. *272*, 23859–23864.

Norris, F. A., Auetghavekiat, V. and Majerus, P. W. (1995). The isolation and characterization of cDNA encoding human and rat brain inositol polyphosphate 4-phosphatase. J. Biol. Chem. *270*, 16128–16133.

Norris, F. A. and Majerus, P. W. (1994). Hydrolysis of phosphatidylinositol 3,4-bisphosphate by inositol polyphosphate 4-phosphatase isolated by affinity elution chromatography. J. Biol. Chem. *269*, 8716–8720.

Norris, F. A., Wilson, M. P., Wallis, T. S., Galyov, E. E. and Majerus, P. W. (1998). SopB, a protein required for virulence of *Salmonella dublin*, is an inositol phosphate phosphatase. Proc. Natl Acad. Sci. USA *95*, 14057–14059.

Oshima, Y. (1997). The phosphatase system in *Saccharomyces cerevisiae*. Genes Genet. Syst. *72*, 323–334.

Ostanin, K. and Van Etten, R. L. (1993). Asp304 of *Escherichia coli* acid phosphatase is involved in leaving group protonation. J. Biol. Chem. *268*, 20778–20784.

Ostanin, K., Saeed, A. and Van Etten, R. L. (1994). Heterologous expression of human acid phosphatase and site-directed mutagenesis of the enzyme active site. J. Biol. Chem. *269*, 8971–8978.

Palmer, F. B. St. C. (1990). Enzymes that degrade phosphatidylinositol 4-phosphate and phosphatidylinositol 4,5-bisphosphate have different developmental profiles in chicken brain. Biochem. Cell Biol. *68*, 800–803.

Palmer, F. B. St. C., Theolis, R., Cook, H. W. and Byers, D. M. (1994). Purification of two immunologically related phosphatidylinositol-(4,5)-bisphosphate phosphatases from bovine brain cytosol. J. Biol. Chem. *269*, 3403–3410.

Parthasarathy, R., Parthasarathy, L. and Vadnal, R. (1997). Brain inositol mono phosphatase identified as a galactose 1-phosphatase. Brain Res. *778*, 99–106.

Parthasarathy, L., Vadnal, R. E., Ramesh, T. G., Shyamaldevi, C. S. and Parthasarathy, R. (1993). *myo*-Inositol monophosphatase from rat testes: purification and properties. Arch. Biohcem. Biophys. *304*, 94–101.

Pesesse, X., Deleu, S., DeSmedt, F., Drayer, L. and Erneux, C. (1997). Identification of a second SH-2-domain-containing protein closely related to the phosphatidylinositol polyphosphate 5-phosphatase SHIP. Biochem. Biophys. Res. Commun. *239*, 697–700.

Pesesse, X., Dewaste, V., De Smedt, F., Laffargue, M., Giuriato, S., Moreau, C., Payastre, B. and Erneux, C. (2001). The SH2 domain containing inositol 5-phosphatase SHIP2 is recruited to the EGF receptor and dephosphorylates phosphatidylinositol 3,4,5-trisphosphate in EGF stimulated COS-7 cells. J. Biol. Chem. *276*, 28348–28355.

Piddington, C. S., Houston, C. S., Paloheimo, M., Cantrell, M., Miettinen-Oinonen, A., Nevalainen, H. and Rambosck, J. (1993). The cloning and sequencing of the genes encoding phytase (phy) and pH 2.5-optimum acid phosphatase (aph) from *Aspergillus niger* var. *awamori*. Gene *133*, 55–62.

Pike, L. J. and Eakes, A. T. (1987). Epidermal growth factor stimulates the production of phosphatidylinositol monophosphate and the breakdown of polyphosphoinositides in A431 cells. J. Biol. Chem. *262*, 1644–1651.

Rameh, L. E., Tolias, K. F., Duckworth, B. C. and Cantley, L. C. (1997). A new pathway for synthesis of phosphatidylinositol-4,5-bisphosphate. Nature *390*, 192–196.

Rass Hope, H. M. and Pike, L. J. (1994). Purification and characterization of a polyphosphoinositide phosphatase from rat brain. J. Biol. Chem. *269*, 23648–23654.

Rivera, J., Lozano, M. L., Gonzalez-Canejero, R., Corral, J., De Arriba, F. and Vicente, V. (1998). A radioreceptor assay for mass measurement of inositol (1,4,5)-trisphosphate using saponin-permeabilized outdated human platelets. Anal. Biochem. *256*, 117–121.

Rodriguez, E., Han, Y. and Lei, X. G. (1999). Cloning, sequencing, and expression of an *Escherichia coli* acid phosphatase/phytase gene (appA2) isolated from pig colon. Biochem. Biophys. Res. Commun. *257*, 117–123.

Rohrschneider, L. R., Fuller, J. F., Wolf, I., Liu, Y. and Lucas, D. M. (2000). Structure, function, and biology of DSHIP proteins. Genes Dev. *14*, 505–520.

Romano, P. R., Wang, J., O'Keefe, R. J., Puzas, J. E., Rosier, R. N. and Reynolds, P. R. (1998). HiPER1, a phosphatase of the endoplasmic reticulum with a role in chondrocyte maturation. J. Cell Sci. *111*, 803–813.

Ross, T. S., Jefferson, A. B., Mitchell, C. A. and Majerus, P. W. (1991). Cloning and expression of human 75-kDa inositol polyphosphate 5-phosphatase. J. Biol. Chem. *266*, 20283–20289.

Safrany, S. T., Caffrey, J. J., Yang, X., Bembenek, M., Moyer, M. B., Burkhart, W. A. and Shears, S. B. (1998). A novel context for the "MutT" module, a guardian of cell integrity, in a diphosphoinositol polyphosphate phosphohydrolase. EMBO J. *17*, 6599–6607.

Safrany, S. T. and Shears, S. B. (1998). Turnover of bis-diphosphoinositol tetrakisphosphate in a smooth muscle cell line is regulated by β_2-adrenergic receptors through a cAMP-mediated, A-kinase-independent mechanism. EMBO J. *17*, 1710–1716.

Sakisaka, T., Itoh, T., Miura, K. and Takenawa, T. (1997). Phosphatidylinositol 4, 5-bisphosphate phosphatase regulates the rearrangement of actin filaments. Mol. Cell. Biol. *17*, 3841–3849.

Schorr, M., Then, A., Tahirovic, S., Hug, N. and Mayinger, P. (2001). The phosphoinositide phosphatase Sac1p controls trafficking of the yeast Chs3p chitin synthase. Curr. Biol. *11*, 1421–1426.

Seet, L-F., Cho, S., Hessel, A. and Dumont, D. J. (1998). Molecular cloning of multiple isoforms of synaptojanin 2 and assignment of the gene to mouse chromosome 17A2-3.1. Biochem. Biophys. Res. Commun. *247*, 116–122.

Serunian, L. A., Haber, M. T., Fukui, T., Kim, J. W., Rhee, S. G., Lowenstein, J. M. and Cantley, L. C. (1989). Polyphosphoinositides produced by phosphatidylinositol 3-kinase are poor substrates for phospholipases C from rat liver and bovine brain. J. Biol. Chem. *264*, 19911–19915.

Sha, B., Phillips, S. E., Bankaitis, B. A. and Luo, M. (1998). Crystal structure of the *Saccharomyces cerevisiae* phosphatidylinositol transfer protein. Nature *391*, 506–510.

Shears, S. B. (1992). Metabolism of inositol phosphates. Adv. Second Messenger Phosphoprotein Res. *63*, 63–91.

Shears, S. B. (1998). The versatility of inositol phosphates as cellular signals. Biochim. Biophys. Acta *1436*, 49–67.

Shears, S. B. (2001). Assessing the omnipotence of inositol hexakisphosphate. Cell Signal. *13*, 151–158.

Shears, S. B., Ali, N., Craxton, A. and Bembenek, M. E. (1995). Synthesis and metabolism of bis-diphospoinositol tetrakisphosphate *in vitro* and *in vivo*. J. Biol. Chem. *270*, 10489–10497.

Sherman, W. R., Ackermann, K. E., Berger, R. A., Gish, B. G. and Zinbo, M. (1986). Analysis of inositol mono- and polyphosphates by gas chromatography/mass spectrometry and fast atom bombardment. Biomed. Environ. Mass Spectrom. *13*, 333–341.

Singh, A. K. (1992). Quantitative analysis of inositol lipids and inositol phosphates in synaptosomes and microvessels by column chromatography: comparison of the mass analysis and the radiolabeling methods. J.Chromatogr. *581*, 1–10.

Spencer, C. E. L., Stephens, L. R. and Irvine, R. F. (1990). Separation of higher inositol phosphates by polyethyleneimine-cellulose thin-layer chromatography and by Dowex chloride column chromatography. In: Methods in Inositide Research (Irvine, R. F., ed.). Raven Press, New York, pp. 39–43.

Stack, J. H., DeWald, D. B., Takegawa, K. and Emr, S. D. (1995). Vesicle mediated protein transport: regulatory interactions between the Vps15 protein kinase and the Vps34 PtdIns 3-kinase essential for protein sorting to the vacuole in yeast. J. Cell Biol. *129*, 321–334.

Stec, B., Yang, H., Johnson, K. A. and Roberts, M. F. (2000). MJ0109 is an enzyme that is both an inositol monophosphatase and the "missing" archael fructose-1,6-bisphosphatase. Nat. Struct. Biol. *7*, 1046–1050.

Stein, J. M., Smith, G. and Luzio, J. P. (1990). Quantification of polypho sphoinositides by acetylation with [³H]acetic anhydride. In: Methods of Inositide Research (Irvine, R. F., ed.). Raven Press Ltd, New York, pp. 167–177.

Stephens, L. R. (1990). Preparation and separation of inositol tetrakisphosphates and inositol pentakisphosphates and the establishment of enantiomeric configurations by the use of L-iditol dehydrogenase. In: Methods in Inositide Research (Irvine, R. F., ed.). Raven Press Ltd, New York, pp. 9–30.

Stephens, L. R., Hughes, K. T. and Irvine, R. F. (1991). Pathway of phosphatidylinositol (3,4,5)-trisphosphate synthesis in activated neutrophils. Nature *351*, 33–39.

Stephens, L. R., Radenberg, T., Thiel, U., Vogel, G., Khoo, K-H., Dell, A., Jackson, T. R., Hawkins, P. T. and Mayr, G. (1993). The detection, purification, structural characterization and metabolism of diphosphoinositol pentakisphosphate(s) and bisdiphosphoinositol tetrakisphosphate(s). J. Biol. Chem. *268*, 4009–4015.

Stolz, L. E., Huynh, C. V., Thorner, J. and York, J. D. (1998a). Identification and characterization of an essential family of inositol polyphosphate 5-phosphatases (INP51, NP52 and INP53 gene products) in the yeast *Saccharomyces cerevisiae*. Genetics *148*, 1715–1729.

Stolz, L. E., Kuo, W. J., Longchamps, J., Sekhon, M. K. and York, J. D. (1998b). INP51, a yeast inositol polyphosphate 5-phosphatase required for phosphatidylinositol 4,5-bisphosphate homeostasis and whose absence confers a cold resistant phenotype. J. Biol. Chem. *273*, 11852–11861.

Storey, D. J., Shears, S. B., Kirk, C. J. and Michell, R. H. (1984). Stepwise enzymatic dephosphorylation of inositol 1,4,5-trisphosphate to inositol in liver. Nature *312*, 374–376.

Stricker, R., Adelt, S., Vogel, G. and Reiser, G. (1999). Translocation between membranes and cytosol of p42^{IP4}, a specific inositol 1,3,4,5-tetrakisphosphate/ phosphatidylinositol 3,4,5-trisphosphate-receptor protein from brain, is induced by inositol 1,3,4,5-tetrakisphosphate and regulated by a membrane-associated 5-phosphatase. Eur. J. Biochem. *265*, 815–824.

Suchy, S. F., Olivos-Glander, I. M. and Nussabaum, R. L. (1995). Lowe syndrome, a deficiency of phosphatidylinositol 4,5-bisphosphate 5-phosphatase in the Golgi apparatus. Hum. Mol. Genet. *4*, 2245–2250.

Sun, G. Y. and Lin, T. N. (1990). Separation of phosphoinositides and other phospholipids by high performance thin-layer chromatography. In: Methods in Inositide Research (Irvine, R. F., ed.). Raven Press Ltd, New York, pp. 153–158.

Sun, G. Y., Lin, T. N., Premkumar, N., Carter, S. and MacQuarries, R. A. (1990). Separation and quantification of isomers of inositol phosphates by ion chromatography. In: Methods in Inositide Research (Irvine, R. F., ed.). Raven Press Ltd, New York, NY, pp. 135–143.

Takazawa, K., Passareiro, H., Dumont, J. E. and Erneux, C. (1988). Ca^{2+}/ calmodulin sensitive inositol 1,4,5-trisphosphate 3-kinase in rat and bovine brain tissues. Biochem. Biophys. Res. Commun. *153*, 632–641.

Taylor, G. S., Liu, Y., Baskerville, C. and Charbonneau, H. (1997). The activity of Cdc14p, an oligomeric dual specificity protein phosphatase from *Saccharomyces cerevisiae*, is required for cell cycle progression. J. Biol. Chem. *272*, 24054–24063.

Taylor, G. S., Maehama, T. and Dixon, J. E. (2000a). Myotubularin, a protein tyrosine phosphatase mutated in myotubular myopathy, dephosphorylates the lipid second messenger, phosphatidyl 3-phosphatase. Proc. Natl Acad. Sci. USA *97*, 8910–8915.

Taylor, V., Wong, M., Brandts, C., Reilly, L., Dean, N. M., Cowsert, L. M., Moodie, S. and Stokoe, D. (2000b). 5′-Phospholipid phosphatase SHIP-2 causes protein kinase B inactivation and cell cycle arrest in glioblastoma cells. Mol. Cell Biol. *20*, 6860–6871.

Traynor-Kaplan, A. E., Harris, A. L., Thompson, B. L., Taylor, P. and Sklar, L. A. (1988). An inositol tetrakisphosphate-containing phospholipid in activated neutrophils. Nature (London) *334*, 353–356.

Tsujishita, Y., Guo, S., Stolz, L. E., York, J. D. and Hurley, J. H. (2001). Specificity determinants in phosphoinositide dephosphorylation: crystal structure of an archetypal inositol polyphosphate 5-phosphatase. Cell *105*, 379–389.

Ullah, A. H. J., Cummins, B. J. and Dischinger, H. C., Jr (1991). Cyclohexadione modification of arginine at the active site of *Aspergillus ficuum* phytase. Biochem. Biophys. Res. Commun. *178*, 45–53.

Van der Kaay, J. and Van Haastert, P. J. (1995). Stereospecificity of inositol hexakisphosphate dephosphorylation by *Paramecium* phytase. Biochem. J. *312*, 907–910.

Van der Kaay, J., Batty, I. H., Cross, D. A. E., Watt, P. W. and Downes, C. P. (1997). A novel, rapid, and highly sensitive mass assay for phosphatidyl-inositol 3,4,5-trisphosphate [PtdIns(3,4,5)P$_3$] and its application to measure insulin stimulated PtdIns(3,4,5)P$_3$ production in rat skeletal muscle *in vivo*. J. Biol. Chem. *272*, 5477–5481.

Van der Kaay, J., Culle, P. J. and Downes, C. P. (1998). Phosphatdidylinositol(3, 4,5)trisphosphate [PtdIns(3,4,5)P$_3$] mass measurement using a radioligand and displacement assay. Methods Mol. Biol. *105*, 109–125.

Van Dijken, P., Bergsma, J. C., Hiemstra, H. S., De Vries, B., Van der Kaay, J. and Van Haastert, P. J. (1996). *Dictyostelium discoideum* contains three inositol monophosphatase activities with different substrate specificities and sensitivities to lithium. Biochem. J. *314*, 491–495.

Vanhaesebroeck, B., Leevers, S. J., Madi, K., Timms, J., Katso, R., Driscoll, P. C., Woscholski, R., Parker, P. J. and Waterfield, M. D. (2001). Synthesis and function of 3-phosphorylated inositol lipids. Annu. Rev. Biochem. *70*, 535–602.

Van Lookeren Campagne, M., Erneux, C., Van Eijk, R. and Van Haastert, P. J. M. (1988). Two dephosphorylation pathways of inositol 1,4,5-trisphosphate in homogenates of the cellular slime mould of *Dictiostelium discoideum*. Biochem. J. *254*, 343–350.

Verjans, B., Lecocq, R., Moreau, C. and Erneux, C. (1992). Purification of bovine brain inositol-1,4,5-trisphosphate 5-phosphatase. Eur. J. Biochem. *204*, 1083–1087.

Verjans, B., Moreau, C. and Erneux, C. (1994a). The control of intracellular signal molecules at the level of their hydrolysis: the example of inositol 1,4,5-trisphosphate 5-phosphatase. Mol. Cell. Endocrinol. *98*, 167–171.

Verjans, B., De Smedt, F., Lecocq, R., Vanweyenberg, V., Moreau, C. and Erneux, C. (1994b). Cloning and expression in *Escherichia coli* of a dog thyroid cDNA encoding a novel inositol 1,4,5-trisphosphate 5-phosphatase. Biochem. J. *300*, 85–90.

Vyas, P., Norris, F. A., Joseph, R., Majerus, P. W. and Orkin, S. H. (2000). Inositol polyphosphate 4-phosphatase type I regulates cell growth downstream of transcription factor GATA-1. Proc. Natl Acad. Sci. USA *97*, 13696–13701.

Walker, D. M., Urbe, S., Dove, S. K., Tenza, D., Raposo, G. and Clague, M. J. (2001a). Characterization of MTMR3: an inositol lipid 3-phosphatase with novel substrate specificity. Curr. Biol. *11*, 1600–1605.

Walker, S. M., Downes, C. P. and Leslie, N. R. (2001b). TPIP: a novel phosphoinositide 3-phosphatase. Biochem. J. *360*, 277–283.

Walsh, J. P., Caldwell, K. K. and Majerus, P. W. (1991). Formation of phosphatidylinositol 3-phosphate by isomerization from phosphatidylinositol 4-phosphate. Proc. Natl Acad. Sci. USA *88*, 9184–9187.

Wilson, M. P. and Majerus, P. W. (1996). Isolation of inositol 1,3,4-trisphosphate 5/6-kinase, cDNA cloning and expression of the recombinant enzyme. J. Biol. Chem. *271*, 11904–11910.

Wisniewski, D., Strife, A., Swendeman, S., Erdjument-Bromage, H., Geromanos, S., Kavanaugh, W. M., Tempst, P. and Clarkson, B. (1999). A novel SH2-containing phosphatidylinositol 3,4,5-trisphospahe 5-phosphatase (SHIP2) is constitutively tyrosine phosphorylated and associated with src homologous and collagen gene (SHC) in chronic myelogenous leukemia progenitor cells. Blood *93*, 2707–2720.

Wodzinski, R. J. and Ullah, A. H. J. (1996). Phytase. Adv. Appl. Microbiol. *42*, 263–302.

Woscholski, R. and Parker, P. J. (1997). Inositol lipid 5-phosphatases—traffic signals and signal traffic. Trends Biochem. Sci. *22*, 427–431.

Woscholski, R. and Parker, P. J. (2000). Inositol phosphatases—constructive destruction of phosphoinositides and inositol phosphates. In: Biology of Phosphoinositides (Cockcroft, S., eds.), Vol. 27. Oxford Univ. Press, London, pp. 320–338.

Woscholski, R., Waterfield, M. D. and Parker, P. J. (1995). Purification and biochemical characterization of a mammalian phosphatidylinositol 3,4,5-trisphosphate 5-phosphatase. J. Biol. Chem. *52*, 31001–31007.

Woscholski, R., Finan, P. M., Radley, E., Totty, N. F., Sterling, A. E., Hsuan, J. J., Waterfield, M. D. and Parker, P. J. (1997). Synaptojanin is the major constitutively active phosphatdylinositol-3,4,5-trisphosphate 5-phosphatase in rodent brain. J. Biol. Chem. *272*, 9625–9628.

Wyss, M., Brugger, R., Kronenberger, A., Remy, R., Fimbel, R., Oesterhelt, G., Lehmann, M. and van Loon, A. P. G. M. (1999). Biochemical characterization of fungal phytases (*myo*-inositol hexakisphosphate phosphohydrolases): catalytic properties. Appl. Environ. Microbiol. *65*, 367–373.

York, J. D. and Majerus, P. W. (1990). Isolation and heterologous expression of a cDNA encoding bovine inositol polyphosphate 1-phosphatase. Proc. Natl Acad. Sci. USA *87*, 9548–9552.

York, J. D., Chen, Z-W., Pnder, J. W., Chauhan, A. K., Mathews, F. R. S. and Majerus, P. W. (1994). Crystallization and initial X-ray crystallographic characterization of recombinant bovine inositol polyphosphate 1-phosphatase produced in *Spodoptera frugiperda* cells. J. Mol. Biol. *236*, 584–589.

York, J. D., Ponder, J. W. and Majerus, P. W. (1995). Definition of a metal-dependent/Li$^+$-inhibited phosphomonoesterase protein family based upon a conserved three-dimensional core structure. Proc. Natl Acad. Sci. USA *92*, 5149–5153.

Zhang, L. and Buxton, I. L. (1998). Measurement of phosphatidylinositols and phosphoinositides using high performance liquid chromatography flow detection. In: Phospholipid Signaling ProtocolsMethods Mol. Biol., (Bird, I. M., eds.), Vol. 105. Humana Press, Totowa, NJ, pp. 47–63.

Zhang, X., Jefferson, A. B., Auethavekiat, V. and Majerus, P. W. (1995). The protein deficient in Lowe syndrome is a phosphatidylinositol 4,5-bisphosphate 5-phosphatase. Proc. Natl Acad. Sci. USA *92*, 4853–5856.

Zhao, R., Qi, Y. and Zhao, Z. J. (2000). FIVE DSP1, a dual-specificity protein phosphatase containing an FYVE domain. Biochem. Biophys. Commun. *270*, 222–229.

Chapter 9

Ackerman, E. J. and Dennis, E. A. (1995). Mammalian calcium-independent phospholipase A$_2$. Biochim. Biophys. Acta *1259*, 125–136.

Ameratunga, R. V., Hawkins, R., Prestidge, R. and Marbrook, J. (1995). A high efficiency method for purification and assay of bee venom phospholipase A$_2$. Pathology *27*, 157–160.

Ananthanarayanan, B., Das, S., Rhee, S. G., Murray, D. and Cho, W. (2002). Membrane targeting of C2 domains of phospholipase C-δ isoforms. J. Biol. Chem. *277*, 3568–3575.

Armah, D. A. and Mensa-Wilmot, K. (1999). S-Myristoylation of a glycosylphosphatidylinositol-specific phospholipase C in *Trypanosoma brucei*. J. Biol. Chem. *274*, 5931–5938.

Armah, D. A. and Mensa-Wilmot, K. (2000). Tetramerization of glycosylphosphatidylinositol-specific phospholipase C from *Trypanosoma brucei*. J. Biol. Chem. *275*, 19334–19342.

Auger, K. R., Serunian, L. A. and Cantley, L. C. (1990). Separation of novel polyphosphatides. In: Methods in Inositide Research (Irvine, R. F., ed.). Raven Press, New York, pp. 159–166.

Aurich, I., Hirche, F. and Ulbrich-Hofmann, R. (1999). The determination of phospholipase D activity in emulsion systems. Anal. Biochem. *268*, 337–342.

Bae, Y. S. D., Cantley, L. G., Chen, C.-S., Kim, S.-R., Kwon, K.-S. and Rhee, S. G. (1998b). Activation of phospholipase C-γ by phosphatidylinositol 3,4,5-trisphosphate. J. Biol. Chem. *273*, 4465–4469.

Bae, C. D., Min, D. S., Fleming, I. N. and Exton, J. H. (1998a). Determination of interaction sites on the small G protein RhoA for phospholipase D. J. Biol. Chem. *273*, 11596–11604.

Balboa, M. A., Balsinde, J. and Dennis, E. A. (2000). Phosphorylation of cytosolic group IV phospholipase A_2 is necessary but not sufficient for arachidonic acid release in P388D macrophages. Biochem. Biophys. Res. Commun. *267*, 145–148.

Balsinde, J., Balboa, M. A. and Dennis, E. A. (1997). Antisense inhibition of group VI Ca^{2+}-independent phospholipase A_2 blocks phospholipid fatty acid remodeling in murine P388D1. J. Biol. Chem. *272*, 29317–29321.

Balsinde, J., Balboa, M. A., Insel, P. A. and Dennis, E. A. (1999). Regulation and inhibition of phospholipase A_2. Annu. Rev. Pharmacol. Toxicol. *39*, 175–189.

Balsinde, J., Balboa, M. A., Li, W. H., Llopis, J. and Dennis, E. A. (2000). Cellular regulation of cytosolic group IV phospholipase A_2 by phosphatidylinositol bisphosphate levels. J. Immunol. *164*, 5398–5402.

Balsinde, J., Dietz, E., Fernandez, B. and Mollinedo, F. (1989). Biochemical characterization of phospholipase D activity from human neutrophils. Eur. J. Biochem. *186*, 717–724.

Bandoh, K., Aoki, J., Hosono, H., Kobayashi, S., Kobayashi, T., Murakami-Murofushi, K., Tsujimoto, M., Arai, H. and Inoue, K. (1999). J. Biol. Chem. *274*, 27776–27785.

Banno, Y., Fujita, H., Ono, Y., Nakashima, S., Ito, Y., Kuzumaki, N. and Nozawa, Y. (1999). Differential phospholipase D activation by bradykinin and sphingosine 1-phosphate in NIH 3T3 fibroblasts overexpressing gelsolin. J. Biol. Chem. *274*, 27385–27391.

Banno, Y., Ito, Y., Ojio, K., Kanoh, H., Nakashima, S. and Nozawa, Y. (1996). Membrane-associated phospholipase D activity in neural cell line PC12. J. Lipid Mediators Cell Signal. *14*, 237–243.

Banno, Y. and Nozawa, Y. (1987). Characterization of partially purified phospholipase C from human platelet membranes. Biochem. J. *248*, 95–101.

Banno, Y., Yada, Y. and Nozawa, Y. (1988). Purification and characterization of membrane-bound phospholipase C specific for phosphoinositides from human platelets. J. Biol. Chem. *263*, 11459–11465.

Bayburt, T., Yu, B.-Z., Lin, H.-K., Browning, J., Jain, M. K. and Gelb, M. H. (1993). Human nonpancreatic secreted phospholipase A_2: interfacial parameters, substrate specificities and competitive inhibitors. Biochemistry *32*, 573–582.

Bayburt, T., Yu, B. Z., Street, I., Ghomashchi, F., Laliberte, F., Perrier, H., Wang, Z., Homan, R., Jain, M. K. and Gelb, M. H. (1995). Continuous, vesicle based fluorometric assays of 14- and 85-kDa phospholipases A_2. Anal. Biochem. *232*, 7–23.

Becker, M., Spohn, U. and Ulbrich-Hofmann, R. (1997). Detection and characterization of phospholipase D by flow injection analysis. Anal. Biochem. *244*, 55–61.

Berridge, M. J. (1987). Inositol trisphosphate and diacylglycerol: two interacting second messengers. Annu. Rev. Biochem. *56*, 159–193.

Bertello, L., Alves, M. J. M., Colli, W. and de Lederkremer, R. M. (2000). Evidence for phospholipases from *Trypanosoma cruzi* active on phosphatidylinositol and inositol phosphoceramide. Biochem. J. *345*, 77–84.

Bezzine, S., Koduri, R. S., Valentin, E., Murakami, M., Kudo, I., Ghomashchi, F., Sadilek, M., Lambeau, G. and Gelb, M. H. (2000). Exogenously added human group X secreted phospholipase A_2 but not the group IB, IIA, and V enzymes efficiently release arachidonic from adherent mammalian cells. J. Biol. Chem. *275*, 3179–3191.

Bingham, C. O., III, Fijneman, R. J. A., Friend, D. S., Goddeau, R. P., Rogers, R. A., Austen, K. F. and Arm, J. P. (1999). Low molecular weight group IIA and group V phospholipase A_2 enzymes have different intracellular locations in mouse bone marrow-derived mast cells. J. Biol. Chem. *274*, 31476–31484.

Bligh, E. G. and Dyer, W. J. (1959). A rapid method of total lipid extraction and purification. Can. J. Biochem. Physiol. *37*, 911–917.

Bollag, W. B. (1998). Measurement of phospholipase D activity. Methods Mol. Biol. *105*, 151–160.

Bredt, D. S., Mourey, R. J. and Snyder, S. H. (1989). A simple, sensitive, and specific radioreceptor assay for inisitol-1,4,5-trisphosphate in biological tissues. Biochem. Biophys. Res. Commun. *159*, 976–982.

Broad, L. M., Braun, F.-J., Lievremont, J.-P., Bird, G. St. J., Kurosaki, T. and Putney, J. W., Jr. (2001). Role of the phospholipase C-inositol 1,4,5-

trisphosphate pathway in calcium release-activated calcium current and capacitative calcium entry. J. Biol. Chem. *276*, 15945–15952.

Brodbeck, U. (1998). Signaling properties of glycosylphosphatidylinositols and their regulated release from membranes in the turnover of glycosylphosphatidylinositol-anchored proteins. Biol. Chem. *378*, 1041–1044.

Bromann, A., Boetticher, E. E. and Lomasney, J. W. (1997). A single amino acid substitution in the pleckstrin homology domain of phospholipase Cδ1 enhances the rate of substrate hydrolysis. J. Biol. Chem. *272*, 16240–16246.

Brown, H. A., Gutowski, S., Moomaw, C. R., Slaughter, C. and Sternweis, P. C. (1993a). ADP-ribosylation factor, a small GTP-dependent regulatory protein, stimulates phospholipase D activity. Cell *75*, 1137–1144.

Brown, H. A., Gutowski, S., Moomaw, C. R., Slaughter, C. and Sternweis, P. (1993b). ADP-ribosylation factor, a small GTP-dependent regulatory protein, stimulates phospholipase D activity. Cell *75*, 1137–1140.

Brown, H. A., Gutowski, S., Kahn, R. A. and Sternweis, P. C. (1995). Partial purification and characterization of Arf-sensitive phospholipase D from porcine brain. J. Biol. Chem. *270*, 14935–14943.

Brunner, G., Metz, C. N., Nguyen, H., Gabrilove, J., Patel, S. R., Davitz, M. A., Rifkin, D. B. and Wilson, E. L. (1994). An endogenous glycosylphosphatidylinositol-specific phospholipase D releases basic fibroblast growth factor-heparan sulfate proteoglycan complexes from human bone marrow cultures. Blood *83*, 2115–2125.

Bruzik, K. S., Morocho, A. M., Jhon, D.-Y., Rhee, S. G. and Tsai, M.-D. (1992). Phospholipids chiral phosphoprus: stereochemical mechanism for the formation of inositol 1-phosphate catalyzed by phosphatidylinositol-specific phospholipase C. Biochemistry *31*, 5183–5193.

Bruzik, K. S., Hakeem, A. A. and Tsai, M. D. (1994). Are D- and L-*chiro*-phosphoinositides substrates of phosphatidylinositol-specific phospholipase C? Biochemistry *33*, 8367–8374.

Bublitz, R., Armesto, J., Hoffmann-Blume, E., Schulze, M., Rhode, H., Horn, A., Aulwurm, S., Hannappel, E. and Fischer, W. (1993). Heterogeneity of glycosyl-phosphatidyl-inositol-anchored alkaline phosphatase of calf intestine. Eur. J. Biochem. *217*, 199–207.

Cai, S. and Exton, J. H. (2001). Determination of interaction sites of phospholipase D1 for RhoA. Biochem. J. *355*, 779–785.

Caramelo, J. J., Florin-Christensen, J., Florin-Christensen, M. and Delfino, J. M. (2000). Mapping the catalytic pocket of phospholipases A$_2$ and C using a novel set of phosphatidylcholines. Biochem. J. *346*, 679–690.

Chalifour, R. and Kanfer, J. N. (1982). Fatty acid activation and temperature perturbation of rat brain microsomal phospholipase. D. J. Neurochem. *39*, 299–305.

Chen, Y. and Dennis, E. A. (1998). Expression and characterization of human group V phospholipase A_2. Biochim. Biophys. Acta *1394*, 57–64.

Chilton, F. H. (1991). Assays for measuring arachidonic acid release from phospholipids. Methods Enzymol. *197*, 166–182.

Chilton, F. H. and Murphy, R. C. (1986). Remodeling of arachidonate-containing phosphoglyceride within the human neutrophil. J. Biol. Chem. *261*, 7771–7777.

Ching, T.-T., Wang, D.-S., Hsu, A.-L., Lu, P.-J. and Chen, C.-S. (1999). Identification of multiple phosphoinositide-specific phospholipases D as new regulatory enzymes for phosphatidylinositol 3,4,5-trisphosphate. J. Biol. Chem. *274*, 8611–8617.

Cho, W., Wu, S. K., Yoon, E. and Lichtenbergova, L. (1999). Fluorometric phospholipase assays based on polymerized liposome substrates. Methods Mol. Biol. *109*, 7–17.

Chung, J. K., Sekiya, F., Kang, H. S., Lee, C., Han, J. S., Kim, S. R., Bae, Y. S., Morris, A. J. and Rhee, S. G. (1997). Synaptojanin inhibition of phospholipase D activity by hydrolysis of phosphatidylinositol 4,5-bisphosphate. J. Biol. Chem. *272*, 15980–15985.

Cifuentes, M. E., Honkanen, L. and Rebecchi, M. J. (1993). Proteolytic fragments of phosphoinositide-specific phospholipase C-δ 1. Catalytic and membrane binding properties. J. Biol. Chem. *268*, 11586–11593.

Civenni, G., Butikofer, P., Stadelmann, B. and Brodbeck, U. (1999). In vitro phosphorylation of purified glycosylphosphatidylinositol-specific phospholipase D. Biol. Chem. *380*, 585–588.

Clark, J. D., Lin, L.-L., Kriz, R. W., Ramesha, Ch. S., Sultzman, L. A., Lin, A. Y., Milona, N. and Knopf, J. L. (1991). A novel arachidonic acid-selective cytosolic PLA_2 contains a Ca^{2+}-dependent translocation domain with homology to pkc and gap. Cell *65*, 1043–1051.

Clark, J. D., Schievella, A. R., Nalefski, E. A. and Lin, L.-L. (1995). Cytosolic phospholipase A_2. A review. J. Lipid Mediators Cell Signal. *12*, 83–118.

Clark, M. A., Shorr, R. G. L. and Bomalski, J. S. (1986). Antibodies prepared to Bacillus cereus phospholipase C crossreact with a phosphatidylcholine preferring phospholipase C in mammalian cells. Biochem. Biophys. Res. Commun. *140*, 114–119.

Cockcroft, S. (2001). Signaling roles of mammalian phospholipase D1 and D2. Cell Mol. Life Sci. *58*, 1674–1687.

Cockcroft, S., Baldwin, J. M. and Alla, D. (1984) The Ca^{2+}-activated polyphosphoinositide phosphodiesterase of human and rabbit neutrophyl membranes. Biochem. J. *221*, 477–482.

Cockcroft, S. and Thomas, G. M. H. (1992). Inositol-lipid-specific phospholipase C isozymes and their differential regulation by receptors. Biochem. J. *288*, 1–14.

Colley, W. C., Altshuller, Y. M., Sue-Lng, C. K., Copeland, N. G., Gilbert, D. J., Jenkins, N. A., Branch, K. D., Tsirka, S. E., Bollag, R. J., Bollag, W. B. and Frohman, M. A. (1997). Cloning and expression analysis of murine phospholipase D1. Biochem. J. *326*, 745–753.

Colley, W. C., Sung, T. C., Roll, R., Jenco, J., Hammond, S. M., Altshuller, Y., Bar-Sagi, D., Morris, A. J. and Frohman, M. A. (1997). Phospholipase D2, a distinct phospholipase D isoform with novel regulatory properties that provokes cytoskeletal reorganization. Curr. Biol. *7*, 191–201.

Conde-Frieboes, K., Reynolds, L. J., Lio, Y., Hale, M., Wasserman, H. H. and Dennis, E. A. (1996). J. Am. Chem. Soc. *118*, 5519–5525.

Cordoso de Almeida, M. L., Turner, M. J., Stambuk, B. U. and Schenkman, S. (1988). Identification of an acid-lipase in human serum which is capable of solubilizing glycosylphosphatidylinositol-anchored proteins. Biochem. Biophys. Res. Commun. *150*, 476–482.

Crooke, S. T. and Bennett, C. F. (1989). Mammalian phosphoinositide-specific phospholipase C isoenzymes. Cell Calcium *10*, 309–323.

Cubitt, A. B., Dharmawardhane, S. and Firtel, R. A. (1993). Developmentally regulated changes in 1,2-diacylglycerol in *Dictyostelium*. Regulation by light and G proteins. J. Biol. Chem. *268*, 17431–17439.

Cupillard, L., Koumanov, K., Mattei, M.-G., Lazdunski, M. and Lambeau, G. (1997). Cloning, chromosomal mapping, and expression of a novel human secretory phospholipase A_2. J. Biol. Chem. *272*, 15745–15752.

Cussac, D., Newman-Tancredi, A., Quentric, Y., Carpentier, N., Poissonnet, G., Parmentier, J. G., Goldstein, S. and Millan, M. J. (2002). Characterization of phospholipase C activity at h5-HT(2C) compared with h5-HT(2B) receptors: influence of novel ligands upon membrane-bound levels of [(3)H]phosphatidylinositols. Naunyn Schmiedebergs Arch. Pharmacol. *365*, 242–252.

Darnell, J. C. and Saltiel, A. R. (1991). Coenzyme A-dependent, ATP-independent acylation of 2-acyllysophosphatidylinositol in rat liver microsomes. Biochim. Biophys. Acta *1084*, 292–299.

Davis, L. L., Maglio, J. J. and Horwitz, J. (1998). Phospholipase D hydrolyzes short-chain analogs of phosphatidylcholine in the absence of detergent. Lipids *33*, 223–227.

Davitz, M. A., Hom, J. and Schenkman, S. (1989). Purification of a glycosyl phosphatidylinositol-specific phospholipase D from human plasma. J. Biol. Chem. *264*, 13760–13764.

Dawson, R. M. C., Irvine, R. F., Hemington, N. L. and Kirasawa, H. (1983). The alkaline phospholipase A1 of rat liver cytosol. Biochem. J. *209*, 865–873.

Deeg, M. A. and Davitz, M. A. (1994). Structure and function of the glycosyl phosphatidylinositol-specific phospholipase D. In: Signal-Activated Phospholipases (Liscovitch, M., eds.). Landes, Austin, TX, pp. 125–138.

Deeg, M. A. and Davitz, M. A. (1995). Glycosylphosphatidylinositol-phospholipase D: a tool for glycosylphosphatidylinositol structural analysis. Methods Enzymol. *250*, 630–640.

Deeg, M. A. and Verchere, C. B. (1997). Regulation of glycosylphosphatidylinositol-specific phospholipase D secretion from beta TC3 cells. Endocrinology *138*, 819–826.

Deng, J. T., Hoylaerts, M. F., De broe, M. E. and Van Hoof, V. O. (1996). Hydrolysis of membrane-bound liver alkaline phosphatase by GPI-PLD requires bile salts. Am. J. Physiol. *271*, G655–G663.

Dennis, E. A. (1994). Diversity of group types, regulation, and function of phospholipase A_2. J. Biol. Chem. *269*, 13057–13060.

Dennis, E. A. (1997). The growing phospholipase A_2 superfamily of signal transduction enzymes. Trends Biochem. Sci. *253*, 1–2.

Diagne, A., Mitjavila, S., Fauvel, J., Chap, H. and Douste-Blazy, L. (1987). Intestinal absorption of ester and ether glycerophospholipids in guinea pig. Role of a phospholipase A_2 from brush border membrane. Lipids *22*, 33–40.

Diez, E., Chilton, F. H., Stroup, G., Mayer, R. J., Winkler, J. D. and Fonteh, A. N. (1994). Fatty acid and phospholipid selectivity of different phospholipase A_2 enzymes studied by using a mammalian membrane as a substrate. Biochem. J. *301*, 721–726.

Diez, E., Louis-Flamberg, P., Hall, R. H. and Mayer, R. J. (1992). Substrate specificities and properties of human phospholipase A_2 in a mixed vesicle model. J. Biol. Chem. *267*, 18342–18348.

Divecha, N., Banfic, H. and Irvine, R. F. (1991). The polyphosphoinositide cycle exists in the nuclei of Swiss 3T3 cells under the control of a receptor (for IGF-I) in the plasma membrane and stimulation of the cycle increases nuclear diacylglycerol and apparently induces translocation of protein kinase C to the nucleus. EMBO J. *10*, 3207–3214.

Divecha, N., Roefs, M., Halstead, J. R., Dandrea, S., Fernandez-Borga, M., Oomen, L., Saqib, K., Wakelam, M. J. O. and D'Santos, C. (2000). Interaction of the Type 1α PIPkinase with phospholipase D: a role for the local generation of phosphatidylinositol 4,5-bisphosphate in the regulation of PLD2 activity. EMBO J. *19*, 5440–5449.

Dole, V. P. and Meinertz, H. (1960). Microdetermination of long-chain fatty acids in plasma and tissues. J. Biol. Chem. *235*, 2595–2599.

Downes, C. P., Dibner, M. D. and Hanley, M. R. (1983). Sympathetic denervation impairs agonist stimulated phosphatidylinositol metabolism in rat parotid glands. Biochem. J. *214*, 865–870.

Downes, C. P. and Mitchell, R. H. (1981). The polyphosphoinositide phosphodiesterase of erythrocyte membranes. Biochem. J. *198*, 133–140.

Du, X., Cai, J., Zhou, J.-Z., Stevens, V. L. and Low, M. G. (2002b). Tolerance of glycosylphosphatidylinositol (GPI)-specific phospholipase D overexpression by Chinese hamster ovary cell mutants with aberrant GPI biosynthesis. Biochem. J. *361*, 113–118.

Du, X. and Low, M. G. (2001). Down regulation of glycosylphosphatidylinositol-specific phospholipase D induced by lipopolysaccharide and oxidative stress in the murine monocyte-macrophage cell line RAW 264.7. Infect. Immun. *69*, 3214–3223.

Du, G., Morris, A. J., Scorra, V. A. and Frohman, M. A. (2002a). G-protein-coupled receptor regulation of phospholipase D. Methods Enzymol. *345*, 265–274.

Ella, K. M., Dolan, J. W. and Meier, K. E. (1995). Characterization of a regulated form of phospholipase D in the yeast *Saccharomyces cerevisiae*. Biochem. J. *307*, 799–805.

Ellis, M. V., Carne, A. and Katan, M. (1993). Structural requirements of phosphatidylinositol-specific phospholipase C δ1 for enzyme activity. Eur. J. Biochem. *213*, 339–347.

Ellis, M. V., James, S. R., Perisic, O., Downes, C. P., Williams, R. L. and Katan, M. (1998). Catalytic domain of phosphoinositide specific phospholipase C (PLC). Mutational analysis of residues within the active site and hydrophobic ridge of plcδ1. J. Biol. Chem. *273*, 11650–11659.

Essen, L. O., Perisic, O., Cheung, R., Katan, M. and Williams, R. L. (1996). Crystal structure of mammalian phosphoinositide-specific phospholipase C δ. Nature *380*, 595–602.

Evellin, S., Nolte, J., Tysack, K., vom Dorp, F., Thiel, M., Oude Weernink, P. A., Jakobs, K. H., Webb, E., Lomasney, J. W. and Schmidt, M. (2002). Stimulation of phospholipase C-epsilon by the M3 muscarinic acetylcholine receptor mediated by cyclic AMP and the GTPase Rap2B. J. Biol. Chem. *277*, 16805–16813.

Exton, J. H. (1994). Phosphoninositide phospholipases and G proteins in hormone action. Annu. Rev. Physiol. *56*, 349–369.

Exton, J. H. (1996). Annu. Rev. Pharmacol. Toxicol. *36*, 481–509.

Exton, J. H. (1997). Phospholipase D: enzymology, mechanisms of regulation, and function. Physiol. Rev. 77, 303–320.

Exton, J. H. (1998). Phospholipase D. Biochim. Biophys. Acta 1436, 105–115.

Exton, J. H. (1999). Regulation of phospholipase D. Biochim. Biophys. Acta 1439, 121–133.

Exton, J. H. (2000). Phospholipase D. Ann. NY Acad. Sci. 905, 61–68.

Faenza, I., Matteucci, A., Manzoli, L., Billi, A. M., Aluigi, M., Peruzzi, D., Vitale, M., Castorina, S., Suh, P.-G. and Cocco, L. (2000). A role for nuclear phospholipase Cβ1 in cell cycle control. J. Biol. Chem. 275, 30520–30524.

Falasca, M., Carvelli, A., Iursci, C., Qiu, R. G., Symons, M. H. and Corda, D. (1997). Fast receptor-induced formation of glycerophosphoinositol-4-phosphate, a putative novel intracellular messenger in the Ras pathway. Mol. Biol. Cell 8, 443–453.

Falasca, M., Iurisci, C., Carvelli, A., Sacchetti, A. and Corda, D. (1998). Release of the mitogen lysophosphatidylinositol from H-Ras-transformed fibroblasts; a possible mechanism of autocrine control of cell proliferation. Oncogene 16, 2357–2365.

Falasca, M., Logan, S. K., Lehto, V. P., Baccante, G., Lemmon, M. A. and Schlessinger, J. (1998). Activation of phospholipase C gamma by PI 3-kinase-induced domain-mediated membrane targeting. EMBO J. 17, 414–422.

Ferguson, M. A. J. (1999). The structure, biosynthesis and functions of glycosylphosphatidylinositol anchors, and the contributions of trypanosome research. J. Cell Sci. 112, 2799–2809.

Ferguson, K. M., Lemmon, M. A., Schlessinger, J. and Sigler, P. B. (1995). Structure of the high affinity complex of inositol trisphosphate with a phospholipase C pleckstrin homology domain. Cell 83, 1037–1046.

Ferguson, M. A. J. and Williams, A. F. (1988). Cell surface anchoring of proteins via glycosylphosphatidylinositol structures. Ann. Rev. Biochem. 57, 285–320.

Flick, J. S. and Thorner, J. (1993). Genetic and biochemical characterization of a phosphatidylinositol-specific phospholipase C in Saccharomyces cerevisiae. Mol. Cell. Biol. 13, 5861–5876.

Folch, J., Lees, M. and Sloane-Stanley, G. H. (1957). A simple method for the isolation and purification of total lipids from animal tissues. J. Biol. Chem. 226, 497–509.

Franken, P. A., Berg, L. V. D., Huang, J., Gunyuzlu, P., Lugtigheid, R. B., Verheij, H. M. and De Haas, G. H. (1992). Purification and characterization of a mutant human platelet phospholipase A2 expressed in E. coli. Eur. J. Biochem. 203, 89–98.

Frohman, M. A., Kanaho, Y., Zhang, Y. and Morris, A. J. (2000). Regulation of phospholipase D activity by Rho GTPases. Methods Enzymol. 325, 177–189.

Frohman, M. A., Sung, T. C. and Morris, A. J. (1999). Mammalian phospholipase D structure and regulation. Biochim. Biophys. Acta *1439*, 175–186.

Furuya, T., Kashuba, C., Docampo, R. and Moreno, S. N. J. (2000). A novel phosphatidylinositol-phospholipase C of *Trypanosoma cruzi* that is lipid modified and activated during trypomastigote to amastigote dfferentiation. J. Biol. Chem. *275*, 6428–6438.

Garcia, P., Gupta, R., Shah, S., Morris, A. J., Rudge, S. A., Scarlata, S., Petrova, V., McLaughlin, S. and Rebecchi, M. J. (1995). The pleckstrin homology domain of phospholipase C-δ1 binds with high affinity to phosphatidylinositol 4,5-bisphosphate in bilayer membranes. Biochemistry *34*, 16228–16234.

Gascard, P., Sauvage, M., Sulpice, J.-C. and Giraud, F. (1993). Characterization of structural and functional phosphoinositide domains in human erythrocyte membranes. Biochemistry *32*, 5941–5948.

Gascard, P., Tran, D., Sauvage, M., Sulpice, J.-C., Fukami, K., Takenawa, T., Claret, M. and Giraud, F. (1991). Asymmetric distribution of phosphoinositides and phosphatidic acid in the human erythrocyte membrane. Biochim. Biophys. Acta *1069*, 27–36.

Gassama-Diagne, A., Fauvel, J. and Chap, H. (1991). Phospholipase A_1 activity of guinea pig pancreatic lipase. Methods Enzymol. *197*, 316–325.

Gassama-Diagne, A., Rogalle, P., Fauvel, J., Wilson, M., Klaebe, A. and Chap, H. (1992). Substrate specificity of phospholipase B from guinea pig intestine: a glycerol ester lipase with broad specificity. J. Biol. Chem. *267*, 13418–13424.

Gehrhardt, S., Blue, E., Cumme, G. A., Bublitz, R., Rhode, H. and Horn, A. (2000). Gel chromatographic characterization of the hydrophobic interaction of glycosylphosphatidylinositol-alkaline phosphatase with detergents. Biol. Chem. *381*, 161–172.

Gelb, M. H., Jain, M. K., Hanel, A. M. and Berg, O. G. (1995). Interfacial enzymology of glycerolipid hydrolases: lessons from secreted phospholipase A_2. Annu. Rev. Biochem. *64*, 653–688.

Ghomashchi, F., Lin, Y., Hixon, M. S., Yu, B.-Z., Annand, R., Jain, M. K. and Gelb, M. H. (1998). Interfacial recognition by bee venom phospholipase A_2: insights into nonelectrostatic molecular determinants by charge reversal mutagenesis. Biochemistry *37*, 6697–6710.

Ghomashchi, F., Yu, B.-Z., Berg, O., Jain, M. K. and Gelb, M. H. (1991). Interfacial catalysis by phospholipase A_2: substrate specificity in vesicles. Biochemistry *30*, 7318–7329.

Goetzl, E. J., An, S. and Smith, W. L. (1995). Specificity of expression and effects of eicosanoid mediators in normal physiology and human diseases. FASEB J. *9*, 1051–1058.

Gomez-Munoz, A., O'Brien, L., Hundal, R. and Steinbrecher, U. P. (1999). Lysophosphatidylcholine stimulates phospholipase D activity in mouse peritoneal macrophages. J. Lipid Res. 40, 988–993.

Gray, N. C. C. and Strickland, K. P. (1982). The purification and characterization of a phospholipase A_2 activity from the 106 000 g pellet (microsomal fraction) of bovine brain acting on phosphatidylinositol. Can. J. Biochem. 60, 108–117.

Griffith, O. H. and Ryan, M. (1999). Bacterial phosphatidylinositol-specific phospholipase C: structure, function, and interaction with lipids. Biochim. Biophys. Acta 1441, 237–254.

Gross, W., Yang, W. and Boss, W. F. (1992). Release of carrot membrane-associated phosphatidylinositol kinase by phospholipase A2 and activation of a 70 kDa protein. Biochim. Biophys. Acta. 1134, 73–80.

Hagishita, T., Nishikawa, M. and Hatanaka, T. (1999). A spectrophotometric assay for the transphosphatidylation activity of phospholipase D enzyme. Anal. Biochem. 276, 161–165.

Hammond, S. M., Altshuller, Y. M., Sung, T. C., Rudge, S. A., Rose, K., Engebrecht, J., Morris, A. J. and Frohman, M. A. (1995). Human ADP-ribosylation factor-activated phosphatidylcholine-specific phospholipase D defines a new and highly conserved gene family. J. Biol. Chem. 270, 29640–29643.

Hammond, S. M., Jenco, J. M., Nakashima, S., Cadwallader, K. G. Q., Cook, S., Nozawa, Y., Prestwich, G. D., Frohman, M. A. and Morris, A. J. (1997). Characterization of two alternatively spliced forms of phospholipase D1. J. Biol. Chem. 272, 3860–3868.

Hanasaki, K., Ono, T., Saiga, A., Morioka, Y., Ikeda, M., Kawamoto, K., Higashino, K.-I., Nakano, K., Yamada, K., Ishizaki, J. and Arita, H. (1999). Purified group X secretory phospholipase A_2 induced prominent release of arachidonic acid from human myeloid leukemia cells. J. Biol. Chem. 274, 34203–34211.

Hanel, A. M., Schuttel, S. and Gelb, M. H. (1993). Processive interfacial catalysis by mammalian 85-kilodalton phospholipase A_2 enzymes on product-containing vesicles: application to the determination of substrate preferences. Biochemistry 32, 5949–5958.

Hanley, M. R., Poyner, D. R. and Hawkins, P. T. (1991). Inositol phospholipids for investigation of intact cell phospholipase C substrates and products. Methods Enzymol. 197, 149–158.

Hawkins, P. T., Stephens, L. R. and Downes, C. P. (1986). Rapid formation of inositol 1,3,4,5-tetrakisphosphate and inositol 1,3,4-trisphosphate in rat parotid glands may both result indirectly from receptor stimulated release of inositol 1,

4,5-trisphosphate from phosphatidylinositol-4,5-bisphosphate. Biochem. J. *238*, 507–516.

Heinz, D. W., Ryan, M., Bullock, T. L. and Griffith, O. H. (1995). Crystal structure of the phosphatidylinositol-specific phospholipase C from *Bacillus cereus* in complex with *myo*-inositol. EMBO J. *14*, 3855–3863.

Hendrickson, H. S. and Hendrickson, E. K. (1999). Binding of phosphatidyl-inositol-specific phospholipase C to phospholipid interfaces, determined by fluorescence resonance energy transfer. Biochim. Biophys. Acta *1440*, 107–117.

Hereld, D., Krakow, J. L., Bangs, J. D., Hart, G. W. and Englund, P. T. (1986). A phospholipase C from *Trypanosoma brucei* which selectively cleaves the glycolipid of the variant surface protein. J. Biol. Chem. *261*, 13813–13819.

Hergenrother, P. J. and Martin, S. F. (1997). Determination of the kinetic parameters for phospholipase C (*Bacillus cereus*) on different phospholipid substrates using a chromogenic assay based on the quantitation of inorganic phosphate. Anal. Biochem. *251*, 45–49.

Hergenrother, P. J., Spaller, M. R., Haas, M. K. and Martin, S. F. (1995). Chromatographic assay for phospholipase C from *Bacillus cereus*. Anal. Biochem. *229*, 313–316.

Higgs, H. N. and Glomset, J. A. (1994). Identification of a phosphatidic acid-preferring phospholipase from bovine brain and testis. Proc. Natl Acad. Sci. USA *91*, 9574–9578.

Higgs, H. N. and Glomset, J. A. (1996). Purification and properties of a phosphatidic acid-preferring phospholipase A_1 from bovine testis. J. Biol. Chem. *271*, 10874–10883.

Higgs, H. N., Han, M. H., Johnson, G. E. and Glomset, J. A. (1998). Cloning of a phosphatidic acid-preferring phospholipase A_1 from bovine testis. J. Biol. Chem. *273*, 5468–5477.

Hirasawa, K., Irvine, R. F. and Dawson, R. M. C. (1981). The catabolism of phosphatidylinositol by an EDTA-insensitive phospholipase A_1 and calcium-dependent phosphatidylinositol phosphodiesterase in rat brain. Eur. J. Biochem. *120*, 53–58.

Hodgkin, M. N., Clark, J. M., Rose, S., Saqib, K. and Wakelam, M. J. (1999). Characterization of the regulation of phospholipase D activity in the detergent-insoluble fraction of HL60 cells by protein kinase C and small G-proteins. Biochem. J. *339*, 87–93.

Hodgkin, M. N., Masson, M. R., Powner, D., Saqib, K. M., Ponting, C. P. and Wakelman, M. J. (2000). PhospholipaseD regulation and localization is dependent upon phosphatidylinositol 4,5-bisphosphate-specific PH domain. Curr. Biol. *10*, 43–46.

Hoerner, M. C. and Brodbeck, U. (1992). Phosphatidylinositol glycan-anchor-specific phospholipase D is an amphiphilic glycoprotein that in serum is associated with high density lipoproteins. Eur. J. Biochem. *206*, 747–757.

Hofmann, S. L. and Marjerus, P. W. (1982). Identification and properties of two distinct phosphatidylinositol-specific phospholipase C enzymes from sheep seminal vesicular glands. J. Biol. Chem. *257*, 6461–6469.

Holub, B. J. and Kuksis, A. (1978). Metabolism of molecular species of diacylglycerophospholipids. Adv. Lipid Res. *16*, 1–125.

Holub, B. J., Kuksis, A. and Thompson, W. (1970). Molecular species of mono-, di-, and triphosphoinositides of bovine brain. J. Lipid Res. *11*, 558–564.

Homma, Y. and Emori, Y. (1997). In: Signaling by Inositides (Shears, S., ed.). Oxford University Press, Oxford, pp. 99–116.

Homma, Y., Imaki, J., Nakanishi, O. and Takenawa, T. (1988). Isolation and characterization of two different forms of inositol phospholipid specific phospholipase C from rat brain. J. Biol. Chem. *263*, 6592–6598.

Hondal, R. J., Zhao, Z., Kravchuk, A. V., Liao, H., Riddle, S. R., Yue, X., Bruzik, K. S. and Tsai, M.-D. (1998). Mechanism of phosphatidylinositol-specific phospholipase C: a unified view of the mechanism of catalysis. Biochemistry *37*, 4568–4580.

Horwitz, J. and Davis, L. L. (1993). The substrate specificity of brain microsomal phospholipase D. Biochem. J. *295*, 793–798.

Hostetler, K. Y. and Gardner, M. F. (1991). Purification of rat kidney lysosomal phospholipase A₁. Methods Enzymol. *197*, 325–330.

Hostetler, K. Y., Gardner, M. F. and Aldern, K. A. (1991). Assay of phospholipases C and D in the presence of other lipid hydrolases. Methods Enzymol. *197*, 125–134.

Hough, E., Hansen, L. K., Birkness, B., Jynge, K., Hansen, S., Hordvik, A., Little, C., Dodson, E. and Derewenda, Z. (1989). High resolution (1.5 A) crystal strucutre of phospholipase C from *Bacilus cereus*. Nature *338*, 357–360.

Huang, Z., Laliberte, F., Tremblay, N. M., Weech, P. K. and Street, I. P. (1994). A continuous fluorescence-based assay for human high-molecular weight cytosolic phospholipase A₂. Anal. Biochem. *222*, 110–115.

Huang, K. S., Li, S., Fung, W. J., Hulmes, J. D., Reik, L., Pan, Y. C. and Low, M. G. (1990). Purification and characterization of glycosyl-phosphatidyl-inositol-specific phospholipase D. J. Biol. Chem. *265*, 17738–17745.

Ikezawa, H. (1991). Bacterial PIPLCs—unique properties and usefulness in studies on GPI anchors. 15, 1115–1131.

Illenberger, D., Stephan, I., Gierschik, P. and Schwald, F. (2000). Stimulation of phospholipase C-β2 by Rho GTPases. Methods Enzymol. *325*, 16–177.

Irvine, R. F., Letcher, A. J. and Dawson, R. M. C. (1984). Phosphatidylinositol-4,5-bisphosphate phosphodiesterase and phosphomonoesterase activities of rat brain. Some properties and possible control mechanisms. Biochem. J. *218*, 177–185.

Ishizaki, J., Suzuki, N., Higashino, K.-I., Yokota, Y., Ono, T., Kawamoto, K., Fujii, N., Arita, H. and Nakasaki, K. (1999). Cloning and characterization of novel mouse and human secretory phospholipase A_2s. J. Biol. Chem. *274*, 24973–24979.

Jain, H. M. and Gelb, M. H. (1991). Phospholipase A_2-catalyzed hydrolysis of vesicles: uses of interfacial catalysis in the scooting mode. Methods Enzymol. *197*, 112–125.

James, S. R. and Downes, C. P. (1997). Structure and mechanistic features of phospholipase C: effectors of inositol phospholipid mediators signal transduction. Cell. Signal. *9*, 329–336.

Jenco, J. M., Rawlingson, A., Daniels, B. and Morris, A. J. (1998). Regulation of phospholipase D2: selective inhibition of mammalian D isozymes of α and β-synucleins. Biochemistry *37*, 4901–4909.

Johnson, S. C., Dahl, J., Shih, T.-L., Schedler, D. J. A., Anderson, L., Benjamin, T. L. and Baker, D. C. (1993). Synthesis and evaluation of 3-modified 1D-*myo*-inositols as inhibitors and substrates of phosphatidylinositol synthase and inhibitors of *myo*-inositol uptake by cells. J. Med. Chem. *36*, 3628–3635.

Kafoury, R. M., Pryor, W. A., Squadrito, G. L., Salgo, M. G., Zou, X. and Friedman, M. (1998). Lipid ozonization products activate phospholipase A_2, C, and D. Toxicol. Appl. Pharmacol. *150*, 338–349.

Kam, Y. and Exton, J. H. (2002). Dimerization of phospholipase D isozymes. Biochem. Biophys. Res. Commun. *290*, 375–380.

Katan, M. (1998). Families of phosphoinositide-specific phospholipase C: structure and function. Biochim. Biophys. Acta *1436*, 5–17.

Kim, K. P., Han, S. K., Hing, M. and Cho, W. (2000). The molecular basis of phosphatidylcholine preference of human group-V phospholipase A_2. Biochem. J. *348*, 643–647.

Kim, D.-K., Kudo, I. and Inoue, K. (1991). Purification and characterization of rabbit platelet cytosolic phospholipase A_2. Biochim. Biophys. Acta *1083*, 80–88.

Kim, M. J., Min, D. S., Ryu, S. H. and Suh, P.-G. (1998). A cytosolic, $G\alpha_q$- and βγ-insensitive spliced variant of phospholipase C-β4. J. Biol. Chem. *273*, 3618–3624.

Kobayashi, H., Homma, S., Nakahata, N. and Ohizumi, Y. (2000). Involvement of phosphatidylcholine-specific phospholipase C in thromboxane A_2-induced activation of mitogen-activated protein kinase in astrocytoma cells. J. Neurochem. 74, 2167–2173.

Kobayashi, M. and Kanfer, J. N. (1987). Phosphatidylethanol formation via transphosphatidylation by rat brain synaptosomal phospholipase D. J. Neurochem. 48, 1597–1603.

Kodaki, T. and Yamashita, S. (1997). Cloning, expression, and characterization of a novel phospholipase D complementary DNA from rat brain. J. Biol. Chem. 272, 11408–11413.

Koonin, E. V. (1996). A duplicate catalytic motif in a new superfamily of phosphohydrolases and phospholipid synthases that includes poxvirus envelope proteins. Trends Biochem. Sci. 21, 242–243.

Kramer, R. M. and Sharp, J. D. (1995). Novel Approaches to Anti-inflammatory Therapy (Pruzanski, W. and Vadas P., eds.). Birkhaeuser Verlag, Basel, pp. 65–76.

Kudo, I., Murakami, M., Hara, S. and Inoue, K. (1993). Mammalian non-pancreatic phospholipases A_2. Biochim. Biophys. Acta 117, 217–231.

Kuksis, A. and Myher, J. J. (1990). Mass analysis of molecular species of diradylglycerols. In: Methods in Inositide Research (Irvine, R. F., ed.). Raven Press, New York, pp. 187–216.

Laine, J., Bourgoin, S., Bourassa, J. and Morisett, J. (2000). Subcellular distribution and characterization of rat pancreatic phospholipase D isoforms. Pancreas 20, 323–336.

LeBoeuf, R. C., Caldwell, M., Guo, Y., Metz, C., Davitz, M. A., Olson, L. K. and Deeg, M. A. (1998). Mouse glycosylphosphatidylinositol-specific phospholipase D (Gpld1) characterization. Mamm. Genome 9, 710–714.

Lee, C., Kim, S. R., Chung, J. K., Frohman, M. A., Kilimann, M. W. and Rhee, S. G. (2000). Inhibition of phospholipase D by amphiphysins. J. Biol. Chem. 275, 18751–18758.

Lee, J. Y., Lee, H. J., Kim, M. R., Myung, P. K. and Sok, D. E. (1999). Regulation of brain glycosylphosphatidylinositol-specific phospholipase D by natural amphiphiles. Neurochem. Res. 12, 1577–1583.

Lehto, M. T. and Sharom, F. J. (2002). PI-specific phospholipase C cleavage of a reconstituted GPI-anchored protein: modulation by the lipid bilayer. Biochemistry 41, 1398–1408.

Leslie, C. C. (1997). Properties and regulation of cytosolic phospholipase A_2. J. Biol. Chem. 272, 16709–16712.

Leslie, C. and Channon, J. Y. (1990). Anionic phospholipids stimulate arachidonoyl-hydrolyzing phospholipase A_2 from macrophages and reduce the calcium requirement for activity. Biochim. Biophys. Acta *1045*, 261–270.

Lemmon, M. A., Ferguson, K. M., O'Brien, R., Sigler, P. B. and Schlesinger, J. (1995). Proc. Natl Acad. Sci. USA *92*, 10472–10476.

Lewis, K. A., Bian, J., Sweeney, A. and Roberts, M. F. (1990). Asymmetric short-chain phosphatidylcholines: defining chain binding constraints in phospholipases. Biochemistry *29*, 9962–9970.

Lewis, K. A., Garigapati, V. R., Zhou, C. and Roberts, M. F. (1993). Substrate requirements of bacterial phosphatidylinositol-specific phospholipase C. Biochemistry *32*, 8836–8841.

Li, L. and Fleming, N. (1999). Aluminum fluoride inhibition of cabbage phospholipase D by a phosphate-mimicking mechanism. FEBS Lett. *461*, 1–5.

Lierheimer, R., Kunz, B., Vogt, L., Savoca, R., Brodbeck, U. and Sonderegger, P. (1997). The neuronal cell-adhesion molecule axonin-1 is specifically released by an endogenous glycosylphosphatidylinositol-specific phospholipase. Eur. J. Biochem. *243*, 502–510.

Lin, Q., Higgs, H. N. and Glomset, J. A. (2000). Membrane lipids have multiple effects on interfacial catalysis by a phosphatidic acid-preferring phospholipase A_1 from bovine testis. Biochemistry *39*, 9335–9344.

Liscovitch, M. (1996). Phospholipase D: role in signal transduction and membrane traffic. J. Lipid Med. Cell Signal. *14*, 215–221.

Liscovitch, M., Chalifa, V., Pertile, P., Chen, C.-S. and Cantley, L. C. (1994). Novel function of phosphatidylinositol 4,5-bisphosphate as a cofactor for brain membrane phospholipase D. J. Biol. Chem. *269*, 21403–21406.

Liscovitch, M., Czarny, M., Fiucci, G., Lavie, Y. and Tang, X. (1999). Localization and possible functions of phospholipase D isozymes. Biochim. Biophys. Acta *1439*, 245–263.

Liscovitch, M., Czarny, M., Fiucci, G. and Tang, X. (2000). Phospholipase D: molecular and cell biology of a novel gene family. Biochem. J. *345*, 401–415.

Lomasney, J. W., Cheng, H. F., Roffler, S. R. and King, K. (1999). Activation of phospholipase C-δ1 through C2 domain by a Ca^{2+}-enzyme-phosphatidylserine ternary complex. J. Biol. Chem. *274*, 21995–22001.

Lomasney, J. W., Cheng, H.-F., Wang, L.-P., Kuan, Y., Liu, S., Fesik, S. W. and King, K. (1996). Phosphatidylinositol 4,5-bisphosphate binding to the pleckstrin homology domain of phospholipase C-δ1 enhances enzyme activity. J. Biol. Chem. *271*, 25316–25326.

Lopez, I., Arnold, R. S. and Lambeth, J. D. (1998). Cloning and initial characterization of a human phospholipase D2 (hPLD2). ADP-ribosylation factor regulates hPLD2. J. Biol. Chem. *273*, 12846–12852.

Lopez, I., Mak, E. C., Ding, J., Hamm, H. E. and Lomasney, W. (2001). A novel bifunctional phospholipase C that is regulated b Galpha12 and stimulates the Ras/mitogen-activated protein kinase pathway. J. Biol. Chem. *276*, 2758–2765.

Low, M. (1981). Phosphatidylinositol-specific phospholipase C from *Staphylococcus aureus*. Methods Enzymol. *71*, 741–746.

Low, M. G., Carroll, R. C. and Weglicki, W. B. (1984). Multiple forms of phosphoinositide specific phospholipase C of different relative molecular masses in animal tissues. Biochem. J. *221*, 813–820.

Low, M. and Prasand, A. R. S. (1988). A phospholipase D specific for the phosphatidylinositol anchor of cell-surface proteins is abundant in plasma. Proc. Natl Acad. Sci. USA *85*, 980–984.

Low, M. G. and Saltiel, A. R. (1988). Structural and functional roles of glycosylphosphatidylinositol in membranes. Science *239*, 268–275.

Ma, Z., Wang, X., Nowatzke, W., Ramachandran, S. and Turk, J. (1999). Human pancreatic islets express mRNA species encoding two distinct catalytically active isoforms of group IV phospholipase A$_2$ (iPLA$_2$) that arise from an exon-skipping mechanism of alternative splicing of the transcript from the iPLA$_2$ gene on chromosome 22q13. J. Biol. Chem. *274*, 9607–9616.

Martin, S. F., DeBlanc, R. L. and Hergenrother, P. J. (2000a). Determination of the substrate specificity of the phospholipase D from *Streptomyces chromofuscus* via an inorganic phosphate quantitation assay. Anal. Biochem. *278*, 106–110.

Martin, S. F., Follows, B. C., Hergenrother, P. J. and Trotter, B. K. (2000b). The choline binding site of phospholipase C (*Bacillus cereus*): insights into substrate specificity. Biochemistry *39*, 3410–3415.

Martin, S. F. and Hergenrother, P. J. (1998). Enzymatic synthesis of a modified phospholipid and its evaluation as a substrate for *B. cereus* phospholipase C. Biorg. Med. Chem. Lett. *8*, 593–596.

Martin, T. W. and Wysolmerski, R. B. (1987). Ca^{2+}-dependent and Ca^{2+}-independent pathways for release of arachidonic acid from phosphatidyl-inositol in endothelial cells. J. Biol. Chem. *262*, 13086–13092.

Massenburg, D., Han, J.-S., Liyanage, M., Patton, W. A., Rhee, S. G., Moss, J. and Vaughan, M. (1994). Activation of rat brain phospholipase D by ADP-ribosylation factors 1,5, and 6: separation of ADP-ribosylation factor-dependent and oleate-dependent enzymes. Proc. Natl. Acad. Sci. U. S. A. *91*, 11718–11722.

Masterson, W. J., Raper, J., Doering, T. L., Hart, G. W. and Englund, P. T. (1990). Fatty acid remodeling: a novel reaction sequence in the biosynthesis of *Trypanosome* glycosyl phosphatidylinositol membrane anchors. Cell *62*, 73–80.

McDonald, L. J. and Mamrack, M. D. (1988). Aluminum affects phosphoinositide hydrolysis by phosphoinositidase C. Biochem. Biophys. Res. Commun. *155*, 203–208.

Meier, K. E. and Gibbs, T. (1999). Signal Transduction: A Practical Approach 2nd ed., IRL Press, Oxford, pp. 301.

Mensa-Wilmot, K., Morris, J. C., Al-Qahtani, A. and Englund, P. (1995). Purification and use of recombinant glycosylphosphatidylinositol-phospholipase C. Methods Enzymol. *250*, 641–655.

Metz, C. N., Brunner, G., Choi-Muira, N. H., Nguyen, H., Gabrilove, J., Caras, I. W., Alzuler, N., Rifkin, D. B., Weilson, E. L. and Davitz, M. A. (1994). Release of GPI-anchored membrane proteins by a cell associated GPI-specific phospholipase D. EMBO J. *13*, 1741–1751.

Min, D. S., Park, S.-K. and Exton, J. H. (1998). Characterization of a rat brain phospholipase D isozyme. J. Biol. Chem. *273*, 7044–7051.

Mitchell, C. J., Kelly, M. M., Blewitt, M., Wilson, J. R. and Biden, T. J. (2001). Phospholipase C-γ mediates the hydrolysis of phosphatidylinositol but not of phosphatidylinositol 4,5-bisphosphate, in carbamylcholine-stimulated islets of Langerhans. J. Biol. Chem. *276*, 19072–19077.

Mizenina, O., Musatkina, E., Yanushevic, Y., Rodina, A., Krasilnikov, M., de Gunzburg, J., Camonis, J. H., Tavitian, A. and Tatosyan, A. (2001). A novel group IIA phospholipase A_2 interacts with v-Src oncoprotein from RSV-transformed hamster cells. J. Biol. Chem. *276*, 34006–34012.

Morris, A. J., Frohman, M. A. and Engebrecht, J. (1997). Measurement of phospholipase D activity. Anal. Biochem. *252*, 1–9.

Mosior, M., Six, D. A. and Dennis, E. A. (1998). Group IV cytosolic phospholipase A_2 binds with high affinity and specificity to phosphatidylinositol 4,5-bisphosphate resulting in dramatic increases in activity. J. Biol. Chem. *273*, 2184–2191.

Munnik, T., Irine, R. F. and Musgrave, A. (1998). Phospholipid signalling in plants. Biochim. Biophys. Acta *1389*, 222–272.

Murakami, M., Kambe, T., Shimbara, S., Higashino, K.-I., Hanasaki, K., Arita, H., Horiguchi, M., Arita, M., Arai, H., Inoue, K. and Kudo, I. (1999). Different functional aspects of the group II subfamily (Types IIA and V) and Type X secretory phospholipase A_2s in regulating arachidonic acid release and prostaglandin generation. J. Biol. Chem. *274*, 31435–31444.

Murakami, M., Matsumoto, R., Urade, Y., Austen, K. T. and Arm, J. P. (1995). c-Kit ligand mediates increased expression of cytosolic phospholipase A_2, prostaglandin endoperoxide synthase-1, and initiates aortic prostaglandin D_2 synthase and increased IgF-dependent prostaglandin D_2 generation in immature mouse mast cells. J. Biol. Chem. *270*, 3239–3246.

Murakami, M., Nakatani, Y., Atsumi, G., Inoue, K. and Kudo, I. (1997). Regulatory functions of phospholipase A_2. Crit. Rev. Immunol. *17*, 225–283.

Myher, J. J. and Kuksis, A. (1984). Molecular species of plant phosphatidylinositol with selective cytotoxicity towards tumor cells. Biochim. Biophys. Acta *795*, 85–90.

Nakagawa, Y., Setaka, M. and Nojima, S. (1991). Detergent-resistant phospholipase A_1 from *Escherichia coli* membranes. Methods Enzymol. *197*, 309–315.

Nakahata, N., Takano, H. and Ohizumi, Y. (2000). Thromboxane A_2, receptor-mediated tonic contraction is attributed to an activation of phosphatidylcholine-specific phospholipase C in rabbit aortic muscles. Life Sci. *66*, 71–76.

Nakamura, S., Kiyohara, Y., Jinnai, H., Hityomi, T., Ogino, C., Yoshida, K. and Nishizuka, Y. (1996). Mammalian phospholipase D: phosphatidylethanolamine as an essential component. Proc. Natl Acad. Sci. USA *93*, 4300–4304.

Nalbone, G. and Hostetler, K. Y. (1985). Subcellular localization of the phospholipase A of rat heart: evidence for a cytosolic phospholipase A_1. J. Lipid Res. *26*, 104–114.

O'Brien, K. D., Pineda, C., Chiu, W. S., Bowen, R. and Deeg, M. A. (1999). Glycosylphosphatidylinositol-specific phospholipase D is expressed by macrophages in human atherosclerosis and colocalizes with oxidation epitopes. Circulation *99*, 2876–2882.

Ogino, C., Negi, Y., Daido, H., Kanemasu, M., Kondo, A., Kuroda, S., Tanizawa, K., Shimizu, N. and Fukuda, H. (2001). Identification of novel membrane-bound phospholipase D from *Streptoverticillium cinnamoneum*, possessing only hydrolytic activity. Biochim. Biophys. Acta *1530*, 23–31.

Ogino, C., Negi, Y., Matsumiya, T., Nakaoka, K., Kondo, A., Kuroda, S., Tokuyama, S., Kikkawa, U., Yamane, T. and Fukuga, H. (1999). Purification, characterization, and sequence determination of phospholipase D secreted by *Streptoverticillium ciannamoneum*. J. Biochem (Tokyo) *125*, 263–269.

Oh, S. O., Hong, J. H., Kim, Y. R., Yoo, H. S., Lee, S. H., Lim, K., Hwang, B. D., Exton, J. H. and Park, S. K. (2000). Regulation of phospholipase D2 by H_2O_2 in PC12 cells. J. Neurochem. *75*, 2445–2454.

Okamura, S. and Yamashita, S. (1994). Purification and characterization of phosphatidylcholine phospholipase D from pig lung. J. Biol. Chem. *269*, 31207–31213.

Pappan, K. and Wang, X. (1999). Molecular and biochemical properties and physiological roles of plant phospholipase D. Biochim. Biophys. Acta *1439*, 151–166.

Pappan, K., Qin, W., Dyer, J. H., Zheng, L. and Wang, X. (1997a). Molecular cloning and functional analysis of polyphosphoinositide-dependent phospholipase D, PLDβ, from *Arabidopsis*. J. Biol. Chem. *272*, 7055–7061.

Pappan, K., Zheng, S. and Wang, X. (1997b). Identification and characterization of a novel plant phospholipase D that requires polyphosphoinositides and submicromolar calcium for activity in *Arabidopsis*. J. Biol. Chem. *272*, 7048–7054.

Park, J. B., Kim, J. H., Kim, Y., Ha, S. H., Yoo, J. S., Du, G., Frohman, M. A., Suh, P. G. and Ryu, S. H. (2000). Cardiac phospholipase D2 localizes to sarcolemmal membranes and is inhibited by α actin in an AFR-reversible manner. J. Biol. Chem. *275*, 21295–21301.

Park, S. K., Provost, J. J., Bae, C. D., Ho, W. T. and Exton, J. H. (1997). Cloning and characterization of phospholipase D from rat brain. J. Biol. Chem. *272*, 29263–29271.

Pasquet, J.-M., Quek, L., Stevens, C., Bobe, R., Huber, M., Duronio, V., Krystal, G. and Watson, S. P. (2000). Phosphatidylinositol 3,4,5-trisphosphate regulates Ca^{2+} entry via Btk in platelets and megakaryocytes without increasing phospholipase C activity. EMBO J. *19*, 2793–2800.

Pertile, P., Liscovitch, M., Chalifa, V. and Cantley, L. C. (1995). Phosphatidylinositol 4,5-bisphosphate synthesis is required for activation of phospholipase D in U937 cells. J. Biol. Chem. *270*, 5130–5135.

Pete, M. J., Ross, A. H. and Exton, J. H. (1994). Purification and properties of phospholipase A1 from bovine brain. J. Biol. Chem. *269*, 19494–19500.

Pettitt, T. R., McDermott, M., Aqib, K. M., Shimwell, N. and Wakelam, M. J. (2001). Phospholipase D1b and D2a generate structurally identical phosphatidic acid species in mammalian cells. Biochem. J. *360*, 707–715.

Pind, S. and Kuksis, A. (1987). Isolation of purified brush border membranes from rat jejunum containing a Ca^{2+}-independent phospholipase A_2 activity. Biochim. Biophys. Acta *901*, 78–87.

Pind, S. and Kuksis, A. (1988). Solubilization and assay of phospholipase A_2 activity for rat jejunal brush border membranes. Biochim. Biophys. Acta *938*, 211–221.

Pind, S. and Kuksis, A. (1989). Association of the intestinal brush-border membrane phospholipase A_2 and lysophospholipase activities (phospholipase B) with a stalked membrane protein. Lipids *24*, 357–362.

Pind, S. and Kuksis, A. (1991). Further characterization of a novel phospholipase B (phospholipase A_2-lysophospholipase) from intestinal brush border membranes. Biochem. Cell Biol. *69*, 346–357.

Ponting, C. P. and Kerr, I. D. (1997). A novel family of phospholipase D homologues that includes phospholipid synthases and putative endonucleases:

identification of duplicated repeats and potential active site residues. Protein Sci. *5*, 914–922.

Portilla, D., Morrissey, J. and Morrison, A. R. (1988). Bradykinin-activated membrane-associated phospholipase C in Madin–Darby canine kidney cells. J. Clin. Invest. *81*, 1896–1902.

Preiss, J., Loomis, C. R., Bishop, W. R., Stein, R., Niedel, J. E. and Bell, R. M. (1986). Quantitative measurement of *sn*-1,2-diacylglycerols present in platelets, hepatocytes and *ras*- and *sis*-transformed normal rat kidney cells. J. Biol. Chem. *261*, 8597–8600.

Pruzanski, W. and Vadas, P. (1991). Phospholipase A_2-mediator between proximal and distal effectors of inflammation. Immunol. Today *12*, 143–146.

Qin, W., Pappan, K. and Wang, X. (1997). Molecular heterogeneity of phospholipase D (PLD). Cloning of PLDγ and regulation of plant PLDγ, -β and -α by polyphosphoinositides. J. Biol. Chem. *272*, 28267–28273.

Rameh, L. E. and Cantley, L. C. (1999). The role of phosphoinositide 3-kinase lipid products in cell function. J. Biol. Chem. *274*, 8347–8350.

Ravandi, A. and Kuksis, A. (2000). Phospholipids of plasma lipoproteins, red blood cells and atheroma, analysis of. In: Encyclopedia of Analytical Chemistry, (Myers, R. A., eds.). John Wiley & Sons Ltd, Chichester, pp. 1531–1570.

Razzini, G., Brancaccio, A., Lemmon, M. A., Guarnieri, S. and Falasca, M. (2000). The role of the pleckstrin homology domain in membrane targeting and activation of phospholipase Cβ1. J. Biol. Chem. *275*, 14873–14881.

Reynolds, L. J., Hughes, L. L. and Dennis, E. A. (1992). Analysis of human synovial fluid phospholipase A_2 on short-chain phosphatidylcholine-micelles: development of a spectrophotometric assay suitable for microtiter plate reader. Anal. Biochem. *204*, 190–197.

Reynolds, L. J., Washburn, W. N., Deems, R. A. and Dennis, E. A. (1991). Assay strategies and methods for phospholipases. Methods Enzymol. *197*, 3–23.

Rhee, S. G. (2001). Regulation of phosphoinositide-specific phospholipase C. Annu. Rev. Biochem. *70*, 281–312.

Rhee, S. G. and Bae, Y. S. (1997). Regulation of phosphoinositide-specific phospholipase C isozymes. J. Biol. Chem. *272*, 15045–15048.

Rhee, S. G., Suh, P. G., Ryu, S. H. and Lee, S. Y. (1989). Studies of inositol phospholipid-specific phospholipase C. Science *244*, 546–550.

Rhode, H., Hoffmann-Blume, E., Schilling, K., Gehrhardt, S., Goehlert, A., Buettner, A., Bublitz, R., Cumme, G. A. and Horn, A. (1995). Glycosylphosphatidylinositol-alkaline phosphatase from calf intestine as substrate for glycosylphosphatidyl-inositol-specific phospholipases—microassay using hydrophobic chromatography in pipette tips. Anal. Biochem. *231*, 99–108.

Rhode, H., Lopatta, E., Schulze, M., Pascual, C., Schulze, H. P., Schubert, K., Schubert, H., Reinhart, K. and Horn, A. (1999). Glycosylphosphatidylinositol-specific phospholipase D in blood serum: is the liver the only source of the enzyme? Clin. Chim. Acta *281*, 127–145.

Rhode, H., Schulze, M., Cumme, G. A., Bublitz, R., Schilling, K. and Horn, A. (2000). Glycosylphosphatidylinositol-specific phospholipase D of human serum—activity modulation by naturally occurring amphiphiles. Biol. Chem. *379*, 471–485.

Ries, U., Fleer, E. A., Unger, C. and Eibl, H. (1992). Synthetic phospholipids as substrates for phospholipase C from *Bacillus cereus*. Biochim. Biophys. Acta *1125*, 160–170.

Rigtrup, K. M., Kakkad, B. and Ong, D. E. (1994). Purification and partial characterization of a retinyl ester hydrolase from the brush border of rat small intestine mucosa: probable identity with brush border phospholipase B. Biochemistry *33*, 2661–2666.

Roberts, M. F. (1996). Phospholipases: structural and functional motifs for working at an interface. FASEB J. *10*, 1159–1172.

Rose, K., Rudge, S. A., Frohman, M. A., Morris, A. J. and Engebrecht, J. (1995). Phospholipase D signaling is essential for meiosis. Proc. Natl Acad. Sci. USA *92*, 12151–12155.

Ryu, S. H., Cho, K. S., Lee, K.-Y., Suh, P.-G. and Rhee, S. G. (1986). Two forms of phosphatidylinositol-specific phospholipase C from bovine brain. Biochem. Biophys. Res. Commun. *141*, 137–144.

Ryu, S. H., Cho, K. S., Lee, K.-Y., Suh, P.-G. and Rhee, S. G. (1987a). Purification and characterization of two immunologically distinct phosphoinositide-specific phospholipase C from bovine brain. J. Biol. Chem. *262*, 12511–12518.

Ryu, S. H., Suh, P. G., Cho, K. S., Lee, K. Y. and Rhee, S. G. (1987b). Bovine brain cytosol contains three immunologically distinct forms of inositol phospholipid-specific phospholipase C. Proc. Natl Acad. Sci. USA *84*, 6649–6653.

Saito, M. and Kanfer, J. N. (1975). Phosphatidohydrolase activity in a solubilized preparation from rat brain particulate fraction. Arch. Biochem. Biophys. *169*, 318–323.

Saito, K. and Kates, M. (1974). Substrate specificity of highly purified phospholipase B from *Penicillium notatum*. Biochim. Biophys. Acta *369*, 245–253.

Saito, K., Sugatani, J. and Okumura, T. (1991). Phospholipase B from *Penicillium notatum*. Methods Enzymol. *197*, 446–456.

Salto, M. L., Furuya, T., Moreno, S. N., Docampo, R. and de Lederkremer, R. M. (2002). The phosphatidylinositol-phospholipase C from *Trypanosoma cruzi* is active on inositophosphoceramide. Mol. Biochem. Parasitol. *119*, 131–133.

Sawai, H., Okomoto, Y., Luberto, C., Mao, C., Bielawska, A., Domae, N. and Hannun, Y. A. (2000). Identification of ISC1 (YERO019w) as inositol phosphosphingolipid phospholipase C in *Saccharomyces cerevisiae*. J. Biol. Chem. *275*, 39793–39798.

Scallon, B. J., Fung, W.-J. C., Tsang, T. C., Li, S., Kado-Fong, H., Huang, K.-S. and Kochan, J. P. (1991). Science *252*, 446–448.

Scharenberg, A. M., El-Hillal, O., Fruman, D. A., Beitz, L. O., Li, Z., Lin, S., Gout, I., Cantley, L. C., Rawlings, D. J. and Kinet, J.-P. (1998). Phosphatidylinositol-3,4,5-trisphosphate (PtdIns-3,4,5,-P3)/Tec kinase-dependent calcium signaling pathway: a target for SHIP-mediated inhibitory signals. EMBO J. *17*, 1961–1972.

Schmidt, M., Evellin, S., Weernink, P. A., van Dorp, F., Rehmann, H., Lomasney, J. W. and Jacobs, K. H. (2001). A new phospholipase C calcium signaling pathway mediated by cyclic AMP and Rap GTPase. Nat. Cell Biol. *3*, 1020–1024.

Sciorra, V. A., Rudge, S. A., Prestwich, G. D., Frohman, M. A., Engebrecht, J. and Morris, A. J. (1999). Identification of a phosphoinositide binding motif that mediates activation of mammalian and yeast phospholipase D isoenzymes. EMBO J. *20*, 5911–5921.

Serhan, C. N., Haeggstrom, J. Z. and Leslie, C. C. (1996). FASEB J. *10*, 1147–1158.

Serunian, L. A., Haber, M. T., Fukui, T., Kim, J. W., Rhee, S. G., et al. (1989). Polyphosphoinositides produced by phosphatidylinositol 3-kinase are poor substrates for phospholipases C from rat liver and bovine brain. J. Biol. Chem. *264*, 17809–17815.

Shears, S. B. (1989). Metabolism of the inositol phosphates produced upon receptor activation. Biochem. J. *260*, 313–324.

Shibatohge, M., Kariya, K.-I., Liao, Y., Hu, C.-D., Atari, Y., Goshima, M., Shia, F. and Kataoka, T. (1998). The identification of PLC210, a *Caenorhabditis elegans* phospholipase C, as putative effector of Ras. J. Biol. Chem. *273*, 6218–6222.

Shikano, M., Masuzawa, Y., Yazawa, K., Takayama, K., Kudo, L. and Inoue, K. (1994). Complete discrimination of docosahexaenoate from arachidonate by 85-kDa cytosolic phospholipase A_2 during the hydrolysis of diacyl- and alkenylacylglycero-phosphoethanolamine. Biochim. Biophys. Acta *1212*, 211–216.

Siddiqi, A. R., Smith, J. L., Ross, A. H., Qiu, R.-G., Symons, M. and Exton, J. H. (1995). Regulation of phospholipase D in HL60 cells. Evidence for cytosolic phospholipase D. J. Biol. Chem. *270*, 8466–8473.

Singer, W. D., Brown, H. A. and Sternweis, P. C. (1997). Regulation of eukaryotic phosphatidylinositol-specific phospholipase C and phospholipase D. Annu. Rev. Biochem. *66*, 475–509.

Singh, I. N., Stromberg, L. M., Bourgoin, S. G., Sciorra, V. A., Morris, A. J. and Brindley, D. N. (2001). Ceramide inhibition of mammalian phospholipase D1 and D2 activities is antagonized by phosphatidylinositol-4,5-bisphosphate. Biochemistry *40*, 11227–11233.

Six, D. A. and Dennis, E. A. (2000). The expanding superfamily of phospholipase A_2 enzymes: classification and characterization. Biochim. Biophys. Acta *1488*, 1–19.

Shukla, S. D. (1982). Phosphatidylinositol specific phospholipase C. Life Sci. *30*, 1323–1335.

Slotboom, A. J., De Haas, G. H., Bonsen, P. P. M., Burbach-Westerhuis, G. J. and Van Deenen, L. L. M. (1970). Hydrolysis of phosphoglycerides by purified lipase preparations. I. Substrate-, positional- and stereospecificity. Chem. Phys. Lipids *4*, 15–29.

Smith, T. K., Milne, F. C., Sharma, D. K., Crossman, A., Brimacombe, J. S. and Ferguson, M. A. J. (1997a). Early steps in glycosylphosphatidylinositol biosynthesis in Leishmania major. Biochem. J. *326*, 393–400.

Smith, T. K., Sharma, D. K., Crossman, A., Dix, A., Brinacombe, J. S. and Ferguson, M. A. J. (1997b). Parasite and mammalian GPI biosynthetic pathways can be distinguished using synthetic substrate analogues. EMBO J. *16*, 6667–6675.

Song, C., Hu, C.-D., Masago, M., Kariya, K.-I., Yamawaki-Kataoka, Y., Shibatohge, M., Wu, D., Satoh, T. and Kataoka, T. (2001). Regulation of a novel human phospholipase C, PLCε, through membrane targeting by Ras. J. Biol. Chem. *276*, 2752–2757.

Stadelmann, B., Butikofer, P., Konig, A. and Brodbeck, U. (1997). The C-terminus of glycosylphosphatidylinositol-specific phospholipase D is essential for biological activity. Biochim. Biophys. Acta *1355*, 107–113.

Stafforini, D. NM., Prescott, S. M., Zimmerman, G. A. and McIntyre, T. M. (1991). Platelet-activating factor acetylhydrolase activity in human tissues and blood cells. Lipids *26*, 979–985.

Stambuk, B. U. and Curdoso de Almeida, L. C. (1996). An assay for glycosylphos-phatidylinositol-anchoring degrading phospholipases. J. Biochem. Biophys. Methods in press.

Stephens, L. R., Hawkins, P. T., Carter, N. G., Chahwala, S., Morris, A. J., Whetton, A. D. and Downes, C. P. (1988). L-*myo*-Inositol-1,4,5,6-tetrakis phosphate is present in both mammalian and avian cells. Biochem. J. *249*, 271–282.

Stephens, L., Hawkins, P. T. and Downes, C. P. (1989). Metabolic and structural evidence for the existence of a third species of polyphosphoinositide in cells: D-phosphatidyl-*myo*-inositol 3-phosphate. Biochem. J. *259*, 267–276.

Sternfeld, L., Thevenod, F. and Schulz, I. (2000). fMLP-induced arachidonic acid release in db-cAMP-differentiated HL-60 cells is independent of phosphatidylinositol-4,5-bisphosphate-specific phospholipase C activation and cytosolic phospholipase A$_2$ a. activation. Arch. Biochem. Biophys. *378*, 246–258.

Takemori, H, Zolotaryov, F. N., Ting, M., Urbain, L., Komatsubara, T., Hatano, O., Okamoto, M. and Tojo, H. (1998). Identification of functional domains of rat intestinal phospholipase B/Lipase. J. Biol. Chem. *273*, 2222–2231.

Tall, E., Dorman, G., Garcia, P., Runnels, L., Shah, S., Chen, J., Profit, A. Gu, Q.-M., Chaudhary, A., Prestwich, G. D and Rebecchi, M. J. (1997). Phosphoinositide binding specificity among phospholipase C isozymes as determined by photo-cross-linking to novel substrate and product analogs. Biochemistry *36*, 7239–7248.

Tan, C. A. and Roberts, M. F. (1996). Vanadate is a potent competitive inhibitor of phospholipase C from *Bacillus cereus*. Biochim. Biophys. Acta *1298*, 58–68.

Thompson, W. (1969). Positional distribution of fatty acids in brain polyphosphoinositides. Biochim. Biophys. Acta *187*, 150–153.

Tishfield, J. A. (1997). A reassessment of the low molecular weight phospholipase A$_2$ gene family in mammals. J. Biol. Chem. *272*, 17247–17250.

Tojo, H., Ichida, T. and Okamoto, M. (1998). Purification and characterization of a catalytic domain of rat intestinal phospholipase B/lipase assembled with brush border membrane. J. Biol. Chem. *273*, 2214–2221.

Treumann, A., Lifely, M. R., Schneider, P. and Ferguson, M. A. J. (1995). Primary structure of CD52. J. Biol. Chem. *270*, 6088–6099.

Tsujioka, H., Misumi, Y., Takami, N. and Ikehara, Y. (1998). Posttranslational modification of glycosylinositol (GPI)-specific phospholipase D and its activity in cleavage of GPI anchors. Biochem. Biophys. Res. Commun. *251*, 737–743.

Tsujioka, H., Takami, N., Misumi, Y. and Ikehara, Y. (1999). Intracellular cleavage of glycosylphosphatidylinositol by phospholipase D induces activation of protein kinase Cα. Biochem. J. *343*, 449–455.

Tsujita, Y., Asaoka, Y. and Nishizuka, Y. (1994). Regulation of phospholipase A$_2$ in human leukemia cell lines: its implication for intracellular signaling. Proc. Natl Acad. Sci. USA *91*, 6274–6278.

Tysnes, O.-B., Aarbakke, G. M., Verhoeven, A. J. M. and Holmsen, H. (1985). Thin-layer chromatography of polyphosphoinositides from platelet extracts: interference by an unknown phospholipid. Thromb. Res. *40*, 329–338.

Ueda, H., Kobayashi, T., Kishimoto, M., Tsutsumi, T., Watanabe, S. and Okuyama, H. (1993). The presence of Ca^{2+}-independent phospholipase A_1 highly specific for phosphatidylinositol in bovine brain. Biochem. Biophys. Res. Commun. *195*, 1272–1279.

Ueki, J., Morioka, S., Komari, T. and Kumashiro, T. (1995). Purification and characterization of phospholipase D (PLD) from rice (*Oryza sativa* L.) and cloning of cDNA for PLD from rice and maize (*Zea mays* L.). Plant Cell Physiol. *36*, 903–914.

Van Deenen, L. L. M. and DeHaas, G. H. (1964). The synthesis of phosphoglycerides and some biochemical applications. Adv. Lipid res. *2*, 167–234.

Virto, C., Svensson, I. and Adlercreutz, P. (2000). Hydrolytic and transphosphatidylation activities of phospholipase D from Savoy cabbage towards lysophosphatidylcholine. Chem. Phys. Lipids *106*, 41–51.

Volwerk, J. J., Filhuth, E., Griffith, O. H. and Jain, M. K. (1994). Phosphatidylinositol-specific phospholipase C from *Bacillus cereus*: interfacial binding, catalysis, and activation. Biochemistry *33*, 3464–3474.

Volwerk, J. J., Koke, J. A., Wetherwax, P. B. and Griffith, O. H. (1989). Functional characteristics of phosphatidylinositol-specific phospholipase C from *Bacillus cereus* and *Bacillus thyringiensis*. FEMS Microbiol. Lett. *61*, 237–242.

Volwerk, J. J., Shashidhar, M. S., Kuppe, A. and Griffith, O. H. (1990). Phosphatidylinoisitol-specific phospholipase C from *Bacillus cereus* combines intrinsic phosphotransferase and cyclic phosphodiesterase activities: a ^{31}P NMR study. Biochemistry *29*, 8056–8062.

Wakelam, M. J., Hodgkin, M. and Martin, A. (1995). The measurement of phospholipase D-linked signaling in cells. Methods Mol. Biol. *41*, 271–278.

Wakelam, M. J., Hodgkin, M. N., Martin, A. and Sqib, K. (1997). Phospholipase D. Semin. Cell Dev. Biol. *3*, 305–310.

Waksman, M., Eli, Y., Liscovitch, M. and Gerst, J. E. (1996). Identification and characterization of a gene encoding phospholipase D activity in yeast. J. Biol. Chem. *271*, 2361–2364.

Wakelam, M. J., Martin, A., Hodgkin, M. N., Brown, F., Pettitt, T. R., Cross, M. J., De Takats, P. G. and Reynolds, J. L. (1997). Role and regulation of phospholipase D activity in normal and cancer cells. Adv. Enzyme Regul. *37*, 29–34.

Wang, X. (2000). Multiple forms of phospholipase D in plants: the gene family, catalytic and regulatory properties, and cellular functions. Progr. Lipid Res. *39*, 109–149.

Wang, T., Dowal, L., El-Maghrabi, M. R., Rebecchi, M. and Scarlata, S. (2000). The pleckstrin homology domain of phospholipase C-β_2 links the binding of Gβγ to activation of the catalytic core. J. Biol. Chem. *275*, 7466–7469.

Wang, X. M., Dyer, J. H. and Zheng, L. (1993). Purification and immunological analysis of phospholipase D from castor bean endosperm. Arch. Biochem. Biophys. *306*, 486–494.

Wang, A., Lo, R., Chen, Z. and Dennis, E. A. (1997). Regiospecificity and catalytic triad of lysophospholipase I. J. Biol. Chem. *272*, 22030–22036.

Wang, T., Pentyla, S., Rebecchi, M. J. and Scarlata, S. (1999). Differential association of the pleckstrin homology domains of phospholipases C-β1, C-β2, and C-δ1 with lipid bilayers and the βγ subunits of heterotrimeric G proteins. Biochemistry *38*, 1517–1524.

Wang, X., Xu, L. and Zheng, L. (1994). Cloning and expression of phosphatidylcholine-hydrolyzing phospholipase D from *Ricinus communis* L. J. Biol. Chem. *269*, 20312–20317.

Wilhelm, O. G., Wilhelm, S., Escott, G. M., Lutz, V., Magdolen, V., Schmitt, M., Rifkin, D. B., Wilson, E. L., Graeff, H. and Brunner, G. (1999). Cellular glycosylphosphatidyl-inositol-specific phospholipase D regulates urokinase receptor shedding and cell surface expression. J. Cell. Physiol. *180*, 225–235.

Williams, R. L. (1999). Mammalian phosphoinositide-specific phospholipase C. Biochim. Biophys. Acta *1441*, 255–267.

Williams, S. D., Hsu, F.-F. and Ford, D. A. (2000). Electrospray ionization mass spectrometry analyses of nuclear membrane phospholipid loss after reperfusion of ischemic myocardium. J. Lipid Res. *41*, 1585–1595.

Williams, R. L. and Katan, M. (1996). Structural views of phosphoinositide-specific phospholipase C: signalling the way ahead. Structure *4*, 1387–2126.

Wissing, J. B., Grabo, P. and Kornak, B. (1996). Plant Sci. *117*, 17–31.

Wolf, R. A. and Gross, R. W. (1985). Semi-synthetic approach for the preparation of homogeneous plasmenylethanolamine utilizing phospholipase D from *Streptomyces chromofuscus*. J. Lipid Res. *26*, 629–633.

Wu, S. K. and Cho, W. (1994). A continuous fluorescence assay for phospholipases using polymerized mixed liposomes. Anal. Biochem. *221*, 152–159.

Xie, Z., Ho, W. T. and Exton, J. H. (2001). Requirements and effects of palmitoylation of rat PLD1. J. Biol. Chem. *276*, 9383–9391.

Xie, Z., Ho, W.-T., Spellman, R., Cai, S. and Exton, J. H. (2002a). Mechanisms of regulation of phospholipase D1 and D2 by the heterotrimeric G proteins G_{13} and G_q. J. Biol. Chem. *277*, 11979–11986.

Xie, Z., Kim, H. K. and Exton, J. H. (2002b). Expression and characterization of rat brain phospholipase D. Methods Enzymol. *345*, 255–264.

Xie, Y. and Meier, K. E. (2002). Assays for phospholipase D reaction products. Methods Enzymol. *345*, 294–305.

Xu, Z., Byers, D. M., Palmer, F. B., Spence, M. W. and Cook, H. W. (1993). Limited metabolic interaction of serine with ethanolamine and choline in the turnover of phosphatidylserine, phosphatidylethanolamine and plasmalogens in cultured glioma cells. Biochim. Biophys. Acta *1168*, 167–174.

Yagisawa, H., Sakuma, K., Paterson, H. F., Cheung, R., Allen, V., Hirata, H., Watanabe, Y., Hirata, M., Williams, R. L. and Katan, M. (1998). Replacements of single basic amino acids in the pleckstrin homology domain of phospholipase C-δ1 alter the ligand binding, phospholipase activity, and interaction with the plasma membrane. J. Biol. Chem. *273*, 417–424.

Yamamoto, T.-A., Takeuchi, H., Kanematsu, T., Allen, V., Yagisawa, H., Kikkawa, U., Watanabe, Y., Kasima, A., Katan, M. and Hirata, M. (1999). Involvement of EF hand motifs in the Ca^{2+}-dependent binding of the pleckstrin homology domain to phosphoinositides. Eur. J. Biochem. *265*, 481–490.

Yang, S. F., Freer, S. and Benson, A. A. (1967). Transphosphatidylation by phospholipase D. J. Biol. Chem. *242*, 477–484.

Yang, H.-Ch., Mosior, M., Johnson, Ch. A., Chen, Y. and Dennis, E. A. (1999). Group-specific assays that distinguish between the four major types of mammalian phospholipase A_2. Anal. Biochem. *269*, 278–288.

York, J. D., Odom, A. R., Murphy, R., Ives, E. B. and Wente, S. R. (1999). A phospholipase C-dependent inositol polyphosphate kinase pathway required for efficient messenger RNA export. Science *285*, 96–100.

Yoshida, H., Tsujishita, Y., Hullin, F., Yoshida, K., Nakamura, S., Kikkawa, U. and Asaoka, Y. (1998). Isolation and properties of a novel phospholipase A from rat brain that hydrolyses fatty acids at *sn*-1 and *sn*-2 positions. Ann. Clin. Biochem. *35*, 295–301.

Zhang, Y. Y., Deems, R. A. and Dennis, E. A. (1991). Lysophospholipases I and II from P388D1 macrophage—like cell line. Methods Enzymol. *197*, 456–468.

Zhao, Y., Stuckey, J. A., Lohse, D. L. and Dixon, J. E. (1997). Expression, characterization and crystallization of a member of the novel phospholipase D family of phosphodiesterases. Protein Sci. *6*, 2655–2658.

Zhou, C., Garigapati, V. and Roberts, M. F. (1997a). Short-chain phosphatidyl-inositol conformation and its relevance to phosphatidylinositol-specific phospholipase C. Biochemistry *36*, 15925–15931.

Zhou, C., Wu, Y. and Roberts, M. F. (1997b). Activation of phosphatidylinositol-specific phospholipase C toward inositol 1,2-(cyclic)-phosphate. Biochemistry *36*, 347–355.

Chapter 10

Adelt, S., Plettenburg, O., Dallmann, G., Ritter, F. P., Shears, S. B., Altenbach, H-J. and Vogel, G. (2001). Regiospecific phosphohydrolases from *Dictyostelium* as tools for the chemoenzymatic synthesis of the enantiomers D-*myo*-inositol 1,2,4,-trisphosphate and D-*myo*-inositol 2,3,6-trisphosphate: non-physiological, potential analogues of biologically active D-*myo*-inositol 1, 3,4-trisphosphate. Bioorg. Med. Chem. Letters *11*, 2705–2708.

Adelt, S., Plettenburg, O., Stricker, R., Reiser, G., Altenbach, H.-J. and Vogel, G. (1999). Enzyme-assisted total synthesis of the optical antipodes D-*myo*-inositol 3,4,5-trisphosphate and D-*myo*-inositol 1,5,6-trisphosphate: Aspects of their structure–activity relationship to biologically active inositol phosphates. J. Med. Chem. *42*, 1262–1273.

Aquilo, A., Martin-Lomas, M. and Penades, S. (1992). The regioselective synthesis of enantiomerically pure *myo*-inositol derivatives. Efficient synthesis of *myo*-inositol 1,2,4,5-tetrakisphosphate. Tetrahedron Lett. *33*, 401–404.

Auger, K. R., Serunian, L. A. and Cantley, L. C. (1990). Separation of novel polyphosphoinositides. In: Methods in Phosphatide Research (Irvine, R. F., ed.). Raven Press, New York, pp. 159–166.

Auger, K. R., Serunian, L. A., Soltoff, S. P., Libby, P. and Cantley, L. C. (1989). PDGF-dependent tyrosine phosphorylation stimulates production of novel polyphosphoinositides in intact cells. Cell *57*, 167–175.

Balla, T., Guillemette, G., Baukal, A. J. and Catt, K. J. (1987). Metabolism of inositol 1,3,4-trisphosphate to a new tetrakisphosphate isomer in angiotensin-stimulated adrenal glomerulosa cells. J. Biol. Chem. *262*, 9952–9955.

Ballou, C. E. and Pizer, L. I. (1960). The absolute configuration of the *myo*-inositol 1-phosphates and a confirmation of the bornesitol configurations. J. Am. Chem. Soc. *82*, 3333–3335.

Batty, I. R., Nahorski, S. R. and Irvine, R. F. (1985). Rapid formation of inositol 1,3,4,5-tetrakisphosphate following muscarinic receptor stimulation of rat cerebral cortical slices. Biochem. J. *232*, 211–215.

Batty, I. A., Letcher, A. J. and Nahorski, S. R. (1989). Accumulation of inositol polyphosphate isomers in agonist-stimulated cerebral cortex slices. Biochem. J. *258*, 23–32.

Baudin, G., Glanzer, B. I., Swaminathan, K. S. and Vasella, A. (1988). A synthesis of 1D- and 1L-*myo*-inositol 1,2,3,5-tetrakisphosphate. Helv. Chim. Acta *71*, 1367–1378.

Bird, I. M. (1998). Preparation of [^{3}H]phosphoinositol standards and conversion of [^{3}H]phosphoinositides to [^{3}H]phosphoinositols. Methods Mol. Biol. *105*, 65–76.

Bird, I. M., Sadler, I. H., Williams, B. C. and Walker, S. W. (1989). The preparation of *myo*-inositol 1,4-bisphosphate and D-*myo*-inositol 1,4,5-trisphosphate in milligram quantities from a readily available starting material. Mol. Cell. Endocrinol. 66, 215–229.

Brown, D. M. and Stewart, J. C. (1966). The structure of triphospoinositide from rat brain. Biochim. Biophys. Acta 125, 413–421.

Burgess, G. M., Godfrey, P. P., McKinney, J. S., Berridge, M. J., Irvine, R. F. and Putney, J. W. (1984b). The second messenger linking receptor activation to internal Ca^{++} release in liver. Nature 309, 63–66.

Carter, A. N. and Downes, C. P. (1992). Phosphatidylinositol 3-kinase is activated by nerve growth factor and epidermal growth factor in PC12 cells. J. Biol. Chem. 267, 14563–14567.

Carter, A. N., Huang, R., Sorisky, A., Downes, C. P. and Rittenberg, S. E. (1994). Phosphatidylinositol 3,4,5-trisphosphate is formed from phosphatidylinositol 4,5-bisphosphate in thrombin-stimulated platelets. Biochem. J. 301, 415–420.

Chen, J., Profit, A. A. and Prestwich, G. D. (1996). Synthesis of photoactivatable 1,2-diacyl-*sn*-glycerol derivatives of 1-L-phosphatidyll-D-*myo*-inositol 4,5-bisphosphate (PtdInsP$_2$) and 3,4,5-trisphosphate (PtdInsP$_3$). J. Org. Chem. 61, 6305–6312.

Chen, L., Zhou, C., Yang, H. and Roberts, M. F. (2000). Inositol-1-phosphate synthase from *Archaeglobus fulgidus* is a class II aldolase. Biochemistry 39, 12415–12423.

Chung, S. K., Shin, B. G., Chang, Y. T., Suh, B. C. and Kim, K. T. (1998). Syntheses of D- and L-*myo*-inositol 1,2,4,5-tetrakisphosphate and stereoselectivity of the I(1,4,5)P$_3$ receptor binding. Bioorg. Med. Chem. Lett. 17, 659–662.

Clarke, N. G. and Dawson, R. M. C. (1981). Alkaline O–N-transacylation. A new method for the quantitative deacylation of phospholipids. Biochem. J. 195, 301–306.

Cosgrove, D. J. (1969). Ion exchange chromatography of inositol polyphosphates. Ann. N.Y. Acad. Sci. 165, 677–686.

Cosgrove, D. J. (1980). In: Inositol Phosphates (Cosgrove, B. J., ed.). Elsevier, Amsterdam, Chapter 12.

Cronholm, T., Viestam-Rains, M. and Sjovall, J. (1992). Decreased content of arachidonoyl species of phosphatidylinositol phosphates in pancreas of rats fed on an ethanol-containing diet. Biochem. J. 287, 925–928.

Dean, N. M. and Moyer, J. D. (1987). Separation of multiple isomers of inositol phosphates formed in GH3 cells. Biochem. J. 242, 361–366.

Dove, S. K., Cooke, F. T., Douglas, M. R., Sayers, L. G., Parker, P. J. and Michell, R. H. (1997). Osmotic stress activates phosphatidylinositol-3,5-bisphosphate synthesis. Nature 390, 187–192.

Downes, C. P., Hawkins, P. T. and Irvine, R. F. (1986). Inositol 1,3,4,5-tetrakis-phosphate and not phosphatidylinositol 3,4-bisphosphate is the probable precursor of inositol 1,3,4-trisphosphate in agonist-stimulated parotid gland. Biochem. J. *238*, 501–506.

Downes, C. P., Mussat, M. C. and Michell, R. H. (1982). The inositol trisphosphate phosphomonoesterase of the human erythrocyte membrane. Biochem. J. *203*, 169–177.

Eisenberg, F., Jr. and Parthasarathy, R. (1987). Measurement of biosynthesis of myo-inositol from glucose 6-phosphate. Methods Enzymol. *141*, 127–143.

Falck, J. R. and Abdali, A. (1993). Total synthesis of D-*myo*-inositol 3,4,5-trisphosphate and 1,3,4,5-tetrakisphosphate. Bioorg. Med. Chem. Lett. *3*, 717–720.

Filthuth, E. and Eibl, H. (1992). Synthesis of enantiomerically pure lysophos-phatidylinositols and alkylphosphoinositols. Chem. Phys. Lipids *60*, 253–261.

Folch, J. (1949). Complete fractionation of brain cephalin: isolation from it of phosphatidylserine, phosphatidylethanolamine and diphosphoinositide. J. Biol. Chem. *177*, 497–504.

Gero, S. G., Mercier, D. and Barrett, J. E. (1972). 1L-MYO-Inositol 1-phosphate. Methods Carbohydr. Chem. *6*, 403–408.

Grado, C. and Ballou, C. E. (1961). *Myo*-inositol phosphates obtained by alkaline hydrolysis of beef brain phospholipids. J. Biol. Chem. *236*, 54–60.

Gunnarsson, T., Ekblad, L., Karlsson, A., Michelsen, P., Odham, G. and Jergil, B. (1997). Separation of polyphosphoinositides using normal-phase high-performance liquid chromatography and evaporative light scattering detection or electrospray mass spectrometry. Anal. Biochem. *254*, 293–296.

Hansen, C. A., Dahl, S., Huddell, B. and Williams, J. R. (1987). Characterization of inositol 1,3,4-tris-phosphate phosphorylation in rat liver. FEBS Lett. *236*, 53–56.

Hawkins, P. T., Reynolds, D. J. M., Poyner, D. R. and Hanley, M. R. (1990). Identification of a novel inositol phosphate recognition site: specific [3H]inositol hexakisphosphate binding to brain regions and cerebellar membranes. Biochem. Biophys. Res. Commun. *167*, 819–827.

Hendrickson, H. S. and Ballou, C. E. (1964). Ion exchange chromatography of intact brain phosphoinositides on diethylaminoethyl cellulose by gradient salt elution in a mixed solvent system. J. Biol. Chem. *239*, 1369–1373.

Hirata, M., Kimura, Y., Ishimatsu, T., Yanaga, F., Shuto, T., Sasaguri, T., Koga, T., Watanabe, Y. and Ozaki, S. (1991). Synthetic inositol 1,3,4,5-tetrakisphos-phate analogues. Biochem. J. *276*, 333–336.

Horne, G. and Potter, B. V. L. (2001). Synthesis of the enantiomers of 6-deoxy-*myo*-inositol 1,3,4,5-tetrakisphosphate, structural analogues of *myo*-inositol 1,3,4,5-tetrakisphosphate. Chemistry *7*, 80–87.

Hsu, F.-F. and Turk, J. (2000). Characterization of phosphatidylinositol, phosphatidylinositol-4-phosphate, and phosphatidylinositol-4,5-bisphosphate by electrospray ionization tandem mass spectrometry: a mechanistic study. J. Am. Soc. Mass Spectrom. *11*, 986–999.

Irvine, R. F. (1986). The structure, metabolism and analysis of inositol lipids and inositol phosphates. In: Phosphoinositides and Receptor Mechanisms (Putney, J. W., Jr., ed.). Alan Liss, New York, pp. 89–107.

Irvine, R. F., Anggard, E. E., Letcher, A. J. and Downes, C. P. (1985). Metabolism of inositol 1,4,5-trisphosphate and inositol 1,3,4-trisphosphate in rat parotid glands. Biochem. J. *229*, 505–511.

Irvine, R. F., Letcher, A. J. and Dawson, R. M. C. (1979). Fatty acid stimulation of membrane phosphatidylinositol hydrolysis by brain phosphatidylinositol phosphodiesterase. Biochem. J. *178*, 497–500.

Irvine, R. F., Letcher, A. J., Heslop, J. P. and Berridge, M. J. (1986a). The inositol tris/tetrakisphosphate pathway—demonstration of Ins(1,4,5)P$_3$ 3-kinase activity in animal tissues. Nature *320*, 631–634.

Irvine, R. F., Letcher, A. J., Lander, D. J. and Berridge, M. J. (1986b). Specificity of inositol phosphate-stimulated Ca^{2+} mobilization from Swiss-mouse 3T3 cells. Biochem. J. *240*, 301–304.

Irvine, R. F., Letcher, A. J., Lander, D. J. and Downes, C. P. (1984). Inositol trisphosphates in carbachol-stimulated rat parathyroid glands. Biochem. J. *223*, 237–243.

Irvine, R. F., Letcher, A. J., Lander, D. J., Heslop, J. P. and Berridge, M. J. (1987). Inositol(3,4)bisphosphate and inositol(1,3)bisphosphate in GH4 cells— evidence for complex breakdown of inositol(1,3,4)trisphosphate. Biochem. Biophys. Res. Commun. *143*, 353–359.

Irving, G. C. J. and Cosgrove, D. J. (1972). Inositol phosphates phosphatases of microbiological origin: the inositol pentaphosphate products of *Aspergillus ficuum* phytases. J. Bacteriol. *112*, 434–438.

Jiang, T., Sweeney, G., Rudolf, M. T., Klip, A., Traynor-Kaplan, A. and Tsien, R. Y. (1998). Membrane-permeant esters of phosphatidylinositol 3,4,5-trisphosphate. J. Biol. Chem. *273*, 11017–11024.

Johnson, L. F. and Tate, M. E. (1969). The structure of 'phytic acids'. Can. J. Chem. *47*, 63–73.

Kieley, D. E., Abruscato, G. J. and Bubaro, V. (1974). A synthesis of *myo*-inositol 1-phosphate. Carbohydr. Res. *34*, 307–313.

Kuksis, A. and Myher, J. J. (1990). Mass analysis of molecular species of diradylglycerols. In: Methods in Phosphatide Research (Irvine, R. F., ed.). Raven Press, New York, pp. 187–216.

Laussmann, T., Eujen, R., Weisshuhn, C. M., Thiel, U. and Vogel, G. (1996). Structures of diphospho-*myo*-inositol pentakisphosphate and bisdiphospho-

myo-inositol tetrakisphosphate from *Dictyostelium* resolved by NMR analysis. Biochem. J. *315*, 715–720.

Laussmann, T., Reddy, K. M., Reddy, K. K., Falck, J. R. and Vogel, G. (1997). Diphospho-*myo*-inositol phosphates from *Dictyostelium* identified as D-6-diphospho-*myo*-inositol pentakisphosphate and D-5,6-bisdiphospho-*myo*-inositol tetrakisphosphate. Biochem. J. *322*, 31–33.

Letcher, A. J., Stephens, L. R. and Irvine, R. F. (1990). Preparation of ^{32}P-Labeled inositol 1,4,5-trisphosphate and ^{14}C-labeled inositol 1,4-bisphosphate. In: Methods in Phosphatide Research (Irvine, R. F., ed.). Raven Press, New York, pp. 31–37.

Leung, L. W. and Bittman, R. (1997). A convenient synthesis of D-*myo*-inositol 1, 4,5-trisphosphate (Ins(1,4,5)P$_2$) and L-*myo*-inositol 1,4,5-trisphosphate (Ins(3,5,6)P$_3$. Carbohydr. Res. *305*, 171–179.

Li, W., Llopis, J., Whitney, M., Zlokarnik, G. and Tsien, R. Y. (1998). Cell-permeant caged InsP3 ester shows that a2 + spike frequency can optimize gene expression. Nature *392*, 863–866.

Li, W., Schultz, C., Llopis, J. and Tsien, R. Y. (1997). Tetrahedron *53*, 12017–12040.

Lim, P. E. and Tate, M. E. (1973). The phytases. II. Biochim. Biophys. Acta *302*, 316–328.

Ling, I. and Okazaki, S. (1994). A chemoenzymatic synthesis of D-*myo*-inositol 1,4,5-trisphosphate. Carbohydr. Res. *256*, 49–58.

Lips, D. L., Majerus, P. W., Gorga, F. R., Young, A. T. and Benjamin, T. L. (1989). Phosphatidylinositol 3-phosphate is present in normal and transformed fibroblasts and is resistant to hydrolysis by bovine brain phospholipase C II. J. Biol. Chem. *264*, 8759–8763.

Liu, Y.-C. and Chen, C.-S. (1989). An efficient synthesis of optically active D-*myo*-inositol 1,4,5-trisphosphate. Tetrahedron Lett. *30*, 1617–1620.

Low, M. G. (1990). Purification of phosphatidylinositol 4-phosphate and phosphatidylinositol 4,5-bisphosphate by column chromatography. In: Methods in Phosphatide Research (Irvine, R. F., ed.). Raven Press, New York, pp. 145–151.

Manifava, M., Thuring, W. J. F., Lim, Z.-Y., Packman, L., Holmes, A. B. and Ktistakis, N. T. (2001). Differential binding of traffic-related proteins to phosphatidic acid- or phosphatidylinositol(4,5)bisphosphate-coupled affinity reagents. J. Biol. Chem. *276*, 8987–8994.

Mayr, G. W. (1988). A novel metal-dye detection system permits picomolar-range hplc analysis of inositol polyphosphates from non-radioactively labeled cell or tissue specimens. Biochem. J. *254*, 585–591.

Mayr, G. W. (1990). Mass determination of inositol phosphates by high-performance liquid chromatography with postcolumn complexometry (Metal-Dye Detection). In: Methods in Inositide Research (Irvine, R. F., ed.). Raven press, New York, pp. 83–108.

Mayr, G. W., Radenberg, T., Thiel, U., Vogel, G. and Stephens, L. D. (1992). Carbohydr. Res. *234*, 247–262.

McConnell, F. M., Stephens, L. R. and Shears, S. B. (1991). Multiple isomers of inositol pentakisphosphate in Epstein-Bar virus-transformed (T5-1) B-lymphocytes. Identification of inositol 1,3,4,5,6-pentakisphosphate, D-inositol 1,2,4,5,6-pentakisphosphate and L-inositol 1,2,4,5,6-pentakisphosphate. Biochem. J. *280*, 323–329.

Menniti, F. S., Miller, R. N., Putney, J. W., Jr. and Shears, S. B. (1993). Turnover of inositol polyphosphates in pancreatoma cells. J. Biol. Chem. *268*, 3850–3856.

Niggli, V. (2000). A membrane-permeant ester of phosphatidylinositol 3,4,5-trisphosphate (PIP$_3$) is an activator of human neutrophil migration. FEBS Lett. *473*, 217–221.

Ongusaha, P. P., Hughes, P. J., Davey, J. and Michell, R. H. (1998). Inositol hexakisphosphate in *Schizosaccharomyces pombe*: synthesis from Ins(1,4,5)P$_3$ and osmotic regulation. Biochem. J. *335*, 671–679.

Ozaki, S., DeWald, D. B., Shope, J. C., Chen, J. and Prestwich, G. D. (2000). Intracellular delivery of phosphoinositides and inositol phosphates using polyamine carriers. Proc. Natl Acad. Sci. USA *97*, 11286–11291.

Ozaki, S., Ling, L., Ogasawara, T., Watanabe, Y. and Masato, H. (1994). A convenient chemoenzymatic synthesis of D- and L-*myo*-inositol 1,4,5,6-tetrakisphosphate. Carbohydr. Res. *259*, 307–310.

Palmer, S., Hughes, K. T., Lee, D. Y. and Wakelam, M. J. O. (1989). Development of a novel Ins(1,4,5)P$_3$-specific binding assay; its use to determine the intracellular concentration of Ins(1,4,5)P$_3$ in unstimulated and vasopressin-stimulated rat hepatocytes. Cell. Signalling *1*, 147–156.

Phillippy, B. Q. and Bland, J. M. (1988). Gradient ion chromatography of inositol phosphates. Anal. Biochem. *175*, 162–166.

Phillippy, B. Q., White, K. D., Johnson, M. R., Tao, S. H. and Fox, M. R. S. (1987). Preparation of inositol phosphates from sodium phytate by enzymatic and non-enzymatic hydrolysis. Anal. Biochem. *162*, 115–121.

Pietrusiewicz, K. M., Salamonczyk, G. M., Bruzik, K. S. and Wieczork, W. (1992). The synthesis of homochiral inositol phosphates from *myo*-inositol. Tetrahedron Lett. *48*, 5523–5542. Corrigendum. Pietrosiewicz, K. M., Salamonczyk, G. M. and Bruzik, K. S. (1994). Tetrahedron *50*, 573–574.

Pizer, F. L. and Ballou, C. E. (1959). Studies on *myo*-inositol phosphates of natural origin. J. Am. Chem. Soc. *81*, 915–921.

Qiao, L., Nan, F., Kunkel, M., Gallegos, A., Powis, G. and Kozikowski, A. P. (1998). 3-Deoxy-D-*myo*-inositol 1-phosphate, 1-phosphonate, and ether analogues as inhibitors of phosphatidylinositol-3-kinase signaling and cell growth. J. Med. Chem. *41*, 3303–3306.

Rameh, L. E., Tolias, K. F., Duckworth, B. C. and Cantley, L. C. (1997). A new pathway for synthesis of phosphatidylinositol 4,5-bisphosphate. Nature *390*, 192–196.

Reddy, K. K., Saady, M., Falck, J. R. and Whited, G. (1995). J. Org. Chem. *60*, 3385–3390.

Romer, S., Stadler, C., Rudolf, M. T., Jasdorf, B. and Schultz, C. (1996). Membrane permeant analogues of the putative second-messenger *myo*-inositol 3,4,5,6-tetrakisphosphate. J. Chem. Soc. Perkin Trans. 1, 1683–1694.

Romer, S., Stadler, C., Rudolf, M. T. and Schultz, C. (1995). Synthesis of D-*myo*-Inositol 3,4,5,6- and 1,4,5,6-tetrakisphosphates analogues and their membrane permeant derivatives. J. Chem. Soc., Chem. Commun., 411–412.

Rudolf, M. T., Li, W. H., Wolfson, N., Traynor-Kaplan, A. E. and Schultz, C. (1998). 2-Deoxy derivative is a partial agonist of the intracellular messenger 3, 4,5,6-tetrakisphosphate in the epithelial cell line T_{84}. J. Med. Chem. *41*, 3635–3644.

Salamonczyk, G. M. and Pietrusiewicz, K. M. (1991). Expedient synthesis of D-*myo*-inositol 1,4,5-trisphosphate and D-*myo*-inositol 1,4-bisphosphate. Tetrahedron Lett. *32*, 6167–6170.

Schiller, J., Arnhold, J., Bernard, S., Muller, M., Reichl, S. and Arnold, K. (1999). Lipid analysis by matrix assisted laser desorption and ionization mass spectrometry: A methodological approach. Anal. Biochem. *267*, 46–56.

Schultz, C., Burmester, A. and Stadler, C. (1996). Synthesis, separation and identification of different inositol phosphates. Subcell. Biochem. *26*, 371–413.

Serunian, L. A., Auger, K. R. and Cantley, L. C. (1991). Identification and quantification of polyphosphoinositides produced in response to platelet-derived growth factor stimulation. Methods Enzymol. *198*, 78–89.

Sharpes, E. S. and McCarl, R. L. (1982). A high performance liquid chromatographic method to measure ^{32}P incorporation into phosphorylated metabolites in cultured cells. Anal. Biochem. *124*, 421–424.

Shears, S. B. (1989). The pathway of *myo*-inositol 1,3,4-triphosphate phosphorylation in liver. J. Biol. Chem. *264*, 19879–19886.

Shears, S. B. (1996). Inositol pentakis- and hexakisphosphate metabolism adds versatility to the action of inositol polyphosphates. Novel effects on ion channels and protein traffic. In: Subcellular Biochemistry, Vol. 26: *myo*-Inositol Phosphates, Phosphoinositides, and Signal Transduction (Biwas, B. B. and Biwas S., eds.). Plenum Press, New York, pp. 187–225.

Shears, S. B., Ali, N., Craxton, A. and Bembenek, M. E. (1995). Synthesis and metabolism of bis-diphosphoinositol tetrakisphosphate *in vitro* and *in vivo*. J. Biol. Chem. *270*, 10489–10497.

Shears, S. B., Parry, J. B., Tang, E. Y. K., Irvine, R. F., Michell, R. H. and Kirk, C. J. (1987). Metabolism of D-*myo*-inositol-1,3,4,5-tetrakisphosphate by rat liver homogenates including synthesis of a novel isomer of myo-inositol tetrakisphosphate. Biochem. J. 246, 139–147.

Shvets, V. I., Klyashchitskii, B. A., Stepanov, A. E. and Evstigneeva, R. P. (1973). Resolution of asymmetrically substituted myo-inositols into optical antipodes. Tetrahedron Lett. 29, 331–340.

Stephens, L. R. (1990). Preparation and separation of inositol tetrakisphosphates and inositol pentakisphosphates and the establishment of enantiomeric configurations by the use of L-iditol dehydrogenase. In: Methods in Phosphatide Research (Irvine, R. F., ed.). Raven Press, New York, pp. 9–30.

Stephens, L. R., Berrie, C. P. and Irvine, R. F. (1990). Agonist-stimulated inositol phosphate metabolism in avian erythrocytes. Biochem. J. 269, 65–72.

Stephens, L. R. and Downes, C. P. (1990). Precursor–product relationships amongst inositol polyphosphates. Biochem. J. 265, 435–452.

Stephens, L. R., Hawkins, P. T., Barker, C. J. and Downes, C. P. (1988a). Synthesis of myo-inositol 1,3,4,5,6-pentakisphsphate from inositol phosphates generated by receptor activation. Biochem. J. 253, 721–723.

Stephens, L. R., Hawkins, P. T., Carter, N., Chahwala, S. B., Morris, A. J., Whetton, A. D. and Downes, C. P. (1988b). L-myo-inositol 1,4,5,6-tetrakisphosphate is present in both mammalian and avian cells. Biochem J. 249, 271–282.

Stephens, L. R., Hawkins, P. T., Morris, A. J. and Downes, C. P. (1988c). L-myo-inositol 1,4,5,6-tetrakis phosphate (3-hydroxyl)kinase. Biochem. J. 249, 283–292.

Stephens, L. R., Hawkins, P. T. and Downes, C. P. (1989a). Metabolic and structural evidence for the existence of a third species of polyphosphoinositide in cells: D-phosphatidyl-myo-inositol 3-phosphate. Biochem. J. 259, 267–276.

Stephens, L. R., Hawkins, P. T. and Downes, C. P. (1989b). Analysis of myo-[3H]inositol trisphosphates found in myo-[3H]inositol prelabeled avian erythrocytes. Biochem. J. 262, 727–737.

Stephens, L. R., Hawkins, P. T., Stanley, A. F., Moore, T., Poyner, D. R., Morris, P. J., Hanley, M. R., Kay, R. R. and Irvine, R. F. (1991). myo-Inositol pentakisphosphates. Biochem. J. 275, 485–499.

Stephens, L. R. and Irvine, R. F. (1990) Stepwise phosphorylation of myo-inositol leading to myo-inositol hexakisphosphate in Dictyostelium.

Stephens, L., Radenberg, T., Thiel, U., Vogel, G., Khoo, K.-H., Dell, A., Jackson, T. R., Hawkins, P. T. and Mayr, G. W. (1993). The detection, purification, structural characterization, and metabolism of diphosphoinositol pentakisphosphate(s) and bisdiphosphoinositol tetrakisphosphate(s). J. Biol. Chem. 268, 4009–4015.

Tate, M. E. (1968). Separation of myo-inositol pentakisphosphates by moving paper electrophoresis. Anal. Biochem. 23, 141–149.

Tolias, K. F., Couvillon, A. D., Cantley, L. C. and Carpenter, C. L. (1998a). Characterization of Rac1- and RhoGD1-associated lipid kinases of signaling complex. Mol. Cell Biol. *18*, 762–770.

Tolias, K. F., Rameh, L. E., Ishihara, H., Shibasaki, Y., Chen, J., Prestwich, G. L. C. and Carpenter, C. L. (1998). Type 1 phosphatidylinositol-4-phosphate 5-kinase synthesizes the novel lipids phosphatidylinositol 3,5-bisphosphate and phosphatidylinositol 5-phosphate. J. Biol. Chem. *273*, 18040–18046.

Tomlinson, R. V. and Ballou, C. E. (1962). *Myo*-inositol polyphosphate intermediates in the dephosphorylation of phytic acid by phytase. Biochemistry *1*, 166–171.

Traynor-Kaplan, A. E., Thompson, B. L., Harris, A. L., Taylor, P., Omann, G. M. and Sklar, L. A. (1989). Transient increase in phosphatidylinositol 3,4-bisphosphate and phosphatidylinositol trisphosphate during activation of human neutrophils. J. Biol. Chem. *264*, 15668–15673.

Tuominen, E. K. J., Holopainen, J. M., Chen, J., Prestwich, G. D., Bachiller, P. R., Kinnunen, P. K. J. and Janmey, P. A. (1999). Fluorescent phosphoinositide derivatives reveal specific binding of gelsolin and other actin regulatory proteins to mixed lipid bilayers. Eur. J. Biochem. *263*, 85–92.

Vajanaphanich, M., Schultz, C., Rudolph, M. T., Wasserman, M., Enyedi, P., Craxton, A., Shears, S., Tsien, R. Y., Barrett, K. E. and Traynor-Kaplan, A. E. (1994). Long-term uncoupling of chloride secretion from intracellular calcium levels by Ins(3,4,5,6)P$_4$. Nature *371*, 711–714.

Vaskovsky, V. E. and Kostetsky, E. Y. (1968). Modified spray for the detection of phospholipids on thin-layer chromatograms. J. Lipid Res. *9*, 396–397.

Walker, D. M., Urbe, S., Dove, S. K., Tenza, D., Raposo, G. and Clague, M. J. (2001). Characterization of MTMR3: an inositol lipid 3-phosphatase with novel substrate specificity. Curr. Biol. *11*, 1600–1605.

Walsh, J. P., Caldwell, K. K. and Majerus, P. W. (1991). Formation of phosphatidylinositol 3-phosphate by isomerization from phosphatidylinositol 4-phosphate. Proc. Natl Acad. Sci. *88*, 9184–9187.

Watanabe, Y., Mitani, M., Morita, T. and Ozaki, S. (1989). Highly efficient protection by the tetraisopropyldisiloxane-1,3-diyl group in the synthesis of *myo*-inositol phosphates as inositol 1,3,4,6-tetrakisphosphate. J. Chem. Soc. Chem. Commun. *1989*, 482–483.

Watanabe, Y., Shinohara, T., Fujimoto, T. and Ozaki, S. (1990). A short step practical synthesis of *myo*-inositol 1,3,4,5-tetrakisphosphate. Chem. Pharm. Bull. *38*, 562–563.

Whiteford, C. C., Brearley, C. A. and Ulug, E. T. (1997). Phosphatidylinositol 3,5-bisphosphate defines a novel PI 3-kinase pathway in resting mouse fibroblasts. Biochem. J. *323*, 597–601.

Whitman, M., Downes, C. P., Keeler, M., Keller, T. and Cantley, L. (1988). Type I phosphatidylinositol kinase makes a novel inositol phospholipid, phosphatidylinositol-3-phosphate. Nature *332*, 644–646.

Whitman, M., Kaplan, D. R., Schaffhausen, B., Cantley, L. and Roberts, T. M. (1985). Association of phosphatidylinositol kinase activity with polyoma middle-T competent transformation. Nature *315*, 239–242.

Wilcox, R. A., Safrany, S. T., Lampe, D., Mills, S. J., Nahorski, S. R. and Potter, B. V. (1994). Modification at C_2 of *myo*-inositol 1,4,5-trisphosphate produces trisphosphates and tretrakisphosphates with potential biological activity. Eur. J. Biochem. *223*, 115–124.

Wiley, M. G., Przetakiewicz, M., Takahashi, M. and Lowenstein, J. M. (1992). An extended method for separating and quantitating molecular species of phospholipids. Lipids *27*, 295–301.

Zhu, X. and Eichberg, J. (1993). Molecular species composition of glycerophospholipids in rat sciatic nerve and its alteration in stretozotocin-induced diabetes. Biochim. Biophys. Acta *1168*, 1–12.

Chapter 11

Abdullah, M., Hughes, P. J., Craxton, A., Gigg, R., Desai, T., Marecek, J. F., Prestwich, G. D. and Shears, S. B. (1992). Purification and characterization of inositol-1,3,4-trisphosphate 5/6-kinase from liver using an inositol hexakisphosphate affinity column. J. Biol. Chem. *267*, 22340–22345.

Abreu, M. T., Arnold, E. T., Chow, J. Y. C. and Barrett, K. E. (2001). Phosphatidylinositol 3-kinase-dependent pathways oppose fas-induced apoptosis and limit chloride secretion in human intestinal epithelial cells. J. Biol. Chem. *276*, 47563–47574.

Acehan, D., Jiang, X., Morgan, D. G., Heuser, J. E., Wang, X. and Akey, C. W. (2002). Three-dimensional structure of the apoptosome: implications for assembly, procaspase-9 binding, and activation. Mol. Cell *9*, 423–432.

Albert, C., Safrany, S. T., Bembenek, M. E., Reddy, K. M., Reddy, K., Falck, J., Brocker, M., Shears, S. B. and Mayer, G. W. (1997). Biological variability in the structures of diphosphoinositol polyphosphates in *Dictyostelium discoideum* and mammalian cells. Biochem. J. *327*, 553–560.

Alessi, D. R. (2001). Discovery of PDK1, one of the missing links in insulin signal transduction. Colworth Medal lecture. Biochem. Soc. Trans. *29*, 1–14.

Alessi, D. R. and Downes, C. P. (1998). The role of PI 3-kinase in insulin action. Biochim. Biophys. Acta *1436*, 151–164.

Ali, N., Craxton, A. and Shears, S. B. (1993). HepaticIns(1,3,4,5)P4 3-phosphatase is compartmentalized inside endoplasmic reticulum. J. Biol. Chem. *268*, 6161–6167.

Amano, M., Ito, M., Kimura, K., Fukata, Y., Chihara, K., et al. (1996). Phosphorylation and activation of myosin by Rho-associated kinase. Curr. Biol. *7*, 776–789.

Anderson, R. A., Boronenkov, I. V., Doughman, S. D., Kunz, J. and Loijens, J. C. (1999). Phosphatidylinositol phosphate kinases, a multifaceted family of signaling enzymes. J. Biol. Chem. *274*, 9907–9910.

Andjelkovic, M., Alessi, D. R., Meier, R., Fernandez, A., Lamb, N. J., Frech, M., Cron, P., Cohen, P., Lucocq, J. M. and Hemmings, B. A. (1997). Role of translocation in the activation and function of protein kinase B. J. Biol. Chem. *272*, 31515–31524.

Arcaro, A., Zvelebil, M. J., Wallasch, C., Ullrich, A., Waterfield, M. D. and Domin, A. (2000). Class II phosphoinositide 3-kinases are downstream targets of activated polypeptide growth factor receptors. Mol. Cell. Biol. *20*, 3817–3830.

Attree, O., Olivos, I. M., Okabe, I., Bailey, L. C., Nelson, D. L., Lewis, R. A., McInnes, R. R. and Nussbaum, R. L. (1992). The Lowe's oculocerebrorenal syndrome gene encodes a protein highly homologous to inositol polyphosphate-5-phosphatase. Nature *358*, 239–242.

Auger, K. R., Serunian, L. A. and Cantley, L. C. (1990). Separation of novel polyphosphoinositides. In: Methods in Inositide Research (Irvine, R. F., ed.). Raven press, Ltd, New York, pp. 159–166.

Azuma, T., Koths, K., Flanagan, L. and Kwiatkowski, D. (2000). Gelsolin in complex with phosphatiodylinositol 4,5-bisphosphate inhibits caspase -3 and -9 to retard apoptotic progression. J. Biol. Chem. *275*, 3761–3766.

Banfic, H., Tang, H-W., Batty, I. H., Downes, C. P., Chen, C. and Rittenhouse, S. E. (1998). A novel integrin-activated pathway forms PKB/Akt-stimulatory phosphatidylinositol 3,4-bisphosphate via phosphatidylinositol 3-phosphate in platelets. J. Biol. Chem. *273*, 13–19.

Baraldi, E., Carugo, K. D., Hyvonen, M., Surdo, P., Riley, A. M., Potter, B. V. L., O'Brien, R., Ladbury, J. E. and Saraste, M. (1999). Structure of the pH domain from Bruton's tyrosine kinase in complex with inositol 1,3,4,5-tetrakisphosphate. Structure Fold Des. *7*, 449–460.

Barber, A. J., Nakamura, M., Wolpert, E. B., Reiter, C. E. N., Seigel, G. M., Antonetti, D. A. and Gardner, T. W. (2001). Insulin rescues retinal neurons from apoptosis by a phosphatidylinositol 3-kinase/Akt-mediated mechanism that reduces the activation of caspase-3. J. Biol. Chem. *276*, 32814–32821.

Barylko, B., Bins, D., Lin, K., Atkinson, M. A. L., Jameson, D. M., Lin, H. L. and Albanesi, J. P. (1998). Synergistic activation of dynamin GTPase by Grb2 and phosphoinositides. J. Biol. Chem. *273*, 3791–3797.

Batty, I. H. and Downes, C. P. (1996). Thrombin receptors modulate insulin-stimulated phosphatidylinositol 3,4,5-trisphosphate accumulation in 1321N1 astrocytoma cells. Biochem. J. *317*, 347–351.

Batty, I. H., Letcher, A. J. and Nahorski, S. R. (1989). Accumulation of inositol polyphosphate isomers in agonist-stimulated cerebral-cortex slices. Biochem. J. *258*, 23–32.

Batty, J. P., Nahorski, S. R. and Irvine, R. F. (1985). Rapid formation of inositol 1, 3,4,5-tetrakisphosphate following muscarinic receptor stimulation of rat cerebral cortical slices. Biochem. J. *232*, 211–215.

Berditchevski, F., Tolias, K. F., Wong, K., Carpenter, C. L. and Hemler, M. E. (1997). A novel link between integrins, transmembrane 4 superfamily proteins and phosphatidylinositol 4-kinase. J. Biol. Chem. *272*, 2595–2598.

Berridge, M. J. (1993). Inositol trisphosphate and calcium signaling. Nature (London) *361*, 315–325.

Berridge, M. J. (1995). Capacitative calcium entry. Biochem. J. *312*, 1–11.

Berridge, M. J. and Irvine, R. F. (1984). Inositol trisphosphate, a novel second messenger in cellular signal transduction. Nature *312*, 315–321.

Berridge, M. J. and Irvine, R. F. (1989). Inositol phosphates and cell signaling. Nature *341*, 197–205.

Berridge, M. J., Lipp, P. and Bootman, M. D. (2000). The calcium entry pas de deux. Science *287*, 1604–1605.

Bird, G. S. and Putney, J. W. Jr. (1996). Effect of inositol 1,3,4,5-tetrakisphosphate on inositol trisphosphate-activated Ca^{2+} signaling in mouse lacrimal acinar cells. J. Biol. Chem. *271*, 6766–6770.

Bird, I. M., Williams, B. C. and Walker, W. W. (1992). Identification and metabolism of phosphoinositol species formed on angiotensin II stimulation of zona fasciculate-reticularis cells from the bovine adrenal cortex. Mol. Cell. Endocrinol. *83*, 29–38.

Blatt, M. R. and Grabov, A. (1997). Physiol. Plant. *100*, 481–490.

Boggon, T. J., Shan, W-S., Santagata, S., Myers, S. C. and Shapiro, L. (1999). Implication of Tubby proteins as transcription factors by structure-based functional analysis. Science *286*, 2119–2125.

Bokkala, S. and Joseph, S. K. (1997). Angiotensin II-induced down-regulation of inositol trisphosphate receptors in WB rat liver epithelial cells. J Biol. Chem. *272*, 12454–12461.

Bony, C., Roche, S., Shuichi, U., Sasaki, T., Crackower, M. A., Penninger, J., Mano, H. and Puceat, M. (2001). A specific role of phosphatidylinositol 3-kinase. A reglation of autonomic Ca^{2+} oscillations in cardiac cells. J. Cell Biol. *152*, 717–727.

Boronenkov, I. V. and Anderson, R. A. (1995). The sequence of phosphatidyl-inositol 4-phosphate 5-kinase defines a novel family of lipid kinases. J. Biol. Chem. *270*, 2881–2884.

Bottomley, J. R., Reynolds, J. S., Lockyer, P. J. and Cullen, P. J. (1998a). Structural and functional analysis of the putative inositol 1,3,4,5-tetrakispho-sphate receptors GAP1[IP4BP] and GAP1[m]. Biochem. Biophys. Res. Commun. *250*, 143–149.

Bottomley, M. J., Salim, K. and Panayotou, G. (1998b). Phospholipid-binding domains. Biochim. Biophys. Acta *1436*, 165–183.

Boulay, G., Brown, D. M., Qin, N., Jiang, M., Dietrich, A., Zhu, M. X., Chen, Z., Birnbaumer, M., Mikoshiba, K. and Birnbaumer, L. (1999). Proc. Natl. Acad. Sci. USA *96*, 14955–14960.

Bradford, P. G. and Rubin, R. P. (1986). Quantitative changes in inositol 1,4,5-trisphosphate in chemoattractant-stimulated neutrophils. J. Biol. Chem. *261*, 15644–15647.

Brown, F. D., Rozelle, A. L., Yin, H. L., Balla, T. and Donaldson, J. D. (2001). PI(4, 5)P$_2$ and ARF6-regulated membrane traffic. J. Cell Biol. *154*, 1007–1017.

Bryant, N. J., Govers, R. and James, D. E. (2002). Regulated transport of the glucose transporter GLUT4. Nature Revs. Mol. Cell Biol. *3*, 267–277.

Burd, C. G. and Emr, S. D. (1998). Phosphatidylinositol(3)phosphate signaling mediated by specific binding to RING FYVE domains. Mol. Cell *2*, 157–162.

Camina, J. P., Casabiell, X. and Casanueva, F. F. (1999). Inositol 1,4,5-trisphosphate-independent Ca^{2+} mobilization triggered by a lipid factor isolated from vitreous body. J. Biol. Chem. *274*, 28134–28144.

Cantley, L. C. (2001). Translocating Tubby. Science *292*, 2019–2023.

Cantley, L. C. (2002). The phosphoinositide 3-kinase pathway. Science *296*, 1655–1657.

Carew, M. A., Yang, X., Schultz, C. and Shears, S. B. (2000). *myo*-Inositol 3,4,5, 6-tetrakisphosphate inhibits an apical calcium-activated chloride conductance in polarized monolayers of a cystic fibrosis cell line. J. Biol. Chem. *275*, 26906–26913.

Caroni, P. (2001). Actin cytoskeleton regulation through modulation of PI(4,5)P$_2$ rafts. EMBO J. *20*, 4332–4336.

Carpenter, C. L. and Cantley, L. C. (1996a). Phosphoinositide 3-kinase and the regulation of cell growth. Biochim. Biophys. Acta *1288*, M11–M16.

Carpenter, C. L. and Cantley, L. C. (1996b). Phosphoinositide kinases. Curr. Opin. Cell Biol. *8*, 153–158.

Carpenter, C. L., Tolias, K. F., Van Vugt, A. and Hartwig, J. (1999). Lipid kinases are novel effectors of the GTPase Rac1. Adv. Enzyme Regul. *39*, 299–312.

Castellino, A. M., Paker, G. J., Boronenkov, I. V., Anderson, R. A. and Chao, M. V. (1997). A novel interaction between the juxtamembrane region of the p55 tumor necrosis factor receptor and phosphatidylinositol 4-phosphate 5-kinase. J. Biol. Chem. *272*, 5861–5870.

Chalifour, R. J. and Kanfer, J. N. (1980). Microsomal phospholipase D of rat brain and lung tissues. Biochem. Biophys. Res. Commun. *96*, 742–747.

Challis, R. A., Batty, I. H. and Nahorski, S. R. (1988). Mass measurements of inositol(1,4,5)trisphosphate in rat cerebral cortex slices using a radioreceptor assay: effects of neurotransmitters and depolarization. Biochem. Biophys. Res. Commun. *157*, 684–691.

Challis, R. A., Blackledge, M. J. and Radda, G. K. (1988). Spatial heterogeneity of metabolism in skeletal muscle in vivo studied by ^{31}P-NMR spectroscopy. Am. J. Physiol. *254*, C417–C422.

Challis, R. A. and Nahorski, S. R. (1993). Measurement of inositol 1,4,5-trisphosphate, inositol 1,3,4,5-tetrakisphosphate, and phosphatidylinositol 4,5-bisphosphate in brain. In: Lipid Metabolism in Signaling Systems, Methods in Neurosciences, (Fain, J. N., eds.) Vol. 18. Academic Press, New York, pp. 224–244.

Challiss, R. A., Jones, J. A., Owen, P. J. and Boarder, M. R. (1991). Changes in inositol 1,4,5-trisphosphate and inositol 1,3,4,5-tetrakisphosphate mass accumulation in cultured adrenal chromaffin cells in response to bradykinin and histamine. J. Neurochem. *56*, 1083–1086.

Challiss, R. A., Willcocks, A. L., Mulloy, B., Potter, B. V. and Nahorski, S. R. (1991). Characterization of inositol 1,4,5-trisphosphate- and inositol 1,3,4,5-tetrakisphosphate-binding sites in rat cerebellum. Biochem. J. *274*, 861–867.

Chang, H. W., Aoki, M., Fruman, D., Auger, K. R., Bellacosa, A., Tsichlis, P. N., Cantley, L. C., Roberts, T. M. and Vogt, P. K. (1997). Transformation of chicken cells by the gene encoding the catalytic subunit of PI 3-kinase. Science *276*, 1848–1850.

Cheatham, B., Vlahos, C. J., Cheatham, L., Wang, B. J. and Kahn, C. R. (1994). Phosphatidylinositol 3-kinase activation is required for insulin stimulation of pp70 S6 kinase, DNA synthesis, and glucose transported translocation. Mol. Cell Biol. *14*, 4902–4911.

Cheever, M. L., Sato, T. K., de Beer, T., Kutaledze, T. G., Emr, S. D. and Overduin, M. (2001). PX domain interaction with PtdIns(3)P targets the Vam7 t-SNARE to vacuole membranes. Nat. Cell Biol. *3*, 613–618.

Chi, T. H. and Crabtree, G. R. (2000). Perspectives: signal transduction. Inositol phosphates in the nucleus. Science *287*, 1937–1939.

Chiang, S-H., Baumann, C. A., Kanzaki, M., Thurmond, D. C., Watson, R. T., Neudauer, C. L., Macara, I. G., Pessin, J. E. and Saltiel, A. R. (2001). Insulionstimulated GLUT4 translocation requires the CAP-dependent activation of TC10. Nature *410*, 944–948.

Ching, T. T., Hsu, A-L., Johnson, A. J. and Chen, C-S. (2001a). Phosphoinositide 3-kinase facilitates antigen-stimulated Ca^{2+} influx in RBL-2H3 mast cells via a phosphatidylinositol 3,4,5-trisphosphate-sensitive Ca^{2+} entry mechanism. J. Biol. Chem. *276*, 14814–14820.

Ching, T. T., Lin, H. P., Yang, C. C., Oliveira, M., Lu, P. J. and Chen, C. S. (2001b). Specific binding of the C-terminal Src homology 2 domain of the p85α subunit of phosphoinositide 3-kinase to phosphatidylinositol 3,4,5-trisphosphate. Localization and engineering of the phosphoinositide-binding motif. J. Biol. Chem. *276*, 43932–43938.

Clapham, D. E. (1995). Calcium signaling. Cell *80*, 259–268.

Clark, S. F., Martin, S., Carozzi, A. J., Hill, M. M. and James, D. E. (1998). Intracellular localization of phosphatidylinositide 3-kinase and insulin receptor substrate-1 in adipocytes: potential involvement of a membrane skeleton. J. Cell Biol. *140*, 1211–1225.

Clement, S., Krause, U., Desmedt, F., Tanti, J-F., Behrends, J., Pesesse, X., Sasaki, T., Penninger, J., Doherty, M., Malaisse, W., Dumont, J. E., Le Marchand-Brustel, Y., Erneux, C., Hue, L. and Schurmans, S. (2001). The lipid phosphatase SHIP2 controls insulin sensitivity. Nature *409*, 92–96.

Cocco, L., Capitani, S., Maraldi, N. M., Mazotti, G., Barnabei, O., Rizzoli, R., Gimour, R. S., Wirtz, K. W., Rhee, S. G. and Manzoli, F. A. (1998). Inositides in the nucleus: taking stock of PLC β1. Adv. Enzyme Regul. *38*, 351–363.

Cocco, L., Maraldi, N. M., Capitani, S., Martelli, A. M. and Manzoli, F. A. (2001). Nuclear localization and signaling activity of inositol lipids. Ital. J. Anat. Embryol. *106*(2 Suppl 1), 31–43.

Cohen, P., Alessi, R. and Cross, D. A. E. (1997). PDK1, one of the missing links in insulin signal transduction? FEBS Lett. *410*, 3–10.

Connelly, T. M., Bansal, V. S., Bross, T. E., Irvine, R. F. and Majerus, P. W. (1987). The metabolism of tris- and tetrakisphosphates of inositol by 5-phosphomonoesterase and 3-kinase enzymes. J. Biol. Chem. *262*, 2146–2149.

Conricode, K. M. (1995). Involvement of phosphatidylinositol 3-kinase in stimulation of glucose transport by growth factors in 3T3-L1 adipocytes. Biochem. Mol. Biol. Int. *36*, 835–843.

Cozier, G. E., Lockyer, P. J., Reynolds, J. S., Kupzig, S., Bottomley, J. R., Millard, T. H., Banting, G. and Cullen, P. J. (2000). GAP1[IP4BP] contains a novel group 1 pleckstrin homology domain that directs constitutive plasma membrane association. J. Biol. Chem. *275*, 28261–28268.

Craxton, A., Erneux, C. and Shears, S. B. (1994). Inositol 1,4,5,6-tetrakisphosphate is phosphorylated in rat liver by a 3-kinase distinct from inositol 1,4,5-trisphosphate 3-kinase. J. Biol. Chem. *269*, 4337–4342.

Cremona, O., DI Paolo, G., Wenk, M. R., Luthi, A., Kim, W. T., Takei, K., Daniell, L., Nemoto, Y., Shears, S. B., Flavell, R. A., McCormick, D. A. and De Camilli, P. (1999). Essential role of phosphoinositide metabolism in synaptic vesicle recycling. Cell *99*, 179–188.

Cross, D. A. E., Watt, P. W., Shaw, M., Kay, J. V. D., Downes, C. P., Holder, J. C. and Cohen, P. (1997). Insulin activates protein kinase B, inhibits glycogen synthase kinase-3 and activates glycogen synthase by rapamycin-insensitive pathways in skeletal muscle and adipose tissue. FEBS Lett. *406*, 211–215.

Cullen, P. J. (1998). Bridging the GAP in inositol 1,3,4,5-tetrakisphosphate signaling. Biochim. Biophys. Acta *1436*, 35–47.

Cullen, P. J. (2002). Integration of calcium and ras signaling. Nature Revs. Mol. Biol. *3*, 339–348.

Cullen, P. J., Cozier, G. E., Banting, G. and Mellor, H. (2001). Modular phosphoinositide-binding domains—their role in signaling and membrane trafficking. Current Biol. *11*, R882–R893.

Cullen, P. J., Dawson, A. P. and Irvine, R. F. (1995b). Purification and characterization of an Ins(1,3,4,5)P$_4$ binding protein from pig platelets: possible identification of a novel non-neuronal Ins(1,3,4,5)P$_4$ receptor. Biochem. J. *305*, 139–143.

Cullen, P. J., Hsuan, J. J., Truong, O., Letcher, A. J., Jackson, T. R., Dawson, A. P. and Irvine, R. F. (1995a). Identification of a specific Ins(1,3,4,5)P$_4$-binding protein as a member of the GAP1 family. Nature *376*, 527–530.

Cullen, P. J., Loomis-Husselbee, J., Dawson, A. P. and Irvine, R. F. (1997). Inositol 1,3,4,5-tetrakisphosphate and Ca^{2+} homeostasis: the role of GAP1IP4BP. Biochem. Soc. Trans. *25*, 991–996.

Czech, M. P. (2003). Dynamics of phosphoinositides in membrane retrieval and insertion. Annu. Rev. Physiol. *65*, 791–815.

D'Santos, C. S., Clarke, J. H. and Divecha, N. (1998). Phospholipid signaling in the nucleus. Review. Biochim. Biophys. Acta *1436*, 201–232.

De Camilli, P., Emr, S. D., McPherson, P. S. and Novick, P. (1996). Phosphoinositides as regulators in membrane traffic. Science *271*, 1533–1539.

Del Peso, L., Gonzales-Garcia, M., Page, C., Herrera, R. and Nunez, G. (1997). Interleukin 3-induced phosphorylation of BAD through the protein kinase Akt. Science *278*, 687–689.

Denker, S. P. and Barber, D. L. (2002). Ion transport proteins anchor and regulate the cytoskeleton. Curr. Opin. Cell Biol. *14*, 214–220.

Dhand, R., Hiles, I., Panayotou, G., Roche, S., Fry, M. J., Gout, I., Totty, N. F., Truong, O., Vicendo, P., Yonezawa, K., Kasuga, M., Courtneidge, S. A. and Waterfield, M. D. (1994). PI 3-kinase is a dual specificity enzyme: autoregulation by an intrinsic protein-serine kinase activity. EMBO J. *13*, 522–533.

Didichenko, S. A., Tilton, B., Hemmings, B. A., Ballmerhofer, K. and Thelen, M. (1996). Constitutive activation of protein -kinase-b and phosphorylation of p47(phox) by membrane-targeted phosphoinositide 3-kinase. Curr. Biol. *6*, 1271–1278.

Dohlman, H. G. and Thorner, J. (2001). Regulation of G protein-initiated signal transduction in yeast: paradigms and principles. Annu. Rev. Biochem. *70*, 703–754.

Domin, J. and Waterfield, M. D. (1997). Using structure to define the function of phosphoinositide 3-kinase family members. FEBS Lett. *410*, 91–95.

Donie, F. and Reiser, G. (1989). A novel, specific binding protein assay for quantification of intracellular inositol 1,3,4,5-tetrakisphosphate (InsP$_4$) using high-affinity InsP$_4$ receptor from cerebellum. FEBS Lett. *254*, 155–158.

Dove, S. K., Cooke, F. T., Douglas, M. R., Sayers, L. G., Parker, P. J. and Mitchell, R. H. (1997). Osmotic stress activates phosphatidylinositol-3,5-bisphosphate synthesis. Nature (London) *390*, 187–192.

Dove, S. K., McEwen, R. K., Cooke, F. T., Parker, P. J. and Michell, R. H. (1999). Phosphatidylinositol 3,5-bisphosphate: a novel lipid that links stress responses to membrane trafficking events. Biochem. Soc. Trans. *27*, 674–677.

Dowler, S., Currie, R. A., Campbell, D. G., Deak, M., Kular, G., Downes, C. P. and Alessi, D. R. (2000). Identification of pleckstrin-homology-domain-containing proteins with novel phosphoinositide-binding specificities. Biochem. J. *351*, 19–31.

Downes, C. P. and Carter, A. N. (1990). Curr. Opin. Cell Biol. *2*, 185–191.

Downward, J. (1998). Mechanisms and consequences of activation of protein kinase B/Akt. Curr. Opin. Cell Biol. *10*, 262–267.

Eckmann, L., Rudolf, M. T., Ptaszni, A., Schultz, C., Jiang, T., Wolfson, N., Tsien, R., Fierer, J., Shears, S. B., Kagnoff, M. F. and Traynor-Kaplan, A. E. (1997). D-*myo*-Inositol 1,4,5,6-tetrakisphosphate produced in human intestinal epithelial cells n response to *Salmonella* invasion inhibits phosphoinositide 3-kinase signaling pathways. Proc. Natl. Acad. Sci. USA *94*, 14456–14460.

Efanov, A. M., Zaltsev, S. V. and Berggren, P. O. (1997). Inositol hexakisphosphate stimulates non-Ca^{2+}-mediated and primes Ca^{2+}-mediated exocytosis of insulin by activation of protein kinase C. Proc. Natl. Acad. Sci. USA *94*, 4435–4439.

Ellson, C. D., Andrews, S., Stephens, L. R. and Hawkins, P. T. (2002). The PX domain: a new phosphoinositide-binding module. J. Cell Sci. *115*, 1099–1105.

Enyedi, P., Brown, E. and Williams, G. H. (1989). Distinct binding sites for Ins(1,4,5)P$_3$ and Ins(1,3,4,5)P$_4$ in bovine parathyroid glands. Biochem. Biophys. Res. Commun. *159*, 200–208.

Feng, Y., Wente, S. R. and Majerus, P. W. (2001). Overexpression of the inositol phosphate SopB in human 293 cells stimulates cellular chloride influx and inhibits nuclear mRNA transport. Proc. Natl. Acad. Sci. USA *98*, 875–879.

Ferguson, K. M., Kavran, J. M., Sankaran, V. G., Fournier, E., Isakoff, S. J., Skolnik, E. Y. and Lemmon, M. A. (2000). Structural basis for discrimination of 3-phosphoinositides by pleckstrin homology domains. Molecular Cell *6*, 373–384.

Ferris, C. D., Huganir, R. L., Supattapone, S. and Snyder, S. H. (1989). Purified inositol 1,4,5-trisphosphate receptor mediates calcium flux in reconstituted lipid vesicles. Nature *342*, 87–89.

Fleischer, B., Xie, J., Mayrleitner, M., Shears, S. B., Palmer, D. J. and Fleischer, S. (1994). Golgi coatomer binds, and forms K(+)-selective channels gated by inositol polyphosphates. J. Biol. Chem. *269*, 17826–17832.

Ford, M. G. J., Pearse, B. M. F., Hggins, M. K., Vallis, Y., Owen, D. J., Gibson, A., Hopkins, C. R., Evans, P. R. and McMahon, H. T. (2001). Simultaneous binding of PtdIns(4,5)P2 and clathrin by AP180 in the nucleation of clathrin lattices on membranes. Science *291*, 1051–1055.

Franke, T. F., Kaplan, D. R. and Cantley, L. C. (1997a). PI3K: downstream AKTion blocks apoptosis. Cell *88*, 435–437.

Franke, T. F., Kaplan, D. R., Cantley, L. C. and Toker, A. (1997b). Direct regulation of the Akt proto-oncogene product by phosphatidylinositol-3,4-bisphosphate. Science *275*, 665–668.

Fratti, R. A., Backer, J. M., Gruenberg, J., Corvera, S. and Deretic, V. (2001). Role of phosphatidylinositol 3-kinase and Rab5 effectors in phagosomal biogenesis and mycobacterial phagosome maturation arrest. J. Cell Biol. *154*, 631–644.

Frech, M., Andjelkovic, M., Inley, E., Reddy, K. K., Falck, J. R. and Hemmings, B. A. (1997). High affinity binding of inositol phosphates and phosphoinositides to the pleckstrin homology domain of RAC/protein kinase B and their influence on kinase activity. J. Biol. Chem. *272*, 8474–8481.

Frevert, E. U. and Kahn, B. B. (1997). Differential effects of constitutively active PI 3-kinase on glucose transport, glycogen synthase activity and DNA synthesis in 3T3-L1 adipocytes. Mol. Cell. Biol. *17*, 190–198.

Fruman, D. A., Meyers, R. E. and Cantley, L. C. (1998). Phosphoinositide kinases. Annu. Rev. Biochem. 67, 481–507.

Fruman, D. A., Rameh, L. R. and Cantley, L. C. (1999). Phosphoinositide binding domains: Embracing 3-phosphate. Cell 97, 817–820.

Fukami, K., Sawada, N., Endo, T. and Takenawa, T. (1996). Identification of phosphatidylinositol 4,5-bisphosphate binding site in chicken skeletal muscle α-actinin. J. Biol. Chem. 271, 2646–2650.

Fukuda, M., Aruga, J., Niinobe, M., Aimoto, S. and Mikoshiba, K. (1994). Inositol-1,3,4,5-tetrakisphosphate binding to C2B domain of IP4BP/synaptotagmin II. J. Biol. Chem. 269, 29206–29211.

Fukuda, M. and Mikoshiba, K. (1996). Structure-function relationships of the mouse Gap1m. Determination of the inositol 1,3,4,5-tetrakisphosphate-binding domain. J. Biol. Chem. 271, 18838–18842.

Fukuda, M., Kojima, T., Kabayama, H. and Mikoshiba, K. (1996). Mutation of the pleckstrin homology domain of Bruton's tyrosine kinase in immunodeficiency impaired inositol 1,3,4,5-tetrakisphosphate binding capacity. J. Biol. Chem. 271, 30303–30306.

Fukumi, K., Matsuoka, K., Nakanishi, O., Yamakawa, A., Kawai, S. and Takenawa, T. (1988). Antibody to phosphatidylinositol 4,5-bisphosphate inhibits oncogene-induced mitogenesis. Proc. Natl. Acad. Sci. USA 85, 9057–9061.

Gaidarov, I., Krupnick, J. G., Falck, J. R., Benovic, J. L. and Keen, J. H. (1999). Arrestin function in G protein-coupled receptor endocytosis requires phosphoinositide binding. EMBO J. 18, 871–881.

Galectic, I. et al. (1999). Mechanism of protein kinase B activation by insulin/insulin-like growth factor revealed by specific inhibitors of phosphoinositide 3-kinase—Significance for diabetes and cancer. Pharmacol. Ther. 82, 409–425.

Gaullier, J.-M., Simonsen, A., D'Arrigo, A., Bremmes, B., Aasland, R. and Stenmark, H. (1998). FYVE fingers bind PtdIns(3)P. Nature 394, 432–433.

Gaullier, J.-M., Simonsen, A., D'Arrigo, A., Bremmes, B. and Stenmark, H. (1999). FYFE finger proteins as effectors of phosphatidylinositol 3-phosphate. Chem. Phys. Lipids 98, 87–94.

Gillooly, D. J., Morrow, I. C., Lindsay, M., Gould, R., Bryant, N. J., Gaulier, J. M., Parton, R. G. and Stenmark, H. (2000). Localization of phosphatidylinositol 3-phosphate in yeast and mammalian cells. EMBO J. 19, 4577–4588.

Gillooly, D. J., Simonsen, A. and Stenmark, H. (2001). Cellular functions of PI(3)P and FYVE domain proteins. Biochem. J. 355, 249–258.

Gilmore, A. P. and Burridge, K. (1996). Regulation of vinculin binding to talin and action by phosphatidylinositol 4,5-bisphosphate. Nature 381, 531–535.

Glaser, M., Wanaski, S., Buser, C. A., Boguslavsky, V., Rashidzada, W., Morris, A., Rebecchi, M., Scarlata, S. F., Runnels, L. W., Prestwich, G. D., Chen, J., Aderem, A., Ahn, J. and McLaughlin, S. (1996). Myristoylated alanine-rich C kinase substrate (MARCKS) produces reversible inhibition of phospholipase C by sequestering phosphatidylinositol 4,5-bisphosphate in lateral domains. J. Biol. Chem. *271*, 26187–26193.

Hall, A. (1998). Rho GTPases and the actin cytoskeleton. Science *279*, 509–514.

Hammond, S. M., Jenco, J. M., Nakashima, S., Cadwallader, K., Gu, Q. et al. (1997). Characterization of two alternatively spliced forms of phospholipase D1. J. Biol. Chem. *272*, 3860–3868.

Hammonds-Odie, L. P., Jackson, T. R., Profit, A. A., Blader, I. J., Turck, C. W., Prestwich, G. D. and Thiebert, A. B. (1996). Identification and cloning of centaurin-α. J. Biol. Chem. *271*, 1885–18868.

Hanahan, D. and Weinberg, R. A. (2000). The hallmarks of cancer. Cell *100*, 57–70.

Hanakahi, L. A., Bartlet-Jones, M., Chappell, C., Pappin, D. and West, S. C. (2000). Binding of inositol phosphate to DNA-PK and stimulation of double-strand break repair. Cell *102*, 721–729.

Harlan, J. E., Hajduk, P. J., Yoon, H. S. and Fesik, S. W. (1994). Pleckstrin homology domains bind to phosphatidylinositol-4,5-bisphosphate. Nature *371*, 168–170.

Harlan, J. E., Yoon, H. S., Hajduk, P. J. and Fesik, S. W. (1995). Structural characterization of the interaction between a pleckstrin homology domain and phosphatidylinositol-4,5-bisphosphate. Biochemistry *34*, 9859–9864.

Hartwig, J. H., Bokoch, G. M., Carpenter, C. L., Janmey, P. A., Taylor, L. A. et al. (1995). Thrombin receptor ligation and activated Rac uncap actin filament barbed ends through phosphoinositide synthesis in permeabilized human platelets. Cell *82*, 643–653.

Haruta, T., Morris, A. J., Rose, D. W., Nelson, J. G., Meuckler, M. and Olefsky, J. M. (1995). Insulin-stimulated GLUT4 translocation is mediated by a divergent intracellular signaling pathway. J. Biol. Chem. *270*, 27991–27994.

Hawkins, P. T., Jackson, T. R. and Stephens, L. R. (1992). Platelet-derived growth factor stimulates synthesis of PtdIns(3,4,5)P$_3$ by activating a PtdIns(4,5)P$_2$ 3-OH kinase. Nature *358*, 157–159.

Hawkins, P. T., Poyner, D. R., Jackson, T. R., Letcher, A. J., Lander, D. A. and Irvine, R. J. (1993). Inhibition of iron-catalyzed hydroxyl radical formation by inositol polyphosphates: a possible physiological function for *myo*-inositol hexakisphosphate. Biochem. J. *294*, 929–934.

Hawkins, P. T., Stephens, L. and Downes, C. P. (1986). Rapid formation of inositol 1,3,4,5-tetrakisphosphate and inositol 1,3,4-trisphosphate in rat parotid glands may both result indirectly from receptor-stimulated release of inositol 1,4,5-trisphosphate from phosphatidylinositol 4,5-bisphosphate. Biochem. J. *238*, 507–516.

Hawkins, P. T., Stephens, L. R. and Piggott, J. R. (1993). Analysis of inositol metabolites produced by *Saccharomyces cerevisiae* in response to glucose stimulation. J. Biol. Chem. *268*, 3374–3383.

Hay, J. C., Fisette, P. L., Jenkins, G. H., Fukami, K., Takenawa, T., Anderson, R. A. and Martin, T. F. (1995). ATP-dependent inositide phosphorylation required for Ca^{2+}-activated exocytosis. Nature *374*, 173–177.

Hay, J. C. and Martin, T. F. J. (1993). Phosphatidylinositol transfer protein required for ATP-dependent priming of Ca^{2+}-activated secretion. Nature *366*, 572–575.

Hemler, M. E., Mannion, B. A. and Berditchevski, F. (1996). Association of TN4SF proteins with integrins: relevance to cancer. Biochim. Biophys. Acta *1287*, 67–71.

Heraud, J-H., Racuud-Sultan, C., Giroincel, D., Albiges-Rizo, C. et al. (1998). Lipid products of phosphoinositide 3-kinase and phosphatidyl $4',5'$-bisphosphate are both required for ADP-dependent platelet spreading. J. Biol. Chem. *273*, 17817–17823.

Hermosura, M. C., Takeuchi, H., Fleig, A., Riley, A. M., Potter, V., Hirata, M. and Penner, R. (2000). InsP$_4$ facilitates store-operated calcium influx by inhibition of InsP$_3$ 5-phosphatase. Nature *408*, 353–356.

Hildebrant, J. P. and Shuttleworth, T. J. (1992). Calcium-sensitivity of inositol 1,4,5-trisphosphate metabolism in exocrine cells from the avian salt gland. Biochem. J. *282*, 703–710.

Hilton, J. M., Plomann, M., Ritter, B., Modregger, J., Freeman, H. N., Falck, J. R., Krishna, U. M. and Tobin, A. B. (2001). Phospohorylation of a synaptic vesicle associated protein by an inositol hexakisphosphate-regulated protein kinase. J. Biol. Chem. In press.

Hinchliffe, K. A., Ciruela, A. and Irvine, R. F. (1998). PIPkins1, their substrates and their products: new functions for old enzymes. Biochim. Biophys. Acta *1436*, 87–104.

Hirata, M. and Kanematsu, T. (1993). Inositol 1,4,5-trisphosphate-binding proteins in rat brain cytosol. Lipid Metabolism in Signaling Systems Methods in Neurosciences, (Fain, J. N., eds.)18. Academic press, New York, pp. 298–311.

Ho, M. W. Y., Carew, M. A., Yang, X. and Shears, S. B. (2000). Regulation of chloride channel conductance by Ins(3,4,5,6)P$_4$: A phosphoinositide-initiated signaling pathway that acts down-stream of Ins(1,4,5)P$_3$. In: Frontiers in Molecular Biology: Biology of Phosphoinositides (Cockroft, S., ed.). Oxford University Press, Oxford, pp. 298–319.

Ho, M. W., Kaetzel, M. A., Armstrong, D. L. and Shears, S. B. (2001). Regulation of a human chloride channel: a paradigm for integrating input from calcium, type II calmodulin-dependent protein kinase, and inositol 3,4,5,6-tetrakisphosphate. J. Biol. Chem. *276*, 18673–18680.

Ho, M. W., Yang, X., Carew, M. A., Zhang, T., Hua, L., Kwon, Y. U., Chung, S. K., Adelt, S., Vogel, G., Riley, A. M., Potter, B. V. and Shears, S. B. (2002). Regulation of Ins(3,4,5,6)P$_4$ signaling by a reversible kinase/phosphatase. Curr. Biol. *12*, 477–482.

Hokin, L. E. (1985). Receptors and phosphoinositide-generated second messengers. Ann. Rev. Biochem. *54*, 205–235.

Holman, G. D. and Kassuga, M. (1997). From receptor to transporter: insulin signaling to glucose transport. Diabetologia *40*, 991–1003.

Holz, R. W., Hlubek, M. D., Sorensen, S. D., Fisher, S. K., Balla, T., Ozaki, S., Prestwich, G. D., Stuenkel, E. L. and Bittner, M. A. (2000). A pleckstrin homology domain specific for phosphatidylinositol 4,5-bisphosphate (PtdIns-4, 5-P$_2$) and fused to green fluorescent protein identifies plasma membrane PtdIns-4,5-P$_2$ as being important in exocytosis. J. Biol. Chem. *275*, 17878–17885.

Honda, A., Nogami, M., Yokezieki, T., Yamazaki, M., Nakamura, H., Watanabe, H. K., Nakayama, K., Morris, A. J., Frohman, M. A. and Kanaho, Y. (1999). PI(4)P 5-kinase α is a downstream effector of the small G protein ARF6 in membrane ruffle formation. Cell *99*, 521–532.

Honda, Z., Takano, T., Hirose, N., Suzuki, T., Muto, A., Kume, S., Mikoshiba, K., Itoh, K. and Shimizu, T. (1995). J. Biol. Chem. *270*, 4840–4844.

Hope, H. R. and Pike, L. J. (1996). Phosphoinositides and phosphoinositide-utilizing enzymes in detergent-insoluble lipid domains. Mol. Biol. Cell *7*, 843–851.

Horiuchi, H., Lippe, R., McBride, H. M., Rubino, M., Woodman, P. et al. (1997). A novel Rab5 GDP/GTP exchange factor complexed to Rabaptin-5 links nucleotide exchange to effector recruitment and function. Cell *90*, 1149–1159.

Hsu, A. L., Ching, T. T., Sen, G., Wang, D. S., Bondada, S., Authi, K. S., Chen, C. S., Ching, T., Sen, G., Wang, D. S., Bondada, S., Authi, K. S. and Chen, C. S. (2000). Novel function of phosphoinositide 3-kinase in T cell Ca^{2+} signaling. A phosphatidylinositol 3,4,5-trisphosphate-mediated Ca^{2+} entry mechanism. J. Biol. Chem. *274*, 16242–16250.

Hu, Q., Klippel, A., Muslin, A. J., Fantl, W. J. and Williams, L. T. (1995). Ras-dependent induction of cellular responses by constitutive active phosphatidyl-inositol 3-kinase. Science *268*, 100–102.

Hughes, P. J., Hughes, A. R., Putney, J. W. Jr. and Shears, S. B. (1989). The regulation of the phosphorylation of inositol 1,3,4-trisposphate in cell-free preparations and its relevance to the formation of inositol 1,3,4,6-tetrakisphosphate in agonist-stimulated rat parotid acinar cells. J. Biol. Chem. *264*, 19871–19878.

Hughes, W. E., Cooke, F. T. and Parker, P. J. (2000). Sac phosphatase domain proteins. Biochem. J. *350*, 337–352.

Hunter, T. (1997). Oncoprotein networks. Cell *88*, 333–346.

Hunter, T. (2000). Signaling –2000 and beyond. Cell *100*, 113–127.

Hurley, J. H., Anderson, D. E., Beach, B., Canagarajah, B., Ho, Y. S. J., Jones, E., Miller, G., Misra, S., Pearson, M., Saidi, L., Suer, S., Trievel, R. and Tsujishita, Y. (2002). Structural genomics and signaling domains. Trends Biochem. Sci. *27*, 48–53.

Hurley, J. H. and Meyer, T. (2001). Subcellular targeting by membrane lipids. Curr. Opin. Cell Biol. *13*, 146–152.

Hyman, J., Chen, H., Di Fiore, P. P., De Camilli, P. and Brunger, A. T. (2000). Apsin 1 undergoes nucleocytoslic shuttling and its Eps 15 interactor NH2-terminal homology (ENTH) domain, structurally similar to armadillo and HEAT repeats, interacts with the transcription factor promyelocytic leukemia Zn^{2+} finger protein (PLZF). J. Cell Biol. *149*, 537–546.

Ijuin, T., Mochizuki, Y., Fkami, K., Funaki, M., Asano, T. and Takenawa, T. (2000). Identification and characterization of a novel inositol polyphosphate 5-phosphatase. J. Biol. Chem. *275*, 10870–10875.

Irvine, R. F. (1991). Inositol tetrakisphosphate as a second messenger: confusions, contradictions, and a potential resolution. Bioassays *13*, 419–427.

Irvine, R. F. (1992a). Is inositol tetrakisphosphate the second messenger that controls Ca^{2+} entry into cells? Adv. Second Messenger Phosphoprotein Res. *26*, 161–185.

Irvine, R. F. (1992b). Inositol phosphates and Ca^{2+} entry: toward a proliferation or a simplification? FASEB J. *6*, 3085–3091.

Irvine, R. F. (1998). Inositol phospholipids: translocation, translocation, translocation. Curr. Biol. *8*, R557–R559.

Irvine, R. F. (2001). Does IP_4 run a protection racket? Curr. Biol. *11*, R173–R174.

Irvine, R. F., Brown, K. D. and Berridge, M. J. (1984). Specificity of inositol trisphosphate-induced calcium release from permeabilized Swiss-mouse 3T3 cells. Biochem. J. *222*, 269–272.

Irvine, R. F., Letcher, A. J., Heslop, J. P. and Berridge, M. J. (1986). The inositol tris/tetrakisphosphate pathway—demonstration of Ins(1,4,5)P3 3-kinase activity in animal tissues. Nature *320*, 631–634.

Irvine, R. F., Moor, R. M., Pollock, W. K., Smith, P. M. and Wreggett, K. A. (1988). Inositol phosphates: proliferation, metabolism and function. Philos. Trans. R. Soc. Lond. B Biol. Sci. *320*, 281–298.

Irvine, R. F., McNulty, T. J. and Schell, M. J. (1999). Inositol 1,3,4,5-tetrakisphosphate as a second messenger—a special role in neurones. Chem. Phys. Lipids *98*, 49–57.

Irvine, R. F. and Schell, M. J. (2001). Back in the water: The return of the inositol phosphates. Nature Mol. Cell Biol. 2, 327–333.

Ishihara, H., Shibasaki, Y., Kizuki, N., Katagiri, H., Yazaki, Y., Asano, T. and Oka, Y. (1996). Cloning of cDNAs encoding two isoforms of 68-kDa type I phosphatidylinositol-4-phosphate 5-kinase. J. Biol. Chem. 271, 23611–23614.

Ishihara, H., Shibasaki, Y., Kizuki, N., Wada, T., Yazaki, Y., Asano, T. and Oka, Y. (1998). Type I phosphatidylinositol-4-phosphate 5-kinase. J. Biol. Chem. 273, 8741–8748.

Ismailov, I. I., et al. (1996). Biologic function for an 'orphan' messenger: D-myo-inositol-3,4,5,6-tetrakisphosphate selectively blocks epithelial calcium-activated chloride channels. Proc. Natl. Acad. Sci. USA 93, 10505–10509.

Itoh, T., Koshiba, S., Kigawa, T., Kikuchi, A., Yokoyama, S. and Takenawa, T. (2001). Role of the ENTH domain in phosphatidylinositol-4,5-bisphosphate binding and endocytosis. Science 291, 1047–1051.

Jacobson, M. D., Weil, M. and Raff, M. C. (1997). Programmed cell death in animal development. Cell 88, 347–354.

James, S. R., Downes, C. P., Gigg, R., Grove, S. J., Holmes, A. B. and Alessi, D. R. (1996). Specific binding of Akt-1 protein kinase to phosphatidylinositol 3,4,5-trisphosphate without subsequent activation. Biochem. J. 315, 709–713.

Janmey, P. A. (1994). Phosphoinositides and calcium as regulators of cellular actin assembly and disassembly. Annu. Revs. Physiol. 56, 169–191.

Janmey, P. A., Lam, J., Allen, P. G. and Matsudaira, P. T. (1992). Phosphoinositide-binding peptides derived from the sequences of gelsolin and vilin. J. Biol. Chem. 267, 11818–11823.

Janmey, P. A., Xian, W. and Flanagan, L. A. (1999). Controlling cytoskeleton structure by phosphoinositide-protein interactions: phosphoinositide binding protein domains and effects of lipid packing. Chem. Phys. Lipids 101, 93–107.

Jenkins, G. H., Fisette, P. L. and Anderson, R. A. (1994). Type I phosphatidylinositol 4-phosphate 5-kinase isoforms are specifically stimulated by phosphatidic acid. J. Biol. Chem. 269, 11547–11554.

Jiang, Z. Y., Chawla, A., Bose, A., Way, M. and Czech, M. P. (2002). A phosphatidylinositol 3-kinase-independent insulin signaling pathway to N-WASP/Arp2/3/F-actin required for GLUT4 glucose transporter recycling. J. Biol. Chem. 277, 509–515.

Jiang, T., Sweeney, G., Rudolf, M. T., Klip, A., Traynor-Kaplan, A. and Tsien, R. Y. (1998). Membrane-permeant esters of phosphatidylinositol 3,4,5-trisphosphate. J. Biol. Chem. 273, 11017–11024.

Jones, D. R., Gonzalez-Garcia, A., Diez, E., Martinez, A. C., Carrera, A. C. and Merida, I. (1999). The identification of phosphatidylinositol 3,5-bisphosphate in T-lymphocytes and its regulation by interleukin-2. J. Biol. Chem. *274*, 18407–18413.

Jones, D. R., Sanjuan, M. A. and Merida, I. (2000). Type Iα phosphatidylinositol 4-phosphate 5-kinase is putative target for increased intracellular phosphatidic acid. FEBS Lett. *476*, 160–165.

Joseph, S. K. (1996). The inositol tris-phosphate receptor family. Cell Signal. *8*, 1–7.

Jost, M., Simpson, F., Kavran, J. M., Lemmon, M. A. and Schmid, S. L. (1998). Phosphatidylinositol-4,5-bisphosphate is required for endocytic coated vesicle formation. Curr. Biol. *8*, 1399–1402.

Joyal, J. L., Burks, D. J., Pons, S., Matter, W. F., Vlahos, C. J., White, M. F. and Sacks, D. B. (1997). Calmodulin activates phosphatidylinositol 3-kinase. J. Biol. Chem. *272*, 28183–28186.

Kaliman, P., Canicio, J., Testar, X., Palacin, M. and Zorzano, A. (1999). Insulin-like growth factor-II, phosphatidylinositol 3-kinase, nuclear factor-κB and inducible nitric-oxide synthase define a common myogenic signaling pathway. J. Biol. Chem. *274*, 17437–17444.

Kanematsu, T., Misumi, Y., Watanabe, Y., Ozaki, S., Koga, T., Iwanaga, S., Ikehara, Y. and Hirata, M. (1996). A new inositol 1,4,5-trisphosphate binding protein similar to phospholipase C-δ1. Biochem. J. *313*, 319–325.

Kapeller, R. and Cantley, L. C. (1994). Phosphatidylinositol 3-kinase. Bioassays *16*, 565–576.

Kavran, J. M., Klein, D. E., Lee, A., Falasca, M., Isakoff, S. J., Skolnik, E. Y. and Lemmon, M. A. (1998). Specificity and promiscuity in phosphoinositide binding by pleckstrin homology domains. J. Biol. Chem. *273*, 30497–30508.

Kay, B. K., Yamabhai, M., Wendland, B. and Emr, S. D. (1999). Identification of a novel domain shared by putative components of the endocytic and cytoskeletal machinery. Protein Sci. *8*, 435–438.

Kimura, K., Ito, M., Amano, M., Chihara, K., Fukata, Y., et al. (1996). Regulation of myosin phosphatase by Rho and Rho-associated kinase. Science *273*, 245–248.

Kinuta, M., Yamada, H., Abe, T., Watanabe, M., Li, S-A., Kamitani, A., Yasuda, T., Matsukawa, T., Kumon, H. and Takei, K. (2002). Phosphatidylinositol 4, 5-bisphosphate stimulates vesicle formation from liposomes by brain cytosol. Proc. Natl. Acad. Sci. USA *99*, 2842–2847.

Kirchhausen, T., Bonifacino, J. S. and Riezman, H. (1997). Linking cargo to vesicle formation: receptor tail interactions with coat proteins. Curr. Opin. Cell Biol. *9*, 488–495.

Kiselyov, K., Mignery, G. A., Zhu, M. X. and Muallem, S. (1999). The N-terminal domain of the IP3 receptor gated store-operated hTrp3 channels. Mol. Cell *4*, 123–132.

Kisseleva, M. V., Cao, L. and Majerus, P. W. (2002). Phosphoinositide-specific inositol polyphosphate 5-phosphatase IV inhibits Akt/protein kinase B phosphorylation and leads to apoptotic cell death. J. Biol. Chem. 277, 6266–6272.

Klarlund, J. K., Guilherme, A., Holik, J. J., Virbasius, J. V., Chawla, J. V. A. and Czech, M. P. (1997). Signaling by phosphoinositide-3,4,5-trisphosphate through proteins containing pleckstrin and Sec7 homology domains. Science 275, 1927–1930.

Klarlund, J. K., Tsiaras, W., Holik, J. J., Chawla, A. and Czech, M. P. (2000). Distinct polyphosphoinositide binding selectivity for pleckstrin homology domains of GRP1-like proteins based on diglycine versus triglycine motifs. J. Biol. Chem. 275, 32816–32821.

Klippel, A., Kavanaugh, W. M., Pot, D. and Williams, L. T. (1997a). A specific product of phosphatidylinositol 3-kinase directly activates the protein kinase Akt through its pleckstrin homology domain. Mol. Cell Biol. 17, 338–344.

Klippel, A., Kavanaugh, W. M., Pot, D. and Williams, L. T. (1997b). A specific product of phosphatidylinositol 3-kinase directly activates the protein kinase Akt through its pleckstrin homology domain. Mol. Cell Biol. 17, 338–344.

Kochendorfer, K. U., Then, A. R., Kearns, B. G., Bankaitis, V. A. and Mayinger, P. (1999). Sac1p plays a crucial role in microsomal ATP transport, which is distinct from its function in Golgi phospholipid metabolism. EMBO J. 18, 1506–1515.

Kohn, A. D., Takeuchi, F. and Roth, R. A. (1996). Akt, a pleckstrin homology domain containing kinase, is activated primarily by phosphorylation. J. Biol. Chem. 271, 21920–21926.

Kojima, T., Fukuda, M., Watanabe, Y., Hamazato, F. and Mikoshiba, K. (1997). Characterization of the pleckstrin homology domain of Btk as an inositol polyphosphate and phosphoinositide binding domain. Biochem. Biophys. Res. Commun. 236, 333–339.

Koppler, P., Mersel, M., Humbert, J. P., Vignon, J., Vincendon, G. and Malviya, A. N. (1996). High affinity inositol 1,3,4,5-tetrakisphosphate receptor from rat liver nuclei: purification, characterization, and amino-terminal sequence. Biochemistry 35, 5481–5487.

Kothakota, S., Azuma, T., Reinhard, C., Klippel, A., Tang, J., Chu, K., McGarry, T. J., Kirschner, M. W., Koths, K., Kwiatkowski, D. J. and Williams, L. T. (1997). Caspase-3-generated fragment of gelsolin: effector of morphological change in apoptosis. Science 278, 294–298.

Koya, R. C., Fujita, H., Shimizu, S., Ohtsu, M., Takimoto, M., Tsujimoto, Y. and Kuzumaki, N. (2000). Gelsolin inhibits apoptosis by blocking mitochondrial membrane potential loss and cytochrome C release. J. Biol. Chem. *275*, 15343–15349.

Krasilnikov, M. A. (2000). Phosphatidylinositol-3 kinase dependent pathways: the role in control of cell growth, survival, and malignant transformation. Biochemistry (Moscow) *65*, 59–67.

Ktistakis, N. T., Brown, H. A., Aters, M. G., Sternwes, P. C. and Roth, M. G. (1996). Evidence that phospholipase D mediates ADP ribosylation factor-dependent formation of Golgi coated vesicles. J. Cell Biol. *134*, 295–306.

Kunz, J., Fuelling, A., Kolbe, L. and Anderson, R. A. (2002). Stereo-specific substrate recognition by phosphatidylinositol phosphate kinases is swapped by changing a single amino acid residue. J. Biol. Chem. *277*, 5611–5619.

Kunz, J., Wilson, M. P., Kisseleva, M., Hurley, J. H., Majerus, P. W. and Anderson, R. A. (2000). The activation of loop of phosphatidylinositol phosphate kinase determines signalling specificity. Mol. Cell *5*, 1–11.

Kurzchalia, T. V. and Parton, R. G. (1999). Membrane microdomains and caveolae. Curr. Opin. Cell Biol. *11*, 424–431.

Larsson, O., Barker, C. J., Sjoholm, A., Carlqvist, H., Michell, R. H., Bertorello, A., Nilsson, T., Honkanen, R. E., Mayr, G. W., Zwiler, J. and Berggren, P.-O. (1997). Inhibition of phosphatases and increased Ca^{2+} channel activity by inositol hexakisphosphate. Science *278*, 471–474.

Laussmann, T., Hansen, A., Reddy, K. M., Reddy, K. K., Falck, J. R. and Vogel, G. (1998). Diphospho-*myo*-inositol phosphates in *Dictyostelium* and *Polyspondylium*: identification of a new bisdiphospho-*myo*-inositol tetrakisphosphate. FEBS Lett. *426*, 145–150.

Laussmann, T., Reddy, K. M., Reddy, K. K., Falck, J. R. and Vogel, G. (1997). Diphospho-*myo*-inositol phosphates from *Dictyostelium* identified as D-6-diphospho-*myo*-inositol pentakisphosphate and D-5,6-bisdiphospho-*myo*-inositol tetrakisphosphate. Biochem. J. *322*, 31–33.

Lemmon, M. A. (1999). Phospholipids: regulators of membrane traffic and signaling. Biochem. Soc. Trans. *27*, 617–624.

Lemmon, M. A. and Ferguson, K. M. (2000). Signal-dependent membrane targeting by pleckstrin homology (PH) domains. Biochem. J. *350*, 1–18.

Lemtiri-Chlieh, F., MacRobbie, E. A. C. and Brearley, C. A. (2000). Inositol hexakisphosphate is a physiological signal regulating the K + -inward rectifying conductance in guard cells. Proc. Natl. Acad. Sci. USA *97*, 8687–8692.

Li, G., D'Souza-Schorey, C., Barbieri, M. A., Roberts, R. L., Klippel, A., Williams, L. T. and Stahl, P. D. (1995). Evidence for phosphatidylinositol 3-kinase as a regulator of endocytosis via activation of Rab5. Proc. Natl. Acad. Sci. USA *92*, 10207–10211.

Lietzke, S. E., Bose, S., Cronin, T., Klarlund, J., Chawla, A., Czech, M. P. and Lambright, D. G. (2000). Structural basis of 3-phosphoinositide recognition by pleckstrin homology domains. Molecular Cell *6*, 385–394.

Lips, D. L. and Majerus, P. W. (1989). The discovery of a 3-phosphomonoesterase that hydrolyzes phosphatidylinositol 3-phosphate in NIH 3T3 cells. J. Biol. Chem. *264*, 19911–19915.

Lips, D. L., Majerus, P. W., Gorga, F. R., Young, A. T. and Benjamin, T. L. (1989). Phosphatidylinositol 3-phosphate is present in normal and transformed fibroblasts and is resistant to hydrolysis by bovine brain phospholipase C II. J. Biol. Chem. *264*, 8759–8763.

Liscovitch, M. and Cantley, L. C. (1995). Signal transduction and membrane traffic: the PITP/phosphoinositide connection. Cell *81*, 659–662.

Liscovitch, M., Chalifa, V., Pertile, P., Chen, C-S. and Cantley, L. C. (1994). Novel functions of phosphatidylinositol 4,5-bisphosphate as a cofactor for brain membrane phospholipase D. J. Biol. Chem. *269*, 21403–21406.

Lizcano, J. M. and Alessi, D. R. (2002). The insulin signaling pathway. Curr. Biol. *12*, R236–R238.

Lockyer, P. J., Wennstrom, S., Kupzig, S., Venkatesvarlu, K., Downward, J. and Cullen, P. J. (1999). Identification of the ras GTPase-activating protein GAP1^m as phosphatidylinositol-3,4,5-trisphosphate-binding protein in vivo. Curr. Biol. *9*, 265–268.

Lohi, O. and Lehto, V. P. (1998). VHS domain marks a group of proteins involved in endocytosis and vascular trafficking. FEBS Lett. *440*, 255–257.

Loyet, K. M., Kowalchyk, J. A., Chaudhary, A., Chen, J., Prestwich, G. D. and Martin, T. F. J. (1998). Specific binding of phosphatidylinositol 4,5-bisphosphate to CAPS, a potential phosphoinositide effector protein for regulated exocytosis. J. Biol. Chem. *273*, 8337–8343.

Lu, P.-J. and Chen, C.-S. (1997). Selective recognition of phosphatidylinositol 3, 4,5-trisphosphate by a synthetic peptide. J. Biol. Chem. *272*, 466–472.

Luckhoff, A. and Clapham, D. E. (1992). Inositol 1,3,4,5-tetrakisphosphate activates an endothelial Ca^{2+}-permeable channel. Nature *355*, 356–358.

Luo, H. R., Saiardi, A., Yu, H., Nagata, E., Ye, K. and Snyder, S. H. (2002). Inositol pyrophosphates are required for DNA hyperrecombination in protein kinase C1 mutant yeast. Biochemistry *41*, 2509–2515.

Ma, Y. and Lieber, M. R. (2002). Binding of inositol hexakisphosphate (IP_6) to Ku but not to DNA-PK$_{cs}$. J. Biol. Chem. *277*, 10756–10759.

Ma, H-T., Venkatachlam, K., Li, H-S., Montell, C., Kurosaki, T., Patterson, R. L. and Gill, D. L. (2001). Asessment of the role of the inositol 1,4,5-trisphosphate receptor in the activation of transient receptor potential channels and store-operated Ca^{2+} entry channels. J. Biol Chem. *276*, 18888–18986.

MacLennan, D. H. (2000). Ca^{2+} signaling and muscle disease. Eur. J. Biochem. *267*, 5291–5297.

Magun, R., Burgering, B. M. T., Coffer, P. J., Pardasani, D., Lin, Y., Chabot, J. and Sorisky, A. (1996). Expression of a constitutively activated form of protein-kinase-β (c-akt) in 3T3-I1 predipose cells causes spontaneous differentiation. Endocrinology *137*, 3590–3593.

Majerus, P. W., Kisseleva, M. V. and Norris, F. A. (1999). The role of phosphatase in inositol signaling reactions. J. Biol. Chem. *274*, 10669–10672.

Mao, Y., Nickitenko, A., Duan, X., Lloyd, T. E., Wu, M. N., Bellen, H. and Quiocho, F. A. (2000). Crystal structure of the VHS and FYVE tandem domains of Hrs, a protein involved in membrane trafficking and signal transduction. Cell *100*, 447–456.

Maraldi, N. M., Marmirli, S., Rizzoli, R., Mazzotti, G. and Manzoli, F. A. (1999). Phosphatidylinositol 3-kinase translocation to the nucleus is an early event in the interleukin-1 signalling mechanism inhuman osteosarcoma Saos-2 cells. Adv. Enzyme Regul. *39*, 33–49.

Maraldi, N. M., Zini, N., Santi, S. and Manzoli, F. A. (1999). Topology of inositol lipid signal transduction in the nucleus. J. Cell Physiol. *181*, 203–217.

Maraldi, N. M., Zini, N., Santi, S., Riccio, M., Falconi, M., Capitani, S. and Mazoli, F. A. (2000). Nuclear domains involved in inositol lipid signal transduction. Adv. Enzyme Regul. *40*, 219–253.

Maroun, C. R., Moscatello, D. K., Naujokas, M. A., Holgado-Madrugas, M., Wong, A. J. and Park, M. (1999). A conserved inositol phospholipid binding site within the pleckstrin homology of the docking protein is required for epithelial morphogenesis. J. Biol. Chem. *274*, 31719–31726.

Marte, B. M. and Downward, J. (1997). PKB/Akt: connecting phosphoinositide 3-kinase to cell survival and beyond. TIBS *22* Trends Biochem. Sci. 22, 355–358.

Martin, S. S., Haruta, T., Morris, A. J., Klippel, A., Williams, L. T. and Olefsky, J. M. (1996). Activated phosphatidylinositol 3-kinase is sufficient to mediate actin rearrangement and GLUT4 translocation in adipocytes. J. Biol. Chem. *271*, 17605–17608.

Martin, T. F. J. (1998). Phosphoinositide lipids as signaling molecules: common themes for signal tansduction, cytoskeletal regulation, and membrane trafficking. Annu Rev. Cell Dev. Biol. *14*, 231–264.

Martin, T. F. (2001). PI(4,5)P$_2$ regulation of surface membrane traffic. Curr. Opin. Cell Biol. *13*, 493–499.

Martin, T. F. J., Loyet, K. M., Barry, V. A. and Kowalchyk, J. A. (1997). The role of PtdIns 4,5-bisphosphate in exocytic membrane fusion. Biochem. Soc. Trans. *25*, 113–1141.

Mattingly, R. R., Stephens, L. R., Irvine, R. F. and Garrison, J. C. (1991). Effects of transformation with the v-*src* oncogene on inositol phosphate metabolism in rat-1 fibroblasts. D-myo-Inositol 1,4,5,6-tetrakisphosphate is increased in v-*src* transformed rat-1 fibroblasts and can be synthesized from D-*myo*-inositol 1,3, 4-trisphosphate in cytosolic extracts. J. Biol. Chem. *266*, 15144–15153.

McPherson, P. S., Garcia, E. P., Slepnev, V. I., David, C., Zhang, X., Grabs, D., et al. (1996). A presynaptic inositol 5-phosphatase. Nature *379*, 353–357.

Mehrotra, B., Elliott, J. T., Che, J., Olszewski, J. D., Profit, A. A., Chaudhay, A., Fukuda, M., Mikoshiba, K. and Prestwich, G. D. (1997). Selective photoaffinity labeling of the inositol polyphosphate binding C2B domains of synaptotagmins. J. Biol. Chem. *272*, 4237–4244.

Mehrotra, B., Myszka, D. G. and Prestwich, G. D. (2000). Binding kinetics and ligand specificity for the interactions of the C2B domain of synaptotagmin II with inositol polyphosphates and phosphoinositides. Biochemistry *39*, 9679–9686.

Mejillano, M., Yamamoto, M., Rozelle, A. L., Sun, H-Q., Ang, X. and Yin, H. L. (2001). Regulation of apoptosis by phosphatidylinositol 4,5-bisphosphate inhibition of caspases, and caspase inactivation of phosphatidylinositol phosphate 5-kinases. J. Biol. Chem. *276*, 1865–1872.

Melo, J. and Toczyski, D. (2002). A unified view of the DNA-damage checkpoint. Curr. Opin. Cell Biol. *14*, 237–245.

Menniti, F. S., Oliver, K. G., Putney, J. W. Jr. and Shears, S. B. (1993a). Inositol phosphates and cell signaling: new views of $InsP_5$ and $InsP_6$. Trends Biochem. Sci. *18*, 53–56.

Menniti, F. S., Miller, R. N., Putney, J. W. Jr. and Shears, S. B. (1993b). Turnover of inositol polyphosphate pyrophosphates in pancreatoma cells. J. Biol. Chem. *268*, 3850–3856.

Mignery, G. A. and Sudhof, T. C. (1993). Molecular analysis of inositol 1,4,5-trisphosphate receptors. In: Lipid Metabolism in Signaling SystemsMethods in Neurosciences (Fain, J. N., eds.) Vol. 18. Academic Press, New York, pp. 247–265.

Mikoshiba, K., Fukuda, M., Ibata, K., Kabayama, H. and Mizutani, A. (1999). Role of synaptotagmin, a Ca^{2+} and inositol polyphosphate binding protein, in neurotransmitter release and neurite outgrowth. Chem. Phys. Lipids *98*, 59–67.

Misra, S. and Hurley, J. H. (1999). Crystal structure of a phosphatidylinositol 3-phosphate-specific membrane targeting motif, the FYVE domain of Vps27p. Cell *97*, 657–666.

Mitchell, R. H. (1989). Current Opinion in Cell Biology *1*, 201–205.

Mithieux, G., Daniele, N., Payrastres, B. and Zitoun, C. (1998). Live microsomal glucose-6-phosphatase is competitively inhibited by the lipid products of phosphatidylinositol 3-kinase. J. Biol. Chem. *273*, 17–19.

Molz, L., Chen, Y. W., Hirano, M. and Williams, L. T. (1996). Cpk is a novel class of Drosophila PtdIns 3-kinase containing a C2 domain. J. Biol. Chem. *271*, 13892–13899.

Morrison, B. H., Bauer, J. A., Kalvakolanu, D. V. and Lindner, D. J. (2001). Inositol hexakisphosphate kinase 2 mediates growth suppressive and apoptotic effects of interferon-β in ovarian carcinoma cells. J. Biol. Chem. *276*, 24965–24970.

Munger, S. D., Gleeson, R. A., Aldrich, H. C., Rust, N. C., Ache, B. W. and Greenberg, R. M. (2000). Characterization of a phosphoinositide-mediated odor transduction pathway reveals plasma membrane localization of an inositol 1,4,5-trisphosphate receptor in lobster olfactory receptor neurons. J. Biol. Chem. *275*, 20450–20457.

Munnik, T., Irvine, R. F. and Musgrave, A. (1998). Phospholipid signaling in plants. Review. Biochim. Biophys. Acta *1389*, 222–272.

Murthy, P. P. N. (1996). Inositol phosphates and their metabolism in plants. Subcell. Biochem. *26*, 227–255.

Nahorski, S. R. and Potter, B. V. (1989). Molecular recognition of inositol polyphosphates by intracellular receptors and metabolic enzymes. Trends Pharmacol. Sci. *10*, 139–144.

Nakanishi, H., Brewer, K. A. and Exton, J. H. (1993). Activation of the zeta isozyme of protein kinase C by phosphatidylinositol 3,4,5-trisphosphate. J. Biol. Chem. *268*, 13–16.

Nave, B. T., Haigh, R. J., Hayward, A. C., Siddle, K. and Shepherd, P. R. (1996). Compartment-specific regulation of phosphoinositide 3-kinase by platelet-derived growth factor and insulin in 3T3-L1 adipocytes. Biochem. J. *318*, 55–60.

Nishizuka, Y. (1984). The role of protein kinase C in cell surface signal transduction and tumor promotion. Nature (London) *308*, 693–697.

Nobes, C. D. and Hall, A. (1995). Rho, Rac and Cdc42 GTPases regulate the assembly of multimolecular focal complexes associated with actin stress fibers, lamellipodia and filipodia. Cell *81*, 53–62.

Norris, F. A., Ungewickell, E. and Majerus, P. W. (1995). Inositol hexakisphosphate binds to clathrin assembly protein 3 (AP-3/AP180) and inhibits clathrin cage assembly in vitro. J. Biol. Chem. *270*, 214–217.

Norris, F. A., Wilson, M. P., Wallis, T. S., Galyov, E. E. and Majerus, P. W. (1998). SopB, a protein required for virulence of *Salmonella dublin*, is an inositol phosphate phosphatase. Proc. Natl. Acad. Sci. USA *95*, 14006–14008.

Odom, R. A., Stahlberg, A., Wente, S. R. and York, J. D. (2000). A role for nuclear inositol 1,4,5-trisphosphate kinase in transcriptional control. Science *287*, 2026–2029.

Ohtsu, M., Sakai, N., Fujita, H., Kashiwagi, M., Gasa, S., Shimizu, S., Eguchi, Y., Tsujimoto, Y., Sakiyama, Y., Kobayashi, K. and Kuzumaki, N. (1997). Inhibition of apoptosis by the actin-regulatory protein gelsolin. EMBO J. *16*, 4650–4656.

Okada, T., Hazeki, O., Ui, M. and Katada, T. (1996). Synergistic activation of PtdIns 3-kinase by tyrosine-phosphorylated peptide and beta/gamma-subunits of GTP-binding proteins. Biochem. J. *317*, 475–480.

Okada, T., Inoue, R., Yamazaki, K., Maeda, A., Kurosaki, T., Yamakuni, T., Tanaka, I., Shimizu, S., Ikenaka, K., Imoto, K. and Mori, Y. (1999). J. Biol. Chem. *274*, 27359–27370.

Okada, T., Kawano, Y., Sakakibara, T., Hazeki, O. and Ui, M. (1994). Essential role of phosphatidylinositol 3-kinase in insulin-induced glucose transport and antilipolysis in rat adipocytes. J. Biol. Chem. *269*, 3568–3573.

Oliver, K. G., Putney, J. W. Jr., Obie, J. F. and Shears, S. B. (1992). J. Biol. Chem. *267*, 21528–21534.

Olsson, H., Martinez-Arias, W., Drobak, B. K. and Jergil, B. (1995). Presence of a novel form of phosphatidylinositol 4-kinase in rat liver. FEBS Lett. *361*, 282–286.

Ongusaha, P. P., Hughes, P. J., Davey, J. and Mitchell, R. H. (1998). Inositol hexakisphosphate in Saccharomyces pombe: synthesis from Ins(1,4,5)P$_3$ and osmotic regulation. Biochem. J. *335*, 671–679.

Palmer, R. H., Dekker, L. V., Woscholski, R., LeGood, J. A., Gigg, R. and Parker, P. J. (1995). Activation of PRK1 by phosphatidylinositol 4,5-bisphosphate and phosphatidylinositol 3,4,5-trisphosphate. A comparison with protein kinase C isotypes. J. Biol. Chem. *270*, 22412–22416.

Palmer, S., Hughes, K. T., Lee, D. Y. and Wakelam, M. J. O. (1988). Development of a novel, Ins(1,4,5)P3-specific binding assay. Its use to determined the intracellular concentration of Ins(1,4,5)P3 in unstimulated and vasopressin-stimulated rat hepatocytes. Cell Signalling *1*, 147–154.

Palmer, S. and Wakelam, M. J. O. (1989). Mass measurement of inositol phosphates. Biochim. Biophys. Acta *1014*, 239–246.

Paris, S., Beraud-Dufour, S., Robineau, S., Bigay, J., Antonny, B., et al. (1997). Role of protein-phospholipid interactions in the activation of ARF1 by the guanine nucleotide exchange factor Arno. J. Biol. Chem. *272*, 22221–22226.

Patki, V., Virbasius, J., Lane, W. S., Toh, B. H., Shpetner, H. S. and Corvera, S. (1997). Identification of an early endosomal protein regulated by phosphatidylinositol 3-kinase. Proc. Nat. Acad. Sci. USA *94*, 7326–7330.

Patki, V., Lawe, D. C., Corvera, S., Virbasius, J. V. and Chawla, A. (1998). A functional PtdIns(3)P-binding motif. Nature *394*, 433–434.

Pertile, P. and Cantley, L. C. (1995). Type 2 phosphatidylinositol 4-kinase is recruited to CD4 in response to CD4 cross-linking. Biochim. Biophys. Acta *1248*, 129–140.

Phillips-Mason, P. J., Raben, D. M. and Baldassare, J. J. (2000). Phosphatidylinositol 3-kinase activity regulates alpha-thrombin-stimulated G1 progression by its effect on cyclin D1 expression and cyclin-dependent kinase 4 activity. J. Biol. Chem. *275*, 18046–18053.

Putney, J. W., Jr (1990). Capacitative calcium entry revisited. Cell Calcium *11*, 611–624.

Putney, J. W. Jr. (1999). TRP, inositol 1,4,5-trisphosphate receptors, and capacitative calcium entry. Proc. Natl. Acad. Sci. USA *96*, 14669–14671.

Putney, J. W. Jr., Broad, L. M., Braun, F. J., Lievremont, J. P. and Bird, G. S. (2001). Mechanism of capacitative calcium entry. J. Cell Sci. *114*, 2223–2229.

Putney, J. W. Jr. and McKay, R. R. (1999). Capacitative calcium entry channels. BioEssays *21*, 38–46.

Pyrzynska, B., Serrano, M., Martinez-A, M. and Kaminska, B. (2002). Tumor suppressor p53 mediates apoptotic cell death triggered by cyclosporin A. J. Biol. Chem. *277*, 14102–14108.

Quick, M. W., Lester, H. A., Davidson, N., Simon, M. I. and Aragay, A. M. (1996). Desensitization of inositol 1,4,5-trisphosphate/Ca^{2+}-induced Cl^- currents by prolonged activation of G proteins in *Xenopus* ooctes. J. Biol. Chem. *271*, 32021–32027.

Rameh, L. E., Arvidsson, A., Carraway, K. L. III, Couvillon, A. D., Rathbun, G., Crompton, A., VanRenterghem, B., Czech, M. P., Ravichandran, S., Burakoff, S. J., Wang, D. S., Chen, C. S. and Cantley, L. C. (1997a). A comparative analysis of the phosphoinositide binding specificity of pleckstrin homology domains. J. Biol. Chem. *272*, 22059–22066.

Rameh, L. E., Tolias, K. F., Duckworth, B. C. and Cantley, L. C. (1997b). A new pathway for synthesis of phosphatidylinositol 4,5-bisphosphate. Nature *390*, 192–196.

Rameh, L. E. and Cantley, L. C. (1999). The role of phosphoinositide 3-kinase lipid products in cell function. J. Biol. Chem. *274*, 8347–8350.

Randazzo, P. A. (1997). Functional interaction of ADP-ribosylation factor I with phosphatidylinositol 4,5-bisphosphate. J. Biol. Chem. *272*, 7688–7692.

Rappoport, I., Miyazaki, M., Bll, W., Duckworth, B., Cantley, L. C., et al. (1997). Regulatory interactions in the recognition of endocytic sorting signals by AP-2 complexes. EMBO J. *16*, 2240–2250.

Raucher, D., Stauffer, T., Chen, W., Shen, K., Guo, S., York, J. D., Sheetz, M. P. and Meyer, T. (2000). Phosphatidylinositol 4,5-bisphosphate functions as a second messenger that regulates cytoskeleton-plasma membrane adhesion. Cell *100*, 221–228.

Reaves, B., Bright, N., Mullock, B. and Luzio, J. (1996). The effect of wortmannin on the localization of lysosomal type I integral membrane glycoproteins suggests a role for phosphoinositide 3-kinase activity in regulating membrane traffic late in the endocytic pathway. J. Cell Sci. *109*, 749–762.

Reif, K., Nobes, C. D., Thomas, G., Hall, A. and Cantrell, D. A. (1996). Phosphatidylinositol 3-kinase signals activate a selective subset of Rac/Rho-dependent effector pathways. Curr. Biol. *6*, 1445–1455.

Reiser, G. (1993). High-affinity inositol 1,3,4,5-tetrakisphosphate receptor from cerebellum. In: Lipid metabolism in signaling systems, Methods in Neurosciences, (Fain, J. N., eds.) 18. Academic press, New York, pp. 280–297.

Rhee, S. G. (1991). Inositol phospholipids-specific phospholipase C: interaction of the γ_1 isoform with tyrosine kinase. Trends Biochem. Sci. *16*, 297–301.

Rhee, S. G. and Bae, Y. S. (1997). Regulation of phosphoinositide-specific phospholipase C isozymes. J. Biol. Chem. *272*, 15045–15048.

Rodriguez-Viciana, P., Warne, P. H., Dhand, R., Vanhasebroeck, B., Gout, I., Fry, M. J., Waterfield, M. D. and Downward, J. (1994). Phosphatidylinositol-3-OH kinase as a direct target of Ras. Nature *370*, 527–532.

Rodriguez-Viciana, P., Warne, P. H., Vanhasebroeck, B., Waterfield, M. D. and Downward, J. (1996). Activation of phosphoinositide 3-kinase by interaction with Ras and by point mutation. EMBO J. *15*, 2442–2451.

Rohatgi, R., Nollau, P., Ho, H-Y. H., Kirschner, M. W. and Mayer, B. J. (2001). Nck and phosphatidylinositol 4,5-bisphosphate synergistically activate actin polymerization through the N-WASP-Arp2/3 pathway. J. Biol. Chem. *276*, 26448–26452.

Roth, M. G. and Sternweis, P. C. (1997). The role of lipid signaling in constitutive membrane traffic. Curr. Opin. Cell Biol. *9*, 519–526.

Safrany, S. T., Caffrey, J. J., Yang, X. and Shears, S. B. (1999). Diphosphoinositol polyphosphates: the final frontier for inositide research. Biol. Chem. *380*, 945–951.

Safrany, S. T. and Shears, S. B. (1998). Turnover of bis-diphosphoinositol tetrakisphosphate in a smooth muscle cell regulated by beta2-adrenergic receptors through a cAMP-mediated, A-kinase-independent mechanism. EMBO J. *17*, 1710–1716.

Saiardi, A., Caffrey, J. J., Snyder, S. H. and Shears, S. B. (2000a). Inositol polyphosphate multikinase (ArgRIII) determines nuclear mRNA export in *Saccharomyces cerevisiae*. FEBS Lett *468*, 28–32.

Saiardi, A., Caffrey, J. J., Snyder, S. H. and Shears, S. B. (2000b). The inositol hexakisphosphate kinase family. Catalytic flexibility and function in yeast vacuole biogenesis. J. Biol. Chem. 275, 24686–24692.

Saiardi, A., Erdjument-Bromage, H., Snowman, A. M., Tempst, P. and Snyder, S. H. (1999). Synthesis of diphosphoinositol pentakisphosphate by a newly identified family of higher inositol polyphosphate kinases. Curr. Biol. 9, 1323–1326.

Saito, K., Scharenberg, A. M. and Kinet, J-P. (2001). Interaction between the Btk pH domain and phosphatidylinositol-3,4,5-trisphosphate directly regulates Btk. J. Biol. Chem. 276, 16201–16206.

Sakisaka, T., Itoh, T., Miura, K. and Takenawa, T. (1997). Phosphatidylinositol 4, 5-bisphosphate phosphatase regulates the rearrangement of actin filaments. Mol. Cell. Biol. 17, 3841–3849.

Salim, K., Bottomley, M. J., Querfurth, E., Zvelebil, M. J., Gout, I., Scaife, R., Margolis, R. L., Gigg, R., Smith, C. I., Driscoll, P. C., et al. (1996). Distinct specificity in the recognition of phosphoinositides by the pleckstrin homology domains of dynamin and Bruton's tyrosine kinase. EMBO J. 15, 6241–6250.

Saltiel, A. R. and Pessin, J. E. (2002). Insulin pathways in time and space. Trends Cell Biol. 12, 65–71.

Santagata, S., Boggon, T. J., Baid, C. L., Gomez, C. A., Zhao, J., Shan, W. S., Myszka, D. G. and Shapiro, L. (2001). G-protein signaling through tubby proteins. Science 292, 2041–2050.

Sbrissa, D., Ikonomov, O. C. and Shisheva, A. (1999). PIKfyve, a mammalian ortholog of yeast Fab1p lipid kinase, synthesizes 5-phosphoinositides. Effect of insulin. J. Biol. Chem. 274, 21589–21597.

Sbrissa, D., Ikonomov, O. C. and Shisheva, A. (2000). PIKfyve lipid kinase is a protein kinase: downregulation of 5′-phosphoinositide product formation by autophosphorylation. Biochemistry 39, 15980–15989.

Sbrissa, D., Ikonomov, O. C. and Shisheva, A. (2002). Phosphatidylinositol 3-phosphate-interacting domains in PIKfyve. J. Biol. Chem. 277, 6073–6079.

Schell, M. J., Letcher, A. J., Brearley, C. A., Biber, J., Murer, H. and Irvine, R. F. (1999). PLUS (PI uptake stimulator) is an inositol hexakisphosphate kinase. FEBS Lett. 461, 169–172.

Schell, M. J., Erneux, C. and Irvine, R. F. (2001). Inositol 1,4,5-trisphosphate 3-kinase associates with F-actin and dendritic spines via its N terminus. J. Biol. Chem. 276, 37537–37546.

Schiavo, G., Gu, Q., Prestwich, G. D., Sollner, T. H. and Rothman, J. E. (1996). Calcium-dependent switching of the specificity of phosphoinositide binding to synaptotagmin. Proc. Natl. Acad. Sci. USA 93, 13327–13332.

Schorr, M., Then, A., Tahirovic, S., Hug, N. and Mayinger, P. (2001). The phosphoinositide phosphatase Sac1p controls trafficking of the yeast Chs3p chitin synthase. Curr. Biol. *11*, 1421–1426.

Schu, P. V., Takegawa, K., Fry, M. J., Stack, J. H., Waterfield, M. D. and Emr, S. D. (1993). Phosphatidylinositol 3-kinase encoded by yeast VPS34 gene essential for protein sorting. Science *260*, 88–91.

Sears, R. C. and Nevins, J. R. (2002). Signaling networks that link cell proliferation and cell fate. J. Biol. Chem. *277*, 11617–11620.

Serunian, L. A., Haber, M. T., Fukui, T., Kim, J. W., Hee, S. G., Lowenstein, J. M. and Cantley, L. C. (1989). Polyphosphoinositides produced by phosphatidyl-inositol 3-kinase are poor substrates for phospholipases C from rat liver and bovine brain. J. Biol. Chem. *264*, 17809–17815.

Shears, S. B. (1998). The versatility of inositol phosphates as cellular signals. Biochim. Biophys. Acta *1436*, 49–67.

Shears, S. B. (2001). Assessing the functional omnipotence of inositol hexaki-sphosphate. Cell Signalling *13*, 151–158.

Shepherd, P. R., Reaves, B. J. and Davidson, H. W. (1996). Phosphoinositide 3-kinases and membrane traffic. Trends Cell Biol. *6*, 92–97.

Shepherd, P. R., Withers, D. J. and Siddle, K. (1998). Phosphoinositide 3-kinase: the key switch mechanism in insulin signalling. Biochem. J. *333*, 471–490.

Shiavo, G., Gu, Q.-M., Prestwich, G. D., Soellner, T. H. and Rothman, J. E. (1996). Calcium dependent switching of the specificity of phosphoinositide binding to synaptotagmin. Proc. Natl. Acad. Sci., USA *93*, 13327–13332.

Shi, Y. (2001). A structural view of mitochondria-mediated apoptosis. Nat. Struct. Biol. *8*, 394–401.

Shi, Y. (2002). Mechanisms of caspase activation and inhibition during apoptosis. Mol. Cell *9*, 459–470.

Shibasaki, Y., Ishihara, H., Kizuki, N., Sano, T., Oka, Y. and Yzaki, Y. (1997). Massive actin polymerization induced by phosphatidylinositol 4-phosphate 5-kinase in vivo. J. Biol. Chem. *272*, 7578–7581.

Shisheva, A. (2001). PIKfyve: the road to PtdIns 5-P and PtdIns 3,5-P$_2$. Cell Biol. Int. *25*, 1201–1206.

Shpetner, H., Joly, M., Hartley, D. and Corvera, S. (1996). Potential sites of PI-3 kinase function in the endocytic pathway revealed by the PI-3 kinase inhibitor wortmannin. J. Cell Biol. *132*, 595–605.

Simonsen, A. and Stenmark, H. (2001). PX domains: attracted by phosphoinosi-tides. Nat. Cell Biol. *3*, (8) E179–E182.

Simonsen, A., Wurmser, A. E., Emr, S. D. and Stenmark, H. (2001). The role of phosphoinositides in membrane transport. Curr. Opin. Cell Biol. *13*, 485–492.

Sims, C. E. and Allbritton, N. L. (1998). Metabolism of inositol 1,4,5-trisphosphate and inositol 1,3,4,5-tetrakisphosphate by the oocytes of *Xenopus laevis*. J. Biol. Chem. *273*, 4052–4058.

Smith, C. D. and Chang, K. J. (1989). Regulation of brain phosphatidylinositol 4-phosphate kinase by GTP analogues. J. Biol. Chem. *264*, 310–320.

Smith, F. D. and Scott, J. D. (2002). Signaling complexes: junctions on the intracellular information superhighway. Curr. Biol. *12*, R32–R40.

Smith, P. M., Harmer, A. R., Letcher, A. and Irvine, R. F. (2000). The effect of inositol 1,4,5,-tetrakisphosphate on inositol trisphosphate-induced Ca^{2+} mobilization in freshly isolated and cultured mouse lacrimal acinar cells. Biochem. J. *347*, 77–82.

Smith, C. I., Islam, T. C., Mattson, P. T., Mohamed, A. J., Nore, B. F. and Vihinen, M. (2001). The Tec family of cytoplasmic tyrosine kinases: mammalian Btk, Bmx, Itk, Tec, Txk and homologs in other species. Bioessays *23*, 436–446.

Snyder, J. T., Rossman, K. T., Baumeister, M. A., Pruitt, W. M., Siderovski, D. P., Der, C. J., Lemmon, M. A. and Sondek, J. (2001). Quantitative analysis of the effect of phosphoinositide interactions on the function of Db1 family proteins. J. Biol. Chem. *276*, 45868–45875.

Sohn, R. H., Chen, J., Koblan, K. S., Bray, P. F. and Goldschmidt-Clermont, P. J. (1995). Localization of a binding site for phosphatidylinositol 4,5-bisphosphate on human profilin. J. Biol. Chem. *270*, 21114–21120.

Somwar, R., Niu, W., Kim, D. Y., Sweeney, G., Randhavwa, V. K., Huang, C., Ramlal, T. and Klip, A. (2001). Differential effects of phosphatidylinositol 3-kinase inhibition on intracellular signals regulating GLUT4 translocation and glucose transport. J. Biol. Chem. *276*, 46079–46087.

Specht, K. M. and Shokat, K. M. (2002). The emerging power of chemical genetics. Curr. Opin. Cell Biol. *14*, 155–159.

Sprong, H., van der Sluijs, and Van Meer, G (2001) How proteins move lipids and lipids.

Stack, J. H., Horazdovsky, B. and Er, S. D. (1995). Receptor-mediated protein sorting to the vacuole in yeast. Annu. Rev. Cell Dev. Biol. *11*, 1–33.

Stauffer, T. P., Ahn, S. and Meyer, T. (1998). Receptor-induced transient reduction in plasma membrane $PI(4,5)P_2$ concentration monitored in living cells. Curr. Biol. *8*, 343–346.

Stenmark, H. (2000). Membrane traffic: Cycling lipids. Curr. Biol. *10*, R57–R59.

Stenmark, H. (2000). Phosphatidylinositol 3-kinase and membrane trafficking. In: Biology of Phosphoinositides (Crockcroft, S., ed.). Oxford University Press, Oxford, pp. 239–267.

Stenmark, H. and Aasland, R. (1999). FYVE-finger proteins—effectors of an inositol lipid. J. Cell Science *112*, 4175–4183.

Stenmark, H., Aasland, R., Toh, B. H. and D'Arrigo, A. (1996). Endosomal localization of the autoantigen EEA1 is mediated by a zinc-binding FYVE finger. J. Biol. Chem. *271*, 24048–24054.

Stephens, L. R., Berrie, C. P. and Irvine, R. F. (1990). Agonist-stimulated inositol phosphate metabolism in avian erythrocytes. Biochem. J. *269*, 65–72.

Stephens, L. R., Cooke, F. T., Walters, R., Jackson, T., Volinia, S., et al. (1994). Characterization of a phosphatidylinositol-specific phosphoinositide 3-kinase from mammalian cells. Curr. Biol. *4*, 203–214.

Stephens, L., Eguinoa, A., Corey, S., Jackson, T. and Hawkins, P. T. (1993). Receptor stimulated accumulation of phosphatidylinositol (3,4,5)-trisphosphate b G-protein mediated pathways in human myeloid derived cells. EMBO J. *12*, 2265–2273.

Stephens, L. R., Eguinoa, A., Erdjument-Bromage, H., Lui, M., Cooke, F., Coadwell, J., Smrcka, A. S., Thelen, M., Cadwallader, K., Tempst, P. and Hawkins, P. T. (1997). The G beta gamma sensitivity of a PI3K is dependent upon a tightly associated adaptor, p101. Cell *89*, 105–114.

Stephens, L., Anderson, K., Stokoe, D., Erdjument-Bromage, H., Painter, G. F., Holmes, A. B., Gaffney, P. R. J., Reese, C. B., McCormick, F., Tempst, P., Coldwell, J. and Hawkins, P. T. (1998). Protein kinase B kinases that mediate phosphatidylinositol 3,4,5-trisphosphate-dependent activation of protein kinase B. Science *279*, 710–714.

Stephens, L., Ellson, C. and Hawkins, P. (2002). Roles of PI3Ks in leukocyte chemotaxis and phagocytosis. Curr. Opin. Cell Biol. *14*, 203–213.

Stephens, L. R., Hawkins, P. T. and Downes, C. P. (1989a). Metabolic and structural evidence for the existence of a third species of polyphosphoinositide in cells: D-phosphatidyl-*myo*-inositol 3-phosphate. Biochem. J. *259*, 267–276.

Stephens, L. R., Hawkins, P. T. and Downes, C. P. (1989b). An analysis of *myo*-[^{3}H]inositol trisphosphates founding *myo*-[^{3}H]inositol prelabelled avina erythrocytes. Biochem. J. *262*, 727–737.

Stephens, L. R., Hawkins, P. T., Stanley, A. F., Moore, T., Poyner, D. R., Morris, P. J., Hanley, M. R., Kay, R. R. and Irvine, R. F. (1991). *myo*-Inositol pentakisphosphates. Structure, biological occurrence and phosphorylation to *myo*-inositol hexakisphosphate. Biochem. J. *275*, 485–499.

Stephens, L., McGregor, A. and Hawkins, P. (2000). In: Biology of Phosphoinositides, (Cockcroft, S., ed.). Oxford University Press, Oxford, UK, pp. 32–107.

Sternweis, P. C. and Smrcka, A. V. (1993). G proteins in signal transduction: the regulation of phospholipase C. Ciba Found. Symp. *176*, 96–106.

Stevenson, J. M., Perera, I. Y. and Boss, W. F. (1998). A phosphatidylinositol 4-kinase pleckstrin homology that binds phosphatidylinositol 4-monophospate. J. Biol. Chem. *273*, 22761–22767.

Stokoe, D., Stephens, L. R., Copeland, T., Gaffney, P. R., Reese, C. B., Painter, G. F., Holmes, A. B., McCormick, F. and Hawkins, P. T. (1997). Dual role of phosphatidylinositol-3,4,5-trisphosphate in the activation of protein kinase B. Science *277*, 567–570.

Stolz, L. E., Kuo, W. J., Longchamps, J., Sekhon, M. and York, J. D. (1998). INP51, a yeast inositol polyphosphate 5-phosphatase required for phosphatidylinositol 4,5-bisphosphate homeostasis and whose absence confers a cold-resistant phenotype. J. Biol. Chem. *273*, 11852–11861.

Stoyanov, B., Volinia, S., Hanck, T., Rubio, I., Loubtchenko, M., Malek, D., Stoyanova, S., Vanhaesebroeck, B., Dhand, R., Nurnberg, B., et al. (1995). Cloning and characterization of a G protein-activated human phosphoinositide 3-kinase. Science *269*, 690–693.

Streb, H., Irvine, R. F., Berridge, M. J. and Schulz, I. (1983). Release of Ca^{2+} from a nonmitochondrial intracellular store in pancreatic acinar cells by inositol-1,3,5-trisphosphate. Nature *306*, 67–68.

Stricker, R., Hulser, E., Fischer, J., Jarchau, T., Walter, U., Lottspeich, F. and Reiser, R. (1997). cDNA cloning of porcine p42IP4, a membrane-associated and cytosolic 42 kDa inositol(1,3,4,5)tetrakisphosphate receptor from pig brain with similarly high affinity for phosphatidylinositol (3,4,5)P₃. FEBS Lett. *405*, 229–236.

Stubbs, E. B., Kelleher, J. A. and Sun, G. Y. (1988). Phosphatidylinositol kinase, phosphatidylinositol 4-phosphate kinase and diacylglycerol kinase activities in rat brain subcellular fractions. Biochim. Biophys. Acta *958*, 247–254.

Sudhof, T. C. (1995). The synaptic vesicle cycle: a cascade of protein–protein interactions. Nature *375*, 645–653.

Supattapone, S., Danoff, S. K., Theibert, A., Joseph, S. K., Steiner, J. and Snyder, S. H. (1988). Cyclic AMP-dependent phosphorylation of a brain inositol trisphosphate receptor decreases its release of calcium. Proc. Natl. Acad. Sci. USA *85*, 8747–8750.

Sutton, R. B., Davletov, B. A., Berghuis, A. M., Sudhof, T. C. and Sprang, S. R. (1995). Structure of the first C2 domain of synaptotagmin I: a novel Ca^{2+}/ phospholipid-binding fold. Cell *80*, 929–938.

Takenawa, T. and Itoh, T. (2001). Phosphoinositides, key molecules for regulation of actin cytoskeletal organization and membrane traffic from the plasma membrane. Biochim. Biophys. Acta *1533*, 190–206.

Tan, Z., Bruzik, K. S. and Shears, S. B. (1997). Properties of the inositol 3,4,5,6-tetrakisphosphate 1-kinase purified from rat liver. J. Biol. Chem. *272*, 2285–2290.

Tanaka, K., Imajoh-Ohmi, S., Sawada, T., Shirai, R., Hashimoto, Y., Iwasaki, S., Kaibuchi, K., Kanaho, Y., Shirai, T., Terada, Y., et al. (1997). A target of phosphatidylinositol 3,4,5-trisphosphate with a zinc finger motif similar to that of the ADP-ribosylation-factor GTPase-activating protein and two pleckstrin homology domains. Eur. J. Biochem. 245, 512–519.

Tang, X., Downes, C. P., Whetton, A. D. and Owen-Lynch, P. J. (2000). Role of phosphatidylinositol 3-kinase and specific protein kinase B isoforms in the suppression of apoptosis mediated by the Abelson protein-tyrosine kinase. J. Biol. Chem. 275, 13142–13148.

Tapon, N. and Hall, A. (1997). Rho, rac and Cdc42 GTPases regulate the organization of the actincytoiskeleton. Curr. Opin. Cell Biol. 9, 86–92.

Taylor, C. W. (1998). Inositol trisphosphate receptors: Ca^{2+}-modulated intracellular Ca^{2+} channels. Biochim. Biophys. Acta 1436, 19–33.

Theibert, A. B., Estevez, V. A., Mourey, R. J., Maracek, J. F., Barrow, R. K., Prestwich, G. D. and Snyder, S. H. (1992). Photoaffinity labeling and characterization of isolated inositol 1,3,4,5-tetrakisphosphate- and inositol hexakisphosphate-binding proteins. J. Biol. Chem. 267, 9071–9079.

Theibert, A. B., Prestwich, G. D., Jackson, T. R. and Hammonds-Odie, L. P. (1997). The purification and assay of inositide-binding proteins. In: Signalling by Inositides: A Practical Approach (Shears, S., ed.). Oxford Press, New York, pp. 117–150.

Thomsen, P., Roepstorff, K., Stahlhut, M. and van Deurs, B. (2002). Caveolae are highly immobile plasma membrane microdomains, which are not involved in constitutive endocytic trafficking. Mol. Biol. Cell 13, 238–250.

Toker, A. (1998). The synthesis and cellular roles of phosphatidylinositol 4,5-bisphosphate. Curr. Opin. Cell Biol. 10, 254–261.

Toker, A. and Cantley, L. C. (1997). Signaling through the lipid products of phosphoinositide-3-OH kinase. Nature (London) 387, 673–676.

Toker, A., Meyer, M., Reddy, K. K., Flack, J. R., Aneda, R., Aneda, S., Parra, A., Burns, D. J., Ballas, L. M. and Cantley, L. C. (1994). J. Biol. Chem. 269, 32358–32367.

Tolias, K. F. and Carpenter, C. L. (2000). Enzymes involved in Hormone signaling via G-protein: regulation of phosphatidylinositol 4,5-bisphosphate hydrolysis by Gq. Philos. Trans (R. Soc. Lond.) B Biol. Sci. 336, 35-41 the synthesis of PtdIns(4,5)P_2 and their regulation: PdIns kinases and PtdInsP kinases. In Biology of Phosphoinositides (S. Crockcroft, ed.) Oxford University Press, pp. 109–130.

Tolias, K. F., Cantley, L. C. and Carpenter, C. L. (1995). Rho family GTPases bind to phosphoinositide kinases. J. Biol. Chem. 270, 17656–17659.

Tolias, K. F., Rameh, L. E., Ishihara, H., Shibasaki, Y., Chen, J., Prestwich, G. D., Cantley, L. C. and Carpenter, C. L. (1998). Type I phosphatidylinositol-

4-phosphate 5-kinases synthesize the novel lipids phosphatidylinositol 3, 5-bisphosphate and phosphatidylinositol 5-phosphate. J. Biol. Chem. *273*, 18040–18046.

Tolias, K. F., Hartwig, J. H., Ishihara, M., Shibasaki, Y., Cantley, L. C. and Carpenter, C. L. (2000). Type 1α phosphatidylinositol-4-phosphate 5-kinase mediates Rac-dependent actin assembly. Curr. Biol. *10*, 153–156.

Tran, D., Gascard, P., Berthon, B., Fukami, K., Takenawa, T., Giraud, F. and Claret, M. (1993). Cellular distribution of polyphosphoinositides in rat hepatocytes. Cell. Signal. *5*, 565–581.

Tsakiridis, T., McDowell, H. E., Walker, T., Downes, C. P., Hundal, H. S., Vranic, M. and Klip, A. (1995). Multiple roles of phosphatidylinositol 3-kinase in regulation of glucose transport, amino acid transport, and glucose transporters in L6 skeletal muscle cells. Endocrinology *136*, 4315–4322.

Tsakiridis, T., Taha, C., Grinstein, S. and Klip, A. (1996). Insulin activates a p21-activated kinase in muscle cells via phosphatidylinositol 3-kinase. J. Biol. Chem. *271*, 19664–19667.

Tsunoda, S., Sierralta, J., Sun, Y., Bodner, R., Suzuki, E., Becker, A., Socolich, M. and Zuker, C. S. (1997). A multivalent PDZ-domain protein assembles signaling complexes in a G-protein-coupled cascade. Nature *388*, 243–249.

Tsuruta, F., Masuyama, N. and Gotoh, Y. (2002). The phosphatidylinositol 3-kinase (PI3K)-Akt pathway suppresses Bax translocation to mitochondria. J. Biol. Chem. *277*, 14040–14047.

Vajanaphanich, M., Schultz, C., Rudolf, M. T., Wasserman, M., Enyedi, M. T., Craxton, A., Shears, S. B., Tsien, R. Y., Barrett, E. and Tranor-Kaplan, A. (1994). Long-term uncoupling of chloride secretion from intracellular calcium levels by Ins(3,4,5)P$_4$. Nature *371*, 711–714.

Valhmu, W. B. and Raia, F. J. (2002). *myo*-Inositol 1,4,5-trisphosphate and Ca^{2+}/ calmodulin-dependent factors mediate transduction of compression-induced signals in bovine articular chondrocytes. Biochem. J. *361*, 689–696.

Valius, M. and Kazlauskas, A. (1993). Phospholipase C-gamma 1 and phosphatidylinositol 3 kinase are the downstream mediators of the PDGF receptor's mitogenic signal. Cell *73*, 321–334.

Van der Kaay, J., Batty, I. H., Cross, D. A. E., Watt, P. W. and Downes, C. P. (1997). A novel, rapid, and highly sensitive mass assay for phosphatidyl-inositol 3,4,5-trisphosphate (PtdIns(3,4,5)P3) and its application to measure insulin-stimulated PtdIns (3,4,5)P3 production in rat skeletal muscle in vivo. J. Biol. Chem. *272*, 5477–5481.

Van der Kaay, J., Wesseling, J. and Van Haastert, P. J. (1995). Nucleus-associated phosphorylation of Ins(1,4,5)P$_3$ to InsP$_6$ in *Dictyostelium*. Biochem. J. *312*, 911–917.

Vanhaesebroeck, B., Leevers, S. J., Ahmadi, K., Timms, J., Katso, R., Dsriscoll, R. C., Woscholski, R., Parker, P. J. and Waterfield, M. D. (2001). Synthesis and function of 3-phosphorylated inositol lipids. Annu. Rev. Biochem. 70, 535–602.

Vanhaesebroeck, B., Leevers, S. J., Panayotu, G. and Waterfield, M. D. (1997a). Phosphoinositide 3-kinases: a conserved family of signal transducers. Trends Biochem. Sci. 22, 267–272.

Vanhaesebroeck, B., Welham, M. J., Kotani, K., Sten, R., Warme, P. H., Zvelebil, M. J., Higashi, K., Volinia, S., Downward, J. and Wateffield, M. D. (1997b). p110δ, a novel phosphoinositide 3-kinase in leukocytes. Proc. Natl. Acad. Sci. USA 94, 4330–4335.

Venkateswarlu, K., Oatey, P. B., Tavare, J. M. and Cullen, P. J. (1998). Insulin-dependent translocation of ARNO to the plasma membrane of adipocytes requires PI 3-kinase. Curr. Biol. 8, 463–466.

Vlahos, C. J., Matter, W. F., Hui, K. Y. and Brown, R. F. (1994). A specific inhibitor of phosphatidylinositol 3-kinase, 2-(4-morpholinyl)-8-phenyl-4-H-1-benzopyran-4-one (LY294002). J. Biol. Chem. 269, 5241–5248.

Voglmaier, S. M., et al. (1992). Inositol hexakisphosphate receptor identified as the clathrin assembly protein AP-2. Biochem. Biophys. Res. Commun. 187, 158–163.

Voglmaier, S. M., Bembenek, M. E., Kaplin, A. I., Dorman, G., Olszewski, J. D., Prestwich, G. D. and Snyder, S. H. (1996). Purified inositol hexakisphosphate kinase is an ATP synthase: diphosphoinositol pentakisphosphate as a high energy phosphate donor. Proc. Natl. Acad. Sci. USA 93, 4305–4310.

Voglmaier, S. M., Bembenek, M. E., Kaplin, A. I., Dorman, G., Olszewski, J. D., Prestwich, G. D. and Snyder, S. H. (1997). In: Signaling by Inositides: A Practical Approach (Shears, S. B., ed.). IRL Press, Oxford, pp. 195–201.

Vollenweider, P., et al. (1999). An SH2 domain-containing 5'-inositolphosphatase inhibits insulin-induced Glut4 translocation and growth factor-induced actin filament rearrangement. Mol. Cell. Biol. 19, 1081–1091.

Voorhout, W. F., van Genderen, I. L., Yoshioka, T., Fukami, K., Geuze, H. J. and van Meer, G. (1992). Subcellular localization of glycolipids as revealed by immunoelectromicroscopy. Trends Glycosci. Glycotechnol. 4, 533–546.

Watson, R. T., Shigematsu, S., Chiang, S-H., Mora, S., Kanzaki, M., Macara, I. G., Saltiel, A. R. and Pessin, J. E. (2001). Lipid raft microdomain compartmentalization of TC10 is required for insulin signaling and GLU4 translocation. J. Cell Biol. 154, 829–840.

Watt, S. A., Kular, G., Fleming, I. N., Downes, C. P. and Lucocq, J. M. (2002). Subcellular localization of phosphatidylinositol 4,5-bisphosphate using the pleckstrin homology domain of phoslipase C δ1. Biochem. J. 363, 657–666.

Wendland, B., Steece, K. E. and Emr4, S. D. (1999). Yeast pepsins contain an essential N-terminal ENTH domain, bind clathrin and are required for endocytosis. EMBO J. *18*, 4383–4393.

Whiteford, C. A., Brearley, C. A. and Ulug, E. T. (1997). Phosphatidylinositol 3,5-bisphosphate defines a novel PI 3-kinase pathway in resting mouse fibroblasts. Biochem. J. *323*, 597–601.

Willcocks, A. L., Cooke, A. M., Potter, B. V. and Nahorski, S. R. (1987). Stereospecific recognition sites for [^{3}H]inositol(1,4,5)-trisphosphate in particulate preparations of rat cerebellum. Biochem. Biophys. Res. Commun. *146*, 1071–1078.

Wilson, M. P. and Majerus, P. W. (1996). Isolation of inositol 1,3,4-trisphosphate 5/6 kinase, cDNA cloning, and expression of the recombinant enzyme. J. Biol. Chem. *271*, 11904–11910.

Wishart, M. J., Taylor, G. S. and Dixon, J. E. (2001). Phoxy lipids: revealing PX domains as phosphoinositide binding molecules. Cell *105*, 817–820.

Wojcikiewicz, R. J. H. (1995). Type I, II and III inositol 1,4,5-trisphosphate receptors are uniquely susceptible to down-regulation and are expressed in markedly different proportions in different cell types. J. Biol. Chem. *270*, 11678–11683.

Wojcikiewicz, R. J. and Nahorski, S. R. (1991). Prolonged stimulation of SH-SY5Y cells with carbachol inhibits inositol 1,4,5-trisphosphate (InsP3)binding and InsP3-induced Ca^{2+} mobilization. Biochem. Soc. Trans. *19*, 97S.

Worley, P. F., Baraban, J. M., Mc Carren, M., Snyder, S. H. and Alger, B. E. (1987). Cholinergic phosphatidylinositol modulation of inhibitory, G protein-linked neurotransmitter actions: electrophysiological studies in rat hippocampus. Proc. Natl. Acad. Sci. USA *84*, 3467–3471.

Worley, P. F., Baraban, J. M., Colvin, J. S. and Snyder, S. H. (1987). Inositol trisphosphate receptor localization in brain: variable stoichiometry with protein kinase C. Nature *325*, 159–161.

Worley, P. F., Baraban, J. M., Supattapone, S., Wilson, V. S. and Snyder, S. H. (1987). Characterization of inositol trisphosphate receptor binding in brain. Regulation by pH and calcium. J. Biol. Chem. *262*, 12132–12136.

Woscholski, R. and Parker, P. J. (1997). Inositol lipid 5-phosphatases—traffic signals and signal traffic. Trends Biochem. Sci. *22*, 427–431.

Wurmser, A. E. and Emr, S. D. (1998). Phosphoinositide signaling and turnover: PtdIns(3)P, a regulator of membrane traffic is transported to the vacuole and degraded by a process that requires lumenal vacuolar hydrolase activities. EMBO J. *17*, 4930–4942.

Wurmser, A. E., Gary, J. D. and Emr, S. D. (1999). Phosphoinositide 3-kinases and their FYVE domain-containing effectors as regulators of vacuolar/lysosomal membrane trafficking pathways. J. Biol. Chem. *274*, 9129–9132.

Xie, W., Kaetzel, M. A., Bruzik, K. S., Dedman, J. R., Shears, S. B. and Nelson, D. J. (1996). Inositol 3,4,5,6-tetrakisphosphate inhibits the calmodulin-dependent protein kinase II-activated chloride conductance in T84 colonic epithelial cells. J. Biol. Chem. *271*, 14092–14097.

Xu, Y., Horstman, H., Seet, L., Wong, S. H. and Hong, W. (2001). SNX3 regulates endosomal function through its PX domain-mediated interaction with PtdIns(3)P. Nat. Cell Biol. *3*, 658–666.

Yamamoto, T-A., Takeuchi, H., Kanematsu, T., Allen, V., Yagisawa, H., Kikkawa, U., Watanabe, Y., Nakasima, A., Katan, M. and Hirata, M. (1999). Involvement of EF hand motifs in the Ca^{2+}-dependent binding of the pleckstrin homology domain to phosphoinositides. Eur. J. Biochem. *265*, 481–490.

Yang, X., Rudolph, M., Yoshida, M., Carew, M. A., Riley, A. M., Chung, S-K., Bruzik, K. S., Potter, B. V. L., Schultz, C. and Shears, S. B. (1999). Inositol 1, 3,4-trisphosphate acts in vivo as a specific regulator of cellular signaling by inositol 3,4,5,6-tetrakisphosphate. J. Biol Chem. *274*, 18973–18980.

Yang, X. and Shears, S. B. (2000). Multitasking in signal transduction by a promiscuous human Ins(3,4,5,6)P$_4$1-kinase/Ins(1,3,4)P$_3$ 5/6 kinase. Biochem. J. *351*, 551–555.

Yauch, R. L. and Hemler, M. E. (2000). Specific interactions among transmembrane 4 superfamily (TM4SF) proteins and phosphoinositide 4-kinase. Biochem. J. *351*, 629–637.

Ye, W., Ali, N., Bembenek, M. E., Shears, S. B. and Lafer, E. M. (1995). Inhibition of clathrin assembly by high affinity binding of specific inositol polyphosphates to the synapse-specific clathrin assembly protein AP-3. J. Biol. Chem. *270*, 1564–1568.

Yeh, J., Gulve, E. A., Rameh, L. and Birnbaum, M. J. (1995). The effects of wortmannin on rat skeletal muscle. Dissociation of signaling pathways for insulin- and contraction-activated hexose transport. J. Biol. Chem. *270*, 2107–2111.

Yonezawa, N., Homma, Y., Yahara, I., Sakai, H. and Nishida, E. (1991). A short sequence responsible for both phosphoinositide binding and actin binding activities of cofilin. J. Biol. Chem. *266*, 17218–17221.

York, J. D., Xiong, J. P. and Spiegelberg, B. (1998). Nuclear inositol signalling: a structural and functional approach. Adv. Enzyme Regul. *38*, 365–374.

York, J. D., Odom, R. A., Murphy, R., Ives, E. B. and Wente, S. R. (1999). A phospholipase C-dependent inositol phosphate kinase pathway required for efficient messenger RNA export. Science *285*, 96–100.

York, J. D., Guo, S., Odom, A. R., Spiegelberg, B. D. and Stolz, L. E. (2001). An expanded view of inositol signaling. Adv. Enzyme Regul. *41*, 57–71.

Yoshida, Y. and Imai, S. (1997). Structure and function of inositol 1,4,5-trisphosphate receptor. Review. Jpn. J. Pharmacol. *74*, 125–137.

Yu, F.-X., Sun, H.-Q., Janmey, P. A. and Yin, H. L. (1992). Identification of a polyphosphoinositide-binding sequence in an actin monomer-binding domain of gelsolin. J. Biol. Chem. 267, 14616–14621.

Yue, G., Malik, B., Yue, G. and Eaton, D. C. (2002). Phosphatidylinositol 4, 5-bisphosphate (PIP$_2$) stimulates epithelial sodium channel activity in A6 cells. J. Biol. Chem. 277, 11965–11969.

Zhang, L. (1998). Inositol 1,4,5-trisphosphate mass assay. In: Phospholipid Signaling Protocols (Bird, I. M., ed.). Humana Press, Totowa, NJ, pp. 77–87.

Zhang, X., Loijens, J. C., Boronenkov, I. V., Parker, G. J., Norris, F. A., Chen, J., Thum, O., Prestwich, G. D., Majerus, P. W. and Anderson, R. A. (1997). Phosphatidylinositol-4-phosphate 5-kinase isozymes catalyze the synthesis of 3-phosphate-containing phosphatidylinositol signaling molecules. J. Biol. Chem. 272, 17756–17761.

Zheng, L., Shan, J., Krishnamoorthi, R. and Wang, X. (2002). Activation of plant phospholipase Dβ by phosphatidylinositol 4,5-bisphosphate: characterization of binding site and mode of action. Biochemistry 41, 4546–4553.

Zhou, D., Chen, L. M., Hernandez, L., Hears, S. B. and Galan, J. E. (2001). A Salmonella inositol polyphosphatase acts in conjunction with other bacterial effectors to promote host cell actin cytoskeleton rearrangements and bacterial internalization. Mol. Microbiol. 39, 248–260.

Zhu, C-C., Furuichi, T., Mikoshiba, K. and Wojcikiewicz, R. J. H. (1999). Inositol 1,4,5-trisphosphate receptor down-regulation is activated directly by inositol 1, 4,5-trisphosphate binding. J. Bio. Chem. 274, 3476–3484.

Zhu, D-M., Teke, E., Huang, C. Y. and Chock, P. B. (2000). Inositol tetrakisphosphate as frequency regulator in calcium oscillations in HeLa cells. J. Biol. Chem. 275, 6063–6066.

Zubov, A. I., Kaznacheeva, E. V., Nikolaev, A. V., Alexeenko, V. A., Kiselyov, K., Muallem, S. and Mozhayeva, G. N. (1999). Regulation of the miniature plasma membrane Ca^{2+} channel I(min) by inositol 1,4,5-trisphosphate receptors. J. Biol. Chem. 274, 25983–25985.

Subject Index

A isoforms 703–704
AcChE *see* acetylcholine esterase
acetic acid 42
acetolysis 297
acetylation 172
acetylcholine esterase (AcChE) 305, 306
acetylcholine esterase glycosyl-phosphatidylinositols 323
acid extracts 176–179
acid phosphatases 539
actin 479–480, 729–730, 754
acyl esters 384–385
acyl group exchange 345–348
adipocytes 362
ADP-ribosylation factor (ARF) 615, 740, 744–747
ADP/ATP elimination 190
adrenals 240
affinity assays 607–612
affinity chromatography 127–128
    see also high performance liquid chromatography; high performance thin-layer chromatography
affinity reagents 688
Agaricus spp. 10, 329
Akt/PKB 756–757, 759, 762–763
alkaline hydrolysis 50
alkaline phosphatases 537
alkenylacyl moieties 152
alkylacylglycerols 152, 316–320
alkylglycerol composition 303–304
altritol enantiomers 226, 227, 228
aluminium chloride 589

AM esters 687
amino columns 122
analytical evidence 691–696, 720–722
anchors *see* glycosyl phosphatidyl-inositol-anchored proteins
anion exchange cartridge chromatography 47–48, 124–125, 180, 182–185
    see also strong anion exchange; weak anion exchange
AP-2/AP-3 *see* clathrin adaptor proteins
APCI *see* atmospheric pressure chemical ionization
apoptosis 756–762
Arabidopsis spp. 351, 413, 432, 449, 733
arabinogalactans 331–334
arabitols 231
arachidonic acid 552, 553, 554, 555
ARF *see* ADP-ribosylation factor
Aspergillus spp. 532–533
assay methods 577–579, 591–599, 604–606, 619–626, 629–632
atmospheric pressure chemical ionization (APCI) 72
autoradiography 165–167
B isoforms 703–704
Bacillus spp.
    B. cereus 56, 63, 83, 561–564, 588
    B. thuringiensis 56, 294, 296, 586, 590
    phytases 537, 539, 540

bacterial phosphatidylinositol-PLC 581, 586–588
bee venom phospholipase A_2 576–577, 578–579
benzoate derivatives 89–93
benzyl esters 299
binding 695, 696–702, 726–734
biological effects 9–14, 740–764
biosynthesis
  glucosyl PtdIns glycolipids 401
  glycoinositol phospholipids 401–402
  GPtdIns 376–402
  GPtdIns protein anchors 377–398
  inositol phospholipids 335–402
  protein-free GPtdIns 398–399
  PtdIns 336–352
  PtdIns bisphosphates 365–373
  PtdIns monophosphates 353–365
  PtdIns phosphates 352–376
  PtdIns trisphosphates 373–376
  PtdIns(3,4,5)P_3 373–376
  PtdIns(3,4)P_2 369–371
  PtdIns(3,5)P_2 371–373
  PtdIns(3)P 359–363
  PtdIns(4,5)P_2 366–368
  PtdIns(4)P 354–359
  PtdIns(5)P 363–365
bis-diphosphoinositol tetrakisphosphate 3
bis-pyrophosphate-inositol tetrakisphosphate 462
BODIPY-PtdCho 620–621
bound glycosyl phosphatidylinositol anchors 263–267
bovine adrenals 241–242
bovine brain 123, 145, 146, 150, 159–161, 357, 442
bovine liver 150

brain
  Ca^{2+}/CaM-sensitive InsP$_3$ 3-kinase 442
  InsP$_5$ hydroxy-kinase activity 455
  InsPs identification 192–193
  InsPs isolation 177
  PP-InsP$_5$ kinase 462
  PtdIns phospholipases 617, 618
  PtdIns/PtdInsPs major molecular species 159, 160, 161
  PtdIns(4)P synthesis 357
  PtdInsP mass spectra 145, 146, 150
  PtdInsPs HPLC 122
  PtdInsPs isolation 123
Btk 727
n-butanol 40

calcium activated chloride channels 700, 701, 705, 706, 713, 716
calcium activation 703–705
calcium mobilization 693, 695, 697–698, 700, 701, 711–715
calcium signaling 176
calcium-independent phospholipases 546–547, 551–553, 560
calcium/calmodulin sensitivity 443–444
calf cerebellum 241–242
calmodulin 441, 442, 443, 444, 701–703, 703–705, 753
capacitative Ca^{2+} entry hypothesis 712
CAPS 744
cartridge extraction 47–48, 124–125, 180, 182–185
catalytic assay 608, 611–612
Catharathus roseus 572
CD52 antigen 292–294
CD52-glycosyl phosphatidylinositol-peptide 276
cell differentiation 747–751

cell growth 747–751
cell lysates assay 411–412
cell motility 752–756
cellular proteins assays 420–422
cellular signaling 689–764
  evidence for 691–696
  Ins phospholipids 335–336
  InsP–protein interaction 696–702
  InsP$_6$ receptors 702
  InsPs 691–719
  ion channel physiology 711–715
  membrane dynamics 715–718
  pathways 10–13
  PLC 689–691
  PtdInsPs 719–764
ceramides 30, 327, 328, 391, 394, 614
cerebellum 241–242
CF-LS-IMS *see* continuous flow liquid
  secondary ion mass
  spectrometry
CHAPS detergent 698–699
charcoal 178
chemical analysis 129–132, 207–211,
  326
chemical families 179–191
chemical labeling 262–263
chemical structure determination
  50–54, 207–223, 267–307
chiro-inositol phosphatides 586
chiro-inositols 6, 20–23, 28–29,
  52–54, 57, 95–96
chloride channels 700, 701, 705, 706,
  713, 716
chloroform/methanol 38–40
choline release 621–622
chromatography
  see also gas chromatography;
  gas–liquid chromatography;
  high performance liquid
  chromatography; high
  performance thin-layer

chromatography; thin-layer
  chromatography
diradylglycerols 315–322
GPtdIns 311–322
GPtdIns anchors 324–326
PLC assay 591–594, 597
PtdInsPs 135–142
CID *see* collision induced dissociation
Clarke and Dawson's deacylation
  reagent 662
Classes of phosphatidylinositol
  3-kinases 404–406
clathrin adaptor proteins 696
collision induced dissociation (CID)
  59, 73, 283–285
commercial inositol(1,4,5)trisphos-
  phate radioreceptor assay
  kits 240–242
conformational inositol phosphate
  diagrams 208
continuous flow liquid secondary ion
  mass spectrometry
  (CF-LSIMS) 77–78, 80
COS-7, HEK-293 cells 624–625
cPLA$_1$ *see* cytosolic phospholipase A$_1$
cPLA$_2$ *see* cytosolic phospholipase A$_2$
cPLD *see* cytosolic PLD
CTP *see* cytidinetriphosphate
cultured cells 109–110
cycles *see* metabolic cycles
cysteine phosphatases *see* acid
  phosphatases
cytidinediphosphate-diacylglycerol
  336–345
cytidinediphosphate-diacylglycerol
  synthetase 337, 338, 339,
  340, 341–345
cytidinetriphosphate (CTP) 341–345
cytidylyltransferase 340
cytoskeletal organization 752–756
cytosolic phospholipase D (cPLD) 615

cytosolic phospholipase A₁ (cPLA₁)
    545
cytosolic phospholipase A₂ (cPLA₂)
    551, 553–554, 606–608
cytosolic phospholipase A₂-IV assay
    558
cytosomal phospholipase A₂ 627

DAF *see* decay accelerating factor
DAG *see* sn-1,2-diacylglycerol;
    diacylglycerol
Dbi homology (DH) domain 733
de novo phosphatidylinositol synthesis
    336–345
deacetylation 382–383
deacylated phospholipids 126–127,
    130
deacylation reagent 662
decay accelerating factor (DAF) 277
deglyceration 130–132
deoxy-*myo*-inositol phosphates
    685–686
dephosphorylation 132
desalting process 237
detergent-resistant membrane (DRM)
    fragments 260
DFH *see* Dbi homology domain
DG *see* diacylglycerol
di-C₈-GroPInsP₃/AM 687
di-C₁₂-GroPInsP₃/AM 687
diacylglycerol acetates 295
sn-1,2-diacylglycerol (DAG) 11, 13,
    587
diacylglycerol kinase 629
diacylglycerols
    assay 566–568
    benzoate derivatives 92–93
    enantiomers 300, 301
    GPtdIns protein anchors
        remodeling 391–393
    HPLC and LC/MS 87, 88–89

PLC 94
PtdInsPs 158–159
quantification 99–101
Dictyostelium spp. 6, 8, 202, 289
    GIns phosphoceramides 327, 328
    Ins hexakisphosphates 205
    Ins(1,3,4,5,6)P₅ 2-kinase
        453–457
    Ins(3)P phosphatases 502
    InsP₄ 5-phosphatases 527
    InsP₆ biosynthesis 694
    InsP₆/InsP₅ cycle 708
    InsPP phosphatases 534
    InsPP standards 683–685
    InsPs NMR 213, 215, 216, 221
    InsPs stereochemical structure 230,
        232–233
    PLD activity 568
    PP cycle 710
    PP-InsP₅ kinase 234–235
    PtdIns 3-kinase 407
dinitrophenylurethane (DNPU)
    derivatives 94, 95,
        300–301
diphosphoinositol pentakisphosphate
    3, 4, 683–684
    kinase 234–235
diphosphoinositol polyphosphate
    phosphohydrolase (DIPP)
        535–536
diphosphoinositol tetrakisphosphate
    3, 4
    kinase 459–460
diradyl GroPIns 312–313
diradyl GroPOH 313–315
diradylglycerobenzoate derivatives
    321–322
diradylglycerols 83–95, 294–302,
    315–322
discovery of phospholipids 1–6
DNA synthesis 762–764

DNPU *see* dinitrophenylurethane
Dole procedure 558, 560
DRM *see* detergent-resistant
    membrane fragments
Drosophila spp. 337

EI spectrum 218–219
electron spray ionisation mass
    spectrometry (ESI/MS) 61,
    98, 276
electron spray ionisation tandem mass
    spectrometry (ESI/MS/MS)
    61, 143–148
electrophoresis 190
enantiomers
    see also stereochemistry
    arabitols 231
    chiro-inositols 6, 20–23, 28–29,
        52–54, 57, 95–96
    diacylglycerols 300, 301
    Ins pentakisphosphates 200–205
    Ins resolution 141
    InsPs 208, 213, 215, 223–235
endosome-to-vacuole transport 746
Entamoeba histolytica 31, 215
enzyme proteins 723–725
enzymic methods 54–59, 132–135,
    250
epithelial cells 706
epsin amino-terminal homology
    (ETNH) domain 726
erythrocytes
    acetylcholine esterase
        glycosyl-PtdIns 323
    diradylglycerobenzoate derivatives
        321
    diradylglycerol benzoates 90–91
    diradylglycerols 85–87
    GPtdIns anchor 305–306
    InsPs anion exchange cartridge
        chromatography 183

PtdIns(4)P 5-kinase 367–368
PtdInsPs 117
Escherichia coli 54, 337, 531, 532,
    544–545
ESI/MS *see* electron spray ionisation
    mass spectrometry
ESI/MS/MS *see* electron spray
    ionisation tandem mass
    spectrometry
ET-18-OCH$_3$ *see* 1-*O*-octadecyl-2-*O*-
    methyl-*rac*-glycero-3-
    phosphocholine
ethanolamine 260–262, 271–278
ETNH *see* epsin amino-terminal
    homology domain
eukaryotic cells 253–334
evidence, signaling 691–696,
    720–725
exocytic fusion 745
exoglycosidase digestion 280–282
extracellular phospholipase A$_2$
    see secretory
    phospholipase A$_2$
extraction, free GPtdIns-anchors
    256–258

fast atom bombardment/mass
    spectrometry (FAB/MS)
    InsPs 219–222
    intact PtdIns 71–72
    intact PtdInsPs 153–155
    proteolysed GPtdIns-anchored
        proteins 274–275, 277, 283
    PtdIns 59, 71–72
    PtdInsPs 142–143, 153–155
    PtdOH 81–82
fast atom bombardment/tandem mass
    spectrometry (FAB/MS/MS)
    82
fatty acids
    cPLA$_2$ selectivity 627

GPtdIns protein anchors
  remodeling 391–393
PLD 612
positional distribution in PtdInsPs
  159, 161–162
PtdIns 51–52, 66–70, 346–347
quantification 101
fatty chains 302–306
FERM domain 725
flow injection ES/MS 72–73
fluorescent compounds 251,
  686–688
free glycosyl phosphatidylinositol-
  anchors 255, 256–258
FYVE domain 725, 733

galactosamine 271–272
GAP1^{IP4BP} 698–700
GAP1^{IPA4BP} 705–706
gas chromatography/mass spectrometry (GC/MS) 84–85,
  222–223, 247–248
gas–liquid chromatography (GLC)
  84–87, 223
gelsolin 618–619, 729, 757–758
gene expression 464, 602–603
GIPLs see glycoinositolphospholipids
Glc6Pase see glucose-6-phosphatase
GlcN-phosphatidylinositol 382–383
GlcN-phosphatidylinositol acyl esters
  384–385
GlcN-phosphatidylinositol
  mannosides 385–388
GlcN-phosphatidylinositol(Man)$_3$EtnP
  388–389
GlcNAc-phosphatidylinositol
  379–382
glucitols 229
glucosamine 271–272
glucosamine-inositol linkage 279–280
glucose-6-phosphatase (GlcPase) 760

glucosyl phosphatidylinositol
  glycolipids 401
glycan moiety 280–288
glycerol content 51
L-α-glycerophospho-D-myo-inositol-
  4,5-bisphosphate (GPMI-P$_2$)
  731
glycerophosphoinositol
  lipid species 142–143
  phosphates 136–140, 372–373,
    638–639, 647–648
glycerophospholipids 542–543
glycoinositolphospholipids (GIPLs)
  268, 401
glycosyl phosphatidylinositol specific
  phospholipase D (GPI-PLD)
  313
glycosyl phosphatidylinositol
  transamidase complex 271
glycosyl phosphatidylinositol-
  anchored proteins 323–334
  Bacillus thuringiensis PtdIns-PLC
    cleavage 590
  biosynthesis 379–398
  biosynthetic transfer to protein
    390–398
  bound 263–267
  CD52 antigens 292–294
  chemical labeling 262–263
  chemical structure determination
    267–307
  exoglycosidase digestion
    280–282
  glucosamine-inositol linkage
    279–280
  glycan moiety 280–288
  isolation 259–260
  metabolic labeling 260–262
  protein-ethanolamine bridge
    271–278
  release of bound 263–267

remodeling 391–398
Saccharomyces cerevisiae Golgi
    apparatus 395–397
species-specific 254
glycosyl phosphatidylinositol-
    phospholipase D
    (GPtdIns-PLD) 601–603,
    604, 605
glycosyl phosphatidylinositol-specific
    phosphatidylinositol
    phospholipases
    575–606
glycosyl phosphatidylinositols
    (GPtdIns) 253–334,
    376–402, 601
  chromatography/mass
    spectrometry 311–322
  diradylglycerol moiety chemical
    structure 294–302
  discovery 3, 6
  extraction 256–258
  fatty chains chemical structure
    302–306
  glycan moiety 280–288
  Ins moiety 307
  isolation 254–255
  molecular species determination
    307–322
  natural occurrence 8–9,
    254–255
  PtdIns moieties 288–294
  purification 258–259
  quantification 322–326
  structure 5
glycosylated phosphatidylinositols
    (GPIs) 268
glycosylinositol phosphoceramides
    326–330
Golgi apparatus 356, 395–397
GPIs *see* glycosylated phosphatidyl-
    inositols

GPMI-P$_2$ *see* L-α-glycerophospho-
    D-*myo*-inositol-4,5-
    bisphosphate
GPtdIns *see* glycosyl phosphatidyl-
    inositols
GPtdIns-anchored proteins *see* glyco-
    syl phosphatidylinositol-
    anchored proteins
GPtdIns-PLD *see* glycosyl
    phosphatidylinositol-
    phospholipase D
GroPIns *see* glycerophosphoinositol
group X sPLA$_2$ *see* X sPLA$_2$
GTPase-activating protein 699–700
GTPases 434–435, 438
guard cells 708–709, 715

'hard' ionization methods 143
hDIPP *see* human diphosphoinositol
    polyphosphate phospho-
    hydrolases
He-La cells 599
head-group exchange 348–352
heptafluorobutyric anhydride (HFBA)
    249
hexane/isopropanol 40, 41
HFBA *see* heptafluorobutyric
    anhydride
high performance liquid
    chromatography (HPLC)
  desalting process 237
  diradylglycerols 88–95
  InsPs 185–190, 245, 246
  normal phase 73–76, 82–83
  PLC assay 596
  polyols 209–211
  prior to MS 62
  PtdIns extraction 45, 47
  PtdInsPs 121–126, 138–139,
    140
  reversed phase 76–79, 82, 89

high performance liquid chromatography metal–dye detector
(HPLC-MDD) 198, 201
high performance thin-layer
chromatography (HPTLC)
42, 44–45, 48, 111–121,
181–182, 592–593
high-pH anion chromatography
(HPAEC) 287
HPLC *see* high performance liquid
chromatography
HPTLC *see* high performance
thin-layer chromatography
human diphosphoinositol polyphosphate phosphohydrolases
(hDIPP) 536
human erythrocyte phosphatidylinositol 4-kinase 638
human erythrocyte plasma membrane
139
human plasma lipoproteins 99–100
human platelets 108, 122–123, 172,
637, 742
hydrofluoric acid 267

identification, PtdIns 48–96
IDH *see* myo-inositol dehydrogenase
L-iditol dehydrogenase 225, 226, 228
iditols 227
immunoprecipitates assay 408–411
in vitro PLD assay 619–624
in vivo PLD assay 624–626
inositol 20–23, 52, 95–96, 101–104,
162, 307
inositol 5-phosphatase catalytic
(IPP5C) domain 522
inositol bisphosphate 195–196,
655–660
1-phosphatase 505–506
3-phosphatases 506–508

4-kinases 441
4-phosphatases 508–510
5-kinases 441
5-phosphatases 510
6/3 kinase 441–442
inositol glycerophospholipids
15–19
inositol hexakisphosphates 205, 207,
682–683, 694, 708–710,
717, 719
3-phytase 532
5-kinase 458–461
5-phytase 532–533
6-phytase 531–532
InsP$_6$/Ins$_5$ cycle 708–710
kinases 457–462
phosphatases 531–533
receptors 702
inositol monophosphates 194–195,
654–655
kinases 439
phosphatases 497–504
inositol pentakisphosphates 200–205,
230, 676–682, 708–710
1-kinases 453–454
1-phosphatases 528
2-kinases 454–457
3-kinases 457
3-phosphatases 528–530
5/6 kinases 457
kinases 453–457
phosphatases 528–531
pyrophosphokinase 457
inositol phosphates 141,
175–252
1-kinases 440
6-kinases 441
anion exchange cartridge
chromatography 182–185
arabitols 231
Ca^{2+} signaling 176

chemical structure 207–223
conformational diagrams 208
electrophoresis 190
FAB/MS 219–222
HPLC 185–190
HPTLC 181–182
hydrolysis 1, 3
InsP–protein interaction 696–702
isolation 176–179
isotope studies 235–243
kinases 232–233, 702–710
mass assays 239, 243–252
natural occurrence 7–8
NMR 212–217
nomenclature 19–30
nucleotides removal 178
optical rotation 224
phosphatases 233–235, 463–465,
    497–536
phytases 8
polyol dehydrogenases 224–231
positional isomers 191–207
quantification 235–252
radio-receptor assays 239–243
signaling 689–764
standard preparation 653–683
stereochemical structure 223–235
TLC 180–182
inositol phospholipids 335–402, 464
inositol pyrophosphates 8, 27, 30–31,
    205, 207, 683–685
    phosphatases 534–536
inositol sphingophospholipids 30
inositol tetrakisphosphates 198–200,
    668–676, 699–700,
    705–707
    1-kinases 450–451
    2/4 kinases 452
    3-kinases 452
    3-phosphatases 526
    4-kinases 452

4-phosphatases 526
5-kinases 452
5-phosphatases 527–528
kinases 450–452
phosphatases 526–528
inositol trisphosphates 196–198,
    660–668, 692–694, 695,
    697
    1-kinases 442
    1-phosphatase 511–513
    3-kinases 442–445
    3-phosphatases 513–514
    4-phosphatases 514–516
    5-phosphatases 516–526
    6/3 kinases 445–448
    6/5 kinase 448–450
    kinases 442–450
    phosphatases 510–526
inositol(1,2,3,4,5)pentakisphosphate
    679–680
inositol(1,2,3,4,6)pentakisphosphate
    682
D/L-inositol(1,2,3,4,6)pentakis-
    phosphate 5/6-kinase 457
inositol(1,2,3,4)tetrakisphosphate 668
inositol(1,2,3,5,6)pentakisphosphate
    680
inositol(1,2,4,5,6)pentakisphosphate
    679–681
    3-kinase 457
inositol(1,2,4,5)tetrakisphosphate
    669
inositol(1,2)bisphosphate 656
inositol(1,2cyc4,5)trisphosphate 667
inositol(1,3,4,5,6)pentakisphosphate
    681–682
    2-kinase 454–457
    6-phosphatases 531
Ins(1,4,5,6)P$_4$ cycle 707–708
Ins(3,4,5,6)P$_4$ cycle 705–707
pyrophosphokinase 457

inositol(1,3,4,5)tetrakisphosphate 669–671
  receptors 699–700
inositol(1,3,4,6)tetrakisphosphate 671–673
inositol(1,3,4)trisphosphate 660–661
  6/5-kinase 448–450
inositol(1,3,5,6)tetrakisphosphate 2/4-kinase 452
inositol(1,3,5)trisphosphate 662
  6-kinase 450
inositol(1,3,6)trisphosphate 662
inositol(1,3)bisphosphate 656–658
  5-kinase 441
inositol(1,4,5,6)tetrakisphosphate 673, 707–708
inositol(1,4,5)trisphosphate 662–666, 692–694, 695, 697
  3-kinase 442–445
  6/3 kinase 445–448
  Ins(1,3,4,5)P$_4$ cycle 703–705
  receptors 697–699
inositol(1,4,6)trisphosphate 666
inositol(1,4)bisphosphate 658–659
inositol(1,5)bisphosphate 659
inositol(1)phosphate 654
  phosphatases 497–502
inositol(2,3,4,5,6)pentakisphosphate 1-kinase 453–454
inositol(2,5,6)trisphosphate 3-kinase 445
inositol(2,6)bisphosphate 5-kinase 441
inositol(2)phosphate 654
  6-kinases 440
inositol(3,4,5,6)tetrakisphosphate 674–676
  1-kinase 450–451
  receptors 700–701
inositol(3,4,5)trisphosphate 666
inositol(3,4,6)trisphosphate 666
  1-kinase 442

inositol(3,4)bisphosphate 656–658
inositol(3,5)bisphosphate 659
inositol(3,6)bisphosphate 4-kinases 441
inositol(3)phosphate 655
  1-kinase 439
  phosphatases 502–504
inositol(4,5,6)trisphosphate 667
inositol(4,5)bisphosphate 659–660
inositol(4)phosphate 655
Ins see inositol
InsP see inositol phosphate
InsP$_7$ see diphosphoinositol pentakisphosphate
InsP$_8$ see bis-diphosphoinositol tetrakisphosphate
InsPs see inositol phosphates
insulin 747–751, 753
intact glycosyl phosphatidylinositol 310–312
intact phosphatidylinositol 70–72, 77, 98–99
  phosphates 135–136, 153–158
intact protein 308–309
intracellular phospholipase A$_2$ 551
ion channel physiology 711–715
ion exchange HPLC 126–127
ion-pair chromatography 189–190
IPP5C see inositol 5-phosphatase catalytic domain
isolation
  GPtdIns 255
  GPtdIns-anchored proteins 259–260
  InsPs 176–179, 206
  PtdIns 37–48
  PtdInsPs 107
isopropanol, PtdIns extraction 40, 41
isotope studies
  arabitols 231
  inositol pentakisphosphates 203

Ins kinases 232–233
InsP$_5$ isomers 230
InsPs quantification 235–243
InsPs as signaling molecules
    691–693
InsPs stereochemistry 224–231
PLC assay 591–592, 594, 596
PLD substrate specificity
    622–623
PtdInsPs 136, 155, 169
isozymes 580, 581–582, 584,
    614–615, 618, 628

kinases 11–12, 31–33, 133–134
    see also individual kinases
Kinetoplastida spp. 9

lamellapodia formation 752
LC/ES/MS, intact PtdIns
    quantification 98
Leishmania spp. 9, 268, 269
    GPtdIns biosynthesis 399–400
    GPtdIns-related structures 398
    lipoarabinomannans 333
    lipophosphoglycans in GPtdIns
        biosynthesis 398–399
    lipophosphophoglycans 331
lipases *see* phosphatidylinositol
    phospholipases
lipoarabinomannans 331–334
lipophosphoglycans 330–331,
    398–399
liposomes 547–548
liquid chromatography/mass
    spectrometry (LC/MS)
    88–95, 98
liquid–liquid partition 549
lithium 511
LY294002 inhibitor 740
lysophospholipases 549–550

McCorkindale-Edson rules 226
Malachite method 501, 539
malate dehydrogenase (MDH) 249
MALDI-TOF-MS *see* matrix-assisted
    laser desorption and ion-
    isation time-of-flight-mass-
    spectrometry
mammalian cells
    GlcN-PtdIns acyl esters synthesis
        383
    GlcN-PtdIns mannosides synthesis
        386
    GPtdIns protein anchors synthesis
        377–378
    PtdIns 4-kinases 412–413
    PtdIns-PLC 579–581
    PtdIns(3,5)P$_2$ synthesis 371–372
    PtdInsPs 107
    PtdInsPs biosynthetic pathways
        summary 354, 355–357
Man α1-4GlcNα1-6*myo*-inositol-1-
    PO$_4$-DG structural motif
    see glycosylphosphatidyl-
    inositols
Manα1-4GlcNH$_2$α1-6 *myo*-inositol
    core region 278–279
mannosides 385–387
mass assays
    Ins(1)P phosphatases 501–502
    Ins(3)P phosphatases 503–504
    InsP$_2$
        1-phosphatase 506
        3-phosphatases 508
        4-phosphatases 509–510
    InsP$_3$
        1-phosphatase 512–513
        3-phosphatases 514
        4-phosphatases 516
        5-phosphatases 526
    InsP$_4$ 5-phosphatases
        527–528

InsP$_5$ 1-phosphatases 528
InsP$_6$ 5-phytase 533–534
InsPs 239, 243–252, 692
PPInsP$_5$ pyrophosphatases
536–537
PtdIns(4)P 4-phosphatase 474
PtdInsP 471–472
PtdInsP$_2$
3-phosphatases 475
4-phosphatases 477–478
5-phosphatases 483
PtdInsP$_3$
3-phosphatases 488–489
4-phosphatases 490
mass spectrometry
diradylglycerols 315–322
GPtdIns anchors 323–324
GPtdIns molecular species
308–311
InsPs 217
PtdIns structure 59–62
PtdInsPs 141–151
mass spectrometry/collision-induced
dissociation/mass
spectrometry (MS/CID/MS)
analysis 284–286
matrix-assisted laser desorption and
ionisation time-of-flight-
mass-spectrometry
(MALDI-TOF/MS) 62,
148–151, 155–158, 165
MDD *see* metal–dye-detection
membrane dynamics 715–718, 737
membrane permeators 686–687
membrane preparations 620
membrane ruffling 752
membrane trafficking 741–747
membrane-bound PLC 581, 595
metabolic cycles 702–710
metabolic evidence 693–694,
722–723

metabolic intermediates 13–14
metabolic labeling 260–262
metal–dye-detection (MDD)
Ins trisphosphates 197–198
InsP 198, 201
kinases 232
mass assays 244–245
phosphatases 233–235
InsP$_3$
1-phosphatase 512
5-phosphatases 526
InsP$_4$ 5-phosphatases 527
Methanococcus jannaschii 500
methanol 38–40
micelles 416–417
MIDH *see myo*-inositol
dehydrogenase
MIKE spectra 290–291
Minpp1 530, 533
MIPP *see* multiple inositol
polyphosphate phosphatase
modular structures 405
molecular signaling evidence
695–696, 723–725
molecular species
AcChE GPtdIns 323
diradylglycerols 83–95
GPtdIns 307–322
PtdIns 70–96
PtdInsPs 151–162
mouse fibroblasts 372–373
MTMR3 *see* myotubularin-related
protein 3
multiple inositol polyphosphate
phosphatase (MIPP)
530–531, 694, 707
mung beans 230
Mycobacterium smegmatis 500
myo-inositols
dehydrogenase (MIDH) 249

hexakisphosphate (phytate)
531–533
history 1, 3
D-*myo*-inositol-1-phosphate 129
myo-inositol(1,4,5)trisphosphate
665
nomenclature 20–23, 26–30
occurrence 6, 95
phosphatases 497–498
phosphates 227, 228, 229, 231
PtdIns incorporation mechanisms
348–352
quantitative analysis 102–104, 222
myo-phosphatidylinositol 337–339,
340, 341
myocytes 183
myotubularin 467–468
myotubularin-related protein 3
(MTMR3) 469, 470, 474

N-acetylglucosaminyl-phosphatidyl-
inositol 379–382
N-deacetylase 382–383
N-WASP protein 754–756
NAD/NADH 102, 104, 225, 228, 229,
249, 250
naphthylethyl urethanes 88, 89
naphthylurethanes 299
natural occurrence 6–9, 254–255
negative chemical ionization (NCI) 93
negative ion CF-LSIMS 80
negative-ion electrospray mass spectra
144–146
neo-inositol 20–21, 30
neo-inositol phosphates 215, 217
neomycin columns 127–128
neurogranin 727
neurospora 40
neurotransmitters 743–744
neutral glycans 280–288
neutral losses 218

neutralization of acid extracts
178–179
NICI GC/MS 96, 97
NIE-115 cells 197
NIH-3T3 fibroblasts 419
nitrous acid deamination 266–267,
290–291
nomenclature 14–33
InsGPLs 15–19
InsPPs 27, 30–31
InsPs 19–30
kinases 31–33
numbering of phosphate groups
24–30
phosphatases 31–33
phospholipases 31–33
positional isomerism 19
PtdInsPs 19–30
stereoisomerism 16–19,
20–23
non-cyclic *myo*-inositol phosphates
229, 231
non-polar capillary GLC and GC/MS
84, 85, 86
normal phase HPLC and LC/MS
73–76, 82–83
nuclear magnetic resonance
spectrosopy (NMR) 62–65,
212–217, 585
nuclear signaling 718–719
nucleotide removal 178

occurrence, natural 6–9, 254–255
OCRL *see* oculocerebrorenal
1-*O*-octadecyl-2-*O*-methyl-*rac*-
glycero-3-phosphocholine
(ET-18-OCH₃) 761–762
oculocerebrorenal (OCRL) syndrome
479, 518, 519, 521
one-dimensional thin-layer

chromatography 42–43,
112, 113
optical rotation 224
organic phosphorus 96
ox brain 88, 159, 160, 161

p53 tumor suppressor 761
pancreatic lipase 546
papain digestion 271–273
paper chromatography 191
PAPs *see* purple acid phosphatases
PAR *see* 4-(2-pyridylazo) resorcinol
Paramecium primaurelia 10, 327,
398
PARP *see* procyclic acidic repetitive
protein
PCA *see* perchloric acid
PCI *see* positive chemical ionization
PDGF *see* platelet-derived growth
factor
Penicillium notatum 546
perchloric acid (PCA) 176
periodate oxidation 278–279
PH *see* pleckstrin homology domains
phagosomal maturation 746
phosphatases 11–12, 31–33, 133,
463–540
see also individual phosphatases
phosphatidic acids 158
phosphatidylcholine (PtdCho)
630–631
phosphatidylinositol 3-kinase 407
apoptosis 756–762
cell growth 747–751
cell motility 752–756
cytoskeletal organization
752–756
DNA synthesis 762–764
Ca²⁺ signaling 715
PtdIns(3)P biosynthesis 359–363
regulation 734–735

phosphatidylinositol 4-kinase
354–358, 412–416,
737–738
phosphatidylinositol 5-kinase
363–365, 419–423,
738–740
phosphatidylinositol bisphosphates 1,
2, 3, 365–373, 645–651,
729–732
3-phosphatases 474–475
4-phosphatases 475–478
5-phosphatases 478–483
-dependent PLD activity 622
kinases 434–439
phosphatases 474–483
phosphatidylinositol kinases 404–423
FYVE 738–739
D-3/D-4 phosphatidylinositol lipids
116, 118–119
phosphatidylinositol monophosphates
353–365, 635–645
kinases (PIPKINS) 423–434
phosphatases 465–474
phosphatidylinositol phosphates
107–173, 732–734
3-kinases 424–427
4-kinases 427–430
5-kinases 430–434
affinity chromatography 127–128
analytical evidence 720–722
biosynthesis 352–376
cellular signaling 719–764
chemical analyses 129–132
chromatographic analysis 135–142
deacylation 130
deglyceration 130–132
dephosphorylation 132
discovery 1
enzymic structure analysis methods
132–135
ES/MS/MS 143–148

FAB/MS 142–143
fatty acids positional distribution
    159, 161–162
HPLC 121–126
ion exchange HPLC 126–127
isolation 107
kinases 734–740
MALDI-TOF/MS 148–151
mass spectrometry 141–151
membrane/vesicular trafficking
    741–747
molecular species 151–162
natural occurrence 7
nomenclature 19–30
phosphatases 463–497
protein interactions 725–734
quantification 162–173
radioreceptor assay 167–171
signaling 689–764
solid phase extraction 111–128
solvent extraction 107–111
specific biological effects 740–764
standard preparation 634–653
structure determination 128–151
TLC and HPTLC 111–121
phosphatidylinositol trisphosphates
    373–376, 651–653,
    726–729
3-phosphatases 484–489
4-phosphatases 489–490
5-phosphatases 490–497
phosphatases 483–497
phosphatidylinositol-nonspecific
    phospholipases 542–575
phosphatidylinositol-specific enzymes
    phosphatidylinositol
        phospholipases 575–606
    phospholipase C 294–297,
        582–591
    phospholipase D 629–632
phosphatidylinositol(3,4,5)

trisphosphate 373–376, 633,
    651–653
-activated phospholipases 626–632
phosphatidylinositol(3,4)bisphosphate
    369–371, 633, 649–651
5-kinase 437–439
phosphatidylinositol(3,5)bisphosphate
    371–373, 648–649
4-kinase 437
phosphatidylinositol(3)phosphate
    359–363, 633, 639–643
3-phosphatase 465–472
4-kinase 369, 427–428
5-kinase 430–432
phosphatidylinositol(4,5)bisphosphate
    366–368, 645–648
3-kinase 434–437
-activated phospholipases 606–626
phosphatidylinositol(4)phosphate
    354–359, 635–639
3-kinase 425–426
4-phosphatase 472–474
5-kinase 366–368, 420, 422–423,
    432–434
phosphatidylinositol(5)phosphate
    363–365, 420–422,
    644–645
3-kinase 426–427
4-kinase 367, 420, 428–430
5-phosphatase 474
phosphatidylinositols 37–106
biosynthesis 336–352
cartridge extraction 47–48,
    124–125, 180, 182–185
chemical structure determination
    50–54
de novo synthesis 336–345
diacylglycerols quantification
    99–101
enzymic structure determination
    54–59

FAM/MS 59, 71–72
fatty acids 66–70, 101, 346–347
flow injection ES/MS 72–73
head-group exchange 348–352
history 1, 6
identification 48–96
inositol isomers 95–96
inositols content 101–104
isolation 37–48
mass spectrometry 59–62
moieties 288–294, 311
molecular species determination
    70–96
NMR studies 62–65
normal phase HPLC and LC/MS
    73–76
phospholipases 541–632
phosphorus content 105–106
PLC two-step hydrolysis 587
quantification 96–106
remodeling 345–352
reversed phase HPLC and LC/MS
    76–79
solid phase extraction 42–48
solvent extraction 37–42
structure determination 49–65
synthase 337–339, 340
TLC and HPTLC extraction 42–45
phospholipase C γ (PLC γ), Ca^{2+}
    signaling 714
phospholipase C (PLC) 542–543,
    561–568
activation mechanisms 627–629
affinity assays 609–612
assay methods 565–568, 591–599,
    611–612
catalytic assay 611–612
characteristics 579–599,
    609–612
diacylglycerols from PtdInsPs 158
distribution 581

Ins(1,4,5)trisphosphate receptors
    698
isozymes 580, 582–584, 629
NMR analysis 585
PH domains 580, 583–584
PtdIns 3-kinase activity 749
PtdIns diacylglycerols 94
PtdIns and GPtdIns-specific
    579–599
PtdIns structure 55–57, 63
PtdIns-specific PLC substrate
    specificity 582–591
signaling 689, 690, 691
substrate specificity 561–564,
    582–591
phospholipase C_{Bc} (PLC_{Bc}) 561–564,
    565–566
phospholipase D (PLD) 542–543,
    568–575
assay methods 572–575, 604–606,
    629–632
characteristics 599–606, 612,
    614–626
GPtdIns 599–601
GPtdIns-PLD 601–603
isozymes 614, 618
phosphatidic acids from PtdInsPs
    158
PtdIns structure 57–59
PtdInsPs in membrane trafficking
    744–745
regulation 616–619
substrate specificity 570–572,
    603–604
superfamily 615
transphosphatidylation 349–352
in vitro assay 619–624
in vivo assay 624–626
phospholipase A (PLA) 54
phospholipase A_1 (PLA_1) 5–54,
    542–550, 576–579

phospholipase A_2 (PLA$_2$) 542–543,
    550–560
  assay methods 556–560
  characteristics 606–608, 613,
    626–627
  PtdIns structure 55
  PtdIns/GptIns-specific 576–579
  sn-1-position 133
  substrate specificity 553–556
  Vipera ammodytes 119
phospholipases
  assay strategies 541–542
  nomenclature 31–33
  PtdIns 541–632
  PtdIns and GPtdIns-specific
    575–606
  PtdIns-nonspecific 542–575
  PtdINs(3,4,5)P$_3$ activated
    626–632
  PtdIns(4,5)P$_2$ activated 606–626,
    630
  PtdInsPs analysis 132–133
phosphomonoesterases 537, 539
phosphorus content 51, 105–106
phosphorylation 602
Phreatamoeba balamuthi 31, 215
phytases 8, 531–533, 537, 539–540
pig cerebellum 444–445
PIPKINS see phosphatidylinositol
    monophosphate kinases
PKB see protein kinase B
PLA$_1$ see phospholipase A$_1$
PLA$_2$ see phospholipase A$_2$
placenta 518
plants
  see also myo-inositols
  guard cells 708–709, 715
  PLD specificity 603–604
  PtdIns 4-kinases 415
  PtdIns de novo synthesis 344–345
  PtdIns(4)P biosynthesis 357

tissue assays 418
plasmalogens 38, 39, 108
Plasmodium spp. 5, 257, 287, 377
platelet-derived growth factor (PDGF)
    747, 749, 750, 762–763
platelets 108, 122–123, 172, 637, 742
PLC see phospholipase C
PLD see phospholipase D
pleckstrin homology (PH) domain
  Ins(1,4,5,6)P$_4$ 3-kinase 707
  InsP$_3$ 5-phosphatases 518, 520
  PLC isozymes 580, 582, 583–584
  PtdInsP$_2$ binding 729
  PtdInsP$_3$ binding 726–727
  PtdInsPs 171, 723–724, 725
polar capillary GLC and GC/MS
    85–87
polyol dehydrogenases 134–135,
    224–231
polyols 207, 209, 210, 211
polyoma virus-transformed cells
    369–370
positional isomers
  Ins bisphosphates 195–196
  Ins hexakisphosphates 205, 207
  Ins monophosphates 194–195
  Ins pentakisphosphates 200–205
  Ins tetrakisphosphates 198–200
  Ins trisphosphates 196–198
  InsPs 191–207
  nomenclature 19
positive chemical ionization (PCI) 93
potassium current 709, 715
PP-InsP$_4$ see diphosphoinositol
    tetrakisphosphate
PP-InsP$_5$ see diphosphoinositol
    pentakisphosphate
[PP]$_2$-InsP$_4$ see bis-diphosphoinositol
    tetrakisphosphate
procyclic acidic repetitive protein
    (PARP) 289, 292, 294

protein bound glycosyl phosphatidyl-
    inositol-anchors 263–267
protein kinase B (PKB) 724, 756
protein kinase B/Akt (PKB/Akt)
    756–757, 758–759
protein–ethanolamine bridge
    271–278
protein–phospholipid interactions 723
protein-free GPtdIns 398–399
proteins *see* binding proteins
proteolysed glycosyl phosphatidyl-
    inositol-anchored proteins
    274–275, 277, 283
proteolysis 264–266
protozoal GPtdIns-related structures
    399–400
PtdCho *see* phosphatidylcholine
PtdIns *see* phosphatidylinositol
PtdIns-PLC *see* phosphatidylinositol
    phospholipase C
PtdInsP$_2$ *see* phosphatidylinositol
    bisphosphates
PtdInsPs *see* phosphatidylinositol
    phosphates
PtdOH, molecular species
    determination 79–83
PTEN tumor suppressor gene
    InsP$_2$ 3-phosphatases 506
    InsP$_3$ 3-phosphatases 513
    InsP$_5$ 1-phosphatases 528
    PtdIns(3)P 469
    PtdInsP$_2$ 3-phosphatases 474–475
    PtdInsP$_3$ 484–486
purple acid phosphatases (PAPs) 537
PX domain 732–733
4-(2-pyridylazo) resorcinol (PAR) 245
Pyrococcus woes 215
pyrophosphates
    see also inositol pyrophosphates
    cycle 710
Pyrus communis 334

quantification
    diacylglycerols 99–101
    fatty acids 101
    GPtdIns 322–326
    Ins content 101–104
    InsPs 235–252
    PtdIns 96–106
    PtdInsPs 162–173

radio-receptor assays 167–171,
    239–243, 692
radioisotope assays 236–238
    see also isotope studies
    Ins(1,3,4,5,6)P$_5$ 6-phosphatases
    531
    Ins(1)P phosphatases 500–501
    Ins(3)P phosphatases 502–503
    InsP$_2$
        1-phosphatase 505–506
        3-phosphatases 506–508
        4-phosphatases 508–510
    InsP$_3$
        1-phosphatase 511–513
        3-phosphatases 513–514
        4-phosphatases 516
        5-phosphatases 523–526
    PPInsP$_5$ pyrophosphatases 536
    PtdIns(3)P 3-phosphatase 469–471
    PtdIns(4)P 4-phosphatase 472–474
    PtdInsP$_2$
        3-phosphatases 475
        4-phosphatases 476–477
        5-phosphatases 480–482
    PtdInsP$_3$
        3-phosphatases 486–488
        4-phosphatases 490
        5-phosphatases 493–497
Ras 406, 407, 590
rat brain 177, 192–193, 443, 457, 462
rat liver microsomes 346–347

rat pancreas 125, 153–154
recombinant proteins 417–418, 422–423
regulation
   InsP kinases 702–710
   PLD 616–619
   PtdIns 3-kinases 735–736
   PtdIns 4-kinases 737–738
   PtdIns 5-kinases 738–740
   PtdIns kinases FYVE 738–739
   PtdInsP kinases 734–740
relative fluorescence units (RFU) 251
remodeling 345–352, 391–398
resazurin 251
resorufin 251
reversed phase HPLC 76–79, 82, 89
R_f values 117
RFU *see* relative fluorescence units
Rho/Rac-binding proteins 754–755

Sac domain 472, 473, 480
Saccharomyces cerevisiae
   CDP-diacylglycerol synthetase 337
   Golgi apparatus 394–396
   GPtdIns
      protein anchors biosynthesis 380
      PtdIns moiety 289
      transamidase complex 271
   Ins(1,3,4,5,6)P_5 2-kinase 454
   InsP$_6$ 5-kinase 458
   myotubularin-related protein 3 468
   phosphatases 463
   Vps34p 405
Salmonella spp. 447, 513, 529
SAX *see* strong anion exchange
Schizosaccharomyces pombe 52
scooting assay 542, 547–548, 556–557

scyllo-inositols 6, 20–23, 28–29, 95
scyllo-phosphatidylinositols 337–339, 340, 341, 349
second messengers 235, 335
   Ins(1,4,5)P_3 689, 693–694, 695, 697–699, 715
   receptor 176
   Ins(3,4,5,6)P_4 714
   molecular basis 695–696
secretory phospholipase A$_2$ (sPLA$_2$) 550–551, 555–556, 558, 559–560
Sep-Pak columns 124–125, 184, 185
serine phosphatases *see* alkaline phosphatases
SHIP 492–493, 519, 522
signaling *see* cellular signaling
skeletal muscle and kidney enriched inositol phosphatase (SKIP) 521
SKIP *see* skeletal muscle and kidney enriched inositol phosphatase
small GTPases 434–435, 438
sn-1-position 66–68, 69, 74–75, 94, 133
sn-2-position 54, 67, 68–70, 94
snake venoms 550
SNARE proteins 741, 743
sn-1,2-diacylglycerol (DAG) 11, 13, 586, 587
'soft' ionization methods 143
solid phase extraction 42–48, 111–128
solvent extraction 37–42, 107–111
solvent systems, TLC 42–47
SopB *Salmonella* virulence protein 513–514, 529–530, 707–708
soybeans, intact PtdInsPs 155
species-specific glycosyl phosphatidylinositol anchors 254

specific biological effects, PtdInsPs 740–764
specificity *see* substrate specificity
sPLA$_2$ *see* secretory phospholipase A$_2$
Sporothrix schenckii 10, 330
standards preparation 633–688
Staphylococcus aureus 266, 294, 296
sn-1-stearoyl-2-arachidonoyl GroPIns(3,4,5)P$_3$ 728
stereochemical structure 16–19, 20–23, 223–235
    see also enantiomers
Streptomyces chromofuscus 570, 572
strong anion exchange (SAX) columns 136–140, 185, 186, 188, 199, 200, 205, 237–238
structural features
    Man α1–4GlcNα1–6*myo*-inositol-1-PO$_4$-DG 253
    PLC isozymes 584
    PtdIns 49–65
    PtdIns 4-kinases 416
    PtdIns(4)P 5-kinases 420
    PtdIns(5)P 4-kinases 420
    PtdInsPs 128–151
STT4 355
substrate specificity 234–235, 577, 582–591, 603–604
SUV 579
Swiss mouse 3T3 cells 171, 374–375
synaptojanin 519
    Ins 5-phosphatase catalytic domain 522
    InsP$_3$ 5-phosphatases 519
    PtdIns(4,5)P$_2$ hydrolysis 479–480
    PtdIns(4)P 4-phosphatase 472
    PtdInsP$_3$ 5-phosphatases 490
    PtdInsPs in membrane trafficking 742–743
synaptotagmin 743, 746–747

T5-1 cells 186, 187
tandem mass spectrometry (MS/MS) 61, 82, 143–148, 163–165
TBAHS *see* tetrabutylammonium hydrogen sulfate
TBDMS ethers 84, 87, 93, 298–299
TCA *see* trichloroacetic acid
TEAB *see* triethylammonium bicarbonate
tetrabutylammonium hydrogen sulfate (TBAHS) 189
Tetrahymena spp. 65
Thermotoga maritime 499–500
thin-layer chromatography (TLC)
    see also high performance thin-layer chromatography
    InsPs 180–182
    PLA$_1$ assay 548–549
    PLC assay 592–594
    prior to MS 62
    PtdIns extraction 42–45, 48
    PtdInsPs 111–121
TM4SF *see* transmembrane 4 superfamily
TMS ethers 298–299
TNF *see* tumor necrosis factor
Torpedo marmorata acetylcholinesterase 152
Toxoplasma gondii 398, 401–402
TPIPα 469, 475
transmembrane 4 superfamily (TM4SF) 737
transphosphatidylation 349–352, 570–572, 573, 575
trichloroacetic acid (TCA) 176
triethylammonium bicarbonate (TEAB) 184, 185
TRP proteins 712
Trypanosoma brucei PtdIns-specific PLC 588

trypanosomes 9
  exoglycosidase sequencing 282
  GlcN-PtdIns(Man)$_3$EtnP
    biosynthesis 387
  glycan moiety 280
  glycosyl PtdIns anchors 285
  GPtdIns remodeling 391
  inositol acylation 384
  natural PtdIns structures
    29–30
  PARP 289, 292
  PLA$_2$ fatty acid remodeling 560
  PtdIns-PLC 564
TUBBY proteins 730–732
tumor necrosis factor (TNF) 429,
  738
two-dimensional HPTLC 44–45
two-dimensional TLC 112–113

vacuolar protein sorting mutants 405
venoms 550, 576–577, 578–579
vesicular trafficking 741–747
VHS domain 726
Vipera ammodytes 119
Vps34p *see* vacuolar protein sorting
  mutant

weak anion exchange (WAX)
    columns 186, 199, 202, 203
wortmannin 408, 742, 748

X sPLA$_2$ 555–556, 559–560
Xenopus 459

yeast 40, 363, 377–378, 383, 386
yttrium 245

Printed and bound by CPI Group (UK) Ltd, Croydon, CR0 4YY

07/10/2024

01041904-0001